Lecture Notes in Computer Science 9294

Commenced Publication in 1973
Founding and Former Series Editors:
Gerhard Goos, Juris Hartmanis, and Jan van Leeuwen

Advanced Research in Computing and Software Science

Subline of Lecture Notes in Computer Science

T0171917

More information about this series at http://www.springer.com/series/7407

Nikhil Bansal · Irene Finocchi (Eds.)

Algorithms – ESA 2015

23rd Annual European Symposium
Patras, Greece, September 14–16, 2015
Proceedings

 Springer

Editors
Nikhil Bansal
University of Technology
Eindhoven
The Netherlands

Irene Finocchi
Sapienza University of Rome
Rome
Italy

ISSN 0302-9743
Lecture Notes in Computer Science
ISBN 978-3-662-48349-7
DOI 10.1007/978-3-662-48350-3

ISSN 1611-3349   (electronic)

ISBN 978-3-662-48350-3   (eBook)

Library of Congress Control Number: 2015948726

LNCS Sublibrary: SL1 – Theoretical Computer Science and General Issues

Printed on acid-free paper

Springer-Verlag GmbH Berlin Heidelberg is part of Springer Science+Business Media
(www.springer.com)

Preface

This volume contains the extended abstracts selected for presentation at ESA 2015, the 23rd European Symposium on Algorithms, held in Patras, Greece, September 14–16, 2015, as part of ALGO 2015. The ESA symposia are devoted to fostering and disseminating the results of high-quality research on algorithms and data structures. ESA seeks original algorithmic contributions for problems with relevant theoretical and/or practical applications and aims at bringing together researchers in the computer science and operations research communities. Ever since 2002, it has had two tracks, the Design and Analysis Track (Track A), intended for papers on the design and mathematical analysis of algorithms, and the Engineering and Applications Track (Track B), for submissions dealing with real-world applications, engineering, and experimental analysis of algorithms. Information on past symposia, including locations and LNCS volume numbers, is maintained at{http://esa-symposium.org}.

In response to the call for papers, ESA 2015 attracted a record number of 320 submissions, 261 for Track A and 59 for Track B. Paper selection was based on originality, technical quality, and relevance. Considerable effort was devoted to the evaluation of the submissions, with at least three reviews per paper. With the help of more than 980 expert reviews and more than 514 external reviewers, the two committees selected 86 papers for inclusion in the scientific program of ESA 2015, 71 in Track A and 15 in Track B, yielding an acceptance rate of about 26%. In addition to the accepted contributions, the symposium featured two invited lectures by Rasmus Pagh (IT University of Copenhagen, Denmark) and Paul Spirakis (University of Liverpool, UK and CTI & University of Patras, Greece). Abstracts of the invited lectures are also included in this volume.

The European Association for Theoretical Computer Science (EATCS) sponsored a best paper award and a best student paper award. A submission was eligible for the best student paper award if all authors were doctoral, master, or bachelor students at the time of submission. This award was shared by two papers: one by Sascha Witt for his contribution on "Trip-Based Public Transit Routing" and the other by Meirav Zehavi for her contribution entitled "Mixing Color Coding-Related Techniques." The best paper award went to Christina Boucher, Christine Lo, and Daniel Lokshtanov for their paper entitled "Consensus Patterns (Probably) Has No EPTAS." Our warmest congratulations to all of them for these achievements!

We wish to thank all the authors who submitted papers for consideration, the invited speakers, the members of the Program Committees for their hard work, and all the external reviewers who assisted the Program Committees in the evaluation process. Special thanks go to Giuseppe Italiano for answering our many questions along the way, and to Christos Zaroliagis, who helped with the local organization of the conference.

July 2015

Nikhil Bansal
Irene Finocchi

Organization

Program Committee

Per Austrin	KTH Stockholm, Sweden
Nikhil Bansal	Eindhoven University of Technology, The Netherlands
Maike Buchin	Ruhr Univ. Bochum, Germany
Benjamin Doerr	Ecole Polytechnique, France
Christoph Dürr	University of Pierre et Marie Curie, France
Esther Ezra	Georgia Tech, USA and New York Univ., USA
Rolf Fagerberg	University of Southern Denmark, Denmark
Sandor Fekete	University of Technology, Braunschweig, Germany
Irene Finocchi	Sapienza University of Rome, Italy
Tobias Friedrich	Friedrich Schiller Univ. Jena, Germany
Kasper Green Larsen	Aarhus Univ., Denmark
Riko Jacob	ETH Zurich, Switzerland
Moshe Lewenstein	Bar Ilan Univ., Israel
Alex Lopez-Ortiz	University of Waterloo, Canada
Brendan Lucier	Microsoft Research, USA
Andrew McGregor	University of Massachusetts, USA
Viswanath Nagarajan	University of Michigan, USA
Thomas Pajor	Microsoft Research, USA
Seth Pettie	University of Michigan, USA
Cynthia Phillips	Sandia National Laboratories, USA
Marcin Pilipczuk	University of Warsaw, Poland and University of Warwick, UK
Tomasz Radzik	King's College London, UK
Thomas Rothvoss	University of Washington, USA
Marie-France Sagot	INRIA Grenoble Rhone-Alpes, France
Luciana Salete Buriol	Federal University of Rio Grande do Sul, Brazil
Piotr Sankowski	University of Warsaw, Poland
Saket Saurabh	Institue of Mathematical Sciences, India
Roy Schwartz	Princeton Univ., USA
Anastasios Sidiropoulos	Ohio State Univ., USA
Francesco Silvestri	University of Padova, Italy
Kunal Talwar	Google, USA

Philippas Tsigas	Chalmers University of Technology, Sweden
Charalampos Tsourakakis	Harvard Univ., USA
Leo van Iersel	CWI, The Netherlands
Rob van Stee	University of Leicester, UK
Sergei Vassilvitskii	Google, USA
Andreas Wiese	MPI for Informatics, Germany
Virginia Williams	Stanford Univ., USA

Additional Reviewers

Abboud, Amir
Abo Khamis, Mahmoud
Abraham, Ittai
Ackermann, Marcel R.
Afshani, Peyman
Ahn, Hee-Kap
Ailon, Nir
Ajwani, Deepak
Akutsu, Tatsuya
Alekseyev, Max
Alon, Noga
Álvarez-Miranda,
    Eduardo
Anagnostopoulos, Aris
Andrade, Ricardo
Angelopoulos, Spyros
Annamalai,
    Chidambaram
Anshelevich, Elliot
Antoniadis, Antonios
Arnosti, Nick
Ashok, Pradeesha
Aziz, Haris
Backurs, Arturs
Balkanski, Eric
Bampis, Evripidis
Bannai, Hideo
Barbay, Jérémy
Barman, Siddharth
Bateni,
    Mohammadhossein
Bauer, Ulrich
Bavarian, Mohammad
Becker, Ruben
Belazzougui, Djamal

Bender, Michael
Berenbrink, Petra
Berkholz, Christoph
Bhaskara, Aditya
Bhattacharya, Sayan
Bilardi, Gianfranco
Bille, Philip
Bilò, Davide
Birmele, Etienne
Bodwin, Greg
Bohler, Cecilia
Bonifaci, Vincenzo
Bonsma, Paul
Borassi, Michele
Borodin, Alan
Braverman, Vladimir
Bringmann, Karl
Brodal, Gerth Stølting
Buchbinder, Niv
Buchin, Kevin
Bus, Norbert
Błasiok, Jarosław
Cabello, Sergio
Cadek, Martin
Cai, Yang
Calinescu, Gruia
Campelo, Manoel
Cao, Yixin
Cardinal, Jean
Carmi, Paz
Chakaravarthy,
    Venkatesan
Chakrabarti, Amit
Chalermsook, Parinya
Chambers, Erin

Chan, Siu On
Chan, Timothy M.
Chang, Hsien-Chih
Chauve, Cedric
Chechik, Shiri
Chekuri, Chandra
Chen, Lin
Chen-Xu, Shen
Cheng, Siu-Wing
Cheung, Wang Chi
Chlamtac, Eden
Chuzhoy, Julia
Cicalese, Ferdinando
Clarkson, Ken
Clifford, Raphael
Cormode, Graham
Croitoru, Cosmina
Cseh, Ágnes
Cygan, Marek
Dadush, Daniel
Das, Gautam K.
Das, Syamantak
de Freitas, Rosiane
De Keijzer, Bart
De Oliveira Oliveira,
    Mateus
de Rezende, Susanna F.
de Zeeuw, Frank
Derka, Martin
Despotakis, Stelios
Devillers, Olivier
Dey, Tamal
Dibbelt, Julian
Dietzfelbinger, Martin
Ding, Bolin

Disser, Yann
Dobzinski, Shahar
Drange, Pål Grønås
Drechsler, Rolf
Dregi, Markus Sortland
Driemel, Anne
Duetting, Paul
van Duijn, Ingo
Dujmovic, Vida
Durand, Arnaud
Dutta, Kunal
Dziembowski, Stefan
Efentakis, Alexandros
Eisenstat, Sarah
Elffers, Jan
Elsässer, Robert
Emiris, Ioannis
Ene, Alina
Englert, Matthias
Epasto, Alessandro
Eppstein, David
Epstein, Leah
Erdos, Peter
Erikson, Jeff
Erlebach, Thomas
Even, Guy
Faenza, Yuri
Feldman, Moran
Fellows, Michael
Felsner, Stefan
Feng, Qilong
Fernau, Henning
Filmus, Yuval
Fischer, Johannes
Fomin, Fedor
Friedrich, Tobias
Frieze, Alan
Friggstad, Zachary
Fulek, Radoslav
Gaertner, Bernd
Gagie, Travis
Ganian, Robert
Gao, Jie
Gaspers, Serge
Ge, Rong

Gemsa, Andreas
Giakkoupis, George
Gibson, Matt
Glisse, Marc
Goldberg, Paul
Golovach, Petr
Goswami, Mayank
Gouleakis, Themistoklis
Gouveia, João
Goyal, Vineet
Grandoni, Fabrizio
Gravin, Nick
Grigoriev, Alexander
Grossi, Roberto
Gudmundsson, Joachim
Guha, Sudipto
Guo, Jiong
Gupta, Sushmita
Guttmann, Tony
Gálvez, Waldo
Hagerup, Torben
Haghpanah, Nima
Hansen, Thomas
   Dueholm
Har-Peled, Sariel
Hariharan, Ramesh
Harris, David
Hassidim, Avinatan
Hassin, Refael
Hastad, Johan
He, Meng
Heeringa, Brent
Heggernes, Pinar
Hemmer, Michael
Hermelin, Danny
Heydrich, Sandy
Hoefer, Martin
Hoffmann, Michael
Hong, Seok-Hee
Horel, Thibaut
Hosseini, Kaave
Huang, Chien-Chung
Huang, Sangxia
Huang, Zengfeng
Hume, Thomas

Iacono, John
Ileri, Atalay Mert
Im, Sungjin
Immorlica, Nicole
Inenaga, Shunsuke
Inostroza-Ponta, Mario
Irving, Robert
Italiano, Giuseppe F.
Jansen, Bart M.P.
Jansen, Klaus
Jansen, Thomas
Jeż, Łukasz
Jones, Mark
Jurdzinski, Marcin
Kaibel, Volker
Kamali, Shahin
Kamiyama, Naoyuki
Kammer, Frank
Karczmarz, Adam
Karrenbauer, Andreas
Kaski, Petteri
Kawamura, Akitoshi
Kelk, Steven
Kerenidis, Iordanis
Kesselheim, Thomas
Khan, Arindam
Khanna, Sanjeev
Kindermann, Philipp
King, Valerie
Klein, Rolf
Kleinberg, Robert
Kliemann, Lasse
Knauer, Christian
Knudsen, Mathias Bæk
   Tejs
Kobayashi, Yusuke
Kociumaka, Tomasz
Koebler, Johannes
Koenemann, Jochen
Kogan, Kirill
Koivisto, Mikko
Kolay, Sudeshna
Kollias, Konstantinos
Kolpakov, Roman
Komusiewicz, Christian

Konrad, Christian
Kontogiannis, Spyros
Kopelowitz, Tsvi
Korula, Nitish
Kostitsyna, Irina
Kotrbcik, Michal
Kowalik, Lukasz
Kozik, Marcin
Kratsch, Stefan
Krishnaswamy,
   Ravishankar
Krithika, R.
Krohmer, Anton
Krysta, Piotr
Kshemkalyani, Ajay
Ku, Jason S.
Kuhnert, Sebastian
Kulkarni, Janardhan
Kumar, Amit
Kumar, Nirman
Kumar, Ravi
Kurokawa, David
Kurpisz, Adam
Kwon, O-Joung
Kärkkäinen, Juha
Köhler, Ekkehard
Künnemann, Marvin
Łącki, Jakub
Lattanzi, Silvio
Laue, Soeren
van Leeuwen, Erik Jan
Leniowski, Dariusz
Leung, Vitus
Levit, Vadim
Li, Fei
Li, Jian
Li, Shi
Liaghat, Vahid
Lidbetter, Thomas
Lin, Bertrand
Lingas, Andrzej
Linz, Simone
Lokshtanov, Daniel
Lopez-Ortiz, Alejandro
Lotker, Zvi

Löffler, Maarten
M.S., Ramanujan
Maftuleac, Daniela
Maheshwari, Anil
Makarychev, Konstantin
Makarychev, Yury
Mallmann-Trenn,
   Frederik
Manlove, David
Manthey, Bodo
Markakis, Evangelos
Marx, Dániel
Mastrolilli, Monaldo
Matuschke, Jannik
Maurer, Olaf
Mavronicolas, Marios
Mccauley, Samuel
Megow, Nicole
Meir, Reshef
Melo, Emerson
Mengel, Stefan
Mertens, Stephan
Mertzios, George
Mestre, Julian
Mihalák, Matúš
Miksa, Mladen
Miltersen, Peter Bro
Misra, Pranabendu
Misra, Neeldhara
Mitchell, Joseph
Mnich, Matthias
Moitra, Ankur
Monaco, Gianpiero
Morgenstern, Jamie
Morin, Pat
Moruz, Gabriel
Moseley, Benjamin
Mouawad, Amer
Mozes, Shay
Mulzer, Wolfgang
Munro, Ian
Musco, Christopher
Mustafa, Nabil
Mömke, Tobias
Müller-Hannemann,

Matthias
Nachmanson, Lev
Naor, Seffi
Nastos, James
Nayyeri, Amir
Nederlof, Jesper
Newman, Alantha
Ngo, Hung
Nguyen, Huy
Nguyen, Kim Thang
Niazadeh, Rad
Nicholson, Patrick K.
Nichterlein, André
Nickerson, Bradford
Niedermann, Benjamin
Niedermeier, Rolf
Nielsen, Jesper Sindahl
Nikolov, Aleksandar
Nilsson, Bengt J.
Nordstrom, Jakob
O'Hara, Stephen
Ochoa, Carlos
Olver, Neil
Onak, Krzysztof
Ordóñez Pereira, Alberto
Orlin, James
Otachi, Yota
Ott, Sebastian
Ozkan, Ozgur
Pachocki, Jakub
Paes Leme, Renato
Pagh, Rasmus
Panigrahi, Debmalya
Panolan, Fahad
Parekh, Ojas
Parter, Merav
Pascual, Fanny
Patel, Amit
Patel, Viresh
Patt-Shamir, Boaz
Paul, Saurabh
Pedrosa, Lehilton L.C.
Peng, Richard
Penna, Paolo
Pettie, Seth

Pfetsch, Marc
Pham, Ninh
Phillips, Cynthia
Pietracaprina, Andrea
Pilipczuk, Michał
Piperno, Adolfo
Pisanti, Nadia
Pontecorvi, Matteo
Pouget-Abadie, Jean
Price, Eric
Pritchard, David
Proietti, Guido
Pruhs, Kirk
Prädel, Lars
Pröger, Tobias
Pukelsheim, Friedrich
Pérez-Lantero, Pablo
Quedenfeld, Frank
Quimper, Claude-Guy
Raichel, Benjamin
Rajaraman, Rajmohan
Raman, Venkatesh
Ramon, Jan
Rav, Mathias
Rawitz, Dror
Ray, Saurabh
Raz, Orit E.
Regev, Oded
Reidl, Felix
Roditty, Liam
Roeloffzen, Marcel
Romero, Jazmín
Rotenberg, Eva
Ruskey, Frank
Sabharwal, Yogish
Sach, Benjamin
Sadakane, Kunihiko
Saha, Barna
Salvagnin, Domenico
Sanders, Peter
Sanita, Laura
Sarpatwar, Kanthi
Satti, Srinivasa Rao
Sau, Ignasi
Saumell, Maria

Schabanel, Nicolas
Schaudt, Oliver
Schieber, Baruch
Schill Collberg, Adam
Schlauch, Wolfgang
Schlipf, Lena
Schmid, Andreas
Schmid, Stefan
Schmidt, Jens M.
Schrijvers, Okke
Schulz, André
Schutt, Andreas
Schweitzer, Pascal
Scornavacca, Celine
Scquizzato, Michele
Segal, Ilya
Segev, Danny
Shapira, Asaf
Shi, Cong
Singer, Yaron
Sitchinava, Nodari
Sitters, Rene
Sivan, Balasubramanian
Skala, Matthew
Skutella, Martin
Smet, Pieter
Smid, Michiel
Smorodinsky, Shakhar
Snoeyink, Jack
Sohler, Christian
Sommer, Christian
Sorge, Manuel
Soto, José A.
Stefankovic, Daniel
Stiller, Sebastian
Stojakovic, Milos
Storandt, Sabine
Strasser, Ben
Stubbs, Daniel
Stöckel, Morten
Sudholt, Dirk
Suomela, Jukka
Sutton, Andrew
Svensson, Ola
Swamy, Chaitanya

Syrgkanis, Vasilis
Takeda, Masayuki
Tamir, Tami
Tamura, Takeyuki
Tan, Xuehou
Tang, Pingzhong
Thaler, Justin
Thatte, Bhalchandra
Thilikos, Dimitrios
Thorup, Mikkel
Tokuyama, Takeshi
Toth, Csaba
Tsichlas, Kostas
Tzamos, Christos
Ullman, Jonathan
Urrutia, Sebastián
Vaz, Daniel
Ventre, Carmine
Verschae, José
Vigneron, Antoine
Vinga, Susana
Vinyals, Marc
Vondrak, Jan
Vosoughpoor, Hamide
Wahlström, Magnus
Wakabayashi, Yoshiko
Walczak, Bartosz
Wang, Yusu
Ward, Justin
Wegrzycki, Karol
Weimann, Oren
Weinberg, S. Matthew
Wenk, Carola
Werneck, Renato
White, Colin
Wieder, Udi
Wiese, Andreas
Wilkens, Christopher
Wilkinson, Bryan T.
Woeginger, Gerhard J.
Wong, Prudence W.H.
Wood, David R.
Woodruff, David
Wrochna, Marcin
Xie, Ning

Yaniv, Jonathan
Yaroslavtsev, Grigory
Ye, Deshi
Yi, Ke
Yoshida, Yuichi
Yu, Huacheng

Zehavi, Meirav
Zeng, Bo
Zenklusen, Rico
Zhang, Qiang
Zhang, Qin
Zwick, Uri

Zych, Anna
Živný, Stanislav
van Zuylen, Anke

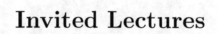

Invited Lectures

Correlated Locality-Sensitive Hashing*

Rasmus Pagh

IT University of Copenhagen, Denmark

After an introduction to the area, we consider a new construction of locality-sensitive hash functions for Hamming space that is *covering* in the sense that is it guaranteed to produce a collision for every pair of vectors within a given radius r. The construction is *efficient* in the sense that the expected number of hash collisions between vectors at distance cr, for a given $c > 1$, comes close to that of the best possible data independent LSH without the covering guarantee, namely, the seminal LSH construction of Indyk and Motwani (FOCS '98). The efficiency of the new construction essentially *matches* their bound if $\log(n)/(cr)$ is integer, where n is the number of points in the data set, and differs from it by at most a factor $\ln(4) < 1.4$ in the exponent for larger values of cr. As a consequence, LSH-based similarity search in Hamming space can avoid the problem of false negatives at little or no cost in efficiency.

---

* The research leading to these results has received funding from the European Research Council under the European Union's Seventh Framework Programme (FP7/2007-2013) / ERC grant agreement no. 614331.

On the Discrete Dynamics of Probabilistic (Finite) Population Protocols

Paul Spirakis[1,2,3]

[1]Department of Computer Science, University of Liverpool, UK
P.Spirakis@liverpool.ac.uk
[2]Department of Computer Engineering and Informatics, University of Patras,
26504 Patras, Greece
[3]Computer Technology Institute and Press "Diophantus",
N. Kazantzaki Str., Patras University Campus, 26504 Patras, Greece

Population Protocols are a recent model of computation that captures the way in which complex behavior of systems can emerge from the underlying local interactions of agents. Agents are usually anonymous and the local interaction rules are scalable (independent of the size, n, of the population). Such protocols can model the antagonism between members of several "species" and relate to evolutionary games.

In the recent past the speaker was involved in joint research studying the discrete dynamics of cases of such protocols for finite populations. Such dynamics are, usually, probabilistic in nature , either due to the protocol itself or due to the stochastic nature of scheduling local interactions. Examples are (a) the generalized Moran process (where the protocol is evolutionary because a fitness parameter is crucially involved) (b) the Discrete Lotka-Volterra Population Protocols (and associated Cyclic Games) and (c) the Majority protocols for random interactions.

Such protocols are usually discrete time transient Markov Chains. However the detailed states description of such chains is exponential in size and the state equations do not facilitate a rigorous approach. Instead, ideas related to filtering, stochastic domination and Potentials (leading to Martingales) help in understanding the dynamics of the protocols.

In the talk we discuss such rigorous approaches and their techniques. We examine the question of fast (in time polynomial in the population size) convergence (to an absorbing state). We also discuss the question of most probable eventual state of the protocols (and the computation of the probability of such states). Several aspects of such discrete dynamics are wide open and it seems that the algorithmic thought can contribute to the understanding of this emerging subfield of science.

Contents

Improved Approximation Algorithms for Stochastic Matching*

Marek Adamczyk[1], Fabrizio Grandoni[2], and Joydeep Mukherjee[3]

[1] Department of Computer, Control, and Management Engineering,
Sapienza University of Rome, Italy
adamczyk@dis.uniroma1.it
[2] IDSIA, University of Lugano, Switzerland
fabrizio@idsia.ch
[3] Institute of Mathematical Sciences, CIT, India
joydeepm@imsc.res.in

Abstract. In this paper we consider the *Stochastic Matching* problem, which is motivated by applications in kidney exchange and online dating. We are given an undirected graph in which every edge is assigned a probability of existence and a positive profit, and each node is assigned a positive integer called *timeout*. We know whether an edge exists or not only after probing it. On this random graph we are executing a process, which one-by-one probes the edges and gradually constructs a matching. The process is constrained in two ways: once an edge is taken it cannot be removed from the matching, and the timeout of node v upper-bounds the number of edges incident to v that can be probed. The goal is to maximize the expected profit of the constructed matching.

For this problem Bansal et al. [4] provided a 3-approximation algorithm for bipartite graphs, and a 4-approximation for general graphs. In this work we improve the approximation factors to 2.845 and 3.709, respectively.

We also consider an online version of the bipartite case, where one side of the partition arrives node by node, and each time a node b arrives we have to decide which edges incident to b we want to probe, and in which order. Here we present a 4.07-approximation, improving on the 7.92-approximation of Bansal et al. [4].

The main technical ingredient in our result is a novel way of probing edges according to a random but non-uniform permutation. Patching this method with an algorithm that works best for large probability edges (plus some additional ideas) leads to our improved approximation factors.

1 Introduction

In this paper we consider the *Stochastic Matching* problem, which is motivated by applications in kidney exchange and online dating. Here we are given an

---

* This work was partially done while the first and last authors were visiting IDSIA. The first and second authors were partially supported by the ERC StG project NEWNET no. 279352, and the first author by the ERC StG project PAAl no. 259515. The third author was partially supported by the ISJRP project Mathematical Programming in Parameterized Algorithms.

© Springer-Verlag Berlin Heidelberg 2015
N. Bansal and I. Finocchi (Eds.): ESA 2015, LNCS 9294, pp. 1–12, 2015.
DOI: 10.1007/978-3-662-48350-3_1

undirected graph $G = (V, E)$. Each edge $e \in E$ is labeled with an (existence) probability $p_e \in (0, 1]$ and a weight (or profit) $w_e > 0$, and each node $v \in V$ with a *timeout* (or *patience*) $t_v \in \mathbb{N}^+$. An algorithm for this problem probes edges in a possibly adaptive order. Each time an edge is probed, it turns out to be *present* with probability p_e, in which case it is (irrevocably) included in the matching under construction and provides a profit w_e. We can probe at most t_u edges among the set $\delta(u)$ of edges incident to node u (independently from whether those edges turn out to be present or absent). Furthermore, when an edge e is added to the matching, no edge $f \in \delta(e)$ (i.e., incident on e) can be probed in subsequent steps. Our goal is to maximize the expected weight of the constructed matching. Bansal et al. [4] provide an LP-based 3-approximation when G is bipartite, and via reduction to the bipartite case a 4-approximation for general graphs (see also [3]).

We also consider the *Online Stochastic Matching with Timeouts* problem introduced in [4]. Here we are given in input a bipartite graph $G = (A \cup B, A \times B)$, where nodes in B are *buyer types* and nodes in A are *items* that we wish to sell. Like in the offline case, edges are labeled with probabilities and profits, and nodes are assigned timeouts. However, in this case timeouts on the item side are assumed to be unbounded. Then a second bipartite graph is constructed in an online fashion. Initially this graph consists of A only. At each time step one random buyer $\tilde{b}$ of some type b is sampled (possibly with repetitions) from a given probability distribution. The edges between $\tilde{b}$ and A are copies of the corresponding edges in G. The online algorithm has to choose at most t_b unmatched neighbors of $\tilde{b}$, and probe those edges in some order until some edge $a\tilde{b}$ turns out to be present (in which case $a\tilde{b}$ is added to the matching and we gain the corresponding profit) or all the mentioned edges are probed. This process is repeated n times, and our goal is to maximize the final total expected profit[1].

For this problem Bansal et al. [4] present a 7.92-approximation algorithm. In his Ph.D. thesis Li [8] claims an improved 4.008-approximation. However, his analysis contains a mistake [9]. By fixing that, he still achieves a 5.16-approximation ratio improving over [4].

1.1   Our Results

Our main result is an approximation algorithm for bipartite Stochastic Matching which improves the 3-approximation of Bansal et al. [4] (see Section 2).

Theorem 1. *There is an expected 2.845-approximation algorithm for Stochastic Matching in bipartite graphs.*

Our algorithm for the bipartite case is similar to the one from [4], which works as follows. After solving a proper LP and rounding the solution via a rounding technique from [7], Bansal et al. probe edges in uniform random order. Then they show that every edge e is probed with probability at least $x_e \cdot g(p_{max})$, where

---

[1] As in [4], we assume that the probability of a buyer type b is an integer multiple of $1/n$.

x_e is the fractional value of e, $p_{max} := \max_{f \in \delta(e)} \{p_f\}$ is the largest probability of any edge incident to e (e excluded), and $g(\cdot)$ is a decreasing function with $g(1) = 1/3$.

Our idea is to rather consider edges in a carefully chosen *non-uniform* random order. This way, we are able to show (with a slightly simpler analysis) that each edge e is probed with probability $x_e \cdot g(p_e) \geq \frac{1}{3} x_e$. Observe that we have the same function $g(\cdot)$ as in [4], but depending on p_e rather than p_{max}. In particular, according to our analysis, small probability edges are more likely to be probed than large probability ones (for a given value of x_e), regardless of the probabilities of edges incident to e. Though this approach alone does not directly imply an improved approximation factor, it is not hard to patch it with a simple greedy algorithm that behaves best for large probability edges, and this yields an improved approximation ratio altogether.

We also improve on the 4-approximation for general graphs in [4]. This is achieved by reducing the general case to the bipartite one as in prior work, but we also use a refined LP with blossom inequalities in order to fully exploit our large/small probability patching technique.

Theorem 2. *There is an expected* 3.709-*approximation algorithm for Stochastic Matching in general graphs.*

Similar arguments can also be successfully applied to the online case. By applying our idea of non-uniform permutation of edges we would get a 5.16-approximation (the same as in [8], after correcting the mentioned mistake). However, due to the way edges have to be probed in the online case, we are able to finely control the probability that an edge is probed via *dumping factors*. This allows us to improve the approximation from 5.16 to 4.16. Our idea is similar in spirit to the one used by Ma [10] in his neat 2-approximation algorithm for correlated non-preemptive stochastic knapsack. Further application of the large/small probability trick gives an extra improvement down to 4.07 (see Section 3).

Theorem 3. *There is an expected* 4.07-*approximation algorithm for Online Stochastic Matching with Timeouts.*

1.2 Related Work

The Stochastic Matching problem falls under the framework of adaptive stochastic problems presented first by Dean et al. [6]. Here the solution is in fact a process, and the optimal one might even require larger than polynomial space to be described.

The Stochastic Matching problem was originally presented by Chen et al. [5] together with applications in kidney exchange and online dating. The authors consider the unweighted version of the problem, and prove that a greedy algorithm is a 4-approximation. Adamczyk [1] later proved that the same algorithm is in fact a 2-approximation, and this result is tight. The greedy algorithm does not provide a good approximation in the weighted case, and all known algorithms for this case are LP-based. Here, Bansal et al. [4] showed a 3-approximation for the bipartite case. Via a reduction to the bipartite case, Bansal et al. [4] also obtain a 4-approximation algorithm for general graphs (see also [3]).

2   Stochastic Matching

2.1   Bipartite Graphs

Let us denote by OPT the optimum probing strategy, and let $\mathbb{E}[OPT]$ denote its expected outcome. Consider the following LP:

$$\max \sum_{e \in E} w_e p_e x_e \qquad \qquad \text{(LP-BIP)}$$

$$\text{s.t.} \sum_{e \in \delta(u)} p_e x_e \leq 1, \qquad \qquad \forall u \in V; \qquad (1)$$

$$\sum_{e \in \delta(u)} x_e \leq t_u, \qquad \qquad \forall u \in V; \qquad (2)$$

$$0 \leq x_e \leq 1, \qquad \qquad \forall e \in E. \qquad (3)$$

The proof of the following Lemma is already quite standard [3,4,6] — just note that $x_e = \mathbb{P}[OPT \text{ probes } e]$ is a feasible solution of LP-BIP.

Lemma 1. *[4] Let LP_{bip} be the optimal value of* LP-BIP. *It holds that $LP_{bip} \geq \mathbb{E}[OPT]$.*

Our approach is similar to the one of Bansal et al. [4] (see also Algorithm 2.1 in the figure). We solve LP-BIP: let $x = (x_e)_{e \in E}$ be the optimal fractional solution. Then we apply to x the rounding procedure by Gandhi et al. [7], which we shall call just GKPS. Let $\hat{E}$ be the set of rounded edges, and let $\hat{x}_e = 1$ if $e \in \hat{E}$ and $\hat{x}_e = 0$ otherwise. GKPS guarantees the following properties of the rounded solution:

1. (Marginal distribution) For any $e \in E$, $\mathbb{P}[\hat{x}_e = 1] = x_e$.
2. (Degree preservation) For any $v \in V$, $\sum_{e \in \delta(v)} \hat{x}_e \leq \lceil \sum_{e \in \delta(v)} x_e \rceil \leq t_v$.
3. (Negative correlation) For any $v \in V$, any subset $S \subseteq \delta(v)$ of edges incident to v, and any $b \in \{0,1\}$, it holds that $\mathbb{P}[\wedge_{e \in S}(\hat{x}_e = b)] \leq \prod_{e \in S} \mathbb{P}[\hat{x}_e = b]$.

Our algorithm sorts the edges in $\hat{E}$ according to a random permutation and probes each edge $e \in \hat{E}$ according to that order, but provided that the endpoints of e are not matched already. It is important to notice that, by the degree preservation property, in $\hat{E}$ there are at most t_v edges incident to each node v. Hence, the timeout constraint of v is respected even if the algorithm probes all the edges in $\delta(u) \cap \hat{E}$.

Our algorithm differs from [4] and subsequent work in the way edges are randomly ordered. Prior work exploits a random uniform order on $\hat{E}$. We rather use the following, more complex strategy. For each $e \in \hat{E}$ we draw a random

---

Algorithm 1. Approximation algorithm for bipartite Stochastic Matching.

1. Let $(x_e)_{e \in E}$ be the solution to LP-BIP.
2. Round the solution $(x_e)_{e \in E}$ with GKPS; let $(\hat{x}_e)_{e \in E}$ be the rounded 0-1 solution, and $\hat{E} = \{e \in E | \hat{x}_e = 1\}$.
3. For every $e \in \hat{E}$, sample a random variable Y_e distributed as $\mathbb{P}[Y_e \leq y] = \frac{1 - e^{-yp_e}}{p_e}$.
4. For every $e \in \hat{E}$ in increasing order of Y_e:
   (a) If no edge $f \in \hat{\delta}(e) := \delta(e) \cap \hat{E}$ is yet taken, then probe edge e

---

variable Y_e distributed on the interval $\left[0, \frac{1}{p_e} \ln \frac{1}{1-p_e}\right]$ according to the following cumulative distribution: $\mathbb{P}[Y_e \leq y] = \frac{1}{p_e}(1 - e^{-p_e y})$. Observe that the density function of Y_e in this interval is e^{-yp_e} (and zero otherwise). Edges of $\hat{E}$ are sorted in increasing order of the Y_e's, and they are probed according to that order. We next let $Y = (Y_e)_{e \in \hat{E}}$.

Define $\hat{\delta}(v) := \delta(v) \cap \hat{E}$. We say that an edge $e \in \hat{E}$ is *safe* if, at the time we consider e for probing, no other edge $f \in \hat{\delta}(e)$ is already taken into the matching. Note that the algorithm can probe e only in that case, and if we do probe e, it is added to the matching with probability p_e.

The main ingredient of our analysis is the following lower-bound on the probability that an arbitrary edge e is safe.

Lemma 2. *For every edge e it holds that* $\mathbb{P}\left[e \text{ is safe} | e \in \hat{E}\right] \geq g(p_e)$, *where*

$$g(p) := \frac{1}{2+p}\left(1 - \exp\left(-(2+p)\frac{1}{p}\ln\frac{1}{1-p}\right)\right).$$

Proof. In the worst case every edge $f \in \hat{\delta}(e)$ that is before e in the ordering can be probed, and each of these probes has to fail for e to be safe. Thus

$$\mathbb{P}\left[e \text{ is safe} | e \in \hat{E}\right] \geq \mathbb{E}_{\hat{E}\backslash e, Y}\left[\prod_{f \in \hat{\delta}(e): Y_f < Y_e}(1 - p_f)\middle| e \in \hat{E}\right].$$

Now we take expectation on Y only, and using the fact that the variables Y_f are independent, we can write the latter expectation as

$$\mathbb{E}_{\hat{E}\backslash e}\left[\int_0^{\frac{1}{p_e}\ln\frac{1}{1-p_e}}\left(\prod_{f \in \hat{\delta}(e)}(\mathbb{P}[Y_f \leq y](1 - p_f) + \mathbb{P}[Y_f > y])\right)e^{-p_e \cdot y}dy\middle| e \in \hat{E}\right].$$

$$(4)$$

Observe that $\mathbb{P}[Y_f \leq y](1 - p_f) + \mathbb{P}[Y_f > y] = 1 - p_f \mathbb{P}[Y_f \leq y]$. When $y > \frac{1}{p_f}\ln\frac{1}{1-p_f}$, then $\mathbb{P}[Y_f \leq y] = 1$, and moreover, $\frac{1}{p_f}(1 - e^{-p_f \cdot y})$ is an increasing function of y. Thus we can upper-bound $\mathbb{P}[Y_f \leq y]$ by $\frac{1}{p_f}(1 - e^{-p_f \cdot y})$ for any

$y \in [0, \infty]$, and obtain that $1 - p_f \mathbb{P}[Y_f \leq y] \geq 1 - p_f \frac{1}{p_f}(1 - e^{-p_f \cdot y}) = e^{-p_f \cdot y}$.
Thus (4) can be lower bounded by

$$\mathbb{E}_{\hat{E} \backslash e}\left[\int_0^{\frac{1}{p_e} \ln \frac{1}{1-p_e}} e^{-\sum_{f \in \hat{\delta}(e)} p_f \cdot y - p_e \cdot y} dy \,\Big|\, e \in \hat{E}\right]$$

$$= \mathbb{E}_{\hat{E} \backslash e}\left[\frac{1}{\sum_{f \in \hat{\delta}(e)} p_f + p_e}\left(1 - e^{-\left(\sum_{f \in \hat{\delta}(e)} p_f + p_e\right)\frac{1}{p_e} \ln \frac{1}{1-p_e}}\right)\,\Big|\, e \in \hat{E}\right].$$

From the negative correlation and marginal distribution properties we know
that $\mathbb{E}_{\hat{E} \backslash e}\left[\hat{x}_f \,|\, e \in \hat{E}\right] \leq \mathbb{E}_{\hat{E} \backslash e}[\hat{x}_f] = x_f$ for every $f \in \delta(e)$, and therefore
$\mathbb{E}_{\hat{E} \backslash e}\left[\sum_{f \in \hat{\delta}(e)} p_f \,\Big|\, e \in \hat{E}\right] \leq \sum_{f \in \delta(e)} p_f x_f \leq 2$, where the last inequality follows
from the LP constraints. Consider function $f(x) := \frac{1}{x+p_e}\left(1 - e^{-(x+p_e)\frac{1}{p_e} \ln \frac{1}{1-p_e}}\right)$.
This function is decreasing and convex. From Jensen's inequality we know that
$\mathbb{E}[f(x)] \geq f(\mathbb{E}[x])$. Thus

$$\mathbb{E}_{\hat{E} \backslash e}\left[f\left(\sum_{f \in \hat{\delta}(e)} p_f\right)\,\Big|\, e \in \hat{E}\right] \geq f\left(\mathbb{E}_{\hat{E} \backslash e}\left[\sum_{f \in \hat{\delta}(e)} p_f \,\Big|\, e \in \hat{E}\right]\right)$$

$$\geq f(2) = \frac{1}{2 + p_e}\left(1 - e^{-(2+p_e)\frac{1}{p_e} \ln \frac{1}{1-p_e}}\right) = g(p_e). \qquad \square$$

From Lemma 2 and the marginal distribution property, the expected contribution of edge e to the profit of the solution is

$$w_e p_e \cdot \mathbb{P}\left[e \in \hat{E}\right] \cdot \mathbb{P}\left[e \text{ is safe}\,|\, e \in \hat{E}\right] \geq w_e p_e x_e \cdot g(p_e) \geq w_e p_e x_e \cdot g(1) = \frac{1}{3} w_e p_e x_e.$$

Therefore, our analysis implies a 3 approximation, matching the result in [4].
However, by playing with the probabilities appropriately we can do better.

Patching with Greedy. We next describe an improved approximation algorithm,
based on the patching of the above algorithm with a simple greedy one. Let $\delta \in (0, 1)$ be a parameter to be fixed later. We define E_{large} as the (*large*) edges with
$p_e \geq \delta$, and let E_{small} be the remaining (*small*) edges. Recall that LP_{bip} denotes
the optimal value of LP-BIP. Let also LP_{large} and LP_{small} be the fraction of
LP_{bip} due to large and small edges, respectively; i.e., $LP_{large} = \sum_{e \in E_{large}} w_e p_e x_e$
and $LP_{small} = LP_{bip} - LP_{large}$. Define $\gamma \in [0, 1]$ such that $\gamma LP_{bip} = LP_{large}$.
By refining the above analysis, we obtain the following result.

Lemma 3. *Algorithm 2.1 has expected approximation ratio $\frac{1}{3}\gamma + g(\delta)(1 - \gamma)$.*

Proof. The expected profit of the algorithm is at least:

$$\sum_{e \in E} w_e p_e x_e \cdot g(p_e) \geq \sum_{e \in E_{large}} w_e p_e x_e \cdot g(1) + \sum_{e \in E_{small}} w_e p_e x_e \cdot g(\delta)$$

$$= \frac{1}{3} LP_{large} + g(\delta) LP_{small} = \left(\frac{1}{3}\gamma + g(\delta)(1 - \gamma)\right) LP_{bip}. \qquad \square$$

Consider the following greedy algorithm. Compute a maximum weight matching M_{grd} in G with respect to edge weights $w_e p_e$, and probe the edges of M_{grd} in any order. Note that the timeout constraints are satisfied since we probe at most one edge incident to each node (and timeouts are strictly positive by definition and w.l.o.g.).

Lemma 4. *The greedy algorithm has expected approximation ratio $\delta\gamma$.*

Proof. It is sufficient to show that the expected profit of the obtained solution is at least $\delta \cdot LP_{large}$. Let $x = (x_e)_{e \in E}$ be the optimal solution to LP-BIP. Consider the solution $x' = (x'_e)_{e \in E}$ that is obtained from x by setting to zero all the variables corresponding to edges in E_{small}, and by multiplying all the remaining variables by δ. Since $p_e \geq \delta$ for all $e \in E_{large}$, x' is a feasible fractional solution to the following matching LP:

$$\max \sum_{e \in E} w_e p_e z_e \qquad\qquad \text{(LP-MATCH)}$$

$$\text{s.t.} \sum_{e \in \delta(u)} z_e \leq 1, \qquad\qquad \forall u \in V;$$

$$0 \leq z_e \leq 1, \qquad\qquad \forall e \in E. \qquad (5)$$

The value of x' in the above LP is $\delta \cdot LP_{large}$ by construction. Let LP_{match} be the optimal profit of LP-MATCH. Then $LP_{match} \geq \delta \cdot LP_{large}$. Given that the graph is bipartite, LP-MATCH defines the matching polyhedron, and we can find an integral optimal solution to it. But such a solution is exactly a maximum weight matching according to weights $w_e p_e$, i.e. $\sum_{e \in M_{grd}} w_e p_e = LP_{match}$. The claim follows since the expected profit of the greedy algorithm is precisely the weight of M_{grd}. $\qquad\square$

The overall algorithm, for a given δ, simply computes the value of γ, and runs the greedy algorithm if $\gamma\delta \geq \left(\frac{1}{3}\gamma + g(\delta)(1 - \gamma)\right)$, and Algorithm 2.1 otherwise[2].

The approximation factor is given by $\max\{\frac{\gamma}{3} + (1 - \gamma)g(\delta), \gamma\delta\}$, and the worst case is achieved when the two quantities are equal, i.e., for $\gamma = \frac{g(\delta)}{\delta + g(\delta) - \frac{1}{3}}$, yielding an approximation ratio of $\frac{\delta \cdot g(\delta)}{\delta + g(\delta) - \frac{1}{3}}$. Maximizing (numerically) the latter function in δ gives $\delta = 0.6022$, and the final 2.845-approximation ratio claimed in Theorem 1.

2.2 General Graphs

For general graphs, we consider the linear program LP-GEN which is obtained from LP-BIP by adding the following *blossom inequalities*:

$$\sum_{e \in E(W)} p_e x_e \leq \frac{|W| - 1}{2} \qquad\qquad \forall W \subseteq V, |W| \text{ odd}. \qquad (6)$$

---

[2] Note that we cannot run both algorithms, and take the best solution.

Here $E(W)$ is the subset of edges with both endpoints in W. We remark that, using standard tools from matching theory, we can solve LP-GEN in polynomial time despite its exponential number of constraints; see the book of Schrijver for details [11]. Also in this case $x_e = \mathbb{P}[OPT$ probes $e]$ is a feasible solution of LP-GEN, hence the analogue of Lemma 1 still holds.

Our Stochastic Matching algorithm for the case of a general graph $G = (V, E)$ works via a reduction to the bipartite case. First we solve LP-GEN; let $x = (x_e)_{e \in E}$ be the optimal fractional solution. Second we randomly split the nodes V into two sets A and B, with E_{AB} being the set of edges between them. On the bipartite graph $(A \cup B, E_{AB})$ we apply the algorithm for the bipartite case, but using the fractional solution $(x_e)_{e \in E_{AB}}$ induced by LP-GEN rather than solving LP-BIP. Note that $(x_e)_{e \in E_{AB}}$ is a feasible solution to LP-BIP for the bipartite graph $(A \cup B, E_{AB})$.

The analysis differs only in two points w.r.t. the one for the bipartite case. First, with $\hat{E}_{AB}$ being the subset of edges of E_{AB} that were rounded to 1, we have now that $\mathbb{P}\left[e \in \hat{E}_{AB}\right] = \mathbb{P}[e \in E_{AB}] \cdot \mathbb{P}\left[e \in \hat{E}_{AB}\middle| e \in E_{AB}\right] = \frac{1}{2}x_e$. Second, but for the same reason, using again the negative correlation and marginal distribution properties, we have

$$\mathbb{E}\left[\sum_{f \in \hat{\delta}(e)} p_f \middle| e \in \hat{E}_{AB}\right] \leq \sum_{f \in \delta(e)} p_f \mathbb{P}\left[f \in \hat{E}_{AB}\right] = \sum_{f \in \delta(e)} \frac{p_f x_f}{2} \leq \frac{2 - 2p_e x_e}{2} \leq 1.$$

Repeating the steps of the proof of Lemma 2 and including the above inequality we get the following.

Lemma 5. *For every edge e it holds that* $\mathbb{P}\left[e \text{ is safe}\middle| e \in \hat{E}_{AB}\right] \geq h(p_e)$, *where*

$$h(p) := \frac{1}{1 + p}\left(1 - \exp\left(-(1 + p)\frac{1}{p}\ln\frac{1}{1 - p}\right)\right).$$

Since $h(p_e) \geq h(1) = \frac{1}{2}$, we directly obtain a 4-approximation which matches the result in [4]. Similarly to the bipartite case, we can patch this result with the simple greedy algorithm (which is exactly the same in the general graph case). For a given parameter $\delta \in [0, 1]$, let us define γ analogously to the bipartite case. Similarly to the proof of Lemma 3, one obtains that the above algorithm has approximation factor $\frac{\gamma}{4} + \frac{1-\gamma}{2}h(\delta)$. Similarly to the proof of Lemma 4, the greedy algorithm has approximation ratio $\gamma\delta$ (here we exploit the blossom inequalities that guarantee the integrality of the matching polyhedron). We can conclude similarly that in the worst case $\gamma = \frac{h(\delta)}{2\delta + h(\delta) - 1/2}$, yielding an approximation ratio of $\frac{\delta \cdot h(\delta)}{2\delta + h(\delta) - 1/2}$. Maximizing (numerically) this function over δ gives, for $\delta = 0.5580$, the 3.709 approximation ratio claimed in Theorem 2.

3   Online Stochastic Matching with Timeouts

Let $G = (A \cup B, A \times B)$ be the input graph, with items A and buyer types B. We use the same notation for edge probabilities, edge profits, and timeouts as in Stochastic Matching. Following [4], we can assume w.l.o.g. that each buyer type is sampled uniformly with probability $1/n$. Consider the following linear program:

$$\max \sum_{a\in A, b\in B} w_{ab}p_{ab}x_{ab} \qquad\qquad \text{(LP-ONL)}$$

$$\text{s.t.} \sum_{b\in B} p_{ab}x_{ab} \le 1, \qquad\qquad \forall a \in A$$

$$\sum_{a\in A} p_{ab}x_{ab} \le 1, \qquad\qquad \forall b \in B$$

$$\sum_{a\in A} x_{ab} \le t_b, \qquad\qquad \forall b \in B$$

$$0 \le x_{ab} \le 1, \qquad\qquad \forall ab \subset E.$$

The above LP models a bipartite Stochastic Matching instance where one side of the bipartition contains exactly one buyer per buyer type. In contrast, in the online case several buyers of the same buyer type (or none at all) can arrive, and the optimal strategy can allow many buyers of the same type to probe edges. Still, that is not a problem since the following lemma from [4] allows us just to look at the graph of buyer types and not at the actual realized buyers.

Lemma 6. *([4], Lemmas 9 and 11) Let $\mathbb{E}[OPT]$ be the expected profit of the optimal online algorithm for the problem. Let LP_{onl} be the optimal value of LP-ONL. It holds that $\mathbb{E}[OPT] \le LP_{onl}$.*

We will devise an algorithm whose expected outcome is at least $\frac{1}{4.07} \cdot LP_{onl}$, and then Theorem 3 follows from Lemma 6.

The Algorithm. We initially solve LP-ONL and let $(x_{ab})_{ab\in A\times B}$ be the optimal fractional solution. Then buyers arrive. When a buyer of type b is sampled, then 1) if a buyer of the same type b was already sampled before we simply discard her, do nothing, and wait for another buyer to arrive, 2) if it is the first buyer of type b, then we execute the following *subroutine for buyers*. Since we take action only when the first buyer of type b comes, we shall denote such a buyer simply by b, as it will not cause any confusion.

Subroutine for Buyers. Let us consider the step of the online algorithm in which the first buyer of type b arrived, if any. Let A_b be the items that are still available when b arrives. Our subroutine will probe a subset of at most t_b edges ab, $a \in A_b$. Consider the vector $(x_{ab})_{a\in A_b}$. Observe that it satisfies the constraints $\sum_{a\in A_b} p_{ab}x_{ab} \le 1$ and $\sum_{a\in A_b} x_{ab} \le t_b$. Again using GKPS, we round this vector in order to get $(\hat{x}_{ab})_{a\in A_b}$ with $\hat{x}_{ab} \in \{0,1\}$, and satisfying the marginal distribution, degree preservation, and negative correlation properties[3]. Let $\hat{A}_b$ be the

---

[3] Actually in this case we have a bipartite graph where one side has only one vertex, and here GKPS reduces to Srinivasan's rounding procedure for level-sets [12].

set of items a such that $\hat{x}_{ab} = 1$. For each ab, $a \in \hat{A}_b$, we independently draw a random variable Y_{ab} with distribution: $\mathbb{P}\left[Y_{ab} < y\right] = \frac{1}{p_{ab}}\left(1 - \exp\left(-p_{ab} \cdot y\right)\right)$ for $y \in \left[0, \frac{1}{p_{ab}} \ln \frac{1}{1-p_{ab}}\right]$. Let $Y = (Y_{ab})_{a \in \hat{A}_b}$.

Next we consider items of $\hat{A}_b$ in increasing order of Y_{ab}. Let $\alpha_{ab} \in [\frac{1}{2}, 1]$ be a *dumping factor* that we will define later. With probability α_{ab} we probe edge ab and as usual we stop the process (of probing edges incident to b) if ab is present. Otherwise (with probability $1 - \alpha_{ab}$) we *simulate* the probe of ab, meaning that with probability p_{ab} we stop the process anyway — like if edge ab were probed and turned out to be present. Note that we do not get any profit from the latter simulation since we do not really probe ab.

Dumping Factors. It remains to define the dumping factors. For a given edge ab, let

$$\beta_{ab} := \mathbb{E}_{\hat{A}_b \setminus a, Y}\left[\prod_{a' \in A_b : Y_{a'b} < Y_{ab}} (1 - p_{a'b}) \middle| a \in \hat{A}_b\right].$$

Using the inequality $\sum_{a \in A_b} p_{ab} x_{ab} \leq 1$, by repeating the analysis from Section 2 we can show that

$$\beta_{ab} \geq h(p_{ab}) = \frac{1}{1 + p_{ab}}\left(1 - \exp\left(-(1 + p_{ab})\frac{1}{p_{ab}} \ln \frac{1}{1 - p_{ab}}\right)\right) \geq \frac{1}{2}.$$

Let us assume for the sake of simplicity that we are able to compute β_{ab} exactly. We set $\alpha_{ab} = \frac{1}{2\beta_{ab}}$. Note that α_{ab} is well defined since $\beta_{ab} \in [1/2, 1]$.

Analysis. Let us denote by $\mathcal{A}_b$ the event that at least one buyer of type b arrives. The probability that an edge ab is probed can be expressed as:

$$\mathbb{P}\left[\mathcal{A}_b\right] \cdot \mathbb{P}\left[\text{no } b' \text{ takes } a \text{ before } b \middle| \mathcal{A}_b\right] \cdot \mathbb{P}\left[b \text{ probes } a \middle| \mathcal{A}_b \wedge a \text{ is not yet taken}\right].$$

The probability that b arrives is $\mathbb{P}\left[\mathcal{A}_b\right] = 1 - \left(1 - \frac{1}{n}\right)^n \geq 1 - \frac{1}{e}$. We shall show first that

$$\mathbb{P}\left[b \text{ probes } a \middle| \mathcal{A}_b \wedge a \text{ is not yet taken}\right]$$

is exactly $\frac{1}{2}x_{ab}$, and later we shall show that $\mathbb{P}\left[\text{no } b' \text{ takes } a \text{ before } b \middle| \mathcal{A}_b\right]$ is at least $\frac{1}{1 + \frac{1}{2}\left(1 - \frac{1}{e}\right)}$. This will yield that the probability that ab is probed is at least

$$\left(1 - \frac{1}{e}\right)\frac{1}{1 + \frac{1}{2}\left(1 - \frac{1}{e}\right)} \cdot \frac{1}{2}x_{ab} = \frac{e - 1}{3e - 1}x_{ab} > \frac{1}{4.16}x_{ab}.$$

Consider the probability that some edge $a'b$ appearing before ab in the random order *blocks* edge ab, meaning that ab is not probed because of $a'b$. Observe that each such $a'b$ is indeed considered for probing in the online model, and the probability that $a'b$ blocks ab is therefore $\alpha_{a'b}p_{a'b} + (1 - \alpha_{a'b})p_{a'b} = p_{a'b}$. We can conclude that the probability that ab is not blocked is exactly β_{ab}.

Due to the dumping factor α_{ab}, the probability that we actually probe edge $ab \in \hat{A}_b$ is exactly $\alpha_{ab} \cdot \beta_{ab} = \frac{1}{2}$. Recall that $\mathbb{P}\left[a \in \hat{A}_b\right] = x_{ab}$ by the marginal distribution property. Altogether

$$\mathbb{P}\left[b \text{ probes } a \mid \mathcal{A}_b \wedge a \text{ is not yet taken}\right] = \frac{1}{2}x_{ab}. \tag{7}$$

Next let us condition on the event that buyer b arrived, and let us lower bound the probability that ab is not blocked on the a's side in such a step, i.e., that no other buyer has taken a already. The buyers, who are first occurrences of their type, arrive uniformly at random. Therefore, we can analyze the process of their arrivals as if it was constructed by the following procedure: every buyer b' is given an independent random variable $Y_{b'}$ distributed exponentially on $[0, \infty]$, i.e., $\mathbb{P}[Y_{b'} < y] = 1 - e^y$; buyers arrive in increasing order of their variables $Y_{b'}$. Once buyer b' arrives, it probes edge ab' with probability (exactly) $\alpha_{ab'}\beta_{ab'}x_{ab'} = \frac{1}{2}x_{ab'}$ — these probabilities are independent among different buyers. Thus, conditioning on the fact that b arrives, we obtain the following expression for the probability that a is safe at the moment when b arrives:

$$\mathbb{P}\left[\text{no } b' \text{ takes } a \text{ before } b \mid \mathcal{A}_b\right]$$

$$\geq \mathbb{E}\left[\prod_{b' \in B \setminus b : Y_{b'} < Y_b} (1 - \mathbb{P}\left[\mathcal{A}_{b'} \mid \mathcal{A}_b\right] \mathbb{P}\left[b' \text{ probes } ab' \mid \mathcal{A}_{b'}\right] p_{ab'}) \,\Big|\, \mathcal{A}_b\right]$$

$$= \int_0^\infty \prod_{b' \in B \setminus b} (1 - \mathbb{P}\left[\mathcal{A}_{b'} \mid \mathcal{A}_b\right] \cdot \mathbb{P}\left[Y_{b'} < y \mid \mathcal{A}_{b'}\right] \cdot \mathbb{P}\left[b' \text{ probes } ab' \mid \mathcal{A}_{b'}\right] p_{ab'}) e^{-y}dy.$$

Now let us upper-bound each of the probability factors in the above product. First of all $\mathbb{P}\left[\mathcal{A}_{b'} \mid \mathcal{A}_b\right] = 1 - \left(1 - \frac{1}{n}\right)^{n-1} \leq 1 - \frac{1}{e}$. Second, $\mathbb{P}\left[Y_{b'} < y \mid \mathcal{A}_{b'}\right] = 1 - e^{-y}$ just by definition[4]. Third, from (7) we have that $\mathbb{P}\left[b' \text{ probes } ab' \mid \mathcal{A}_{b'}\right] = \frac{x_{ab'}}{2}$.

Thus the above integral can be lower bounded by

$$\int_0^\infty \prod_{b' \in B \setminus b} \left(1 - \left(1 - \frac{1}{e}\right)\left(1 - e^{-y}\right) \cdot \frac{1}{2}x_{ab'} \cdot p_{ab'}\right) e^{-y}dy$$

$$\geq \int_0^\infty \prod_{b' \in B \setminus b} \exp\left(-\left(1 - \frac{1}{e}\right)\frac{1}{2}x_{ab'} \cdot p_{ab'} \cdot y\right) e^{-y}dy$$

$$= \frac{1}{1 + \left(1 - \frac{1}{e}\right)\frac{1}{2}\left(\sum_{b' \in B \setminus b} p_{ab'} \cdot x_{ab'}\right)} \geq \frac{1}{1 + \frac{1}{2}\left(1 - \frac{1}{e}\right)} = \frac{2e}{3e - 1}.$$

Above in the first inequality we used the fact that $1 - c(1 - e^{-y}) \geq e^{-cy}$ for $c \in [0, 1]$ and any $y \in \mathbb{R}$: here $c = \left(1 - \frac{1}{e}\right)\frac{1}{2}x_{ab'} \cdot p_{ab'}$. In the first equality we used $\int_0^\infty e^{-ax}dx = \frac{1}{a}$. In the last inequality we used the LP constraint $\sum_{b' \in B \setminus b} p_{ab'} \cdot x_{ab'} \leq 1$.

---

[4] The $\mathcal{A}_{b'}$ event in the condition simply indicates that $Y_{b'}$ was drawn.

Altogether, as anticipated earlier,

$$\mathbb{P}\left[ab \text{ is probed}\right] \geq \left(1 - \frac{1}{e}\right) \frac{x_{ab}}{2} \cdot \frac{2e}{3e - 1} = x_{ab} \cdot \frac{e - 1}{3e - 1} > \frac{1}{4.16} \cdot x_{ab}.$$

Technical Details. Recall that we assumed that we are able to compute the quantities β_{ab}, hence the desired dumping factors α_{ab}. Indeed, for our goals it is sufficient to estimate them with large enough probability and with sufficiently good accuracy. This can be done by simulating the underlying random process a polynomial number of times. This way the above probability can be lower bounded by $(\frac{e-1}{3e-1} + \varepsilon)x_e$ for an arbitrarily small constant $\varepsilon > 0$. In particular, by choosing a small enough ε the factor 4.16 is still guaranteed. The approximation factor can be further reduced to 4.07 via the technique based on small and big probabilities that we introduced before. The omitted technical details will be given in the full version of the paper (see also [2]). Theorem 3 follows.

References

1. Adamczyk, M.: Improved analysis of the greedy algorithm for stochastic matching. Information Processing Letters 111, 731–737 (2011)
2. Adamczyk, M., Grandoni, F., Mukherjee, J.: Improved approximation algorithms for stochastic matching. CoRR, abs/1505.01439 (2015)
3. Adamczyk, M., Sviridenko, M., Ward, J.: Submodular stochastic probing on matroids. In: STACS 2014, pp. 29–40 (2014)
4. Bansal, N., Gupta, A., Li, J., Mestre, J., Nagarajan, V., Rudra, A.: When LP is the cure for your matching woes: Improved bounds for stochastic matchings. Algorithmica 63(4), 733–762 (2012)
5. Chen, N., Immorlica, N., Karlin, A.R., Mahdian, M., Rudra, A.: Approximating matches made in heaven. In: Albers, S., Marchetti-Spaccamela, A., Matias, Y., Nikoletseas, S., Thomas, W. (eds.) ICALP 2009, Part I. LNCS, vol. 5555, pp. 266–278. Springer, Heidelberg (2009)
6. Dean, B.C., Goemans, M.X., Vondrák, J.: Approximating the stochastic knapsack problem: The benefit of adaptivity. Math. Oper. Res. 33(4), 945–964 (2008)
7. Gandhi, R., Khuller, S., Parthasarathy, S., Srinivasan, A.: Dependent rounding and its applications to approximation algorithms. Journal of the ACM 53(3), 324–360 (2006)
8. Li, J.: Decision making under uncertainty. PhD thesis, University of Maryland (2011)
9. Li, J.: Private communication (2015)
10. Ma, W.: Improvements and generalizations of stochastic knapsack and multi-armed bandit approximation algorithms: Extended abstract. In: SODA 2014, pp. 1154–1163 (2014)
11. Schrijver, A.: Combinatorial Optimization - Polyhedra and Efficiency. Springer (2003)
12. Srinivasan, A.: Distributions on level-sets with applications to approximation algorithms. In: FOCS 2001, pp. 588–597 (2001)

Sorting and Permuting
without Bank Conflicts on GPUs

Peyman Afshani[1,*] and Nodari Sitchinava[2]

[1] MADALGO, Aarhus University, Denmark
[2] University of Hawaii – Manoa, HI, USA

Abstract. In this paper, we look at the complexity of designing algorithms without any bank conflicts in the shared memory of Graphical Processing Units (GPUs). Given input of size n, w processors and w memory banks, we study three fundamental problems: sorting, permuting and w-way partitioning (defined as sorting an input containing exactly n/w copies of every integer in $[w]$).

We solve sorting in optimal $O(\frac{n}{w} \log n)$ time. When $n \geq w^2$, we solve the partitioning problem optimally in $O(n/w)$ time. We also present a general solution for the partitioning problem which takes $O(\frac{n}{w} \log^3_{n/w} w)$ time. Finally, we solve the permutation problem using a randomized algorithm in $O(\frac{n}{w} \log \log \log_{n/w} n)$ time. Our results show evidence that when working with banked memory architectures, there is a separation between these problems and the permutation and partitioning problems are not as easy as simple parallel scanning.

1 Introduction

Graphics Processing Units (GPUs) over the past decade have been transformed from special-purpose graphics rendering co-processors, to a powerful platform for general purpose computations, with runtimes rivaling best implementations on many-core CPUs. With high memory throughput, hundreds of physical cores and fast context switching between thousands of threads, they became very popular among computationally intensive applications. Instead of citing a tiny subset of such papers, we refer the reader to `gpgpu.org` website [11], which lists over 300 research papers on this topic.

Yet, most of these results are experimental and the theory community seems to shy away from designing and analyzing algorithms on GPUs. In part, this is probably due to the lack of a simple theoretical model of computation for GPUs. This has started to change recently, with introduction of several theoretical models for algorithm analysis on GPUs.

A Brief Overview of GPU Architecture. A modern GPU contains hundreds of physical cores. To implement such a large number of cores, a GPU is designed hierarchically. It consists of a number of *streaming multiprocessors*

---

* Work supported in part by the Danish National Research Foundation grant DNRF84 through Center for Massive Data Algorithmics (MADALGO).

N. Bansal and I. Finocchi (Eds.): ESA 2015, LNCS 9294, pp. 13–24, 2015.
DOI: 10.1007/978-3-662-48350-3_2

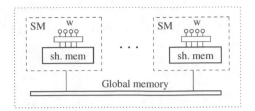

Fig. 1. A schematic of GPU architecture

(SMs) and a *global memory* shared by all SMs. Each SM consists of a number of cores (for concreteness, let us parameterize it by w) and a *shared memory* of limited size which is shared by all the cores within the SM but is inaccessible by other SMs. With computational power of hundreds of cores, latency of accessing memory becomes non-negligible and GPUs take several approaches to mitigate the problem.

First, they support massive hyper-threading, i.e., multiple logical threads may run on each physical core with light context switching between the threads. Thus, when a thread stalls on a memory request, the other threads can continue running on the same core. To schedule all these threads efficiently, groups of w threads, called *warps*, are scheduled to run on w physical cores simultaneously in *single instructions, multiple data (SIMD)* [9] fashion.

Second, there are limitations on how data is accessed in memory. Accesses to global memory are most efficient if they are *coalesced*. Essentially, it means that w threads of a warp should access contiguous w addresses of global memory. On the other hand, shared memory is partitioned into w *memory banks* and each memory bank may service at most one thread of a warp in each time step. If more than one thread of a warp requests access to the same memory bank, a *bank conflict* occurs and multiple accesses to the same memory bank are sequentialized. Thus, for optimal utilization of processors, it is recommended to design algorithms that perform coalesced accesses to global memory and incur no bank conflicts in shared memory [19].

Designing GPU Algorithms. Several papers [13, 17, 18, 22] present theoretical models that incorporate the concept of coalesced accesses to global memory into the performance analysis of GPU algorithms. In essence, all of them introduce a complexity metric that counts the number of blocks transferred between the global and internal memory (shared memory or registers), similar to the I/O-complexity metric of sequential and parallel external memory and cache-oblivious models on CPUs [2, 3, 5, 10]. The models vary in what other features of GPUs they incorporate and, consequently, in the number of parameters introduced into the model.

Once the data is in shared memory, the usual approach is to implement standard parallel algorithms in the PRAM model [14] or interconnection networks [16]. For example, sorting data in shared memory, is usually implemented using sorting networks, e.g. Batcher's odd-even mergesort [4] or bitonic merge-sort [4]. Even though these sorting networks are not asymptotically optimal, they provide good empirical runtimes because they exhibit small constant factors, they fit well in the SIMD execution flow within a warp, and for small inputs asymptotic optimality is irrelevant. On very small inputs (e.g. on w items – one per memory bank) they also cause no bank conflicts.

However, as the input sizes for shared memory algorithms grow, bank conflicts start to affect the running time.

The first paper that studies bank conflicts on GPUs is by Dotsenko et al. [8]. The authors view shared memory as a two-dimensional matrix with w rows, where each row represents a separate memory bank. Any one-dimensional array $A[0..n-1]$ will be laid out in this matrix in column-major order in $\lceil n/w \rceil$ contiguous columns. Note, that *strided* parallel access to data, that is each thread t accessing array entries $A[wi + t]$ for integer $0 \leq i < \lceil n/w \rceil$, does not incur bank conflicts because each thread accesses a single row. The authors also observed that the *contiguous* parallel access to data, that is each thread scanning a contiguous section of $\lceil n/w \rceil$ items of the array, also incurs no bank conflicts if $\lceil n/w \rceil$ is co-prime with w. Thus, with some extra padding, contiguous parallel access to data can also be implemented without any bank conflicts.

Instead of adding padding to ensure that $\lceil n/w \rceil$ is co-prime with w, another solution to bank-conflict-free contiguous parallel access is to convert the matrix from column-major layout to row-major layout and perform strided access. This conversion is equivalent to in-place transposition of the matrix. Catanzaro et al. [6] study this problem and present an elegant bank-conflict-free parallel algorithm that runs in $\Theta(n/w)$ time, which is optimal.

Sitchinava and Weichert [22] present a strong correlation between bank conflicts and the runtime for some problems. Based on the matrix view of shared memory, they developed a sorting network that incurred no bank conflicts. They show that although compared to Batcher's sorting networks their solution incurs extra $\Theta(\log n)$ factor in parallel time and work, it performs better in practice because it incurs no bank conflicts.

Nakano [18] presents a formal definition of a parallel model with the matrix view of shared memory, which is also extended to model memory access latency hiding via hyper-threading.[1] He calls his model *Discrete Memory Model (DMM)* and studies the problem of *offline permutation*, which we define in detail later.

The DMM model is probably the simplest abstract model that captures the important aspects of designing bank-conflict-free algorithms for GPUs. In this paper we will work in this model. However, to simplify the exposition, we will assume that each memory access incurs no latency (i.e. takes a unit time) and we have exactly w processors. This simplification still captures the key algorithmic challenges of designing bank-conflict-free algorithms without the added complexity of modeling hyper-threading. We summarize the key features of the model below, and for more details refer the reader to [18].

---

[1] Nakano's DMM exposition actually swapped the rows and columns and viewed memory banks as columns of the matrix and the data laid out in row-major order.

Model of Computation. Data of size n is laid out in memory as a matrix $\mathcal{M}$ with dimensions $w \times m$, where $m = \lceil n/w \rceil$. As in the PRAM model, w processors proceed synchronously in discrete time steps, and in each time step perform access to data (a processor may skip accessing data in some time step). Every processor can access any item within the matrix. However, an algorithm must ensure that in each time step at most one processor accesses data within a particular row.[2] Performing computation on a constant number of items takes unit time and, therefore, can be completed within a single time step. The parallel time complexity (or simply *time*) is the number of time steps required to complete the task. The work complexity (or simply *work*) is the product of w and the time complexity of an algorithm.[3]

Although the DMM model allows each processor to access any memory bank (i.e. any row of the matrix), to simplify the exposition, it is helpful to think of each processor fixed to a single row of the matrix and the transfer of information between the processors being performed via "message passing", where at each step, processor i may send a message (constant words of information) to another processor j (e.g., asking to read or write a memory location within row j). Next, the processor j can respond by sending constant words of information back to row i. Crucially and to avoid bank conflicts, we demand that at each parallel computation step, at most one message is received by each row; we call this the "Conflict Avoidance Condition" or CAC.

Note that this view of interprocessor communication via message passing is only done for the ease of exposition, and algorithms can be implemented in the DMM model (and in practice) by processor i directly reading or writing the contents of the target memory location from the appropriate location in row j. Finally, CAC is equivalent to each memory bank being accessed by at most one processor in each access request made by a warp.

Problems of Interest. Using the above model, we study complexity of developing bank conflict free algorithms for the following fundamental problems:

• *Sorting*: The matrix $\mathcal{M}$ is populated with items from a totally ordered universe. The goal is to have $\mathcal{M}$ sorted in row-major order.[4]

• *Partition*: The matrix $\mathcal{M}$ is populated with labeled items. The labels form a permutation that contains m copies of every integer in $[w]$ and an item with label i needs to be sent to row i.

• *Permutation*: The matrix $\mathcal{M}$ is populated with labeled items. The labels form a permutation of tuples $[w] \times [m]$. And an item with label (i, j) needs to be sent to the j-th memory location in row i.

---

[2] This is analogous to how EREW PRAM model requires the algorithms to be designed so that in each time step at most one processor accesses any memory address.

[3] Work complexity is easily computed from time complexity and number of processors, therefore, we don't mention it explicitly in our algorithms. However, we mention it here because it is a useful metric for efficiency, when compared to the runtime of the optimal sequential algorithms.

[4] The final layout within the matrix (row-major or column-major order) is of little relevance, because the conversion between the two layouts can be implemented efficiently in time and work required to simply read the input [6].

While the sorting problem is natural, we need to mention a few remarks regarding the permutation and the partition problems. Often these two problems are equivalent and thus there is little motivation to separate them. However rather surprisingly, it turns out that in our case these problems are in fact very different. Nonetheless, the permutation problem can be considered an "abstract" sorting problem where all the comparisons have been resolved and the goal is to merely send each item to its correct location. The partition problem is more practical and it appears in scenarios where the goal is to split an input into many subproblems. For example, consider a multi-way quicksort algorithm with pivots, $p_0 = -\infty, p_1, \cdots, p_{w-1}, p_w = \infty$. In this case, row i would like to send all the items that are greater than p_{j-1} but less than p_j to row j but row i would not know the position of the items in the final sorted matrix. In other words, in this case, for each element in row i, we only know the destination row rather than the destination row and the rank.

Prior Work. Sorting and permutation are among the most fundamental algorithmic problems. The solution to the permutation problem is trivial in the random access memory models – $\mathcal{O}(n)$ work is required in both RAM and PRAM models of computation – while sorting is often more difficult (the $\Omega(n \log n)$ comparison-based lower bound is a very classical result). However, this picture changes in other models. For example, in both the sequential and parallel external memory models [2, 3], which model hierarchical memories of modern CPU processors, existing lower bounds show that permutation is as hard as sorting (for most realistic parameters of the models) [2, 12].

In the context of GPU algorithms, matrix transposition was studied as a special case of the permutation problem by Catanzaro et al. [6] and they showed that one can transpose a matrix in-place without bank conflicts in $\mathcal{O}(n)$ work. While they didn't explicitly mentioned the DMM model, their analysis holds trivially in the DMM model with unit latency. Nakano [18] studied the problem of performing arbitrary permutations in the DMM model *offline*, where the permutation is known in advance and we are allowed to pre-compute some information before running the algorithm. The time to perform the precomputation is not counted toward the complexity of performing the permutation. Offline permutation is useful if the permutation is fixed for a particular application and we can encode the precomputation result in the program description. Common examples of fixed permutations include matrix transposition, bit-reversal permutations, and FFT permutations. Nakano showed that any offline permutation can be implemented in linear work. The required pre-computation in Nakano's algorithm is coloring of a regular bipartite graph, which seems very difficult to adapt to the *online* permutation problem.

As mentioned earlier, Sitchinava and Weichert [22] presented the first algorithm for sorting $w \times w$ matrix which incurs no bank conflicts. They use Shearsort [21], which repeatedly sorts columns of the matrix in alternating order and rows in increasing order. After $\Theta(\log w)$ repetitions, the matrix is sorted in column-major order. Rows of the matrix can be sorted without bank conflicts. And since matrix transposition can be implemented without bank conflicts, the

Fig. 2. Consider one row that is being converted to column-major layout. The elements that will be put in the same column are bundled together. It is easily seen that each row will create at most one dirty column.

columns can also be sorted without bank conflicts via transposition and sorting of the rows. The resulting runtime is $\Theta(t(w) \log w)$, where $t(w)$ is the time it takes to sort an array of w items using a single thread.

Our Contributions. In this paper we present the following results.

- *Sorting*: We present an algorithm that runs in $O(m \log(mw))$ time, which is optimal in the context of comparison based algorithms.
- *Partition*: We present an optimal solution that runs in $O(m)$ time when $w \leq m$. We generalize this to a solution that runs in $O(m \log_m^3 w)$ time.
- *Permutation*: We present a randomized algorithm that runs in expected $O(m \log \log \log_m w)$ time. Even though this is a rather technical solution (and thus of theoretical interest), it strongly hints at a possible separation between the partition and permutation problems in the DMM model.

2   Sorting and Partitioning

In this section we improve the sorting algorithm of Sitchinava and Weichert [22] by removing a $\Theta(\log w)$ factor. We begin with our base case, a "short and wide" $w \times m$ matrix $\mathcal{M}$ where $w \leq \sqrt{m}$. The algorithm repeats the following twice and then sorts the rows in ascending order.

1. Sort rows in alternating order (odd rows ascending, even rows descending).
2. Convert the matrix from row-major to column-major layout (e.g., the first w elements of the first row form the first column, the next w elements form the second column and so on).
3. Sort rows in ascending order.
4. Convert the matrix from column-major to row-major layout (e.g., the first m/w columns will form the first row).

Lemma 1. *The above algorithm sorts the matrix correctly in $O(m \log m)$ time.*

Proof. We would like to use the *0-1 principle* [15] but since we are not working with a sorting network, we simply reprove the principle. Observe that our algorithm is essentially oblivious to the values in the matrix and only takes into account the relative order of them. Pick a parameter i, $1 \leq i \leq mw$, that we call the *marking value*. Based on the above observation, we attach a mark of "0" to any element that has rank less than i and "1" to the remaining elements. These marks are symbolic and thus invisible to the algorithm. We say a column (or row) is *dirty* if it contains both elements with mark "0" and "1". After the first execution of steps 1 and 2, the number of dirty columns is reduced to at most w (Fig. 2). After the next execution of steps 3 and 4, the number of dirty rows is reduced to two. Crucially, the two dirty rows are adjacent. After one more execution of steps 1 and 2, there would be two dirty columns, however, since the rows were ordered in alternating order in step 1, one of the dirty rows will have "0"s toward the top and "1"s toward the bottom, while the other will have them in reverse order. This means, after step 3, there will be only one dirty column and only one dirty row after step 4. The final sorting round will sort the elements in the dirty row and thus the elements will be correctly sorted by their mark.

As the algorithm is oblivious to marks, the matrix will be sorted by marks regardless of our choice of the marking value, meaning, the matrix will be correctly sorted. Conversions between column-major and row-major (and vice versa) can be performed in $O(m)$ time using [6] while respecting CAC. Sorting rows is the main runtime bottleneck and the only part that needs more than $O(m)$ time. □

Corollary 1. *The partition problem can be solved in $O(m)$ time if $w \leq \sqrt{m}$.*

Proof. Observe that for the partition problem, we can "sort" each row in $O(m)$ time (e.g., using radix sort).   □

Using the above as a base case, we can now sort a square matrix efficiently.

Theorem 1. *If $w = m$, then the sorting problem can be solved in $O(m \log m)$ time. The partition problem can be solved in $O(m)$ time.*

Proof. Once again, we use the idea of the 0-1 principle and assume the elements are marked with "0" and "1" bits, invisible to the algorithm. To sort an $m \times m$ matrix, partition the rows into groups of $\sqrt{m}$ adjacent rows. Call each group a *super-row*. Sort each super-row using Lemma 1. Each super-row has at most one dirty row so there are at most $\sqrt{m}$ dirty rows in total. We now sort the columns of the matrix in ascending order by transposing the matrix, sorting the rows, and then transposing it again. This will place all the dirty rows in adjacent rows. We sort each super-row in alternating order (the first super-row ascending, the next descending and so on). After this, there will be at most two dirty rows left, sorted in alternating order. We sort the columns in ascending order, which will reduce the number of dirty rows to one. A final sorting of the rows will have the matrix in the correct sorted order.

The running time is clearly $O(m \log m)$. In the partition problem, we use radix sort to sort each row and thus the running time will be $O(m)$.   □

For non-square matrices, the situation is not so straightforward and thus more interesting. First, we observe that by using the $O(\log w)$ time EREW PRAM algorithm [7] to sort w items using w processors, the above algorithm can be generalized to non-square matrices:

Theorem 2. *A $w \times m$ matrix $\mathcal{M}$, $w \geq m$, can be sorted in $O(m \log w)$ time.*

Proof. As before assume "0" and "1" marks have been placed on the elements in $\mathcal{M}$. First, we sort the rows. Then we sort the columns of $\mathcal{M}$ using the EREW PRAM algorithm [7]. Next, we convert the matrix from column-major to row-major layout, meaning, in the first column, the first m elements will form the first row, the next m elements the next row and so on. This will leave at most m dirty rows. Once again, we sort the columns, which will place the dirty rows in adjacent rows. We sort each $m \times m$ sub-matrix in alternating order, using Theorem 1, which will reduce the number of dirty rows to two. The rest of the argument is very similar to the one presented for Theorem 1.   □

The next result might not sound very strong, but it will be very useful in the next section. It also reveals some differences between the partition and sorting problems. Note that the result is optimal as long as w is polynomial in m.

Lemma 2. *The partition problem on a $w \times m$ matrix $\mathcal{M}$, $w \geq m > 2\sqrt{\log w}$, can be solved in $O(m \log_m^3 w)$ time. The same bound holds for sorting $O(\log n)$-bit integers.*

Proof. We use the idea of 0-1 principle, combined with radix sort, which allows us to sort every row in $O(m)$ time. The algorithm uses the following steps.

Balancing. Balancing has $\lceil \log_m w \rceil$ rounds; for simplicity, we assume $\log_m w$ is an integer. In the zeroth round, we create w/m sub-matrices of dimensions $m \times m$ by partitioning the rows into groups of m adjacent rows and then sort each sub-matrix in column-major order (i.e., the elements marked "0" will be to the left and the elements marked "1" to the right). At the beginning of each subsequent round i, we have w/m^i sub-matrices with dimensions $m^i \times m$. We partition the sub-matrices into groups of m adjacent sub-matrices; from every group, we create m^i square $m \times m$ matrices: we pick the j-th row of every sub-matrix in the group, for $1 \leq j \leq m^i$, to create the j-th square matrix in the group. We sort each $m \times m$ matrix in column-major order. Now, each group will form an $m^{i+1} \times m$ sub-matrix for the next round. Observe that balancing takes $O(m \log_m w)$ time in total.

Convert and Divide (C&D). With a moment of thought, it can be proven that after *balancing*, each row will have between $\frac{w_0}{w} - \log_m w$ and $\frac{w_0}{w} + \log_m w$ (resp. $\frac{w_1}{w} - \log_m w$ and $\frac{w_1}{w} + \log_m w$) elements marked "0" (resp. "1") where w_0 (resp. w_1) is the total number of elements marked "0" (resp. "1"). This means, that there are at most $d = 2 \log_m w$ dirty columns. We convert the matrix from column-major to row-major layout, which leaves us with $\frac{w}{md}$ adjacent dirty rows (each column is placed in $\frac{w}{m}$ rows after conversion). Now, we divide $\mathcal{M}$ into md smaller matrices of dimension $\frac{w}{md} \times m$. Each matrix will be a new subproblem.

The Algorithm. The algorithm repeatedly applies *balancing* and *C&D* steps: after the i-th execution of these two steps, we have $(md)^i$ subproblems where each subproblem is a $\frac{w}{(md)^i} \times m$ matrix, and in the next round, *balancing* and *C&D* operate locally within each subproblem (i.e., they work on $\frac{w}{(md)^i} \times m$ matrices). Note that before the divide step, each subproblem has at most $\frac{w}{(md)^i}$ adjacent dirty rows. After the divide step, these dirty rows will be sent to at most two different sub-problems, meaning, at the depth i of the recursion, all the dirty rows will be clustered into at most 2^i groups of adjacent rows, each containing at most $\frac{w}{(md)^i}$ rows. Since, $m > 2\sqrt{\log_m w}$, after $2\log_m w$ steps of recursion, the subproblems will have at most m rows and thus each subproblem can be sorted in $O(m)$ time, leaving at most one dirty row per sub-problem. Thus, we will end up with only $2^{2\log_m w}$ dirty rows in the whole matrix $\mathcal{M}$.

We now sort the columns of the matrix $\mathcal{M}$: we fit each column of $\mathcal{M}$ in a $\frac{w}{m} \times m$ submatrix and recurse on the submatrices using the above algorithm, then convert each submatrix back to a column. We will end up with a matrix with $2^{2\log_m w}$ dirty rows but crucially, these rows will be adjacent. Equally important is the observation that $2^{2\log_m w} \le m/2$ since $m > 2\sqrt{\log w}$. To sort these few dirty rows, we decompose the matrix into square matrices of $m \times m$, and sort each square matrix. Next, we shift this decomposition by $m/2$ rows and sort them again. In one of these decompositions, an $m \times m$ matrix will fully contain all the dirty rows, meaning, we will end up with a fully sorted matrix. If $f(w, m)$ is the running time on a $w \times m$ matrix, then we have the recursion $f(w, m) = f(w/m, m) + O(m \log_m^2 w)$ which gives our claimed bound.   □

3   A Randomized Algorithm for Permutation

In this section we present an improved algorithm for the permutation problem that beats our best algorithm for partitioning for when the matrix is "tall and narrow" (i.e., $w \gg m$). We note that the algorithm is only of theoretical interest as it is a bit technical and it uses randomization. However, at least from this theoretical point of view, it hints that perhaps in the GPU model, the permutation problem and the partitioning problem are different and require different techniques to solve.

Remember that we have a matrix $\mathcal{M}$ which contains a set of elements with labels. The labels form a permutation of $[w] \times [m]$, and an element with label (i, j) needs to be placed at the j-th memory location of row i. From now on, we use the term "label" to both refer to an element and its label.

To gain our speed up, we use two crucial properties: first, we use randomization and second we use the existence of the second index, j, in the labels.

Intuition and Summary. Our algorithm works as follows. It starts with a preprocessing phase where each row picks a random word and these random words are used to shuffle the labels into a more "uniform" distribution. Then, the main body of the algorithm begins. One row picks a random hash function and communicates it to the rest of the rows. Based on this hash function, all the rows compute a common coloring of the labels using m colors $1, \ldots, m$ such that the following property holds: for each index i, $1 \le i \le w$, and among the labels $(i, 1), (i, 2), \ldots, (i, m)$, there is exactly one label with color j, for every $1 \le j \le m$. After establishing such a coloring, in the upcoming j-th step, each row will send one label of color j to its correct destination. Our coloring guarantees that CAC is not violated. At the end of the m-th step, a significant majority of the labels are placed at their correct position and thus we are left only with a few "leftover" labels. This process is repeated until the number of "leftover" labels is a $1/\log_m^3 w$ fraction of the original input. At this point, we use Lemma 2 which we show can be done in $O(m)$ time since very few labels are left. Next, we address the technical details that arise.

Preprocessing Phase. For simplicity, we assume w is divisible by m. Each row picks a random number r between 1 and m and shifts its labels by that amount, i.e., places a label at position j into position $j + r$ (mod m). Next, we group the rows into matrices of size $m \times m$ which we transpose (i.e., the first m rows create the first matrix and so on).

The Main Body and Coloring. We use a result by Pagh and Pagh [20].

Theorem 3. *[20] Consider the universe $[w]$ and a set $S \subset [w]$ of size m. There exists an algorithm in the word RAM model that (independent of S) can select a family H of hash functions from $[w]$ to $[v]$ in $O(\log m(\log v)^{O(1)})$ time and using $O(\log m + \log \log w)$ bits, such that:*

- *H is m-wise independent on S with probability $1 - m^{-c}$ for a constant c*
- *Any member of H can be represented by a data structure that uses $O(m \log v)$ bits and the hash values can be computed in constant time. The construction time of the structure is $O(m)$.*

The first row builds H then picks a hash function h from the family. This function, which is represented as a data structure, is communicated to the rest of the rows in $O(\log w + m)$ time as follows: assume the data structure consumes $S_m = O(m)$ space. In the first S_m steps, the first row sends the j-th word in the data structure to row j. In the subsequent $\log(w)$ rounds, the j-th word is communicated to any row j' with $j' \equiv j$ (mod S_m), using a simple broadcasting strategy that doubles the number of rows that have received the j-th word. This boils down the problem of transferring the data structure to the problem of distributing S_m random words between S_m rows which can be easily done in S_m steps while respecting CAC.

Coloring. A label (i, j) is assigned color k where $j \equiv h(i) + k$ (mod m).

Communication. Each row computes the color of its labels. Consider a row i containing some labels. The communication phase has αm steps, where α is a large enough constant. During step k, $1 \le k \le m$, the row i picks a label of color k. If no such label exists, then the row does nothing so let's assume a label (i_k, j_k) has assigned color k. The row i sends this label to the destination row i_k. After performing these initial m steps, the algorithm repeats these m steps $\alpha - 1$ times for a total of αm steps. We claim the communication step respects CAC: assume otherwise that another label (i'_k, j'_k) is sent to row i_k during step k; clearly, we must have $i'_k = i_k$ but this contradicts our coloring since it implies $j'_k \equiv h(i_k) + k \equiv j_k$ (mod m) and thus $j'_k = j_k$. Note that the communication phase takes $O(m)$ time as α is a constant.

Synchronization. Each row computes the number of labels that still need to be sent to their destination, and in $O(\log w + m)$ time, these numbers are summed and broadcast to all the rows. We repeat the main body of the algorithm as long as this number is larger than $\frac{wm}{\log_m^3 w}$.

Finishing. We present the details in the full version of the paper [1] but roughly speaking, we do the following: first, we show that we can pack the remaining elements into a $w \times O(\frac{m}{\log_m^3 w})$ matrix in $O(m)$ time (this is not trivial as the matrix can have rows with $\Omega(m)$ labels left). Then, we use Lemma 2 to sort the matrix in $O(m)$ time. Finishing off the sorted matrix turns out to be relatively easy.

Correctness and Running Time. Correctness is trivial as at the end all rows receive their labels and the algorithm is shown to respect CAC. Thus, the main issue is the running time. The *finishing* and *preprocessing* steps clearly takes $O(m)$ time. The running time of each repetition of *coloring*, *communication* and *synchronization* is $O(\log w + m)$ and thus to bound the total running time we simply need to bound how many times the synchronization steps are executed. To do this, we need the following lemma (proof is in the full paper [1]).

Lemma 3. *Consider a row ℓ containing a set S_ℓ of k labels, $k \leq m$. Assume the hash function h is m-wise independent on S_ℓ. For a large enough constant α, let C_{heavy} be the set of colors that appear more than α times in ℓ, and S_{heavy} be the set of labels assigned colors from C_{heavy}. Then $\mathbb{E}[|S_{heavy}|] = m(k/2m)^\alpha$ and the probability that $|S_{heavy}| = \Omega(m(k/2m)^{\alpha/10})$ is at most $2^{-\sqrt{m}(k/2m)^{\alpha/10}}$.*

Consider a row ℓ and let k_i be the number of labels in the row after the i-th iteration of the *synchronization* ($k_0 = m$). We call ℓ a *lucky* row if the hash function is m-wise independent on ℓ during all the i iterations. Assume ℓ is lucky. During the *communication* step, we process all the labels except those in the set S_{heavy}. By the above lemma, it follows that with high probability, we will have $O(m(k_i/2m)^{a/10})$ labels left for the next iteration on ℓ, meaning, with high probability, $k_{i+1} = O(m(k_i/2m)^{a/10})$.

Observe that if $m \geq \log^{O(1)} w$ (for an appropriate constant in the exponent) and $k_i > m/\log_m^3 w$, then $\sqrt{m}\left(\frac{k_i}{2m}\right)^{\alpha/10} > 3\log w$ which means the probability that Lemma 3 "fails" (i.e., the "high probability" fails) is at most $1/w^3$. Thus, with probability at least $1 - 1/w$, Lemma 3 never fails for any lucky row and thus each lucky row will have at most

$$O\left(m\left(\frac{k_0}{2m}\right)^{(\alpha/10)^i}\right)$$

labels left after the i-th iteration. By Theorem 3, the expected number of unlucky rows is at most n/m^c at each iteration which means after $i = O(\log \log \log_m w)$ iterations, there will be $O(\frac{mw}{\log_m^3 w} + \frac{mwi}{m^c}) = O(mw\log_m^3 w)$ labels left. So we proved that the algorithm repeats the *synchronization* step at most $O(\log \log \log_m n)$ times, giving us the following theorem.

Theorem 4. *The permutation problem on an $w \times m$ matrix can be solved in $O(m \log \log \log_m w)$ expected time, assuming $m = \Omega(\log^{O(1)} w)$. Furthermore, the total number of random words used by the algorithm is $w + O(m \log \log \log_m w)$.*

References

1. Afshani, P., Sitchinava, N.: Sorting and permuting without bank conflicts on GPUs. CoRR abs/1507.01391 (2015), http://arxiv.org/abs/1507.01391
2. Aggarwal, A., Vitter, J.S.: The input/output complexity of sorting and related problems. Commun. ACM 31, 1116–1127 (1988)
3. Arge, L., Goodrich, M.T., Nelson, M.J., Sitchinava, N.: Fundamental parallel algorithms for private-cache chip multiprocessors. In: 20th ACM Symposium on Parallelism in Algorithms and Architectures (SPAA), pp. 197–206 (2008)
4. Batcher, K.E.: Sorting networks and their applications. In: AFIPS Spring Joint Computer Conference, pp. 307–314
5. Blelloch, G.E., Chowdhury, R.A., Gibbons, P.B., Ramachandran, V., Chen, S., Kozuch, M.: Provably good multicore cache performance for divide-and-conquer algorithms. In: 19th ACM-SIAM Symp. on Discrete Algorithms, pp. 501–510 (2008)
6. Catanzaro, B., Keller, A., Garland, M.: A decomposition for in-place matrix transposition. In: 19th ACM SIGPLAN Principles and Practices of Parallel Programming (PPoPP), pp. 193–206 (2014)
7. Cole, R.: Parallel merge sort. In: 27th IEEE Symposium on Foundations of Computer Science. pp. 511–516 (1986)
8. Dotsenko, Y., Govindaraju, N.K., Sloan, P.P., Boyd, C., Manfedelli, J.: Fast Scan Algorithms on Graphics Processors. In: 22nd International Conference on Supercomputing, pp. 205–213 (2008)
9. Flynn, M.: Some computer organizations and their effectiveness. IEEE Transactions on Computers C 21(9), 948–960 (1972)
10. Frigo, M., Leiserson, C.E., Prokop, H., Ramachandran, S.: Cache-oblivious algorithms. In: 40th IEEE Symp. on Foundations of Comp. Sci., pp. 285–298 (1999)
11. GPGPU.org: Research papers on gpgpu.org, http://gpgpu.org/tag/papers
12. Greiner, G.: Sparse Matrix Computations and their I/O Complexity. Dissertation, Technische Universität München, München (2012)
13. Haque, S., Maza, M., Xie, N.: A many-core machine model for designing algorithms with minimum parallelism overheads. In: High Performance Computing Symposium (2013)
14. JáJá, J.: An Introduction to Parallel Algorithms. Addison Wesley (1992)
15. Knuth, D.E.: The Art of Computer Programming, Volume III: Sorting and Searching. Addison-Wesley (1973)
16. Leighton, F.T.: Introduction to Parallel Algorithms and Architectures: Arrays, Trees, and Hypercubes. Morgan-Kaufmann, San Mateo (1991)
17. Ma, L., Agrawal, K., Chamberlain, R.D.: A memory access model for highly-threaded many-core architectures. Future Generation Computer Systems 30, 202–215 (2014)
18. Nakano, K.: Simple memory machine models for gpus. In: 26th IEEE International Parallel and Distributed Processing Symposium Workshops & PhD Forum (IPDPSW), pp. 794–803 (2012)
19. NVIDIA Corp.: CUDA C Best Practices Guide. Version 7.0 (March 2015)
20. Pagh, A., Pagh, R.: Uniform hashing in constant time and optimal space. SIAM Journal on Computing 38(1), 85–96 (2008)
21. Sen, S., Scherson, I.D., Shamir, A.: Shear Sort: A True Two-Dimensional Sorting Techniques for VLSI Networks. In: International Conference on Parallel Processing, pp. 903–908 (1986)
22. Sitchinava, N., Weichert, V.: Provably efficient GPU algorithms. CoRR abs/1306.5076 (2013), http://arxiv.org/abs/1306.5076

Approximating Minimum-Area Rectangular and Convex Containers for Packing Convex Polygons*

Helmut Alt[1], Mark de Berg[2], and Christian Knauer[3]

[1] Institute of Computer Science, Freie Universität Berlin,
Takustr. 9, 14195 Berlin, Germany
alt@mi.fu-berlin.de
[2] Department of Computing Science, TU Eindhoven, P.O. Box 513,
5600 MB Eindhoven, The Netherlands
mdberg@win.tue.nl
[3] Universität Bayreuth, Institut für Angewandte Informatik, AG Algorithmen und
Datenstrukturen (AI VI), 95440 Bayreuth, Germany
christian.knauer@uni-bayreuth.de

Abstract. We investigate the problem of finding a minimum-area container for the disjoint packing of a set of convex polygons by translations. In particular, we consider axis-parallel rectangles or arbitrary convex sets as containers. For both optimization problems which are NP-hard we develop efficient constant factor approximation algorithms.

1 Introduction

Algorithms for efficiently packing objects into containers have important applications: In two dimensions, the problem occurs, e.g., in the context of cutting out a given set of patterns from a given large piece of material minimizing waste, typically in apparel fabrication or sheet metal processing. In three dimensions the problem occurs naturally in minimizing storage space or container space for transportation.

The problem has numerous variants. The most basic one is the decision problem whether a set of given objects can be packed into a given container. In sheet metal and apparel processing, mostly the problem of *strip packing* [12,13,5] occurs, i.e., a given set of objects needs to be packed inside a strip of a given fixed width using as short a piece of the strip as possible. Here, we will consider the problem of minimizing the area of the container.

Moreover, the shape of objects to be packed is significant, e.g., in two dimensions arbitrary rectangles, axis-parallel rectangles, convex polygons, or simple polygons. Furthermore, the allowable transformations for placing the objects play an important role: Is it allowed to rotate them, i.e., apply *rigid motions*, or

---

* Research was partially carried out at the International INRIA-McGill-Victoria Workshop on Problems in Computational Geometry, Barbados 2015. Research by Mark de Berg was also supported by the Netherlands' Organisation for Scientific Research (NWO) under project no. 024.002.003.

© Springer-Verlag Berlin Heidelberg 2015
N. Bansal and I. Finocchi (Eds.): ESA 2015, LNCS 9294, pp. 25–34, 2015.
DOI: 10.1007/978-3-662-48350-3_3

are only *translations* allowed? We will consider packing by translations, which is an important variant in practice. For example, in apparel production there are usually patterns of weaving or texture on the material so that the position where a piece should be cut out cannot be rotated arbitrarily.

Already the most simple versions of the problem are NP-hard, e.g., packing a given set of axis-parallel rectangles into a given axis-parallel container by translation, as can easily be seen by reduction from PARTITION (see, e.g., [4]). So, only for a constant number of objects polynomial-time algorithms are known, see [1,2,14,3,8,9,10,7]. Therefore, numerous approximation algorithms and heuristics have been investigated, mostly in the operations research and combinatorial optimization communities and mostly on axis-parallel rectangles under translation for a given container or strip. For a survey see [11].

Not much work has been done so far on approximation algorithms for the problem of finding *minimum area containers*, either convex ones or axis-parallel boxes. For objects that are axis-parallel, rectangles under translation, and arbitrary rectangles or convex polygons under rigid motion, approximation algorithms can be developed from the known ones for strip packing [13]. In fact, in [15] it is shown that packing axis-parallel rectangles under translation with an approximation factor of 2, and packing convex polygons under rigid motions with an approximation factor of 4 is possible efficiently.

Thus the status of the packing problem in the translational case can be summarized as follows: all known algorithms that produce optimal results or have provable approximation ratios fall into one of two categories: they either pack only a very special kind of objects like line-segments or axis-parallel boxes, or they pack a constant number of polygons. Since the problem for a non-constant number of objects is NP-hard, this leaves the following question: can the packing problem for n objects that are not axis-parallel be efficiently approximated? We answer this question affirmatively for packing a set P of convex polygons into a minimum-area rectangular container. Using this result, we also show how to approximate the minimum-area convex container for P. Our algorithms run in $O(n \log n)$ time.

We remark that the restriction to translational packing makes the problem harder. Indeed, if we allow rotations then we can compute a minimum-area oriented bounding box for each input polygon, rotate those boxes so that they become axis-parallel, and then pack the boxes using a known algorithm for axis-parallel boxes. Since the minimum-area oriented bounding box has area at most twice the area of the polygon itself, this can give a good approximation ratio. When we are not allowed to rotate the polygons this approach fails, since the area of the axis-parallel bounding box of a polygon can be arbitrarily much larger than the area of the polygon itself. Our result shows that we can still get a constant-factor approximation in the translational case. The approximation factors we obtain are fairly large, but to the best of our knowledge this is the first proof that these NP-hard optimization problems can be approximated at all.

2   The Algorithm

Let $P := \{p_1, \ldots, p_k\}$ be a set of convex polygons with n vertices in total. We call an axis-aligned box into which we can pack all polygons (without rotating them) a *container* for P. Our goal is to find a container for P of minimum area. Let b_{opt} be a minimum-area container for P, and let OPT be its area. Below we present an algorithm that finds a container of area at most $17.45 \cdot \text{OPT}$.

Define the *height* of a polygon p, denoted height(p), to be the difference between its maximum and minimum y-coordinates, and let $h_{\max} := \max_{p \in P} \text{height}(p)$. Furthermore, define the *width* of a polygon p, denoted width(p), to be the difference between its maximum and minimum x-coordinates, and let $w_{\max} := \max_{p \in P} \text{width}(p)$. We partition P into *height classes* using a parameter α, with $0 < \alpha < 1$ which is determined later to give the optimal approximation factor. More precisely, P is partitioned into subsets $P_0, P_1, \ldots$ according to the height: Polygons with height between $h_{\max}$ and $\alpha h_{\max}$ go into P_0, polygons with height between $\alpha h_{\max}$ and $\alpha^2 h_{\max}$ go into P_1, and so on. More precisely, P_i contains all polygons $p \in P$ such that $h_{i+1} < \text{height}(p) \leqslant h_i$, where $h_i = \alpha^i h_{\max}$. Our general strategy is now as follows:

1. Pack each height class P_i separately into a container B_i of height h_i.
2. Replace each nonempty container B_i by a collection of axis-aligned *mini-containers* that are not too wide. Pack all mini-containers into a single container B.

Next we explain each of these steps in more detail.

Step 1: Packing Polygons from One Height Class. Consider the height class P_i, which contains all polygons whose height lies in the range $(\alpha h_i, h_i]$. Let $\sigma := [0, \infty) \times [0, h_i]$ be a semi-infinite strip of height h_i. We place the polygons from P_i into σ in a greedy manner, as follows: For a polygon p, let $s(p)$ be the segment connecting the lowest vertex of p to the highest vertex of p. (If p has horizontal edges, then $s(p)$ connects the bottom-left vertex to the top-right vertex.) We call $s(p)$ the *spine* of p. We sort the polygons in P_i according to the slopes of their spines and then we place them one by one into σ, where each polygon is pushed as far to the left as possible—that is, until it hits a previously placed polygon or the left edge of σ—while keeping its lowest vertex on the bottom edge of σ. Fig. 1 illustrates the process. After we have placed all polygons, we close the container B_i; the right edge of B_i is defined by the vertical line through the rightmost vertex of any of the placed polygons.

Lemma 1. *The area of the container B_i computed for P_i satisfies*

$$\text{area}(B_i) \leqslant 2/\alpha \cdot \sum_{p \in P_i} \text{area}(p) + 2h_i \cdot \max_{p \in P_i} \text{width}(p).$$

Proof. Recall that we push each polygon (in order of the slope of the spines) to the left until either it hits the left edge of σ or until it hits a previously placed polygon. We define a polygon p to be *relevant* if

Fig. 1. Example of a packing produced by our algorithm for a single height class

(i) it is the last polygon that hits the left edge of σ, or

(ii) there is a previously placed relevant polygon r' such that p is the last polygon that hits r'.

By definition, there is a relevant polygon touching the left edge of B_i. There must also be a relevant polygon touching the right edge of B_i. Indeed, assume otherwise and let p be the relevant polygon placed last. Then either the right edge of B_i is determined by the rightmost vertex of some irrelevant polygon p' placed before p or there are polygons placed after p. In the first case, p' would block p from reaching the left side of B_i by some chain of relevant polygons. In the second case, none of the polygons placed after p can touch p, because otherwise one of them would become relevant, as well. Let p' be the one placed directly after p. On its left, p' must touch some (irrelevant) polygon p'' placed before p. But then, the union of p' and p'' would separate p from the left edge of B_i. Hence, p cannot be relevant, a contradiction.

We conclude that in our analysis we can safely restrict our attention to the relevant polygons. Now let P_i^* be the set of relevant polygons in P_i. We will prove that

$$\text{area}(B_i) \leqslant 2/\alpha \cdot \sum_{p \in P_i^*} \text{area}(p) + 2h_i \cdot \max_{p \in P_i^*} \text{width}(p), \tag{1}$$

which obviously implies the lemma.

Let $p_1, p_2, \ldots, p_t$ be the polygons in P_i^*, ordered according to the slope of their spines, and let $S := \{s(p_1), \ldots, s(p_t)\}$ be the set of spines of the relevant polygons. Imagine placing each spine with its lower endpoint in the origin. Let $B(S)$ be the box of height h_i and minimum width containing all the spines placed in this manner, see Figure 2. Note that

$$\text{area}(B(S)) \leqslant 2h_i \cdot \max_{p \in P_i^*} \text{width}(p). \tag{2}$$

Partition $B(S)$ into pieces by extending the spines until they hit the boundary of $B(S)$. Let $\Delta_0, \Delta_1, \ldots, \Delta_t$ be the resulting set of pieces, where the numbering is in clockwise order around the origin.

To bound the area of B_i, we first partition B_i into regions (which are either triangles, quadrilaterals, or 5-gons) by extending the spines of the polygons in P_i^*

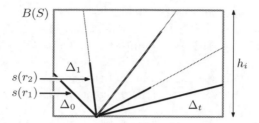

Fig. 2. Illustration for the proof of Lemma 1

until they hit the boundary of B_i. Let $Q_0, Q_1, \ldots, Q_t$ be these regions ordered from left to right. Thus Q_j lies between spines $s(p_j)$ and $s(p_{j+1})$, except for the first and last region which lie before the first spine and after the last spine, respectively. We claim (and will prove below) that

$$\text{area}(Q_j) \leqslant 2/\alpha \cdot \left(\text{area}(p_j^r) + \text{area}(p_{j+1}^l)\right) + \text{area}(\Delta_j). \qquad (3)$$

where p^l and p^r are the right and left hand side, respectively, into which a polygon p is split by its spine, and, with a slight abuse of notation (since p_0 and p_{t+1} do not exist) we define $\text{area}(p_0) = \text{area}(p_{t+1}) = 0$. From the claim we derive

$$
\begin{aligned}
\text{area}(B_i) &= \sum_j \text{area}(Q_j) \\
&\leqslant \sum_j \left(2/\alpha \cdot (\text{area}(p_j^r) + \text{area}(p_{j+1}^l)) + \text{area}(\Delta_j)\right) \qquad (4) \\
&\leqslant 2/\alpha \cdot \sum_{p \in P_i^*} \text{area}(p) + \text{area}(B(S)),
\end{aligned}
$$

which, using (2), proves the lemma.

It remains to prove claim (3). To simplify the presentation we assume $0 < j < t$; it is easily verified that a similar argument applies when $j = 0$ and when $j = t$. To get a bound on the area of Q_j, let s_p be the segment parallel to $s(p_j)$ that splits Q_j and passes through the point p where p_j touches p_{j+1}. Let q be the lower endpoint of s_p. There are two cases, as illustrated in Fig. 3.

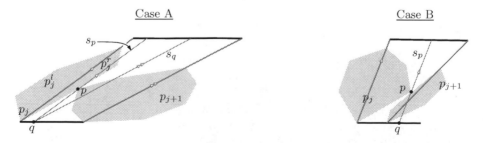

Fig. 3. The two cases in the proof of Lemma 1

Case A: The point q lies to the left of the lower endpoint of $s(p_{j+1})$. Now connect q to the opposite boundary of B_i with a segment, s_q, which is parallel to $s(p_{j+1})$. The segments s_p and s_q partition Q_j into three pieces Q_j^1, Q_j^2, and Q_j^3. First consider the left piece, Q_j^1, which is bounded from the left by $s(p_j)$ and from the right by s_p. Let A be the parallelogram between the lines spanned by $s(p_j)$, s_p, and the upper and lower edges of B_i. Because height$(p_j) \geqslant \alpha h_i$, the area of the triangle T defined by $s(p_j)$ and p is at least area$(A) \cdot \alpha/2$ which is at least area$(Q_j^1) \cdot \alpha/2$. (Q_j^1 can be smaller than A, because unlike A it can be bounded by the right or left edge of B_i.) Since also the area of T is at most area(p_j^r), we have

$$\text{area}(Q_j^1) \leqslant 2/\alpha \cdot \text{area}(p_j^r). \tag{5}$$

A similar argument shows that the area of Q_j^3, the rightmost piece, is at most $2/\alpha \cdot$ area(p_{j+1}^l). The middle piece, Q_j^2, is bounded by $s(p_j)$ and $s(p_{j+1})$, so area$(Q_j^2) \leqslant$ area(Δ_j). This finishes the claim for Case A.

Case B: the point q lies to the right of or coincides with the lower endpoint of $s(p_{j+1})$. As before we can bound the area between $s(p_j)$ and s_p by $2/\alpha \cdot$ area(p_j^r). (This area may not be completely contained in Q_j, but this does not matter.) The only part of Q_j we have not accounted for is the part between s_p and the line through $s(p_{j+1})$ above the intersection point of s_p and $s(p_{j+1})$. The area of this part is bounded by the area of Δ_j. This finishes the proof of claim (3) for Case B and, hence, finishes the proof of the lemma.   □

Step 2: Generating and Packing Mini-Containers. Step 1 results in a collection of containers B_i of various lengths l_i, each containing all polygons from the height class P_i. We replace each B_i by mini-containers of equal lengths as follows. Recall that $w_{\max}$ is the maximum width of any polygon in P. First, partition B_i into boxes of width $cw_{\max}$—we will determine a suitable value for c later—and height h_i, except for the last box which may have width smaller than $w_{\max}$. Now assign each polygon $p \in P_i$ to the box b containing its leftmost point. (If the leftmost point lies on the boundary between two boxes we assign it to the righthand box.) We now generate a mini-container from each box b by extending b to the right until its width is exactly $(c + 1)w_{\max}$. Note that (the extended) b contains all polygons assigned to b. This results in a collection $\overline{R}_i$ of at most $l_i/(cw_{max}) + 1$ mini-containers each having width exactly $(c+1)w_{\max}$. Since the height of B_i and of each mini-container is h_i, we have

$$\sum_{b \in \overline{R}_i} \text{area}(b) \leqslant (1 + \frac{1}{c}) \cdot \text{area}(B_i) + (c + 1)w_{\max} h_i. \tag{6}$$

Let $\overline{R} := \bigcup \overline{R}_i$ be the collection of all mini-containers obtained in this manner. Packing these mini-containers can trivially be done without any loss of area: since all mini-containers in $\overline{R}$ have the same width we can simply stack them on top of each other to obtain our final container B.

We can now state our result about packing polygons.

Theorem 1. *Let P be a set of polygons in the plane with n vertices in total. We can pack P in $O(n \log n)$ time into an axis-aligned rectangular container B such that $\mathrm{area}(B) \leqslant 17.45 \cdot \mathrm{OPT}$, where OPT is the minimum area of any axis-aligned rectangular container for P.*

Proof. Let B be the container computed by our algorithm. Observe that $\mathrm{OPT} \geqslant \sum_{p \in P} \mathrm{area}(p)$ and $\mathrm{OPT} \geqslant w_{\max} h_{\max}$. We have

$$\mathrm{area}(B) = \sum_{b \in \overline{R}} \mathrm{area}(b)$$

$$\leqslant \sum_i \left\{ (1 + \tfrac{1}{c}) \cdot \mathrm{area}(B_i) + (c+1) w_{\max} h_i \right\} \qquad \text{by (6)}$$

$$\leqslant (1 + \tfrac{1}{c}) \sum_i \left\{ 2/\alpha \cdot \sum_{p \in P_i} \mathrm{area}(p) + 2 h_i \cdot \max_{p \in P_i} \mathrm{width}(p) \right\}$$
$$+ \tfrac{1}{1-\alpha} \cdot (c+1) w_{max} h_{max}$$

$$\text{(by Lemma 1 and because } h_i = \alpha^i h_{\max})$$

$$\leqslant (1 + \tfrac{1}{c}) \cdot \left(2/\alpha \cdot \sum_{p \in P} \mathrm{area}(p) + 2/(1-\alpha) \cdot w_{\max} h_{\max} \right)$$
$$+ \tfrac{1}{1-\alpha}(c+1) \cdot \mathrm{OPT}$$

$$\leqslant \left((1 + \tfrac{1}{c})(\tfrac{2}{\alpha} + \tfrac{2}{1-\alpha}) + \tfrac{c+1}{1-\alpha} \right) \cdot \mathrm{OPT}.$$

$$(7)$$

The term before OPT simplifies to

$$f(c, \alpha) := \left(1 + \frac{1}{c} \right) \cdot \frac{2 + c\alpha}{\alpha - \alpha^2} \qquad (8)$$

In order to minimize the approximation factor, we determine the optimal values for c and α by setting the partial derivatives to zero: $\frac{\partial f}{\partial c} = 0$ and $\frac{\partial f}{\partial \alpha} = 0$.

The first equation yields the identity $\alpha = 2/c^2$. Using this in the second equation gives that c is obtained by $c^3 - 4c - 2 = 0$ which has the solution $c = 2.214..$ which gives $\alpha = 0.407...$ and the approximation factor $f(c, \alpha) = 17.449....$

In order to get the desired runtime, we first observe that partitioning the polygons in P into height classes takes linear time and sorting the ones in each class by the slopes of their spines takes a total time of $O(n \log n)$.

In order to pack each polygon p efficiently in Step 1, we maintain a balanced binary search tree. It contains, ordered by y-coordinate, those vertices of the set P' of polygons already packed which are visible from the right, see the dashed lines in Figure 4. Thus, in order to find the point where p hits first a polygon of P' when moved to the left, we search for the y-coordinates of the vertices of p visible from the left finding the corresponding candidate edges in P' in $O(\log n)$ time (dotted arrows). Vice versa, for all vertices of P' in the y-range of p we find the corresponding edges of p (dashed arrows) in time $O(\log n)$ assuming that the vertices of p are available sorted clockwise in an array. Observe that this operation is done altogether at most once per vertex, since afterwards it is not visible from the right any more and removed from the data structure. Instead, the vertices of p visible from the right must be inserted. Consequently, each vertex is inserted into and deleted from the data structure at most once, taking $O(n \log n)$ time in total.

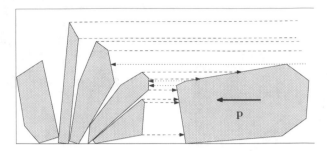

Fig. 4. The data structure for inserting polygons

□

3 Convex Containers

Next, we consider the problem of finding an arbitrary convex container of minimal area for a set of convex polygons . In other words, we want to pack the input polygons by translation such that the area of the convex hull of their union is minimized. We will give an approximation algorithm based on the one for minimum enclosing boxes from before.

The idea is to find a suitable orientation $\phi^* \in S^1$, determine the approximately optimal bounding box B with that orientation using the algorithm from Section 2 and then return B as the approximate solution.

Given a set P of polygons, we choose ϕ^* to be that orientation minimizing $h_{max}(\phi)w_{max}(\phi)$, where $h_{max}(\phi)$ is the maximal extent of any polygon in direction perpendicular to ϕ and $w_{max}(\phi)$ the extent in direction ϕ. The orientation ϕ^* can be determined in $O(n \log n)$ time by using rotating calipers simultaneously around all the polygons from P.

To see this, observe that the functions $h_{max}(\phi)$ and $w_{max}(\phi)$ are composed piecewise of functions of the form $a \sin(\phi + b)$ for some constants $a, b \in \mathbb{R}$. The number of these functions equals the number of pairs of antipodal vertices in all polygons, so it is $O(n)$. Two of these functions can intersect at most once, so by an algorithm of Hershberger [6] their upper envelope can be constructed in $O(n \log n)$ time and consists of $O(n\alpha(n))$ pieces of the functions above. The upper envelope of $h_{max}(\phi)w_{max}(\phi)$ is a piecewise trigonometric function as well where the pieces result from merging the pieces of the upper envelopes of $h_{max}(\phi)$ and $w_{max}(\phi)$. Therefore, it also consists of $O(n\alpha(n))$ pieces and it can be constructed and its minimum determined in $O(n \log n)$ time.

To get an estimate on the quality of the solution, let us consider the optimal solution which is some convex polygon C_{opt}. Consider a bounding box B_{opt} of C_{opt} that has a side parallel to a longest line segment $\overline{pq}$ contained in C_{opt}. We claim that

$$\text{area}(B_{opt}) \leq 2 \cdot \text{area}(C_{opt}). \tag{9}$$

To see this, observe that B_{opt} is partitioned into two rectangles B_1 and B_2 (one of which can be empty) by $\overline{pq}$. The triangle in B_1 formed by $\overline{pq}$ and the opposite tangent point has half the area of B_1. The same holds for the corresponding triangle in B_2. Since the union of both triangles is contained in C_{opt}, claim (9) follows.

Now, let h_{opt} and w_{opt} be the height and width of B_{opt}, respectively. Then by the choice of ϕ^*,

$$h_{max}(\phi^*)w_{max}(\phi^*) \leqslant h_{opt}w_{opt} = \text{area}(B_{opt}) \leqslant 2 \cdot \text{OPT}. \tag{10}$$

Observe that now OPT denotes the area of the smallest enclosing convex container rather than bounding box. Therefore, using (10) the derivation in (7) can be replaced by

$$\text{area}(B) = \sum_{b \in \overline{R}} \text{area}(b)$$
$$\leqslant \sum_i \left\{ (1 + \tfrac{1}{c}) \cdot \text{area}(B_i) + (c+1)w_{max}h_i \right\}$$
$$\leqslant (1 + \tfrac{1}{c}) \sum_i \left\{ 2/\alpha \cdot \sum_{p \in P_i} \text{area}(p) + 2h_i \cdot \max_{p \in P_i} \text{width}(p) \right\}$$
$$+ 1/(1-\alpha) \cdot (c+1)w_{max}h_{max} \tag{11}$$
$$\leqslant (1 + \tfrac{1}{c}) \cdot \left(2/\alpha \cdot \sum_{p \in P} \text{area}(p) + 2/(1-\alpha) \cdot w_{max}h_{max} \right)$$
$$+ 2/(1-\alpha) \cdot (c+1) \cdot \text{OPT}$$
$$\leqslant \tfrac{c+1}{c} \left(\tfrac{2}{\alpha} + \tfrac{4}{1-\alpha} + \tfrac{2c}{1-\alpha} \right) \cdot \text{OPT}.$$

Let $f(c, \alpha)$ denote the factor before OPT in the last line, then the partial derivatives are:

$$\tfrac{\partial f}{\partial c} = \tfrac{-1}{c^2} \cdot \left(\tfrac{2}{\alpha} + \tfrac{4+2c}{1-\alpha} \right) + \tfrac{c+1}{c} \cdot \tfrac{2}{1-\alpha}$$
$$\tfrac{\partial f}{\partial \alpha} = \tfrac{c+1}{c} \cdot \left(-\tfrac{2}{\alpha^2} + \tfrac{4+2c}{(1-\alpha)^2} \right)$$

As easily can be verified, both expressions evaluate to zero for $c = 2$ and $\alpha = 1/3$, so we obtain the optimal approximation factor $f(2, 1/3) = 27$, if we choose these values in our algorithm. We obtain:

Theorem 2. *Let P be a set of convex polygons in the plane with n vertices in total. We can pack P in $O(n \log n)$ time into a convex polygon B such that* $\text{area}(B) \leqslant 27 \cdot \text{OPT}$, *where* OPT *is the minimum area of any convex container for P.*

Acknowledgement. We would like to thank an anonymous referee for a useful hint by which we could improve the approximation factors.

References

1. Ahn, H.-K., Alt, H., Bae, S.W., Park, D.: Bundling Three Convex Polygons to Minimize Area or Perimeter. In: Dehne, F., Solis-Oba, R., Sack, J.-R. (eds.) WADS 2013. LNCS, vol. 8037, pp. 13–24. Springer, Heidelberg (2013)

2. Ahn, H.-K., Cheong, O.: Aligning two convex figures to minimize area or perimeter. Algorithmica 62(1-2), 464–479 (2012)
3. Alt, H., Hurtado, F.: Packing convex polygons into rectangular boxes. In: JCDCG, pp. 67–80 (2000)
4. Fowler, R.J., Paterson, M., Tanimoto, S.L.: Optimal packing and covering in the plane are np-complete. Inf. Process. Lett. 12(3), 133–137 (1981)
5. Harren, R., Jansen, K., Prädel, L., van Stee, R.: A $(5/3 + \epsilon)$-approximation for strip packing. Comput. Geom. 47(2), 248–267 (2014)
6. Hershberger, J.: Finding the upper envelope of n line segments in o(n log n) time. Inf. Process. Lett. 33(4), 169–174 (1989)
7. Lee, H.-C., Woo, T.C.: Determining in linear time the minimum area convex hull of two polygons. IIE Transactions 20(4), 338–345 (1988)
8. Milenkovic, V.: Translational polygon containment and minimal enclosure using linear programming based restriction. In: STOC, pp. 109–118 (1996)
9. Milenkovic, V.: Multiple translational containment. part ii: Exact algorithms. Algorithmica 19(1/2), 183–218 (1997)
10. Milenkovic, V.: Rotational polygon containment and minimum enclosure. In: Symposium on Computational Geometry, pp. 1–8 (1998)
11. Scheithauer, G.: Zuschnitt- und Packungsoptimierung: Problemstellungen, Modellierungstechniken, Lösungsmethoden. Studienbücher Wirtschaftsmathematik. Vieweg+Teubner Verlag (2008)
12. Schiermeyer, I.: Reverse-fit: A 2-optimal algorithm for packing rectangles. In: van Leeuwen, J. (ed.) ESA 1994. LNCS, vol. 855, pp. 290–299. Springer, Heidelberg (1994)
13. Steinberg, A.: A strip-packing algorithm with absolute performance bound 2. SIAM J. Comput. 26(2), 401–409 (1997)
14. Tang, K., Wang, C.C.L., Chen, D.Z.: Minimum area convex packing of two convex polygons. Int. J. Comput. Geometry Appl. 16(1), 41–74 (2006)
15. von Niederhäusern, L.: Packing polygons. Master's thesis, EPFL Lausanne, FU Berlin (2014)

Primal-Dual and Dual-Fitting Analysis of Online Scheduling Algorithms for Generalized Flow Time Problems

Spyros Angelopoulos[1,2,*], Giorgio Lucarelli[3,*,**], and Kim Thang Nguyen[4,***]

[1] Sorbonne Universités, UPMC Univ Paris 06, Paris, France
[2] CNRS, Paris, France
[3] LIG, University Grenoble-Alpes, France
[4] IBISC, University of Evry Val d'Essonne, France

Abstract. We study online scheduling problems on a single processor that can be viewed as extensions of the well-studied problem of minimizing total weighted flow time. In particular, we provide a framework of analysis that is derived by duality properties, does not rely on potential functions and is applicable to a variety of scheduling problems. A key ingredient in our approach is bypassing the need for "black-box" rounding of fractional solutions, which yields improved competitive ratios.

We begin with an interpretation of Highest-Density-First (HDF) as a primal-dual algorithm, and a corresponding proof that HDF is optimal for total fractional weighted flow time (and thus scalable for the integral objective). Building upon the salient ideas of the proof, we show how to apply and extend this analysis to the more general problem of minimizing $\sum_j w_j g(F_j)$, where w_j is the job weight, F_j is the flow time and g is a non-decreasing cost function. Among other results, we present improved competitive ratios for the setting in which g is a concave function, and the setting of same-density jobs but general cost functions. We further apply our framework of analysis to online weighted completion time with general cost functions as well as scheduling under polyhedral constraints.

1 Introduction

We consider online scheduling problems in which a set of jobs $\mathcal{J}$ arriving over time must be executed on a single processor. In particular, each job $j \in \mathcal{J}$ is characterized by its *processing time* $p_j > 0$ and its *weight* $w_j > 0$, which become known after its *release time* $r_j \geq 0$. The *density* of j is $\delta_j = w_j/p_j$, whereas, given a schedule, its *completion time*, C_j, is defined as the first time $t \geq r_j$ such that p_j units of j have been processed. The *flow time* of j is then defined as $F_j = C_j - r_j$,

---

* Research supported in part by project ANR-11-BS02-0015 "New Techniques in Online Computation–NeTOC".
** Research supported by the ANR project Moebus.
*** Research supported by the FMJH program Gaspard Monge in optimization and operations research and by EDF.

© Springer-Verlag Berlin Heidelberg 2015
N. Bansal and I. Finocchi (Eds.): ESA 2015, LNCS 9294, pp. 35–46, 2015.
DOI: 10.1007/978-3-662-48350-3_4

and represents the time elapsed after the release of job j and up to its completion. A natural optimization objective is to design schedules that minimize the *total weighted flow time*, namely the sum $\sum_{j \in \mathcal{J}} w_j F_j$ of all processed jobs. A related objective is to minimize the weighted sum of completion times, as given by the expression $\sum_{j \in \mathcal{J}} w_j C_j$. We assume that *preemption* of jobs is allowed.

Im *et al.* [14] studied a generalization of the total weighted flow time problem, in which jobs may incur non-linear contributions to the objective. More formally, they defined the *Generalized Flow Time Problem* (GFP) in which the objective is to minimize the sum $\sum_{j \in \mathcal{J}} w_j g(F_j)$, where $g : \mathbb{R}^+ \to \mathbb{R}^+$ is a given non-decreasing cost function with $g(0) = 0$. This extension captures many interesting and natural variants of flow time with real-life applications. Moreover, it is an appropriate formulation of the setting in which we aim to simultaneously optimize several objectives. We define the *Generalized Completion Time Problem* (GCP) along the same lines, with the only difference being the objective function, which equals to $\sum_{j \in J} w_j g(C_j)$. A further generalization of the above problems, introduced in [6], associates each job j with a non-decreasing cost function $g_j : \mathbb{R}^+ \to \mathbb{R}^+$ and $g_j(0) = 0$; in the *Job-Dependent Generalized Flow Time Problem* (JDGFP), the objective is to minimize the sum $\sum_{j \in \mathcal{J}} w_j g_j(F_j)$.

Very recently, Im *et al.* [11] introduced and studied a general scheduling problem called *Packing Scheduling Problem* (PSP). Here, at any time t, the scheduler may assign rates $\{x_j(t)\}$ to each job $j \in \mathcal{J}$. In addition, we are given a matrix B of non-negative entries. The goal is to minimize the total weighted flow time subject to packing constraints $\{Bx \leq 1, x \geq 0\}$. This formulates applications in which each job j is associated with a resource-demand vector $\mathbf{b}_j = (b_{1j}, b_{2j}, \ldots, b_{Mj})$ so that it requires an amount b_{ij} of the i-th resource.

In this paper, we present a general framework based on LP-duality principles, for online scheduling with generalized flow time objectives. Since no online algorithm even for total weighted flow time is constant competitive [4], we study the effect of *resource augmentation*, introduced by Kalyanasundaram and Pruhs [15]. More precisely, given some optimization objective (e.g. total flow time), an algorithm is said to be α-speed β-competitive if it is β-competitive with respect to an offline optimal scheduling algorithm of speed $1/\alpha$ (here $\alpha \geq 1$).

Related Work. It is well-known that the algorithm Shortest Remaining Processing Time (SRPT) is optimal for online total (unweighted) flow time. Becchetti *et al.* [7] showed that the natural algorithm Highest-Density-First (HDF) is $(1 + \epsilon)$-speed $\frac{1+\epsilon}{\epsilon}$-competitive for total weighted flow time. At each time, HDF processes the job of highest density. Concerning the online GFP, Im *et al.* [14] showed that HDF is $(2 + \epsilon)$-speed $O(\frac{1}{\epsilon})$-competitive algorithm for general non decreasing functions g. On the negative side, they showed that no *oblivious* algorithm is $O(1)$-competitive with speed augmentation $2 - \epsilon$, for any $\epsilon > 0$ (an oblivious algorithm does not know the function g). In the case in which g is a twice-differentiable, concave function, they showed that the algorithm Weighted Late Arrival Processor Sharing (WLAPS) is $(1+\epsilon)$-speed $O(\frac{1}{\epsilon^2})$-competitive. For equal-density jobs and general cost functions [14] prove that FIFO is $(1+\epsilon)$-speed $\frac{4}{\epsilon^2}$-competitive. Fox *et al.* [9] studied the problem of convex cost functions in the *non-clairvoyant* variant providing a $(2+\epsilon)$-speed $O(\frac{1}{\epsilon})$-competitive algorithm; in this variant, the scheduler learns the processing time of a job only when the job is completed. Bansal and Pruhs [5] considered a special class of convex functions, namely the weighted ℓ_k norms of flow time, with $1 < k < \infty$, and they showed that HDF is $(1 + \epsilon)$-speed $O(\frac{1}{\epsilon^2})$-competitive. Moreover, they showed how to transform this result in order to obtain an 1-speed $O(1)$-competitive algorithm for the weighted ℓ_k norms of completion time.

Most of the above works rely to techniques based on amortized analysis (see also [13] for a survey). More recently, techniques based on LP duality have been applied in the context of online scheduling for generalized flow time problems. Gupta *et al.* [10] gave a primal-dual algorithm for a class of non-linear load balancing problems. Devanur and Huang [8] used a duality approach for the problem of minimizing the sum of energy and weighted flow time on unrelated machines. Of particular relevance to our paper is the work of Antoniadis *et al.* [3], which gives an optimal *offline* energy and fractional weighted flow trade-off schedule for a speed-scalable processor with discrete speeds, and uses an approach based on primal-dual properties (similar geometric interpretations arise in the context of our work, in the online setting). Anand *et al.* [1] were the first to propose an approach to online scheduling by linear/convex programming and dual fitting. Nguyen [16] presented a framework based on Lagrangian duality for online scheduling problems beyond linear and convex programming. Im *et al.* [11] applied dual fitting in the context of PSP. For the weighted flow time objective, they gave a non-clairvoyant algorithm that is $O(\log n)$-speed $O(\log n)$-competitive, where n denotes the number of jobs. They also showed that for any constant $\epsilon > 0$, any $O(n^{1-\epsilon})$-competitive algorithm requires speed augmentation compared to the offline optimum.

We note that a common approach in obtaining a competitive, resource-augmented scheduling algorithm for flow time and related problems is by first deriving an algorithm that is competitive for the *fractional* objective [7,13]. An informal interpretation of the fractional objective is that a job contributes to the objective proportionally to the amount of its remaining work (see Section 2 for a formal definition). It is known that any α-speed β-competitive algorithm for fractional GFP can be converted, in "black-box" fashion, to a $(1+\epsilon)\alpha$-speed $\frac{1+\epsilon}{\epsilon}\beta$-competitive algorithm for (integral) GFP, for $0 < \epsilon \leq 1$ [9]. Fractional objectives are often considered as interesting problems in their own (as in [3]).

Contribution. We present a framework for the design and analysis of algorithms for generalized flow time problems that is based on primal-dual and dual-fitting techniques. Our proofs are based on intuitive geometric interpretations of the primal/dual objectives; in particular, we do not rely on potential functions. An interesting feature in our primal-dual approach, that differs from previous ones, is that when a new job arrives, we may update the dual variables for jobs that already have been scheduled without affecting the past portion (primal solution) of the schedule. Another important ingredient of our analysis consists in relating, in a direct manner, the primal integral and fractional dual objectives, without passing through the fractional primal. This allows us to bypass the "black-box" transformation of fractional to integral solutions which has been the canonical approach up to now. As a result, we obtain an improvement to the competitive ratio by a factor of $O(\frac{1}{\epsilon})$, for $(1 + \epsilon)$-speed.

In Section 3 we begin with an interpretation of HDF as a primal-dual algorithm for total weighted flow time. Our analysis, albeit significantly more complicated than the known combinatorial one [7], yields insights about more complex problems. Note that our approach differs from [8] (in which the objective is to minimize the sum of energy and weighted flow time), even though the two settings are seemingly similar. More precisely, the relaxation considered in [8] consists only of covering constraints, whereas for minimizing weighted flow time, one has to consider both covering and packing constraints in the primal LP.

In Sections 4 and 5 we expand the salient ideas behind the above analysis of HDF and derive a framework which is applicable to more complicated objectives. More precisely, we show that HDF is $(1 + \epsilon)$-speed $\frac{1+\epsilon}{\epsilon}$-competitive for GFP with concave functions, improving the $(1+\epsilon)$-speed $O(\frac{1}{\epsilon^2})$-competitive analysis of WLAPS [14], and removing the assumption that g is twice-differentiable. For GFP with general cost functions and jobs of the same density, we show that FIFO is $(1 + \epsilon)$-speed $\frac{1+\epsilon}{\epsilon}$-competitive, which improves again the analysis in [14] by a factor of $O(\frac{1}{\epsilon})$ in the competitive ratio. For the special case of GFP with equal-density jobs and convex (resp. concave) cost functions we show that FIFO (resp. LIFO) are fractionally optimal, and $(1 + \epsilon)$-speed $\frac{1+\epsilon}{\epsilon}$-competitive for the integral objective. In addition, we apply our framework to the following problems: i) online GCP: here, we show that HDF is optimal for the fractional objective, and $(1 + \epsilon)$-speed $\frac{1+\epsilon}{\epsilon}$-competitive for the integral one; and ii) online PSP assuming a matrix B of strictly positive elements: here, we derive an adaptation of HDF which we prove is 1-competitive and which requires resource augmentation $\max_j \frac{B_j}{b_j}$, with $B_j = \max_i b_{ij}$ and $b_j = \min_i b_{ij}$.

Last, we extend ideas of [12], using, in addition, the Lagrangian relaxation of a non-convex formulation for the online JDGFT problem. We thus obtain a non-oblivious $(1+\epsilon)$-speed $\frac{4(1+\epsilon)^2}{\epsilon^2}$-competitive algorithm, assuming each function g_j is concave and differentiable (this result can be entirely found in the full version [2] of this paper since it does not rely on our framework).

Complete proofs, that are omitted or sketched, can be found in [2].

Notation. Let z be a job that is released at time τ. For a given scheduling algorithm, we denote by P_τ the set of pending jobs at time τ (i.e., jobs released up to and including τ but not yet completed), and by $C_{\max}^\tau$ the last completion time among jobs in P_τ, assuming no jobs are released after τ. We also define R_τ as the set of all jobs released up to and included τ and $\mathcal{J}_\tau$ as the set of all jobs that have been completed up to time τ.

2   Linear Programming Relaxation

In order to give a linear programming relaxation of GFP, we pass through the corresponding fractional variant. Formally, let $q_j(t)$ be the remaining processing time of job j at time t (in a schedule). The *fractional remaining weight* of j at time t is defined as $w_j q_j(t)/p_j$. The fractional objective of GFP is now defined as $\sum_j \int_{r_j}^\infty w_j \frac{q_j(t)}{p_j} g(t - r_j)$. An advantage of fractional GFP is that it admits a linear-programming formulation (in fact, the same holds even for the stronger problem JDGFP). Let $x_j(t) \in [0,1]$ be a variable that indicates the execution rate of $j \in \mathcal{J}$ at time t. The primal and dual LPs are:

$$\min \sum_{j \in \mathcal{J}} \delta_j \int_{r_j}^\infty g(t - r_j) x_j(t) dt \quad (P) \qquad \max \sum_{j \in \mathcal{J}} \lambda_j p_j - \int_0^\infty \gamma(t) dt \qquad (D)$$

$$\int_{r_j}^\infty x_j(t) dt \geq p_j \quad \forall j \in \mathcal{J} \quad (1) \qquad \lambda_j - \gamma(t) \leq \delta_j g(t - r_j) \qquad \forall j \in \mathcal{J}, t \geq r_j \quad (3)$$

$$\sum_{j \in \mathcal{J}} x_j(t) \leq 1 \quad \forall t \geq 0 \quad (2) \qquad \lambda_j, \gamma(t) \geq 0 \qquad \forall j \in \mathcal{J}, \forall t \geq 0$$

$$x_j(t) \geq 0 \quad \forall j \in \mathcal{J}, t \geq 0$$

In this paper, we avoid the use of the standard transformation from fractional to integral GFP. However, we always consider the fractional objective as a lower bound for the integral one. Specifically, we will prove the performance of an algorithm by comparing its integral objective to that of a feasible dual solution (D). Note that by weak duality the latter is upper-bounded by the optimal solution of (P), which is a lower bound of the optimum solution for integral GFP.

Moreover, we will analyze algorithms that are α-speed β-competitive. In other words, we compare the performance of our algorithm to an offline optimum with speed $1/\alpha$ ($\alpha, \beta > 1$). In turn, the cost of this offline optimum is the objective of a variant of (P) in which constraints (2) are replaced by constraints $\sum_{j \in \mathcal{J}} x_j(t) \leq 1/\alpha$ for all $t \geq 0$. The corresponding dual is the same as (D), with the only difference that the objective is equal to $\sum_{j \in \mathcal{J}} \lambda_j p_j - \frac{1}{\alpha} \int_0^\infty \gamma(t) dt$. We denote these modified primal and dual LP's by (P_α) and (D_α), respectively. In order to prove that the algorithm is α-speed β-competitive, it will then be sufficient to show that there is a feasible dual solution to (D_α) for which the algorithm's cost is at most β times the objective of the solution.

3   A Primal-Dual Interpretation of HDF for $\sum_j w_j F_j$

In this section we give an alternative statement of HDF as a primal-dual algorithm for the total weighted flow time problem. We begin with an intuitive understanding of the complementary slackness (CS) conditions. In particular, the primal CS condition states that for a given job j and time t, if $x_j(t) > 0$, i.e., if the algorithm were to execute job j at time t, then it should be that $\gamma(t) = \lambda_j - \delta_j(t - r_j)$. We would like then the dual variable $\gamma(t)$ to be such that we obtain some information about which job to schedule at time t. To this end, for any job $j \in \mathcal{J}$, we define the line $\gamma_j(t) = \lambda_j - \delta_j(t - r_j)$, with domain $[r_j, \infty)$.

The slope of this line is equal to the negative density of the job, i.e, $-\delta_j$. Our algorithm will always choose $\gamma(t)$ to be equal to $\max\{0, \max_{j \in \mathcal{J}:r_j \leq t}\{\gamma_j(t)\}\}$ for every $t \geq 0$. We say that at time t the line γ_j (or the job j) is *dominant* if $\gamma_j(t) = \gamma(t)$. We can thus restate the primal CS condition as a *dominance* condition: if a job j is executed at time t, then γ_j must be dominant at t.

We will consider a class of scheduling algorithms, denoted by $\mathcal{A}$, that comply to the following rules: i) the processor is never idle if there are pending jobs; and ii) if at time τ a new job z is released, the algorithm will first decide an ordering on the set P_τ of all pending jobs at time τ. Then for every $t \geq \tau$, it schedules all jobs in P_τ according to the above ordering, unless a new job arrives after τ.

We now proceed to give a primal-dual algorithm in the class $\mathcal{A}$ (which will turn out to be identical to HDF). The algorithm will use the dominance condition so as to decide how to update the dual variables λ_j, and, on the primal side, which job to execute at each time. Note that once we define the λ_j's, the lines γ_j's as well as $\gamma(t)$ are well-defined, as we emphasized earlier. In our scheme we change the primal and dual variables only upon arrival of a new job, say at time τ. We also modify the dual variables for jobs in $\mathcal{J}_\tau$, i.e., jobs that have already completed in the past (before time τ) without however affecting the primal variables of the past, so as to comply with the online nature of the problem.

By induction, suppose that the primal-dual algorithm $A \in \mathcal{A}$ satisfies the dominance condition up to time τ, upon which a new job z arrives. Let q_j be the remaining processing time of each job $j \in P_\tau$ at time τ and $|P_\tau| = k$. To satisfy CS conditions, each line γ_j must be defined such that to be dominant for a total period of time at least q_j, in $[\tau, \infty)$. If a line γ_j is dominant at times t_1, t_2, it must also be dominant in the entire interval $[t_1, t_2]$. This implies that for two jobs $j_1, j_2 \in P_\tau$, such that j_1 (resp. j_2) is dominant at time t_1 (resp. t_2), if $t_1 < t_2$ then the slope of γ_{j_1} must be smaller than the slope of γ_{j_2} (i.e., $-\delta_{j_1} \leq -\delta_{j_2}$). We derive that A must make the same decisions as HDF. Consequently, the algorithm A orders the jobs in P_τ in non-decreasing order of the slopes of the corresponding lines γ_j. For every job $j \in P_\tau$, define $C_j = \tau + \sum_{j' \prec j} q_{j'}$, where the precedence is according to the above ordering of A. These are the completion times of jobs in P_τ in A's schedule, if no new jobs are released after time τ; so we set the primal variables $x_j(t) = 1$ for all $t \in (C_{j-1}, C_j]$. Procedure 1 formalizes the choice of λ_j for all $j \in P_\tau$; intuitively, it ensures that if a job $j \in P_\tau$ is executed at time $t > \tau$ then γ_j is dominant at t (see Figure 1 for an illustration).

Procedure 1. Assignment of dual variables λ_j for all $j \in P_\tau$

1: Consider the jobs in P_τ in increasing order of completion times $C_1 < C_2 < \ldots < C_k$
2: Choose λ_k such that $\gamma_k(C_k) = 0$
3: **for** each pending job $j = k - 1$ to 1 **do**
4:    Choose λ_j such that $\gamma_j(C_j) = \gamma_{j+1}(C_j)$

The following lemma shows that if no new jobs were to be released after time τ, HDF would guarantee the dominance condition for all times $t \geq \tau$.

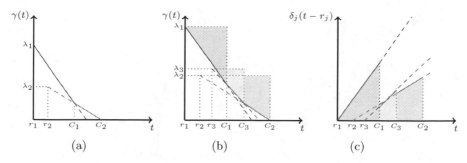

Fig. 1. Figure (a) depicts the situation right before τ: the two lines γ_1, γ_2 correspond to two pending jobs prior to the release of z. In addition, $\gamma(t)$ is the upper envelope of the two lines. Figure (b) illustrates the situation after the release of a third job z at time $\tau = r_3$; the area of the shaded regions is the dual objective. In Figure (c), the area of the shaded regions is the primal fractional objective for the three jobs of Figure (b).

Lemma 1 (future dominance). *For λ_j's as defined by Procedure 1, and $A \equiv HDF$, if job $j \in P_\tau$ is executed at time $t \geq \tau$, then γ_j is dominant at t.*

Observe that Procedure 1 modifies (increases) the λ_j variables of all jobs pending at time τ. In turn, this action may violate the dominance condition prior to τ. We thus need a second procedure that will rectify the dominance condition for $t \leq \tau$. We consider again the jobs in P_τ in increasing order of their completion times, i.e., $C_1 < C_2 < \ldots < C_k$, with $k = |P_\tau|$. We partition $\mathcal{R}_\tau$ into k disjoint sets $S_1, S_2, \ldots, S_k$. Each set S_j is initialized with the job $j \in P_\tau$, which is called the *representative* element of S_j (we use the same index to denote the set and its representative job). Informally, the set S_j will be constructed in such a way that it will contain all jobs $a \in \mathcal{J}_\tau$ whose corresponding variable λ_a will be increased by the same amount in the procedure. This amount is equal to the increase, say Δ_j, of λ_j, due to Procedure 1 for the representative job of S_j. We then define Procedure 2 that increases the dual variables for jobs in $\mathcal{J}_\tau$.

Procedure 2. Updating of dual variables λ_j for all jobs $j \in \mathcal{J}_\tau$

1: **for** $j = 1$ to k **do**
2:     Add j in S_j
3: **for** each job $a \in \mathcal{J}_\tau$ in decreasing order of completion times **do**
4:     Let b be the job such that $\gamma_a(C_a) = \gamma_b(C_a)$
5:     Let S_j be the set that contains b
6:     Add a in S_j
7: **for** each set S_j, $1 \leq j \leq k$, **do**
8:     Let Δ_j be the increase of λ_j, due to Procedure 1, for the representative of S_j
9:     Increase λ_a by an amount of Δ_j for all $a \in S_j \setminus \{j\}$

Geometrically, the update operation is a vertical translation of the line $\gamma(t)$ for $t < \tau$. The following lemma shows that, if a line γ_j was dominant for a time

$t < \tau$ prior to the arrival of the new job at time τ, then it will remain dominant after the application of Procedures 1 and 2.

Lemma 2 (past dominance). *For λ_j's as defined by both Procedure 1 and Procedure 2, and $A \equiv HDF$, if job $j \in \mathcal{J}_\tau \cup P_\tau$ is executed at time $t < \tau$, then γ_j is dominant at t.*

The following lemma states that the dual variable $\gamma(t)$ has been defined in such a way that it is zero for all $t > C_{\max}^\tau$. This will be required in order to establish that the primal and dual solutions have the same objective value.

Lemma 3 (completion). *For λ_j's defined by Procedures 1 and 2, we have that $\gamma(t) = 0$ for every $t > C_{\max}^\tau$.*

The proof of the following theorem is based on Lemmas 1, 2 and 3, and it is a simplified case of the proof of Theorem 2 which is given in the next section.

Theorem 1. *The primal-dual algorithm $A \equiv HDF$ is an optimal online algorithm for the total fractional weighted flow time and a $(1 + \epsilon)$-speed $\frac{1+\epsilon}{\epsilon}$-competitive algorithm for the total (integral) weighted flow time.*

4   A Framework for Primal-Dual Algorithms

Building on the primal-dual analysis of HDF for total weighted flow time, we can abstract the essential properties that we need to satisfy in order to obtain online algorithms for other similar problems. For the problems we consider, the primal solution is generated by an online primal-dual algorithm $A \in \mathcal{A}$ which may not necessarily be HDF. In addition, each job j will now correspond to a *curve* γ_j (for the total weighted flow time problem, γ_j is a line), and we will also have a dual variable $\gamma(t)$ that will be set equal to $\max\{0, \max_{j \in \mathcal{J}: r_j \leq t}\{\gamma_j(t)\}\}$ for every $t \geq 0$. Finally, the crux is in maintaining dual variables λ_j, upon release of a new job z at time τ, such that the following properties are satisfied: ($\mathcal{P}$1) *Future dominance*: if the algorithm A executes job j at time $t \geq \tau$, then γ_j is dominant at t; ($\mathcal{P}$2) *Past dominance*: if the algorithm A executes job j at time $t < \tau$, then γ_j remains dominant at t. In addition, the primal solution (i.e., the algorithm's scheduling decisions) for $t < \tau$ does not change due to the release of z; and ($\mathcal{P}$3) *Completion*: $\gamma(t) = 0$ for all $t > C_{\max}^\tau$. Essentially properties ($\mathcal{P}$1), ($\mathcal{P}$2) and ($\mathcal{P}$3) reflect that the statements of Lemmas 1, 2 and 3 are not tied exclusively to the total weighted flow time problem.

Theorem 2. *Any algorithm that satisfies the properties ($\mathcal{P}$1), ($\mathcal{P}$2) and ($\mathcal{P}$3) with respect to a feasible dual solution is an optimal online algorithm for fractional GFP and a $(1 + \epsilon)$-speed $\frac{1+\epsilon}{\epsilon}$-competitive algorithm for integral GFP.*

Proof. The feasibility of the solution is directly implied by the fact that $\lambda_j \geq 0$ and from our definition of $\gamma(t)$ which implies that the constraints (3) are satisfied and $\gamma(t) \geq 0$. Let $C_{\max}$ be the completion time of the last job. We will assume, without loss of generality, that at time $t \leq C_{\max}$ there is at least one pending job in the schedule; otherwise, there are idle times in the schedule and we can apply the same type of analysis for jobs scheduled between consecutive idle periods.

We will first show that the primal and the dual objectives are equal. Consider a job j and let $[t_1, t_2], [t_2, t_3], \ldots, [t_{k-1}, t_k]$ be the time intervals during which j is executed. Note that $x_j(t) = 1$ for every t in these intervals (and $x_{j'}(t) = 0$ for $j' \neq j$). Hence, the contribution of j to the primal (fractional) objective is $\sum_{i=1}^{k-1} \delta_j \int_{t_i}^{t_{i+1}} g(t - r_j)dt$. By properties $(\mathcal{P}1)$ and $(\mathcal{P}2)$, the line γ_j is dominant during the same time intervals. Thus, the contribution of job j to the dual is

$$\lambda_j p_j - \sum_{i=1}^{k-1} \int_{t_i}^{t_{i+1}} \gamma(t)dt = \sum_{i=1}^{k-1} \delta_j \int_{t_i}^{t_{i+1}} g(t - r_j)dt$$

since $\sum_{i=1}^{k-1} \int_{t_i}^{t_{i+1}} \lambda_j dt = \lambda_j \sum_{i=1}^{k-1} \int_{t_i}^{t_{i+1}} x_j(t)dt = \lambda_j p_j$. The first part of the theorem follows by summing over all jobs j, and by accounting for the fact that $\int_{C_{\max}}^{\infty} \gamma(t) = 0$ (from property $(\mathcal{P}3)$).

For the second part of the theorem, consider again the time intervals during which a job j is executed. The contribution of j to the integral objective is $w_j g(C_j - r_j) = \delta_j g(C_j - r_j)p_j$. By properties $(\mathcal{P}1)$ and $(\mathcal{P}2)$, for any $t \in \bigcup_{i=1}^{k-1}[t_i, t_{i+1}]$ we have that $\gamma_j(t) \geq 0$. In particular, it holds for $t_k = C_j$, that is $\lambda_j \geq \delta_j g(C_j - r_j)$. Therefore, the contribution of j to the integral objective is at most $\lambda_j p_j$. Since we consider the speed augmentation case, we will use as lower bound of the optimal solution the dual program that uses a smaller speed as explained in Section 2. By properties $(\mathcal{P}1)$ and $(\mathcal{P}2)$, we have $\gamma(t) = \lambda_j - \delta_j g(t - r_j) \leq \lambda_j$ during the time intervals where the job j is executed. Thus, the contribution of j to the dual objective is at least

$$\lambda_j p_j - \frac{1}{1+\epsilon} \sum_{i=1}^{k-1} \int_{t_i}^{t_{i+1}} \gamma(t)dt \geq \lambda_j p_j - \frac{1}{1+\epsilon} \sum_{i=1}^{k-1} \int_{t_i}^{t_{i+1}} \lambda_j dt = \frac{\epsilon}{1+\epsilon} \lambda_j p_j,$$

since $\sum_{i=1}^{k-1} \int_{t_i}^{t_{i+1}} \lambda_j dt = \lambda_j \sum_{i=1}^{k-1} \int_{t_i}^{t_{i+1}} x_j(t)dt = \lambda_j p_j$. From property $(\mathcal{P}3)$ we have $\int_{C_{\max}}^{\infty} \gamma(t)dt = 0$. Summing up over all jobs, the theorem follows.   □

We can apply the proposed framework to three different settings and we get the following results (see [2]).

Theorem 3. *The primal-dual algorithm $A \equiv HDF$ is an optimal algorithm for fractional GCP and a $(1 + \epsilon)$-speed $\frac{1+\epsilon}{\epsilon}$-competitive algorithm for integral GCP.*

Theorem 4. *The primal-dual algorithm $A \equiv FIFO$ (resp. $A \equiv LIFO$) is an optimal online algorithm for fractional GFP and a $(1 + \epsilon)$-speed $\frac{1+\epsilon}{\epsilon}$-competitive algorithm for integral GFP, when we consider convex (reps. concave) cost functions and jobs of equal density.*

Theorem 5. *For the online PSP problem with constraints $Bx \leq 1$ and $b_{ij} > 0$ for every i, j, an adaptation of HDF is $\max_j\{B_j/b_j\}$-speed 1-competitive for fractional weighted flow time and $\max_j\{(1+\epsilon)B_j/b_j\}$-speed $(1+\epsilon)/\epsilon$-competitive for integral weighted flow time.*

5   A Generalized Framework Using Dual-Fitting

In this section we relax certain properties in order to generalize our framework and apply it to the integral variant of more problems. Our analysis here is based on the dual-fitting paradigm, since the analysis of Section 3 provides us with intuition about the geometric interpretation of the primal and dual objectives. We consider, as concrete applications, GFP for given cost functions g. We again associate with each job j the curve γ_j and set $\gamma(t) = \max\{0, \max_{j \in \mathcal{J}: r_j \le t}\{\gamma_j(t)\}\}$. Then, we need to define how to update the dual variables λ_j, upon release of a new job z at time τ, such that the following properties are satisfied: $(\mathcal{Q}1)$: if the algorithm A schedules job j at time $t \ge \tau$ then $\gamma_j(t) \ge 0$ and $\lambda_j \ge \gamma_{j'}(t)$ for every other pending job j' at time t; $(\mathcal{Q}2)$: if the algorithm A schedules job j at time $t < \tau$, then $\gamma_j(t) \ge 0$ and $\lambda_j \ge \gamma_{j'}(t)$ for every other pending job j' at time t. In addition, the primal solution for $t < \tau$ is not affected by the release of z; and $(\mathcal{Q}3)$: $\gamma(t) = 0$ for all $t > C_{\max}^\tau$.

Note that $(\mathcal{Q}1)$ is relaxed with respect to property $(\mathcal{P}1)$ of Section 4, since it describes a weaker dominance condition. Informally, $(\mathcal{Q}1)$ guarantees that for any time t the job that is scheduled at t does not have negative contributions in the dual. On the other hand, property $(\mathcal{Q}2)$ is the counterpart of $(\mathcal{Q}1)$, for times $t < \tau$ (similar to the relation between $(\mathcal{P}1)$ and $(\mathcal{P}2)$). Finally, note that even though the relaxed properties do not guarantee anymore the optimality for the fractional objectives, the following theorem (Theorem 6) establishes exactly the same result as Theorem 2 for the *integral* objectives. This is because in the second part of the proof of Theorem 2 we only require that when j is executed at time t then $\lambda_j \ge \gamma_{j'}(t)$ for every other pending job j' at t, which is in fact guaranteed by properties $(\mathcal{Q}1)$ and $(\mathcal{Q}2)$. Therefore, the proof of the following theorem is identical with the one of Theorem 2.

Theorem 6. *Any algorithm that satisfies the properties $(\mathcal{Q}1)$, $(\mathcal{Q}2)$ and $(\mathcal{Q}3)$ with respect to a feasible dual solution is a $(1+\epsilon)$-speed $\frac{1+\epsilon}{\epsilon}$-competitive algorithm for integral GFP with general cost functions g.*

5.1   Online GFP with General Cost Functions and Equal-Density Jobs

We will analyze the FIFO algorithm using dual fitting. We will use a single procedure, namely Procedure 3, for the assignment of the λ_j variables for each job j released by time τ. We denote this set of jobs by R_τ, and $k = |R_\tau|$.

Procedure 3. Assignment and updating of λ_j's for the set R_τ of all jobs released by time τ.

1: Consider jobs in R_τ in increasing order of completion times $C_1 < C_2 < \ldots < C_k$
2: Choose λ_k such that $\gamma_k(C_k) = 0$
3: **for** $j = k - 1$ to 1 **do**
4:     Choose the maximum possible λ_j such that for every $t \ge C_j$, $\gamma_j(t) \le \gamma_{j+1}(t)$
5:     **if** $\gamma_j(C_j) < 0$ **then**
6:         Choose λ_j such that $\gamma_j(C_j) = 0$

We can show that the dual solution created by Procedure 3 satisfies the properties ($Q1$), ($Q2$) and ($Q3$). Hence, the following theorem is an immediate consequence of Theorem 6.

Theorem 7. *FIFO is a* $(1 + \epsilon)$*-speed* $\frac{1+\epsilon}{\epsilon}$*-competitive for integral GFP with general cost functions and equal-density jobs.*

5.2 Online GFP with Concave Cost Functions

We will analyze the HDF algorithm using dual fitting. As in Section 3, we will employ two procedures for maintaining the dual variables λ_j. The first one is Procedure 4 which updates the λ_j's for $j \in P_\tau$. The second procedure updates the λ_j's for $j \in \mathcal{J}_\tau$; this procedure is identical to Procedure 2 of Section 3.

The intuition behind Procedure 4 is to ensure property ($Q1$) that is, $\gamma_j(t) \geq 0$ and $\lambda_j \geq \gamma_{j'}(t)$ for all $j' \in P_\tau$, which in some sense is the "hard" property to maintain. Specifically, for given job j there is a set of jobs A (initialized in line 4) for which the property does not hold. The while loop in the procedure decreases the λ values of jobs in A so as to rectify this situation (see line 6(ii)). However, this decrement may, in turn, invalidate this property for some jobs b (see line 6(i)). These jobs are then added in the set of "problematic" jobs A and we continue until no problematic jobs are left.

Procedure 4. Assignment of dual variables λ_j for all $j \in P_\tau$.

1: Consider the jobs in P_τ in increasing order of completion times $C_1 < C_2 < \ldots < C_k$

2: For every $1 \leq j \leq k$ choose λ_j such that $\gamma_j(C_k) = 0$
3: **for** $j = 2$ to k **do**
4:     Define $A := \{$jobs $1 \leq a \leq j - 1 : \gamma_a(C_{j-1}) > \lambda_j\}$
5:     **while** $A \neq \emptyset$ **do**
6:         Continuously reduce λ_a by the same amount for all jobs $a \in A$ until:
        (i) $\exists\, a \in A$ and $b \in P_\tau \setminus A$ with $b < a$ s.t. $\lambda_a = \gamma_b(C_{a-1})$; then $A \leftarrow A \cup \{b\}$
        (ii) $\exists\, a \in A$ s.t. $\gamma_a(C_{j-1}) = \lambda_j$; then $A \leftarrow A \setminus \{a\}$

We can show that the dual solution maintained by Procedures 3 and 2 satisfies the properties ($Q1$), ($Q2$) and ($Q3$). Hence, the following theorem is an immediate consequence of Theorem 6.

Theorem 8. *HDF is a* $(1 + \epsilon)$*-speed* $\frac{1+\epsilon}{\epsilon}$*-competitive for integral GFP with concave cost functions.*

6 Conclusion

A promising direction for future work is to apply our framework to *non-clairvoyant* problems. It would be very interesting to obtain a primal-dual analysis of Shortest Elapsed Time First (SETF) which is is known to be scalable [15]; moreover, this algorithm has been analyzed in [9] in the context of the online GFP with convex/concave cost functions. Interestingly, one can use duality to argue that

SETF is the non-clairvoyant counterpart of HDF; more precisely, one can derive SETF as a primal-dual algorithm in a similar manner as the discussion of HDF in Section 3. It remains to bound the primal and dual objectives, which appears to be substantially harder than in the clairvoyant setting. A further open question is extending the results of this paper to multiple machines; here, one potentially needs to define the dual variable $\gamma(t)$ with respect to as many curves per job as machines. Last, we would like to further relax the conditions of the current framework in order to allow for algorithms that are not necessarily scalable.

References

1. Anand, S., Garg, N., Kumar, A.: Resource augmentation for weighted flow-time explained by dual fitting. In: SODA, pp. 1228–1241 (2012)
2. Angelopoulos, S., Lucarelli, G., Thang, N.K.: Primal-dual and dual-fitting analysis of online scheduling algorithms for generalized flow-time problems. CoRR, abs/1502.03946 (2015)
3. Antoniadis, A., Barcelo, N., Consuegra, M., Kling, P., Nugent, M., Pruhs, K., Scquizzato, M.: Efficient computation of optimal energy and fractional weighted flow trade-off schedules. In: STACS. LIPIcs, vol. 25, pp. 63–74 (2014)
4. Bansal, N., Chan, H.-L.: Weighted flow time does not admit $o(1)$-competitive algorithms. In: SODA, pp. 1238–1244 (2009)
5. Bansal, N., Pruhs, K.R.: Server scheduling in the weighted ℓ_p norm. In: Farach-Colton, M. (ed.) LATIN 2004. LNCS, vol. 2976, pp. 434–443. Springer, Heidelberg (2004)
6. Bansal, N., Pruhs, K.: The geometry of scheduling. In: FOCS, pp. 407–414 (2010)
7. Becchetti, L., Leonardi, S., Marchetti-Spaccamela, A., Pruhs, K.: Online weighted flow time and deadline scheduling. J. Discrete Algorithms 4(3), 339–352 (2006)
8. Devanur, N.R., Huang, Z.: Primal dual gives almost optimal energy efficient online algorithms. In: SODA, pp. 1123–1140 (2014)
9. Fox, K., Im, S., Kulkarni, J., Moseley, B.: Online non-clairvoyant scheduling to simultaneously minimize all convex functions. In: Raghavendra, P., Raskhodnikova, S., Jansen, K., Rolim, J.D.P. (eds.) RANDOM 2013 and APPROX 2013. LNCS, vol. 8096, pp. 142–157. Springer, Heidelberg (2013)
10. Gupta, A., Krishnaswamy, R., Pruhs, K.: Online primal-dual for non-linear optimization with applications to speed scaling. In: Erlebach, T., Persiano, G. (eds.) WAOA 2012. LNCS, vol. 7846, pp. 173–186. Springer, Heidelberg (2013)
11. Im, S., Kulkarni, J., Munagala, K.: Competitive algorithms from competitive equilibria: Non-clairvoyant scheduling under polyhedral constraints. In: STOC, pp. 313–322 (2014)
12. Im, S., Kulkarni, J., Munagala, K., Pruhs, K.: Selfishmigrate: A scalable algorithm for non-clairvoyantly scheduling heterogeneous processors. In: FOCS, pp. 531–540 (2014)
13. Im, S., Moseley, B., Pruhs, K.: A tutorial on amortized local competitiveness in online scheduling. SIGACT News 42(2), 83–97 (2011)
14. Im, S., Moseley, B., Pruhs, K.: Online scheduling with general cost functions. SIAM Journal on Computing 43(1), 126–143 (2014)
15. Kalyanasundaram, B., Pruhs, K.: Speed is as powerful as clairvoyance. Journal of the ACM 47(4), 617–643 (2000)
16. Nguyen, K.T.: Lagrangian duality in online scheduling with resource augmentation and speed scaling. In: Bodlaender, H.L., Italiano, G.F. (eds.) ESA 2013. LNCS, vol. 8125, pp. 755–766. Springer, Heidelberg (2013)

Buffer Management
for Packets with Processing Times

Yossi Azar* and Oren Gilon

Blavatnik School of Computer Science, Tel-Aviv University, Israel
azar@tau.ac.il, oren.gilon@gmail.com

Abstract. We discuss the well known job scheduling problem with release times and deadlines, alongside an extended model - buffer management for packets with processing requirements. For job scheduling,
an $\Omega(\sqrt{\frac{\log \kappa}{\log \log \kappa}})$ lower bound for any randomized preemptive algorithm
was shown by Irani and Canetti (1995), where κ is the the maximum job
duration or the maximum job value (the minimum is assumed to be 1).
The proof of this well-known result is fairly elaborate and involved. In
contrast, we show a significantly improved lower bound of $\Omega(\log \kappa)$ using
a simple proof. Our result matches the easy upper bound and closes a
gap which was supposedly open for 20 years.

We also discuss an interesting extension of job scheduling (for tight
jobs). We discuss the problem of handling a FIFO buffer of a limited
capacity, where packets arrive over time and may be preempted. Most of
the work in buffer management considers the case where each packet has
unit processing requirement. We consider a model where packets require
some number of processing cycles before they can be transmitted. We aim
to maximize the value of transmitted packets. We show an $\Omega(\frac{\log \kappa}{\log \log \kappa})$
lower bound on the competitive ratio of randomized algorithms in this
setting. We also present bounds for several special cases. For packets with
unit values we also show a $\varphi \approx 1.618$ lower bound on the competitive
ratio of deterministic algorithms, and a 2-competitive algorithm for this
problem. For the case of packets with constant densities we present a
4-competitive algorithm.

Keywords: Competitive analysis, buffer management, job scheduling,
online algorithms, deadlines.

1 Introduction

We discuss the job scheduling problem with release times, deadlines and values.
Jobs arrive over time at a server. At each time step, the server may choose some
job to process. The server gains the value of a job if it is fully processed before
its deadline. This model has been discussed in detail by Canetti and Irani [5].
To phrase their results we add the following definition:

---

* Supported in part by the Israel Science Foundation (grant No. 1404/10) and by the
Israeli Centers of Research Excellence (I-CORE) program (Center No. 4/11).

© Springer-Verlag Berlin Heidelberg 2015
N. Bansal and I. Finocchi (Eds.): ESA 2015, LNCS 9294, pp. 47–58, 2015.
DOI: 10.1007/978-3-662-48350-3_5

Definition 1. *Let κ be the minimum between the maximum job value V, the maximum job duration T and the maximum job value-density ρ. The minimum value and duration is assumed to be 1.*

In [5] an $\Omega(\sqrt{\frac{\log \kappa}{\log \log \kappa}})$ lower bound for randomized preemptive algorithms is shown. We improve this bound to $\Omega(\log \kappa)$, and close a gap which was open for 20 years.

We then discuss a buffer management problem. Unit-sized packets arrive at a server. The server has a FIFO buffer of size B. Each packet has a processing time requirement and a value associated with it. The server may process any packet from its buffer at each time slot. Once the processing requirement is met the packet may be transmitted subject to the buffer's FIFO nature, i.e. only packets at the head of the buffer can be transmitted. A packet may be preempted from the buffer before transmission, but such a packet is lost. We aim to maximize the value of successfully transmitted packets (full description appears in 'Model Description' paragraph).

Our Contributions - Job Scheduling. We discuss the problem of scheduling jobs with release times, deadlines and values. There we show the following result:

- An $\Omega(\log \kappa)$ lower bound for any randomized preemptive algorithm for the job scheduling problem, for κ defined above.

This result is complementary to the upper bound provided by a 'Classify and Randomly Select' algorithm, randomly choosing between possible job values or durations, similar to that presented in [5]. This algorithm is $O(\log \kappa)$-competitive and constitutes an upper bound matching our lower bound.

Our Contributions - Buffer Management. Our results for the buffer management problem include:

- For packets with unit value: A φ lower bound on the competitive ratio of any deterministic algorithm, where $\varphi \approx 1.618$ is the golden ratio; A deterministic 2-competitive algorithm called $SRPTB$.
- For packets with constant densities: A deterministic 4-competitive algorithm called $KeepPackets$.
- For packets with arbitrary values and processing requirements: An $\Omega(\frac{\log \kappa}{\log \log \kappa})$ lower bound on the competitive ratio of any randomized algorithm, for κ defined above.

For packets with unit values our algorithm is based on the Shortest-Remaining-Processing-Time scheduler. We show that this algorithm is between 2 and $(2 - \frac{1}{B})$ competitive for maximizing throughput. An independent upper bound of 2 for this problem was shown by Kogan et al [18]. Our φ lower bound is the first lower bound for the unit value model, and previous lower bounds where only known for the case that packets have values [6].

For packets with constant densities the *KeepPackets* (or *KP*) algorithm is between 4 and $(4 - \frac{1}{B})$ competitive. For its admission control, the algorithm prefers more valuable packets, which are at least twice as valuable as the least valuable packet in the buffer.

Our randomized lower bound shows that for arbitrary values the problem is made significantly more difficult. Interestingly the lower bound holds for both a FIFO and a non-FIFO buffer. An upper bound of $O(\log \kappa)$ can be achieved by a 'Classify and Randomly Select' algorithm, randomly choosing between different possible processing times.

We also show an $\Omega(\min(V, \sqrt{T}))$ lower bound on the competitive ratio of all deterministic algorithms in this setting, for V, T defined above.

Related Work. Online scheduling is a widely researched field, with many different variations of the problem studied. The simplest job scheduling problem with release times and deadlines was discussed by Canetti and Irani [5]. They show a poly-logarithmic lower bound on the competitive ratio of all randomized algorithms. A quadratic gap has existed between their lower bound and upper bound. There are various simplifying assumptions that can be made to the basic model described such that a constant competitive deterministic algorithm exists. If all job's densities are constant there exists a 4-competitive algorithm, as shown by Koren et al [21]. Alternatively, if all job durations are constant a 1.828-competitive algorithm was presented by Englert et al [9]. For jobs that have a window length that is at least α times longer than their duration there exists an $(\frac{\alpha}{\alpha-1})$-competitive algorithm [7, 10]. A different approach for relaxing this problem is through resource augmentation. In this relaxation, if we give the algorithm processors which are faster by a factor of $(1 + \epsilon)$ than those of the optimum, a $(1 + \frac{1}{\epsilon})$-competitive algorithm was found by Kalyanasundaram et al [12]. All these modifications show that there are many variations to the model, all of which result in the logarithmic bound collapsing to a constant competitive algorithm. Extensive surveys of the job scheduling problem can be found in [24, 25].

The buffer management problem with unit processing times has been well researched. The research in this field was initiated in [15, 22]. Kesselman et al [15] analyzed the performance of the greedy algorithm in the bounded-buffer model for packets with values. In this model, they show that the greedy algorithm is 2-competitive. The model where packets have values has also been considered in [1, 8]. A model where packets have dependencies between them was studied in [16, 23]. The multi-queue model has also been considered, for example in [2–4, 13, 14].

Buffer management for a bounded buffer with processing time requirements has been researched in numerous articles. Kogan et al [18] showed a 2-competitive deterministic algorithm for this variation of the scheduling problem. In [17] resource augmentation is studied. It is shown that by moderately increasing the speed of the processor, the gap between an algorithm's performance and that of the optimal algorithm can be closed. In [20], packets that have sizes in addition to required processing time is researched, for non-FIFO buffers. Chuprikov et al [6] discuss a model where packets have values in addition to processing requirements. Surveys of this field can be found in [11, 19].

Model Description. In Section 2, we discuss the well-known job scheduling problem for single or multiple processors. In this model, jobs arrive over time at a server. Each job has some duration for which it must be processed, some deadline, and a value attained by successfully processing the packet before its deadline. Jobs may be preempted and may migrate between different processors. The goal is to maximize the value of completed jobs.

In Section 3 we consider the problem of a server managing a FIFO buffer of limited capacity. We denote the buffer's capacity as B. Packets arrive sequentially and must be handled. The packets are unit-sized, meaning each packet p occupies a single unit of space in the buffer. It also has a certain amount of processing time $r(p) \in \{1, \ldots, T\}$, and a value $v(p) \in \{1, \ldots, V\}$ associated with it. We denote the packet's arrival time by $a(p)$. For a given algorithm processing the input, we denote by $r_t(p)$ the residual processing time of a packet p at time t. Note that $r_{a(p)}(p) = r(p)$. At each time step t three tasks must be performed sequentially:

i *Scheduling*: one of the packets in the buffer, p, that has $r_{t-1}(p) > 0$, is chosen for processing. The residual processing time of the chosen packet is then reduced by 1, i.e. $r_t(p) = r_{t-1}(p) - 1$

ii *Transmission*: all packets p at the head of the buffer that have $r_t(p) = 0$ are transmitted and leave the buffer. Note that more than one packet can be transmitted at a given time-step.

iii *Buffer Management*: new packets arrive at the server and are handled. At this stage, some packets that are already in the buffer may be *preempted*. This means those packets are lost, but all trailing packets are pushed forward in the buffer. Newly arriving packets can be either *accepted* or *rejected*. Accepted packets are placed at the end of the buffer, while rejected packets are lost forever.

The goal is to maximize the value of transmitted packets. It is easy to see that for a unit sized buffer, this is equivalent to the job scheduling problem. In Section 3.1 we analyze packets that have unit values, i.e. each packet p has $v(p) = 1$. In Section 3.2 we analyze packets that have a *density* of 1, i.e. each packet p has $v(p) = r(p)$. In Section 3.3 we discuss the most general model where a packet's processing time and value are arbitrary. There we also discuss a modification of this model where the buffer is not FIFO. This means that packets can be transmitted from any place in the buffer, regardless of the order of their arrival.

2   Job Scheduling

We proceed to discuss the job scheduling problem for m processors. As mentioned before, a $O(\log \kappa)$ upper bound is provided by the 'Classify and Randomly Select' algorithm. We provide a tight lower bound improving on the previously known lower bound of $\Omega(\sqrt{\frac{\log \kappa}{\log \log \kappa}})$ shown in [5].

Theorem 1. *Any randomized preemptive algorithm for the job scheduling problem with deadlines has a competitive ratio which is $\Omega(\log \kappa)$ for κ defined above.*

Proof. Let ALG be some randomized algorithm solving the job scheduling problem. If we let ALG transmit packets fractionally, clearly a 1-competitive algorithm exists for tight jobs. Nevertheless, in our analysis, we give ALG *some* additional power: we assume ALG can fractionally transmit jobs that have been continuously processed since their arrival, i.e. for a job with value v and duration d that arrived at time t_0, after being processed for some $t < d$ time by time $t_0 + t$, ALG gains $\frac{t}{d}v$ value from the job. We define $\frac{\log \kappa}{2}$ job types, where the i'th job type has a duration of 4^i and a value of 2^i for each $0 \leq i < \frac{\log \kappa}{2}$. We build an input sequence that is composed of a series of *phases*. At the beginning of each phase, m jobs of each type are sent to the server. We define p_i to be the expected number of i-type jobs accepted by ALG. Note that since all $m\frac{\log \kappa}{2}$ arrived at the same time and they are all tight, at most m of them can be accepted by ALG. This means that $\sum_i p_i \leq m$. We define $r_i = \sum_{j \leq i} \frac{p_j}{2^{i-j}} + \sum_{j > i} \frac{p_j}{2^{j-i}}$. We note that

$$\sum_i r_i = \sum_i \sum_{j \leq i} \frac{p_j}{2^{i-j}} + \sum_i \sum_{j > i} \frac{p_j}{2^{j-i}} = \sum_i \left(p_i \sum_{j \geq i} 2^{i-j}\right) + \sum_i \left(p_i \sum_{j < i} 2^{j-i}\right)$$

$$= \sum_i 2p_i + \sum_i p_i \leq 3m$$

Hence, there exists some i such that $r_i \leq \frac{3m}{\frac{\log \kappa}{2}} = \frac{6m}{\log \kappa}$. OPT processes the jobs of type i during this phase and gains a value of $2^i m$. ALG's expected gain during this phase is $\sum_{j \leq i} 2^j p_j + \sum_{j > i} 2^i \frac{2^{j+1}}{4^{j+1}} p_j$. This is since ALG may be semi-fractional. Denote the k'th phase as σ_k. Then

$$\frac{ALG(\sigma_k)}{OPT(\sigma_k)} = \frac{1}{m}\left(\sum_{j \leq i} 2^{j-i} p_j + \sum_{j > i} \frac{2^{j+1}}{4^{j+1}} p_j\right) \leq \frac{1}{m}\left(\sum_{j \leq i} 2^{j-i} p_j + \sum_{j > i} 2^{-j-1} p_j\right)$$

$$\leq \frac{1}{m}\left(\sum_{j \leq i} 2^{j-i} p_j + \sum_{j > i} 2^{i-j} p_j\right) \leq \frac{r_i}{m} \leq \frac{6}{\log \kappa}$$

We begin a new phase immediately when OPT completes processing its jobs. We repeat this process N times, for some large N. Note that due to ALG being semi-fractional, it can only improve its situation by replacing a partially processed job of type i with the job of type i that arrives at the beginning of the new phase. Hence we can assume that ALG preempts the currently processed job immediately before the start of a new phase. Thus the analysis holds for all phases but the last one, where ALG has a gain of at most mV. We denote the full input sequence by σ. Using the claims above, we get that

$$\frac{OPT(\sigma)}{ALG(\sigma)} = \frac{\sum_i OPT(\sigma_i)}{mV + \sum_i ALG(\sigma_i)} \geq \frac{\frac{\log \kappa}{6}\left(\sum_i ALG(\sigma_i)\right)}{mV + \sum_i ALG(\sigma_i)}$$

Thus as the number of phases N tends to ∞, the ratio tends to $\frac{\log \kappa}{6}$ which implies the lower bound. $\square$

3   Buffer Management

In this section, we discuss the buffer management problem for packets with processing requirements. We begin by discussing the case that arriving packets have unit values. We present a 2-competitive deterministic algorithm called $SRPTB$, and a φ lower bound on the competitive ratio of all deterministic algorithms for this problem. We then discuss packets with constant densities, and show a 4-competitive deterministic algorithm called KP. Finally, we present a lower bound on the competitive ratio of any randomized algorithm for packets with arbitrary values and processing requirements.

3.1   Packets with Unit Values

We begin by defining the $SRPTB$'s scheduling policy and its buffer management policy:

i *Scheduling*: always process the packet with the shortest remaining processing time (SRPT). Break ties by processing the packet closer to the head of the buffer.

ii *Buffer Management*: accept a packet p if there is room in the buffer, or if the buffer is full and the packet in the buffer with the greatest remaining processing time, q, has more processing time remaining than p (in this case, preempt q). The algorithm also maintains a counter of *completed* packets, i.e. packets whose processing time has reached 0 (including both transmitted packets and those still in the buffer). When this counter reaches B, the buffer is *cleared*, i.e. all packets that have positive remaining processing time are preempted, and then all packets still remaining in the buffer (if any), all of which have 0 remaining processing time, are transmitted. The counter is reset once the buffer is cleared.

Theorem 2. *$SRPTB$ is at most 2-competitive.*

The following lemma shows that the analysis of $SRPTB$'s competitive ratio is tight:

Lemma 1. *$SRPTB$ is at least $(2 - \frac{1}{B})$-competitive.*

We proceed to show a lower bound on the competitive ratio of any deterministic algorithm for this problem. The lower bound is shown using a simple choice between one of two possible input sequences.

Theorem 3. *Any deterministic online algorithm ALG is at least $(\varphi - \frac{1}{B})$-competitive, where $\varphi \approx 1.618$ is the golden ratio.*

Proof. Given a deterministic online algorithm ALG, we design an input sequence σ on which the number of packets transmitted by OPT is at least $(\varphi - \frac{1}{B})$ times more than those transmitted by ALG. At time 0, we send $B - \lfloor \frac{B}{\varphi} \rfloor$ packets whose required processing time is B. We call these *type-a* packets. Immediately thereafter, we send $\lfloor \frac{B}{\varphi} \rfloor$ packets whose required processing time is 1. We call these *type-b* packets. We then wait until time $t = \lfloor \frac{B}{\varphi} \rfloor$. At this time there are two options for the remainder of the input sequence:

i No packets were transmitted by ALG by this time. In this case, we send B type-b packets. We then wait for ALG to finish processing and transmitting its buffer. Regardless of what ALG decided to do with the new type-b packets, the maximum number of packets that ALG can transmit is B, the number of packets it can have in its buffer. OPT, on the other hand, could have rejected all type-a packets. It could then have transmitted all $\lfloor \frac{B}{\varphi} \rfloor$ initial type-b packets before time t. It could then transmit all B new type-b packets that arrived at time t. Thus we get

$$\frac{OPT(\sigma)}{ALG(\sigma)} \geq \frac{B + \lfloor \frac{B}{\varphi} \rfloor}{B} \geq 1 + \frac{1}{\varphi} - \frac{1}{B} = \varphi - \frac{1}{B}$$

ii Some packets were transmitted by ALG during this time. In this case, we send no more new packets. All the packets transmitted by ALG must have been type-b. This is since $t < B$, so no type-a packet could have been fully processed. Due to the FIFO nature of the buffer, all type-a packets must have been preempted or rejected. This means that ALG's maximum gain is $\lfloor \frac{B}{\varphi} \rfloor$, the total number of type-b packets. OPT can simply process and transmit all B packets. This yields

$$\frac{OPT(\sigma)}{ALG(\sigma)} \geq \frac{B}{\lfloor \frac{B}{\varphi} \rfloor} \geq \varphi$$

In either case, we get that ALG's competitive ratio on the input sequence is $\frac{OPT(\sigma)}{ALG(\sigma)} \geq \varphi - \frac{1}{B}$, as desired. By repeating the sequence, we conclude that the lower bound holds even if we allow the competitive ratio an additive constant.

$\square$

3.2   Packets with Constant Density

In this section we limit ourselves only to packets that have a constant density. We present an algorithm called $KeepPackets$, or KP, which is 4-competitive in this setting. We first define the algorithm:

i *Scheduling*: greedily process the packet at the head of the buffer.
ii *Buffer Management*: accept a packet if there is room in the buffer, or if the buffer is full and the packet in the buffer with the smallest value, q, has $v(q) \leq \frac{1}{2} v(p)$ (in this case, preempt q).

We prove the following upper bound on the competitive ratio of KP:

Theorem 4. *The KP algorithm is at most 4-competitive.*

Proof. We assume WLOG that KP's buffer is empty only at the beginning and the end of its processing of the input sequence σ. Otherwise, we could simply divide σ into phases $\sigma_1, \ldots, \sigma_k$, where our assumption holds in each σ_i. We could then analyze KP's operation on each σ_i independently. Let σ be some input sequence. As explained above, we assume that during KP's operation on σ its buffer is only empty at times 0 and t_{max}, where after time t_{max} no more packets arrive. Let $0 < t \le t_{max}$ be some time step. If OPT doesn't schedule a packet at this time, then it obviously has no gain at this time step. Otherwise, at this time, OPT schedules some packet p. If p is not eventually transmitted by OPT then it gains nothing from processing p, and otherwise over the $r(p)$ time steps that OPT takes to process p, it gains $v(p) = r(p)$. Thus, it can be said that OPT has a gain of at most 1 at time step t, and we see that OPT's gain until t_{max}, $OPT_{t \le t_{max}}(\sigma) \le t_{max}$. We now analyze KP's gain until time t_{max}. KP's buffer is not empty, so at any time it schedules some packet, q. We inspect q's preemption chain, $q_1, q_2, \ldots, q_j = q, \ldots, q_n$, such that q_i was preempted at the arrival of q_{i+1} and q_n was accepted by KP. By the buffer management policy, $v(q_i) \le \frac{1}{2} v(q_{i+1})$. This means that

$$\sum_{i=1}^{n} v(q_i) \le 2v(q_n)$$

We denote by c_i the i'th preemption chain, and by n_i its length. We denote the packets of the i'th preemption chain by $c_{i,1}, \ldots, c_{i,n_i}$. We see that the value gained by KP until t_{max}, which is precisely all of KP's gain, holds:

$$KP(\sigma) = \sum_i v(c_{i,n_i}) \ge \frac{1}{2} \sum_i \left(\sum_{j=1}^{n_i} v(c_{i,j}) \right) \ge \frac{1}{2} t_{max} \ge \frac{1}{2} OPT_{t \le t_{max}}(\sigma) \qquad (1)$$

The final inequality holds since $\{c_{i,j}\}$ contains at least all the packets scheduled by KP until t_{max}, and as of such their total duration (and their total value) is at least t_{max}.

It remains to see what happens after time t_{max}. By the definition of t_{max}, no new packets arrive after this time. This means that OPT can gain at most the total value stored in its buffer at time t_{max}. We match each packet p in OPT's buffer to some packet q transmitted by KP, in a manner such that the sum of the values of the packets in OPT's buffer is at most twice the sum of the values gained by KP from the packets in the matched packets' preemption chains. This will mean that the value in OPT's buffer at time t_{max} is at most $2KP(\sigma)$. The matching scheme is as follows:

i We first match all packets in OPT's buffer at time t_{max} that where accepted by KP. For each such packet p we match it with itself. Clearly this matching is one-to-one. The sum of values of all such packets is clearly at most twice the value of the packets at the head of the matched preemption chains, which is precisely the value gained by KP from these preemption chains.

ii We now match the remaining packets. Let p be a packet in OPT's buffer at time t_{max} that was rejected by KP, and let $t = a(p)$. As p is rejected by KP, KP's buffer is full at time t. As we are matching at most B packets, there exists some packet q in KP's buffer at time t such that no packet in q's preemption chain is already matched (each packet in KP's buffer at a given time belongs to a different preemption chain). We then match p with the packet q' at the head of q's preemption chain. As p was rejected, $v(q') \geq v(q) > \frac{1}{2}v(p)$.

This matching gives us that

$$OPT_{t>T}(\sigma) \leq 2KP(\sigma) \tag{2}$$

We conclude the proof. By combining inequalities (1) and (2), we get that

$$\frac{OPT(\sigma)}{KP(\sigma)} = \frac{OPT_{t \leq t_{max}}(\sigma) + OPT_{t > t_{max}}(\sigma)}{KP(\sigma)} \leq \frac{2KP(\sigma) + 2KP(\sigma)}{KP(\sigma)} = 4$$

$\square$

The following lemma shows that the analysis of KP's competitive ratio is tight:

Lemma 2. KP is at least $(4 - \frac{1}{B})$-competitive.

3.3 Arbitrary Value and Processing Requirement

We begin by providing a lower bound on the competitive ratio of any deterministic algorithm for packets with general values:

Lemma 3. *Any deterministic algorithm for this model is $\Omega(\min(V, \sqrt{T}))$-competitive, for V and T defined above.*

The above lower bound shows than no constant-competitive deterministic algorithm exists for the model with general values. Thus we inspect randomized algorithms and show a lower bound for them. This lower bound shows us that even randomized algorithms for this problem cannot be constant-competitive. This means that the model with added values is substantially more difficult than that previously described. The lower bound we provide holds even for the non-FIFO setting, where completed packets can be transmitted from any place in the buffer. Note that for the non-FIFO case, one can easily find 1-competitive algorithms for packets with unit values or unit processing times. The construction of this lower bound is similar to that in Theorem 1.

Theorem 5. *Any randomized preemptive algorithm in this setting has a competitive ratio which is $\Omega(\frac{\log \kappa}{\log \log \kappa})$ for κ defined above, even if the buffer is not FIFO.*

Proof. We define $\gamma = \frac{\log \kappa}{\log \log \kappa}$. We define $\frac{\gamma}{2}$ packet types, where the i'th packet type has a processing requirement of γ^{2i} and a value of γ^i for each $0 \leq i < \frac{\gamma}{2}$. Note that $\gamma^\gamma \leq \kappa$, and as of such all defined packet types have legal processing requirements and values. We build an input sequence that is composed of a series of *phases*. At the beginning of each phase, B packets of each type are sent to the server. We define e_i to be the expected number of type-i packets that ALG accepts. Note that since all $B\frac{\gamma}{2}$ packets arrive at the same time, at most B of them can be accepted by ALG. This means that $\sum_i e_i \leq B$. As there are $\frac{\gamma}{2}$ packet types, there exists some i such that $e_i \leq \frac{2B}{\gamma}$. We assume that OPT processes the B type-i packets during this phase. This means that during the phase, OPT has a gain of $\gamma^i B$ over a time span of $\gamma^{2i} B$. ALG's expected gain during this phase has 3 components:

i Gain from type-i packets: this is clearly at most $\gamma^i e_i \leq 2\gamma^{i-1}B$.

ii Gain from smaller packets: there are at most B smaller packets in ALG's buffer, each with value at most γ^{i-1}. Thus the gain from these packets is at most $\gamma^{i-1}B$.

iii Gain from larger packets: there are at most B larger packets in ALG's buffer. The total time for which they are processed is at most $\gamma^{2i}B$. Notice that each of these has a density of at most $\frac{\gamma^{i+1}}{\gamma^{2i+2}} = \frac{1}{\gamma^{i+1}}$. Thus the gain from these packets is at most $\frac{\gamma^{2i}B}{\gamma^{i+1}} = \gamma^{i-1}B$.

If we denote the n'th phase as σ_n, notice that this means that

$$\frac{ALG(\sigma_n)}{OPT(\sigma_n)} \leq \frac{2\gamma^{i-1}B + \gamma^{i-1}B + \gamma^{i-1}B}{\gamma^i B} \leq \frac{4}{\gamma}$$

We begin a new phase once OPT finishes processing its packets. In the above calculation, we gave ALG gain proportional to the fractional part of packets it has processed, regardless of their position in the buffer. As of such, ALG can only improve its situation by replacing a partially processed packet of type i with a packet of type i that arrives at the beginning of the new phase. Hence we can assume that ALG preempts all currently processed packets immediately before the start of a new phase. We repeat this process N times, for some large N. Thus the above analysis holds for all phases but the last one, where ALG has a gain of at most VB. We denote the full input sequence by σ. Using the claims

$$\frac{OPT(\sigma)}{ALG(\sigma)} = \frac{\sum_i OPT(\sigma_i)}{VB + \sum_i ALG(\sigma_i)} \geq \frac{\frac{\gamma}{4}(\sum_i ALG(\sigma_i))}{VB + \sum_i ALG(\sigma_i)}$$

Thus as the number of phases N tends to ∞, the ratio tends to $\frac{\gamma}{4} = \Omega(\frac{\log \kappa}{\log \log \kappa})$. This means that the input sequence σ gives the required lower bound. $\square$

References

1. Aiello, W.A., Mansour, Y., Rajagopolan, S., Rosén, A.: Competitive queue policies for differential services. In: Proceedings of IEEE INFOCOM, pp. 431–440 (2000)
2. Albers, S., Schmidt, M.: On the performance of greedy algorithms in packet buffering. SIAM Journal on Computing 35(2), 278–304 (2005)
3. Azar, Y., Litichevskey, A.: Maximizing throughput in multi-queue switches. In: Albers, S., Radzik, T. (eds.) ESA 2004. LNCS, vol. 3221, pp. 53–64. Springer, Heidelberg (2004)
4. Azar, Y., Richter, Y.: An improved algorithm for cioq switches. ACM Transactions on Algorithms 2(2), 282–295 (2006)
5. Canetti, R., Irani, S.: Bounding the power of preemption in randomized scheduling. In: Proceedings of the 27th Annual ACM Symposium on Theory of Computing, pp. 606–615. ACM (1995)
6. Chuprikov, P., Nikolenko, S., Kogan, K.: Priority queueing with multiple packet characteristics (2015)
7. DasGupta, B., Palis, M.A.: Online real-time preemptive scheduling of jobs with deadlines. In: Jansen, K., Khuller, S. (eds.) APPROX 2000. LNCS, vol. 1913, pp. 96–107. Springer, Heidelberg (2000)
8. Englert, M., Westermann, M.: Lower and upper bounds on FIFO buffer management in qoS switches. In: Azar, Y., Erlebach, T. (eds.) ESA 2006. LNCS, vol. 4168, pp. 352–363. Springer, Heidelberg (2006)
9. Englert, M., Westermann, M.: Considering suppressed packets improves buffer management in qos switches. In: Proceedings of the 18th Annual ACM-SIAM Symposium on Discrete Algorithms, pp. 209–218. Society for Industrial and Applied Mathematics (2007)
10. Garay, J.A., Naor, J., Yener, B., Zhao, P.: On-line admission control and packet scheduling with interleaving. In: Proceedings of 21st Annual Joint Conference of the IEEE Computer and Communications Societies, vol. 1, pp. 94–103. IEEE (2002)
11. Goldwasser, M.: A survey of buffer management policies for packet switches. ACM SIGACT News 41(1), 100–128 (2010)
12. Kalyanasundaram, B., Pruhs, K.: Speed is as powerful as clairvoyance. Journal of the ACM 47(4), 617–643 (2000)
13. Kesselman, A., Kogan, K., Segal, M.: Packet mode and qos algorithms for buffered crossbar switches with fifo queuing. Distributed Computing 23(3), 163–175 (2010)
14. Kesselman, A., Kogan, K., Segal, M.: Improved competitive performance bounds for cioq switches. Algorithmica 63(1-2), 411–424 (2012)
15. Kesselman, A., Lotker, Z., Mansour, Y., Patt-Shamir, B., Schieber, B., Sviridenko, M.: Buffer overflow management in qos switches. SIAM Journal on Computing 33(3), 563–583 (2004)
16. Kesselman, A., Patt-Shamir, B., Scalosub, G.: Competitive buffer management with packet dependencies. In: Proceedings of the IEEE International Symposium on Parallel & Distributed Processing, pp. 1–12. IEEE (2009)
17. Kogan, K., López-Ortiz, A., Nikolenko, S., Sirotkin, A.: Multi-queued network processors for packets with heterogeneous processing requirements. In: 5th International Conference on Communication Systems and Networks, pp. 1–10. IEEE (2013)
18. Kogan, K., López-Ortiz, A., Nikolenko, S., Sirotkin, A.V., et al.: A taxonomy of semi-fifo policies. In: IEEE 31st International Performance Computing and Communications Conference (IPCCC), pp. 295–304. IEEE (2012)

19. Kogan, K., Nikolenko, S.: Single and multiple buffer processing (2014)
20. Kogan, K., Nikolenko, S., López-Ortiz, A., Scalosub, G., Segal, M.: Balancing work and size with bounded buffers. In: COMSNETS, pp. 1–8 (2014)
21. Koren, G., Shasha, D.: D over; an optimal on-line scheduling algorithm for overloaded real-time systems. In: Real-Time Systems Symposium, pp. 290–299. IEEE (1992)
22. Mansour, Y., Patt-Shamir, B., Lapid, O.: Optimal smoothing schedules for real-time streams. In: Proceedings of the 19th Annual ACM Symposium on Principles of Distributed Computing, pp. 21–29. ACM (2000)
23. Mansour, Y., Patt-Shamir, B., Rawitz, D.: Overflow management with multipart packets. Computer Networks 56(15), 3456–3467 (2012)
24. Pruhs, K.: Competitive online scheduling for server systems. ACM SIGMETRICS Performance Evaluation Review 34(4), 52–58 (2007)
25. Pruhs, K., Sgall, J., Torng, E.: Online scheduling. In: Handbook of Scheduling: Algorithms, Models, and Performance Analysis, pp. 115–124 (2003)

A Triplet-Based Exact Method
for the Shift Minimisation Personnel Task
Scheduling Problem

Davaatseren Baatar[1], Mohan Krishnamoorthy[2], and Andreas T. Ernst[3]

[1] School of Mathematical Sciences, Monash University, Clayton, Australia
[2] Department of Mechanical and Aerospace Engineering, Monash University
[3] CSIRO, Clayton, Australia

Abstract. In this paper we describe a new approach for solving the shift minimisation personnel task scheduling problem. This variant of fixed job scheduling problems arises when tasks with fixed start and end times have to be assigned to personnel with shift time constraints. We present definitions, formulations and briefly discuss complexity results for the variant that focuses on minimising the number of machines (or workers) that are required to schedule all jobs. We first develop some mathematical properties of the problem and subsequently, the necessary and sufficient conditions for feasibility. These properties are used to develop a new branch and bound scheme, which is used in conjunction with two column generation based approaches and a heuristic algorithm to create an efficient solution procedure. We present extensive computational results for large instances and thereby, empirically demonstrate the effectiveness of our new approach.

1 Introduction

There are many situations in which tasks or jobs with fixed start and end times have to be scheduled on multiple machines. Such problems are variously called the *fixed job scheduling problem* (FJSP), *interval scheduling problem* or *task scheduling problem*. We come across fixed start and end times for tasks (or activities) when there is an externally imposed timetable on tasks and/or machines. For example, airline schedules imply that cleaning, maintenance, baggage handling and other tasks at airports have fixed times.

Recently the FJSP with spread time constraints has received considerable attention in the literature (see for example [1, 3, 12, 15, 19–22]) In these problems, there is a limit on the maximum time between the start of the first task and the end of the last task assigned to a machine or worker. Another type of constraint is a working time constraint that limits the total duration of all jobs assigned to a machine irrespective of the amount of idle time in between (see [21]). The variant presented in [12] has the machines different in the speed at which they process tasks, so that intervals always have the same start time but different finishing time depending on the machine that they are assigned to. Alternatively machines may differ in the cost for completing tasks or in terms of whether or not they are, in some sense, qualified to carry out tasks. The *operational* variant

© Springer-Verlag Berlin Heidelberg 2015
N. Bansal and I. Finocchi (Eds.): ESA 2015, LNCS 9294, pp. 59–70, 2015.
DOI: 10.1007/978-3-662-48350-3_6

of these problems aims to maximise the number or value of tasks completed
with a fixed number of machines. The *tactical* variant minimises the number of
machines required to complete all of a given list of tasks.

These problems are also referred to as *interval scheduling* (e.g. [11]). The bus
driver scheduling discussed in [4] is an example application of this problem.
Surveys of the interval scheduling literature can be found in [7, 8].

The motivation for our work in this paper is to solve the Shift Minimization
Personnel Task Scheduling Problem (SMPTSP). This variant of the personnel
task scheduling problem [9] (PTSP) was first introduced in [10]. The SMPTSP
is a tactical fixed job scheduling problem with (machine) availability and (machine to job) qualification constraints. A heuristic approach that is provided in
[10] solves large instances of the SMPTSP. A metaheuristic approach for solving
the SMPTSP is published in [17] and a more recent heuristic for this problem
is described in [13]. The SMPTSP, which is derived from a personnel scheduling
application, is not limited to rostering problems faced in large organizations.
It can be extended to solve a wide variety of applications: in the telecommu-
nications industry, in production centers with high product variety, car rental
applications, luxury cottage rental applications and so on.

The SMPTSP consists of assigning a set of *tasks* with specific start and end
times to *workers* or *shifts* who have specific skill sets and availability intervals.
The qualification constraints specify which workers are qualified to perform each
task. The objective is to find a feasible assignment of all of the tasks that minimises the number of workers used. This objective occurs, for example, when
casual workers are used to complete the tasks on a given day and it is advisable
to minimise the total number of staff that are employed.

In this paper, we provide a novel exact solution approach for the SMPTSP.
We first present an integer programming (IP) formulation and provide basic notation and definitions. Using mathematical properties of the problem developed
in this paper we create a unique branching scheme for this problem. This is
used in conjunction with two column generation formulations. We then present
computational results and analysis of the algorithmic alternatives.

2   IP Formulation and Complexity

Throughout the paper we use the following notation

$J = \{1, \ldots, n\}$ The set of tasks or jobs,

$W = \{1, \ldots, m\}$ The set of workers that can perform tasks,

s_j, f_j The start time and finish time of the tasks, respectively,

P_j The set of staff or workers that can perform task j,

T_w The set of tasks that worker $w \in W$ can perform.

The aim is to solve a minmax problem of completing as many tasks as possible
with the minimum number of people The sets P_j and T_w may also be defined
in terms of the skills or preferences of workers, although they may also take
into account start and end times of availability (e.g. shift start and end times).
Multi–tasking is not allowed. This leads us to define the following set of tasks
which is the fundamental block for the formulation of the problem.

Definition 1. *A set of tasks $K \subseteq J$ is called a clique if all tasks overlap for some interval of time in the planning time horizon. A* **maximal clique** *is a clique that cannot be extended further, that is for which no additional task in $J \setminus K$ overlaps with the tasks in the clique.*

Obviously, for any clique at most one task can be assigned to the same worker. Let $C = \{K_1, \ldots, K_p\}$ be the set of all *maximal cliques* defined on the set of tasks J. The set of maximal cliques can be found in polynomial time (see for example [6]). We can find the maximal clique set C^w, for each worker w, with respect to the tasks T_w.

To define the problem let $x_{jw} = 1$ if task $j \in J$ is assigned to worker $w \in W$. Let $y_w = 1$ if worker $w \in W$ is used and zero otherwise. Also, let $u_j = 1$ if task j is unallocated. The SMPTSP can then be formulated as the following IP:

Problem $\mathbb{P}$: $\min \quad \sum_{w \in W} y_w + \sum_{j \in J} M\, u_j$

$$s.t. \quad \sum_{w \in P_j} x_{jw} + u_j = 1 \qquad \forall\, j \in J \tag{1}$$

$$\sum_{j \in K} x_{jw} \leq y_w \qquad \forall\, w \in W,\ K \in C^w \tag{2}$$

$$x_{jw} \in \{0,1\},\ y_w \in [0,1],\ u_j \in [0,1] \quad \forall j \in J,\ w \in W$$

where M is a sufficiently large number.

Problem $\mathbb{P}$ is always feasible, since $u_j = 1$, for all $j \in J$, is a trivial solution to the problem. Moreover, for the instances of the problem $\mathbb{P}$ where all tasks can be assigned to machines without multi–tasking and preemption, we could fix all variables $u_j \equiv 0$. For those instances ($\mathbb{P}$) is shown to be a *NP–hard* problem – see [10]. Therefore, obviously, ($\mathbb{P}$) is *NP–hard* as well.

However, some special cases of the problem can be solved in polynomial time. In [10] it is shown that if all workers are identical in terms of 'qualification', $T^w = P$ and $P_j = W$ for all w and j, then ($\mathbb{P}$) can be solved in $O(n \log n)$ with a 'sorting of jobs' dominating the algorithm. This can be achieved by assigning the tasks, in order of its start time, to an available worker, with a preference to available workers who have been used in the solution before (already).

As a special case of the polynomially solvable variant with no qualification constraints, consider the case where there is only a single worker. In what follows we are interested in solving a variant of the single worker problem which we refer to as the *packing problem*. Consider a worker w and the following problem of assigning the highest value set of tasks to the worker given a set of prices π.

Problem $Pack(\pi, T_w)$: $\max_{x_j \in \{0,1\}} \sum_{j \in T_w} \pi_j x_j$ s.t. $\sum_{j \in K} x_j \leq 1 \quad \forall\, K \in C^w.$ (3)

From each maximal clique at most one task is assigned to the worker such that the objective function is maximized. Hereafter we refer to this problem as the *packing problem* and denote it as $Pack(\pi, T_w)$. In [10] it is shown that the packing problem can be solved in polynomial time and it was suggested to use a longest path algorithm, that is a dynamic programming approach on the acyclic network, to solve the packing problem.

3   Mathematical Foundations of the Problem

This section investigates the mathematical properties of the problem $\mathbb{P}$ in order to identify necessary and sufficient conditions for feasibility. To do so, we summarize some relevant results on totally unimodular matrices (see [14, 16]).

For any totally unimodular (TU) matrix A and for an integral vector b, the polyhedron $P := \{x | Ax \leq b\}$ is integral (i.e. every vertex of P is integral). Hence we can find an integer solution by solving the LP relaxation of the corresponding IP. It is well known that consecutive ones matrices (C1 matrices) are TU. A $C1$ matrix is one in which there is permutation of the columns (rows) so that all ones in each row (or column) are consecutive. The following proposition follows immediately from Ghouila–Houri's characterisation of TU matrices [5]:

Proposition 1. *Let A, B be $\{0, 1, -1\}$ matrices and B has at most one non-zero entry in each column. Then A is TU if and only if the matrix $[A\ B]$ is TU.*

3.1   Unimodularity in Submatrices of $\mathbb{P}$

Let $C = \{K_1, \ldots, K_p\}$ be the set of all *maximal cliques* defined on the set of tasks J. Consider a matrix $A_{p \times n}$ defined by C, where rows and columns represent maximal cliques and tasks respectively, i.e. $a_{ij} = 1$ if task $j \in K_i$ and $a_{ij} = 0$ otherwise. We refer to this matrix as the (maximal) *clique–task incidence matrix*. Sorting cliques by time gives consecutive ones in each column, hence:

Proposition 2. *The clique incidence matrix is a C1's matrix.*

Corollary 1. *The clique incidence matrix is totally unimodular (TU).*

Let A^w be the maximal clique incidence matrix defined on the set T_w for worker w. Due to Proposition 2 each A^w is TU. This provides another proof that $Pack(\lambda^w, T_w)$ can be solved in polynomial time. Without loss of generality, we assume that 1's are consecutive in each column of A^w.

Consider the coefficient matrix corresponding to variables of the problem $\mathbb{P}$

$$\begin{pmatrix} B^1 & B^2 & B^3 & \ldots & B^m & 0 & I \\ A^1 & 0 & 0 & \ldots & 0 & D^1 & 0 \\ 0 & A^2 & 0 & \ldots & 0 & D^2 & 0 \\ \ldots & \ldots & \ldots & \ldots & \ldots & \ldots & \ldots \\ 0 & 0 & 0 & \ldots & A^m & D^m & 0 \end{pmatrix} \tag{4}$$

B^w is a $n \times |T_w|$ binary matrix where columns represent tasks that can be performed by the worker and the rows represent the set of tasks J; and each row and column has a single non–zero element representing the task.

I is the identity matrix,

A^w is the maximal clique–task incidence matrix of the worker w.

D^w represents the coefficients corresponding to variable y_w thus has 0's everywhere except w–th column, where all elements are 1.

In general, the matrix (4) is not TU. However, some parts (submatrices) of the matrix are TU. We will exploit this structure to develop new solution approaches for $\mathbb{P}$.

Theorem 1. *Let the coefficient matrix of $\mathbb{P}$ be given in the form as in (4). Then*

1. *each block matrix is a C1's matrix with 1's consecutively in a single block,*
2. *each column partition in (4) is a TU matrix.*

Proof. The proof immediately follows from the definition of the matrices A^w, B^w, D^w and I; and the Corollary 1. $\qquad\square$

Moreover, certain types of TU sub–matrices can be obtained from the matrix (4). Let us consider the following block matrix:

$$\begin{pmatrix} \bar{B}^1 & \bar{B}^2 & \bar{B}^3 & \dots & \bar{B}^m & I \\ A^1 & 0 & 0 & \dots & 0 & 0 \\ 0 & \bar{A}^2 & 0 & \dots & 0 & 0 \\ \dots & \dots & \dots & \dots & \dots & \dots \\ 0 & 0 & 0 & \dots & \bar{A}^m & 0 \end{pmatrix} \tag{5}$$

This was obtained from (4) by removing columns corresponding to some of the x_{jw} and the entire column partition corresponding to y_w variables. Some rows of the matrices $\bar{A}^w$ may no longer represent maximal cliques but they all represent cliques (not necessarily unique so that $\bar{A}^w$ may have multiple identical rows). Note that the $\bar{A}^w$ are a column-wise C1's matrices.

Theorem 2. *For any worker w and for any two cliques K_p and K_q, represented by the rows of the matrix $\bar{A}^w$, the matrix (5) is TU if*

$$rK_p \cap K_q \neq \emptyset \quad \text{implies either} \quad K_p \subseteq K_q \text{ or } K_q \subseteq K_p \tag{6}$$

3.2   Existence of Non–trivial Solutions

Problem $\mathbb{P}$ has the nice property that certain fractional solutions of the LP relaxation of the problem ($\mathbb{LP}$) have an associated integer solution to $\mathbb{P}$.

Let us consider a feasible solution $(\bar{x},\ \bar{y},\ \bar{u})$ of the $\mathbb{LP}$.

Definition 2. *The solution $(\bar{x},\ \bar{y},\ \bar{u})$ has a **conflicting triplet** if for some $w \in W$ there exits a triplet $(\bar{x}_{j_1w}, \bar{x}_{j_2w}, \bar{x}_{j_3w}) > 0$ such that no clique $K \in C^w$ contains all three tasks j_1, j_2 and j_3, but there exist cliques K_p and K_q in C^w such that $j_1, j_2 \in K_p$ and $j_2, j_3 \in K_q$.*

For a conflicting triplet jobs j_1 and j_3 do not overlap but j_2 overlaps with both, and all three jobs have a non-zero allocation to the same worker. We refer to a feasible $\mathbb{LP}$ solution as *triplet free* if it doesn't contain any conflicting triplets.

Note that, any integer solution to $\mathbb{P}$ is a triplet free solution to $\mathbb{LP}$ due to the constraints (2). The converse is also true. If we only consider a triplet free set of possible task allocations then each task only appears in one of the constraints (2) for each shift. Hence $\mathbb{P}$ becomes a linear assignment assignment problem and $\bar{x}$ is a fractional assignment. It is well known that feasible set of the linear assignment problem is integral. Thus, using simplex methods we can get an integral solution which, obviously, must be not worse than the fractional solution." Thus triplet free solutions of $\mathbb{LP}$ are integer and any integer solution is a triplet free solution. This property can be used to develop a novel branching scheme to solve $\mathbb{P}$ as described in the next section.

4   A Triplet-Based Branch and Bound Algorithm

In this section we describe a branching strategy based on the notion of conflicting triplets. If at the current node, the LP relaxation $\mathbb{LP}$ has a conflicting triplet then we branch on x variables. For the chosen conflicting triplet $(\bar{x}_{j_1w}, \bar{x}_{j_2w}, \bar{x}_{j_3w}) > 0$ we create two descendant nodes. If $j_1, j_2 \in K_p$ and $j_2, j_3 \in K_q$, for some cliques $K_p, K_q \in C^w$, then we branch on x_{j_2w} fixing its value to 0 and 1. Note that, for all descendant nodes of the current node the triplet $(x_{j_1w}, x_{j_2w}, x_{j_3w})$ is conflict free for any feasible solution to $\mathbb{LP}$.

In the case that the solution of $\mathbb{LP}$ at the current node is triplet free, then we can always find an integer optimal solution $(\tilde{x}, \tilde{u})$ to the auxiliary problem $\mathbb{P}_{aux}$ and an integer solution to $\mathbb{P}$ is constructed as follows:

$$\tilde{y}_w = \begin{cases} 1 \text{ if } \exists\ j \in T_w \text{ s.t. } \tilde{x}_{jw} > 0 \\ 0 \text{ otherwise} \end{cases} \tag{7}$$

Note that the integer solution $(\tilde{x}, \tilde{y}, \tilde{u})$ to $\mathbb{P}$ is not necessarily an optimal solution for $\mathbb{P}$ at the current node. Thus, in order to find a better integer solution, we branch on y_w variables. We add a descendant node for each worker w used in the integer solution by fixing it to 0, unless it has a fixed assignment. In other words, if $\tilde{y}_w > 0$ and none of the variables x_{jw} are fixed to 1 due to branching then we add a node with $y_w = 0$.

An initial upper bound is obtained at the root node using the heuristic described in [10]. At each node of the branch and bound (B&B) tree, the branching variable is chosen in the following way:

1. If the solution of the LP relaxation has a conflicting triplet then the task j which is most split (i.e. has the greatest number of positive x_{jw} values) and has a conflicting triplet is chosen. For the chosen task j we choose the worker w for which $\bar{x}_{jw}$ has the largest value and is a member of a conflicting triplet. We create two branches $x_{jw} = 0$ and $x_{jw} = 1$.

2. If the solution of the LP relaxation is triplet free then an integer solution is obtained as described above. If the objective value is worse than the objective value of the LP relaxation, then for each $\tilde{y}_w = 1$ we add a branch $y_w = 0$ (unless one of the variables x_{jw} is fixed to 1).

The node selection of the B&B algorithm uses the following hierarchical comparison to determine the next node. Node A is processed before node B if

1. the lower bound at node A is strictly smaller than the one at node B;
2. node A is branching on a y variable and node B is on an x variable;
3. both of them branch on y variables but (a) A is in a deeper position (in terms of level); or in case of ties (b) A has larger node number;
4. node A has branched $x_{j_1 w_1} = 1$ and node B has $x_{j_2 w_2} = 0$;
5. A is in a deeper position;
6. A has larger node number (was generated later) in the case of ties.

From our numerical tests we have observed that $\mathbb{LP}$ provides a tight lower bound. Thus, we always pursue the node with lowest lower bound. If there are several nodes with an identical lower bound then a node branching on variable y is chosen. That gives us a chance to make the upper bound tighter quickly. If there is no node branching on variable y then we consider branches with $x_{jw} = 1$, which yields $\mathbb{LP}$ with a smaller number of variables than the branch with $x_{jw} = 0$. Moreover it increases the chance of getting a fractional solution with a smaller number of conflicting triplets. If there are several candidates we choose the node which is deeper in the branch and bound tree. Overall this processing order is designed to maximise the chance of finding a good integer solution as early as possible in the tree search.

As described above, the B&B method is used to explore triplet free solutions that could lead to an optimal solution of the problem $\mathbb{P}$. We refer to this approach as *triplet based branching*. We also tried different branching strategies to solve $\mathbb{P}$ such as branching on the x_{jw} variables where j is the task split across the most workers and w is the worker with largest fractional value. We implemented and tested this branching strategy, only to find out that with this approach we can solve only small sized problem instances. Moreover, the computation times were much worse than using the triplet based branching.

5   Column Generation Approaches

In the B&B approach we solve the LP relaxation $\mathbb{LP}$ at each node of the tree. As the problem size increases solving the underlying linear programs consumes most of the computation time (especially at root node) and reduces the ability to explore enough nodes in the B&B procedure to get the exact optimal solution within a fixed time limit. This motivated us to consider a column generation approach to $\mathbb{P}$ (see [2] for an introduction to column generation concepts).

A typical approach for column generation would be to use a Dantzig-Wolfe decomposition where variables z_{wP} represent a pattern (or complete schedule)

$P \subseteq T_w$ for worker w. This means that the problem can be written as a set-partitioning formulation one pattern per worker, and each task covered by one of these patterns (or assigned to u_j). Here $Pack(\pi, T_w)$ solves the pricing problem.

However we only had limited success with this formulation as the solutions are often highly fractional. We can apply column generation directly to the IP formulation $\mathbb{P}$. This approach is based on C1's property of the matrices A^w and hence will be referred to as CG–$C1$. Recall, that in each column of A^w 1's occur in a block. In other words, for each variable x_{jw} the non–zero coefficients corresponding to the constraints (2) occur in a single block. That can be achieved since, as we mentioned before, the Maximum Clique Algorithm extracts the maximal cliques by the order of their occurrence. Therefore, while extracting the maximal cliques we can get the start and end positions of the consecutive 1's block for each variable x_{jw}. We denote the start and end positions by ℓ_{jw} and r_{jw}, respectively. Since we know the start and end positions for 1's blocks the generation of a single column with negative reduced cost is quite straight forward.

However, to add a single column for each worker in each iteration is not efficient from computational point of view. Therefore, for each worker we added several columns, with negative reduced costs and non-overlapping tasks, at the same time. Recall, that if two tasks are not overlapping then there is no maximal clique that contains both of them. Therefore, one can interpret this column generation procedure as a packing problem of the tasks in T_w with new start and end times i.e.

$$\max \sum_{j \in T^w} \pi_j x_j \ \text{ s.t. } \sum_{j \in K} x_j \leq 1 \ \forall \ K \in \tilde{C}^w, \text{ and } x_j \in \{0,1\} \ \forall \ j \in J.$$

Here the maximal clique set $\tilde{C}^w$ is defined with respect to the set of tasks T_w, where each task j has start and end times ℓ_{jw} and r_{jw}, respectively, which represent the start and end position of the consecutive 1's. Therefore, new columns can be generated by solving the packing problem.

6   Computational Results

We implemented the algorithms using CPLEX 12.5 embedded in a C++ code. The algorithms are tested on benchmark data previously generated by the authors which are available from ORLIB[1]. The benchmark data pool contains 137 problem instances. However many of the small instances are too easy to solve on recent computers so we only provide results for larger instances. We also tested our method on ten instances created by [18][2]. We used $M = 1000$ in all tests. Tests were run on a cluster with nodes that have dual Intel Xeon E5-2680 CPUs (20 cores total) running at 2.8GHz.

Preliminary tests indicated that the C1's based column generation (CG–C1) is more efficient than the pattern formulation. Note that this is somewhat solver

---

[1] http://people.brunel.ac.uk/~mastjjb/jeb/orlib/ptaskinfo.html
[2] http://allserv.kahosl.be/~pieter/smptsp.zip

Table 1. Comparison of our triplet based branching scheme with column generation CG–C1 against CPLEX 12.5 and the heuristic in [13] on data sets from [10] with a time limit of 1800 seconds. * indicates that CPLEX failed to solve the root node, otherwise lower bounds for CPLEX and the triplet method are identical. For large instances CPLEX produced no solution in 1800sec (UB=—).

Dataset			CPLEX			Triplet		[13]	Dataset			Triplet			[13]
No.	|W|	|J|	LB	UB	secs	UB	secs	UB	No.	|W|	|J|	LB	UB	secs	UB
080	112	691	99	**99**	714	**99**	487	99	109	205	1115	157	161	1809	**159**
081	97	692	80	**80**	909	**80**	6	**80**	110	183	1143	155	159	1800	**157**
082	89	697	80	**80**	941	80	1345	81	111	155	1211	139	143	1802	**142**
083	222	700	180	—	1801	**180**	17	**180**	112	200	1213	169	**169**	21	171
084	136	718	120	**120**	1356	120	4	**120**	113	141	1221	110	**110**	25	**110**
085	217	720	180	**180**	1053	**180**	15	**180**	114	157	1227	138	**138**	22	141
086	178	721	140	**140**	371	**140**	6	**140**	115	228	1257	177	**177**	26	178
087	203	735	170	**170**	392	**170**	254	**170**	116	205	1262	176	**176**	64	**176**
088	137	777	120	—	1800	**120**	10	**120**	117	192	1285	149	**149**	14	150
089	88	788	70	**70**	963	**70**	15	**70**	118	180	1302	147	**147**	34	150
090	157	791	139	—	1800	**139**	13	**139**	119	236	1335	188	**188**	31	189
091	147	851	118	—	1800	**118**	22	**118**	120	228	1341	187	**187**	30	190
092	126	856	98	—	1800	**98**	24	99	121	147	1345	120	**120**	16	123
093	141	856	119	—	1800	**119**	15	120	122	422	1358	348	**348**	93	**348**
094	93	881	80	—	1800	**80**	372	82	123	187	1376	159	**159**	23	160
095	204	882	170	—	1800	**170**	7	**170**	124	198	1383	158	**158**	30	161
096	98	886	80	—	1800	**80**	10	**80**	125	157	1448	130	**130**	21	132
097	383	895	300	**300**	952	**300**	7	**300**	126	193	1462	167	**167**	38	170
098	91	896	80	—	1801	80	283	82	127	192	1472	170*	**170**	55	**170**
099	176	956	160	—	1801	**160**	16	**160**	128	207	1542	178	**178**	91	**178**
100	194	956	160	—	1801	**160**	16	**160**	129	233	1546	178	**178**	1271	180
101	166	997	140	—	1800	**140**	22	**140**	130	176	1562	140	**140**	49	**140**
102	179	997	138	—	1800	**139**	1809	**139**	131	415	1610	348	**348**	60	**348**
103	348	1024	300	—	1801	**300**	21	**300**	132	216	1645	186	**186**	206	189
104	181	1057	146	—	1801	**146**	111	150	133	211	1647	185	**185**	89	189
105	173	1075	150	—	1800	**150**	28	151	134	184	1776	160	**160**	62	**160**
106	121	1096	100	—	1800	**102**	1801	104	135	213	1988	179	182	1800	**180**
107	114	1112	100	—	1800	**103**	1801	**103**	136	216	2000	179*	**179**	100	**179**
108	162	1115	128	—	1800	**128**	21	132	137	245	2105	190	**190**	60	191

dependent – earlier experiments with an older versions of CPLEX concluded that the pattern formulation is useful as a method for the root node to give solutions more quickly. However in the results reported here we only use CG–C1.

Tests were run with a time limit of 1800 seconds. Note that the CPLEX implementation has an advantage on the parallel computing architecture used as it uses a fairly sophisticated parallel B&B implementation, while we only run the heuristic used for the initial solution in parallel, while the main triple based B&B algorithm and column generation are single threaded.

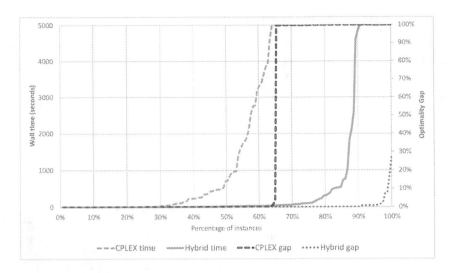

Fig. 1. Cumulative results for instances with a 5000 second elapsed time limit. Shows percentage of instances that could be solved with the indicated gap or elapsed time.

In Table 1, *LB* is the final lower bound obtained when terminating the B&B, *UB* is the best upper bound found, and *secs* the total computational elapsed time in seconds We also compare against heuristic methods from the literature. Results for the three phase heuristic method in [13] were taken as reported in the paper for runs of 1800 seconds. Clearly our exact approach outperforms [13] producing equal or better solutions in all but 4 of the 58 instance reported.

The cumulative performance of both CPLEX and our approach for the instances in both tables is summarized in Figure 1 where we also included the computational time when running the branch and bound for 5000 seconds. This clearly shows that the superior performance of our method is not dependent on the time limit. There are only a few instances that cannot be solved with the longer computational times and in all cases the upper bound is very close to the lower bound. By contrast there are many instances where CPLEX cannot find any integer solution (instances where *UB* =—). There are also two rare instances where CPLEX fails to complete solving the root node in 1800 seconds (marked with ∗) demonstrating the value of the column generation approach in producing good lower bounds quickly.

The much smaller collection of instances presented in Table 2 were specifically designed by the authors to be more challenging. In this table we have also included the results from the heuristic presented in [18]. Note that this is indicative only as the results are simply quoted from the paper and were run on a different machine. It is interesting that while these tend to be harder for both methods than equivalent sized problems in Table 1, this trend is not universal and depends on the method. For example *smet01* is "easy" for CPLEX but "hard" for our method. Conversely *smet07* and *smet08* cannot be solved by CPLEX but are not too challenging for our method. While there are no clear winners the

Table 2. Results on datasets by Smet & van der Berghe [18]

Dataset			CPLEX			Triplet			Heur. [18]		CPLEX		Triplet	
			1800 sec limit								5000 sec limit			
	$|W|$	$|J|$	LB	UB	secs	LB	UB	secs	UB_{avg}	secs	UB	secs	UB	secs
smet01	50	258	40	**40**	5	40	41	1800	40.8	641	**40**	5	41	5000
smet02	44	510	40	**40**	54	40	**40**	61	41.2	683	**40**	53	**40**	63
smet03	102	525	77	**77**	492	77	**77**	198	77.4	939	77	474	**77**	180
smet04	113	647	98	**98**	439	98	**98**	324	98.0	163	98	435	**98**	340
smet05	77	777	59	**59**	1694	59	**59**	495	59.8	1615	**59**	1709	59	474
smet06	135	777	116	117	1801	116	121	1801	**116.9**	1800	116	3210	117	5000
smet07	70	781	59	61	1801	59	**59**	147	61.5	1800	61	5009	**59**	150
smet08	88	1022	79	81	1800	79	**79**	1286	80.5	1800	80	5001	**79**	1014
smet09	125	1308	98	—	1800	98	106	1801	**101.9**	1800	—	5000	106	5002
smet10	153	1577	116	—	1813	116	128	1806	**123.2**	1800	—	5000	126	5004

exact methods of CPLEX and CG–C1 appear to be competitive with the heuristic approach from [18]. Where both exact methods solve instances our approach tends to be faster, in some cases significantly so.

7   Conclusions

In this paper we have developed a comprehensive approach for a fixed job scheduling problem in which machines have fixed start and end times and qualifications; jobs require start times, end times and require qualifications to perform these. The particular version of the problem of interest to us was one in which the jobs are allocated to machines and number of machines that are used is minimised. The effectiveness of our approach relies on the combination of:

(a) A novel branching scheme that exploits mathematical properties of our problem and steers the tree search towards areas where the LP relaxation will produce integer solutions.

(b) An alternative column generation approach formulation that solves more efficiently than the naive pattern formulation.

We also use an existing heuristic to ensure we obtain upper bounds in parallel to the branch and bound search.

Computational results demonstrate that our approach performs better than just using a standard commercial MILP solver (CPLEX). While the proposed method is significantly better than the standard solver, we also found that it starts to fail for some larger instances. However there is no clear pattern to this, with many large instances with well over 1000 (and in fact even over 2000) tasks being solved relatively easily, while the smallest unsolved instance for our method has only 258 tasks. Further research will need to concentrate on solving these much larger instances much more effectively, perhaps by concentrating more on the heuristic search as the lower bounds produced by our method are tight and produced very quickly.

References

1. Bekki, O., Azizoglu, M.: Operational fixed interval scheduling problem on uniform parallel machines. Int. J. Prod. Econ., 756–768 (2008)
2. Desrosiers, J., Lübbecke, M.E.: A primer in column generation. Springer (2005)
3. Eliiyi, D., Azizoglu, M.: Working time constraints in operational fixed job scheduling. Int. J. Prod. Res., 6211–6233 (2010)
4. Fischetti, M., Martello, S., Toth, P.: The fixed job schedule problem with working-time constraints. Operations Research 37(3), 395–403 (1989)
5. Ghouila-Houri, A.: Charactérisations des matrices totalement unimodulaires. Comptes Rendus de l'Acadmie des Sciences 254, 1192–1194 (1962)
6. Gupta, U.I., Lee, D.T., Leung, J.Y.T.: Efficient algorithms for interval graphs and circular-arc graphs. Networks 12(4), 459–467 (1982)
7. Kolen, A.W.J., Lenstra, J.K., Papadimitriou, C.H., Spieksma, F.C.R.: Interval scheduling: A survey. Naval Research Logistics 54(5), 530–543 (2007)
8. Kovalyov, M., Ng, C., Cheng, T.: Fixed interval scheduling: models, applications, computational complexity and algorithms. European J. of O.R., 331–342 (2007)
9. Krishnamoorthy, M., Ernst, A.T.: The personnel task scheduling problem. In: Yang, X., Teo, K.L., Caccetta, L. (eds.) Optimization Methods and Applications, pp. 343–368. Kluwer Academic Publishers (2001)
10. Krishnamoorthy, M., Ernst, A.T., Baatar, D.: Algorithms for large scale shift minimisation personnel task scheduling problems. European Journal of Operational Research 219, 34–48 (2012)
11. Kroon, L.G., Salomon, M., Van Wassenhove, L.N.: Exact and approximation algorithms for the tactical fixed interval scheduling problem. Operations Research 45(4), 624–638 (1997)
12. Krumke, S.O., Thielen, C., Westphal, S.: Interval scheduling on related machines. Computers & Operations Research 38(12), 1836–1844 (2011)
13. Lin, S.W., Ying, K.C.: Minimizing shifts for personnel task scheduling problems: A three-phase algorithm. European J. of Op. Research 237(1), 323–334 (2014)
14. Nemhauser, G., Wolsey, L.: Integer and Combinatorial Optimization. John Wiley & Sons, Inc. (1988)
15. Rossi, A., Singh, A., Sevaux, M.: A metaheuristic for the fixed job scheduling problem under spread time constraints. Computers & O.R. 37(6), 1045–1054 (2010)
16. Schrijver, A.: Theory of Linear and Integer Programming. Wiley & Sons (1986)
17. Smet, P., Berghe, G.V.: A matheuristic approach to the shift minimisation personnel task scheduling problem. In: Proc. of the 9th Int. Conf. on the Practice and Theory of Automated Timetabling (PATAT), Son, Norway, pp. 145–160 (August 2012)
18. Smet, P., Wauters, T., Mihaylov, M., Berghe, G.V.: The shift minimisation personnel task scheduling problem: A new hybrid approach and computational insights. Omega 46, 64–73 (2014)
19. Solyali, O., Ozpeynirci, O.: Operational fixed job scheduling problem Under spread time constraints: a branch-and-price algorithm. Int. J. Prod. Res., 1877–1893 (2009)
20. Solyali, O., Özpeynirci, O.: Operational fixed job scheduling problem under spread time constraints: a branch-and-price algorithm. International Journal of Production Research 47(7), 1877–1893 (2009)
21. Türsel Eliiyi, D., Azizolu, M.: Heuristics for operational fixed job scheduling problems with working and spread time constraints. International Journal of Production Economics 132(1), 107–121 (2011)
22. Zhou, S., Zhang, X., Chen, B., Velde, S.v.d.: Tactical fixed job scheduling with spread-time constraints. Computers & Operations Research, 53–60 (2014)

Exact Minkowski Sums of Polygons With Holes

Alon Baram[1], Efi Fogel[1], Dan Halperin[1], Michael Hemmer[2],
and Sebastian Morr[2]

[1] School of Computer Science, Tel Aviv University, Israel
{alontbst,efifogel}@gmail.com, danha@post.tau.ac.il
[2] Dept. of Computer Science, TU Braunschweig, Germany
mhsaar@gmail.com, sebastian@morr.cc

Abstract. We present an efficient algorithm that computes the Minkowski sum of two polygons, which may have holes. The new algorithm is based on the convolution approach. Its efficiency stems in part from a property for Minkowski sums of polygons with holes, which in fact holds in any dimension: Given two polygons with holes, for each input polygon we can fill up the holes that are relatively small compared to the other polygon. Specifically, we can always fill up all the holes of at least one polygon, transforming it into a simple polygon, and still obtain exactly the same Minkowski sum. Obliterating holes in the input summands speeds up the computation of Minkowski sums.

We introduce a robust implementation of the new algorithm, which follows the Exact Geometric Computation paradigm and thus guarantees exact results. We also present an empirical comparison of the performance of Minkowski sum construction of various input examples, where we show that the implementation of the new algorithm exhibits better performance than several other implementations in many cases.

The software is available as part of the *2D Minkowski Sums* package of CGAL (Computational Geometry Algorithms Library), starting from Release 4.7. Additional information and supplementary material is available at our project page http://acg.cs.tau.ac.il/projects/rc.

1 Introduction

Let P and Q be two point sets in $\mathbb{R}^d$. The Minkowski sum of P and Q is defined as $P \oplus Q = \{p + q \mid p \in P, q \in Q\}$. In this paper we focus on the computation of Minkowski sums of general polygons in the plane, that is, polygons that may have holes. However, some of our results also apply to higher dimensions. Minkowski sums are ubiquitous in many fields and applications including robot motion planning [16], assembly planning [7], computer aided design [5], and collision detection in general [19].

[1] Work by E.F. and D.H. has been supported in part by the Israel Science Foundation (grant no. 1102/11), by the German-Israeli Foundation (grant no. 1150-82.6/2011), and by the Hermann Minkowski–Minerva Center for Geometry at Tel Aviv University.
[2] Work by S.M. and M.H. has been supported by Google Summer of Code 2014.

© Springer-Verlag Berlin Heidelberg 2015
N. Bansal and I. Finocchi (Eds.): ESA 2015, LNCS 9294, pp. 71–82, 2015.
DOI: 10.1007/978-3-662-48350-3_7

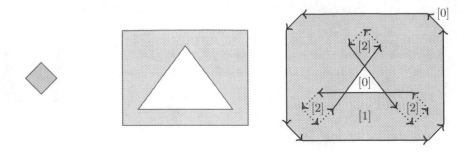

Fig. 1. The convolution of a convex polygon and a non-convex polygon; winding numbers are indicated in brackets; dotted edges are left out for the reduced convolution.

1.1 Terminology and Related Work

During the last four decades many algorithms to compute the Minkowski sum of polygons or polyhedra were introduced. For exact two-dimensional solutions see, e.g., [8]. For approximate solutions see, e.g., [13] and [15]. For exact and approximate three-dimensional solutions see, e.g., [11], [17], [18], and [24].

Computing the Minkowski sum of two convex polygons P and Q is rather easy. As $P \oplus Q$ is a convex polygon bounded by copies of the edges of P and Q ordered according to their slope, the Minkowski sum can be computed using an operation similar to merging two sorted lists of numbers. If the polygons are not convex, it is possible to use one of the two following general approaches:

Decomposition. Algorithms that follow the decomposition approach decompose P and Q into two sets of convex sub-polygons. Then, they compute the pairwise sums using the simple procedure described above. Finally, they compute the union of the pairwise sums. This approach was first proposed by Lozano-Pérez [20]. The performance of this approach heavily depends on the method that computes the convex decomposition of the input polygons. Flato et al. [1] described an implementation of the first exact and robust version of the decomposition approach, which handles degeneracies. They also tried different decomposition methods, but none of them handles polygons with holes.

Ghosh [9] introduced *slope diagrams*—a data structure that was used later on by some of us to construct Minkowski sums of bounded convex polyhedra in 3D [6]. Hachenberger [11] constructed Minkowski sums of general polyhedra in 3D. Both implementations are based on the Computational Geometry Algorithms Library (CGAL) and follow the Exact Geometric Computation (EGC) paradigm.

Convolution. Let $V_P = (p_0, \ldots, p_{m-1})$ and $V_Q = (q_0, \ldots, q_{n-1})$ denote the sequence of vertices in counter-clockwise order along the boundaries of the input polygons P and Q, respectively. Assume that their boundaries wind in a counterclockwise order around their interiors. The *convolution* of these two polygons, denoted $P * Q$, is a collection of line segments of the form[3] $[p_i + q_j, p_{i+1} + q_j]$, where the vector $\overrightarrow{p_i p_{i+1}}$ lies counterclockwise in between $\overrightarrow{q_{j-1} q_j}$ and $\overrightarrow{q_j q_{j+1}}$ and, symmetrically, of segments of the form $[p_i + q_j, p_i + q_{j+1}]$, where the vector $\overrightarrow{q_j q_{j+1}}$ lies counterclockwise in between $\overrightarrow{p_{i-1} p_i}$ and $\overrightarrow{p_i p_{i+1}}$.

According to the *Convolution Theorem* stated in 1983 by Guibas et al. [10], the convolution $P * Q$ of two polygons P and Q is a superset of the boundary of the Minkowski sum $P \oplus Q$. The segments of the convolution form a number of closed (possibly self-intersecting) polygonal curves called *convolution cycles*. The set of points having a nonzero winding number with respect to the convolution cycles comprise the Minkowski sum $P \oplus Q$.[4] However, this theorem has not been completely proven. Though, in the introduction of the thesis of Ramkumar [22], there are some statements about the correctness of the *Convolution Theorem*.

Wein [25] implemented the standard convolution algorithms for simple polygons. He computed the winding number for each face in the arrangement induced by the convolution cycles and used it to determine whether the face is part of the Minkowski sum or not; see Figure 1. Wein's implementation is available in CGAL [26], and as such, it follows the EGC paradigm.

Kaul et al. [14] observed that a segment $[p_i + q_j, p_{i+1} + q_j]$ (resp. $[p_i + q_j, p_i + q_{j+1}]$) cannot possibly contribute to the boundary of the Minkowski sum if q_j (resp. p_i) is a reflex vertex (see dotted edges in Figure 1). The remaining subset of convolution segments, the *reduced convolution*, is still a superset of the Minkowski sum boundary, but the idea of winding numbers can not be applied any longer as there are no closed cycles anymore. Instead, Behar and Lien [2], first identify faces in the arrangement of the reduced convolution that may represent holes (based on proper orientation of all boundary edges of the face). Thereafter, they check whether such a face is indeed a proper hole by selecting a point x inside the face and performing a collision detection of P and $x \oplus -Q$. Their implementation exhibits faster running time than Wein's implementation. However, although it uses advanced multi-precision arithmetic, it does not handle some degenerate cases. The method was also extended to three dimensions [18].

Milenkovic and Sacks [21] defined the *Monotonic Convolution*, which is another superset of the Minkowski sum boundary. They show that this set defines cycles and induces winding numbers, which are positive only in the interior of the Minkowski sum.

1.2   Our Results

We present an efficient algorithm that computes the Minkowski sum of two polygons, which may have holes. The new algorithm is a variant of the algorithm proposed by Behar and Lien [2], which computes the reduced convolution set. In our new algorithm, the initial set of filters proposed in [2] is enhanced by

---

[3] Addition of vertex indices is carried out modulo m for P and modulo n for Q.

[4] Informally, the winding number of a point $p \in \mathbb{R}^2$ with respect to some planar curve γ is an integer number counting how many times does γ wind in a counterclockwise orientation around p.

the removal of complete holes in the input. This enhancement reduces the size of the reduced convolution set even further. The enhancement is backed up by a theorem, the proof of which is also presented; see Section 2. Moreover, we show that at least one of the input polygons can always be made simple (before applying the convolution). These latter results are applicable to any dimension and are independent of the used approach. In addition, roughly speaking, we show that every boundary cycle of the Minkowski sum is induced by exactly one boundary cycle of each summand; see Section 2. It implies that we can compute the convolution of each pair of boundary cycles of the summands separately in order to obtain the correct boundary cycles of the final Minkowski sum. This result is also applicable to any dimension and it is independent of the used approach.

We introduce an implementation of the new algorithm. We also introduce implementations of two new convex decomposition methods that handle polygons with holes as input—one is based on vertical decomposition and the other is based on triangulation. These two methods can be directly applied to compute the Minkowski sum of polygons with holes via decomposition. All our implementations are robust and handle degenerate cases.

We present an empirical comparison of all the implementations above and existing implementations; see Section 4. We show that the implementation of our new algorithm that computes the reduced convolution after filling up some holes in the input exhibits better performance than all other implementations in many cases.

2   Filtering Out Holes

The fundamental observation of the convolution theorem is that only points on the boundary of P and Q can contribute to the boundary of $P \oplus Q$. Specifically, the union of the segments in the convolution $P * Q$, as a point set, is a super-set of the union of the segments of the boundary of $P \oplus Q$.

The idea behind the reduced convolution method is to filter out segments of $P * Q$ that can not possibly contribute to the boundary of $P \oplus Q$ using a local criterion; see Section 1.1. In this section we introduce a global criterion. We show that if a hole in one polygon is relatively small compared to the other polygon, the hole is irrelevant for the computation of $P \oplus Q$; see Figure 2 for an illustration. Thus, we can ignore all segments in $P * Q$ that are induced by the hole when computing $P \oplus Q$. It implies that the hole can be removed (that is, filled up) before the main computation starts, regardless of the approach that one uses to compute the Minkowski sum.

Definition 1. *A hole H of polygon P leaves a trace in $P \oplus Q$, if there exists a point $r = p + q \in \partial(P \oplus Q)$, such that $p \in \partial H$ and $q \in \partial Q$. We say that r is a trace of H. Conversely, we say that a hole H is irrelevant for the computation of $P \oplus Q$ if it does not leave a trace at all.*

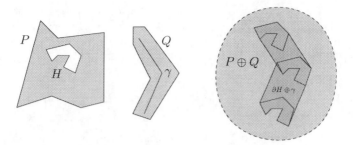

Fig. 2. A small hole H is irrelevant for the computation of $P \oplus Q$ as adding ∂H and $\gamma \subset Q$ fills up any potential hole in $P \oplus Q$ related to H

Lemma 1. *If H leaves a trace in $P \oplus Q$ at a point r, then r is on the boundary of a hole $\tilde{H}$ in $P \oplus Q$.*

Proof. Consider the point $r = p + q$, which is on the boundary of $P \oplus Q$, such that $p \in \partial H$ and $q \in \partial Q$. Since the polygons are closed, for every neighborhood of r there exists a point $r' \notin P \oplus Q$, see Figure 3. Consequently, its corresponding point $p' = r' - q$, which is in the neighborhood of p must be in H. Thus, r' must be enclosed by $\partial H \oplus q$, implying that r' is inside a hole of $P \oplus Q$. $\square$

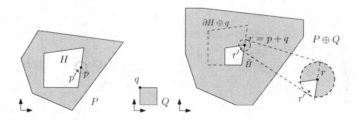

Fig. 3. Hole H leaves a trace in $P \oplus Q$ at point r, which must be on the boundary of some hole $\tilde{H}$ in $P \oplus Q$; see Lemma 1

Lemma 2. *Let $\tilde{H}$ be a hole in $P \oplus Q$ that contains a point $r = p + q \in \partial\tilde{H}$, such that $p \in \partial H$ and $q \in \partial Q$; that is, r is a trace of H. Then $\forall z \in \tilde{H}$ and $\forall y \in Q$, it must hold that $z \in H \oplus y$. In other words, $\tilde{H} \subseteq \bigcap_{\forall y \in Q} H \oplus y$.*

Proof. As in Lemma 1, there is a point r' in the neighborhood of r, which is enclosed by $\partial H \oplus q$. Furthermore, there exists a continuous path from $r' \in \tilde{H}$ to any $z \in \tilde{H}$, which means that every $z \in \tilde{H}$ is also enclosed by $\partial H \oplus q$, or in other words: $z \in H \oplus q$.

Now, assume for contradiction that there is a point $y_0 \in Q$, for which z is not in $H \oplus y_0$. Consider the continuous path γ that connects q and y_0 within Q. Observe that $z \in H \oplus q$ and $z \notin H \oplus y_0$ are equivalent to $z - q \in H$ and

$z - y_0 \notin H$, respectively. This means that $z - y_0$ is either in the unbounded face, or in some other hole in P. Now observe that the path $z \oplus (-\gamma)$ connects $z - q$ and $z - y_0$. Thus, since γ is continuous, there must be a point $y_0' \in \gamma \subset Q$, for which $z - y_0' \in P$. Hence, $z \in P \oplus y_0'$, which implies $z \in P \oplus Q$—a contradiction. $\qquad \square$

Corollary 1. *Let $\tilde{H}$ be a hole in $P \oplus Q$ with $r \in \partial \tilde{H}$ being a trace of H. Then $\forall s \in \partial \tilde{H}$ it holds that s is a trace of H.*

Proof. Consider an arbitrary point $s \in \partial \tilde{H}$, and assume by contradiction that $s = x + y$, where $y \in \partial Q$ and $x \in \partial H'$ is on a boundary cycle of P different than ∂H. By Lemma 2, $\tilde{H} \subseteq \bigcap_{\forall y \in Q} H \oplus y$, it also holds that $s = x + y$, where $x \in H$, which implies that x is in two different holes—a contracdiction. $\qquad \square$

Theorem 1. *Let H be a closed hole in polygon P. H is irrelevant for the computation of $P \oplus Q$ iff there is a path contained in polygon Q that does not fit under any translation in $-H$.*

Proof. We first show that H is irrelevant for the computation of $P \oplus Q$ if there is a path $\gamma \subset Q$ that does not fit under any translation in $-H$. Assume for contradiction that H leaves a trace in $P \oplus Q$; that is, there is an $r = p + q \in \partial(P \oplus Q)$, such that $p \in \partial H$ and $q \in \partial Q$. By Lemma 1, the point r is on the boundary of a hole $\tilde{H}$ in $P \oplus Q$. By Lemma 2, for any point $x \in \tilde{H}$ it must hold that $x \in H \oplus y$ $\forall y \in Q$. Specifically, it must hold $\forall y \in \gamma \subset Q$. This is equivalent to $y \in (x \oplus -H)$ for all $y \in \gamma$, stating that γ fits into $-H$ under some translation—a contradiction.

Conversely, if there is no path that does not fit into $-H$ then all paths contained in Q fit in $-H$. Thus, also Q itself fits in $-H$ under some translation x with $x \oplus Q \subseteq -H$. In this case $x + q \in -H$ for all $q \in Q$, which is equivalent to $-x \in H \oplus q$ for all $q \in Q$. This implies that $-x \notin P \oplus Q$, whereas $-x \in (P \cup H) \oplus Q$, that is, H is relevant for $P \oplus Q$. $\qquad \square$

Corollary 2. *If the closed axis-aligned bounding box B_Q of Q does not fit under any translation in the open axis-aligned bounding box $\mathring{B}_H$ of a hole H in P, then H does not have a trace in $P \oplus Q$.*

Proof. W.l.o.g. assume that B_Q does not fit into $\mathring{B}_H$ with respect to the x-direction. Consider the two extreme points on ∂Q in that direction and connect them by a closed path γ, which obviously does not fit into $-H$, as it does not fit into $\mathring{B}_H$. $\qquad \square$

Theorem 2. *Let P and Q be two polygons with holes and let P' and Q' be their filtered versions, that is, with holes filled up according to Corollary 2 with $P \oplus Q = P' \oplus Q'$. Then, at least P' or Q' is a simple polygon.*

Proof. Note that if B_Q does not fit in the open axis-aligned bounding box $\mathring{B}_P$ of P, it cannot fit in the bounding box of any hole in P, implying that all holes of P can be ignored. Since for any two bounding boxes either $B_Q \not\subset \mathring{B}_P$ or $B_P \not\subset \mathring{B}_Q$ holds, we need to consider the holes of at most one polygon. $\qquad \square$

Consequently, we can remove all holes in one polygon whose bounding boxes are, in x- or y-direction, smaller than, or as large as, the bounding box of the other polygon, as an initial phase of all methods. With fewer holes, convex decomposition results in fewer pieces. Moreover, when all holes of a polygon become irrelevant, one can choose a decomposition method that handles only simple polygons instead of a decomposition method that handles polygons with holes, which is typically more costly. As for the convolution approach, the intermediate arrangements become smaller, speeding up the computation.

3   Implementation

The software has been developed as part of the *2D Minkowski Sums* package of CGAL [26], and it uses other components of CGAL [23]. As such, it is written in C++ and rigorously adheres to the generic-programming paradigm and the EGC paradigm. In the following we provide some details about each one of the new implementations.

3.1   Reduced Convolution

We compute the reduced convolution set of segments filtering out features that cannot possibly contribute to the boundary of the Minkowski sum (see Section 1.1) and in particular complete holes (see Section 2). Then, we construct the arrangement induced by the reduced convolution set.[5] Finally, we traverse the arrangement and extract the boundary of the Minkowski sum. We apply two different filters to identify valid holes in the Minkowski sum: (i) We ignore any face in the arrangement the outer boundary of which forms a cycle that is not properly oriented, as suggested in [2]. (ii) We ignore any face f, such that $(-P \oplus x)$ and Q collide, where $x \in f$ is a sampled point inside f, as suggested in [14]. We use axis-aligned bounding box trees to expedite the collision tests. After applying these two filters, only segments that constitute the Minkowski sum boundary remain.

3.2   Decomposition

Vertical decomposition [12] (a.k.a. trapezoidal decomposition) and triangulation [3] have been extensively used ever since they have been independently introduced a long time ago. We provide a brief overview of these two structures for completeness and explain how they are used in our implementations.

Vertical decomposition for a planar subdivisions is the partition of the (already subdivided) plane into a finite collection of pseudo trapezoids. Each pseudo trapezoid is either a trapezoid that has vertical sides, or a triangle (which is a

---

[5] Currently, we use a single arrangement and do not separate segments that originate from different boundary cycles in the summands (exploiting Corollary 1). We plan to apply this enhancement in the near future.

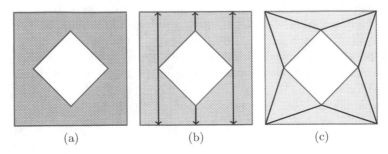

Fig. 4. Convex decomposition. (a) A polygon with holes. (b) Vertical decomposition of the polygon in (a). (c) Triangulation of the polygon in (a).

degenerate trapezoid). Given a polygon with holes, we obtain the decomposition as follows: At every vertex of the polygon, we extend a ray upward if it does not escape the polygon, until either another vertex or an edge is hit. Similarly, we extend a ray downward; see Figure 4b. In our implementation we exploit the vertical decomposition functionality provided by the CGAL package *2D Arrangements* [27].

A Delaunay triangulation for a set of points in a plane is the partition of the plane into triangles, such that no point in the input is inside the circumcircle of any triangle in the triangulation. A constrained Delaunay triangulation is a generalization of the Delaunay triangulation that forces certain required segments into the triangulation. Given a polygon with holes we obtain the constrained Delaunay triangulation confined to the given polygon and provide the polygon edges as constraints; see Figure 4c. In our implementation, we use the *2D Triangulations* [28] CGAL package.

4   Experiments

We have conducted our experiments on families of randomly generated simple and general polygons from AGPLib [4]; these polygons are depicted in Figure 5a and 5b, respectively. All experiments were run on an *Intel Core i7-4770* CPU clocked at 3.4 GHz with 16 GB of RAM, a high-class desktop CPU which would be a good fit for CAD applications. For each instance size the diagrams in the figures show an average over 30 runs on different input. Every run was allowed 20 minutes of CPU time and aborted when it did not finish within this limit.

First, we compared the running time of the implementations of all methods for simple polygons available in CGAL (for details, see [8, Section 9.1.2]), the new implementations, and Behar and Lien's implementation; see Figure 7a. The reduced convolution method consumed about ten times less time than the full convolution method for large instances, whereas the decomposition methods were the fastest for instances larger than 225 vertices.

Secondly, we compared the running time of the implementations of the three new methods (i.e., the reduced convolution method (RC), the triangular based decomposition method (TD), and the vertical decomposition based method (VD))

(a)   (b)

Fig. 5. Randomly generated polygons: (a) simple polygon with 200 vertices, and (b) general with 200 vertices and 20 holes

and Behar and Lien's implementation on instances of general polygons with n vertices and $n/10$ holes; see Figure 7b. For each pair of polygons, one was scaled down by a factor of 1000, to avoid the effect of the hole filter in this experiment. For all executions, the reduced convolution method consumed significantly less time than the two decomposition methods. Behar and Lien's implementation generally performs worse than our reduced convolution method.

In order to demonstrate the effect of the hole filter, we compared the running time of the implementations above fed with a circle with 32 vertices of varying size (see the horizontal axis in Figure 7c and 7d) and with randomly generated polygons having 2000 vertices and 200 holes. Without the hole filter the running time of the reduced convolution method increases as the circle grows due to an increase of the complexity of the intermediate arrangement. Behar and Lien's implementation exhibited constant running time, as it performs pairwise intersection testing. When applying the hole filter to our methods, the reduced convolution method consumed less time than all other methods. The two diagrams clearly show the impact of filtering holes.

(a)   (b)

Fig. 6. Letters from the font *Tangerine* used for the real-world benchmark, displayed with their offset versions. (a) Lowest-resolution "M" with 75 vertices (b) Highest-resolution "A" with 8319 vertices.

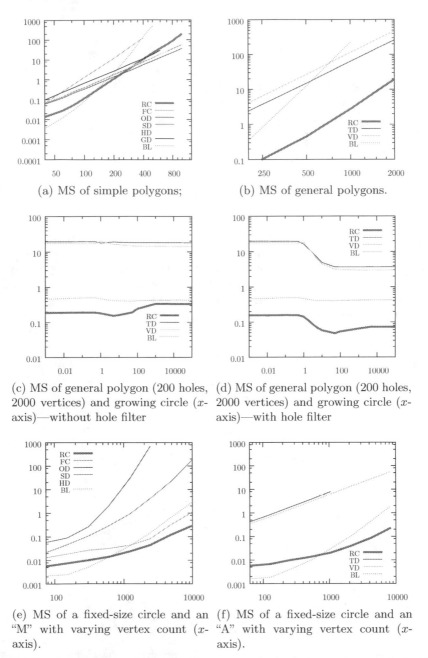

(a) MS of simple polygons;

(b) MS of general polygons.

(c) MS of general polygon (200 holes, 2000 vertices) and growing circle (x-axis)—without hole filter

(d) MS of general polygon (200 holes, 2000 vertices) and growing circle (x-axis)—with hole filter

(e) MS of a fixed-size circle and an "M" with varying vertex count (x-axis).

(f) MS of a fixed-size circle and an "A" with varying vertex count (x-axis).

Fig. 7. Time consumption of Minkowski sum construction using various methods. The y-axis indicates the time measured in seconds; the x-axis indicates the number of vertices of each input polygon, unless otherwise stated. Legend: (RC) reduced convolution; (FC) full convolution; (TD) constrained triangulation decomposition; (VD) vertical decomposition; (SD) small-side angle-bisector decomposition; (OD) optimal convex decomposition; (HD) Hertel-Mehlhorn decomposition; (GD) Greene decomposition; (BL) Behar and Lien's reduced convolution.

Note that the polygons used for the benchmarks above do not represent a real-world case. Instead, the complex shapes essentially constitute the worst-case, as most segments intersections are inside the Minkowski sum anyway. For a more realistic scenario, consider a text, which we would like to offset (for example, for printing stickers). In Figure 7e, we show the running times of the methods available for simple polygons when calculating the Minkowski sum of a glyph of the letter "M" (Figure 6) with a varying amount of vertices and a circle with 128 vertices. In Figure 7f, we show the running times of the methods available for general polygons when calculating the Minkowski sum of a glyph of the letter "A" and the same circle. For both letters, our implementation of the reduced convolution is faster than all other methods for large n.

5   Conclusion

All implementations introduced in this work are available as part of the *2D Minkowski Sums* package of CGAL, which now also supports polygons with holes. The decomposition approaches that handle only simple polygons outperform the new reduced convolution method (which, naturally handles also simple polygons) for instances of random simple polygons with more than 150 vertices. However, these rather chaotic polygons somewhat constitute the worst case scenario for the reduced convolution method. In all other scenarios, the reduced convolution method with hole filter outperforms all other methods by a factor of at least 5. Consequently, starting with CGAL version 4.7, this is the new default method to compute Minkowski sums for simple polygons as well as polygons with holes.

References

1. Agarwal, P.K., Flato, E., Halperin, D.: Polygon decomposition for efficient construction of Minkowski sums. Comput. Geom. Theory Appl. 21, 39–61 (2002)
2. Behar, E., Lien, J.-M.: Fast and robust 2D Minkowski sum using reduced convolution. In: Proc. IEEE Conf. on Intelligent Robots and Systems (2011)
3. Bern, M.: Triangulations and mesh generation. In: Goodman, J.E., O'Rourke, J. (eds.) Handb. Disc. Comput. Geom., ch. 25, 2nd edn., pp. 529–582. Chapman & Hall/CRC, Boca Raton (2004)
4. Couto, M.C., de Rezende, P.J., de Souza, C.C.: Instances for the Art Gallery Problem (2009), http://www.ic.unicamp.br/~cid/Problem-instances/Art-Gallery
5. Elber, G., Kim, M.-S.: Offsets, sweeps, and Minkowski sums. Comput. Aided Design 31(3), 163 (1999)
6. Fogel, E., Halperin, D.: Exact and efficient construction of Minkowski sums of convex polyhedra with applications. In: Proc. 8th Workshop Alg. Eng. Experiments, pp. 3–15 (2006)
7. Fogel, E., Halperin, D.: Polyhedral assembly partitioning with infinite translations or the importance of being exact. IEEE Trans. on Automation Sci. and Eng. 10, 227–241 (2013)
8. Fogel, E., Halperin, D., Wein, R.: CGAL Arrangements and Their Applications, A Step by Step Guide. Springer, Heidelberg (2012)

9. Ghosh, P.K.: A unified computational framework for Minkowski operations. Comput. & Graphics 17(4), 357–378 (1993)
10. Guibas, L.J., Ramshaw, L., Stolfi, J.: A kinetic framework for computational geometry. In: Proc. 24th Annu. IEEE Symp. Found. Comput. Sci., pp. 100–111 (1983)
11. Hachenberger, P.: Exact Minkowksi sums of polyhedra and exact and efficient decomposition of polyhedra into convex pieces. Algorithmica 55(2), 329–345 (2009)
12. Halperin, D.: Arrangements. In: Goodman, J.E., O'Rourke, J. (eds.) Handb. Disc. Comput. Geom., ch. 24, 2nd edn., pp. 529–562. Chapman & Hall/CRC, Boca Raton (2004)
13. Hartquist, E.E., Menon, J., Suresh, K., Voelcker, H.B., Zagajac, J.: A computing strategy for applications involving offsets, sweeps, and Minkowski operations. Comput. Aided Design 31, 175–183 (1999)
14. Kaul, A., O'Connor, M., Srinivasan, V.: Computing Minkowski sums of regular polygons. In: Proc. 3rd Canadian Conf. on Comput. Geom., pp. 74–77 (1991)
15. Kavraki, L.E.: Computation of configuration-space obstacles using the fast fourier transform. In: Proc. IEEE Int. Conf. on Robotics & Automation, pp. 255–261 (1993)
16. Latombe, J.-C.: Robot Motion Planning. Kluwer Academic Publishers, Norwell (1991)
17. Li, W., McMains, S.: A GPU-based voxelization approach to 3D Minkowski sum computation. In: Proc. 2010 ACM Symp. Solid Phys. Model., pp. 31–40. ACM Press (2010)
18. Lien, J.-M.: A simple method for computing minkowski sum boundary in 3D using collision detection. In: Chirikjian, G.S., Choset, H., Morales, M., Murphey, T. (eds.) Algorithmic Foundation of Robotics VIII. STAR, vol. 57, pp. 401–415. Springer, Heidelberg (2009)
19. Lin, M.C., Manocha, D.: Collision and proximity queries. In: Goodman, J.E., O'Rourke, J. (eds.) Handb. Disc. Comput. Geom., ch. 35, 2nd edn., pp. 787–807. Chapman & Hall/CRC, Boca Raton (2004)
20. Lozano-Pérez, T.: Spatial planning: A configuration space approach. IEEE Trans. on Comput. C-32, 108–120 (1983)
21. Milenkovic, V., Sacks, E.: A monotonic convolution for Minkowski sums. Int. J. of Comput. Geom. Appl. 17(4), 383–396 (2007)
22. Ramkumar, G.: Tracings and Their Convolutions: Theory and Application. Phd thesis, Stanford, California (1998)
23. The Cgal Project. Cgal User and Reference Manual. Cgal Editorial Board, 4.6 edn. (2015), http://doc.cgal.org/latest/Manual/index.html
24. Varadhan, G., Manocha, D.: Accurate Minkowski sum approximation of polyhedral models. Graphical Models 68(4), 343–355 (2006)
25. Wein, R.: Exact and efficient construction of planar Minkowski sums using the convolution method. In: Proc. 14th Annu. Eur. Symp. Alg., pp. 829–840 (2006)
26. Wein, R.: 2D Minkowski sums. In: CGAL User and Reference Manual. CGAL Editorial Board, 4.6 edn. (2015). http://doc.cgal.org/latest/Manual/packages.html#PkgMinkowskiSum2Summary.
27. Wein, R., Berberich, E., Fogel, E., Halperin, D., Hemmer, M., Salzman, O., Zukerman, B.: 2D arrangements. In CGAL User and Reference Manual. CGAL Editorial Board, 4.6 edn. (2015). http://doc.cgal.org/latest/Manual/packages.html#PkgArrangement2Summary.
28. Yvinec, M.: 2D triangulations. In: CGAL User and Reference Manual. CGAL Editorial Board, 4.6 edn. (2015). http://doc.cgal.org/latest/Manual/packages.html#PkgTriangulation2Summary.

λ > 4*

Gill Barequet[1], Günter Rote[2], and Mira Shalah[1]

[1] Dept. of Computer Science
Technion—Israel Institute of Technology
Haifa 32000, Israel
{barequet,mshalah}@cs.technion.ac.il
[2] Institut für Informatik
Freie Universität Berlin
Takustraße 9, D-14195 Berlin, Germany
rote@inf.fu-berlin.de

Abstract. A *polyomino* ("lattice animal") is an edge-connected set of squares on the two-dimensional square lattice. Counting polyominoes is an extremely hard problem in enumerative combinatorics, with important applications in statistical physics for modeling processes of percolation and collapse of branched polymers. We investigated a fundamental question related to polyominoes, namely, what is their growth constant, the asymptotic ratio between $A(n + 1)$ and $A(n)$ when $n \to \infty$, where $A(n)$ is the number of polyominoes of size n. This value is also known as "Klarner's constant" and denoted by λ. So far, the best lower and upper bounds on λ were roughly 3.98 and 4.65, respectively, and so not even a single decimal digit of λ was known. Using extremely high computing resources, we have shown (still rigorously) that $\lambda > 4.00253$, thereby settled a long-standing problem: proving that the leading digit of λ is 4.

Keywords: Polyominoes, lattice animals, growth constant.

1 Introduction

1.1 What Is λ?

The universal constant λ appears in the study of three seemingly completely unrelated fields: combinatorics, percolation, and branched polymers. In combinatorics, the analysis of *self-avoiding walks* (SAWs, non-self-intersecting lattice paths starting at the origin, counted by lattice units), *simple polygons* or *self-avoiding polygons* (SAPs, closed SAWs, counted by either perimeter or area), and *polyominoes* (SAPs possibly with holes, edge-connected sets of lattice squares, counted by area), are all related. In statistical physics, SAWs and SAPs play a significant role in percolation processes and in the collapse transition which branched polymers undergo when being heated. A recent collection edited by A. J. Guttmann [11] provides an excellent review of all these topics and the

---

* We acknowledge the support of the facilities and staff of the HPI Future SOC Lab in Potsdam, where this project has been carried out.

© Springer-Verlag Berlin Heidelberg 2015
N. Bansal and I. Finocchi (Eds.): ESA 2015, LNCS 9294, pp. 83–94, 2015.
DOI: 10.1007/978-3-662-48350-3_8

connections between them. In this paper we describe our effort to prove that the growth constant of polyominoes is strictly greater than 4. To this aim we exploited to the maximum possible computer resources which were available to us, designing and implementing carefully the algorithm and the required data structures. Eventually we obtained a computer-generated proof which was verified by other programs implemented independently. Let us start with a brief description of the history of λ and the three research areas, then describe the method and computation of the lower bound on λ.

1.2   Brief History

Determining the exact value of λ (or even setting good bounds on it) is a hard problem in enumerative combinatorics. In 1967, Klarner [13] showed that the limit $\lim_{n\to\infty} \sqrt[n]{A(n)}$ exists and denoted it by λ. Since then, λ has been called "Klarner's constant." Only in 1999, Madras [16] proved the stronger statement that the asymptotic growth rate in the sense of the limit $\lambda = \lim_{n\to\infty} A(n + 1)/A(n)$ exists.

By using interpolation methods, Sykes and Glen [21] *estimated* in 1976 that $\lambda = 4.06 \pm 0.02$. This estimate was sharpened several times, the most accurate $(4.0625696 \pm 0.0000005)$ given by Jensen [12] in 2003. Before carrying out this project, the best *proven* bounds on λ were roughly 3.9801 from below [4] and 4.6496 from above [14]. Thus, λ has always been an elusive constant, of which not even a single significant digit was (rigorously) known. Our goal was to raise the lower bound on λ over the barrier of 4, and thus reveal its first decimal digit and proving that $\lambda \neq 4$. The current improvement of the lower bound on λ to 4.0025 also cuts the difference between the known lower bound and the estimated value of λ by about 25% (from 0.0825 to 0.0600).

1.3   Enumerative Combinatorics

A *polyomino* is a connected set of cells on the planar square lattice, where connectivity is along sides but not through corners of the cells. Polyominoes were made popular by the pioneering book of Golomb [10] and by Martin Gardner's columns in Scientific American, and counting polyominoes by size became a popular fascinating combinatorial problem. The size of a polyomino is the number of its cells. In this article we consider "fixed" polyominoes; two such polyominoes are considered identical if one can be obtained from the other by a translation, while rotations and flipping are not allowed. In the mathematical literature, the number of polyominoes of size n is usually denoted as $A(n)$, but no formula is known yet for it. Researchers have suggested efficient back-tracking [19,20] and transfer-matrix [6,12] algorithms for computing $A(n)$ for a given value of n. The latter algorithm was adapted in [4] and also in this work for twisted cylinders. To-date, the sequence $A(n)$ has been determined up to $n = 56$ by a parallel computation on an HP server cluster using 64 processors [12]. The exact value of the growth constant of this sequence, $\lambda = \lim_{n\to\infty} A(n + 1)/A(n)$, has also been elusive for many years. It has been interesting to know whether this value is smaller or greater than the *connective constant* of this lattice. This latter constant is simply the number of neighbors each cell of the lattice has, which is, in this case, 4. In this work we reveal the leading decimal digit of λ: It is 4.

1.4   Percolation Processes

In physics, chemistry, and materials science, percolation theory deals with the movement and filtering of fluids through porous materials. Giving it a mathematical model, the theory describes the behavior of connected clusters in random graphs. Suppose that a unit of liquid L is poured on top of some porous material M. What is the chance that L makes its way through M and reach the bottom? An idealized mathematical model of this process is a two- or three-dimensional grid of vertices ("sites") connected with edges ("bonds"), where each bond is independently open (or closed) for liquid flow with some probability p. Broadbent and Hammersley [5] asked in 1957, for a fixed value of p and for the size of the grid tending to infinity, what is the probability that a path consisting of open bonds exists from the top to the bottom. They essentially investigated solute diffusing through solvent and molecules penetrating a porous solid, representing space as a lattice with two distinct types of cells.

In the literature of statistical physics, fixed polyominoes are usually called "strongly-embedded lattice animals," and there, the analogue of the growth rate of polyominoes is the growth constant of lattice animals. The terms high and low *temperature* mean high and low *density* of clusters, respectively, and the term *free energy* corresponds to the natural logarithm of the growth constant. Lattice animals were used for computing the mean cluster density in percolation processes (Gaunt et al. [9]), in particular those of fluid flow in random media. Sykes and Glen [21] were the first to observe that $A(n)$, the total number of connected clusters of size n, grows asymptotically like $C\lambda^n n^\theta$, where λ is Klarner's constant and C, θ are two other fixed values.

1.5   Collapse of Branched Polymers

Another important topic in statistical physics is the existence of a collapse transition of branched polymers in dilute solution at a high temperature. In physics, a *field* is an entity each of whose points has a value which depends on location and time. Lubensky and Isaacson [15] developed a field theory of branched polymers in the dilute limit, using statistics of (bond) lattice animals (which are important in the theory of percolation) to imply when a solvent is good or bad for a polymer. Derrida and Herrmann [7] investigated two-dimensional branched polymers by looking at lattice animals on a square lattice and studying their free energy. Flesia et al. [8] made the connection between collapse processes to percolation theory, relating the growth constant of strongly-embedded site animals to the free energy in the processes. Madras et al. [17] considered several models of branched polymers in dilute solution, proving bounds on the growth constants for each such model.

2 Twisted Cylinders

A "twisted cylinder" is a half-infinite wrap-around spiral-like square lattice, as is shown in Fig. 1. We denote the perimeter (or "width") of the twisted cylinder by

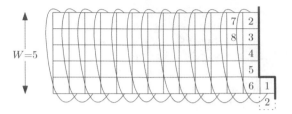

Fig. 1. A twisted cylinder of perimeter $W = 5$

the symbol W. Like in the plane, one can count polyominoes on a twisted cylinder of width W and study their asymptotic growth constant, λ_W. It was proven that the sequence $(\lambda_W)_{W=1}^{\infty}$ is monotone increasing [4] and that it converges to λ [2]. Thus, the bigger W is, the better (higher) the lower bound λ_W on λ is.

It turns out that analyzing the growth rate of polyominoes is more convenient on a twisted cylinder than in the plane. The reason is that we want to build up polyominoes incrementally by considering one square at a time. On a twisted cylinder, this can be done in a uniform way, without having to jump to a new row from time to time. Imagine that we walk along the spiral order of squares, and at each square decide whether or not to add it to the polyomino. Naturally, the size of a polyomino is the number of positive decisions we make on the way. The crucial observation is that no matter how big polyominoes are, they can be characterized in a *finite* number of ways that depends only on W. This is because all one needs to remember is the structure of the last W squares of the twisted cylinder (the "boundary"), and how they are inter-connected through cells that were considered before the boundary. This provides enough information for the continuation of the process: whenever a new square is considered, and a decision is taken about whether or not to add it to the polyomino, the boundary is updated accordingly. Thus, the growth of polyominoes on a twisted cylinder can be modeled by a finite-state automaton whose states are all possible boundaries. Every state in this automaton has two outgoing edges that correspond to whether or not the next square is added to the polyomino.

The number of states in the automaton that models the growth of polyominoes on a twisted cylinder of perimeter W is large [3,4]: it is the $(W{+}1)$st Motzkin number M_{W+1}. The nth Motzkin number, M_n, counts the number of Motzkin paths of length n (see Fig. 2 for an illustration): Such a path connects the integer grid points $(0,0)$ and $(n,0)$ with n steps, consisting only of steps taken from $\{(1,1),(1,0),(1,-1)\}$, and not going under the x axis. Motzkin numbers can also be defined in a variety of other ways [1]. Asymptotically, $M_n \sim 3^n n^{-3/2}$, thus, M_W increases roughly by a factor of 3 when W is incremented by 1.

The number of polyominoes with n cells that have state s as the boundary configuration is equal to the number of paths that the automaton can take from the starting state to the state s, paths which involve n transitions in which a cell is added to the polyomino. We compute these numbers by using a dynamic-programming recursion.

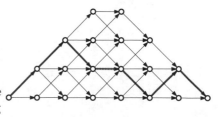

Fig. 2. A Motzkin path of length 7

See Appendix A for more details about Motzkin paths, and Appendix B for a small example automaton and more explanations of this topic.

3   Method

In 2004, a sequential program that computes λ_W for any perimeter was developed by Ms. Ares Ribó as part of her Ph.D. thesis under the supervision of G. Rote. The program first computes the endpoints of the outgoing edges from all states of the automaton and saves them in two long arrays $succ0$ and $succ1$, which correspond to adding an empty or an occupied cell. Both arrays are of length $M := M_{W+1}$. Two successive iteration vectors (which contain the number of polyominoes corresponding to each boundary) are stored as two arrays y^{old} and y^{new} of floating point numbers, also of length M. The four arrays are indexed from 0 to $M-1$. After initializing $y^{\text{old}} := (1, 0, 0, \dots)$, each iteration computes the new version of y by performing the following simple loop.

$$y^{\text{new}}[0] := 0;$$
$$\textbf{for } s := 1, \dots, M-1:$$
$$(*) \qquad y^{\text{new}}[s] := y^{\text{new}}[succ0[s]] + y^{\text{old}}[succ1[s]];$$

As mentioned, the pointer arrays $succ0[\,]$ and $succ1[\,]$ are computed beforehand. The pointer $succ0[s]$ may be null, in which case the corresponding zero entry ($y^{\text{new}}[0]$) is used.

As explained above, each index s represents a state. The states are encoded by Motzkin paths, and these paths can be bijectively mapped to numbers s between 0 and $M-1$. In the iteration $(*)$, the vector y^{new} depends on itself, but this does not cause any problem because $succ0[s]$, if it is non-null, is always less than s. Therefore, there are no circular references and each entry is set before it is used. In fact, the states can be partitioned into groups $G_1, G_2, \dots, G_W$: The group G_i contains the states corresponding to boundaries in which i is the smallest occupied cell, or, in other words, boundaries that start with $i-1$ empty cells. The dependence between the entries of the groups is schematically shown in Fig. 3: $succ0[s]$ of an entry $s \in G_i$ (for $1 \le i \le W-1$), if it is non-null, belongs to G_{i+1}. Naturally, $succ0$ of the single state in G_W is null since a boundary with all cells empty is invalid.

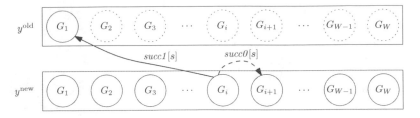

Fig. 3. The dependence between the different groups of y^{new} and y^{old}

At the end, y^{new} is moved to y^{old} to start the new iteration. It was proven [4] that after every iteration, we have the following interval for bounding λ_W:

$$\min_s \frac{y^{\mathrm{new}}[s]}{y^{\mathrm{old}}[s]} \leq \lambda_W \leq \max_s \frac{y^{\mathrm{new}}[s]}{y^{\mathrm{old}}[s]} \tag{1}$$

In this procedure, the two bounds converge (by the Perron-Frobenius Theorem) to λ_W, and y^{old} converges to a corresponding eigenvector. The vector y^{old} is normalized after every few iterations in order to prevent overflow. The scale of the vector is irrelevant to the process. The program terminates when the two bounds are close enough. The left-hand side of (1) is a lower bound on λ_W, which in turn is a lower bound on λ, and this is our real goal.

4   Sequential Runs

In 2004 we obtained good approximations of λ_W up to $W = 22$. The program required extremely high resources in terms of main memory (RAM) by the standards of that time. The computation of $\lambda_{22} \approx 3.9801$ (with a single processor) took about 6 hours on a machine with 32 GB of RAM. (Today, the same program runs in 20 minutes on a regular workstation.) We extrapolated the first 22 values of (λ_W) (see Fig. 4) and estimated that only when we reach $W = 27$ we would break the mythical barrier of 4.0. However, as mentioned above, the required storage is proportional to M_W, which increases roughly by a factor of 3 when W is incremented by 1. With this exponential growth of both memory and running time, the goal of breaking the barrier seemed then to be out of reach.

5   Computing λ_{27}

Environment. The computation of λ_{27} was performed on a Hewlett Packard ProLiant DL980 G7 server of HPI (Hasso Plattner Institute) Future SOC Lab in Potsdam, Germany. It consists of 8 Intel Xeon X7560 nodes (Intel64 architecture), each having eight physical 2.26 GHz processors (16 virtual cores), for a total of 64 processors (128 virtual cores). Hyperthreading was used to allow processes to run on the physical cores of a node while sharing certain resources, thus yielding twice as many virtual processor cores as physical processors. Each

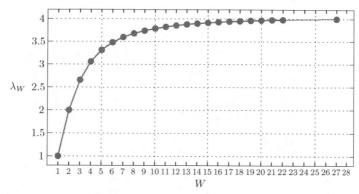

Fig. 4. Extrapolating the sequence λ_W

node was equipped with 256 GiB of RAM (and 24 MiB of cache memory), for a total of **2 TiB of RAM**. Simultaneous access by all processors to the shared main memory was crucial to the success of the project. Distributed memory would incur a severe penalty in running time. The machine was run with the Ubuntu version of Gnu/Linux. Compilation was done using the gcc C compiler with OpenMP 2.0 directives for parallel processing.

Programming Improvements. Since for $W = 27$ the finite automaton has $M_{28} \approx 2.1 \cdot 10^{11}$ states, we initially estimated that we would need memory for two 8-byte arrays (for storing $succ0$ and $succ1$) and two 4-byte arrays (for storing y^{old} and y^{new}), all of length M_{28}, for a total of $24 \cdot 2.1 \cdot 10^{11} \approx 4.6$ TiB of RAM, which was certainly out of reach, even with the available supercomputer. Apparently, a combination of parallelization, storage-compression techniques, and a few other enhancements and tricks allowed us to push the lower bound on λ above 4.0.

1. **Parallelization.** Since the set of states G of the automaton could be partitioned into groups $G_1, \ldots, G_W$, such that $succ0[s]$ for an element $s \in G_i$ belongs to G_{i+1}, the groups $G_W, \ldots, G_1$ had to be processed sequentially (in this order) but all elements in one group could be computed in parallel. The size of the groups is exponentially decreasing; in fact, G_1 comprises more than half of all states, and G_W contains only a single state. Therefore, for the bulk of the work (the iterative computation of y^{new}), we easily achieved coarse-grained parallelization and a distribution of the work on all the 128 available cores, requiring concurrent read but no concurrent write operations. We also parallelized the preprocessing phase (computing the $succ$ arrays) and various house-keeping tasks (e.g., rescaling the y vectors after every 10th iteration). Tests with different numbers of processors revealed indeed a speed-up close to linear.
2. **Elimination of Unreachable States.** A considerable portion of the states of the automaton (about 11% asymptotically) are unreachable, i.e., there is no binary string leading to these states. This happens because not all seemingly legal states can be realized by a valid boundary. These states do not affect the correctness of the computation, and there was no harm

in leaving them, apart from the effect on the performance of the iteration. We were able to characterize the unreachable states fairly easily in terms of their Motzkin paths. After eliminating these states, we had to modify the bijection between the Motzkin paths representing the remaining states and the successive integers.

3. **Bit-streaming of the *succ0/1* Arrays.** Instead of storing each entry of the *succ0/1* arrays in a full word (8 bytes, once the number of states exceeded 2^{32}), we allocated to each entry exactly the number of required bits and stored all entries consecutively in a packed manner. Since the *succ0/1* entries were only accessed sequentially, there was only a small overhead in running time for unpacking the resulting bit sequence. In addition, since we knew *a priori* to which group G_i each pointer belonged, we needed only $\lceil \log_2 |G_i| \rceil$ bits per pointer, for all entries in G_i (plus a negligible amount of bits required to delimit between the different sets G_i). On top of that, the *succ0*-pointer was often null because the choice of not adding the next cell to the polyomino caused a connected component of the polyomino to lose contact with the boundary. By spending one extra indicator bit per pointer, we eliminated altogether these illegal pointers, which comprised about 11% of all *succ0* entries.

4. **Storing Higher Groups Only Once.** For states s not in the group G_1, $y^{\mathrm{old}}[s]$ is not needed in the recursion (∗). Thus, we did not need to keep two separate arrays in memory. The quotient $y^{\mathrm{new}}[s]/y^{\mathrm{old}}[s]$ could still be computed before overwriting $y^{\mathrm{old}}[s]$ by $y^{\mathrm{new}}[s]$, and thus the minimum and maximum of these quotients, which give the bounds (1) on λ, could be accumulated as we scanned the states.

5. **Recomputing *succ0*.** Instead of storing the *succ0* array, we computed its entries on-the-fly whenever they were needed, and thus saved completely the memory needed to store these pointers. Naturally, this required more running time. Streamlined computation of the pointers accelerated the successor computation (see below). This variation has also benefited from parallelization since each processor could do the pointer computations independently. Since the elimination of the *succ0* pointers was sufficient to get the program running with $W = 27$, we did not pursue the option of eliminating the *succ1* array in an analogous way.

6. **Streamlining the Conversion from Motzkin Paths to Integers.** Originally, we represented a Motzkin path by a sequence of $W+1$ integer numbers taking values from $\{-1, 0, +1\}$. However, we compressed the representation into a sequence of $(W+1)$ 2-bit items, each one encoding one step of the path, which we could store in one 8-byte word (since $W \leq 31$). This compact storage opened up the possibility of word-level operations. For converting paths to numbers, we could process several symbols at a time, using look-up tables.

Execution. After 120 iterations, the program announced the lower bound 4.00064 on λ_{27}, thus breaking the 4 barrier. We continued to run the program for a few more days. Then, after 290 iterations, the program reached the stable situation (observed in a few successive tens of iterations) $4.002537727 \leq \lambda_{27} \leq 4.002542973$, establishing the new record $\lambda > 4.00253$. The total running time for the computations leading to this result was about 36 hours. In total, we used a few dozens of hours of exclusive use of the server spread over several weeks.

6 Validity and Certification

Our proof depends heavily on computer calculations. This raises two issues about its validity: (a) Elaborate calculations on a large computer are hard to reproduce, and in particular when a complicated parallel computer program is involved, one should be skeptical. (b) We performed the computations with 32-digit floating-point numbers. We address these issues in turn.

(a) What our program tries to compute is an eigenvalue of a matrix. The amount and length of the computations are irrelevant to the fact that eventually we have a witness array of floating-point numbers (the "proof"), about 450 GB in size, which is a good approximation of the eigenvector corresponding to λ_{27}. This array provides rigorous bounds on the true eigenvalue λ_{27}, because the relation (1) holds for *any* vector y^{old} and its successor vector y^{new}. To check the proof and evaluate the bounds (1), one only has to read the approximate eigenvector y^{old} and carry out one iteration ($*$). This approach of providing simple certificates for the result of complicated computations is the philosophy of *certifying algorithms* [18]. We ran two different programs for the checking task. The code for the only technically challenging part of the algorithm, the successor computation, was based on programs written independently by two people who used different state representations. Both programs ran in a purely sequential manner, and the running time was about 20 hours each.

(b) Regarding the accuracy of the calculations, one can look how the recurrence ($*$) produces y^{new} from y^{old}. One finds that each term in the lower bound (1) results from the input data (the approximate eigenvector y^{old}) through at most 26 additions of positive numbers for computing $y^{\text{new}}[s]$, plus one division, all in single-precision `float`. The final minimization is error-free. Since we made sure that no denormalized floating-point numbers occurred, the magnitude of the numerical errors is comparable to the accuracy of floating-point numbers, and the accumulated error is much smaller than the gap that we opened above 4. By bounding the floating-point error, we obtain 4.00253176 as a certified lower bound on λ. Thus, in particular, we now know that the leading digit of λ is 4.

7 Conclusion

In this project we computed λ_{27} and set a new lower bound on λ, which is greater than 4. By this we also excluded the possibility that $\lambda = 4$. We believe that with some more effort, it will be feasible to run the program for $W = 28$. This would probably require (a) To eliminate also the storage for the *succ1*-successors and compute them along with the *succ0*-successors; (b) To eliminate all groups $G_2, \ldots, G_W$ and keep only group G_1; and (c) To implement a customized floating-point storage format for the numbers y. (With a total of 2 TiB of RAM, we can only afford 27 bits per entry.) We anticipate that this would increase the lower bound on λ to about 4.0065.

References

1. Aigner, M.: Motzkin Numbers. European J. of Combinatorics 19, 663–675 (1998)
2. Aleksandrowicz, G., Asinowski, A., Barequet, G., Barequet, R.: Formulae for polyominoes on twisted cylinders. In: Dediu, A.-H., Martín-Vide, C., Sierra-Rodríguez, J.-L., Truthe, B. (eds.) LATA 2014. LNCS, vol. 8370, pp. 76–87. Springer, Heidelberg (2014)
3. Barequet, G., Moffie, M.: On the complexity of Jensen's algorithm for counting fixed polyominoes. J. of Discrete Algorithms 5, 348–355 (2007)
4. Barequet, G., Moffie, M., Ribó, A., Rote, G.: Counting polyominoes on twisted cylinders. INTEGERS: Elec. J. of Comb. Number Theory 6(A22), 1–37 (2006)
5. Broadbent, S.R., Hammersley, J.M.: Percolation processes: I. Crystals and mazes. Proc. Cambridge Philosophical Society 53, 629–641 (1957)
6. Conway, A.: Enumerating 2D percolation series by the finite-lattice method: Theory. J. of Physics, A: Mathematical and General 28, 335–349 (1995)
7. Derrida, B., Herrmann, H.J.: Collapse of branched polymers. J. de Physique 44, 1365–1376 (1983)
8. Flesia, S., Gaunt, D.S., Soteros, C.E., Whittington, S.G.: Statistics of collapsing lattice animals. J. of Physics, A: Mathematical and General 27, 5831–5846 (1994)
9. Gaunt, D.S., Sykes, M.F., Ruskin, H.: Percolation processes in d-dimensions. J. of Physics A: Mathematical and General 9, 1899–1911 (1976)
10. Golomb, S.W.: Polyominoes, 2nd edn. Princeton Univ. Press, Princeton (1994)
11. Guttmann, A.J. (ed.): Polygons, Polyominoes and Polycubes. Lecture Notes in Physics, vol. 775. Springer, Heidelberg (2009)
12. Jensen, I.: Counting polyominoes: A parallel implementation for cluster computing. In: Sloot, P.M.A., Abramson, D., Bogdanov, A.V., Gorbachev, Y.E., Dongarra, J., Zomaya, A.Y. (eds.) ICCS 2003, Part III. LNCS, vol. 2659, pp. 203–212. Springer, Heidelberg (2003)
13. Klarner, D.A.: Cell growth problems. Canad. J. of Mathematics 19, 851–863 (1967)
14. Klarner, D.A., Rivest, R.L.: A procedure for improving the upper bound for the number of n-ominoes. Canadian J. of Mathematics 25, 585–602 (1973)
15. Lubensky, T.C., Isaacson, J.: Statistics of lattice animals and dilute branched polymers. Physical Review A 20, 2130–2146 (1979)
16. Madras, N.: A pattern theorem for lattice clusters. Annals of Combinatorics 3, 357–384 (1999)
17. Madras, N., Soteros, C.E., Whittington, S.G., Martin, J.L., Sykes, M.F., Flesia, S., Gaunt, D.S.: The free energy of a collapsing branched polymer. J. of Physics, A: Mathematical and General 23, 5327–5350 (1990)
18. McConnell, R.M., Mehlhorn, K., Näher, S., Schweitzer, P.: Certifying algorithms. Computer Science Review 5, 119–161 (2011)
19. Mertens, S., Lautenbacher, M.E.: Counting lattice animals: A parallel attack. J. of Statistical Physics 66, 669–678 (1992)
20. Redelmeier, D.H.: Counting polyominoes: Yet another attack. Discrete Mathematics 36, 191–203 (1981)
21. Sykes, M.F., Glen, M.: Percolation processes in two dimensions: I. Low-density series expansions. J. of Physics, A: Mathematical and General 9, 87–95 (1976)

Appendix A: Representing Boundaries as Motzkin Paths

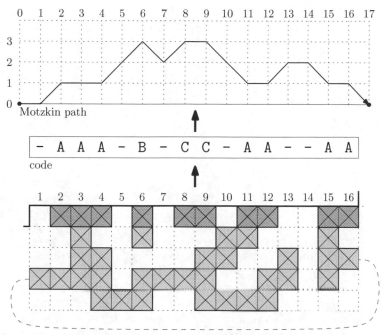

The figure above illustrates the representation of boundaries of polyominoes on twisted cylinders as Motzkin paths. The figure should be read from bottom to top. The bottom of the figure shows a partially constructed polyomino on a twisted cylinder of width 16. The dashed line indicates two adjacent cells which are connected "around the cylinder," where this is not immediately apparent. The boundary cells (top row) are shown darker. The light-gray cells away from the boundary need not be remembered individually; what matters is the connectivity among the boundary cells that they provide. This is indicated in a symbolic *code* -AAA-B-CC-AA--AA. Boundary cells in the same component are represented by the same letter, and the character '-' denotes an empty cell. However, this code was not used in our program. Instead, we represented a boundary as a Motzkin path, as shown in the top part of the figure, because this representation allows for a convenient bijection to successive integers and therefore for a compact storage of the boundary in a vector. Intuitively, the Motzkin path follows the movements of a stack when reading the code from left to right. Whenever a new component starts, like component A in position 2 or component B in position 6, the path moves *up*. Whenever a component is temporarily interrupted, such as component A in position 5, the path also moves *up*. The path moves *down* when an interrupted component is resumed (e.g., component A in positions 11 and 15) or when a component is completed (positions 7, 10, and 17). The crucial property is that components cannot cross, i.e., a pattern like ...A...B...A...B... cannot occur. As a consequence of these rules, the occupied cells correspond to odd levels in the path, and the free cells correspond to even levels.

Appendix B: Automata for Modeling Polyominoes on Twisted Cylinders

A finite automaton is a very convenient tool for representing the growth of polyominoes. Below is the automaton for width $W = 3$. The starting state is A--. The states A-B and --A have no 0-successors.

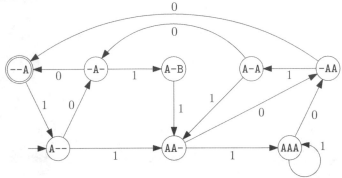

We associate the labels '1' and '0' with the edges, corresponding to whether or not a cell was added to the polyomino in the corresponding step. The construction of a polyomino is modeled by tracing the path in the automaton while processing an input word of 1s and 0s, where the number of occurrences of '1' is the size of the polyomino (minus 1, since the starting state contains already one cell). We want to accept only legal polyominoes, that is, polyominoes that are composed of *connected* squares. It is sufficient to consider a single *accepting state* for the automaton (--A in the example) because this state can always be reached by adding enough empty cells. The number of polyominoes on a twisted cylinder is equal to the number of binary words recognized by the automaton, that is, whose processing by the automaton terminates in an accepting state.

Automata theory and linear algebra give us strong tools to analyze the behavior of a finite automaton. First, we can represent the automaton as an $M \times M$ 0/1 transfer matrix B, where M is the number of states of the automaton, and the (ij)th entry of B is 1 if an edge leads from the ith to the jth state of the automaton. One can derive from B (through its characteristic polynomial) the generating function of the sequence enumerating polyominoes on the twisted cylinder, and a linear recurrence formula satisfied by this sequence. This has been carried out [2] up to width $W = 10$.

Second, this matrix has a few interesting properties. It is proven [4] that the largest eigenvalue (in absolute value) of B is exactly the desired growth constant λ_W. Moreover, B is a primitive and irreducible matrix, and, hence, λ_W is the only positive eigenvalue of B. Under these conditions, the Perron-Frobenius theorem provides an effective method for computing this eigenvalue: Start from any positive vector (e.g., the vector in which all entries are 1) and repeatedly multiply it by B. In the limit, this process converges to the eigenvector, and the ratios between successive vectors in this process converge to the desired eigenvalue, which is the desired growth constant.

Revenue Maximization for Selling Multiple Correlated Items*

MohammadHossein Bateni[1], Sina Dehghani[2],
MohammadTaghi Hajiaghayi[2], and Saeed Seddighin[2]

[1] Google Research
[2] University of Maryland

Abstract. We study the problem of selling n items to a single buyer
with an additive valuation function. We consider the valuation of the
items to be correlated, i.e., desirabilities of the buyer for the items are
not drawn independently. Ideally, the goal is to design a mechanism
to maximize the revenue. However, it has been shown that a revenue
optimal mechanism might be very complicated and as a result inapplicable to real-world auctions. Therefore, our focus is on designing a simple mechanism that achieves a constant fraction of the optimal revenue.
Babaioff et al. [3] propose a simple mechanism that achieves a constant
fraction of the optimal revenue for independent setting with a single
additive buyer. However, they leave the following problem as an open
question: *"Is there a simple, approximately optimal mechanism for a
single additive buyer whose value for n items is sampled from a common base-value distribution?"* Babaioff et al. show a constant approximation factor of the optimal revenue can be achieved by either selling
the items separately or as a whole bundle in the independent setting.
We show a similar result for the correlated setting when the desirabilities of the buyer are drawn from a common base-value distribution. It
is worth mentioning that the core decomposition lemma which is mainly
the heart of the proofs for efficiency of the mechanisms does not hold
for correlated settings. Therefore we propose a modified version of this
lemma which is applicable to the correlated settings as well. Although
we apply this technique to show the proposed mechanism can guarantee
a constant fraction of the optimal revenue in a very weak correlation,
this method alone can not directly show the efficiency of the mechanism
in stronger correlations. Therefore, via a combinatorial approach we reduce the problem to an auction with a weak correlation to which the
core decomposition technique is applicable. In addition, we introduce a
generalized model of correlation for items and show the proposed mechanism achieves an $O(\log k)$ approximation factor of the optimal revenue
in that setting.

1 Introduction

Suppose an auctioneer wants to sell n items to a single buyer. The buyer's
valuation for a particular item comes from a known distribution, and the his

---

* Supported in part by NSF CAREER award 1053605, NSF grant CCF-1161626, ONR
YIP award N000141110662, and a DARPA/AFOSR grant FA9550-12-1-0423.

N. Bansal and I. Finocchi (Eds.): ESA 2015, LNCS 9294, pp. 95–105, 2015.
DOI: 10.1007/978-3-662-48350-3_9

values are assumed to be additive (i.e., value of a set of items for the buyer is equal to the summation of the values of the items in the set). The buyer is considered to be strategic, that is, he is trying to maximize $v(S) - p(S)$, where S is the set of purchased items, $v(S)$ is the value of these items to the buyer and $p(S)$ is the price of the set. Knowing that the valuation of the buyer for item j is drawn from a given distribution D_j, what is a revenue optimal mechanism for the auctioneer to sell the items? Myerson [19] solves the problem for a very simple case where we only have a single item and a single buyer. He shows that in this special case the optimal mechanism is to set a fixed reserved price for the item. Despite the simplicity of the revenue optimal mechanism for selling a single item, this problem becomes quite complicated when it comes to selling two items even when we have only one buyer. Hart and Reny [15] show an optimal mechanism for selling two independent items is much more subtle and may involve randomization.

Though there are several attempts to characterize the properties of a revenue optimal mechanism of an auction, most approaches seem to be too complex and as a result impractical to real-world auctions [1, 2, 4, 5, 7, 8, 9, 12, 10, 13, 16]. Therefore, a new line of investigation is to design simple mechanisms that are approximately optimal. In a recent work of Babaioff, Immorlica, Lucier, and Weinberg [3], it is shown that we can achieve a constant factor approximation of the optimal revenue by selling items either separately or as a whole bundle in the independent setting. However, they leave the following important problem as an open question:

– ***Open Problem 3**. Is there a simple, approximately optimal mechanism for a single additive buyer whose value for n items is sampled from a common base-value distribution? What about other models of limited correlation?*

Hart and Nisan [14] show there are instances with correlated valuations in which neither selling items separately nor as a whole bundle can achieve any approximation of the optimal revenue. This holds, even when we have only two times. Therefore, it is essential to consider limited models of correlation for this problem. As an example, Babaioff et al. propose to study common base-value distributions. This model has also been considered by Chawla, Malec, and Sivan [11] to study optimal mechanisms for selling multiple items in a unit-demand setting.

In this work we study the problem for the case of correlated valuation functions and answer the above open question. In addition we also introduce a generalized model of correlation between items. Suppose we have a set of items and want to sell them to a single buyer. The buyer has a set of features in his mind and considers a value for each feature which is randomly drawn from a known distribution. Furthermore, the buyer formulates his desirability for each item as a linear combination of the values of the features. More precisely, the buyer has l distributions $F_1, F_2, \ldots, F_l$ and an $l \times n$ matrix M (which are known in advance) such that the value of feature i, denoted by f_i, is drawn from F_i and the value of item j is calculated by $V_f \cdot M_j$ where $V_f = \langle f_1, f_2, \ldots, f_l \rangle$ and M_j is the j-th row of matrix M.

This model captures the behavior of the auctions especially when the items have different features that are of different value to the buyers. Note that every common base-value distribution is a special case of this general correlation where we have $n + 1$ features $F_1, F_2, \ldots, F_n, B$ and the value of item j is determined by $v_j + b$ where v_j is drawn from F_j and b is equal for all items which is drawn from distribution B.

2   Related Work

As mentioned earlier, the problem originates from the seminal work of Myerson [19] in 1981 which characterizes a revenue optimal mechanism for selling a single item to a single buyer. This result was important in the sense that it was simple and practical while promising the maximum possible revenue. In contrast to this result, it is known that designing an optimal mechanism is much harder for the case of multiple items. There has been some efforts to find a revenue optimal mechanism for selling two heterogeneous items [20] but, unfortunately, so far too little is known about the problem even for this simple case.

Hardness of this problem is even more highlighted when Hart and Reny [15] observed randomization is necessary for the case of multiple items. This reveals the fact that even if we knew how to design an optimal mechanism for selling multiple items, it would be almost impossible to implement the optimal strategy in a real-world auction. Therefore, so far studies are focused on finding simple and approximately optimal mechanisms.

Speaking of simple mechanisms, it is very natural to think of selling items separately or as a whole bundle. The former mechanism is denoted by SRev and the latter is referred to by BRev. Hart and Nissan [13] show SRev mechanism achieves at least an $\Omega(1/\log^2 n)$ approximation of the optimal revenue in the independent setting and BRev mechanism yields at least an $\Omega(1/\log n)$ approximation for the case of identically independent distributions. Later on, this result was improved by the work of Li and Yao, that prove an $\Omega(1/\log n)$ approximation factor for SRev and a constant factor approximation for BRev for identically independent distributions [17]. These bounds are tight up to a constant factor. Moreover, it is shown BRev can be $\theta(n)$ times worse than the revenue of an optimal mechanism in the independent setting. Therefore in order to achieve a constant factor approximation mechanism we should think of more non-trivial strategies.

The seminal work of Babaioff et al. [3] shows despite the fact that both strategies SRev and BRev may separately result in a bad approximation factor, $\max\{$SRev, BRev$\}$ always has a revenue at least $\frac{1}{6}$ of an optimal mechanism. They also show we can determine which of these strategies has more revenue in polynomial time which yields a deterministic simple mechanism that can be implemented in polynomial time. However, there has been no significant progress in the case of correlated items, as [3] leave it as an open question.

In addition to this, they posed two more questions which became the subject of further studies. In the first question, they ask if there exists a simple mechanism which is approximately optimal in the case of multiple additive buyers?

This question is answered by Yao [22] via proposing a reduction from k-item n-bidder auctions to k-item auctions. They show, as a result of their reduction, a deterministic mechanism achieves a constant fraction of the optimal revenue by any randomized mechanism. In the second question, they ask if the same result can be proved for a mechanism with a single buyer whose valuation is k-demand? This question is also answered by a recent work of Rubinstein and Weinberg [21] which presents a positive result. They show the same mechanism that either sells the items separately or as a whole bundle, achieves a constant fraction of the optimal revenue even in the sub-additive setting with independent valuations. They, too, use the core decomposition technique as their main approach. Their work is very similar in spirit to ours since we both show the same mechanism is approximately optimal in different settings.

Another line of research investigated optimal mechanism for selling n items to a single unit-demand buyer. Briest et al. [6] show how complex the optimal strategies can become by proving that the gap between the revenue of deterministic mechanisms and that of non-deterministic mechanisms can be unbounded even when we have a constant number of items with correlated values. This highlights the fact that when it comes to general correlations, there is not much that can be achieved by deterministic mechanisms. However, Chawla et al. [11] study the problem with a mild correlation known as the common base-value correlation and present positive results for deterministic mechanisms in this case.

3   Results and Techniques

We study the mechanism design for selling n items to a single buyer with additive valuation function when desirabilities of each buyer for items are correlated. The main result of the paper is $\max\{\mathsf{SRev}, \mathsf{BRev}\}$, that is, the revenue we get by the better of selling items separately or as a whole bundle achieves a constant approximation of the optimal revenue when we have only one buyer and the distribution of valuations for this buyer is a common base-value distribution. This problem was left open in [3]. Our method for proving the effectiveness of the proposed mechanism is consisted of two parts. In the first part, we consider a very weak correlation between the items, which we call semi-independent correlation, and show the same mechanism achieves a constant fraction of the optimal revenue in this setting. To this end, we use the core decomposition technique which has been used by several similar works [17, 3, 21]. The second part, however, is based on a combinatorial reduction which reduces the problem to an auction with a semi-independent valuation function.

Theorem 1. *For an auction with one seller, one buyer, and a common base-value distribution of valuations we have* $\max\{\mathsf{SRev}(D), \mathsf{BRev}(D)\} \geq \frac{1}{12} \times \mathsf{Rev}(D)$.

Furthermore, we consider a natural model of correlation in which the buyer has a number of features and scores each item based on these features. The valuation of each feature for the buyer is realized from a given distributions which is known in advance. The value of each item to the buyer is then determined by a linear formula in terms of the values of the features. This can also be seen as a generalization of the common base-value correlation since a common base-value correlation can be though of as a linear correlation with $n + 1$ features. We show that if all of the features have the same distribution then $\max\{\mathsf{SRev}(D), \mathsf{BRev}(D)\}$ is at least a $\frac{1}{O(\log k)}$ fraction of $\mathsf{Rev}(D)$ where k is the maximum number of features that determine the value of each item.

Theorem 2. *In an auction with one seller, one buyer, and a linear correlation with i.i.d distribution of valuations for the features* $\max\{\mathsf{SRev}, \mathsf{BRev}\} \geq O(\frac{\mathsf{Rev}}{\log k})$ *where the value of each item depends on at most k features.*

Our approach is as follows: First we study the problem in a setting which we call *semi-independent*. In this setting, the valuation of the items are realized independently, but each item can have many copies with the same value. More precisely, each pair of items are either similar or different. In the former case, they have the same value for the buyer in each realization whereas in the latter case they have independent valuations.

Inspired by [3], we show $\max\{\mathsf{SRev}(D), \mathsf{BRev}(D)\} \geq \frac{\mathsf{Rev}(D)}{6}$ for every semi-independent distribution D. To do so, we first modify the core decomposition lemma to make it applicable to the correlated settings. Next, we apply this lemma to the problem and prove $\max\{\mathsf{SRev}(D), \mathsf{BRev}(D)\}$ achieves a constant fraction of the optimal revenue.

Given $\max\{\mathsf{SRev}(D), \mathsf{BRev}(D)\}$ is optimal up to a constant factor in the semi-independent setting, we analyze the behavior of $\max\{\mathsf{SRev}, \mathsf{BRev}\}$ in each of the settings by creating another auction in which each item of the original auction is split into several items and the distributions are semi-independent. We show that the maximum achievable revenue in the secondary auction is no less than the optimal revenue of the original auction and also selling all items together has the same revenue in both auctions. Finally, we bound the revenue of SRev in the original auction by a fraction of the revenue that SRev achieves in the new auction and by putting all inequalities together we prove an approximation factor for $\max\{\mathsf{SRev}, \mathsf{BRev}\}$. In contrast to the prior methods for analyzing the efficiency of mechanism, our approach in this part is purely combinatorial.

Although the main contribution of the paper is analyzing $\max\{\mathsf{SRev}, \mathsf{BRev}\}$ in common base-value and linear correlations, we show the following as auxiliary lemmas which might be of independent interest.

- One could consider a variation of independent setting, wherein each item has a number of copies and the value of all copies of an item to the buyer is always the same. We show in this setting $\max\{\mathsf{SRev}, \mathsf{BRev}\}$ is still a constant fraction of Rev.
- A natural generalization of i.i.d settings, is a setting in which the distributions of valuations are not exactly the same, but are the same up to scaling. We show, in the independent setting with such valuation functions BRev is at least an $O(\frac{1}{\log n})$ fraction of Rev.

Due to the space constraints proofs are omitted in this version. The reader can find the full version of the paper on arXiv for a formal discussion.

4 Preliminaries

Throughout this paper we study the optimal mechanisms for selling n items to a risk-neutral, quasi-linear buyer. The items are considered to be indivisible and not necessarily identical i.e. the buyer can have different distributions of desirabilities for different items. In our setting, distributions are denoted by $D = \langle D_1, D_2, \ldots, D_n \rangle$ where D_j is the distribution for item j. Moreover, the buyer has a valuation vector $V = \langle v_1, v_2, \ldots, v_n \rangle$ which is randomly drawn from D specifying the values he has for the items. Note that, values may be correlated. Once a mechanism is set for selling items, the buyer purchases a set S_V of the items that maximizes $v(S_V) - p(S_V)$, where $v(S_V)$ is the desirability of S_V for the buyer and $p(S_V)$ is the price that he pays. The revenue achieved by a mechanism is equal to $\sum \mathbb{E}\big[p(S_V)\big]$ where V is randomly drawn from D. The following terminology is used in [3] in order to compare the performance of different mechanisms. In this paper we use similar notations.

- Rev(D): Maximum possible revenue that can be achieved by any truthful mechanism.
- SRev(D): The revenue that we get when selling items separately using Myerson's optimal mechanism for selling each item.
- BRev(D): The revenue that we get when selling all items as a whole bundle using Myerson's optimal mechanism.

We refer to the expected value and variance of a one-dimensional distribution D by Val(D) and Var(D) respectively. We say an n-dimensional distribution D of the desirabilities of a buyer is independent over the items if for every $a \neq b$, v_a and v_b are independent variables when $V = \langle v_1, v_2, \ldots, v_n \rangle$ is drawn from D. Furthermore, we define the semi-independent distributions as follows.

Definition 1. *Let D be a distribution of valuations of a buyer over a set of items. We say D is semi-independent iff the valuations of every two different items are either always equal or completely independent. Moreover, we say two items a and b are similar in a semi-independent distribution D if for every $V \sim D$ we have $v_a = v_b$.*

Moreover, we define the common base-value distributions as follows.

Definition 2. *We say a distribution D is common base-value, if there exist independent distributions $F_1, F_2, \ldots, F_n, B$ such that for $V = \langle v_1, v_2, \ldots, v_n \rangle \sim D$ and every $1 \leq j \leq n$, $v_j = f_j + b$ where f_j comes from distribution F_j and b is drawn from B which is equal for all items.*

A natural generalization of common base-value distributions are distributions in which the valuation of each item is determined by a linear combination of k independent variables which are the same for all items. More precisely, we define the linear distributions as follows.

Definition 3. *Let D be a distribution of valuations of a buyer for n items. We say D is a linear distribution if there exist independent desirability distributions $F_1, F_2, \ldots, F_k$ and a $k \times n$ matrix M with non-negative rational values such that $V = \langle v_1, v_2, \ldots, v_n \rangle \sim D$, can be written as $W \times M$ where $W = \langle w_1, w_2, \ldots, w_k \rangle$ is a vector such that w_i is drawn from F_i.*

5   The Core Decomposition Technique

Most of the results in this area are mainly achieved by the core decomposition technique which was first introduced in [17]. Using this technique we can bound the revenue of an optimal mechanism without taking into account the complexities of the revenue optimal mechanism. The underlying idea is to split distributions into two parts: the core and the tail. If for each realization of the values we were to know in advance for which items the valuations in the core part will be and for which items the valuations in the tail part will be, we would achieve at least the optimal revenue achievable without such information. This gives us an intuition which we can bound the optimal revenue by the total sum of the revenues of 2^n auctions where in each auction we know which valuation is in which part. The tricky part then would be to separate the items whose valuations are in the core part from the items whose valuations are in the tail and sum them up separately. We use the same notation which was used in [3] for formalizing our arguments as follows.

- D_i: The distribution of desirabilities of the buyer for item i.
- D_A: (A is a subset of items): The distribution of desirabilities of the buyer for items in A.
- r_i: The revenue that we get by selling item i using Myerson's optimal mechanism.
- r: The revenue we get by selling all of the items separately using Myerson's optimal mechanism which is equal to $\sum r_i$.
- t_i: A real number separating the core from the tail for the distribution of item i. we say a valuation v_i for item i is in the core if $0 \leq v_i \leq r_i t_i$ and is in the tail otherwise.
- p_i: A real number equal to the probability that $v_i > r_i t_i$ when v_i is drawn from D_i.
- p_A: (A is a subset of items): A real number equal to the probability that $\forall i \notin A, v_i \leq r_i t_i$ and $\forall i \in A, v_i > r_i t_i$.
- D_i^C: A distribution of valuations of the i-th item that is equal to D_i conditioned on $v_i \leq r_i t_i$.
- D_i^T: A distribution of valuations of the i-th item for the buyer that is equal to D_i conditioned on $v_i > r_i t_i$.
- D_A^C: (A is a subset of items): A distribution of valuations of the items in $[N] - A$ for the buyer that is equal to $D_{[N]-A}$ conditioned on $\forall i \notin A, v_i \leq r_i t_i$.
- D_A^T: (A is a subset of items): A distribution of valuations of the items in A for the buyer that is equal to D_A conditioned on $\forall i \in A, v_i > r_i t_i$.

– D^A: A distribution of valuations for all items which is equal to D conditioned on both $\forall i \notin A, v_i \leq r_i t_i$ and $\forall i \in A, v_i > r_i t_i$.

In Lemma 2 we provide an upper bound for p_i. Next we bound $\mathsf{Rev}(D_i^C)$ and $\mathsf{Rev}(D)$ in Lemmas 3 and 4 and finally in Lemma 6 which is known as Core Decomposition Lemma we prove an upper bound for $\mathsf{Rev}(D)$. All these lemmas are proved in [3] for the case of independent setting.

Lemma 1. *For every $A \subset [N]$, if the valuation of items in A are independent of items in $[N] - A$ then we have $\mathsf{Rev}(D) \leq \mathsf{Rev}(D_A) + \mathsf{Val}(D_{[N]-A})$.*

Lemma 2. $p_i \leq \frac{1}{t_i}$.

Lemma 3. $\mathsf{Rev}(D_i^C) \leq r_i$.

Lemma 4. $\mathsf{Rev}(D_i^T) \leq r_i/p_i$.

Lemma 5. $\mathsf{Rev}(D) \leq \sum_A p_A \mathsf{Rev}(D^A)$.

For independent setting we can apply Lemma 1 to Lemma 5 and finally with application of some algebraic inequalities come up with the following inequality

$$\mathsf{Rev}(D) \leq \mathsf{Val}(D_\emptyset^C) + \sum_A p_A \mathsf{Rev}(D_A^T).$$

Unfortunately this does not hold for correlated settings since in Lemma 1 we assume valuation of items of A are independent of the items of $[N]-A$. Therefore, we need to slightly modify this lemma such that it becomes applicable to the correlated settings as well. Thus, we add the following restriction to the valuation of items: For each subset A such that p_A is non-zero, the valuation of items in A are independent of items of $[N] - A$.

Lemma 6. *If for every A with $p_A > 0$ the values of items in A are drawn independent of the items in $[N] - A$ we have $\mathsf{Rev}(D) \leq \mathsf{Val}(D_\emptyset^C) + \sum_A p_A \mathsf{Rev}(D_A^T)$.*

6   Semi-independent Distributions

In this section we show the better of selling items separately and as a whole bundle is approximately optimal for the semi-independent correlations. To do so, we first show $k \cdot \mathsf{SRev}(D) \geq \mathsf{Rev}(D)$ where we have n items divided into k types such that items of each type are similar. Next we leverage this lemma in order to prove $\max\{\mathsf{SRev}(D), \mathsf{BRev}(D)\}$ achieves a constant-factor approximation of the revenue of an optimal mechanism. We start by stating the following lemma which is proved in [18].

Lemma 7. *In an auction with one seller, one buyer, and multiple similar items we have $\mathsf{Rev}(D) = \mathsf{SRev}(D)$.*

We also need Lemma 8 proved in [13] and [3] which bounds the revenue when we have a sub-domain S two independent value distributions D and D' over disjoint sets of items. Moreover we use Lemma 9 as an auxiliary lemma in the proof of Lemma 10.

Lemma 8. ("Marginal Mechanism on Sub-Domain [13, 3]") *Let D and D' be two independent distributions over disjoint sets of items. Let S be a set of values of D and D' and s be the probability that a sample of D and D' lies in S, i.e. $s = Pr[(v, v') \sim D \times D' \in S]$. $\mathsf{sRev}(D \times D'|(v, v') \in S) \leq s\mathsf{Val}(D|(v, v') \in S) + \mathsf{Rev}(D')$.*

Lemma 9. *In a single-seller mechanism with m buyers and n items with a semi-independent correlation between the items in which there are at most k non-similar items we have $\mathsf{Rev}(D) \leq mk \cdot \mathsf{SRev}(D)$.*

Next, we show $\max\{\mathsf{SRev}(D), \mathsf{BRev}(D)\} \geq \frac{1}{6} \cdot \mathsf{Rev}(D)$. The proof is very similar in spirit to the proof of Babaioff et al. for showing $\max\{\mathsf{SRev}(D), \mathsf{BRev}(D)\}$ achieves a constant approximation factor of the revenue optimal mechanism in independent setting [3]. In this proof, we first apply the core decomposition lemma with $t_i = r/(r_i n_i)$ and break down the problem into two sub-problems. In the first sub-problem we show $\sum_A p_A \mathsf{Rev}(D_A^T) \leq 2\mathsf{SRev}(D)$ and in the second sub-problem we prove $4 \max\{\mathsf{SRev}(D), \mathsf{BRev}(D)\} \geq \mathsf{Val}(D_\emptyset^C)$. Having these two bounds together, we apply the core decomposition lemma to imply $\max\{\mathsf{SRev}(D), \mathsf{BRev}(D)\} \geq \frac{1}{6} \cdot \mathsf{Rev}(D)$.

Lemma 10. *Let D be a semi-independent distribution of valuations for n items in single buyer setting. In this problem we have $\max\{\mathsf{SRev}(D), \mathsf{BRev}(D)\} \geq \frac{1}{6} \cdot \mathsf{Rev}(D)$.*

7   Common Base-Value Distributions

In this section we study the same problem with a common base-value distribution. Recall that in such distributions desirabilities of the buyer are of the form $v_j = f_j + b_i$ where f_j is drawn from a known distribution F_j and b_i is the same for all items and is drawn from a known distribution B. Again, we show $\max\{\mathsf{SRev}, \mathsf{BRev}\}$ achieves a constant factor approximation of Rev when we have only one buyer. Note that, this result answers an open question raised by Babaioff et al. in [3].

Theorem 3. *For an auction with one seller, one buyer, and a common base-value distribution of valuations we have $\max\{\mathsf{SRev}(D), \mathsf{BRev}(D)\} \geq \frac{1}{12} \times \mathsf{Rev}(D)$.*

Proof. Let I be an instance of the auction. We create an instance $\mathsf{Cor}(I)$ of an auction with $2n$ items such that the distribution of valuations is a semi-independent distribution D' where $D_i' = F_i$ for $1 \leq i \leq n$ and $D_i' = B$ for $n + 1 \leq i \leq 2n$. Moreover, the valuations of the items $n + 1, n + 2, \ldots, 2n$ are always equal and all other valuations are independent. Thus, by the definition, D' is a semi-independent distribution of valuations and by Lemma 10 we have

$$\max\{\mathsf{SRev}(D'), \mathsf{BRev}(D')\} \geq \frac{1}{6} \times \mathsf{Rev}(D'). \tag{1}$$

Since every mechanism for selling the items of D can be mapped to a mechanism for selling the items of D' where items i and $n+i$ are considered as a single package containing both items, we have

$$\mathsf{Rev}(D) \leq \mathsf{Rev}(D'). \tag{2}$$

Moreover, since in the bundle mechanism we sell all of the items as a whole bundle, the revenue achieved by bundle mechanism is the same in both auctions. Hence,

$$\mathsf{BRev}(D) = \mathsf{BRev}(D'). \tag{3}$$

Note that, we can consider $\mathsf{SRev}(D)$ as a mechanism for selling items of $\mathsf{Cor}(\mathsf{I})$ such that items are packed into partitions of size 2 (item i is packed with item $n+i$) and each partition is priced with Myerson's optimal mechanism. Since for every two independent distributions F_i, F_{i+n} we have $\mathsf{SRev}(F_i \times F_{n+i}) \leq 2 \cdot \mathsf{BRev}(F_i \times F_{n+i})$ we can imply

$$\mathsf{SRev}(D) = \sum_{i=1}^{n} \mathsf{BRev}(F_i \times F_{n+i}) \geq \sum_{i=1}^{n} \frac{\mathsf{SRev}(F_i \times F_{i+n})}{2} = \frac{\mathsf{SRev}(D')}{2}. \tag{4}$$

According to Inequalities (1),(2), and (3) we have

$$\max\{\mathsf{SRev}(D), \mathsf{BRev}(D)\} \geq \max\{\mathsf{SRev}(D')/2, \mathsf{BRev}(D')\} \geq$$

$$\max\{\mathsf{SRev}(D'), \mathsf{BRev}(D')\}/2 \geq \mathsf{Rev}(D')/12 \geq \mathsf{Rev}(D)/12.$$

References

[1] Alaei, S., Fu, H., Haghpanah, N., Hartline, J., Malekian, A.: Bayesian optimal auctions via multi-to single-agent reduction. arXiv preprint arXiv:1203.5099 (2012)
[2] Alaei, S., Fu, H., Haghpanah, N., Hartline, J.: The simple economics of approximately optimal auctions. In: 2013 IEEE 54th Annual Symposium on Foundations of Computer Science (FOCS), pp. 628–637. IEEE (2013)
[3] Babaioff, M., Immorlica, N., Lucier, B., Matthew Weinberg, S.: A simple and approximately optimal mechanism for an additive buyer. arXiv preprint arXiv:1405.6146 (2014)
[4] Bhalgat, A., Gollapudi, S., Munagala, K.: Optimal auctions via the multiplicative weight method. In: Proceedings of the Fourteenth ACM Conference on Electronic Commerce, pp. 73–90. ACM (2013)
[5] Bhattacharya, S., Goel, G., Gollapudi, S., Munagala, K.: Budget constrained auctions with heterogeneous items. In: Proceedings of the Forty-Second ACM Symposium on Theory of Computing, pp. 379–388. ACM (2010)

[6] Briest, P., Chawla, S., Kleinberg, R., Matthew Weinberg, S.: Pricing randomized allocations. In: Proceedings of the Twenty-First Annual ACM-SIAM Symposium on Discrete Algorithms, pp. 585–597. Society for Industrial and Applied Mathematics (2010)

[7] Cai, Y., Daskalakis, C., Matthew Weinberg, S.: An algorithmic characterization of multi-dimensional mechanisms. In: Proceedings of the Forty-Fourth Annual ACM Symposium on Theory of Computing, pp. 459–478. ACM (2012a)

[8] Cai, Y., Daskalakis, C., Weinberg, M.: Optimal multi-dimensional mechanism design: Reducing revenue to welfare maximization. In: 2012 IEEE 53rd Annual Symposium on Foundations of Computer Science (FOCS), pp. 130–139. IEEE (2012b)

[9] Cai, Y.: Constantinos Daskalakis, and S Matthew Weinberg. Reducing revenue to welfare maximization: Approximation algorithms and other generalizations. In: Proceedings of the Twenty-Fourth Annual ACM-SIAM Symposium on Discrete Algorithms, pp. 578–595. SIAM (2013)

[10] Chawla, S., Hartline, J.D., Malec, D.L., Sivan, B.: Multi-parameter mechanism design and sequential posted pricing. In: Proceedings of the Forty-Second ACM Symposium on Theory of Computing, pp. 311–320. ACM (2010a)

[11] Chawla, S., Malec, D.L., Sivan, B.: The power of randomness in bayesian optimal mechanism design. In: Proceedings of the 11th ACM Conference on Electronic Commerce, pp. 149–158. ACM (2010b)

[12] Daskalakis, C., Deckelbaum, A., Tzamos, C.: The complexity of optimal mechanism design. In: SODA, pp. 1302–1318. SIAM (2014)

[13] Hart, S., Nisan, N.: Approximate revenue maximization with multiple items. arXiv preprint arXiv:1204.1846 (2012)

[14] Hart, S., Nisan, N.: The menu-size complexity of auctions. arXiv preprint arXiv:1304.6116 (2013)

[15] Hart, S., Reny, P.J.: Maximal revenue with multiple goods: Nonmonotonicity and other observations. Center for the Study of Rationality (2012)

[16] Kleinberg, R., Weinberg, S.M.: Matroid prophet inequalities. In: Proceedings of the Forty-Fourth Annual ACM Symposium on Theory of Computing, pp. 123–136. ACM (2012)

[17] Li, X., Yao, A.C.-C.: On revenue maximization for selling multiple independently distributed items. Proceedings of the National Academy of Sciences 110(28), 11232–11237 (2013)

[18] Maskin, E., Riley, J., Hahn, F.: Optimal multi-unit auctions. The Economics of Missing Markets, Information, and Games (1989)

[19] Myerson, R.B.: Optimal auction design. Mathematics of Operations Research 6(1), 58–73 (1981)

[20] Pavlov, G.: Optimal mechanism for selling two goods. The BE Journal of Theoretical Economics 11(1) (2011)

[21] Rubinstein, A., Matthew Weinberg, S.: Simple mechanisms for a combinatorial buyer and applications to revenue monotonicity. arXiv preprint arXiv:1501.07637 (2015)

[22] Yao, A.C.-C.: An n-to-1 bidder reduction for multi-item auctions and its applications. arXiv preprint arXiv:1406.3278 (2014)

Efficient Implementation of a Synchronous Parallel Push-Relabel Algorithm

Niklas Baumstark[1], Guy Blelloch[2], and Julian Shun[2]

[1] Institute of Theoretical Informatics, Karlsruhe Institute of Technology, Germany
niklas.baumstark@gmail.com
[2] Computer Science Department, Carnegie Mellon University, USA
{guyb,jshun}@cs.cmu.edu

Abstract. Motivated by the observation that FIFO-based push-relabel algorithms are able to outperform highest label-based variants on modern, large maximum flow problem instances, we introduce an efficient implementation of the algorithm that uses coarse-grained parallelism to avoid the problems of existing parallel approaches. We demonstrate good relative and absolute speedups of our algorithm on a set of large graph instances taken from real-world applications. On a modern 40-core machine, our parallel implementation outperforms existing sequential implementations by up to a factor of 12 and other parallel implementations by factors of up to 3.

1 Introduction

The problem of computing the maximum flow in a network plays an important role in many areas of research such as resource scheduling, global optimization and computer vision. It also arises as a subproblem of other optimization tasks like graph partitioning. There exist near-linear approximate algorithms for the problem [20], but exact solutions can in practice be found even for very large instances using modern algorithms. It is only natural to ask how we can exploit readily available multi-processor systems to further reduce the computation time. While a large fraction of the prior work has focused on distributed and parallel implementations of the algorithms commonly used in computer vision, fewer publications are dedicated to finding parallel algorithms that solve the problem for other graph families.

To assess the practicality of existing algorithms, we collected a number of benchmark instances. Some of them are taken from a common benchmark suite for maximum flow and others we selected specifically to represent various applications of maximum flow. Our experiments suggest that Goldberg's *hi_pr* program (a highest label-based push-relabel implementation) which is often used for comparison in previous publications is not optimal for most of the graphs that we studied. Instead, push-relabel algorithms processing active vertices in first-in-first-out (FIFO) order seems to be better suited to these graphs, and at the same time happen to be amenable for parallelization. We proceeded to design and implement our own shared memory-based parallel algorithm for the

© Springer-Verlag Berlin Heidelberg 2015
N. Bansal and I. Finocchi (Eds.): ESA 2015, LNCS 9294, pp. 106–117, 2015.
DOI: 10.1007/978-3-662-48350-3_10

maximum flow problem, inspired by an old algorithm and optimized for modern shared-memory platforms. In contrast to previous parallel implementations we try to keep the usage of atomic CPU instructions to a minimum. We achieve this by employing coarse-grained synchronization to rebalance the work and by using a parallel version of global relabeling instead of running it concurrently with the rest of the algorithm.

We are able to demonstrate good speedups on the graphs in our benchmark suite, both compared to the best sequential competitors, where we achieve speedups of up to 12 with 40 threads, and to the most recent parallel solver, which we often outperform by a factor of three or more with 40 threads.

2 Preliminaries and Related Work

We consider a directed graph G with vertices V, together with a designated source s and sink t, where $s \neq t \in V$ as well as a capacity function $c : V \times V \to \mathbb{R}_{\geq 0}$. The set of edges is $E = \{(v, w) \in V \times V \mid c(v, w) > 0\}$. We define $n = |V|$ and $m = |E|$. A *flow* in the graph is a function $f : E \to \mathbb{R}$ that is bounded from above by the capacity function and respects the *flow conservation* and *asymmetry* constraints

$$\forall w \in V : \qquad \sum_{(v,w) \in E, v \neq w} f(v, w) = \sum_{(w,x) \in E, w \neq x} f(w, x) \qquad (1)$$

$$\forall v, w \in V : \qquad f(v, w) = -f(w, v) \qquad (2)$$

We define the *residual graph* G_f with regard to a specific flow f using the residual weight function $c_f(v, w) = c(v, w) - f(v, w)$. The set of residual edges is just $E_f = \{(v, w) \in V \times V \mid c_f(v, w) > 0\}$. The *reverse residual graph* G_f^R is the same graph with each edge inverted.

A *maximum flow* in G is a flow that maximizes the *flow value*, i.e. the sum of flow on edges out of the source. It so happens that a flow is maximum if and only if there is no path from s to t in the residual graph G_f [9, Corollary 5.2]. The maximum flow problem is the problem of finding such a flow function. It is closely related to the minimum cut problem, which asks for a disjoint partition $(S \subset V, T \subset V)$ of the graph with $s \in S, t \in T$ that minimizes the cumulative capacity of edges that cross from S to T. It can be shown that the value of a maximum flow is equal to the value of a minimum cut and a minimum cut can be easily computed from a given maximum flow in linear time as the set of vertices reachable from the source in the residual graph [9, §5].

2.1 Sequential Max-Flow and Min-Cut Computations

Existing work related to the maximum flow problem is generally split into two categories: work on algorithms specific to computer vision applications and work on general-purpose algorithms. Most of the algorithms that work well for the type of grid graphs found in computer vision tend to be inferior for other graph families and vice versa [11, Concluding Remarks]. In this paper we aim to design a general-purpose algorithm that performs reasonably well on all sorts of graphs.

Traditional algorithms for the maximum flow problem typically fall into one of two categories: *Augmenting path-based* algorithms directly apply the *Ford–Fulkerson theorem* [9, Corallary 5.2] by incrementally finding augmenting paths from s to t in the residual graph and increasing the flow along them. They mainly differ in their methods of finding augmenting paths. Modern algorithms for minimum cuts in computer vision applications such as [11] belong to this family. *Preflow-based* algorithms do not maintain a valid flow during their execution and instead allow for vertices to have more incoming than outgoing flow. The difference in flow on in-edges and out-edges of a vertex is called *excess*. Vertices with positive excess are called *active*. A prominent member of this family is the classical *push-relabel* algorithm due to Goldberg and Tarjan [13]. It maintains vertex labels that estimate the minimal number of edges on a path to the sink. Excess flow can be *pushed* from a vertex to a neighbor by increasing the flow value on the connecting edge. Pushes can only happen along *admissible* residual edges to vertices of lower label. When none of the edges out of an active vertex are admissible for a push, the vertex gets *relabeled* and to a higher label. It is crucial for practical performance of push-relabel that the labels estimate the sink distance as accurately as possible. A simple way to keep them updated is to regularly run a BFS in the reverse residual graph to set them to the exact distance. This optimization is called *global relabeling*.

The more recent *pseudoflow* algorithm due to Hochbaum [16] does not need global relabeling and uses specialized data structures that allow for pushes along more than one edge. Implementations of push-relabel algorithms and Hochbaum's algorithm differ mainly in the order in which they process active vertices. *Highest label*-based implementations process active vertices in order of decreasing labels, while FIFO-based implementations select active vertices in queue order. Goldberg's *hi_pr* program [10] uses the former technique and is considered one of the fastest generic maximum flow solvers. It is often used for comparison purposes in related research. For push-relabel and Hochbaum's algorithm, it is beneficial to compute merely a *maximum preflow* that maximizes the cumulative flow on in-edges of the sink, rather than a complete flow assignment. In the case where we are looking only for a minimum cut this is already enough. In all the other cases, computing a valid flow assignment for a given maximum preflow can be achieved using a greedy decomposition algorithm that tends to take much less time than the computation of a preflow [6].

2.2 Parallel and Distributed Approaches to the Problem

Parallel algorithms for the maximum flow problem date back to 1982, where Shiloach *et al.* propose a work-efficient parallel algorithm in the PRAM model, based on blocking flows [23]. Most of the more recent work however is based on the push-relabel family of algorithms. With regard to parallelization, it has the fundamental, distinct advantage that its primitive operations are inherently local and thus largely independent.

As far as we know, Anderson and Setubal give the first implementation of a practical parallel algorithm for the maximum flow problem [1]. In their algorithm, a global queue of active vertices approximates the FIFO selection order. A fixed number of threads fetch vertices from the queue for processing and add newly activated vertices to the queue using locks for synchronization. The authors report speedups over a sequential FIFO push-relabel implementation of up to a factor of 7 with 16 processors. The authors also describe a concurrent version of global relabeling that works in parallel to the asynchronous processing of active vertices. We will refer to this technique as *concurrent global relabeling*. Bader and Sachdeva [2] modify the approach by Anderson and Setubal and introduce the first parallel algorithm that approximates the highest-label vertex selection order used by *hi_pr*.

Hong [17] proposes an asynchronous implementation that completely removes the need for locking. Instead it makes use of atomic instructions readily available in modern processors. Hong and He later present an implementation of the algorithm that also incorporates concurrent global relabeling [18]. Good speedups over a FIFO-based sequential solver and an implementation of Anderson and Setubal's algorithm are reported. There is also a GPU-accelerated implementation of the algorithm [15].

Pmaxflow [25] is a parallel, asynchronous FIFO-based push-relabel implementation. It does not use the concurrent global relabeling proposed by [1] and instead regularly runs a parallel breadth-first search on all processors. They report speedups of up to 3 over *hi_pr* with 32 threads.

3 A Synchronous Parallel Implementation of Push-Relabel

The parallel algorithms mentioned in subsection 2.2 are exclusively implemented in an asynchronous manner and differ mainly in the load-balancing schemes they use and in how they resolve conflicts between adjacent vertices that are processed concurrently. We believe the motivation for using asynchronous methods this is that in the tested benchmark instances, often there is only a handful of active vertices available for concurrent processing at a given point in time. In this work we try to also consider larger instances, where there is an obvious need for accelerated processing and where it might not be possible to solve multiple independent instances concurrently, due to memory limitations. With a higher number of active vertices per iteration a synchronous approach becomes more attractive because less work is wasted on distributing the load.

From initial experiments with sequential flow push-relabel algorithms, we learned the following things: As expected, the average number of active vertices increases with the size of the graph for a fixed family of inputs. Also, on almost all of the graphs we tested, a FIFO-based solver outperformed the highest label-based *hi_pr* implementation. This is somewhat surprising as *hi_pr* is clearly superior on the standard DIMACS benchmark [6]. These observations led us to an initial design of a simple synchronous parallel algorithm, inspired by an algorithm proposed in the original push-relabel article [13]. After the standard initialization, where all edges adjacent to the source are saturated, it proceeds in a series of iterations, each iteration consisting of the following steps:

1. All of the active vertices are processed in parallel. For each such vertex, its edges are checked sequentially for admissibility. Possible pushes are performed, but the excess changes are only applied to a copy of the old excess values. The final values are copied back in step 4.
2. New temporary labels are computed in parallel for vertices that have been processed in step 1 but are still active.
3. The new labels are applied by iterating again over the set of active vertices in parallel and setting the distance labels to the values computed in step 2.
4. The excess changes from step 1 are applied by iterating over the new set of active vertices in parallel.

These steps are repeated until there are no more active vertices with a label smaller than n. The algorithm is deterministic in that it requires the same amount of work regardless of the number of threads, which is a clear advantage over other parallel approaches that exhibit a considerable increase in work when adding more threads [18]. As soon as there are no more active vertices, we have computed a maximum preflow and can determine a minimum cut immediately or proceed to reconstruct a maximum flow assignment using a sequential greedy decomposition.

It is important to note that in step 1 we modify shared memory from multiple threads concurrently. To ensure correctness, we use atomic fetch-and-add instructions here to update the excess values of neighbor vertices (or rather, copies thereof). Contention on these values is typically low, so overhead caused by cache coherency mechanisms is not a problem. To collect the new set of active vertices for the next iteration we use atomic test-and-set instructions that resolve conflicts when a vertex is activated simultaneously by multiple neighbors, a situation that occurs only very rarely. We want to point out that synchronization primitives are kept to a minimum by design, which to our knowledge constitutes a significant difference to the state-of-the-art.

Instead of running global relabeling concurrently with the rest of the algorithm as done by [1] and [18], we regularly insert a global relabeling step in between certain iterations. The work threshold we use to determine when to do this is the same as the one used by *hi_pr*.[1] The global relabeling is implemented as a simple parallel reverse breadth-first search from the sink. Atomic compare-and-swap primitives are used during the BFS to test whether adjacent vertices have already been discovered. Apart from global relabeling, we also experimented with other heuristics such as *gap relabeling*, described in [6, Chapter 3], but could not achieve speedups by applying them to the parallel case.

---

[1] Global relabeling is performed approximately after every $12n + 2m$ edge scans.

3.1   Improving the Algorithm

We implemented the above algorithm in C++ with OpenMP extensions. For common parallel operations like prefix sums and filter, we used library functions from our *Problem Based Benchmark Suite* [24] for parallel algorithms. Even with this very simple implementation, we could measure promising speedups compared to sequential solvers. However, we conjectured that the restriction of doing at most one relabel per vertex in each iteration has some negative consequences: For one, it hinders the possible parallelism: A low-degree vertex can only activate so many other vertices before getting relabeled. It would be preferrable to imitate the sequential algorithms and completely discharge each active vertex in one iteration by alternating push and relabel operations until its excess is zero. Also, the per-vertex work is small. As we parallelize on a vertex level, we want to maximize the work per vertex to improve multi-threaded performance in the common case that only few vertices are active.

To be able to relabel a vertex more than once during one iteration, we need to allow for non-determinism and develop a scheme to resolve conflicts between adjacent vertices when both are active in the same iteration. We experimented with several options here, including the lock-free vertex discharge routine introduced by Hong and He [18]. Another approach turned out to be more successful and works without any additional synchronization. In the case where two adjacent vertices v and w are both active, a deterministic winning criterion is used to decide which one of the vertices owns the connecting edges during the current iteration. We say that v wins the competition if $d(v) < d(w) - 1$ or $d(v) = d(w) + 1$ or $v < w$ (the latter condition is a tie-breaker in the case where $d(v) = d(w)$). In this case, v is allowed to push along the edge (v, w) but w is not allowed to push along the edge (w, v). The discharge of w is thus aborted if (w, v) is the last remaining admissible edge. The particular condition is chosen such that one of the vertices can get relabeled past the other, to ensure progress. There is an edge case to consider where two adjacent vertices v and w are active, v owns the connecting edge but w is still relabeled because the residual capacity $c_f(w, v)$ is zero. We allow this scenario, but apply relabels only to a copy of the distance function d, called d'. The new admissibility condition for an edge $(x, y) \in E_f$ becomes $d'(x) = d(y) + 1$, i.e. the old distance of y is considered. The new labels are applied at the end of the iteration.

By using this approach, we ensure that for each sequence of push and relabeling operations in our algorithm during one iteration, there exists an equivalent sequence of pushes and relabels that is valid with regard to the original admissibility conditions from [13]. Thus the algorithm is correct as per the correctness proof for the push-relabel algorithm.

The resulting algorithm works similar to the simple algorithm stated above, but mixes steps 1 and 2, to enable our changes. We will refer to our own implementation of this algorithm as *prsn* in the remainder of this document. Pseudocode can be found in the longer version of our paper [4].

4   Evaluation

4.1   A Modern Benchmark Suite for Flow Computations

Traditionally, instance families from the twenty-year-old DIMACS implementation challenge [19] are used to compare the performance of maximum flow algorithms. Examples of publications that use primarily these graph families are [1, 2, 5, 6, 12, 18]. We believe that the instance families from the DIMACS benchmark suite do not accurately represent the flow and cut problems that are typically found today. Based on different applications where flow computations occur as subproblems, we compiled a benchmark suite for our experiments that we hope will give us better insight into which approaches are the most successful in practice.

Saito *et al.* describe how minimum cut techniques can be used for spam detection on the internet [21]: They observe that generally spam sites link to "good" (non-spam) sites a lot while the opposite is rarely the case. Thus the sink partition of a minimum cut between a seed set of good and spam sites is likely to contain mostly spam sites. We used their construction on a graph of pay level domains provided by a research group at the University of Mannheim with edges of capacity 1 between domains that have at least one hyperlink.[2] A publicly accessible spam list[3] and a list of major internet sites[4] helped us build good and bad seed sets of size 100 each, resulting in the *pld_spam* graph.

Very similar constructions can be used for community detection in social networks [8] It is known that social networks, the Web and document graphs like Wikipedia share a lot of common characteristics, in particular sparsity and low diameter. Halim *et al.* include in their article a comprehensive collection of references that observe these properties for different classes of graphs [14]. Based on this we believe that *pld_spam* is representative of a more general class of applications that involve community detection in such graphs.

Graph partitioning software such as *KaHIP* due to Sanders and Schulz commonly use flow techniques internally [22]. The KaHIP website[5] provides an archive of flow instances for research purposes which we used as part of our test suite. We included multiple instances from this suite, because the structure of the flow graphs is very close to the structure of the input graphs and those cover a wide range of practical applications.

The input graphs for KaHIP are taken from the 10th DIMACS graph partitioning implementation challenge [3]: *delaunay* is a family of graphs representing the Delaunay triangulations of randomly generated sets of points in the plane. *rgg* is a family of random geometric graphs generated from a set of random points in the unit square. Points are connected via an edge if their distance is smaller than $0.55 \cdot \frac{\ln n}{n}$. *europe.osm* is the largest amongst a set of street map graphs. *nlp-kkt240* is the graph representation of a large sparse matrix arising in non-linear

---

[2] http://webdatacommons.org/hyperlinkgraph/
[3] http://www.joewein.de/sw/blacklist.htm
[4] https://www.quantcast.com/top-sites
[5] http://algo2.iti.kit.edu/documents/kahip/

optimization. For the cases where graphs of different sizes are available (*delaunay* and *rgg*), we included the largest instance whose internal representation fits into the main memory of our test machine.

As a third application, in computer vision a lot of different problems reduce to minimum cut: For reference, Fishbain and Hochbaum [7, Section 3.2] describe various examples of applications. We included *BL06-camel-lrg*, an instance of multi-view reconstruction from the vision benchmark suite of the University of Western Ontario.[6]

For completeness, we also included instances of two of the harder graph families from the DIMACS maximum flow challenge, *rmf_wide_4* and *rlg_wide_16*, which are described for example by [6]. Table 1 shows the complete list of graphs we used in our benchmarks, together with their respective vertex and edge counts, as well as the maximum edge capacities.

Table 1. Properties of our benchmark graph instances. The maximum edge capacity is excluding source or sink adjacent edges.

graph name	num. vertices	num. edges	max. edge capacity
rmf_wide_4	1,048,576	5,160,960	10000
rlg_wide_16	4,194,306	12,517,376	30000
delaunay_28	161,061,274	966,286,764	1
rgg_27	80,530,639	1,431,907,505	1
europe.osm	15,273,606	32,521,077	1
nlpkkt240	8,398,082	222,847,493	1
pld_spam	42,889,802	623,056,513	1
BL06-camel-lrg	18,900,002	93,749,846	16000

4.2   Comparison and Testing Methodology

Our aim was to compare the practical efficiency of our algorithm to the sequential and parallel state-of-the-art on a common Intel architecture. For comparison with sequential implementations, we selected the publicly available *f_prf*[7], *hi_pr* and *hpf*[8] programs, implementing FIFO and highest label-based push-relabel and Hochbaum's pseudoflow algorithm, respectively. For *hi_pr*, we did not find a canonical URL for the most recent 3.7 version of the code and instead used the copy embedded in a different project.[9] Our results show that *hpf* is the best sequential solver for our benchmark suite, only outperformed by *f_prf* on the *pld_spam* graph. The most recent parallel algorithm is the asynchronous lock-free algorithm by Hong and He [18]. Since it has no public implementation, we implemented their algorithm based on the pseudocode description. We will refer to it as *hong_he* in the remainder of this document. Our own implementation of the algorithm described in subsection 3.1 is called *prsn*. We also experimented with a parallel push-relabel implementation that is part of the Galois project.[10]

---

[6] http://vision.csd.uwo.ca/data/maxflow/

[7] http://www.avglab.com/soft.html

[8] http://riot.ieor.berkeley.edu/Applications/Pseudoflow/maxflow.html

[9] https://code.google.com/p/pmaxflow/source/browse/trunk/goldberg/hipr

[10] http://iss.ices.utexas.edu/?p=projects/galois/benchmarks/preflow_push

Although their code is competitive on certain small inputs, it did not complete within a reasonable amount of time on larger instances.

To eliminate differences due to graph representation and initialization overhead, we modified the algorithms to use the same internal graph representation, namely adjacency arrays with each edge also storing the residual capacity of its reverse edge, as described in [2]. For all five algorithms we only measured the time to compute the maximum preflow, not including the data structure initialization time. The reconstruction of the complete flow function from there is the same in every case and takes only a negligible fraction (less than 3 percent) of the total sequential computation time for all of our input graphs. We measured each combination of algorithm and input at least five times.

We carried out the experiments on a NUMA Intel Nehalem machine. It hosts four Xeon E7-8870 sockets clocked at 2.4 GHz per core, making for a total of 40 physical and 80 logical cores. Every socket has 64 GiB of RAM associated with it, making for a total of 256 GiB.

4.3   Results

The longer version of our paper contains comprehensive tables of the absolute timings we collected in all our experiments [4]. *rgg_27*, *delaunay_28* and *nlpkkt240* are examples of graphs where an effective parallel solution is possible: Figure 1 shows that both *hong_he* and *prsn* outperform *hpf* in the case of *rgg_27*; furthermore *prsn* is three times faster than *hong_he* with 32 threads. The speedup plot for *delaunay_28* looks almost identical to the one for *rgg_27*. In the case of *nlpkkt240*, we can tell from Figure 2 that *prsn* outperforms *hpf* with four threads and achieves a speedup of 5.7 over *hpf* with 32 threads. *hong_he* does not achieve any absolute speedup even with 40 threads. *prsn* does remarkably well on our spam detection instance *pld_spam*: Even with one thread, our implementation outperforms *hpf* and *hi_pr* and is on par with *f_prf*. Figure 3 shows that with 40 threads, an absolute speedup of 12 is achieved over the best sequential run. We noticed here that the algorithm spends most of the time performing a small number of iterations on a very large number of active vertices, which is very advantageous for parallelization. Note that *hong_he* did not finish on the *pld_spam*

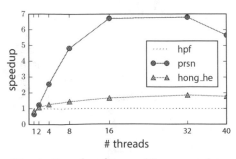

Fig. 1. Speedup for *rgg_27* compared to the best single-threaded timing

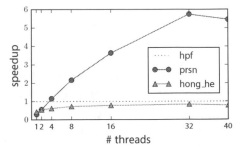

Fig. 2. Speedup for *nlpkkt240* compared to the best single-threaded timing

benchmark after multiple hours of run time. We conjecture that this is because the algorithm is simply very inefficient for this particular instance and were able to confirm that this is the case by reimplementing the same vertex discharge in the sequential *f_prf* program. Before each push, it scans all the edges of a vertex to find the neighbor with the lowest label. This is necessary in *hong_he* because the algorithm does not maintain the label invariant $d(x) \leq d(y) + 1$ for all $(x, y) \in E_f$. The modified *f_prf* also did not finish solving the benchmark instance within a reasonable time frame.

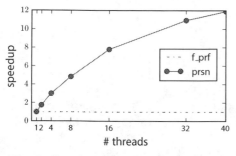

Fig. 3. Speedup for *pld_spam* compared to the best timing. *hong_hc* did not finish in our experiments

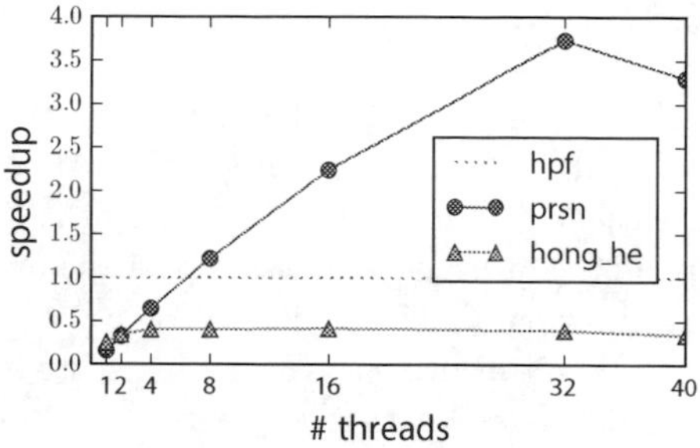

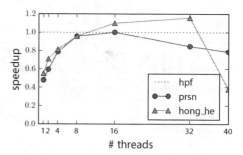

Fig. 4. Speedup for *BL06-camel-lrg* compared to the best single-threaded timing

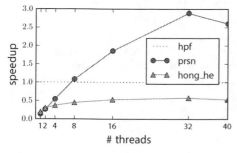

Fig. 5. Speedup for *rlg_wide_16* compared to the best sequential timing

Fig. 6. Speedup for *europe.osm* compared to the best single-threaded timing

BL06-camel-lrg is a benchmark from computer vision. Figure 4 shows that *prsn* is able to outperform *hpf* with 8 threads and achieves a speedup of almost four with 32 threads. *hpf* has in turn been shown to perform almost as well as the specialized BK algorithm on this benchmark [7].

As we can tell from Figure 5, in the case of *rlg_wide_16*, *prsn* requires eight threads to outperform *hpf* and achieves an absolute speedup of about three with 32 threads. *europe.osm* appears to be a hard instance for the parallel algorithms, as shown in Figure 6: Only *hong_he* achieves a small speedup with 32 threads. Both parallel algorithms fail to outperform the best sequential algorithm in the case of the *rmf_wide_4* graph. In all cases, making use of hyper-threading by running the algorithms with 80 threads did not yield any performance improvements, which we attribute to the fact that the algorithm is mostly memory bandwidth-bound.

Overall, different graph types lead to different behaviour of the tested algorithms. We have shown that especially for large, sparse graphs of low diameter, our algorithm can provide significant speedups over existing sequential and parallel maximum flow solvers.

5   Conclusion

In this paper, we presented a new parallel maximum flow implementation and compared it with existing state-of-the-art sequential and parallel implementations on a variety of graphs. Our implementation uses coarse-grained synchronization to avoid the overhead of fine-grained locking and hardware-level synchronization used by other parallel implementations. We showed experimentally that our implementation outperforms the fastest existing parallel implementation and achieves good speedup over existing sequential implementations on different graphs. Therefore, we believe that our algorithm can considerably accelerate many flow and cut computations that arise in practice. To evaluate the performance of our algorithm, we identified a new set of benchmark graphs representing maximum flow problems occuring in practical applications. We believe this contribution will help in evaluating maximum flow algorithms in the future.

Acknowledgments. We want to thank Professor Dr. Peter Sanders from Karlsruhe Institute of Technology for contributing initial ideas and Dr. Christian Schulz from KIT for preparing flow instances for us.

This work is partially supported by the National Science Foundation under grant CCF-1314590, and by the Intel Labs Academic Research Office for the Parallel Algorithms for Non-Numeric Computing Program.

References

1. Anderson, R., Setubal, J.: A Parallel Implementation of the Push–Relabel Algorithm for the Maximum Flow Problem. Journal of Parallel and Distributed Computing (1995)
2. Bader, D., Sachdeva, V.: A Cache-aware Parallel Implementation of the Push–Relabel Network Flow Algorithm and Experimental Evaluation of the Gap Relabeling Heuristic (2006)
3. Bader, D.A., Meyerhenke, H., Sanders, P., Wagner, D. (eds.): Graph Partitioning and Graph Clustering - 10th DIMACS Implementation Challenge Workshop, Georgia Institute of Technology, Atlanta, GA, USA, February 13-14, 2012. Proceedings, Contemporary Mathematics, vol. 588. American Mathematical Society (2013)
4. Baumstark, N., Blelloch, G., Shun, J.: Efficient Implementation of a Synchronous Parallel Push-Relabel Algorithm. CoRR abs/1507.01926 (2015)
5. Chandran, B.G., Hochbaum, D.S.: A Computational Study of the Pseudoflow and Push–Relabel Algorithms for the Maximum Flow Problem. Operations Research 57(2), 358–376 (2009)

6. Cherkassky, B.V., Goldberg, A.V.: On Implementing Push–Relabel Method for the Maximum Flow Problem. Algorithmica 19(4), 390–410 (1997)
7. Fishbain, B., Hochbaum, D.S., Mueller, S.: A Competitive Study of the Pseudoflow Algorithm for the Minimum st Cut Problem in Vision Applications. Journal of Real-Time Image Processing (2013)
8. Flake, G.W., Lawrence, S., Giles, C.L.: Efficient Identification of Web Communities. In: KDD 2000, pp. 150–160. ACM, New York (2000)
9. Ford, L., Fulkerson, D.: Flows in Networks (1962)
10. Goldberg, A.: hi_pr Maximum Flow Solver, http://www.avglab.com/soft.html
11. Goldberg, A.V., Hed, S., Kaplan, H., Tarjan, R.E., Werneck, R.F.: Maximum Flows by Incremental Breadth-first Search. In: Demetrescu, C., Halldórsson, M.M. (eds.) ESA 2011. LNCS, vol. 6942, pp. 457–468. Springer, Heidelberg (2011)
12. Goldberg, A.: Two-level Push–Relabel Algorithm for the Maximum Flow Problem. Algorithmic Aspects in Information and Management (2009)
13. Goldberg, A., Tarjan, R.: A New Approach to the Maximum-Flow Problem. Journal of the ACM (JACM) 35(4), 921–940 (1988)
14. Halim, F., Yap, R., Wu, Y.: A MapReduce-Based Maximum-Flow Algorithm for Large Small-World Network Graphs. In: ICDCS 2011, pp. 192–202 (June 2011)
15. He, Z., Hong, B.: Dynamically Tuned Push–Relabel Algorithm for the Maximum Flow Problem on CPU-GPU-Hybrid Platforms. In: IPDPS 2010 (April 2010)
16. Hochbaum, D.S.: The Pseudoflow Algorithm: A New Algorithm for the Maximum-Flow Problem. Operations Research 56(4), 992–1009 (2008)
17. Hong, B.: A Lock-Free Multi-Threaded Algorithm for the Maximum Flow Problem. In: MTAAP 2008 (April 2008)
18. Hong, B., He, Z.: An Asynchronous Multithreaded Algorithm for the Maximum Network Flow Problem with Nonblocking Global Relabeling Heuristic. IEEE Transactions on Parallel and Distributed Systems 22(6), 1025–1033 (2011)
19. Johnson, D.S., McGeoch, C.C. (eds.): Network Flows and Matching: First DIMACS Implementation Challenge. American Mathematical Society, Boston (1993)
20. Kelner, J.A., Lee, Y.T., Orecchia, L., Sidford, A.: An Almost-linear-time Algorithm for Approximate Max Flow in Undirected Graphs, and Its Multicommodity Generalizations. In: SODA 2014, pp. 217–226. SIAM (2014)
21. Saito, H., Toyoda, M., Kitsuregawa, M., Aihara, K.: A Large-Scale Study of Link Spam Detection by Graph Algorithms. In: AIRWeb 2007 (2007)
22. Sanders, P., Schulz, C.: Engineering multilevel graph partitioning algorithms. In: Demetrescu, C., Halldórsson, M.M. (eds.) ESA 2011. LNCS, vol. 6942, pp. 469–480. Springer, Heidelberg (2011)
23. Shiloach, Y., Vishkin, U.: An $O(n^2 \log n)$ Parallel Max-Flow Algorithm. J. Algorithms 3(2), 128–146 (1982)
24. Shun, J., Blelloch, G.E., Fineman, J.T., Gibbons, P.B., Kyrola, A., Simhadri, H.V., Tangwongsan, K.: Brief Announcement: The Problem Based Benchmark Suite. In: SPAA 2012, pp. 68–70. ACM (2012)
25. Soner, S., Ozturan, C.: Experiences with Parallel Multi-threaded Network Maximum Flow Algorithm. Partnership for Advanced Computing in Europe (2013)

Towards Tight Lower Bounds
for Scheduling Problems

Abbas Bazzi and Ashkan Norouzi-Fard

School of Computer and Communication Sciences, EPFL
{abbas.bazzi,ashkan.norouzifard}@epfl.ch

Abstract. We show a close connection between structural hardness for
k-partite graphs and tight inapproximability results for scheduling problems with precedence constraints. Assuming a natural but nontrivial generalisation of the bipartite structural hardness result of [1], we obtain a hardness of $2 - \epsilon$ for the problem of minimising the makespan for scheduling precedence-constrained jobs with preemption on identical parallel machines. This matches the best approximation guarantee for this problem [6,4]. Assuming the same hypothesis, we also obtain a super constant inapproximability result for the problem of scheduling precedence-constrained jobs on related parallel machines, making progress towards settling an open question in both lists of ten open questions by Williamson and Shmoys [17], and by Schuurman and Woeginger [14].

The study of structural hardness of k-partite graphs is of independent interest, as it captures the intrinsic hardness for a large family of scheduling problems. Other than the ones already mentioned, this generalisation also implies tight inapproximability to the problem of minimising the weighted completion time for precedence-constrained jobs on a single machine, and the problem of minimising the makespan of precedence-constrained jobs on identical parallel machine, and hence unifying the results of Bansal and Khot[1] and Svensson [15], respectively.

Keywords: hardness of approximation, scheduling problems, unique game conjecture.

1   Introduction

The study of scheduling problems is motivated by the natural need to efficiently allocate limited resources over the course of time. While some scheduling problems can be solved to optimality in polynomial time, others turn out to be NP-hard. This difference in computational complexity can be altered by many factors, from the machines model that we adopt, to the requirements imposed on the jobs, as well as the optimality criterion of a feasible schedule. For instance, if we are interested in minimising the completion time of the latest job in a schedule (known as the maximum makespan), then the scheduling problem is NP-hard to approximate within a factor of $3/2 - \epsilon$, for any $\epsilon > 0$, if the machines are *unrelated*, whereas it admits a Polynomial Time Approximation Scheme (PTAS) for

© Springer-Verlag Berlin Heidelberg 2015
N. Bansal and I. Finocchi (Eds.): ESA 2015, LNCS 9294, pp. 118–129, 2015.
DOI: 10.1007/978-3-662-48350-3_11

the case of *identical parallel* machines [8]. Adopting a model in between the two, in which the machines run at different speeds, but do so uniformly for all jobs (known as *uniform parallel* machines), also leads to a PTAS for the scheduling problem [9].

Although this somehow suggests a similarity in the complexity of scheduling problems between identical parallel machines and uniform parallel machines, our hopes for comparably performing algorithms seem to be shattered as soon as we add precedence requirements among the jobs. On the one hand, we know how to obtain a 2-approximation algorithm for the problem where the parallel machines are identical [6,4] (denoted as P|prec |C_{max} in the language of [7]), whereas on the other hand the best approximation algorithm known to date for the uniform parallel machines case (denoted as Q|prec |C_{max}), gives a $\log(m)$-approximation guarantee [3,2], m being the number of machines. In fact obtaining a constant factor approximation algorithm for the latter, or ruling out any such result is a major open problem in the area of scheduling algorithms. Perhaps as a testament to that, is the fact that it is listed by Williamson and Shmoys [17] as Open Problem 8, and by Schuurman and Woeginger [14] as Open Problem 1.

Moreover, our understanding of scheduling problems even on the same model of machines does not seem to be complete either. On the positive side, it is easy to see that the maximum makespan of any feasible schedule for P|prec |C_{max} is at least $\max\{L, n/m\}$, where L is the length of the longest chain of precedence constraints in our instance, and n and m are the number of jobs and machines respectively. The same lower bound still holds when we allow preemption, i.e., the scheduling problem P|prec, pmtn|C_{max}. Given that both 2-approximation algorithms of [6] and [4] rely in their analysis on the aforementioned lower bound, then they also yield a 2-approximation algorithm for P|prec, pmtn|C_{max}. However, on the negative side, our understanding for P|prec, pmtn|C_{max} is much less complete. For instance, we know that it is NP-hard to approximate P|prec |C_{max} within any constant factor strictly better than 4/3 [10], and assuming (a variant of) the UNIQUE GAMES Conjecture, the latter lower bound is improved to 2 [15]. However for P|prec, pmtn|C_{max}, only NP-hardness is known. It is important to note here that the hard instances yielding the $(2 - \epsilon)$ hardness for P|prec |C_{max} are easy instances for P|prec, pmtn|C_{max}. Informally speaking, the hard instances for P|prec |C_{max} can be thought of as k-partite graphs, where each partition has $n + 1$ vertices that correspond to $n + 1$ jobs, and the edges from a layer to the layer above it emulate the precedence constraints. The goal is to schedule these $(n+1)k$ jobs on n machines. If the k-partite graph is *complete*, then any feasible schedule has a makespan of at least $2k$, whereas if the graph was a collection of *perfect matchings* between each two consecutive layers, then there exists a schedule whose makespan is $k+1$[1]. However, if we allow preemption, then it is easy to see that even if the k-partite graph is complete, one can nonetheless find a feasible schedule whose makespan is $k + 1$.

---

[1] In fact, the gap is between k-partite graphs that have nice structural properties in the completeness case, and behave like node expanders in the soundness case.

The effort of closing the inapproximability gap between the best approximation guarantee and the best known hardness result for some scheduling problems
was successful in recent years; two of the results that are of particular interest for
us are [1] and [15]. Namely, Bansal and Khot studied in [1] the scheduling prob-
lem $1|\mathsf{prec}\ |\sum_j w_j C_j$, the problem of scheduling precedence constrained jobs on
a single machine, with the goal of minimsing the weighted sum of completion
time, and proved tight inapproximability results for it, assuming a variant of
the UNIQUE GAMES Conjecture. Similarly, Svensson proved in [15] a hardness of
$2 - \epsilon$ for $P|\mathsf{prec}\ |C_{max}$, assuming the same conjecture. In fact, both papers relied
on a structural hardness result for bipartite graphs, first introduced in [1], by
reducing a bipartite graph to a scheduling instance which leads to the desired
hardness factor.

Our Results. We propose a natural but non-trivial generalisation of the structural hardness result of [1] from bipartite to k-partite graphs, that captures
the intrinsic hardness of a large family of scheduling problems. Concretely, this
generalisation yields

1. A super constant hardness for $Q|\mathsf{prec}\ |C_{max}$, making progress towards re-
   solving an open question by [17,14]
2. A hardness of $2 - \epsilon$ for $P|\mathsf{prec},\ \mathsf{pmtn}|C_{max}$, even for the case where the
   processing time of each jobs is 1, denote by $P|\mathsf{prec},\ \mathsf{pmtn},\ p_j = 1|C_{max}$, and
   hence closing the gap for this problem.

Also, the results of [1] and [15] will still hold for $1|\mathsf{prec}\ |\sum_j w_j C_j$ and $P|\mathsf{prec}$
$|C_{max}$, respectively, under the same assumption.

On the one hand, our generalisation rules out any constant factor polynomial
time approximation algorithm for the scheduling problem $Q|\mathsf{prec}\ |C_{max}$. On the
other hand, one may speculate that the preemption flexibility when added to
the scheduling problem $P|\mathsf{prec}\ |C_{max}$ may render this problem easier, especially
that the hard instances of the latter problem become easy when preemption is
allowed. Contrary to such speculations, our generalisation to k-partite graphs
enables us to prove that it is NP-hard to approximate the scheduling problem
$P|\mathsf{prec},\ \mathsf{pmtn},\ p_j = 1|C_{max}$ within any factor strictly better than 2. Formally, we
prove the following:

Theorem 1. *Assuming Hypothesis 5, it is NP-hard to approximate the scheduling
problems* $P|prec,\ pmtn,\ p_j = 1|C_{max}$ *within any constant factor strictly better than
2, and* $Q|prec\ |C_{max}$ *within any constant factor.*

This suggests that the intrinsic hardness of a large family of scheduling problems
seems to be captured by structural hardness results for k-partite graphs. For the
case of $k = 2$, our hypothesis coincides with the structure bipartite hardness
result of [1], and yields the following result:

Theorem 2. *Assuming a variant of the* UNIQUE GAMES *Conjecture, it is NP-
hard to approximate the scheduling problem* $P|prec,\ pmtn,\ p_j = 1|C_{max}$ *within
any constant factor strictly less than 3/2.*

In fact, the $3/2$ lower bound holds even if we only assume that $1|\text{prec}\,|\sum_j w_j C_j$ is NP-hard to approximate within any factor strictly better than 2, by noting the connection between the latter and a certain bipartite ordering problem. This connection was observed and used by Svensson [15] to prove tight hardness of approximation lower bounds for $\text{P}|\text{prec}\,|C_{max}$, and this yields a somehow stronger statement; even if the UNIQUE GAMES Conjecture turns out to be false, $1|\text{prec}\,|\sum_j w_j C_j$ might still be hard to approximate to within a factor of $2 - \epsilon$, and our result for $\text{P}|\text{prec, pmtn}, p_j = 1|C_{max}$ will still hold as well. Formally,

Corollary 1. *For any $\epsilon > 0$, and $\eta \geq \eta(\epsilon)$, where $\eta(\epsilon)$ tends to 0 as ϵ tends to 0, if $1|\text{prec}\,|\sum_j w_j C_j$ has no $(2 - \epsilon)$-approximation algorithm, then $\text{P}|\text{prec}, \text{pmtn}, p_j = 1|C_{max}$ has no $(3/2 - \eta)$-approximation algorithm.*

Although we believe that Hypothesis 5 holds, the proof is still eluding us. Nonetheless, understanding the structure of k-partite graphs seems to be a very promising direction to understanding the inapproximability of scheduling problems, due to its manifold implications on the latter problems. As mentioned earlier, a similar structure for bipartite graphs was proved assuming a variant of the UNIQUE GAMES Conjecture in [1] (see Theorem 4), and we show in the full version how to extend it to k-partite graphs, while maintaining a somehow similar structure. However the resulting structure does not suffice for our purposes, i.e., does not satisfy the requirement for Hypothesis 5. Informally speaking, a bipartite graph corresponding to the completeness case of Theorem 4, despite having a *nice* structure, contains some noisy components that we cannot fully control. This follows from the fact that these graphs are derived from UNIQUE GAMES *PCP-like* tests, where the resulting noise is either intrinsic to the UNIQUE GAMES instance (i.e., from the non-perfect completeness of the UNIQUE GAMES instance), or artificially added by the test. Although we can overcome the latter, the former prohibits us from replicating the structure of the bipartite graph to get a k-partite graph with an *equally nice* structure.

Further Related Work. The scheduling problem $\text{P}|\text{prec, pmtn}, p_j = 1|C_{max}$ was first shown to be NP-hard by Ullman [16]. However, if we drop the precedence rule, the problem can be solved to optimality in polynomial time [11]. Similarly, if the precedence constraint graph is a *tree*[12,13,5] or the number of machines is 2 [12,13], the problem also becomes solvable in polynomial time. Yet, for an arbitrary precedence constraints structure, it remains open whether the problem is polynomial time solvable when the number of machines is a constant greater than or equal to 3 [17]. A closely related problem to $\text{P}|\text{prec, pmtn}|C_{max}$ is $\text{P}|\text{prec}\,|C_{max}$, in which preemption is not allowed. In fact the best 2-approximation algorithms known to date for $\text{P}|\text{prec, pmtn}|C_{max}$ were originally designed to approximate $\text{P}|\text{prec}\,|C_{max}$ [6,4], by noting the common lower bound for a makespan to any feasible schedule for both problems. As mentioned earlier, [10] and [15] prove a $4/3 - \epsilon$ NP-hardness, and $2 - \epsilon$ UGC-hardness respectively for $\text{P}|\text{prec}\,|C_{max}$, for any $\epsilon > 0$. However, to this date, only NP-hardness is known for the $\text{P}|\text{prec, pmtn}, p_j = 1|C_{max}$ scheduling problem. Although one may speculate that allowing preemption might enable us to get better approximation guarantees, no substantial progress has been made in this direction since [6] and [4].

One can easily see that the scheduling problem P|prec |C_{max} is a special case of Q|prec |C_{max}, since it corresponds to the case where the speed of every machine is equal to 1, and hence the $(4/3 - \epsilon)$ NP-hardness of [10] and the $(2 - \epsilon)$ UGC-hardness of [15] also apply to Q|prec |C_{max}. Nonetheless, no constant factor approximation for this problems is known; a $\log(m)$-approximation algorithm was designed by Chudak and Shmoys [3], and Chekuri and Bender [2] independently, where m is the number of machines.

Outline. We start in Section 2 by defining the UNIQUE GAMES problem, along with the variant of the UNIQUE GAMES Conjecture introduced in [1]. We then state in Section 3 the structural hardness result for bipartite graphs proved in [1], and propose our new hypothesis for k-partite graphs (Hypothesis 5) that will play an essential role in the hardness proofs of Section 4. Namely, we use it in Section 4.1 to prove a super constant inapproximability result for the scheduling problem Q|prec |C_{max}, and $2 - \epsilon$ inapproximability for P|prec, pmtn, $p_j = 1|C_{max}$. The reduction for the latter problem can be seen as replicating a certain scheduling instance $k - 1$ times, and hence we note that if we settle for one copy of the instance, we can prove an inapproximability of $3/2$, assuming the variant of the UNIQUE GAMES Conjecture of [1]. In the full version of the paper, we prove a structural hardness result for k-partite graphs which is similar to Hypothesis 5, although not sufficient for our scheduling problems of interest. We also note in the full version that the integrality gap instances for the natural Linear Programming (LP) relaxation for P|prec, pmtn, $p_j = 1|C_{max}$, have a very similar structure to the instances yielding the hardness result.

2   Preliminaries

In this section, we start by introducing the UNIQUE GAMES problem, along with a variant of Khot's UNIQUE GAMES conjecture as it appears in [1], and then we formally define the scheduling problems of interest.

Definition 1. *A* UNIQUE GAMES *instance* $\mathcal{U}(G = (V, W, E), [R], \Pi)$ *is defined by a bipartite graph* $G - (V, W, E)$ *with bipartitions* V *and* W *respectively, and edge set* E. *Every edge* $(v, w) \in E$ *is associated with a bijection map* $\pi_{v,w} \in \Pi$ *such that* $\pi_{v,w} : [R] \mapsto [R]$, *where* $[R]$ *is the label set. The goal of this problem is find a labeling* $\Lambda : V \cup W \mapsto [R]$ *that maximises the number of satisfied edges in* E, *where an edge* $(u, v) \in E$ *is satisfied by* Λ *if* $\pi_{v,w}(\Lambda(w)) = \Lambda(v)$.

Bansal and Khot [1] proposed the variant of the UNIQUE GAMES Conjecture in Hypothesis 3, and used it to (implicitly) prove the structural hardness result for bipartite graphs in Theorem 4.

Hypothesis 3. *[Variant of the UGC[1]] For arbitrarily small constants* $\eta, \zeta,$ $\delta > 0$, *there exists an integer* $R = R(\eta, \zeta, \delta)$ *such that for a* UNIQUE GAMES *instance* $\mathcal{U}(G = (V, W, E), [R], \Pi)$, *it is NP-hard to distinguish between:*

- *(YES Case:)* There are sets $V' \subseteq V$, $W' \subseteq W$ such that $|V'| \geq (1 - \eta)|V|$ and $|W'| \geq (1 - \eta)|W|$, and a labeling $\Lambda : V \cup W \mapsto [R]$ such that all the edges between the sets (V', W') are satisfied.
- *(NO Case:)* No labeling to $\mathcal{U}$ satisfies even a ζ fraction of edges. Moreover, the instance satisfies the following expansion property. For every $S \subseteq V$, $T \subseteq W$, $|S| = \delta|V|$, $|T| = \delta|W|$, there is an edge between S and T.

Theorem 4. [Section 7.2 in [1]] *For every $\epsilon, \delta > 0$, and positive integer Q, the following problem is NP-hard assuming Hypothesis 3: given an n-by-n bipartite graph $G = (V, W, E)$, distinguish between the following two cases:*

- *YES Case:* V *can be partitioned into* $V_0, \ldots, V_{Q-1}$ *and* W *can be partitioned into* $W_0, \ldots, W_{Q-1}$, *such that*
  - *There is no edge between V_i and W_j for all $0 \leq j < i < Q$.*
  - $|V_i| \geq \frac{(1-\epsilon)}{Q}n$ and $|W_i| \geq \frac{(1-\epsilon)}{Q}n$, for all $i \in [Q]$.
- *NO Case: For any $S \subseteq V$, $T \subseteq W$, $|S| = \delta n$, $|T| = \delta n$, there is an edge between S and T.*

In the scheduling problems that we consider, we are given a set $\mathcal{M}$ of machines and a set $\mathcal{J}$ of jobs with precedence constraints, and the goal is find a feasible schedule in a way to minimise the makespan, i.e., the maximum completion time. We will be interested in the following two variants of this general setting:

P|prec, pmtn $|\mathbf{C}_{max}$: In this model, the machines are assumed to be be parallel and identical, i.e., the processing time of a job $J_j \in \mathcal{J}$ is the same on any machine $M_i \in \mathcal{M}$ ($p_{i,j} = p_j$ for all $M_i \in \mathcal{M}$). Furthermore, preemption is allowed, and hence the processing of a job can be paused and resumed at later stages, not necessarily on the same machine.

Q|prec $|\mathbf{C}_{max}$: In this model, the machines are assumed to be parallel and uniform, i.e., each machine $M_i \in \mathcal{M}$ has a speed s_i, and the time it takes to process job $J_j \in \mathcal{J}$ on this machine is p_j/s_i.

Before we proceed we give the following notations that will come in handy in the remaining sections of the paper. For a positive integer Q, $[Q]$ denotes the set $\{0, 1, \ldots, Q - 1\}$. In a scheduling context, we say that a job J_i is a predecessor of a job J_j, and write it $J_i \prec J_j$, if in any feasible schedule, J_j cannot start executing before the completion of job J_i. Similarly, for two *sets* of jobs $\mathcal{J}_i$ and $\mathcal{J}_j$, $\mathcal{J}_i \prec \mathcal{J}_j$ is equivalent to saying that all the jobs in $\mathcal{J}_j$ are successors of all the jobs in $\mathcal{J}_i$.

3   Structured k-partite Problem

We propose in this section a natural but nontrivial generalisation of Theorem 4 to k-partite graphs. Assuming hardness of this problem, we can get the following hardness of approximation results:

1. It is NP-hard to approximate Q|prec $|\mathbf{C}_{max}$ within any constant factor.
2. It is NP-hard to approximate P|prec, pmtn, $p_j = 1|\mathbf{C}_{max}$ within a $2 - \epsilon$ factor.

3. It is NP-hard to approximate $1|\text{prec} |\sum_j w_j C_j$ within a $2 - \epsilon$ factor.
4. It is NP-hard to approximate $P|\text{prec} |C_{max}$ within a $2 - \epsilon$ factor.

The first and second result are presented in Section 4.1 and 4.2, respectively. Moreover, one can see that the reduction presented in [1] for the scheduling problem $1|\text{prec} |\sum_j w_j C_j$ holds using the hypothesis for the case that $k = 2$. The same holds for the reduction in [15] for the scheduling problem $P|\text{prec} |C_{max}$. This suggests that this structured hardness result for k-partite graphs somehow unifies a large family of scheduling problems, and captures their common intrinsic hard structure.

Hypothesis 5. *[k-partite Problem] For every $\epsilon, \delta > 0$, and constant integers $k, Q > 1$, the following problem is NP-hard: given a k-partite graph $G = (V_1, ..., V_k, E_1, ..., E_{k-1})$ with $|V_i| = n$ for all $1 \leq i \leq k$ and E_i being the set of edges between V_i and V_{i+1} for all $1 \leq i < k$, distinguish between following two cases:*

- *YES Case: every V_i can be partitioned into $V_{i,0}, ..., V_{i,Q-1}$, such that*
  - *There is no edge between V_{i,j_1} and V_{i-1,j_2} for all $1 < i \leq k, j_1 < j_2 \in [Q]$.*
  - *$|V_{i,j}| \geq \frac{(1-\epsilon)}{Q}n$, for all $1 \leq i \leq k, j \in [Q]$.*
- *NO Case: For any $1 < i \leq k$ and any two sets $S \subseteq V_{i-1}, T \subseteq V_i, |S| = \delta n, |T| = \delta n$, there is an edge between S and T.*

This says that if the k-partite graph $G = (V_1, ..., V_k, E_1, ..., E_{k-1})$ satisfies the YES Case, then for every $1 \leq i \leq k-1$, the induced subgraph $\tilde{G} = (V_i, V_{i+1}, E_i)$ behaves like the YES Case of Theorem 4, and otherwise, every such induced subgraph corresponds to the NO case. Moreover, if we think of G as a directed graph such that the edges are oriented from V_i to V_{i-1}, then all the partitions in the YES case are consistent in the sense that a vertex $v \in V_{i,j}$ can only have paths to vertices $v' \in V_{i',j'}$ if $i' < i \leq k$ and $j' \leq j \leq Q - 1$.

We can prove that assuming the previously stated variant of the UNIQUE GAMES Conjecture, Hypothesis 5 holds for $k = 2$. Also we can extend Theorem 4 to a k-partite graph using a perfect matching approach which results in the following theorem. We delegate its proof to the full version of the paper.

Theorem 6. *For every $\epsilon, \delta > 0$, and constant integers $k, Q > 1$, the following problem is NP-hard: given a k-partite graph $G = (V_1, ..., V_k, E_1, ..., E_{k-1})$ with $|V_i| = n$ and E_i being the set of edges between V_i and V_{i+1}, distinguish between following two cases:*

- *YES Case: every V_i can be partitioned in to $V_{i,0}, ..., V_{i,Q-1}, V_{i,err}$, such that*
  - *There is no edge between V_{i,j_1} and V_{i-1,j_2} for all $1 < i \leq k, j_1 \neq j_2 \in [Q]$.*
  - *$|V_{i,j}| \geq \frac{(1-\epsilon)}{Q}n$ for all $1 \leq i \leq k, j \in [Q]$.*
- *NO Case: For any $1 < i \leq k$ and any two sets $S \subseteq V_i, T \subseteq V_{i-1}, |S| = \delta n, |T| = \delta n$, there is an edge between S and T.*

Note that in the YES Case, the induced subgraphs on $\{V_{i,j}\}$ for $1 \leq i \leq k$, $0 \leq j \leq Q - 1$, have the perfect structure that we need for our reductions to scheduling problems. However, we do not get the required structure between the noise partitions (i.e., $\{V_{i,err}\}$ for $1 \leq i \leq k$), which will prohibit us from getting the desired gap between the YES and NO Cases when performing a reduction from this graph to our scheduling instances of interest. The structure of the noise that we want is that the vertices in the noise partition are only connected to the vertices in the noise partition of the next layer.

4   Lower Bounds for Scheduling Problems

In this section, we show that, assuming Hypothesis 5, there is no constant factor approximation algorithm for the scheduling problem Q|prec |C_{max}, and there is no c-approximation algorithm for the scheduling problem P|prec, pmtn, $p_j = 1$|C_{max}, for any constant c strictly better than 2. We also show that, assuming a special case of Hypothesis 5, i.e., $k = 2$ which is equivalent to (a variant) of UNIQUE GAMES Conjecture (Hypothesis 3), there is no approximation algorithm better than $3/2 - \epsilon$ for P|prec, pmtn, $p_j = 1$|C_{max}, for any $\epsilon > 0$.

4.1   Q|prec |C_{max}

In this section, we reduce a given k-partite graph G to an instance $\mathcal{I}(k)$ of the scheduling problem Q|prec |C_{max}, and show that if G corresponds to the YES Case of Hypothesis 5, then the maximum makespan of $\mathcal{I}(k)$ is roughly n, whereas a graph corresponding to the NO Case leads to a scheduling instance whose makespan is roughly the number of vertices in the graph, i.e., nk. Formally, we prove the following theorem.

Theorem 7. *Assuming Hypothesis 5, it is NP-hard to approximate the scheduling problem* Q|prec |C_{max} *within any constant factor.*

Reduction. We present a reduction from a k-partite graph $G = (V_1, ..., V_k, E_1, ..., E_{k-1})$ to an instance $\mathcal{I}(k)$ of the scheduling problem Q|prec |C_{max}. The reduction is parametrised by a constant k, a constant $Q \gg k$ such that Q divides n, and a large enough value $m \gg nk$.

- For each vertex in $v \in V_i$, let $\mathcal{J}_{v,i}$ be a set of $m^{2(k-i)}$ jobs with processing time m^{i-1}, for every $1 \leq i \leq k$.
- For each edge $e = (v, w) \in E_i$, we have $\mathcal{J}_{v,i} \prec \mathcal{J}_{w,i+1}$, for $1 \leq i < k$.
- For each $1 \leq i \leq k$ we create a set $\mathcal{M}_i$ of $m^{2(k-i)}$ machines with speed m^{i-1}.

Completeness. We show that if the given k-partite graph satisfies the properties of the YES Case, then there exist a schedule with makespan $(1 + \epsilon_1)n$ for some small $\epsilon_1 > 0$. Towards this end, assume that the given k-partite graph satisfies the properties of the YES Case and let $\{V_{i,j}\}$ for $1 \leq i \leq k$ and $0 \leq j \leq Q-1$ be the claimed partitioning of Hypothesis 5.

The partitioning of the vertices naturally induces a partitioning $\{\tilde{\mathcal{J}}_{i,j}\}$ for the jobs for $1 \leq i \leq k$ and $0 \leq j \leq Q - 1$ in the following way:

$$\tilde{\mathcal{J}}_{i,j} = \bigcup_{v \in V_{i,j}} \mathcal{J}_{v,i}$$

Consider the schedule where for each $1 \leq i \leq k$, all the jobs in a set $\tilde{\mathcal{J}}_{i,0}, \ldots, \tilde{\mathcal{J}}_{i,Q-1}$ are scheduled on the machines in $\mathcal{M}_i$. Moreover, we start the jobs in $\tilde{\mathcal{J}}_{i,j}$ after finishing the jobs in both $\tilde{\mathcal{J}}_{i-1,j}$ and $\tilde{\mathcal{J}}_{i,j-1}$ (if such sets exist). In other words, we schedule the jobs as follows:

- For each $1 \leq i \leq k$, we first schedule the jobs in $\tilde{\mathcal{J}}_{i,0}$, then those in $\tilde{\mathcal{J}}_{i,1}$ and so on up until $\tilde{\mathcal{J}}_{i,Q-1}$. The scheduling of the jobs on machines in $\mathcal{M}_0$ starts at time 0 in the previously defined order.
- For each $2 \leq i \leq k$, we start the scheduling of jobs $\tilde{\mathcal{J}}_{i,0}$ right after the completion of the jobs in $\tilde{\mathcal{J}}_{i-1,0}$.
- To respect the remaining precedence requirements, we start scheduling the jobs in $\tilde{\mathcal{J}}_{i,j}$ right after the execution of jobs in $\tilde{\mathcal{J}}_{i,j-1}$ and as soon as the jobs in $\tilde{\mathcal{J}}_{i-1,j}$ have finished executing, for $2 \leq i \leq k$ and $1 \leq j \leq Q - 1$.

By the aforementioned construction of the schedule, we know that the precedence constraints are satisfied, and hence the schedule is feasible. That is, since we are in YES Case, we know that vertices in $V_{i',j'}$ might only have edges to the vertices in $V_{i,j}$ for all $1 \leq i' < i \leq k$ and $1 \leq j' \leq j < Q$, which means that the precedence constraints may only be from the jobs in $\tilde{\mathcal{J}}_{i',j'}$ to jobs in $\tilde{\mathcal{J}}_{i,j}$ for all $1 \leq i' < i \leq k$ and $0 \leq j' \leq j < Q$. Therefore the precedence constraints are satisfied.

Moreover, we know that there are at most $m^{2(k-i)}n(1 + \epsilon)/Q$ jobs of length m^{i-1} in $\tilde{\mathcal{J}}_{i,j}$, and $m^{2(k-i)}$ machines with speed m^{i-1} in each $\mathcal{M}_i$ for all $1 \leq i \leq k$, $j \in [Q]$. This gives that it takes $(1 + \epsilon)n/Q$ time to schedule all the jobs in $\tilde{\mathcal{J}}_{i,j}$ on the machines in $\mathcal{M}_i$ for all $1 \leq i \leq k$, $j \in [Q]$, which in turn implies that we can schedule all the jobs in a set $\tilde{\mathcal{J}}_{i,j}$ between time $(i + j - 1)(1 + \epsilon)n/Q$ and $(i + j)(1 + \epsilon)n/Q$. This gives that the makespan is at most $(k + Q)(1 + \epsilon)n/Q$ which is equal to $(1 + \epsilon_1)n$, by the assumption that $Q \gg k$.

Soundness. We shall now show that if the k-partite graph G corresponds to the NO Case of Hypothesis 5, then any feasible schedule for $\mathcal{I}(k)$ must have a makespan of at least cnk, where $c := (1 - 2\delta)(1 - k^2/m)$ can be made arbitrary close to one.

Lemma 1. *In a feasible schedule σ for $\mathcal{I}(k)$ such that the makespan of σ is at most nk, the following is true: for every $1 \leq i \leq k$, at least a $(1 - k^2/m)$ fraction of the jobs in $\mathcal{L}_i = \cup_{v \in V_i} \mathcal{J}_{v,i}$ are scheduled on machines in $\mathcal{M}_i$.*

Proof. We first show that no job in $\mathcal{L}_i$ can be scheduled on machines in $\mathcal{M}_j$, for all $1 \leq j < i \leq k$. This is true, because any job $J \in \mathcal{J}_i$ has a processing time of m^{i-1}, whereas the speed of any machine $M \in \mathcal{M}_j$ is m^{j-1} by construction, and hence scheduling the job J on the machine M would require m^{i-1}/m^{j-1} $\geq m$ time steps. But since $m \gg nk$, this contradicts the assumption that the makespan is at most nk.

We now show that at most k^2/m fraction of the jobs in $\mathcal{L}_i$ can be scheduled on the machines in $\mathcal{M}_j$ for $1 \leq i < j \leq k$. Fix any such pair i and j, and assume that all the machines in $\mathcal{M}_j$ process the jobs in $\mathcal{L}_i$ during all the $T \leq nk$ time steps of the schedule. This accounts for a total $T\frac{m^{2(k-j)}m^{j-1}}{m^{i-1}} \leq m^{2k-j-i}nk$ jobs processed from $\mathcal{L}_i$, which constitutes at most $\frac{m^{2k-j-i}nk}{nm^{2(k-i)}} \leq \frac{k}{m}$ fraction of the total number of jobs in $\mathcal{L}_i$.

Let σ be a schedule whose makespan is at most nk, and fix $\gamma > k^2/m$ to be a small constant. From Lemma 1 we know that for every $1 \leq i \leq k$, at least an $(1 - \gamma)$ fraction of the jobs in $\mathcal{L}_i$ is scheduled on machines in $\mathcal{M}_i$. From the structure of the graph in the NO Case of the k-partite Problem, we know that we cannot start more than δ fraction of the jobs in $\mathcal{L}_i$ before finishing $(1 - \delta)$ fraction of the jobs in $\mathcal{L}_{i-1}$, for all $2 \leq i \leq k$. Hence the maximum makespan of any such schedule σ is at least $(1 - 2\delta)(1 - \gamma)nk$.

4.2   P|prec, pmtn, $p_j = 1$|C$_{max}$

We present in this section a reduction from a k-partite graph to an instance of the scheduling problem P|prec, pmtn, $p_j = 1$|C$_{max}$, and prove a tight inapproximability result for the latter, assuming Hypothesis 5. Formally, we prove the following result:

Theorem 8. *Assuming Hypothesis 5, it is NP-hard to approximate the scheduling problem* P|prec, pmtn, $p_j = 1$|C$_{max}$ *within any constant factor strictly better than 2.*

To prove this, we first reduce a k-partite graph $G = (V_1, ..., V_k, E_1, ..., E_{k-1})$ to a scheduling instance $\tilde{\mathcal{I}}(k)$, and then show that

1. If G satisfies the YES Case of Hypothesis 5, then $\tilde{\mathcal{I}}(k)$ has a feasible schedule whose makespan is roughly $kQ/2$.
2. if G satisfies the NO Case of Hypothesis 5, then any schedule for $\tilde{\mathcal{I}}(k)$ must have a makespan of roughly kQ.

Reduction. The reduction has three parameters: an odd integer k, an integer Q such that $Q \gg k$ and n divides Q, and a real $\epsilon \gg 1/Q^2 > 0$.

Given a k-partite graph $G = (V_1, ..., V_k, E_1, ..., E_{k-1})$, we construct an instance $\tilde{\mathcal{I}}(k)$ of the scheduling problem P|prec, pmtn, $p_j = 1$|C$_{max}$ as follows:

- For each vertex $v \in V_{2i-1}$ and every $1 \leq i \leq (k + 1)/2$, we create a set $\mathcal{J}_{2i-1,v}$ of $Qn - (Q - 1)$ jobs.
- For each vertex $v \in V_{2i}$ and every $1 \leq i < (k + 1)/2$, we create a chain of length $Q - 1$ of jobs, i.e., a set $\mathcal{J}_{2i,v}$ of $Q - 1$ jobs

$$\mathcal{J}_{2i,v} = \{J_{2i,v}^1, J_{2i,v}^2, \ldots, J_{2i,v}^{Q-1}\}$$

where we have $J_{2i,v}^l \prec J_{2i,v}^{l+1}$ for all $l \in \{1, 2, \ldots, Q-2\}$.
- For each edge $e = (v, w) \in E_{2i-1}$ and every $1 \le i < (k+1)/2$, we have $J_{2i-1,v} \prec J_{2i,w}^1$.
- For each edge $e = (v, w) \in E_{2i}$ and every $1 \le i < (k+1)/2$, we have $J_{2i,v}^{Q-1} \prec J_{2i+1,w}$.

Finally the number of machines is $(1 + Q\epsilon)n^2$.

The following lemma concludes the proof of Theorem 8.

Lemma 2. *Scheduling instance $\tilde{\mathcal{I}}(k)$ has the following two properties.*

1. *If G satisfies the YES Case of Hypothesis 5, then $\tilde{\mathcal{I}}(k)$ has a feasible schedule whose makespan is $(1 + \epsilon)kQ/2$, where ϵ can be arbitrary close to zero.*
2. *if G satisfies the NO Case of Hypothesis 5, then any feasible schedule for $\tilde{\mathcal{I}}(k)$ must have a makespan of $(1 - \epsilon)kQ$, where ϵ can be arbitrary close to zero.*

Although not formally defined, one can devise a similar reduction for the case of $k = 2$, and prove a $3/2$-inapproximability result for $\mathrm{P}|\mathrm{prec}, \mathrm{pmtn}, p_j = 1|\mathrm{C}_{max}$, assuming the variant of the UNIQUE GAMES Conjecture in [1]. We illustrate this in the full version of the paper and prove the following result:

Theorem 9. *For any $\epsilon > 0$, it is NP-hard to approximate $\mathrm{P}|\mathrm{prec}, \mathrm{pmtn}, p_j = 1|\mathrm{C}_{max}$ within a factor of $3/2 - \epsilon$, assuming (a variant of) the UNIQUE GAMES Conjecture.*

5   Discussion

We proposed in this paper a natural but nontrivial generalisation of Theorem 4, that seems to capture the hardness of a large family of scheduling problems with precedence constraints. It is interesting to investigate whether this generalisation also illustrates potential intrinsic hardness of other scheduling problems, for which the gap between the best known approximation algorithm and the best known hardness result persists.

On the other hand, a natural direction would be to prove Hypothesis 5; we show in the full version how to prove a *less-structured* version of it using the bipartite graph resulting from the variant of the UNIQUE GAMES Conjecture in [1]. One can also tweak the dictatorship $T_{\epsilon,t}$ of [1], to yield a k-partite graph instead of a bipartite one. However, composing this test with a UNIQUE GAMES instance adds a noisy component to our k-partite graph, that we do not know how to control, since it is due to the non-perfect completeness of the UNIQUE GAMES instance. One can also try to impose (a variant of) this dictatorship test on d-to-1 GAMES instances, and perhaps prove the hypothesis assuming the d-to-1 Conjecture, although we expect the size of the partitions to deteriorate as k increases.

Acknowledgments. The authors are grateful to Ola Svensson for inspiring discussions and valuable comments that influenced this work. We also wish to thank Hyung Chan An, Laurent Feuilloley, Christos Kalaitzis and the anonymous reviewers for several useful comments on the exposition.

References

1. Bansal, N., Khot, S.: Optimal long code test with one free bit. In: Proc. FOCS 2009, FOCS 2009, pp. 453–462. IEEE Computer Society, Washington, DC (2009)
2. Chekuri, C., Bender, M.: An efficient approximation algorithm for minimizing makespan on uniformly related machines. Journal of Algorithms 41(2), 212–224 (2001)
3. Chudak, F.A., Shmoys, D.B.: Approximation algorithms for precedence-constrained scheduling problems on parallel machines that run at different speeds. Journal of Algorithms 30(2), 323–343 (1999)
4. Gangal, D., Ranade, A.: Precedence constrained scheduling in (2-7/(3p+1)) optimal. Journal of Computer and System Sciences 74(7), 1139–1146 (2008)
5. Gonzalez, T.F., Johnson, D.B.: A new algorithm for preemptive scheduling of trees. Journal of the ACM (JACM) 27(2), 287–312 (1980)
6. Graham, R.L.: Bounds for certain multiprocessing anomalies. Bell System Technical Journal 45(9), 1563–1581 (1966)
7. Graham, R.L., Lawler, E.L., Lenstra, J.K., Rinnooy Kan, A.H.G.: Optimization and approximation in deterministic sequencing and scheduling: a survey. Annals of Discrete Mathematics 5(2), 287–326 (1979)
8. Hochbaum, D.S., Shmoys, D.B.: Using dual approximation algorithms for scheduling problems theoretical and practical results. Journal of the ACM (JACM) 34(1), 144–162 (1987)
9. Hochbaum, D.S., Shmoys, D.B.: A polynomial approximation scheme for scheduling on uniform processors: Using the dual approximation approach. SIAM Journal on Computing 17(3), 539–551 (1988)
10. Lenstra, J.K., Kan, A.R.: Computational complexity of discrete optimization problems. Annals of Discrete Mathematics 4, 121–140 (1979)
11. McNaughton, R.: Scheduling with deadlines and loss functions. Management Science 6(1), 1–12 (1959)
12. Muntz, R.R., Coffman Jr., E.G.: Optimal preemptive scheduling on two-processor systems. IEEE Transactions on Computers 100(11), 1014–1020 (1969)
13. Muntz, R.R., Coffman Jr., E.G.: Preemptive scheduling of real-time tasks on multiprocessor systems. Journal of the ACM (JACM) 17(2), 324–338 (1970)
14. Schuurman, P., Woeginger, G.J.: Polynomial time approximation algorithms for machine scheduling: Ten open problems. Journal of Scheduling 2(5), 203–213 (1999)
15. Svensson, O.: Hardness of precedence constrained scheduling on identical machines. SIAM Journal on Computing 40(5), 1258–1274 (2011)
16. Ullman, J.D.: Complexity of sequencing problems. In: Coman Jr., E.G. (ed.) Computer and Job-Shop Scheduling Theory (1976)
17. Williamson, D.P., Shmoys, D.B.: The design of approximation algorithms. Cambridge University Press (2011)

1-Planar Graphs have Constant Book Thickness

Michael A. Bekos[1], Till Bruckdorfer[1], Michael Kaufmann[1], Chrysanthi Raftopoulou[2]

[1] Wilhelm-Schickard-Institut für Informatik, Universität Tübingen, Germany
{bekos,bruckdor,mk}@informatik.uni-tuebingen.de
[2] School of Applied Mathematics and Physical Science, NTUA, Greece
crisraft@mail.ntua.gr

Abstract. In a book embedding the vertices of a graph are placed on the "spine" of a book and the edges are assigned to "pages", so that edges on the same page do not cross. In this paper, we prove that every 1-planar graph (that is, a graph that can be drawn on the plane such that no edge is crossed more than once) admits an embedding in a book with constant number of pages. To the best of our knowledge, the best non-trivial previous upper-bound was $O(\sqrt{n})$, where n is the number of vertices of the graph.

1 Introduction

A *book embedding* is a special type of a graph embedding in which (i) the vertices of the graph are restricted to a line along the *spine* of a book and (ii) the edges are assigned to the *pages* of the book such that edges of the same page are *not in conflict*, i.e., they do not cross. The minimum number of pages required for such an embedding is known as *book thickness* or *page number* of a graph. An upper bound on the page number of an n-vertex graph is $\lceil n/2 \rceil$, which is tight for complete graphs [4]. Book embeddings have a long history of research dating back to early seventies [19], see e.g., [5,20].

For the class of planar graphs, a central result is due to M. Yannakakis [22], who in the late eighties proved that planar graphs have page number at most four. It remains however unanswered whether the known bound of four is tight. Heath [11] for example proves that all planar 3-trees are 3-page book embeddable. For more restricted subclasses of planar graphs, Bernhart and Kainen [4] show that the graphs with page number one are the outerplanar graphs, while the class of two-page embeddable graphs coincides with the class of subhamiltonian graphs (recall that a graph is *subhamiltonian* if and only if it is subgraph of a planar Hamiltonian graph). Testing whether a graph is subhamiltonian is NP-complete [21]. However, several graph classes are known to be subhamiltonian (and therefore two-page book embeddable), see e.g., [2,8,12,13,18]

In this paper, we go a step beyond planar graphs. In particular, we prove that 1-planar graphs can be embedded in 39 pages. Recall that a graph is 1-*planar*, if it admits a drawing in which each edge is crossed at most once. To the best of our knowledge, the only (non-trivial) upper bound on the page number of 1-planar graphs on n vertices is $O(\sqrt{n})$. This is due to two known results. First, graphs with m edges have page number $O(\sqrt{m})$ [16]. Second, 1-planar graphs with n vertices have at most $4n - 8$ edges [6]. Minor-closed graphs (e.g., graphs of constant treewidth [9] or genus [15]) have constant page number [17]. However, 1-planar graphs are not closed under minors [17].

© Springer-Verlag Berlin Heidelberg 2015
N. Bansal and I. Finocchi (Eds.): ESA 2015, LNCS 9294, pp. 130–141, 2015.
DOI: 10.1007/978-3-662-48350-3_12

In the remainder, we will assume that a simple 1-planar drawing $\Gamma(G)$ of the input 1-planar graph G is also specified as part of the input. Recall that the problem of determining whether a graph is 1-planar is NP-hard [10,14], even if the deletion of a single edge makes the input graph planar [7]. We also assume biconnectivity, as the page number of a graph equals to the maximum page number of its biconnected components [4].

2 Definitions and Yannakakis' Algorithm

Let G be a simple topological graph, that is, undirected and drawn in the plane. Unless otherwise specified, we consider *simple* drawings, that is, no edge crosses itself, no two edges meet tangentially and no two edges cross more than once. A drawing uniquely defines the cyclic order of the edges incident to each vertex and therefore specifies a *combinatorial embedding*. A 1-planar topological graph is called *planar-maximal* or simply *maximal*, if the addition of a non-crossed edge is not possible. The following lemma shows that two crossing edges induce a K_4 (see, e.g., Lemma 1 in [1]).

Lemma 1. *In a maximal 1-planar topological graph, the endpoints of two crossing edges are pairwise adjacent, that is, they induce a K_4.*

The basis of our approach is the simple version of Yannakakis' algorithm, which embeds any (internally-triangulated) plane graph in a book of five pages [22] (not four). This algorithm is based on a "peeling" into *levels* approach: (i) vertices on the outerface are at level zero, (ii) vertices that are on the outerface of the graph induced by deleting all vertices of levels $\leq i - 1$ are at level i, (iii) edges between vertices of the same (different, resp.) level are called *level* (*binding*, resp.) edges (see Figure 1).

Let $G = (V, E)$ be a graph consisting of two levels, say L_0 and L_1 (it is also assumed that L_0 has no chords). The vertices, say $u_1, \ldots, u_k$, of L_0 are called *outer* and appear in this order along the clockwise traversal of the outerface of G. The remaining vertices are called *inner* (and obviously belong to L_1). The graph induced by all outer vertices is biconnected. The biconnected components (or *blocks*), say $B_1, \ldots, B_m$, of the graph induced by the inner vertices form a tree (in the absence of chords in L_0). It is assumed that the block tree is rooted at block, say w.l.o.g. B_1, that contains the so-called *first inner vertex*, which is uniquely defined as the third vertex of the bounded face containing the outer vertices u_1 and u_k. Given a block B_i, an outer vertex is said to be *adjacent* to B_i if it is adjacent to a vertex of it. The set of outer vertices adjacent to B_i is denoted by $N(B_i)$, $i = 1, 2, \ldots, m$. Furthermore, a vertex w is said to *see* an edge (x, y), if w is adjacent to x and y and the triangle x, y, w is a face. An outer vertex sees a block if it sees an edge of it.

The *leader* $\ell(B_i)$ of a block B_i is the first vertex of B_i that is encountered in any path in L_1 from the first inner vertex to block B_i. An inner vertex that belongs to only one block is *assigned* to that block. One that belongs to more than one blocks is assigned to the "highest block" in the block tree that contains it. Given an inner vertex $v \in L_1$, we denote by $B(v)$ the block that v is assigned to. The *dominator* of a block B is the first outer vertex that is adjacent to a vertex assigned to B and is denoted by $dom(B)$.

Let B be a block of level L_1 and assume that $v_0, v_1, \ldots, v_t$ are the vertices of B as they appear in a counterclockwise traversal of the boundary of B starting from

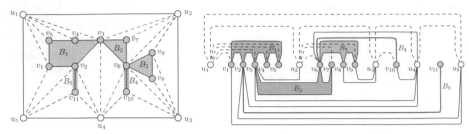

(a) An internally-triangulated graph.   (b) A book embedding in three pages taken from [22].

Fig. 1. (a) Outer (inner) vertices are colored white (gray). Level (binding) edges are solid (dashed). Blocks are highlighted in gray. The first inner vertex is v_1. So, the root of the block tree is B_1. $N(B_3) = \{u_2, u_3\}$. Vertex u_2 sees (v_3, v_7) and so sees B_2. The leaders of B_1, B_2, B_3, B_4 and B_5 are v_1, v_3, v_6, v_6 and v_2, resp. The dominators of B_1, B_2, B_3, B_4 and B_5 are u_1, u_2, u_2, u_3 and u_4, resp. The red edges indicate that $u_f(B_2) = u_2$ and $u_l(B_2) = u_4$. Hence, $P[u_f(B_2) \to u_l(B_2)] = u_2 \to u_3 \to u_4$. (b) Linear order and assignment of edges to pages.

$v_0 = \ell(B)$. Denote by $u_f(B)$ and $u_l(B)$ the smallest- and largest-indexed vertices of level L_0 that see edges (v_0, v_k) and (v_0, v_1), respectively. Equivalently, $u_f(B)$ and $u_l(B)$ are defined as the smallest- and largest-indexed vertices of $N(B)$. Note that $u_f(B) = dom(B)$. The path on level L_0 from $u_f(B)$ to $u_l(B)$ in clockwise direction along L_0 is denoted by $P[u_f(B) \to u_l(B)]$.

The *linear order* of the vertices along the spine is computed as follows. First, the outer vertices are embedded in the order $u_1, u_2, \ldots, u_k$. For $j = 1, 2, \ldots, k$, the blocks dominated by the outer vertex u_j are embedded right next to u_j one after the other in the top-to-bottom order of the block tree. The vertices that belong to block B_i are ordered along the spine in the order that appear in the counterclockwise traversal of the boundary of B_i starting from $\ell(B_i)$, $i = 1, 2, \ldots, m$ (which is already placed).

The edges are assigned to pages as follows. All level edges of L_0 are assigned to the first page. Level edges of L_1 are assigned either to the second or to the third page based on whether they belong to a block that is in an odd or even distance from the root of the block tree, respectively. Binding edges are further classified as *forward* or *back*. A binding edge is forward if the inner vertex precedes the outer vertex. Otherwise it is back (recall that a binding edge connects an outer and an inner vertex). All back edges are assigned to the first page. A forward edge incident to a block B_i is assigned to the second page, if B_i is on the third page. Otherwise to the third page, for $i = 1, 2, \ldots, m$.

In the case where more than two levers are present, the algorithm is as follows. Possible chords in level L_0 are assigned to the first page. Note however that in the presence of such chords the blocks of level L_1 form a forest in general (i.e., not a single tree). Therefore, each block tree of the underlying forest must be embedded according to the rules described above. Graphs with more than two layers are embedded by "recycling" the remaining available pages. More precisely, consider a block B of level $i - 1$ and let B' be a block of level i that is in the interior of B in the peeling order. Let $\{p_1, \ldots, p_5\}$ be a permutation of $\{1, \ldots, 5\}$ and assume w.l.o.g. that the boundary of block B is assigned to page p_1, while the boundary of all blocks in its interior (including B')

are assigned to pages p_2 and p_3. Then, the boundary of all blocks of level $i + 1$ that are in the interior of B' in the peeling order will be assigned to pages p_4 and p_5. In the following we present properties that we use in the remainder of the paper.

Lemma 2 (Yannakakis [22]). *Let G be a planar graph consisting of two levels L_0 and L_1. Let B be a block of level L_1 and let $v_0, \ldots, v_t$ be the vertices of B in a counter-clockwise order along the boundary of B starting from $v_0 = \ell(B)$. Then: (i) Vertices $v_1, \ldots, v_t$ are consecutive along the spine. (ii) $u_f(B) \neq u_l(B)$. (iii) If $u_i = u_f(B)$ and $u_j = u_l(B)$ for some $i < j$, then vertices $v_1, \ldots, v_t, u_{i+1}, \ldots, u_j$ appear in this order from left to right along the spine. (iv) Let $G[B]$ be the subgraph of G in the interior of cycle $P[u_f(B) \to u_l(B)] \to \ell(B) \to u_f(B)$. Then, a block $B' \in G[B]$ if and only if B is an ancestor of B', that is, B' belongs to the block subtree rooted at B.*

Lemma 3 (Yannakakis [22]). *Let G be a planar graph consisting of two levels L_0 and L_1 and assume that (u_i, u_j), $i < j$, is a chord of L_0. Denote by H the subgraph of G in the interior of the cycle $P[u_i \to u_j] \to u_i$. Then: (i) Vertices u_i and u_j form a separation pair in G. (ii) All vertices of H lie between u_i and u_j along the spine. (iii) If there is a vertex between u_i and u_j that does not belong to H, then this vertex belongs to a block B dominated by u_i. In addition, all vertices of H, except for u_i are to the right of B along the spine.*

3 An Upper Bound on the Page Number of 1-Planar Graphs

Let $G = (V, E)$ be a 1-planar graph and $\Gamma(G)$ be a 1-planar drawing of G. Initially, we consider the case where $\Gamma(G)$ contains no crossings incident to its unbounded face. To simplify the presentation, we further assume that G satisfies the K_4-*emptiness property*, that is, G is internally maximal 1-planar and in $\Gamma(G)$ the interior of all K_4's of Lemma 1 are free of vertices and edges. In order to guarantee this property, first we planarize G by replacing each crossing in $\Gamma(G)$ with a so-called *crossing vertex*. The planarized graph is then triangulated (only in its interior), so that no new edge is incident to a crossing vertex. Note that the latter restriction may lead to a non-simple graph (containing multiedges), as we will see in Section 3.2. However, if we treat all crossing vertices as actual crossings, then the implied augmented graph satisfies the K_4-emptiness property (at the cost of non-simplicity).

First, we consider the case where graph G is simple, internally-maximal 1-planar, satisfies the K_4-emptiness property and has no crossings at its unbounded face (i.e., no multiedges are created during the procedure described above). We prove that if there are only two levels, then such a graph fits in 16 pages. Otherwise, 39 pages suffice. Finally, we show how to cope with multiedges and crossings on the unbounded face of $\Gamma(G)$.

Let G be a simple and internally-maximal 1-planar graph that satisfies the K_4-emptiness property and has no crossings at its unbounded face. Then, (i) vertices on the outerface of G are at level zero, (ii) vertices that are at distance i from the level zero vertices are at level i. Similarly to Yannakakis' naming scheme, edges that connect vertices of the same (different, resp.) level are called *level* (*binding*, resp.) edges.

If we remove one edge from each pair of crossing edges, then the result is an internally-triangulated plane graph (which we call *underlying planar structure*). For a

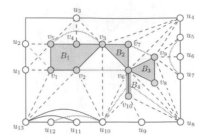

Fig. 2. Level (binding) edges are solid (dashed). The planar structure G_P is colored black. Gray colored edges do not belong to G_P. Edges (u_5, u_7), (u_8, u_{10}) and (u_{10}, u_{12}) are outer crossing chords that cross a binding edge, a bridge-block and a chord of G_P, resp. Edges (v_3, v_5), (v_7, v_9) and (v_2, v_6) are 2-hops crossing binding edges. Edge (u_2, v_1) is forward.

pair of binding crossing edges or for a pair of level crossing edges, we choose arbitrarily one to remove (we will shortly adjust this choice for two special cases). However, for a pair of crossing edges consisting of a binding edge and a level edge, we always choose to remove the level edge. This approach allow us to define the blocks, the leaders and the dominators of the blocks as Yannakakis does. Also observe that if all removed edges are plugged back to the graph, then a binding edge cannot cross a block.

3.1   The Two-Level Case

We first consider the base case where the given graph consists of two levels L_0 and L_1. We also assume that there is no level edge of L_1 which by the combinatorial embedding is strictly in the interior of a block of L_1. In addition, G is simple, internally maximal 1-planar satisfies the K_4-emptiness property and has no crossings on its unbounded face. We proceed to obtain a 3-page book embedding of the underlying planar structure G_P using Yannakakis' algorithm (see Section 2). We argue that we can embed the removed edges in the linear order implied by the book embedding of G_P using 13 more pages.

We first introduce an important notion useful in "eliminating" possible crossing situations. We say that two edges e_1 and e_2 of G *form a strong pair* if (i) they are both assigned to the same page, say p, and (ii) if an edge e, that is assigned also to page p, is in conflict with e_i, then it is also in conflict with e_j, where $i \neq j \in \{1, 2\}$. Suppose that $e_1 \notin E[G_P]$ and $e_2 \in E[G_P]$ form a strong pair of edges. If $e_3 \in E[G_P]$, then e_3 can cross neither e_1 nor e_2 (due to Yannakakis' algorithm). On the other hand, if $e_3 \notin E[G_P]$ and forms a strong pair with another edge $e_4 \in E[G_P]$, then again e_3 can cross neither e_1 nor e_2, as otherwise e_4 would also be involved in a crossing with e_1 or e_2, contradicting the correctness of Yannakakis' algorithm as $e_4 \in E[G_P]$.

We now describe six types of crossings that occur when the removed edges are plugged back to G (see Figure 2). Level edges of L_0 that do not belong to the G_P are called *outer crossing chords*. Such chords may be involved in crossings with (i) other chords of L_0 that belong to G_P or (ii) binding edges (between levels L_0 and L_1), or, (iii) degenerated blocks (so-called *block-bridges*) of level L_1 that are simple edges.

Level edges of L_1 that do not belong to G_P are called *inner crossing chords* or simply 2-*hops* (since it can "bypass" only one vertex along the boundary of the block tree). We claim that 2-hops do not cross with each other. Assume to the contrary that $e = (u, v)$ and $e' = (u', v')$ are two 2-hops that cross and say w.l.o.g. that u, u', v and v' appear in this order along the boundary of the block tree of G_P. Since G is maximal 1-planar, by Lemma 1 and the K_4-emptiness property it follows that (u, v') belongs to G_P. However, in the presence of this edge both vertices u' and v are not anymore at the boundary of the block tree of level L_1 of G_P, which is a contradiction as e and e' are both level edges of L_1. Hence, 2-hops are involved in crossings only with binding edges. Since level edges of different levels cannot cross, the only type of crossings that we have not reported are those between binding edges.

Recall that binding edges are of two types, forward and back. For a pair of crossing binding edges, say $e = (u_i, v_j)$ and $e' = (u_{i'}, v_{j'})$, where $u_i, u_{i'} \in L_0$ and $v_i, v_{i'} \in L_1$, we mentioned that we can arbitrarily choose which one is assigned to G_P. Here we adjust this choice. Edge e is assigned to G_P if and only if vertex u_i is lower-indexed than $u_{i'}$ in L_0, that is $i < i'$. Hence, e' is always forward and no two back edges cross.

Similarly, for a pair of crossing level edges of level L_0 we adjust our initial choice as follows. If a level edge is incident to $u_1 \in L_0$, then it is necessarily assigned to G_P.

From the above, it follows that for a pair of crossing edges, say $e \in E[G_P]$ and $e' \notin E[G_P]$, we have the following crossing situations each of which is separately treated in the following lemmas (except for the last one which is more demanding).

C.1: e' is an outer crossing chord and e is a chord of L_0 that belongs to G_P.
C.2: e' is an outer crossing chord and e is a binding edge.
C.3: e' is an outer crossing chord and e is a block-bridge of L_1.
C.4: e' is a forward edge and e is a forward edge.
C.5: e' is a forward edge and e is a back edge.
C.6: e' is a 2-hop and e is a binding edge.

Theorem 1. *Any simple internally-maximal 1-planar graph G with 2 levels, that satisfies the K_4-emptiness property and has no crossings at its unbounded face admits a book embedding on 16 pages.*

Proof. The underlying planar structure can be embedded in three pages. Case C.1 requires one extra page (due to Lemma 5.i). The crossing edges that fall into Cases C.2 and C.3 can be accommodated on the same pages used for the underlying planar structure (see Lemma 5.ii). Case C.4 requires two extra pages due to Lemma 6. Case C.5 requires three extra pages due to Lemma 7. Finally, Case C.6 requires seven more pages due to Lemma 8. Summing up the above yields 16 pages in total. □

We start by investigating the case where e' is an outer crossing chord of G.

Lemma 4. *Let $u_1, \ldots, u_k$ be the vertices of level L_0 in clockwise order along its boundary. Let $c = (u_i, u_j)$ and $c' = (u_{i'}, u_{j'})$ be two chords of L_0, such that $i < i' < j < j'$. Then, exactly one of c and c' is an outer crossing chord.*

Proof. Since $c = (u_i, u_j)$ and $c' = (u_{i'}, u_{j'})$ are chords of L_0 with $i < i' < j < j'$, chords c and c' cross in the 1-planar drawing $\Gamma(G)$ of G. So, one of them would belong to G_P and the other one would be an outer crossing chord. □

By Lemma 4, outer crossing chords can be placed on one page. However, in order to achieve more flexibility for the multi-level case, we chose to place some of them on a separate page (see Lemma 5). We call this particular page *universal*, because it contains outer crossing chords from all levels. Recall that we use three pages for G_P: p_1 (for level edges of L_0 and back edges), p_2 and p_3 (for level edges of L_1 and forward edges).

Lemma 5 (Cases C.1 - C.3). *Let $e = (u, v) \in E(G_P)$ and $e' = (u', v') \notin E(G_P)$ be two edges of G that are involved in a crossing, where e' is an outer crossing chord of L_0. (i) If e is a chord of L_0, then e' is placed on a universal page denoted by up_c (Case C.1). (ii) If e is a binding or a block-bridge of L_1, then e' is assigned to page p_1, that is, the page used for level edges of L_0 and back edges of G_P (Cases C.2 and C.3).*

Proof. (i) Since e is chord of level L_0, e is placed on page p_1 (recall that $e \in E(G_P)$) and e' is placed on the universal page up_c. Since up_c contains only outer crossing chords of G, by Lemma 4 they do not cross with each other.

(ii) If e is a binding or a block-bridge of level L_1, then e' is assigned to page p_1. Suppose that e' is in conflict with another edge, say e'', of page p_1. By Lemma 4, edge e'' is not an outer crossing chord, that is, e'' belongs to the underlying planar structure G_P of G. So, $e'' \in E[G_P]$ and it is either: (a) a level edge of level L_0 or (b) a back edge of G_P. In the first case, the endpoints of e'' cannot be consecutive vertices of level L_0, since that would not lead to a crossing situation. Hence, e'' must be a chord of level L_0. However, if e' is involved in such a crossing, then e' is assigned to page up_c, a contradiction. In the second case, e'' is back edge of G_P. So, edge e'' is nested by a level edge of level L_0 that is not a chord of L_0 and therefore if e' crosses e'', then e' must also cross this particular level edge of level L_0, which is not possible.   □

Lemma 6 (Case C.4). *All forward edges that are involved in crossings with forward edges of the underlying planar structure can be assigned to 2 new pages.*

Proof. Observe that for a pair of crossing forward edges, the choice of the edge that will be assigned to G_P affects neither the decomposition into blocks nor the choice of dominators and leaders of blocks. Therefore, it does not affect the linear order of the vertices along the spine. This ensures that two new pages suffice.   □

We proceed with Case C.5, where the back edge $e = (u, v) \in E[G_P]$ crosses the forward edge $e' = (u', v') \notin E[G_P]$. Let P be the block containing (v, v') and let $v_0, \ldots, v_t$ be the vertices of P in counterclockwise order around P starting from $v_0 = \ell(P)$. Since e is back, it follows that $u = u_f(P)$. By definition of $u_f(P)$, u sees (v_i, v_{i+1}), $\ldots$, (v_{t-1}, v_t), (v_t, v_0) of P, for some $1 \le i \le t$. Hence, edges (u, v_0), (u, v_t), $\ldots$, (u, v_i) exist and are back. This implies that either $v = v_i$ and $v' = v_{(i+1) \bmod t}$ or $v = v_0$ and $v' = v_t$. In the latter case and since u' is to the right of u on the spine, P is a root-block. In both cases (u', v) is forward.

Lemma 7 (Case C.5). *Let $e = (u, v)$ and $e' = (u', v')$ a back and a forward edge of G that cross. Let $v_0, \ldots, v_t$ be the vertices of block P in counterclockwise order around P starting from $v_0 = \ell(P)$, where P contains (v, v'). Finally, let i be the minimum s.t. $u = u_f(P)$ sees (v_i, v_{i+1}), $\ldots$, (v_{t-1}, v_t), (v_t, v_0). Then, we use three new pages p'_1, p'_2 and p'_3 as follows. (i) If $v = v_i$ and $v' = v_{(i+1) \bmod t}$, then edge e' is placed on a new*

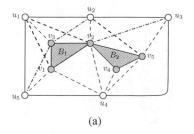
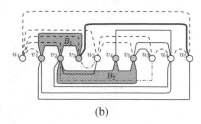

(a)   (b)

Fig. 3. (a) Red edges indicate forward edges involved in crossings. (b) Linear order and assignment of edges to pages. The fat edge is assigned to p_1'. The dashed-dotted ones to p_2' and p_3'.

page p_j' if and only if forward edges incident to block $B(v')$ are assigned to page p_j, $j = 2, 3$. (ii) If $v = v_0$ and $v' = v_t$, then edge e' is placed on a page p_1'.

Proof. (i) We prove that if the forward edges incident to $B(v')$ are on p_j, then e' can also be on p_j. If this is true, the lemma follows, as one can always split one page into two. We distinguish two cases based on whether $v = v_t$ (i.e., $i = t$) or $v = v_i$ for $i < t$.

First, assume that $v = v_t$ and $v' = v_0$ (refer to $e = (u_2, v_5)$ and $e' = (u_3, v_2)$ in Figure 3a). Let $B = B(v)$ and $B' = B(v')$. It follows that $B = P$ and B' is the parent-block of B. W.l.o.g. assume that the boundary of B' is on p_2. We claim that e' can be placed on page p_3 (together with forward edges of B'). To prove it, we show that $e' = (u', v_0)$ and (v_0, v_t) form a strong pair. First, observe that (v_0, v_t) is on page p_3. Indeed, B' is on page p_2. So, B is on p_3 and (v_0, v_t) is an edge of B. By Lemma 2.iii, vertices $v_1, v_2, \ldots, v_t$ and u' appear in the same order from left to right along the spine, so (v_0, v_t) is nested by e'. If v_t and u' are consecutive along the spine, then e' and (v_0, v_t) clearly form a strong pair. Otherwise, (u, u') is chord and by Lemma 3.iii, again e' and (v_0, v_t) form a strong pair.

In the case where $v = v_i$ and $v' = v_{i+1}$ for some $i < t$ (refer to $e = (u_1, v_2)$ and $e' = (u_2, v_3)$ in Figure 3a), we have that $B = B' = P$. Suppose w.l.o.g. that P is on p_2. We claim that $e' = (u', v_{i+1})$ and (u', v_i) form a strong pair. By Lemma 1 and the K_4-emptiness property, (u', v_i) exists and is forward. So, it is on page p_3. Since vertices v_i and v_{i+1} are consecutive along the spine, edges e' and (u', v_i) form a strong pair.

(ii) In this case (refer to $e = (u_1, v_1)$ and $e' = (u_5, v_3)$ in Figure 3a), P is a root-block. Let $e_1' = (u_1', v_1')$ and $e_2' = (u_2', v_2')$ be two edges that are assigned to the new page p_1'. Since $P_1 = B(v_1')$ and $P_2 = B(v_2')$ are both root blocks, it follows that P_1 and P_2 are separated by a chord of L_0. So, by Lemma 3, e_1' is not in conflict with e_2'.   □

Finally, we consider Case C.6 where $e' = (x, y)$ is a 2-hop of level L_1 and $e = (u, z)$ is a binding edge of G_P, where $x, y, z \in L_1$ and $u \in L_0$. Let x, z and y be assigned to blocks B_x, B_z and B_y, resp., that are not necessarily distinct. By Lemma 1 and the K_4-emptiness property, $x \to z \to y$ is a path in L_1. So, B_x and B_y are at distance at most two on the block tree of G. If x and y are assigned to the same block (that is, $B_x = B_z = B_y$), then e is called *simple 2-hop* (see Figure 4a). Suppose w.l.o.g. that B_x precedes B_y in the pre-order traversal of the block tree of G. Then, there exist two cases depending on whether B_x is an ancestor of B_y on the block tree. If this is not

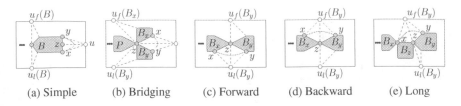

| (a) Simple | (b) Bridging | (c) Forward | (d) Backward | (e) Long |

Fig. 4. Different types of 2-hops (drawn in gray).

the case, then B_x and B_y have the same parent-block, say B_p. In this case, e' is called *bridging* 2-*hop* (see Figure 4b). Suppose now that B_x is an ancestor of B_y. Then, the path $x \to z \to y$ contains the leader of B_y, which is either x or z. By Lemma 1 and the K_4-emptiness property, (u, x) (u, z) and (u, y) exist in G. So, u is either $u_l(B_y)$ or $u_f(B_y)$. In the first subcase, e' is called *forward* 2-*hop* (see Figure 4c). In the second subcase, since B_x is ancestor of B_y and the two blocks are at distance at most two, if B_x is the parent-block of B_y, then e' is called *backward* 2-*hop* (see Figure 4d). Finally, if B_x is the grand-parent-block of B_y, then e' is called *long* 2-*hop* (see Figure 4e).

Lemma 8 (Case C.6). *All crossing 2-hops can be assigned to seven pages in total.*

Sketch of Proof. In high level description, one can prove that all simple 2-hops can be embedded in any page that contains 2-hops, all bridging 2-hops can be embedded in two new pages, forward 2-hops can be embedded in one new page, and finally, backward and long 2-hops can be embedded in two new pages each. Summing up the above yields a total of seven pages for all 2-hops. The detailed proof can be found in [3].   □

3.2 The Multi-Level Case

We now consider the more general case according to which the given 1-planar graph G consists of more than two levels, say $L_0, L_1, \ldots, L_\lambda, \lambda \geq 2$.

Lemma 9. *Any simple internally-maximal 1-planar graph G with $\lambda \geq 2$ levels, that satisfies the K_4-emptiness property and has no crossings at its unbounded face admits a book embedding on 34 pages.*

Proof. We first embed in 5 pages the underlying planar structure G_P of G using the algorithm of Yannakakis [22]. This implies that all vertices of a block of level i, except possibly for its leader, are between two consecutive vertices of level $i-1, i = 1, \ldots, \lambda$. So, for outer crossing chords that are involved in crossings with level edges of G_P (Case C.1), one universal page (denoted by up_c in Lemma 5.i) suffices, since such chords are not incident to block-leaders.

Next, we consider the outer crossing chords that are involved in crossings with binding edges or bridge-blocks of G (Cases C.2 and C.3). Such a chord $c_{i,j} = (v_i, v_j)$ of a block B is on the same page as the boundary and the non-crossing chords of B. The path $P[v_i \to v_j]$ on the boundary of B joins the endpoints of the crossing chord. Hence, if another edge of the same page crosses with $c_{i,j}$, then it must also cross with an edge of B, a contradiction. Therefore, such chords do not require additional pages.

For binding edges of Case C.4, 5 pages in total suffice (one page for each page of G_P). For binding edges of Case C.5, we argue differently. Since a binding edge between levels L_{i+2} and L_{i+1} cannot cross with a binding edge between levels L_{i-1} and L_{i-2}, $i = 2, \ldots, \lambda - 2$, it follows that binding edges that bridge pairs of levels at distance at least 3 are independent. So, for binding edges of Case C.5 we need a total of 9 pages.

Similarly, all blocks of level $i + 1$ that are in the interior of a certain block of level i are always between two consecutive vertices of level $i - 1$, $i = 1, 2, \ldots, \lambda - 1$. Hence, 2-hops that are by at least two levels apart in the peeling order are independent, which implies that for 2-hops we need a total of $2 * 7 = 14$ pages (Case C.6). Summing up we need $5 + 1 + 5 + 9 + 14 = 34$ pages for G.   □

Coping with Multiedges. At the beginning of the algorithm, one must augment the input 1-planar graph, in order to guarantee the K_4-emptiness property. The augmentation, however, may introduce multiedges. On the positive side, we can assume that all multiedges are crossing-free. Indeed, if a multiedge contains an edge that is involved in a crossing, then this particular edge can be safely removed from the graph, as it can be "replaced" by any of the corresponding crossing-free ones.

Let (v, w) be a double edge of G. Denote by $G_{in}[(v, w)]$ the so-called *interior subgraph* of G w.r.t. (v, w) bounded by the double edge (v, w) in $\Gamma(G)$. By $G_{ext}[(v, w)]$ we denote the so-called *exterior subgraph* of G w.r.t. (v, w) derived from G by substituting $G_{in}[(v, w)]$ by a single edge (see Figure 5). Clearly, $G_{ext}[(v, w)]$ stays internally-maximal 1-planar, satisfies the K_4-emptiness property, has no crossings at its unbounded face and simultaneously has fewer multiedges than G. So, it can be recursively embedded. The base of the recursion is a graph that can be embedded based on Lemma 9.

On the other hand, we cannot assure that the interior subgraph has fewer multiedges than G. Our aim is to modify it appropriately, so as to reduce the number of its multiedges by one. To do so, we will "remove" the multiedge (v, w) that defines the boundary of $G_{in}[(v, w)]$, so as to be able to recursively embed it (again we seek to employ Lemma 9 in the base of the recursion). Let $e_i(v)$ ($e_i(w)$, resp.) be the i-th edge incident to vertex v (w, resp.) in clockwise direction and between the two edges that form the double edge (v, w). We replace vertex v (w, resp.) by a path of $d(v)$ ($d(w)$, resp.) vertices, say $v_1, v_2, \ldots, v_{d(v)}$ ($w_1, w_2, \ldots, w_{d(w)}$, resp.), such that vertex v_i (w_i, resp.) is the endpoint of edge $e_i(v)$ ($e_i(w)$, resp.). Let $\overline{G}_{in}[(v, w)]$ be the implied graph. Since $\overline{G}_{in}[(v, w)]$ has no new crossings, it can be augmented to internally-maximal 1-planar, that satisfies the K_4-emptiness property, has no crossings at its unbounded face and has fewer multiedges than $G_{in}[(v, w)]$. So, $\overline{G}_{in}[(v, w)]$ can be embedded recursively.

We now describe how to plug the embedding of $\overline{G}_{in}[(v, w)]$ to the one of $G_{ext}[(v, w)]$. Suppose that (v, w) of $G_{ext}[(v, w)]$ is on page p. Clearly, p is one of the pages used to embed the planar structure of $G_{ext}[(v, w)]$, since (v, w) is not involved in crossings in $G_{ext}[(v, w)]$. Let the boundary of $\overline{G}_{in}[(v, w)]$ be also on page p. Since (v, w) is present in the embedding of $G_{ext}[(v, w)]$, it suffices to plug in the embedding of $G_{ext}[(v, w)]$ only the interior of $\overline{G}_{in}[(v, w)]$, which is the same as the one of $G_{in}[(v, w)]$. Suppose w.l.o.g. that in the embedding of $G_{ext}[(v, w)]$ v appears before w. Then, we place the interior subgraph of $\overline{G}_{in}[(v, w)]$ to the right of v. The edges connecting the interior of $\overline{G}_{in}[(v, w)]$ with v (w, resp.) are assigned to page p (a new page p' which is in correspondence to p, resp.). In this way, we create 5 new pages.

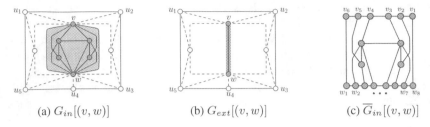

(a) $G_{in}[(v, w)]$ (b) $G_{ext}[(v, w)]$ (c) $\overline{G}_{in}[(v, w)]$

Fig. 5. Illustration of the decomposition in case of multiedges.

Next, we prove that no crossings are introduced. Since the boundary of $\overline{G}_{in}[(v, w)]$ is on page p, all edges incident to v towards $\overline{G}_{in}[(v, w)]$ become back edges of $\overline{G}_{in}[(v, w)]$. So, edges that join v with vertices in the interior of $\overline{G}_{in}[(v, w)]$ do not cross with other edges in the interior of $\overline{G}_{in}[(v, w)]$. Since $\overline{G}_{in}[(v, w)]$ is placed next to v, edges incident to v do not cross edges of $G_{ext}[(v, w)]$ on page p. Similarly, we argue that edges incident to w towards the interior of $\overline{G}_{in}[(v, w)]$ do not cross other edges in the interior of $\overline{G}_{in}[(v, w)]$ on page p'. It remains to prove that edges incident to w do not cross edges of $G_{ext}[(v, w)]$ on page p'. Such a crossing can only be in the presence of another double-edge (v', w') also of page p in $G_{ext}[(v, w)]$. Say w.l.o.g. that v' is to the left of w' (recall that v is to the left of w). First, consider the case where $v \neq v'$. If a conflict occurs because of $\overline{G}_{in}[(v, w)]$ and $\overline{G}_{in}[(v', w')]$, then (v, w) and (v', w'), which belong to the planar structure, must cross. If on the other hand $v = v'$ (and w.l.o.g. w to the left of w'), it suffices to place $\overline{G}_{in}[(v, w)]$ before $\overline{G}_{in}[(v', w')]$.

Coping with Crossings on Graph's Unbounded Face. If there exist crossings incident to the unbounded face of G, then, when we augment G in order to ensure the K_4-emptiness property, we must also triangulate the unbounded face of the planarized graph implied by replacing all crossings of G with crossing vertices (recall the first step of our algorithm). This procedure may lead to a situation where the unbounded face is a double edge, say (v, w). In this case, $G_{ext}[(v, w)]$ consists of two vertices and a single edge between them. $\overline{G}_{in}[(v, w)]$ is treated as described above.

Theorem 2. *Any 1-planar graph admits a book embedding in a book of 39 pages.*

4   Conclusions and Open Problems

In this paper, we proved that 1-planar graphs have constant page number. To keep the description simple, we decided not to "slightly" reduce the page number by more complicated arguments. A reasonable question is whether the page number can be further reduced, e.g., to less than 20. This question is of importance even for *optimal 1-planar graphs*, i.e., graphs with n vertices and exactly $4n-8$ edges. Other classes of non-planar graphs that fit in books with constant number of pages are also of interest. Finding 1-planar graphs that need a certain number of pages (e.g., ≥ 4) is also important.

Acknowledgement. We thank S. Kobourov and J. Toenniskoetter for useful discussions. We also thank David R. Wood for pointing out an error in the first version of this paper.

References

1. Alam, M. J., Brandenburg, F.J., Kobourov, S.G.: Straight-line grid drawings of 3-connected 1-planar graphs. In: Wismath, S., Wolff, A. (eds.) GD 2013. LNCS, vol. 8242, pp. 83–94. Springer, Heidelberg (2013)
2. Bekos, M., Gronemann, M., Raftopoulou, C.: Two-page book embeddings of 4-planar graphs. In: STACS. LIPIcs, vol. 25, pp. 137–148. Schloss Dagstuhl (2014)
3. Bekos, M.A., Bruckdorfer, T., Kaufmann, M., Raftopoulou, C.N.: The book thickness of 1-planar graphs is constant. CoRR, abs/1503.04990 (2015)
4. Bernhart, F., Kainen, P.: The book thickness of a graph. Combinatorial Theory 27(3), 320–331 (1979)
5. Bilski, T.: Embedding graphs in books: a survey. IEEE Proceedings Computers and Digital Techniques 139(2), 134–138 (1992)
6. Bodendiek, R., Schumacher, H., Wagner, K.: Über 1-optimale graphen. Mathematische Nachrichten 117(1), 323–339 (1984)
7. Cabello, S., Mohar, B.: Adding one edge to planar graphs makes crossing number and 1-planarity hard. SIAM Journal on Computing 42(5), 1803–1829 (2013)
8. Cornuéjols, G., Naddef, D., Pulleyblank, W.: Halin graphs and the travelling salesman problem. Mathematical Programming 26(3), 287–294 (1983)
9. Dujmovic, V., Wood, D.: Graph treewidth and geometric thickness parameters. Discrete and Computational Geometry 37(4), 641–670 (2007)
10. Grigoriev, A., Bodlaender, H.: Algorithms for graphs embeddable with few crossings per edge. Algorithmica 49(1), 1–11 (2007)
11. Heath, L.: Embedding planar graphs in seven pages. In: FOCS, pp. 74–83. IEEE (1984)
12. Heath, L.: Algorithms for Embedding Graphs in Books. PhD thesis, University of North Carolina (1985)
13. Kainen, P., Overbay, S.: Extension of a theorem of whitney. AML 20(7), 835–837 (2007)
14. Korzhik, V., Mohar, B.: Minimal obstructions for 1-immersions and hardness of 1-planarity testing. Journal of Graph Theory 72(1), 30–71 (2013)
15. Malitz, S.: Genus g graphs have pagenumber $O(\sqrt{g})$. Journal of Algorithms 17(1), 85–109 (1994)
16. Malitz, S.: Graphs with E edges have pagenumber $O(\sqrt{E})$. Journal of Algorithms 17(1), 71–84 (1994)
17. Nesetril, J., Ossona de Mendez, P.: Sparsity: Graphs, Structures, and Algorithms. Algorithms and Combinatorics, vol. 28. Springer, New York (2012)
18. Nishizeki, T., Chiba, N.: Planar Graphs: Theory and Algorithms. In: Hamiltonian Cycles, ch. 10. Dover Books on Mathematics, pp. 171–184. Courier Dover Publications (2008)
19. Ollmann, T.: On the book thicknesses of various graphs. In: Proceedings of the 4th Southeastern Conference on Combinatorics, Graph Theory and Computing. Congressus Numerantium, vol. VIII, p. 459 (1973)
20. Overbay, S.: Graphs with small book thickness. Missouri Journal of Mathematical Science 19(2), 121–130 (2007)
21. Wigderson, A.: The complexity of the hamiltonian circuit problem for maximal planar graphs. Technical Report TR-298, EECS Department, Princeton University (1982)
22. Yannakakis, M.: Embedding planar graphs in four pages. Journal of Computer and System Science 38(1), 36–67 (1989)

Access, Rank, and Select
in Grammar-compressed Strings*

Djamal Belazzougui[1], Patrick Hagge Cording[2], Simon J. Puglisi[1],
and Yasuo Tabei[3]

[1] Department of Computer Science,
University of Helsinki Helsinki, Finland
{belazzou,puglisi}@cs.helsinki.fi

[2] DTU Compute, Technical University of Denmark
Kgs. Lyngby, Denmark
phaco@dtu.dk

[3] PRESTO, Japan Science and Technology Agency
Saitama, Japan
tabei.y.aa@m.titech.ac.jp

Abstract. Given a string S of length N on a fixed alphabet of σ symbols, a grammar compressor produces a context-free grammar G of size n that generates S and only S. In this paper we describe data structures to support the following operations on a grammar-compressed string: $\mathsf{access}(S, i, j)$ (return substring $S[i, j]$), $\mathsf{rank}_c(S, i)$ (return the number of occurrences of symbol c before position i in S), and $\mathsf{select}_c(S, i)$ (return the position of the ith occurrence of c in S). Our main result for access is a method that requires $O(n \log N)$ bits of space and $O(\log N + m/\log_\sigma N)$ time to extract $m = j - i + 1$ consecutive symbols from S. Alternatively, we can achieve $O(\log_\tau N + m/\log_\sigma N)$ query time using $O(n\tau \log_\tau(N/n) \log N)$ bits of space, matching a lower bound stated by Verbin and Yu for strings where N is polynomially related to n when $\tau = \log^\epsilon N$. For rank and select we describe data structures of size $O(n\sigma \log N)$ bits that support the two operations in $O(\log N)$ time. We also extend our other structure to support both operations in $O(\log_\tau N)$ time using $O(n\tau\sigma \log_\tau(N/n) \log N)$ bits of space. When $\tau = \log^\epsilon N$ the query time is $O(\log N/\log \log N)$ and we provide a hardness result showing that significantly improving this would imply a major breakthrough on a hard graph-theoretical problem.

1 Introduction

Given a string S of length N, a grammar compressor [5] produces a context-free grammar G that generates S and only S. The size of the grammar refers to the total length of the right-hand sides of all rules.

* This research is supported by Academy of Finland grants 258308 and 284598 (CoECGR), JSPS grant KAKENHI 24700140, and by the JST PRESTO program.

© Springer-Verlag Berlin Heidelberg 2015
N. Bansal and I. Finocchi (Eds.): ESA 2015, LNCS 9294, pp. 142–154, 2015.
DOI: 10.1007/978-3-662-48350-3_13

It is well known that when the data to be compressed has a high degree of repetition (so called highly repetitive data, see, e.g. [20]), grammar compressors (and their close relative LZ77 [5]) can achieve compression significantly better than statistical compressors, whose performance is expressed in terms of $H_k(S)$, the kth-order empirical entropy [18]. $H_k(S)$ is a lower bound on the bits-per-symbol compression achievable by any statistical compressor that models each symbol's probability as a function of the k symbols preceding it in the S.

In this paper we consider support for three basic operations on grammar-compressed strings:

access(S, i, j) = return substring $S[i, j]$, the symbols in S between i and j inclusive;

rank$_c(S, i)$ = number of occurrences of symbol $c \in \Sigma$ among the first i symbols in S;

select$_c(S, j)$ = position in S of the jth occurrence of symbol $c \in \Sigma$.

The access operation allows one to process areas of interest in the compressed string without full decompression of all the symbols prior to $S[i, j]$. This is important, for example, in index-directed approximate pattern matching, where an index data structure first finds "seed" sites at which approximate matches to a query string may occur, before more expensive alignment of the pattern to the surrounding text (obtained via the access operation). Bille, Landau, Raman, Sadakane, Rao, and Weimann [3] show how, given a grammar of size n, it is possible to build a data structure of size $O(n \log N)$ bits that supports access to any substring $S[i, j]$ in time $O(\log N + (j - i))$.

Operations rank and select are of great importance on regular (uncompressed) strings, where they serve as building blocks for fast pattern matching indexes [22], wavelet trees [11], and document retrieval methods [12,21,23]. On binary strings, efficient rank and select support has been the germ for the now busy field of succinct data structures [19]. Although many space-efficient data structures supporting rank and select operations have been presented [10,11,24,25], they are generally not able to compress S beyond its statistical entropy.

In a recent full version of Bille et al.'s [3] paper it has been shown how to support rank and select in $O(\log N)$ time using $O(n \log N)$ bits of space for binary alphabets. Two other related results exist for rank and select in grammar-compressed strings. Navarro and Ordóñez [23] investigated practical methods for the rank operation in the context of indexed pattern matching. Their results pertain to grammars that are *height balanced* and in Chomsky normal form (i.e. straight-line programs [17]). Because their grammars are balanced their rank algorithm takes $O(\log N)$ time. The results we describe in this paper are faster and work for any context-free grammar, balanced or not. Recently, Bille, Cording, and Gørtz [2], used a weak form of select query, called *select-next*[1] as part of their compressed subsequence matching algorithm. A select-next(S,i,c) query returns the smallest $j > i$ such that $S[j] = c$.

Our Contribution. This paper provides the following results:

---

[1] These queries are referred to as *labeled successor queries* in [2].

1. We show how to support access to m consecutive symbols in $O(\log N + m/\log_\sigma N)$ time using $O(n \log N)$ bits of space. This is an improvement over the $O(\log N + m)$ time solution of Bille et al., within the same space. Our result is also significantly more powerful than the scheme of Ferragina and Venturini [7], which has access time as $O(1 + m/\log_\sigma N)$, but only achieves k^{th} order entropy compression (for some $k = o(\log_\sigma N)$).

2. We then show that by increasing the space usage slightly, we obtain a data structure that supports access to m consecutive symbols in $O(\log_\tau N + m/\log_\sigma N)$ time and $O(n\tau \log_\tau(N/n) \log N)$ bits of space. When setting τ to $\log^\epsilon N$ for some constant $\epsilon > 0$ our upper bound for access matches the lower bound of Verbin and Yu [27] who have shown that for "not-so-compressible" strings — those that have a grammar of length n such that $N \leq n^{1+\epsilon}$ for some constant ϵ — the query time cannot be better that $O(\log n/\log \log n) = O(\log N/\log \log N)$ if the used space is not more than $O(n \log^c n)$ for some constant c.

   Our data structure also easily extends to supporting computing Karp-Rabin fingerprints [16] of substrings. This improves the time to compute the longest common common prefix of two suffixes of S from $O(\log N \log \ell)$ time (due to Bille et al. [4]) to $O(\log_\tau N \log \ell)$ time, where ℓ is the length of the longest common prefix.

3. We describe data structures supporting rank and select operations in $O(\log N)$ time and $O(n\sigma \log N)$ bits of space, or $O(\log_\tau N)$ time and $O(n\tau\sigma \log_\tau(N/n) \log N)$ bits of space.

4. The above schemes for rank and select are fairly straightforward augmentations to our access data structures, but our final result suggests that it is probably difficult to do much better. In particular, we show a reduction between rank and select operations in grammar compressed strings and the problem of counting the number of distinct paths between two nodes in a directed acyclic graph — an old problem in graph theory. No significant progress has been made so far, even on the seemingly easier problem of reachability [6] (just returning whether the number of paths is non-zero).

2   Notation and Preliminaries

We consider a string S of total length N over an integer alphabet $[1..\sigma]$. The string S is generated by a grammar that contains exactly n *non-terminals* (or *variables*) and σ *terminals* (corresponding to each character in the alphabet of S). We assume that the grammar is in Chomsky normal form (CNF). That is, each grammar rule is either of the form $R_0 = R_1 R_2$, where R_1 and R_2 are non-terminals, or of the form $R_0 = c$, where c is a terminal. The grammar has thus exactly σ terminals. We denote by $|R|$ the length of the string generated by the non-terminal R. In what follows, we only consider grammars in CNF, since every grammar of size n with σ terminals, can be transformed into a grammar in CNF of size $O(n)$ with σ terminals.

We will often use a directed acyclic graph (DAG) representation of the grammar with one source (corresponding to the unique non-terminal that generates the whole string) and σ sinks (corresponding to the terminals). The height of a grammar is the maximal distance between the source and one of the sinks. The grammar is said to be balanced if its height is $O(\log N)$. Further, it is said to be AVL-balanced if the height of R_1 and R_2 differ by at most one.

Throughout we assume the RAM model with word length at least $\log N$ bits, and all the usual arithmetic and logic operations require constant time each.

3   Improved Access Time with Rank and Select Support

We now extend Bille et al.'s access scheme [3] so that it uses the same space, $O(n \log N)$ bits (for a grammar of length n that is not necessarily smallest possible), but allows access to m consecutive symbols in time $O(\log N + m/\log_\sigma N)$ instead of $O(\log N + m)$. To support rank and select queries we also extend Bille et al.'s data structure. However, due to lack of space the details of this will appear in a full version of the paper.

We start by reviewing the method of Bille et al.

3.1   The Method of Bille et al.

The method of Bille et al. uses heavy-weight paths in the DAG to support the access operation. More precisely, suppose the non-terminal that generates the whole string is R_0. Given a position x, the method of Bille et al. generates a sequence of triplets $(R_1, s_1, e_1), (R_2, s_2, e_2), ..., (R_t, s_t, e_t)$, where $t \leq \log N$, $(\sum_{1 \leq i \leq t} (s_i - 1)) + 1 = x$ and R_t is a non-terminal that generates a single character c. Note that we have $t \leq \log N$ because $|R_{i+1}| \leq |R_i|/2$ and $R_0 \leq N$. Thus c is the character that is at position x in the substring generated by R_0.

Each triplet (R_i, s_i, e_i) indicates that a substring generated by the non-terminal R_i starts at position s_i and ends at position e_i inside the substring generated by the non-terminal R_{i-1}. The non-terminal R_i is found using the heavy path that starts at the non-terminal R_{i-1} (further explained below).

To every non-terminal R_i, there is an associated position p_i, which is the rank of the leaf in the heavy path that starts from R_i among all leaves in the subtree of R_i. We call p_i the *center* of R_i.

3.2   Heavy Path Decomposition of a DAG

Given a non-terminal R, the heavy path starting from the variable $R = P_0$ is the sequence of non-terminals $P_1, P_2, \ldots, P_t$, such that:

1. For all $i \in [0, t-1]$, either $P_i = P_{i+1}Q_{i+1}$ or $P_i = Q_{i+1}P_{i+1}$
2. $|P_i| \geq |Q_i|$.
3. The non-terminal P_t generates a single character c.

Informally speaking, the heavy path starting from a non-terminal P_0 is the sequence of non-terminals $P_1, P_2, \ldots, P_t$ such that every non-terminal P_{i+1} is the heaviest among the two non-terminals in the right-hand side of the non-terminal P_i and the variable P_t generates a single character. We associate with each non-terminal R a center point $p = \sum_{1 \leq i \leq k} |Q_{i_j}| + 1$ where the i_j is the sequence of indices in $[1..t]$ such that $P_{i_j-1} = Q_{i_j+1}P_{i_j+1}$ (that is, Q_{i_j} is the left non-terminal in the right-hand side of the non-terminal P_{i_j-1}). The character at position p in the string generated by R is precisely the character generated by the non-terminal P_t.

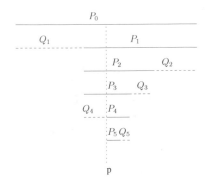

Fig. 1. Illustration of the heavy path starting from variable P_0 and its center point p.

3.3 Biased Skip Trees

The main data structure used by Bille et al. is a forest of trees built as follows. Each forest has as a root a node that corresponds to a sink in the DAG representation of the grammar. A node x is a child of another node y in the forest if and only if y is the heavy child of x in the original DAG. Thus the forest can be thought of as being built by reversing the original DAG (upside-down) and keeping only the edges that correspond to heavy paths. Then each of the resulting trees is represented using a biased skip tree. The representation allows us to find the sequence of triplets $(R_1, s_1, e_1), (R_2, s_2, e_2), \ldots, (R_t, s_t, e_t)$, given a position x in the string S. Suppose that we are given a non-terminal R whose center point is p and heavy path is $P_1, P_2, \ldots, P_m$. Given a point $x \neq p$ inside the substring generated by R, we wish to find inside the heavy path decomposition of R the non-terminal P_i such that either:

1. $P_i = P_{i+1}Q_{i+1}$ with $x - p > |P_{i+1}| - p_{i+1}$, where p_{i+1} is the center of the non-terminal P_{i+1}.
2. $P_i = Q_{i+1}P_{i+1}$ with $p - x > p_{i+1} - 1$, where p_{i+1} is the center the non-terminal P_{i+1}.

Informally speaking, the non-terminal P_i is the last non-terminal in the heavy path that contains the point x and the non-terminal Q_{i+1} hangs from the heavy-path, either from the right (first case above) of from the left (second case above). The biased skip tree allows to find the non-terminal P_i in time

$O(\log(|R|/|Q_{i+1}|))$. Then, the algorithm produces the triplet (R_1, s_1, e_1) by setting $R_1 = Q_{i+1}$, $s_1 = p - p_i + 1$ and $e_1 = s_1 + |R_1| - 1$ and starts the same procedure above by replacing R by R_1, which results in the triplet (R_2, s_2, e_2). The algorithm continues in this way until it gets the triplet (R_t, s_t, e_t). The total running time of the procedure is $O(\log N)$, since $|R| = N$ and the successive running times $O(\log(|R|/|R_1|)), O(\log(|R_1|/|R_2|)), \ldots, O(\log(|R_{t-1}|/|R_t|))$ add up to $O(\log N)$ time by a telescoping argument.

3.4   Improved Access Time

The above scheme can be extended to allow decompression of an arbitrary substring that covers positions $[x, x']$ in time $O(m + \log N)$, where $m = x' - x + 1$ is the length of the decompressed substring. The decompression works as follows. We first find the sequence of triplets $(R_1, s_1, e_1), (R_2, s_2, e_2), \ldots, (R_t, s_t, e_t)$ corresponding to the point x. We then find the sequence of triplets $(R'_1, s'_1, e'_1), (R'_2, s'_2, e'_2), \ldots, (R'_{t'}, s'_{t'}, e'_{t'})$ corresponding to the point x'. We let $(R_i, s_i, e_i) = (R'_i, s'_i, e'_i)$ be the last common triplet between the two sequences. Without loss of generality, assume that R_{i+1} hangs at a higher point than R'_{i+1} and that P_i is the last non-terminal on the heavy path of R_i that contains point x (note that $P_i = R_{i+1}P_{i+1}$). Then the non-terminal P_{i+1} still contains the point x' and we need to decompress all the non-terminals that hang on the left of the heavy path that starts at P_{i+1} down to the non-terminal from which the non-terminal R'_{i+1} hangs. Afterwards, we just need to 1) decompress all the non-terminals that hang on the right of the heavy path that starts at R_j down to the non-terminal from which R_{j+1} hangs for all $j \in [i+1, t-1]$, and then 2) symmetrically decompress all the non-terminals that hang on the left of the heavy path that starts at R'_j down to the non-terminal from which R'_{j+1} hangs for all $j \in [i+1, t'-1]$ [2]. This whole procedure takes $O(m + \log N)$ time. The main point of the procedure is to be able to decompress all the non-terminals that hang on the right or on the left of the portion of some heavy path that starts at some non-terminal P_i down to some non-terminal P_{i+1} inside that heavy path.

In what follows, we show how to reduce the time to just $O(m/\log_\sigma N + \log N)$. For every non-terminal X, we will store the following additional fields, which occupy $O(\log N)$ bits:

1. The $\log N/\log \sigma$ leftmost and $\log N/\log \sigma$ rightmost characters in the substring generated by X.
2. Three jump pointers. Each jump pointer is a pair of the form (R, p), where R is non-terminal and p is a position inside the substring generated by X.

The three jump pointers are called left, right and central (any of them may be empty). The jump pointers will allow us to accelerate the extraction of characters. The central jump pointer allows to fully decompress any given non-terminal that generates a string of length m in time $O(1 + m/\log_\sigma N)$.

---

[2] In addition if R_{j+1} (R'_{j+1}) hangs on the left (right) of $R_j(R'_j)$, we will need to decompress the right (left) child of the node from which R_{j+1} (R'_{j+1}) hangs

Lemma 1. *Suppose that we are given a grammar of size n that generates a string S of length N over an alphabet of size σ is of size n. Then we can build a data structure that uses $O(n \log N)$ bits so that decompressing the string generated by any non-terminal takes $O(\log N + m/\log_\sigma N)$ time, where m is the length of the string generated by the non-terminal.*

The proof will appear in the full version of the paper. The right and left jump pointers will allow us to jump along the heavy paths, avoiding the traversal of all the non-terminals in the heavy path. Their use is shown next.

Decompression of Arbitrary Substrings. We now show how to decompress an arbitrary substring that is not necessarily aligned on a non-terminal.

Recall that the core procedure for decompression is as follows. We are given a non-terminal R and another non-terminal Q that hangs from the heavy path down from R and we want to decompress what hangs from the left (respectively right) of the heavy path down to the point where Q hangs from the heavy path. To accelerate the decompression we will use the right and left jump pointers. Since right and left decompression are symmetric, we only describe the left case.

Before describing the decompression itself, we first describe how the left jump pointer for non-terminal R is set. We will keep a counter C (initially set to zero), assume $P_0 = R$ and inspect the sequence $P_1, P_2, \ldots, P_t$. For increasing i starting from 1 such that $P_{i-1} = Q_i P_i$, we increment C by $|Q_i|$ and stop when $C + |Q_i| > \log N/\log \sigma$. Then the left jump pointer will point to the non-terminal $P_{i-1} = L$ along with its starting point p_L inside R. If $P_0 = Q_1 P_1$ and $|Q_1| > \log N/\log \sigma$ or $C + |Q_i|$ never exceeds $\log N/\log \sigma$, then we do not store a left jump pointer at all.

The decompression of the left of a heavy path is done as follows. We are given the non-terminal Q and its starting position p_Q inside P. We first check whether $p_Q \leq \log N/\log \sigma$, in which case everything to the left of Q inside P is already in the left substring of P and it can be decompressed in time $O(1 + p_Q/\log_\sigma N)$ and we are done.

If $p_Q > \log N/\log \sigma$, then we have two cases:

1. If P has a left jump pointer (L, p_L) then $p_Q \geq p_L$ and we decompress the first $p_L \leq \log N/\log \sigma$ characters of the string generated by P (from the left substring of P), then replace P by L and recurse on L and Q.
2. If P does not have a left jump pointer, then we necessarily have that $P = Q_1 P_1$ with $|Q_1| > \log N/\log \sigma$ and $p_Q > |Q_1|$, we just decompress Q_1 (using the procedure shown above for fully decompressing non-terminals using central jump pointers). replace P by P_1 and recurse on P_1 and Q.

It remains to show that the bound for the procedure is $O(1 + y/\log_\sigma N)$, where y is the total length of the decompressed string. Analyzing the recursion, it can easily be seen that when we follow two successive left jump pointers, we are decompressing at least $\log_\sigma N$ characters from left substrings.

Otherwise, if we do not follow a jump pointer, then we are either decompressing a non-terminal of length at least $\log_\sigma N$ characters in optimal time or we terminate by decompressing at most $\log_\sigma N$ characters. We thus have shown the following theorem.

Theorem 1. *Suppose that we are given a grammar of size n that generates a string S of length N over an alphabet of size σ is of size n. Then we can build a data structure that uses $O(n \log N)$ bits that supports the access to m consecutive characters in time $O(\log N + m/\log_\sigma N)$ time.*

4   Optimal Access Time for Not-so-compressible Strings

Theorem 2. *Given a balanced grammar of size n generating a string S of length N over an alphabet of size σ, we can build a data structure of $O(n \log^{1+\epsilon} N)$ bits (for any constant ϵ), that supports random access to any character of S in $O(\log N/\log \log N)$ time, and access to m consecutive characters in $O(\log N/\log \log N + m/\log_\sigma N)$ time. Furthermore, we can build a data structure of $O(n\sigma \log^{1+\epsilon} N)$ bits that supports rank and select in $O(\log N/\log \log N)$ time.*

For every variable R we generate a new right-hand side that contains at most $\log^\epsilon N$ variables by iteratively expanding the right-hand side of R down $\epsilon \log \log N$ levels (or less if we reach a terminal). We then store in a fusion tree [8,9] the prefix-sum of the lengths of the strings generated by each variable in the right-hand side. More precisely, assuming that the expanded right-hand side of R is $R_1 R_2 \ldots R_t$ with $t \leq 2^{\epsilon \log \log n}$, the prefix sums are $s_1, s_2 \ldots s_t$, where $s_i = \sum_{j=1}^{i} |R_j|$. The height of the resulting grammar is $O(\log N/\log \log N)$ since the original grammar is balanced and thus have height $O(\log N)$ [5,26]. Every fusion tree uses $O(\log^{1+\epsilon} N)$ bits of space and the expanded grammar has $O(n)$ nodes, so we use $O(n \log^{1+\epsilon} N)$ bits of space.

To access a specific character $S[i]$ we traverse the grammar top-down and use the prefix-sums for navigation. Suppose we have reached a rule R producing the string $S[i', j']$ (where $i' \leq i \leq j'$). To find the child of R to continue the search from, we ask for the predecessor of $i - i'$ among the prefix-sums stored in R. The values for i' and j' are not unique to a node, so we must keep a counter indicating what i' is at every step. The counter is initially zero and when the search exits a node R in child R_k, we add s_{k-1} to the counter (or 0 if $k = 1$).

The fusion tree allows predecessor searches on a set of t integers of w bits in $O(\log t/\log w)$. Since in our case, we have $t = \log^\epsilon N$ and $w \geq \log N$, the query time is constant. The traversal therefore takes $O(\log N/\log \log N)$ time.

The data structure can be extended to support access(i, j) queries in $O(\log N + m/\log_\sigma N)$ time, and rank/select queries in $O(\log N/\log \log N)$ time (multiplying the space by a factor σ). For lack of space the details are deferred to the full version of the paper.

4.1   Fast Queries for Unbalanced Grammars

For unbalanced grammars we may use the results of Charikar et al. [5] or Rytter [26] to generate a balanced grammar that produces the same string as the unbalanced, but is of size larger by a factor $O(\log(N/n))$. This immediately gives a data structure that uses $O(n \log(N/n) \log^{1+\epsilon} N)$ bits (for any constant ϵ) supporting access to m consecutive characters in $O(\log N/\log\log N + m/\log_\sigma N)$ time, and a data structure using $O(n\sigma \log(N/n) \log^{1+\epsilon} N)$ bits of space that supports rank and select in $O(\log N/\log\log N)$ time. In this section we show the following theorem.

Theorem 3. *Given any grammar of size n generating a string S of length N over an alphabet of size σ, we can build a data structure that uses $O(n\tau \log_\tau \frac{N}{n} \log N)$ bits that supports random access to any character of S in $O(\log_\tau N)$ time, and access to m consecutive characters in $O(\log_\tau N + m/\log_\sigma N)$ time. Furthermore, we can build a data structure of $O(n\tau\sigma \log_\tau \frac{N}{n} \log N)$ bits that supports rank and select in $O(\log_\tau N)$ time. τ can be any value between 2 and $\log^\epsilon N$ for some constant ϵ.*

Our data structure uses Rytter's algorithm to generate a balanced grammar and then uses the data structure that we described in the previous section. One of the key observations in the analysis of Rytter's algorithm is that when joining two AVL-balanced grammars we add only a number of new rules proportional to the difference in their heights. Taking a closer look at the algorithm will reveal the following useful property. The proof is omitted due to lack of space.

Lemma 2. *Let S be an arbitrary grammar of size n and S' an AVL-balanced grammar generated by Rytter's algorithm producing the same string as S. The number of non-terminals with height h in S' is $O(n)$ for any h.*

Suppose we are given a grammar balanced by Rytter's algorithm. We want to expand the right-hand sides of rules to be of size $O(\tau)$. Because the grammar is AVL-balanced, we may find a set of rules where each rule has height $h - 1$ or h for some h such that S can be partitioned into substrings of size $\frac{1.6^{h-1}}{\sqrt{5}} < \tau \le 2^h$ produced by these rules. We expand the right-hand sides of these rules and proceed to a higher level. We then have the following corollary.

Corollary 1. *Given a string of size N compressed by a grammar of size n. After applying Rytter's algorithm to balance the grammar and expanding the right-hand sides of rules to size $\frac{1.6^{h-1}}{\sqrt{5}} < \tau \le 2^h$ for some h, the resulting grammar has $O(n \log_\tau N)$ rules.*

Proof. The grammar is balanced by Rytter's algorithm so after it is expanded its height is $O(\log N/(h - 1))$ which in terms of τ is $O(\log_\tau N)$. We select rules with a height difference of at most one in every iteration, so from Lemma 2 we know that $O(n)$ rules is selected at each level. Therefore the resulting grammar has $O(n \log_\tau N)$ rules. □

We now have seen how to obtain an expanded, balanced grammar with $O(n \log_\tau N)$ rules. To get Theorem 3 we build a fusion tree in each node using $O(\tau \log N)$ bits of space each totalling to $O(n\tau \log_\tau N \log N)$ bits of space. A full proof of Theorem 3 that also shows how to get the $\log_\tau(N/n)$-factor instead of $\log_\tau N$ will appear in a full version of the paper.

This scheme also extends to computing the Karp-Rabin fingerprint [16] of a substring of S in the same time as it takes to access a single character.

Corollary 2. *Given a grammar of size n generating a string S of length N, we can build a data structure using $O(n\tau \log_\tau N \log N)$ bits of space that supports finding the fingerprint of a substring $S[i, j]$ in $O(\log_\tau N)$ time.*

This improves upon the $O(\log N)$ query time of [4]. By adapting the exponential search algorithm from [4] it further follows that we can compute the longest common prefix of two suffixes of S in $O(\log N \log \ell / \log \log N)$ time, where ℓ is the length of the longest common prefix.

5   Hardness of Rank/Select in Grammar-compressed Strings

We will now show a reduction from the problem of path counting in DAGs to rank and select in grammar-compressed strings.

Suppose that we are given a DAG with m nodes and n edges that has β nodes with indegree 0 (sources) and σ nodes with outdegree 0 (sinks) and later for any node u and any sink v we want to be able to compute the number of distinct paths that connect u to v. We allow multi-edges. Let N be the total number of distinct paths that connect the β sources to the σ sinks. We can show that we can construct a grammar-compressed string of (uncompressed) length N having $O(n)$ non-terminals and σ terminals and such that answering the above queries on the DAG reduces to answering rank queries on the compressed string.

We modify the DAG such that it contains a single source and all nodes it contains have outdegree either 0 or 2:

1. For every node v of outdegree 1 do the following. If the successor of v is w, then for every edge uv create a new edge uw and remove edge uv. Node v now has indegree 0. We remove v and keep a structure mapping v to w. Since any path starting at v must pass through w, we will know that counting the number of paths starting at v and ending at a node x is the same as the number starting at w and ending at x. Note that all paths that went through v are preserved. Thus the count of the number of paths is unchanged.
2. If the number of nodes having indegree 0 is $t \geq 2$, then create a new root and connect the root to all the nodes of indegree 0 by creating $t - 2$ intermediate nodes. The root and the newly created nodes will have outdegree 2.
3. For every node v of outdegree $d \geq 3$, we will add exactly $d - 2$ intermediate nodes of outdegree 2 that connect the original nodes with the destination and modify v so that it has outdegree 2.

Clearly, the constructed DAG will have $O(m)$ nodes and will generate a string of length exactly N, where N is the total number of distinct paths between one of the original β sources and one of the original σ sinks.

For every non-terminal, we will store two pointers that delimit the leftmost occurrence of the rule in the text. This array occupies $O(n)$ words of space. Then, in time $T(n, \sigma, N)$, we build a data structure of size $S(n, \sigma, N)$ that answers rank queries on the string generated by the grammar in time $t(n, \sigma, N)$. To answer a query that counts the number of paths between a node u and a designated sink v, we will find the non-terminal R that corresponds to u and the terminal c that corresponds to v. We then find (using the array of pointers) the two positions i and j that correspond to the leftmost occurrences of R in the text. Finally, the count is returned by doing two rank queries for symbol c at positions i and j.

We have proved the following.

Theorem 4. *Suppose there exists a scheme that can preprocess a grammar of size n with σ non-terminals that generates a string of length N in $T(n, \sigma, N)$ time and produces a data structure of size $S(n, \sigma, N)$ that answers to rank queries on the string generated by the grammar in time $t(n, \sigma, N)$. Then given a DAG with m nodes, n edges (possibly with multiedges), β sources and α sinks, we can, after preprocessing time $O(n + T(n, \sigma, N))$, produce a data structure of size $O(n + S(n, \sigma, N))$ that can count the number of distinct paths from any node of the DAG to one of the σ sinks in time $O(t(n, \sigma, N))$, where N is the number of distinct paths that connect the β sources to the α sinks.*

6   Concluding Remarks

Perhaps the most interesting open question we raise is whether our results for rank and select are optimal. As we have shown, proving this one way or the other would lead to progress on path counting and reachability in DAGs, an old and interesting problem in graph theory. Perhaps similar approaches will be fruitful in judging the hardness of other problems on grammar-compressed strings, many solutions to which currently seem to be loose upperbounds [2,1,15,13,14].

Our result for access closes the gap between Bille et al.'s random access result [3] and the lowerbound of Verbin and Yu [27] for the (large) set of strings whose grammar-compressed size n is polynomially related to N. We leave closing the gap for the remaining strings as an open problem.

References

1. Bannai, H., Gagie, T.I., Inenaga, S., Landau, G.M., Lewenstein, M.: An efficient algorithm to test square-freeness of strings compressed by straight-line programs. Information Processing Letters 112(19), 711–714 (2012)
2. Bille, P., Cording, P.H., Gørtz, I.L.: Compressed subsequence matching and packed tree coloring. In: Kulikov, A.S., Kuznetsov, S.O., Pevzner, P. (eds.) CPM 2014. LNCS, vol. 8486, pp. 40–49. Springer, Heidelberg (2014)
3. Bille, P., Landau, G.M., Raman, R., Sadakane, K., Satti, S.R., Weimann, O.: Random access to grammar-compressed strings. In: Proc. 22nd SODA, pp. 373–389. SIAM (2011)
4. Bille, P., Cording, P.H., Gørtz, I.L., Sach, B., Vildhøj, H.W., Vind, S.: Fingerprints in compressed strings. In: Dehne, F., Solis-Oba, R., Sack, J.-R. (eds.) WADS 2013. LNCS, vol. 8037, pp. 146–157. Springer, Heidelberg (2013)

5. Charikar, M., Lehman, E., Liu, D., Panigrahy, R., Prabhakaran, M., Sahai, A., Shelat, A.: The smallest grammar problem. IEEE Transactions on Information Theory 51(7), 2554–2576 (2005)

6. Cohen, E., Halperin, E., Kaplan, H., Zwick, U.: Reachability and distance queries via 2-hop labels. SIAM Journal on Computing 32(5), 1338–1355 (2003)

7. Ferragina, P., Venturini, R.: A simple storage scheme for strings achieving entropy bounds. Theoretical Computer Science 372(1), 115–121 (2007)

8. Fredman, M.L., Willard, D.E.: Blasting through the information theoretic barrier with fusion trees. In: Proceedings of the Twenty-Second Annual ACM Symposium on Theory of Computing, pp. 1–7. ACM (1990)

9. Fredman, M.L., Willard, D.E.: Surpassing the information theoretic bound with fusion trees. Journal of Computer and System Sciences 47(3), 424–436 (1993)

10. Golynski, A., Munro, J.I., Rao, S.S.: Rank/select operations on large alphabets: a tool for text indexing. In: Proc. 17th SODA, pp. 368–373. SIAM (2006)

11. Grossi, R., Gupta, A., Vitter, J.S.: High-order entropy-compressed text indexes. In: Proc. 14th SODA, pp. 841–850. SIAM (2003)

12. Hon, W.K., Patil, M., Shah, R., Thankachan, S.V., Vitter, J.S.: Indexes for document retrieval with relevance. In: Brodnik, A., López-Ortiz, A., Raman, V., Viola, A. (eds.) Ianfest-66. LNCS, vol. 8066, pp. 351–362. Springer, Heidelberg (2013)

13. I, T., Matsubara, W., Shimohira, K., Inenaga, S., Bannai, H., Takeda, M., Narisawa, K., Shinohara, A.: Detecting regularities on grammar-compressed strings. In: Chatterjee, K., Sgall, J. (eds.) MFCS 2013. LNCS, vol. 8087, pp. 571–582. Springer, Heidelberg (2013)

14. I, T., Nakashima, Y., Inenaga, S., Bannai, H., Takeda, M.: Faster lyndon factorization algorithms for SLP and LZ78 compressed text. In: Kurland, O., Lewenstein, M., Porat, E. (eds.) SPIRE 2013. LNCS, vol. 8214, pp. 174–185. Springer, Heidelberg (2013)

15. Inenaga, S., Bannai, H.: Finding characteristic substrings from compressed texts. International Journal of Foundations of Computer Science 23(2), 261–280 (2012)

16. Karp, R.M., Rabin, M.O.: Efficient randomized pattern-matching algorithms. IBM Journal of Research and Development 31(2), 249–260 (1987)

17. Karpinski, M., Rytter, W., Shinohara, A.: An efficient pattern-matching algorithm for strings with short descriptions. Nordic Journal of Computing 4, 172–186 (1997)

18. Manzini, G.: An analysis of the Burrows-Wheeler transform. Journal of the ACM 48(3), 407–430 (2001)

19. Munro, J.I.: Tables. In: Chandru, V., Vinay, V. (eds.) FSTTCS 1996. LNCS, vol. 1180, pp. 37–42. Springer, Heidelberg (1996)

20. Navarro, G.: Indexing highly repetitive collections. In: Smyth, B. (ed.) IWOCA 2012. LNCS, vol. 7643, pp. 274–279. Springer, Heidelberg (2012)

21. Navarro, G.: Spaces, trees and colors: The algorithmic landscape of document retrieval on sequences. ACM Computing Surveys 46(4), article 52, 47 pages (2014)

22. Navarro, G., Mäkinen, V.: Compressed full-text indexes. ACM Computing Surveys 39(1), article 2 (2007)

23. Navarro, G., Ordóñez, A.: Grammar compressed sequences with rank/Select support. In: Moura, E., Crochemore, M. (eds.) SPIRE 2014. LNCS, vol. 8799, pp. 31–44. Springer, Heidelberg (2014)

24. Okanohara, D., Sadakane, K.: Practical entropy-compressed rank/select dictionary. In: Proc. 9th ALENEX, pp. 60–70. SIAM (2007)
25. Raman, R., Raman, V., Rao, S.S.: Succinct indexable dictionaries with applications to encoding k-ary trees, prefix sums and multisets. ACM Transactions on Algorithms 3(4) (2007)
26. Rytter, W.: Application of Lempel-Ziv factorization to the approximation of grammar-based compression. Theor. Comp. Sci. 302(1–3), 211–222 (2003)
27. Verbin, E., Yu, W.: Data structure lower bounds on random access to grammar-compressed strings. In: Fischer, J., Sanders, P. (eds.) CPM 2013. LNCS, vol. 7922, pp. 247–258. Springer, Heidelberg (2013)

Fully-Dynamic Approximation
of Betweenness Centrality

Elisabetta Bergamini and Henning Meyerhenke

Institute of Theoretical Informatics
Karlsruhe Institute of Technology (KIT), Germany
{elisabetta.bergamini,meyerhenke}@kit.edu

Abstract. Betweenness is a well-known centrality measure that ranks
the nodes of a network according to their participation in shortest paths.
Since an exact computation is prohibitive in large networks, several approximation algorithms have been proposed. Besides that, recent years
have seen the publication of dynamic algorithms for efficient recomputation of betweenness in evolving networks. In previous work we proposed
the first semi-dynamic algorithms that recompute an *approximation* of
betweenness in connected graphs after batches of edge insertions.

In this paper we propose the first fully-dynamic approximation algorithms (for weighted and unweighted undirected graphs that need not to
be connected) with a provable guarantee on the maximum approximation error. The transfer to fully-dynamic and disconnected graphs implies
additional algorithmic problems that could be of independent interest.
In particular, we propose a new upper bound on the vertex diameter for
weighted undirected graphs. For both weighted and unweighted graphs,
we also propose the first fully-dynamic algorithms that keep track of
this upper bound. In addition, we extend our former algorithm for semi-
dynamic BFS to batches of both edge insertions and deletions.

Using approximation, our algorithms are the first to make in-memory
computation of betweenness in fully-dynamic networks with millions of
edges feasible. Our experiments show that they can achieve substantial
speedups compared to recomputation, up to several orders of magnitude.

Keywords: betweenness centrality, algorithmic network analysis, fully-
dynamic graph algorithms, approximation algorithms, shortest paths.

1 Introduction

The identification of the most central nodes of a network is a fundamental
problem in network analysis. *Betweenness centrality* (BC) is a well-known index that ranks the importance of nodes according to their participation in
shortest paths. Intuitively, a node has high BC when it lies on many shortest
paths between pairs of other nodes. Formally, BC of a node v is defined as
$c_B(v) = \frac{1}{n(n-1)} \sum_{s \neq v \neq t} \frac{\sigma_{st}(v)}{\sigma_{st}}$, where n is the number of nodes, σ_{st} is the number of shortest paths between two nodes s and t and $\sigma_{st}(v)$ is the number of these
paths that go through node v. Since it depends on *all* shortest paths, the exact

© Springer-Verlag Berlin Heidelberg 2015
N. Bansal and I. Finocchi (Eds.): ESA 2015, LNCS 9294, pp. 155–166, 2015.
DOI: 10.1007/978-3-662-48350-3_14

computation of BC is expensive: the best known algorithm [5] is quadratic in the
number of nodes for sparse networks and cubic for dense networks, prohibitive
for networks with hundreds of thousands of nodes. Many graphs of interest,
however, such as web graphs or social networks, have millions or even billions
of nodes and edges. For this reason, approximation algorithms [6,9,1] must be
used in practice. In addition, many large graphs of interest evolve continuously,
making the efficient recomputation of BC a necessity. In a previous work, we
proposed the first two approximation algorithms [4] (IA for unweighted and IAW
for weighted graphs) that can efficiently recompute the approximate BC scores
after batches of edge insertions or weight decreases. IA and IAW are the only
semi-dynamic algorithms that can actually be applied to large networks. The
algorithms build on RK [19], a static algorithm with a theoretical guarantee on
the quality of the approximation, and inherit this guarantee from RK. However,
IA and IAW target a relatively restricted configuration: only connected graphs
and edge insertions/weight decreases.

Our Contributions. In this paper we present the first fully-dynamic algorithms
(handling edge insertions, deletions and arbitrary weight updates) for BC approximation in weighted and unweighted undirected graphs. Our algorithms extend the semi-dynamic ones we presented in [4], while keeping the theoretical
guarantee on the maximum approximation error. The transfer to fully-dynamic
and disconnected graphs implies several additional problems compared to the restricted case we considered previously [4]. Consequently, we present the following
intermediate results, all of which could be of independent interest. (i) We propose a new upper bound on the vertex diameter VD (i. e. number of nodes in the
shortest path(s) with the maximum number of nodes) for weighted undirected
graphs. This can improve significantly the one used in the RK algorithm [19] if
the network's weights vary in relatively small ranges (from the size of the largest
connected component to at most twice the vertex diameter times the ratio between the maximum and the minimum edge weights). (ii) For both weighted
and unweighted graphs, we present the first fully-dynamic algorithm for updating an approximation of VD, which is equivalent to the diameter in unweighted
graphs. (iii) We extend our previous semi-dynamic BFS algorithm [4] to batches
of both edge insertions and deletions. In our experiments, we compare our algorithms to recomputation with RK on both synthetic and real dynamic networks.
Our results show that our algorithms can achieve substantial speedups, often
several orders of magnitude on single-edge updates and are always faster than
recomputation on batches of more than 1000 edges.

2   Related Work

2.1   Overview of Algorithms for Computing BC

The best static exact algorithm for BC (BA) is due to Brandes [5] and requires
$\Theta(nm)$ operations for unweighted graphs and $\Theta(nm + n^2 \log n)$ for graphs with
positive edge weights. The algorithm computes a single-source shortest path

(SSSP) search from every node s in the graph and adds to the BC score of each node $v \neq s$ the fraction of shortest paths that go through v. Several static approximation algorithms have been proposed that compute an SSSP search from a set of randomly chosen nodes and extrapolate the BC scores of the other nodes [6,9,1]. The static approximation algorithm by Riondato and Kornaropoulos (RK) [19] samples a set of shortest paths and adds a contribution to each node in the sampled paths. This approach allows a theoretical guarantee on the quality of the approximation and will be described in Section 2.2. Recent years have seen the publication of a few dynamic exact algorithms [15,11,13,12,17,10]. Most of them store the previously calculated BC values and additional information, like the distance of each node from every source, and try to limit the recomputation to the nodes whose BC has actually been affected. All the dynamic algorithms perform better than recomputation on certain inputs. Yet, none of them is in general better than BA. In fact, they all require updating an all-pairs shortest paths (APSP) search, for which no algorithm has an improved worst-case complexity compared to the best static algorithm [20]. Also, the scalability of the dynamic exact BC algorithms is strongly compromised by their memory requirement of $\Omega(n^2)$. To overcome these problems, we presented two algorithms that efficiently recompute an approximation of the BC scores instead of their exact values [4]. The algorithms have shown significantly high speedups compared to recomputation with RK and a good scalability, but they are limited to connected graphs and batches of edge insertions/weight decreases (see Section 2.3).

2.2   RK Algorithm

The static approximation algorithm RK [19] is the foundation for the incremental approach we presented in [4] and our new fully-dynamic approach. RK samples a set $S = \{p_{(1)}, ..., p_{(r)}\}$ of r shortest paths between randomly-chosen source-target pairs (s, t). Then, RK computes the approximated betweenness $\tilde{c}_B(v)$ of a node v as the fraction of sampled paths $p_{(k)} \in S$ that go through v, by adding $\frac{1}{r}$ to v's score for each of these paths. In each of the r iterations, the probability of a shortest path p_{st} to be sampled is $\pi_G(p_{st}) = \frac{1}{n(n-1)} \cdot \frac{1}{\sigma_{st}}$. The number r of samples required to approximate the BC scores with the given error guarantee is $r = \frac{c}{\epsilon^2} \left(\lfloor \log_2 (VD - 2) \rfloor + 1 + \ln \frac{1}{\delta} \right)$, where ϵ and δ are constants in $(0, 1)$ and $c \approx 0.5$. Then, if r shortest paths are sampled according to π_G, with probability at least $1 - \delta$ the approximations $\tilde{c}_B(v)$ are within ϵ from their exact value: $\Pr(\exists v \in V \ s.t. \ |c_B(v) - \tilde{c}_B(v)| > \epsilon) < \delta$. To sample the shortest paths according to π_G, RK first chooses a source-target node pair (s, t) uniformly at random and performs a shortest-path search (Dijkstra or BFS) from s to t, keeping also track of the number σ_{sv} of shortest paths between s and v and of the list of predecessors $P_s(v)$ (i.e. the nodes that immediately precede v in the shortest paths between s and v) for any node v between s and t. Then one shortest path is selected: starting from t, a predecessor $z \in P_s(t)$ is selected with probability $\sigma_{sz} / \sum_{w \in P_s(t)} \sigma_{sw} = \sigma_{sz} / \sigma_{st}$. The sampling is repeated iteratively until node s is reached.

Approximating the Vertex Diameter. RK uses two upper bounds on *VD* that can be both computed in $O(n + m)$. For unweighted undirected graphs, it samples a source node s_i for each connected component of G, computes a BFS from each s_i and sums the two shortest paths with maximum length starting in s_i. The *VD* approximation is the maximum of these sums over all components. For weighted graphs, RK approximates *VD* with the size of the largest connected component, which can be a significant overestimation for complex networks, possibly of orders of magnitude. In this paper, we present a new approximation for weighted graphs, described in Section 3.

2.3   IA and IAW Algorithms

IA and IAW are the incremental approximation algorithms (for unweighted and weighted graphs, respectively) that we presented previously [4]. The algorithms are based on the observation that if only edge insertions are allowed and the graph is connected, *VD* cannot increase, and therefore also the number r of samples required by RK for the theoretical guarantee. Instead of recomputing r new shortest paths after a batch of edge insertions, IA and IAW *replace* each old shortest path $p_{s,t}$ with a new shortest path between the same node pair (s,t). In IAW the paths are recomputed with a slightly-modified T-SWSF [2], whereas IA uses a new semi-dynamic BFS algorithm. The BC scores are updated by subtracting $1/r$ to the BC of the nodes in the old path and adding $1/r$ to the BC of nodes in the new shortest path.

2.4   Batch Dynamic SSSP Algorithms

Dynamic SSSP algorithms recompute distances from a source node after a single edge update or a batch of edge updates. Algorithms for the batch problem have been published [18,8,2] and compared in experimental studies [2,7]. The experiments show that the tuned algorithm T-SWSF presented in [2] performs well on many types of graphs and edge updates. For batches of only edge insertions in unweighted graphs, we developed an algorithm asymptotically faster than T-SWSF [4]. The algorithm is in principle similar to T-SWSF, but has an improved complexity thanks to different data structures.

3   New *VD* Approximation for Weighted Graphs

Let G be an undirected graph. For simplicity, let G be connected for now. If it is not, we compute an approximation for each connected component and take the maximum over all the approximations. Let $T \subseteq G$ be an SSSP tree from any source node $s \in V$. Let p_{xy} denote a shortest path between x and y in G and let p_{xy}^T denote a shortest path between x and y in T. Let $|p_{xy}|$ be the number of nodes in p_{xy} and $d(x, y)$ be the distance between x and y in G, and analogously for $|p_{xy}^T|$ and $d^T(x, y)$. Let $\overline{\omega}$ and $\underline{\omega}$ be the maximum and minimum edge weights, respectively. Let u and v be the nodes with maximum distance from s, i. e. $d(s, u) \geq d(s, v) \geq d(s, x) \; \forall x \in V, x \neq u$.

We define the VD approximation $\tilde{VD} := 1 + \frac{d(s,u)+d(s,v)}{\underline{\omega}}$. Then:

Proposition 1. $VD \leq \tilde{VD} < 2 \cdot \frac{\overline{\omega}}{\underline{\omega}} VD$. *(Proof in extended version [3])*

To obtain the upper bound $\tilde{VD}$, we can simply compute an SSSP search from any node s, find the two nodes with maximum distance and perform the remaining calculations. Notice that $\tilde{VD}$ extends the upper bound proposed for RK [19] for unweighted graphs: When the graph is unweighted and thus $\underline{\omega} = \overline{\omega}$, $\tilde{VD}$ becomes equal to the approximation used by RK. Complex networks are often characterized by a small diameter and in networks like coauthorship, friendship, communication networks, VD and $\frac{\overline{\omega}}{\underline{\omega}}$ can be several order of magnitude smaller than the size of the largest component. This translates into a substantially improved VD approximation.

4   New Fully-Dynamic Algorithms

Overview. We propose two fully-dynamic algorithms, one for unweighted (DA, dynamic approximation) and one for weighted (DAW, dynamic approximation weighted) graphs. Similarly to IA and IAW, our new fully-dynamic algorithms keep track of the old shortest paths and substitute them only when necessary. However, if G is not connected or edge deletions occur, VD can grow and a simple substitution of the paths is not sufficient anymore. Although many real-world networks exhibit a shrinking-diameter behavior [16], to ensure our theoretical guarantee, we need to keep track of $\tilde{VD}$ over time and sample new paths in case $\tilde{VD}$ increases. The need for an efficient update of $\tilde{VD}$ augments significantly the difficulty of the fully-dynamic problem, as well as the necessity to recompute the SSSPs after batches of both edge insertions and deletions. The building block for the BC update are basically two: a fully-dynamic algorithm that updates distances and number of shortest paths from a certain source node (SSSP update) and an algorithm that keeps track of a VD approximation for each connected component of G. The following paragraphs give an overview of such building blocks, which could be of independent interest. The last paragraph outlines the dynamic BC approximation algorithm. **Due to space constraints, a detailed description of the algorithms as well as the pseudocodes and the omitted proofs can be found in the full version of this paper [3].**

SSSP Update in Weighted Graphs. Our SSSP update is based on T-SWSF [2], which recomputes distances from a source node s after a batch β of weight updates (or edge insertions/deletions). For our BC algorithm, we need two extensions of T-SWSF: an algorithm that also recomputes the number of shortest paths between s and the other nodes (updateSSSP-W) and one that also updates a VD approximation for the connected component of s (updateApprVD-W). The VD approximation is computed as described in Section 3. Thus, updateApprVD-W keeps track of the two maximum distances d' and d'' from s and the minimum edge weight $\underline{\omega}$. We call *affected nodes* the nodes whose distance (or also whose number of shortest paths, in updateSSSP-W) from s has changed as a consequence of β. Basically, the idea is to put the set A of affected nodes w into a priority queue Q with priority $p(w)$ equal to the candidate distance of w. When w is extracted, if there is actually a path of length $p(w)$ from s to w, the new distance of w is set to $p(w)$, otherwise w is reinserted into Q with a higher candidate distance. In both cases, the affected neighbors of w are inserted into Q. In updateApprVD-W, d' and d'' are recomputed while updating the distances and $\underline{\omega}$ is updated while scanning β. In updateSSSP-W, the number σ_{sw} of shortest paths of w is recomputed as the sum of the σ_{sz} of the new predecessors z of w.

Let $|\beta|$ represent the cardinality of β and let $||A||$ represent the sum of the nodes in A and of the edges that have at least one endpoint in A. Then, the following complexity derives from feeding Q with the batch and inserting into/extracting from Q the affected nodes and their neighbors.

Lemma 1. *The time required by* updateApprVD-W *(*updateSSSP-W*) to update the distances and* $\tilde{VD}$ *(the number of shortest paths) is* $O(|\beta| \log |\beta| + ||A|| \log ||A||)$.

SSSP update in unweighted graphs. For unweighted graphs, we basically replace the priority queue Q of updateApprVD-W and updateSSSP-W with a list of queues, as the one we used in [4] for the incremental BFS. Each queue represents a *level* from 0 (which only the source belongs to) to the maximum distance d'. The levels replace the priorities and also in this case represent the candidate distances for the nodes. In order not to visit a node multiple times, we use colors to distinguish the unvisited nodes from the visited ones. The replacement of the priority queue with the list of queues decreases the complexity of the SSSP update algorithms for unweighted graphs, that we call updateApprVD-U and updateSSSP-U, in analogy with the ones for weighted graphs.

Lemma 2. *The time required by* updateApprVD-U *(*updateSSSP-U*) to update the distances and* $\tilde{VD}$ *(the number of shortest paths) is* $O(|\beta| + ||A|| + d_{\max})$, *where* $d_{\max}$ *is the maximum distance from* s *reached during the update.*

Fully-dynamic VD approximation. The algorithm keeps track of a *VD* approximation for the whole graph G, i.e. for each connected component of G. It is composed of two phases. In the initialization, we compute an SSSP from a source node s_i for each connected component C_i. During the SSSP search from s_i, we also compute a *VD* approximation $\tilde{VD}_i$ for C_i, as described in Sections 2.2 and 3. In the update, we recompute the SSSPs and the *VD* approximations with updateApprVD-W (or updateApprVD-U). Since components might split or merge, we might need to compute new approximations, in addition to update the old ones. To do this, for each node, we keep track of the number of times it has been visited. This way we discard source nodes that have already been visited and compute a new approximation for components that have become unvisited. The complexity of the update of the *VD* approximation derives from the $\tilde{VD}$ update in the single components, using updateApprVD-W and updateApprVD-U.

Theorem 1. *The time required to update the VD approximation is $O(n_c \cdot |\beta| \log$ $|\beta| + \sum_{i=1}^{n_c} ||A^{(i)}|| \log ||A^{(i)}||)$ in weighted graphs and $O(n_c \cdot |\beta| + \sum_{i=1}^{n_c} ||A^{(i)}|| + d_{max}^{(i)})$ in unweighted graphs, where n_c is the number of components in G before the update and $A^{(i)}$ is the sum of affected nodes in C_i and their incident edges.*

Dynamic BC approximation. Let G be an undirected graph with n_c connected components. Now that we have defined our building blocks, we can outline a fully-dynamic BC algorithm: we use the fully dynamic VD approximation to recompute $\tilde{VD}$ after a batch, we update the r sampled paths with updateSSSP and, if $\tilde{VD}$ (and therefore r) increases, we sample new paths. However, since updateSSSP and updateApprVD share most of the operations, we can "merge" them and update at the same time the shortest paths from a source node s and the VD approximation for the component of s. We call such hybrid function updateSSSPVD. Instead of storing and updating n_c SSSPs for the VD approximation and r SSSPs for the BC scores, we recompute a VD approximation for each of the r samples while recomputing the shortest paths with updateSSSPVD. This way we do not need to compute an additional SSSP for the components covered by r sampled paths (i.e. in which the paths lie), saving time and memory. Only for components that are not covered by any of them (if they exist), we compute and store a separate VD approximation. We refer to such components as R' (and to $|R'|$ as r'). The high-level description of the update after a batch β is shown as Algorithm 1. After changing the graph according to β (Line 1), we recompute the previous r samples and the VD approximations for their components (Lines 2 - 5). Then, similarly to IA and IAW, we update the BC scores of the nodes in the old and in the new shortest paths. Thus, we update a VD approximation for the components in R' (Lines 6 - 8) and compute a new approximation for new components that have formed applying the batch (Lines 9 - 12). Then, we use the results to update the number of samples (Lines 13 - 14). If necessary, we sample additional paths and normalize the BC scores (Lines 15 - 21). The difference between DA and DAW is the way the SSSPs and the VD approximation are updated: in DA we use updateApprVD-U and in DAW updateApprVD-W. Differently from RK and our previous algorithms IA and IAW, in DA and DAW we scan the neighbors every time we need the predecessors instead of storing them. This allows us to use $\Theta(n)$ memory per sample (i.e., $\Theta((r + r')n)$ in total) instead of $\Theta(m)$ per sample, while our experiments show that the running time is hardly influenced. The number of samples depends on ϵ, so in theory this can be as large as $|V|$. However, the experiments conducted in [4] show that relatively large values of ϵ (e.g. $\epsilon = 0.05$) lead to good ranking of nodes with high BC and for such values the number of samples is typically much smaller than $|V|$, making the memory requirements of our algorithms significantly less demanding than those of the dynamic exact algorithms ($\Omega(n^2)$) for many applications.

Theorem 2. *Algorithm 1 preserves the guarantee on the maximum absolute error, i.e. naming $c_B'(v)$ and $\tilde{c}_B'(v)$ the new exact and approximated BC values, respectively, $\Pr(\exists v \in V \ s.t. \ |c_B'(v) - \tilde{c}_B'(v)| > \epsilon) < \delta$.*

Algorithm 1. BC update after a batch β of edge updates

---

1 applyBatch(G, β);
2 **for** $i \leftarrow 1$ **to** r **do**
3 | $\tilde{VD}_i \leftarrow$ updateSSSPVD(s_i, β);
4 | replacePath(s_i, t_i) ;   /* update of BC scores */
5 **end**
6 **foreach** $C_i \in R'$ **do**
7 | $\tilde{VD}_i \leftarrow$ updateApprVD(C_i, β);
8 **end**
9 **foreach** *unvisited* C_j **do**
10 | add C_j to R';
11 | $\tilde{VD}_j \leftarrow$ initApprVD(C_j);
12 **end**
13 $\tilde{VD} \leftarrow \max_{C_i \in R \cup R'} \tilde{VD}_i$;
14 $r_{\text{new}} \leftarrow (c/\epsilon^2)(\lfloor \log_2(\tilde{VD} - 2) \rfloor + \ln(1/\delta))$;
15 **if** $r_{new} > r$ **then**
16 | sampleNewPaths() ;   /* update of BC scores */
17 | **foreach** $v \in V$ **do**
18 | | $\tilde{c}_B(v) \leftarrow \tilde{c}_B(v) \cdot r/r_{\text{new}}$;   /* renormalization of BC scores */
19 | **end**
20 | $r \leftarrow r_{\text{new}}$;
21 **end**
22 **return** $\{(v, \tilde{c}_B(v)) : v \in V\}$

Theorem 3. *Let Δr be the difference between the value of r before and after the batch and let $||A^{(i)}||$ be the sum of affected nodes and their incident edges in the i-th SSSP. The time required for the BC update in unweighted graphs is* $O((r + r')|\beta| + \sum_{i=1}^{r+r'} (||A^{(i)}|| + d_{\max}^{(i)}) + \Delta r(|V| + |E|))$. *In weighted graphs, it is* $O((r + r')|\beta| \log |\beta| + \sum_{i=1}^{r+r'} ||A^{(i)}|| \log ||A^{(i)}|| + \Delta r(|V| \log |V| + |E|))$.

Notice that, if $\tilde{VD}$ does not increase, $\Delta r = 0$ and the complexities are the same as the only-incremental algorithms IA and IAW we proposed in [4]. Also, notice that in the worst case the complexity can be as bad as recomputing from scratch. However, no dynamic SSSP (and so probably also no BC approximation) algorithm exists that is faster than recomputation.

5   Experiments

Implementation and settings. We implement our two dynamic approaches DA and DAW in C++, building on the open-source *NetworKit* framework [22], which also contains the static approximation RK. In all experiments we fix δ to 0.1 and ϵ to 0.05, as a good tradeoff between running time and accuracy [4]. This means that, with a probability of at least 90%, the computed BC values deviate at most 0.05 from the exact ones. In our previous experimental study [4], we showed that

Table 1. Overview of real dynamic graphs used in the experiments.

Graph	Type	Nodes	Edges	Type
repliesDigg	communication	30,398	85,155	Weighted
emailSlashdot	communication	51,083	116,573	Weighted
emailLinux	communication	63,399	159,996	Weighted
facebookPosts	communication	46,952	183,412	Weighted
emailEnron	communication	87,273	297,456	Weighted
facebookFriends	friendship	63,731	817,035	Unweighted
arXivCitations	coauthorship	28,093	3,148,447	Unweighted
englishWikipedia	hyperlink	1,870,709	36,532,531	Unweighted

for such values of ϵ and δ, the ranking error (how much the ranking computed by the approximation algorithm differs from the rank of the exact algorithm) is low for nodes with high betweenness. Since our algorithms simply update the approximation of RK, our accuracy in terms or ranking error does not differ from that of RK (see [4] for details). Also, our experiments in [4] have shown that dynamic exact algorithms are not scalable, because of both time and memory requirements, so we do not include them in our tests. The machine used has 2 x 8 Intel(R) Xeon(R) E5-2680 cores at 2.7 GHz, of which we use only one core, and 256 GB RAM.

Data sets and experiments. We concentrate on two types of graphs: synthetic and real-world graphs with real edge dynamics. The real-world networks are taken from The Koblenz Network Collection (KONECT) [14] and are summarized in Table 1. All the edges of the KONECT graphs are characterized by a time of arrival. In case of multiple edges between two nodes, we extract two versions of the graph: one unweighted, where we ignore additional edges, and one weighted, where we replace the set E_{st} of edges between two nodes with an edge of weight $1/|E_{st}|$. In our experiments, we let the batch size vary from 1 to 1024 and for each batch size, we average the running times over 10 runs. Since the networks do not include edge deletions, we implement additional simulated dynamics. In particular, we consider the following experiments. (i) *Real dynamics.* We remove the x edges with the highest timestamp from the network and we insert them back in batches, in the order of timestamps. (ii) *Random insertions and deletions.* We remove x edges from the graph, chosen uniformly at random. To create batches of both edge insertions and deletions, we add back the deleted edges with probability $1/2$ and delete other random edges with probability $1/2$. (iii) *Random weight changes.* In weighted networks, we choose x edges uniformly at random and we multiply their weight by a random value in the interval $(0, 2)$.

For synthetic graphs we use a generator based on a unit-disk graph model in hyperbolic geometry [21], where edge insertions and deletions are obtained by moving the nodes in the hyperbolic plane. The networks produced by the model were shown to have many properties of real complex networks, like small diameter and power-law degree distribution (see [21] and the references therein). We generate seven networks, with $|E|$ ranging from about $2 \cdot 10^4$ to about $2 \cdot 10^7$ and $|V|$ approximately equal to $|E|/10$.

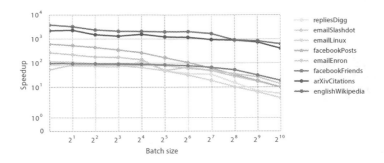

Fig. 1. Speedups of DA on RK in real unweighted networks using real dynamics.

Table 2. Times and speedups of DA on RK in unweighted real graphs under real dynamics and random updates, for batch sizes of 1 and 1024.

	Real				Random			
	Time [s]		Speedups		Time [s]		Speedups	
Graph	$|\beta| = 1$	$|\beta| = 1024$	$|\beta| = 1$	$|\beta| = 1024$	$|\beta| = 1$	$|\beta| = 1024$	$|\beta| = 1$	$|\beta| = 1024$
repliesDigg	0.078	1.028	76.11	5.42	0.008	0.832	94.00	4.76
emailSlashdot	0.043	1.055	219.02	9.91	0.038	1.151	263.89	28.81
emailLinux	0.049	1.412	108.28	3.59	0.051	2.144	72.73	1.33
facebookPosts	0.023	1.416	527.04	9.86	0.015	1.520	745.86	8.21
emailEnron	0.368	1.279	83.59	13.66	0.203	1.640	99.45	9.39
facebookFriends	0.447	1.946	94.23	18.70	0.448	2.184	95.91	18.24
arXivCitations	0.038	0.186	2287.84	400.45	0.025	1.520	2188.70	28.81
englishWikipedia	1.078	6.735	3226.11	617.47	0.877	5.937	2833.57	703.18

Speedups. Figure 1 reports the speedups of DA on RK in real graphs using real dynamics. Although some fluctuations can be noticed, the speedups tend to decrease as the batch size increases. We can attribute fluctuations to two main factors: First, different batches can affect areas of G of varying sizes, influencing also the time required to update the SSSPs. Second, changes in the VD approximation can require to sample new paths and therefore increase the running time of DA (and DAW). Nevertheless, DA is significantly faster than recomputation on all networks and for every tested batch size. The tests with random dynamics lead to similar results, reported in our full version [3]. Table 2 summarizes the running times of DA and its speedups on RK with batches of size 1 and 1024 in unweighted graphs, under both real and random dynamics. Even on the larger graphs (`arXivCitations` and `englishWikipedia`) and on large batches, DA requires at most a few seconds to recompute the BC scores, whereas RK requires about one hour for `englishWikipedia`. The results for weighted graphs are shown in [3]. In both real dynamics and random updates, the speedups vary between ≈ 50 and $\approx 6 \cdot 10^3$ for single-edge updates and between ≈ 5 and ≈ 75 for batches of size 1024. In hyperbolic graphs, the speedups of DA on RK increase with the size of the graph, varying between ≈ 100 and $\approx 3 \cdot 10^5$ for single-edge updates and between ≈ 3 and $\approx 5 \cdot 10^3$ for batches of 1024 edges (see [3] for details). The results show that DA and DAW are faster than recomputation with

RK in all the tested instances, even when large batches of 1024 edges are applied to the graph. With small batches, the algorithms are always orders of magnitude faster than RK, often with running times of fraction of seconds or seconds compared to minutes or hours. Such high speedups are made possible by the efficient update of the sampled shortest paths, which limit the recomputation to the nodes that are actually affected by the batch. Also, processing the edges in batches, we avoid to update multiple times nodes that are affected by several edges of the batch.

6   Conclusions

Betweenness is a widely used centrality measure, yet expensive if computed exactly. In this paper we have presented the first fully-dynamic algorithms for betweenness approximation (for weighted and for unweighted undirected graphs). The consideration of edge deletions and disconnected graphs is made possible by the efficient solution of several algorithmic subproblems (some of which may be of independent interest). Now BC can be approximated with an error guarantee for a much wider set of dynamic real graphs compared to previous work.

Our experiments show significant speedups over the static algorithm RK. In this context it is interesting to remark that dynamic algorithms require to store additional memory and that this can be a limit to the size of the graphs they can be applied to. By not storing the predecessors in the shortest paths, we reduce the memory requirement from $\Theta(|E|)$ per sampled path to $\Theta(|V|)$ – and are still often more than 100 times faster than RK despite rebuilding the paths.

Future work may include the transfer of our concepts to approximating other centrality measures in a fully-dynamic manner, e. g. closeness, and the extension to directed graphs, for which a good VD approximation is the only obstacle. Moreover, making the betweenness code run in parallel will further accelerate the computations in practice. Our implementation will be made available as part of a future release of the network analysis tool suite *NetworKit* [22].

Acknowledgements. This work is partially supported by DFG grant FINCA (ME-3619/3-1) within the SPP 1736 *Algorithms for Big Data*. We thank Moritz von Looz, Matteo Riondato (Brown University) and the numerous contributors to the *NetworKit* project.

References

1. Bader, D.A., Kintali, S., Madduri, K., Mihail, M.: Approximating betweenness centrality. In: Bonato, A., Chung, F.R.K. (eds.) WAW 2007. LNCS, vol. 4863, pp. 124–137. Springer, Heidelberg (2007)
2. Bauer, R., Wagner, D.: Batch dynamic single-source shortest-path algorithms: An experimental study. In: Vahrenhold, J. (ed.) SEA 2009. LNCS, vol. 5526, pp. 51–62. Springer, Heidelberg (2009)
3. Bergamini, E., Meyerhenke, H.: Fully-dynamic approximation of betweenness centrality. CoRR, abs/1504.07091 (2015)

4. Bergamini, E., Meyerhenke, H., Staudt, C.: Approximating betweenness centrality in large evolving networks. In: 17th Workshop on Algorithm Engineering and Experiments, ALENEX 2015, pp. 133–146. SIAM (2015)
5. Brandes, U.: A faster algorithm for betweenness centrality. Journal of Mathematical Sociology 25, 163–177 (2001)
6. Brandes, U., Pich, C.: Centrality estimation in large networks. I. J. Bifurcation and Chaos 17(7), 2303–2318 (2007)
7. D'Andrea, A., D'Emidio, M., Frigioni, D., Leucci, S., Proietti, G.: Experimental evaluation of dynamic shortest path tree algorithms on homogeneous batches. In: Gudmundsson, J., Katajainen, J. (eds.) SEA 2014. LNCS, vol. 8504, pp. 283–294. Springer, Heidelberg (2014)
8. Frigioni, D., Marchetti-Spaccamela, A., Nanni, U.: Semi-dynamic algorithms for maintaining single-source shortest path trees. Algorithmica 22, 250–274 (2008)
9. Geisberger, R., Sanders, P., Schultes, D.: Better approximation of betweenness centrality. In: 10th Workshop on Algorithm Engineering and Experiments, ALENEX 2008, pp. 90–100. SIAM (2008)
10. Goel, K., Singh, R.R., Iyengar, S., Sukrit: A faster algorithm to update betweenness centrality after node alteration. In: Bonato, A., Mitzenmacher, M., Prałat, P. (eds.) WAW 2013. LNCS, vol. 8305, pp. 170–184. Springer, Heidelberg (2013)
11. Green, O., McColl, R., Bader, D.A.: A fast algorithm for streaming betweenness centrality. In: SocialCom/PASSAT, pp. 11–20. IEEE (2012)
12. Kas, M., Carley, K.M., Carley, L.R.: An incremental algorithm for updating betweenness centrality and k-betweenness centrality and its performance on realistic dynamic social network data. Social Netw. Analys. Mining 4(1), 235 (2014)
13. Kourtellis, N., De Francisci Morales, G., Bonchi, F.: Scalable online betweenness centrality in evolving graphs. IEEE Transactions on Knowledge and Data Engineering (99), 1 (2015)
14. Kunegis, J.: KONECT: the koblenz network collection. In: 22nd Int. World Wide Web Conf., WWW 2013, pp. 1343–1350 (2013)
15. Lee, M., Lee, J., Park, J.Y., Choi, R.H., Chung, C.: QUBE: a quick algorithm for updating betweenness centrality. In: 21st World Wide Web Conf. 2012, WWW 2012, pp. 351–360. ACM (2012)
16. Leskovec, J., Kleinberg, J.M., Faloutsos, C.: Graphs over time: densification laws, shrinking diameters and possible explanations. In: 11th Int. Conf. on Knowledge Discovery and Data Mining, pp. 177–187. ACM (2005)
17. Nasre, M., Pontecorvi, M., Ramachandran, V.: Betweenness centrality – incremental and faster. In: Csuhaj-Varjú, E., Dietzfelbinger, M., Ésik, Z. (eds.) MFCS 2014, Part II. LNCS, vol. 8635, pp. 577–588. Springer, Heidelberg (2014)
18. Ramalingam, G., Reps, T.: An incremental algorithm for a generalization of the shortest-path problem. Journal of Algorithms 21, 267–305 (1992)
19. Riondato, M., Kornaropoulos, E.M.: Fast approximation of betweenness centrality through sampling. In: 7th ACM Int. Conf. on Web Search and Data Mining (WSDM 2014), pp. 413–422. ACM (2014)
20. Roditty, L., Zwick, U.: On dynamic shortest paths problems. Algorithmica 61(2), 389–401 (2011)
21. von Looz, M., Staudt, C.L., Meyerhenke, H., Prutkin, R.: Fast generation of complex networks with underlying hyperbolic geometry (2015), http://arxiv.org/abs/1501.03545v2
22. Staudt, C., Sazonovs, A., Meyerhenke, H.: NetworKit: An interactive tool suite for high-performance network analysis (2014), http://arxiv.org/abs/1403.3005

Improved Purely Additive
Fault-Tolerant Spanners[*]

Davide Bilò[1], Fabrizio Grandoni[2], Luciano Gualà[3],
Stefano Leucci[4], and Guido Proietti[4,5]

[1] Dipartimento di Scienze Umanistiche e Sociali, Università di Sassari, Italy
[2] IDSIA, University of Lugano, Switzerland
[3] Dipartimento di Ingegneria dell'Impresa, Università di Roma "Tor Vergata", Italy
[4] DISIM, Università degli Studi dell'Aquila, Italy
[5] Istituto di Analisi dei Sistemi ed Informatica, CNR, Roma, Italy
davide.bilo@uniss.it, fabrizio@idsia.ch, guala@mat.uniroma2.it,
{stefano.leucci,guido.proietti}@univaq.it

Abstract. Let G be an unweighted n-node undirected graph. A β-*additive spanner* of G is a spanning subgraph H of G such that distances in H are stretched at most by an additive term β w.r.t. the corresponding distances in G. A natural research goal related with spanners is that of designing *sparse* spanners with *low* stretch.

In this paper, we focus on *fault-tolerant* additive spanners, namely additive spanners which are able to preserve their additive stretch even when one edge fails. We are able to improve all known such spanners, in terms of either sparsity or stretch. In particular, we consider the sparsest known spanners with stretch 6, 28, and 38, and reduce the stretch to 4, 10, and 14, respectively (while keeping the same sparsity).

Our results are based on two different constructions. On one hand, we show how to augment (by adding a *small* number of edges) a fault-tolerant additive *sourcewise spanner* (that approximately preserves distances only from a given set of source nodes) into one such spanner that preserves all pairwise distances. On the other hand, we show how to augment some known fault-tolerant additive spanners, based on clustering techniques. This way we decrease the additive stretch without any asymptotic increase in their size. We also obtain improved fault-tolerant additive spanners for the case of one vertex failure, and for the case of f edge failures.

1 Introduction

We are given an unweighted, undirected n-node graph $G = (V(G), E(G))$. Let $d_G(s, t)$ denote the shortest path distance between nodes s and t in G. A *spanner* H of G is a spanning subgraph such that $d_H(s, t) \leq \varphi(d_G(s, t))$ for all s, $t \in$

---

[*] This work was partially supported by the Research Grant PRIN 2010 "ARS TechnoMedia", funded by the Italian Ministry of Education, University, and Research, and by the ERC Starting Grant "New Approaches to Network Design".

© Springer-Verlag Berlin Heidelberg 2015
N. Bansal and I. Finocchi (Eds.): ESA 2015, LNCS 9294, pp. 167–178, 2015.
DOI: 10.1007/978-3-662-48350-3_15

$V(G)$, where φ is the so-called *stretch* or *distortion* function of the spanner. In particular, when $\varphi(x) = \alpha x + \beta$, for constants α, β, the spanner is named an (α,β) spanner. If $\alpha = 1$, the spanner is called *(purely) additive* or also *β-additive*. If $\beta = 0$, the spanner is called *α-multiplicative*.

Finding *sparse* (i.e., with a small number of edges) spanners is a key task in many network applications, since they allow for a small-size infrastructure onto which an efficient (in terms of paths' length) point-to-point communication can be performed. Due to this important feature, spanners were the subject of an intensive research effort, aiming at designing increasingly sparser spanners with lower stretch.

However, as any sparse structure, a spanner is very sensitive to possible failures of *components* (i.e., edges or nodes), which may drastically affect its performances, or even disconnect it! Thus, to deal with this drawback, a more robust concept of *fault-tolerant* spanner is naturally conceivable, in which the distortion must be guaranteed even after a subset of components of G fails.

More formally, for a subset F of edges (resp., vertices) of G, let $G - F$ be the graph obtained by removing from G the edges (resp., vertices and incident edges) in F. When $F = \{x\}$, we will simply write $G - x$. Then, an *f-edge fault-tolerant* (*f-EFT*) spanner with distortion (α, β), is a subgraph H of G such that, for every set $F \subseteq E(G)$ of at most f failed edges, we have[1]

$$d_{H-F}(s,t) \le \alpha \cdot d_{G-F}(s,t) + \beta \quad \forall s,t \in V(G).$$

We define similarly an *f-vertex fault-tolerant* (*f-VFT*) spanner. For $f = 1$, we simply call the spanner edge/vertex fault-tolerant (EFT/VFT).

Chechik et al. [10] show how to construct a $(2k-1)$-multiplicative f-EFT spanner of size $O(f \cdot n^{1+1/k})$, for any integer $k \ge 1$. Their approach also works for weighted graphs and for vertex-failures, returning a $(2k-1)$-multiplicative f-VFT spanner of size $\widetilde{O}(f^2 \cdot k^{f+1} \cdot n^{1+1/k})$.[2] This latter result has been finally improved through a randomized construction in [14], where the expected size was reduced to $\widetilde{O}(f^{2-1/k} \cdot n^{1+1/k})$. For a comparison, the sparsest known $(2k-1)$-multiplicative *standard* (non fault-tolerant) spanners have size $O(n^{1+\frac{1}{k}})$ [2], and this is believed to be asymptotically tight due to the girth conjecture of Erdős [15].

Additive fault-tolerant spanners can be constructed with the following approach by Braunshvig et al [8]. Let M be an α-multiplicative f-EFT spanner, and A be a β-additive standard spanner. Then $H = M \cup A$ is a $(2f(2\beta + \alpha - 1) + \beta)$-additive f-EFT spanner. One can exploit this approach to construct *concrete* EFT spanners as follows. We know how to construct 6-additive spanners of size $O(n^{4/3})$ [4], randomized spanners that, w.h.p., have size $\widetilde{O}(n^{7/5})$ and additive distortion 4 [9], and 2-additive spanners of size $O(n^{3/2})$ [1]. By setting $f = 1$ and choosing k properly, this leads to EFT spanners of size $O(n^{4/3})$ with additive distortion 38, size $\widetilde{O}(n^{7/5})$ with additive distortion 28 (w.h.p.), and size

---

[1] Note that in this definition we allow $d_{G-F}(s,t)$ to become infinite (if the removal of F disconnects s from t). In that case we assume the inequality to be trivially satisfied.

[2] The $\widetilde{O}$ notation hides poly-logarithmic factors in n.

Table 1. State of the art and new results on additive EFT spanners. Distortions and sizes marked with "*" hold w.h.p.

State of the art		Our results	
Size	Additive distortion	Size	Additive distortion
$\widetilde{O}(n^{5/3})$	2 [17]	$O(n^{5/3})$	2
$\widetilde{O}(n^{3/2})$	6 [17]	$O(n^{3/2})$	4
$\widetilde{O}(n^{\frac{7}{5}})^*$	28* [8,9]	$\widetilde{O}(n^{\frac{7}{5}})^*$	10*
$O(n^{\frac{4}{3}})$	38 [4,8]	$O(n^{\frac{4}{3}})$	14

$O(n^{3/2})$ with additive distortion 14. Finally, using a different approach, Parter [17] recently presented 2- and 6-additive EFT/VFT spanners of size $\widetilde{O}(n^{5/3})$ and $\widetilde{O}(n^{3/2})$, respectively.

1.1 Our Results

In this paper, we focus on additive EFT spanners, and we improve all the known such spanners in terms of sparsity or stretch (see Table 1). We also present some better results for additive VFT and f-EFT spanners.

In more detail, our improved EFT spanners exploit the following two novel approaches. Our first technique (see Section 2), assumes that we are given an additive *sourcewise* fault-tolerant spanner A_S, i.e., a fault-tolerant spanner that guarantees low distortion only for the distances from a given set S of source nodes. We show that, by carefully choosing S and by augmenting A_S with a conveniently selected *small* subset of edges, it is possible to construct a fault-tolerant spanner (approximately preserving *all* pairwise distances) with a moderate increase of the stretch. This, combined with the sourcewise EFT spanners in [7,18], leads to the first two results in the table. In particular, we reduce the additive stretch of the best-known spanner of size $\widetilde{O}(n^{3/2})$ from 6 [17] to 4 (actually, we also save a polylogarithmic factor in the size here). For the case of stretch 2, we slightly decrease the size from $\widetilde{O}(n^{5/3})$ [17] to $O(n^{5/3})$. This technique also applies to VFT spanners. In particular, we achieve a 2-additive VFT spanner of size $O(n^{5/3})$ rather than $\widetilde{O}(n^{5/3})$ [17], and a 4-additive VFT spanner of size $O(n^{3/2}\sqrt{\log n})$, improving on the 6-additive VFT spanner of size $\widetilde{O}(n^{3/2})$ in [17].

Our second technique (see Section 3) relies on some properties of known additive spanners. We observe that some known additive spanners are based on *clustering techniques* that construct a small-enough number of clusters. Furthermore, the worst-case stretch of these spanners is achieved only in some specific cases. We exploit these facts to augment the spanner $H = M \cup A$ based on the already mentioned construction of [8] with a small number of inter and intra-cluster edges. This allows us to reduce the additive stretch without any asymptotic increase in the number of edges.

Finally, for the case of multiple edge failures, we are able to prove that the construction in [8] has in fact an additive stretch of only $2f(\beta+\alpha-1)+\beta$ (rather than $2f(2\beta+\alpha-1)+\beta$).

Theorem 1. *Let A be a β-additive spanner of G, and let M be an α-multiplicative f-EFT spanner of G. The graph $H = (V(G), E(A)\cup E(M))$ is a $(2f(\beta+\alpha-1)+\beta)$-additive f-EFT spanner of G. In the special case $f = 1$, the additive stretch is at most $2\beta + \alpha - 1$.*

We see this as an interesting result since, to the best of our knowledge, the construction of [8] is the only known approach for building additive spanners withstanding more than a single edge fault. At a high-level, in [8] the shortest path in $G - F$ between two vertices is decomposed into (roughly $2f$) subpaths called *blocks*. The authors then show that it is possible to build a *bypass* (i.e., a fault-free path) in H between the endpoints of each block such that the additive error incurred by using this path is at most $\beta + \alpha - 1$. Actually, in addition to those intra-block bypasses, the spanner H contains some inter-block *shortcuts*, that are exploited in order to prove a better distortion. Due to space limitations, the proof of this result will be given in the full version of the paper.

1.2 Related Work

A notion closely relate to fault-tolerant spanners is the one of *Distance Sensitivity Oracles* (DSO). The goal here is to compute, with a *low* preprocessing time, a *compact* data structure which is able to *quickly* answer distance queries following some component failures (possibly in an approximate way). For recent achievements on DSO, we refer the reader to [5,6,11,16].

Another setting which is very close in spirit to fault-tolerant spanners is the recent work on fault-tolerant *approximate shortest-path trees*, both for unweighted [19] and for weighted [5,7] graphs. In [3] it was introduced the resembling concept of *resilient spanners*, i.e., spanners that approximately preserve the *relative* increase of distances due to an edge failure.

There was also some research (see for example [12,13]) on spanners approximately preserving the distance from a given set of nodes (*sourcewise* spanners), among a given set of nodes (*subsetwise* spanners), or between given pairs of nodes (*pairwise* spanners). In this framework a spanner is called a *preserver* if distances are preserved (in other words, the stretch function is the identity function). In particular, in one of our constructions we exploit a fault-tolerant version of a sourcewise preserver.

1.3 Notation

Given an unweighted, undirected graph G, let us denote by $\pi_G(u, v)$ a shortest path) between u and v in G. When the graph G is clear from the context we might omit the subscript. Given a simple path π in G and two vertices $s, t \in V(\pi)$, we define $\pi[s,t]$ to be the subpath of π connecting s and t. Moreover, we denote by $|\pi|$ the *length* of π, i.e., the number of its edges. When dealing with one or

---

Algorithm 1: Algorithm for computing a fault-tolerant additive spanner of G from a β-additive f-EFT/VFT sourcewise spanner. The parameter p affects the size of the returned spanner and it will be suitably chosen.

---

1 $\text{color}(v) \leftarrow \text{white} \quad \forall v \in V; \text{counter}(v) \leftarrow f+1 \quad \forall v \in V$
2 $S \leftarrow \emptyset; E' \leftarrow \emptyset$

3 **while** $\exists s \in V \setminus S : \delta_{white}(s) \geq p$ **do**
4 $\quad\quad S \leftarrow S \cup \{s\}$ /* Add a new source s */
5 $\quad\quad \text{color}(s) \leftarrow \text{red}$
6 $\quad\quad$ **foreach** $u \in N_{white}(s)$ **do**
7 $\quad\quad\quad\quad \text{counter}(u) \leftarrow \text{counter}(u) - 1$
8 $\quad\quad\quad\quad E' \leftarrow E' \cup \{(s,u)\}$
9 $\quad\quad\quad\quad$ **if** $counter(u) = 0$ **then**
10 $\quad\quad\quad\quad\quad\quad \text{color}(u) \leftarrow \text{black}$

11 $E' \leftarrow E' \cup \{(u,v) \in E : \text{color}(u) = \text{white}\}$
12 $A_S \leftarrow \beta\text{-additive f-EFT/VFT sourcewise spanner w.r.t. sources in } S$
13 **return** $H \leftarrow (V(G), E' \cup E(A_S))$

---

multiple *failed edges*, we say that a path π is *fault-free* if it does not contain any of such edges. Finally, if two paths π and π' are such that the last vertex of π coincides with the first vertex of π', we will denote by $\pi \circ \pi'$ the path obtained by concatenating π with π'.

2 Augmenting Sourcewise Fault-Tolerant Spanners

We next describe a general procedure (see Algorithm 1) to derive purely additive fault-tolerant spanners from sourcewise spanners of the same type.

The main idea of the algorithm is to select a small subset S of *source* vertices of G, which we call **red**. These vertices are used to build a fault-tolerant sourcewise spanner of the graph. The remaining vertices are either **black** or **white**. The former ones are always adjacent to a source, even in the event of f edge/vertex failures. Finally, edges incident to **white** vertices are added to the sought spanner, as their overall number is provably small.

During the execution of the algorithm, we let $\text{color}(u) \in \{\text{white}, \text{black}, \text{red}\}$ denote the current color of vertex u. We define $N_{\text{white}}(u)$ to be the set of neighbors of u which are colored **white**, and we let $\delta_{\text{white}}(u) = |N_{\text{white}}(u)|$. We will also assign a non-negative counter $\text{counter}(u)$ to each vertex. Initially all these counters will be positive, and then they will only be decremented. A vertex u is colored **black** only when $\text{counter}(u)$ reaches 0, and once a **white** vertex is colored either **black** or **red** it will never be recolored **white** again. Therefore, we have that $\text{color}(u) = \text{black}$ implies $\text{counter}(u) = 0$.

We first bound the size of the spanner H.

Lemma 1. *Algorithm 1 computes a spanner of size* $O\left(np + nf + \gamma\left(n, \left\lfloor \frac{(f+1)n}{p} \right\rfloor\right)\right)$, *where $\gamma(n, \ell)$ is the size of the spanner A_S for $|S| = \ell$.*

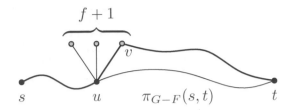

Fig. 1. A case of the proof of Theorem 2. Bold edges are in $H - F$. The black vertex u is adjacent to at least $f + 1$ vertices in S.

Proof. The edges added to E' by line 8 are at most $(f+1)n$, since each time an edge (s, u) is added to H the counter `counter(u)` is decremented, and at most $(f+1)n$ counter decrements can occur.

To bound the edges added to E' by line 11, observe that all the edges in $\{(u, v) \in E : \mathtt{color}(u) = \mathtt{white}\}$ which are incident to a **red** vertex, have already been added to H by line 8, hence we only consider vertices v which are either **white** or **black**. Let v be such a vertex and notice that, before line 11 is executed, we must have $\delta_{\mathtt{white}}(v) < p$, as otherwise v would have been selected as a source and colored **red**. This immediately implies that line 11 causes the addition of at most np edges to H.

It remains to bound the size of A_S. It is sufficient to show that $|S| \leq \left\lfloor \frac{(f+1)n}{p} \right\rfloor$ at the end of the algorithm. Each time a source s is selected, s has at least p white neighbors in G, hence the quantity $\sum_{u \in V(G)} \mathtt{counter}(u)$ decreases by at least p. The claim follows by noticing that $\sum_{u \in V(G)} \mathtt{counter}(u) = n(f+1)$ at the beginning of the algorithm and, when the algorithm terminates, it must be non-negative.   □

We next bound the distortion of H.

Theorem 2. *Algorithm 1 computes a $(\beta + 2)$-additive f-EFT/VFT spanner.*

Proof. Consider two vertices $s, t \in V(G)$ and a set F of at most f failed edges/vertices of G, we will show that $d_{H-F}(s, t) \leq d_{G-F}(s, t) + \beta + 2$. We assume, w.l.o.g., that s and t are connected in $G - F$, as otherwise the claim trivially holds.

If all the vertices in $\pi = \pi_{G-F}(s, t)$ are **white**, then all their incident edges have been added to H (see line 11 of Algorithm 1), hence $d_{H-F}(s, t) = d_{G-F}(s, t)$.

Otherwise, let $u \in V(\pi)$ be the closest vertex to s such that $\mathtt{color}(u) \neq \mathtt{white}$. Notice that, by the choice of u, $H - F$ contains all the edges of $\pi[s, u]$. If $\mathtt{color}(u) = \mathtt{red}$ then:

$$d_{H-F}(s, t) \leq d_{H-F}(s, u) + d_{H-F}(u, t) \leq d_{G-F}(s, u) + d_{A_S-F}(u, t)$$
$$\leq d_{G-F}(s, u) + d_{G-F}(u, t) + \beta = d_{G-F}(s, t) + \beta$$

where we used the fact that $u \in \pi_{G-F}(s, t)$.

Finally, if $\mathtt{color}(u) = \mathtt{black}$ then $\mathtt{counter}(u) = 0$, hence u has at least $f + 1$ red neighbors in H (see Figure 1). As a consequence, there is at least one red vertex v such that $(u, v) \in H - F$ (and hence $(u, v) \in G - F$), therefore:

$$
\begin{aligned}
d_{H-F}(s, t) &\leq d_{H-F}(s, u) + d_{H-F}(u, v) + d_{H-F}(v, t) \\
&\leq d_{G-F}(s, u) + 1 + d_{A_s - F}(v, t) \leq d_{G-F}(s, u) + 1 + d_{G-F}(v, t) + \beta \\
&\leq d_{G-F}(s, u) + 1 + d_{G-F}(v, u) + d_{G-F}(u, t) + \beta \\
&= d_{G-F}(s, t) + \beta + 2.
\end{aligned}
$$

$\square$

Let $S' \subset V(G)$ be a set of *sources*. In [18] it is shown that a sourcewise EFT/VFT *preserver* (i.e, a $(1, 0)$ EFT/VFT sourcewise spanner) of G having size $\gamma(n, |S'|) = O(n\sqrt{n|S'|})$ can be built in polynomial time. Combining this preserver and Algorithm 1 with $p = n^{\frac{2}{3}}$, we obtain the following:

Corollary 1. *There exists a polynomial time algorithm to compute a 2-additive EFT/VFT spanner of size $O(n^{\frac{5}{3}})$.*

Furthermore, we can exploit the following result in [7].[3]

Lemma 2 ([7]). *Given an (α, β)-spanner A and a subset S' of vertices, it is possible to compute in polynomial time a subset of $O(|S'| \cdot n)$ edges E', so that $A \cup E'$ is an (α, β) EFT sourcewise spanner w.r.t. S'. The same result holds for VFT spanners, with E' of size $O(|S'| \cdot n \log n)$.*

Combining the above result with the 2-additive spanner of size $O(n^{3/2})$ in [1], we obtain 2-additive EFT and VFT sourcewise spanners of size $\gamma(n, |S'|) = O(n\sqrt{n} + |S'| \cdot n)$ and $\gamma(n, |S'|) = O(n\sqrt{n} + |S'| \cdot n \log n)$, respectively. By using these spanners in Algorithm 1, with $p = \sqrt{n}$ and $p = \sqrt{n \log n}$ respectively, we obtain the following result.

Corollary 2. *There exists a polynomial time algorithm to compute a 4-additive EFT spanner of size $O(n^{3/2})$, and a 4-additive VFT spanner of size $O(n^{3/2}\sqrt{\log n})$.*

3   Augmenting Clustering-Based Additive Spanners

Most additive spanners in the literature are based on a *clustering* technique. A subset of the vertices of the graph G is partitioned into *clusters*, each containing a special *center* vertex along with some of its neighbors. The distances between these clusters is then reduced by adding a suitable set of edges to the spanner. This technique is used, for example, in [4,9]. We now describe a general technique which can be used to augment such spanners in order to obtain a fault-tolerant additive spanner.

---

[3] Actually, the result in [7] is claimed for the single source case only, but it immediately extends to multiple sources.

Algorithm 2: Algorithm for computing a fault-tolerant additive spanner from multiplicative and clustering-based additive spanners. Here $\{\mathcal{C}, \text{cnt}(\cdot)\}$ denotes the clustering of G while $\delta(C, C')$ is the set of the edges in $E(G)$ with one endpoint in C and the other in C'.

---

1   $E' \leftarrow \emptyset$; $M \leftarrow (\alpha, 0)$ EFT spanner; $A \leftarrow (1, \beta)$ clustering-based spanner
2   **foreach** $C \in \mathcal{C}$ **do**
3   **foreach** $v \in C$ **do**
4   **if** $\exists (v, x) \in E(G) : x \in C \setminus \{cnt(v)\}$ **then**
5         $E' \leftarrow E' \cup \{(v, x)\}$.
6   **foreach** $C, C' \in \mathcal{C} : C \neq C'$ **do**
7   **if** $\exists e, e' \in \delta(C, C') : e$ *and* e' *are vertex-disjoint* **then**
8       $E' \leftarrow E' \cup \{e, e'\}$.
9   **else if** $\exists e, e' \in \delta(C, C') : e \neq e'$ **then**
10      $E' \leftarrow E' \cup \{e, e'\}$.
11   **else**
12      $E' \leftarrow E' \cup \delta(C, C')$ /* $\delta(C, C')$ contains at most one edge */
13 **return** $H \leftarrow (V(G), E' \cup E(M) \cup E(A))$

More formally, a *clustering* of G is a partition $\mathcal{C}$ of a subset of $V(G)$. We call each element $C \in \mathcal{C}$ a *cluster*. We say that a vertex v is *clustered* if it belongs to a cluster, and *unclustered* otherwise. Each cluster $C \in \mathcal{C}$ is associated with a vertex $u \in C$ which is the *center* of C. For each *clustered* vertex v, we denote by $\text{cnt}(v)$ the center of the cluster containing v.

We say that a β-additive spanner A is *clustering-based* if there exists a clustering $\mathcal{C}$ of G such that: (i) A contains all the edges incident to unclustered vertices, (ii) A contains all the edges between every clustered vertex v and $\text{cnt}(v)$, and (iii) the following property holds:

Property 1. For every $u, v \in V(G)$ such that v is a clustered vertex, there exists a path $\tilde{\pi}(u, v)$ in A such that one of the following conditions holds:

(P1) $|\tilde{\pi}(u, v)| \leq d_G(u, v) + \beta - 2$;
(P2) $|\tilde{\pi}(u, v)| = d_G(u, v) + \beta - 1$ and either (i) $v = \text{cnt}(v)$, or (ii) the last edge of $\tilde{\pi}(u, v)$ is $(\text{cnt}(v), v)$.
(P3) $|\tilde{\pi}(u, v)| = d_G(u, v) + \beta$, $v \neq \text{cnt}(v)$, and the last edge of $\tilde{\pi}(u, v)$ is $(\text{cnt}(v), v)$.

Our algorithm works as follows (see Algorithm 2). We add to our spanner H a β-additive clustering-based spanner A, and a α-multiplicative EFT spanner M. Note that so far our construction is the same as in [8], with the extra constraint that A is clustering-based. We then augment H by adding a carefully chosen subset E' of inter and intra-cluster edges.

Let $\{\mathcal{C}, \text{cnt}(\cdot)\}$ be the clustering of A. It is easy to see that E' contains at most $O(n + |\mathcal{C}|^2)$ edges and hence $|E(H)| = O(|E(A)| + |E(M)| + |\mathcal{C}|^2)$. We now prove an useful lemma which is then used to upper-bound the distortion of the spanner H.

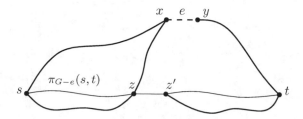

Fig. 2. Decomposition of $\pi_{G-e}(s,t)$ so that all shortest paths from s to z (resp. from z' to t) in A are fault-free. Bold lines denote shortest paths in A.

Lemma 3. *Let A be a spanning subgraph of G, let $e \in E(G)$ be a failed edge and $s, t \in V(G)$ be two vertices satisfying $d_A(s,t) < d_{A-e}(s,t) \neq \infty$. There exist two consecutive vertices z, z' in $V(\pi_{G-e}(s,t))$, with $d_{G-e}(s,z) < d_{G-e}(s,z')$, such that every shortest path in A between s and z (resp. t and z') is fault-free.*

Proof. First of all, notice that $e = (x,y)$ belongs to every shortest path between s and t in A, therefore let $\pi_A(s,t) = \langle s, \ldots, x, y, \ldots, t \rangle$. Consider the vertices of $\pi_{G-e}(s,t)$ from s to t, let $z \in V(\pi)$ be the last vertex such that there exists a shortest path π between z and t in A that contains e (z can possibly coincide with s), and call z' the vertex following z in π (see Figure 2). By the choice of z, we have that no shortest path between z' and t in A can contain e. Moreover, π must traverse e in the same direction as $\pi_A(s,t)$, i.e., $\pi = \langle z, \ldots, x, y, \ldots, t \rangle$. This is true since otherwise we would have $\pi = \langle z, \ldots, y, x, \ldots, t \rangle$ and hence $\pi_A(s,t)[s,x] \circ \pi[x,t]$ would be a fault-free shortest path between s and t in A, a contradiction.

It remains to show that no shortest path between s and z in A can contain e. Suppose this is not the case, then:

$$d_A(s,z) = d_A(s,y) + d_A(y,z) = d_A(s,x) + 1 + 1 + d_A(z,x) = d_A(s,z) + 2$$

which is again a contradiction.   □

We are now ready to prove the main theorem of this section.

Theorem 3. *Algorithm 2 computes a $(2\beta + \max\{2, \alpha - 3\})$-additive EFT-spanner.*

Proof. Choose any two vertices $s, t \in V(G)$ and a failed edge $e \in E(G)$. Suppose that s, t are connected in $G - e$, and that every shortest path between s and t in A contains e (as otherwise the claim trivially holds). We partition $\pi_{G-e}(s,t)$ by finding $z, z' \in V(\pi_{G-e}(s,t))$ as shown by Lemma 3.

The edge (z, z') is in $\pi_{G-e}(s,t)$ and hence cannot coincide with e. Moreover, we suppose $(z, z') \notin E(H)$, as otherwise we would immediately have:

$$d_{H-e}(s,t) \leq d_A(s,z) + d_{H-e}(z,z') + d_A(z',t)$$
$$\leq d_{G-e}(s,z) + \beta + 1 + d_{G-e}(z',t) + \beta + (d_{G-e}(z,z') - 1)$$
$$= d_{G-e}(s,t) + 2\beta - 1.$$

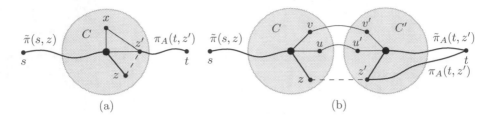

Fig. 3. Cases considered in the proof of Theorem 3 to build a fault-free path between s and t with small additive distortion. Bold lines represent shortest paths/edges in A. Solid lines represent paths/edges in H while the dashed edge (z, z') does not belong to $E(H)$ and cannot coincide with e.

This means that both z and z' must be clustered. Let C (resp. C') be the (unique) cluster that contains z (resp. z').[4]

Let $\gamma := d_A(s, z) - d_{G-e}(s, z)$ and $\gamma' := d_A(t, z') - d_{G-e}(z', t)$. Clearly $0 \leq \gamma, \gamma' \leq \beta$. If $\gamma + \gamma' \leq 2\beta - 2$ then we are done as:

$$
\begin{aligned}
d_{H-e}(s, t) &\leq d_A(s, z) + d_{B-e}(z, z') + d_A(z', t) \\
&\leq d_{G-e}(s, z) + d_{G-e}(z', t) + 2\beta - 2 + \alpha d_{G-e}(z, z') + (d_{G-e}(z, z') - 1) \\
&\leq d_{G-e}(s, t) + 2\beta + \alpha - 3.
\end{aligned}
$$

Next we assume that $\gamma + \gamma' \geq 2\beta - 1$. This means that either (i) γ and γ' are both equal to β, or (ii) exactly one of them is β while the other equals $\beta - 1$. Assume w.l.o.g. that $\gamma = \beta$. This implies that $d_A(s, z) = |\tilde{\pi}(s, z)|$, hence $\tilde{\pi}(s, z)$ is a shortest path between s and z in A, and by Lemma 3, it is fault-free.

In the rest of the proof we separately consider the cases $C = C'$ and $C \neq C'$. In the former case, since $(z, z') \notin E(H)$, we know that, during the execution of the loop in line 2 of Algorithm 2, an edge (z', x) such that $x \in C \setminus \{z, \text{cnt}(z)\}$ has been added to E (see Figure 3 (a)). Since the paths $\langle \text{cnt}(z), z' \rangle$ and $\langle \text{cnt}(z), x, z' \rangle$ are edge-disjoint, at least one of them is fault free, hence: $d_{H-e}(\text{cnt}(z), z') \leq 2 = d_{G-e}(z, z') + 1$. Thus, by (P3) of Property 1:

$$
\begin{aligned}
d_{H-e}(s, t) &\leq d_A(s, \text{cnt}(z)) + d_{H-e}(\text{cnt}(z), z') + d_A(z', t) \\
&\leq d_{G-e}(s, z) + \beta - 1 + d_{G-e}(z, z') + 1 + d_{G-e}(z', t) + \beta \\
&\leq d_{G-e}(s, t) + 2\beta.
\end{aligned}
$$

We now consider the remaining case, namely $C \neq C'$. We have that $(z, z') \notin E(H)$, therefore during the execution of the loop in line 6 of Algoritm 2, two distinct edges $(u, u'), (v, v')$ so that $u, v \in C$ and $u', v' \in C'$ must have been added to E (see Figure 3 (b)).

Notice that u' and v' might coincide, but this would imply that $u \neq v$ and hence $u' = v' = z'$. This, in turn, implies the existence of two edge-disjoint paths

---

[4] Notice that C and C' may coincide.

of length 2 between $\mathrm{cnt}(z)$ and z' in H, namely $\langle \mathrm{cnt}(z), u, z' \rangle$ and $\langle \mathrm{cnt}(z), v, z' \rangle$. As at least one of them must be fault-free. Therefore:

$$\begin{aligned} d_{H-e}(s,t) &\leq d_A(s, \mathrm{cnt}(z)) + d_{H-e}(\mathrm{cnt}(z), z') + d_A(z', t) \\ &\leq d_{G-e}(s,z) + \beta - 1 + d_{G-e}(z, z') + 1 + d_{G-e}(z', t) + \beta \\ &= d_{G-e}(s,t) + 2\beta. \end{aligned}$$

On the other hand, if $u' \neq v'$, we consider the two paths $\pi' = \langle \mathrm{cnt}(z), u, u',$ $\mathrm{cnt}(z') \rangle$ and $\pi'' = \langle \mathrm{cnt}(z), v, v', \mathrm{cnt}(z') \rangle$.[5] Notice that π' and π'' can share at most a single edge, namely $(\mathrm{cnt}(z), z)$ (when $u = v = z$), and that this edge cannot coincide with e as it belongs to $\tilde{\pi}(s, z)$ which is a fault-free shortest path between s and z in A. This implies that at least one of π' and π'' is fault-free and hence $d_{H-e}(\mathrm{cnt}(z), \mathrm{cnt}(z')) \leq 3 = d_{G-e}(z, z') + 2$. If $e = (\mathrm{cnt}(z'), z')$ then, since $|\tilde{\pi}(t, z')| \geq d_A(z', t) \geq d_{G-e}(z', t) + \beta - 1$, either (P2) or (P3) of Property 1 must hold, so we know that $\tilde{\pi}(t, z')[t, \mathrm{cnt}(z')]$ is fault-free and has a length of at most $d_{G-e}(z', t) + \beta - 1$. We have:

$$\begin{aligned} d_{H-e}(s,t) &\leq d_A(s, \mathrm{cnt}(z)) + d_{H-e}(\mathrm{cnt}(z), \mathrm{cnt}(z')) + d_A(\mathrm{cnt}(z'), t) \\ &\leq d_{G-e}(s,z) + \beta - 1 + d_{G-e}(z, z') + 2 + d_{G-e}(z', t) + \beta - 1 \\ &\leq d_{G-e}(s,t) + 2\beta. \end{aligned}$$

Finally, when $e \neq (\mathrm{cnt}(z'), z')$, we have:

$$\begin{aligned} d_{H-e}(s,t) &\leq d_A(s, \mathrm{cnt}(z)) + d_{H-e}(\mathrm{cnt}(z), \mathrm{cnt}(z')) + d_{H-e}(\mathrm{cnt}(z'), z') + d_A(z', t) \\ &\leq d_{G-e}(s,z) + \beta - 1 + d_{G-e}(z, z') + 2 + 1 + d_{G-e}(z', t) + \beta \\ &\leq d_G(s,t) + 2\beta + 2. \end{aligned}$$

This concludes the proof.   □

This result can immediately be applied to the 6-additive spanner of size $O(n^{\frac{4}{3}})$ in [4], which is clustering-based and uses $O(n^{\frac{2}{3}})$ clusters. Using the 5-multiplicative EFT spanner M of size $O(n^{4/3})$ from [10], we obtain:

Corollary 3. *There exists a polynomial time algorithm to compute a 14-additive EFT spanner of size $O(n^{\frac{4}{3}})$.*

We can similarly exploit the clustering-based spanner of [9] which provides, w.h.p., an additive stretch of 4 and a size of $\widetilde{O}(n^{\frac{7}{5}})$ by using $O(n^{\frac{3}{5}})$ clusters.

Corollary 4. *There exists a polynomial time randomized algorithm that computes w.h.p. a 10-additive EFT spanner of G of size $\widetilde{O}(n^{\frac{7}{5}})$.*

---

[5] Some consecutive vertices of π' (resp. π'') might actually coincide. In this case, we ignore all but the first of such vertices and define π' (resp. π'') accordingly.

References

1. Aingworth, D., Chekuri, C., Indyk, P., Motwani, R.: Fast estimation of diameter and shortest paths (without matrix multiplication). SIAM J. Comput. 28(4), 1167–1181 (1999)
2. Althöfer, I., Das, G., Dobkin, D.P., Joseph, D., Soares, J.: On sparse spanners of weighted graphs. Discrete & Computational Geometry 9, 81–100 (1993)
3. Ausiello, G., Franciosa, P.G., Italiano, G.F., Ribichini, A.: On resilient graph spanners. In: Bodlaender, H.L., Italiano, G.F. (eds.) ESA 2013. LNCS, vol. 8125, pp. 85–96. Springer, Heidelberg (2013)
4. Baswana, S., Kavitha, T., Mehlhorn, K., Pettie, S.: Additive spanners and (alpha, beta)-spanners. ACM Transactions on Algorithms 7(1), 5 (2010)
5. Baswana, S., Khanna, N.: Approximate shortest paths avoiding a failed vertex: Near optimal data structures for undirected unweighted graphs. Algorithmica 66(1), 18–50 (2013)
6. Bernstein, A., Karger, D.R.: A nearly optimal oracle for avoiding failed vertices and edges. In: STOC, pp. 101–110 (2009)
7. Bilò, D., Gualà, L., Leucci, S., Proietti, G.: Fault-tolerant approximate shortest-path trees. In: Schulz, A.S., Wagner, D. (eds.) ESA 2014. LNCS, vol. 8737, pp. 137–148. Springer, Heidelberg (2014)
8. Braunschvig, G., Chechik, S., Peleg, D.: Fault tolerant additive spanners. In: Golumbic, M.C., Stern, M., Levy, A., Morgenstern, G. (eds.) WG 2012. LNCS, vol. 7551, pp. 206–214. Springer, Heidelberg (2012)
9. Chechik, S.: New additive spanners. In: SODA, pp. 498–512 (2013)
10. Chechik, S., Langberg, M., Peleg, D., Roditty, L.: Fault-tolerant spanners for general graphs. In: STOC, pp. 435–444 (2009)
11. Chechik, S., Langberg, M., Peleg, D., Roditty, L.: f-sensitivity distance oracles and routing schemes. In: de Berg, M., Meyer, U. (eds.) ESA 2010, Part I. LNCS, vol. 6346, pp. 84–96. Springer, Heidelberg (2010)
12. Coppersmith, D., Elkin, M.: Sparse sourcewise and pairwise distance preservers. SIAM J. Discrete Math. 20(2), 463–501 (2006)
13. Cygan, M., Grandoni, F., Kavitha, T.: On pairwise spanners. In: STACS, pp. 209–220 (2013)
14. Dinitz, M., Krauthgamer, R.: Fault-tolerant spanners: better and simpler. In: PODC, pp. 169–178 (2011)
15. Erdős, P.: Extremal problems in graph theory. In: Theory of Graphs and its Applications, pp. 29–36 (1964)
16. Grandoni, F., Williams, V.V.: Improved distance sensitivity oracles via fast single-source replacement paths. In: FOCS, pp. 748–757 (2012)
17. Parter, M.: Vertex fault tolerant additive spanners. In: Kuhn, F. (ed.) DISC 2014. LNCS, vol. 8784, pp. 167–181. Springer, Heidelberg (2014)
18. Parter, M., Peleg, D.: Sparse fault-tolerant BFS trees. In: Bodlaender, H.L., Italiano, G.F. (eds.) ESA 2013. LNCS, vol. 8125, pp. 779–790. Springer, Heidelberg (2013)
19. Parter, M., Peleg, D.: Fault tolerant approximate BFS structures. In: SODA, pp. 1073–1092 (2014)

Subexponential Time Algorithms for Finding Small Tree and Path Decompositions

Hans L. Bodlaender[1,*] and Jesper Nederlof[2,**]

[1] Utrecht University and Technical University Eindhoven, The Netherlands
H.L.Bodlaender@uu.nl
[2] Technical University Eindhoven, The Netherlands
j.nederlof@tue.nl

Abstract. The Minimum Size Tree Decomposition (MSTD) and Minimum Size Path Decomposition (MSPD) problems ask for a given n-vertex graph G and integer k, what is the minimum number of bags of a tree decomposition (respectively, path decomposition) of width at most k. The problems are known to be NP-complete for each fixed $k \geq 4$. In this paper we present algorithms that solve both problems for fixed k in $2^{O(n/\log n)}$ time and show that they cannot be solved in $2^{o(n/\log n)}$ time, assuming the Exponential Time Hypothesis.

1 Introduction

In this paper, we consider two bicriteria problems concerning path and tree decompositions, namely, for an integer k, find for a given graph G a path or tree decomposition with the minimum number of bags. For both problems, we give exact algorithms that use $2^{O(n/\log n)}$ time and give a matching lower bound, assuming the Exponential Time Hypothesis. The results have a number of interesting features. To our knowledge, this is the first problem for which a matching upper and lower bound (assuming the ETH) with the running time $2^{\Theta(n/\log n)}$ is known. The algorithmic technique is to improve the analysis of a simple idea by van Bodlaender and van Rooij [3]: a branching algorithm with memorization would use $2^{O(n)}$ time, but combining this with the easy observation that isomorphic subgraphs 'behave' the same, by adding isomorphism tests on the right locations in the algorithm, the savings in time is achieved. Our lower bound proofs use a series of reductions; the intermediate problems in the reductions seem quite useful for showing hardness for other problems.

Bicriteria problems are in many cases more difficult than problems with one criterion that must be optimized. For the problems that we consider in this paper, this is not different: if we just ask for a tree or path decomposition with the minimum number of bags, then the problem is trivial as there always is a tree or path decomposition with one bag. Also, it is well known that the problem

---

* Partially supported by the Networks project, funded by the Dutch Ministry of Education, Culture and Science through NWO.
** Supported by NWO Veni project 639.021.438

© Springer-Verlag Berlin Heidelberg 2015
N. Bansal and I. Finocchi (Eds.): ESA 2015, LNCS 9294, pp. 179–190, 2015.
DOI: 10.1007/978-3-662-48350-3_16

to decide if the treewidth or pathwidth of a graph is bounded by a given number k is fixed parameter tractable. However, recent results show that if we ask to minimize the number of bags of the tree or path decomposition of width at most k, then the problem becomes para-NP-complete (i.e., NP-complete for some fixed k) [4,7].

The problem to find path decompositions with a bound on the width and a minimum number of bags was first studied by Dereniowski et al. [4]. Formulated as a decision problem, the problem MSPD_k is to determine, given a graph $G = (V, E)$ and integer s, whether G has a path decomposition of width at most k and with at most s bags. Dereniowski et al. [4] mention a number of applications of this problem and study the complexity of the problem for small values of k. They show that for $k \geq 4$, the problem is NP-complete, and for $k \geq 5$, the problem is NP-complete for connected graphs. They also give polynomial time algorithms for the MSPD_k problem for $k \leq 3$ and discuss a number of applications of the problem, including the Partner Units problem, problems in scheduling, and in graph searching.

Li et al. [7] introduced the MSTD_k problem: given a graph G and integer ℓ, does G have a tree decomposition of width at most k and with at most ℓ bags. They show the problem to be NP-complete for $k \geq 4$ and for $k \geq 5$ for connected graphs, with a proof similar to that of Dereniowski et al. [4] for the pathwidth case, and show that the problem can be solved in polynomial time when $k \leq 2$.

In this paper, we look at exact algorithms for the MSPD_k and MSTD_k problems. Interestingly, these problems (for fixed values of k) allow for subexponential time algorithms. The running time of our algorithm is of a form that is not frequently seen in the field: for each fixed k, we give algorithms for MSPD and MSTD that use $2^{O(n/\log n)}$ time. Moreover, we show that these results are tight in the sense that there are no $2^{o(n/\log n)}$ time algorithms for MSPD_k and MSTD_k for some large enough k, assuming the Exponential Time Hypothesis.

Our algorithmic technique is a variation and extension of the technique used by Bodlaender and van Rooij [3] for subexponential time algorithms for INTERVALIZING k-COLORED GRAPHS. That algorithm has the same running time as ours; we conjecture a matching lower bound (assuming the ETH) for INTERVALIZING 6-COLORED GRAPHS.

2   Preliminaries

Notation In this paper, we interpret vectors as strings and vice versa whenever convenient, and for clarity use boldface notation for both. When $\mathbf{a}, \mathbf{b} \in \Sigma^\ell$ are strings, we denote $\mathbf{a}||\mathbf{b}$ for the string obtained by concatenating $\mathbf{a}$ and $\mathbf{b}$. We let s^ℓ denote the string repeating symbol s ℓ times. Also, we denote $\mathbf{a} \preceq \mathbf{b}$ to denote that $a_i \leq b_i$ for every $1 \leq i \leq n$ and use $\mathbf{1}$ to denote the vector with each entry equal to 1 (the dimension of $\mathbf{1}$ will always be clear from the context). We also add vectors, referring to component-wise addition.

Tree and Path decompositions. Unless stated otherwise, the graphs we consider in this paper are simple and undirected. We let $n = |V|$ denote the number of vertices of the graph $G = (V, E)$. A *path decomposition* of a graph $G = (V, E)$ is a sequence of subsets of V: $(X_1, \ldots, X_s)$ such that

- $\bigcup_{1 \leq i \leq s} X_i = V$,
- For all edges $\{v, w\} \in E$: there is an i, $1 \leq i \leq s$, with $v, w \in X_i$,
- For all vertices $v \in V$: there are i_v, j_v, such that $i \in [i_v, j_v] \Leftrightarrow v \in X_i$.

The *width* of a path decomposition $(X_1, \ldots, X_s)$ is $\max_{1 \leq i \leq s} |X_i| - 1$; its *size* is s. The *pathwidth* of a graph G is the minimum width of a path decomposition of G. We will refer to X_s as the *last bag* of $(X_1, \ldots, X_s)$. A *tree decomposition* of a graph $G = (V, E)$ is a pair $(\{X_i \mid i \in I\}, T = (I, F))$ with $\{X_i \mid i \in I\}$ a family of subsets of V, and T a rooted tree, such that

- $\bigcup_{i \in I} X_i = V$.
- For all edges $\{v, w\} \in E$: there is an $i \in I$ with $v, w \in X_i$.
- For all vertices $v \in V$: the set $I_v = \{i \in I \mid v \in X_i\}$ induces a subtree of T (i.e., is connected.)

The *width* of a tree decomposition $(\{X_i \mid i \in I\}, T = (I, F))$ is $\max_{i \in I} |X_i| - 1$; its *size* is $|I|$. The *treewidth* of a graph G is the minimum width of a tree decomposition of G. In the definition above, we assume that T is rooted: this does not change the minimum width or size, but makes proofs slightly easier. Elements of I and numbers in $\{1, 2, \ldots, s\}$ and their corresponding sets X_i are called *bags*.

In the full version of this paper, we show the following (loose) estimate. The proof is not very complicated, but generalizes a result by Bodlaender and van Rooij [3].

Lemma 1. *The number of non-isomorphic graphs with n vertices of treewidth at most k is at most $2^{10k^2 n}$.*

A sketch of the proof is as follows. It is known that a graph of treewidth at most k has a 'nice tree decomposition'[1] of size at most $4n$ [6]. For each bag in the nice tree decomposition, there are at most $2^{k^2} + 2^k + k + 1$ 'non-isomorphic possibilities', the number of unlabeled trees on n vertices is at most $O(3^n)$ [9], and a nice tree decomposition can be described by an unlabeled tree and along with one of the 'non-isomorphic possibilities' for each bag.

Usually, nice tree and path decompositions have more than the minimum number of bags. The notions are still very useful for our analysis; in particular, they help to count the number of non-isomorphic graphs of treewidth or pathwidth at most k.

3 Path and Tree Decompositions with Few Bags

3.1 Finding Path Decompositions with Memorization and Isomorphism Tests

In this section, we describe our algorithm for the MSPD problem. Throughout the section, we assume that k is a fixed positive integer and that G has treewidth

---

[1] Some normalized treedecomposition in which each bag behaves in a restricted way.

at most k (note that we can determine this in linear time for fixed k (cf. [2])
and return NO if the treewidth is higher than k). Our branching algorithm is
parameterized by 'a good pair', formalized as follows:

Definition 1. *A* good pair *is a pair of vertex sets* (X, W), *such that*

- $|X| \leq k + 1$,
- $X \cap W = \emptyset$, *and*
- *for all* $v \in W$, $N(v) \subseteq W \cup X$. *Equivalently,* W *is the union of the vertex
  sets of zero or more connected components of* $G[V \setminus X]$.

For a good pair (X, W), let $\mathsf{mstd}_k(X, W)$ $(\mathsf{mspd}_k(X, W))$ *be the minimum* s
such that there is a tree (path) decomposition of $G[X \cup W]$ *of width at most* k,
where X *is the root bag (last bag) of the tree (path) decomposition.*

A recursive formulation for path decompositions. The following lemma gives a
recursive formulation for mspd_k. The formulation is the starting point for our
algorithm, but we will in addition exploit graph isomorphisms, see below.

Lemma 2. *If* $|X| \leq k + 1$, *then* $\mathsf{mspd}_k(X, \emptyset) = 1$. *Otherwise, let* (X, W) *be a
good pair, and* $W \neq \emptyset$. *Then*

$$\mathsf{mspd}_k(X, W) = \min_{\substack{Y \subseteq X \cup W \\ X \neq Y \\ W \cap N(X \setminus Y) = \emptyset}} 1 + \mathsf{mspd}_k(Y, W \setminus Y). \tag{1}$$

Proof. The first part with $|X| \leq k+1$ is trivial: take the only path decomposition
with one bag X.

Otherwise, suppose Y fulfills $W \cap N(X \setminus Y) = \emptyset$. Let $PD = (X_1, \ldots, X_s)$ be a
path decomposition of width at most k of $G[Y \cup W]$ with $X_s = Y$. Now we verify
that $(X_1, \ldots, X_s, X)$ is a path decomposition of width at most k of $G[X \cup W]$.
Since there can be no edges between $X \setminus Y$ and W, all the edges incident to
$X \setminus Y$ are covered in the bag X and all other edges are covered in PD since it
is a path decomposition of $G[Y \cup W]$. Also, a vertex $v \in X \setminus Y$ cannot occur in
PD so the bags containing a particular vertex will still induce a connected part
in the path decomposition.

Conversely, suppose $(X_1, \ldots, X_s, X)$ is a minimal size path decomposition
of width at most k of $G[X \cup W]$. Note that $X = X_s$ contradicts this path
decomposition being of minimal size, so we may assume $X \neq X_s$. Vertices in
$X \setminus X_s$ do not belong to $\bigcup_{1 \leq i \leq s-1} X_i$, by the definition of path decomposition,
so we must have that $W \cap N(X \setminus X_s) = \emptyset$ since otherwise not all edges incident
to X are covered in the path decomposition. Hence, X_s fulfills all the conditions
of the minimization in the recurrence.

We have that, $(X_1, \ldots, X_s)$ is a path decomposition of $G[X_s \cup (W \setminus X_s)]$,
which has at least $\mathsf{mspd}_k(X_{s-1}, W \setminus X_{s-1})$ bags and hence taking $Y = X_s$ shows
that $\mathsf{mspd}_k(X, W) \leq s + 1$. $\qquad\square$

Isomorphism. The following notion will be needed for presenting the used recurrence for tree decompositions and essential for quickly evaluating (1). Intuitively, it indicates $G[X \cup W]$ being isomorphic to $G[Y \cup Z]$ with an isomorphism that maps X to Y. More formally,

Definition 2. *Good pairs* (X, W) *and* (X, Z) *are isomorphic if there is a bijection* $f : X \cup W \leftrightarrow X \cup Z$, *such that*

1. *For all* $v, w \in X \cup W$: $\{v, w\} \in E \Leftrightarrow \{f(v), f(w)\} \in E$, *and*
2. $f(v) = v$ *for all* $v \in X$.

We will use the following obvious fact:

Observation 1. Suppose good pair (X, W) is isomorphic to good pair (X, Z). Then $\mathsf{mspd}_k(X, W) = \mathsf{mspd}_k(X, Z)$ and $\mathsf{mstd}_k(X, W) = \mathsf{mstd}_k(X, Z)$

In our algorithm we use a result by Loksthanov et al. [8] which gives an algorithm that for fixed k maps each graph G of treewidth at most k to a string $\mathsf{can}(G)$ (called its *canonical form*), such that two graphs G and H are isomorphic if and only if $\mathsf{can}(G) = \mathsf{can}(H)$. The result also holds for graphs where vertices (or edges) are labeled with labels from a finite set and the isomorphism should map vertices to vertices of the same label. We can use this result to make canonical forms for good pairs:

Observation 2. An isomorphism class of the good pairs (X, W) can be described by the triple $\mathsf{can}(X, W) := (X, \mathsf{can}(G[X \cup W], f)$ where f is a bijection from X to $|X|$.

Here, $f : X \leftrightarrow X$ can be (for example) be defined as the restriction of π onto X of the lexicographically smallest (with respect to some arbitrary ordering) isomorphism π of $G[X \cup W]$.

A recursive algorithm with memorization. We now give a recursive algorithm PD_k to compute for a given good pair (X, W) the value $\mathsf{mspd}_k(X, W)$. The algorithm uses memorization. In a dynamic map datastructure D, we store values that we have computed. We can use e.g., a balanced tree or a hash table for D, that is initially assumed empty. Confer Algorithm 1.

Algorithm $\mathsf{PD}_k(X, W)$
1: **if** $|X| \leq k + 1$ and $W = \emptyset$ **then return** 1
2: **if** $D(\mathsf{can}(X, W))$ is stored **then return** $D(\mathsf{can}(X, W))$
3: $m \leftarrow \infty$.
4: **for all** $Y \subseteq X \cup W$ such that $Y \neq X$ and $N(X \setminus Y) \subseteq X$ **do**
5:     $m \leftarrow \min\{m, 1 + \mathsf{PD}_k(Y, W \setminus Y)\}$.
6: Store $D(\mathsf{can}(X, W)) \leftarrow m$.
7: **return** m.

Algorithm 1. Finding a small path decompositions of width at most k.

The correctness of this method follows directly from Lemma 2 and Observation 1. The main difference with a traditional evaluation (with memorization) of the recursive formulation of mspd is that we store and lookup values under their canonical form under isomorphism — this simple change is essential for obtaining a subexponential running time. The fact that we work with graphs of bounded treewidth and for these, GRAPH ISOMORPHISM is polynomial [1,8] makes that we can perform this step sufficiently fast.

Equipped with the $\mathsf{PD_k}$ algorithm, we solve the MSPD problem as follows: for all $X \subseteq V$ with $|X| \leq k + 1$, run $\mathsf{PD_k}(X, V \setminus X)$; report the smallest value over all choices of X.

3.2   The Number of Good Pairs

We now will analyze the number of good pairs. This is the main ingredient of the analysis of the running time of the algorithm given above.

Theorem 1. *Let k be a constant. Let G be a graph with n vertices and treewidth at most k. Then G has $2^{O(n/\log n)}$ non-isomorphic good pairs.*

Proof. Let us define a *basic good pair* as a good pair (X, W) where $G[W]$ is connected. The isomorphism classes (with respect to the notion of isomorphism from Definition 2) of good pairs can be described as follows: let X be a set of size at most k. Let $\mathcal{C}_1, \ldots, \mathcal{C}_\ell$ be a partition of the connected components of $G[V \setminus X]$ into basic good pair isomorphism classes, e.g.: we have for two connected components C_a, C_b that $C_a, C_b \in \mathcal{C}_i$ for some i if and only if there exists a bijection $X \cup C_a \leftrightarrow X \cup C_a$ such that for all $v \in X$ we have $f(v) = v$ and for all $v, w \in X \cup C_a : \{v, w\} \in E \Leftrightarrow \{f(v), f(w)\} \in E$. We order the isomorphism classes arbitrarily (e.g., in some lexicographical order).

Then an isomorphism class of *all* good pairs can be described by a triple $(X, \mathbf{s} = \{c_1, \ldots, c_s\}, f)$ where c_i is the number of connected components of $G[V \setminus X]$ in basic pair isomorphism class $\mathcal{C}_i$. Then we have the following bound:

Claim. For a constant k, the number of isomorphism classes of basic good pairs (X, W) with $|W| + |X| \leq \frac{1}{22k^2} \log n$ is at most $\sqrt{n}$.

Proof. By Lemma 1, the number of graph isomorphism classes of $G[X \cup W]$ is at most $2^{\frac{10k^2}{22k^2} \log n}$ since we assumed the treewidth to be at most k (as stated in the beginning of this section).

The isomorphism class of a basic good pair is described by the set X, the permutation of X and the graph isomorphism class of $G[W \cup X]$, thus we have that the number of basic good pair isomorphism classes is at most $k! 2^{\frac{10k^2}{22k^2} \log n} \leq 2^{k^2 + \frac{10k^2}{22k^2} \log n}$ which is $\sqrt{n}$ for large enough n. $\square$

Say an isomorphism class $\mathcal{C}_i$ is *small* if $(X, W) \in \mathcal{C}_i$ implies $|X| + |W| \leq \frac{1}{22k^2} \log n$, and it is *large* otherwise. Assume $\mathcal{C}_1, \ldots, \mathcal{C}_z$ are small. By the above Claim, z is at most $\sqrt{n}$. Thus, since we know $c_i \leq n$, the number of possibilities of $\mathbf{s}$ on the small isomorphism classes is at most $n^{O(\sqrt{n})}$. For the remaining $\ell - z$ isomorphism classes of large connected components, we have that $\sum_{j=z+1}^{\ell} c_i \le 22k^2 n / \log n = O(n / \log n)$. Thus, there are only $2^{O(n / \log n)}$ subsets of the large connected components that can be in W. Combining both bounds gives the upper bound of $2^{O(n / \log n)}$ for the number of non-isomorphic good pairs, as desired. $\square$

3.3   Analysis of the Algorithm

In this section, we analyze the running time of Algorithm 1. First, we note that we have $O(n^{k+1})$ calls of the form $\mathsf{PD}_k(X, V \setminus X)$. Observe that each call to PD_k is with a good pair as parameters, and these good pairs on Line 4 can be enumerated with linear delay. Thus, by Theorem 1, there are $2^{O(n / \log n)}$ calls to PD_k that make recursive calls to PD_k. Within each single call, we have $O(n^{k+1})$ choices for a set Y; computing $s = \mathsf{can}(X, W)$ can be done in $O(n^5)$ time (confer [8]), and thus, the overhead of a single recursive call is bounded by $O(n^{\max\{k+1, 5\}})$. Putting all ingredients together shows that the algorithm uses $2^{O(n / \log n)}$ time.

Theorem 2. *For fixed k, the MSPD problem can be solved in $2^{O(n / \log n)}$ time.*

3.4   Extension to Finding Tree Decompositions

Now we discuss how to extend the algorithm for solving the MSTD problem. Note that, like usual, dealing with tree decompositions instead of path decompositions amounts to dealing with join bags. We have the following analogue of Lemma 2.

Lemma 3. *If $|X| \le k + 1$, then $\mathsf{mstd}_k(X, \emptyset) = 1$. Otherwise, let (X, W) be a good pair, and $W \ne \emptyset$. Then $\mathsf{mstd}_k(X, W) = \min\{\mathsf{extend}, \mathsf{branch}\}$ where*

$$
\mathsf{extend} = \min_{\substack{Y \subset X \cup W \\ X \ne Y \\ W \cap N(X \setminus Y) = \emptyset}} 1 + \mathsf{mstd}_k(Y, W \setminus Y).
$$

$$
\mathsf{branch} = \min_{\substack{W_1 \subseteq W \\ N(W_1) \subseteq W_1 \cup X}} \mathsf{mstd}(X, W_1) + \mathsf{mstd}(X, W \setminus W_1) - 1.
$$

(2)

Proof. The cases extend and branch refer to whether the root bag r with vertex set X has exactly one child, or at least two children. If r has one child, then the same arguments that show (1) can be used to show correctness of the extend case. If r has two or more children, then we can guess the set of vertices $W_1 \subseteq W$ that appear in bags in the subtree rooted by the first child of r. We must have that W_1 is a union of connected components of $G[W]$ by the definition of tree decompositions. Thus, the tree decomposition can be obtained by taking a tree decomposition of $G[X \cup W_1]$ and a tree decomposition of $G[X \cup (W \setminus W_1)]$, both with X as the vertex set of the root bag, and then taking the union, identifying the two root bags. The number of bags thus equals the minimum number of bags for the first tree decomposition (which equals $\mathsf{mstd}(X, W_1)$), plus the minimum number of bags for the second (equally $\mathsf{mstd}(X, W \setminus W_1)$), subtracting one as we counted the bag with vertex set X twice. $\square$

Given Algorithm 1 and (2), the algorithm for computing mstd suggests itself since it is easy to see that again we only need to evaluate $\mathsf{mstd}_k(X, W)$ for good pairs (X, W). This is indeed our approach but there is one small complication, since we cannot compute branch in a naive way because the number of connected components of $G[W]$ could be $\Omega(n)$. We deal with this by even further restricting the set of subsets of W we iterate over, again based on Observation 1.

Algorithm $\mathsf{TD_k}(X,W)$

1: **if** $|X| \leq k + 1$ and $W = \emptyset$ **then return** 1
2: **if** $D(\mathsf{can}(X, W))$ is stored **then return** $D(\mathsf{can}(X, W))$
3: $m \leftarrow \infty$
4: **for all** $Y \subseteq X \cup W$ such that $Y \neq X$ **and** $N(X \setminus Y) \subseteq X$ **do**
5:    $m \leftarrow \min\{m, 1 + \mathsf{TD_k}(Y, W \setminus Y)\}$
6: Let $\mathcal{C}_1, \ldots, \mathcal{C}_\ell$ be the isomorphism classes of the basic good pairs (X, W'), where W' is a connected component of W
7: For $1 \leq i \leq \ell$, let c_i be the number of $(X, W') \in \mathcal{C}_i$ where W' is a connected component of W
8: For $1 \leq i \leq \ell$ and $0 \leq j \leq c_i$, let W_j^i be the union of the j lexicographically first connected components W' such that $(X, W') \in \mathcal{C}_i$
9: **for all** vectors $\mathbf{y} \preceq (c_1, \ldots, c_\ell)$ **do**
10:    $W_1 \leftarrow \bigcup_{i=1}^{\ell} W_{y_i}^i$
11:    $W_2 \leftarrow W \setminus W_1$
12:    $m \leftarrow \min\{m, \mathsf{TD_k}(X, W_1) + \mathsf{TD_k}(X, W_2) - 1\}$
13: Store $D(\mathsf{can}(X, W)) \leftarrow m$
14: **return** m.

Algorithm 2. Extension of Algorithm 1 to find small tree decompositions of width at most k.

We solve the mstd problem in Algorithm 2. Let us first discuss the correctness of this algorithm. Note that similarly as in Algorithm 1, it implements the memorization with the datastructure D. It is easy to see that after Line 5, m equals the quantity extend from (2). By Lemma 3, it remains to show that at Line 13, m equals $\min\{\mathsf{extend}, \mathsf{branch}\}$.

To see this, note that by construction we iterate over a subset of 2^W generating all isomorphism classes that (X, W_1) subject to $N(W_1) \subseteq W_1 \cup X$ can generate, and by Observation 1 this is sufficient to find any optimal partition of W into W_1, W_2.

4   Lower Bound

This section is devoted to the proof of the following theorem:

Theorem 3. *Suppose the Exponential Time Hypothesis holds, then there is no algorithm for MSPD or MSTD for fixed $k \geq 39$ using $2^{o(n/\log n)}$ time.*

We will use a reduction from the following problem:

STRING 3-GROUPS
   Given: Sets $A, B, C \subseteq \{0, 1\}^{O(\log n)}$, with $|A| = |B| = |C| = n$
   Question: Choose n elements from $A \times B \times C$, such that each element
      in A, B, and C appears exactly once in a triple, and if $(\mathbf{a}, \mathbf{b}, \mathbf{c})$ is a
      chosen triple, then $\mathbf{a} + \mathbf{b} + \mathbf{c} \preceq \mathbf{1}$.

This is sufficient to show hardness of the MSPD and MSTD problems by virtue of the following theorem:

Theorem 4. *Suppose the Exponential Time Hypothesis holds. Then there is no algorithm for* STRING 3-GROUPS *using $2^{o(n)}$ time.*

A proof of Theorem 4 appears in the full version of this paper. The result is obtained by a series of reductions, starting a variation of a result on PARTITION INTO TRIANGLES for 4-regular graphs by van Rooij et al [11]. An interesting intermediate result is that 3-DIMENSIONAL MATCHING where each element appears in at most three triples cannot be solved in $2^{o(n)}$ time, unless the Exponential Time Hypothesis does not hold.

Vector gadgets. We will use the following notions extensively:

Definition 3. *The* fingerprint *of a path decomposition $(X_1, \ldots, X_r)$ is the vector $(|X_1|, \ldots, |X_r|)$. A path decomposition is* minimal *if (i) for all path decompositions $(X'_1, \ldots, X'_{r'})$ of G we have $r' > r$ or if $r' = r$, then $(|X_1|, \ldots, |X_r|) \preceq (|X'_1|, \ldots, |X'_r|)$. Graph G k-implements $\mathbf{w} \in \mathbb{N}^{\ell}_{>0}$ if (i) every tree decomposition of G of size r and width $k + 1$ is a path decomposition, (ii) all minimal path decompositions of size r have fingerprint $\mathbf{w}$.*

A *palindrome* is a vector $\mathbf{w} \in \mathbb{N}^r$ such that $(w_1, \ldots, w_r) = (w_r, \ldots, w_1)$. The most important part of our reduction is the gadget summarized by the following lemma:

Lemma 4. *For every integer $k \geq 3$ and palindrome $\mathbf{w} \in \mathbb{N}^r_{>0}$ such that $\lceil 2k/3 \rceil < w_i \leq k$ for all $i \leq r$, we can in polynomial time construct a graph G that k-implements $\mathbf{w}$.*

Proof. Construct G as follows:

- Construct disjoint cliques $C_0, \ldots, C_r$ all of size $\lfloor k/3 \rfloor$ and for $i = 1, \ldots, r$ make all vertices from C_{i-1} and C_i adjacent,
- Construct disjoint cliques $C^p_1, \ldots, C^p_r$ where $|C^p_i| = w_i - 2\lfloor k/3 \rfloor$ for all $i = 1, \ldots, r$ and for all i, make all vertices of C^p_i adjacent with all vertices of C_{i-1} and C_i.

For $i = 1, \ldots, r$, let us denote $M_i = C_{i-1} \cup C_i \cup C_i^p$ for the maximal cliques of G. Since any clique must be contained in a bag of any tree decomposition we have that for every $i = 1, \ldots, r$ some bag must contain a M_i. Since all bags must be of width at most k, the maximal cliques of G are of size w_i for some i and the maximal cliques intersect in only $\lfloor k/3 \rfloor$ vertices, one bag cannot contain two maximal cliques. Hence in a path decomposition of width at most k and size r each bag contains exactly one maximal clique. Let $(\{X_i\}, T)$ be a tree decomposition of width at most k and size at most r, and suppose that X_i is the bag containing M_i. Note that in T, bags X_i and X_{i+1} must be adjacent since they are the only bags that can contain all of C_i. Therefore, we know that T must be a path $X_1, \ldots, X_r$ or the path $X_r, \ldots, X_1$. Also notice that using such a T and setting $X_i = M_i$ gives us two valid tree decompositions that are path decompositions and both have $\mathbf{w}$ as fingerprint since $\mathbf{w}$ is a palindrome. Also, these are the only minimal ones since X_i must contain M_i.   □

Construction. Let A, B, C be an instance of STRING 3-GROUPS. Note that without loss of generality, we way assume that all elements of A, B, C are palindromes: if we change all strings $\mathbf{x} \in A \cup B \cup C$ to $\mathbf{x} || \overleftarrow{\mathbf{x}}$, where $\overleftarrow{\mathbf{x}}$ denotes the reverse of $\mathbf{x}$, we obtain a clearly equivalent instance where all strings are palindromes. Also, by padding zero's we may assume that for the length ℓ of all vectors, we have $\ell = 12 \lceil \log n \rceil + 2$.

Let us now construct a graph G such that G has no tree decomposition with maximum bag size $k = 53$ and size $s = n(\ell + 1)$ if (A, B, C) is a no-instance of STRING 3-GROUPS, and G has a path decomposition with maximum bag size k and size s otherwise.

Let us denote $A = \{\mathbf{a^1}, \ldots, \mathbf{a^n}\}, B = \{\mathbf{b^1}, \ldots, \mathbf{b^n}\}, C = \{\mathbf{c^1}, \ldots, \mathbf{c^n}\}$ for the binary strings in A, B, C. Set $k = 53$, $\ell = (n-1) + 6n \log n$, and construct G as follows

1. Add one graph $G(A)$ 40-implementing $\mathbf{a^1} + \mathbf{27} || 40 || \mathbf{a^2} + \mathbf{27} || 40 || \ldots || 40 || \mathbf{a^n} + \mathbf{27} || 40$,
2. For every $\mathbf{b^i} \in B$, add a graph $G(\mathbf{b^i})$ that 13-implements $\mathbf{b^i} + \mathbf{9}$,
3. For every $\mathbf{c^i} \in C$, add a graph $G(\mathbf{c^i})$ that 4-implements $\mathbf{c^i} + \mathbf{3}$.

Applying Lemma 4 we see that all graphs $G(A)$, $G(\mathbf{b})$ and $G(\mathbf{c})$ exist and can be found in polynomial time since respectively $27 > \frac{2}{3}40, 9 > \frac{2}{3}13, 3 > \frac{2}{3}4$.

Figure 1 gives a schematic intuitive illustration of the construction, and its correctness.

Suppose that the instance of STRING 3-GROUPS is a yes-instance and without loss of generality assume that $\mathbf{a^i} + \mathbf{b^i} + \mathbf{c^i} \preceq \mathbf{1}$ for all $1 \le i \le n$. Let $(A_1, \ldots, A_s)$ be a minimal path decomposition of $G(A)$, for $i = 1, \ldots, n$ let $(B_1^i, \ldots, B_\ell^i)$ be a minimal path decompositions of $G(\mathbf{b_i})$ and $(C_1^i, \ldots, C_\ell^i)$ be a minimal path decompositions of $G(\mathbf{c_i})$. Then it is easy to see that

$$(A_1 \cup B_1^1 \cup C_1^1, \ldots, A_\ell \cup B_\ell^1 \cup C_\ell^1, A_{\ell+1}, A_{\ell+2} \cup B_1^2 \cup C_1^2, \ldots, A_{s-1} \cup B_\ell^n \cup C_\ell^n, A_s),$$

is a valid path decomposition of G of size s. Moreover, all bags have size at most 40: for j being a positive multiple of $(\ell + 1)$ we have $|A_j| = 40$ and otherwise if

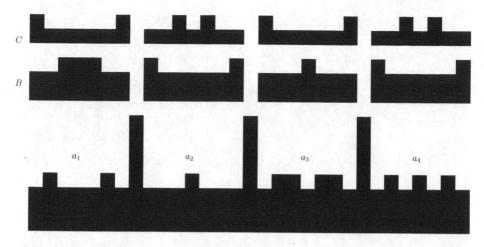

Fig. 1. Schematic illustration for the proof of Theorem 3. The larger object represents all elements in A; the smaller objects each represent one element from B or C. In each 'gap' between two towers, we must fit an element from B and an element from C; $\mathbf{b^{i_1}}$ and $\mathbf{c^{i_2}}$ fit in the gap with $\mathbf{a^{i_3}}$, iff $\mathbf{a^{i_3}} + \mathbf{b^{i_1}} + \mathbf{c^{i_2}} \preceq \mathbf{1}$ — the one by one protruding blocks each represent one vertex, and we can fit at most one such vertex in the respective bag.

$j = g(\ell+1)+i$ for $1 \leq i \leq \ell+1$ then the size of the j'th bag equals $39 + a_i^g + b_i^g + c_i^g$ which is at most 40 by the assumption $\mathbf{a^i} + \mathbf{b^i} + \mathbf{c^i} \preceq \mathbf{1}$ for all $1 \leq i \leq n$.

Suppose that G has a tree decomposition $\mathbb{T}$ of width at most k and size s. Restricted to the vertices of $G(A)$ we see that by the construction of $G(A)$, $\mathbb{T}$ has to be a path decomposition $P_1, \ldots, P_s$ where $A_i \subseteq P_i$ for all i or $A_i \subseteq P_{s-i}$. These cases are effectively the same, so let us assume the first case holds. We have that there are n sets of ℓ consecutive bags that are of size 12 or 13, separated with bags of size 40.

Then, for t being a positive multiple of $\ell+1$ we have that $P_t \cap B_i^j = \emptyset$ for any i, j, and therefore for each j, the bags of $\mathbb{T}$ containing elements of $G(\mathbf{b^j})$ must be a consecutive interval of length at most ℓ. Moreover, since all bags of $\mathbb{T}$ contain at least 27 vertices from $G(A)$, we see that the partial path decomposition induced by $G(\mathbf{b})$ is of size at most ℓ and width at most 13 and hence by construction it must have fingerprint $\mathbf{b}$. Since we have n intervals of consecutive bags in $G(A)$ and n graphs $G(\mathbf{b})$ and no two graphs can be put into the same interval we see that we can reorder $B = \{\mathbf{b^1}, \ldots, \mathbf{b^n}\}$ such that if $(B_1^j, \ldots, B_\ell^j)$ is a minimal path decomposition of $G(\mathbf{b^j})$ then either $B_i^j \subseteq P_{j(\ell+1)+i}$ for each i or $B_i^j \subseteq P_{j(\ell+1)+1-i}$ for each i. Note that in both cases the fingerprint of $\mathbb{T}$ induced by the vertices from $G(A)$ and $G(\mathbf{b})$ for each b is the same.

Focusing on the vertices from $G(c)$, we have that since all bags of $\mathbb{T}$ contain at least 36 vertices of $G(A)$ and $G(\mathbf{b})$ for some b that the path decomposition of $G(c)$ must be of width at most 4, and by construction thus of length at least ℓ. By similar arguments as in the preceding paragraph, we see we may assume that

$C = \{\mathbf{c^1}, \ldots, \mathbf{c^n}\}$ such that either $C_i^j \subseteq P_{j(\ell+1)+i}$ for each i or $C_i^j \subseteq P_{j(\ell+1)+1-i}$ for each i.

By the definitions of $G(A)$, $G(b)$, $G(c)$ and the assumption that $\mathbb{T}$ has width at most 53 we then see that $\mathbf{a^i} + \mathbf{b^i} + \mathbf{c^i} \preceq 1$ for every $1 \leq i \leq n$, as desired.

For the efficiency of the reduction: notice that the graph G has at most $40s = 40((\ell+1)n) \leq 40((\lceil 12 \log n \rceil + 3)n) = O(n \log n)$ vertices. Hence, an $2^{o(n/\log n)}$ algorithm solving mspd or mstd implies by the reduction a $2^{o((n \log n)/(\log n - \log \log n))} = 2^{o(n)}$ algorithm for STRING 3-GROUPS, which violates the ETH by Theorem 4.

Acknowledgements. We thank the anonymous referees for their detailed and helpful comments.

References

1. Bodlaender, H.L.: Polynomial algorithms for graph isomorphism and chromatic index on partial k-trees. Journal of Algorithms 11, 631–643 (1990)
2. Bodlaender, H.L.: A linear time algorithm for finding tree-decompositions of small treewidth. SIAM Journal on Computing 25, 1305–1317 (1996)
3. Bodlaender, H.L., van Rooij, J.M.M.: Exact algorithms for intervalizing colored graphs. In: Marchetti-Spaccamela, A., Segal, M. (eds.) TAPAS 2011. LNCS, vol. 6595, pp. 45–56. Springer, Heidelberg (2011)
4. Dereniowski, D., Kubiak, W., Zwols, Y.: Minimum length path decompositions. ArXiv e-prints 1302.2788 (2013)
5. Impagliazzo, R., Paturi, R., Zane, F.: Which problems have strongly exponential complexity? Journal of Computer and System Sciences 63, 512–530 (2001)
6. Kloks, T.: Treewidth. LNCS, vol. 842. Springer, Heidelberg (1994)
7. Li, B., Moataz, F.Z., Nisse, N.: Minimum size tree-decompositions. In: 9th International Colloquium on Graph Theory and Combinatorics, ICGT, number hal-01023904, Grenoble, France (2013)
8. Lokshtanov, D., Pilipczuk, M., Pilipczuk, M., Saurabh, S.: Fixed-parameter tractable canonization and isomorphism test for graphs of bounded treewidth. In: Proceedings of the 55th Annual Symposium on Foundations of Computer Science, FOCS 2014, pp. 186–195 (2014)
9. Otter, R.: The number of trees. Annals of Mathematics 49(3), 583–599 (1948)
10. Schaefer, T.J.: The complexity of satisfiability problems. In: Proceedings of the 10th Annual Symposium on Theory of Computing, STOC 1978, pp. 216–226 (1978)
11. van Rooij, J.M.M., van Kooten Niekerk, M.E., Bodlaender, H.L.: Partition into triangles on bounded degree graphs. Theory Comput. Syst. 52(4), 687–718 (2013)

Enumeration of 2-Level Polytopes

Adam Bohn[1], Yuri Faenza[2], Samuel Fiorini[1], Vissarion Fisikopoulos[1],
Marco Macchia[1], and Kanstantsin Pashkovich[3]

[1] Université libre de Bruxelles, Brussels, Belgium
{adam.bohn,sfiorini,vfisikop,mmacchia}@ulb.ac.be
[2] Ecole Polytechnique Fédérale de Lausanne (EPFL), Lausanne, Switzerland
yuri.faenza@epfl.ch
[3] C & O Department, University of Waterloo, Waterloo, Canada
kanstantsin.pashkovich@gmail.com

Abstract. We propose the first algorithm for enumerating all combinatorial types of 2-level polytopes of a given dimension d, and provide complete experimental results for $d \leqslant 6$. Our approach is based on the notion of a simplicial core, that allows us to reduce the problem to the enumeration of the closed sets of a discrete closure operator, along with some convex hull computations and isomorphism tests.

Keywords: Polyhedral computation, Optimization, Formal concept analysis

1 Introduction

A (convex) polytope $P \subseteq \mathbb{R}^d$ is said to be 2-*level* if for every facet-defining hyperplane H, there exists another hyperplane H' parallel to H which contains all the vertices of P that are not contained in H.

There are a number of alternative ways to define 2-level polytopes. For example, a polytope P is said to be *compressed* if every pulling triangulation of P is unimodular with respect to the lattice generated by its vertices [18,11,5]. In [20] this property is shown to be equivalent to 2-levelness. Given a finite set $V \subseteq \mathbb{R}^d$ and a positive integer k, the k-th *theta body* of V is a tractable convex relaxation of the convex hull of V. The *theta rank* of V is defined as the smallest k such that this relaxation is exact. These notions were introduced in [8] in a more general context in which V can be the set of real solutions of any finite system of real polynomials. The authors of [8] show that a finite set has theta rank 1 if and only if it is the vertex set of a 2-level polytope.

Families of 2-level polytopes appear in a number of different combinatorial contexts: Birkhoff polytopes, Hanner polytopes [12], stable set polytopes of perfect graphs [4], Hansen polytopes [13], order polytopes [19] and spanning tree polytopes of series-parallel graphs [10] all have the 2-level property. Because they appear in such a wide variety of contexts, 2-level polytopes are interesting objects. However, our understanding of them remains relatively poor. In this paper we study the problem of enumerating all combinatorial types of 2-level polytopes of a fixed dimension.

© Springer-Verlag Berlin Heidelberg 2015
N. Bansal and I. Finocchi (Eds.): ESA 2015, LNCS 9294, pp. 191–202, 2015.
DOI: 10.1007/978-3-662-48350-3_17

Since every 2-level polytope is affinely equivalent to a 0/1-polytope, one might think to compute all those of a given dimension simply by enumerating all 0/1-polytopes of that dimension and discarding those which are not 2-level polytopes. However, the complete enumeration of d-dimensional 0/1-polytopes has been implemented only for $d \leqslant 5$ [1]. The same author has enumerated all those 6-dimensional 0/1-polytopes having up to 12 vertices, but the complete enumeration even for this low dimension is not expected to be feasible: the output of the combinatorial types alone is so huge that it is not currently possible to store it or search it efficiently [22]. Thus for all but the lowest dimensions, there is no hope of working with a pre-existing list of 0/1-polytopes, and it is desirable to find an efficient algorithm which computes 2-level polytopes from scratch.

1.1   Contribution and Outline

We present the first algorithm to enumerate all combinatorial types of 2-level polytopes of a given dimension d. The algorithm uses new structural results on 2-level polytopes which we develop here.

Our starting point is a pair of full-dimensional embeddings of a given 2-level d-polytope that are related to each other via some $d \times d$ unimodular, lower-triangular 0/1-matrix. This is explained in Section 3. In one embedding, which we refer to as the $\mathcal{H}$-embedding, the facets have 0/1-coefficients. In the other – the $\mathcal{V}$-embedding – the vertices have 0/1-coordinates. The $\mathcal{H}$- and $\mathcal{V}$-embeddings are determined by a structure, which we call a simplicial core (see Section 3.2)

Our algorithm is described in detail in Section 4. It computes a complete list L_d of non-isomorphic 2-level d-polytopes, from a similar list L_{d-1} of 2-level $(d-1)$-polytopes. In these lists, each polytope is stored via its slack matrix (see Section 3.1).

For some polytope $P_0 \in L_{d-1}$, define $L(P_0)$ to be the collection of all 2-level polytopes that have P_0 as a facet. Then the union of these collections $L(P_0)$ over all polytopes $P_0 \subset L_{d-1}$ is our desired set L_d, because every facet of a 2-level polytope is 2-level. We proceed as follows: given some $P_0 \in L_{d-1}$, we realize it in the hyperplane $\{x \in \mathbb{R}^d \mid x_1 = 0\} \simeq \mathbb{R}^{d-1}$. We compute a collection $\mathcal{A} \subseteq \{x \in \mathbb{R}^d \mid x_1 = 1\}$ of point sets, such that for each 2-level polytope $P \in L(P_0)$, there exists $A \in \mathcal{A}$ with $P \simeq \text{conv}(P_0 \cup \{e_1\} \cup A)$. For each $A \in \mathcal{A}$, we compute $P = \text{conv}(P_0 \cup \{e_1\} \cup A)$ and, in case it is 2-level and not isomorphic to any of the polytopes already generated by the algorithm, we add P to the list L_d. The efficiency of this approach depends greatly on how the collection $\mathcal{A}$ is chosen. Here, we exploit the pair of embeddings to define a proxy for the notion of 2-level polytopes in terms of closed sets with respect to a certain discrete closure operator, and use this proxy to construct a suitable collection $\mathcal{A}$. This turns out to provide a significant speedup in the computations.

We implemented this algorithm and ran it to obtain L_d for $d \leqslant 6$. The outcome of our experiments is discussed in Section 5. We found that the number of combinatorial types of 2-level d-polytopes is surprisingly small for low dimensions d. Moreover, low-dimensional 2-level polytopes can be used to understand the structure of higher-dimensional 2-level polytopes. For instance, they show which polytopes can appear as low-dimensional faces of higher-dimensional 2-level polytopes.

We conclude the paper by discussing one conjecture inspired by our experiments, and some ideas for future work (see Section 6).

1.2 Previous Work

The problem closest to the one which we study here is that of enumerating 0/1-polytopes, see [1,22]. In our approach, we use techniques from formal concept analysis, in particular we use a previously existing algorithm to enumerate all concepts of a relation, see [6,15]. Some general properties of 2-level polytopes are established e.g., in [20] and [8].

2 Preliminaries

We list here a number of definitions and properties used throughout the paper. For basic notions on polytopes that do not appear here, we refer the reader to [21]. Given a positive integer d, we set $[d] := \{1, \ldots, d\}$. A d-polytope is a polytope of dimension d. For $x \in \mathbb{R}^d$ and $E \subseteq [d]$, we let $x(E) := \sum_{i \in E} x_i$.

While in general two polytopes can be combinatorially isomorphic without being affinely isomorphic, for 2-level polytopes these two notions coincide. This is not difficult to see but requires some definitions, so we defer it to Section 3.1 (see Lemma 1). We then say that two 2-level polytopes are *isomorphic* if and only if they are combinatorially isomorphic. A condition stronger than isomorphism is congruency: two polytopes are *congruent* if there is an isometry mapping one to the other.

The *f-vector* of a d-polytope P is the d-dimensional vector whose i-th entry is the number of $(i-1)$-dimensional faces of P. Thus $f_0(P)$ gives the number of vertices of P, and $f_{d-1}(P)$ the number of facets of P. We use vert(P) to denote the vertex set of polytope P.

3 Embeddings

3.1 Slack Matrices and Slack Embeddings

The *slack matrix* of a polytope $P \subseteq \mathbb{R}^d$ with m facets $F_1, \ldots, F_m$ and n vertices $v_1, \ldots, v_n$ is the $m \times n$ nonnegative matrix $S = S(P)$ such that S_{ij} is the *slack* of the vertex v_j with respect to the facet F_i, that is, $S_{ij} = g_i(v_j)$ where $g_i : \mathbb{R}^d \to \mathbb{R}$ is any affine form such that $g_i(x) \geqslant 0$ is valid for P and $F_i = \{x \in P \mid g_i(x) = 0\}$. The slack matrix of a polytope is defined up to scaling its rows by positive reals.

The slack matrix provides a canonical way to embed any polytope, which we call the *slack embedding*. This embedding maps every vertex v_j to the corresponding column $S^j \in \mathbb{R}^m_+$ of the slack matrix $S = S(P)$. Every polytope is affinely isomorphic to the convex hull of the columns of its slack matrix.

Due to definition a polytope P is 2-level if and only if $S(P)$ can be scaled to be 0/1. Given a 2-level polytope, we henceforth always assume that its facet-defining inequalities are scaled so that the slacks are 0/1. Thus, the slack embedding of a 2-level polytope depends only on the *support* of its slack matrix, which only depends on its combinatorial structure. The next lemma follows from this observation.

Lemma 1. *Two 2-level polytopes are affinely isomorphic if and only if they have the same combinatorial type.*

3.2   Simplicial Cores

A *simplicial core* for a d-polytope P is a $(2d+2)$-tuple $(F_1, \ldots, F_{d+1}; v_1, \ldots, v_{d+1})$ of facets and vertices of P such that each facet F_i does not contain vertex v_i but contains vertices $v_{i+1}, \ldots, v_{d+1}$.

Every d-polytope P admits a simplicial core and this fact can be proved by a simple induction on the dimension, see, e.g., [9, Proposition 3.2]. Actually, simplicial cores for P correspond to $(d+1) \times (d+1)$ submatrices of $S(P)$ that are invertible and lower-triangular, for some ordering of rows and columns.

Notice that, for each i, the affine hull of F_i contains v_j for $j > i$, but does not contain v_i; thus the vertices of a simplicial core are affinely independent. That is, $v_1, \ldots, v_{d+1}$ form the vertices of a d-simplex contained in P.

3.3 $\mathcal{H}$- and $\mathcal{V}$-Embeddings

Although canonical, the slack embedding is never full-dimensional, which can be a disadvantage. To remedy this, we use simplicial cores to define two types of embeddings that are full-dimensional. Let P be a 2-level d-polytope with m facets and n vertices, and let $\Gamma := (F_1, \ldots, F_{d+1}; v_1, \ldots, v_{d+1})$ be a simplicial core for P.

From now on, we assume that the rows and columns of the slack matrix $S(P)$ are ordered compatibly with the simplicial core, so that the i-th row of $S(P)$ corresponds to facet F_i for $1 \leqslant i \leqslant d+1$ and the j-th column of $S(P)$ corresponds to vertex v_j for $1 \leqslant j \leqslant d+1$.

The $\mathcal{H}$-*embedding* with respect to Γ is defined by mapping each v_j to the unit vector e_j of $\mathbb{R}^d$ for $1 \leqslant j \leqslant d$, and v_{d+1} to the origin. In the $\mathcal{H}$-embedding of P, facet F_i for $1 \leqslant i \leqslant m$ is defined by the inequality $\sum_{j \in [d], S_{ij}=1} x_j \geqslant 0$ if $v_{d+1} \in F_i$ and by $\sum_{j \in [d], S_{ij}=0} x_j \leqslant 1$ if $v_{d+1} \notin F_i$.

In the $\mathcal{V}$-*embedding* of P with respect to Γ, vertex v_j is the point of $\mathbb{R}^d$ whose i-th coordinate is S_{ij}, for $1 \leqslant j \leqslant n$ and $1 \leqslant i \leqslant d$. Equivalently, the $\mathcal{V}$-embedding can be defined via the transformation $x \mapsto Mx$, where $M = M(\Gamma)$ is the top left $d \times d$ submatrix of $S(P)$ and $x \in \mathbb{R}^d$ is a point in the $\mathcal{H}$-embedding. We stick to this convention for the rest of the paper. The next lemma summarizes the discussion.

Lemma 2. *Let P be a 2-level d-polytope and let $(F_1, \ldots, F_{d+1}; v_1, \ldots, v_{d+1})$ be a simplicial core for P. In the corresponding $\mathcal{H}$-embedding, all the facets of P are of the form $x(E) \leqslant 1$ or $x(E) \geqslant 0$ for some nonempty $E \subseteq [d]$. Moreover, in the corresponding $\mathcal{V}$-embedding i-th coordinate is the slack with respect to facet F_i. In particular, in the $\mathcal{V}$-embedding, all the vertices of P have 0/1-coordinates.*

We call the submatrix $M := M(\Gamma)$ of $S(P)$ the *embedding matrix* of Γ. Note that every embedding matrix M is unimodular. Indeed, M is an invertible, lower-triangular, 0/1-matrix. Thus $\det(M) = 1$. The next lemma is the key to our approach.

Lemma 3. *In the $\mathcal{H}$-embedding P of a 2-level d-polytope with respect to any simplicial core Γ, the vertex set of P equals $P \cap M^{-1} \cdot \{0, 1\}^d \subseteq \mathbb{Z}^d$, where $M = M(\Gamma)$ is the embedding matrix of Γ.*

For a hypergraph $H = (V, \mathcal{E})$ with $V = [d]$, let $P(H) := \{x \in \mathbb{R}^d \mid 0 \leqslant x(E) \leqslant 1 \text{ for each } E \in \mathcal{E}\}$. We refer to a pair of inequalities $0 \leqslant x(E) \leqslant 1$ as a pair of *hyperedge constraints* where E is a *hyperedge*. It follows from Lemma 2 that any $\mathcal{H}$-embedding of a 2-level d-polytope is of the form $P(H)$ for some hypergraph H such that $P(H)$ is integral. Conversely, each $P(H)$ that is integral is a 2-level polytope.

4   Algorithm

4.1   Closed Sets

An operator $\mathrm{cl} : 2^{\mathcal{X}} \to 2^{\mathcal{X}}$ over a ground set $\mathcal{X}$ is a *closure operator* if it is idempotent, $\mathrm{cl}(\mathrm{cl}(A)) = \mathrm{cl}(A)$; extensive, $A \subseteq \mathrm{cl}(A)$; and monotone, $A \subseteq B \implies \mathrm{cl}(A) \subseteq \mathrm{cl}(B)$. A set $A \subseteq \mathcal{X}$ is said to be *closed* with respect to cl if $\mathrm{cl}(A) = A$. In [6], Ganter and Reuter provided a polynomial delay algorithm for enumerating all the closed sets of a given closure operator.

Below, the ground set $\mathcal{X}$ will be a finite subset of points in $\mathbb{R}^d$. Let $\mathcal{F} \subseteq \mathbb{R}^d$ be another finite set of points that is disjoint from $\mathcal{X}$. For $A \subseteq \mathcal{X}$, define $\mathcal{E}_{\mathcal{F}}(A)$ to be the set of all hyperedges whose pair of hyperedge constraints is verified by $A \cup \mathcal{F}$:

$$\mathcal{E}_{\mathcal{F}}(A) := \{E \subseteq [d] \mid 0 \leqslant x(E) \leqslant 1 \text{ for every } x \in A \cup \mathcal{F}\}.$$

Our first closure operator is parametrized by $(\mathcal{X}, \mathcal{F})$ and is defined as:

$$\mathrm{cl}_{(\mathcal{X},\mathcal{F})}(A) := \{x \in \mathcal{X} \mid 0 \leqslant x(E) \leqslant 1 \text{ for every } E \in \mathcal{E}_{\mathcal{F}}(A)\}$$

for $A \subseteq \mathcal{X}$. In other words, $\mathrm{cl}_{(\mathcal{X},\mathcal{F})}(A)$ is the subset of $\mathcal{X}$ verifying all hyperedge inequalities that are satisfied by $A \cup \mathcal{F}$.

To obtain a 2-level d-polytope P, we fix one of its possible facets, i.e. we choose a 2-level $(d-1)$-polytope P_0 and an embedding matrix M_{d-1} of P_0. Afterwards, we extend M_{d-1} to an embedding matrix M_d of P so that P_0 is embedded in $\{x \in \mathbb{R}^d \mid x_1 = 0\} \simeq \mathbb{R}^{d-1}$ via the corresponding $\mathcal{H}$-embedding.

Then the algorithm enumerates all 2-level d-polytopes P such that M_d is an embedding matrix and P_0 is the facet defined by $x_1 \geqslant 0$ in the $\mathcal{H}$-embedding of P.

A first insight to achieve this goal is that $A := \mathrm{vert}(P) \cap \mathcal{X}$ is closed with respect to $\mathrm{cl}_{(\mathcal{X},\mathcal{F})}$, where $\mathcal{X} := \left(M_d^{-1} \cdot (\{1\} \times \{0,1\}^{d-1}) \right) \setminus \{e_1\}$ and $\mathcal{F} := \mathrm{vert}(P_0) \cup \{e_1\}$. Hence, to enumerate the possible 2-level d-polytopes P with a prescribed facet P_0 and embedding matrix M_d, it suffices to enumerate the closed sets $A \subseteq \mathcal{X}$ with respect to $\mathrm{cl}_{(\mathcal{X},\mathcal{F})}$.

A second insight is that the closure operator $\mathrm{cl}_{(\mathcal{X},\mathcal{F})}$ can be improved by recalling that each facet of $P_0 \subseteq \{x \in \mathbb{R}^d \mid x_1 = 0\}$ extends uniquely to a facet of P distinct from P_0. Since each facet of P should satisfy the 2-level property, certain choices of pairs of points of $\mathcal{X}$ are forbidden. To model this, we introduce an *incompatibility graph* $G = G(P_0, M_d)$ on $\mathcal{X}$. We declare two points $u, v \in \mathcal{X}$ *incompatible* whenever there exists a facet F_0 of P_0 such that u, v and e_1 lie on three different translates of $\mathrm{aff}(F_0)$. The nodes $u, v \in \mathcal{X}$ of G are connected by an edge if and only if they are incompatible.

Next, we define the closure operator cl_G on $\mathcal{X}$ such that, for every $A \subseteq \mathcal{X}$, $\mathrm{cl}_G(A) := A$ if A is a stable set in G and $\mathrm{cl}_G(A) := \mathcal{X}$ otherwise. It can be easily checked that the composed operator $\mathrm{cl}_G \circ \mathrm{cl}_{(\mathcal{X},\mathcal{F})}$ is a closure operator over $\mathcal{X}$. This is the closure operator that we use in our enumeration algorithm.

4.2   The Enumeration Algorithm

We now provide a detailed description of our algorithm. We start with the list L_{d-1} of combinatorial types of 2-level $(d-1)$-polytopes. Each combinatorial type is stored as a slack matrix together with a simplicial core. As before, we may assume that the simplicial core is formed by the facets and vertices indexing the first $(d-1)+1$ rows and columns of the slack matrix, respectively. The algorithm below then generates the list L_d of all combinatorial types of 2-level d-polytopes, each with a simplicial core.

Theorem 1. *Algorithm 1 outputs the list of all combinatorial types of 2-level d-polytopes, each with a simplicial core.*

Proof. Consider a 2-level d-polytope P. In the rest of the proof, we consider P only as a combinatorial structure. Later on, P will be embedded in $\mathbb{R}^d$ via a $\mathcal{H}$-embedding. To simplify notation, we use the same letters for both the abstract polytope P and its realization in $\mathbb{R}^d$. We use also this convention for facets of P. We prove that a $\mathcal{H}$-embedding of P is obtained at some point by the algorithm and is added to the list L_d.

Let P_0 be any facet of P. Thus P_0 is a 2-level $(d-1)$-polytope, and hence P_0 is stored in L_{d-1} together with a simplicial core $\Gamma_0 := (F_2', \ldots, F_{d+1}'; v_2, \ldots, v_{d+1})$. Extend Γ_0 to a simplicial core $\Gamma = (F_1, \ldots, F_{d+1}; v_1, \ldots, v_{d+1})$ for P by defining v_1 to be a vertex of F_2 not contained in F_1, and defining F_1 to be P_0 and F_i for $2 \leqslant i \leqslant d+1$ to be a unique facet of P such that $F_i' = F_i \cap P_0$. Observe that the embedding matrix $M_d := M(\Gamma)$ is of the form (1) for some $b = (b_1, \ldots, b_{d-2}) \in \{0,1\}^{d-2}$ and for $M_{d-1} := M(\Gamma_0)$.

Algorithm 1: Enumeration algorithm

---

1 Set $L_d := \varnothing$;
2 **foreach** $P_0 \in L_{d-1}$ *with simplicial core* $\Gamma_0 := (F'_2, \ldots, F'_{d+1}; v_2, \ldots, v_{d+1})$ **do**
3   Construct the $\mathcal{H}$-embedding of P_0 in $\{0\} \times \mathbb{R}^{d-1} \simeq \mathbb{R}^{d-1}$ w.r.t. Γ_0;
4   Let $M_{d-1} := M(\Gamma_0)$;
5   **foreach** *bit vector* $b \in \{0,1\}^{d-2}$ **do**
6   Complete M_{d-1} to a $d \times d$ matrix in the following way:

$$M_d := \begin{pmatrix} 1 & 0 \cdots 0 \\ 0 & \\ b_1 & \\ \vdots & M_{d-1} \\ b_{d-2} & \end{pmatrix} \tag{1}$$

7   Let $\mathcal{F} := \mathrm{vert}(P_0) \cup \{e_1\}$ and $\mathcal{X} := M_d^{-1} \cdot (\{1\} \times \{0,1\}^{d-1}) \smallsetminus \{e_1\}$;
8   Let G be the incompatibility graph on $\mathcal{X}$ w.r.t. P_0 and M_d;
9   Using the Ganter-Reuter algorithm [6], compute the list $\mathcal{A}$ of closed sets of the closure operator $\mathrm{cl}_G \circ \mathrm{cl}_{(\mathcal{X},\mathcal{F})}$;
10   **foreach** $A \in \mathcal{A}$ **do**
11   Let $P := \mathrm{conv}(A \cup \mathcal{F})$;
12   **if** P *is 2-level and not isomorphic to any polytope in* L_d **then**
13   Let $F_1 := P_0$ and $v_1 := e_1$;
14   **for** $i = 2, \ldots, d+1$ **do**
15   Let F_i be the facet of P distinct from F_1 s.t. $F_i \supseteq F'_i$;
16   **end**
17   Add P to L_d with $\Gamma := (F_1, \ldots, F_{d+1}; v_1, \ldots, v_{d+1})$;
18   **end**
19   **end**
20   **end**
21 **end**

Now, consider the $\mathcal{H}$-embedding of P defined by Γ. The vertices $v_2, \ldots, v_{d+1}$ are mapped to $e_2, \ldots, e_d$ and the origin, and v_1 is mapped to e_1. In this realization of P, the facet P_0 is embedded in $\{x \in \mathbb{R}^d \mid x_1 = 0\}$. In fact, P_0 is the facet of P defined by $x_1 \geqslant 0$.

As in the algorithm, take $\mathcal{F} := \mathrm{vert}(P_0) \cup \{e_1\}$ and $\mathcal{X} := M_d^{-1} \cdot (\{1\} \times \{0,1\}^{d-1}) \smallsetminus \{e_1\}$. Let $A := \mathrm{vert}(P) \smallsetminus (\mathrm{vert}(P_0) \cup \{e_1\})$. We claim that A is closed for $\mathrm{cl}_{(\mathcal{X},\mathcal{F})}$.

By Lemma 3, $A = \mathrm{vert}(P) \cap \mathcal{X}$, thus $A \subseteq \mathcal{X}$. By Lemma 2, P can be described by the linear system $\{x \in \mathbb{R}^d \mid 0 \leqslant x(E) \leqslant 1 \text{ for every } E \in \mathcal{E}_\mathcal{F}(A)\}$. Hence $\mathrm{vert}(P) = \mathcal{F} \cup \{x \in \mathcal{X} \mid 0 \leqslant x(E) \leqslant 1 \text{ for every } E \in \mathcal{E}_\mathcal{F}(A)\}$. Since $A = \mathrm{vert}(P) \cap \mathcal{X}$ and $\mathcal{X} \cap \mathcal{F} = \varnothing$, we see that $A = \mathrm{cl}_{(\mathcal{X},\mathcal{F})}(A)$. This proves the claim.

Finally, consider the incompatibility graph $G = G(P_0, M_d)$. If A were not a stable set of G then, among the facets of P adjacent to P_0, there would exist a facet that violates the 2-level property. Thus $\mathrm{cl}_G(\mathrm{cl}_{(\mathcal{X},\mathcal{F})}(A)) = A$, i.e. A is closed also with respect to $\text{cl}_G \circ \text{cl}_{(\mathcal{X}, \mathcal{F})}$. It follows that the combinatorial type of P is added at some point by the algorithm to the list L_d.

Clearly, L_d contains at most one member for each combinatorial type of 2-level d-polytope, because a 2-level polytope is added to L_d only if it is not isomorphic to any other polytope in the list.   □

4.3   Implementation

We implement the algorithm presented in Section 4.2 in `Perl`. We use `polymake` [7] for the geometric computations, such as congruence and isomorphism tests, convex hull and f-vector computations, and general linear algebra operations.

Isomorphism testing is in general a harder problem than congruence testing for polytopes given by sets of vertices, as it involves a convex hull computation. For this reason, before computing the convex hull in Step 11, we first ascertain whether or not there is an existing congruent polytope in L_d by testing the corresponding sets of vertices. For congruence tests, `polymake` uses the reduction of the congruence problem for arbitrary point sets to the graph isomorphism problem [2]. For isomorphism tests, the problem is reduced to graph isomorphism of the vertex-facet incidence graphs. For the 2-level test in Step 12 we check if every facet inequality of P computed by a convex hull algorithm in Step 11 attains two values when evaluated on vertices of P.

As part of our code, we implement the Ganter-Reuter algorithm [6]. The sets are represented by bit vectors and all the operations we need—such as order test between two sets, and closed set computations—are implemented by bit operations. For these we use the `Perl` library `Bit::Vector` [3].

The choice of the closed sets enumeration algorithm is not crucial for our problem: experiments indicate that more than 99% of the enumeration time is spent in geometric computation (i.e. convex hull and isomorphism tests) and the rest is spent computing the next closed set from the current one.

Since convex hull computation is crucial for our enumeration algorithm, we perform experiments on the performance of 4 state-of-the-art convex hull implementations: `beneath_beyond` (`bb`), which implements the incremental beneath and beyond algorithm; `lrs`, which implements the reverse search algorithm; and `cdd`, `ppl`, which implement the double description method. In $d = 6$ without redundancy removal the fastest implementation is `bb`, `cdd`, `lrs`, `ppl` for 224, 23, 3, 879 polytopes respectively and with redundancy removal for 28, 45, 376,

Table 1. Numbers of non-isomorphic 2-level polytopes, equivalence classes (isomorphic and congruent) and closed sets computed by the algorithm.

d	closed sets	2-level	isomorphic	congruent	closed sets/2-level
4	277	19	203	42	0.95
5	10963	106	7669	621	0.77
6	1908641	1150	414714	42076	0.24

Table 2. Numbers of combinatorially equivalent 0/1 polytopes, 2-level polytopes and sub-classes; 2L: 2-level polytopes, Δ-f: with one simplicial facet, STAB: stable sets of perfect graphs, polar: 2-level polytopes whose polar is 2-level, CS: centrally symmetric, Birk: Birkhoff polytope faces from [16], '-': exact numbers are unknown.

d	2L	Δ-f	STAB	polar	CS	Birk	0/1
3	5	4	4	4	2	4	8
4	19	12	11	12	4	11	192
5	106	41	33	42	13	33	1,048,576
6	1150	248	148	276	45	129	-
7	-	-	-	-	238	661	-

700 polytopes respectively. Interestingly, `lrs` and `bb` exchange roles in these two cases. We conclude that `ppl` is the most efficient implementation in most of the cases and thus the one we choose for our implementation. Note that since we know that the input points are always in convex position we can avoid redundancy removal thus earn a $5\times$ speed-up in dimension 6.

5   Experimental Results

5.1   Outcome of the Experiments

In dimension 4, the set of 2-level polytopes is computed by our algorithm in 20 seconds, while for $d = 5$ it takes 12 minutes to enumerate 106 2-level polytopes on an `Intel(R) Core(TM) i7-4700HQ CPU @ 2.40GHz`. For $d = 6$ we exploited one property of our algorithm: its straightforward parallelization. We created one job for each branch of commands in the two outer for loops of the algorithm and submitted these jobs to a cluster[1]. In particular, we created one job for each 2-level 5-polytope and each b vector, i.e. 1696 jobs. The total computation lasted 1 day (the sequential time is estimated in 4.5 days).

We illustrate the attained speed-up gained by using $\mathrm{cl}_G \circ \mathrm{cl}_{(\mathcal{X},\mathcal{F})}$ instead of $\mathrm{cl}_{(\mathcal{X},\mathcal{F})}$. In $d = 6$ the use of the first leads to $\sim 1.9 \cdot 10^6$ closed sets while the later to $\sim 10^8$ closed sets.

Not surprisingly, as the dimension increases, more computation time is consumed in testing polytopes that are not 2-level as depicted in Table 1.

To understand the current limits of computation note that in $d = 7$ we have to create 36800 jobs, while experiments show there are jobs that need more than 5 days to terminate.

Table 2 summarizes our results regarding the number of 2-level polytopes and interesting subclasses. Our main result is the number of isomorphism classes of 2-level polytopes in $d = 6$. Additionally, we make use of properties of 2-level centrally symmetric polytopes to enumerate all of them in $d = 7$.

---

[1] Hydra balanced cluster: `https://cc.ulb.ac.be/hpc/hydra.php`

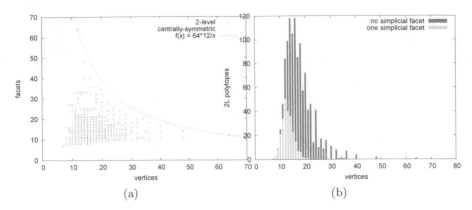

Fig. 1. (a) The relation between the number of facets and the number of vertices of 2-level 6-polytopes; (b) the number of 2-level 6-polytopes and the class with the ones with a simplicial facet as a function of the number of vertices.

The computed polytopes in `polymake` format as well as more information on the experiments and data are available online[2]. Taking advantage of the computed data we perform a number of statistical tests to understand the structure and properties of 2-level polytopes.

We experimentally study the number of 2-level polytopes as a function of the number of vertices in dimension 6 (see Fig. 1(b)). Interestingly, most of the polytopes, namely 1048 (i.e. more than 90%) have 10 to 24 vertices. The number of polytopes with a simplicial facet is maximum when the number of vertices is 12 and the extreme cases are the simplex (7 vertices) and the hypersimplex $\Delta_6(2)$ (21 vertices) [21].

The relation between the number of vertices and the number of facets in $d = 6$ is depicted in Fig. 1(a). Experiments show that the bound $f_0(P)f_{d-1}(P) \leqslant d2^{d+1}$ holds for all 2-level d-polytopes up to $d = 6$ and for the centrally symmetric 2-level polytopes in $d = 7$. Note that $f_0(P)f_{d-1}(P) = d2^{d+1}$ when P is the cube or its polar.

Our experiments show that all 2-level centrally symmetric polytopes, up to dimension 7, validate Kalai's 3^d conjecture [14] (note that for general centrally symmetric polytopes, Kalai's conjecture is known to be true only up to dimension 4 [17]). Dimension 5 is the lowest dimension in which we found centrally symmetric polytopes that are not Hanner nor Hansen (e.g. one with f-vector $(12, 60, 120, 90, 20)$). In dimension 6 we found a 2-level centrally symmetric polytope with f-vector $(20, 120, 290, 310, 144, 24)$, for which therefore $f_0 + f_4 = 44$; this offers a stronger counterexample to the conjecture B of [14] than the one presented in [17] having $f_0 + f_4 = 48$.

---

[2] `http://homepages.ulb.ac.be/~vfisikop/data/2-level.html`

Note that the stored polytopes are in a slightly different format than described in the algorithm, i.e. we store an $\mathcal{H}$-embedding without the slack matrix and the simplicial core.

6   Discussion

We think that the experimental evidence we gathered will lead to interesting research questions. As a sample, we propose the following question motivated by Fig. 1(a): is it true that $f_0(P)f_{d-1}(P) \leqslant d2^{d+1}$ for all 2-level d-polytopes P? And if yes, is equality attained only by the cube and cross-polytope? It is known that $f_0(P) \leqslant 2^d$ with equality if and only if P is a cube and $f_{d-1}(P) \leqslant 2^d$ with equality if and only if P is a cross-polytope [8]. In these cases, $f_0(P)f_{d-1}(P) = d2^{d+1}$.

One way to decrease the computation time of the algorithm is to exploit the symmetries of the embedding matrix M_d and reduce the possible choices for the bit vector b. Given M_{d-1} two vectors b are equivalent if the resulting matrices M_d can be transformed from one to the other by swapping columns and rows. Therefore, only one b for each equivalent class should be considered by the algorithm.

Acknowledgments. We acknowledge support from the following research grants: ERC grant *FOREFRONT* (grant agreement no. 615640) funded by the European Research Council under the EU's 7th Framework Programme (FP7/2007-2013), *Ambizione* grant PZ00P2 154779 *Tight formulations of 0-1 problems* funded by the Swiss National Science Foundation, the research grant *Semidefinite extended formulations* (Semaphore 14620017) funded by F.R.S.-FNRS, and the *ARC* grant AUWB-2012-12/17-ULB2 *COPHYMA* funded by the French community of Belgium. We also thank the anonymous referees for their comments which helped us improve the presentation.

References

1. Aichholzer, O.: Extremal properties of 0/1-polytopes of dimension 5. In: Ziegler, G., Kalai, G. (eds.) Polytopes - Combinatorics and Computation, pp. 111–130. Birkhäuser (2000)

2. Akutsu, T.: On determining the congruence of point sets in d dimensions. Computational Geometry 9(4), 247–256 (1998)

3. Beyer, S.: Comprehensive perl archive network: Bit-vector-7.4 (2014). http://search.cpan.org/~stbey/Bit-Vector-7.4

4. Chvátal, V.: On certain polytopes associated with graphs. J. Combinatorial Theory Ser. B 18, 138–154 (1975)

5. De Loera, J., Rambau, J., Santos, F.: Triangulations. Algorithms and Computation in Mathematics, vol. 25 (2010)

6. Ganter, B., Reuter, K.: Finding all closed sets: A general approach. Order 8(3), 283–290 (1991) (English)

7. Gawrilow, E., Joswig, M.: Polymake: an approach to modular software design in computational geometry. In: International Symposium on Computational Geometry (SOCG), pp. 222–231. ACM Press (2001)

8. Gouveia, J., Parrilo, P., Thomas, R.: Theta bodies for polynomial ideals. SIAM Journal on Optimization 20(4), 2097–2118 (2010)

9. Gouveia, J., Robinson, R., Thomas, R.: Polytopes of minimum positive semidefinite rank. Discrete & Computational Geometry 50(3), 679–699 (2013)

10. Grande, F., Sanyal, R.: Theta rank, levelness, and matroid minors, arXiv preprint, arXiv:1408.1262 (2014)
11. Haase, C.: Lattice polytopes and unimodular triangulations, Ph.D. thesis, Technical University of Berlin (2000)
12. Hanner, O.: Intersections of translates of convex bodies. Mathematica Scandinavica 4, 65–87 (1956)
13. Hansen, A.: On a certain class of polytopes associated with independence systems. Mathematica Scandinavica 41, 225–241 (1977)
14. Kalai, G.: The number of faces of centrally-symmetric polytopes. Graphs and Combinatorics 5(1), 389–391 (1989)
15. Kuznetsov, S., Obiedkov, S.: Comparing performance of algorithms for generating concept lattices. Journal of Experimental and Theoretical Artificial Intelligence 14, 189–216 (2002)
16. Paffenholz, A.: Faces of Birkhoff polytopes. Electronic Journal of Combinatorics 22 (2015)
17. Sanyal, R., Werner, A., Ziegler, G.: On Kalai's conjectures concerning centrally symmetric polytopes. Discrete & Computational Geometry 41(2), 183–198 (2009)
18. Stanley, R.: Decompositions of rational convex polytopes. Annals of Discrete Mathematics, 333–342 (1980)
19. Stanley, R.: Two poset polytopes. Discrete & Computational Geometry 1(1), 9–23 (1986)
20. Sullivant, S.: Compressed polytopes and statistical disclosure limitation. Tohoku Mathematical Journal, Second Series 58(3), 433–445 (2006)
21. Ziegler, G.: Lectures on polytopes, vol. 152, Springer Science & Business Media (1995)
22. Ziegler, G.: Lectures on 0/1-polytopes. In: Kalai, G., Ziegler, G. (eds.) Polytopes - Combinatorics and Computation. DMV Seminar, vol. 29, pp. 1–41. Springer, Birkhäuser, Basel (2000)

Upper and Lower Bounds for Online Routing on Delaunay Triangulations

Nicolas Bonichon[1,2,*], Prosenjit Bose[3,**], Jean-Lou De Carufel[3], Ljubomir Perković[4], and André van Renssen[5,6]

[1] Univ. Bordeaux, LaBRI, UMR 5800, F-33400 Talence, France
[2] CNRS, LaBRI, UMR 5800, F-33400 Talence, France
bonichon@labri.fr
[3] School of Computer Science, Carleton University, Ottawa, Canada
{jit,jdecaruf}@scs.carleton.ca
[4] School of Computing, DePaul University, USA
lperkovic@cs.depaul.edu
[5] National Institute of Informatics, Tokyo, Japan
andre@nii.ac.jp
[6] JST, ERATO, Kawarabayashi Large Graph Project, Tokyo, Japan

Abstract. Consider a weighted graph G whose vertices are points in the plane and edges are line segments between pairs of points whose weight is the Euclidean distance between its endpoints. A routing algorithm on G sends a message from any vertex s to any vertex t in G. The algorithm has a *competitive ratio* of c if the length of the path taken by the message is at most c times the length of the shortest path from s to t in G. It has a *routing ratio* of c if the length of the path is at most c times the Euclidean distance from s to t. The algorithm is *online* if it makes forwarding decisions based on (1) the k-neighborhood in G of the message's current position (for constant $k > 0$) and (2) limited information stored in the message header.

We present an online routing algorithm on the Delaunay triangulation with routing ratio less than 5.90, improving the best known routing ratio of 15.48. Our algorithm makes forwarding decisions based on the 1-neighborhood of the current position of the message and the positions of the message source and destination only.

We present a lower bound of 5.7282 on the routing ratio of our algorithm, so the 5.90 upper bound is close to the best possible. We also show that the routing (resp., competitive) ratio of any deterministic k-local algorithm is at least 1.70 (resp., 1.23) for the Delaunay triangulation and 2.70 (resp., 1.23) for the L_1-Delaunay triangulation. In the case of the L_1-Delaunay triangulation, this implies that even though there always exists a path between s and t whose length is at most $2.61|[st]|$, it is not always possible to route a message along a path of length less than $2.70|[st]|$ using only local information.

Keywords: Delaunay triangulation, online routing, routing ratio, competitive ratio.

* Partially supported by ANR grant JCJC EGOS ANR-12-JS02-002-01 and LIRCO.
** Partially supported by NSERC.

N. Bansal and I. Finocchi (Eds.): ESA 2015, LNCS 9294, pp. 203–214, 2015.
DOI: 10.1007/978-3-662-48350-3_18

1   Introduction

Navigation is a fundamental problem with a long history [13]. Navigation is encountered in many forms and in a variety of fields including geographic information systems [17], architecture and urban planning [14], robotics [7], and communication networks [16], to name a few. Navigation often occurs in a geometric setting that can be modeled using a *geometric graph*, defined as a weighted graph G whose vertices are points in the plane and whose edges are line segments. The weight of each edge is the Euclidean distance between its endpoints. Navigation is then the problem of finding a path in G–preferably short–from source vertex s to target vertex t. When complete information about the graph is available, classic shortest path algorithms can be applied (e.g., Dijkstra's algorithm [11]). The problem is more challenging when navigation decisions can only use information available locally. To illustrate this, we consider a particular navigation application: the problem of routing a message from a source s to destination t. The goal of a routing algorithm is to select a path for the message.

A routing algorithm on a geometric graph G has a *competitive ratio* of c if the length of the path produced by the algorithm from any vertex s to any vertex t is at most c times the length of the shortest path from s to t in G. If the length of the path is at most c times the Euclidean distance from s to t, we say that the routing algorithm has a *routing ratio* of c. The routing algorithm is *online* (or k-local) if it makes forwarding decisions based on (1) the k-neighborhood in G (for some integer constant $k > 0$) of the current position of the message and (2) limited information stored in the message header.

In this paper we consider online routing on Delaunay triangulations. The (classic) Delaunay triangulation on a set of points P is a geometric graph G such that there is an edge between two vertices u and v if and only if there exists a *circle* with u and v on its boundary that contains no other vertex of P in its interior.[1] If we replace *circle* with *square* in the definition then a different triangulation is defined: the L_1- or the L_∞-Delaunay triangulation, depending on the orientation of the square. If *circle* is replaced with *equilateral triangle*, then yet another triangulation is defined: the TD-Delaunay triangulation. Related previous work on online routing in Delaunay triangulations include [4,5,6,8].

Delaunay triangulations are known to be geometric spanners. A geometric graph G on a set of points P in the plane is a κ-*spanner* for some constant κ (or has a *spanning ratio* of κ) if for any pair of vertices s and t of P, there exists a path in G from s to t with length at most κ times the Euclidean distance between s and t (see [3,15]).

In the mid-1980s, it was not known that Delaunay triangulations were actually spanners. In his seminal 1986 paper, Chew [9] showed that the L_1-Delaunay triangulation is a $\sqrt{10}$-spanner. In fact, he did something stronger: he found a 1-local, online routing algorithm with competitive and routing ratios of $\sqrt{10} \approx 3.162$. In a recent development, Bonichon et al. [1] proved that the L_1- and

---

[1] This definition assumes that points are in general position; we discuss this restriction further in Section 2.

the L_∞-Delaunay triangulations are $\sqrt{4 + 2\sqrt{2}} \approx 2.61$-spanners and that the constant is also tight. This result opens up the question of whether there exists an online routing algorithm with routing ratio of 2.61.

In another significant result, Chew proved that the TD-Delaunay triangulation is a 2-spanner and that the constant 2 is tight [10]. Chew's construction does not lead to an online routing algorithm. In fact, an algorithm with routing ratio of 2 is not even possible: Bose et al. [5] recently showed surprising lower bounds of $\frac{5}{\sqrt{3}}$ on the routing ratio and $\frac{5}{3}$ on the competitive ratio of online routing on TD-Delaunay triangulations. They also found an online routing algorithm that has competitive and routing ratios of $\frac{5}{\sqrt{3}}$.

The results on TD-Delaunay triangulations show that an online routing algorithm cannot provide the same guarantees as an algorithm that selects the routing path based on full knowledge of the graph. A natural, fundamental question is to analyze whether similar gaps exist for Delaunay triangulations other than the TD-Delaunay triangulation.

The (classic) Delaunay triangulation was first shown to be a constant spanner by Dobkin et al. [12]. The current best known upper bound on its spanning ratio is 1.998 by Xia [18] and the best lower bound, by Xia et al. [19], is 1.593. Very recently, Bose et al. developed an online routing algorithm on Delaunay triangulations that has competitive and routing ratios of 15.48 [4].

Our Results. We present an online routing algorithm on Delaunay triangulations that has competitive and routing ratios of 5.90 (Theorem 1). We also show our algorithm has a routing ratio greater than 5.7282 (Theorem 2) so the 5.90 upper bound is close to best possible. The algorithm is a generalization of the deterministic 1-local routing algorithm by Chew on the L_1-Delaunay triangulation [9]. Our algorithm makes forwarding decisions based on the 1-neighborhood of the current position of the message and the positions of the message source and destination. This last requirement is an improvement over the best known online routing algorithms on the Delaunay triangulation [2,4] which require the header of a message to also contain partial sums of distances along the routing path. Although the generalization of Chew's routing algorithm to Delaunay triangulation is natural, the analysis of its routing ratio is non-trivial.

We also show that the routing (resp., competitive) ratios of any deterministic k-local algorithm is at least 1.70 (resp., 1.23) for the Delaunay triangulation and 2.70 (resp., 1.23) for the L_1-Delaunay triangulation (Theorems 3 and 4). In the case of the L_1-Delaunay triangulations, this implies the existence of a gap between the spanning ratio (2.61) and the routing ratio (2.70). See Table 1 for a summary of these results.

2   A Generalization of Chew's Routing Algorithm

In this section we present our online routing algorithm which is a natural adaptation to Delaunay triangulations of Chew's routing algorithm originally designed for L_1-Delaunay triangulations [9] and subsequently adapted for TD-Delaunay triangulations [10].

Table 1. Upper and lower bounds on the spanning and routing ratios on Delaunay triangulations defined using different shapes. We also provide lower bounds on the competitiveness of k-local deterministic routing algorithms on Delaunay triangulations.

Shape	triangle	square	circle
spanning ratio UB	2 [10]	2.61 [1]	1.998 [18]
spanning ratio LB	2 [10]	2.61 [1]	1.593 [19]
routing ratio UB	$5/\sqrt{3} \approx 2.89$ [5]	$\sqrt{10} \approx 3.16$ [9]	$1.185 + 3\pi/2 \approx 5.90$ (Thm 1)
routing ratio LB	$5/\sqrt{3} \approx 2.89$ [5]	2.707 (Thm 4)	1.701 (Thm 3)
competitiveness LB	$5/3 \approx 1.66$ [5]	1.1213 (Thm 4)	1.2327 (Thm 3)

We consider the Delaunay triangulation defined on a finite set of points P in the plane. We assume that the points in P are not all collinear (otherwise the Delaunay triangulation would not be well-defined). To simplify our arguments, we also assume that the points are in *general position* which for us means that no four points of P are cocircular.

In this paper, we denote the source of the routing path by $s \in P$ and its destination by $t \in P$. We assume an orthogonal coordinate system consisting of a horizontal x-axis and a vertical y-axis and we denote by $x(p)$ and $y(p)$ the x- and y-coordinates of any point p in the plane. We denote the line supported by two points p and q by pq, and the line segment with endpoints p and q by $[pq]$. Without loss of generality, we assume that $y(s) = y(t) = 0$ and $x(s) < x(t)$.

When routing from s to t, we consider only (the vertices and edges of) the triangles of the Delaunay triangulation that intersect $[st]$. Without loss of generality, if a vertex (other than s and t) is on $[st]$, we consider it to be slightly above st. Therefore, the triangles that intersect $[st]$ and do not have t as a vertex can be ordered from left to right. Notice that all vertices (other than s and t) from this ordered set of triangles belong to at least 2 of these triangles.

Our Online Routing Algorithm. When the message is at vertex p_i (initially $p_0 = s$), if there is an edge from p_i to t, forward the message to t. Otherwise, let T_i be the rightmost triangle that has p_i as a vertex. Let C_i be the circle circumscribing T_i. Let O_i be the center of C_i, let w_i (w as in *west*) be the leftmost point of C_i, and let r_i be the rightmost intersection of C_i with $[st]$ (see, for example, circle C_3 of Fig. 1). The line segment $[w_i r_i]$ splits C_i into two arcs: the *upper* one, defined by the clockwise walk along C_i from w_i to r_i and the *lower* one, defined by the counterclockwise walk along C_i from w_i to r_i. Both arcs include points w_i and r_i. The forwarding decision at p_i is made as follows:

- If p_i belongs to the upper arc, forward the message to the first vertex of T_i encountered on a clockwise walk along C_i starting at p_i. In Fig. 1, this forwarding decision is made at s, p_2, p_3, and p_4.
- Otherwise, forward the message to the first vertex of T_i encountered on a counter-clockwise walk along C_i starting at p_i. In Fig. 1, this forwarding decision is made at p_1.

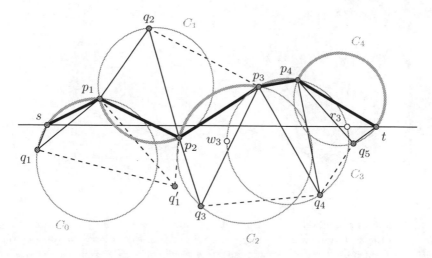

Fig. 1. Illustration of our routing algorithm. The triangles of the Delaunay triangulation that intersect segment $[st]$ are shown using solid and dashed segments; the edges of the *rightmost triangles* are solid and the remaining edges are dashed. Only circles circumscribing the rightmost triangles are shown. The edges of the routing path and the associated arcs are also shown in bold.

Once we reach p_{i+1}, unless $p_{i+1} = t$ we repeat the process. Fig. 1 shows an example of a route computed by this algorithm. We can conclude that the following results by Chew from [9] extend to Delaunay triangulations:

Lemma 1. *The triangles used $(T_0, T_1 \ldots, T_k)$ are ordered along $[st]$. Although not all Delaunay triangulation triangles intersecting $[st]$ are used, those used appear in their order along $[st]$.*

Corollary 1. *The algorithm terminates, producing a path from s to t.*

3 Routing Ratio

In this section, we prove the main theorem of this paper.

Theorem 1. *Our routing algorithm on the Delaunay triangulation has a routing ratio of at most $(1.185043874 + 3\pi/2) \approx 5.89743256$.*

The bound on the routing ratio of our algorithm is close to best possible because, as we show in Sect. 4 and illustrate in Fig. 2, our algorithm has a routing ratio of at least 5.7282.

3.1 Preliminaries

We start by introducing additional definitions, notation, and structural results about our routing algorithm. We denote by $||[pq]||$ the Euclidean length of the

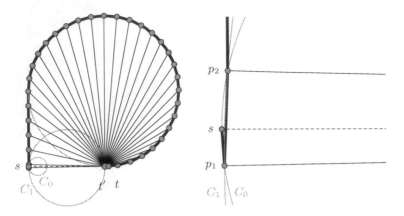

Fig. 2. The lower bound on the routing ratio of our algorithm (left). The right image zooms in on the situation around point s. These images illustrate how the routing ratio of the algorithm must be a bit larger than $1 + 3\pi/2$.

line segment $[pq]$, and by $|\mathcal{P}|$ the length of a path $\mathcal{P}$ in the plane. Given a path $\mathcal{P}$ from p to q and a path $\mathcal{Q}$ from q to r, $\mathcal{P} + \mathcal{Q}$ denotes the concatenation of $\mathcal{P}$ and $\mathcal{Q}$. We say that the path $\mathcal{P}$ from p to q is *inside* a path $\mathcal{Q}$ that also goes from p to q if the path $\mathcal{P}$ is inside the bounded region delimited by $\mathcal{Q} + [qp]$. Note that if $\mathcal{P}$ is convex and inside $\mathcal{Q}$ then $|\mathcal{P}| \leq |\mathcal{Q}|$. Given a path $\mathcal{P}$ and two points p and q on $\mathcal{P}$, we denote by $\mathcal{P}\langle p, q \rangle$ the sub-path of $\mathcal{P}$ from p to q.

Let $s = p_0, p_1, \ldots, p_k = t$ be the sequence of vertices visited by our routing algorithm. If some p_i other than s or t lies on the segment $[st]$, we can separately analyze the routing ratio of the paths from s to p_i and from p_i to t. We assume, therefore, that no p_i, other than $s = p_0$ and $t = p_k$, lies on segment $[st]$.

For every edge (p_i, p_{i+1}), there is a corresponding oriented arc of C_i used by the algorithm which we refer to as $\mathcal{A}_i\langle p_i, p_{i+1} \rangle$: the orientation of $\mathcal{A}_i\langle p_i, p_{i+1} \rangle$ is clockwise if p_i belongs to the upper arc of C_i and counterclockwise if p_i belongs to the lower arc. The arcs are shown in Fig. 1 and Fig. 3. Let $\mathcal{A}$ be the union of these arcs. We call $\mathcal{A}$ the routing path from s to t. The length of the path $s = p_0, p_1, \ldots, p_{k-1}, p_k = t$ along the edges of the Delaunay triangulation is smaller than the length of $\mathcal{A}$.

In order to bound the length of $\mathcal{A}$, we will work with *worst case* circles C_i' defined, for $i = 0, 1, \ldots, k - 1$, as follows. C_i' is a circle that goes through p_i and p_{i+1} and whose center O_i' is obtained by starting at O_i and moving it along the perpendicular bisector of $[p_i p_{i+1}]$ in the direction of arc $\mathcal{A}_i\langle p_i, p_{i+1} \rangle$ until either (1) C_i' is tangent to line st or (2) p_i is the leftmost point of C_i', whichever occurs first. Figure 3 shows the circles C_i' on the example of Fig. 1. Note that (2) holds for C_1' (and so $w_1' = p_1$) and (1) holds for C_3' (and $w_3' \neq p_3$).

By the construction, the circles C_i' intersect st. Let w_i' be the leftmost point of C_i'; if $[p_i p_{i+1}]$ crosses $[st]$, then C_i' must satisfy condition (2), i.e., $p_i = w_i'$. We find it useful to categorize the circles C_i' into three mutually exclusive types:

- Type A_1: $p_i \neq w'_i$, $[p_i p_{i+1}]$ does not cross $[st]$, and C'_i is tangent to $[st]$.
- Type A_2: $p_i = w'_i$ and $[p_i p_{i+1}]$ does not cross $[st]$.
- Type B: $p_i = w'_i$ and $[p_i p_{i+1}]$ crosses $[st]$.

In Fig. 3, $C'_0, C'_1 \ldots C'_4$ are respectively of type A_2, B, B, A_1, A_2. We use the expression "type A" instead of "type A_1 or A_2".

Given two points p, q on C'_i, let $\mathcal{A}'_i \langle p, q \rangle$ be the arc on C'_i from p to q whose orientation (clockwise or counterclockwise) is the same as the orientation of $\mathcal{A}_i \langle p_i, p_{i+1} \rangle$ around C_i. Note that $|\mathcal{A}_i \langle p_i, p_{i+1} \rangle| \leq |\mathcal{A}'_i \langle p_i, p_{i+1} \rangle|$. In fact, C'_i is defined so that $\mathcal{A}'_i \langle p_i, p_{i+1} \rangle$ is the longest possible arc between p_i and p_{i+1} in a route computed by our algorithm if only the points s, t, p_i, p_{i+1} are known. This restriction, in turn, provides enough structure to enable us to bound the length of the union of the arcs $\mathcal{A}'_i \langle p_i, p_{i+1} \rangle$ for $i = 0, \ldots, k-1$ and, in turn, the length of the routing path $\mathcal{A}$.

By the definition of Delaunay triangulations, no point of P is contained inside the circles C_i. This property does not hold for circles C'_i, but the following weaker property does:

Lemma 2. *No point of P lies inside the region bounded by the closed curve $[p_i p_{i+1}] + \mathcal{A}'_i \langle p_{i+1}, p_i \rangle$, for every $i = 0, \ldots, k-1$.*

Proof. The region is part of the region inside circle C_i. $\qquad\square$

Lemma 3. *Let $\angle w'_{i-1} O'_{i-1} p_i$ and $\angle w'_i O'_i p_i$ be the angles defined using the orientations of arcs $\mathcal{A}'_{i-1} \langle p_{i-1}, p_i \rangle$ and $\mathcal{A}'_i \langle p_i, p_{i+1} \rangle$, respectively. Then, for every $i = 1, \ldots, k-1$:*

$$0 \leq \angle w'_i O'_i p_i \leq \angle w'_{i-1} O'_{i-1} p_i \leq 3\pi/2.$$

Let f_i be the first point p_j after p_i such that $[p_i p_j]$ intersects st. Notice that $f_{k-1} = t$. We also set $f_k = t$. In Fig. 3, $f_0 = p_1$, $f_1 = p_2$, $f_2 = p_3$ and $f_3 = f_4 = f_5 = t$.

Lemma 4. *For all $0 < i \leq k$:*

$$x(w'_{i-1}) \leq x(w'_i) \leq x(f_{i-1}) \leq x(f_i). \tag{1}$$

3.2   Proof of Theorem 1

In this subsection, we introduce a key lemma and use it to prove our main theorem. Given two points p and q such that $x(p) < x(q)$ and $y(p) = y(q)$, we define the path $\mathcal{S}_{p,q}$ as follows. Let C be the circle above pq that is tangent to pq at q and tangent to the line $x = x(p)$ at a point that we denote by p'. The path $\mathcal{S}_{p,q}$ consists of line segment $[pp']$ together with the clockwise arc from p' to q on C. We call $\mathcal{S}_{p,q}$ the *snail curve* from p to q. Note that $|\mathcal{S}_{p,q}| = (1 + 3\pi/2)(x(q) - x(p))$.

Let points $\overline{f}_i = (x(f_i), 0)$ and $\overline{w}'_i = (x(w'_i), 0)$ be the orthogonal projections of points f_i and w'_i onto line st. Finally, we define the path $\mathcal{P}_i$ to be $[\overline{w}'_i w'_i] + \mathcal{A}'_i \langle w'_i, p_i \rangle$, for $0 \leq i \leq k-1$ (see Fig. 3).

We start with a simple lemma that motivates these definitions.

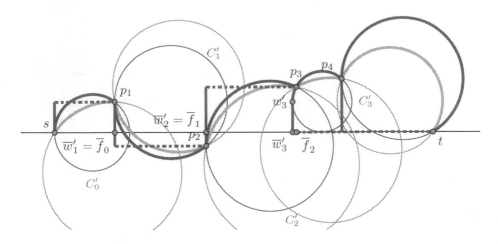

Fig. 3. Illustration of our notation and Lemma 6 on the example of Fig. 1. The unlabeled circles are the empty circles of the Delaunay triangulation and the thick arcs on those circles form the routing path $\mathcal{A}$. Circles C'_i, the paths $\mathcal{P}_i$, arcs $\mathcal{A}'_i \langle p_i, p_{i+1} \rangle$, and segments of height $|y(f_i)|$ are shown in a darker shade. Lengths $|[\overline{f}_{i-1}\overline{f}_i]|$ are represented by dashed horizontal segments.

Lemma 5. $|\mathcal{A}'_{k-1} \langle p_{k-1}, t \rangle| \leq |\mathcal{S}_{\overline{w}'_{k-1}, t}| - |\mathcal{P}_{k-1}|.$

Proof. This follows from the fact that path $\mathcal{P}_{k-1} + \mathcal{A}'_{k-1} \langle p_{k-1}, t \rangle$ from $\overline{w}'_{k-1}$ to t is convex and inside $\mathcal{S}_{\overline{w}'_{k-1}, t}$. $\square$

The following lemma is the key to proving Theorem 1.

Lemma 6. *For all* $0 < i < k$ *and* $\delta = 0.185043874$,

$$|\mathcal{A}'_{i-1} \langle p_{i-1}, p_i \rangle| \leq |\mathcal{P}_i| - |\mathcal{P}_{i-1}| + |\mathcal{S}_{\overline{w}'_{i-1}, \overline{w}'_i}| + |y(f_i)| - |y(f_{i-1})| + \delta|[\overline{f}_{i-1}\overline{f}_i]|. \quad (2)$$

This lemma is illustrated in Fig. 3. We first show how to use Lemma 6 to prove Theorem 1, and then we prove Lemma 6.

Proof of Theorem 1. By Lemma 4, $\sum_{i=1}^{k-1} |[\overline{f}_{i-1}\overline{f}_i]| < |[st]|$ and $\sum_{i=1}^{k} |\mathcal{S}_{\overline{w}'_{i-1}, \overline{w}'_i}| = |\mathcal{S}_{s,t}|$. Since $f_{k-1} = t$, $y(f_{k-1}) = 0$. Therefore, by summing the $k-1$ inequalities from Lemma 6 and the inequality from Lemma 5, we get

$$|\mathcal{A}| \leq \sum_{i=1}^{k} \mathcal{A}'_{i-1} \langle p_{i-1} p_i \rangle < |\mathcal{S}_{s,t}| + \delta|[st]| \leq (1.185043874 + 3\pi/2)|[st]|,$$

which completes the proof. $\square$

3.3 Proof of the Key Lemma

In this subsection, we prove Lemma 6.

Proof of Lemma 6. We consider three cases depending on the types of circles C'_{i-1} and C'_i. Note that if C'_{i-1} is of type A, then $f_{i-1} = f_i$. Hence, in this case, it is sufficient to prove

$$|\mathcal{A}'_{i-1}\langle p_{i-1}, p_i\rangle| \leq |\mathcal{P}_i| - |\mathcal{P}_{i-1}| + |\mathcal{S}_{\overline{w}'_{i-1}, \overline{w}'_i}|$$

or

$$|\mathcal{P}_{i-1} + \mathcal{A}'_{i-1}\langle p_{i-1}, p_i\rangle| \leq |\mathcal{S}_{\overline{w}'_{i-1}, \overline{w}'_i} + \mathcal{P}_i|. \tag{3}$$

This inequality is what we will show in the first two cases of the proof.

- C'_{i-1} **is of type** A **and** C'_i **is of type** A_2 **or** B. *(omitted due to space constraints)*
- C'_{i-1} **is of type** A **and** C'_i **is of type** A_1. We first observe that with the position of p_i and $\overline{w}'_{i-1}$ fixed, $\mathcal{P}_{i-1} + \mathcal{A}'_{i-1}\langle p_{i-1}, p_i\rangle$ is longest if C'_{i-1} is of type A_1 (see Fig. 4).

Hence, we assume that C'_{i-1} is of type A_1. Let b_{i-1} and b_i be the intersections of st with C'_{i-1} and C'_i, respectively. Then

$$|\mathcal{P}_{i-1} + \mathcal{A}'_{i-1}\langle p_{i-1}, p_i\rangle| + |\mathcal{A}'_i\langle p_i, b_i\rangle| =$$
$$|\mathcal{S}_{\overline{w}'_{i-1}, b_{i-1}}| + |\mathcal{S}_{\overline{w}'_i, b_i}| - |\mathcal{A}'_{i-1}\langle p_i, b_{i-1}\rangle| - |\mathcal{P}_i|. \tag{4}$$

If $x(b_{i-1}) \leq x(\overline{w}'_i)$, the right-hand side of (4) is at most $|\mathcal{S}_{\overline{w}'_{i-1}, b_{i-1}}| + |\mathcal{S}_{\overline{w}'_i, b_i}| \leq |\mathcal{S}_{\overline{w}'_{i-1}, b_i}|$. If, however, $x(b_{i-1}) > x(\overline{w}'_i)$ (as is the case in Fig. 4), because the curve $\mathcal{S}_{\overline{w}'_i, b_{i-1}}$ is convex and inside $\mathcal{P}_i + \mathcal{A}'_{i-1}\langle p_i, b_{i-1}\rangle$, it follows that the right-hand side of (4) is at most $|\mathcal{S}_{\overline{w}'_{i-1}, b_{i-1}}| + |\mathcal{S}_{\overline{w}'_i, b_i}| - |\mathcal{S}_{\overline{w}'_i, b_{i-1}}| = |\mathcal{S}_{\overline{w}'_{i-1}, b_i}|$. Either way, we have that

$$|\mathcal{P}_{i-1}| + |\mathcal{A}'_{i-1}\langle p_{i-1}, p_i\rangle| + |\mathcal{A}'_i\langle p_i, b_i\rangle| \leq |\mathcal{S}_{\overline{w}'_{i-1}, b_i}| = |\mathcal{S}_{\overline{w}'_{i-1}, \overline{w}'_i}| + |\mathcal{S}_{\overline{w}'_i, b_i}|.$$

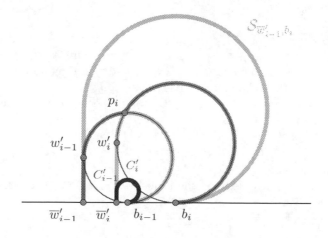

Fig. 4. Illustration of the case when C'_{i-1} is of type A and C'_i is of type A_1. Curve $\mathcal{P}_{i-1} + \mathcal{A}'_{i-1}\langle p_{i-1}, p_i\rangle + \mathcal{A}'_i\langle p_i, b_i\rangle$ is shown in dark gray whereas curves $\mathcal{S}_{\overline{w}'_i, b_i}, \mathcal{P}_i$, and $\mathcal{A}'_{i-1}\langle p_i, b_{i-1}\rangle$ are shown in light gray. Curve $\mathcal{S}_{\overline{w}'_i, b_{i-1}}$ is shown in black.

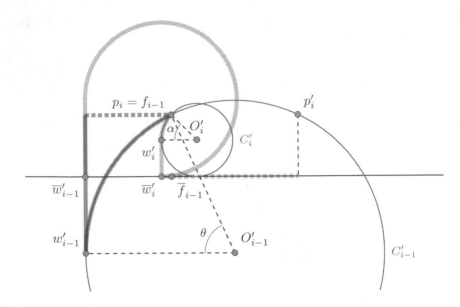

Fig. 5. Illustration of the case of when C'_{i-1} is of type B and $|y(p_{i-1}) - y(p_i)| \geq |y(p_{i-1}) - y(f_i)|$. The dark gray path represents the sum $|\mathcal{P}_{i-1}| + |\mathcal{A}'_{i-1}\langle p_{i-1}p_i\rangle| + |y(f_{i-1})|$. The light gray path represents the sum $|\mathcal{S}_{\overline{w}'_{i-1}, \overline{w}'_i}| + |\mathcal{P}_i| + \delta|[\overline{f}_{i-1}\overline{f}_i]|$ (the last term is represented by the dashed line).

If we subtract $|\mathcal{A}'_i\langle p_i, b_i\rangle|$ from both sides, we get (3).

• C'_{i-1} **is of type** B **and** C'_i **is of type** A **or** B. In this case, $w'_{i-1} = p_{i-1}$ and $f_{i-1} = p_i \neq f_i$. We consider two subcases.

  • **subcase** $|y(p_{i-1}) - y(p_i)| < |y(p_{i-1}) - y(f_i)|$ *(omitted due to space constraints)*

  • **subcase** $|y(p_{i-1}) - y(p_i)| \geq |y(p_{i-1}) - y(f_i)|$ *(refer to Fig. 5)* If we assume that p_i lies above st, then f_i must lie below st. By Lemma 4, $x(p_i) = x(f_{i-1}) \leq x(f_i)$, and by Lemma 2, the region inside C'_{i-1} and to the right of line $x = x(p_i)$ contains no point of P. Therefore, f_i is outside of C'_{i-1}.

  Recall that by Lemma 3, if $\theta = \angle w'_{i-1}O'_{i-1}p_i$ and $\alpha = \angle w'_iO'_ip_i$, then $0 \leq \alpha \leq \theta \leq 3\pi/2$. Without loss of generality, assume that the radius of C'_{i-1} is 1. Let $R = y(w'_i)$. If C'_i is of type A_1, then its radius is R and we have $|\mathcal{P}_i| = (1 + \alpha)R$ and $[\overline{w}'_i\overline{f}_{i-1}] = (1 - \cos(\alpha))R$. If C'_i is of type A_2 or B then $\alpha = 0$, $p_i = w'_i$, and $\overline{w}'_i = \overline{f}_{i-1}$, and we also have that $|\mathcal{P}_i| = (1 + \alpha)R$ and $[\overline{w}'_i\overline{f}_{i-1}] = (1 - \cos(\alpha))R$.

  Let D be the difference between the right-hand side and the left-hand side of inequality (2). Then

$$D = |\mathcal{S}_{\overline{w}'_{i-1}, \overline{w}'_i}| + |\mathcal{P}_i|$$
$$+ \delta|[\overline{f}_{i-1}\overline{f}_i]| + |y(f_i)| - |\mathcal{P}_{i-1}| - |\mathcal{A}'_{i-1}\langle p_{i-1}p_i\rangle| - |y(f_{i-1})|$$
$$= (1 + 3\pi/2)(1 - \cos(\theta) - (1 - \cos(\alpha))R) + (1 + \alpha)R$$
$$+ \delta|[\overline{f}_{i-1}\overline{f}_i]| + |y(f_i)| - \theta - \sin(\theta) \tag{5}$$
$$= R[1 + \alpha - (1 + 3\pi/2)(1 - \cos(\alpha))] + (1 + 3\pi/2)(1 - \cos(\theta))$$
$$+ \delta|[\overline{f}_{i-1}\overline{f}_i]| + |y(f_i)| - \theta - \sin(\theta).$$

It remains to prove that $D \geq 0$.

We first consider the case when $\theta \leq \pi/4$, which, by Lemma 3, implies that $\alpha < \pi/4$ as well. Let p'_i be the intersection, other than p_i, of circle C'_{i-1} with the horizontal line through p_i. Since $\theta \leq \pi/4$, we have $x(p'_i) > x(O'_i)$.

Since $|y(p_{i-1}) - y(p_i)| < |y(p_{i-1}) - y(p)|$ for all points p outside of C'_{i-1} such that $x(p_i) \leq x(p) \leq x(p'_i)$, it follows that $x(f_i) \geq x(p'_i)$. Note that $\angle w'_{i-1}O'_{i-1}p'_i = \pi - \theta$. Since $|[\overline{f}_{i-1}\overline{f}_i]| \geq |[f_{i-1}p'_i]| = 2\cos(\theta)$ (recall that $p_i = f_{i-1}$), we have

$$D \geq R[1 + \alpha - (1 + 3\pi/2)(1 - \cos(\alpha))] + (1 + 3\pi/2)(1 - \cos(\theta))$$
$$+ 2\delta\cos(\theta) - \theta - \sin(\theta).$$

Let $g(\alpha) = 1 + \alpha - (1 + 3\pi/2)(1 - \cos(\alpha))$. There exists an $\alpha_0 > \pi/4$ such that $g(\alpha_0) = 0$ and $g(\alpha_0) \geq 0$ for all $\alpha \in [0, \alpha_0]$. Therefore, to prove that $D \geq 0$ (and therefore that inequality (2) holds), it is sufficient to prove that

$$(1 + 3\pi/2)(1 - \cos(\theta)) + 2\delta\cos(\theta) - \theta - \sin(\theta) \geq 0.$$

If we take $\delta = 0.185043874$, we can show that this inequality is true using elementary calculus arguments.

To complete the proof, it remains to consider the case when $\theta \in [\pi/4, \pi]$. If $\alpha \leq \alpha_0$, from (5) we have that $D \geq (1 + 3\pi/2)(1 - \cos(\theta)) - \theta - \sin(\theta)$, which is positive for all $\theta \in [\pi/4, \pi]$. If $\alpha \in (\alpha_0, \pi]$, $g(\alpha)$ is negative and decreasing. Thus, since $\alpha \leq \theta$ and $R < 1$, we obtain

$$D \geq 1 + \theta - (1 + 3\pi/2)(1 - \cos(\theta)) + (1 + 3\pi/2)(1 - \cos(\theta))$$
$$+ \delta|[\overline{f}_{i-1}\overline{f}_i]| + |y(f_i)| - \theta - \sin(\theta)$$
$$\geq 1 - \sin(\theta) + \delta|[\overline{f}_{i-1}\overline{f}_i]| + |y(f_i)|.$$

This lower bound is trivially positive, hence inequality (2) holds in all cases.   □

4   Lower Bounds

In this section, we provide several lower bounds on the routing and competitive ratios of online routing on Delaunay triangulations. Due to space limitation, proofs have been skipped.

Theorem 2. *The routing ratio of our routing algorithm on a Delaunay triangulation can be greater than 5.7282.*

Theorem 3. *There exists no deterministic k-local routing algorithm on Delaunay triangulations with routing ratio at most 1.7018 or that is 1.2327-competitive.*

Theorem 4. *There exists no deterministic k-local routing algorithm for the L_1- and L_∞-Delaunay triangulations that has a routing ratio less than $(2 + \sqrt{2}/2) \approx 2.7071$ or that is $\frac{2+\sqrt{2}/2}{1+\sqrt{2}} \approx 1.1213$-competitive.*

References

1. Bonichon, N., Gavoille, C., Hanusse, N., Perković, L.: The stretch factor of L_1- and L_∞-Delaunay triangulations. In: Epstein, L., Ferragina, P. (eds.) ESA 2012. LNCS, vol. 7501, pp. 205–216. Springer, Heidelberg (2012)
2. Bose, P., Morin, P.: Online routing in triangulations. SIAM J. Comp. 33(4), 937–951 (2004)
3. Bose, P., Smid, M.: On plane geometric spanners: A survey and open problems. Comput. Geom. 46(7), 818–830 (2013)
4. Bose, P., De Carufel, J.L., Durocher, S., Taslakian, P.: Competitive online routing on Delaunay triangulations. In: SWAT, pp. 98–109 (2014)
5. Bose, P., Fagerberg, R., van Renssen, A., Verdonschot, S.: Competitive routing in the half-θ_6-graph. In: SODA, pp. 1319–1328 (2012)
6. Bose, P., Fagerberg, R., van Renssen, A., Verdonschot, S.: Competitive routing on a bounded-degree plane spanner. In: CCCG, pp. 299–304 (2012)
7. Braunl, T.: Embedded Robotics: Mobile Robot Design and Applications with Embedded Systems. Springer (2006)
8. Broutin, N., Devillers, O., Hemsley, R.: Efficiently navigating a random Delaunay triangulation. In: AofA, pp. 49–60 (2014)
9. Chew, L.P.: There is a planar graph almost as good as the complete graph. In: SoCG, pp. 169–177 (1986)
10. Chew, L.P.: There are planar graphs almost as good as the complete graph. J. Comp. System Sci. 39(2), 205–219 (1989)
11. Dijkstra, E.W.: A note on two problems in connexion with graphs. Numer. Math. 1, 269–271 (1959)
12. Dobkin, D.P., Friedman, S.J., Supowit, K.J.: Delaunay graphs are almost as good as complete graphs. Discrete & Comput. Geom. 5(4), 399–407 (1990). doi:10.1007/BF02187801
13. Hofmann-Wellenhof, B., Legat, K., Wieser, M.: Navigation: Principles of Positioning and Guidance. Springer (2003)
14. Murgante, B., Borruso, G., Lapucci, A.: Geocomputation and Urban Planning. In: Murgante, B., Borruso, G., Lapucci, A. (eds.) Geocomputation and Urban Planning. SCI, vol. 176, pp. 1–17. Springer, Heidelberg (2010)
15. Narasimhan, G., Smid, M.H.M.: Geometric spanner networks. Cambridge University Press (2007)
16. Stojmenovic, I.: Handbook of Wireless Networks and Mobile Computing. Wiley-Interscience (2002)
17. Worboys, M.F., Duckham, M.: GIS: A Computing Perspective, 2nd edn. CRC Press (2004)
18. Xia, G.: The stretch factor of the Delaunay triangulation is less than 1.998. SIAM J. Comput. 42(4), 1620–1659 (2013)
19. Xia, G., Zhang, L.: Toward the tight bound of the stretch factor of Delaunay triangulations. In: CCCG, pp. 175–180 (2011)

On Computing the Hyperbolicity
of Real-World Graphs*

Michele Borassi[1], David Coudert[2], Pierluigi Crescenzi[3], and Andrea Marino[4]

[1] IMT Institute for Advanced Studies Lucca, Italy
[2] Inria, France
[3] Dipartimento di Ingegneria dell'Informazione, Università di Firenze
[4] Dipartimento di Informatica, Università di Pisa

Abstract. The (Gromov) hyperbolicity is a topological property of a graph, which has been recently applied in several different contexts, such as the design of routing schemes, network security, computational biology, the analysis of graph algorithms, and the classification of complex networks. Computing the hyperbolicity of a graph can be very time consuming: indeed, the best available algorithm has running-time $\mathcal{O}(n^{3.69})$, which is clearly prohibitive for big graphs. In this paper, we provide a new and more efficient algorithm: although its worst-case complexity is $\mathcal{O}(n^4)$, in practice it is much faster, allowing, for the first time, the computation of the hyperbolicity of graphs with up to 200,000 nodes. We experimentally show that our new algorithm drastically outperforms the best previously available algorithms, by analyzing a big dataset of real-world networks. Finally, we apply the new algorithm to compute the hyperbolicity of random graphs generated with the Erdös-Renyi model, the Chung-Lu model, and the Configuration Model.

1 Introduction

In recent years, the analysis of complex networks has provided several significant results, with a huge amount of applications in sociology, biology, economics, statistical physics, electrical engineering, and so on. These results are based on the analysis of very big real-world networks, now made available by improvements in computer technology and by Internet. One of the major challenges in this field is to understand which properties distinguish these networks from other kinds of graphs, like random graphs [25], and which properties distinguish networks of different kinds [16], in order to classify general and particular behavior.

In this context, a significant role is played by the hyperbolic structure underlying a complex network, that is usually not present in random graphs [24,6]. For instance, if we draw points from a hyperbolic space and we connect nearby points,

---

* This work has been supported in part by the Italian Ministry of Education, University, and Research under PRIN 2012C4E3KT national research project AMANDA (Algorithmics for MAssive and Networked Data), and by ANR project Stint under reference ANR-13-BS02-0007, ANR program "Investments for the Future" under reference ANR-11-LABX-0031-01.

N. Bansal and I. Finocchi (Eds.): ESA 2015, LNCS 9294, pp. 215–226, 2015.
DOI: 10.1007/978-3-662-48350-3_19

we obtain a graph that shares many properties with real-world networks [21]. Furthermore, the Internet graph can be embedded in the hyperbolic space, preserving some metric properties [26,4].

Consequently, researchers have tried to measure this hyperbolic structure of complex networks, using Gromov's definitions of hyperbolicity [15], which works in any metric space, and does not rely on complicated structures not available in graphs (geodesics, connections, and so on). Intuitively, this parameter reflects how the metric space (distances) of a graph is close to the metric space of a tree. In particular, given an undirected graph $G = (V, E)$ (in this paper, all graphs will be undirected), the Gromov hyperbolicity of a quadruple of nodes $\delta(x, y, v, w)$ is defined as half the difference between the biggest two of the three sums $d(x, y) + d(v, w)$, $d(x, v) + d(y, w)$, and $d(x, w) + d(y, v)$, where $d(\cdot, \cdot)$ denotes the distance function between two nodes, that is, the lenght of the shortest path connecting the two nodes. The hyperbolicity of G is $\delta(G) = \max_{x,y,v,w \in V} \delta(x, y, v, w)$ (the smaller this value, the more hyperbolic the space is).

Several network properties are connected to the value of the hyperbolicity: here we will just recall some of them. In [8], it is shown that a small hyperbolicity implies the existence of efficient distance and routing labeling schemes. In [23], the authors observe that a small hyperbolicity, that is, a negative curvature of an interconnection network, implies a faster congestion within the core of the network itself, and in [18] it is suggested that this property is significant in the context of network security and can, for example, mitigate the effect of distributed denial of service attacks. In [12], instead, the hyperbolicity is used to implement a distance between trees, successively applied to the estimation of phylogenetic trees. From a more algorithmic point of view, it has been shown that several approximation algorithms for problems related to distances in graphs (such as diameter and radius computation [7], and minimum ball covering [9]) have an approximation ratio which depends on the hyperbolicity of the input graph. Moreover, some approximation algorithms with constant approximation factor rely on a data-structure whose size is proportional to the hyperbolicity of the input graph [20]. More in general, the hyperbolicity is connected to other important graph quantities, like treelength [7] and chordality [29]. In the field of the analysis of complex networks, the hyperbolicity and its connection with the size and the diameter of a network has been used in [2] in order to classify networks into three different classes, that is, strongly hyperbolic, hyperbolic, and non-hyperbolic, and to apply this classification to a small dataset of small biological networks (a more extensive analysis of the hyperbolicity of real-world networks has been also recently done in [5]). In general, it is still not clear whether the hyperbolicity value is small in all real-world networks (as it seems from [19,2]), or it is a characteristic of specific networks (as it seems from [1]). Finally, the hyperbolicity of random graphs has been analyzed in the case of several random graph models, such as the Erdös-Renyi model [24] and the Kleinberg model [6]. Moreover, in this latter paper, it is stated that the design of more efficient exact algorithms for the computation of the hyperbolicity would be of interest.

Indeed, it is clear that the hyperbolicity computation problem is polynomial-time solvable by the trivial algorithm that computes $\delta(x, y, v, w)$ for each quadruple of nodes. However, the running-time is $\mathcal{O}(n^4)$, where n is the number of nodes, which is prohibitive for real-world networks. The best known algorithm uses fast (max,min)-matrix multiplication algorithm to obtain a running time $\mathcal{O}(n^{3.69})$ [14], and it has been shown that hyperbolicity cannot be computed in $\mathcal{O}(n^{3.05})$ time, unless there exists a faster algorithm for (max,min)-matrix multiplication than currently known. Such running times are prohibitive for analysing large-scale graphs with more than 10,000 nodes.

Recently, new algorithms have been developed [11,10]. Although these algorithms have worst-case running time $\mathcal{O}(n^4)$, they perform well in practice, making it possible to compute the hyperbolicity of graphs with up to 50,000 nodes.

In this paper, we propose a new algorithm to compute the hyperbolicity of a graph, taking some ideas from the algorithm in [11]. The new algorithm heavily improves the performances through significant speed-ups in the most time-consuming part. The speed-ups will be so efficient that the running-time of the new algorithm will be dominated by the pre-processing part, which needs time $\mathcal{O}(mn)$, where m is the number of edges (we assume the input graph to be connected, and consequently $m + n = \mathcal{O}(m)$). This way, the $\mathcal{O}(n^4)$ bottleneck is almost removed, at least in practical instances. For this reason, we will be able for the first time to compute the hyperbolicity of graphs with up to 200,000 nodes. We will experimentally show these claims by analyzing a big dataset of real-world networks of different kinds. Finally, we apply our algorithm to the computation of the hyperbolicity of random graphs. In particular, in the Chung-Lu model, we compute the hyperbolicity of graphs with up to 200,000 nodes, improving previous experiments that stop at 1,100 nodes [13].

In Section 2, we will sketch the main features of the algorithm in [11], in order to make the paper self-contained. Section 3 will explain how to modify that algorithm, in order to obtain significant improvements, and Section 4 contains our experimental results. Finally, in Section 5, we apply our algorithm to the analysis of the hyperbolicity of random graphs, as suggested in [6], and Section 6 concludes the paper.

2   CCL: The Currently Best Available Algorithm

In this section, we will sketch the algorithm proposed in [11], whose main ideas and lemmas will also be used in the next section. This algorithm improves the trivial algorithm by analyzing quadruples in a specific order, and by cutting the exploration of the quadruples as soon as some conditions are satisfied. We will name this algorithm CCL, from the initials of the surnames of the authors. In particular, for each quadruple (p, q, r, s) of nodes, CCL computes $\tau(p, q; r, s)$ as defined below, instead of computing $\delta(p, q, r, s)$.

$$\tau(p, q; r, s) = \frac{d(p,q) + d(r,s) - \max\{d(p,r) + d(q,s), d(p,s) + d(q,r)\}}{2}.$$

Algorithm 1: Hyperbolicity algorithm proposed in [11], CCL.

Let $P = (\{x_1, y_1\}, \ldots, \{x_N, y_N\})$ be the list of far apart pairs, in decreasing order of distance.

$\delta_L \leftarrow 0$;

for $i \in [1, N]$ **do**

    if $d(x_i, y_i) \leq 2\delta_L$ **then**

        $\lfloor$ **return** δ_L;

    for $j \in [1, i-1]$ **do**

        $\lfloor$ $\delta_L \leftarrow \max(\delta_L, \tau(x_i, y_i; x_j, y_j))$;

return δ_L;

Note that $\delta(G) = \max_{p,q,r,s \in V} \tau(p, q; r, s)$, because if $d(p, q) + d(r, s)$ is the maximum sum, then $\tau(p, q; r, s) = \delta(p, q, r, s)$, otherwise $\tau(p, q; r, s) < 0$.

Lemma 1 (Lemma 3.2 of [11]). *For any quadruple (p, q, r, s) of nodes,* $2\tau(p, q; r, s) \leq \min(d(p, q), d(r, s))$.

In order to exploit this lemma, CCL stores all the $N = \frac{n(n-1)}{2}$ pairs of nodes inside a sequence $P = (\{x_1, y_1\}, \ldots, \{x_N, y_N\})$, in decreasing order of distance (that is, if $d(x_i, y_i) > d(x_j, y_j)$, then $i < j$). For each i, CCL iterates over all pairs $\{x_j, y_j\}$ with $j < i$, and computes $\tau(x_i, y_i; x_j, y_j)$, storing the maximum value found in a variable δ_L (clearly, δ_L is a lower bound for $\delta(G)$). Even if iterating over the whole sequence P would lead us to the trivial algorithm, by applying Lemma 1 we may cut the exploration as soon as $d(x_i, y_i) \leq 2\delta_L$, because the τ value of all remaining quadruples is at most $d(x_i, y_i)$.

A further improvement is provided by the following lemma.

Lemma 2 ([28]). *Let x, y, v, w be four nodes, and let us assume that there exists an edge (x, x') such that $d(x', y) = d(x, y) + 1$. Then, $\tau(x, y; v, w) \leq \tau(x', y; v, w)$.*

Definition 1. *A pair $\{x, y\}$ is far apart if there is no edge (x, x') such that $d(x', y) = d(x, y) + 1$ and no edge (y, y') such that $d(x, y') = d(x, y) + 1$.*

By Lemma 2, CCL only needs to analyze far apart pairs, and, hence, in the following we will denote by P (respectively, N) the list (number) of far apart pairs. The pseudo-code of CCL is provided in Algorithm 1.

Other improvements of this algorithm involve pre-processing the graph: first of all, we may analyze each biconnected component separately [11, Section 2], then, we may decompose the graph by modular decomposition, split decomposition [28], and clique decomposition [10].

3   HYP: The New Algorithm

In this section, we propose a new algorithm, that we will call HYP, that improves upon Algorithm 1 by further reducing the number of quadruples to consider.

Algorithm 2: The new algorithm, HYP

Let P $= (\{x_1, y_1\}, \ldots, \{x_N, y_N\})$ be the ordered list of far apart pairs.
$\delta_L \leftarrow 0$;
mate$[v] \leftarrow \emptyset$ for each v;
for $i \in [1, N]$ **do**
    if $d(x_i, y_i) \leq 2\delta_L$ **then**
        $\lfloor$ **return** δ_L;
    (acceptable , valuable) $\leftarrow$ computeAccVal ();
    for $v \in$ valuable **do**
        for $w \in$ mate$[v]$ **do**
            if $w \in$ acceptable **then**
                $\lfloor$ $\delta_L \leftarrow \max(\delta_L, \tau(x_i, y_i; v, w))$;
    add y_i to mate$[x_i]$;
    add x_i to mate$[y_i]$;
return δ_L

3.1 Overview

The new algorithm HYP speeds-up the inner for loop in Algorithm 1, by decreasing the number of pairs to be analyzed. In particular, let us fix a pair (x_i, y_i) in the outer for loop and a lower bound δ_L: a node v is (i, δ_L)-*skippable* or simply *skippable* if, for any w, $\tau(x_i, y_i; v, w) \leq \delta_L$. It is clear that if a node v is skippable, the algorithm could skip the analysis of all quadruples containing x_i, y_i, and v. Even if it is not easy to compute the set of skippable nodes, we will define easy-to-verify conditions that imply that a node v is skippable (Section 3.2): a node not satisfying any of these conditions will be named (i, δ_L)-*acceptable* or *acceptable* . Our algorithm will then discard all quadruples (x_i, y_i, v, w) where either v or w is not acceptable.

Furthermore, we will define another condition such that if $\tau(x_i, y_i; v, w) > \delta_L$, then either v or w must satisfy this condition (an acceptable node also satisfying this condition will be defined (i, δ_L)-*valuable* or *valuable*). Hence, our algorithm will not only discard all quadruples (x_i, y_i, v, w) where either v or w is not acceptable, but also all quadruples where both v and w are not valuable.

In order to apply these conditions, when analyzing a pair (x_i, y_i), HYP computes the set of acceptable and valuable nodes in time $\mathcal{O}(n)$ (actually, several nodes are skipped, thanks to implementation tricks, so that the time might be much smaller). Then, for each valuable node v, it analyzes pairs (v, w) preceding (x_i, y_i) such that w is acceptable. For this latter loop, we record for each node v the list mate$[v]$ of previously seen pairs (v, w), and then test each time if w is acceptable. The pseudo-code for HYP is provided by Algorithm 2.

Lemma 3. *The algorithm is correct.*

Proof. First of all, $\delta_L \leq \delta(G)$ during the whole algorithm, so we only have to rule out the possibility that the output is strictly smaller than $\delta(G)$. Let x, y, v, w be a quadruple such that $\tau(x, y; v, w) = \delta(G)$. We may assume without loss of generality that $\{x, y\}$ and $\{v, w\}$ are far-apart (otherwise, we change the pairs using Lemma 2), and that $\{v, w\}$ is before $\{x, y\}$ in the ordering of pairs (otherwise, we swap the pairs). By Lemma 1, $d(x, y) \geq 2\delta(G) \geq 2\delta_L$ at any step of the algorithm: if $2\delta_L = d(x, y) \geq 2\delta(G)$ at some step, the algorithm is correct because δ_L never decreases. Otherwise, the pair $\{x, y\}$ is analyzed at some step i, v and w will be (i, δ_L)-acceptable, and either v or w will be (i, δ_L)-valuable (by definition of acceptable and valuable). Hence, in the inner loop, $\tau(x, y; v, w)$ is computed, and afterwards $\delta_L = \tau(x, y; v, w) = \delta(G)$.   □

It remains to define which nodes are acceptable and which nodes are valuable, which is the topic of the following section.

3.2   Acceptable and Valuable Nodes

First of all, let us fix i and δ_L, since in this section they play the role of parameters. Moreover, for the sake of clarity, we will denote x_i and y_i simply by x and y. The following lemmas will provide conditions implying that v is skippable, that is, there is no pair $\{v, w\}$ appearing in P before $\{x, y\}$ such that $\tau(x, y; v, w) > \delta_L$. An acceptable node must not satisfy these conditions. The first lemma holds by definition of skippable.

Lemma 4. *If v does not belong to any (far-apart) pair $\{v, w\}$ before $\{x, y\}$ in P, then v is skippable.*

A second possibility to prove that a node is skippable is given by a simple corollary of the following lemma.

Lemma 5 ([11]). *For each quadruple of nodes (x, y, v, w), $\tau(x, y; v, w) \leq \min_{a,b \in \{x,y,v,w\}} d(a, b)$.*

Corollary 1. *If $d(x, v) \leq \delta_L$ or $d(y, v) \leq \delta_L$, then v is skippable.*

Proof. If the assumptions are satisfied, for each w, $\tau(x, y; v, w) \leq d(x, v) \leq \delta_L$, or $\tau(x, y; v, w) \leq d(y, v) \leq \delta_L$.

The next lemmas make use of the notion of the eccentricity $e(v)$ of a node, defined as $\max_{w \in V} d(v, w)$.

Lemma 6. *If $2e(v) - d(x, v) - d(y, v) < 4\delta_L + 2 - d(x, y)$, then v is skippable.*

Proof. By contradiction, let us suppose that there exists a node w such that $\delta_L < \tau(x, y; v, w)$. Then, $2\delta_L + 1 \leq 2\tau(x, y; v, w) = d(x, y) + d(v, w) - \max(d(x, v) + d(y, w), d(x, w) + d(y, v)) \leq d(x, y) + d(v, w) - \frac{1}{2}(d(x, v) + d(y, w) + d(x, w) + d(y, v)) \leq d(x, y) + e(v) - \frac{1}{2}(d(x, v) + d(y, v)) - \frac{1}{2}d(x, y)$. By rearranging this inequality, we would contradict the hypothesis.   □

Lemma 7. *If $e(v) + d(x, y) - 3\delta_L - \frac{3}{2} < \max\{d(x, v), d(y, v)\}$, then v is skippable.*

Proof. By contradiction, let us suppose that there exists a node w such that $\delta_L < \tau(x,y;v,w)$. By Corollary 1, $d(y,w) > \delta_L$, that is, $d(y,w) \geq \delta_L + \frac{1}{2}$. Consequently, $2\delta_L + 1 \leq 2\tau(x,y;v,w) = d(x,y) + d(v,w) - \max(d(x,v) + d(y,w), d(x,w) + d(y,v)) \leq d(x,y) + d(v,w) - d(x,v) - d(y,w) \leq d(x,y) + e(v) - d(x,v) - \delta_L - 1/2$. By exchanging the roles of x and y, we obtain $2\delta_L + 1 \leq d(x,y) + e(v) - d(y,v) - \delta_L - \frac{1}{2}$. These two inequalities contradict the hypothesis. □

Definition 2. *A node is acceptable if it does not satisfy the assumptions of Lemmas 4, 6 and 7 and Corollary 1.*

Remark 1. Lemma 4 can be verified "on the fly", by keeping track of already-seen nodes. The other items are clearly verifiable in time $\mathcal{O}(1)$ for each node, and consequently the running-time of this operation is $\mathcal{O}(|\{v \in V : \exists\{v,w\} < \{x,y\}\}|)$, which is less than or equal to $\mathcal{O}(n)$.

Remark 2. A variation of HYP verifies on the fly Lemma 7 and not Lemma 4. At the beginning of the algorithm, for each node x, we pre-compute a list of all nodes v in decreasing order of $e(v) - d(x,v)$ (in time $\mathcal{O}(n^2 \log n)$). Then, when computing acceptable nodes, we scan the list corresponding to x, and we stop as soon as we find a node v such that $e(v) + d(x,y) - 3\delta_L - \frac{3}{2} < d(x,v)$. In this case, the running-time of this operation is $\mathcal{O}(|\{v \in V : e(v) + d(x,y) - 3\delta_L - \frac{3}{2} \geq d(x,v)\}|)$. Since we may swap the roles of x and y, at each step, our algorithm chooses between x and y the less central node, according to *closeness centrality* measure [3].

The two remarks above correspond to two versions of our algorithm HYP, that we will call HYP1 and HYP2, respectively. Now we need to define valuable nodes, using the following lemma, which involves a given node c (formally, we would need to write c-valuable instead of valuable). All choices of c are feasible, but if c is "central", the running-time improves. We decided to set c as the most central node according to *closeness centrality* measure [3].

Lemma 8. *Let c be any fixed node, and, for any node z, let $f_c(z) := \frac{1}{2}(d(x,y) - d(x,z) - d(z,y)) + d(z,c)$. Then, for any two nodes v and w, we have $2\tau(x,y;v,w) \leq f_c(v) + f_c(w)$.*

Proof. We have that, $2\tau(x,y;v,w) = d(x,y) + d(v,w) - \max(d(x,v) + d(y,w), d(x,w) + d(y,v)) \leq d(x,y) + d(v,c) + d(c,w) - (d(x,v) + d(y,w) + d(x,w) + d(y,v))/2 = f_c(v) + f_c(w)$. The lemma is thus proved. □

As a consequence, if $2\tau(x,y;v,w) > 2\delta_L$, either $f_c(v) > \delta_L$ or $f_c(w) > \delta_L$. This justifies the following definition.

Definition 3. *An acceptable node v is c-valuable or valuable if $f_c(v) > \delta_L$.*

Hence, if $\tau(x,y;v,w) > \delta_L$, then at least one of v and w must be valuable.

Remark 3. It is possible to compute if an acceptable node is valuable in time $\mathcal{O}(1)$, so there is no time overhead for the computation of valuable nodes.

4 Experimental Results

In this section, we compare the best algorithm available until now [11] (CCL, whose pseudo-code is Algorithm 1), with the two versions of our new algorithm, denoted as HYP1 and HYP2 (using Remark 1 and Remark 2, respectively). Other available algorithms are the trivial algorithm, which is significantly outperformed by CCL in [11], and the algorithm in [14]. This latter is not practical, because it is based on fast matrix multiplication: indeed using $\mathcal{O}(n^3)$ matrix multiplication implementation we get the same time of the trivial algorithm. As far as we know, no other competitors are available.

Both CCL and our algorithm, in both versions HYP1 and HYP2, share the following preprocessing (see [11]):

- compute biconnected components to treat them separately;
- computing the distances between all pairs of nodes;
- computing and sorting the list P of all far-apart pairs.

All the operations above need time $\mathcal{O}(m \cdot n)$ and they will be not part of the comparison since they are common to all three algorithms. Our tests were performed on an AMD Opteron(TM) Processor 6276, running Fedora release 21. Our source code has been written in C and compiled with gcc 4.9.2 with optimization level 3. The code is available at `piluc.dsi.unifi.it/lasagne`.

We have collected a dataset composed by 62 graphs (available with the code) of different kinds: social, peer-to-peer, autonomous systems, citation networks, and so on. The networks were selected from the well-known SNAP dataset (`http://snap.stanford.edu/`), and from CAIDA (`http://www.caida.org`). The number of nodes varies between 4,039 and 265,009 (1,396 and 50,219 after the preprocessing).

Number of quadruples. The first comparison analyzes how many quadruples are processed before the hyperbolicity is computed - note that HYP1 and HYP2 analyze the same number of quadruples, since the only difference between them is how acceptable and valuable nodes are computed. The results are summarized in Figure 1a, which plots the number of quadruples processed by the new algorithms with respect to CCL. More precisely, for each graph G, we draw a point in position (x, y) if CCL analyzed x quadruples and both HYP1 and HYP2 analyzed y quadruples to compute the hyperbolicity of G. We observe that the new algorithm analyzes a much smaller number of quadruples, ranging from one hundred to few millions, drastically outperforming CCL, which often analyzes millions of millions of quadruples, and even billions of millions. Of course, the new algorithm is never outperformed, because the quadruples analyzed by HYP1 and HYP2 are always a subset of the quadruples analyzed by CCL.

Running time. Since the computation of acceptable and valuable nodes has a non-negligible impact on the total running time, for a more fair comparison, we have also considered the running time of the algorithms. In Figure 1b we report the time used by HYP1 and HYP2 with respect to the time used by CCL. Also in this case, both HYP1 and HYP2 drastically outperform CCL: the running-time is lower in most of the graphs, and the only cases where CCL is faster

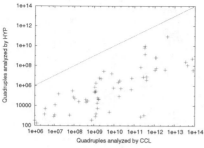

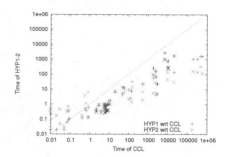

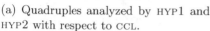

(a) Quadruples analyzed by HYP1 and HYP2 with respect to CCL.

(b) Time used by HYP1 and HYP2 with respect to CCL.

Fig. 1. Comparisons of quadruples analyzed and running-time of HYP1, HYP2, and CCL. The line $y = x$ separates the region where CCL is better (above) from the region where HYP1 and HYP2 are better (below).

need a very small amount of time (a few seconds at most). On the other hand, the new algorithms are much faster when the total time is big: for instance, on input `as-20120601.caida`, CCL needs at least one week (this lower bound was computed from the actual hyperbolicity and all the distances between the nodes), while HYP1 is 367 times faster, needing less than half an hour, and HYP2 is more than 5,000 times faster, needing less than two minutes. Similar results hold in all graphs where the total running-time is big. This does not only mean that we have improved upon CCL, but also that the improvement is concentrated on inputs where the running-time is high. Furthermore, we observe that on all graphs the total running time of algorithm HYP2 is less than half an hour: this means that, even if the worst-case complexity of this algorithm is $\mathcal{O}(n^4)$, in practice, the time used by the second part is comparable to the preprocessing time, which is $\mathcal{O}(m \cdot n)$. Hence, from a practical point of view, since real-world graphs are usually sparse, the algorithm may be considered quadratic.

5   Synthetic Graphs

Recently, a different line of research has tried to compute asymptotic values for the hyperbolicity of random graphs, when the number of nodes n tends to infinity. The simplest model considered is the Erdös-Renyi random graph $G_{n,m}$, that is, we choose a graph with n nodes and m edges uniformly at random. In this model, it has been proved that the hyperbolicity tends to infinity [24], and, if m is "much bigger than n", exact asymptotics for δ have been computed [22]. Instead, the hyperbolicity of sparse Erdös-Renyi graphs is not known, and it is mentioned as an open problem in [22]. Among the other possible models, the Chung-Lu model and the Configuration Model stand out for their simplicity (for more background, we refer to [17]). On these models, as far as we know, it was only proved [27] that the hyperbolicity of a graph generated through the

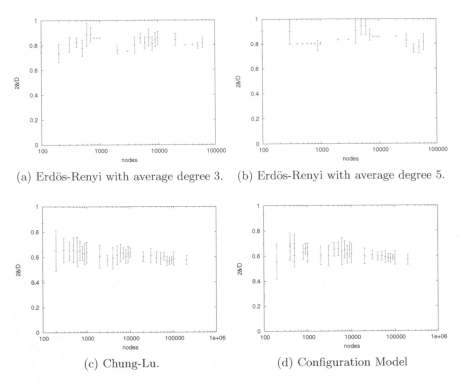

(a) Erdös-Renyi with average degree 3.   (b) Erdös-Renyi with average degree 5.

(c) Chung-Lu.   (d) Configuration Model

Fig. 2. Mean and Standard Deviation of the ratio $\frac{2\delta}{D}$ at growing values of n.

Chung-Lu model tends to infinity if the maximum and minimum degree are "close to each other" (meaning that their ratio is smaller than $2^{\frac{1}{3}}$). Other models were analyzed in [6]: also in that paper, the estimation of the hyperbolicity of random graphs of different kind is mentioned as an open problem.

Following [6], we use our algorithm to shed some light on the behavior of these random graphs, at least experimentally, in order to help formulating sensible conjectures on possible asymptotics. In particular, we have restricted our attention to four examples, chosen among the models where exact asymptotics have not been proved: Erdös-Renyi random graphs with $m = 3n$ and $m = 5n$, and graphs generated through the Chung-Lu and the Configuration Model, with power-law degree distribution with exponent 2.5 (similar to the degree distribution of several real-world networks [25]). For each number of nodes $n = k \cdot 10^i$ where $k < 10$ and $i \geq 2$, we have generated 10 graphs and we have computed their hyperbolicity. More precisely, we have computed the value $\frac{2\delta}{D}$, where D is the diameter, which is always between 0 and 1 because of Lemma 1: this value might be more interesting than the plain hyperbolicity value, since, for most models, asymptotics for the diameter are known. We believe that this ratio can then be used to formulate sensible conjectures. Figure 2 shows the average value of $\frac{2\delta}{D}$ and the corresponding standard error over the 10 measures performed.

We have been able to compute the hyperbolicity of Erdös-Renyi graphs with up to 60,000 nodes, and graphs generated with the Configuration Model or the Chung-Lu model with up to 200,000 nodes. In all models considered, it is quite evident that the ratio $\frac{2\delta}{D}$ does not tend to 0, and consequently $\delta = \Theta(D)$. Furthermore, the ratio in Erdös-Renyi graphs is not very far from 1, even if the results are not precise enough to discriminate between $\delta = \frac{D}{2}$ or $\delta = cD$ for some $c < \frac{1}{2}$. Instead, in graphs generated through the Configuration Model or the Chung-Lu model, this ratio seems to tend to a value between 0.5 and 0.7.

6   Conclusion and Open Problems

In this paper, we have provided a new and more efficient algorithm to compute the hyperbolicity of a graph: even if the running time is $\mathcal{O}(n^4)$ in the worst case, it turns out to be $\mathcal{O}(m \cdot n)$ in practice. As an example of application, we have studied the hyperbolicity of random graphs. The space requirement of the algorithm, as well as of its predecessors, is $\mathcal{O}(n^2)$: in our case this is needed to store all distances and the list of far-apart pairs. It would be nice to better deal with memory usage (for instance, working on the disk) or avoiding the computation and storage of all pairwise distances by using lower and upper bounds instead. Furthermore, this algorithm may be parallelized, by analyzing at the same time different nodes v, or different pairs (x, y). An open issue is determining how parallelization can improve performances. The algorithm can be adapted to deal with weighted graphs. On the other hand, a widely accepted definition of hyperbolicity for directed graphs is still missing. Finally, it would be nice to prove more precise asymptotics for the hyperbolicity of random graphs.

References

1. Albert, R., DasGupta, B., Mobasheri, N.: Topological implications of negative curvature for biological and social networks. Physical Review E 89(3), 32811 (2014)
2. Alrasheed, H., Dragan, F.F.: Core-Periphery Models for Graphs based on their delta-Hyperbolicity: An Example Using Biological Networks. Studies in Computational Intelligence 597, 65–77 (2015)
3. Bavelas, A.: Communication patterns in task-oriented groups. Journal of the Acoustical Society of America 22, 725–730 (1950)
4. Boguna, M., Papadopoulos, F., Krioukov, D.: Sustaining the Internet with Hyperbolic Mapping. Nature Communications 62 (October 2010)
5. Borassi, M., Chessa, A., Caldarelli, G.: Hyperbolicity Measures "Democracy" in Real-World Networks. Preprint on arXiv, pp. 1–10 (March 2015)
6. Chen, W., Fang, W., Hu, G., Mahoney, M.: On the hyperbolicity of small-world and treelike random graphs. Internet Mathematics, pp. 1–40 (2013)
7. Chepoi, V., Dragan, F.F., Estellon, B., Habib, M., Vaxès, Y.: Notes on diameters, centers, and approximating trees of δ-hyperbolic geodesic spaces and graphs. Electronic Notes in Discrete Mathematics 31, 231–234 (2008)
8. Chepoi, V., Dragan, F.F., Estellon, B., Habib, M., Vaxès, Y., Xiang, Y.: Additive Spanners and Distance and Routing Labeling Schemes for Hyperbolic Graphs. Algorithmica 62(3-4), 713–732 (2012)

9. Chepoi, V., Estellon, B.: Packing and covering δ-hyperbolic spaces by balls. In: Charikar, M., Jansen, K., Reingold, O., Rolim, J.D.P. (eds.) RANDOM 2007 and APPROX 2007. LNCS, vol. 4627, pp. 59–73. Springer, Heidelberg (2007)
10. Cohen, N., Coudert, D., Ducoffe, G., Lancin, A.: Applying clique-decomposition for computing Gromov hyperbolicity. Research Report RR-8535, HAL (2014)
11. Cohen, N., Coudert, D., Lancin, A.: On computing the Gromov hyperbolicity. ACM J. Exp. Algor. (2015)
12. Dress, A., Huber, K., Koolen, J., Moulton, V., Spillner, A.: Basic Phylogenetic Combinatorics. Cambridge University Press, Cambridge (2011)
13. Fang, W.: On Hyperbolic Geometry Structure of Complex Networks. Master's thesis, MPRI at ENS and Microsoft Research Asia (2011)
14. Fournier, H., Ismail, A., Vigneron, A.: Computing the Gromov hyperbolicity of a discrete metric space. Information Processing Letters 115(6-8), 576–579 (2015)
15. Gromov, M.: Hyperbolic groups. In: Essays in Group Theory, Springer, Heidelberg (1987)
16. Havlin, S., Cohen, R.: Complex networks: structure, robustness and function. Cambridge University Press, Cambridge (2010)
17. Hofstad, R.V.D.: Random Graphs and Complex Networks (2014)
18. Jonckheere, E.A., Lohsoonthorn, P.: Geometry of network security. In: American Control Conference, vol. 2, pp. 976–981. IEEE, Boston (2004)
19. Kennedy, W.S., Narayan, O., Saniee, I.: On the Hyperbolicity of Large-Scale Networks. CoRR, abs/1307.0031:1–22 (2013)
20. Krauthgamer, R., Lee, J.R.: Algorithms on negatively curved spaces. In: IEEE Symposium on Foundations of Computer Science (FOCS), pp. 119–132 (2006)
21. Krioukov, D., Papadopoulos, F., Kitsak, M., Vahdat, A.: Hyperbolic Geometry of Complex Networks. Physical Review E 82(3), 36106 (2010)
22. Mitche, D., Pralat, P.: On the hyperbolicity of random graphs. The Electronic Journal of Combinatorics 21(2), 1–24 (2014)
23. Narayan, O., Saniee, I.: The Large Scale Curvature of Networks. Physical Review E 84, 66108 (2011)
24. Narayan, O., Saniee, I., Tucci, G.H.: Lack of Hyperbolicity in Asymptotic Erdös–Renyi Sparse Random Graphs. Internet Mathematics, 1–10 (2015)
25. Newman, M.E.J.: The Structure and Function of Complex Networks. SIAM Review 45(2), 167–256 (2003)
26. Papadopoulos, F., Krioukov, D., Boguna, M., Vahdat, A.: Greedy forwarding in scale-free networks embedded in hyperbolic metric spaces. ACM SIGMETRICS Performance Evaluation Review 37(2), 15–17 (2009)
27. Shang, Y.: Non-Hyperbolicity of Random Graphs with Given Expected Degrees. Stochastic Models 29(4), 451–462 (2013)
28. Soto Gómez, M.A.: Quelques propriétés topologiques des graphes et applications à internet et aux réseaux. PhD thesis, Univ. Paris Diderot, Paris 7 (2011)
29. Wu, Y., Zhang, C.: Hyperbolicity and chordality of a graph. The Electronic Journal of Combinatorics 18(1), 1–22 (2011)

Towards Single Face Shortest Vertex-Disjoint Paths in Undirected Planar Graphs

Glencora Borradaile, Amir Nayyeri, and Farzad Zafarani

School of Electrical Engineering and Computer Science,
Oregon State University, Corvallis, Oregon, US
{glencora,nayyeria,zafaranf}@eecs.oregonstate.edu

Abstract. Given k pairs of terminals $\{(s_1, t_1), \ldots, (s_k, t_k)\}$ in a graph G, the min-sum k vertex-disjoint paths problem is to find a collection $\{Q_1, Q_2, \ldots, Q_k\}$ of vertex-disjoint paths with minimum total length, where Q_i is an s_i-to-t_i path between s_i and t_i. We consider the problem in planar graphs, where little is known about computational tractability, even in restricted cases. Kobayashi and Sommer propose a polynomial-time algorithm for $k \leq 3$ in undirected planar graphs assuming all terminals are adjacent to at most two faces. Colin de Verdière and Schrijver give a polynomial-time algorithm when all the sources are on the boundary of one face and all the sinks are on the boundary of another face and ask about the existence of a polynomial-time algorithm provided all terminals are on a common face.

We make progress toward Colin de Verdière and Schrijver's open question by giving an $O(kn^5)$ time algorithm for undirected planar graphs when $\{(s_1, t_1), \ldots, (s_k, t_k)\}$ are in counter-clockwise order on a common face.

1 Introduction

Given k pairs of terminals $\{(s_1, t_1), \ldots, (s_k, t_k)\}$, the k *vertex-disjoint* paths problem asks for a set of k disjoint paths $\{Q_1, Q_2, \ldots, Q_k\}$, in which Q_i is a path between s_i and t_i for all $1 \leq i \leq k$. As a special case of the multi-commodity flow problem, computing vertex disjoint paths has found several applications, for example in VLSI design [KvL84], or network routing [ORS93, SM05]. It is one of Karp's NP-hard problems [Kar74] even for undirected planar graphs if k is part of the input [MP93]. However, there are polynomial time algorithms if k is a constant for general undirected graphs [RS95, KW10]. In general directed graphs, the k-vertex-disjoint paths problem is NP-hard even for $k = 2$ [FHW80] but is fixed parameter tractable with respect to parameter k in directed planar graphs [Sch94, CMPP13].

Surprisingly, much less is known for the optimization variant of the problem, *minimum-sum k vertex-disjoint paths problem (k-min-sum)*, where a set of disjoint paths with *minimum total length* is desired. For example, the 2-min-sum problem and the 4-min-sum problem are open in directed and undirected planar graphs, respectively, even when the terminals are on a common face;

© Springer-Verlag Berlin Heidelberg 2015
N. Bansal and I. Finocchi (Eds.): ESA 2015, LNCS 9294, pp. 227–238, 2015.
DOI: 10.1007/978-3-662-48350-3_20

neither polynomial-time algorithms nor hardness results are known for these problems [KS10]. Bjorklund and Husfeldt gave a randomized polynomial time algorithm for the min-sum two vertex-disjoint paths problem in general undirected graphs [BH14]. Kobayashi and Sommer provide a comprehensive list of similar open problems (Table 2 [KS10]).

One of a few results in this context is due to Colin de Verdière and Schrijver [VS11]: a polynomial time algorithm for the k-min-sum problem in a (directed or undirected) planar graph, given all sources are on one face and all sinks are on another face [VS11]. In the same paper, they ask about the existence of a polynomial time algorithm provided all the terminals (sources and sinks) are on a common face. If the sources and sinks are ordered so that they are in the order $s_1, s_2, \ldots, s_k, t_k, t_{k-1}, \ldots, t_1$ around the boundary, then the k-min-sum problem can be solved by finding a min-cost flow from $s_1, s_2, \ldots, s_k$ to $t_k, t_{k-1}, \ldots, t_1$. For $k \leq 3$ in undirected planar graphs with the terminals in arbitrary order around the common face, Kobayashi and Sommer give an $O(n^4 \log n)$ algorithm [KS10][1]. In this paper, we give the first polynomial-time algorithm for an arbitrary number of terminals on the boundary of a common face, which we call F, so long as the terminals alternate along the boundary. Formally, we prove:

Theorem 1. *There exists an $O(kn^5)$ time algorithm to solve the k-min-sum problem, provided that the terminals $s_1, t_1, s_2, t_2, \ldots, s_k, t_k$ are in counter-clockwise order on the boundary of the graph.*

Definitions and assumptions. We use standard notation for graphs and planar graphs; see full version for details. For simplicity, we assume that the terminal vertices are distinct. One could imagine allowing $t_i = s_{i+1}$; our algorithm can be easily modified to handle this case. We also assume that the shortest path between any two vertices of the input graph is unique as it significantly simplifies the presentation of our result; this assumption can be enforced using a perturbation technique [MVV87].

2   Structural Properties

In this section, we present fundamental properties of the optimum solution that we exploit in our algorithm. To simplify the exposition, we search for pairwise disjoint walks rather than simple paths and refer to a set of pairwise disjoint walks connecting corresponding pair of terminals as a *feasible* solution. Indeed, in an optimal solution, the walks are simple paths.

Let $\{Q_1, Q_2, \ldots, Q_k\}$ be an optimal solution, where Q_i is a s_i-to-t_i path and let $\{P_1, P_2, \ldots, P_k\}$ be the set of shortest paths, where P_i is the s_i-to-t_i shortest path. These shortest paths together with the boundary of the graph, ∂G, define internally disjoint regions of the plane. Specifically, we define R_i to be the subset

---

[1] Kobayashi and Sommer also describe algorithms for the case where terminals are on two different faces, and $k = 3$.

of the plane bounded by the cycle $P_i \cup C_i$, where C_i is the s_i-to-t_i subpath of ∂G that does not contain other terminal vertices. The following lemmas constrain the behavior of the optimal paths. In the interest of space, the proofs of these lemmas are in the full version of our paper.

Lemma 1. *For all $1 \leq i \leq k$, the path Q_i is inside R_i.*

We take the vertices of P_i and Q_i to be ordered along these paths from s_i to t_i.

Lemma 2. *For $u, v \in Q_i \cap P_i$, u precedes v in P_i if and only if u precedes v in Q_i.*

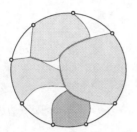 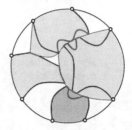

Fig. 1. (left) A 4-min sum instance; regions are shaded and borders are green. (right) A feasible solution; Type I and Type II subpaths are blue and red, respectively.

We call $R_i \cap R_j$ the *border* of R_i and R_j and denote it $B_{i,j}$. Note that a border can be a single vertex. Since we assume shortest paths are unique, $B_{i,j}$ is a single (shortest) path. Figure 1 illustrates borders for a 4-min-sum instance. The following lemma bounds the total number of borders.

Lemma 3. *There are $O(k)$ border paths.*

Consider a region R_i and consider the borders along P_i, $B_{i,i_1}, B_{i,i_2}, \ldots, B_{i,i_t}$. Observe that the intersections of the regions $R_{i_1}, R_{i_2}, \ldots, R_{i_t}$ with ∂G must be in a *clockwise* order. Let $\iota_1, \ldots, \iota_\ell$ be the subsequence of $i_1, \ldots, i_t$ of indices to regions that intersect Q_i. For $j \in \{\iota_1, \ldots, \iota_\ell\}$, let x_j and y_j be the first and last vertex of Q_i in $B_{i,j}$. Additionally, define $y_0 = s_i$ and $x_{\ell+1} = t_i$. We partition Q_i into a collection of subpaths of two types as follows.

Type I : For $h = 0, \ldots, \ell$, $Q_i[y_h, x_{h+1}]$ is a *Type I* subpath in region R_i.
Type II : For $h = 1, \ldots, \ell-1$, $Q_i[x_h, y_h]$ is a *Type II* subpath in region R_i. We say that $Q_i[x_h, y_h]$ is on the border $B_{i,j}$ containing x_h and y_h.

By this definition, all Type I paths are internally disjoint from all borders. By Lemma 2, each Type II path is internally disjoint from all borders except possibly the border that contains its endpoints, with which it may have several intersecting points. See Figure 1 for an illustration of Type I and II paths.

The following lemma demonstrates a key property of Type I paths, implying that (given their endpoints) they can be computed efficiently via a shortest path computation:

Lemma 4. *Let α be a Type I subpath in region R_i. Then α is the shortest path between its endpoints in R_i that is internally disjoint from all borders.*

A Type II path has a similar property if it is the only Type II path on the border that contains its endpoints.

Lemma 5. *Let β be a Type II subpath in region R_i on border $B_{i,j}$. Suppose there is no Type II path on $B_{i,j}$ inside R_j. Then β is the subpath of $B_{i,j}$ between its endpoints.*

The following lemma reveals a relatively more sophisticated structural property of Type II paths on shared borders.

Lemma 6. *Let β and γ be Type II subpaths in R_i and R_j on $B_{i,j}$, respectively, let x_i and y_i be the endpoints of β, and let x_j and y_j be the endpoints of γ. Then, $\{\beta, \gamma\}$ is the pair of paths with minimum total length with the following properties:*

(1) β is an x_i-to-y_i path inside R_i, and it is internally disjoint from all borders except possibly $B_{i,j}[x_i, y_i]$.
(2) γ is an x_j-to-y_j path inside R_j, and it is internally disjoint from all borders except possibly $B_{i,j}[x_j, y_j]$.

3   Algorithmic Toolbox

In this section, we describe algorithms to compute paths of Type I and II for given endpoints. These algorithms are key ingredients of our strongly polynomial time algorithm described in the next section. More directly, they imply an $n^{O(k)}$ time algorithm via enumerating the endpoints, which is sketched at the end of this section.

Each Type I path can be computed in linear time using the algorithms of Henzinger et al. [HKRS97]; they can also be computed in bulk in $O(n \log n)$ time using the multiple-source shortest path algorithm of Klein [Kle05] (although other parts of our algorithms dominate the shortest path computation). Similarly, a Type II path on a border $B_{i,j}$ can be computed in linear time provided it is the only path on $B_{i,j}$. Computing pairs of Type II paths on a shared border is slightly more challenging. To achieve this, our algorithm reduces this problem into a 2-min sum problem that can be solved in linear time via a reduction to the minimum-cost flow problem. The following lemma is implicit in the paper of Kobayashi and Sommer [KS10].

Lemma 7. *There exists a linear time algorithm to solve the 2-min sum problem on an undirected planar graph, provided the terminals are on the outer face.*

We reduce the computation of Type II paths to 2-min sum. The following lemma is a slightly stronger form of this reduction, which finds application in our strongly polynomial time algorithm.

Lemma 8. *Let R_i and R_j be two regions with border $B_{i,j}$ and let $x_i, y_i \in P_i$ and $x_j, y_j \in P_j$. A pair of paths (β, γ) with total minimum length and with the following properties can be computed in linear time.*

1. *β is an x_i-to-y_i path inside R_i, and it is internally disjoint from all borders except possibly $P_i[x_i, y_i] \cap B_{i,j}$.*
2. *γ is an x_j-to-y_j path inside R_j, and it is internally disjoint from all borders except possibly $P_j[x_j, y_j] \cap B_{i,j}$.*

3.1   An $n^{O(k)}$ Time Algorithm

The properties of Type I and II paths imply a naïve $n^{O(k)}$ time algorithm, which we sketch here. An optimal solution defines the endpoints of Type I and Type II paths, so we can simply enumerate over which borders contain endpoints of Type I and II paths and then enumerate over the choices of the endpoints. Consequently, there are zero, two, or four (not necessarily distinct) endpoints of Type I and II paths on $B_{i,j}$, or

$$1 + \binom{\ell(B_{i,j})}{2} + \binom{\ell(B_{i,j})}{4}$$

possibilities, which is $O(n^4)$ since $\ell(B_{i,j}) = O(n)$. Since there are $O(k)$ borders (Lemma 3), there are $n^{O(k)}$ endpoints to guess. Given the set of endpoints, we compute a feasible solution composed of the described Type I and II paths or determines that no such solution exists. Since Type I and II paths can be computed in polynomial time, the overall algorithm runs in $n^{O(k)}$ time.

4   A Fully Polynomial Time Algorithm

We give an $O(kn^5)$-time algorithm via dynamic programming over the regions. For two regions R_i and R_j that have a shared border $B_{i,j}$, R_i and R_j separate the terminal pairs into two sets: those terminals s_ℓ, t_ℓ for $\ell = i+1, \ldots, j-1$ and s_m, t_m for $m = j+1, \ldots, k, 1, \ldots, i-1$ (for $i < j$). Any s_ℓ-to-t_ℓ path that is in region R_ℓ cannot touch any s_m-to-t_m path that is in region R_m since R_ℓ and R_m are vertex disjoint. Therefore any influence the s_ℓ-to-t_ℓ path has on the s_m-to-t_m path occurs indirectly through the s_i-to-t_i and s_j-to-t_j paths. Our dynamic program is indexed by the shared borders $B_{i,j}$ and pairs of vertices on (a subpath of) P_i and (a subpath of) P_j; the vertices on P_i and P_j will indicate a *last point* on the boundary of R_i and R_j that a (partial) feasible solution uses.

We use a tree to order the shared borders for processing by the dynamic program. Since there are $O(k)$ borders (Lemma 3), the dynamic programming table has $O(kn^2)$ entries. We formally define the dynamic programming table below and show how to compute each entry in $O(n^3)$ time.

4.1   Dynamic Programming Tree

Let $\mathcal{R} = \{R_i\}_{i=1}^k$ and $\mathcal{B}$ be the set of all borders between all pairs of regions. We assume, without loss of generality, that $\mathcal{R}$ is connected, otherwise we split the problem into independent subproblems, one for each connected component of $\mathcal{R}$.

We define a graph T (that we will argue is a tree) whose edges are the shared borders $\mathcal{B}$ between the regions $\mathcal{R}$. Two distinct borders $B_{i,j}$ and $B_{h,\ell}$ are incident in this graph if there is an endpoint x of $B_{i,j}$ and y of $B_{h,\ell}$ that are connected by an x-to-y curve in $\mathbb{R}^2 \setminus F$ that does not touch any region $\mathcal{R}$ except at its endpoints x and y; see Figure 2. Note that this curve may be trivial (i.e. $x = y$). The vertices of T (in a non-degenerate instance) correspond one-to-one with components of $\mathbb{R}^2 \setminus (F \cup \mathcal{R})$ (plus some additional trivial components if three or more regions intersect at a point), or *non-regions*. The edges of T cannot form a cycle, since by the Jordan Curve Theorem this would define a disk that is disjoint from ∂F; an edge $B_{i,j}$ in the cycle bounds two regions, one of which would be contained by the disk, contradicting that each region shares a boundary with ∂F. Therefore T is indeed a tree. We use an embedding of T that is derived naturally from the embedding of G according to this construction. We use this tree to guide the dynamic program.

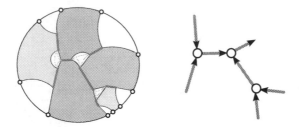

Fig. 2. (left) Thick green segments are borders, thin green curves are in $\mathbb{R}^2 \setminus (F \cup \mathcal{R})$ connecting borders that are incident in T. (right) The directed tree T used for dynamic programming.

By the correspondence of the vertices of T to non-regions, we have:

Observation 2. *The borders $B_{i,j}, B_{i,\ell}, \dots$ along a given region R_i form a path in T. The order of the borders from s_i to t_i along P_i is the same as in the path in T.*

Consider two edges $B_{i,j}$ and $B_{h,\ell}$ that are incident to the same vertex v of T and that are consecutive in a cyclic order of the edges incident to v in T's embedding. By Observation 2 and the embedding of T, there is a labeling of i, j, h, ℓ such that:

Observation 3. *Either $j = h$ or $B_{i,j}$ is the last border of R_j along P_j and $B_{h,\ell}$ is the first border of R_h along P_h.*

Root T at an arbitrary leaf. Since $\mathcal{R}$ is connected, the leaf of T is non-trivial; that is, it has an edge $B_{i,j}$ incident to it. By Observation 2, $B_{i,j}$ is (w.l.o.g.) the last border of R_i along P_i and the first border of R_j along P_j. By the correspondence of the vertices of T to non-regions, and the connectivity of $\mathcal{R}$, either $j = i + 1$ or $i = 1$ and $j = k$. For ease of notation, we assume that the terminals are numbered so that $i = 1$ and $j = k$. We get:

Observation 4. *Every non-root leaf edge of T corresponds to a border $B_{i,i+1}$.*

We consider the borders to be both paths in G and edges in T. In T we orient the borders toward the root. In G, this gives a well defined start $a_{i,j}$ and endpoint $b_{i,j}$ of the corresponding path $B_{i,j}$ (note that $a_{i,j} = b_{i,j}$ is possible). By our choice of terminal numbering and orientation of the edges of T, from s_i to t_i along P_i, $b_{i,j}$ is visited before $a_{i,j}$, and from s_j to t_j along P_j, $a_{i,j}$ is visited before $b_{i,j}$.

4.2   Dynamic Programming Table

We populate a dynamic programming table $D_{i,j}$ for each border $B_{i,j}$. $D_{i,j}$ is indexed by two vertices x and y: x is a vertex of $P_i[b_{i,j}, t_i]$ and y is a vertex of $P_j[s_j, b_{i,j}]$. $D_{i,j}[x, y]$ is defined to be the minimum length of a set of vertex-disjoint paths that connect:

$$x \text{ to } t_i, \ s_j \text{ to } y, \text{ and } s_h \text{ to } t_h \text{ for every } h = i + 1, \ldots, j - 1$$

These paths are illustrated in Figure 3. We interpret y as the last vertex of $P_j[s_j, b_{i,j}]$ that is used in this sub-solution and we interpret x as the first vertex of $P_i[b_{i,j}, t_i]$ that can be used in this sub-solution (or, more intuitively, the last vertex of the reverse of $P_i[b_{i,j}, t_i]$). By Lemma 1, each of the paths defining $D_{i,j}[x, y]$ are contained by their respective region.

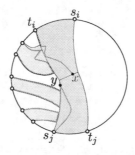

Fig. 3. An illustration of the paths defined by $D_{ij}[x, y]$.

Optimal solution. Given $D_{1,k}$, we can compute the value of the optimal solution. By Lemma 4, Q_1 and Q_k contain a shortest (possibly trivial) path from s_1 to a vertex x on P_1, and from a vertex y on P_k to t_k, respectively. Let y be the last vertex of $P_k[s_k, b_{1,k}]$ that Q_k contains and let x be the first vertex of $P_1[b_{1,k}, t_1]$ that Q_1 contains. Then, by Lemma 4, the optimal solution has length $D_{1,k}[x, y] + \ell(\alpha(s_1, x)) + \ell(\alpha(y, t_k))$ where α is a Type I path between the given vertices. The optimal solution can be computed in $O(n^2)$ time by enumerating over all choices of x and y. Computing all such Type I paths takes $O(n^2)$ since there are $O(n)$ such paths to compute, and each path can be found using the linear-time shortest paths algorithm for planar graphs [HKRS97].

Base case: Leaf edges of **T**. Consider a non-root leaf edge of T, which, by Observation 4, is $B_{i,i+1}$ for some i. Then $D_{i,i+1}[x, y]$ is the length of minimum vertex disjoint x-to-t_i and s_{i+1}-to-y paths in $R_i \cup R_{i+1}$. By Lemma 8, $D_{i,i+1}[x, y]$ can be computed in $O(n)$ time for any x and y and so $D_{i,i+1}$ can be populated in $O(n^2)$ time.

4.3   Non-base Case of the Dynamic Program

Consider a border $B_{i,j}$ and consider the edges of T that are children of $B_{i,j}$. These edges considered counter-clockwise around their common node of T correspond to borders $B_{i_1,j_1}, B_{i_2,j_2}, \ldots, B_{i_t,j_t}$ where $i \leq i_1 \leq j_1 \leq \cdots \leq i_t \leq j_t \leq j$. For simplicity of notation, we additionally let $j_0 = i$ and $i_{t+1} = j$. Then, by Observation 3, either $j_\ell = i_{\ell+1}$ or B_{i_ℓ,j_ℓ} is the last border on P_{j_ℓ} and $B_{i_{\ell+1},j_{\ell+1}}$ is the first border on $P_{i_{\ell+1}}$ for $\ell = 0, \ldots, t$.

An acyclic graph H to piece together sub-solutions. To populate $D_{i,j}$ we create a directed acyclic graph H with sources corresponding to vertices of $P_i[b_{i,j}, t_i]$ and sinks corresponding to $P_j[s_j, b_{i,j}]$. A source-to-sink (u-to-v) path in H will correspond one-to-one with vertex disjoint paths from:

$$u \text{ to } t_i, \ s_j \text{ to } v, \text{ and } s_h \text{ to } t_h \text{ for every } h = i+1, \ldots, j-1$$

Here u and v do not correspond to the vertices x and y that index $D_{i,j}$; to these vertex disjoint paths, we will need to append vertex disjoint x-to-u and v-to-y paths (which can be found using a minimum cost flow computation by Lemma 8).

The arcs of H are of two types: (a) Type I arcs and (b) sub-problem arcs. Directed paths in H alternate between these two types of arcs. The Type I arcs correspond to Type I paths and the endpoints of the Type I arcs correspond to the endpoints of the Type I paths. Sub-problem arcs correspond to the sub-solutions from the dynamic programming table and the endpoints of the sub-problem arcs correspond to the indices of the dynamic programming table (and so are the endpoints of the *incomplete* paths represented by the table). Note that vertices of a border may appear as either the first or second index to the dynamic programming table; in H, two copies of the border vertices are included so the endpoints of the resulting sub-solution arcs are distinct. Formally:

- Type I arcs go from vertices of P_{j_ℓ} to vertices of $P_{i_{\ell+1}}$ for $\ell = 0, \ldots, t$. Consider regions R_{j_ℓ} and $R_{i_{\ell+1}}$. There are two cases depending on whether or not $R_{j_\ell} = R_{i_{\ell+1}}$.
  - If $R_{j_\ell} = R_{i_{\ell+1}}$, then for every vertex x of $P_{j_\ell}[s_{j_\ell}, b_{i_\ell,j_\ell}]$ and every vertex y of $P_{j_\ell}[b_{i_{\ell+1},j_{\ell+1}}, t_{j_\ell}]$, we define a Type I arc from x to y with length equal to the length of the x-to-y Type I path.
  - If $R_{j_\ell} \neq R_{i_{\ell+1}}$, then for every vertex x of $P_{j_\ell}[s_{j_\ell}, b_{i_\ell,j_\ell}]$ and every vertex y of $P_{i_{\ell+1}}[b_{i_{\ell+1},j_{\ell+1}}, t_{j_\ell}]$, we define a Type I arc from x to y with length equal to the sum of the lengths of the x-to-t_{j_ℓ} and $s_{i_{\ell+1}}$-to-y Type I paths.
- Sub-problem arcs go from vertices of P_{i_ℓ} to vertices of P_{j_ℓ} for $\ell = 1, \ldots, t$. For every $\ell = 1, \ldots, t$ and every vertex x of P_{i_ℓ} and vertex y of P_{j_ℓ} (that are not duplicates of each other), we define a sub-problem arc from x to y with length equal to $D_{i_\ell,j_\ell}[x, y]$.

Shortest paths in H. By construction of H and the definition of D_{i_ℓ,j_ℓ}, for a source u and sink v, we have:

Observation 5. *There is a u-to-v path in H with length L if and only if there are vertex disjoint paths of total length L from u to t_i, s_j to v, and s_h to t_h for every $h = i+1, \ldots, j-1$.*

See Figure 4 for an illustration of the paths in G that correspond to a source-to-sink path in H. Let $H[u, v]$ denote the shortest u-to-v path in H (for a source u and a sink v). We will need to compute $H[u, v]$ for every pair of sources and sinks. Since every vertex in G appears at most twice in H, the size of H is $O(n^2)$ and for a given sink and for all sources, the shortest source-to-sink paths can be found in time linear in the size of H using dynamic programming. Repeating for all sinks results in an $O(n^3)$ running time to compute $H[u, v]$ for every pair of sources and sinks.[2]

Handling vertices that appear in more than two regions. As indicated, a vertex c may appear in more than two regions; this occurs when two or more borders share an endpoint. In the construction above, if c appears in only two regions, then, c can only be used as the endpoint of two sub-paths (whose endpoints meet to form a part of an s_i-to-t_i path in the global solution). However, suppose for example that c appears as an endpoint of both $B_{i,j}$ and $B_{i',j'}$ and so 4 copies of c are included in H (two copies for each of these borders). On the other hand, one need only *guess* which s_i-to-t_i path c should belong to first and construct H accordingly. There are only k possibilities to try.

Unfortunately, there may be $O(k)$ shared vertices among the borders B_{i_1,j_2}, $B_{i_2,j_2}, \ldots, B_{i_t,j_t}$ involved in populating $D_{i,j}$. It seems that for each of these $O(k)$ shared vertices, one would need to guess which s_i-to-t_i path it belongs to, resulting in an exponential dependence on k.

---

[2] Computing the length of the Type I paths is dominated by $O(n^3)$, but can be improved to $O(n \log n)$ time by running Klein's boundary shortest path algorithm [Kle05] in all regions, resulting in an $O(n^2)$ time to construct H.

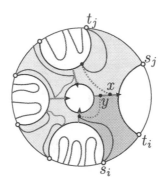

Fig. 4. Mutually disjoint walks represented by a directed path in H for a set of incident borders (green). The blue arcs correspond to the walks represented in sub-problems and the solid red paths correspond to the Type I paths represented by Type I arcs. The dotted red paths represent vertex disjoint u-to-x and v-to-y paths that will be added via a min-cost flow computation.

Here we recall the structure of T: the nodes of T correspond to *non-regions*: disks (or points) surrounded by regions. If there are several shared vertices among the borders, then there is an order of these vertices around the boundary of the non-region. That is, for a vertex shared by a set of borders, these borders must be contiguous subsets of $B_{i_1,j_2}, B_{i_2,j_2}, \dots, B_{i_t,j_t}$. In terms of the construction of H, there is a contiguous set of levels that a given shared vertex appears in and distinct shared vertices participate in non-overlapping sets of levels. For one set of these levels, we can create different copies of the corresponding section of H. In each copy we modify the directed graph to reflect which s_i-to-t_i path the corresponding shared vertex may belong to (see Figure 5). As we have argued, since distinct shared vertices participate in non-overlapping sets of levels, this may safely be repeated for every shared vertex. The resulting graph has size $O(kn^2)$ since there are $O(k)$ borders and shared vertices are shared by borders. The resulting running time for computing all source-to-sink shortest paths in the resulting graph is then $O(kn^3)$.

Computing $D_{i,j}$ from H. To compute $D_{i,j}[x,y]$, we consider all possible u on $P_i[x,t_i]$ and v on $P_j[s_j,y]$ and compute the minimum-length vertex disjoint u-to-x path and v-to-y path that only use vertices that are interior to $R_i \cup R_j$ (that is vertices of $B_{i,j}$ may be used); by Lemma 8, these paths can be computed in linear time. Let $M[u,v]$ be the cost of these paths. Then

$$D_{i,j}[x,y] = \min_{u \in P_i[x,t_i],\, v \in P_j[s_j,y]} M[u,v] + H[u,v].$$

As there are $O(n^2)$ choices for u and v and $M[u,v]$ can be computed in linear time, $D_{i,j}[x,y]$ can be computed in $O(n^3)$ time given that distances in H have been computed.

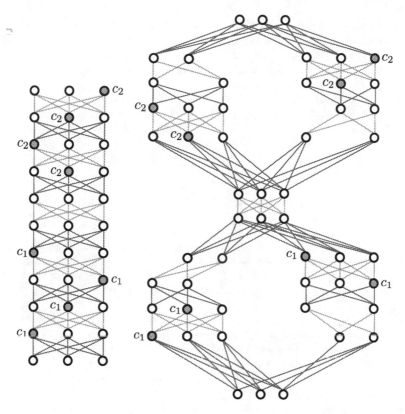

Fig. 5. (left) H constructed without handling the fact that vertices c_1 and c_2 (black) may appear more than twice. The arcs are all directed upwards (arrows are not shown); green arcs are sub-problem arcs and red arcs are Type I arcs. (right) The levels that c_1 appears in are duplicated, and only one pair of copies of c is kept in each copy of these levels. The vertex c_1 may only be visited twice now on a source to sink path.

Overall running time. For each border $B_{i,j}$, H is constructed and shortest source-to-sink paths are computed in $O(kn^3)$ time. For each $x, y \in B_{i,j}$, $D_{i,j}[x, y]$ is computed in $O(n^3)$ time. Since there are $O(n^2)$ pairs of vertices in $B_{i,j}$, $D_{i,j}$ is computed in $O(n^5)$ time (dominating the time to construct and compute shortest paths in H). Since there are $O(k)$ borders (Lemma 3), the overall time for the dynamic program is $O(kn^5)$.

Acknowledgements. This material is based upon work supported by the National Science Foundation under Grant No. CCF-1252833.

References

[BH14]   Björklund, A., Husfeldt, T.: Shortest two disjoint paths in polynomial time. In: Esparza, J., Fraigniaud, P., Husfeldt, T., Koutsoupias, E. (eds.) ICALP 2014. LNCS, vol. 8572, pp. 211–222. Springer, Heidelberg (2014)

[CMPP13] Cygan, M., Marx, D., Pilipczuk, M., Pilipczuk, M.: The planar directed k-vertex-disjoint paths problem is fixed-parameter tractable. In: Proceedings of the 2013 IEEE 54th Annual Symposium on Foundations of Computer Science, FOCS 2013, pp. 197–206. IEEE Computer Society, Washington, DC (2013)

[FHW80] Fortune, S., Hopcroft, J., Wyllie, J.: The directed subgraph homeomorphism problem. Theoretical Computer Science 10(2), 111–121 (1980)

[HKRS97] Henzinger, M.R., Klein, P., Rao, S., Subramanian, S.: Faster shortest-path algorithms for planar graphs. J. Comput. Syst. Sci. 55(1), 3–23 (1997)

[Kar74] Karp, R.: On the computational complexity of combinatorial problems. Networks 5, 45–68 (1974)

[Kle05] Klein, P.N.: Multiple-source shortest paths in planar graphs. In: Proceedings of the Sixteenth Annual ACM-SIAM Symposium on Discrete Algorithms, SODA 2005, Philadelphia, PA, USA, pp. 146–155. Society for Industrial and Applied Mathematics (2005)

[KS10] Kobayashi, Y., Sommer, C.: On shortest disjoint paths in planar graphs. Discrete Optimization 7(4), 234–245 (2010)

[KvL84] Kramer, M.R., van Leeuwen, J.: The complexity of wire-routing and finding minimum area layouts for arbitrary vlsi circuits. Advances in Computing Research 2, 129–146 (1984)

[KW10] Kawarabayashi, K.-I., Wollan, P.: A shorter proof of the graph minor algorithm: The unique linkage theorem. In: Proceedings of the Forty-Second ACM Symposium on Theory of Computing, STOC 2010, pp. 687–694. ACM, New York (2010)

[MP93] Middendorf, M., Pfeiffer, F.: On the complexity of the disjoint paths problem. Combinatorica 13(1), 97–107 (1993)

[MVV87] Mulmuley, K., Vazirani, V., Vazirani, U.: Matching is as easy as matrix inversion. Combinatorica 7(1), 345–354 (1987)

[ORS93] Ogier, R.G., Rutenburg, V., Shacham, N.: Distributed algorithms for computing shortest pairs of disjoint paths. IEEE Transactions on Information Theory 39(2), 443–455 (1993)

[RS95] Robertson, N., Seymour, P.D.: Graph minors. xiii. the disjoint paths problem. Journal of Combinatorial Theory, Series B 63(1), 65–110 (1995)

[Sch94] Schrijver, A.: Finding k disjoint paths in a directed planar graph. SIAM Journal on Computing 23(4), 780–788 (1994)

[SM05] Srinivas, A., Modiano, E.: Finding minimum energy disjoint paths in wireless ad-hoc networks. Wireless Networks 11(4), 401–417 (2005)

[VS11] C. De Verdière, É., Schrijver, A.: Shortest vertex-disjoint two-face paths in planar graphs. ACM Transactions on Algorithms (TALG) 7(2), 19 (2011)

Consensus Patterns (Probably) Has no EPTAS

Christina Boucher[1], Christine Lo[2], and Daniel Lokshantov[3]

[1] Department of Computer Science, Colorado State University in Fort Collins, USA
christina.boucher@colostate.edu
[2] Department of Computer Science and Engineering,
University of California at San Diego, USA
cylo@cs.ucsd.edu
[3] Department of Informatics, University of Bergen, Norway
daniello@ii.uib.no

Abstract. Given n length-L strings $S = \{s_1, \ldots, s_n\}$ over a constant size alphabet Σ together with an integer ℓ, where $\ell \leq L$, the objective of *Consensus Patterns* is to find a length-ℓ string s, a substring t_i of each s_i in S such that $\sum_{\forall i} d(t_i, s)$ is minimized. Here $d(x, y)$ denotes the Hamming distance between the two strings x and y. *Consensus Patterns* admits a PTAS [Li et al., JCSS 2002] is fixed parameter tractable when parameterized by the objective function value [Marx, SICOMP 2008], and although it is a well-studied problem, improvement of the PTAS to an EPTAS seemed elusive. We prove that *Consensus Patterns* does not admit an EPTAS unless FPT=W[1], answering an open problem from [Fellows et al., STACS 2002, Combinatorica 2006]. To the best of our knowledge, *Consensus Patterns* is the first problem that admits a PTAS, and is fixed parameter tractable when parameterized by the value of the objective function but does not admit an EPTAS under plausible complexity assumptions. The proof of our hardness of approximation result combines parameterized reductions and gap preserving reductions in a novel manner.

1 Introduction

Lanctot et al. [16] initiated the study of *distinguishing string selection problems* in bioinformatics, where we seek a representative string satisfying some distance constraints from each of the input strings. The *Consensus Patterns* problem falls within this broad class of stringology problems. Given n length-L strings $S = \{s_1, \ldots, s_n\}$ over a constant size alphabet Σ together with an integer ℓ, where $\ell \leq L$, the objective of *Consensus Patterns* is to find a length-ℓ string s, a length-ℓ substring t_i of each s_i in S such that $\sum_{\forall i} d(t_i, s)$ is minimized. Here $d(x, y)$ denotes the Hamming distance between the two strings x and y. One specific application of *Consensus Patterns* in bioinformatics is the problem of finding transcription factor binding sites [16,23]. Transcription factors are proteins that bind to promoter regions in the genome and have the effect of regulating the expression of one or more genes. Hence, the region where a transcription factor

© Springer-Verlag Berlin Heidelberg 2015
N. Bansal and I. Finocchi (Eds.): ESA 2015, LNCS 9294, pp. 239–250, 2015.
DOI: 10.1007/978-3-662-48350-3_21

binds is very well-conserved, and the problem of detecting such regions can be extrapolated to the problem of finding the substrings $\{t_1, \ldots, t_n\}$.

Consensus Patterns is NP-hard even when the alphabet is binary [17], so we do not expect a polynomial-time algorithm for the problem. On the other hand, the problem admits a *polynomial time approximation scheme* (PTAS), which finds a solution that is at most a factor $(1 + \epsilon)$ worse than the optimum [17] in $n^{O(\frac{1}{\epsilon^2} \log \frac{1}{\epsilon})}$-time. While a superpolynomial dependence of the running time on $\frac{1}{\epsilon}$ is implied by the NP-hardness of *Consensus Patterns*, there is still room for faster approximation schemes for the problem and so a significant effort has been invested in attempting on proving tighter bounds on the running time of the PTAS [5,6]. If the exponent of the polynomial in the running time of a PTAS is independent of ϵ then the PTAS is called an *efficient PTAS* (EPTAS). An interesting question, posed by Fellows et al. [10] is whether *Consensus Patterns* admits an EPTAS.

The difference in running time of a PTAS and an EPTAS can be quite dramatic. For instance, running a $O(2^{1/\epsilon}n)$-time algorithm is reasonable for $\epsilon = \frac{1}{10}$ and $n = 1000$, whereas running a $O(n^{1/\epsilon})$-time algorithm is infeasible on this same input. Hence, considerable effort has been devoted to improving PTASs to EPTASs, and showing that such an improvement is unlikely for some problems. For example, Arora [2] gave a $n^{O(1/\epsilon)}$-time PTAS for *Euclidean TSP*, which was then improved to a $O(2^{O(1/\epsilon^2)}n^2)$-time algorithm in the journal version of the paper [3]. On the other hand *Independent Set* admits a PTAS on unit disk graphs [15] but Marx [19] showed that it does not admit an EPTAS assuming FPT$\neq$W[1]—a widely believed assumption from parameterized complexity. Many more examples of PTASs that have been improved to EPTASs, and problems for which there exists a PTAS but the existence of an EPTAS has been ruled out under the assumption that FPT$\neq$W[1] can be found in the survey of Marx [20]. In this paper we show that assuming FPT$\neq$W[1], *Consensus Patterns* does not admit an EPTAS, resolving the open problem of Fellows et al. [10]. Since *Consensus Patterns* has a PTAS and is FPT, standard methods for ruling out an EPTAS cannot be applied. We discuss this in more details in Section 1.1. Our proof avoids this obstacle by combining gap preserving reductions and parameterized reductions in a novel manner.

1.1   Methods

Our lower bounds are proved under the assumption FPT$\neq$W[1], a standard assumption in parameterized complexity that we will briefly discuss here. In a parameterized problem every instance $\mathcal{I}$ comes with a *parameter* k. A parameterized problem is said to be *fixed parameter tractable* (FPT) if there is an algorithm solving instances of the problem in time $f(k)|\mathcal{I}|^{O(1)}$ for some function f depending only on k and not on $|\mathcal{I}|$. The class of all fixed parameter tractable problems is denoted by FPT. The class W[1] of parameterized problems is the basic class for fixed parameter intractability, FPT $\subseteq$ W[1] and the containment is believed to be proper. A parameterized problem Π with the property that an

FPT algorithm for Π would imply that FPT=W[1] is called W[1]-hard. Thus demonstrating W[1]-hardness of a parameterized problem implies that it is unlikely that the problem is FPT. We refer the reader to the textbooks [7,9,12,22] for a more thorough discussion of parameterized complexity.

W[1]-hardness is frequently used to rule out EPTAS's for optimization problems, since an EPTAS for an optimization problem automatically yields a FPT algorithm for the corresponding decision problem parameterized by the value of the objective function [4,20]. More specifically, if we set $\epsilon = \frac{1}{2\alpha}$, where α is the value of the objective function, then a $(1 + \epsilon)$-approximation algorithm would distinguish between "yes" and "no" instances of the problem. Hence, an EPTAS could be used to solve the problem in $O(f(\epsilon)n^{O(1)}) = O(g(\alpha)n^{O(1)})$-time. Hence, if a problem is W[1]-hard when parameterized by the value of the objective function then the corresponding optimization problem does not admit an EPTAS unless FPT=W[1]. To the best of our knowledge, *all* known results ruling out EPTASs for problems for which a PTAS is known use this approach. However, this approach cannot be used to rule out an EPTAS for *Consensus Patterns* because *Consensus Patterns* parameterized by d has been shown to be FPT by Marx [18]. Thus, different methods are required to rule out an EPTAS for *Consensus Patterns*.

In his survey, Marx [20] introduces a hybrid of FPT reductions and gap preserving reductions and argues that it is conceivable that such a reduction could be used to prove that a problem that has a PTAS and is FPT parameterized by the value of the objective function does not admit an EPTAS unless FPT=W[1]. We show that *Consensus Patterns* does not admit an EPTAS unless FPT=W[1], giving the first example of this phenomenon.

Preliminaries

A PTAS for a minimization problem finds a $(1+\epsilon)$-approximate solution in time $|\mathcal{I}|^{f(1/\epsilon)}$ for some function f. An approximation scheme where the exponent of $|\mathcal{I}|$ in the running time is independent of ϵ is called an *efficient* polynomial time approximation scheme (EPTAS). Formally, an EPTAS is a PTAS whose running time is $f(1/\epsilon)^{O(1)}|\mathcal{I}|^{O(1)}$.

Let $L, L' \subseteq \sum^* \times \mathbb{N}$ be two parameterized problems. We say that L *fpt-reduces* to L' if there are functions $f, g : \mathbb{N} \to \mathbb{N}$, and an algorithm that given an instance $(\mathcal{I}, k)$ runs in time $f(k)|\mathcal{I}|^{f(k)}$ and outputs an instance $(\mathcal{I}', k')$ such that $k' \leq g(k)$ and $(\mathcal{I}, k) \in L \iff (\mathcal{I}', k') \in L'$. These reductions work as expected; if L fpt-reduces to L' and L' is FPT then so is L'. Furthermore, if L fpt-reduces to L' and L is W[1]-hard then so is L'.

Let s be a string over the alphabet Σ. We denote the length of s as $|s|$, and the jth character of s as $s[j]$. Hence, $s = s[1]s[2]\ldots s[|s|]$. For a set S of strings of the same length we denote by $S[i]$ as $\{s[i] : s \in S\}$. Thus, if the same character appears at position i in several strings it is counted several times in $S[i]$. For an interval $P = \{i, i+1, \ldots, j-1, j\}$ of integers, define $s[P]$ to be the substring $s[i]s[i+1]\ldots s[j]$ of s. For a set S of strings and interval P define $S[P]$ to be the (multi)set $\{s[P] : s \in S\}$. For a set S of length-ℓ strings we define the *consensus string* of S, denoted as $c(S)$, as the sequence where $c(S)[i]$ is the most-frequent character in $S[i]$ for all $i \leq \ell$. Ties are broken by selecting the lexicographically first such character, however, we note that the tie-breaking will not affect our arguments.

We denote the sum Hamming distance between a string, s, and a set of strings, S, as $d(S,s)$. Observe that the consensus string $c(S)$ minimizes $d(S,c(S))$—implying that no other string x is closer to S than $c(S)$. However, some $x \neq c(S)$ could achieve $d(S,x) = d(S,c(S))$ and we refer to such strings as *majority strings* because they are obtained by picking a most-frequent character at every position with ties broken arbitrarily.

We will use standard concentration bounds for sums of independent random variables. In particular, the following variant of the Hoeffding's bound [14] given by Grimmett and Stirzaker [13, p. 476] will be needed.

Proposition 1 (Hoeffding's bound). *Let $X_1, X_2, ...X_n$ be independent random variables such that $a_i \leq X_i \leq b_i$ for all i. Let $X = \Sigma_i X_i$ and the expected value of X be $E[X]$ then it follows that:* $\Pr[X - E[X] \geq t] \leq \exp\left(\frac{-2t^2}{\Sigma_{i=1}^n (b_i - a_i)^2}\right)$.

2   Hardness of Approximating Colored Consensus String with Outliers

To show that *Consensus Patterns* does not admit an EPTAS we will first demonstrate hardness the following problem, that we call *Colored Gap-Consensus String with Outliers*. When defining parameterized gap problems, we follow the notation of Marx [20].

In the *Colored Gap-Consensus String with Outliers* (CCWSO) problem the input consists of a (multi)set of n length-L strings $S = \{s_1, \ldots, s_n\}$ over a finite alphabet Σ, an integer $n^* \leq n$, a partitioning of S into n^* sets $S = S_1 \cup S_2 \ldots S_{n^*}$, a rational ϵ and two integers D_{yes} and D_{no} with $D_{no} \geq D_{yes}(1 + \epsilon)$ with the following property. Either (a) there exists a set S^* such that $|S^* \cap S_i| = 1$ for every i and $d(S^*, c(S^*)) \leq D_{yes}$ or (b) for every S^* such that $|S^* \cap S_i| = 1$ for every i we have $d(S^*, c(S^*)) \geq D_{no}$. The task is to determine which one of the two cases holds. In particular, the task is to determine whether there is an S^* such that $|S^* \cap S_i| = 1$ for every $i \leq n^*$ and $d(S^*, c(S^*)) \leq D_{yes}$. The parameter of this instance is $\lceil 1/\epsilon \rceil$. Therefore, algorithms with running time $f(\epsilon)(nL)^{O(1)}$ are considered fixed parameter tractable. The aim of this section is to prove the following lemma.

Lemma 1. *Gap-Colored Consensus String with Outliers is W[1]-hard.*

The proof of Lemma 1 is by reduction from the *MultiColored Clique (MCC)* problem. Here input is a graph G, an integer k and a partition of $V(G)$ into $V_1 \uplus V_2 \ldots \uplus V_k$ such that for each i, $G[V_i]$ is an independent set. The task is to determine whether G contains a clique C of size k. Observe that such a clique must contain exactly one vertex from each V_i, since for each i we have $C \cap V_i \leq 1$. It is well-known that MCC is W[1]-hard [11].

Given an instance (G, k) of MCC we produce in $f(k)n^{O(1)}$-time an instance $(S_1, S_2, \ldots S_{n^*})$ of *Colored Gap-Consensus String with Outliers*. We will say that a subset S^* of S such that $|S^* \cap S_i| = 1$ for every $i \leq n^*$ is a *potential solution* to the CCWSO instance. Our constructed instance will have the following property. If G has a k-clique then there exists a potential solution S^* such that $d(S^*, c(S^*)) \leq D_{yes}$. On the other hand, if no k-clique exists in G then for each potential solution S^* we have $d(S^*, c(S^*)) \geq D_{no}$. The values of D_{yes} and D_{no} will be chosen later in the proof, however, we note that the crucial point of the construction is that $D_{no} \geq \left(1 + \frac{1}{h(k)}\right) D_{yes}$. Hence, a $f(\epsilon)(nL)^{O(1)}$-time algorithm for *Gap-Consensus String with Outliers* could be used to solve to solve the MCC problem in time $g(k)n^{O(1)}$ by setting $\epsilon = \frac{1}{2h(k)}$. Thus, the reduction is a parameterized, gap-creating reduction where the size of the gap decreases as k increases but the decrease is a function of k only.

Construction. We describe how the instance $(S_1, S_2, \ldots, S_{n^*}, D_{yes}, D_{no})$ is constructed from (G, k). Our construction is randomized, and will succeed with probability $\frac{2}{3}$. To prove Lemma 1 we have to change the construction to make it deterministic but for now let us not worry about that.

We start by considering the instance (G, k) and let $E(G) = \{e_1, e_2, \ldots e_m\}$. In the reduction we will create one string s_i for every edge $e_i \in E(G)$. We partition the edge set $E(G)$ into sets $\binom{k}{2}$ sets $E_{\{p,q\}}$ where $1 \leq p, q \leq k$ as follows; $e_i \in E_{p,q}$ if e_i has one endpoint in V_p and the other in V_q. The edge $e_i \in E_{p,q}$ has two endpoints, one in V_p and the other in V_q. The string s_i is inserted into the set $S_{\{p,q\}}$ and the set S of strings in the instance of *Gap-Colored Consensus String with Outliers* will be exactly $S = \bigcup_{p \neq q} S_{\{p,q\}}$.

We set $n^* = \binom{k}{2}$, and use exactly the partition of S into the sets $S_{\{p,q\}}$ as the partition into n^* sets in the instance. Thus, picking a potential solution S^* corresponds to picking a set of edges with exactly one edge from each of the sets $E_{\{p,q\}}$.

There are $K = k \cdot (k-1) \cdot (k-2)$ ordered triples of integers from 1 to k. Consider the lexicographic ordering of such triples. As an example, if $k = 3$ this ordering is $(1, 2, 3), (1, 3, 2), (2, 1, 3), (2, 3, 1), (3, 1, 2), (3, 2, 1)$.

For each i from 1 to K, let $\sigma(i)$ be the i'th triple in this ordering. Thus, for $k = 3$, we have that $\sigma(4) = (2, 3, 1)$. The functions σ^1, σ^2 and σ^3 return the first, second and third entry of the triple returned by σ. Continuing our example for the case that $k = 3$, we have $\sigma^1(4) = 2$, $\sigma^2(4) = 3$ and $\sigma^3(4) = 1$.

Based on G and k, we select an integer ℓ. The exact value of ℓ will be discussed later in the proof, for now the reader may think of ℓ as some function of k times $\log n$. We construct a set $Z = z_1, z_2, \ldots z_m$ of strings, Z will act as a "pool of random bits" in our construction. For each edge $e_i \in E(G)$ we make a string z_i as follows.

$$z_i = \overline{a}_i^{\sigma(1)} \circ \overline{a}_i^{\sigma(2)} \ldots \circ \overline{a}_i^{\sigma(K)}$$

For every $i \leq m$ and $p \leq K$, the strings $\tilde{a}_i$, $\tilde{a}'_i$ and $\overline{a}_i^{\sigma(p)}$ are random binary strings of length ℓ. For each $p \leq K$ and vertex $u \in V_{\sigma_1(p)}$ we make an identification string $id^p(u)$ of length ℓ. Let i be the smallest integer such that the edge e_i is incident to u. We set $id^p(u) = \overline{a}_i^{\sigma(p)}$. Notice that the other endpoint of e_i is a vertex not in V_p. Thus, for any other vertex $v \in V_p$ distinct from u we have that $id^p(v) = \overline{a}_j^{\sigma(p)}$ for some integer $j \neq i$.

We now make the set S of strings in our instance. For each edge $e_i \in E(G)$ we make a string s_i as follows.

$$s_i = a_i^{\sigma(1)} \circ a_i^{\sigma(2)} \ldots \circ a_i^{\sigma(K)}$$

For each $x \leq K$ we define a_i^x using the following rules. Let $e_i = uv$ with $u \in V_p$ and $v \in V_q$. If $\sigma^1(x) = p$ and $\sigma^2(x) = q$ or $\sigma^1(x) = p$ and $\sigma^3(x) = q$, we set $a_i^{\sigma(x)} = id^x(u)$. If $\sigma^1(x) = q$ and $\sigma^2(x) = p$ or $\sigma^1(x) = q$ and $\sigma^3(x) = p$, we set $a_i^{\sigma(x)} = id^x(v)$. Otherwise we set $a_i^{\sigma(x)} = \overline{a}_i^{\sigma(x)}$.

For $1 \leq p \leq K$ we define $B_p = \{(p-1)\ell + 1, (p-1)\ell + 2, \ldots (p-1)\ell + \ell\}$, and will refer to B_p as the p'th *block* of the instance. Notice that for every $i \leq m$ and $p \leq K$ we have $s_i[B_p] = a_i^{\sigma(p)}$. We set $L = K \cdot \ell$ and $N = |S| = m$, this concludes the construction. Recall that n^* is the size of the solution S^* sought for and observe that L is the length of the constructed strings in S.

Analysis. We consider the constructed strings s_i as random variables, and for every j the character $s_i[j]$ is also a random variable which takes value 1 with probability $1/2$ and 0 with probability $1/2$. Observe that for any two positions j and j' such that $j \neq j'$ and any i and i' the random variables $s_i[j]$ and $s_{i'}[j']$ are independent. On the other hand $s_i[j]$ and $s_{i'}[j]$ could be dependent. However, if $s_i[j]$ and $s_{i'}[j]$ are dependent then, by construction $s_i[j] = s_{i'}[j]$.

Let $S^* \subseteq S$ be a potential solution. Here we consider S^* as a set of random string variables, rather than a set of strings. We are interested in studying $d(S^*, c(S^*))$ for different choices of the set S^*. We can write out $d(S^*, c(S^*))$ as

$$d(S^*, c(S^*)) = \sum_{p=1}^{K} d(S^*[B_p], c(S^*)[B_p]) \tag{1}$$

and

$$d(S^*[B_p], c(S^*)[B_p]) = \sum_{j \in B_p} d(S^*[j], c(S^*)[j]).$$

Thus, for each $p \leq K$ we have that $d(S^*[B_p], c(S^*)[B_p])$ is a sum of ℓ independent random variables, each taking values from 0 to n^*. Hence, when ℓ is large enough $d(S^*[B_p], c(S^*)[B_p])$ is sharply concentrated around its expected value. Using a union bound (over the choices of p) we can show that $d(S^*, c(S^*))$ is sharply concentrated around its expectation as well.

We turn our attention to $E[d(S^*, c(S^*))]$ for different choices of S^*. The two main cases that we distinguish between is whether S^* corresponds to the set of edges of a clique in G or not. Note that a potential solution S^* corresponds to a

set E^* of edges with exactly one edge $e_{\{p,q\}} \in E_{\{p,q\}}$ for every (unordered) pair
p,q. *In the remainder of this section S^* is a potential solution and E^* is the edge
set corresponding to S^*. For each pair p,q of integers, $e_{\{p,q\}}$ is the unique edge
in $E^* \cap \in E_{\{p,q\}}$.* We will determine whether E^* is the set of edges of a clique
using the following observation, whose proof is obvious and hence omitted.

Observation 1. *E^* is the edge set of a clique in G if and only if for every
ordered triple (a, b, c) of distinct integers between 1 and k the edge $e_{\{a,b\}}$ and the
edge $e_{\{a,c\}}$ are incident to the same vertex in V_a.*

In the constructed instance the block B_p such that $\sigma(p) = (a, b, c)$ is responsible
for performing the check for the triple (a, b, c).

Before proceeding we need to define a class of variables relating to random
walks. For two integers i and r consider a random walk starting at $x = i$ and
doing r steps. In each step the walk either changes x to $x+1$ or to $x-1$, uniformly
at random. We define $X_{r,0}^i$ as the final value of x of such a walk. Now, for an
integer $t \leq r/2$ we modify the walk such that in t of the steps the walk either
changes x to $x+2$ or to $x-2$, and in $r-2t$ of the steps the walk either changes
x to $x+1$ or to $x-1$. The random variable distributed as the final position of
x after such a walk is denoted by $X_{r,t}^i$. We set $x_{r,t}^i = E[|X_{r,t}^i|]$.

The next lemma characterizes the expectation of $d(S^*[B_p], c(S^*)[B_p])$, subject
to the case distinction on whether the solution S^* passes or fails the test of
Observation 1 for the triple $\sigma(p)$.

Lemma 2. $[\star]$[1] *Let $p \leq K$ and let $\sigma(p) = (a, b, c)$. If $e_{\{a,b\}}$ and $e_{\{a,c\}}$ are
incident to the same vertex in V_a, then $E[d(S^*[B_p], c(S^*)[B_p])] = \ell \cdot (n^*/2 - x_{n^*,1}^0)$. If $e_{\{a,b\}}$ and $e_{\{a,c\}}$ are not incident to the same vertex in V_a, then
$E[d(S^*[B_p], c(S^*)[B_p])] = \ell \cdot (n^*/2 - x_{n^*,0}^0)$.*

We now define E_{yes} as follows.

$$E_{yes} = K \cdot \ell \cdot (n^*/2 - x_{n^*,1}^0) \tag{2}$$

Observe that Equation 1, Lemma 2 and linearity of expectation immedeately
implies that if E^* is the set of edges of a clique then $E[d(S^*, c(S^*))] = E_{yes}$.
Furthermore, By Lemma 2 each triple (a, b, c) of distinct integers from 1 to k
such that the edge $e_{\{a,b\}}$ and $e_{\{a,c\}}$ are not incident to the same vertex in V_a will
contribute exactly $\ell \cdot (n^*/2 - x_{n^*,0}^0)$ instead of $\ell \cdot (n^*/2 - x_{n^*,1}^0)$ to the expectation
$E[d(S^*, c(S^*))]$. This proves the following lemma.

Lemma 3. *Let t be the number of ordered triples (a, b, c) of distinct integers
from 1 to k such that the edge $e_{\{a,b\}}$ and $e_{\{a,c\}}$ are not incident to the same
vertex in V_a. Then $E[d(S^*, c(S^*))] = E_{yes} + t \cdot \ell \cdot (x_{n^*,1}^0 - x_{n^*,0}^0)$*

To conclude the analysis we need to show that as the number of triples t that
fail the test of Observation 1 increases, so does the expected value of $d(S^*, c(S^*))$.
To that end, all we need to prove is that $x_{n^*,1}^0 - x_{n^*,0}^0 > 0$. We will prove this
by "differentiating" $x_{n^*,t}^0$ with respect to t.

---

[1] Proofs of statements labelled with $\star$ are omitted and may be found in the full version.

Lemma 4. [⋆] $x^0_{n^*,0} < x^0_{n^*,1}$. *Furthermore we can compute $x^0_{n^*,0}$ and $x^0_{n^*,1}$ in time polynomial in n^*.*

We now define Δ as follows $\Delta = x^0_{n^*,1} - x^0_{n^*,0}$, and note that Lemma 4 implies that $\Delta > 0$. Furthermore, note that Δ depends only on $n^* = \binom{k}{2}$, so Δ is a computable function of k. Define

$$E_{no} = E_{yes} + \Delta \cdot \ell \tag{3}$$

Observe that $E_{no}/E_{yes} \geq 1 + \frac{2\Delta}{K \cdot n^*}$, and that therefore $E_{no}/E_{yes} \geq 1 + 1/h(k)$ for a function h depending only on k. Lemma 3, Lemma 4 and the definition of E_{no} implies the following lemma, which summarizes the analysis up until now.

Lemma 5. *If E^* is the edge set of a clique in G, then $E[d(S^*[B_p], c(S^*)[B_p])] = E_{yes}$. Otherwise $E[d(S^*[B_p], c(S^*)[B_p])] \geq E_{no}$.*

From the definitions of E_{yes} and E_{no} it follows that there exist constants κ_{yes} and κ_{no} depending only on k such that $E_{yes} = \kappa_{yes}\ell$ and $E_{no} = \kappa_{no}\ell$. Furthermore, $\kappa_{yes} < \kappa_{no}$ and the value of κ_{yes} and κ_{no} can be computed in time $f(k)$ for some function f. Set $\kappa'_{yes} = (2\kappa_{yes} + \kappa_{no})/3$ and $\kappa'_{no} = (\kappa_{yes} + 2\kappa_{no})/3$. Then $\kappa_{yes} < \kappa'_{yes} < \kappa'_{no} < \kappa_{no}$. We set $D_{yes} = \kappa'_{yes}\ell$ and $D_{no} = \kappa'_{no}\ell$. Notice that

$$\kappa'_{yes} - \kappa_{yes} = \kappa_{no} - \kappa'_{no}.$$

In the full version of the paper we give a proof of a randomized analogue of Lemma 1 before proceeding to the proof of Lemma 1. This provides useful insights on how the construction works, but is not strictly necessary to obtain Lemma 1, and is therefore omitted in this short version.

A Deterministic Construction. In order to prove Lemma 1 we need to make the construction deterministic. We only used randomness to construct the set Z, all other steps are deterministic. We now show how Z can be computed deterministically instead of being selected at random, preserving the properties of the reduction. For this, we need the concept of near p-wise independence defined by Naor and Naor [21]. The original definition of near p-wise independence is in terms of sample spaces, we define near p-wise independence in terms of collections of binary strings. This is only a notational difference, and one may freely translate between the two variants.

Definition 1 ([21]). *A set $C = \{c_1, c_2, \ldots c_t\}$ of length ℓ binary strings is (ϵ, p)-independent if for any subset C' of C of size p, if a position $i \leq t$ is selected uniformly at random, then $\sum_{\alpha \in \{0,1\}^p} |P[C'[i] = \alpha] - 2^{-p}| \leq \epsilon$.*

Naor and Naor [21] and Alon et al. [1] give determinsitic constructions of small nearly k-wise independent sample spaces. Reformulated in our terminology, Alon et al. [1] prove a slightly stronger version of the following theorem.

Theorem 1 ([1]). *For every t, p, and ϵ there is a (ϵ, p)-independent set $C = \{c_1, c_2, \ldots c_t\}$ of binary strings of length ℓ, where $\ell = O(\frac{2^k \cdot k \log t}{\epsilon})$. Furthermore, C can be computed in time $O(|C|^{O(1)})$.*

We use Theorem 1 to construct the set Z. We set $\epsilon = \frac{\kappa'_{yes} - \kappa_{yes}}{K \cdot n^*}$ and construct an (ϵ, n^*)-independent set C of $2m$ strings. These strings have length $\ell = f \cdot \log(n)$ for some f depending only on k, and C can be constructed in time $O(gn^{O(1)})$ for some g depending only on k. i we set

$$z_i = c_i \circ c_i \circ \ldots \circ c_i,$$

where we used K copies of c_i such that z_i is a string of length L. That is, in the construction of z_i we set $\overline{a}_i^{\sigma(p)} = c_i$ for all $p \leq K$. The remaining part of the construction, i.e the construction of S from Z remains unchanged. To distinguish between the deterministically constructed S and the randomized construction, we refer to the deterministically constructed S as S_{det}. We now prove that for every potential solution $S^*_{det} \subseteq S_{det}$, if S^* is the set of strings in the randomized construction that corresponds to the same edges as S^*_{det}, then $d(S^*_{det}, c(S^*_{det}))$ is almost equal to $E[d(S^*, c(S^*))]$. When considering $E[d(S^*, c(S^*))]$ we consider the randomized construction, but with the same choice of ℓ as in the construction of S_{det}, so that the strings in S and S_{det} have the same length.

For a subset I of $\{1, 2, \ldots, m\}$ define $S^*(I) = \{s_i \in S : i \in I\}$ and $S^*_{det}(I) = \{s_i \in S_{det} : i \in I\}$. The construction of S_{det} (and S) from Z implies that for every $x < K$, there exists a function $f_x : \mathbb{N} \to \mathbb{N}$ such that for any $i \leq m$, $s_i[B_x] = z_{f(i)}[B_x]$. For any $I \subseteq \{1, 2, \ldots, m\}$ and $x \leq K$ we define $Z^*(I, x)$ to be an arbitrarily chosen subset of Z of size n^* such that $\{z_{f_x(i)} : i \in I\} \subseteq Z^*(I, x)$. The reason we did not define $Z^*(I, x)$ as exactly $\{z_{f_x(i)} : i \in I\}$ is that the function f_x is not injective, and we want to ensure $|Z^*(I, x)| = n^*$. The definition of $Z^*(I, x)$ ensures that for every $I \subseteq \{1, 2, \ldots, m\}$ of size n^*, the string sets $S^*(I)[B_x]$ and $S^*_{det}(I)[B_x]$ are functions of $Z^*(I, x)[B_x]$. Even stronger, for every $j \in B_x$ we have that the *strings* $S^*(I)[j]$ and $S^*_{det}(I)[j]$ are functions of $Z^*(I, x)[j]$. Strictly speaking $S^*(I)[j]$, $S^*_{det}(I)[j]$ and $Z^*(I, x)[j]$ are multi-sets of characters, but we can think of them as strings by, for example, reading the characters in $S^*(I)[j]$ as $s_i[j]$ for all $i \in I$ in increasing order. Since the deterministic and randomized constructions are identical (except for the construction of Z) the strings $S^*(I)[j]$ and $S^*_{det}(I)[j]$ depend on $Z^*(I, x)[j]$ in exactly the same way.

An immediate implication of the fact that $S^*(I)[B_x]$ and $S^*_{det}(I)[B_x]$ are functions of $Z^*(I, x)[B_x]$, is that the distances $d(S^*(I)[j], c(S^*(I)[j]))$ and $d(S^*_{det}(I)[j], c(S^*_{det}(I)[j]))$ are also functions of $Z^*(I, x)[j]$. We now give these functions a name. For every set $I \subseteq \{1, 2, \ldots, m\}$ of size n^* and integer $x < K$ define $d_x^I : \{0, 1\}^{n^*} \to \{0, 1, \ldots, n^*\}$ to be a function such that for any $j \in B_x$, if $Z^*(I)[j] = \alpha$ then $d(S^*(I)[j], c(S^*(I)[j])) = d_x^I(\alpha)$ and $d(S^*_{det}(I)[j], c(S^*_{det}(I)[j])) = d_x^I(\alpha)$.

For every set $I \subseteq \{1, 2, \ldots, m\}$ of size n^* and integer $x \leq K$ we have the following expression for $d(S^*(I)_{det}[B_x], c(S^*(I)_{det}[B_x]))$.

$$d\left(S^*_{det}(I)[B_x], c(S^*_{det}(I)[B_x])\right) = \ell \cdot \sum_{\alpha \in \{0,1\}^{n^*}} P[Z^*(I)[j] = \alpha] \cdot d_j^I(\alpha) \quad (4)$$

Here the probability $P[Z^*(I)[j] = \alpha]$ is taken over random selections of j from B_x. For the randomized construction we have that $P[Z^*(I)[j] = \alpha] = \frac{1}{2^{n^*}}$, which yields the following expression.

$$E\left[d\left(S^*(I)[B_x], c(S^*(I)[B_x])\right)\right] = \ell \cdot \sum_{\alpha \in \{0,1\}^{n^*}} \frac{1}{2^{n^*}} \cdot d_j^I(\alpha) \tag{5}$$

Combining Equations 4 and 5 yields the following bound.

$$\left| d\left(S_{det}^*(I)[B_x], c(S_{det}^*(I)[B_x])\right) - E[d(S^*(I)[B_x], c(S^*(I)[B_x]))] \right|$$

$$= \ell \cdot \left| \sum_{\alpha \in \{0,1\}^{n^*}} \left(P[Z^*(I)[p] = \alpha] - \frac{1}{2^{n^*}} \right) \cdot d_j^I(\alpha) \right| \leq \ell \cdot \epsilon \cdot n^* \tag{6}$$

Summing Equation 6 over $1 \leq x \leq K$ yields the desired bound for every $I \subseteq \{1, 2, \ldots, 2m\}$ of size n^*.

$$\left| d\left(S_{det}^*(I), c(S_{det}^*(I))\right) - E[d(S^*(I), c(S^*(I)))] \right| \leq \ell \cdot K \cdot \epsilon \cdot n^* \leq \ell \cdot (\kappa_{yes'} - \kappa_{yes}) \tag{7}$$

Equation 7 allows us to finish the proof of Lemma 1. For any potential solution S^* that corresponds to a clique in G, we have that $E[d(S^*(I), c(S^*(I)))] = E_{yes} = \ell\kappa_{yes}$, and so by Equation 7, $d\left(S_{det}^*(I), c(S_{det}^*(I))\right) \leq \ell\kappa'_{yes} = D_{yes}$. For any potential solution S^* of size n^* that does not correspond to a clique in G, we have that $E[d(S^*(I), c(S^*(I)))] \geq E_{no} = \ell\kappa_{no}$, and so by Equation 7, $d\left(S_{det}^*(I), c(S_{det}^*(I))\right) \geq \ell\kappa'_{no} = D_{no}$. Since $\frac{D_{no}}{D_{yes}} \geq 1 + \delta$ for some δ depending only on k, the construction is an fpt-reduction from MCC to *Gap-Consensus String With Outliers*, completing the proof of Lemma 1. $\qquad\square$

3   Hardness of Approximating Consensus Patterns

To show that *Consensus Patterns* does not have an EPTAS unless $FPT = W[1]$ we introduce the following gap variant of the problem. In the *Gap-Consensus Patterns* problem, input consists of a set $S = \{s_1, \ldots, s_n\}$ of length-L strings over a constant size alphabet Σ, an integer ℓ, where $\ell \leq L$, a rational ϵ and intgers D_{yes} and D_{no} with $D_{no} \geq D_{yes}(1 + \epsilon)$ such that the following holds. Either there is a length-ℓ substring t_i of each s_i in S such that $\sum_{\forall i} d(t_i, s) \leq D_{yes}$ or for every collection $t_1, \ldots t_n$ such that t_i is a length-ℓ substring s_i we have $\sum_{\forall i} d(t_i, s) \geq D_{no}$. The task is to determine whether there is a length-ℓ substring t_i of each s_i in S such that $\sum_{\forall i} d(t_i, s) \leq D_{yes}$. The parameter of the instance is $\lceil 1/\epsilon \rceil$, and so algorithms with running time $f(\epsilon)(nL)^{O(1)}$ are considered FPT.

We will now give a fpt-reduction from *Gap-Colored Consensus String with Outliers* to *gap-Consensus Patterns*. The main ingredient in our reduction is a gadget string w. The string w has length L_1 (to be determined later), and for every $i \geq 1$, $w[i] = 1$ if $i = j^2$ for an integer j and $w[i] = 0$ otherwise. We will say that an integer i is a *square* if $i = j^2$ for some integer j. Thus $w[i]$ is 1 if and only if i is a square.

Lemma 6. [⋆] *For positive integers x, y and z such that $z \geq \frac{L_1}{4}$, $x < y$ and $y + z \leq L_1$ we have $d(w[\{x, x+1, \ldots, x+z\}], w[\{y, y+1, \ldots, y+z\}]) \geq \lfloor \frac{\sqrt{L_1}}{16} \rfloor$*

Given an instance n^*, $S = S_1 \uplus S_2 \uplus \ldots \uplus S_{n^*}$ of *Gap-Colored Consensus String with Outliers* we construct an instance of *Gap-Cosensus Patterns* as follows. First we ensure that all of the (multi) sets S_i contain the same number of strings; if $|S_i| < |S_j|$ for some i, j we can make duplicates of strings in S_i until equality is obtained. This does not affect any other aspects of the instance, since a solution S^* has to pick one string from each S_i.

Let ℓ be the length of all the strings in S. We choose L_1 such that $\lfloor \frac{\sqrt{L_1}}{16} \rfloor > n^* \cdot \ell$ and construct a gadget string w of length L_1. For every $i \leq n^*$ we make a string $\hat{s}_i$ from the set S_i. Let $S_i = s_i^1, s_i^2, \ldots, s_i^t$. We define

$$\hat{s}_i = w \circ s_i^1 \circ w \circ s_i^2 \circ w \ldots \circ w \circ s_i^t.$$

and set $L = L_1 + \ell$. We keep the values of D_{yes} and D_{no}. This concludes the construction.

Lemma 7. *For every $S^* = \{s_1^*, \ldots, s_{n^*}^*\} \subset S$ such that $s_i^* \in S_i$ for all i there is a collection $T^* = t_1^*, \ldots t_{n^*}^*$ such that t_i^* is a length L substring of $\hat{s}_i$ and $d(c(T^*), T^*) \leq d(C(S^*), S^*)$.*

Proof. For every i, set $t_1^* = w \circ s_i^*$. Since $s_i^* \in S_i$ we have that t_1^* is a length L substring of $\hat{s}_i$. Set $c = w \circ c(S^*)$, we have that $d(c(T^*), T^*) \leq d(c, T^*) \leq d(C(S^*), S^*)$.

Lemma 8. [⋆] *For every collection $T^* = t_1^*, \ldots t_{n^*}^*$ such that t_i^* is a length L substring of $\hat{s}_i$ and $d(c(T^*), T^*) \leq n^* \cdot \ell$ there is a subset $S^* = \{s_1^*, \ldots, s_{n^*}^*\} \subseteq S$ such that $s_i^* \in S_i$ for all i and $d(C(S^*), S^*) \leq d(c(T^*), T^*)$.*

The construction, together with Lemmata 1, 7 and 8 yield that *Gap-Consensus Patterns* is W[1]-hard. Since an EPTAS for *Consensus Patterns* could be used to solve *Gap-Consensus Patterns* in time $f(\epsilon)(nL)^{O(1)}$, this yields our main result.

Theorem 2. Consensus Patterns *does not have an EPTAS unless FPT=W[1]*.

4 Conclusions and Future Work

We have shown that *Consensus Patterns* does not admit an EPTAS unless FPT=W[1]. Our result rules out the possibility of a $(1 + \epsilon)$ approximation algorithms with running time $f(1/\epsilon)n^{O(1)}$, while the best PTAS for *Consensus Patterns* has running time $n^{O(1/\epsilon^4)}$. Hence there is still a significant gap between the known upper and lower bounds, and obtaining tighter bounds warrants further investigation.

References

1. N. Alon, O. Goldreich, J. Håstad and R. Peralta, Simple Construction of Almost k-wise Independent Random Variables. *Random Struct. Algor.*, 3(3): 289–304, 1992.
2. S. Arora. Polynomial Time Approximation Schemes for Euclidean TSP and Other Geometric Problems. *Proc of 37th FOCS*, pages 2-11, 1996.
3. S. Arora. Polynomial Time Approximation Schemes for Euclidean Traveling Salesman and other Geometric Problems. *J. ACM*, 45, 5:753–782, 1998.
4. C. Bazgan Schémas d'approximation et complexité paramétrée. Technical report, Université Paris Sud, 1995.
5. B. Brejová, D.G. Brown, I.M. Harrower, and T. Vinar. New Bounds for Motif Finding in Strong Instances. *Proc. of 17th CPM*, pages 94–105, 2006.
6. B. Brejová, D.G. Brown, I.M. Harrower, A. López-Ortiz and T. Vinar. Sharper Upper and Lower Bounds for an Approximation Scheme for Consensus-Pattern. *Proc. of 16th CPM*, pages 1–10, 2005.
7. M. Cygan, F. V. Fomin, D. Lokshtanov, L. Kowalik, D. Marx, M. Pilipczuk, M. Pilipczuk, and S. Saurabh. *Parameterized Algorithms.* Springer, 2015.
8. C. Lo, B. Kakaradov, D. Lokshtanov, and C. Boucher. SeeSite: Efficiently Finding Co-occurring Splice Sites and Exon Splicing Enhancers. arXiv:1206.5846v1.
9. R. G. Downey and M. R. Fellows. *Fundamentals of Parameterized Complexity.* Texts in Computer Science. Springer, 2013.
10. M.R. Fellows, J. Gramm, and R. Niedermeier. On the parameterized intractability of motif search problems. *Combinatorica*, 26:141–167, 2006.
11. M.R. Fellows, D. Hermelin, F.A. Rosamond, and S. Vialette. On the parameterized complexity of multiple-interval graph problems. *Theor. Comput. Sci.*, 410(1):53–61, 2009
12. J. Flum, and M. Grohe. Parameterized Complexity Theory. *Springer-Verlag*, 2006.
13. F. Grimmett, and D. Stirzaker. Probability and random processes. *Oxford University Press*, 3 edition, 2001.
14. W. Hoeffding. Probability Inequalities for Sums of Bounded Random Variables. *J. Amer. Statistical Assoc.*, 58(301): 13–30, 1963.
15. H.B. Hunt III, M.V. Marathe, V. Radhakrishnan, S.S. Ravi, D.J. Rosenkrantz, and R.E. Stearns, NC-Approximation Schemes for NP- and PSPACE-Hard Problems for Geometric Graphs. *J. Algorithms*, 26(2):238–274, 1998
16. J.K. Lanctot, M. Li, B. Ma, S. Wang, and L. Zhang. Distinguishing string selection problems. *Inform. Comput.*, 185(1):41–55, 2003.
17. M. Li, B. Ma, and L. Wang. Finding similar regions in many sequences. *J. Comput. System Sci.*, 65(1):73–96, 2002.
18. D. Marx. Closest Substring Problems with Small Distances. *SIAM J. Comput.*, 38(4):1283–1410, 2008.
19. D. Marx. Efficient Approximation Schemes for Geometric Problems? *Proc. of 13th ESA*, 51(1): 448–459, 2005.
20. D. Marx. Parameterized complexity and approximation algorithms. *Comput. J.*, 51(1): 60–78, 2008.
21. J. Naor and M. Naor. Small-Bias Probability Spaces: Efficient Constructions and Applications. *SIAM J. Comput.*, 22(4): 838–856, 1993.
22. R. Niedermeier. Invitation to Fixed-Parameter Algorithms. *Oxford University Press*, 2006.
23. P. Pevzner and S. Sze. Combinatorial approaches to finding subtle signals in DNA strings. In *Proc. of the 8th ISMB*, pages 269–278, 2000.

Fast Quasi-Threshold Editing*

Ulrik Brandes[1], Michael Hamann[2], Ben Strasser[2], and Dorothea Wagner[2]

[1] Computer and Information Science, University of Konstanz, Germany
ulrik.brandes@uni-konstanz.de
[2] Faculty of Informatics, Karlsruhe Institute of Technology, Germany
{michael.hamann,strasser,dorothea.wagner}@kit.edu

Abstract. We introduce Quasi-Threshold Mover (QTM), an algorithm to solve the quasi-threshold (also called trivially perfect) graph editing problem with a minimum number of edge insertions and deletions. Given a graph it computes a quasi-threshold graph which is close in terms of edit count, but not necessarily closest as this edit problem is NP-hard. We present an extensive experimental study, in which we show that QTM performs well in practice and is the first heuristic that is able to scale to large real-world graphs in practice. As a side result we further present a simple linear-time algorithm for the quasi-threshold recognition problem.

1 Introduction

Quasi-Threshold graphs, also known as *trivially perfect* graphs, are defined as the P_4- and C_4-free graphs, i.e., the graphs that do not contain a path or cycle of length 4 as node-induced subgraph [20]. They can also be character- ized as the transitive closure of rooted forests [19], as illus- trated in Fig. 1. These forests can be seen as skeletons of quasi-threshold graphs. Further a constructive character- ization exists: Quasi-threshold graphs are the graphs that are closed under disjoint union and the addition of isolated nodes and nodes connected to every existing node [20].

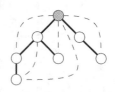

Fig. 1. Quasi-thres. graph with thick skeleton, grey root and dashed transitive closure.

Linear time quasi-threshold recognition algorithms were proposed in [20] and in [9]. Both construct a skeleton if the graph is a quasi-threshold graph. Further, [9] also finds a C_4 or P_4 if the graph is no quasi-threshold graph.

Nastos and Gao [14] observed that components of quasi-threshold graphs have many features in common with the informally defined notion of communities in social networks. They propose to find a quasi-threshold graph that is close to a given graph in terms of edge edit distance in order to detect the communities of that graph. Motivated by their insights we study the quasi-threshold graph editing problem in this paper. Given a graph $G = (V, E)$ we want to find a quasi- threshold graph $G' = (V, E')$ which is closest to G, i.e., we want to minimize the number k of edges in the symmetric difference of E and E'. Figure 2 illustrates

* This work was supported by the DFG under grants BR 2158/6-1, WA 654/22-1, and BR 2158/11-1.

N. Bansal and I. Finocchi (Eds.): ESA 2015, LNCS 9294, pp. 251–262, 2015.
DOI: 10.1007/978-3-662-48350-3_22

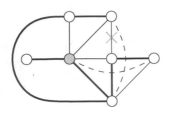

Fig. 2. Edit example with solid input edges, dashed inserted edges, a crossed deleted edge, a thick skeleton with grey root.

an edit example. Unfortunately, the quasi-threshold graph editing problem is NP-hard [14]. However, the problem is fixed parameter tractable (FPT) in k as it is defined using forbidden subgraphs [7]. A basic bounded search tree algorithm which tries every of the 6 possible edits of a forbidden subgraph has a running time in $O(6^k \cdot (|V| + |E|))$. In [11] a polynomial kernel of size $O(k^7)$ was introduced. Unfortunately, our experiments show that real-world social networks have a prohibitively large amount of edits. We prove lower bounds on real-world graphs for k on the scale of 10^4 and 10^5. A purely FPT-based algorithm with the number of edits as parameter can thus not scale in practice. The only heuristic we are aware of was introduced by Nastos and Gao [14]. It greedily picks edits that result in the largest decrease in the number of induced C_4 and P_4 in the graph. Unfortunately, it examines all $\Theta(|V|^2)$ possible edits in each step and thus needs $\Omega(k \cdot |V|^2)$ running time. Even though this running time is polynomial it is still prohibitive for large graphs. In this paper we fill this gap by introducing Quasi-Threshold Mover (QTM), the first scalable quasi-threshold editing heuristic. The final aim of our research is to determine whether quasi-threshold editing is a useful community detection algorithm. Designing an algorithm able to solve the quasi-threshold editing problem on large real-world graphs is a first step in this direction.

1.1 Our Contribution

Our main contribution is Quasi-Threshold Mover (QTM), a scalable quasi-threshold editing algorithm. We provide an extensive experimental evaluation on synthetic as well as a variety of real-world graphs. We further propose a simplified certifying quasi-threshold recognition algorithm. QTM works in two phases: An initial skeleton forest is constructed by a variant of our recognition algorithm, and then refined by moving one node at a time to reduce the number of edits required. The running time of the first phase is dominated by the time needed to count the number of triangles per edge. The best current triangle counting algorithms run in $O(|E|\alpha(G))$ [8,15] time, where $\alpha(G)$ is the arboricity. These algorithms are efficient and scalable in practice on the considered graphs. One round of the second phase needs $O(|V| + |E| \log \Delta)$ time, where Δ is the maximum degree. We show that four rounds are enough to achieve good results.

1.2 Outline

Our paper is organized as follows: We begin by describing how we compute lower bounds on the number of edits. We then introduce the simplified recognition algorithm and the computation of the initial skeleton. The main algorithm is described in Sect. 4. The remainder of the paper is dedicated to the experimental evaluation. An extended version of this paper is available on arXiv [6].

1.3 Preliminaries

We consider simple, undirected graphs $G = (V, E)$ with $n = |V|$ nodes and $m = |E|$ edges. For $v \in V$ let $N(v)$ be the adjacent nodes of v. Let $d(v) := |N(v)|$ for $v \in V$ be the degree of v and Δ the maximum degree in G. Whenever we consider a skeleton forest, we denote by $p(u)$ the parent of a node u.

2   Lower Bounds

A lot of previous research has focused on FPT-based algorithms. To show that no purely FPT-based algorithm parameterized in the number of edits can solve the problem we compute lower bounds on the number of edits required for real-world graphs. The lower bounds used by us are far from tight. However, the bounds are large enough to show that any algorithm with a running time superpolynomial in k can not scale.

To edit a graph we must destroy all forbidden subgraphs H. For quasi-threshold editing H is either a P_4 or a C_4. This leads to the following basic algorithm: Find forbidden subgraph H, increase the lower bound, remove all nodes of H, repeat. This is correct as at least one edit incident to H is necessary. If multiple edits are needed then accounting only for one is a lower bound. We can optimize this algorithm by observing that not all nodes of H have to be removed. If H is a P_4 with the structure $A - B - C - D$ it is enough to remove the two central nodes B and C. If H is a C_4 with nodes A, B, C, and D then it is enough to remove two adjacent nodes. Denote by B and C the removed nodes. This optimization is correct if at least one edit incident to B or C is needed. Regardless of whether H is a P_4 or a C_4 the only edit not incident to B or C is inserting or deleting $\{A, D\}$. However, this edit only transforms a P_4 into a C_4 or vice versa. A subsequent edit incident to B or C is thus necessary.

H can be found using the recognition algorithm. However, the resulting running time of $O(k(n + m))$ does not scale to the large graphs. In the extended version [6] we describe a running time optimization to accelerate computations.

3   Linear Recognition and Initial Editing

The first linear time recognition algorithm for quasi-threshold graphs was proposed in [20]. In [9], a linear time certifying recognition algorithm based on lexicographic breadth first search was presented. However, as the authors note, sorted node partitions and linked lists are needed, which result in large constants behind the big-O. We simplify their algorithm to only require arrays but still provide negative and positive certificates. Further we only need to sort the nodes once to iterate over them by decreasing degree. Our algorithm constructs the forest skeleton of a graph G. If it succeeds G is a quasi-threshold graph and outputs for each node v a parent node $p(v)$. If it fails it outputs a forbidden subgraph H.

To simplify our algorithm we start by adding a super node r to G that is connected to every node and obtain G'. G is a quasi-threshold graph if and only if G' is one. As G' is connected its skeleton is a tree. A core observation is that higher nodes in the tree must have higher degrees, i.e., $d(v) \leq d(p(v))$. We therefore know that r must be the root of the tree. Initially we set $p(u) = r$ for every node u. We process all remaining nodes ordered decreasingly by degree. Once a node is processed its position in the tree is fixed. Denote by u the node that should be processed next. We iterate over all non-processed neighbors v of u and check whether $p(u) = p(v)$ holds and afterwards set $p(v)$ to u. If $p(u) = p(v)$ never fails then G is a quasi-threshold graph as for every node x (except r) we have that by construction that the neighborhood of x is a subset of the one of $p(x)$. If $p(u) \neq p(v)$ holds at some point then a forbidden subgraph H exists. Either $p(u)$ or $p(v)$ was processed first. Assume without lose of generality that it was $p(v)$. We know that no edge $(v, p(u))$ can exist because otherwise $p(u)$ would have assigned itself as parent of v when it was processed. Further we know that $p(u)$'s degree can not be smaller than u's degree as $p(u)$ was processed before u. As v is a neighbor of u we know that another node x must exist that is a neighbor of $p(u)$ but not of u, i.e., (u, x) does not exist. The subgraph H induced by the 4-chain $v - u - p(u) - x$ is thus a P_4 or C_4 depending on whether the edge (v, x) exists. We have that u, v and $p(u)$ are not r as $p(v)$ was processed before them and r was processed first. As x has been chosen such that (u, x) does not exist but (u, r) exist $x \neq r$. H therefore does not use r and is contained in G.

From Recognition to Editing. We modify the recognition algorithm to construct a skeleton for arbitrary graphs. This skeleton induces a quasi-threshold graph Q. We want to minimize Q's distance to G. Note that all edits are performed implicitly, we do not actually modify the input graph for efficiency reasons. The only difference between our recognition and our editing algorithm is what happens when we process a node u that has a non-processed neighbor v with $p(u) \neq p(v)$. The recognition algorithm constructs a forbidden subgraph H, while the editing algorithm tries to resolve the problem. We have three options for resolving the problem: we ignore the edge $\{u, v\}$, we set $p(v)$ to $p(u)$, or we set $p(u)$ to $p(v)$. The last option differs from the first two as it affects all neighbors of u. The first two options are the decision if we want to make v a child of u even though $p(u) \neq p(v)$ or if we want to ignore this potential child. We start by determining a preliminary set of children by deciding for each non-processed neighbor of u whether we want to keep or discard it. These preliminary children elect a new parent by majority. We set $p(u)$ to this new parent. Changing u's parent can change which neighbors are kept. We therefore reevaluate all the decisions and obtain a final set of children for which we set u as parent. Then the algorithm simply continues with the next node.

What remains to describe is when our algorithm keeps a potential child. It does this using two edge measures: The number of triangles $t(e)$ in which an edge

```
1  foreach v_m-neighbor u do
2  |  push u;
3  while queue not empty do
4  |    u ← pop;
5  |    determine child_close(u) by DFS;
6  |    x ← max over score_max of reported u-children;
7  |    y ← Σ over child_close of close u-children;
8  |    if u is v_m-neighbor then
9  |    |    score_max(u) ← max{x, y} + 1;
10 |    else
11 |    |    score_max(u) ← max{x, y} − 1;
12 |    if child_close(u) > 0 or score_max(u) > 0 then
13 |    |    report u to p(u);
14 |    |    push p(u);
15 Best v_m-parent corresponds to score_max(r);
```

(a) Pseudo-Code for moving v_m (b) Moving v_m example

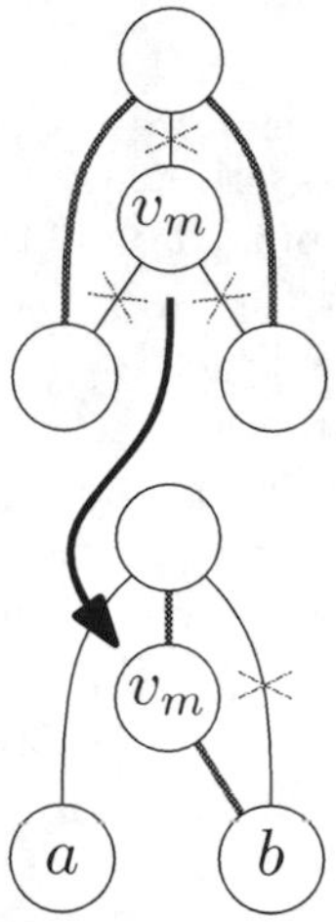

Fig. 3. In Fig. 3b the drawn edges are in the skeleton. By moving v_m, crossed edges are removed and thick blue edges are inserted. a is not adopted while b is.

e participates and a pseudo-C_4-P_4-counter $p_c(e)$, which is the sum of the number of C_4 in which e participates and the number of P_4 in which e participates as central edge. Computing $p_c(x, y)$ is easy given the number of triangles and the degrees of x and y as $p_c(\{x, y\}) = (d(x) - 1 - t(\{x, y\})) \cdot (d(y) - 1 - t(\{x, y\}))$ holds. Having a high $p_c(e)$ makes it likely that e should be deleted. We keep a potential child only if two conditions hold. The first is based on triangles. We know by construction that both u and v have many edges in G towards their current ancestors. Keeping v is thus only useful if u and v share a large number of ancestors as otherwise the number of induced edits is too high. Each common ancestor of u and v results in a triangle involving the edge $\{u, v\}$ in Q. Many of these triangles should also be contained in G. We therefore count the triangles of $\{u, v\}$ in G and check whether there are at least as many triangles as v has ancestors. The other condition uses $p_c(e)$. The decision whether we keep v is in essence the question of whether $\{u, v\}$ or $\{v, p(v)\}$ should be in Q. We only keep v if $p_c(\{u, v\})$ is not higher than $p_c(\{v, p(v)\})$. The details of the algorithm can be found in the extended version [6]. The time complexity of this editing heuristic is dominated by the triangle counting algorithm as the rest is linear.

4   The Quasi-Threshold Mover Algorithm

The Quasi-Threshold Mover (QTM) algorithm iteratively increases the quality of a skeleton T using an algorithm based on local moving. Local moving is a technique that is successfully employed in many heuristic community detection algorithms [2,12,16]. As in most algorithm based on this principle, our algorithm works in rounds. In each round it iterates over all nodes v_m in random order and tries to move v_m. In the context of community detection, a node is moved to a neighboring community such that a certain objective function is increased. In our setting we want to minimize the number of edits needed to transform the input graph G into the quasi-threshold graph Q implicitly defined by T. We need to define the set of allowed moves for v_m in our setting. Moving v_m consists of moving v_m to a different position within T and is illustrated in Fig. 3b. We need to chose a new parent u for v_m. The new parent of v_m's old children is v_m's old parent. Besides choosing the new parent u we select a set of children of u that are *adopted* by v_m, i.e., their new parent becomes v_m. Among all allowed moves for v_m we chose the move that reduces the number of edits as much as possible. Doing this in sub-quadratic running time is difficult as v_m might be moved anywhere in G. By only considering the neighbors of v_m in G and a few more nodes per neighbor in a bottom-up scan in the skeleton, our algorithm has a running time in $O(n + m \log \Delta)$ per round. While our algorithm is not guaranteed to be optimal as a whole we can prove that for each node v_m we choose a move that reduces the number of edits as much as possible. Our experiments show that given the result of the initialization heuristic our moving algorithm performs well in practice. They further show that in practice four rounds are good enough which results in a near-linear total running time.

Basic Idea. Our algorithm starts by isolating v_m, i.e., removing all incident edges in Q. It then finds a position at which v_m should be inserted in T. If v_m's original position was optimal then it will find this position again. For simplicity we will assume again that we add a virtual root r that is connected to all nodes. Isolating v_m thus means that we move v_m below the root r and do not adopt any children. Choosing u as parent of v_m requires Q to contain edges from all ancestors of u to v_m. Further if v_m adopts a child w of u then Q must have an edge from every descendant of w to v_m. How good a move is depends on how many of these edges already exist in G and how many edges incident to v_m in G are not covered. To simplify notation we will refer to the nodes incident to v_m in G as v_m-*neighbors*. We start by identifying which children a node should adopt. For this we define the *child closeness* $\text{child}_{\text{close}}(u)$ of u as the number of v_m-neighbors in the subtree of u minus the non-v_m-neighbors. A node u is a *close child* if $\text{child}_{\text{close}}(u) > 0$. If v_m chooses a node u as new parent then it should adopt all close children. A node can only be a close child if it is a neighbor of v_m or when it has a close child. Our algorithm starts by computing all close children and their closeness using many short DFS searches in a bottom up fashion. Knowing which nodes are good children we can identify which nodes are good parents for v_m. A potential parent must have a close child or must be a neighbor of v_m. Using the set of close children we can easily derive a set of parent candidates and an optimal selection of adopted children for every potential parent. We need to determine the candidate with the fewest edits. We do this in a bottom-up fashion. To implement the described moving algorithm we need to put $O(d_G(v_m))$ elements into a priority queue. The running time is thus amortized $O(d_G(v_m) \log d_G(v_m))$ per move or $O(n + m \log \Delta)$ per round.

We analyze the running time complexity using tokens. Initially only the v_m-neighbors have tokens. The tokens are consumed by the short DFS searches and the processing of parent nodes. The details of the analysis are complex and are described in the extended version [6].

Close Children. To find all close children we attach to each node u a DFS instance that explores the subtree of u. Note that every DFS instance has a constant state size and thus the memory consumption is still linear. u is close if this DFS finds more v_m-neighbors than non-v_m-neighbors. Unfortunately we can not fully run all these searches as this requires too much running time. Therefore a DFS is aborted if it finds more non-v_m-neighbors than v_m-neighbors. We exploit that close children are v_m-neighbors or have themselves close children. Initially we fill a queue of potential close children with the neighbors of v_m and when a new close child is found we add its parent to the queue. Let u denote the current node removed from the queue. We run u's DFS and if it explores the whole subtree then u is a close child. We need to take special care that every node is visited only by one DFS. A DFS therefore looks at the states of the DFS of the nodes it visits. If one of these other DFS has run then it uses their state information to skip the already explored part of the subtree. To avoid that a DFS is run after its state was inspected we organize the queue as priority queue ordered by tree depth. If the DFS of u starts by first inspecting the wrong children then it can get stuck because it would see the v_m-neighbors too late. The DFS must first visit the close children of u. To assure that u knows which children are close every close child must report itself to its parent when it is detected. As all children have a greater depth they are detected before the DFS of their parent starts.

Potential Parents. Consider the subtree T_u of u and a potential parent w in T_u. Let X_w be the set of nodes given by w, the ancestors of w, the close children of w and the descendants of the close children of w. Moving v_m below w requires us to insert an edge from v_m to every non-v_m-neighbor in X_w. Likewise, not including v_m-neighbors in X_w requires us to delete an edge for each of them. We therefore want X_w to maximize the number of v_m-neighbors minus the number of non-v_m-neighbors. This value gives us a score for each potential parent in T_u. We denote by $\text{score}_{\max}(u)$ the maximum score over all potential parents in T_u. Note that $\text{score}_{\max}(u)$ is always at least -1 as we can move v_m below u and not adopt any children. We determine in a bottom-up fashion all $\text{score}_{\max}(u)$ that are greater than 0. Whether $\text{score}_{\max}(u)$ is -1 or 0 is irrelevant because isolating v_m is never worse. The final solution will be in $\text{score}_{\max}(r)$ of the root r as its "subtree" encompasses the whole graph. $\text{score}_{\max}(u)$ can be computed recursively. If u is a best parent then the value of $\text{score}_{\max}(u)$ is the sum over the closenesses of all of u's close children ± 1. If the subtree T_w of a child w of u contains a best parent then $\text{score}_{\max}(u) = \text{score}_{\max}(w) \pm 1$. The ± 1 depends on whether w is a v_m-neighbor. Unfortunately not only potential parents u have a $\text{score}_{\max}(u) > 0$. However, we know that every node u with $\text{score}_{\max}(u) > 0$ is a v_m-neighbor or has a child w with $\text{score}_{\max}(w) > 0$. We can therefore process all $\text{score}_{\max}$ values in a similar bottom-up way using a tree-depth ordered priority queue as we used to compute child$_{close}$. As both bottom-up procedures have the same structure we can interweave them as optimization and use only a single queue. The algorithm is illustrated in Fig. 3a in pseudo-code form.

5   Experimental Evaluation

We evaluated the QTM algorithm on the small instances used by Nastos and Gao [14], on larger synthetic graphs and large real-world social networks and web graphs. We measured both the number of edits needed and the required running time. For each graph we also report the lower bound b of necessary edits that we obtained using our lower bound algorithm. We implemented the algorithms in C++ using NetworKit [17]. All experiments were performed on an Intel Core i7-2600K CPU with 32GB RAM. We ran all algorithms ten times with ten different random node id permutations.

Comparison with Nastos and Gao's Results. Nastos and Gao [14] did not report any running times, we therefore re-implemented their algorithm. Our implementation of their algorithm has a complexity of $O(m^2 + k \cdot n^2 \cdot m)$, the details can be found in the extended version [6]. Similar to their implementation we used a simple exact bounded search tree (BST) algorithm for the last 10 edits. In Table 1 we report the minimum and average number of edits over ten runs. Our implementation of their algorithm never needs more edits than they reported[1]. For two of the graphs (dolphins and lesmis) our implementation needs slightly less edits due to different tie-breaking rules.

For all but one graph QTM is at least as good as the algorithm of Nastos and Gao in terms of edits. QTM needs only one more edit than Nastos and Gao for the grass_web graph. The QTM algorithm is much faster than their algorithm, it needs at most 2.5 milliseconds while the heuristic of Nastos and Gao needs up to 6 seconds without bounded search tree and almost 17 seconds with bounded search tree. The number of iterations necessary is at most 5. As the last round only checks whether we are finished four iterations would be enough.

Large Graphs. For the results in Table 2 we used two Facebook graphs [18] and five SNAP graphs [13] as social networks and four web graphs from the 10th DIMACS Implementation Challenge [1,3,4,5]. We evaluate two variants of QTM. The first is the standard variant which starts with a non-trivial skeleton obtained by the heuristic described in Section 3. The second variant starts with a trivial skeleton where every node is a root. We chose these two variants to determine which part of our algorithm has which influence on the final result. For the standard variant we report the number of edits needed before any node is moved. With a trivial skeleton this number is meaningless and thus we report the number of edits after one round. All other measures are straightforward and are explained in the table's caption.

---

[1] Except on Karate, where they report 20 due to a typo. They also need 21 edits.

Table 1. Comparison of QTM and [14]. We report n and m, the lower bound b, the number of edits (as minimum, mean and standard deviation), the mean and maximum of number of QTM iterations, and running times in ms.

Name	n	m	b	Algorithm	Edits			Iterations		Time [ms]	
					min	mean	std	mean	max	mean	std
				QTM	72	74.1	1.1	2.7	4.0	0.6	0.1
dolphins	62	159	24	NG w/ BST	73	74.7	0.9	-	-	15 594.0	2 019.0
				NG w/o BST	73	74.8	0.8	-	-	301.3	4.0
				QTM	251	254.3	2.7	3.5	4.0	2.5	0.4
football	115	613	52	NG w/ BST	255	255.0	0.0	-	-	16 623.3	3 640.6
				NG w/o BST	255	255.0	0.0	-	-	6 234.6	37.7
				QTM	35	35.2	0.4	2.0	2.0	0.5	0.1
grass_web	86	113	10	NG w/ BST	34	34.6	0.5	-	-	13 020.0	3 909.8
				NG w/o BST	38	38.0	0.0	-	-	184.6	1.2
				QTM	21	21.2	0.4	2.0	2.0	0.4	0.1
karate	34	78	8	NG w/ BST	21	21.0	0.0	-	-	9 676.6	607.4
				NG w/o BST	21	21.0	0.0	-	-	28.1	0.3
				QTM	60	60.5	0.5	3.3	5.0	1.4	0.3
lesmis	77	254	13	NG w/ BST	60	60.8	1.0	-	-	16 919.1	3 487.7
				NG w/o BST	60	77.1	32.4	-	-	625.0	226.4

Even though for some of the graphs the mover needs more than 20 iterations to terminate, the results do not change significantly compared to the results after round 4. In practice we can thus stop after 4 rounds without incurring a significant quality penalty. It is interesting to see that for the social networks the initialization algorithm sometimes produces a skeleton that induces more than m edits (e.g. in the case of the "Penn" graph) but still the results are always slightly better than with a trivial initial skeleton. This is even true when we do not abort moving after 4 rounds. For the web graphs, the non-trivial initial skeleton does not seem to be useful for some graphs. It is not only that the initial number of edits is much higher than the finally needed number of edits, also the number of edits needed in the end is slightly higher than if a trivial initial skeleton was used. This might be explained by the fact that we designed the initialization algorithm with social networks in mind. Initial skeleton heuristics built specifically for web graphs could perform better. While the QTM algorithm needs to edit between approximately 50 and 80% of the edges of the social networks, the edits of the web graphs are only between 10 and 25% of the edges. This suggests that quasi-threshold graphs might be a good model for web graphs while for social networks they represent only a core of the graph that is hidden by a lot of noise. Concerning the running time one can clearly see that QTM is scalable and suitable for large real-world networks.

As we cannot show for our real-world networks that the edit distance that we get is close to the optimum we generated synthetic graphs by generating quasi-threshold graphs and applying random edits to these graphs. The details of the

Table 2. Results for large real-world and synthetic graphs. Number of nodes n and edges m, the lower bound b and the number of edits are reported in thousands. Column "I" indicates whether we start with a trivial skeleton or not. • indicates an initial skeleton as described in Section 3 and ○ indicates a trivial skeleton. Edits and running time are reported for a maximum number of 0 (respectively 1 for a trivial initial skeleton), 4 and ∞ iterations. For the latter, the number of actually needed iterations is reported as "It". Edits, iterations and running time are the average over the ten runs.

	Name	n [K] m [K]	b [K]	I	Edits [K] 0/1	4	∞	It ∞	Time [s] 0/1	4	∞
Social Networks	Caltech	0.77 16.66	0.35	• ○	15.8 12.6	11.6 11.7	11.6 11.6	8.5 9.4	0.0 0.0	0.0 0.0	0.1 0.1
	amazon	335 926	99.4	• ○	495 433	392 403	392 403	7.2 8.9	0.3 1.3	5.5 4.9	9.3 10.7
	dblp	317 1 050	53.7	• ○	478 444	415 424	415 423	7.2 9.0	0.4 1.4	5.8 5.2	9.9 11.5
	Penn	41.6 1 362	19.9	• ○	1 499 1 174	1 129 1 133	1 127 1 129	14.4 16.2	0.6 1.0	4.2 3.7	13.5 14.4
	youtube	1 135 2 988	139	• ○	2 169 2 007	1 961 1 983	1 961 1 983	9.8 10.0	1.4 7.1	31.3 28.9	73.6 72.7
	lj	3 998 34 681	1 335	• ○	32 451 26 794	25 607 25 803	25 577 25 749	18.8 19.9	23.5 58.3	241.9 225.9	1 036.0 1 101.3
	orkut	3 072 117 185	1 480	• ○	133 086 106 367	103 426 103 786	103 278 103 507	24.2 30.2	115.2 187.9	866.4 738.4	4 601.3 5 538.5
Web Graphs	cnr-2000	326 2 739	48.7	• ○	1 028 502	409 410	407 409	11.2 10.7	0.8 3.2	12.8 11.8	33.8 30.8
	in-2004	1 383 13 591	195	• ○	2 700 1 909	1 402 1 392	1 401 1 389	11.0 13.5	7.9 16.6	72.4 65.0	182.3 217.6
	eu-2005	863 16 139	229	• ○	7 613 4 690	3 917 3 919	3 906 3 910	13.7 14.5	6.9 22.6	90.7 85.6	287.7 303.5
	uk-2002	18 520 261 787	2 966	• ○	68 969 42 193	31 218 31 092	31 178 31 042	19.1 22.3	200.6 399.8	1 638.0 1 609.6	6 875.5 8 651.8
Synthetic	Gen. 160K	100 930	42	• ○	200 193	158 158	158 158	4.6 6.1	0.2 1.0	3.5 3.3	4.1 4.9
	Gen. 0.4K	1 000 10 649	0.391	• ○	1.161 182	0.395 5.52	0.395 5.52	3.0 6.1	3.3 15.9	43.8 52.9	43.8 78.8

generation process are described in the extended version [6]. In Table 2 we report the results of two of these graphs with 400 and 160 000 random edits. In both cases the number of edits the QTM algorithm finds is below or equal to the generated editing distance. If we start with a trivial skeleton, the resulting edit distance is sometimes very high, as can be seen for the graph with 400 edits. This shows that the initialization algorithm from Section 3 is necessary to achieve good quality on

graphs that need only few edits. As it seems to
be beneficial for most graphs and not very bad
for the rest, we suggest to use the initialization
algorithm for all graphs.

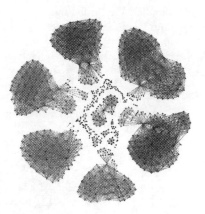

Case Study: Caltech. The main application
of our work is community detection. While a
thorough experimental evaluation of its usefulness in this context is future work we want
to give a promising outlook. Figure 4 depicts the edited Caltech university Facebook
network from [18]. Nodes are students and
edges are Facebook-friendships. The dormitories of most students are known. We colored
the graph accordingly. The picture clearly
shows that our algorithm succeeds at identifying most of this structure.

Fig. 4. Edited Caltech network,
edges colored by dormitories of
endpoints.

6   Conclusion

We have introduced Quasi-Threshold Mover (QTM), the first heuristic algorithm
to solve the quasi-threshold editing problem in practice for large graphs. As a side
result we have presented a simple certifying linear-time algorithm for the quasi-
threshold recognition problem. A variant of our recognition algorithm is also used
as initialization for the QTM algorithm. In an extensive experimental study with
large real world networks we have shown that it scales very well in practice. We
generated graphs by applying random edits to quasi-threshold graphs. QTM
succeeds on these random graphs and often even finds other quasi-threshold
graphs that are closer to the edited graph than the original quasi-threshold
graph. A surprising result is that web graphs are much closer to quasi-threshold
graphs than social networks, for which quasi-threshold graphs were introduced
as community detection method. A logical next step is a closer examination of
the detected quasi-threshold graphs and the community structure they induce.
Further our QTM algorithm might be adapted for the more restricted problem
of threshold editing which is NP-hard as well [10] or extended with an improved
initialization algorithm, especially for web graphs.

Acknowledgment. We thank James Nastos for helpful discussions.

References

1. Bader, D.A., Meyerhenke, H., Sanders, P., Wagner, D.: Graph Partitioning and
Graph Clustering: 10th DIMACS Implementation Challenge, vol. 588. American
Mathematical Society (2013)

2. Blondel, V., Guillaume, J.L., Lambiotte, R., Lefebvre, E.: Fast unfolding of communities in large networks. Journal of Statistical Mechanics: Theory and Experiment 2008(10) (2008)

3. Boldi, P., Codenotti, B., Santini, M., Vigna, S.: Ubicrawler: A scalable fully distributed web crawler. Software - Practice and Experience 34(8), 711–726 (2004)

4. Boldi, P., Rosa, M., Santini, M., Vigna, S.: Layered label propagation: A multiresolution coordinate-free ordering for compressing social networks. In: Proceedings of the 20th International Conference on World Wide Web (WWW 2011), pp. 587–596. ACM Press (2011)

5. Boldi, P., Vigna, S.: The WebGraph framework I: Compression techniques. In: Proceedings of the 13th International Conference on World Wide Web (WWW 2004), pp. 595–602. ACM Press (2004)

6. Brandes, U., Hamann, M., Strasser, B., Wagner, D.: Fast quasi-threshold editing (2015), http://arxiv.org/abs/1504.07379

7. Cai, L.: Fixed-parameter tractability of graph modification problems for hereditary properties. Information Processing Letters 58(4), 171–176 (1996)

8. Chiba, N., Nishizeki, T.: Arboricity and subgraph listing algorithms. SIAM Journal on Computing 14(1), 210–223 (1985)

9. Chu, F.P.M.: A simple linear time certifying lbfs-based algorithm for recognizing trivially perfect graphs and their complements. Information Processing Letters 107(1), 7–12 (2008)

10. Drange, P.G., Dregi, M.S., Lokshtanov, D., Sullivan, B.D.: On the threshold of intractability. In: Proceedings of the 23rd Annual European Symposium on Algorithms (ESA 2015). LNCS. Springer (2015)

11. Drange, P.G., Pilipczuk, M.: A polynomial kernel for trivially perfect editing. In: Proceedings of the 23rd Annual European Symposium on Algorithms (ESA 2015). LNCS. Springer (2015)

12. Görke, R., Kappes, A., Wagner, D.: Experiments on density-constrained graph clustering. ACM Journal of Experimental Algorithmics 19, 1.6:1.1–1.6:1.31 (2014)

13. Leskovec, J., Krevl, A.: Snap datasets: Stanford large network dataset collection (June 2014), http://snap.stanford.edu/data

14. Nastos, J., Gao, Y.: Familial groups in social networks. Social Networks 35(3), 439–450 (2013)

15. Ortmann, M., Brandes, U.: Triangle listing algorithms: Back from the diversion. In: Proceedings of the 16th Meeting on Algorithm Engineering and Experiments (ALENEX 2014), pp. 1–8. SIAM (2014)

16. Rotta, R., Noack, A.: Multilevel local search algorithms for modularity clustering. ACM Journal of Experimental Algorithmics 16, 2.3:2.1–2.3:2.27 (2011)

17. Staudt, C., Sazonovs, A., Meyerhenke, H.: Networkit: An interactive tool suite for high-performance network analysis (2014), http://arxiv.org/abs/1403.3005

18. Traud, A.L., Mucha, P.J., Porter, M.A.: Social structure of facebook networks. Physica A: Statistical Mechanics and its Applications 391(16), 4165–4180 (2012)

19. Wolk, E.S.: A note on "the comparability graph of a tree". Proceedings of the American Mathematical Society 16(1), 17–20 (1965)

20. Yan, J.H., Chen, J.J., Chang, G.J.: Quasi-threshold graphs. Discrete Applied Mathematics 69(3), 247–255 (1996)

Sublinear Estimation of Weighted Matchings in Dynamic Data Streams[*]

Marc Bury and Chris Schwiegelshohn

Efficient Algorithms and Complexity Theory, TU Dortmund, Germany
{firstname.lastname}@tu-dortmund.de

Abstract. This paper presents an algorithm for estimating the weight of a maximum weighted matching by augmenting any estimation routine for the size of an unweighted matching. The algorithm is implementable in any streaming model including dynamic graph streams. We also give the first constant estimation for the maximum matching size in a dynamic graph stream for planar graphs (or any graph with bounded arboricity) using $\tilde{O}(n^{4/5})$ space which also extends to weighted matching. Using previous results by Kapralov, Khanna, and Sudan (2014) we obtain a polylog(n) approximation for general graphs using polylog(n) space in random order streams, respectively. In addition, we give a space lower bound of $\Omega(n^{1-\varepsilon})$ for any randomized algorithm estimating the size of a maximum matching up to a $1 + O(\varepsilon)$ factor for adversarial streams.

1 Introduction

Large graph structures encountered in social networks or the web-graph have become focus of analysis both from theory and practice. To process such large input, conventional algorithms often require an infeasible amount of running time, space or both, giving rise to other models of computation. Much theoretical research focuses on the streaming model where the input arrives one by one with the goal of storing as much information as possible in small, preferably polylogarithmic, space. Streaming algorithms on graphs were first studied by Henzinger et al. [17], who showed that even simple problems often admit no solution with such small space requirements. The semi-streaming model [14] where the stream consists of the edges of a graph and the algorithm is allowed $O(n \cdot \text{polylog}(n))$ space and allows few (ideally just one) passes over the data relaxes these requirements and has received considerable attention. Problems studied in the semi-streaming model include sparsification, spanners, connectivity, minimum spanning trees, counting triangles and matching, for an overview we refer to a recent survey by McGregor [27]. Due to the fact that graphs motivating this research are dynamic structures that change over time there has recently been research on streaming algorithms supporting deletions. We now review the literature on streaming algorithms for matching and dynamic streams.

[*] Supported by Deutsche Forschungsgemeinschaft, grant BO 2755/1-2 and within the Collaborative Research Center SFB 876, project A2.

© Springer-Verlag Berlin Heidelberg 2015
N. Bansal and I. Finocchi (Eds.): ESA 2015, LNCS 9294, pp. 263–274, 2015.
DOI: 10.1007/978-3-662-48350-3_23

Matching. Maintaining a 2 approximation to the maximum matching (MM) in an insertion-only stream can be straightforwardly done by greedily maintaining a maximal matching [14]. Improving on this algorithm turns out to be difficult as Goel et al. [16] showed that no algorithm using $\tilde{O}(n)$ space can achieve an approximation ratio better than $\frac{3}{2}$ which was improved by Kapralov to $\frac{e}{e-1}$ [19]. Konrad et al. [22] gave an algorithm using $\tilde{O}(n)$ space with an approximation factor of 1.989 if the edges are assumed to arrive in random order. For weighted matching (MWM), a series of results have been published [14,26,11,32,12] with the current best bound of $4 + \varepsilon$ being due to Crouch and Stubbs [10].

To bypass the natural $\Omega(n)$ bound required by any algorithm maintaining an approximate matching, recent research has begun to focus on estimating the size of the maximum matching. Kapralov et al. [20] gave a polylogrithmic approximate estimate using polylogarithmic space for random order streams. For certain sparse graphs including planar graphs, Esfandiari et al. [13] describe how to obtain a constant factor estimation using $\tilde{O}(n^{2/3})$ space in a single pass and $\tilde{O}(\sqrt{n})$ space using two passes or assuming randomly ordered streams. The authors also gave a lower bound of $\Omega(\sqrt{n})$ for any approximation better than $\frac{3}{2}$.

Dynamic Streams. In the turnstile model, the stream consists of a sequence of additive updates to a vector. Problems studied in this model include numerical linear algebra problems such as regression and low-rank approximation, and maintaining certain statistics of a vector like frequency moments, heavy hitters or entropy. Linear sketches have proven to be the algorithmic technique of choice and might as well be the only algorithmic tool able to efficiently do so, see Li, Nguyen and Woodruff [24]. Dynamic graphs as introduced and studied by Ahn, Guha and McGregor [1,2,3,4] are similar to, but weaker than turnstile updates. Though both streaming models assume update to the input matrix, there usually exists a consistency assumption for streams, i.e. at any given time the multiplicity of an edge is either 0 or 1 and edge weights cannot change arbitrarily but are first set to 0 and then reinserted with the desired weight. The authors extend some of the aforementioned problems such as connectivity, sparsification and minimum spanning trees to this setting. Recent results by Assadi et al. [5] showed that approximating matchings in dynamic streams is hard by providing a space lower bound of $\Omega(n^{2-3\varepsilon})$ for approximating the maximum matching within a factor of $\tilde{O}(n^{\varepsilon})$. Simultaneously, Konrad [21] showed a similar but slightly weaker lower bound of $\Omega(n^{3/2-4\varepsilon})$. Both works presented an algorithm with an almost matching upper bound on the space complexity of $\tilde{O}(n^{2-2\varepsilon})$ [21] and $\tilde{O}(n^{2-3\varepsilon})$ [5]. Chitnis et al. [7] gave a streaming algorithm using $\tilde{O}(k^2)$ space that returns an exact maximum matching under the assumption that the size is at most k. It is important to note that all these results actually compute a matching. In terms of estimating the size of the maximum matching, Chitnis et al. [7] extended the estimation algorithms for sparse graphs from [13] to the settings of dynamic streams using $\tilde{O}(n^{4/5})$ space. A bridge between dynamic graphs and the insertion-only streaming model is the sliding window model studied by Crouch et al. [9]. The authors give a $(3 + \varepsilon)$-approximation algorithm for maximum matching.

The p-Schatten norm of a matrix A is defined as the ℓ_p-norm of the vector of singular values. It is well known that computing the maximum matching size is equivalent to computing the rank of the Tutte matrix [29,25] (see also Section 2.1). Estimating the maximum matching size therefore is a special case of estimating the rank or 0-Schatten norm of a matrix. Li, Nguyen and Woodruff gave strong lower bounds on the space requirement for estimating Schatten norms in dynamic streams [23]. Any estimation of the rank within any constant factor is shown to require $\Omega(n^2)$ space when using bi-linear sketches and $\Omega(\sqrt{n})$ space for general linear sketches.

Table 1. Results for estimating the size (weight) of a maximum (weighted) matching in data streams.

	Reference	Graph class	Streaming model	Approx. factor	Space
MM:	Greedy	General	Adversarial	2	$O(n)$
	[20]	General	Random	polylog(n)	polylog(n)
	[13]	Trees	Adversarial	$2 + \varepsilon$	$\tilde{O}(\sqrt{n})$
	[13]	Bounded arboricity	Adversarial	$O(1)$	$\tilde{O}(n^{2/3})$
	here	Trees	Dynamic	$2 + \varepsilon$	$O(\frac{\log^2 n}{\varepsilon^2})$
	here	Bounded arboricity	Dynamic	$O(1)$	$\tilde{O}(n^{4/5})$
	[13]	Forests	Adversarial	$\frac{3}{2} - \varepsilon$	$\Omega(\sqrt{n})$
	here	General	Adversarial	$1 + O(\varepsilon)$	$\Omega\left(n^{1-\varepsilon}\right)$
MWM:	[10]	General	Adversarial	$4 + \varepsilon$	$O(n \log^2 n)$
	here	General	Random	polylog(n)	polylog(n)
	here	Bounded arboricity	Dynamic	$O(1)$	$\tilde{O}(n^{4/5})$

Techniques and Contribution. Table 1 gives an overview of our results in comparison to previously known algorithms and lower bounds. Our first main result (Section 2) is an approximate estimation algorithm for the maximum weight of a matching. We give a generic procedure using any unweighted estimation as black box. In particular:

Theorem 1 (informal version). *Given a λ-approximate estimation using S space, there exists an $O(\lambda^4)$-approximate estimation algorithm for the weighted matching problem using $O(S \cdot \log n)$ space.*

The previous algorithms for weighted matchings in insertion only streams analyzed in [14,26,11,32] extend the greedy approach by a charging scheme. If edges are mutually exclusive, the new edge will be added if the weight of the matching increases by a given threshold, implicitly partitioning the edges into sets of geometrically increasing weights. We use a similar scheme, but with a twist: Single edge weights cannot be charged to an edge with larger weight as estimation routines do not necessarily give information on distinct edges. However, entire matchings can be charged as the contribution of a specific range of weights r can only be large if these edges take up a significant part of any maximum matching in the subgraph containing only the edges of weight at least r. For analysis, we use a result on parallel algorithms by Uehara and Chen [30]. We show that the weight outputted by our algorithm is close to the weight of the matching computed by the authors, implying an approximation to the maximum weight.

We can implement this algorithm in dynamic streams although at submission, we were unaware of any estimations for dynamic streams. Building on the work by Esfandiari et al. [13], we give a constant estimation on the matching size in bounded arboricity graphs. The main obstacle to adapt their algorithms for bounded arboricity graphs is that they maintain a small size matching using the greedy algorithm which is hard for dynamic streams. Instead of maintaining a matching, we use the Tutte matrix to get a 1-pass streaming algorithm using $\tilde{O}(n^{4/5})$ space, which immediately extends to weighted matching. Similar bounds have been obtained independently by Chitnis et al. [7].

Our lower bound (Section 3) is proven via reduction from the Boolean Hidden Hypermatching problem introduced by Verbin and Yu [31]. In this setting, two players Alice and Bob are given a binary n-bit string and a perfect t-hypermatching on n nodes, respectively. Bob also gets a binary string w. The players are promised that the parity of bits corresponding to the nodes of the i-th hypermatching either are equal to w_i for all i or equal to $1 - w_i$ for all i and the task is to find out which case holds using only a single round of communication. We construct a graph consisting of a t-clique for each hyperedge of Bob's matching and a single edge for each bit of Alice's input that has one node in common with the t-cliques. Then we show that approximating the matching size within a factor better than $1 + O(1/t)$ can also solve the Boolean Hidden Hypermatching instance. Using the lower bound of $\Omega(n^{1-1/t})$ from [31] we have

Theorem 2 (informal version). *Any 1-pass streaming algorithm approximating the size of the maximum matching matching up to an $(1 + O(\varepsilon))$ factor requires $\Omega(n^{1-\varepsilon})$ bits of space.*

This lower bound also implies an $\Omega(n^{1-\varepsilon})$ space bound for $1 + O(\varepsilon)$ approximating the rank of a matrix in data streams which also improves the $\Omega(\sqrt{n})$ bound by Li, Nguyen, and Woodruff [23] for linear sketches.

1.1   Preliminaries

We use $\tilde{O}(f(n))$ to hide factors polylogarithmic in $f(n)$. Any randomized algorithm succeeding with high probability has at least $1 - 1/n$ chance of success. Graphs are denoted by $G(V, E, w)$ where V is the set of n nodes, E is the set of edges and $w : E \to \mathbb{R}^+$ is a weight function. Our estimated value $\widehat{M}$ is a λ-approximation to the size of the maximum matching M if $\widehat{M} \leq |M| \leq \lambda\widehat{M}$.

2   Weighted Matching

We start by describing the parallel algorithm by Uehara and Chen [30] which we call the *partitioning algorithm*. Let $\gamma > 1$ and $k > 0$ be constant. We partition the edge set by t ranks where all edges e in rank $i \in \{1, \ldots, t\}$ have a weight $w(e) \in \left(\gamma^{i-1} \cdot \frac{w_{max}}{kN}, \gamma^i \cdot \frac{w_{max}}{kN}\right]$ where w_{max} is the maximal weight in G. Let $G' = (V, E, w)$ be equal to G but each edge e in rank i has weight $r_i := \gamma^i$ for all $i = 1, \ldots, t$. Starting with $i = t$, we compute an unweighted maximal matching M_i considering only edges in rank i (in G') and remove all edges incident to a matched node. Continue with $i - 1$. The weight of the matching $M = \bigcup M_i$ is $w(M) = \sum_{i=1}^t r_i \cdot |M_i|$ and satisfies $w_G(M^*) \geq w_{G'}(M) \geq \frac{1}{2\gamma} \cdot w_G(M^*)$ where M^* is an optimal weighted matching in G. The previous algorithms [14,26,11,32,10] for insertion-only streams use a similar partitioning of edge weights. Since these algorithms are limited to storing one maximal matching (in case of [10] one maximal matching per rank), they cannot compute residual maximal matchings in each rank. However, by charging the smaller edge weights into the higher ones, the resulting approximation factor can be made reasonably close to that of Uehara and Chen. Since these algorithms maintain matchings, they cannot have sublinear space in an insertion-only stream and they need at least $\Omega(n^{2-3\varepsilon})$ in a dynamic stream even when the maintained matching is only a $O(n^\varepsilon)$ approximation ([5]). Though the complexity for unweighted estimating unweighted matchings is not settled for any streaming model, there exist graph classes for which one can improve on these algorithms wrt space requirement. Therefore, we assume the existence of a black box λ-approximate matching estimation algorithm.

Algorithm and Analysis. The partitioning of Uehara and Chen can be constructed almost obliviously in a stream. Let $(e_0, w(e_0))$ be the first inserted edge. Then an edge e belongs to rank i iff $2^{i-1} \cdot w(e_0) < w(e) \leq 2^i \cdot w(e_0)$ for some $i \in \mathbb{N}$. Note that we can assume that the weights are greater than 0. Then the number of sets is $O(\log \frac{w_{max}}{w_{min}})$. For the sake of simplicity, we assume that the edge weights are in $[1, W]$. Further details can be found in the full version of the paper.

We now introduce a bit of notation we will use in the algorithm and throughout the proof. We partition the edge set $E = \bigcup_{i=0}^t E_i$ by $t + 1 = O(\log W)$ ranks where the set E_i contains all edges e with weight $w(e) \in [2^i, 2^{i+1})$. Wlog we assume $E_t \neq \emptyset$ (otherwise let t be the largest rank with $E_t \neq \emptyset$). Let $G' = (V, E, w')$ be equal to G but each edge $e \in E_i$ has weight $w'(e) := r_i := 2^i$ for all $i = 0, \ldots, t$. Let $M = \bigcup_{i=0}^t M_i$ be the matching computed by the partitioning algorithm and S be a $(t + 1)$-dimensional vector with $S_i = \sum_{j=i}^t |M_j|$.

Algorithm 1 now proceeds as follows: For every $i \in \{0, \ldots t\}$ the size of a maximum matching in $(V, \bigcup_{j=i}^t E_j)$ and S_i differ by only a constant factor. Conceptually, we set our estimator $\widehat{S_i}$ of S_i to be the approximation of the size of the maximum matching of $(V, \bigcup_{j=i}^t E_i)$ and the estimator of the contribution of the edges in E_i to the weight of an optimal weighted matching is $\widehat{R_i} = \widehat{S_i} - \widehat{S_{i+1}}$. The estimator $\widehat{R_i}$ is crude and generally not a good approximation to $|M_i|$. What helps us is that if the edges M_i have a significant contribution to $w(M)$, then $|M_i| \gg \sum_{j=i+1}^t |M_j| = S_{i+1}$. In order to detect whether the matching M_i has a significant contribution to the objective value, we introduce two parameters T and c. The first matching M_t is always significant (and the simplest to approximate by setting $\widehat{R_t} = \widehat{S_t}$). For all subsequent matchings $i < t$, let j be the most

---

Algorithm 1. Weighted Matching Approximation

Require: Graph $G = (V, \bigcup_{i=0}^{t} E_i)$ with weights r_i for edges in E_i
Ensure: Estimator of the weighted matching

> **for** $i = t$ **to** 0 **do**
>> $\widehat{S}_i = \widehat{R}_i = 0$
>
> $weight = 0,\ last = t$
> $\widehat{R}_t = \widehat{S}_t = $ **Unweighted Matching Estimation**(V, E_t)
> **for** $i = t - 1$ **to** 0 **do**
>> $\widehat{S}_i = $ **Unweighted Matching Estimation**$(V, \bigcup_{j=i}^{t} E_j)$
>> **if** $\widehat{S}_i > \widehat{S_{last}} \cdot T$ **then**   ▷ Add current index i to I_{good}
>>> **if** $\widehat{S}_i - \widehat{S_{last}} \geq c \cdot \widehat{R}_{last}$ **then**   ▷ Add current index i to I_{sign}
>>>> $\widehat{R}_i = \widehat{S}_i - \widehat{S_{last}}$
>>>> $last = i$
>>
>> **else**
>>> $\widehat{S}_i = 0$
>
> **return** $\frac{2}{5} \sum_{i=0}^{t} r_i \cdot \widehat{R}_i$

---

recent matching which we deemed to be significant. We require $\widehat{S}_i \geq T \cdot \widehat{S}_j$ and $\widehat{R}_i \geq c \cdot \widehat{R}_j$. If both criteria are satisfied, we use the estimator $\widehat{R}_i = \widehat{S}_i - \widehat{S}_j$ and set i to be the now most recent, significant matching, otherwise we set $\widehat{R}_i = 0$. The final estimator of the weight is $\sum_{i=0}^{t} r_i \cdot \widehat{R}_i$. The next definition gives a more detailed description of the two sets of ranks which are important for the analysis.

Definition 1 (Good and Significant Ranks). *Let $\widehat{S}$ and $\widehat{R}$ be the vectors at the end of Algorithm 1. An index i is called to be a* good rank *if $\widehat{S}_i \neq 0$ and i is a* significant rank *if $\widehat{R}_i \neq 0$. We denote the set of good ranks by I_{good} and the set of significant ranks by I_{sign}, i.e., $I_{good} := \left\{ i \subseteq \{0, \ldots t\}\ |\widehat{S}_i \neq 0 \right\}$ and $I_{sign} := \left\{ i \subseteq \{0, \ldots t\}\ |\widehat{R}_i \neq 0 \right\}$. We define I_{good} and I_{sign} to be in descending order and we will refer to the ℓ-th element of I_{good} and I_{sign} by $I_{good}(\ell)$ and $I_{sign}(\ell)$, respectively. That means $I_{good}(1) > I_{good}(2) > \ldots > I_{good}(|I_{good}|)$ and $I_{sign}(1) > I_{sign}(2) > \ldots > I_{sign}(|I_{sign}|)$. We slightly abuse the notation and set $I_{sign}(|I_{sign}| + 1) = 0$. Let $D_1 := |M_t|$ and for $\ell \in \{2, \ldots, |I_{sign}|\}$ we define the sum of the matching sizes between two significant ranks $I_{sign}(\ell)$ and $I_{sign}(\ell-1)$ where the smaller significant rank is included by $D_\ell := \sum_{i=I_{sign}(\ell)}^{I_{sign}(\ell-1)-1} |M_i|$.*

In the following, we subscript indices of significant ranks by s and of good ranks by g. We state a simple property of I_{good} and I_{sign}.

Fact 1. $I_{good}(1) = I_{sign}(1) = t$ *and* $I_{sign} \subseteq I_{good}$.

Now, we have the necessary notations and properties of good and significant ranks to proof our main theorem.

Theorem 1. *Let $G = (V, E, w)$ be a weighted graph where the weights are from $[1, W]$. Let A be an algorithm that returns an λ-estimator $\widehat{M}$ for the size of a maximum matching M of a graph with $1/\lambda \cdot |M| \leq \widehat{M} \leq |M|$ with failure probability at most δ and needs space S. If we partition the edge set into sets $E_0, \ldots, E_t$ with $t = \lfloor \log W \rfloor$ where E_i consists of all edges with weight in $[2^i, 2^{i+1})$, set $r_i = 2^i$, and use A as the unweighted matching estimator in Algorithm 1, then there are parameters T and c depending on λ such that the algorithm returns an $O(\lambda^4)$-estimator $\widehat{W}$ for the weight of the maximum weighted matching with failure probability at most $\delta \cdot (t + 1)$ using $O(S \cdot t)$ space, i.e. there is a constant c such that $\frac{1}{c\lambda^4} \cdot w(M^*) \leq \widehat{W} \leq w(M^*)$ where M^* is an optimal weighted matching.*

Proof (sketch). In the following we condition on the event that all calls to the unweighted estimation routine succeed, which happens with probability at least $1 - \delta \cdot (t + 1)$. The estimator returned by Algorithm 1 can be written as $\sum_{\ell=1}^{|I_{sign}|} r_{I_{sign}(\ell)} \cdot \widehat{R_{I_{sign}(\ell)}}$. Using similar arguments as found in Lemma 4 of [30], we have $\frac{1}{8} \cdot w(M^*) \leq \sum_{i=0}^{t} r_i |M_i| \leq w(M^*)$. Thus, it is sufficient to show that $\sum_{\ell=1}^{|I_{sign}|} r_{I_{sgin}(\ell)} \cdot \widehat{R_{I_{sign}(\ell)}}$ is a good estimator for $\sum_{i=0}^{t} r_i |M_i|$. We first consider the problem of estimating D_ℓ, and then how to charge the matching sizes.

(1) Estimation of D_ℓ. Since $\bigcup_{j=i}^{t} M_j$ is a maximal matching in $\bigcup_{j=i}^{t} E_j$, $\widehat{S_i}$ is a good estimator for S_i:

Lemma 1. *For all $i \in \{0, \ldots, t\}$ we have $\frac{1}{\lambda} \cdot S_i \leq \widehat{S_i} \leq 2 \cdot S_i$.*

Next, we show that for an index $i_g \in I_{good}$ the difference $\widehat{S_{i_g}} - \widehat{S_{I_{sign}(\ell)}}$ to the last significant rank is a good estimator for $\sum_{i=i_g}^{I_{sign}(\ell)-1} |M_i|$.

Lemma 2. *For all $i_g \in I_{good}$ with $I_{sign}(\ell + 1) \leq i_g < I_{sign}(\ell)$ for some $\ell \in \{1, \ldots, |I_{sign}|\}$ and $T = 8\lambda^2 - 2\lambda$,*

$$\frac{1}{2\lambda} \cdot \sum_{i=i_g}^{I_{sign}(\ell)-1} |M_i| < \widehat{S_{i_g}} - \widehat{S_{I_{sign}(\ell)}} < \frac{5}{2} \cdot \sum_{i=i_g}^{I_{sign}(\ell)-1} |M_i|$$

and $\frac{1}{\lambda}|M_t| \leq \widehat{S_t} \leq 2|M_t|$.

From Fact 1 we know that $I_{sign} \subseteq I_{good}$ which together with the last Lemma 2 implies that $\widehat{R_{I_{sign}(\ell)}}$ is a good estimator for D_ℓ.

Corollary 1. *For $\ell \in \{1, \ldots, |I_{sign}|\}$, $\frac{1}{2\lambda} \cdot D_\ell \leq \widehat{R_{J(\ell)}} \leq \frac{5}{2} \cdot D_\ell$. Furthermore, if $c > 5\lambda$ then the values of the D_ℓ are exponentially increasing:*

$$D_1 \leq \frac{5\lambda}{c} D_2 \leq \ldots \leq \left(\frac{5\lambda}{c}\right)^{|I_{sign}|-1} D_{|I_{sign}|-1}.$$

(2) The Charging Argument. We show that the sum of the matching sizes between two significant ranks $I_{sign}(\ell+1)$ and $I_{sign}(\ell)$ is bounded by $O(\lambda \cdot T \cdot D_\ell) = O\left(\lambda \cdot T \cdot \sum_{i=I_{sign}(\ell)}^{I_{sign}(\ell-1)+1} |M_i|\right)$.

Lemma 3. *Setting* $c = \frac{2}{5} \cdot T + 5\lambda$ *in Algorithm 1. Then for* $\ell \in \{1, \ldots, |I_{sign}| - 1\}$,

$$\sum_{i=I_{sign}(\ell+1)+1}^{I_{sign}(\ell)-1} |M_i| \leq (2\lambda \cdot T + 25\lambda^2) \cdot D_\ell \ and \quad \sum_{i=0}^{I_{sign}(|I_{sign}|)-1} |M_i| \leq (2\lambda \cdot T + 25\lambda^2) \cdot$$

$D_{|I_{sign}|}$ *if* $0 \notin I_{sign}$.

Proof. For the proof of the first inequality, let $i_g \in I_{good}$ be minimal such that $I_{sign}(\ell+1) < i_g < I_{sign}(\ell)$ for $\ell \in \{1, \ldots, |I_{sign}| - 1\}$. If such a good rank does not exist, set $i_g = -1$. We distinguish between two cases. Note that $c = \frac{5}{2} \cdot T + 5\lambda > 5\lambda$.

Case 1: $i_g = I_{sign}(\ell+1) + 1$. For the sake of simplicity, we abuse the notation and set $\widehat{S_{I_{sign}(0)}} = 0$ such that $\widehat{R_{I_{sign}(\ell)}} = \widehat{S_{I_{sign}(\ell)}} - \widehat{S_{I_{sign}(\ell-1)}}$ also holds for $\ell = 1$. Using Lemma 2 we have

$$\sum_{i=I_{sign}(\ell+1)+1}^{I_{sign}(\ell)-1} |M_i| \quad = \quad \sum_{i=i_g}^{I_{sign}(\ell)-1} |M_i| \underset{\text{Lem. 2}}{\leq} 2\lambda \cdot \left(\widehat{S_{i_g}} - \widehat{S_{I_{sign}(\ell)}}\right)$$

$$\underset{i_g \notin I_{sign}}{\leq} 2\lambda c \cdot \widehat{R_{I_{sign}(\ell)}} = 2\lambda \cdot c \cdot \left(\widehat{S_{I_{sign}(\ell)}} - \widehat{S_{I_{sign}(\ell-1)}}\right)$$

$$\underset{\text{Lem. 2}}{\leq} 5\lambda \cdot c \cdot \sum_{i=I_{sign}(\ell)}^{I_{sign}(\ell-1)-1} |M_i| = 5 \cdot c \cdot D_\ell \qquad (1)$$

Case 2: $i_g \neq I_{sign}(\ell+1) + 1$. In this case $\widehat{S_{I_{sign}(\ell+1)+1}} \leq T \cdot \widehat{S_{I_{sign}(\ell)}}$. Thus

$$\sum_{i=I_{sign}(\ell+1)+1}^{I_{sign}(\ell)-1} |M_i| \quad \leq \quad S_{I_{sign}(\ell+1)+1} \underset{\text{Lem. 1}}{\leq} \lambda \cdot \widehat{S_{I_{sign}(\ell+1)+1}}$$

$$\leq \quad \lambda \cdot T \cdot \widehat{S_{I_{sign}(\ell)}} \underset{\text{Lem. 1}}{\leq} 2\lambda \cdot T \cdot S_{I_{sign}(\ell)} = 2\lambda \cdot T \cdot \sum_{i=1}^{\ell} D_i$$

$$\underset{\text{Cor. 1}}{\leq} 2\lambda \cdot T \cdot D_\ell \cdot \sum_{i=1}^{\ell} \left(\frac{5\lambda}{c}\right)^i \leq 2\lambda \cdot T \cdot D_\ell \cdot \frac{1}{1 - \frac{5\lambda}{c}} \qquad (2)$$

Combining the inequalities 1 and 2, we have $\sum_{i=I_{sign}(\ell+1)+1}^{I_{sign}(\ell)-1} |M_i| \leq \max\left\{5\lambda \cdot c, \frac{2\lambda \cdot T}{1 - \frac{5\lambda}{c}}\right\} \cdot D_\ell$ which simplifies to

$$\sum_{i=I_{sign}(\ell+1)+1}^{I_{sign}(\ell)-1} |M_i| \leq (2\lambda \cdot T + 25\lambda^2) \cdot D_\ell \qquad \text{for } \ell \in \{1, \ldots, |I_{sign}| - 1\}.$$

If $0 \notin I_{sign}$ we can do the same arguments to bound $\sum_{i=0}^{I_{sign}(|I_{sign}|)-1} |M_i|$ by $(2\lambda \cdot T + 25\lambda^2) \cdot D_{|I_{sign}|}$.   □

We use Lemma 3 to show that $w(M)$ is bounded in terms of $\sum_{\ell=1}^{|I_{sign}|} r_{I_{sign}(\ell)} \cdot D_\ell$:

$$\sum_{i=0}^{t} r_i \cdot |M_i| \geq \sum_{\ell=1}^{|I_{sign}|} r_{I_{sign}(\ell)} \cdot D_\ell \tag{3}$$

$$\sum_{i=0}^{t} r_i \cdot |M_i| \leq (1 + 2\lambda \cdot T + 25\lambda^2) \cdot \sum_{\ell=1}^{|I_{sign}|} r_{I_{sign}(\ell)} \cdot D_\ell. \tag{4}$$

Putting Everything Together. Using Corollary 1 we have $\frac{1}{2\lambda} \cdot D_\ell \leq \widehat{R_{I_{sign}(\ell)}} \leq \frac{5}{2} \cdot D_\ell$ for all $\ell \in \{1, \ldots, |I_{sign}|\}$ which with (3) and (4) gives $\frac{1}{2\lambda \cdot (1+2\lambda \cdot T + 25\lambda^2)} \cdot w(M) \leq \sum_{\ell=1}^{|I_{sign}|} r_{I_{sign}(\ell)} \cdot \widehat{R_{I_{sign}(\ell)}} \leq \frac{5}{2} \cdot w(M)$. Recall that we set $T = 8\lambda^2 - 2\lambda$. Now, folding in the factor of $\frac{1}{8}$ from the partitioning and rescaling the estimator gives an $O(\lambda^4)$-estimation on the weight of an optimal weighted matching.

2.1   Applications

Since every edge insertion and deletion supplies the edge weight, it is straightforward to determine the rank for each edge upon every update. Using the following results for unweighted matching, we can obtain estimates with similar approximation guarantee and space bounds for weighted matching.

Random Order Streams. For an arbitrary graph whose edges are streamed in random order, Kapralov, Khanna and Sudan [20] gave an algorithm with polylog n approximation guarantee using polylog n space with failure probability $\delta = 1/\text{polylog } n$. Since this probability takes the randomness of the input permutation into account, we cannot easily amplify it, though for $\log W \leq \delta$, the extension to weighted matching still succeeds with at least constant probability.

Adversarial Streams. The arboricity of a graph G is defined as $\max_{U \subseteq V} \left\lceil \frac{|E(U)|}{|U|-1} \right\rceil$. Examples of graphs with constant arboricity include planar graphs and graphs with constant degree. For graphs of bounded arboricity ν, Esfandiari et al. [13] gave an algorithm with an $O(\nu)$ approximation guarantee using $\tilde{O}(\nu \cdot n^{2/3})$ space.

Dynamic Streams. Matching in trees can be easily sketched by counting the number of distinct elements of the degree vector initialized to $-\mathbf{1}^n$. We briefly sketch how to extend the algorithm by Esfandiari et al. [13] to dynamic streams: The only part that is not straightforwardly adapted is the small matching up to some threshold k maintained by the greedy algorithm in insertion-only streams. For this, we summarize entries of the adjacency matrix as well as entries of the Tutte-matrix [29], where each non-zero entry (i, j) of the adjacency matrix is replaced by the variable x_{ij} if $i < j$ and $-x_{ij}$ if $i > j$. Lovász [25] showed that the maximum rank of T over all choices of the indeterminates is twice the size of the maximum matching. Furthermore, Lovász also noted that T has maximum rank with high probability if the indeterminates are chosen independently and uniformly at random from $\{1, \ldots, \text{poly}(n)\}$. This shows that computing the size of the maximum matching is a special case of computing the rank of a matrix; we simply maintain the Tutte-matrix with randomly chosen values for each entry. Uniformly random bits would require $O(n^2)$ space, which can be averted by using Nisan's pseudorandom generator for bounded space computation [28,18].

Given a positive integer k and a stream over updates to a matrix A, an algorithm for the *rank decision problem* outputs 1 if $\text{rank}(A) \geq k$ and 0 otherwise. Clarkson and Woodruff [8] proposed an algorithm operating in fully dynamic streams using $O(k^2 \log n)$ space. Setting $k = n^{2/5}$ then gives us the following.

Theorem 3. *Let G be a graph with bounded arboricity ν. Then there exists an algorithm estimating the size of the maximum matching in G within an $O(\nu)$-factor in the dynamic streaming model using a single pass over the data and $\tilde{O}(\nu \cdot n^{4/5})$ space or two passes over the data and $\tilde{O}(\nu \cdot n^{2/3})$ space.*

3   Lower Bound

Esfandiari et al. [13] showed a space lower bound of $\Omega(\sqrt{n})$ for any estimation better than $3/2$. Their reduction (see below) uses the Boolean Hidden Matching Problem introduced by Bar-Yossef et al. [6], and further studied by Gavinsky et al. [15]. We will use the following generalization due to Verbin and Yu [31].

Definition 2 (Boolean Hidden Hypermatching Problem [31]). *In the Boolean Hidden Hypermatching Problem $BHH_{t,n}$ Alice gets a vector $x \in \{0,1\}^n$ with $n = 2kt$ and $k \in \mathbb{N}$ and Bob gets a perfect t-hypermatching M on the n coordinates of x, i. e., each edge has exactly t coordinates, and a string $w \in \{0,1\}^{n/t}$. We denote the vector of length n/t given by $(\bigoplus_{1 \leq i \leq t} x_{M_{1,i}}, \ldots, \bigoplus_{1 \leq i \leq t} x_{M_{n/t,i}})$ by Mx where $(M_{1,1}, \ldots, M_{1,t}), \ldots, (M_{n/t,1}, \ldots, M_{n/t,t})$ are the edges of M. The problem is to return 1 if $Mx \oplus w = 1^{n/t}$ and 0 if $Mx \oplus w = 0^{n/t}$, otherwise the algorithm may answer arbitrarily.*

Verbin and Yu [31] showed a lower bound of $\Omega(n^{1-1/t})$ for the randomized one-way communication complexity for $BHH_{t,n}$. For our reduction we require $w = 0^{n/t}$ and thus $Mx = 1^{n/t}$ or $Mx = 0^{n/t}$. We denote this problem by $BHH_{t,n}^0$. We can show that this does not reduce the communication complexity.

Lemma 4. *The communication complexity of $BHH_{t,4n}^0$ is lower bounded by the communication complexity of $BHH_{t,n}$.*

Let us now sketch the reduction from $BHH_{2,n}^0$ to approximate maximum matching to get the idea how to extend it to the general bound. Let x, M be the input for Alice and Bob. They construct a graph consisting of $2n$ nodes denoted by

$v_{1,i}$ and $v_{2,i}$, for $i \in \{1, \ldots, n\}$. For each bit x_i of $x \in \{0,1\}^n$, Alice adds an edge $\{v_{1,i}, v_{2,i}\}$ iff $x_i = 1$ and sends a message to Bob. Bob adds an edge between $v_{2,i}$ and $v_{2,j}$ for each edge $\{x_i, x_j\} \in M$ and approximates the size of the matching. If all parities are 1 then the size of the maximum matching is $n/2$. If the parities are all 0 then the size is $3n/4$. Every streaming algorithm that approximates better than $3/2$ can distinguish between these two cases. The first observation is that the size of the matching is lower bounded by the number of ones in x. The second observation is that the added edges by Bob increase the matching iff the parities of all pairs are 0 and only the edges between the two 0 input bits of Alice increase the matching. Since it is promised that all parities are equal and the number of ones is exactly $n/2$ we can calculate the number of $(0,0)$ pairs. For our lower bound we show that this calculation is still possible if Bob adds a t-clique between the corresponding nodes of the hyperedge.

Theorem 2. *Any randomized streaming algorithm that approximates the maximum matching size within a $1 + \frac{1}{3t/2-1}$ factor for $t \geq 2$ needs $\Omega(n^{1-1/t})$ space.*

Finally, constructing the Tutte-matrix with randomly chosen entries gives us

Corollary 2. *Any randomized streaming algorithm that approximates $\mathrm{rank}(A)$ of $A \in \mathbb{R}^{n \times n}$ within a $1 + \frac{1}{3t/2-1}$ factor for $t \geq 2$ requires $\Omega(n^{1-1/t})$ space.*

References

1. Ahn, K., Guha, S.: Graph sparsification in the semi-streaming model. In: Albers, S., Marchetti-Spaccamela, A., Matias, Y., Nikoletseas, S., Thomas, W. (eds.) ICALP 2009, Part II. LNCS, vol. 5556, pp. 328–338. Springer, Heidelberg (2009)
2. Ahn, K., Guha, S., McGregor, A.: Analyzing graph structure via linear measurements. In: SODA, pp. 459–467 (2012)
3. Ahn, K., Guha, S., McGregor, A.: Graph sketches: sparsification, spanners, and subgraphs. In: PODS, pp. 5–14 (2012)
4. Ahn, K., Guha, S., McGregor, A.: Spectral sparsification in dynamic graph streams. In: Raghavendra, P., Raskhodnikova, S., Jansen, K., Rolim, J.D.P. (eds.) RANDOM 2013 and APPROX 2013. LNCS, vol. 8096, pp. 1–10. Springer, Heidelberg (2013)
5. Assadi, S., Khanna, S., Li, Y., Yaroslavtsev, G.: Tight bounds for linear sketches of approximate matchings. CoRR, abs/1505.01467 (2015)
6. Bar-Yossef, Z., Jayram, T.S., Kerenidis, I.: Exponential separation of quantum and classical one-way communication complexity. SIAM J. Comput. 38(1), 366–384 (2008)
7. Chitnis, R., Cormode, G., Esfandiari, H., Hajiaghayi, M., McGregor, A., Monemizadeh, M., Vorotnikova, S.: Kernelization via sampling with applications to dynamic graph streams. CoRR, abs/1505.01731 (2015)
8. Clarkson, K., Woodruff, D.: Numerical linear algebra in the streaming model. In: STOC, pp. 205–214 (2009)
9. Crouch, M., McGregor, A., Stubbs, D.: Dynamic graphs in the sliding-window model. In: Bodlaender, H.L., Italiano, G.F. (eds.) ESA 2013. LNCS, vol. 8125, pp. 337–348. Springer, Heidelberg (2013)
10. Crouch, M., Stubbs, D.: Improved streaming algorithms for weighted matching, via unweighted matching. In: APPROX/RANDOM 2014, pp. 96–104 (2014)

11. Epstein, L., Levin, A., Mestre, J., Segev, D.: Improved approximation guarantees for weighted matching in the semi-streaming model. SIAM J. Discrete Math. 25(3), 1251–1265 (2011)
12. Epstein, L., Levin, A., Segev, D., Weimann, O.: Improved bounds for online preemptive matching. In: STACS, pp. 389–399 (2013)
13. Esfandiari, H., Hajiaghayi, M., Liaghat, V., Monemizadeh, M., Onak, K.: Streaming algorithms for estimating the matching size in planar graphs and beyond. In: SODA, pp. 1217–1233 (2015)
14. Feigenbaum, J., Kannan, S., McGregor, A., Suri, S., Zhang, J.: On graph problems in a semi-streaming model. Theor. Comput. Sci. 348(2-3), 207–216 (2005)
15. Gavinsky, D., Kempe, J., Kerenidis, I., Raz, R., de Wolf, R.: Exponential separation for one-way quantum communication complexity, with applications to cryptography. SIAM J. Comput. 38(5), 1695–1708 (2008)
16. Goel, A., Kapralov, M., Khanna, S.: On the communication and streaming complexity of maximum bipartite matching. In: SODA, pp. 468–485 (2012)
17. Henzinger, M., Raghavan, P., Rajagopalan, S.: Computing on data streams (1998)
18. Indyk, P.: Stable distributions, pseudorandom generators, embeddings and data stream computation. In: FOCS, pp. 189–197 (2000)
19. Kapralov, M.: Better bounds for matchings in the streaming model. In: SODA, pp. 1679–1697 (2013)
20. Kapralov, M., Khanna, S., Sudan, M.: Approximating matching size from random streams. In: SODA, pp. 734–751 (2014)
21. Konrad, C.: Maximum matching in turnstile streams. CoRR, abs/1505.01460 (2015)
22. Konrad, C., Magniez, F., Mathieu, C.: Maximum matching in semi-streaming with few passes. In: Gupta, A., Jansen, K., Rolim, J., Servedio, R. (eds.) APPROX 2012 and RANDOM 2012. LNCS, vol. 7408, pp. 231–242. Springer, Heidelberg (2012)
23. Li, Y., Nguyen, H., Woodruff, D.: On sketching matrix norms and the top singular vector. In: SODA, pp. 1562–1581 (2014)
24. Li, Y., Nguyen, H., Woodruff, D.: Turnstile streaming algorithms might as well be linear sketches. In: STOC, pp. 174–183 (2014)
25. Lovász, L.: On determinants, matchings, and random algorithms. In: FCT, pp. 565–574 (1979)
26. McGregor, A.: Finding graph matchings in data streams. In: Chekuri, C., Jansen, K., Rolim, J.D.P., Trevisan, L. (eds.) APPROX 2005 and RANDOM 2005. LNCS, vol. 3624, pp. 170–181. Springer, Heidelberg (2005)
27. McGregor, A.: Graph stream algorithms: a survey. SIGMOD Record 43(1), 9–20 (2014)
28. Nisan, N.: Pseudorandom generators for space-bounded computation. Combinatorica 12(4), 449–461 (1992)
29. Tutte, W.: The factorization of linear graphs. J. London Math. Soc. 22, 107–111 (1947)
30. Uehara, R., Chen, Z.: Parallel approximation algorithms for maximum weighted matching in general graphs. Inf. Process. Lett. 76(1-2), 13–17 (2000)
31. Verbin, E., Yu, W.: The streaming complexity of cycle counting, sorting by reversals, and other problems. In: SODA, pp. 11–25. SIAM (2011)
32. Zelke, M.: Weighted matching in the semi-streaming model. Algorithmica 62(1-2), 1–20 (2012)

An Improved Approximation Algorithm
for Knapsack Median Using Sparsification

Jaroslaw Byrka[1,*], Thomas Pensyl[2,**], Bartosz Rybicki[1,***], Joachim Spoerhase[1],
Aravind Srinivasan[3,†], and Khoa Trinh[2,**]

[1] Institute of Computer Science, University of Wroclaw, Poland
{jby,bry}@cs.uni.wroc.pl, joachim.spoerhase@uni-wuerzburg.de
[2] Department of Computer Science, University of Maryland, College Park, MD 20742
{tpensyl,khoa}@cs.umd.edu
[3] Department of Computer Science and Instute for Advanced Computer Studies,
University of Maryland, College Park, MD 20742
srin@cs.umd.edu

Abstract. Knapsack median is a generalization of the classic k-median problem in which we replace the cardinality constraint with a knapsack constraint. It is currently known to be 32-approximable. We improve on the best known algorithms in several ways, including adding randomization and applying sparsification as a preprocessing step. The latter improvement produces the first LP for this problem with bounded integrality gap. The new algorithm obtains an approximation factor of 17.46. We also give a 3.05 approximation with small budget violation.

Keywords: approximation algorithm, combinatorial optimization, randomized algorithm, facility-location problems.

1 Introduction

k-MEDIAN is a classic problem in combinatorial optimization. Herein, we are given a set of clients $\mathcal{C}$, facilities $\mathcal{F}$, and a symmetric distance metric c on $\mathcal{C} \cup \mathcal{F}$. The goal is to open k facilities such that we minimize the total connection cost (distance to nearest open facility) of all clients. A natural generalization of k-MEDIAN is KNAPSACK MEDIAN (KM), in which we assign nonnegative weight w_i to each facility $i \in F$, and instead of opening k facilities, we require that the sum of the open facility weights be within some budget B.

While KM is not known to be harder than k-MEDIAN, it has thus far proved more difficult to approximate. k-MEDIAN was first approximated within constant factor $6\frac{2}{3}$ in 1999 [2], with a series of improvements leading to the current best-known factor of 2.674 [1][1]. KM was first studied in 2011 by Krishnaswamy et. al. [4], who gave a

---

* Supported by the Polish National Science Centre grant DEC-2012/07/B/ST6/01534.
** Supported in part by NSF Awards CNS 1010789 and CCF 1422569.
*** Research supported by NCN 2012/07/N/ST6/03068 grant.
† Supported in part by NSF Awards CNS 1010789 and CCF 1422569, and by a research award from Adobe, Inc.
[1] The paper claims 2.611, but a very recent correction changes this to 2.674.

© Springer-Verlag Berlin Heidelberg 2015
N. Bansal and I. Finocchi (Eds.): ESA 2015, LNCS 9294, pp. 275–287, 2015.
DOI: 10.1007/978-3-662-48350-3_24

bicriteria $16 + \epsilon$ approximation which slightly violated the budget. Then Kumar gave the first true constant factor approximation for KM with factor 2700 [5], subsequently reduced to 34 by Charikar&Li [3] and then to 32 by Swamy [8].

This paper's algorithm has a flow similar to Swamy's: we first get a half-integral solution (except for a few 'bad' facilities), and then create pairs of half-facilities, opening one facility in each pair. By making several improvements, we reduce the approximation ratio to 17.46. The first improvement is a simple modification to the pairing process so that every half-facility is guaranteed either itself or its closest neighbor to be open (versus having to go through two 'jumps' to get to an open facility). The second improvement is to randomly sample the half-integral solution, and condition on the probability that any given facility is 'bad'. The algorithm can be derandomized with linear loss in the runtime.

The third improvement deals with the bad facilities which inevitabley arise due to the knapsack constraint. All previous algorithms used Kumar's bound from [5] to bound the cost of nearby clients when bad facilities must be closed. However, we show that by using a sparsification technique similar in spirit to - but distinct from - that used in [6], we can focus on a subinstance in which the connection costs of clients are guaranteed to be evenly distributed throughout the instance. This allows for a much stronger bound than Kumar's, and also results in an LP with bounded integrality gap, unlike previous algorithms.

Another alternative is to just open the few bad facilities and violate the budget by some small amount, as Krishnaswamy et. al. did when first introducing KM. By preprocessing, we can ensure this violates the budget by at most ϵB. We show that the bipoint solution based method from [6] can be adapted for KM using this budget-violating technique to get a 3.05 approximation.

1.1 Preliminaries

Let $n = |\mathcal{F}| + |\mathcal{C}|$ be the size of the instance. For the ease of analysis, we assume that each client has unit demand. (Indeed, our algorithm easily extends to the general case.) For a client j, the connection cost of j, denoted as $\text{cost}(j)$, is the distance from j to the nearest *open* facility in our solution. The goal is to *open* a subset $\mathcal{S} \subseteq \mathcal{F}$ of facilities such that the total connection cost is minimized, subject to the *knapsack constraint* $\sum_{i \in \mathcal{S}} w_i \leq B$.

The natural LP relaxation of this problem is as follows.

$$
\begin{aligned}
\text{minimize} \quad & \sum_{i \in \mathcal{F}, j \in \mathcal{C}} c_{ij} x_{ij} \\
\text{subject to} \quad & \sum_{i \in \mathcal{F}} x_{ij} = 1 \qquad \forall j \in \mathcal{C} \\
& x_{ij} \leq y_i \qquad \qquad \forall i \in \mathcal{F}, j \in \mathcal{C} \\
& \sum_{i \in \mathcal{F}} w_i y_i \leq B \\
& 0 \leq x_{ij}, y_i \leq 1 \ \forall i \in \mathcal{F}, j \in \mathcal{C}
\end{aligned}
$$

In this LP, x_{ij} and y_i are indicator variables for the event client j is connected to facility i and facility i is open, respectively. The first constraint guarantees that each client is connected to some facility. The second constraint says that client j can only connect to facility i if it is open. The third one is the knapsack constraint.

In this paper, given a KM instance $\mathcal{I} = (B, \mathcal{F}, \mathcal{C}, c, w)$, let $\text{OPT}_{\mathcal{I}}$ and OPT_f be the cost of an optimal integral solution and the optimal value of the LP relaxation, respectively. Suppose $\mathcal{S} \subseteq \mathcal{F}$ is a solution to $\mathcal{I}$, let $\text{cost}_{\mathcal{I}}(\mathcal{S})$ denote cost of $\mathcal{S}$. Let (x, y) denote the optimal (fractional) solution of the LP relaxation. Let $C_j := \sum_{i \in \mathcal{F}} c_{ij} x_{ij}$ be the fractional connection cost of j. Given $\mathcal{S} \subseteq \mathcal{F}$ and a vector $v \in \mathbb{R}^{|\mathcal{F}|}$, let $v(\mathcal{S}) := \sum_{i \in \mathcal{S}} v_i$. From now on, let us fix any optimal integral solution of the instance for the analysis.

2 An Improved Approximation Algorithm for Knapsack Median

2.1 Kumar's Bound

The main technical difficulty of KM is related to the unbounded integrality gap of the LP relaxation. It is known that this gap remains unbounded even when we strengthen the LP with knapsack cover inequalities [4]. All previous constant-factor approximation algorithms for KM rely on Kumar's bound from [5] to get around the gap. Specifically, Kumar's bound is useful to bound the connection cost of a group of clients via some *cluster center* in terms of $\text{OPT}_{\mathcal{I}}$ instead of OPT_f. We now review this bound, and will improve it later.

Lemma 1. *For each client j, we can compute (in polynomial time) an upper-bound U_j on the connection cost of j in the optimal integral solution (i.e. $\text{cost}(j) \leq U_j$) such that*

$$\sum_{j' \in \mathcal{C}} \max\{0, U_j - c_{jj'}\} \leq \text{OPT}_{\mathcal{I}}.$$

We can slightly strengthen the LP relaxation by adding the constraints: $x_{ij} = 0$ for all $c_{ij} > U_j$. (Unfortunately, the integrality gap is still unbounded after this step.) Thus we may assume that (x, y) satisfies all these constraints.

Lemma 2 (Kumar's bound). *Let S be a set of clients and $s \in S$, where $c_{js} \leq \beta C_j$ for all $j \in S$ and some constant $\beta \geq 1$, then*

$$|\mathcal{S}| U_s \leq \text{OPT}_{\mathcal{I}} + \beta \sum_{j \in \mathcal{S}} C_j.$$

Proof.

$$|\mathcal{S}| U_s = \sum_{j \in \mathcal{S}} U_s = \sum_{j \in \mathcal{S}} (U_s - c_{js}) + \sum_{j \in \mathcal{S}} c_{js} \leq \text{OPT}_{\mathcal{I}} + \beta \sum_{j \in \mathcal{S}} C_j,$$

where we use the property of U_s from Lemma 1 for the last inequality.   □

This bound allows one to bound the cost of clients which rely on the bad facility.

2.2 Sparse Instances

Kumar's bound can only be tight when the connection cost in the optimal solution is highly concentrated around a single client. However, if this were the case, we could guess the client for which this occurs, along with its optimal facility, which would give us a large advantage. On the other hand, if the connection cost is evenly distributed, we can greatly strengthen Kumar's bound. This is the idea behind our definition of sparse instances below.

Let $\mathrm{CBall}(j, r) := \{k \in \mathcal{C} : c_{jk} \leq r\}$ denote the set of clients within radius of r from client j. Let λ_j be the connection cost of j in the optimal integral solution. Also, let $i(j)$ denote the facility serving j in the optimal solution.

Definition 1. *Given some constants $0 < \delta, \epsilon < 1$, we say that a knapsack median instance $\mathcal{I} = \{B, \mathcal{F}, \mathcal{C}, c, w\}$ is (δ, ϵ)-sparse if, for all $j \in \mathcal{C}$,*

$$\sum_{k \in \mathrm{CBall}(j, \delta\lambda_j)} (\lambda_j - c_{jk}) \leq \epsilon \mathrm{OPT}_{\mathcal{I}}.$$

We will show that the integrality gap is bounded on these sparse instances. We also give a polynomial-time algorithm to *sparsify* any knapsack median instance. Moreover, the solution of a sparse instance can be used as a solution of the original instance with only a small loss in the total cost.

Lemma 3. *Given some knapsack median instance $\mathcal{I}_0 = (B, \mathcal{F}, \mathcal{C}_0, c, w)$ and $0 < \delta, \epsilon < 1$, there is an efficient algorithm that outputs $O(n^{2/\epsilon})$ pairs of $(\mathcal{I}, \mathcal{F}')$, where $\mathcal{I} = (B, \mathcal{F}, \mathcal{C}, c, w)$ is a new instance with $\mathcal{C} \subseteq \mathcal{C}_0$, and $\mathcal{F}' \subseteq \mathcal{F}$ is a partial solution, such that at least one of these instances is (δ, ϵ)-sparse.*

The following theorem says that if we have an approximate solution to a sparse instance, then its cost on the original instance can be blown up by a small constant factor.

Theorem 1. *Let $\mathcal{I} = (B, \mathcal{F}, \mathcal{C}, c, w)$ be a (δ, ϵ)-sparse instance obtained from $\mathcal{I}_0 = (B, \mathcal{F}, \mathcal{C}_0, c, w)$ and $\mathcal{F}'$ be the corresponding partial solution. If $S \supseteq \mathcal{F}'$ is any approximate solution to $\mathcal{I}$ (including those open facilities in $\mathcal{F}'$) such that*

$$cost_{\mathcal{I}}(S) \leq \alpha \mathrm{OPT}_{\mathcal{I}},$$

then

$$cost_{\mathcal{I}_0}(S) \leq \max\left\{\frac{1+\delta}{1-\delta}, \alpha\right\} \mathrm{OPT}_{\mathcal{I}_0}.$$

Note that our notion of sparsity differs from that of Li and Svensson in several ways. It is client-centric, and removes clients instead of facilities from the instance. On the negative side, removed clients' costs blow up by $\frac{1+\delta}{1-\delta}$, so our final approximation cannot guarantee better. From now on, assume that we are given some arbitrary knapsack median instance $\mathcal{I}_0 = (B, \mathcal{F}, \mathcal{C}_0, c, w)$. We will transform $\mathcal{I}_0$ into a (δ, ϵ)-sparse instance $\mathcal{I}$ and use Theorem 1 to bound the real cost at the end.

2.3 Improving Kumar's Bound and Modifying the LP Relaxation

We will show how to improve Kumar's bound in sparse instances. Recall that, for all $j \in \mathcal{C}$, we have

$$\sum_{k \in \mathrm{CBall}(j, \delta \lambda_j)} (\lambda_j - c_{jk}) \leq \epsilon \mathrm{OPT}_{\mathcal{I}}.$$

Then, as before, we can guess $\mathrm{OPT}_{\mathcal{I}}$ and take the maximum U_j such that

$$\sum_{k \in \mathrm{CBall}(j, \delta U_j)} (U_j - c_{jk}) \leq \epsilon \mathrm{OPT}_{\mathcal{I}}.$$

(Observe that the LHS is an increasing function of U_j.) Now the constraints $x_{ij} = 0$ for all $i \in \mathcal{F}, j \in \mathcal{C} : c_{ij} > U_j$ are valid and we can add these into the LP. We also add the following constraints: $y_i = 1$ for all facilities $i \in F'$. From now on, assume that (x, y) is the solution of this new LP, satisfying all the mentioned constraints.

Lemma 4. *Let s be any client in sparse instance $\mathcal{I}$ and $\mathcal{S}$ be a set of clients such that $c_{js} \leq \beta C_j$ for all $j \in \mathcal{S}$ and some constant $\beta \geq 1$. Then*

$$|\mathcal{S}|U_s \leq \epsilon \mathrm{OPT}_{\mathcal{I}} + \frac{\beta}{\delta} \sum_{j \in \mathcal{S}} C_j.$$

2.4 Filtering Phase

We will apply the standard filtering method for facility-location problems (see [3, 8]). Basically, we choose a subset $\mathcal{C}' \subseteq \mathcal{C}$ such that clients in $\mathcal{C}'$ are *far* from each other. After assigning each facility to the closest client in $\mathcal{C}'$, it is possible to lower-bound the opening volume of each cluster. Each client in $\mathcal{C}'$ is called a cluster center.

Filtering Algorithm: Initialize $\mathcal{C}' := \mathcal{C}$. For each client $j \in \mathcal{C}'$ in increasing order of C_j, we remove all other clients j' such that $c_{jj'} \leq 4C_j = 4\max\{C_{j'}, C_j\}$ from $\mathcal{C}'$.

For each $j \in \mathcal{C}'$, define $F_j = \{i \in \mathcal{F} : c_{ij} = \min_{k \in \mathcal{C}'} c_{ik}\}$, breaking ties arbitrarily. Let $F'_j = \{i \in F_j : c_{ij} \leq 2C_j\}$ and $\gamma_j = \min_{i \notin F_j} c_{ij}$. Then define $G_j = \{i \in F_j : c_{ij} \leq \gamma_j\}$. We also reassign $y_i := x_{ij}$ for $i \in G_j$ and $y_i := 0$ otherwise. For $j \in \mathcal{C}'$, let M_j be the set containing j and all clients removed by j in the filtering process.

Lemma 5. *We have the following properties:*

- *All sets G_j are disjoint,*
- *$1/2 \leq y(F'_j)$ and $y(G_j) \leq 1$ for all $j \in \mathcal{C}'$.*
- *$F'_j \subseteq G_j$ for all $j \in \mathcal{C}'$.*

It is clear that for all $j, j' \in \mathcal{C}', c_{jj'} \geq 4\max\{C_{j'}, C_j\}$. Moreover, for each $j \in \mathcal{C} \setminus \mathcal{C}'$, we can find $j' \in \mathcal{C}'$, where j' causes the removal of j, or, in other words, $C_{j'} \leq C_j$ and $c_{jj'} \leq 4C_j$. Assuming that we have a solution $\mathcal{S}$ for the instance $\mathcal{I}' = (B, \mathcal{F}, \mathcal{C}', c, w)$ where each client j in $\mathcal{C}'$ has demand $d_j = |M_j|$ (i.e. there are $|M_j|$ copies of j), we can transform it into a solution for $\mathcal{I}$ as follows. Each client $j \in \mathcal{C} \setminus \mathcal{C}'$ will be served by the facility of j' that removed j. Then $\text{cost}(j) = c_{jj'} + \text{cost}(j') \leq \text{cost}(j') + 4C_j$. Therefore,

$$\text{cost}_{\mathcal{I}}(\mathcal{S}) = \sum_{j \in \mathcal{C}'} \text{cost}(j) + \sum_{j \in \mathcal{C} \backslash \mathcal{C}'} \text{cost}(j)$$

$$\leq \sum_{j \in \mathcal{C}'} \text{cost}(j) + \sum_{j \in \mathcal{C} \backslash \mathcal{C}'} \left(\text{cost}(j'(j)) + 4C_j \right)$$

$$\leq \text{cost}_{\mathcal{I}'}(\mathcal{S}) + 4\text{OPT}_f.$$

where, in the second line, $j'(j)$ is the center in $\mathcal{C}'$ that removed j.

2.5   A Basic $(23.09 + \epsilon)$-Approximation Algorithm

In this section, we describe a simple randomized $(23.09 + \epsilon)$-approximation algorithm. In the next section, we will derandomize it and give more insights to further improve the approximation ratio to $17.46 + \epsilon$.

High-Level Ideas: We reuse Swamy's idea from [8] to first obtain an *almost* half integral solution $\hat{y}$. This solution $\hat{y}$ has a very nice structure. For example, each client j only (fractionally) connects to at most 2 facilities, and there is at least a half-opened facility in each G_j. We shall refer to this set of 2 facilities as a bundle. In [8], the author applies a standard clustering process to get disjoint bundles and round $\hat{y}$ by opening at least one facility per bundle. The drawback of this method is that we have to pay extra cost for bundles removed in the clustering step. In fact, it is possible to open at least one facility per bundle without filtering out any bundle. The idea here is inspired by the work of Charikar et. al [2]. In addition, instead of picking $\hat{y}$ deterministically, sampling such a half integral extreme point will be very helpful for the analysis.

We consider the following polytope.

$$\mathcal{P} = \{ v \in [0,1]^{|\mathcal{F}|} : v(F'_j) \geq 1/2, \ v(G_j) \leq 1, \ \forall j \in \mathcal{C}'; \ \sum_{i \in \mathcal{F}} w_i v_i \leq B \}.$$

Lemma 6 ([8]). *Any extreme point of $\mathcal{P}$ is almost half-integral: there exists at most 1 cluster center $s \in \mathcal{C}'$ such that G_s contains variables $\notin \{0, \frac{1}{2}, 1\}$. We call s a fractional client.*

Notice by Lemma 5 that $y \in \mathcal{P}$. By Carathéodory's theorem, y is a convex combination of at most $t = |\mathcal{F}| + 1$ extreme points of $\mathcal{P}$. Moreover, there is an efficient algorithm based on the ellipsoid method to find such a decomposition (e.g., see [7]). We apply this algorithm to get extreme points $y^{(1)}, y^{(2)}, \ldots, y^{(t)} \in \mathcal{P}$ and coefficients $0 \leq p_1, \ldots, p_t \leq 1, \sum_{i=1}^{t} p_i = 1$, such that

$$y = p_1 y^{(1)} + p_2 y^{(2)} + \ldots + p_t y^{(t)}.$$

This representation defines a distribution on t extreme points of $\mathcal{P}$. Let $Y \in [0,1]^{\mathcal{F}}$ be a random vector where $\Pr[Y = y^{(i)}] = p_i$ for $i = 1, \ldots, t$. Observe that Y is *almost* half-integral. Let s be the fractional client in Y. (We assume that s exists; otherwise, the cost will only be smaller.)

Defining primary and secondary facilities: For each $j \in C'$,

- If $j \neq s$, let $i_1(j)$ be the half-integral facility in F'_j. Else ($j = s$), let $i_1(j)$ be the smallest-weight facility in F'_j with $Y_{i_1(j)} > 0$.
- If $Y(i_1(j)) = 1$, let $i_2(j) = i_1(j)$.
- If $Y(G_j) < 1$, then let $\sigma(j)$ be the nearest client to j in C'. Define $i_2(j) = i_1(\sigma(j))$.
- If $Y(G_j) = 1$, then
  - If $j \neq s$, let $i_2(j)$ be the other half-integral facility in G_j.
  - Else ($j = s$), let $i_2(j)$ be the smallest-weight facility in G_j with $Y_{i_2(j)} > 0$.
- We call $i_1(j), i_2(j)$ the primary facility and the secondary facility of j, respectively.

Constructing the neighborhood graph: Initially, construct the directed graph G on clients in C' such that there is an edge $j \to \sigma(j)$ for each $j \in C' : Y(G_j) < 1$. Note that all vertices in G have outdegree ≤ 1. If $Y(G_j) = 1$, then vertex j has no outgoing edge. In this case, we replace j by the edge $i_1(j) \to i_2(j)$, instead. Finally, we relabel all other nodes in G by its primary facility. Now we can think of each client $j \in C'$ as an edge from $i_1(j)$ to $i_2(j)$ in G.

Lemma 7. *Without loss of generality, we can assume that all cycles of G (if any) are of size 2. This means that G is bipartite.*

We are now ready to describe the main algorithm.

Algorithm 1. ROUND(Y)

1: Construct the neighborhood graph G based on Y
2: Let C_1, C_2 be independent sets which partition G.
3: Let W_1, W_2 be the total weight of the facilities in C_1, C_2 respectively.
4: **if** $W_1 \leq W_2$ **then**
5:     **return** C_1
6: **else**
7:     **return** C_2

Theorem 2. *Algorithm 2 returns a feasible solution S where*

$$\mathrm{E}[\, cost_{\mathcal{I}_0}(S)] \leq \max \left\{ \frac{1+\delta}{1-\delta}, 10 + 12/\delta + 3\epsilon \right\} \mathrm{OPT}_{\mathcal{I}_0}.$$

In particular, the approximation ratio is at most $(23.087 + 3\epsilon)$ when setting $\delta :=$ 0.916966.

Algorithm 2. BASICALGORITHM(δ, ϵ, $\mathcal{I}_0$)

1: Generate $O(n^{2/\epsilon})$ pairs $(\mathcal{I}, \mathcal{F}')$ using the algorithm in the proof of Lemma 3.
2: $\mathcal{S} \leftarrow \emptyset$
3: **for** each pair $(\mathcal{I}, \mathcal{F}')$ **do**
4:     Let (x, y) be the optimal solution of the modified LP relaxation in Section 2.3.
5:     Apply the filtering algorithm to get $\mathcal{I}'$.
6:     Use $\mathcal{F}, \mathcal{C}'$ to define the polytope $\mathcal{P}$.
7:     Sample a random extreme point Y of $\mathcal{P}$ as described above.
8:     Let $\mathcal{S}' \leftarrow$ ROUND(Y)
9:     If $\mathcal{S}'$ is feasible and its cost is smaller than the cost of $\mathcal{S}$ then $\mathcal{S} \leftarrow \mathcal{S}'$.
10: **return** $\mathcal{S}$

2.6 A $(17.46 + \epsilon)$-Approximation Algorithm via Conditioning on the Fractional Cluster Center

Recall that the improved Kumar's bound for the fractional client s is

$$|M_s|U_s \leq \epsilon\text{OPT}_{\mathcal{I}} + (4/\delta) \sum_{j \in M_s} C_j.$$

In Theorem 2, we upper-bound the term $\sum_{j \in M_s} C_j$ by OPT_f. However, if this is tight, then the fractional cost of all other clients not in M_s must be zero and we should get an improved ratio.

To formalize this idea, let $u \in \mathcal{C}'$ be the client such that $\sum_{j \in M_u} C_j$ is maximum. Let $\alpha \in [0, 1]$ such that $\sum_{j \in M_u} C_j = \alpha\text{OPT}_f$, then

$$|M_s|U_s \leq \epsilon\text{OPT}_{\mathcal{I}} + (4/\delta)\alpha\text{OPT}_f. \tag{1}$$

The following bound follows immediately by replacing the Kumar's bound by (1) in the proof of Theorem 2.

$$\text{E}[\text{cost}_{\mathcal{I}}(\mathcal{S})] \leq (10 + 12\alpha/\delta + 3\epsilon)\text{OPT}_{\mathcal{I}}. \tag{2}$$

In fact, this bound is only tight when u happens to be the fractional client after sampling Y. If u is not "fractional", the second term in the RHS of (1) should be at most $(1 - \alpha)\text{OPT}_f$. Indeed, if u is *rarely* a fractional client, we should obtain a strictly better bound. To this end, let $\mathcal{E}$ be the event that u is the fractional client after the sampling phase. Let $p = \Pr[\mathcal{E}]$. We get the following lemma.

Lemma 8. *Algorithm 2 returns a solution $\mathcal{S}$ with*

$$\text{E}[\text{cost}_{\mathcal{I}}(\mathcal{S})] \leq (10 + \min\{12\alpha/\delta, (12/\delta)(p\alpha + (1 - p)(1 - \alpha))\} + 3\epsilon)\text{OPT}_{\mathcal{I}}.$$

Finally, conditioning on the event $\mathcal{E}$, we are able to combine certain terms and get the following improved bound.

Lemma 9. *Algorithm 2 returns a solution $\mathcal{S}$ with*

$$\text{E}[\text{cost}_{\mathcal{I}}(\mathcal{S})|\mathcal{E}] \leq (\max\{6/p, 12/\delta\} + 4 + 3\epsilon)\text{OPT}_{\mathcal{I}}.$$

Algorithm 3. DETERMINISTICALGORITHM(δ, ϵ, $\mathcal{I}_0$)

1: Generate $O(n^{2/\epsilon})$ pairs $(\mathcal{I}, \mathcal{F}')$ using the algorithm in the proof of Lemma 3.
2: $\mathcal{S} \leftarrow \emptyset$
3: **for** each pair $(\mathcal{I}, \mathcal{F}')$ **do**
4:     Let (x, y) be the optimal solution of the modified LP relaxation in Section 2.3.
5:     Apply the filtering algorithm to get $\mathcal{I}'$.
6:     Use $\mathcal{F}, \mathcal{C}'$ to define the polytope $\mathcal{P}$.
7:     Decompose y into a convex combination of extreme points $y^{(1)}, y^{(2)}, \ldots, y^{(t)}$ of $\mathcal{P}$.
8:     **for** each $Y \in \{y^{(1)}, y^{(2)}, \ldots, y^{(t)}\}$ **do**
9:         Let $\mathcal{S}' \leftarrow$ ROUND(Y)
10:        If $\mathcal{S}'$ is feasible and its cost is smaller than the cost of $\mathcal{S}$ then $\mathcal{S} \leftarrow \mathcal{S}'$.
11: **return** $\mathcal{S}$

Now we have all the required ingredients to get an improved approximation ratio. Algorithm 3 is a derandomized version of Algorithm 2.

Theorem 3. *Algorithm 3 returns a feasible solution $\mathcal{S}$ where*

$$cost_{\mathcal{I}_0}(\mathcal{S}) \leq (17.46 + 3\epsilon)\mathrm{OPT}_{\mathcal{I}_0},$$

when setting $\delta = 0.891647$.

Note that in [8], Swamy considered a slightly more general version of KM where each facility also has an opening cost. It can be shown that Theorem 3 also extends to this variant.

3 A Bi-Factor 3.05-Approximation Algorithm for Knapsack Median

In this section, we develop a bi-factor approximation algorithm for KM that outputs a pseudo-solution of cost at most $3.05\mathrm{OPT}_{\mathcal{I}}$ and of weight bounded by $(1 + \epsilon)B$. This is a substantial improvement upon the previous comparable result, which achieved a factor of $16 + \epsilon$ and violated the budget additively by the largest weight $w_{\max}$ of a facility. It is not hard to observe that one can also use Swamy's algorithm [8] to obtain an 8-approximation that opens a constant number of extra facilities (exceeding the budget B). Our algorithm works for the original problem formulation of KM where all facility costs are zero. Our algorithm is inspired by a recent algorithm of Li and Svensson [6] for the k-median problem, which beat the long standing best bound of $3 + \epsilon$. The overall approach consists in computing a so-called *bi-point solution*, which is a convex combination $a\mathcal{F}_1 + b\mathcal{F}_2$ of two integral pseudo solutions $\mathcal{F}_1$ and F_2 for appropriate factors $a, b \geq 0$ with $a + b = 1$, and then rounding this bi-point solution to an integral one.

Depending on the value of a, Li and Svensson apply three different bi-point rounding procedures. We extend two of them to the case of KM. The rounding procedures of Li and Svensson have the inherent property of opening $k + c$ facilities where c is a constant. Li and Svensson find a way to preprocess the instance such that any pseudo approximation algorithm for k-median that opens $k + c$ facilities can be turned into a (proper) approximation algorithm by paying only an additional ϵ in the approximation ratio. We did not find a way to prove a similar result also for KM and therefore our algorithms violate the facility budget by a factor of $1 + \epsilon$.

3.1 Pruning the Instance

The bi-factor approximation algorithm that we will describe in Section 3.2 has the following property. It outputs a (possibly infeasible) pseudo-solution of cost at most $\alpha \mathrm{OPT}_{\mathcal{I}}$ such that the budget B is respected when we remove the two heaviest facilities from this solution. This can be combined with a simple reduction to the case where the weight of any facility is at most ϵB. This ensures that our approximate solution violates the budget by a factor at most $1 + 2\epsilon$ while maintaining the approximation factor α.

Lemma 10. *Let* $\mathcal{I} = (B, \mathcal{F}, \mathcal{C}, c, w)$ *be any KM instance. Assume there exists an algorithm A which computes for instance $\mathcal{I}$ a solution S which consists of a feasible solution and two additional facilities, and which has cost at most $\alpha \mathrm{OPT}_{\mathcal{I}}$. Then there exists for any $\epsilon > 0$ a bi-factor approximation algorithm A' which computes a solution of weight $(1 + \epsilon)B$ and of cost at most $\alpha \mathrm{OPT}_{\mathcal{I}}$.*

3.2 Computing and Rounding a Bi-Point Solution

Extending a similar result for the k-median [9], we can compute a bi-point solution, which is a convex combination of two integral pseudo-solutions. See Appendix for more details.

Theorem 4. *We can compute in polynomial time two sets $\mathcal{F}_1$ and $\mathcal{F}_2$ of facilities and factors $a, b \geq 0$ such that $a + b = 1$, $w(\mathcal{F}_1) \leq B \leq w(\mathcal{F}_2)$, $a \cdot w(\mathcal{F}_1) + b \cdot w(\mathcal{F}_2) \leq B$, and $a \cdot \mathrm{cost}_{\mathcal{I}}(\mathcal{F}_1) + b \cdot \mathrm{cost}_{\mathcal{I}}(\mathcal{F}_2) \leq 2 \cdot \mathrm{OPT}_{\mathcal{I}}$.*

We will now give an algorithm which for a given KM instance $\mathcal{I} = (B, \mathcal{F}, \mathcal{C}, d, w)$ returns a pseudo-solution as in Lemma 10 with cost $3.05 \mathrm{OPT}_{\mathcal{I}}$.

We use Theorem 4 to obtain a bi-point solution of cost $2\,\mathrm{OPT}_{\mathcal{I}}$. We will convert it into a pseudo-solution of cost 1.523 times bigger than the bi-point solution. Let $a\mathcal{F}_1 + b\mathcal{F}_2$ be the bi-point solution where $a + b = 1$, $w(\mathcal{F}_1) \leq B < w(\mathcal{F}_2)$ and $aw(\mathcal{F}_1) + bw(\mathcal{F}_2) = B$. For each client $j \in \mathcal{C}$ the closest elements in sets $\mathcal{F}_1$ and $\mathcal{F}_2$ are denoted by $i_1(j)$ and $i_2(j)$, respectively. Moreover, let $d_1(j) = c_{i_1(j)j}$ and $d_2(j) = c_{i_2(j)j}$. Then the (fractional) connection cost of j in our bi-point solution is $ad_1(j) + bd_2(j)$. In a similar way let $d_1 = \sum_{j \in \mathcal{C}} d_1(j)$ and $d_2 = \sum_{j \in \mathcal{C}} d_2(j)$. Then the bi-point solution has cost $ad_1 + bd_2$.

We consider two candidate solutions. In the first we just pick $\mathcal{F}_1$ which has cost bounded by $\frac{d_1}{ad_1 + bd_2} \leq \frac{1}{a + br_D}$, where $r_D = \frac{d_2}{d_1}$. This, multiplied by 2, gives our approximation factor.

To obtain the second candidate solution we use the concept of *stars*. For each facility $i \in \mathcal{F}_2$ define $\pi(i)$ to be the facility from set $\mathcal{F}_1$ which is closest to i. For a facility

$i \in \mathcal{F}_1$ define star $\mathcal{S}_i$ with root i and leafs $S_i = \{i' \in \mathcal{F}_2 | \pi(i') = i\}$. Note that by the definition of stars, we have that any client j with $i_2(j) \in S_i$ has $c_{i_2(j)i} \leq c_{i_2(j)i_1(j)} = d_2(j) + d_1(j)$ and therefore $c_{ji} \leq c_{ji_2(j)} + c_{i_2(j)i} \leq 2d_2(j) + d_1(j)$.

The idea of the algorithm is to open for each star either its root or all of its leaves so that in total the budget is respected. We formulate this subproblem by means of an auxiliary LP. For any star S_i let $\delta(S_i) = \{j \in \mathcal{C} \mid i_2(j) \in S_i\}$. Consider a client $j \in \delta(S_i)$. If we open the root of S_i the connection cost of j is bounded by $2d_2(j) + d_1(j)$, but if we open the leaf $i_2(j) \in S_i$ we pay only $d_2(j)$ for connecting j. Thus, we save in total an amount of $\sum_{j \in \delta(S_i)} d_2(j) + d_1(j)$ when we open all leaves of S in comparison to opening just the root i. This leads us to the following linear programming relaxation where we introduce for each star S_i a variable x_i indicating whether we open the leaves of this star ($x_i = 1$) or its root

$$\max \sum_{i \in \mathcal{F}_1} \sum_{j \in \delta(S_i)} (d_1(j) + d_2(j))x_i \quad \text{subject to} \tag{3}$$

$$\sum_{i \in \mathcal{F}_1} (w(S_i) - w_i)x_i \leq B - w(\mathcal{F}_1)$$

$$0 \leq x_i \leq 1 \quad \forall i \in \mathcal{F}_1.$$

Now observe that this is a knapsack LP. Therefore, any optimum extreme point $\mathbf{x}$ solution to this LP has at most one fractional variable. Note that if we set $x_i = b$ for all $i \in \mathcal{F}_1$ we obtain a feasible solution to the above LP. Therefore the objective value of the above LP is lower bounded by $b(d_1 + d_2)$. We now open for all stars $\mathcal{S}_i$ with integral x_i either its root ($x_i = 0$) or all of its leaves ($x_i = 1$) according to the value of x_i. For the (single) star $\mathcal{S}_i$ where x_i is fractional we apply the following rounding procedure.

We always open i, the root of S_i. To round the leaf set S_i, we set up another auxiliary knapsack LP similar to LP (3). In this LP, each leaf $i' \in S_i$ has a variable $\hat{x}_{i'}$ indicating if the facility is open ($\hat{x}_{i'} = 1$) or not ($\hat{x}_{i'} = 0$). The details of this LP can be found in the appendix. As a result of solving this LP, all leaves except possibly one obtain an integral value $\hat{x}_{i'}$. We open all $i' \in S_i$ with $\hat{x}_{i'} = 1$ and also the only fractional leaf. As a result, the overall set of opened facilities consists of a feasible solution and two additional facilities (namely the root i and the fractional leaf in S_i).

We will now analyze the cost of this solution. Both of the above knapsack LPs only reduce the connection cost in comparison to the original bipoint solution (or equivalently increase the saving with respect to the quantity $d_1 + 2d_2$), the total connection cost of the solution can be upper bounded by $d_1 + 2d_2 - b(d_1 + d_2) = (1 + a)d_2 + ad_1$.

The cost increase of the second algorithm with respect to the bi-point solution is at most

$$\frac{(1 + a)d_2 + ad_1}{(1 - a)d_2 + ad_1} = \frac{(1 + a)r_D + a}{(1 - a)r_D + a},$$

We always choose the better of the solutions of the two algorithms described above. Our approximation ratio is upper bounded by

$$\max_{\substack{r_D \geq 0 \\ a \in [0,1]}} \min \left\{ \frac{(1 + a)r_D + a}{(1 - a)r_D + a}, \frac{1}{a + r_D(1 - a)} \right\} \leq 1.523$$

This, multiplied by 2 gives our overall approximation ratio of 3.05.

Theorem 5. *For any $\epsilon > 0$, there is a bi-factor approximation algorithm for KM that computes a solution of weight $(1 + \epsilon)B$ and has a cost $3.05\mathrm{OPT}_{\mathcal{I}}$.*

Proof. As argued above our algorithm computes a pseudo solution S of cost at most $3.05\mathrm{OPT}_{\mathcal{I}}$. Moreover, S consists of a feasible solution and two additional facilities. Hence, Lemma 10 implies the theorem. ☐

4   Discussion

The proof of Theorem 3 implies that for every (ϵ, δ)-sparse instance $\mathcal{I}$, there exists a solution $\mathcal{S}$ such that $\mathrm{cost}_{\mathcal{I}}(\mathcal{S}) \leq (4+12/\delta)OPT_f + 3\epsilon OPT_{\mathcal{I}}$. Therefore, the integrality gap of $\mathcal{I}$ is at most $\frac{4+12/\delta}{1-3\epsilon}$. Unfortunately, our client-centric sparsification process inflates the approximation factor to at least $\frac{1+\delta}{1-\delta}$, so we must choose some $\delta < 1$ which balances this factor with that of Algorithm 3. In contrast, the facility-centric sparsification used in [6] incurs only a $1 + \epsilon$ factor in cost. We leave it as a open question whether the facility-centric version could also be used to get around the integrality gap of KM.

Our bi-factor approximation algorithm achieves a substantially smaller approximation ratio at the expense of slightly violating the budget by opening two extra facilities. We leave it as an open question, to obtain a pre- and postprocessing in the flavor of Li and Svensson to turn this into an approximation algorithm. It seems even interesting to turn any bi-factor approximation into an approximation algorithm by losing only a constant factor in the approximation ratio. We also leave it as an open question to extend the third bi-point rounding procedure of Li and Svensson to knapsack median, which would give an improved result.

References

1. Byrka, J., Pensyl, T., Rybicki, B., Srinivasan, A., Trinh, K.: An improved approximation for k-median, and positive correlation in budgeted optimization. In: Proceedings of the Annual ACM-SIAM Symposium on Discrete Algorithms (SODA), pp. 737–756 (2015)
2. Charikar, M., Guha, S., Tardos, É., Shmoys, D.B.: A constant-factor approximation algorithm for the k-median problem. In: ACM Symposium on Theory of Computing (STOC), pp. 1–10. ACM (1999)
3. Charikar, M., Li, S.: A dependent lp-rounding approach for the k-median problem. In: Czumaj, A., Mehlhorn, K., Pitts, A., Wattenhofer, R. (eds.) ICALP 2012, Part I. LNCS, vol. 7391, pp. 194–205. Springer, Heidelberg (2012)
4. Krishnaswamy, R., Kumar, A., Nagarajan, V., Sabharwal, Y., Saha, B.: The matroid median problem. In: Proceedings of the Annual ACM-SIAM Symposium on Discrete Algorithms (SODA), pp. 1117–1130. SIAM (2011)
5. Kumar, A.: Constant factor approximation algorithm for the knapsack median problem. In: Proceedings of the Annual ACM-SIAM Symposium on Discrete Algorithms (SODA), pp. 824–832. SIAM (2012)
6. Li, S., Svensson, O.: Approximating k-median via pseudo-approximation. In: STOC, pp. 901–910 (2013)

7. Schrijver, A.: Combinatorial optimization: polyhedra and efficiency, vol. 24. Springer Science & Business Media (2003)
8. Swamy, C.: Improved approximation algorithms for matroid and knapsack median problems and applications. In: APPROX/RANDOM 2014, vol. 28, pp. 403–418 (2014)
9. Williamson, D.P., Shmoys, D.B.: The Design of Approximation Algorithms. Cambridge University Press (2011)

Output-Sensitive Algorithms for Enumerating the Extreme Nondominated Points of Multiobjective Combinatorial Optimization Problems

Fritz Bökler[*] and Petra Mutzel

Department of Computer Science, TU Dortmund, Germany
{fritz.boekler,petra.mutzel}@tu-dortmund.de

Abstract. This paper studies output-sensitive algorithms for enumeration problems in multiobjective combinatorial optimization (MOCO). We develop two methods for enumerating the extreme points of the Pareto-frontier of MOCO problems. The first method is based on a dual variant of Benson's algorithm, which has been originally proposed for multiobjective linear optimization problems. We prove that the algorithm runs in output polynomial time for every fixed number of objectives if the weighted-sum scalarization can be solved in polynomial time. Hence, we propose the first algorithm which solves this general problem in output polynomial time. We also propose a new lexicographic version of the dual Benson algorithm that runs in incremental polynomial time in the case that the lexicographic optimization variant can be solved in polynomial time. As a consequence, the extreme points of the Pareto-frontier of the multiobjective spanning tree problem as well as the multiobjective global min-cut problem can be computed in polynomial time for a fixed number of objectives. Our computational experiments show the practicability of our improved algorithm: We present the first computational study for computing the extreme points of the multiobjective version of the assignment problem with five and more objectives. We also empirically investigate the running time behavior of our new lexicographic version compared to the original algorithm.

1 Introduction

In practical optimization, we often deal with problems having more than one objective. Unlike in single objective optimization, there usually does not exist a single optimal value and we usually have many solutions which are incomparable. If we agree that we will always prefer a solution over all solutions that are worse in all objectives then we can use this relation as a partial order on the solutions, called the *Pareto-dominance*. More precisely, we say a vector $a \in \mathbb{R}^d$ *dominates* a vector $b \in \mathbb{R}^d$ or $a \preceq b$ if $a \leq b$ (componentwise) and $a \neq b$. Analogously, for

---

[*] The author has been supported by the Bundesministerium für Wirtschaft und Energie (BMWi) within the research project "Bewertung und Planung von Stromnetzen" (promotional reference 03ET7505) and by DFG GRK 1855 (DOTS).

N. Bansal and I. Finocchi (Eds.): ESA 2015, LNCS 9294, pp. 288–299, 2015.
DOI: 10.1007/978-3-662-48350-3_25

two solutions $x, y \in \mathcal{X}$ and an objective function $c : \mathcal{X} \to \mathbb{R}^d$ we say $x \preceq y$ if $c(x) \preceq c(y)$. The set of all minimal solutions with respect to this partial order is called the set of *efficient* or *Pareto-optimal* solutions; the corresponding image under the objective function is called the set of *nondominated points* or *Pareto-frontier*.

In this spirit, we can define *multiobjective combinatorial optimization (MOCO) problems* consisting of a base set $\mathcal{A}$, a set of feasible solutions $\mathcal{S} \subseteq 2^{\mathcal{A}}$ and an objective function $c : \mathcal{A} \to \mathbb{Q}^d$. The cost of a solution $x \in \mathcal{S}$ thus is $c(x) := \sum_{a \in x} c(a)$. The goal is to find for all points y of the Pareto-frontier one solution x such that $c(x) = y$. This notion is motivated by the multicriteria decision making community: Two solutions which are mapped to the same image by the objective function are supposed to be essentially the same object.

One serious computational issue of many MOCO problems is that the size of the Pareto-frontier can be large in the worst case, i.e., exponential in the input size. Then again, the output being exponential is also true for many other enumeration problems and has been addressed by the theory of output-sensitive complexity. Johnson, Papadimitriou and Yannakakis [13] provide a summary of these complexity notions: An algorithm runs in *output polynomial time* (originally referred to as *polynomial total time*) if its running time can be bounded by a polynomial in the input and the output size. Besides the total running time, a usually more interesting property is the delay of an enumeration algorithm. Let N be the number of elements to output. We say the 0-th delay is the running time prior to the first output of a solution, the k-th delay is the running time between the output of the k-th and $(k + 1)$-th solution, and the N-th delay is the running time after the last output until the termination of the algorithm. An enumeration algorithm runs in *polynomial (time) delay* if all delays are bounded by a polynomial in the input size. To relax this notion a bit, an algorithm runs in *incremental polynomial time*, if the k-th delay is bounded by a polynomial in the input and k. One purpose of this paper is to encourage a line of research which applies these complexity notions to multiobjective optimization problems and— surprisingly enough—there is only one other paper which considers this [14].

Moreover, in a recent work by Röglin and Brunsch [5], it is shown that in a smoothed analysis setting, the expected size of the Pareto-frontier of a MOCO problem of n variables and d objectives is $\mathcal{O}(n^{2d}\Phi^d)$ for every perturbation parameter $\Phi \geq 1$. This result gives reason to believe that pursuing the goal to find the entire Pareto-frontier might still be practical as long as the number of objectives is not too large.

A very promising approach to solve MOCO problems is the *two-phase method* (cf. e.g., [7]). In this method, the Pareto-frontier is partitioned into two sets: *extreme* nondominated and *nonextreme* nondominated points. (Exact definitions of these sets will be given later.) In a first phase, we compute the extreme nondominated points. In the second phase we can exploit this knowledge to find the remaining nondominated points. The two-phase method is motivated by the observation that it is often **NP**-hard even to decide if there exists a solution which dominates a given point $y \in \mathbb{Q}^d$. But at least in the biobjective case, if the lexicographic variant can be solved in polynomial time, then we can enumerate the set of extreme nondominated points in incremental polynomial time by using standard methods like the dichotomic approach [3]. Although the problem of finding the extreme nondominated points of a MOCO problem seems to be solved in the biobjective case, it is an open problem to enumerate these points and corresponding solutions for problems with an arbitrary number of objectives in reasonable time. Ehrgott and Gandibleux address this problem in their survey paper from 2005 [7] as a "first step to an application of the two phases method in three or more criteria MOCO". The main concern of this paper will be how to compute the set of extreme nondominated points in output polynomial running time in theory and practice.

For $d = 2$, finding the extreme points of the Pareto-frontier is equivalent to the combinatorial parametric optimization problem. See, e.g., [1] for a parametric version of the minimum spanning tree problem. Thus, the problem we consider here is a natural generalization of this very well known class of problems.

Previous Work. There exist only a few other methods to enumerate the extreme nondominated points for MOCO problems with more than two objectives. Przybylski, Gandibleux and Ehrgott [17] propose a method to find these points for general multiobjective integer linear programming problems for which the ideal point exists. The algorithm is theoretically capable of computing the extreme nondominated points for an arbitrary number of objectives and the authors provide deep improvements in the case of three objectives. They also conducted a set of experiments on three-objective assignment and knapsack problems showing a very decent running time. Özpeynirci and Köksalan [15] suggest a method to compute the set of extreme nondominated points based on the dichotomic approach which is usually utilized in the biobjective case. The authors also present experiments on several MOCO problems with three and four objectives. Both of the above works do not provide running time guarantees for their algorithms.

In [14], the authors propose an algorithm that enumerates all efficient spanning trees which are a solution to a weighted-sum of the objectives with polynomial delay. The algorithm bases on the reverse search by Avis and Fukuda [4] to output the efficient vertices of the spanning tree polytope. A note is here in order, because the algorithm follows a different model where all efficient solutions are sought. It is possible that there exist many spanning trees which are mapped to the same point in the objective space. Consider a complete graph with n vertices where all edges are mapped to the 1-vector in $\mathbb{R}^2$ by the objective function. Then each of the n^{n-2} spanning trees is efficient, but the Pareto-frontier consists of only one point.

Another branch of research which tackles the problem of an exponential sized Pareto-frontier of MOCO problems was raised by a paper by Papadimitriou and Yannakakis [16]. In this work and in many papers that followed, an ε-Pareto set, i.e., a subset S of the solution set such that for each point y of the Pareto-frontier there is one solution $x \in S$ such that $c(x) \leq (1 + \varepsilon)y$, is computed. In [16], it is also proven that for each $\varepsilon > 0$ and fixed number of objectives there exist such an ε-Pareto set of size polynomial in the input size and $\frac{1}{\varepsilon}$.

Our Contributions. A well known method to find extreme nondominated points (or more general supported points, see below) in the first phase of the two-phase method for a MOCO problem $(\mathcal{A}, \mathcal{S}, c)$ is to optimize the weighted-sum scalarization: $(P_1(\ell)) : \min\{\ell^T c(x) \mid x \in \mathcal{S}\}$ for some $\ell \in \mathbb{Q}^d, \ell > 0$. On one hand, this problem is as easy as the single objective version of the problem, as long as the encoding lengths of the components of ℓ are not too large. (Considering that ℓ is not part of the original input.) On the other hand, we do not easily know how these ℓ need to be chosen to find all obtainable points of the Pareto-frontier.

The set of solutions which can be obtained by this weighted-sum method are called *supported (efficient) solutions*. A point of the Pareto-frontier which corresponds to a supported solution is called a *supported (nondominated) point*. A geometric view on the supported part of the Pareto-frontier can be acquired by looking at the convex hull of $\mathcal{Y} := c(\mathcal{S})$. A point y of the Pareto-frontier is thus supported iff y is a point on the boundary of conv $\mathcal{Y}$. Moreover, we call a supported point y an *extreme (nondominated) point*, if y is an extreme point of conv $\mathcal{Y}$. The set of extreme nondominated points will be denoted as $\mathcal{Y}_X$. Our concern in this paper will thus be to enumerate the set $\mathcal{Y}_X$.

In general, we can compute the extreme nondominated points of conv $\mathcal{Y}$ by enumerating the extreme nondominated points of the *multiobjective linear program*

$$(\text{MOLP}) : \min\{c(x) \mid x \in \text{conv}(\chi(\mathcal{S}))\}, \tag{1}$$

where $\chi(\mathcal{S})$ denotes the characteristic vectors of the solutions of our MOCO problem for a fixed ordering of the variables. In Sec. 3, we will conduct a running time analysis of a recently proposed algorithm for MOLP. We will prove that it can efficiently find the extreme points of the Pareto-frontier of an MOLP if the ideal point exists. Luckily, in the case in which we derive an MOLP from a MOCO problem, the ideal point exists iff the problem has a solution. But indeed, it might be hard to construct the MOLP, the number of facets of the feasible set might be large or it might have a large encoding length. We will also show that it suffices to have a polynomial time algorithm for problem $(P_1(\ell))$ without constructing the MOLP explicitly.

Theorem 1. *For every MOCO problem P with a fixed number of objectives, the set of extreme nondominated points of P can be enumerated in output polynomial time if we can solve the weighted-sum scalarization of P in polynomial time.*

Subsequently in Sec. 4, we will also suggest an improvement of this algorithm to get a better running time at the expense of needing to solve the lexicographic version of the MOCO problem $(\text{lex-}P_1(\ell)) : \text{lexmin}\{(\ell^T c(x), c_1(x), \ldots, c_d(x)) \mid x \in \mathcal{S}\}$:

Theorem 2. *For every MOCO problem P with a fixed number of objectives, the set of extreme nondominated points of P can be enumerated in incremental polynomial time if we can solve the lexicographic version of P in polynomial time.*

For many classical problems in combinatorial optimization this is not a restriction. For example in the case of shortest path, minimum spanning tree or the assignment problems the lexicographic variant can be solved in polynomial time. In general, if we have a compact LP formulation for a weighted-sum scalarization, then we can solve the lexicographic variant in polynomial time. This is a direct consequence of Lemma 6 in Sec. 4.

If we apply Theorem 1 to the multiobjective spanning tree problem, we immediately obtain an algorithm with a polynomial upper bound on the running time with respect only to the input size to find the extreme nondominated points for each fixed number of objectives. This is well known in the biobjective (or parametric) case, but it is a new result for the general case. It bases on the well known fact that for this problem there exists at most $\mathcal{O}(m^{2(d-1)})$ extreme nondominated points [10]. We obtain a similar result for the multiobjective global min-cut problem, as it has been shown in [2] that the parametric complexity is polynomial in the input size for any fixed number of objectives. To show that these results are not mere theoretical, we implemented the algorithm and a lexicographic oracle for the multiobjective version of the assignment (or minimum weight bipartite matching) problem. The results of this study and a comparison to the method from [15] are presented in Sec. 5. They show that our algorithm is capable of finding the extreme nondominated points of moderately sized instances with up to six objectives. To the best of our knowledge this is the first computational study for five and six objectives.

Moreover, we compare the lexicographic variant of the algorithm to the original algorithm. Since in theory we are able to improve the running time bound, we investigate if this is also true in practice.

2   Theoretical Preliminaries

By $[n]$ we denote the set $\{1, \ldots, n\}$ and $||x||_1$ will be the 1-*norm* of $x \in \mathbb{R}^d$, i.e., $||x||_1 := \sum_{i=1}^d |x_i|$. The nondominated subset of a set of points $M \subseteq \mathbb{R}^d$ is $\min M := \{x \in M \mid x \text{ is nondominated in } M\}$. The *Minkowski sum (product)* of two sets $A, B \subseteq \mathbb{R}^d$ is the set $A + B := \{a + b \mid a \in A, b \in B\}$ ($A \cdot B := \{a \cdot b \mid a \in A, b \in B\}$).

The *multiobjective linear programming problem* (MOLP) is the problem to find for matrices $A \in \mathbb{Q}^{m \times n}, C \in \mathbb{Q}^{d \times n}$ and a vector $b \in \mathbb{Q}^m$ an extreme point representation of the Pareto-frontier $\min\{Cx \mid Ax \geq b\}$. To work on MOLP, it is convenient to define the *feasible set* $P := \{x \in \mathbb{R}^n \mid Ax \geq b\}$ and the *upper image* $\mathcal{P} := C \cdot P + \mathbb{R}^d_{\geq}$, where $\mathbb{R}^d_{\geq} := \{x \in \mathbb{R}^d \mid x \geq 0\}$. It is well known that the extreme points of $\mathcal{P}$ are exactly the extreme points of the Pareto-frontier of the corresponding MOLP (cf. [8]). We will write vert M to denote the set of extreme points of a polyhedral set M and thus, our concern will be to enumerate vert $\mathcal{P}$, i.e., the extreme points of $\mathcal{P}$.

We define the *normalized weighting vectors* to be the set $W_d^0 := \{\ell \in \mathbb{R}^d_{\geq} \mid ||\ell||_1 = 1\}$. The *weighted-sum linear program w.r.t.* $\ell \in W_d^0$ of a given MOLP is the parametric linear program $P_1(\ell) : \min\{\ell^T Cx \mid Ax \geq b\}$. The *ideal point* of an MOLP is then defined as being the point $y^I := (\min P_1(e_i))_{i \in [d]}$, where e_i denotes the i-th unit vector in $\mathbb{R}^d$. Note that an ideal point does not exist for every MOLP.

3   Dual Variant of Benson's Algorithm

The dual variant of Benson's algorithm solves MOLP in the sense that it computes the extreme points of a polyhedron which is a geometric dual polyhedron to $\mathcal{P}$. The algorithm in its original version requires the existence of an ideal point, which we will assume in the following and is always the case in MOCO problems. It was only recently proposed in [8] and got its background theory from [12]. In the next section, we will consider the algorithm as it was given in [8, 11]. We will first give some background theory about the geometric dual polyhedron the algorithm computes and subsequently, describe the algorithm in two levels of detail. The section concludes with its running time analysis in the sense of output-sensitive complexity.

The Geometric Dual Polyhedron. In [12], Heyde and Löhne define a dual polyhedron, or *lower image*, $\mathcal{D}$ to the upper image $\mathcal{P}$ of an MOLP which we will define now.

Since our weight vectors ℓ of the weighted-sum problem are always normalized, i.e., $||\ell||_1 = 1$, it suffices to consider $\ell_1, \ldots, \ell_{d-1}$ and calculate ℓ_d when needed. For ease of notation we define for $v \in \mathbb{R}^d : \lambda(v) := (v_1, \ldots, v_{d-1}, 1 - \sum_{i=1}^{d-1} v_i)$. Then, we consider the dual problem of the weighted-sum LP, which is $(D_1(\ell))$: $\max\{b^T u \mid u \in \mathbb{R}^m_{\geq}, A^T u = C^T \ell\}$. The dual polyhedron $\mathcal{D}$ now consists for all possible vectors $\ell \in W_d^0$ and solutions u to $D_1(\ell)$ of the vectors $(\ell_1, \ldots, \ell_{d-1}, b^T u)$. Thus, $\mathcal{D} := \{(\ell_1, \ldots, \ell_{d-1}, b^T u) \in \mathbb{R}^d \mid \ell \in W_d^0, u \in \mathbb{R}^m_{\geq}, A^T u = C^T \ell\}$. Following LP duality theory, for each point y on the upper boundary of this polyhedron y_d is also the optimal value of $P_1(\lambda(y))$. To take the notion of the upper boundary to a more formal level, we define the $\mathcal{K}_d$-maximal subset of a set $M \subseteq \mathbb{R}^d$, where $\mathcal{K}_d := \{(0, \ldots, 0, y) \in \mathbb{R}^d \mid y \geq 0\}$: A point $y \in M$ is said to be $\mathcal{K}_d$-*maximal* in M if $(y + \mathcal{K}_d) \cap M = \{y\}$. The subset of $\mathcal{K}_d$-maximal points of M is written as $\max_{\mathcal{K}_d} M$.

In [12], it is proven, that the dual polyhedron can be characterized by $\mathcal{D} = \{x \in \mathbb{R}^d \mid \forall y \in \text{vert}\,\mathcal{P} : \psi(y)^T x \geq -y_d, \lambda(x) \geq 0\}$, where $\psi(y) := (v_1 - v_d, \ldots, v_{d-1} - v_d, -1)$. In other words, apart from the inequalities $\lambda(x) \geq 0$, we can describe the polyhedron as an intersection of halfspaces $\{x \in \mathbb{R}^d \mid \psi(y)^T x \geq -y_d\}$ for each extreme point y of the upper image $\mathcal{P}$. Further, Heyde and Löhne also prove that each of these inequalities defines a facet. Thus, we can solve the original MOLP by enumerating the facets of $\mathcal{D}$. While the dual algorithm originally enumerates the extreme points of $\mathcal{D}$, we change the exposition accordingly.

Algorithm Description. We will follow [8] in describing the geometric dual algorithm and use ideas of [11]. Proofs of correctness and finiteness can be found in both places. A formal description of the entire algorithm can be found in

---

Algorithm 1. Dual Variant of Benson's Outer Approximation Algorithm

---

Require: Matrices A, C, and vector $b : P \neq \emptyset$ and $\exists y \in \mathbb{R}^d : y + C \cdot P \subseteq \mathbb{R}^d_{\geq}$
Ensure: List R of pairs (x, y) for all $y \in \text{vert}\, \mathcal{P}$ and some $x \in P$ such that $Cx = y$
1: Find solution x of $P_1(e_1)$ and set $y \leftarrow Cx$
2: $L \leftarrow \{x \in \mathbb{R}^d \mid \lambda(x) \geq 0, \psi(y)^T x \geq -y_d\}$ ▷ Initial polyhedron
3: $M \leftarrow$ Extreme points of L ▷ Perform a vertex enumeration
4: **while** $M \neq \emptyset$ **do**
5:      pick one $v \in M$, $M \leftarrow M \backslash \{v\}$
6:      $x \leftarrow$ optimal solution to $P_1(\lambda(v))$ and $y \leftarrow Cx$ ▷ Shoot ray straight down
7:      **if** $\lambda(v)^T y < v_d$ **then**   ▷ Not an extreme point of $\mathcal{D}$
8:          $L \leftarrow L \cap \{x \in \mathbb{R}^d \mid \psi(y)^T x \geq -y_d\}$ ▷ Add new inequality
9:          $M \leftarrow$ Extreme points of L ▷ Perform a vertex enumeration
10:         $R \leftarrow L \cup \{(x, y)\}$ ▷ Add new candidate extreme point of $\mathcal{P}$
11: Remove redundant entries from R

Alg. 1. From a high-level perspective, the algorithm works as follows: First, the algorithm constructs a polyhedron containing $\mathcal{D}$ (lines 1 to 3). Then, in each iteration, it picks one new extreme point v of the current intermediate polyhedron and shoots a ray into the polyhedron $\mathcal{D}$ (lines 5 and 6). We can shoot a ray in the direction of $-\mathcal{K}_d$ by finding an optimal solution x of $P_1(\lambda(v))$ with value vector $y = Cx$. Either, we discover that v is an extreme point of $\mathcal{D}$, if $\lambda(v)^T y = v_d$ and we proceed to the next iteration. Or, v is not an extreme point, which we see if $\lambda(v)^T y < v_d$ or in other words, the weighted optimal value is smaller than the value represented by v. Then, the algorithm computes a face defining inequality which separates v from $\mathcal{D}$. Because of geometric duality, we can use the new inequality $\psi(y)^T x \geq -y_d$. In lines 8 and 9, the algorithm intersects the current polyhedron with the halfspace corresponding to this inequality. Additionally, it saves y as a candidate for an extreme nondominated point in line 10. This repeats until all extreme points have been confirmed to be part of $\mathcal{D}$. In the end, we still have to remove redundant pairs from the set of candidate extreme nondominated points (see line 11).

Running Time Analysis. The key insight to the running time of the algorithm is that the vertex enumeration steps are performed in the ordinarily much smaller domain of the polyhedron $\mathcal{D}$, which is of dimension d. Additionally, the number of inequalities we enumerate in the process of the algorithm is at most the number of $\mathcal{K}$-maximal faces of $\mathcal{D}$ which—by the geometric duality theorem of [12]—is exactly the number of faces of $\mathcal{P}$. The number of faces of $\mathcal{P}$ can be bounded by reducing the polyhedron to a polytope and using the asymptotic upper bound theorem [18]. Let thus v_e be the number of extreme points of $\mathcal{P}$.

Lemma 3. *Let d be fixed, the number of faces of $\mathcal{P}$ is at most $\mathcal{O}(v_e^{\lfloor \frac{d}{2} \rfloor})$.*

This shows that the number of faces of $\mathcal{D}$ and thus the number of inequalities the algorithm computes does not exceed $\mathcal{O}(v_e^{\lfloor \frac{d}{2} \rfloor})$. To compute the extreme points of the intermediate polyhedra, we can use the asymptotic optimal algorithm for fixed d by Chazelle [6].

Regarding encoding length, there is a subtle problem we need to address. The running times of the known polynomial-time algorithms to compute a solution to the weighted-sum LPs, e.g., the ellipsoid method, depend on the largest number in the input. One potential problem is that the weighted-sum LP not only consists of numbers of the input MOLP, but also of the weight vector ℓ which is recomputed in the process of the algorithm. We can prove that these lengths are not too large by using the fact that the weights are computed from the intermediate extreme points we get. These points are solutions to linear systems of the face defining inequalities we find, which in turn are computed from the extreme points of the upper image $\mathcal{P}$. These extreme points of $\mathcal{P}$ are again linear images of extreme points in the decision space and it is well known that the encoding length of these extreme points can be bounded by $\mathcal{O}(\mathrm{poly}(n, L))$, where L is the encoding length of the largest number in A and thus independent of ℓ.

Lemma 4. *The encoding length of an intermediate extreme point is bounded by* $\mathcal{O}(\mathrm{poly}(d, n, L))$.

This concludes the running time analysis and we can give the running time in the following theorem. We will be able to significantly improve on the d^2 in the exponent in the next section.

Theorem 5. *Let v_e be the number of vertices of $\mathcal{P}$ and d be fixed. Then, Algorithm 1 has a running time bounded by* $\mathcal{O}(v_e^{\lfloor \frac{d}{2} \rfloor}(\mathrm{poly}(n, m, L) + v_e^{\frac{d^2}{4}} \log v_e) + \mathrm{poly}(n, v_e, L))$.

To arrive at Theorem 1, we can solve problem (1). Problem $(P_1(\ell))$ is then equivalent to solving a single linear objective over the feasible set. Instead of constructing the LP explicitly, we can solve the weighted-sum scalarization of the combinatorial optimization problem. Since the encoding length of the weights we use can be bounded by a polynomial in the original input, an algorithm running in weakly polynomial time also suffices to prove the claim.

4   Lexicographic Dual Benson

One serious drawback of the algorithm is that we might enumerate many redundant supporting hyperplanes. This is especially a problem since the number of vertex enumeration steps depends on the number of supporting inequalities we find. Also the number of redundant extreme points we compute depends heavily on this quantity. Hence, it is very much desirable to enumerate facet supporting hyperplanes only. This is the motivation to propose a lexicographic variant of the dual Benson algorithm which we call the lexicographic dual Benson algorithm and will be the subject of this section.

As already stated, in [12], Heyde and Löhne prove that every inequality $\psi(y)^T x \geq -y_d$ is facet defining iff y is an extreme point of $\mathcal{P}$. So it suffices to find only extreme points of $\mathcal{P}$ in lines 1 and 6. We observe that this can be accomplished by computing an optimal solution x of $P_1(\lambda(v))$ with lexicographic minimal $y := Cx$: The set of optimal solutions M of $P_1(\lambda(v))$ is mapped by the objective function to a face F of $\mathcal{P}$ such that $C \cdot M = F$. If we search for a lexicographic minimal Cx for $x \in M$, we will find a lexicographic minimal point of F. Since F is a polyhedron and assuming the ideal point of the MOLP exists, we will always arrive at an extreme point of F and thus of $\mathcal{P}$.

We define lex-$P_1(\ell) : \mathrm{lexmin}\{(\ell^T Cx, c_1 x, \ldots, c_d x) \mid Ax \geq b\}$, where $c_1, \ldots, c_d$ are the rows of C and change lines 1 and 6 accordingly. Moreover, the points we add in line 10 are already vertices of $\mathcal{P}$ and we can skip line 11 and do not need to remove redundant entries anymore. We still have to prove that we can solve this lexicographic LP in polynomial time, but we omit the proof because of space restrictions.

Lemma 6. *A solution to a lexicographic LP can be computed in polynomial time.*

Both of these observations are a game changer in this running time analysis and we arrive at the following theorem.

Theorem 7. *Let v_e be the number of vertices of $\mathcal{P}$ and d be fixed. The lexicographic dual Benson algorithm has a running time of $\mathcal{O}(v_e^{\lfloor \frac{d}{2} \rfloor}(\mathrm{poly}(n, m, L) + v_e \log v_e))$.*

This is a significant improvement over Theorem 5, since we eliminate the term having d^2 in the exponent. Moreover, we are now able to bound the delay of the algorithm. In the original algorithm this was not possible since it could take a large number of iterations until the $(d + 1)$-th extreme point is found.

Theorem 8. *Let d be fixed. For the lexicographic dual Benson algorithm the k-th delay is in $\mathcal{O}(k^{\lfloor \frac{d}{2} \rfloor} \mathrm{poly}(n, m, L))$.*

Analogously to Theorem 1, we want to use this algorithm to find extreme non-dominated points of MOCO problems. Instead of constructing the corresponding lexicographic LP, which is equivalent to solving a lexicographic objective function over the set of feasible points of problem (1), we can solve the lexicographic version of the MOCO problem. Hence, exchanging the lexicographic LP oracle by an algorithm which computes a lexicographic optimal solution of the MOCO problem suffices to prove Theorem 2.

5   Computational Study

We investigated the practical aspects of the lexicographic dual Benson algorithm. To achieve this, we conducted a comparison with the method from [15], which is to the best of our knowledge the only practical method to compute the extreme points of MOCO problems with four and more objectives. Both implementations were tested on instances of the multiobjective assignment (or minimum weight bipartite matching) (MO-A) problem. The MO-A problem is often used as a benchmark problem in the biobjective case, but is also used in the computational studies of the approaches existing for three and four objectives [15, 17].

Table 1. Computational results on multiobjective assignment instances with d objectives and n resources. Instances with an * were taken from [15].

| | | Lexicographic Dual Benson | | | | OK10 implementation | | | | |
| | | Running Time [s] | | $|\mathcal{Y}_X|$ | | Running Time [s] | | $|\mathcal{Y}_X|$ | | no. solved |
d	n	Median	MAD	Median	MAD	Median	MAD	Median	MAD	#
3	10*	0.01	0.002	31.5	4.448	0.03	0.010	31.5	4.448	20
	20*	**0.11**	0.014	150.5	17.050	6.31	1.491	150.5	17.050	20
	30*	**0.70**	0.120	368.5	51.150	160.24	57.858	368.5	51.150	20
	40	**2.57**	0.195	709.0	51.150	1660.63	542.646	709.0	51.150	20
	80	67.22	4.442	2819.0	185.325	—	—	—	—	0 (0/20)
	150	1350.28	87.874	9626.0	374.357	—	—	—	—	0 (20/0)
4	10*	**0.06**	0.019	102.5	26.687	3.29	2.447	102.5	26.687	20
	15	0.33	0.079	453.5	97.852	(253.35)	(105.681)	(347.0)	(14.826)	7 (13/0)
	30	12.97	1.824	3646.0	444.039	—	—	—	—	0 (20/0)
	70	1978.68	269.907	48667.0	5777.692	—	—	—	—	0 (20/0)
5	8	0.22	0.131	125.5	34.100	(29.91)	(36.378)	(124.0)	(31.876)	18 (2/0)
	14	33.30	16.416	1228.0	313.570	—	—	—	—	0 (20/0)
	20	1055.82	252.966	5052.5	699.787	—	—	—	—	0 (20/0)
6	6	0.16	0.094	75.0	20.015	(39.01)	(50.977)	(73.0)	(17.791)	18 (2/0)
	8	1.73	1.096	228.0	54.856	(159.25)	(45.658)	(164.5)	(20.015)	2 (18/0)
	12	213.95	159.990	1798.0	549.303	—	—	—	—	0 (20/0)

Though, the algorithm in [15], to which we will refer to as OK10 algorithm, is not easily implemented efficiently. Instead of using points with large encoding length on the axes, we use projective points at infinity to reduce the numerical inconsistencies which occurred in the experimental setting in [15]. Nevertheless, the implementation still misses some extreme nondominated points, but does find more than the original implementation. To compute optimal assignment solutions in the OK10 algorithm, we use an implementation of the Hungarian method.

The experiments were performed on an Intel Core i7-3770, 3.4 GHz and 16 GB of memory running Ubuntu Linux 12.04. The algorithms were implemented in C++ using double precision arithmetic and compiled using LLVM 3.3. To compute the extreme points of the intermediate polyhedra, we implemented a version of the Double Description method with full adjacency information in the case of $d \in \{3, 4\}$ and for $d > 4$, we used the CDD library [9]. To find lexicographic minimal assignments for the lexicographic dual Benson algorithm, we implemented a lexicographic version of the Hungarian method.

The computational study for computing the set of all extreme nondominated points of MO-A instances in [15] uses a series of 20 randomly generated instances. The sizes of the instances vary from 10 up to 30 resources and the integer objective function coefficients were uniformly drawn in $[1, 20]$. For our experiments, we have taken the instances from [15] for $d \in \{3, 4\}$ and additionally generated similar instances with more resources and objectives.[1] We enforced two kinds of limits on the computational experiments. That is, a memory limit of 16 GB and a computation time limit of one hour should not be exceeded.

---

[1] All instances are available at https://ls11-www.cs.tu-dortmund.de/staff/boekler/moco-instances

Table 2. Comparison of the dual Benson implementation with and without lexicographic oracle on multiobjective assignment instances with $d = 5$

	Running Times $\mathcal{T}_{\text{lex}}/\mathcal{T}$						Points found lex / no lex	
	Hungarian algorithm		VE		Total			
n	Median	MAD	Median	MAD	Median	MAD	Mean	σ
20	1.114	0.0301	1.000	0.0091	1.002	0.0086	1.000	0.0003
22	1.118	0.0221	1.000	0.0046	1.000	0.0045	1.000	0.0002
24	1.130	0.0110	0.999	0.0055	1.000	0.0055	1.000	<0.0001
26	1.126	0.0170	0.999	0.0044	0.999	0.0045	1.000	<0.0001

In Table 1 we can see some selected results including the median and median absolute deviation (MAD) of the running times and numbers of extreme points found. We observe that the lexicographic dual Benson implementation is able to solve all instances in the given limits. The implementation of the OK10 algorithm is only able to solve very small instances with three and four objectives. In all other cases there are instance classes where the implementation can not solve all of the instances, prohibiting a statistical analysis. Nevertheless, we give the median and MAD in parentheses if the OK10 implementation was not able to solve all instances.

In the last column of Table 1, we can see the number of instances the OK10 implementation was able to solve. In parentheses we see the number of instances that could not be solved due to the memory and time limit, respectively. We see that in most cases memory was the limiting factor. This is not surprising, since for every new point d new states are introduced and many survive the pruning steps. In the cases where the OK10 implementation is able to solve all instances, we see that the lexicographic dual Benson implementation is up to a factor of 640 times faster.

In the second set of experiments, we compare the practical performance of the dual Benson algorithm when using our theoretical improvements from Sec. 4 to the original variant. The same instances as in the previous experiments were used; we present the experiments with five objectives. Table 2 displays the medians and MADs of the quotients of the lexicographic variant over the non-lexicographic variant. The table shows these statistics for the cumulated running time of the Hungarian algorithm, the vertex enumeration (VE) and the total time. In addition, the last column of Table 2 displays the mean and standard deviation of the quotient of the number of points found by both algorithms. Median and MAD are in all cases 1 and 0, respectively.

We observe that the running times are very similar. The quotients of the total running time medians are very close to 1. On one hand, the vertex enumeration is only slightly faster when using a lexicographic oracle. On the other hand, the cumulated time of the lexicographic oracle is always slower than the time of the original Hungarian method. Of course, the vertex enumeration dominates the total running time, but we also observe that it does not happen too often that redundant inequalities are found.

We can also observe that the medians of the total running time quotients seem to shrink when increasing the number of ressources. In order to observe if this trend continues, we need to test much larger instances which is not possible with the current implementation, because of running times of already more than 12 hours on instances with 26 ressources.

References

1. Agarwal, P.K., Eppstein, D., Guibas, L.J., Henzinger, M.R.: Parametric and kinetic minimum spanning trees. In: IEEE FoCS, pp. 596–605 (1998)
2. Aissi, H., Mahjoub, A.R., McCormick, S.T., Queyranne, M.: A strongly polynomial time algorithm for multicriteria global minimum cuts. In: Lee, J., Vygen, J. (eds.) IPCO 2014. LNCS, vol. 8494, pp. 25–36. Springer, Heidelberg (2014)
3. Aneja, Y.P., Nair, K.P.K.: Bicriteria transportation problem. Management Science 25(1), 73–78 (1979)
4. Avis, D., Fukuda, K.: A pivoting algorithm for convex hulls and vertex enumeration of arrangements of polyhedra. Discrete and Computational Geometry 8(1), 295–313 (1992)
5. Brunsch, T., Röglin, H.: Improved smoothed analysis of multiobjective optimization. In: ACM SToC, pp. 407–426 (2012)
6. Chazelle, B.: An optimal convex hull algorithm in any fixed dimension. Discrete Computational Geometry 10, 377–409 (1993)
7. Ehrgott, M., Gandibleux, X.: A survey and annotated bibliography of multiobjective combinatorial optimization. OR Spektrum 22, 425–460 (2000)
8. Ehrgott, M., Löhne, A., Shao, L.: A dual variant of Benson's "outer approximation algorithm" for multiple objective linear programming. Journal of Global Optimization 52, 757–778 (2012)
9. Fukuda, K., Prodon, A.: Double description method revisited. In: Deza, M., Manoussakis, I., Euler, R. (eds.) CCS 1995. LNCS, vol. 1120, pp. 91–111. Springer, Heidelberg (1996)
10. Ganley, J.L., Golin, M.J., Salowe, J.S.: The multi-weighted spanning tree problem. In: Li, M., Du, D.-Z. (eds.) COCOON 1995. LNCS, vol. 959, pp. 141–150. Springer, Heidelberg (1995)
11. Hamel, A.H., Löhne, A., Rudloff, B.: Benson type algorithms for linear vector optimization and applications. arXiv:1302.2415 [math.OC] (July 2013)
12. Heyde, F., Löhne, A.: Geometric duality in multiple objective linear programming. SIAM Journal of Optimization 19(2), 836–845 (2008)
13. Johnson, D.S., Yannakakis, M., Papadimitriou, C.H.: On generating all maximal independent sets. Information Processing Letters 27, 119–123 (1988)
14. Okamoto, Y., Uno, T.: A polynomial-time-delay and polynomial-space algorithm for enumeration problems in multi-criteria optimization. In: Tokuyama, T. (ed.) ISAAC 2007. LNCS, vol. 4835, pp. 609–620. Springer, Heidelberg (2007)
15. Özpeynirci, Ö., Köksalan, M.: An exact algorithm for finding extreme supported nondominated points of multiobjective mixed integer programs. Management Science 56(12), 2302–2315 (2010)
16. Papadimitriou, C.H., Yannakakis, M.: On the approximability of trade-offs and optimal access of web sources. In: IEEE FoCS, pp. 86–92 (2000)
17. Przybylski, A., Gandibleux, X., Ehrgott, M.: A recursive algorithm for finding all nondominated extreme points in the outcome set of a multiobjective integer programme. INFORMS Journal on Computing 22(3), 371–386 (2010)
18. Seidel, R.: The upper bound theorem for polytopes: an easy proof of its asymptotic version. Computational Geometry 5, 115–116 (1995)

Self-Adjusting Binary Search Trees:
What Makes Them Tick?

Parinya Chalermsook[1], Mayank Goswami[1], László Kozma[2],
Kurt Mehlhorn[1], and Thatchaphol Saranurak[3,*]

[1] Max-Planck Institute for Informatics, Saarbrücken, 66123, Germany
[2] Department of Computer Science, Saarland University, Saarbrücken, 66123, Germany
[3] KTH Royal Institute of Technology, Stockholm, 11428, Sweden

Abstract. Splay trees (Sleator and Tarjan [11]) satisfy the so-called *access lemma*. Many of the nice properties of splay trees follow from it. *What makes self-adjusting binary search trees (BSTs) satisfy the access lemma?* After each access, self-adjusting BSTs replace the search path by a tree on the same set of nodes (the after-tree). We identify two simple combinatorial properties of the search path and the after-tree that imply the access lemma. Our main result

(i) implies the access lemma for *all* minimally self-adjusting BST algorithms for which it was known to hold: splay trees and their generalization to the class of *local algorithms* (Subramanian [12], Georgakopoulos and Mc-Clurkin [7]), as well as Greedy BST, introduced by Demaine et al. [5] and shown to satisfy the access lemma by Fox [6],

(ii) implies that BST algorithms based on "strict" depth-halving satisfy the access lemma, addressing an open question that was raised several times since 1985, and

(iii) yields an extremely short proof for the $O(\log n \log \log n)$ amortized access cost for the path-balance heuristic (proposed by Sleator), matching the best known bound (Balasubramanian and Raman [2]) to a lower-order factor.

One of our combinatorial properties is *locality*. We show that any BST-algorithm that satisfies the access lemma via the sum-of-log (SOL) potential is necessarily local. The other property states that the sum of the number of leaves of the after-tree plus the number of side alternations in the search path must be at least a constant fraction of the length of the search path. We show that a weak form of this property is necessary for sequential access to be linear.

1 Introduction

The binary search tree (BST) is a fundamental data structure for the dictionary problem. Self-adjusting BSTs rearrange the tree in response to data accesses, and are thus able to adapt to the distribution of queries. We consider the class of *minimally self-adjusting* BSTs: algorithms that rearrange only the search path during each access and make the accessed element the root of the tree. Let s be the element accessed and let P be the search path to s. Such an algorithm can be seen as a mapping from the search path P

---

* Work done while at Saarland University.

N. Bansal and I. Finocchi (Eds.): ESA 2015, LNCS 9294, pp. 300–312, 2015.
DOI: 10.1007/978-3-662-48350-3_26

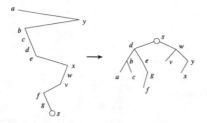

Fig. 1. The search path to s is shown on the left, and the after-tree is shown on the right. The search path consists of 12 nodes and contains four edges that connect nodes on different sides of s ($z = 4$ in the language of Theorem 1). The after-tree has five leaves. The left-depth of a in the after-tree is three (the path from the root to a goes left three times) and the right-depth of y is two. The set $\{a, c, f, v, y\}$ is subtree-disjoint. The sets $\{d, e, g\}$, $\{b, f\}$, $\{x, y\}$, $\{w\}$ are monotone.

(called "before-path" in the sequel) to a tree A with root s on the same set of nodes (called "after-tree" in the sequel). Observe that all subtrees that are disjoint from the before-path can be reattached to the after-tree in a unique way governed by the ordering of the elements. In the BST model, the cost of the access plus the cost of rearranging is $|P|$, see Figure 1 for an example.

Let T be a binary search tree on $[n]$. Let $w : [n] \to \mathbb{R}_{>0}$ be a positive weight function, and for any set $S \subseteq [n]$, let $w(S) = \sum_{a \in S} w(a)$. Sleator and Tarjan defined the sum-of-log (SOL) potential function $\Phi_T = \sum_{a \in [n]} \log w(T_a)$, where T_a is the subtree of T rooted at a. We say that an algorithm $\mathcal{A}$ satisfies the *access lemma (via the SOL potential function)* if for all T' that can be obtained as a rearrangement done by algorithm $\mathcal{A}$ after some element s is accessed, we have

$$|P| \leq \Phi_T - \Phi_{T'} + O(1 + \log \frac{W}{w(s)}),$$

where P is the search path when accessing s in T and $W = w(T)$. The access lemma is known to hold for the splay trees of Sleator and Tarjan [11], for their generalizations to *local algorithms* by Subramanian [12] and Georgakopoulos and McClurkin [7], as well as for Greedy BST, an online algorithm introduced by Demaine et al. [5] and shown to satisfy the access lemma by Fox [6]. For minimally self-adjusting BSTs, the access lemma implies *logarithmic amortized cost*, *static optimality*, and the *static finger* and *working set* properties.

Theorem 1. *Let $\mathcal{A}$ be a minimally self-adjusting BST algorithm. If (i) the number of leaves of the after-tree is $\Omega(|P| - z)$ where P is the search path and z is the number of "side alternations[1]" in P and (ii) for any element $t > s$ (resp. $t < s$), the right-depth of t (left-depth of t) in the after-tree is $O(1)$, then $\mathcal{A}$ satisfies the access lemma.*

Note that the conditions in Theorem 1 are purely combinatorial conditions on the before-paths and after-trees. In particular, the potential function is completely hidden. The theorem directly implies the access lemma for all BST algorithms mentioned above and some new ones.

Corollary 2. *The following BST algorithms satisfy the access lemma: (i) Splay tree, as well as its generalizations to local algorithms (ii) Greedy BST, and (iii) new heuristics based on "strict" depth-halving.*

---

[1] z is the number of edges on the search path connecting nodes on different sides of s. The right-depth of a node is the number of right-going edges on the path from the root to the node.

The third part of the corollary addresses an open question raised by several authors [12,2,7] about whether some form of depth reduction is sufficient to guarantee the access lemma. We show that a strict depth-halving suffices.

For the first part, we formulate a global view of splay trees. We find this new description intuitive and of independent interest. The proof of (i) is only a few lines.

We also prove a partial converse of Theorem 1.

Theorem 3 (Partial Converse). *If a BST algorithm satisfies the access lemma via the SOL-potential function, the after-trees must satisfy condition (ii) of Theorem 1.*

We call a BST algorithm *local* if the transformation from before-path to after-tree can be performed in a bottom-up traversal of the path with a buffer of constant size. Nodes outside the buffer are already arranged into subtrees of the after-tree. We use Theorem 3 to show that BST-algorithms satisfying the access lemma (via the SOL-potential) are necessarily local.

Theorem 4 (Characterization Theorem). *If a minimally self-adjusting BST algorithm satisfies the access lemma via the SOL-potential, then it is local.*

The theorem clarifies, why the access lemma was shown only for local BST algorithms.

In the following, we introduce our main technical tools: subtree-disjoint and monotone sets in § 2, and zigzag sets in § 3. Bounding the potential change over these sets leads to the proof of Theorem 1 in § 3. Corollary 2(i) is also proved in § 3. Corollary 2(iii) is the subject of § 4. In § 5.1 we show that condition (ii) of Theorem 1 is necessary (Theorem 3), and in § 5.2 we argue that a weaker form of condition (i) must also be fulfilled by any reasonably efficient algorithm. We prove Theorem 4 in § 6. For lack of space, in this version of the paper we omit some of the proofs (e.g. Corollary 2(ii)). We refer the reader to the preprint version [3] for full details.

Notation: We use T_a or $T(a)$ to denote the subtree of T rooted at a. We use the same notation to denote the set of elements stored in the subtree. The set of elements stored in a subtree is an interval of elements. If c and d are the smallest and largest elements in $T(a)$, we write $T(a) = [c, d]$. We also use open and half-open intervals to denote subsets of $[n]$, for example $[3, 7)$ is equal to $\{3, 4, 5, 6\}$. We frequently write Φ instead of Φ_T and Φ' instead of $\Phi_{T'}$.

2 Disjoint and Monotone Sets

Let $\mathcal{A}$ be any BST algorithm. Consider an access to s and let T and T' be the search trees before and after the access. The main task in proving the access lemma is to relate the potential difference $\Phi_T - \Phi_{T'}$ to the length of the search path. For our arguments, it is convenient to split the potential into parts that we can argue about separately. For a subset X of the nodes, define a partial potential on X as $\Phi_T(X) = \sum_{a \in X} \log w(T(a))$.

We start with the observation that the potential change is determined only by the nodes on the search path and that we can argue about disjoint sets of nodes separately.

Proposition 5. *Let P be the search path to s. For $a \notin P$, $T(a) = T'(a)$. Therefore, $\Phi_T - \Phi_{T'} = \Phi_T(P) - \Phi_{T'}(P)$. Let $X = \dot{\bigcup}_{i=1}^{k} X_i$ where the sets X_i are pairwise disjoint. Then $\Phi_T(X) - \Phi_{T'}(X) = \sum_{i=1}^{k}(\Phi_T(X_i) - \Phi_{T'}(X_i))$.*

We introduce three kinds of sets of nodes, namely subtree-disjoint, monotone, and zigzag sets, and derive bounds for the potential change for each one of them. A subset X of the search path is *subtree-disjoint* if $T'(a) \cap T'(a') = \emptyset$ for all pairs $a \neq a' \in X$; remark that subtree-disjointness is defined w.r.t. the subtrees after the access. We bound the change of partial potential for subtree-disjoint sets. The proof of the following lemma was inspired by the proof of the access lemma for Greedy BST by Fox [6].

Lemma 6. *Let X be a subtree-disjoint set of nodes. Then*

$$|X| \leq 2 + 8 \cdot \log \frac{W}{w(T(s))} + \Phi_T(X) - \Phi_{T'}(X).$$

Proof. We consider the nodes smaller than s and greater or equal to s separately, i.e. $X = X_{<s} \dot{\cup} X_{\geq s}$. We show $|X_{\geq s}| \leq 1 + \Phi_T(X_{\geq s}) - \Phi_{T'}(X_{\geq s}) + 4 \log \frac{W}{w(T(s))}$, and the same holds for $X_{<s}$. We only give the proof for $X_{\geq s}$.

Denote $X_{\geq s}$ by $Y = \{a_0, a_1, \ldots, a_q\}$ where $s \leq a_0 < \ldots < a_q$. Before the access, s is a descendant of a_0, a_0 is a descendant of a_1, and so on. Let $T(a_0) = [c, d]$. Then $[s, a_0] \subseteq [c, d]$ and $d < a_1$. Let $w_0 = w(T(a_0))$. For $j \geq 0$, define σ_j as the largest index ℓ such that $w([c, a_\ell]) \leq 2^j w_0$. Then $\sigma_0 = 0$ since weights are positive and $[c, d]$ is a proper subset of $[c, a_1]$. The set $\{\sigma_0, \ldots\}$ contains at most $\lceil \log(W/w_0) \rceil$ distinct elements. It contains 0 and q.

Now we upper bound the number of i with $\sigma_j \leq i < \sigma_{j+1}$. We call such an element a_i *heavy* if $w(T'(a_i)) > 2^{j-1} w_0$. There can be at most 3 heavy elements as otherwise $w([c, a_{j+1}]) \geq \sum_{\sigma_j \leq k < \sigma_{j+1}} w(T'(a_k)) > 4 \cdot 2^{j-1} w_0$, a contradiction.

Next we count the number of light (= non-heavy) elements. For each such light element a_i, we have $w(T'(a_i)) \leq 2^{j-1} w_0$. We also have $w(T(a_{i+1})) \geq w([c, a_{i+1}]) > w([c, a_{\sigma_j}])$ and thus $w(T(a_{i+1})) > 2^j w_0$ by the definition of σ_j. Thus the ratio $r_i = w(T(a_{i+1}))/w(T'(a_i)) \geq 2$ whenever a_i is a light element. Moreover, for any $i = 0, \ldots, q-1$ (for which a_i is not necessarily light), we have $r_i \geq 1$. Thus,

$$2^{\text{number of light elements}} \leq \prod_{0 \leq i \leq q-1} r_i = \left(\prod_{0 \leq i \leq q} \frac{w(T(a_i))}{w(T'(a_i))} \right) \cdot \frac{w(T'(a_q))}{w_0}.$$

So the number of light elements is at most $\Phi_T(Y) - \Phi_{T'}(Y) + \log(W/w_0)$. Putting the bounds together, we obtain, writing L for $\log(W/w_0)$:

$$|Y| \leq 1 + 3(\lceil L \rceil - 1) + \Phi_T(Y) - \Phi_{T'}(Y) + L \leq 1 + 4L + \Phi_T(Y) - \Phi_{T'}(Y).$$

$\square$

Now we proceed to analyze our second type of subsets, that we call *monotone sets*. A subset X of the search path is *monotone* if all elements in X are larger (smaller) than s and have the same right-depth (left-depth) in the after-tree.

Lemma 7. *Assume $s < a < b$ and that a is a proper descendant of b in P. If $\{a,b\}$ is monotone, $T'(a) \subseteq T(b)$.*

Proof: Clearly $[s, b] \subseteq T(b)$. The smallest item in $T'(a)$ is larger than s, and, since a and b have the same right-depth, b is larger than all elements in $T'(a)$.   □

Lemma 8. *Let X be a monotone set of nodes. Then*

$$\Phi(X) - \Phi'(X) + \log \frac{W}{w(s)} \geq 0.$$

Proof: We order the elements in $X = \{a_1, \ldots, a_q\}$ such that a_i is a proper descendant of a_{i+1} in the search path for all i. Then $T'(a_i) \subseteq T(a_{i+1})$ by monotonicity, and hence

$$\Phi(X) - \Phi'(X) = \log \frac{\prod_{a \in X} w(T(a))}{\prod_{a \in X} w(T'(a))} = \log \frac{w(T(a_1))}{w(T'(a_q))} + \sum_{i=1}^{q-1} \log \frac{w(T(a_{i+1}))}{w(T'(a_i))}.$$

The second sum is nonnegative. Thus $\Phi(X) - \Phi'(X) \geq \log \frac{w(T(a_1))}{w(T'(a_q))} \geq \log \frac{w(s)}{W}$.   □

Theorem 9. *Suppose that, for every access to an element s, we can partition the elements on the search path P into at most k subtree-disjoint sets D_1 to D_k and at most ℓ monotone sets M_1 to M_ℓ. Then*

$$\sum_{i \leq k} |D_i| \leq \Phi_T(S) - \Phi_{T'}(S) + 2k + (8k + \ell) \log \frac{W}{w(s)}.$$

The proof of Theorem 9 follows immediately from Lemma 6 and 8. We next give some easy applications.

Path-Balance: The path-balance algorithm maps the search path P to a balanced BST of depth $c = \lceil \log_2(1 + |P|) \rceil$ rooted at s. Then

Lemma 10. $|P| \leq \Phi(P) - \Phi'(P) + O((1 + \log|P|)(1 + \log(W/w(s))))$.

Proof: We decompose P into sets P_0 to P_c, where P_k contains the nodes of depth k in the after-tree. Each P_k is subtree-disjoint. An application of Theorem 9 completes the proof.   □

Theorem 11. *Path-Balance has amortized cost at most $O(\log n \log \log n)$.*

Proof: We choose the uniform weight function: $w(a) = 1$ for all a. Let c_i be the cost of the i-th access, $1 \leq i \leq m$, and let $C = \sum_{1 \leq i \leq m} c_i$ be the total cost of the accesses. Note that $\prod_i c_i \leq (C/m)^m$. The potential of a tree with n items is at most $n \log n$. Thus $C \leq n \log n + \sum_{1 \leq i \leq m} O((1 + \log c_i)(1 + \log n)) = O((n + m) \log n) + O(m \log n) \cdot \log(C/m)$ by Lemma 10. Assume $C = K(n + m) \log n$ for some K. Then $K = O(1) + O(1) \cdot \log(K \log n)$ and hence $K = O(\log \log n)$.   □

Greedy BST: The Greedy BST algorithm was introduced by Demaine et al. [5]. It is an online version of the offline greedy algorithm proposed independently by Lucas and Munro [9,8]. The definition of Greedy BST requires a geometric view of BSTs. Our notions of subtree-disjoint and monotone sets translate naturally into geometry, and this allows us to derive the following theorem.

Theorem 12. *Greedy BST satisfies the (geometric) access lemma.*

The geometric view of BSTs and the proof of Theorem 12 are omitted here. We refer to the preprint version [3] for details. We remark that once the correspondences to geometric view are defined, the proof of Theorem 12 is almost immediate.

3 Zigzag Sets

Let s be the accessed element and let $a_1, \ldots, a_{|P|-1}$ be the reversed search path without s. For each i, define the set $Z_i = \{a_i, a_{i+1}\}$ if a_i and a_{i+1} lie on different sides of s, and let $Z_i = \emptyset$ otherwise. The zigzag set Z_P is defined as $Z_P = \bigcup_i Z_i$. In words, the number of non-empty sets Z_i is exactly the number of "side alternations" in the search path, and the cardinality of Z_P is the number of elements involved in such alternations.

Rotate to Root: We first analyze the rotate-to-root algorithm (Allen, Munro [1]), that brings the accessed element s to the root and arranges the elements smaller (larger) than s so the ancestor relationship is maintained, see Figure 2 for an illustration.

Lemma 13. $|Z| \leq \Phi(Z_P) - \Phi'(Z_P) + O(1 + \log \frac{W}{w(T(s))})$.

Proof: Because s is made the root and ancestor relationships are preserved otherwise, $T'(a) = T(a) \cap (-\infty, s)$ if $a < s$ and $T'(a) = T(a) \cap (s, \infty)$ if $a > s$. We first deal with a single side alternation.

Claim. $2 \leq \Phi(Z_i) - \Phi'(Z_i) + \log \frac{w(T(a_{i+1}))}{w(T(a_i))}$.

Proof: This proof is essentially the proof of the zig-zag step for splay trees. We give the proof for the case where $a_i > s$ and $a_{i+1} < s$; the other case is symmetric. Let a' be the left ancestor of a_{i+1} in P and let a'' be the right ancestor of a_i in P. If these elements do not exist, they are $-\infty$ and $+\infty$, respectively. Let $W_1 = w((a', 0))$, $W_2 = w((0, a''))$, and $W' = w((a_{i+1}, 0))$. In T, we have $w(T(a_i)) = W' + w(s) + W_2$ and $w(T(a_{i+1})) = W_1 + w(s) + W_2$, and in T', we have $w(T'(a_i)) = W_2$ and $w(T'(a_{i+1})) = W_1$.

Thus $\Phi(Z_i) - \Phi'(Z_i) + \log \frac{W_1 + w(s) + W_2}{W' + w(s) + W_2} \geq \log(W_1 + w(s) + W_2) - \log W_1 + \log(W_2 + w(s) + W') - \log W_2 + \log \frac{W_1 + w(s) + W_2}{W' + w(s) + W_2} \geq 2\log(W_1 + W_2) - \log W_1 - \log W_2 \geq 2$, since $(W_1 + W_2)^2 \geq 4W_1 W_2$ for all positive numbers W_1 and W_2. □

Let Z_{even} (Z_{odd}) be the union of the Z_i with even (odd) indices. One of the two sets has cardinality at least $|Z_P|/2$. Assume that it is the former; the other case is symmetric. We sum the statement of the claim over all i in Z_{even} and obtain

$$\sum_{i \in Z_{\text{even}}} \left(\Phi(Z_i) - \Phi'(Z_i) + \log \frac{w(T(a_{i+1}))}{w(T(a_i))} \right) \geq 2 |Z_{\text{even}}| \geq |Z_P|.$$

The elements in $Z_P \setminus Z_{even}$ form two monotone sets and hence $\Phi(Z_P \setminus Z_{even}) - \Phi'(Z_P \setminus Z_{even}) + 2\log(W/w(s)) \geq 0$. This completes the proof. $\square$

The following theorem combines all three tools we have introduced: subtree-disjoint, monotone, and zigzag sets.

Theorem 14. *Suppose that, for every access we can partition $P \setminus s$ into at most k subtree-disjoint sets D_1 to D_k and at most ℓ monotone sets M_1 to M_ℓ. Then*

$$\sum_{i \leq k} |D_i| + |Z_P| \leq \Phi(P) - \Phi'(P) + O((k + \ell)(1 + \log \frac{W}{w(s)})).$$

Proof: We view the transformation as a two-step process, i.e., we first rotate s to the root and then transform the left and right subtrees of s. Let Φ'' be the potential of the intermediate tree. By Lemma 13, $|Z_P| \leq \Phi(P) - \Phi''(P) + O(1 + \log \frac{W}{w(T(s))})$. By Theorem 9, $\sum_{i \leq k} |D_i| \leq \Phi''(P) - \Phi'(P) + O((k + \ell)(1 + \log \frac{W}{w(T(s))}))$. $\square$

We next derive an easy to apply corollary from this theorem. For the statement, we need the following proposition that follows directly from the definition of monotone set.

Proposition 15. *Let S be a subset of the search path consisting only of elements larger than s. Then S can be decomposed into ℓ monotone sets if and only if the elements of S have only ℓ different right-depths in the after-tree.*

Theorem 16 (Restatement of Theorem 1). *Suppose the BST algorithm $\mathcal{A}$ rearranges a search path P that contains z side alternations, into a tree A such that (i) s, the element accessed, is the root of A, (ii) the number of leaves of A is $\Omega(|P| - z)$, (iii) for every element x larger (smaller) than s, the right-depth (left-depth) of x in A is bounded by a constant. Then $\mathcal{A}$ satisfies the access lemma.*

Proof: Let B be the set of leaves of T and let $b = |B|$. By assumption (ii), there is a positive constant c such that $b \geq (|T| - z)/c$. Then $|T| \leq cb + z$. We decompose $P \setminus s$ into B and ℓ monotone sets. By assumption (iii), $\ell = O(1)$. An application of Theorem 14 with $k = 1$ and $\ell = O(1)$ completes the proof. $\square$

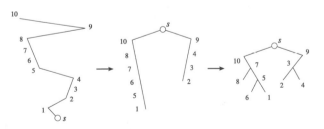

Fig. 2. A global view of splay trees. The transformation from left to the middle illustrates rotate-to-root. The transformation from the left to the right illustrates splay trees.

Splay: Splay extends rotate-to-root: Let $s = v_0, v_1, \dots v_k$ be the reversed search path. We view splaying as a two step process, see Figure 2. We first make s the root and split the search path into two paths, the path of elements smaller than s and the path of elements larger than s. If v_{2i+1} and v_{2i+2} are on the same side of s, we rotate them, i.e., we remove v_{2i+2} from the path and make it a child of v_{2i+1}.

Proposition 17. *The above description of splay is equivalent to the Sleator-Tarjan description.*

Theorem 18. *Splay satisfies the access lemma.*

Proof: There are $|P|/2 - 1$ odd-even pairs. For each pair, if there is no side change, then splay creates a new leaf in the after-tree. Thus

$$\text{\# of leaves} \geq |P|/2 - 1 - \text{\# of side changes.}$$

Since right-depth (left-depth) of elements in the after-tree of splay is at most 2, an application of Theorem 16 finishes the proof.   □

4 New Heuristics: Depth Reduction

Already Sleator and Tarjan [11] formulated the belief that *depth-halving* is the property that makes splaying efficient, i.e. the fact that every element on the access path reduces its distance to the root by a factor of approximately two. Later authors [12,2,7] raised the question, whether a suitable *global* depth-reduction property is sufficient to guarantee the access lemma. Based on Theorem 16, we show that a strict form of depth-halving suffices to guarantee the access lemma.

Let x and y be two arbitrary nodes on the search path. If y is an ancestor of x in the search path, but not in the after-tree, then we say that x has *lost* the ancestor y, and y has lost the descendant x. Similarly we define *gaining* an ancestor or a descendant. We stress that only nodes on the search path (resp. the after-tree) are counted as descendants, and not the nodes of the pendent trees. Let $d(x)$ denote the depth (number of ancestors) of x in the search path. We give a sufficient condition for a good heuristic, stated below. The proof is omitted.

Theorem 19. *Let $\mathcal{A}$ be a minimally self-adjusting BST algorithm that satisfies the following conditions: (i) Every node x on the search path loses at least $(\frac{1}{2} + \epsilon) \cdot d(x) - c$ ancestors, for fixed constants $\epsilon > 0, c > 0$, and (ii) every node on the search path, except the accessed element, gains at most d new descendants, for a fixed constant $d > 0$. Then $\mathcal{A}$ satisfies the access lemma.*

We remark that in general, splay trees do not satisfy condition (i) of Theorem 19. One may ask how tight are the conditions of Theorem 19. If we relax the constant in condition (i) from $(\frac{1}{2} + \epsilon)$ to $\frac{1}{2}$, the conditions of Theorem 16 are no longer implied. There exist rearrangements in which every node loses a $\frac{1}{2}$-fraction of its ancestors, gains at most two ancestors or descendants, yet both the number of side alternations and the number of leaves created are $O(\sqrt{|P|})$, where P is the before-path (details can be found in [3]). If we further relax the ratio to $(\frac{1}{2} - \epsilon)$, we can construct an example where the number of alternations and the number of leaves created are only $O(\log |P|/\epsilon)$.

Allowing more gained descendants and limiting instead the number of gained ancestors is also beyond the strength of Theorem 16. It is possible to construct an example [3] in which every node loses an $(1 - o(1))$-fraction of ancestors, yet the number of leaves created is only $O(\sqrt{|P|})$ (while having no alternations in the before-path).

Finally, we observe that depth-reduction alone is likely not sufficient: one can restructure the access path in such a way that every node reduces its depth by a constant factor, yet the resulting after-tree has an anti-monotone path of linear size [3]. Based on Theorem 20, this means that if such a restructuring were to satisfy the access lemma in its full generality, the SOL potential would not be able to show it.

5   Necessary Conditions

5.1   Necessity of $O(1)$ Monotone Sets

In this section we show that condition (ii) of Theorem 1 is necessary for any minimally self-adjusting BST algorithm that satisfies the access lemma via the SOL potential function.

Theorem 20. *Consider the transformations from before-path P to after-tree A by algorithm $\mathcal{A}$. If $A \setminus s$ cannot be decomposed into constantly many monotone sets, then $\mathcal{A}$ does not satisfy the access lemma with the SOL potential.*

Proof: We may assume that the right subtree of A cannot be decomposed into constantly many monotone sets. Let $x > s$ be a node of maximum right depth in A. By Lemma 15, we may assume that the right depth is $k = \omega(1)$. Let $a_{i_1}, \ldots, a_{i_k}$ be the elements on the path to x where the right child pointer is used. All these nodes are descendants of x in the before-path P.

We now define a weight assignment to the elements of P and the pendent trees for which the access lemma does not hold with the SOL potential. We assign weight zero to all pendent trees, weight one to all proper descendants of x in P and weight K to all ancestors of x in P. Here K is a big number. The total weight W then lies between K and $|P| K$.

We next bound the potential change. Let $r(a_i) = w(T'(a_i))/w(T(a_i))$ be the ratio of the weight of the subtree rooted at a_i in the after-tree and in the before-path. For any element a_{i_j} at which a right turn occurs, we have $w(T(a_{i_j})) \leq |P|$ and $w(T'(a_{i_j})) \geq K$. So $r(a_{i_j}) \geq K/|P|$. Consider now any other a_i. If it is an ancestor of x in the before-path, then $w(T(a_i)) \leq W$ and $w(T'(a_i)) \geq K$. If it is a descendant of x, then $w(T(a_i)) \leq |P|$ and $w(T'(a_i)) \geq 1$. Thus $r(a_i) \geq 1/|P|$ for every a_i. We conclude

$$\Phi'(T) - \Phi(T) \geq k \cdot \log \frac{K}{|P|} - |P| \log |P|.$$

If $\mathcal{A}$ satisfies the access lemma with the SOL potential function, then we must have $\Phi'(T) - \Phi(T) \leq O(\log \frac{W}{w(s)} - |P|) = O(\log(K |P|))$. However, if K is large enough and $k = \omega(1)$, then $k \cdot \lg \frac{K}{|P|} - |P| \lg |P| \gg O(\log(K |P|))$. $\square$

5.2   Necessity of Many Leaves

In this section we study condition (i) of Theorem 1. We show that some such condition is necessary for an efficient BST algorithm: if a local algorithm consistently creates only few leaves, it cannot satisfy the sequential access theorem, a natural efficiency condition known to hold for several BST algorithms [13,6].

Definition 21. *A self-adjusting BST algorithm $\mathcal{A}$ satisfies the* sequential access theorem *if starting from an arbitrary initial tree T, it can access the elements of T in increasing order with total cost $O(|T|)$.*

Theorem 22. *If for all after-trees A created by algorithm $\mathcal{A}$ executed on T, it holds that (i) A can be decomposed into $O(1)$ monotone sets, and (ii) the number of leaves of A is at most $|T|^{o(1)}$, then $\mathcal{A}$ does not satisfy the sequential access theorem.*

The rest of the section is devoted to the proof of Theorem 22.

Let R be a BST over $[n]$. We call a maximal left-leaning path of R a *wing* of R. More precisely, a wing of R is a set $\{x_1, \ldots, x_k\} \subseteq [n]$, with $x_1 < \cdots < x_k$, and such that x_1 has no left child, x_k is either the root of R, or the right child of its parent, and x_i is the left child of x_{i+1} for all $1 \le i < k$. A wing might consist of a single element. Observe that the wings of R partition $[n]$ in a unique way, and we call the set of wings of R the *wing partition* of R, denoted as $wp(R)$. We define a potential function ϕ over a BST R as follows: $\phi(R) = \sum_{w \in wp(R)} |w| \log(|w|)$.

Let T_0 be a left-leaning path over $[n]$ (i.e. n is the root and 1 is the leaf). Consider a minimally self-adjusting BST algorithm $\mathcal{A}$, accessing elements of $[n]$ in sequential order, starting with T_0 as initial tree. Let T_i denote the BST after accessing element i. Then T_i has i as the root, and the elements yet to be accessed (i.e. $[i+1, n]$) form the right subtree of the root, denoted R_i. To avoid treating T_0 separately, we augment it with a "virtual root" 0. This node plays no role in subsequent accesses, and it only adds a constant one to the overall access cost.

Using the previously defined potential function, we denote $\phi_i = \phi(R_i)$. We make the following easy observations: $\phi_0 = n \log n$, and $\phi_n = 0$.

Next, we look at the change in potential due to the restructuring after accessing element i. Let $P_i = (x_1, x_2, \ldots, x_{n_i})$ be the access path when accessing i in T_{i-1}, and let n_i denote its length, i.e. $x_1 = i-1$, and $x_{n_i} = i$. Observe that the set $P_i' = P_i \setminus \{x_1\}$, is a wing of T_{i-1}.

Let us denote the after-tree resulting from rearranging the path P_i as A_i. Observe that the root of A_i is i, and the left child of i in A_i is $i-1$. We denote the tree $A_i \setminus \{i-1\}$ as A_i', and the tree $A_i' \setminus \{i\}$, i.e. the right subtree of i in A_i, as A_i''.

The crucial observation of the proof is that for an arbitrary wing $w \in wp(T_i)$, the following holds: (i) either w was not changed when accessing i, i.e. $w \in wp(T_{i-1})$, or (ii) w contains a portion of P_i', possibly concatenated with an earlier wing, i.e. there exists some $w' \in wp(A_i')$, such that $w' \subseteq w$. In this case, we denote $\text{ext}(w')$ the *extension* of w' to a wing of $wp(T_i)$, i.e. $\text{ext}(w') = w \setminus w'$, and either $\text{ext}(w') = \emptyset$, or $\text{ext}(w') \in wp(T_{i-1})$.

Now we bound the change in potential $\phi_i - \phi_{i-1}$. Wings that did not change during the restructuring (i.e. those of type (i)) do not contribute to the potential difference. Also note, that i contributes to ϕ_{i-1}, but not to ϕ_i. Thus, we have for $1 \le i \le n$, assuming that $0 \log 0 = 0$, and denoting $f(x) = x \log(x)$:

$$\phi_i - \phi_{i-1} = \sum_{w' \in wp(A_i'')} \left(f(|w'| + |\text{ext}(w')|) - f(|\text{ext}(w')|) \right) - f(n_i - 1).$$

By simple manipulation, for $1 \leq i \leq n$:

$$\phi_i - \phi_{i-1} \geq \sum_{w' \in wp(A_i'')} f(|w'|) - f(n_i - 1).$$

By convexity of f, and observing that $|A_i''| = n_i - 2$, we have

$$\phi_i - \phi_{i-1} \geq |wp(A_i'')| \cdot f\left(\frac{n_i - 2}{|wp(A_i'')|}\right) - f(n_i - 1) = (n_i - 2) \cdot \log \frac{n_i - 2}{|wp(A_i'')|} - f(n_i - 1).$$

Lemma 23. *If R has right-depth m, and k leaves, then $|wp(R)| \leq mk$.*

Proof. For a wing w, let $\ell(w)$ be any leaf in the subtree rooted at the node of maximum depth in the wing. Clearly, for any leaf ℓ there can be at most m wings w with $\ell(w) = \ell$. The claim follows. □

Thus, $|wp(A_i'')| \leq n^{o(1)}$. Summing the potential differences over i, we get $\phi_n - \phi_0 = -n \log n \geq -\sum_{i=1}^{n} n_i \log\left(n^{o(1)}\right) - O(n)$. Denoting the total cost of algorithm $\mathcal{A}$ on the sequential access sequence as C, we obtain $C = \sum_{i=1}^{n} n_i = n \cdot \omega(1)$.

This shows that $\mathcal{A}$ does not satisfy the sequential access theorem.

6   Small Monotonicity-Depth and Local Algorithms

In this section we define a class of minimally self-adjusting BST algorithms that we call *local*. We show that an algorithm is local exactly if all after-trees it creates can be decomposed into constantly many monotone sets. Our definition of local algorithm is inspired by similar definitions by Subramanian [12] and Georgakopoulos and Mc-Clurkin [7]. Our locality criterion subsumes both previous definitions, apart from a technical condition not needed in these works: we require the transformation to bring the accessed element to the root. We require this (rather natural) condition in order to simplify the proofs. We mention that it can be removed at considerable expense in technicalities. Apart from this point, our definition of locality is more general: while existing local algorithms are oblivious to the global structure of the after-tree, our definition of local algorithm allows external global advice, as well as non-determinism.

Consider the before-path P and the after-tree A. A *decomposition* of the transformation $P \to A$ is a sequence of BSTs $(P = Q_0 \xrightarrow{P_0} Q_1 \xrightarrow{P_1} \ldots \xrightarrow{P_{k-1}} Q_k = A)$, such that for all i, the tree Q_{i+1} can be obtained from the tree Q_i, by rearranging a path P_i contained in Q_i into a tree T_i, and linking all the attached subtrees in the unique way given by the element ordering. Clearly, every transformation has such a decomposition, since a sequence of rotations fulfills the requirement. The decomposition is *local* with window-size w, if it satisfies the following conditions:

(i) (start) $s \in P_0$, where s is the accessed element in P,
(ii) (progress) $P_{i+1} \setminus P_i \neq \emptyset$, for all i,
(iii) (overlap) $P_{i+1} \cap P_i \neq \emptyset$, for all i,
(iv) (no-revisit) $(P_i - P_{i+1}) \cap P_j = \emptyset$, for all $j > i + 1$,
(v) (window-size) $|P_i| \leq w$, for some constant $w > 0$.

We call a minimally self-adjusting algorithm $\mathcal{A}$ *local*, if all the before-path $\rightarrow$ after-tree transformations performed by $\mathcal{A}$ have a local decomposition with constant-size window. The following theorem shows that local algorithms are exactly those that respect condition (ii) of Theorem 1 (proof omitted).

Theorem 24. *Let $\mathcal{A}$ be a minimally self-adjusting algorithm. (i) If $\mathcal{A}$ is local with window size w, then all the after-trees created by $\mathcal{A}$ can be partitioned into $2w$ monotone sets. (ii) If all the after-trees created by $\mathcal{A}$ can be partitioned into w monotone sets, then $\mathcal{A}$ is local with window-size w.*

Due to the relationship between monotone sets and locality of algorithms, we have

Theorem 25. *If a minimally self-adjusting BST algorithm $\mathcal{A}$ satisfies the access lemma with the SOL potential, then $\mathcal{A}$ can be made local.*

Open Questions: Does the family of algorithms described by Theorem 16 satisfy other efficiency-properties not captured by the access lemma? Properties studied in the literature include sequential access [13], deque [13,10], dynamic finger [4], or the elusive dynamic optimality [11].

One may ask whether locality is a necessary feature of all efficient BST algorithms. We have shown that some natural heuristics (e.g. path-balance or depth reduction) do not share this property, and thus do not satisfy the access lemma with the (rather natural) sum-of-logs potential function. It remains an open question, whether such "truly nonlocal" heuristics are necessarily bad, or if a different potential function could show that they are good.

Acknowledgement. The authors thank Raimund Seidel for suggesting the study of depth-reducing heuristics and for useful insights about BSTs and splay trees.

References

1. Allen, B., Munro, J.I.: Self-organizing binary search trees. J. ACM 25(4), 526–535 (1978)
2. Balasubramanian, R., Venkatesh, R.: Path balance heuristic for self-adjusting binary search trees. In: Proceedings of FSTTCS, pp. 338–348 (1995)
3. Chalermsook, P., Goswami, M., Kozma, L., Mehlhorn, K., Saranurak, T.: Self-adjusting binary search trees: What makes them tick? CoRR, abs/1503.03105 (2015)
4. Cole, R.: On the dynamic finger conjecture for splay trees. part ii: The proof. SIAM Journal on Computing 30(1), 44–85 (2000)
5. Demaine, E.D., Harmon, D., Iacono, J., Kane, D.M., Patrascu, M.: The geometry of binary search trees. In: SODA 2009, pp. 496–505 (2009)
6. Fox, K.: Upper Bounds for Maximally Greedy Binary Search Trees. In: Dehne, F., Iacono, J., Sack, J.-R. (eds.) WADS 2011. LNCS, vol. 6844, pp. 411–422. Springer, Heidelberg (2011)
7. Georgakopoulos, G.F., McClurkin, D.J.: Generalized template splay: A basic theory and calculus. Comput. J. 47(1), 10–19 (2004)
8. Lucas, J.M.: Canonical forms for competitive binary search tree algorithms. Tech. Rep. DCS-TR-250, Rutgers University (1988)
9. Munro, J.I.: On the competitiveness of linear search. In: Paterson, M. (ed.) ESA 2000. LNCS, vol. 1879, pp. 338–345. Springer, Heidelberg (2000)

10. Pettie, S.: Splay trees, Davenport-Schinzel sequences, and the deque conjecture. In: SODA 2008, pp. 1457–1467 (2008)
11. Sleator, D.D., Tarjan, R.E.: Self-adjusting binary search trees. J. ACM 32(3), 652–686 (1985)
12. Subramanian, A.: An explanation of splaying. J. Algorithms 20(3), 512–525 (1996)
13. Tarjan, R.E.: Sequential access in splay trees takes linear time. Combinatorica 5(4), 367–378 (1985)

On Element-Connectivity Preserving Graph Simplification

Chandra Chekuri[1], Thapanapong Rukkanchanunt[2], and Chao Xu[1]

[1] Dept. of Computer Science, Univ. of Illinois at Urbana-Champaign, Urbana, IL, USA
{chekuri,chaoxu3}@illinois.edu
[2] Chiang Mai University, Chiang Mai, Thailand
thapanapong.r@cmu.ac.th

Abstract. The notion of *element-connectivity* has found several important applications in network design and routing problems. We focus on a reduction step that preserves the element-connectivity [18,4,3], which when applied repeatedly allows one to reduce the original graph to a simpler one. This pre-processing step is a crucial ingredient in several applications. In this paper we revisit this reduction step and provide a new proof via the use of setpairs. Our main contribution is algorithmic results for several basic problems on element-connectivity including the problem of achieving the aforementioned graph simplification. We utilize the underlying submodularity properties of element-connectivity to derive faster algorithms.

Keywords: Element-connectivity, Gomory-Hu tree, reduction, bisubmodular.

1 Introduction

Let $G = (V, E)$ be an undirected simple graph. The edge-connectivity between two distinct nodes u, v, denoted by $\lambda_G(u, v)$, is the maximum number of edge-disjoint paths between u and v in G. The vertex-connectivity between u and v, denoted by $\kappa_G(u, v)$ is the maximum number of internally vertex-disjoint paths between u and v. These two connectivity measures are classical and extensively studied in graph theory, combinatorial optimization and algorithms. The notion of *element-connectivity* is more recent and is defined as follows. Let $T \subseteq V$ be a set of terminals; vertices in $V \setminus T$ are referred to as non-terminals. For any two distinct terminals $u, v \in T$, element-connectivity between u and v is the maximum number of u-v paths in G that are pairwise "element"-disjoint where elements consist of edges and non-terminals. Note that element-disjoint paths need not be disjoint in terminals. We use $\kappa'_G(u, v)$ to denote the element-connectivity between u and v. Via Menger's theorem one can characterize edge, vertex and element connectivity in an equivalent way via cuts. We use $\kappa'_G(T) = \min_{u,v \in T} \kappa'_G(u, v)$ to denote the global element-connectivity, the minimum number of elements whose deletion separates some pair of terminals. See Fig. 1 for example.

Element-connectivity can be seen to generalize edge-connectivity by letting $T = V$. At the same time, element-connectivity is also closely related to vertex-connectivity. If T is an independent set then $\kappa'_G(u, v)$ is the maximum number of paths from u to v that are disjoint in non-terminals. In particular, if T contains exactly two vertices s

© Springer-Verlag Berlin Heidelberg 2015
N. Bansal and I. Finocchi (Eds.): ESA 2015, LNCS 9294, pp. 313–324, 2015.
DOI: 10.1007/978-3-662-48350-3_27

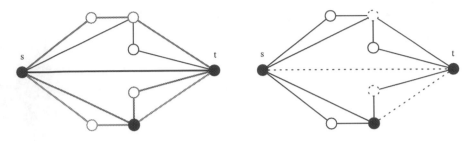

Fig. 1. The black vertices are the terminals. The left image shows 4 element-disjoint st-paths. The right image shows removing 4 elements disconnects s and t. $\kappa'(s,t) = 4$

and t, then $\kappa_G(s,t) = \kappa'_G(s,t)$. Element-connectivity has found several applications in network design, routing and related problems, some of which we will discuss later. Several of these applications rely on an interesting graph reduction operation shown first by Hind and Oellermann [18]. To describe their result we use the notation G/pq to denote the graph obtained from G by contracting the edge pq, and $G - pq$ to denote the graph with edge pq deleted.

Theorem 1 (Hind & Oellermann [18]). *Let $G = (V, E)$ be an undirected graph and $T \subseteq V$ be a terminal-set such that $\kappa'_G(T) \geq k$. Let pq be any edge where $p, q \in V \setminus T$. Then $\kappa'_{G_1}(T) \geq k$ or $\kappa'_{G_2}(T) \geq k$ where $G_1 = G - pq$ and $G_2 = G/pq$.*

Chekuri and Korula generalized the theorem to show that the same reduction operation also preserves the *local* element-connectivity of every pair of terminals.

Theorem 2 (Chekuri & Korula [3]). *Let $G = (V, E)$ be an undirected graph and $T \subseteq V$ be a terminal-set. Let pq be any edge where $p, q \in V \setminus T$ and let $G_1 = G - pq$ and $G_2 = G/pq$. Then one of the following holds: (i) $\forall u, v \in T$, $\kappa'_{G_1}(u, v) = \kappa'_G(u, v)$ (ii) $\forall u, v \in T$, $\kappa'_{G_2}(u, v) = \kappa'_G(u, v)$.*

We refer to the preceding theorem as the reduction lemma following the usage from [3]. By repeatedly applying the reduction lemma, as observed in prior work, we obtain the following corollary.

Corollary 1. *Given a graph $G = (V, E)$ and a terminal set $T \subseteq V$ there is a minor $H = (V', E')$ of G such that (i) $T \subseteq V'$ and (ii) $V' \setminus T$ is an independent set in H and (iii) $\kappa'_H(u, v) = \kappa'_G(u, v)$ for all $u, v \in T$. In particular, if T is an independent set in G then H is a bipartite graph with bipartition $(T, V' \setminus T)$.*

The minor H in the previous corollary is called a *reduced graph* of G. A graph is *reduced* if there are no edges between non-terminals.

Remark 1. A reduced graph $G = (V, E)$ where the terminals T form an independent set can be interpreted as a hypergraph $H = (T, E')$. H contains an edge e_v for every non-terminal v in G, where e_v is the set of neighbors of v in G. Element-connectivity of terminals T in G is equivalent to hypergraph edge-connectivity in H. st-edge connectivity on hypergraphs are defined the same way as graphs : the maximum number of edge disjoint st-paths.

	Min Cut	Min Cut(WHP)	All-pair	All-pair (WHP)	Reduce
λ	$O(m)$[21]	$\tilde{O}(m)$[21]	$\tilde{O}(n^{27/8})$[7]	$\tilde{O}(nm)$[16]	-
κ'	$O(\lvert T\rvert \, \mathrm{MF}(n,m))$	same as all-pair	$O(\lvert T\rvert \, \mathrm{MF}(n,m))$	$\tilde{O}(\lvert T\rvert n^{\omega})$[14],$O(m^{\omega})$[6]	$O(\lvert T\rvert nm)$
κ	$O(n^{7/4}m)$[12]	$\tilde{O}(nm)$[17]	$O(n^{9/2})$[8]	$\tilde{O}(n^{2+\omega})$[14]	-

Fig. 2. The running time for various algorithms for a graph with n vertices and m edges and terminal vertices $\lvert T\rvert$. The row for κ' is our result. $\mathrm{MF}(n,m)$ is the running time for a maximum flow on unit capacity directed graph with n vertices and m edges, which is known to be $O(\sqrt{n}m)$[8]. WHP indicates with high probability bounds for randomized algorithms. ω is the matrix multiplication constant. $\tilde{O}$ notation suppresses poly-logarithmic factors.

Our Results: In this paper we revisit Theorem 2 and Corollary 1 from two related view points. The proofs of Theorems 1 and 2 use elementary arguments on flows and cuts in graphs. We obtain an alternative proof of Theorem 2 using submodularity and super-modularity property of setpairs. Although these properties have been used routinely in network design papers that deal with element-connectivity, they have not been applied in the context of the reduction lemma itself. Second, we examine algorithmic aspects of element-connectivity, and this is our main contribution. Very basic questions have not been explored in contrast to substantial literature on edge and vertex connectivity. For instance, how fast can $\kappa'_G(T)$ be computed? How fast can the graph H promised by Corollary 1 be computed? We obtain several results which are summarized in Fig. 2. In particular, given a graph on n nodes with m edges we obtain an algorithm that outputs the reduced graph for a given set of terminals T in $O(\lvert T\rvert nm)$ time. The key observation that underlies the algorithms is that a symmetric submodular function can be defined over the terminal set T that corresponds to the element-connectivity between then. This in turn allows us to compute and exploit a Gomory-Hu tree for this function.

Applications: Element-connectivity has found important applications in three areas: network design, packing element-disjoint Steiner trees and forests, and more recently in routing for node-disjoint paths and related applications (we omit detailed references due to lack of space). Our algorithmic improvements most directly affect the second application, namely the problem of packing element-disjoint Steiner trees and forests. We briefly describe the simpler case of packing element-disjoint Steiner trees which was the original motivation for the graph reduction step [18]. Here we are given a graph $G = (V, E)$ and terminal set T and the goal is to find the maximum number of Steiner trees for T that are pairwise element-disjoint. It is known that in general graphs one can find $\Omega(k/\log\lvert T\rvert)$ trees where $k = \kappa'_G(T)$ and there are examples where this bound is tight [4]. In planar graphs one can find $\Omega(k)$ trees [1,3]. Algorithms for these first need to compute k, and then work with the reduced graph. Computing the reduced graph is the bottleneck and our result thus implies an $O(\lvert T\rvert nm)$-time algorithm for these packing problems; the previous bounds is $O(k\lvert T\rvert^2 m^2)$.

Discussion: There is a vast amount of literature on algorithms for computing edge and vertex connectivity in graphs, and related problems on flows and cuts. As the table in Fig 2 shows, the edge connectivity versions have faster algorithms and are much better understood. This is not surprising since edge-connectivity has additional structure that can be exploited, including the existence of a Gomory-Hu tree. In contrast, vertex-connectivity does not admit even a weaker notion of flow-trees [2]. Element-connectivity is an interesting notion that has helped bridge edge and vertex connectivity in various applications, and we believe that studying its computational aspects in more depth is an fruitful direction of research. Our work here can be seen as a first step in exploiting the basic properties of element-connectivity to obtain faster algorithms. In this context we mention the the splitting-off operation to preserve edge-connectivity introduced by Lovász [24] and strengthened by Mader [25]. The algorithmic aspects of splitting-off have been exploited in several papers on edge-connectivity including some recent ones [16,23] and we hope similar ideas may bear fruit for element-connectivity.

2   Preliminaries

A set function $f : 2^V \to \mathbb{R}$ is submodular if for all $A, B \subseteq V, f(A) + f(B) \geq f(A \cap B) + f(A \cup B)$. f is non-negative if $f(A) \geq 0$ for all $A \subset V$. f is symmetric if $f(A) = f(V - A)$ for all $A \subset V$. For a non-negative symmetric submodular function $f : 2^V \to \mathbb{R}$, we define the f-connectivity between $s, t \in V$ as:

$$\alpha_f(s,t) = \min_{U \subset V, |U \cap \{s,t\}|=1} f(U). \tag{1}$$

Note that $\alpha_f(s,t) = \alpha_f(t,s)$ for symmetric f. A capacitated spanning tree (R,c) on V is called a *Gomory-Hu tree (cut tree)* of f if for all $st \in E(R), f(A) = \alpha_f(s,t) = c(st)$, where A is a component of $R - st$. $\alpha_f(s,t)$ for all $s, t \in V$ can be read off from the Gomory-Hu tree as the smallest capacity on the unique path between s and t. A Gomory-Hu tree always exists when f is non-negative, symmetric, and submodular (see [26]).

Given an undirected graph $G = (V, E)$, for any $W \subseteq V$, we denote by $\delta_G(W)$ the set of edges with exactly one end point in W. The cut function f, defined by $f(A) = |\delta_G(A)|$, is a symmetric submodular function. For $A, B \subseteq V$, $E(A, B)$ denotes the set of edges with one end point in A and the other in B.

Set Pairs, Bisubmodularity and Bisupermodularity: Given a finite ground set V, a *setpair* $W = (W_t, W_h)$ is an ordered pair of disjoint subsets of V; W_t or W_h may be empty. Let S be the set of all setpairs of V and denote the set of edges with one endpoint in W_t and the other in W_h by $\delta(W) = E(W_t, W_h)$. Given two setpairs $W = (W_t, W_h)$ and $Y = (Y_t, Y_h)$, let $W \otimes Y$ denote the setpair $(W_t \cup Y_t, W_h \cap Y_h)$, and let $W \oplus Y$ denote the setpair $(W_t \cap Y_t, W_h \cup Y_h)$. Additionally, we define $\overline{(W_t, W_h)} = (W_h, W_t)$. Note that $\delta(W) = \delta(\overline{W})$ for an undirected graph and $\overline{W \otimes Y} = \overline{W} \oplus \overline{Y}$.

A function $f : S \to \mathbb{R}$ is called *bisubmodular* if $f(W) + f(Y) \geq f(W \otimes Y) + f(W \oplus Y)$ for any two setpairs W and Y. For *bisupermodular* and *bimodular* functions, we change $\geq$ to $\leq$ and $=$ respectively. f is *symmetric* if $f(W) = f(\overline{W})$ for all $W \in S$.

Definition 1. *Given a ground set V, a non-negative function g on S is called skew-bisupermodular if for any two setpairs W and Y, one of the following holds:*

$$g(W) + g(Y) \leq g(W \otimes Y) + g(W \oplus Y) \tag{2}$$
$$g(W) + g(Y) \leq g(\overline{W} \otimes Y) + g(\overline{W} \oplus Y) \tag{3}$$

3 Element-Connectivity and Connections to Submodularity

It is natural to work with cut-functions that capture element-connectivity. We define a cut function $C_G : 2^T \to \mathbb{R}_+$ over the terminals as follows: for $U \subset T$, $C_G(U)$ is the minimum number of elements whose removal disconnects U from $T \setminus U$ in G. To formally define $C_G(U)$ we consider vertex tri-partitions (A, Z, B) where $U \subseteq A$ and $(T \setminus U) \subseteq B$; among all such tri-partitions the one with minimum $|Z| + |E(A, B)|$ defines $C_G(U)$. $Z \cup E(A, B)$ is called a *cut-set*.

Theorem 3 (Menger's theorem for element-connectivity). *For a graph G with terminal vertices T, for all $s, t \in T$, $\kappa'_G(s, t) = \min\{C_G(U) : U \subset T, |U \cap \{s, t\}| = 1\}$.*

If T is a independent set in the previous theorem, then the cut-set can be taken to contain no edges.

For our purposes a crucial observation is that C_G is a non-negative symmetric submodular function.

Theorem 4. *Let G be a graph with terminal vertices T. C_G is a non-negative symmetric submodular function over T.*

The preceding theorem implies that C_G admits a Gomory-Hu tree.

4 Algorithmic Aspects of Element-Connectivity

In this section we describe our algorithmic contributions to element-connectivity. In particular we describe how the running times in the second row of the table in Fig. 2 can be realized. Our main contribution is a faster algorithm for graph reduction. In the entire section, we are always working with a graph $G = (V, E)$ with n vertices, m edges and terminal vertices T.

Equivalent Directed Graph: One view of element-connectivity that greatly helps with computation is to define a flow problem. One can see that $\kappa'_G(s, t)$ is the maximum s-t-flow in G with unit capacities on the edges *and* non-terminal vertices (terminal vertices have no capacity constraint). This prompts us to define a equivalent directed graph, which we get from applying the standard vertex split operation for a graph when there are vertex capacities.

Let $N = V \setminus T$ be the set of non-terminals. Let $N^- = \{v^-|v \in N\}$ and $N^+ = \{v^+|v \in N\}$. The *equivalent directed graph* of G, denoted by $\tilde{G} = (\tilde{V}, \tilde{E})$, where $\tilde{V} = N^- \cup N^+ \cup T$ and the arc set $\tilde{E}$ is obtained from G as follows:

1. For every $v \in N$, $(v^-, v^+) \in \tilde{E}$.
2. For every $uv \in E$ where $u, v \in N$, $(u^+, v^-), (v^+, u^-) \in \tilde{E}$.
3. For every $uv \in E$ where $u \in T, v \in N$, $(u, v^-), (v^+, u) \in \tilde{E}$.
4. For every $uv \in E$ where $u, v \in T$, $(u, v), (v, u) \in \tilde{E}$.

All the arcs in $\tilde{G}$ implicitly have unit capacity. Any maximum integral acyclic s-t-flow f_{st} in $\tilde{G}$ corresponds to a set of maximum element-disjoint s-t-paths in G. Hence we do not distinguish between maximum flows in $\tilde{G}$ and maximum element-disjoint paths in G. A flow in $\tilde{G}$ contains a vertex v (edge e) in G to mean the corresponding element-disjoint path contains vertex v (edge e).

Lemma 1. $\lambda_{\tilde{G}}(s,t) = \kappa'_G(s,t)$ *for all* $s,t \in T$.

4.1   Computing Element-Connectivity

Single pair element-connectivity The equivalent directed graph allows us to compute local element-connectivity by running a single maximum flow on a unit capacity directed graph.

Lemma 2. $\kappa'_G(s,t)$ *can be computed in* $O(\mathrm{MF}(n,m))$ *time.*

Note that if $T = \{s,t\}$ then $\kappa'_G(s,t) = \kappa_G(s,t)$ and moreover maximum bipartite matching can be reduced to $\kappa_G(s,t)$. Thus, improving the time to compute $\kappa'_G(s,t)$ is not feasible without corresponding improvements to other classical problems.

All-Pair Element Connectivity. To compute $\kappa'_G(s,t)$ for all pairs $s,t \in T$, we can compute the Gomory-Hu tree that we know exists from Theorem 4. Unlike the single-pair case where element-connectivity behaves like vertex-connectivity, in the all-pair case it is closer to edge-connectivity, and in particular there are at most $|T| - 1$ distinct element-connectivity values. A Gomory-Hu tree (R,c) representing the all-pair element-connectivities can be built recursively by solving $|T| - 1$ minimum element cut computations, which correspond to maximum flow computations in the equivalent directed graph. Hence all-pair element-connectivity takes $|T| - 1$ maximum flows. For dense graphs, if we allow randomization, maximum flow can be solved in $\tilde{O}(n^\omega)$ time[14], where ω is the matrix multiplication constant.

There is an alternative approach for sparse graphs using network coding. Cheung *et al.* describe a randomized algorithm that computes the edge-connectivity in a *directed* graph between *every* pair of vertices in $O(m^\omega)$ time with high probability [6]. Since $\kappa'_G(s,t) = \lambda_{\tilde{G}}(s,t)$ for all $s,t \in T$, all-pair element-connectivity can also be computed in $O(m^\omega)$ time with high probability.

Global Element Connectivity. The global element-connectivity $\kappa'_G(T)$ can be easily obtained from the all-pair problem.

Theorem 5. $\kappa'_G(T)$ *can be computed in* $O(|T| \, \mathrm{MF}(n,m))$ *time.*

A different algorithm that results $|T| - 1$ maximum flows computations can be obtained via an approach similar to that of Hao-Orlin's algorithm [15]. We defer the details to full version of the paper. It is a very interesting open problem to improve the running time for computing $\kappa'_G(T)$. Global edge-connectivity admits a near-linear time algorithm [20,21], but global vertex connectivity algorithms are much slower.

Remark 2. For the special case when G is already a reduced graph, an $O(mn)$ time algorithm exists via the problem of computing a minimum cut in a hypergraph [22].

4.2   Computing a Reduced Graph

This section highlights the main result of the paper, an $O(|T|nm)$ time algorithm to find a reduced graph. For G a graph with terminals T, H is called a *reduction* of G if it can be obtained from G by a sequence of reduction operations.

The reduction lemma suggests a simple algorithm: pick any edge incident to two nonterminal vertices, check which one of the two operations preserves element-connectivity, reduce and repeat. For a graph on n vertices and m edges, compute all-pair element-connectivity from scratch. The naive scheme when combined with the non-obvious $O(|T| \, \mathrm{MF}(n,m))$ algorithm for all-pair case would take $O(|T|m \, \mathrm{MF}(n,m))$ time. Cheriyan and Salavatipour [4] described exactly this algorithm where they in fact computed all-pair connectivity using $|T|^2$ max-flow computations; their focus was not in improving the running time. We obtain the following

Theorem 6. *For a graph G with n vertices, m edges and terminals T, a reduced graph of G can be computed in $O(|T|nm)$ time.*

The two high-level ideas to obtain an improved run-time are the following.

1. The algorithm initially computes and then maintains a Gomory-Hu tree (R, c) for the element-connectivity of T. For each edge $st \in E(R)$ it maintains a corresponding maximum flow between s and t in $\tilde{G}$ as it evolves with reduction operations.
2. Instead of considering reduction operations on an edge by edge basis in an ad-hoc manner we consider all edges incident to a non-terminal vertex v and process them as a batch.

The first idea alone would give us a run-time of $O(|T|m^2)$. The second idea gives a further improvement to $O(|T|nm)$.

Reduction by Vertex Elimination: We call an edge pq between two non-terminals a *reducible* edge. For a given non-terminal v let $D(v)$ be the set of all reducible edges incident to v. We say v is *active* if $D(v) \neq \emptyset$. An *elimination* operation on an active vertex v either contracts an edge in $D(v)$ or removes *all* edges in $D(v)$. If the graph is not reduced, there is always an elimination operation that preserves element-connectivity. Indeed, if $D(v)$ cannot be removed, then consider the edges in $D(v)$ in an arbitrary but fixed sequence and apply the reduction operation. At some point there is an edge e such that removing it reduces the element-connectivity. By reduction lemma, we can contract e. An elimination reduces the number of active vertices by at least 1. There can only be $O(n)$ eliminations. Moreover, crucially, we can implement a vertex elimination operation in the same amount of time as an edge reduction operation.

Our goal is to decide quickly whether an active vertex v can be eliminated (that is, $D(v)$ can be removed) and if not which of the edges in $D(v)$ can be contracted. For this purpose we define a weighting of edges of E as follows. First we order the edges in $D(v)$ arbitrarily as $e_1, e_2, \ldots, e_h$ where $h = |D(v)|$. We define a weight function $w : E \to \{1, 2, \ldots, h+1\}$ where $w(e_i) = i$ and $w(e) = h+1$ for all $e \in E \setminus D(v)$.

Given a set of weights ρ on the edges, for each pair (s, t) of terminals we define $\beta_\rho(s, t)$ as the maximum weight a such that the element-connectivity between s and t remains the same even if we remove all edges with weight less than a. We call $\beta_\rho(s, t)$ as the *bottleneck weight* for (s, t). Suppose we used the weight function as defined above based on the numbering for $D(v)$ then v can be eliminated iff $\beta_w(s, t) > h$ for all pairs of terminals (s, t). In fact we can also obtain information on which of the edges in $D(v)$ can be contracted if v cannot be eliminated. Even further, we need to check only the terminal pairs that correspond to edges of a Gomory-Hu tree for the element-connectivity of T. This is captured in the following theorem which forms the crux of our algorithm.

Theorem 7. *Let (R, c) be a Gomory-Hu tree for the element-connectivity of terminal set T in G. Consider an active non-terminal v and the weight function w and let $\ell = \max_{st \in E(R)} \beta_w(s, t)$. Define G' as G/e_ℓ if $\ell < |D(v)| + 1$ and $G - D(v)$ otherwise. Then $\kappa'_{G'}(u, v) = \kappa'_G(u, v)$ for all terminal pairs (u, v).*

Proof. Recall that $D(v) = \{e_1, e_2, \ldots, e_h\}$ where $h = |D(v)|$. Let $S = \{e_1, \ldots, e_{\ell-1}\}$. Since $\beta_w(s, t) \geq \ell$ for all $st \in E(R)$ it follows that the element-connectivity does not change for any pair $st \in E(R)$ if we delete the edges in S from G. From Lemma 5 it follows in fact that the element-connectivity of all pairs remains the same in $G - S$ as in G. Thus all edges in S are deletable. If $\ell = h+1$ then $S = D(v)$ and $G' = G - S$ and we have the desired property. If $\ell \leq h$ then there is at least one $st \in E(R)$ such that $\kappa'_{G-S}(s, t) = \kappa'_G(s, t)$ but $\kappa'_{G-S-e_\ell}(s, t) < \kappa'_G(s, t)$. From the reduction lemma applied to $G - S$ we see that e_ℓ is not deletable, and hence by the reduction lemma $(G - S)/e_\ell$ preserves all the element-connectivities of the terminals. This also implies that G/e_ℓ preserves all element-connectivities. $\square$

Computing $\beta_w(s, t)$ is relatively fast if we already have an existing maximum flow from s to t. This is captured by the lemma below.

Lemma 3. *Given a maximum s-t-flow f_{st} in $\tilde{G}$, an active non-terminal v and a weighting w, we can find $\beta_w(s, t)$ and a corresponding flow in $O(m)$ time.*

Proof. Consider a maximum s-t flow f_{st} in $\tilde{G}$. In a flow decomposition of f_{st} there is at most one flow path that uses the non-terminal v. We can find such a flow path in $O(m)$ time and reduce the flow by at most one unit to obtain a new flow f'_{st} which does not have any flow through v. Not that f'_{st} is non-zero only on edges e with $w(e) = |D(v)| + 1$. If the value of f'_{st} is the same as that of f_{st} then $\beta_w(s, t) = |D(v)| + 1$ and we are done. Otherwise, we claim that $\beta_w(s, t) = \ell$ iff the maximum bottleneck weight for a path from s to t in the residual graph of f'_{st} in $\tilde{G}$ is ℓ. Assuming that the claim is true we can find $\beta_w(s, t)$ by a maximum bottleneck path computation in the residual graph in $O(m)$ time since the edges are sorted by weight (the algorithm is quite simple but we refer the reader to [13]).

Now we prove the claim. If there is a path of maximum bottleneck weight ℓ in the residual graph we can augment f'_{st} by one unit to obtain a maximum flow f''_{st} that uses only edges with weight ℓ or greater and hence $\beta_w(s, t) \geq \ell$. Suppose $\beta_w(s, t) = \ell'$. Remove the edges $\{e_1, \ldots, e_{\ell'-1}\}$ from G and their corresponding arcs from $\tilde{G}$. There is a maximum s-t flow of value $\kappa'_G(s, t)$ in this new graph H. f'_{st} is a flow of value $\kappa'_G(s, t) - 1$ in H and hence there must be an augmenting path in the residual graph of f'_{st} in H and this augmenting path has bottleneck weight at least ℓ' and is also a valid path in residual graph of f'_{st} in $\tilde{G}$. Thus $\ell \geq \ell'$. Thus $\beta_w(s, t) = \ell$ as desired. $\square$

Theorem 7 and Lemma 3 lead to an algorithm as follows. We initially compute and then maintain a Gomory-Hu tree (R, c) for C_G on the terminals. For each edge $st \in E(R)$ we also maintain a maximum flow f_{st} in the current graph $\tilde{G}$. In each iteration we do an elimination procedure on an active vertex using Theorem 7. Each iteration either reduces the number of active vertices by one or contracts an edge. Thus the number of iterations is $O(n)$. Algorithm 1 gives a formal description. Note that the tree (R, c)

Algorithm 1: Reduce a graph G with terminal vertices T

   Input: undirected graph G, terminals T
   `// Preprocessing`
1 $(R, c) \leftarrow$ Gomory-Hu tree of C_G
2 **foreach** $st \in E(R)$ **do**
3 | $f_{st} \leftarrow$ maximum st-flow in $\tilde{G}$
   `// Sequence of eliminations`
4 **while** *there exists an active non-terminal vertex v* **do**
5     $w \leftarrow$ assign weights to $D(v)$
6   **foreach** $st \in E(R)$ **do**
7 | compute $\beta_w(s,t)$ in $O(m)$ time using f_{st}
8     $\ell \leftarrow \min\{\beta_w(s,t) \mid st \in E(R)\}$
9   **if** $\ell > |D(v)|$ **then**
10 | $G \leftarrow G - D(v)$
11   **else**
12 | $G \leftarrow G/e_\ell$
13   **foreach** $st \in E(R)$ **do**
14 | update f_{st} in $O(m)$ time
15 **return** G

---

does not change throughout the algorithm but the flows f_{st} for each $st \in E(R)$ get updated in each iteration. We need to clarify how we update these flows, analyze the overall running time, and argue about the correctness of the algorithm.

Updating the Flows: Each elimination changes the graph. The algorithm updates the maximum flows f_{st} for each $st \in E(R)$ to reflect this change. If the new graph is $G - D(v)$, then no flow need updating since the computation of $\beta_w(s,t)$ already finds a new flow that avoids all edges in $D(v)$. We address the case when we contract an edge from $D(v)$.

Lemma 4. *Let G be a graph with m edges with terminals T. Let $H = G/e$ for some reducible edge e. If f_{st} is a maximum s-t flow in $\tilde{G}$, then we can find f'_{st}, a maximum s-t-flow in $\tilde{H}$ in $O(m)$ time.*

Proof. Let $e = pq$. We delete flow paths in f_{st} that use p or q. This removes at most 2 unit of flow (since each non-terminal has unit capacity) and the reduced flow is a valid flow in $\tilde{H}$. In two augmentations we can find a maximum flow in $\tilde{H}$. Each step can be easily implemented in $O(m)$ time.

Analysis of Running Time: The time spent to build the Gomory-Hu tree (R, c) and the initial maximum flows for each edge $st \in E(R)$ take $|T| - 1$ maximum flow computations. Thus the time for this is $O(|T| \, \text{MF}(n, m))$

As we argued there are $O(n)$ iterations of the while loop. In each iteration we need to compute $\beta_w(st)$ for each $st \in E(R)$. Lemma 3 shows that each such computation can be done in $O(m)$ time using the stored maximum flow; thus the total time is $O(|T|m)$. Updating the maximum flows also takes $O(|T|m)$ time using Lemma 4. Thus the overall running time for all the iterations of the while loop is $O(|T|nm)$ which dominates the time to compute the initial Gomory-Hu tree.

Correctness: Theorem 7 shows the correctness of the elimination procedure. It remains to argue that the Gomory-Hu tree (R, c) computed in the preprocessing step remains valid throughout the algorithm as the graph G changes and gets simplified. A similar idea is used implicitly by Gabow for preserving local edge-connectivity while applying split-off operations [11]. The following simple lemma gives a justification.

Lemma 5. *Let* (R, c) *be a Gomory-Hu tree for* C_G *with terminal vertices* T *in* G. *Let* H *be a reduction of* G. *If* $\kappa'_G(s, t) = \kappa'_H(s, t)$ *for all* $st \in E(R)$ *then* (R, c) *is also a Gomory-Hu tree for* C_H.

5   Proof of Reduction Lemma via Setpairs

In this section we give a new proof of Theorem 2 via the use of setpairs. We will assume without loss of generality that T is an independent set by sub-dividing each edge between terminals. First we introduce some functions over setpairs over the vertex set V. These functions and their properties have been explored in papers on network design for element connectivity [19,9,5] and in particular we refer the reader to [5] for relevant proofs.

Let $\mathcal{S}$ be set of all setpairs over vertex set V of the given graph G. Recall that T is the set of terminals. We define g, ℓ, f on $\mathcal{S}$ as follows. For a set pair $W = (W_t, W_h)$,

$$g(W) = \max\{\kappa'_G(u, v) : u \in W_t \cap T, v \in W_h \cap T\}$$

Here the max follows the convention $\max \emptyset = 0$.

$$\ell(W) = |V - W_t - W_h|$$

$$f(W) = g(W) - \ell(W)$$

The lemmas below are known from prior work; see [9,5].

Lemma 6. *The following properties hold: (i)* g *is symmetric skew-bisupermodular, (ii)* ℓ *is symmetric bimodular, (iii)* f *is symmetric skew-bisupermodular, and (iv)* $|\delta(W)|$ *is symmetric bi-submodular.*

A setpair W is *tight* if $f(W) = |\delta(W)|$. The lemma below follows from skew-bisupermodularity of f and symmetric bi-submodularity of $|\delta()|$.

Lemma 7 (Lemma 3.1 [5]). *If* W *and* Y *are tight, then one of the following holds:*

1. $W \oplus Y$ *and* $W \otimes Y$ *are tight.*
2. $\overline{W} \oplus Y$ *and* $\overline{W} \otimes Y$ *are tight.*

We are now ready to prove the Reduction Lemma.

Proof. Consider any edge pq between non-terminals and graphs $G_1 = G - pq$ and $G_2 = G \setminus pq$. Suppose the lemma is false. There must be distinct[1] terminal pairs (s, t) and (x, y) such that $\kappa'_{G_1}(s, t) < \kappa'_G(s, t)$ and $\kappa'_{G_2}(x, y) < \kappa'_G(x, y)$.

---

[1] It is not hard to see that for any (s, t), $\kappa'_G(s, t) = \kappa'_{G_1}(s, t)$ or $\kappa'_G(s, t) = \kappa'_{G_2}(s, t)$.

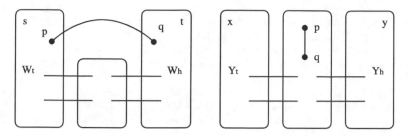

Fig. 3. (left) $\kappa'_{G_1}(s,t) = \kappa'_G(s,t)$ when pq is removed. (right) $\kappa'_{G_2}(x,y) = \kappa'_G(x,y) - 1$ when pq is contracted. Recall that T is an independent set.

For the pair (s,t), since the connectivity decreased when pq was deleted, there exists a setpair W such that $s, p \in W_t, t, q \in W_h, \ell(W) = \kappa'_G(s,t) - 1$ and $\delta(W) = \{pq\}$; see Fig. 3. If we delete pq, then $V - W_t - W_h$ becomes a valid cut-set that separates s and t with smaller value. The cut sets contain only non-terminals because T is an independent set. For the pair (x,y), since connectivity decreased when pq was contracted, there exists a setpair Y such that $x \in Y_t, y \in Y_h, p, q \in V - Y_t - Y_h, \ell(Y) = \kappa'_G(x,y)$ and $\delta(Y) = \emptyset$; see Fig. 3. If we contract pq to form p', $V - Y_t - Y_h - p'$ becomes a smaller cut-set.

We have $f(W) = g(W) - \ell(W) = \kappa'_G(s,t) - (\kappa'_G(s,t) - 1) = 1$, and $f(Y) = g(Y) - \ell(Y) - \kappa'_G(x,y) - \kappa'_G(x,y) = 0$. Because $|\delta(W)| = 1$ and $|\delta(Y)| = 0$, W and Y are tight. We now use Lemma 7 to consider two cases.

Suppose $W \oplus Y$ and $W \otimes Y$ are tight. We have $f(W \oplus Y) = |\delta(W \oplus Y)|$ and $f(W \otimes Y) = |\delta(W \otimes Y)|$. If $\delta(W \oplus Y)$ is not empty, then either $\delta(W) - \{pq\}$ or $\delta(Y)$ must not be empty, which cannot be since $|\delta(W) - pq| = 0$ and $|\delta(Y)| = 0$. (We exclude pq from $\delta(W)$ because $pq \notin \delta(W \oplus Y)$.) Therefore, $\delta(W \oplus Y)$ is empty. Similarly, $\delta(W \otimes Y)$ must be empty. Therefore, $f(W) + f(Y) = 1 + 0 > 0 + 0 = f(W \oplus Y) + f(W \otimes Y)$, which is a contradiction since f is bisupermodular. A similar argument can be applied to the case when $\overline{W} \oplus Y$ and $\overline{W} \otimes Y$ are tight by exchanging W with $\overline{W}$.

Acknowledgements. This work was supported in part by Chandra Chekuri's NSF grants CCF-1319376 and CCF-1526799, and Jeff Erickson's NSF grant CCF-0915519.

References

1. Aazami, A., Cheriyan, J., Jampani, K.: Approximation Algorithms and Hardness Results for Packing Element-Disjoint Steiner Trees in Planar Graphs. Algorithmica 63(1–2), 425–456 (2012)
2. Benczúr, A.A.: Counterexamples for directed and node capacitated cut-trees. SIAM Journal on Computing 24(3), 505–510 (1995)
3. Chekuri, C., Korula, N.: A graph reduction step preserving element-connectivity and packing steiner trees and forests. SIAM Journal on Discrete Mathematics 28(2), 577–597 (2014)
4. Cheriyan, J., Salavatipour, M.: Packing Element-Disjoint Steiner Trees. ACM Transactions on Algorithms 3(4) (2007)

5. Cheriyan, J., Vempala, S., Vetta, A.: Network Design via Iterative Rounding of Setpair Relaxations. Combinatorica 26(3), 255–275 (2006)
6. Cheung, H.Y., Lau, L.C., Leung, K.M.: Graph connectivities, network coding, and expander graphs. In: Proceedings of the 2011 IEEE 52nd Annual Symposium on Foundations of Computer Science, FOCS 2011, pp. 190–199 (2011)
7. Duan, R.: Breaking the $O(n^{2.5})$ Deterministic Time Barrier for Undirected Unit-Capacity Maximum Flow. In: Proceedings of ACM-SIAM SODA, pp. 1171–1179 (2013)
8. Even, S., Tarjan, R.: Network flow and testing graph connectivity. SIAM Journal on Computing 4(4), 507–518 (1975)
9. Fleischer, L., Jain, K., Williamson, D.: Iterative Rounding 2-Approximation Algorithms for Minimum-Cost Vertex Connectivity Problems. Journal of Computer and System Sciences 72(5), 838–867 (2006)
10. Frank, A., Ibaraki, T., Nagamochi, H.: On sparse subgraphs preserving connectivity properties. Journal of Graph Theory 17(3), 275–281 (1993)
11. Gabow, H.N.: Efficient splitting off algorithms for graphs. In: Proceedings of the Twenty-Sixth Annual ACM Symposium on Theory of Computing, STOC 1994, pp. 696–705. ACM, New York (1994)
12. Gabow, H.N.: Using expander graphs to find vertex connectivity. In: Proc. 41st Annual IEEE Symposium on Foundations of Computer Science, pp. 410–420 (2000)
13. Gabow, H.N., Tarjan, R.E.: Algorithms for Two Bottleneck Optimization Problems. Journal of Algorithms 9(3), 411–417 (1988)
14. Gabow, H., Sankowski, P.: Algebraic algorithms for b-matching, shortest undirected paths, and f-factors. In: 2013 IEEE 54th Annual Symposium on Foundations of Computer Science (FOCS), pp. 137–146 (October 2013)
15. Hao, J., Orlin, J.B.: A faster algorithm for finding the minimum cut in a graph. In: Proceedings of the Third Annual ACM-SIAM Symposium on Discrete Algorithms, SODA 1992, pp. 165–174. Society for Industrial and Applied Mathematics, Philadelphia (1992)
16. Hariharan, R., Kavitha, T., Panigrahi, D., Bhalgat, A.: An $\tilde{O}(mn)$ Gomory-Hu tree construction algorithm for unweighted graphs. In: Proceedings of the Thirty-Ninth Annual ACM Symposium on Theory of Computing, STOC 2007, pp. 605–614. ACM (2007)
17. Henzinger, M.R., Rao, S., Gabow, H.N.: Computing vertex connectivity: New bounds from old techniques. Journal of Algorithms 34(2), 222–250 (2000)
18. Hind, H., Oellermann, O.: Menger-Type Results for Three or More Vertices. Congressus Numerantium, pp. 179–204 (1996)
19. Jain, K., Măndoiu, I., Vazirani, V., Williamson, D.: A Primal-Dual Schema Based Approximation Algorithm for the Element Connectivity Problem. In: Proceedings of the 10th Annual ACM-SIAM Symposium on Discrete Algorithms (SODA), pp. 484–489 (1999)
20. Karger, D.: Random Sampling in Graph Optimization Problems. Ph.D. thesis, Stanford University (1994)
21. Kawarabayashi, K., Thorup, M.: Deterministic Global Minimum Cut of a Simple Graph in Near-Linear Time. In: Proceedings of ACM STOC, pp. 665–674 (2015)
22. Klimmek, R., Wagner, F.: A simple hypergraph min cut algorithm. Tech. Rep. B 96-02, Bericht FU Berlin Fachbereich Mathematik und Informatik (1996)
23. Lau, L.C., Yung, C.K.: Efficient edge splitting-off algorithms maintaining all-pairs edge-connectivities. SIAM Journal on Computing 42(3), 1185–1200 (2013)
24. Lovász, L.: On Some Connectivity Properties of Eulerian Graphs. Acta Mathematica Hungarica 28(1), 129–138 (1976)
25. Mader, W.: A Reduction Method for Edge-Connectivity in Graphs. Annals of Discrete Mathematics 3, 145–164 (1978)
26. Schrijver, A.: Combinatorial Optimization: Polyhedra and Efficiency. Springer, Heidelberg (2003)

On Randomized Algorithms
for Matching in the Online Preemptive Model

Ashish Chiplunkar[1,*], Sumedh Tirodkar[2], and Sundar Vishwanathan[2]

[1] Amazon Development Centre, Banglore
ashish.chiplunkar@gmail.com
[2] Department of Computer Science and Engineering, IIT Bombay
{sumedht,sundar}@cse.iitb.ac.in

Abstract. We investigate the power of randomized algorithms for the maximum cardinality matching (MCM) and the maximum weight matching (MWM) problems in the online preemptive model. In this model, the edges of a graph are revealed one by one and the algorithm is required to always maintain a valid matching. On seeing an edge, the algorithm has to either accept or reject the edge. If accepted, then the adjacent edges are discarded. The complexity of the problem is settled for deterministic algorithms [7,9].

Almost nothing is known for randomized algorithms. A lower bound of 1.693 is known for MCM with a trivial upper bound of two. An upper bound of 5.356 is known for MWM. We initiate a systematic study of the same in this paper with an aim to isolate and understand the difficulty. We begin with a primal-dual analysis of the deterministic algorithm due to [7]. All deterministic lower bounds are on instances which are trees at every step. For this class of (unweighted) graphs we present a randomized algorithm which is $\frac{28}{15}$-competitive. The analysis is a considerable extension of the (simple) primal-dual analysis for the deterministic case. The key new technique is that the distribution of primal charge to dual variables depends on the "neighborhood" and needs to be done after having seen the entire input. The assignment is asymmetric: in that edges may assign different charges to the two end-points. Also the proof depends on a non-trivial structural statement on the performance of the algorithm on the input tree.

The other main result of this paper is an extension of the deterministic lower bound of Varadaraja [9] to a natural class of randomized algorithms which decide whether to accept a new edge or not using *independent* random choices. This indicates that randomized algorithms will have to use *dependent* coin tosses to succeed. Indeed, the few known randomized algorithms, even in very restricted models follow this.

We also present the best possible $\frac{4}{3}$-competitive randomized algorithm for MCM on paths.

---

* This work was done when the author was a student at IIT Bombay.

© Springer-Verlag Berlin Heidelberg 2015
N. Bansal and I. Finocchi (Eds.): ESA 2015, LNCS 9294, pp. 325–336, 2015.
DOI: 10.1007/978-3-662-48350-3_28

1   Introduction

Matching has been a central problem in combinatorial optimization. Indeed, algorithm design in various models of computations, sequential, parallel, streaming, etc., have been influenced by techniques used for matching. We study the maximum cardinality matching (MCM) and the maximum weight matching (MWM) problems in the online preemptive model. In this model, edges $e_1, \ldots,$ e_m of a graph, possibly weighted, are presented one by one. An algorithm is required to output a matching M_i after the arrival of each edge e_i. This model constrains an algorithm to accept/reject an edge as soon as it is revealed. If accepted, the adjacent edges, if any, have to be discarded from M_i.

An algorithm is said to have a *competitive ratio* α if the cost of the matching maintained by the algorithm is at least $\frac{1}{\alpha}$ times the cost of the offline optimum over all inputs. The deterministic complexity of this problem is settled. For maximum cardinality matching (MCM), it is an easy exercise to prove a tight bound of two.

The weighted version (MWM) is more difficult. Improving an earlier result of Feigenbaum et al. [5], McGregor [7] gave a deterministic algorithm together with an ingenious analysis to get a competitive ratio of $3 + 2\sqrt{2} \approx 5.828$. Later, this was proved to be optimal by Varadaraja [9].

Very little is known on the power of randomness for this problem. Recently, Epstein et al. [4] proved a lower bound of $1 + \ln 2 \approx 1.693$ on the competitive ratio of randomized algorithms for MCM. This is the best lower bound known even for MWM. Epstein et al. [4] also give a 5.356-competitive randomized algorithm for MWM.

In this paper, we initiate a systematic study of the power of randomness for this problem. Our main contribution is perhaps to throw some light on where lies the difficulty. We first give an analysis of McGregor's algorithm using the traditional Primal-Dual framework (see Appendix A in [3]). All lower bounds for deterministic algorithms (both for MCM and MWM) employ *growing trees*. That is, the input graph is a tree at every stage. It is then natural to start our investigation for this class of inputs. For this class, we give a randomized algorithm (that uses two bits of randomness) that is $\frac{28}{15}$ competitive. While this result is modest, already the analysis is considerably more involved than the traditional primal dual analysis. In the traditional primal dual analysis of the matching problem, the primal charge (every selected edge contributes one to the charge) is distributed (perhaps equally) to the two end-points. In the online case, this is usually done as the algorithm proceeds. Our assignment depends on the structure of the final tree, so this assignment happens at the end. Our charge distribution is *not* symmetric. It depends on the position of the edge in the tree (we make this clear in the analysis) as also the behavior of neighboring edges. The main technical lemma shows that the charge distribution will depend on a neighborhood of distance at most four. We also note that these algorithms are (restricted versions of) randomized greedy algorithms even in the offline setting. Obtaining an approximation ratio less than two for general graphs, even in the offline setting is a notorious problem. See [8,2] for a glimpse of the difficulty.

The optimal maximal matching algorithm for MCM, and McGregor's [7] optimal deterministic algorithm for MWM are both local algorithms. The choice of whether a new edge should accepted or rejected is based only on the weight of the new edge and the weight of the conflicting edges, if any, in the current matching.

It is natural to add randomness to such local algorithms, and to ask whether they do better than the known deterministic lower bounds. An obvious way to add randomness is to accept/reject the new edge with certain probability, which is only dependent on the new edge and the conflicting edges in the current matching. The choice of adding a new edge is independent of the previous coin tosses used by the algorithm. We call such algorithms *randomized local algorithms*. We show that randomized local algorithms cannot do better than optimal deterministic algorithms. This indicates that randomized algorithms may have to use dependent coin tosses to get better approximation ratios. Indeed, the algorithm by Epstein et al. does this. So do our randomized algorithms.

The randomized algorithm of Epstein et al. [4] works as follows. For a parameter θ, they round the weights of the edges to powers of θ randomly, and then they update the matching using a deterministic algorithm. The weights get distorted by a factor $\frac{\theta \ln \theta}{\theta - 1}$ in the rounding step, and the deterministic algorithm has a competitive ratio of $2 + \frac{2}{\theta - 2}$ on θ-*structured graphs*, i.e., graphs with edge weights being powers of θ. The overall competitive ratio of the randomized algorithm is $\frac{\theta \ln \theta}{\theta - 1} \cdot \left(2 + \frac{2}{\theta - 2} \right)$ which is minimized at $\theta \approx 5.356$. A natural approach to reducing this competitive ratio is to improve the approximation ratio for θ structured graphs. However, we prove that the competitive ratio $2 + \frac{2}{\theta - 2}$ is tight for θ-structured graphs, as long as $\theta \geq 4$, for deterministic algorithms.

One (minor) contribution of this paper is a randomized algorithms for MCM on paths, that achieves a competitive ratio of $\frac{4}{3}$, with a matching lower bound.

2   Barely Random Algorithms for MCM

In this section, we present barely random algorithms, that is, algorithms that use a constant number of random bits, for MCM on growing trees.

The ideal way to read the paper, for a reader of leisure, is to first read our analysis of McGregor's algorithm (presented in Appendix A in [3], then the analysis of the algorithm for trees with maximum vertex degree three (presented in Appendix B.2 in [3]) and then this section. The dual variable management which is the key contribution gets progressively more complicated. It is local in the first two cases. Here are the well known Primal and Dual formulations of the matching problem.

Primal LP	Dual LP
$\max \sum_e x_e$	$\min \sum_v y_v$
$\forall v : \sum_{v \in e} x_e \leq 1$	$\forall e : y_u + y_v \geq 1$
$x_e \geq 0$	$y_v \geq 0$

2.1   Randomized Algorithm for MCM on Growing Trees

In this section, by using only two bits of randomness, we beat the deterministic lower bound of 2 for MCM on growing trees.

Algorithm 1. Randomized Algorithm for Growing Trees

---

1. The algorithm maintains four matchings: M_1, M_2, M_3, and M_4.
2. On receipt of an edge e, the processing happens in two phases.
   (a) **The augment phase.** The new edge e is added to each M_i in which there are no edges adjacent to e.
   (b) **The switching phase.** For $i = 2, 3, 4$, in order, e is added to M_i (if it was not added in the previous phase) and the conflicting edge is discarded, provided it decreases the quantity $\sum_{i,j \in [4], i \neq j} |M_i \cap M_j|$.
3. Output matching M_i with probability $\frac{1}{4}$.

We begin by assuming (we justify this below) that all edges that do not belong to any matching are leaf edges. This helps in simplifying the analysis. Suppose that there is an edge e which does not belong to any matching, but is not a leaf edge. By removing e, the tree is partitioned into two subtrees. The edge e is added to the tree in which it has 4 neighboring edges. (There must be such a subtree, see next para.) Each tree is analysed separately.

We will say that a vertex(/an edge) is *covered* by a matching M_i if there is an edge in M_i which is incident on(/adjacent to) the vertex(/edge). We also say that an edge is *covered* by a matching M_i if it belongs to M_i. We begin with the following observations.

– After an edge is revealed, its end points are covered by all 4 matchings.
– An edge e that does not belong to any matching has 4 edges incident on one of its end points such that each of these edges belong to a distinct matching. This holds when the edge is revealed, and does not change subsequently.

An edge is called *internal* if there are edges incident on both its end points. An edge is called *bad* if its end points are covered by only 3 matchings.

We begin by proving some properties about the algorithm. The key structural lemma that keeps "influences" of bad edges local is given below. The two assertions in the Lemma have to be proved together by induction.

Lemma 1. *1. An internal edge is covered by at least four matchings (counted with multiplicities). It is not necessary that these four edges be in distinct matchings.*

2. If p, q and r are three consecutive vertices on a path, then bad edges cannot be incident on all 3 of these vertices, (as in figure 1).

The proof of this lemma is in the Appendix B.4 in [3].

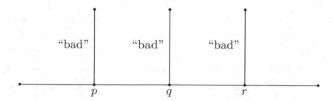

Fig. 1. Forbidden Configuration

Theorem 1. *The randomized algorithm for finding MCM on growing trees is $\frac{28}{15}$-competitive.*

A local analysis like the one in Appendix B.2 in [3] will not work here. For a reason, see Appendix B.3 in [3]. The analysis of this algorithm proceeds in two steps. Once all edges have been seen, we impose a partial order on the vertices of the tree and then with the help of this partial order, we distribute the primal charge to the dual variables, and use the primal-dual framework to infer the competitive ratio. If every edge had four adjacent edges in some matching (counted with multiplicities) then the distribution of dual charge is easy. However we do have edges which have only three adjacent edges in matchings. We would like the edges in matchings to contribute more to the end-points of these edges. Then, the charge on the other end-point would be less and we need to balance this through other edges. Details follow.

Ranks: Consider a vertex v. Let $v_1, \ldots, v_k$ be the neighbors of v. For each i, let d_i denote the maximum distance from v to any leaf if there was no edge between v and v_i. The rank of v is defined as the minimum of all the d_i. Observe that the rank of v is one plus the second highest rank among the neighbors of v. Thus there can be at most one neighbor of vertex v which has rank at least the rank of v. All leaves have rank 0. Rank 1 vertices have at most one non-leaf neighbor.

Lemma 2. *There exists an assignment of the primal charge amongst the dual variables such that the dual constraint for each edge $e \equiv (u, v)$ is satisfied at least $\frac{15}{28}$ in expectation, i.e. $\mathbb{E}[y_u + y_v] \geq \frac{15}{28}$.*

Proof. Consider an edge $e \equiv (u, v)$ where rank of u is i and rank of v is j. We will show that $y_u + y_v \geq 2 + \epsilon$ for such an edge, when summed over all four matchings. The value of ϵ is chosen later. The proof is by induction on the lexicographic order of $< j, i >, j \geq i$.

Dual Variable Management: Consider an edge e from a vertex of rank i to a vertex of rank j, such that $i \leq j$. This edge will distribute its primal weight between its end-points. The exact values are discussed in the proof of the claim below. In general, we look to transfer all of the primal charge to the higher ranked vertex. But this does not work and we need a finer strategy. This is detailed below.

- If e does not belong to any matching, then it does not contribute to the value of dual variables.

- If e belongs to a single matching then, depending on the situation, one of 0, ϵ or 2ϵ of its primal charge will be assigned to the rank i vertex and rest will be assigned to the rank j vertex. The small constant ϵ is determined later.
- If e belongs to two matchings, then at most 3ϵ of its primal charge will be assigned to the rank i vertex as required. The rest is assigned to the rank j vertex.
- If e belongs to three or four matchings, then its entire primal charge is assigned to the rank j vertex.

The analysis breaks up into six cases.

Case 1. Suppose e does not belong to any matching. Then it must be a leaf edge. Hence, $i = 0$. There must be 4 edges incident on v besides e, each belonging to a distinct matching. Of these 4, at least 3 say e_1, e_2, and e_3, must be from lower ranked vertices to the rank j vertex v. The edges e_1, e_2, and e_3, each assign a charge of $1 - 2\epsilon$ to y_v. Therefore, $y_u + y_v \geq 3 - 6\epsilon \geq 2 + \epsilon$.

Case 2. Suppose e is a bad edge that belongs to a single matching. Since no internal edge can be a bad edge, $i = 0$. This implies (Lemma 1) that, there is an edge e_1 from a rank $j - 1$ vertex to v, which belongs to a single matching. Also, there is an edge e_2, from v to a higher ranked vertex, which also belongs to a single matching. The edge e assigns a charge of 1 to y_v. If e_1 assigns a charge of 1 (or $1 - \epsilon$) to y_v, then e_2 assigns ϵ (or 2ϵ respectively) to y_v. In either case, $y_u + y_v = 2 + \epsilon$. The key fact is that e_1 could not have assigned 2ϵ to a lower ranked vertex. Since, then, by Lemma 1, e cannot be a bad edge.

Case 3. Suppose e is not a bad edge, and it belongs to a single matching.
Case 3(a). $i = 0$. There are two sub cases.

- There is an edge e_1 from some rank $j - 1$ vertex to v which belongs to 2 matchings, or there are two other edges e_2 and e_3 from some lower ranked vertices to v, each belonging to separate matchings. The edge e assigns a charge of 1 to y_v. Either e_1 assigns a charge of at least $2 - 3\epsilon$ to y_v, or e_2 and e_3 assign a charge of at least $1 - 2\epsilon$ each, to y_v. In either case, $y_u + y_v \geq 3 - 4\epsilon \geq 2 + \epsilon$.
- There is one edge e_1, from a rank $j - 1$ vertex to v, which belongs to a single matching, and there is one edge e_2, from v to a higher ranked vertex, which belongs to 2 matchings. The edge e assigns a charge of 1 to y_v. If e_1 assigns a charge of 1 (or $1 - \epsilon$ or $1 - 2\epsilon$) to y_v, then e_2 assigns ϵ (or 2ϵ or 3ϵ respectively) to y_v. In either case, $y_u + y_v = 2 + \epsilon$.

Case 3(b). $i > 0$. There are two sub cases.

- There are at least two edges e_1 and e_2 from lower ranked vertices to u, and one edge e_3 from v to a higher ranked vertex. Each of these edges are in one matching only (not necessarily the same matching).
- There is one edge e_4 from a vertex of lower rank to u, at least one edge e_5 from a lower ranked vertex to v, and one edge e_6 from v to a vertex of higher rank. All these edges belong to a single matching (not necessarily the same).

The edge e assigns a charge of 1 among y_u and y_v. If e_1 and e_2 assign a charge of at least $1 - 2\epsilon$ each, to y_u, then $y_u + y_v \geq 3 - 4\epsilon \geq 2 + \epsilon$. Similarly, if e_4 assigns a charge of at least $1 - 2\epsilon$ to y_u, and e_5 assigns a charge of at least $1 - 2\epsilon$ to y_v, then $y_u + y_v \geq 3 - 4\epsilon \geq 2 + \epsilon$.

Case 4. Suppose e is a bad edge that belongs to two matchings. Then $i = 0$. This implies that there is an edge e_1, from v to a vertex of higher rank which belongs to a single matching. The edge e assigns a charge of 2 to y_v, and the edge e_1 assigns a charge of ϵ to y_v. Thus, $y_u + y_v = 2 + \epsilon$.

Case 5. Suppose e is not a bad edge and it belongs to two matchings. This means that either there is an edge e_1 from a lower ranked vertex to u, which belongs to at least one matching, or there is an edge from some lower ranked vertex to v that belongs to at least one matching, or there is an edge from v to some higher ranked vertex which belongs to two matchings. The edge e assigns a charge of 2 among y_u and y_v. The neighboring edges assign a charge of ϵ to y_u or y_v (depending on which vertex it is incident), to give $y_u + y_v \geq 2 + \epsilon$.

Case 6. Suppose, e belongs to 3 or 4 matchings, then trivially $y_u + y_v \geq 2 + \epsilon$. From the above conditions, the best value for the competitive ratio is obtained when $\epsilon = \frac{1}{7}$, yielding $\mathbb{E}[y_u + y_v] \geq \frac{15}{28}$. $\square$

Lemma 2 implies that the competitive ratio of the algorithm is at most $\frac{28}{15}$.

3   Lower Bounds

3.1   Lower Bound for MWM

In this section, we prove a lower bound on the competitive ratio of a natural class of randomized algorithms in the online preemptive model for MWM. The algorithms in this class, which we call *local* algorithms, have the property that their decision to accept or to reject a new edge is completely determined by the weights of the new edge and the conflicting edges in the matching maintained by the algorithm. Indeed, the optimal deterministic algorithm by McGregor [7] is a local algorithm. The notion of locality can be extended to randomized algorithms as well. In case of *randomized local algorithms*, the event that a new edge is accepted is independent of all such previous events, given the current matching maintained by the algorithm. Furthermore, the probability of this event is completely determined by the weight of the new edge and the conflicting edges in the matching maintained by the algorithm. Given that the optimal $(3+2\sqrt{2})$-competitive deterministic algorithm for MWM is a local algorithm, it is natural to ask whether randomized local algorithms can beat the deterministic lower bound of $(3 + 2\sqrt{2})$ by Varadaraja [9]. We answer this question in the negative, and prove the following theorem.

Theorem 2. *No randomized local algorithm for the MWM problem can have a competitive ratio less than $\alpha = 3 + 2\sqrt{2} \approx 5.828$.*

Note that the randomized algorithm by Epstein et al. [4] does not fall in this category, since the decision of accepting or rejecting a new edge is also dependent on the outcome of the coins tossed at the beginning of the run of the algorithm. (For details, see Section 3 of [4].) In order to prove Theorem 2, we will crucially use the following lemma, which is a consequence of Section 4 of [9].

Lemma 3. *If there exists an infinite sequence $(x_n)_{n\in\mathbb{N}}$ of positive real numbers such that for all n, $\beta x_n \geq \sum_{i=1}^{n+1} x_i + x_{n+1}$, then $\beta \geq 3 + 2\sqrt{2}$.*

Characterization of Local Randomized Algorithms. Suppose, for a contradiction, that there exists a randomized local algorithm $\mathcal{A}$ with a competitive ratio $\beta < \alpha = 3 + 2\sqrt{2}$, $\beta \geq 1$. Define the constant γ to be

$$\gamma = \frac{\beta\left(1 - \frac{1}{\alpha}\right)}{\left(1 - \frac{\beta}{\alpha}\right)} = \frac{\beta(\alpha - 1)}{\alpha - \beta} \geq 1 > \frac{1}{\alpha}$$

For $i = 0, 1, 2$, if w is the weight of a new edge and it has i conflicting edges, in the current matching, of weights $w_1, \ldots, w_i$, then $f_i(w_1, \ldots, w_i, w)$ gives the probability of switching to the new edge. The behavior of $\mathcal{A}$ is completely described by these three functions. We need the following key lemma to state our construction of the adversarial input.

The lemma states (informally) that given an edge of weight w_1, there exists weights x and y, close to each other such that if an edge of weight x (respective y) is adjacent to an edge of weight w_1, the probability of switching is at most (respectively at least) δ.

Lemma 4. *For every $\delta \in (0, 1/\alpha)$, $\epsilon > 0$, and w_1, there exist x and y such that $f_1(w_1, x) \geq \delta$, $f_1(w_1, y) \leq \delta$, $x - y \leq \epsilon$, and $w_1/\alpha \leq y \leq x \leq \gamma w_1$.*

The proof of this lemma can be found in Appendix C in [3].

The Adversarial Input. The adversarial input is parameterized by four parameters: $\delta \in (0, 1/\alpha)$, $\epsilon > 0$, m, and n, where m and n determine the graph and δ and ϵ determine the weights of its edges.

Define the infinite sequences $(x_i)_{i\in\mathbb{N}}$ and $(y_i)_{i\in\mathbb{N}}$, as functions of ϵ and δ, as follows. $x_1 = 1$, and for all i, having defined x_i, let x_{i+1} and y_i be such that $f_1(x_i, x_{i+1}) \geq \delta$, $f_1(x_i, y_i) \leq \delta$, $x_{i+1} - y_i \leq \epsilon$, and $x_i/\alpha \leq y_i \leq x_{i+1} \leq \gamma x_i$. Lemma 4 ensures that such x_{i+1} and y_i exist. Furthermore, by induction on i, it is easy to see that for all i,

$$1/\alpha^i \leq y_i \leq x_{i+1} \leq \gamma^i \tag{1}$$

These sequences will be the weights of the edges in the input graph.

Given m and n, the input graph contains several layers of vertices, namely $A_1, A_2, \ldots, A_{n+1}, A_{n+2}$ and $B_1, B_2, \ldots, B_{n+1}$; each layer containing m vertices. The vertices in the layer A_i are named $a_1^i, a_2^i, \ldots, a_m^i$, and those in layer B_i are named analogously. We have a complete bipartite graph J_i between layer A_i

and A_{i+1} and an edge between a_j^i and b_j^i for every i, j (that is, a matching M_i between A_i and B_i).

For $i = 1$ to n, the edges $\{(a_j^i, a_{j'}^{i+1}) | 1 \leq j, j' \leq m\}$, in the complete bipartite graph between A_i and A_{i+1}, have weight x_i, and the edges $\{(a_j^i, b_j^i) | 1 \leq j \leq m\}$, in the matching between A_i and B_i, have weight y_i. The edges in the complete graph J_{n+1} have weight x_n, and those in the matching M_{n+1} have weight y_n. Note that weights x_i and y_i depend on ϵ and δ, but are independent of m and n. Clearly, the weight of the maximum weight matching in this graph is bounded from below by the weight of the matching $\bigcup_{i=1}^{n+1} M_i$. Since $y_i \geq x_{i+1} - \varepsilon$, we have

$$\text{OPT} \geq m \left(\sum_{i=1}^{n} y_i + y_n \right) \geq m \left(\sum_{i=2}^{n+1} x_i + x_{n+1} - (n+1)\epsilon \right) \qquad (2)$$

The edges of the graph are revealed in $n+1$ phases. In the ith phase, the edges in $J_i \cup M_i$ are revealed as follows. The phase is divided into m sub phases. In the jth sub phase of the ith phase, edges incident on a_j^i are revealed, in the order $(a_j^i, a_1^{i+1}), (a_j^i, a_2^{i+1}), \ldots, (a_j^i, a_m^{i+1}), (a_j^i, b_j^i)$.

Analysis of the Lower Bound. The overall idea of bounding the weight of the algorithm's matching is as follows. In each phase i, we will prove that as many as $m - O(1)$ edges of J_i and only $\delta m + O(1)$ edges of M_i are picked by the algorithm. Furthermore, in the $i + 1^{\text{th}}$ phase, since $m - O(1)$ edges from J_{i+1} are picked, all but $O(1)$ edges of the edges picked from J_i are discarded. Thus, the algorithm ends up with $\delta m + O(1)$ edges from each M_i, and $O(1)$ edges from each J_i, except possibly J_n and J_{n+1}. The algorithm can end up with at most m edges from $J_n \cup J_{n+1}$, since the size of the maximum matching in $J_n \cup J_{n+1}$ is m. Thus, the weight of the algorithm's matching is at most mx_n plus a quantity that can be neglected for large m and small δ.

Let X_i (resp. Y_i) be the set of edges of J_i (resp. M_i) held by the algorithm at the end of input. Then we have,

Lemma 5. *For all $i = 1$ to n*

$$E[|Y_i|] \leq \delta m + \frac{1 - \delta}{\delta}$$

Lemma 6. *For all $i = 1$ to $n - 1$*

$$E[|X_i|] \leq \frac{1 - \delta}{\delta}$$

Lemma 7.

$$E[|Y_{n+1}|] \leq \delta m + \frac{1 - \delta}{\delta}$$

The proof of the above lemmas can be found in Appendix C in [3].

We are now ready to prove Theorem 2. The expected weight of the matching held by $\mathcal{A}$ is

$$E[\text{ALG}] \leq \sum_{i=1}^{n} y_i E[|Y_i|] + y_n E[|Y_{n+1}|] + \sum_{i=1}^{n-1} x_i E[|X_i|] + x_n E[|X_n \cup X_{n+1}|]$$

Using Lemmas 5, 7, 6, and the facts that $y_i \leq x_{i+1}$ for all i and $E[|X_n \cup X_{n+1}|] \leq m$ (since $X_n \cup X_{n+1}$ is a matching in $J_n \cup J_{n+1}$), we have

$$E[\text{ALG}] \leq \left(\delta m + \frac{1-\delta}{\delta} \right) \left(\sum_{i=2}^{n+1} x_i + x_{n+1} \right) + \frac{1-\delta}{\delta} \sum_{i=1}^{n-1} x_i + m x_n$$

Since the algorithm is β-competitive, for all n, m, δ and ϵ we must have $E[\text{ALG}] \geq \text{OPT}/\beta$. From the above and equation (2), we must have

$$\begin{aligned} \left(\delta m + \frac{1-\delta}{\delta} \right) \left(\sum_{i=2}^{n+1} x_i + x_{n+1} \right) \\ + \frac{1-\delta}{\delta} \sum_{i=1}^{n-1} x_i + m x_n \end{aligned} \geq \frac{m}{\beta} \left(\sum_{i=2}^{n+1} x_i + x_{n+1} - (n+1)\epsilon \right)$$

Since the above holds for arbitrarily large m, ignoring the terms independent of m (recall that x_i's are functions of ϵ and δ only), we have for all δ and ϵ,

$$\delta \left(\sum_{i=2}^{n+1} x_i + x_{n+1} \right) + x_n \geq \frac{1}{\beta} \left(\sum_{i=2}^{n+1} x_i + x_{n+1} - (n+1)\epsilon \right)$$

that is,

$$x_n \geq \frac{1}{\beta} \left(\sum_{i=2}^{n+1} x_i + x_{n+1} - (n+1)\epsilon \right) - \delta \left(\sum_{i=2}^{n+1} x_i + x_{n+1} \right)$$

Taking limit inferior as $\delta \to 0$ in the above inequality, and noting that limit inferior is super-additive we get for all ϵ,

$$\begin{aligned} \liminf_{\delta \to 0} x_n \geq &\frac{1}{\beta} \left(\sum_{i=2}^{n+1} \liminf_{\delta \to 0} x_i + \liminf_{\delta \to 0} x_{n+1} - (n+1)\epsilon \right) \\ &- \limsup_{\delta \to 0} \delta \left(\sum_{i=2}^{n+1} x_i + x_{n+1} \right) \end{aligned}$$

Recall that x_i's are functions of ϵ and δ, and that from equation (1), $1/\alpha^i \leq x_{i+1} \leq \gamma^i$, where the bounds are independent of δ. Thus, all the limits in the above inequality exist. Moreover, $\lim_{\delta \to 0} \delta \left(\sum_{i=2}^{n+1} x_i + x_{n+1} \right)$ exists and is 0, for all ϵ. This implies $\limsup_{\delta \to 0} \delta \left(\sum_{i=2}^{n+1} x_i + x_{n+1} \right) = 0$ and we get for all ε,

$$\liminf_{\delta \to 0} x_n \geq \frac{1}{\beta} \left(\sum_{i=2}^{n+1} \liminf_{\delta \to 0} x_i + \liminf_{\delta \to 0} x_{n+1} - (n+1)\epsilon \right)$$

Again, taking limit inferior as $\epsilon \to 0$, and using super-additivity,

$$\liminf_{\epsilon \to 0} \liminf_{\delta \to 0} x_n \geq \frac{1}{\beta} \left(\sum_{i=2}^{n+1} \liminf_{\epsilon \to 0} \liminf_{\delta \to 0} x_i + \liminf_{\epsilon \to 0} \liminf_{\delta \to 0} x_{n+1} \right)$$

Note that the above holds for all n. Finally, let $\overline{x_n} = \liminf_{\epsilon \to 0} \liminf_{\delta \to 0} x_{n+1}$. Then we have the infinite sequence $(\overline{x_n})_{n \in \mathbb{N}}$ such that for all n, $\beta \overline{x_n} \geq \sum_{i=1}^{n+1} \overline{x_i} + \overline{x_{n+1}}$. Thus, by Lemma 3, we have $\beta \geq 3 + 2\sqrt{2}$.

3.2   Lower Bound for θ Structured Graphs

Recall that an edge weighted graph is said to be θ-structured if the weights of the edges are powers of θ. The following bound applies to any deterministic algorithm for MWM on θ-structured graphs.

Theorem 3. *No deterministic algorithm can have a competitive ratio less than* $2 + \frac{2}{\theta-2}$ *for MWM on θ-structured graphs, for $\theta \geq 4$.*

The proof of the above theorem can be found in Appendix D in [3].

4   Randomized Algorithm for Paths

When the input graph is restricted to be a collection of paths, then every new edge that arrives connects two (possibly empty) paths. Our algorithm consists of several cases, depending on the lengths of the two paths.

Algorithm 2. Randomized Algorithm for Paths

1: $M = \emptyset$. {M is the matching stored by the algorithm.}
2: **for** each new edge e **do**
3:    Let $L_1 \geq L_2$ be the lengths of the two (possibly empty) paths P_1, P_2 that e connects.
4:    If $L_1 > 0$ (resp. $L_2 > 0$), let e_1 (resp. e_2) be the edge on P_1 (resp. P_2) adjacent to e.
5:    **if** e is a disjoint edge $\{L_1 = L_2 = 0\ \}$ **then**
6:        $M = M \cup \{e\}$.
7:    **else if** e is revealed on a disjoint edge e_1 $\{L_1 = 1, L_2 = 0.\ e_1 \in M\}$ **then**
8:        with probability $\frac{1}{2}$, $M = M \setminus \{e_1\} \cup \{e\}$.
9:    **else if** e is revealed on an end point of path of length > 1 $\{L_1 > 1, L_2 = 0\}$ **then**
10:       if $e_1 \notin M$, $M = M \cup \{e\}$.
11:   **else if** e joins two disjoint edges $\{L_1 = L_2 = 1.\ e_1, e_2 \in M\}$ **then**
12:       with probability $\frac{1}{2}$, $M = M \setminus \{e_1, e_2\} \cup \{e\}$.
13:   **else if** e joins a path and a disjoint edge $\{L_1 > 1, L_2 = 1.\ e_2 \in M\}$ **then**
14:       if $e_1 \notin M$, $M = M \setminus \{e_2\} \cup \{e\}$.
15:   **else if** e joins two paths of length $> 1 \{L_1 > 1, L_2 > 1\}$ **then**
16:       if $e_1 \notin M$ and $e_2 \notin M$, $M = M \cup \{e\}$.
17:   **end if**
18:   Output M.
19: **end for**

The following simple observations can be made by looking at the algorithm:

– All isolated edges belong to M with probability one.

– The end vertex of any path of *length* > 1 is covered by M with probability $\frac{1}{2}$, and this is independent of the end vertex of any other path being covered.
– For a path of length $2, 3$, or 4, each maximal matching is present in M with probability $\frac{1}{2}$.

Theorem 4. *The randomized algorithm for finding MCM on path graphs is $\frac{4}{3}$-competitive.*

The proof of above theorem can be found in Appendix E in [3].

References

1. Buchbinder, N., Naor, J.: The Design of Competitive Online Algorithms via a Primal-Dual Approach. Foundations and Trends in Theoretical Computer Science 3(2-3), 93–263 (2009)
2. Chan, T.H., Chen, F., Wu, X., Zhao, Z.: Ranking on Arbitrary Graphs: Rematch via Continuous LP with Monotone and Boundary Condition Constraints. In: Proceedings of the Twenty-Fifth Annual ACM-SIAM Symposium on Discrete Algorithms, SODA 2014, Portland, Oregon, USA, January 5-7, pp. 1112–1122 (2014)
3. Chiplunkar, A., Tirodkar, S., Vishwanathan, S.: On Randomized Algorithms for Matching in the Online Preemptive Model. CoRR abs/1412.8615 (2014). http://arxiv.org/abs/1412.8615
4. Epstein, L., Levin, A., Segev, D., Weimann, O.: Improved Bounds for Online Preemptive Matching. In: 30th International Symposium on Theoretical Aspects of Computer Science, STACS 2013, Kiel, Germany, pp. 389–399 (2013)
5. Feigenbaum, J., Kannan, S., McGregor, A., Suri, S., Zhang, J.: On Graph Problems in a Semi-streaming Model. Theor. Comput. Sci. 348(2), 207–216 (2005)
6. Karp, R.M., Vazirani, U.V., Vazirani, V.V.: An Optimal Algorithm for On-line Bipartite Matching. In: Proceedings of the Twenty-Second Annual ACM Symposium on Theory of Computing, STOC 1990, pp. 352–358. ACM, New York (1990)
7. McGregor, A.: Finding Graph Matchings in Data Streams. In: Chekuri, C., Jansen, K., Rolim, J.D.P., Trevisan, L. (eds.) APPROX 2005 and RANDOM 2005. LNCS, vol. 3624, pp. 170–181. Springer, Heidelberg (2005)
8. Poloczek, M., Szegedy, M.: Randomized Greedy Algorithms for the Maximum Matching Problem with New Analysis. In: 53rd Annual IEEE Symposium on Foundations of Computer Science, FOCS 2012, New Brunswick, NJ, USA, pp. 708–717 (2012)
9. Badanidiyuru Varadaraja, A.: Buyback Problem - Approximate Matroid Intersection with Cancellation Costs. In: Aceto, L., Henzinger, M., Sgall, J. (eds.) ICALP 2011, Part I. LNCS, vol. 6755, pp. 379–390. Springer, Heidelberg (2011)
10. Yao, A.C.C.: Probabilistic computations: Toward a unified measure of complexity. In: 18th Annual Symposium on Foundations of Computer Science, FOCS 1977, pp. 222–227 (October 1977)

A Characterization of Consistent Digital Line Segments in $\mathbb{Z}^2$

Iffat Chowdhury and Matt Gibson

Department of Computer Science
University of Texas at San Antonio
San Antonio, TX USA
iffat.chowdhury@utsa.edu, gibson@cs.utsa.edu

Abstract. Our concern is the digitalization of line segments in $\mathbb{Z}^2$ as considered by Chun et al. [5] and Christ et al. [4]. The key property that differentiates the research of Chun et al. and Christ et al. from other research in digital line segment construction is that the intersection of any two segments must be connected. Such a system of segments is called a consistent digital line segments system (CDS). Chun et al. give a construction for all segments in $\mathbb{Z}^d$ that share a common endpoint (called consistent digital rays (CDR)) that has asymptotically optimal Hausdorff distance, and Christ et al. give a complete CDS in $\mathbb{Z}^2$ with optimal Hausdorff distance. Christ et al. also give a characterization of CDRs in $\mathbb{Z}^2$, and they leave open the question on how to characterize CDSes in $\mathbb{Z}^2$. In this paper, we answer one of the most important open question regarding CDSes in $\mathbb{Z}^2$ by giving the characterization asked for by Christ et al. We obtain the characterization by giving a set of necessary and sufficient conditions that a CDS must satisfy.

1 Introduction

This paper explores families of digital line segments as considered by Chun et al. [5] and Christ et al. [4]. Consider the unit grid $\mathbb{Z}^2$, and in particular the unit grid graph: for any two points $p = (p^x, p^y)$ and $q = (q^x, q^y)$ in $\mathbb{Z}^2$, p and q are neighbors if and only if $|p^x - q^x| + |p^y - q^y| = 1$. For any pair of grid vertices p and q, we'd like to define a digital line segment $R_p(q)$ from p to q. The collection of digital segments must satisfy the following five properties.

(S1) *Grid path property:* For all $p, q \in \mathbb{Z}^2$, $R_p(q)$ is the points of a path from p to q in the grid topology.

(S2) *Symmetry property:* For all $p, q \in \mathbb{Z}^2$, we have $R_p(q) = R_q(p)$.

(S3) *Subsegment property:* For all $p, q \in \mathbb{Z}^2$ and every $r, s \in R_p(q)$, we have $R_r(s) \subseteq R_p(q)$.

Properties (S2) and (S3) are quite natural to ask for; the subsegment property (S3) is motivated by the fact that the intersection of any two Euclidean line segments is connected. See Fig. 1 (a) for an illustration of a violation of (S3).

© Springer-Verlag Berlin Heidelberg 2015
N. Bansal and I. Finocchi (Eds.): ESA 2015, LNCS 9294, pp. 337–348, 2015.
DOI: 10.1007/978-3-662-48350-3_29

Note that a simple "rounding" scheme of a Euclidean segment commonly used in computer vision produces a good digitalization in isolation, but unfortunately it will not satisfy (S3) when combined with other digital segments, see Fig. 1 (b) and (c).

(S4) *Prolongation property:* For all $p, q \in \mathbb{Z}^2$, there exists $r \in \mathbb{Z}^2$, such that $r \notin R_p(q)$ and $R_p(q) \subseteq R_p(r)$.

The prolongation property (S4) is also a quite natural property to desire with respect to Euclidean line segments. Any Euclidean line segment can be extended to an infinite line, and we would like a similar property to hold for our digital line segments. While (S1)-(S4) form a natural set of axioms for digital segments, there are pathological examples of segments that satisfy these properties which we would like to rule out. For example, Christ et al. [4] describe a CDS where a double spiral is centered at some point in $\mathbb{Z}^2$, traversing all points of $\mathbb{Z}^2$. A CDS is obtained by defining $R_p(q)$ to be the subsegment of this spiral connecting p and q. To rule out these CDSes, the following property was added.

(S5) *Monotonicity property:* For all $p, q \in \mathbb{Z}^2$, if $p^x = q^x = c_1$ for any c_1 (resp. $p^y = q^y = c_2$ for any c_2), then every point $r \in R_p(q)$ has $r^x = c_1$ (resp. $r^y = c_2$).

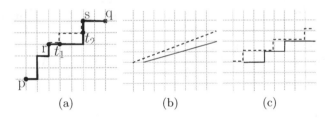

Fig. 1. (a) An illustration of the violation of (S3). The solid segment is $R_p(q)$, and the dashed segment is $R_r(s)$. (b) The dashed line and the solid line denote two different Euclidean line segments. (c) The corresponding digital line segments via a rounding approach.

If a system of digital line segments satisfies the axioms $(S1) - (S5)$, then it is called a *consistent digital line segments system* (CDS). Given such a system, one can easily define digital analogs of various Euclidean objects. For example, a Euclidean object O is convex if for any two points $p, q \in O$ we have that the Euclidean line segment $\overline{pq}$ does not contain any points outside of O. Given a CDS, the natural definition of a digital convex object will satisfy some nice properties. For example, one can see that a digital convex object with respect to a CDS cannot contain any holes (as a result of the prolongation property (S4)). Similarly, one can easily obtain the digital analog of a star-shaped object with a CDS. A Euclidean object S is star-shaped if there is a point $u \in S$ such that for every point $v \in S$ we have the Euclidean line segment $\overline{uv}$ does not contain any points outside of S, and the natural digital generalization easily follows.

Previous Works. Unknown to Chun et al. and Christ et al. when publishing their papers, in 1988 Luby [13] considered *grid geometries* which are equivalent to systems of digital line segments satisfying (S1), (S2), (S5) described in this paper. Let $p, q \in \mathbb{Z}^2$ be such that $p^x \le q^x$. We say the segment $R_p(q)$ has *nonnegative slope* if $p^y \le q^y$ and otherwise has *negative slope*. Luby investigates a property called *smoothness* which uses the following notion of distance between two digital line segments. Consider two digital segments $R_p(q)$ and $R_{p'}(q')$ with nonnegative slope and any point $r \in R_p(q)$. If there is a $s \in R_{p'}(q')$ such that $r^x + r^y = s^x + s^y$, then dist$(R_p(q), R_{p'}(q'), r)$ is defined to be $r^x - s^x$. If there is no such s then we say that dist$(R_p(q), R_{p'}(q'), r)$ is undefined. See Fig. 2 (a). The segments are *smooth* if dist$(R_p(q), R_{p'}(q'), r)$ is either monotonically increasing or monotonically decreasing varying r over its defined domain. See Fig. 2 (b). There is also a symmetric definition of smoothness for pairs of digital segments with negative slope. A grid geometry is said to be smooth if every pair of nonnegative sloped segments and every pair of negative sloped segments are smooth. Luby shows that if a grid geometry is smooth, then it satisfies properties (S3) and (S4) (and therefore is a CDS).

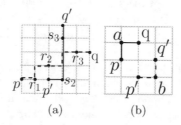

Fig. 2. (a) dist$(R_p(q), R_{p'}(q'), r_1)$ is undefined. dist$(R_p(q), R_{p'}(q'), r_2) = -1$ (using $s_2 \in R_{p'}(q')$), dist$(R_p(q), R_{p'}(q'), r_3) = 1$. (b) An example of segments that are not smooth: dist$(R_p(q), R_{p'}(q'), p) = -1$, dist$(R_p(q), R_{p'}(q'), a) = -2$, and dist$(R_p(q), R_{p'}(q'), q) = -1$.

Chun et al. [5] give an $\Omega(\log n)$ lower bound on the Hausdorff Distance of a CDS where n is the number of points in the segment, and the result even applies to *consistent digital rays* or CDRs (i.e., all segments share a common endpoint). Note that this lower bound is due to property (S3), as it is easy to see that if the requirement of (S3) is removed then digital segments with $O(1)$ Hausdorff distance are easily obtained, for example the trivial "rounding" scheme used in Fig. 1 (c). Chun et al. give a construction of CDRs that satisfy the desired properties (S1)-(S5) with a tight upper bound of $O(\log n)$ on the Hausdorff distance. Christ et al. [4] extend the result to get an optimal $O(\log n)$ upper bound on Hausdorff distance for a CDS in $\mathbb{Z}^2$.

After giving the optimal CDS in $\mathbb{Z}^2$, Christ et al. [4] investigate common patterns in CDSes in an effort to obtain a characterization of CDSes. As a starting point, they are able to give a characterization of CDRs. In their effort to give a characterization, they proved a sufficient condition on the construction of the CDSes but then they give an example of a CDS that demonstrates that their sufficient condition is not necessary. They ask if there are any other interesting examples of CDSes that do not follow their sufficient condition and left open the question on how to characterize the CDSes in $\mathbb{Z}^2$.

Our Contributions. In this paper, we answer one of the most important open question regarding CDSes in $\mathbb{Z}^2$ by giving the characterization asked for by Christ et al. Since Christ et al. have given a characterization of CDRs, we view the construction of a CDS as the assignment of a CDR system to each point in $\mathbb{Z}^2$ such that the union of these CDRs satisfies properties (S1)-(S5). We obtain the characterization by giving a set of necessary and sufficient conditions that the CDRs must satisfy in order to be combined into a CDS. Then to tie together our work with the previous work in CDSes, we analyze the work of Christ et al. and Luby in the context of our characterization.

Motivation and Related Works. Digital geometry plays a fundamental and substantial role in many computer vision applications, for example image segmentation, image processing, facial recognition, fingerprint recognition, and some medical applications. One of the key challenges in digital geometry is to represent Euclidean objects in a digital space so that the digital objects have a similar visual appearance as their Euclidean counterparts. Representation of Euclidean objects in a digital space has been a focus in research for over 25 years, see for example [8,10,15,16,7,2,1].

Digital line segments are particularly important to model accurately, as other digital objects depend on them for their own definitions (e.g. convex and star-shaped objects). In 1986, Greene and Yao [10] gave an interface between the continuous domain of Euclidean line segments and the discrete domain of digital line segments. Goodrich et al. [9] focused on rounding the Euclidean geometric objects to a specific resolution for better computer representation. They gave an efficient algorithm for $\mathbb{R}^2$ and $\mathbb{R}^3$ in the "snap rounding paradigm" where the endpoints or the intersection points of several different line segments are the main concerns. Later, Sivignon et al. [14] also gave some results on the intersection of two digital line segments. In their review paper, Klette et al. [11] discussed the straightness of digital line segments. The characteristics of the subsegment of digital straight line was computed in [12]. Cohen et al. [6] gave a method of converting 3D continuous line segments to discrete line segments based on a voxelization algorithm, but they did not have the requirement that the intersection of two digital line segments should be connected.

2   Preliminaries

Before we describe our characterization, we first need to give some details of the Christ et al. characterization of CDRs. For any point $p \in \mathbb{Z}^2$, let $Q_p^1, Q_p^2, Q_p^3, Q_p^4$ denote the first, second, third, and fourth quadrants of p respectively. Christ et al. show how to construct $R_p(q)$ for $q \in Q_p^1$ from any total order of $\mathbb{Z}$, which we denote $\prec_p^1$. We describe $R_p(q)$ by "walking" from p to q. Starting from p, the segment will repeatedly move either "up" or "right" until it reaches q. Suppose on the walk we are currently at a point $r = (r^x, r^y)$. Then it needs to move to either $(r^x + 1, r^y)$ or $(r^x, r^y + 1)$. Either way, the sum of the two coordinates of the current point is increased by 1 in each step. The segment will move up $q^y - p^y$ times, and it will move right $q^x - p^x$ times. If the line segment is at a point r for which $r^x + r^y$ is among the $q^y - p^y$ greatest integers in the interval $I(p, q) := [p^x + p^y, q^x + q^y - 1]$ according to $\prec_p^1$, the line segment will move up. Otherwise, it will move right. See Fig. 3 (a) for an example. Throughout the paper, when we say that $a < b$, we mean that a is less than b in natural total order and when we say that $a \prec b$, we mean that a is less than b according to total order $\prec$.

Property (S3) is generally the most difficult property to deal with, and we will argue that the segments $R_p(q)$ and $R_p(q')$ will not violate (S3) for any points q and q' in the first quadrant of p. As shown in [4], (S3) is violated if and only if two segments intersect at a point t_1, one segment moves vertically from t_1 while the other moves horizontally from t_1, and the segments later intersect again. Consider two digital segments that "break apart" at some point t_1 in this manner, and suppose they do intersect again. Let t_2 be the first point at which they intersect after "splitting apart". Then we say that (t_1, t_2) is *witness to the violation of (S3)* or a *witness* for short. Therefore, one can show that any two segments satisfy $(S3)$ by showing that they do not have witnesses, and this is how we will prove the segments satisfy $(S3)$ now (and also in our characterization). Consider the segments $R_p(q)$ and $R_p(q')$ generated according to the Christ et al. definition, and suppose for the sake of contradiction that they have a witness (t_1, t_2) as in Fig. 1 (a). One segment moves up at point t_1 and moves right into the point t_2 which implies $(t_2^x + t_2^y - 1) \prec_p^1 (t_1^x + t_1^y)$, and the other segment moves right at point t_1 and moves up into the point t_2 which implies $(t_1^x + t_1^y) \prec_p^1 (t_2^x + t_2^y - 1)$, a contradiction. Therefore $R_p(q)$ and $R_p(q')$ do not have any witnesses and therefore satisfy (S3). Christ et al. [4] show that digital segments in quadrants Q_p^2, Q_p^3, and Q_p^4 can also be generated with total orders $\prec_p^2$, $\prec_p^3$, and $\prec_p^4$ (described formally below), and moreover they establish a one-to-one correspondence between CDRs and total orders. That is, (1) given any total order of $\mathbb{Z}$, one can generate all digital rays in any quadrant of p, and (2) for any set of digital rays R in some quadrant of p, there is a total order that will generate R. This provides a characterization of CDRs.

Given the characterization of CDRs, the problem of constructing a complete CDS can be viewed as assigning total orders to all points in $\mathbb{Z}^2$ so that the segments obtained using these total orders are collectively a CDS. Suppose that for every point $p \in \mathbb{Z}^2$, we assign to p a total order $\prec_p^1$ to generate segments to all points in Q_p^1. Now suppose that we want to define a "third-quadrant segment" $R_p(q)$ to some point $q \in Q_p^3$. Note that $q \in Q_p^3$ implies that $p \in Q_q^1$. Since all first-quadrant segments have been defined, this means $R_q(p)$ has been defined, and the symmetry property (S2) states that $R_p(q) = R_q(p)$. Therefore we do not need to use a total order to generate these third-quadrant segments; we simply

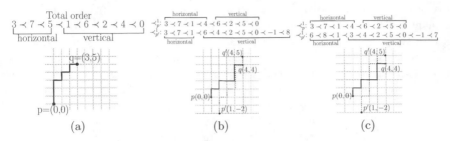

Fig. 3. (a) The digital line segment between $p = (0,0)$ and $q = (3,5)$. According to $\prec_p$, the $q^x - p^x = 3$ smallest integers in $[0,7]$ correspond to the horizontal movements, and the $q^y - p^y = 5$ largest integers in $[0,7]$ correspond to the vertical movements. (b) A choice of $\prec_p^1$ and $\prec_{p'}^1$ that satisfies (S3). (c) A choice of $\prec_p^1$ and $\prec_{p'}^1$ that does not satisfy (S3).

use the corresponding first-quadrant segments which have already been defined. In order to be part of a CDS, $\bigcup_{q \in Q_p^3} R_q(p)$ must be a system of rays in Q_p^3 that satisfies (S1)-(S5). From the characterization of rays, we know that there is an implicit third-quadrant total order $\prec_p^3$ on the integers in the range $(-\infty, p^x + p^y]$ that can be used to generate these rays. This generation is done in a very similar manner as in first quadrant rays. The key differences are: (1) the first quadrant segment $R_q(p)$ uses the interval $[q^x + q^y, p^x + p^y - 1]$ and the third quadrant segment $R_p(q)$ uses the interval $[q^x + q^y + 1, p^x + p^y]$, and (2) the sum of the coordinates of our "current point" *decreases* by 1 each time as we walk from p to q. Note that when considering first-quadrant segments, a horizontal movement (resp. vertical movement) is determined by the sum of the coordinates of the "left" endpoint (resp. "bottom" endpoint), whereas in a third quadrant segment a horizontal movement (resp. vertical movement) is determined by the sum of the coordinates of the "right" endpoint (resp. "top" endpoint). This implies that if the first-quadrant segment made a horizontal (resp. vertical) movement at a point where the sum of the coordinates is a, then the corresponding third-quadrant segment should make a horizontal (resp. vertical) movement at $(a+1)$. For example consider Fig. 3 (a). Since the horizontal movements of this first quadrant segment are at 3, 7, and 5, then the third quadrant segment should make horizontal movements at 4, 8, and 6. Similarly, the third quadrant segment should make vertical movements at 2, 7, 3, 5, and 1. We again state that we do not explicitly construct third quadrant segments using this technique and instead obtain them directly from the corresponding first quadrant segments. But note that if there is no total order $\prec_p^3$ which can be used to generate these third-quadrant segments then the segments necessarily must not satisfy at least one of (S1)-(S5). These implicit third-quadrant total orders will play an important role in the proof of our characterization.

Now consider the definition of segments $R_p(q)$ with negative slope, that is, $R_p(q)$ for which $q \in Q_p^2$ or $q \in Q_p^4$. We "mirror" p and q by multiplying both x-coordinates by -1. Let $m(p) = (-p^x, p^y)$ and $m(q) = (-q^x, q^y)$ denote the mirrored points. Note that if $q \in Q_p^2$, then $m(q) \in Q_{m(p)}^1$, and if $q \in Q_p^4$, then $m(q) \in Q_{m(p)}^3$. Therefore $R_{m(p)}(m(q))$ is a segment with nonnegative slope and can be defined as described above. We compute a second-quadrant segment $R_p(q)$ by making the same sequence of horizontal/vertical movements as $R_{m(p)}(m(q))$ when generated by a second-quadrant total order $\prec_p^2$ on the integers in the range $[-p^x + p^y, \infty)$. Similarly to third quadrant segments, fourth-quadrant segments $R_p(q)$ are set to be the same as $R_q(p)$ and there is an implicit fourth-quadrant total order $\prec_p^4$ on the integers in the range $(-\infty, -p^x + p^y]$.

3   A Characterization of CDSes in $\mathbb{Z}^2$

In a complete CDS, the segments that are adjacent to any point $p \in \mathbb{Z}^2$ can be viewed as a system of CDRs emanating from p, and therefore there is a total order that can be used to generate these segments. Christ et al. show that if the same total order is used by every point in $\mathbb{Z}^2$ to generate its adjacent segments, then the result is a CDS (the analysis follows very closely to the analysis for CDRs shown in the previous section). However, there are some situations in which points can be assigned different total orders and we still get a CDS. To illustrate this, consider Fig. 3 (b). Note that $\prec_p^1$ and $\prec_{p'}^1$ disagree on the relative ordering of 4 and 6, yet the resulting segments $R_p(q)$ and $R_{p'}(q')$ satisfy property (S3). But if we instead use the total orders as shown in Fig. 3 (c), they once again disagree on the ordering of 4 and 6 but this time $R_p(q)$ and $R_{p'}(q')$ do not satisfy property (S3). The issue is then to identify a set of necessary and sufficient properties of the total orders in a CDS.

We are now ready to give our characterization. We assume that we are considering segments $R_p(q)$ with nonnegative slope for the majority of this section, and we give set of necessary and sufficient conditions which $\prec_p^1$ must satisfy for each p in $\mathbb{Z}^2$. To help explain what must happen we first look at the interaction between first quadrant rays and third quadrant rays. Even though our conditions are only with respect to first-quadrant total orders, it will be useful to describe that our condition is necessary by showing that if the condition is not satisfied then there is some point q such that *any* definition of $\prec_q^3$ would generate third-quadrant segments that violate (S3).

Suppose, we have the total order $\prec_{p_1}^1$ for some point $p_1 \in \mathbb{Z}^2$, and let $p_2 \in Q_{p_1}^1$ and recall that we must have $R_{p_2}(p_1) = R_{p_1}(p_2)$ by property (S2). Now consider how $\prec_{p_2}^3$ must be defined so that $R_{p_2}(p_1) = R_{p_1}(p_2)$. Any integer on which $R_{p_2}(p_1)$ moves horizontally should be smaller than any integer on which the path moves vertically with respect to $\prec_{p_2}^3$, otherwise $R_{p_2}(p_1) \neq R_{p_1}(p_2)$. Motivated by this, we say that an integer on which $R_{p_2}(p_1)$ moves vertically has *priority* over an integer on which it moves horizontally. So, for any two integers a and b such that a has priority over b, $(a + 1)$ must be larger than $(b + 1)$ with respect to $\prec_{p_2}^3$.

Now, suppose we have three points p_1, p_2, p_3, where $p_1, p_2 \in Q_{p_3}^3$. $\prec_{p_3}^3$ has a set of priorities induced by $R_{p_1}(p_3)$ and another set of priorities induced by $R_{p_2}(p_3)$. Let a and b be two integers in $I(p_1, p_3) \cap I(p_2, p_3)$. If a has priority over b in $\prec_{p_1}^1$ and b has priority over a in $\prec_{p_2}^1$, then we call this a *conflicting priority*. If we have a conflicting priority then any definition of $\prec_{p_3}^3$ will violate (S2). Indeed, if $(b+1) \prec_{p_3}^3 (a+1)$ then this would imply $R_{p_3}(p_2) \neq R_{p_2}(p_3)$, and if $(a+1) \prec_{p_3}^3 (b+1)$ then this would imply $R_{p_3}(p_1) \neq R_{p_1}(p_3)$. Therefore it is necessary to define $\prec_{p_1}^1$ and $\prec_{p_2}^1$ so that there will not be any conflicting priorities for any choice of $p_3 \in Q_{p_1}^1 \cap Q_{p_2}^1$.

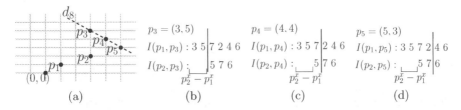

Fig. 4. The layout view of the intervals with $p_1 = (1,1)$ and $p_2 = (3,2)$. (a) The points in the grid. (b) Dividing line for p_3. (c) Dividing line for p_4. (d) Dividing line for p_5.

To help visualize what must happen to avoid these conflicting priorities, we describe a "layout" of the integers in the interval. Consider a point $p_3 \in Q_{p_1}^1 \cap Q_{p_2}^1$ and the intervals $I(p_1, p_3)$ and $I(p_2, p_3)$ that are used to define the segments $R_{p_1}(p_3)$ and $R_{p_2}(p_3)$ respectively, and without loss of generality assume that $p_1^x \leq p_2^x$. We write the intervals in increasing order in a matrix with two rows with $I(p_1, p_3)$ in the top row and $I(p_2, p_3)$ in the bottom row. The first element of $I(p_2, p_3)$ is "shifted" to the right $(p_2^x - p_1^x)$ positions after the first element of $I(p_1, p_3)$. Note that the integers in $I(p_1, p_3)$ and $I(p_2, p_3)$ are determined by the natural total order on the integers, but then are sorted by the total orders $\prec_{p_1}^1$ and $\prec_{p_2}^1$ respectively. The advantage of the layout view is that a single vertical line can break both of the intervals into the horizontal movements portion and vertical movements portion. We call such a line a *dividing line*. The left parts consist of the integers on which the segments make horizontal movements and the right parts consist of the integers on which the segments make vertical movements. We define the *antidiagonal* d_C to be the set of all of the points $p = (p^x, p^y)$ in $\mathbb{Z}^2$ such that $(p^x + p^y) = C$. Note that for any two points $q, q' \in Q_p^1 \cap d_C$, we have $I(p, q) = I(p, q')$, and if we "slide" q up (resp. down) that antidiagonal d_C, then the dividing line that corresponds to q moves to the left (resp. to the right). See Fig. 4. Now, let a and b be two integers in $I(p_1, p_3) \cap I(p_2, p_3)$, and consider these intervals in layout view. Suppose there exists some dividing line ℓ such that in $I(p_1, p_3)$ we have a on the left side of ℓ and b on the right side of ℓ, and simultaneously in $I(p_2, p_3)$ we have b on the left side of ℓ and a on the right side of ℓ. Then we call $\{a, b\}$ a *bad pair*, and we say ℓ *splits* the bad pair. See Fig. 5 (a). We say total orders $\prec_{p_1}^1$ and $\prec_{p_2}^1$ have a bad pair if there is a C satisfying $C \geq (p_1^x + p_1^y)$ and $C \geq (p_2^x + p_2^y)$ such that the interval $[p_1^x + p_1^y, C]$ sorted by $\prec_{p_1}^1$ and the interval $[p_2^x + p_2^y, C]$ sorted by $\prec_{p_2}^1$ in the layout view have a bad pair. Now we have the following lemma.

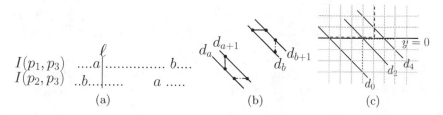

Fig. 5. (a) An illustration of a bad pair. (b) An illustration of conflicting priority. (c) An illustration of that the waterline example is not smooth.

Lemma 1. *If $\prec_{p_1}^1$ and $\prec_{p_2}^1$ have a bad pair, then there exists a $p_3 \in Q_{p_1}^1 \cap Q_{p_2}^1$ such that $I(p_1, p_3)$ and $I(p_2, p_3)$ have a bad pair and the dividing line corresponding to p_3 splits this bad pair.*

Proof. Suppose that $\prec_{p_1}^1$ and $\prec_{p_2}^1$ have a bad pair. Let C where $C \geq (p_1^x + p_1^y)$ and $C \geq (p_2^x + p_2^y)$ be such that there is a bad pair in $[p_1^x + p_1^y, C]$ and $[p_2^x + p_2^y, C]$ in layout view. Let $\{a, b\}$ denote the bad pair in the intervals, and let ℓ denote a dividing line that splits that bad pair. Let p_3 be a point where $p_3^x + p_3^y = C$ and ℓ is the dividing line corresponding with p_3. We complete the proof by showing that $p_3 \in Q_{p_1}^1 \cap Q_{p_2}^1$.

Because $\{a, b\}$ is a bad pair, we can assume without loss of generality that a is to the left of ℓ and b is to the right of ℓ in $I(p_1, p_3)$. This implies that there is at least one horizontal movement and at least one vertical movement to get from p_1 to p_3 (i.e., $p_3^x > p_1^x$ and $p_3^y > p_1^y$). On the other hand, a is to the right of ℓ and b is to the left of ℓ in $I(p_2, p_3)$. So we similarly have $p_3^x > p_2^x$ and $p_3^y > p_2^y$. So, $p_3 \in Q_{p_1}^1 \cap Q_{p_2}^1$, completing the proof. $\qquad\square$

The following lemma implies that it is necessary that any pair of first quadrant total orders do not have any bad pairs.

Lemma 2. *There is a point $p_3 \in Q_{p_1}^1 \cap Q_{p_2}^1$ that has a conflicting priority with respect to $\prec_{p_1}^1$ and $\prec_{p_2}^1$ if and only if there is a bad pair in $\prec_{p_1}^1$ and $\prec_{p_2}^1$.*

Proof. Assume $\prec_{p_1}^1$ and $\prec_{p_2}^1$ have a bad pair. We will show that there is a point $p_3 \in Q_{p_1}^1 \cap Q_{p_2}^1$ that has a conflicting priority.

Let p_3 be a point as described in Lemma 1, and let $\{a, b\}$ denote the bad pair that p_3's dividing line splits. Without loss of generality, $\prec_{p_3}^3$ must give $(a + 1)$ priority over $(b + 1)$ with respect to $R_{p_1}(p_3)$ and must give $(b + 1)$ priority over $(a + 1)$ with respect to $R_{p_2}(p_3)$. Therefore we have a conflicting priority. See Fig. 5 (b).

Now assume that there is a conflicting priority for p_3 with respect to $\prec_{p_1}^1$ and $\prec_{p_2}^1$. We will complete the proof by showing that $I(p_1, p_3)$ and $I(p_2, p_3)$ must have a bad pair. Let a and b denote the integers in the conflicting priority, and let ℓ denote the dividing line with respect to p_3 for $I(p_1, p_3)$ and $I(p_2, p_3)$ in layout view. Then by the definition of conflicting priority we must have a to the left of ℓ and b to the right of ℓ in one interval, and simultaneously we have b

to the left of ℓ and a to the right of ℓ in the other interval, forming a bad pair. Since $I(p_1, p_3)$ and $I(p_2, p_3)$ have a bad pair, we have that $\prec^1_{p_1}$ and $\prec^1_{p_2}$ have a bad pair.   □

Lemma 2 implies that it is necessary for any two points p_i and p_j that $\prec^1_{p_i}$ and $\prec^1_{p_j}$ do not have a bad pair. We now show that this condition is also sufficient.

Lemma 3. *If all pairs of total orders have no bad pairs, then the line segments will satisfy properties (S1)-(S5).*

Proof. It is easy to see that (S1), (S2), (S4) and (S5) are automatically satisfied by construction, and it is only (S3) that we need to prove. We first will show that a segment with nonnegative slope and a segment with non-positive slope will always satisfy (S3). To see this, consider a nonnegative line segment $R_{p'}(q')$ and a non-positive line segment $R_p(q)$. If they violate (S3) then there must be a witness (t_1, t_2) in $R_p(q) \cap R_{p'}(q')$ for two points $t_1 = (t_1^x, t_1^y)$ and $t_2 = (t_2^x, t_2^y)$ such that $t_1^x \neq t_2^x$ and $t_1^y \neq t_2^y$. But we will show that for any two points r_1 and r_2 that satisfy $r_1^x \neq r_2^x$ and $r_1^y \neq r_2^y$, it cannot be that r_1 and r_2 are in both segments. Without loss of generality, assume that $r_1^x < r_2^x$ and $r_1 \in R_{p'}(q') \cap R_p(q)$.

All points after r_1 in $R_{p'}(q')$ have y-coordinate at least r_1^y, and all points after r_1 in $R_p(q)$ have y-coordinate at most r_1^y. Therefore if there is a point z that comes after r_1 in $R_{p'}(q') \cap R_p(q)$ then it must satisfy $z^y = r_1^y$. This implies that if both segments contain r_1 and $R_p(q)$ contains r_2 then $R_{p'}(q')$ cannot contain r_2. Thus $R_p(q)$ and $R_{p'}(q')$ do not have a witness and do not violate (S3).

Now without loss of generality, consider segments $R_{p_1}(q_1)$ and $R_{p_2}(q_2)$ with nonnegative slope. In order to violate (S3), there must be a witness (t_1, t_2) to the violation of (S3). Suppose we have such a witness, and consider the subsegments $R_{p_1}(t_2)$ and $R_{p_2}(t_2)$. If we consider the intervals $I(p_1, t_2)$ and $I(p_2, t_2)$ in layout view, we can see that the dividing line corresponding to t_2 will split the bad pair $\{t_1^x + t_1^y, t_2^x + t_2^y - 1\}$. Therefore if there are no bad pairs, then there cannot be a witness to the violation of (S3), and therefore the segments satisfy (S3).   □

Combining Lemma 2 and 3, we get our characterization.

Theorem 1. *A system of nonnegative sloped line segments in $\mathbb{Z}^2$ is a CDS if and only if we have a total order for the first quadrants for each point such that each pair of total orders have no bad pairs and the third quadrant segments are induced by the corresponding first quadrant segments.*

4   Luby and Christ et al. in the Context of Our Characterization

In an attempt to tie together some of the previous works on CDSes, we now analyze the work of Luby [13] and Christ et al. [4] in the context of our characterization. Chun et al. and Christ et al. were not aware of Luby's work when publishing [4] and [5], although Christ gives a comparison of his work with that of Luby in his thesis [3].

We will first provide an analysis relating smooth grid geometries given by Luby [13]. To do so, we need the following definition. Consider any two points p_1 and p_2 with first quadrant total orders $\prec_{p_1}^1$ and $\prec_{p_2}^1$. We say that $\prec_{p_1}^1$ and $\prec_{p_2}^1$ are *in agreement* if $a \prec_{p_1}^1 b$ if and only if $a \prec_{p_2}^1 b$ for every pair of integers a and b such that antidiagonals d_a and d_b intersect $Q_{p_1}^1 \cap Q_{p_2}^1$. Intuitively, $\prec_{p_1}^1$ and $\prec_{p_2}^1$ are in agreement if they are the same ordering when considering antidiagonals intersecting both first quadrants. We now prove the following lemma about smooth grid geometries. An equivalent lemma was proved by Christ [3] using a different proof technique. We prove the lemma for segments with nonnegative slope, but a symmetric argument holds for segments with negative slope.

Lemma 4. *A CDS is a smooth grid geometry if and only if $\prec_p^1$ and $\prec_q^1$ are in agreement for any pair of points $p, q \in \mathbb{Z}^2$.*

We now turn our attention to analyzing the work of Christ et al. [4] in the context of bad pairs. They give two methods for choosing total orders to construct a CDS. The first method is to assign total orders to points so that all pairs of total orders are in agreement (e.g., assigning the same total order to all points). Note that if two total orders have a bad pair, then there necessarily has to be two integers a and b such that one total order has $a \prec b$ while another has $b \prec a$. But this clearly cannot happen if all total orders are in agreement. Therefore there are no bad pairs and by Theorem 1 it is a CDS.

They also give an example of a CDS constructed using total orders that are not in agreement. Specifically, a point's total order depends on if it is above or below the x-axis. Because of the special role of the x-axis, this example is called the *waterline example*. In the waterline example, every point p such that $p^y \geq 0$ uses the natural total order, that is $\prec_p^1 = (p^x + p^y) \prec (p^x + p^y + 1) \prec \cdots \prec (+\infty)$. For points p such that $p^y < 0$, the total order is a function of its x-coordinate p^x. Specifically, we have $\prec_p^1 = (p^x) \prec (p^x + 1) \prec \cdots \prec (+\infty) \prec (p^x - 1) \prec (p^x - 2) \prec \cdots \prec (-\infty)$. After giving this definition, Christ et al. point out that it is easy to see that the segments form a CDS. We give a formal proof using our characterization.

Lemma 5. *The waterline example is a CDS.*

Note that the waterline example is a CDS that is not smooth. See Fig. 5 (c). The dashed segment is $R_p(q)$ with $p = (0,0)$ and $q = (4,3)$ and the dotted segment is $R_{p'}(q')$ with $p' = (3,-3)$ and $q' = (6,3)$. If we let $z = (3,0)$, then we have that $\text{dist}(R_p(q), R_{p'}(q'), p) = -3$, $\text{dist}(R_p(q), R_{p'}(q'), z) = 0$, and $\text{dist}(R_p(q), R_{p'}(q'), q) = -2$. By definition, these segments are not smooth and therefore the waterline example is not a smooth grid geometry.

Acknowledgement. We would like to thank Dr. Xiaodong Wu for introducing the problem to us, and Dr. Wu and Dr. Kasturi Varadarajan for some valuable discussions.

References

1. Andres, E.: Discrete linear objects in dimension n: the standard model. Graphical Models 65(1-3), 92–111 (2003)
2. Berthé, V., Labbé, S.: An arithmetic and combinatorial approach to three-dimensional discrete lines. In: Discrete Geometry for Computer Imagery - 16th IAPR International Conference, Nancy, France, April 6-8, pp. 47–58 (2011)
3. Christ, T.: Discrete Descriptions of Geometric Objects. PhD thesis, Zürich (2011)
4. Christ, T., Pálvölgyi, D., Stojakovic, M.: Consistent digital line segments. Discrete & Computational Geometry 47(4), 691–710 (2012)
5. Chun, J., Korman, M., Nöllenburg, M., Tokuyama, T.: Consistent digital rays. Discrete & Computational Geometry 42(3), 359–378 (2009)
6. Cohen-Or, D., Kaufman, A.E.: 3d line voxelization and connectivity control. IEEE Computer Graphics and Applications 17(6), 80–87 (1997)
7. Eckhardt, U.: Digital lines and digital convexity. In: Bertrand, G., Imiya, A., Klette, R. (eds.) Digital and Image Geometry. LNCS, vol. 2243, pp. 209–228. Springer, Heidelberg (2002)
8. Franklin, W.R.: Problems with raster graphics algorithm. In: Peters, F.J., Kessener, L.R.A., van Lierop, M.L.P. (eds.) Data Structures for Raster Graphics, Steensel, Netherlands (1985)
9. Goodrich, M.T., Guibas, L.J., Hershberger, J., Tanenbaum, P.J.: Snap rounding line segments efficiently in two and three dimensions. In: Symposium on Computational Geometry, pp. 284–293 (1997)
10. Greene, D.H., Yao, F.F.: Finite-resolution computational geometry. In: 27th Annual Symposium on Foundations of Computer Science, Toronto, Canada, October 27-29, pp. 143–152. IEEE Computer Society (1986)
11. Klette, R., Rosenfeld, A.: Digital straightness - a review. Discrete Applied Mathematics 139(1-3), 197–230 (2004)
12. Lachaud, J.-O., Said, M.: Two efficient algorithms for computing the characteristics of a subsegment of a digital straight line. Discrete Applied Mathematics 161(15), 2293–2315 (2013)
13. Luby, M.G.: Grid geometries which preserve properties of euclidean geometry: A study of graphics line drawing algorithms. In: Earnshaw, R.A. (ed.) Theoretical Foundations of Computer Graphics and CAD, vol. 40, pp. 397–432 (1988)
14. Sivignon, I., Dupont, F., Chassery, J.-M.: Digital intersections: minimal carrier, connectivity, and periodicity properties. Graphical Models 66(4), 226–244 (2004)
15. Sugihara, K.: Robust geometric computation based on topological consistency. In: International Conference on Computational Science (1), pp. 12–26 (2001)
16. van Lierop, M.L.P., van Overveld, C.W.A.M., van de Wetering, H.M.M.: Line rasterization algorithms that satisfy the subset line property. Computer Vision, Graphics, and Image Processing 41(2), 210–228 (1988)

On the Efficiency of All-Pay Mechanisms

George Christodoulou*, Alkmini Sgouritsa, and Bo Tang

University of Liverpool, UK

Abstract. We study the inefficiency of mixed equilibria, expressed as the price of anarchy, of all-pay auctions in three different environments: combinatorial, multi-unit and single-item auctions. First, we consider item-bidding combinatorial auctions where m all-pay auctions run in parallel, one for each good. For fractionally subadditive valuations, we strengthen the upper bound from 2 [22] to 1.82 by proving some structural properties that characterize the mixed Nash equilibria of the game. Next, we design an all-pay mechanism with a randomized allocation rule for the multi-unit auction. We show that, for bidders with submodular valuations, the mechanism admits a unique, 75% efficient, pure Nash equilibrium. The efficiency of this mechanism outperforms all the known bounds on the price of anarchy of mechanisms used for multi-unit auctions. Finally, we analyze single-item all-pay auctions motivated by their connection to contests and show tight bounds on the price of anarchy of social welfare, revenue and maximum bid.

1 Introduction

It is a common economic phenomenon in competitions that agents make irreversible investments without knowing the outcome. *All-pay* auctions are widely used in economics to capture such situations, where all players, even the losers, pay their bids. For example, a lobbyist can make a monetary contribution in order to influence decisions made by the government. Usually the group invested the most increases their winning chances, but all groups have to pay regardless of the outcome. In addition, all-pay auctions have been shown useful to model rent seeking, political campaigns and R&D races. There is a well-known connection between all-pay auctions and *contests* [20]. In particular, the all-pay auction can be viewed as a single-prize contest, where the payments correspond to the effort that players make in order to win the competition.

In this paper, we study the efficiency of mixed Nash equilibria in all-pay auctions with complete information, from a worst-case analysis perspective, using the *price of anarchy* [16] as a measure. As social objective, we consider the *social welfare*, i.e. the sum of the bidders' valuations. We study the equilibria induced from all-pay mechanisms in three fundamental resource allocation scenarios; combinatorial auctions, multi-unit auctions and single-item auctions.

In a combinatorial auction a set of items are allocated to a group of selfish individuals. Each player has different preferences for different subsets of items

---

* This author was supported by EPSRC grants EP/M008118/1 and EP/K01000X/1.

N. Bansal and I. Finocchi (Eds.): ESA 2015, LNCS 9294, pp. 349–360, 2015.
DOI: 10.1007/978-3-662-48350-3_30

and this is expressed via a *valuation set* function. A multi-unit auction can be considered as an important special case, where there are multiple copies of a single good. Hence the valuations of the players are not set functions, but depend only on the number of copies received. Multi-unit auctions have been extensively studied since the seminal work by Vickrey [23]. As already mentioned, all-pay auctions have received a lot of attention for the case of a single item, as they model all-pay contests and procurements via contests.

1.1   Contribution

Combinatorial Auctions. Our first result is on the price of anarchy of simultaneous all-pay auctions with item-bidding that was previously studied by Syrgkanis and Tardos [22]. For fractionally subadditive valuations, it was previously shown that the price of anarchy was at most 2 [22] and at least $e/(e-1) \approx 1.58$ [8]. We narrow further this gap, by improving the upper bound to 1.82. In order to obtain the bound, we come up with several structural theorems that characterize mixed Nash equilibria in simultaneous all-pay auctions.

Multi-unit Auctions. Our next result shows a novel use of all-pay mechanisms to the multi-unit setting. We propose an all-pay mechanism with a randomized allocation rule inspired by Kelly's seminal proportional allocation mechanism [15]. We show that this mechanism admits a *unique*, 75% efficient *pure* Nash equilibrium and no other mixed Nash equilibria exist, when bidders' valuations are submodular. As a consequence, the price of anarchy of our mechanism outperforms all current price of anarchy bounds of prevalent multi-unit auctions including uniform price [18] and discriminatory [14] auctions, with bound $e/(e-1)$.

Single-item Auctions. Finally, we study the efficiency of a single-prize contest that can be modeled as a single-item all-pay auction. We show a tight bound on the price of anarchy for mixed equilibria which is approximately 1.185. By following previous study on the procurement via contest, we further study two other standard objectives, *revenue* and *maximum bid*. We evaluate the performance of all-pay auctions in the prior-free setting, i.e. no distribution over bidders' valuation is assumed. We show that both the revenue and the maximum bid of any mixed Nash equilibrium are at least as high as $v_2/2$, where v_2 is the second highest valuation. In contrast, the revenue and the maximum bid in some mixed Nash equilibrium may be less than $v_2/2$ when using reward structure other than allocating the entire reward to the highest bidder. This result coincides with the optimal crowdsourcing contest developed in [6] for the setting with prior distributions. We also show that in conventional procurements (modeled by first-price auctions), v_2 is exactly the revenue and maximum bid in the worst equilibrium. So procurement via all-pay contests is a 2-approximation to the conventional procurement in the context of worst-case equilibria.

1.2   Related Work

The inefficiency of Nash equilibria in auctions has been a well-known fact (see e.g. [17]). Existence of efficient equilibria of simultaneous sealed bid auctions in full information settings was first studied by Bikhchandani [3]. Christodoulou, Kovács and Schapira [7] initiated the study of the (Bayesian) price of anarchy of simultaneous auctions with item-bidding. Several variants have been studied since then [2,12,11], as well as multi-unit auctions [14,18].

Syrgkanis and Tardos [22] proposed a general smoothness framework for several types of mechanisms and applied it to settings with fractionally subadditive bidders obtaining several upper bounds (e.g., first price auction, all-pay auction, and multi-unit auction). Christodoulou et al. [8] constructed tight lower bounds for first-price auctions and showed a tight price of anarchy bound of 2 for all-pay auctions with subadditive valuations. Roughgarden [19] presented an elegant methodology to provide price of anarchy lower bounds via a reduction from the hardness of the underlying optimization problems.

All-pay auctions and contests have been studied extensively in economic theory. Baye, Kovenock and de Vries [1], fully characterized the Nash equilibria in single-item all-pay auction with complete information. The connection between all-pay auctions and crowdsourcing contests was proposed in [9]. Chawla et al. [6] studied the design of optimal crowdsourcing contest when agents' value are drawn independently from a specific distribution.

2   Preliminaries

In a *combinatorial auction*, n *players* compete on m *items*. Every player (or *bidder*) $i \in [n]$ has a valuation function $v_i : \{0,1\}^m \to \mathbb{R}^+$ which is monotone and normalized, that is, $\forall S \subseteq T \subseteq [m]$, $v_i(S) \leq v_i(T)$, and $v_i(\emptyset) = 0$. The outcome of the auction is represented by a tuple of $(\mathbf{X}, \mathbf{p})$ where $\mathbf{X} = (X_1, \ldots, X_n)$ specifies the allocation of items (X_i is the set of items allocated to player i) and $\mathbf{p} = (p_1, \ldots, p_n)$ specifies the buyers' payments (p_i is the payment of player i for the allocation $\mathbf{X}$). In the *simultaneous item-bidding* auction, every player $i \in [n]$ submits a non-negative bid b_{ij} for each item $j \in [m]$. The items are then allocated by independent auctions, i.e. the allocation and payment rule for item j only depend on the players' bids on item j. In a simultaneous *all-pay* auction the allocation and payment for each player is determined as follows: each item $j \in [m]$ is allocated to the bidder i^* with the highest bid for that item, i.e. $i^* = \arg\max_i b_{ij}$, and each bidder i is charged an amount equal to $p_i = \sum_{j \in [m]} b_{ij}$. It is worth mentioning that, for any bidder profile, there always exists a tie-breaking rule such that mixed equilibria exist [21]. Actually, our results hold for arbitrary tie-breaking rule. For completeness, we specify a tie-breaking rule where the mechanism will allocate the item to a winner picked uniformly from all highest bidders as in [1].

Definition 1 (Valuations). *Let $v : 2^{[m]} \to \mathbb{R}$ be a valuation function. Then v is called a) additive, if $v(S) = \sum_{j \in S} v(\{j\})$; b) submodular, if $v(S \cup T) + v(S \cap T) \leq v(S) + v(T)$; c) fractionally subadditive or XOS, if v is determined by a finite set of additive valuations ξ_k such that $v(S) = \max_k \xi_k(S)$.*

The classes of the above valuations are in increasing order of inclusion.

Multi-unit Auction. In a multi-unit auction, m copies of an item are sold to n bidders. Here, bidder i 's valuation is a function that depends on the number of copies he gets. That is $v_i : \{0, 1, \dots, m\} \to \mathbb{R}^+$ and it is non-decreasing and normalized, with $v_i(0) = 0$. We say a valuation v_i is *submodular*, if it has non-increasing marginal values, i.e. $v_i(s + 1) - v_i(s) \geq v_i(t + 1) - v_i(t)$ for all $s \leq t$.

Nash equilibrium and price of anarchy. We use b_i to denote a pure strategy of player i which might be a single value or a vector, depending on the auction. So, for the case of m simultaneous auctions, $b_i = (b_{i1}, \dots, b_{im})$. We denote by $\mathbf{b}_{-i} = (b_1, \dots, b_{i-1}, b_{i+1}, \dots, b_n)$ the strategies of all players except for i. Any *mixed strategy* B_i of player i is a probability distribution over pure strategies.

For any profile of strategies, $\mathbf{b} = (b_1, \dots, b_n)$, $\mathbf{X}(\mathbf{b})$ denotes the allocation under the strategy profile $\mathbf{b}$. The valuation of player i for the allocation $\mathbf{X}(\mathbf{b})$ is denoted by $v_i(\mathbf{X}(\mathbf{b})) = v_i(\mathbf{b})$. The *utility* u_i of player i is defined as the difference between her valuation and payment: $u_i(\mathbf{X}(\mathbf{b})) = u_i(\mathbf{b}) = v_i(\mathbf{b}) - p_i(\mathbf{b})$.

Definition 2 (Nash equilibria). *A bidding profile* $\mathbf{b} = (b_1, \dots, b_n)$ *forms a pure Nash equilibrium if for every player i and all bids b_i', $u_i(\mathbf{b}) \geq u_i(b_i', \mathbf{b}_{-i})$. Similarly, a mixed bidding profile $\mathbf{B} = \times_i B_i$ is a mixed Nash equilibrium if for all bids b_i' and every player i, $\mathbb{E}_{\mathbf{b} \sim \mathbf{B}}[u_i(\mathbf{b})] \geq \mathbb{E}_{\mathbf{b}_{-i} \sim \mathbf{B}_{-i}}[u_i(b_i', \mathbf{b}_{-i})]$. Clearly, any pure Nash equilibrium is also a mixed Nash equilibrium.*

Our global objective is to maximize the sum of the valuations of the players for their received allocations, i.e., to maximize the *social welfare* $SW(\mathbf{X}) = \sum_{i \in [n]} v_i(X_i)$. So $\mathbf{O}(\mathbf{v}) = \mathbf{O} = (O_1, \dots, O_n)$ is an *optimal allocation* if $SW(\mathbf{O}) = \max_{\mathbf{X}} SW(\mathbf{X})$. In Sect. 5, we also study two other objectives: the *revenue*, which equals the sum of the payments, $\sum_i p_i$, and the *maximum payment*, $\max_i b_i$. We also refer to the maximum payment as the *maximum bid*.

Definition 3 (Price of anarchy). *Let $\mathcal{I}([n], [m], \mathbf{v})$ be the set of all instances, i.e. $\mathcal{I}([n], [m], \mathbf{v})$ includes the instances for every set of bidders and items and any possible valuation functions. The mixed price of anarchy, PoA, of a mechanism is defined as*

$$PoA = \max_{I \in \mathcal{I}} \max_{\mathbf{B} \in \mathcal{E}(I)} \frac{SW(\mathbf{O})}{\mathbb{E}_{\mathbf{b} \sim \mathbf{B}}[SW(\mathbf{X}(\mathbf{b}))]} \quad,$$

where $\mathcal{E}(I)$ is the class of mixed Nash equilibria for the instance $I \in \mathcal{I}$. The pure PoA is defined as above but restricted in the class of pure Nash equilibria.

Let $\mathbf{B} = (B_1, \dots, B_n)$ be a profile of mixed strategies. Given the profile $\mathbf{B}$, we fix the notation for the following *cumulative distribution functions (CDF)*: G_{ij} is the CDF of the bid of player i for item j; F_j is the CDF of the highest bid for item j and F_{ij} is the CDF of the highest bid for item j if we exclude the bid of player i. Observe that $F_j = \prod_k G_{kj}$ and $F_{ij} = \prod_{k \neq i} G_{kj}$. We also use $\varphi_{ij}(x)$ to denote the probability that player i gets item j by bidding x. Then, $\varphi_{ij}(x) \leq F_{ij}(x)$. When we refer to a single item, we may drop the index j. Whenever it is clear from the context, we will use shorter notation for expectations, e.g. we use $\mathbb{E}[u_i(\mathbf{b})]$ instead of $\mathbb{E}_{\mathbf{b} \sim \mathbf{B}}[u_i(\mathbf{b})]$, or even $SW(\mathbf{B})$ to denote $\mathbb{E}_{\mathbf{b} \sim \mathbf{B}}[SW(\mathbf{X}(\mathbf{b}))]$.

3   Combinatorial Auctions

In this section we prove an upper bound of 1.82 for the mixed price of anarchy
of simultaneous all-pay auctions when bidders' valuations are fractionally subadditive (XOS). This result improves over the previously known bound of 2 due
to [22]. We first state our main theorem and present the key ingredients. Then
we prove these ingredients in the following subsections. Due to space limitation,
we give proofs of the lemmas and theorems in the full version.

Theorem 4. *The mixed price of anarchy for simultaneous all-pay auctions with
fractionally subadditive (XOS) bidders is at most 1.82.*

Proof. Given a valuation profile $\mathbf{v} = (v_1, \ldots, v_n)$, let $\mathbf{O} = (O_1, \ldots, O_n)$ be a
fixed optimal solution, that maximizes the social welfare. We can safely assume
that $\mathbf{O}$ is a partition of the items. Since v_i is an XOS valuation, let $\xi_i^{O_i}$ be a
maximizing additive function with respect to O_i. For every item j we denote by
o_j item j's contribution to the optimal social welfare, that is, $o_j = \xi_i^{O_i}(j)$, where
i is such that $j \in O_i$. The optimal social welfare is thus $SW(\mathbf{O}) = \sum_j o_j$. In
order to bound the price of anarchy, we consider only items with $o_j > 0$, as it is
without loss of generality to omit items with $o_j = 0$.

For a fixed mixed Nash equilibrium $\mathbf{B}$, recall that by F_j and F_{ij} we denote the
CDFs of the maximum bid on item j among all bidders, with and without the bid
of bidder i, respectively. For any item $j \in O_i$, let $A_j = \max_{x \geq 0}\{F_{ij}(x)o_j - x\}$.

As a key part of the proof we use the following two inequalities that bound
from below the social welfare in any mixed Nash equilibrium $\mathbf{B}$.

$$SW(\mathbf{B}) \geq \sum_{j \in [m]} \left(A_j + \int_0^{o_j - A_j} (1 - F_j(x))dx \right) , \tag{1}$$

$$SW(\mathbf{B}) \geq \sum_{j \in [m]} \int_0^{o_j - A_j} \sqrt{F_j(x)}dx . \tag{2}$$

Inequality (1) suffices to provide a weaker upper bound of 2 (see [8]). The proof of
(2) is much more involved, and requires a deeper understanding of the equilibria
properties of the induced game. We postpone their proofs in Sect. 3.1 (Lemma 5)
and Sect. 3.2 (Lemma 6), respectively. By combining (1) and (2),

$$SW(\mathbf{B}) \geq \frac{1}{1+\lambda} \cdot \sum_j \left(A_j + \int_0^{o_j - A_j} \left(1 - F_j(x) + \lambda \cdot \sqrt{F_j(x)} \right) dx \right) , \tag{3}$$

for every $\lambda \geq 0$. It suffices to bound from below the right-hand side of (3) with
respect to the optimal social welfare. For any cumulative distribution function
F, and any positive real number v, let

$$R(F, v) \stackrel{\text{def}}{=} A + \int_0^{v-A} (1 - F(x))dx + \lambda \cdot \int_0^{v-A} \sqrt{F(x)}dx ,$$

where $A = \max_{x \geq 0}\{F(x) \cdot v - x\}$. Inequality (3) can then be rewritten as $SW(\mathbf{B}) \geq \frac{1}{1+\lambda}\sum_j R(F_j, o_j)$. Finally, we show a lower bound of $R(F, v)$ that holds for any CDF F and any positive real v.

$$R(F, v) \geq \frac{3 + 4\lambda - \lambda^4}{6} \cdot v \ . \tag{4}$$

The proof of (4) is given in Sect. 3.3 (Lemma 9). Finally, we obtain that for any $\lambda > 0$,

$$SW(\mathbf{B}) \geq \frac{1}{1+\lambda}\sum_j R(F_j, o_j) \geq \frac{3+4\lambda-\lambda^4}{6\lambda+6} \cdot \sum_j o_j = \frac{3+4\lambda-\lambda^4}{6\lambda+6} \cdot SW(\mathbf{O}) \ .$$

By taking $\lambda = 0.56$, we conclude that the price of anarchy is at most 1.82.   □

3.1 Proof of Inequality (1)

This section is devoted to the proof of the following lower bound. Recall that the definition o_j is from the definition of XOS functions.

Lemma 5. $SW(\mathbf{B}) \geq \sum_{j \in [m]} (A_j + \int_0^{o_j - A_j}(1 - F_j(x))dx)$.

Proof. Recall that $A_j = \max_{x_j \geq 0}\{F_{ij}(x)o_j - x_j\}$. We can bound bidder i's utility in the Nash equilibrium $\mathbf{B}$ by $u_i(\mathbf{B}) \geq \sum_{j \in O_i} A_j$. To see this, consider the deviation for bidder i, where he bids only for items in O_i, namely, for each item j, he bids the value x_j that maximizes the expression $F_{ij}(x_j)o_j - x_j$. Since for any obtained subset $T \subseteq O_i$, he has value $v_i(T) \geq \sum_{j \in T} o_j$, and the bids x_j must be paid in any case, the expected utility with these bids is at least $\sum_{j \in O_i} \max_{x_j \geq 0} (F_{ij}(x)o_j - x_j) = \sum_{j \in O_i} A_j$. With $\mathbf{B}$ being an equilibrium, we infer that $u_i(\mathbf{B}) \geq \sum_{j \in O_i} A_j$. By summing up over all bidders,

$$SW(\mathbf{B}) = \sum_{i \in [n]} u_i(\mathbf{B}) + \sum_{i \in [n]}\sum_{j \in [m]} \mathbb{E}[b_{ij}] \geq \sum_{j \in [m]} A_j + \sum_{j \in [m]}\sum_{i \in [n]} \mathbb{E}[b_{ij}]$$

$$\geq \sum_{j \in [m]} (A_j + \mathbb{E}[\max_{i \in [n]}\{b_{ij}\}]) \geq \sum_{j \in [m]} \left(A_j + \int_0^{o_j - A_j}(1 - F_j(x))dx\right) \ .$$

The first equality holds because $\sum_i \mathbb{E}_{\mathbf{b}}[v_i(\mathbf{b})] = \sum_i \mathbb{E}_{\mathbf{b}}[u_i(\mathbf{b}) + \sum_{j \in [m]} b_{ij}]$. The second inequality follows because $\sum_i b_{ij} \geq \max_i b_{ij}$ and the last one is implied by the definition of the expected value of any positive random variable.   □

3.2 Proof of Inequality (2)

Here, we prove the following lemma for any mixed Nash equilibrium $\mathbf{B}$.

Lemma 6. $SW(\mathbf{B}) \geq \sum_{j \in [m]} \int_0^{o_j - A_j} \sqrt{F_j(x)}dx$.

First we show a useful lemma that holds for XOS valuations. We will further use the technical Proposition 8.

Lemma 7. *For any fractionally subadditive (XOS) valuation function v,*

$$v(S) \geq \sum_{j \in [m]} (v(S) - v(S \setminus \{j\})) \ .$$

Proof. Let ξ be a maximizing additive function of S for the XOS valuation v. By definition, $v(S) = \xi(S)$ and for every j, $v(S \setminus \{j\}) \geq \xi(S \setminus \{j\})$. Then, $\sum_{j \in [m]} (v(S) - v(S \setminus \{j\})) \leq \sum_{j \in S} (\xi(S) - \xi(S \setminus \{j\})) = \sum_{j \in S} \xi(j) = v(S)$. $\square$

Proposition 8. *For any integer $n \geq 2$, any positive reals $G_i \leq 1$ and positive reals g_i, for $1 \leq i \leq n$,*

$$\sum_{i=1}^n \frac{g_i}{\sum_{k \neq i} \frac{g_k}{G_k}} \geq \sqrt{\prod_{i=1}^n G_i} \ .$$

We are now ready to prove Lemma 6. We only state a proof sketch here to illustrate the main ideas.

Proof (Sketch of Lemma 6). Recall that G_{ij} is the CDF of the bid of player i for item j. For simplicity, we assume $G_{ij}(x)$ is continuous and differentiable, with $g_{ij}(x)$ being the PDF of player i's bid for item j. First, we define the *expected marginal valuation* of item j w.r.t player i,

$$v_{ij}(x) \stackrel{\text{def}}{=} \mathop{\mathbb{E}}_{\mathbf{b} \sim \mathbf{B}} [v_i(X_i(\mathbf{b}) \cup \{j\}) - v_i(X_i(\mathbf{b}) \setminus \{j\}) | b_{ij} = x] \ .$$

Given the above definition and a careful characterization of mixed Nash equilibria, we are able to show $F_{ij}(x) \cdot v_{ij}(x) = \mathbb{E}[v_i(X_i(\mathbf{b})) - v_i(X_i(\mathbf{b}) \setminus \{j\}) | b_{ij} = x]$ and $\frac{1}{v_{ij}(x)} = \frac{dF_{ij}(x)}{dx}$ for any x in the support of G_{ij}. Let $g_{ij}(x)$ be the derivative of $G_{ij}(x)$. Using Lemma 7, we have

$$SW(\mathbf{B}) = \sum_i \mathbb{E}[v_i(X_i(\mathbf{b}))] \geq \sum_i \sum_j \mathbb{E}[v_i(X_i(\mathbf{b})) - v_i(X_i(\mathbf{b}) \setminus \{j\})]$$

$$\geq \sum_i \sum_j \int_0^{o_j - A_j} \mathbb{E}[v_i(X_i(\mathbf{b})) - v_i(X_i(\mathbf{b}) \setminus \{j\}) | b_{ij} = x] \cdot g_{ij}(x) dx$$

$$\geq \sum_i \sum_j \int_0^{o_j - A_j} F_{ij}(x) \cdot v_{ij}(x) \cdot g_{ij}(x) dx \ ,$$

where the second inequality follows by the law of total probability. By using the facts that $F_{ij}(x) = \prod_{k \neq i} G_{kj}(x)$ and $\frac{1}{v_{ij}(x)} = \frac{dF_{ij}(x)}{dx}$, for any $x > 0$ such that $g_{ij}(x) > 0$ (x is in the support of player i) and $F_j(x) > 0$, we obtain

$$F_{ij}(x) \cdot v_{ij}(x) \cdot g_{ij}(x) = \frac{F_{ij}(x) \cdot g_{ij}(x)}{\frac{dF_{ij}}{dx}(x)} = \frac{\prod_{k \neq i} G_{kj}(x) \cdot g_{ij}(x)}{\sum_{k \neq i} \left(g_{kj} \cdot \prod_{s \neq k \wedge s \neq i} G_{sj} \right)} = \frac{g_{ij}(x)}{\sum_{k \neq i} \frac{g_{kj}(x)}{G_{kj}(x)}} \ .$$

For every $x > 0$, we use Proposition 8 only over the set S of players with $g_{ij}(x) > 0$. After summing over all bidders we get,

$$\sum_{i \in [n]} F_{ij}(x) \cdot v_{ij}(x) \cdot g_{ij}(x) \geq \sum_{i \in S} \frac{g_{ij}(x)}{\sum_{k \neq i, k \in S} \frac{g_{kj}}{G_{kj}}} \geq \sqrt{\prod_{i \in S} G_{ij}(x)} \geq \sqrt{F_j(x)} .$$

The above inequality also holds for $F_j(x) = 0$. Finally, by merging the above inequalities, we conclude that $SW(\mathbf{B}) \geq \sum_{j \in [m]} \int_0^{o_j - A_j} \sqrt{F_j(x)} dx$.   □

3.3 Proof of Inequality (4)

In this section we prove the following technical lemma.

Lemma 9. *For any CDF F and any real $v > 0$, $R(F, v) \geq \frac{3 + 4\lambda - \lambda^4}{6} v$.*

In order to obtain a lower bound for $R(F, v)$ as stated in the lemma, we show first that we can restrict attention to cumulative distribution functions of a simple special form, since these constitute worst cases for $R(F, v)$. In the next lemma, for an arbitrary CDF F we will define a simple piecewise linear function $\hat{F}$ that satisfies the following two properties:

$$\int_0^{v - A} (1 - \hat{F}(x)) dx = \int_0^{v - A} (1 - F(x)) dx \; ; \quad \int_0^{v - A} \sqrt{\hat{F}(x)} dx \leq \int_0^{v - A} \sqrt{F(x)} dx .$$

Once we establish this, it is convenient to lower bound $R(\hat{F}, v)$ for the given type of piecewise linear functions $\hat{F}$.

Lemma 10. *For any CDF F and real $v > 0$, there always exists another CDF $\hat{F}$ such that $R(F, v) \geq R(\hat{F}, v)$ that, for $A = \max_{x \geq 0}\{F(x) \cdot v - x\}$, is defined by*

$$\hat{F}(x) = \begin{cases} 0 & , \text{ if } x \in [0, x_0] \\ \frac{x + A}{v} & , \text{ if } x \in (x_0, v - A] \end{cases} .$$

Now we are ready to proceed with the proof of Lemma 9.

Proof (of Lemma 9). By Lemma 10, for any fixed $v > 0$, we only need to consider the CDF's in the following form: for any positive A and x_0 such that $x_0 + A \leq v$,

$$F(x) = \begin{cases} 0 & , \text{ if } x \in [0, x_0] \\ \frac{x + A}{v} & , \text{ if } x \in (x_0, v - A] \end{cases} .$$

Clearly, $\max_{x \geq 0}\{F(x) \cdot v - x\} = A$. Let $t = \frac{A + x_0}{v}$. Then

$$R(F, v) = A + \int_0^{v - A} (1 - F(x)) dx + \lambda \cdot \int_0^{v - A} \sqrt{F(x)} dx$$

$$= A + v - A - \frac{v}{2} \cdot \left(\frac{x + A}{v}\right)^2 \Big|_{x_0}^{v - A} + \lambda \cdot \frac{2v}{3} \cdot \left(\frac{x + A}{v}\right)^{\frac{3}{2}} \Big|_{x_0}^{v - A}$$

$$= v - \frac{v}{2} \cdot (1 - t^2) + \lambda \cdot \frac{2v}{3} \cdot (1 - t^{\frac{3}{2}}) = v \cdot \left(\frac{1}{2}(1 + t^2) + \frac{2\lambda}{3}(1 - t^{\frac{3}{2}})\right) .$$

By optimizing over t, the above formula is minimized when $t = \lambda^2 \leq 1$. That is,

$$R(F, v) \geq v \cdot \left(\frac{1}{2}(1 + \lambda^4) + \frac{2\lambda}{3}(1 - \lambda^3) \right) = \frac{3 + 4\lambda - \lambda^4}{6} \cdot v \ . \qquad \square$$

4   Multi-unit Auctions

In this section, we propose a randomized all-pay mechanism for the multi-unit setting, where m identical items are to be allocated to n bidders. Markakis and Telelis [18] and de Keijzer et al. [14] have studied the price of anarchy for several multi-unit auction formats. The current best upper bound obtained was 1.58 for both pure and mixed Nash equilibria.

We propose a *randomized* all-pay mechanism that induces a *unique pure* Nash equilibrium, with an improved price of anarchy bound of 4/3. We call the mechanism Random proportional-share allocation mechanism (PSAM), as it is a randomized version of Kelly's celebrated proportional-share allocation mechanism for divisible resources [15]. The mechanism works as follows (illustrated as Mechanism 1).

Each bidder submits a non-negative real b_i to the auctioneer. After soliciting all the bids from the bidders, the auctioneer associates a real number x_i with bidder i that is equal to $x_i = \frac{m \cdot b_i}{\sum_{i \in [n]} b_i}$. Each player pays their bid, $p_i = b_i$. In the degenerate case, where $\sum_i b_i = 0$, then $x_i = 0$ and $p_i = 0$ for all i.

We turn the x_i's to a random allocation as follows. Each bidder i secures $\lfloor x_i \rfloor$ items and gets one more item with probability $x_i - \lfloor x_i \rfloor$. An application of the Birkhoff-von Neumann decomposition theorem guarantees that given an allocation vector $(x_1, x_2, \ldots, x_n)$ with $\sum_i x_i = m$, one can always find a randomized allocation[1] with random variables $X_1, X_2, \ldots, X_n$ such that $\mathbb{E}[X_i] = x_i$ and $\Pr[\lfloor x_i \rfloor \leq X_i \leq \lceil x_i \rceil] = 1$ (see for example [10,4]).

We next show that the game induced by the Random PSAM when the bidders have submodular valuations is *isomorphic* to the game induced by Kelly's mechanism for a single divisible resource when bidders have piece-wise linear concave valuations.

Theorem 11. *Any game induced by the Random PSAM applied to the multi-unit setting with submodular bidders is isomorphic to a game induced from Kelly's mechanism applied to a single divisible resource with piece-wise linear concave functions.*

Proof. For each bidder i's submodular valuation function $f_i : \{0, 1, \ldots, m\} \to R^+$, we associate a concave function $g_i : [0, 1] \to R^+$ such that,

$$\forall x \in [0, m], \ g_i(x/m) = f_i(\lfloor x \rfloor) + (x - \lfloor x \rfloor) \cdot (f_i(\lfloor x \rfloor + 1) - f_i(\lfloor x \rfloor)) \ . \quad (5)$$

---

[1] As an example, assume $x_1 = 2.5, x_2 = 1.6, x_3 = 1.9$. One can define a random allocation such that assignments $(3, 2, 1)$, $(3, 1, 2)$ and $(2, 2, 2)$ occur with probabilities 0.1, 0.4, and 0.5 respectively.

Mechanism 1. Random PSAM

Input: Total number of items m and all bidders' bid $b_1, b_2, \ldots, b_n$

Output: Ex-post allocations $X_1, X_2, \ldots, X_n$ and payments $p_1, p_2, \ldots, p_n$

if $\sum_{i \in [n]} b_i > 0$ then

    foreach *bidder* $i = 1, 2, \ldots, n$ **do**

        $x_i \leftarrow \frac{m \cdot b_i}{\sum_{i \in [n]} b_i}$;

        $p_i \leftarrow b_i$;

    Sample $\{X_i\}_{i \in [n]}$ from $\{x_i\}_{i \in [n]}$ by using Birkhoff-von Neumann decomposition theorem such that $\lfloor x_i \rfloor \leq X \leq \lceil x_i \rceil$ and the expectation of sampling X_i is x_i;

else Set $\mathbf{X} = \mathbf{0}$ and $\mathbf{p} = \mathbf{0}$;

Return X_i and p_i for all $i \in [n]$;

Essentially, g_i is the piecewise linear function that comprises the line segments that connect $f_i(k)$ with $f_i(k+1)$, for all nonnegative integers k. It is easy to see that g_i is concave if f_i is submodular. We use identity functions as the bijections ϕ^i in the definition of game isomorphism. Therefore, it suffices to show that, for any pure strategy profile $\mathbf{b}$, $u_i(\mathbf{b}) = u'_i(\mathbf{b})$, where u_i and u'_i are the bidder i's utility functions in the first and second game, respectively. Let $x_i = \frac{m \cdot b_i}{\sum_i b_i}$, then

$$
\begin{aligned}
u_i(\mathbf{b}) &= (x_i - \lfloor x_i \rfloor) f_i(\lfloor x_i \rfloor + 1) + (1 - x_i + \lfloor x_i \rfloor) f_i(\lfloor x_i \rfloor) - b_i \\
&= f_i(\lfloor x_i \rfloor) + (x_i - \lfloor x_i \rfloor)(f_i(\lfloor x_i \rfloor + 1) - f_i(\lfloor x_i \rfloor)) - b_i \\
&= g_i\left(\frac{x_i}{m}\right) - b_i = g_i\left(\frac{b_i}{\sum_i b_i}\right) - b_i = u'_i(\mathbf{b}) \ .
\end{aligned}
$$

Note that $g_i\left(\frac{b_i}{\sum_i b_i}\right) - b_i$ is player i's utility, under $\mathbf{b}$, in Kelly's mechanism. □

We next show an equivalence between the optimal welfare.

Lemma 12. *The optimum social welfare in the multi-unit setting, with submodular valuations* $\mathbf{f} = (f_1, \ldots, f_n)$, *is equal to the optimal social welfare in the divisible resource allocation setting with concave valuations* $\mathbf{g} = (g_1, \ldots g_n)$, *where* $g_i(x/m) = f_i(\lfloor x \rfloor) + (x - \lfloor x \rfloor) \cdot (f_i(\lfloor x \rfloor + 1) - f_i(\lfloor x \rfloor))$.

Theorem 11 and Lemma 12, allow us to obtain the existence and uniqueness of the pure Nash equilibrium, as well as the price of anarchy bounds of Random PSAM by the corresponding results on Kelly's mechanism for a single divisible resource [13]. Moreover, it can be shown that there are no other mixed equilibria by adopting the arguments of [5] for Kelly's mechanism. The main conclusion of this section is summarized in the following Corollary.

Corollary 13. *Random PSAM induces a unique pure Nash equilibrium when applied to the multi-unit setting with submodular bidders. Moreover, the price of anarchy of the mechanism is exactly 4/3.*

5 Single Item Auctions

In this section, we study mixed Nash equilibria in the single item all-pay auction. First, we measure the inefficiency of mixed Nash equilibria, showing tight results for the price of anarchy. En route, we also show that the price of anarchy is 8/7 for two players. Then we analyze the quality of two other important criteria, the *expected revenue (the sum of bids)* and the quality of the expected *highest submission (the maximum bid)*, which is a standard objective in crowdsourcing contests [6]. For these objectives, we show a tight lower bound of $v_2/2$, where v_2 is the second highest value among all bidders' valuations. In the following, we drop the word expected while referring to the revenue or to the maximum bid.

We quantify the loss of revenue and the highest submission in the worst-case equilibria. We show that the all-pay auction achieves a 2-approximation comparing to the conventional procurement (modeled as the first price auction), when considering worst-case mixed Nash equilibria; we show that the revenue and the maximum bid of the conventional procurement equals v_2 in the worst case. We also consider other structures of rewards allocation and conclude that allocating the entire reward to the highest bidder is the only way to guarantee the approximation factor of 2. Roughly speaking, allocating all the reward to the top prize is the optimal way to maximize the maximum bid and revenue among all the prior-free all-pay mechanisms where the designer has no prior information about the participants.

Due to the lack of space we give the proofs of theorems and lemmas in the full version.

Theorem 14. *The mixed price of anarchy of the single item all-pay auction is 1.185.*

Theorem 15. *In any mixed Nash equilibrium of the single-item all-pay auction, the revenue and the maximum bid are at least half of the second highest valuation.*

Lemma 16. *For any $\epsilon > 0$, there exists a valuation vector $\mathbf{v} = (v_1, \ldots, v_n)$, such that in a mixed Nash equilibrium of the induced single-item all-pay auction, the revenue and the maximum bid is at most $v_2/2 + \epsilon$.*

Finally, the next theorem indicates that allocating the entire reward to the highest bidder is the best choice. In particular a prior-free all-pay mechanism is presented by a probability vector $\mathbf{q} = (q_i)_{i \in [n]}$, with $\sum_{i \in [n]} q_i = 1$, where q_i is the probability that the ith highest bidder is allocated the item, for every $i \leq n$. Note that the reward structure considered here does not depend on the index of the bidder, i.e. the mechanisms are anonymous.

Theorem 17. *For any prior-free all-pay mechanism that assigns the item to the highest bidder with probability strictly less than 1, i.e. $q_1 < 1$, there exists a valuation profile and mixed Nash equilibrium such that the revenue and the maximum bid are strictly less than $v_2/2$.*

References

1. Baye, M.R., Kovenock, D., de Vries, C.G.: The all-pay auction with complete information. Economic Theory 8(2), 291–305 (1996)
2. Bhawalkar, K., Roughgarden, T.: Welfare guarantees for combinatorial auctions with item bidding. In: SODA 2011. SIAM (January 2011)
3. Bikhchandani, S.: Auctions of Heterogeneous Objects. Games and Economic Behavior (January 1999)
4. Cai, Y., Daskalakis, C., Weinberg, S.M.: An algorithmic characterization of multidimensional mechanisms. In: Proceedings of the Forty-Fourth Annual ACM Symposium on Theory of Computing, STOC 2012, pp. 459–478. ACM, New York (2012)
5. Caragiannis, I., Voudouris, A.A.: Welfare guarantees for proportional allocations. In: SAGT 2014 (2014)
6. Chawla, S., Hartline, J.D., Sivan, B.: Optimal crowdsourcing contests. In: SODA 2012, Kyoto, Japan, January 17-19, pp. 856–868 (2012)
7. Christodoulou, G., Kovács, A., Schapira, M.: Bayesian Combinatorial Auctions. In: Aceto, L., Damgård, I., Goldberg, L.A., Halldórsson, M.M., Ingólfsdóttir, A., Walukiewicz, I. (eds.) ICALP 2008, Part I. LNCS, vol. 5125, pp. 820–832. Springer, Heidelberg (2008)
8. Christodoulou, G., Kovács, A., Sgouritsa, A., Tang, B.: Tight bounds for the price of anarchy of simultaneous first price auctions. CoRR abs/1312.2371 (2013)
9. DiPalantino, D., Vojnovic, M.: Crowdsourcing and all-pay auctions. In: EC 2009, pp. 119–128. ACM, New York (2009)
10. Dobzinski, S., Fu, H., Kleinberg, R.D.: Optimal auctions with correlated bidders are easy. In: Proceedings of the Forty-third Annual ACM Symposium on Theory of Computing, STOC 2011, pp. 129–138. ACM, New York (2011)
11. Feldman, M., Fu, H., Gravin, N., Lucier, B.: Simultaneous Auctions are (almost) Efficient. In: STOC 2013 (September 2013)
12. Hassidim, A., Kaplan, H., Mansour, Y., Nisan, N.: Non-price equilibria in markets of discrete goods. In: EC 2011. ACM (June 2011)
13. Johari, R., Tsitsiklis, J.N.: Efficiency loss in a network resource allocation game. Mathematics of Operations Research 29(3), 407–435 (2004)
14. de Keijzer, B., Markakis, E., Schäfer, G., Telelis, O.: On the Inefficiency of Standard Multi-Unit Auctions. In: Bodlaender, H.L., Italiano, G.F. (eds.) ESA 2013. LNCS, vol. 8125, pp. 385–396. Springer, Heidelberg (2013)
15. Kelly, F.: Charging and rate control for elastic traffic. Eur. Trans. Telecomm. 8(1), 33–37 (1997)
16. Koutsoupias, E., Papadimitriou, C.: Worst-case equilibria. In: Meinel, C., Tison, S. (eds.) STACS 1999. LNCS, vol. 1563, p. 404. Springer, Heidelberg (1999)
17. Krishna, V.: Auction Theory. Academic Press (2002)
18. Markakis, E., Telelis, O.: Uniform price auctions: Equilibria and efficiency. In: Serna, M. (ed.) SAGT 2012. LNCS, vol. 7615, pp. 227–238. Springer, Heidelberg (2012)
19. Roughgarden, T.: Barriers to near-optimal equilibria. In: FOCS 2014, Philadelphia, PA, USA, October 18-21, pp. 71–80 (2014)
20. Siegel, R.: All-pay contests. Econometrica 77(1), 71–92 (2009)
21. Simon, L.K., Zame, W.R.: Discontinuous games and endogenous sharing rules. Econometrica: Journal of the Econometric Society, 861–872 (1990)
22. Syrgkanis, V., Tardos, E.: Composable and Efficient Mechanisms. In: STOC 2013: Proceedings of the 45th symposium on Theory of Computing (November 2013)
23. Vickrey, W.: Counterspeculation, auctions, and competitive sealed tenders. The Journal of Finance 16(1), 8–37 (1961)

Dictionary Matching in a Stream

Raphaël Clifford[1], Allyx Fontaine[1], Ely Porat[2],
Benjamin Sach[1], and Tatiana Starikovskaya[1]

[1] University of Bristol, Department of Computer Science, Bristol, UK
[2] Bar-Ilan University, Department of Computer Science, Israel

Abstract. We consider the problem of dictionary matching in a stream. Given a set of strings, known as a dictionary, and a stream of characters arriving one at a time, the task is to report each time some string in our dictionary occurs in the stream. We present a randomised algorithm which takes $O(\log \log(k + m))$ time per arriving character and uses $O(k \log m)$ words of space, where k is the number of strings in the dictionary and m is the length of the longest string in the dictionary.

1 Introduction

We consider the problem of dictionary matching in a stream. Given a set of strings, known as a dictionary, and a stream of characters arriving one at a time, the task is to determine when some string in our dictionary matches a suffix of the growing stream. The dictionary matching problem models the common situation where we are interested in not only a single pattern that may occur but in fact a whole set of them.

The solutions we present will be analysed under a particularly strong model of space usage. We will account for all the space used by our algorithm and will not, for example, even allow ourselves to store a complete version of the input. In particular, we will neither be able to store the whole of the dictionary nor the streaming text. We now define the problem which will be the main object of study for this paper more formally.

Problem 1. In the dictionary-matching problem we have a set of patterns $\mathcal{P}$ and a streaming text $T = t_1 \ldots t_n$ which arrives one character at a time. We must report all positions in T where there exists a pattern in $\mathcal{P}$ which matches exactly. More formally, we output all the positions x such that there exists a pattern $P_i \in \mathcal{P}$ with $t_{x-|P_i|+1} \ldots t_x = P_i$. We must report an occurrence of some pattern in $\mathcal{P}$ as soon as it occurs and before we can process the subsequent arriving character.

If each of the k patterns in the dictionary had the same length m then we could straightforwardly deploy the fingerprinting method of Karp and Rabin [13] to maintain a fingerprint of a window of length m successive characters of the text. We can then compare this for each new character that arrives to a hash table of stored fingerprints of the patterns in the dictionary. In our notation this

© Springer-Verlag Berlin Heidelberg 2015
N. Bansal and I. Finocchi (Eds.): ESA 2015, LNCS 9294, pp. 361–372, 2015.
DOI: 10.1007/978-3-662-48350-3_31

approach would require $O(k + m)$ words of space and constant time per arrival. However if the patterns are not all the same length this technique no longer works.

For a single pattern, Porat and Porat [17] showed that it is possible to perform exact matching in a stream quickly using very little space. To do this they introduced a clever combination of the randomised fingerprinting method of Karp and Rabin and the deterministic and classical KMP algorithm [14]. Their method uses $O(\log m)$ words of space and takes $O(\log m)$ time per arriving character where m is the length of the single pattern. Breslauer and Galil subsequently made two improvements to this method. First, they sped up the method to only require $O(1)$ time per arriving character and they also showed that it was possible to eliminate the possibility of false negatives, which could occur using the previous approach [3].

Our solution takes the single-pattern streaming algorithm of Breslauer and Galil [3] as its starting point. If we were to run this algorithm independently in parallel for each separate string in the dictionary, this would take $O(k)$ time per arriving character and $O(k \log m)$ words of space. Our goal in this paper is to reduce the running time to as close to constant as possible without increasing the total space. Achieving this presents a number of technical difficulties which we have to overcome.

The first such hurdle is how to process patterns of different lengths efficiently. In the method of Breslauer and Galil prefixes of power of two lengths are found until either we encounter a mismatch or a match is found for a prefix of length at least half of the total pattern size. Exact matches for such long prefixes can only occur rarely and so they can afford to check each one of these potential matches to see if it can be extended to a full match of the pattern. However when the number of patterns is large we can no longer afford to inspect each pattern every time a new character arrives.

Our solution breaks down the patterns in the dictionary into three cases: short patterns, long patterns with short periods, long patterns with long periods. A key conceptual innovation that we make is a method to split the patterns into parts in such a way that matches for all of these parts can be found and stitched together at exactly the time they are needed. We achieve this while minimising the total space and taking only $O(\log \log(k + m))$ time per arriving symbol.

A straightforward counting argument tells us that any randomised algorithm with inverse polynomial probability of error requires at least $\Omega(k \log n)$ *bits* of space, see for example [5]. Our space requirements are therefore within a logarithmic factor of being optimal. However, unlike the single-pattern algorithm of Breslauer and Galil, our dictionary matching algorithm can give both false positives and false negatives with small probability.

Throughout the rest of this paper, we will refer to the arriving text character as the arrival. We can now give our main new result which will be proven in the remaining parts of this paper.

Theorem 1. *Consider a dictionary $\mathcal{P}$ of k patterns of size at most m and a streaming text T. The streaming dictionary matching problem can be solved in*

$O(\log \log(k + m))$ *time per arrival and* $O(k \log m)$ *words of space. The probability of error is* $O(1/n)$ *where* n *is the length of the streaming text.*

1.1 Related Work

The now standard offline solution for dictionary matching is based on the Aho-Corasick algorithm [1]. Given a dictionary $\mathcal{P} = \{P_1, P_2, \ldots, P_k\}$, and a text $T = t_1 \ldots t_n$, let occ denote the number of matches and M denote the sum of the lengths of the patterns in $\mathcal{P}$, that is $M = \sum_{i=1}^{k} |P_i|$. The Aho-Corasick algorithm finds all occurrences of elements in $\mathcal{P}$ in the text T in $O(M + n + \text{occ})$ time and $O(M)$ space. Where the dictionary is large, the space required by the Aho-Corasick approach may however be excessive.

There is now an extensive literature in the streaming model. Focusing narrowly on results related to the streaming algorithm of Porat and Porat [17], this has included a form of approximate matching called parameterised matching [12], efficient algorithms for detecting periodicity in streams [11] as well as identifying periodic trends [10]. Fast deterministic streaming algorithms have also been given which provided guaranteed worst case performance for a number of different approximate pattern matching problems [7,8] as well as pattern matching in multiple streams [6].

The streaming dictionary matching problem has also been considered in a weaker model where the algorithm is allowed to store a complete read-only copy of the pattern and text but only a constant number of extra words in working space. Breslauer, Grossi and Mignosi [4] developed a real-time string matching algorithm in this model by building on previous work of Crochemore and Perrin [9]. The algorithm is based on the computation of periods and critical factorisations allowing at the same time a forward and a backward scan of the text.

1.2 Definitions

We will make extensive use of Karp-Rabin fingerprints [13] which we now define along with some useful properties.

Definition 1. *Karp-Rabin fingerprint function* ϕ. *Let* p *be a prime and* r *a random integer in* $\mathbb{F}_p$. *We define the fingerprint function* ϕ *for a string* $S = s_1 \ldots s_\ell$ *such that:*

$$\phi(S) = \sum_{i=1}^{\ell} s_i r^i \mod p.$$

The most important property is that for any two equal length strings U and V with $U \neq V$, the probability that $\phi(U) = \phi(V)$ is at most $1/n^2$ if $p > n^3$. We will also exploit several well known arithmetic properties of Karp-Rabin fingerprints which we give in Lemma 1. All operations will be performed in the word-RAM model with word size $\Theta(\log n)$.

Lemma 1. *Let* U *be a string of size* ℓ *and* V *another string, then:*

- $\phi(UV) = \phi(U) + r^{\ell}\phi(V) \bmod p,$
- $\phi(U) = \phi(UV) - r^{\ell}\phi(V) \bmod p,$
- $\phi(V) = r^{-\ell}(\phi(UV) - \phi(U)) \bmod p.$

For a non-empty string x, an integer p with $0 < p \leq |x|$ is called a *period* of x if $x_i = x_{i+p}$ for all $i \in \{1, \ldots, |x| - p\}$. The period of a non-empty string x is simply the smallest of its periods. We will also assume that all logarithms are base 2 and are rounded to the nearest integer.

We describe three algorithms: $\mathcal{A}_1$ in Section 2 which handles short patterns in the dictionary, and $\mathcal{A}_{2a}$ and $\mathcal{A}_{2b}$ in Section 3 which deal with the long patterns. Theorem 1 is obtained by running all three algorithms simultaneously.

2   Short Patterns

Lemma 2. *There exists an algorithm $\mathcal{A}_1$ which solves the streaming dictionary matching problem and runs in $O(\log \log(k + m))$ time per arrival and uses $O(k \log m)$ space on a dictionary of k patterns whose maximum length is at most $2k \log m$.*

For very short patterns, shorter than $2 \log m$, we can straightforwardly construct an Aho-Corasick automaton [1]. The automaton occupies $O(k \log m)$ space and reports occurrences of short patterns in constant time per arrival. As the input alphabet may not be constant, at each node of the automaton we store the transition function using a static perfect hash table allowing constant time transitions. From now on, we continue under the assumption that all patterns are longer than $2 \log m$.

Our solution partitions each of the patterns into prefix/suffix pairs in multiple ways. For each pattern there is one partition for each $\ell \in [\log m, 2 \log m]$. Each suffix has length ℓ and is referred to as the tail. The prefix makes up the rest of the pattern and is referred to as the head. We partition each pattern into at most $\log m$ head/tail pairs, making a total of at most $k \log m$ heads overall.

The overall idea is to insert all heads into a data structure so that we can find potential matches in the stream efficiently. We will only look for potential matches every $\log m$ arrivals. We use the remaining at least $\log m$ arrivals before a full match can occur both to de-amortise the cost of finding head matches as well as to check whether the relevant tails match as well.

In order to look for matches with heads of patterns efficiently we put them into a compacted trie. To make use of the compacted trie that we build, we will need to be able to find the *exit node* of any string x efficiently. This is the deepest node in the trie which represents a prefix of x. Given a string x and the fingerprints of all its prefixes, the exit node of x can be found using a variant of binary search in $O(\log m)$ time. This variant of binary search is called 2-fattest binary search and was introduced by Djamal Bellazougui et al. [2] for this purpose.

The basic idea is to perform binary search on the length of the longest prefix of x which matches a string in the set using fingerprint comparisons. The problem with this approach is that traditional binary search would require us to store the fingerprint of every prefix of every string corresponding to a node in the trie. The 2-fattest binary search preprocessing avoids this by carefully selecting a single prefix length for each node. These are chosen in such a way that for any x the binary search only makes comparisons with these preselected prefix lengths. Therefore we only store $O(k \log m)$ fingerprints. The details of 2-fattest binary search can be found in Section 4.1 of [2]. In their work a general 'signature function' is supported which in our work is implemented as the Karp-Rabin fingerprint function.

We can now describe Algorithm $\mathcal{A}_1$ assuming that all patterns are longer than $2 \log m$ but no longer than $2k \log m$. As a preprocessing step, we build the compacted trie for the reverse of the at most $k \log m$ heads and preprocess it to allow efficient computation of exit nodes. For regularly spaced indices of the text, we will use the compacted trie to find the longest head that matches at each of these locations.

We will also augment the compacted trie during preprocessing so that we can support a second operation which will allow us to extend head matches into full matches. We mark each node labelled by a head with a colour representing the fingerprint of the corresponding tail. In the end, each node may be marked by several colours, and the total number of colours will be $k \log m$. On top of the trie we build a coloured-ancestor data structure [16]. This occupies $O(k \log m)$ space and supports $\mathsf{Find}(u, c)$ queries in $O(\log \log(k \log m)) = O(\log \log(k + m))$ time, where $\mathsf{Find}(u, c)$ is the lowest ancestor of a node u marked with colour c. We will use the coloured-ancestor queries to extend a matching head into the longest possible match with a whole pattern by using the fingerprints of different tails as queries.

At all times we maintain a circular buffer of size $2k \log m$ which holds the fingerprints of the most recent $2k \log m$ prefixes of the text. Let i be an integer multiple of $\log m$. For each such i, we query the trie with a string $x = t_i \dots t_{i-2k \log m+1}$. Note that for each prefix of x we can compute its fingerprint in $O(1)$ time with the help of the buffer. The query returns the exit node $e(x)$ of x in $O(\log m)$ time, which is used to analyse arrivals in the interval $[i + \log m, i + 2 \log m]$. This exit node corresponds to the longest head that matches ending at index i. The $O(\log m)$ cost of performing the query is de-amortised during the interval $(i, i + \log m]$.

For each arrival t_ℓ, $\ell \in (i + \log m, i + 2 \log m]$ we compute the fingerprint ϕ of $t_{i+1} \dots t_\ell$. This can be done in constant time as we store the last $2k \log m \geq 2 \log m$ fingerprints. If $\mathsf{Find}(e(x), \phi)$ is defined, ℓ is an endpoint of a whole pattern match and we report it. Otherwise, we proceed to the next arrival. The overall time per arrival is therefore dominated by the time to perform the coloured-ancestor queries which is $O(\log \log(k + m))$.

We remark that the algorithm can be extended to permit patterns of length at most $4k \log m$ (instead of $2k \log m$) without affecting the time or space complexity. Moreover, if there are several possible patterns that match for a given arrival, the algorithm reports the longest such pattern. These two properties will be needed when we describe Algorithm $\mathcal{A}_{2b}$ in Section 3.2.

3   Long Patterns

We now assume that all the patterns have length greater than $2k \log m$. We distinguish two cases according to the periodicity of those patterns: those with short period and those with long period. Hereafter, to distinguish the cases, we use the following notation. Let $m_i = |P_i|$ and Q_i be the prefix of P_i such that $|Q_i| = m_i - k \log m$. Let ρ_{Q_i} be the period of Q_i. The remaining patterns are then partitioned in two disjoint groups of patterns, those with $\rho_{Q_i} < k \log m$ and those with $\rho_{Q_i} \geq k \log m$. We describe two algorithms: $\mathcal{A}_{2a}$ and $\mathcal{A}_{2b}$, one for each case respectively. Finally, the overall solution is then to run all three algorithms $\mathcal{A}_1$, $\mathcal{A}_{2a}$, $\mathcal{A}_{2b}$ simultaneously to obtain Theorem 1.

3.1   Algorithm $\mathcal{A}_{2a}$: Patterns with Short Periods

This section gives an algorithm for a dictionary of patterns $\mathcal{P} = P_1, \ldots, P_k$ such that $m_i \geq 2k \log m$ and $\rho_{Q_i} < k \log m$. Recall that Q_i is the prefix of P_i of length $m_i - k \log m$ and ρ_{Q_i} is the period of Q_i. The overall idea for this case is that if we can find enough repeated occurrences of the period of a pattern then we know we have almost found a full pattern match. As the pattern may end with a partial copy of its period we will have to handle this part separately. The main technical hurdle we overcome is how to process different patterns with different length periods in an efficient manner.

We define the tail of a pattern P_i to be its suffix of length $2k \log m$. Observe that a P_i match occurs if and only if there is a match of Q_i followed by a match with the tail of P_i.

Let K_i be the prefix of Q_i of length $k \log m$. Further observe that Q_i can only match if there is a sequence of $\left\lfloor \frac{|Q_i| - |K_i|}{\rho_{Q_i}} + 1 \right\rfloor$ occurrences of K_i in the text, each occurring exactly ρ_{Q_i} characters after the last. This follows immediately from the fact that K_i has length $k \log m$ and Q_i has period $\rho_{Q_i} < k \log m$.

We now describe algorithm $\mathcal{A}_{2a}$ which solves this case. At all times we maintain a circular buffer of size $2k \log m$ which holds the fingerprints of the most recent $2k \log m$ prefixes of the text. That is, if the last arrival is t_ℓ, then the buffer contains the fingerprints $\phi(t_1 \ldots t_{\ell - 2k \log m + 1}), \ldots, \phi(t_1 \ldots t_\ell)$.

To find K_i matches, we store the fingerprint $\phi(K_i)$ of each distinct K_i in a static perfect hash table. By looking up $\phi(t_{\ell - k \log m + 1} \ldots t_\ell)$ we can find whether some K_i matches in $O(1)$ time. For each distinct K_i we maintain a list of recent matches stored as an arithmetic progression. Each time we find a new match with K_i we check whether it is exactly ρ_{Q_i} characters from the last match. If so we include it in the current arithmetic progression. If not, then we delete the current progression and start a new progression containing only the latest match. Note that $K_i = K_j$ implies that $\rho_{Q_i} = \rho_{Q_j}$ and therefore there is no ambiguity in the description.

We store the fingerprint of each tail in another static perfect hash table. For each arrival t_ℓ we use this hash table to check whether $\phi(t_{\ell-2k\log m+1}\cdots t_\ell)$ matches the fingerprint of some tail. This takes $O(1)$ time per arrival.

Assume that the tail of some P_i matched. We will justify below that we can assume that each tail corresponds to a unique P_i. It remains to decide whether this is in-fact a full match with P_i. This is determined by a simple check, that is whether the current arithmetic progression for K_i contains at least $\left\lfloor \frac{|Q_i|-|K_i|}{\rho_{Q_i}} + 1 \right\rfloor$ occurrences.

Lemma 3. *Algorithm $\mathcal{A}_{2a}$ takes $O(1)$ time per character and uses $O(k\log m)$ space.*

Proof. The algorithm stores two hash tables, each containing $O(k\log m)$ fingerprints as well as $O(k)$ arithmetic progressions. The total space is therefore $O(k\log m)$ as claimed. The time complexity of $O(1)$ per character follows by the use of static perfect hash tables (which are precomputed and depend only on $\mathcal{P}$).

We first prove the claim that each tail corresponds to a unique P_i. To this end, we assume in this section that no pattern contains another pattern as a suffix. In particular, any such pattern can be deleted from the dictionary during the preprocessing stage as it does not change the output. This implies the claim that each P_i has a distinct tail because the tail contains a full period of P_i.

The correctness follows almost immediately from the algorithm description via the observation that each Q_i is formed from $\left\lfloor \frac{|Q_i|-|K_i|}{\rho_{Q_i}} + 1 \right\rfloor$ repeats of K_i followed by a prefix of K_i. We check explicitly whether there are sufficient repeats of K_i in the text stream to imply a Q_i match. While we do not check explicitly that either final prefix of K_i is a match or that the full P_i matches, this is implied by the tail match. This is because the tail has length $2k\log m$ and hence includes the final prefix of K_i and the last $k\log m$ characters of P_i (those in P_i but not in Q_i). □

3.2   Algorithm $\mathcal{A}_{2b}$: Patterns with Long Periods

Consider a dictionary $\mathcal{P}$ in which the patterns are such that $m_i \geq 2k\log m$ and $\rho_{Q_i} \geq k\log m$. Let us define k to be number strings in this dictionary. We can now describe Algorithm $\mathcal{A}_{2b}$. Recall that Q_i is the prefix of P_i s.t. $|Q_i| = m_i - k\log m$. For each pattern P_i, we define $P_{i,j}$ to be the prefix of P_i with length 2^j, $1 \leq 2^j \leq m_i - 2k\log m$.

We will first give an overview of an algorithm that identifies $P_{i,j}$ matches in $O(\log m)$ time per arrival. With the help of $\mathcal{A}_1$ and $\mathcal{A}_{2a}$ we will speed it up to achieve an algorithm with $O(\log\log(k+m))$ time per arrival. The algorithm will identify the matches with a small delay up to $k\log m$ arrivals. We then show how to extend $P_{i,j}$ to Q_i matches. This stage will still report the matches after they occur. Finally we show how to find whole pattern matches in the stream using the Q_i matches while also completely eliminating the delay in the reporting of these matches. In other words, any matches for whole patterns will be reported as soon as they occur and before the next arrival in the stream as desired.

$O(\log m)$-time Algorithm. We define a logarithmic number of *levels*. Level j will represent all the matches for prefixes $P_{i,j}$. We store only *active* prefix matches, that still have the potential to indicate the start of full matches of a pattern in the dictionary. This means that any match at level j whose position is more than 2^{j+1} from the current position of an arrival is simply removed. We will use the following well-known fact.

Fact 2 (Lemma 3.2[3]). *If there are at least three matches of a string U of length 2^j in a string V of length 2^{j+1}, then positions of all matches of U in V form an arithmetic progression. The difference of the progression is equal to the length of the period of U.*

It follows that if there are at least three active matches for the same prefix at the same level, we can compactly store them as a *progression* in constant space. Consider a set of distinct prefixes of length 2^j of the patterns in $\mathcal{P}$. For each of them we store a progression that contains:

(1) The position `fp` of the first match;
(2) The fingerprint of $t_1 \ldots t_{\texttt{fp}}$;
(3) The fingerprint of the period ρ of the prefix;
(4) The length of the period ρ of the prefix;
(5) The position `lp` of the last match.

With this information, we can deduce the position and the fingerprint of the text from the start to the position of any active match of the prefix. Moreover, we can add a new match or delete the first match in a progression in $O(1)$ time.

We make use of a perfect hash table $\mathcal{H}$ that stores the fingerprints of all the prefixes of the patterns in $\mathcal{P}$. The keys of $\mathcal{H}$ correspond to the fingerprints of all the prefixes and the associated value indicates whether the prefix from which the key was obtained is a proper prefix of some pattern, a whole pattern itself, or both. Using the construction of [18], for example, the total space needed to store all the fingerprints and their corresponding values is $O(k \log m)$.

When a character t_ℓ of the text arrives, we update the current position and the fingerprint of the current text. The algorithm then proceeds by the progressions over $\log m$ levels. We start at level 0. If the fingerprint $\phi(t_\ell)$ is in $\mathcal{H}$, we insert a new match to the corresponding progression at level 0.

For each level j from 0 to $\log m$, we retrieve the position p of the first match at level j. If p is at distance 2^{j+1} from t_ℓ, we delete the match and check if the fingerprint $\phi(t_p \ldots t_\ell)$ is in $\mathcal{H}$. If it is and the fingerprint is a fingerprint of one of the patterns, we report a match (ending at t_ℓ, the current position of the text). If the fingerprint is in $\mathcal{H}$ and if it is a fingerprint of a proper prefix, then p is a plausible position of a match of a prefix of length 2^{j+1}. We check if it fits in the appropriate progression π at level $j+1$. (Which might not be true if the fingerprints collided). If it does, we insert p to π. If p does not match in π, we discard it and proceed to the next level.

As updating progressions at each level only takes $O(1)$ time, and there are $\log m$ levels, the time complexity of the algorithm is $O(\log m)$ per arrival. The space complexity is $O(k \log m)$.

$O(\log\log(k+m))$-time Algorithm. We will follow the same level-based idea. To speed up the algorithm, we will consider prefixes $P_{i,j}$ with short and long periods separately. The number of matches of the prefixes with short periods can be big, but we will be able to compute them fast with the help of $\mathcal{A}_1$ and $\mathcal{A}_{2a}$. On the other hand, matches of the prefixes with long periods are rare, and we will be able to compute them in a round robin fashion.

Let $\rho_{i,j}$ be the period of $P_{i,j}$. We first build a dictionary D_1 containing at most one prefix for each P_i. Specifically, containing the largest $P_{i,j}$ with the period $\rho_{i,j} < k\log m$ and $2k\log m \leq |P_{i,j}| \leq m_i - 2k\log m$. If no such $P_{i,j}$ exists we do not insert a prefix for P_i. This dictionary is processed using a modification of algorithm $\mathcal{A}_{2a}$ which we described in Section 3.1. The modification is that when a text character t_ℓ arrives, the output of the algorithm identifies the longest pattern in D_1 which matches ending at t_ℓ or 'no match' if no pattern matches. This is in contrast to $\mathcal{A}_{2a}$ as described previously where we only outputted whether some pattern matches. The modification takes advantage of the fact that prefixes in D_1 all have power-of-two lengths and uses a simple binary search approach over the $O(\log m)$ distinct pattern lengths. This increases the run-time of $\mathcal{A}_{2a}$ to $O(\log\log m)$ time per arrival.

Whenever a match is found with some pattern in D_1, we update the match progression of the reported pattern (but not of any of its suffixes that might be in D_1). Importantly, we will still have at most two progressions of active matches per prefix because of the following lemma and corollary.

Lemma 4. *Let $P_{i,j}, P_{i',j'}$ be two prefixes in D_1 and suppose that $P_{i,j}$ is a suffix of $P_{i',j'}$. The periods of $P_{i,j}, P_{i',j'}$ are equal.*

Proof. Assume the contrary. Then $P_{i,j}$ has two periods: $\rho_{i,j}$ and $\rho_{i',j'}$ (because it is a suffix of $P_{i',j'}$). We have $\rho_{i,j} + \rho_{i',j'} < 2k\log m \leq |P_{i,j}|$. By the periodicity lemma (see, e.g., [15]), $\rho_{i,j}$ is a multiple of $\rho_{i',j'}$. But then $P_{i,j}$ is periodic with period $\rho_{i',j'} < \rho_{i,j}$, a contradiction. $\qquad\square$

Corollary 1. *Let $P_{i,j}$, $P_{i',j'}$, and $P_{i'',j''}$ be prefixes in D_1. Suppose that $P_{i,j}$ is a suffix of $P_{i',j'}$ and simultaneously is a suffix of $P_{i'',j''}$. Then $P_{i',j'}$ is a suffix of $P_{i'',j''}$ (or vice versa).*

We now consider any P_i for which we did not find a suitable small period prefix. In this case it is guaranteed that there is a prefix $P_{i,j}$ with period longer than $k\log m$ but length at most $4k\log m$. We build another dictionary D_2 for each of these prefixes. We apply algorithm $\mathcal{A}_1$ and for each arrival t_ℓ return the longest prefix $P_{i,j}$ in D_2 that matches at it in $O(\log\log(k+m))$ time. We then need to update the match progression of $P_{i,j}$ as well as the match progressions of all $P_{i',j'} \in D_2$ that are suffixes of $P_{i,j}$. Fortunately, each of the prefixes in D_2 can match at most once in every $k\log m$ arrivals, because the period of each of them is long, meaning that we can schedule the updates in a round robin fashion to take $O(1)$ time per arrival.

We denote a set of all $P_{i,j}$ such that $\rho_{i,j} \geq k\log m$ by S. Any of these prefixes can have at most one match in $k\log m$ arrivals. Because of that and because

$|S| \leq k \log m$, we will be able to afford to update the matches in a round robin fashion.

We will have two update processes running in parallel. The first process will be updating matches of prefixes $P_{i,j} \in S$ such that $P_{i,j-1} \in S \cup D_2$. We consider one of these prefixes per arrival. If there is a match with $P_{i,j}$ in $[t_\ell - k \log m, t_\ell]$ then there must be a corresponding match with $P_{i,j-1}$ ending in $[t_{\ell-2^{j-1}-k \log m}, t_{\ell-2^{j-1}}]$. As $P_{i,j-1} \in S$, $\rho_{i,j} \geq k \log m$ so there is at most one match. We can determine whether this match can be extended into a $P_{i,j}$ match using a single fingerprint comparison as described in the $O(\log m)$-time algorithm. This is facilitated by storing a circular buffer of the fingerprints of the most recent $k \log m$ text prefixes.

The second process will be updating matches of prefixes $P_{i,j} \in S$ such that $P_{i,j-1} \in D_1$. Again, if there is a match with $P_{i,j}$ in $[t_\ell - k \log m, t_\ell]$ then there must be a corresponding match with $P_{i,j-1}$ ending in $[t_{\ell-2^{j-1}-k \log m}, t_{\ell-2^{j-1}}]$. However, the second process will be more complicated for two reasons. First, $P_{i,j-1}$ has a small period so there could be many $P_{i,j-1}$ matches ending in this interval. Second, the information about $P_{i,j-1}$ matches can be stored not only in the progressions corresponding to $P_{i,j-1}$, but also in the progressions corresponding to prefixes that have $P_{i,j-1}$ as a suffix. The first difficulty can be overcome because of the following lemma.

Lemma 5. *Consider any $P_{i,j}$ such that $\rho_{i,j-1} < k \log m \leq \rho_{i,j}$. Given a match progression for $P_{i,j-1}$, only one match could also correspond to a match with $P_{i,j}$.*

Proof. Let U be the prefix of $P_{i,j-1}$ of length $\rho_{i,j-1}$. That is, the substrings bounded by consecutive matches in the match progression for $P_{i,j-1}$ are equal to U. Suppose that $P_{i,j}$ starts with exactly r copies of U. Then we have $P_{i,j} = U^r V$ for some string V. Note that as $\rho_{i,j-1} < k \log m \leq \rho_{i,j}$, the string V cannot be a prefix of U. Then the only match in the progression which could match with $P_{i,j}$ is the r-th last one.   □

To overcome the second difficulty, we use Corollary 1. It implies that prefixes in D_1 can be organized in chains based on the "being-a-suffix" relationship. We consider prefixes in each chain in a round robin fashion again. We start at the longest prefix, let it be $P_{i,j}$. At each moment we store exactly one progression initialized to the progression of $P_{i,j}$. If the progression intersects with $[t_{\ell-2^{j-1}-k \log m}, t_{\ell-2^{j-1}}]$, we identify the 'interesting' match in $O(1)$ time with the help of Lemma 5 and try to extend it as in the first process. We then proceed to the second longest prefix $P_{i',j'}$. If the stored progression intersects with $[t_{\ell-2^{j'-1}-k \log m}, t_{\ell-2^{j'-1}}]$, we proceed as for $P_{i,j}$. Otherwise, we update the progression to be the progression of $P_{i',j'}$ and repeat the previous steps for it. We continue this process for all prefixes in the chain.

From the description of the processes it follows that the matches for each $P_{i,j}$ (in particular, for the longest $P_{i,j}$ for each i) are outputted in $O(\log \log(k + m))$ time per arrival with a delay of up to $k \log m$ characters (i.e. at most $k \log m$ characters after they occur).

Finding Q_i Matches. We now show how to find Q_i matches using $P_{i,j}$ matches. If there is a match with Q_i in $[t_\ell - k \log m, t_\ell]$, there must be a match with the longest $P_{i,j}$ in $[t_\ell - 2^j - k \log m, t_\ell - 2^j]$. Because $|P_{i,j}| \leq m_i - 2k \log m$, this match has been identified by the algorithm and it is the first match in the progressions. We can determine whether this match can be extended into a Q_i match using a single fingerprint comparison.

Therefore the Q_i matches are outputted in $O(\log \log(k+m))$ time with a delay of up to $k \log m$ characters (i.e. at most $k \log m$ characters after they occur). We can then remove this delay using coloured ancestor queries in a similar manner to algorithm $\mathcal{A}_1$ as described below.

Finding Whole Pattern Matches and Removing the Delay. Up to this point, we have shown that we can find each Q_i match in $O(\log \log(k + m))$ time per arrival with a delay of at most $k \log m$ characters. Further we only report one Q_i match at each time. We will show how to extend these Q_i matches into P_i matches using coloured ancestor queries in $O(\log \log(k+m))$ time per arrival.

Build a compacted trie of the reverse of each string Q_i. The edges labels are not stored. The space used is $O(k)$. For each i we can find the reverse of Q_i in the trie in $O(1)$ time (by storing an $O(k)$ space look-up table).

The tail of each P_i is its $(k \log m)$-length suffix, i.e. the portion of P_i which is not in Q_i. Each distinct tail is associated with a colour. As there are at most $k \log m$ patterns, there are at most $k \log m$ colours. Computing the colour from the tail is achieved using a standard combination of fingerprinting and static perfect hashing. For each node in the tree which represents some Q_i we colour the node with the colour of the tail of P_i.

Whenever we find a Q_i match, we identify the place in the tree where the reverse of Q_i occurs. Recall that these matches may be found after a delay of at most $k \log m$ characters. A Q_i match ending at position $\ell - k \log m$ implies a possible P_i match at position ℓ. We remember this potential match until t_ℓ arrives.

More specifically when t_ℓ arrives we determine the node u in the trie representing the reverse of the longest Q_i which has a match at position $\ell - k \log m$. This can be done in $O(1)$ time by storing a circular buffer of fingerprints.

We now need to decide whether Q_i implies the existence of some P_j match. It is important to observe that as we discarded all but the longest such Q_i, we might find a P_j with $j \neq i$.

For each arrival t_ℓ, we compute the fingerprint ϕ of $t_{\ell - k \log m + 1} \ldots t_\ell$. This can be done in constant time as we store the last $k \log m$ fingerprints. If $\mathsf{Find}(u, \phi)$ is defined, t_ℓ is an endpoint of a pattern match and we report it. Otherwise, we proceed to the next arrival.

Lemma 6. *Algorithm $\mathcal{A}_{2b}$ takes $O(\log \log(k+m))$ time per character. The space complexity of the algorithm is $O(k \log m)$.*

References

1. Aho, A.V., Corasick, M.J.: Efficient string matching: an aid to bibliographic search. Communications of the ACM 18(8), 333–340 (1975)
2. Belazzougui, D., Boldi, P., Pagh, R., Vigna, S.: Monotone minimal perfect hashing: Searching a sorted table with O(1) accesses. In: SODA 2009: Proc. 20th ACM-SIAM Symp. on Discrete Algorithms, pp. 785–794 (2009)
3. Breslauer, D., Galil, Z.: Real-time streaming string-matching. ACM Transactions on Algorithms 10(4), 22 (2014)
4. Breslauer, D., Grossi, R., Mignosi, F.: Simple real-time constant-space string matching. In: Giancarlo, R., Manzini, G. (eds.) CPM 2011. LNCS, vol. 6661, pp. 173–183. Springer, Heidelberg (2011)
5. Broder, A.Z., Mitzenmacher, M.: Survey: Network applications of bloom filters: A survey. Internet Mathematics 1(4), 485–509 (2003)
6. Clifford, R., Jalsenius, M., Porat, E., Sach, B.: Pattern matching in multiple streams. In: Kärkkäinen, J., Stoye, J. (eds.) CPM 2012. LNCS, vol. 7354, pp. 97–109. Springer, Heidelberg (2012)
7. Clifford, R., Sach, B.: Pseudo-realtime pattern matching: Closing the gap. In: Amir, A., Parida, L. (eds.) CPM 2010. LNCS, vol. 6129, pp. 101–111. Springer, Heidelberg (2010)
8. Clifford, R., Sach, B.: Pattern matching in pseudo real-time. Journal of Discrete Algorithms 9(1), 67–81 (2011)
9. Crochemore, M., Perrin, D.: Two-way string matching. J. ACM 38(3), 651–675 (1991)
10. Crouch, M.S., McGregor, A.: Periodicity and cyclic shifts via linear sketches. In: Goldberg, L.A., Jansen, K., Ravi, R., Rolim, J.D.P. (eds.) APPROX and RANDOM 2011. LNCS, vol. 6845, pp. 158–170. Springer, Heidelberg (2011)
11. Ergun, F., Jowhari, H., Sağlam, M.: Periodicity in streams. In: Serna, M., Shaltiel, R., Jansen, K., Rolim, J. (eds.) APPROX and RANDOM 2010. LNCS, vol. 6302, pp. 545–559. Springer, Heidelberg (2010)
12. Jalsenius, M., Porat, B., Sach, B.: Parameterized matching in the streaming model. In: STACS 2013: Proc. 30th Annual Symp. on Theoretical Aspects of Computer Science, pp. 400–411 (2013)
13. Karp, R.M., Rabin, M.O.: Efficient randomized pattern-matching algorithms. IBM Journal of Research and Development 31(2), 249–260 (1987)
14. Knuth, D.E., Morris, J.H., Pratt, V.B.: Fast pattern matching in strings. SIAM Journal on Computing 6, 323–350 (1977)
15. Lothaire, M.: Algebraic Combinatorics on Words. Cambridge University Press (2002)
16. Muthukrishnan, S., Müller, M.: Time and space efficient method-lookup for object-oriented programs. In: SODA 1996: Proc. 7th ACM-SIAM Symp. on Discrete Algorithms, pp. 42–51 (1996)
17. Porat, B., Porat, E.: Exact and approximate pattern matching in the streaming model. In: FOCS 2009: Proc. 50th Annual Symp. Foundations of Computer Science, pp. 315–323 (2009)
18. Ružić, M.: Constructing efficient dictionaries in close to sorting time. In: Aceto, L., Damgård, I., Goldberg, L.A., Halldórsson, M.M., Ingólfsdóttir, A., Walukiewicz, I. (eds.) ICALP 2008, Part I. LNCS, vol. 5125, pp. 84–95. Springer, Heidelberg (2008)

Multicuts in Planar and Bounded-Genus Graphs with Bounded Number of Terminals[*]

Éric Colin de Verdière

Département d'informatique, École normale supérieure, Paris, France, and CNRS
eric.colin.de.verdiere@ens.fr

Abstract. Given an undirected, edge-weighted graph G together with pairs of vertices, called pairs of *terminals*, the *minimum multicut problem* asks for a minimum-weight set of edges such that, after deleting these edges, the two terminals of each pair belong to different connected components of the graph. Relying on topological techniques, we provide a polynomial-time algorithm for this problem in the case where G is embedded on a fixed surface of genus g (e.g., when G is planar) and has a fixed number t of terminals. The running time is a polynomial of degree $O\left(\sqrt{g^2 + gt}\right)$ in the input size.

In the planar case, our result corrects an error in an extended abstract by Bentz [Int. Workshop on Parameterized and Exact Computation, 109–119, 2012]. The minimum multicut problem is also a generalization of the *multiway cut problem*, also known as the *multiterminal cut problem*; even for this special case, no dedicated algorithm was known for graphs embedded on surfaces.

1 Introduction

The minimum cut problem is one of the most fundamental problems in combinatorial optimization, originally introduced in relation to railway transshipment problems during the cold war (see Schrijver [20] for a fascinating historical account). In this context, the railway network is modeled by a planar graph, each edge having a weight (its capacity), and the goal is to compute the minimum-weight set of edges that need to be removed to disconnect two given vertices of the network, the source and destination for a given commodity. While countless generalizations of this problem have been studied, we are interested here in two natural extensions:

1. What if there are several commodities, corresponding to different source and destination pairs? In other words, we are studying an instance of the minimum multicut problem: Given several pairs of source and destination vertices, how to find a minimum-weight set of edges to disconnect every destination vertex from its corresponding source?
2. What if the network is not planar, but includes a few tunnels and bridges? In other words, what happens if the graph is embedded, not in the plane, but on some fixed surface?

[*] Work supported by the French ANR Blanc project ANR-12-BS02-005 (RDAM).

N. Bansal and I. Finocchi (Eds.): ESA 2015, LNCS 9294, pp. 373–385, 2015.
DOI: 10.1007/978-3-662-48350-3_32

More formally, let $G = (V, E)$ be an undirected graph. Furthermore, let T be a subset of vertices of G, called **terminals**, and let R be a set of unordered pairs of vertices in T, called **terminal pairs**. A subset E' of E is a **multicut** (with respect to (T, R)) if for every terminal pair $\{t_1, t_2\} \in R$, the vertices t_1 and t_2 are in different connected components of the graph $(V, E \setminus E')$. In the **minimum multicut problem** (also known as the *minimum multiterminal cut problem*), we assume in addition that G is positively edge-weighted, and the goal is to find a multicut of minimum weight. We prove that this problem is polynomial-time solvable if G is embedded on a fixed surface and the number t of terminals is fixed:

Theorem 1.1. *Assume that G is cellularly embedded on a surface (orientable or not) of Euler genus g. Then the minimum multicut problem can be solved in $O(m \log m) + (g + t)^{O(g+t)} n^{O\left(\sqrt{g^2+gt}\right)}$ time, where $t = |T|$ is the number of terminals, m is the number of edges of G, and n is the number of faces of G.*

This is the first polynomial-time algorithm for this purpose, even when specialized to either the multiway cut problem (see below for details) or the planar version. Moreover, the $n^{O(\sqrt{t})}$ dependence in the number of terminals is unavoidable, assuming the Exponential Time Hypothesis [17], even in these two special cases.

Comparison with Former Work

Many instances of the **minimum multicut** problem are hard, even in very restricted cases. In particular, it is NP-hard, and even APX-hard, in unweighted stars [14, Theorem 3.1]. In the case where the number of pairs of terminals is fixed and at least three, Dahlhaus et al. [10] have proved that the problem is APX-hard in general graphs; nonetheless, it becomes polynomial-time solvable for bounded-treewidth graphs, as proved by Bentz [2, Theorem 1].

In the case where the graph is **planar**, the number of terminals is fixed, and they all lie on the outer face, Bentz [3] has given a polynomial-time algorithm for the minimum multicut problem. More recently [4], he has announced an algorithm for the same case, but removing the condition that the terminals lie on the outer face. Unfortunately, his proof has several flaws, leaving little hope for repair. We give a faster algorithm that also works on arbitrary surfaces.

A special case that is somewhat more tractable is the **multiway cut problem** (also known as the *multiterminal cut problem*); this is the case where each pair of distinct vertices of the set of terminals $T \subset V$ is a terminal pair. In the planar case, Dahlhaus et al. [10] have proved that it is still NP-hard, but Bateni et al. [1] have given a polynomial-time approximation scheme. Again in the planar case, the problem is also polynomial-time solvable if the number of terminals is fixed, as proved in the early 1990s [10]. In stark contrast, the complexity of the multicut problem has remained open until now, although it is a natural generalization of the multiway cut problem (the multicut problem is "dual" to the multicommodity flow problem, largely studied [19, Chapters 70–76]).

More recently, Klein and Marx have shown that the planar multiway cut problem can be solved in $2^{O(t)} n^{O(\sqrt{t})}$ time (where n is the complexity of the graph) [16]; Marx has proved that the $n^{O(\sqrt{t})}$-dependence is the best one could hope for, assuming the Exponential Time Hypothesis (ETH) [17]. Our algorithm is more general since it deals with multicut, not multiway cut, and works on arbitrary surfaces; its running time, for fixed genus, is $t^{O(t)}n^{O(\sqrt{t})}$; while the $t^{O(t)}$ factor is slightly worse than the $2^{O(t)}$ of Klein and Marx, the second factor is the same, and optimal unless ETH is false. Since approximability in the planar case is very different for multicut and multiway cut, our result is surprising, since it shows that, for exact computations, both are (essentially) equally hard.

On the other hand, graph algorithms dedicated to **graphs embedded on a fixed surface** have flourished during the last decade. One reason is that many graphs arising in geometric settings are naturally embedded on a surface; another one is that the family of graphs embeddable on a fixed surface is closed under minor, and such studies can be seen as a first step towards efficient algorithms for minor-closed families. Testing whether a graph of complexity n embeds on a surface of genus g can be done in $2^{O(g)}n$ time [18]. Flows and cuts can be computed more efficiently for graphs embeddable on a fixed surface [6,12,7]; the main tool is homology, which is the appropriate algebraic formalism for dealing with graphs separating two given vertices, but it appears to be insufficient alone in the multicommodity case. Hence, to our knowledge, we present here the first algorithm for the minimum multicut problem (or even the multiway cut problem) for surface-embedded graphs.

Overview and Discussion of Proof Techniques

The strategy for proving Theorem 1.1 is the following. In Section 3, we first show that a multicut corresponds, in a dual sense, to a graph drawn on S that separates all pairs of terminals; such a graph will be called a *multicut dual*. Moreover, if the multicut is minimum, then this multicut dual is as short as possible, when distances on S are measured using the *cross-metric*: namely, the sum of the weights of the edges of G crossed by the multicut dual is minimum. The topological structure of the multicut dual can be described suitably after we cut the surface open into a disk with all terminals on its boundary. We then show that this structure is constrained (Section 4), that we can enumerate its various possibilities (Section 5), and (roughly) that, for each of these topologies, we can compute a shortest multicut dual with that topology efficiently (Section 6).

At a high level, our approach follows a similar pattern to Klein and Marx [16], since they also rely on enumerating the various candidate topologies for the dual solution, and find the optimum solution for each topology separately. This strategy is also present in Dahlhaus et al. [10] and, in a different context, in Chambers et al. [5], which was our initial source of inspiration. The details are, however, rather different.

Indeed, Klein and Marx [16] need a reduction to the biconnected case [16, Section 3], which is shortcut in our approach. Also, the structural properties that we develop for the minimum multicut problem are more involved than the known ones for the multiway cut problem; indeed, the solution, viewed in the dual graph, has less structure in the multicut problem than in the multiway cut problem: in the multiway cut case, it has as many faces as terminals, and thus many cycles, whereas for the minimum multicut problem, the possible topologies are more diverse (e.g., an optimal solution could be a single cycle).

Chambers et al. [5] have developed related techniques for computing a shortest splitting cycle, subsequently reused [13,6]. A key difference, however, is that we extend the method to work with graphs instead of paths or cycles, which makes the arguments more complicated. We need to encode precisely the locations of the vertices and edges of the multicut dual with respect to the cut graph; the cross-metric setting is very convenient for this purpose.

Our approach also requires other techniques from computational topology, in particular, homology techniques developed for the single commodity minimum cut problem [6], homotopy techniques for shortest homotopic paths [9], and treewidth techniques for the surface case [11]. Finally, we remark that we are not aware of any significantly simpler proof for the planar case.

2   Preliminaries

We recall here standard definitions on the topology of surfaces; for more details, we refer to a book [21], a survey [8], or an article [9]. In this article, S is a compact, connected **surface** without boundary; g denotes its Euler genus. For simplicity of exposition, we only consider orientable surfaces in this extended abstract, but the arguments extend almost directly to the non-orientable case. Thus, $g \geq 0$ is even, and S is (homeomorphic to) a sphere with $g/2$ handles attached. We consider paths drawn on S. A **path** p is a continuous map from $[0,1]$ to S; a **loop** is a path p such that its two endpoints coincide: $p(0) = p(1)$. A path is **simple** if it is one-to-one (except, of course, that its endpoints $p(0)$ and $p(1)$ may coincide). We thus emphasize that, contrary to the standard terminology in graph theory, paths may self-intersect.

All the graphs considered in this article may have loops and multiple edges. A **drawing** of a graph G on S maps the vertices of G to points on S and the edges of G to paths on S whose endpoints are the images of the incident vertices. An **embedding** of G is a "crossing-free" drawing: The images of the vertices are pairwise distinct, and the image of each edge is a simple path intersecting the image of no other vertex or edge, except possibly at its endpoints. A **face** of an embedding of G on S is a connected component of S minus (the image of) G. A graph is **cellularly embedded** on S if every face of the graph is an open disk. In particular, a **cut graph** of S is a graph G embedded on S whose unique face is a disk. **Euler's formula** states that, if G is a graph cellularly embedded on S with v vertices, e edges, and f faces, then $v - e + f = 2 - g$.

Algorithmically, we can store graphs cellularly embedded on S by their *combinatorial map*, which essentially records the cyclic ordering of the edges around each vertex; there are efficient data structures for this purpose [11].

3   The Cross-Metric Setting

We will first prove that a minimum multicut corresponds, in an appropriate sense, to a shortest graph satisfying certain properties.

We say that a graph H embedded on S is in **general position** with respect to our input graph G if there are finitely many intersection points between G and H, and each such point corresponds to a crossing between an edge of G and an edge of H. The **length** of H is the sum, over all crossing points between G and H, of the weight of the corresponding edge of G. In other words, G is now seen as a graph that provides a discrete (or *cross-metric*) distance function on S [9]. Algorithmically, we can store a graph H in general position with respect to G by recording the combinatorial map of the *overlay* of G and H, obtained by adding vertices at each intersection point between G and H and subdividing edges of G and H.

*In the following, unless noted otherwise, **all graphs drawn on S will be in general position with respect to G**. Moreover, whenever we consider distances between two points in S (not lying on G) or lengths of paths in S, we implicitly consider them in the above cross-metric sense.* In some clearly mentioned cases below (see Proposition 3.2), we will need to consider paths p that are in general position with respect to G, except that some of their endpoints may lie on G. In such cases, the endpoints of p are not taken into account for determining the length of p.

A **multicut dual** is a graph C embedded on S such that, for every pair $\{t_1, t_2\} \in R$, the vertices t_1 and t_2 are in different faces of C. As the terminology suggests, we have the following easy proposition, which will guide our approach.

Proposition 3.1. *Let C be a shortest multicut dual. Then the set E' of edges of G crossed at least once by C is a minimum multicut.*

As a side remark, it follows that the minimum multicut problem can be seen as a discrete version of the following topological problem: Given a metric surface S with boundary, and a set R of pairs of boundary components, compute a shortest graph on S that separates every pair of boundaries in R. We are exactly solving this problem in the realm of cross-metric surfaces.

Our algorithm starts by computing a particular cut graph K of S passing through all the terminals. If S is the sphere (equivalently, if G is planar), we could take for K a shortest spanning tree of T (with respect to the cross-metric setting), which can be obtained using a simple modification of any algorithm for computing minimum spanning trees in graphs. For the general case, we use a known construction, a so-called *greedy system of arcs* [5]:

Proposition 3.2. *In $O(m \log m + (g+t)n)$ time, we can compute a cut graph K on S, whose $O(g+t)$ vertices contain T, and with $O(g+t)$ edges, each of which is a shortest path on S. Some vertices of K may lie on G (either on vertices or on edges).*

At a high level, the algorithm consists in (1) enumerating all possible "topologies" of the multicut dual with respect to K, (2) for each of these possible topologies, computing a shortest multicut dual with that topology, and (3) returning the overall shortest multicut dual.

4   Structural Properties of a Shortest Multicut Dual

In this section, we prove some structural properties of a shortest multicut dual.

Consider all shortest multicut duals in general position with respect to $K \cup G$. Among all these, let C_0 be one that crosses K a minimum number of times. We can, without loss of generality, assume that C_0 is inclusionwise minimal, in the sense that no edge can be removed from C_0 without violating the fact that it is a multicut dual. Of course, we can assume that C_0 has no isolated vertex. If C_0 has a degree-one vertex, we can "prune" it, by removing it together with its incident edge. If C_0 has a degree-two vertex that is not a loop, we can "dissolve" it, by removing it and identifying the two incident edges. Thus, we can assume that C_0 has minimum degree at least two, and that every degree-two vertex is the vertex of a connected component that is a loop.

4.1   Crossing Bound

We start with an easy consequence of Euler's formula.

Lemma 4.1. C_0 has $O(g + t)$ vertices and edges.

The main structural property of C_0 is isolated in the following lemma:

Lemma 4.2. There are $O(g + t)$ crossings between C_0 and each edge of K.

As in algorithms for other problems using the same approach [16,5,13,6], the proof of this lemma consists of an exchange argument: If C_0 crosses an edge of K too many times, we can replace C_0 with a no longer multicut dual that crosses K fewer times, contradicting the choice of C_0. The proof ultimately boils down to topological considerations. Let us also mention that the only property that we are using on the edges of K is that they are disjoint shortest paths (except possibly at their endpoints).

Proof of Lemma 4.2. We focus on a specific edge e of K crossed by C_0, forgetting about the others. It is convenient to put an *obstacle* close to each of the two endpoints of e (since e is a shortest path, its endpoints are distinct). It is also convenient to temporarily look at the situation differently, by forgetting about G and by modifying C_0 in the vicinity of e by pushing all crossings of C_0 with e to a single point p on e (Figure 1(a, b)). This transforms C_0 into another graph C_0' that has p as a new vertex. To prove the lemma, it suffices to prove that the degree of p in C_0' is $O(g + t)$. Moreover, every non-loop edge of C_0' corresponds to one of the two endpoints of an edge of C_0, and there are $O(g + t)$ of these by Lemma 4.1. Hence, if we let L be the one-vertex subgraph of C_0' made of the loops of C_0' based at p, it suffices to prove that the number of loops in L is $O(g + t)$.

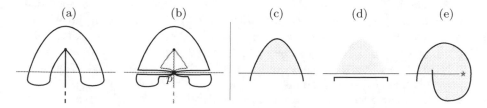

Fig. 1. (a): The part of the multicut dual C_0 close to e (depicted as a horizontal line). (b): its modified version C_0' obtained by pushing all crossings with e to a single point p. The black lines are the loops in L, the grey ones are the other edges of C_0'. (c): The configuration corresponding to a monogon. The disk is shaded. (d): The new configuration to replace (c). (e): This configuration is not a monogon, because of the presence of an obstacle (shown as a star).

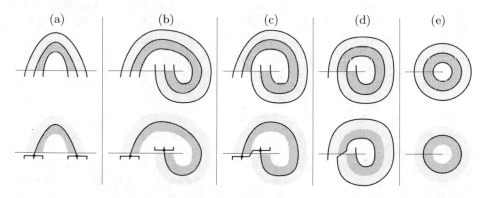

Fig. 2. The exchange argument. The horizontal segment represents edge e of K. The strips are shaded; they represent disks with no terminal, no obstacle, and no piece of C_0 in their interior.

If the sides of the strips are all on the same side of e, there is a single case (a). Replacing the top configuration of C_0 with the bottom configuration (creating two new vertices) still yields a multicut dual (as all pairs of faces that were separated in the top configuration are still separated in the bottom configuration, except possibly for the strips, but these contain no terminal), which is no longer than the original (because e is a shortest path) and has less crossings with K. This is a contradiction with the choice of C_0.

If the sides of the strips are on different sides of e, we need to distinguish according to four cases (b–e), depending on how the sides of the strips overlap. In all cases, the same argument shows that we could find a no longer multicut dual with fewer crossings with K, a contradiction. (We could also remark that case (e) is impossible because it involves closed curves in C_0 without vertex.)

A *monogon*, resp. a *bigon*, is a face of L that is, topologically, an open disk with one, resp. two, copies of p on its boundary and containing in its interior no obstacle, no vertex of C_0, and no terminal. We first claim that no face of L can be a monogon. Otherwise (Figure 1(c)), some edge e' of C_0 crosses e twice consecutively, at points x and y say, such that the pieces of e and e' between x and y bound a disk containing in its interior no obstacle, no vertex of C_0, and no terminal. Since the disk contains no obstacle, the boundary of the disk lies entirely on one side of e, as in Figure 1(c), and other cases such as the one shown in Figure 1(e) cannot occur. Since the disk contains no vertex of C_0, it contains no piece of C_0 in its interior. We can thus replace the piece of e' between x and y with a path that runs along e (Figure 1(d)). This operation does not make e' longer, since e is a shortest path; it removes the two intersection points with e and does not introduce other crossings with K. Moreover, since the disk contains no terminal in its interior, the resulting graph is also a multicut dual. This contradiction with the choice of C_0 proves the claim.

We will prove below that no loop in L can be incident to two bigons. Taking this fact for granted for now, whenever one face of L is a bigon, we remove one of the two incident loops, and iterate until there is no bigon any more. The previous fact implies that these iterations remove at most half of the loops: If L' is the remaining set of loops, we have $|L| \leq 2|L'|$. Furthermore, L' has no monogon or bigon. This latter fact, together with arguments based on Euler's formula, implies that the number of loops in L' is $O(g + t)$ [5, Lemma 2.1], because S has Euler genus g, and the total number of obstacles, vertices of C_0, and terminals (which are the points that prevent a face that is a disk of degree one or two to be a monogon or bigon) is $O(g + t)$ (Lemma 4.1). This implies that $|L| = O(g + t)$, which concludes the lemma.

So there only remains to prove that no loop in L can be incident to two bigons. Assume that such a loop exists. On the original surface S, this corresponds to two "strips" glued together, see Figure 2, top: Each strip is bounded by two pieces of e and two pieces of edges of C_0, and these two strips share a common piece of edge of C_0. Since a bigon contains no obstacle, the sides of the strips contain none of the endpoints of e. Since the interiors of these strips contain no vertex of C_0, they contain no piece of C_0.

Since S is assumed to be orientable, there are five possible cases up to symmetry, see Figure 2, top: (a) is the case where each strip has its two sides on the same side of e, (b–e) are the various cases where each strip has its two sides on opposite sides of e. In each case, we change C_0 by modifying some edges and possibly by adding vertices (see Figure 2, bottom). Since e is a shortest path and the new pieces "run along" e, one can check that the resulting graph is no longer than C_0; moreover, it crosses K fewer times. Also, each replacement may split some faces of the original graph and attach each of the strips to some of the resulting faces, but pairs of terminals that were initially separated by C_0 are still separated by the new graph, which is thus also a multicut dual. This contradicts the choice of C_0.   □

4.2   Some Shortest Multicut Dual is Good

We now give a more precise description of the intersection pattern between C_0 and K, using the properties proved in the previous section. Cutting the surface S along K yields a topological disk $\boldsymbol{D}$. The boundary $\partial \boldsymbol{D}$ of D corresponds to copies of vertices and edges of K; there are $O(g + t)$ of these. The copies of the vertices of K on ∂D are called the **corners** of D, while the copies of the edges of K are called the **sides** of D. The sides of D can be reglued pairwise to obtain S.

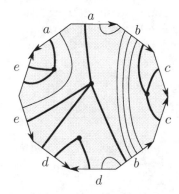

Fig. 3. A view of the graph C after cutting S along K

Let C be a graph on S in general position with respect to $K \cup G$. In particular, C_0 is such a graph. Cutting S along K transforms the overlay of K and C into a graph U drawn in D (Figure 3): Each edge of that graph corresponds to a piece of an edge of K or of C; each vertex of that graph corresponds to a vertex of K, a vertex of C, or a (four-valent) intersection point between an edge of K and an edge of C. We denote by $\bar{C}$ the subgraph of U made of the edges corresponding to pieces of edges of C (thus, $\bar{C}$ lies in the interior of D except possibly for some of its leaves), and by $\bar{K}$ the subgraph of U made of the edges corresponding to pieces of edges of K (thus, the image of $\bar{K}$ is the boundary of D).

Definition 4.3. *(See Figure 3) We say that C is **good** if $\bar{C}$ is the disjoint union of **trees** with at least one vertex of degree at least two, with all their leaves on ∂D (drawn in thick lines), and **arcs**, namely, edges with both endpoints on ∂D, on different sides of ∂D (drawn lin thin lines), and, moreover: there are $O(g + t)$ intersection points between C and each side of ∂D, and the total number of edges of the trees is $O(g + t)$.*

Here is the main result of this section, which follows from Lemmas 4.1 and 4.2:

Proposition 4.4. *Some shortest multicut dual is good.*

5   Enumerating Topologies

Let C be a good graph on S; recall that the union of $\bar{K}$ and $\bar{C}$ forms a connected planar graph U. The **topology of C** is the data of the combinatorial map of U, where the outer face is distinguished, and the sides are paired. Intuitively, it describes combinatorially the relative positions of C and K. More generally, a **topology** is the combinatorial map of a connected, planar graph with a distinguished outer face and a pairing of the sides (these are the subpaths of the outer cycle connecting consecutive degree-two vertices).

Whether C is a multicut dual or not is completely determined by its topology. Hence the following terminology: A topology is **valid** if it is the topology of a good graph that is a multicut dual.

Proposition 5.1. *The number of valid topologies is* $(g+t)^{O(g+t)}$; *these topologies can be enumerated within the same time bound.*

6   Dealing With Each Valid Topology

The strategy for our algorithm is roughly as follows: For each valid topology, we compute a shortest graph C with that topology; then we return the set of edges of G crossed by the overall shortest graph C. Actually, we do not exactly compute an embedding of a shortest graph C; instead, we compute a shortest drawing (possibly with crossings) of the same graph, with some homotopy constraints; that drawing is no longer than the shortest embedding, and we prove that it also corresponds to a minimum multicut. We shall prove:

Proposition 6.1. *Given a valid topology, we can compute, in* $(g+t)^{O(1)} \times n^{O\left(\sqrt{g^2+gt}\right)}$ *time, some multicut whose weight is at most the length of each multicut dual with that topology.*

Let C be a good graph. The **crossing sequence** of an edge e of C is the ordered sequence of edges in K crossed by e, together with the indication of the orientation of the crossing (i.e., whether e crosses the edge of K from left to right or from right to left). Given the topology of C, one can determine the crossing sequence of every edge of C. We say that a drawing C' of the (abstract) graph C has the same **topology** as C if each edge of C' has the same crossing sequence as the corresponding edge of C.

Lemma 6.2. *Let C' be a drawing of a multicut dual C with the same topology as C. Then, the set of edges of G crossed by C' is a multicut.*

Proof. Let $\{t_1, t_2\} \in R$ be a pair of terminals, and let f be the face of C containing t_1. The set of edges of C that are incident exactly once to the face of C containing t_1 forms an *even* subgraph C_1 of C, in which every vertex has even degree. Moreover, C_1 separates t_1 from t_2. Let C_1' be the drawing of the same subgraph in C'. To prove our result, it suffices to prove that C_1' also separates t_1 from t_2, using the fact that the crossing sequences are the same in C_1 and C_1'.

We can assume that C' has a finite number of self-intersection points, each of which is a crossing. Therefore, C_1' can be seen as an even graph embedded on S (by adding a new vertex at each crossing between two edges of C_1'). Our lemma is implied by two results by Chambers et al. [6], reformulated here in our setting:

- if, for every edge e of K, the even graphs C_1 and C_1' cross e with the same parity, then they are homologous (over $\mathbb{Z}/2\mathbb{Z}$) [6, Lemma 3.4]. In our case, since the crossing sequences are equal, C_1 and C_1' are homologous;
- an even graph separates t_1 from t_2 if and only if it is homologous, on the surface $S \setminus \{t_1, t_2\}$, to a small circle around t_1 [6, Lemma 3.1]. Thus, since C_1 separates t_1 from t_2, it is also the case for C_1'.   □

Proof of Proposition 6.1. Lemma 6.2 implies that it suffices to compute (the set of edges of G crossed by) a shortest drawing with the given topology.

Let us first explain how to achieve this, assuming that the locations of the vertices are prescribed. Let C be any multicut dual with that topology and these vertex locations; we do not know C, but know the crossing sequence of its edges (they are determined by the topology). To achieve our goal, it suffices, for each edge e of C, to compute a shortest path with the same crossing sequence as e and the same endpoints. We remark that algorithms for computing a shortest homotopic path in S minus the vertex set of K precisely achieve this goal [9]: In short, we glue copies of the disk D according to the specified crossing sequence, in such a way that edge e "lifts" to that space $\hat{S}_e$; then we compute a shortest path in $\hat{S}_e$ connecting the same endpoints as that lift, and "project" back to S. (We can assign infinitesimal crossing weights to the edges of K to ensure that the crossing sequences of e and e' are the same.) The complexity of D (with its internal structure defined by the edges of G) is $O((g+t)n)$, and the crossing sequences have total length $O(g+t)^2$, so the total complexity of the spaces $\hat{S}_e$ is $O((g+t)^3 n)$. Since $\hat{S}_e$ is planar, and since shortest paths can be computed in linear time in planar graphs [15], this is also the complexity of computing (the set of edges of G crossed by) a shortest graph drawing with a given topology and specified vertex locations.

To compute a shortest drawing with the given topology, over all choices of vertex locations, we can naïvely enumerate all possible locations of the $O(g+t)$ vertices. Note that it is only relevant to consider which face of the overlay of G and K each vertex belongs to, and there are $O((g+t)n)$ such faces. To get a better running time, we use treewidth techniques, also used by Klein and Marx in the planar multiway cut case [16]. The (abstract) graph C defined by the specified topology has $O(g+t)$ vertices and is embedded on a surface with genus g. We can thus, in $(g+t)^{O(1)}$ time, compute a tree decomposition of C of width $O(\sqrt{g^2 + gt})$. We use standard dynamic programming on the tree decomposition: At each node N of the path, we have a table that indicates, for every choice of the locations of the vertices in N, the length of the shortest drawing of the subgraph of C induced by the vertices in N and its descendents, among those that respect the crossing sequence constraints. We can fill in the tables by a bottom-up traversal of the tree decomposition. The claimed running time follows, since each node contains $O(\sqrt{g^2 + gt})$ vertices. $\qquad\square$

We can now conclude the proof of Theorem 1.1:

Proof of Theorem 1.1. We compute the cut graph K in $O(m \log m + gn)$ time (Proposition 3.2), and enumerate all valid topologies in $(g+t)^{O(g+t)}$ time (Proposition 5.1). For each valid topology, we apply the algorithm of Proposition 6.1 in $O(g+t)^{O(1)} n^{O(\sqrt{g^2+gt})}$ time, and return a shortest multicut found. This implies the claimed running time. The correctness is easy: By Proposition 3.1, it suffices to compute a multicut whose weight is at most the length of any multicut dual. By Proposition 4.4, some shortest multicut dual has a valid topology; when this

topology is chosen, Proposition 6.1 guarantees that we have computed a shortest
multicut.   □

Acknowledgments. Thanks to Cédric Bentz and Claire Mathieu for inspiring
discussions, and to the referees of a previous version, one of them suggesting that
an improvement might be possible using treewidth, as in Klein and Marx [16].

References

1. Bateni, M., Hajiaghayi, M., Klein, P.N., Mathieu, C.: A polynomial-time approximation scheme for planar multiway cut. In: Proc. ACM-SIAM Symp. on Discrete
   Algorithms, pp. 639–655 (2012)
2. Bentz, C.: On the complexity of the multicut problem in bounded tree-width graphs
   and digraphs. Disc. Applied Math. 156(10), 1908–1917 (2008)
3. Bentz, C.: A simple algorithm for multicuts in planar graphs with outer terminals.
   Disc. Applied Math. 157, 1959–1964 (2009)
4. Bentz, C.: A polynomial-time algorithm for planar multicuts with few source-sink
   pairs. In: International Workshop on Parameterized and Exact Computation, pp.
   109–119 (2012), Also in arXiv:1206.3999
5. Chambers, E.W., Colin de Verdière, É., Erickson, J., Lazarus, F., Whittlesey, K.:
   Splitting (complicated) surfaces is hard. Comput. Geom.: Theory Appl. 41(1–2),
   94–110 (2008)
6. Chambers, E.W., Erickson, J., Nayyeri, A.: Minimum cuts and shortest homologous
   cycles. In: Proc. Symp. on Computational Geometry, pp. 377–385. ACM (2009)
7. Chambers, E.W., Erickson, J., Nayyeri, A.: Homology flows, cohomology cuts.
   SIAM J. Comput. 41(6), 1605–1634 (2012)
8. Colin de Verdière, É.: Topological algorithms for graphs on surfaces. Habilitation
   thesis, École normale supérieure (2012). http://www.di.ens.fr/~colin/
9. Colin de Verdière, É., Erickson, J.: Tightening nonsimple paths and cycles on
   surfaces. SIAM J. Comput. 39(8), 3784–3813 (2010)
10. Dahlhaus, E., Johnson, D.S., Papadimitriou, C.H., Seymour, P.D., Yannakakis, M.:
    The complexity of multiterminal cuts. SIAM J. Comput. 23(4), 864–894 (1994)
11. Eppstein, D.: Dynamic generators of topologically embedded graphs. In: Proc.
    ACM-SIAM Symp. on Discrete Algorithms, pp. 599–608 (2003)
12. Erickson, J., Nayyeri, A.: Minimum cuts and shortest non-separating cycles via
    homology covers. In: Proc. ACM-SIAM Symp. on Discrete Algorithms, pp. 1166–
    1176 (2011)
13. Erickson, J., Nayyeri, A.: Shortest non-crossing walks in the plane. In: Proc. ACM-
    SIAM Symp. on Discrete Algorithms, pp. 297–308 (2011)
14. Garg, N., Vazirani, V.V., Yannakakis, M.: Primal-dual approximation algorithms
    for integral flow and multicut in trees. Algorithmica 18, 3–20 (1997)
15. Henzinger, M.R., Klein, P., Rao, S., Subramanian, S.: Faster shortest-path algo-
    rithms for planar graphs. J. Comput. System Sciences 55(1, part 1), 3–23 (1997)
16. Klein, P.N., Marx, D.: Solving PLANAR k-TERMINAL CUT in $O(n^{c\sqrt{k}})$ time. In: Proc.
    Int. Coll. on Automata, Languages and Programming, vol. 1, pp. 569–580 (2012)
17. Marx, D.: A tight lower bound for planar multiway cut with fixed number of
    terminals. In: Proc. Int. Coll. on Automata, Languages and Programming, vol. 1,
    pp. 677–688 (2012)

18. Mohar, B.: A linear time algorithm for embedding graphs in an arbitrary surface. SIAM J. Disc. Math. 12(1), 6–26 (1999)
19. Schrijver, A.: Combinatorial optimization. Polyhedra and efficiency. Algorithms and Combinatorics, vol. 24. Springer, Berlin (2003)
20. Schrijver, A.: On the history of combinatorial optimization (till 1960). In: Handbook of Discrete Optimization, pp. 1–68. Elsevier (2005)
21. Stillwell, J.: Classical topology and combinatorial group theory, 2nd edn. Springer, New York (1993)

A Fixed Parameter Tractable Approximation Scheme for the Optimal Cut Graph of a Surface[*]

Vincent Cohen-Addad[1] and Arnaud de Mesmay[2]

[1] Département d'informatique, École normale supérieure, Paris, France
[2] IST Austria, Vienna, Austria

Abstract. Given a graph G cellularly embedded on a surface Σ of genus g, a cut graph is a subgraph of G such that cutting Σ along G yields a topological disk. We provide a fixed parameter tractable approximation scheme for the problem of computing the shortest cut graph, that is, for any $\varepsilon > 0$, we show how to compute a $(1 + \varepsilon)$ approximation of the shortest cut graph in time $f(\varepsilon, g)n^3$.

Our techniques first rely on the computation of a spanner for the problem using the technique of brick decompositions, to reduce the problem to the case of bounded tree-width. Then, to solve the bounded tree-width case, we introduce a variant of the surface-cut decomposition of Rué, Sau and Thilikos, which may be of independent interest.

1 Introduction

Embedded graphs are commonly used to model a wide array of discrete structures, and in many cases it is necessary to consider embeddings into surfaces instead of the plane or the sphere. For example, many instances of network design actually feature some crossings, coming from tunnels or overpasses, which are appropriately modeled by a surface of small genus. In other settings, such as in computer graphics or computer-aided design, we are looking for a discrete model for objects which inherently display a non-trivial topology (e.g., holes), and graphs embedded on surfaces are the natural tool for that. From a more theoretical point of view, the graph structure theorem of Robertson and Seymour showcases a very strong connection between graphs embedded on surfaces and minor-closed families of graphs.

When dealing with embedded graphs, a classical problem, to which a lot of effort has been devoted in the past decade, is to find a *topological decomposition* of the underlying surface, i.e., to cut the surface into simpler pieces so as to simplify its topology, or equivalently to cut the embedded graph into a planar graph, see the recent surveys [19,7]. This is a fundamental operation in algorithm design for surface-embedded graphs, as it allows to apply the vast number of tools

---

[*] The research of the first author was partially supported by ANR RDAM. The research of the second author leading to these results has received funding from the People Programme (Marie Curie Actions) of the European Union's Seventh Framework Programme (FP7/2007-2013) under REA grant agreement n° [291734].

© Springer-Verlag Berlin Heidelberg 2015
N. Bansal and I. Finocchi (Eds.): ESA 2015, LNCS 9294, pp. 386–398, 2015.
DOI: 10.1007/978-3-662-48350-3_33

available for planar graphs to this more general setting. Furthermore, making a graph planar is useful for various purposes in computer graphics and mesh processing, see for example [20]. No matter the application, a crucial parameter is always the length of the topological decomposition: having good control on it ensures that the meaningful features of the embedded graphs did not get too much distorted during the cutting.

In this article, we are interested in the problem of computing a short cut graph: For a graph G with n vertices embedded on a surface Σ of genus g, a **cut graph** of G is a subgraph $C \subseteq G$ such that cutting Σ along C gives a topological disk. The problem of computing the shortest possible cut graph of an embedded graph was introduced by Erickson and Har-Peled [8], who showed that it is NP-hard, provided an $n^{O(g)}$ algorithm to compute it, as well as an $O(g^2 n \log n)$ algorithm to compute a $O(\log^2 g)$ approximation. Now, since in most practical applications, the genus of the embedded graph tends to be quite small compared to the complexity of the graph, it is natural to also investigate this problem through the lens of parametrized complexity, which provides a natural framework to study the dependency of cutting algorithms with respect to the genus. In this direction, Erickson and Har-Peled asked whether computing the shortest cut graph is *fixed-parameter tractable*, i.e. whether it can be solved in time $f(g)n^{O(1)}$ for some function f. This question is, up to our knowledge, still open, and we address here the neighborly problem of devising a good approximation algorithm working in time fixed parameter tractable with respect to the genus; we refer to the survey of Marx [12] for more background on these algorithms at the intersection of approximation algorithms and parametrized complexity.

Our results. In this article, we provide a *fixed-parameter tractable approximation scheme* for the problem of computing the shortest cut graph of an embedded graph.

Theorem 1. *Let G be a weighted graph cellularly embedded on a surface Σ of genus g. For any $\varepsilon > 0$, there exists an algorithm computing a $(1 + \varepsilon)$-approximation of the shortest cut graph of G, which runs in time $f(\varepsilon, g)n^3$ for some function f.*

Our techniques. Our algorithm uses the brick decompositions of Borradaile, Klein and Mathieu [3] for subset-connectivity problems in planar graphs, which have been extended to bounded genus graphs by Borradaile, Demaine and Tazari [2]. Although brick decompositions are now a common tool for optimization problems for embedded graphs, it is to our knowledge the first time they are applied to compute topological decompositions. In a nutshell, the idea is the following:

1. We first compute a *spanner* G_{span} for our problem, namely a subgraph of the input graph containing a $(1+\varepsilon)$-approximation of the optimal cut graph, and having total length bounded by $f(g, \varepsilon)$ times the length of the optimal cut graph, for some function f. This is achieved via *brick decompositions*.

2. Using a result of *contraction-decomposition* of Demaine, Hajiaghayi and Mohar [6], we *contract* a set of edges of controlled length in G_{span}, obtaining a graph G_{tw} of bounded tree-width.

3. We use dynamic programming on G_{tw} to compute its optimal cut graph.

4. We incorporate back the contracted edges, which gives us a subgraph of G cutting the surface into one or more disks. Removing edges so that the complement is a single disk gives our final cut graph.

The first steps of this framework mostly follow from the same techniques as in the article of Borradaile et al. [2], the only difference being that we need a specific structure theorem to show that the obtained graph is indeed a spanner for our problem. However, as the restriction of a cut graph to a brick, i.e., a small disk on the surface, is a forest, this structure theorem is a variation of an existing theorem for the Steiner tree problem [3].

The main difficulty of this approach lies instead in the third step. Since a cut graph is inherently a topological notion, it is key for a dynamic programming approach to work with a tree-decomposition having nice topological properties. An appealing concept has been developed by Rué, Sau and Thilikos [17] for the neighborly (and for our purpose, equivalent) notion of branch-decomposition: they introduced *surface-cut decompositions* with this exact goal of giving a nice topological structure to work with when designing dynamic programs for graph on surfaces (see also Bonsma [1] for a related concept). However, their approach is cumbersome for our purpose when the graph embeddings are not *polyhedral* (we refer to the introduction for precise definitions), as it first relies on computing a *polyhedral decomposition* of the input graph. While dynamic programming over these polyhedral decompositions can be achieved for the class of problems that they consider, it seems unclear how to do it for the problem of computing a shortest cut graph.

We propose two ways to circumvent this issue. In the first one, we observe that the need for polyhedral embeddings in surface-cut decompositions can be traced back exclusively to a theorem of Fomin and Thilikos [17, Lemma 5.1][10, Theorem 1] relating the branch-width of an embedded graph and the carving-width of its medial graph, the proof of which uses crucially that the graph embedding is polyhedral. But another proof of this theorem which does not rely on this assumption was obtained by Inkmann [11, Theorem 3.6.1]. Therefore, the full strength of surface cut decompositions can be used without first relying on polyhedral decompositions.

However, since Inkmann's proof is intricate and has never been published we also propose an alternative, self-contained, solution tailored to our problem. For our purpose, it is enough to make the graph polyhedral at the end of the second step of the framework while preserving a strong bound on the branch-width of the graph, we show that this can be achieved by superposing medial graphs and triangulating with care. With appropriate heavy weights on the new edges, we can ensure that they do not impact the length of the optimal cut graph and that we still obtain a valid solution to our problem.

Finally, both approaches allow us to work with a branch decomposition that possesses a nice topological structure. We then show how to exploit it to write a dynamic program to compute the shortest cut graph in fixed parameter tractable time for graphs of bounded tree-width.

Open problems. One of the main challenges is whether the problem of computing the shortest cut graph can be solved *exactly* in FPT complexity – the recent application of brick decompositions to exact solutions for Steiner problems [15] might help in this direction. In the approximability direction, it is also unknown whether there exists a polynomial time constant factor approximation to this problem, or even a PTAS.

Organization of the paper. We start by introducing the main notions surrounding embedded graphs and brick decompositions in Section 2. We then prove the structure theorem in Section 3, showing that the brick decomposition with portals contains a cut graph which is at most $(1+\varepsilon)$ longer than the optimal one. In Section 4, we show how to combine this structure theorem with the aforementioned framework to obtain our algorithm. This algorithm relies on one to solve the problem when the input graph has bounded tree-width, which is described in Section 5.

Due to space restrictions, many proofs are omitted, but they are included in the full version of this paper [4].

2   Preliminaries

All graphs $G = (V, E)$ in this article are multigraphs, possibly with loops, have n vertices, m edges, are undirected and their edges are weighted with a length $\ell(e)$. These weights induce naturally a length on paths and subgraphs of G.

Graphs on surfaces. We will be using classical notions of graphs embedded on surfaces, for more background on the subject, we refer to the textbook of Mohar and Thomassen [14]. Throughout the article, Σ will denote a compact connected surface of Euler genus g, which we will simply call genus. An **embedding** of G on Σ is a crossing-free drawing of G on Σ, i.e. the images of the vertices are pairwise distinct and the image of each edge is a simple path intersecting the image of no other vertex or edge, except possibly at its endpoints. We will always identify an abstract graph with its embedding. A **face** of the embedding is a connected component of the complement of the graph. A **cellular embedding** is an embedding of a graph where every face is a topological disk. Every embedding in this paper will be assumed to be cellular. A graph embedding is a triangulation if all the faces have degree three. **Euler's formula** states that for a graph G embedded on a surface Σ, we have $n - m + f = 2 - g$, for f the number of faces of the embedding. A **noose** is an embedding of the circle $\mathbb{S}^1$ on Σ which intersects G only at its vertices. An embedding of a graph G on a surface is said to be **polyhedral** if G is 3-connected and the smallest length of a non-contractible noose is at least 3 or if G is a clique and it has at most 3 vertices. In particular, a polyhedral embedding is cellular. If G is a graph embedded on Σ, the surface Σ' obtained by **cutting** Σ along G is the disjoint union of the faces of G, it is a (a priori disconnected) surface with boundary. When we cut a surface along a set of nooses, viewed as a graph, the resulting connected components will be called **regions**. A **combinatorial map** of an embedded graph is the combinatorial description of its embedding, namely the cyclic ordering of the edges around each vertex.

Given an embedded graph G, the **medial graph** M_G is the embedded graph obtained by placing a vertex v_e for every edge e of G, and connecting the vertices v_e and v'_e with an edge whenever e and e' are adjacent on a face of G. The **barycentric subdivision** of an embedded graph G is the embedded graph obtained by adding a vertex on each edge and on each face and an edge between every such face vertex and its adjacent (original) vertices and edge vertices.

For Σ a surface and G a graph embedded on Σ, a **cut graph** of (Σ, G) is a subgraph H of G whose unique face is a disk. The length of the cut graph is the sum of the lengths of the edges of H. Throughout the whole paper, OPT will denote the length of the shortest cut graph of (Σ, G).

We refer the reader to [2,17] for definitions pertaining to **tree decomposition** and **branch decomposition**. A **carving decomposition** of a graph G is the analogue of a branch decomposition with vertices and edges inverted, with the **carving-width** defined analogously. A **bond carving decomposition** is a special kind of carving decomposition where the middle sets always separate the graph in two connected components. Since these concepts only appear sporadically in this paper, we refer to [17] for a precise definition.

Mortar graph and bricks. The framework of mortar graphs and bricks has been developed by Borradaile, Klein and Mathieu [3] to efficiently compute spanners for subset connectivity problems in planar graphs. We recall here the main definitions around mortar graphs and bricks and refer to the articles [2,3] for more background on these objects.

Let G be a graph embedded on Σ of genus g. A path P in a graph G is ε-short in G if for every pair of vertices x and y on P, the distance from x to y along P is at most $(1 + \varepsilon)$ times the distance from x to y in G: $\mathrm{dist}_P(x, y) \leq (1 + \varepsilon)\mathrm{dist}_G(x, y)$. For $\varepsilon > 0$, let $\kappa(g, \varepsilon)$ and $\alpha(g, \varepsilon)$ be functions to be defined later. A **mortar graph** $MG(G, \varepsilon)$ is a subgraph of G such that $\ell(MG) \leq \alpha OPT$, and the faces of MG partition G into **bricks** B that satisfy the following properties:

1. B is planar.
2. The boundary of B is the union of four paths in clockwise order N, E, S, W.
3. N is 0-short in B, and every proper subpath of S is ε-short in B.
4. There exists a number $k \leq \kappa$ and vertices $s_0 \ldots s_k$ ordered from left to right along S such that, for any vertex x of $S[s_i, s_{i+1})$, $\mathrm{dist}_S(x, s_i) \leq \varepsilon \mathrm{dist}_B(x, N)$.

The mortar graph is computed using a slight variant of the procedure in [2, Theorem 4], the idea is the following:

1. Cut Σ along an approximate cut graph, yielding a disk D with boundary ∂D.

2. Find shortest paths between certain vertices of ∂D. This defines the N and S boundaries of the bricks.
3. Find shortest paths between vertices of the previous paths. These paths are called the columns.
4. Take every κth path found in the last step. These paths are called the **su-percolumns** and form the E and W boundaries of the bricks. The constant κ is called the **spacing** of the supercolumns.

This leads to the following theorem to compute the mortar graph in time $O(g^2 n \log n)$. The proof, very similar to the one in [2, Theorem 2], is omitted.

Theorem 2. *Let* $\varepsilon > 0$ *and* G *be a graph embedded on* Σ *of genus* g. *There exists* $\alpha = O(\log^2 g\varepsilon^{-1})$ *such that there is a mortar graph* $MG(G, \varepsilon)$ *of* G *such that* $\ell(MG) \leq \alpha OPT$ *and the supercolumns of* MG *have length* $\leq \varepsilon OPT$ *with spacing* $\kappa = O(\log^2 g\varepsilon^{-3})$. *This mortar graph can be found in* $O(g^2 n \log n)$ *time.*

3   Structure Theorem

In this section, we prove the structure theorem, which shows that there exists an ε-approximation to the optimal cut graph which only crosses the mortar graph at a small subset of vertices called *portals*.

In order to state this theorem, following the literature, we define a *brick-copy* operation B^+ as follows. For each brick B, a subset of θ vertices is chosen as *portals* such that the distance along ∂B between any vertex and the closest portal is at most $\ell(\partial B)/\theta$. For every brick B, embed B in the corresponding face of MG and connect every portal of B to the corresponding vertex of MG with a zero-length *portal edge*; this defines $B^+(MG, \theta)$. The edges originating from MG are called the *mortar edges*.

We note that by construction, $B^+(MG, \theta)$ embeds on the plane in such a way that every brick of $B^+(MG, \theta)$ is included in the corresponding brick of MG. Furthermore, every vertex of G corresponds to a vertex of $B^+(MG, \theta)$ by mapping the insides of bricks to the insides of bricks in $B^+(MG, \theta)$, and the mortar graph to itself, cf. Figure 1. We denote this map by $i : V(G) \to V(B^+(MG, \theta))$.

Moreover, we contract the E and W boundaries of each brick of $B^+(MG, \theta)$ and their copies in the mortar graph. Since the sum of the length of the E and

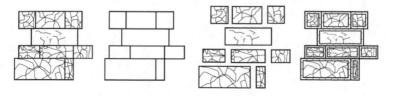

Fig. 1. The different stages of the brick decomposition of a graph G, the mortar graph, the set of bricks and the graph $B^+(MG, \theta)$.

W boundaries is at most εOPT, any solution of length ℓ in $B^+(MG, \theta)$ going through a vertex resulting from a contraction can be transformed into a solution of length at most $\ell + 2\varepsilon OPT$ in $B^+(MG, \theta)$ where no edge is contracted. The structure theorem is then the following:

Theorem 3. *Let G be a graph embedded on Σ of genus g, and $\varepsilon > 0$. Let $MG(G, \varepsilon)$ be a corresponding mortar graph of length at most αOPT and supercolumns of length at most εOPT with spacing κ. There exists a constant $\theta(\alpha, \varepsilon, \kappa)$ depending polynomially on $\alpha, 1/\varepsilon$ and κ such that:*

$$OPT(B^+(MG, \theta)) \le (1 + c\varepsilon)OPT.$$

The proof of this theorem essentially consists in plugging in the structure theorem of [3] and verifying that it fits. Let us first recall the structural theorem of bricks [3]. For a graph H and a path $P \subseteq H$, a *joining vertex* of H with P is a vertex in P that lies on an edge of $H \setminus P$.

Theorem 4. *[3, Theorem 10.7] Let B be a brick, and F be a set of edges of B. There is a forest $\widetilde{F}$ in B with the following properties:*
1. *If two vertices of $N \cup S$ are connected by F, they are also connected by $\widetilde{F}$,*
2. *The number of joining vertices of $\widetilde{F}$ with both N and S is bounded by $\gamma(\kappa, \varepsilon)$,*
3. *$\ell((\widetilde{F})) \le \ell(F)(1 + c\varepsilon)$.*
   In the above, $\gamma(\kappa, \varepsilon) = o(\varepsilon^{-2.5}\kappa)$ and c is a fixed constant.

From this we can deduce the following proposition, and we proceed to the proof of the structure theorem.

Proposition 1. *Let C be a subgraph of G of length OPT. There exists a constant $\theta(\alpha, \varepsilon, \kappa)$ depending polynomially on $\alpha, 1/\varepsilon$ and κ and a subgraph $\widehat{C}$ of $B^+(MG, \theta)$ with the following properties:*
- *$\ell(\widehat{C}) \le (1 + \tilde{c}\varepsilon)\ell(C) = (1 + \tilde{c}\varepsilon)OPT$, where $\tilde{c}$ is a fixed constant.*
- *If we denote by D the closed disk on which MG has been constructed, for any two vertices $s, t \in \partial D$ that are connected by C in D, $i(s)$ and $i(t)$ are connected by $\widehat{C}$ in D as well.*

Proof of Theorem 3. Let C be an optimal cut graph of (Σ, G). We apply Proposition 1 to C, it yields a subgraph $\widehat{C}$ of $B^+(MG, \theta)$ of length $\ell(\widehat{C}) \le (1+\varepsilon)\ell(C)$. We claim that this graph $\widehat{C}$ contains a cut graph of Σ.

Suppose on the contrary that there exists a non-contractible cycle γ in $(\Sigma, B^+(MG, \theta))$ which does not cross $\widehat{C}$. This cycle γ corresponds to a cycle γ' in (Σ, G) by contracting portal edges, and since C is a cut graph, there exists a maximal subpath P of C restricted to D and a maximum subpath P' of γ' in D such that P' crosses P an odd number of times, otherwise, by flipping bigons we could find a cycle homotopic to γ' not crossing C. Denote by (s, t) and (s', t') the intersections of P and P' with ∂D. Then, without loss of generality, s, s', t and t' appear in this order on ∂D. Furthermore, the vertices $i(s)$ and $i(t)$ in $B^+(MG, \theta)$ are connected by $\widehat{C}$ by Proposition 1, since s and t are connected by C. Therefore, γ crosses $\widehat{C}$, and we reach a contradiction. $\qquad\square$

4   Algorithm

We now explain how to apply the spanner framework of Borradaile et al. in [2] to compute an approximation of the optimal cut graph. We start by computing the optimal Steiner tree, for each subset of the portals in every brick by using the algorithm or Erickson et al. [9], and then take the union of all these trees over all bricks, plus the edges of the mortar graph. As this algorithm runs in time $O(nk^3)$, this step takes time $O_{g,\varepsilon}(n)$. This defines the graph G_{span}, which by construction has length $\leq f(g,\varepsilon)OPT$, where $f(g,\varepsilon) = O(2^\theta) = 2^{O(\log^2 g)poly(1/\varepsilon)}$, and contains a $(1+\varepsilon)$ approximation of the optimal cut graph by the structure theorem.

We will use the following theorem of Demaine et al. [6, Theorem 1.1] (the complexity of this algorithm can be improved to $O_g(n)$ [5]).

Theorem 5. *For a fixed genus g, any $k \geq 2$ and any graph G of genus at most g, the edges of G can be partitioned into k sets such that contracting any one of the sets results in a graph of tree-width at most $O(g^2 k)$. Furthermore, the partition can be found in time $O(g^{5/2}n^{3/2}\log n)$ time.*

The four steps of the framework are now the following.
1. Compute the spanner G_{span}.
2. Apply Theorem 5 with $k = f(g,\varepsilon)/\varepsilon$, and contract the edges in the set of the partition with the least weight. The resulting graph G_{tw} has tree-width at most $O(g^2\varepsilon^{-1}f(g,\varepsilon))$.
3. Use the bounded tree-width to compute a cut graph of (Σ, G_{tw}). An algorithm to do this is described in Section 5.
4. Incorporate the contracted edges back. By definition, they have length at most $f(g,\varepsilon)OPT/k = \varepsilon OPT$. Therefore, the final graph we obtain has the desired length. If the resulting graph has more than one face, remove edges until we obtain a cut-graph.

We now analyze the complexity of this algorithm. The spanner is computed in time $O_{g,\varepsilon}(n\log n)$. Using [5], the second step takes time $O_{g,\varepsilon}(n)$. Dynamic programming takes time $O_{g,\varepsilon}(n^3)$ (see thereafter), and the final lifting step takes linear time. Assuming the dynamic programming step described in the next section, this proves Theorem 1.

5   Computing Cut Graphs for Bounded Tree-Width

There remains to prove that computing the optimal cut graph of (Σ, G) is fixed parameter tractable with respect to both the tree-width of G and the genus of Σ as a parameter. Out of convenience, we work with the branch-width instead, which gives the result since they are within a constant factor [16, Theorem 5.1]. As cut graphs are a topological object, we will rely on surface-cut decompositions [17], which are a topological strengthening of branch decompositions. Note that, for reasons which will be clear later, our definition is slightly different from the one of Rué, Sau and Thilikos as it does not rely on polyhedral decompositions.

Given a graph G embedded in a surface Σ of genus G, a *surface-cut decomposition* of G is a branch decomposition (T, μ) of G such that for each edge $e \in E(T)$, the vertices in mid(e) are contained in a set $\mathcal{N}$ of nooses in Σ such that:

- $|\mathcal{N}| = O(g)$ and $\theta(\mathcal{N}) = O(g)$
- The nooses in $\mathcal{N}$ pairwise intersect only at subsets of mid(e)
- $\Sigma \setminus \bigcup \mathcal{N}$ contains exactly two connected components, of which closures contain respectively G_1 and G_2.

where θ is defined as follows: for a point p in Σ, if we denote by $\mathcal{N}(p)$ the number of nooses in $\mathcal{N}$ containing p, and let $P(\mathcal{N}) = \{p \in \Sigma \mid \mathcal{N}(p) \geq 2\}$, we define

$$\theta(\mathcal{N}) = \sum_{p \in P(\mathcal{N})} \mathcal{N}(p) - 1.$$

Rué et al. showed how to compute such a surface-cut decomposition when the input graph G is embedded **polyhedrally** on the surface Σ:

Theorem 6 ([17, Theorem 7.2]). *Given a graph G on n vertices polyhedrally embedded on a surface of genus g and with $bw(G) \leq k$, one can compute a surface-cut decomposition of G of width $O(g + k)$ in time $2^{O(k)} n^3$.*

When the input graph is not polyhedral, Rué et al. propose a more intricate version of surface-cut decompositions relying on *polyhedral decompositions*, but it is unclear how to incorporate these in a dynamic program to compute optimal cut graphs.

Instead, we present two ways to circumvent polyhedral decompositions and use these surface-cut decompositions directly. The first one consists of observing that the difficulties involved with computing surface-cut decompositions of non-polyhedral embeddings can be circumvented by using a theorem of Inkmann [11]. Since Inkmann's theorem has, up to our knowledge, not been published outside of his thesis, and the proof is quite intricate, for the sake of clarity we also provide a different approach, based on modifying the input graph to make it polyhedral.

In both cases, we obtain a branch decomposition with a strong topological structure, which we can then use as a basis for a dynamic program to compute the optimal cut graph.

5.1   A Simpler Version of Surface-Cut Decompositions

The algorithm [17, Algorithm 2] behind the proof of Theorem 6 relies on the following steps. Starting with a polyhedral embedding of G on a surface,

1. Compute a branch decomposition *branch(G)* of G.
2. Transform *branch(G)* into a carving decomposition *carv(G)* of M_G.
3. Transform *carv(G)* into a *bond* carving decomposition *bond(G)* of M_G.
4. Transform *bond(G)* into a branch decomposition of G.

The second step is the only one where where the polyhedrality of the embedding is used, as it relies on the following lemma:

Lemma 1 ([17, **Lemma 5.1**]). *Let G be a polyhedral embedding on a surface Σ of genus g, and M_G be the embedding of the medial graph. Then $bw(G) \leq cw(M_G)/2 \leq 6bw(G) + 4g + O(1)$, and the corresponding carving decomposition of M_G can be computed from the branch decomposition of G in linear time.*

We observe that the following theorem of Inkmann shows that the branch-width of a surface-embedded graph and the carving-width of its medial graph are tightly related, even for non-polyhedral embeddings.

Theorem 7 ([11, **Theorem 3.6.1**]). *For every surface Σ there is a non-negative constant $c(\Sigma)$ such that if G is embedded on Σ with $|E(G)| \geq 2$ and M_G is its medial graph, we have $2bw(G) \leq cw(M) \leq 4bw(G) + c(\Sigma)$.*

Digging into the proof reveals that $c(\Sigma) = O(g^2)$ for Σ of genus g. The idea is therefore that replacing Lemma 5.1 of Thilikos et al. by Theorem 7 allows us to lift the requirement of polyhedral embedding in their construction. One downside is that this theorem does not seem to be constructive, and therefore we need an alternative way to compute the carving decomposition in step 2. This can be achieved in fixed parameter tractable time with respect to the carving-width (and linear in n) using the algorithm of Thilikos et al. [18]. In conclusion, we obtain the following corollary (note that the bottleneck in the complexity is the same as in the one of Rué et al., which is the transformation between a carving and a bond carving decomposition).

Corollary 1. *Given a graph G on n vertices embedded in a surface of genus g with $bw(G) \leq k$, there exists an algorithm running in time $O_k(n^3)$ computing a surface-cut decomposition (T, μ) of G of width at most $O(k + g^2)$.*

5.2 Making a Graph Polyhedral

In this section, we show how to go from an embedded graph to a polyhedral embedding, without increasing the tree-width too much. The construction will be split in several lemmas. The proofs of the three following lemmas are omitted. The proofs of Lemma 3 and 4 rely on a theorem of Mazoit [13] connecting the tree-width of a graph embedded on a surface and the one of its dual.

Lemma 2. *Let G be a graph of tree-width at most $k \geq 2$, embedded on a surface of genus g. Then there exists a triangulation of G of tree-width at most k. Moreover, given a tree-decomposition of width k, one can compute a triangulation of G of tree-width at most k in polynomial time.*

Lemma 3. *Let G be a triangulated graph of tree-width at most k, embedded on a surface of genus g. Then its barycentric subdivision $B(G)$ has tree-width at most $f(k, g)$ for some function f.*

Lemma 4. *Let G be a triangulated graph of tree-width at most k, embedded on a surface of genus g. Let M_G denote the medial graph of G, and G' the superposition of G and M_G. Then the tree-width of G' is bounded by some function of k and g.*

Now, we observe that superposing medial graphs two times increases the length of non-contractible nooses of a graph. Furthermore, if the new edges are weighted heavily enough (e.g., with a weight larger than OPT, which we know how to approximate), they will not change the value of the optimal cut graph[1]. Therefore this allows us to assume that the embedded graph of which we want to compute an optimal cut graph has only non-contractible nooses of length at least three. By subdividing it to remove loops and multiple edges and triangulating it, we can also assume that it is 3-connected (since the link of every vertex of a triangulated simple graph is 2-connected), and therefore that it is polyhedral.

For a polyhedral embedding, our definition of surface-cut decompositions and the one of Thilikos et al. [17] coincide, and therefore we can use their algorithm to compute it.

5.3 Dynamic Programming on Surface-Cut Decompositions

We now show how to compute an optimal cut graph of an embedded graph of bounded branch-width, using surface-cut decompositions. We first recall the following lemma of Erickson et al. which follows from Euler's formula and allows us to bound the complexity of the optimal cut graph. For a graph H embedded on a region R, we define its *reduced graph* to be the embedded graph obtained by repeatedly removing from G the degree 1 vertices which are not on a boundary and their adjacent edges, and contracting each maximal path through degree 2 vertices to a single edge (weighted as the length of the path).

Lemma 5 ([8, Lemma 4.2]). *Let Σ be a surface of genus g. Then any reduced cut graph on Σ has less than $4g$ vertices and $6g$ edges.*

The idea is then to compute in a dynamic programming fashion, for every region R of the surface-cut decomposition, every possible combinatorial map M corresponding to the restriction of a reduced cut graph of Σ to R, and every possible position P of the vertices of the boundary of M on the boundary of R, the shortest reduced graph embedded on R with map M and position P. The bounds on the size of the boundaries of the region (coming from the definition of surface-cut decompositions), as well as Lemma 5 allow us to bound the size of the dynamic table, the proof is omitted.

Theorem 8. *If a graph G of complexity n embedded on a genus g surface has branch-width at most k, an optimal cut graph of G can be computed in time $O_{g,k}(n^3)$.*

---

[1] When an edge is cut in two halves, the weight is spread in half on each sub-edge.

Acknowledgments. We are grateful to Sergio Cabello, Éric Colin de Verdière, Frederic Dorn and Dimitrios M. Thilikos for very helpful remarks at various stages of the elaboration of this article.

References

1. Bonsma, P.: Surface split decompositions and subgraph isomorphism in graphs on surfaces. In: Proc. of the Symp. on Theoretical Aspects of Computer Science, STACS 2012, pp. 531–542 (2012)
2. Borradaile, G., Demaine, E.D., Tazari, S.: Polynomial-time approximation schemes for subset-connectivity problems in bounded-genus graphs. Algorithmica 68(2), 287–311 (2014)
3. Borradaile, G., Klein, P.N., Mathieu, C.: An $O(n \log n)$ approximation scheme for Steiner tree in planar graphs. ACM Trans. on Algorithms 5(3) (2009)
4. Cohen-Addad, V., de Mesmay, A.: A fixed paramater tractable approximation scheme for the optimal cut graph of a surface (2015) (in preparation)
5. Demaine, E.D., Hajiaghayi, M., Kawarabayashi, K.I.: Contraction decomposition in h-minor-free graphs and algorithmic applications. In: Proc. of the ACM Symp. on Theory of Computing, STOC 2011, pp. 441–450. ACM (2011)
6. Demaine, E.D., Hajiaghayi, M., Mohar, B.: Approximation algorithms via contraction decomposition. Combinatorica, 533 552 (2010)
7. Erickson, J.: Combinatorial optimization of cycles and bases. In: Zomorodian, A. (ed.) Computational Topology. Proc. of Symp. in Applied Mathematics, AMS (2012)
8. Erickson, J., Har-Peled, S.: Optimally cutting a surface into a disk. Discrete & Computational Geometry 31(1), 37–59 (2004)
9. Erickson, R.E., Monma, C.L., Veinott Jr., A.F.: Send-and-split method for minimum-concave-cost network flows. Mathematics of Operations Research 12(4), 634–664 (1987)
10. Fomin, F.V., Thilikos, D.M.: On self duality of pathwidth in polyhedral graph embeddings. Journal of Graph Theory 55(1), 42–54 (2007)
11. Inkmann, T.: Tree-based decompositions of graphs on surfaces and applications to the Traveling Salesman Problem. Ph.D. thesis, Georgia Inst. of Technology (2007)
12. Marx, D.: Parameterized complexity and approximation algorithms. The Computer Journal 51(1), 60–78 (2008)
13. Mazoit, F.: Tree-width of hypergraphs and surface duality. J. Comb. Theory, Ser. B 102(3), 671–687 (2012)
14. Mohar, B., Thomassen, C.: Graphs on surfaces. Johns Hopkins Studies in the Mathematical Sciences. Johns Hopkins University Press (2001)
15. Pilipczuk, M., Pilipczuk, M., Sankowski, P., van Leeuwen, E.J.: Network sparsification for steiner problems on planar and bounded-genus graphs. In: Proceedings of Foundations of Computer Science (FOCS), pp. 276–285 (2014)
16. Robertson, N., Seymour, P.: Graph minors. X. obstructions to tree-decomposition. J. Combin. Theory Ser. B 52(2), 153–190 (1991)
17. Rué, J., Sau, I., Thilikos, D.M.: Dynamic programming for graphs on surfaces. ACM Trans. Algorithms 10(2), 8:1–8:26 (2014)

18. Thilikos, D.M., Serna, M., Bodlaender, H.L.: Constructive linear time algorithms for small cutwidth and carving-width. In: Lee, D.T., Teng, S.-H. (eds.) ISAAC 2000. LNCS, vol. 1969, pp. 192–203. Springer, Heidelberg (2000)
19. Colin de Verdière, É.: Topological algorithms for graphs on surfaces (2012), habilitation thesis. http://www.di.ens.fr/~colin/
20. Wood, Z., Hoppe, H., Desbrun, M., Schröder, P.: Removing excess topology from isosurfaces. ACM Transactions on Graphics 23(2), 190–208 (2004)

Explicit Expanding Expanders

Michael Dinitz[1],*, Michael Schapira[2],**, and Asaf Valadarsky[3],**

[1] Johns Hopkins University
mdinitz@cs.jhu.edu
[2] Hebrew University of Jerusalem
schapiram@huji.ac.il
[3] Hebrew University of Jerusalem
asaf.valadarsky@mail.huji.ac.il

Abstract. Deterministic constructions of expander graphs have been an important topic of research in computer science and mathematics, with many well-studied constructions of infinite families of expanders. In some applications, though, an infinite family is not enough: we need expanders which are "close" to each other. We study the following question: Construct an an infinite sequence of expanders $G_0, G_1, \ldots$, such that for every two consecutive graphs G_i and G_{i+1}, G_{i+1} can be obtained from G_i by adding a single vertex and inserting/removing a small number of edges, which we call the *expansion cost* of transitioning from G_i to G_{i+1}. This question is very natural, e.g., in the context of datacenter networks, where the vertices represent racks of servers, and the expansion cost captures the amount of rewiring needed when adding another rack to the network. We present an *explicit* construction of d-regular expanders with expansion cost at most $\frac{5d}{2}$, for any $d \geq 6$. Our construction leverages the notion of a "2-lift" of a graph. This operation was first analyzed by Bilu and Linial [1], who repeatedly applied 2-lifts to construct an infinite family of expanders which double in size from one expander to the next. Our construction can be viewed as a way to "interpolate" between Bilu-Linial expanders with low expansion cost while preserving good edge expansion throughout.

1 Introduction

Expander graphs (aka expanders) have been the object of extensive study in theoretical computer science and mathematics (see e.g. the survey of [2]). Originally introduced in the context of building robust, high-performance communication networks [3], expanders are both very natural from a purely mathematical perspective and play a key role in a host of other applications (from complexity theory to coding). While d-regular random graphs are, in fact, very good expanders [4, 5], many applications require *explicit*, deterministic constructions of

* Supported in part by NSF grant #1464239
** Supported in part by ISF grant 420/12, Israel Ministry of Science Grant 3-9772, Marie Curie Career Integration Grant, the Israeli Center for Research Excellence in Algorithms (I-CORE), Microsoft Research PhD Scholarship

© Springer-Verlag Berlin Heidelberg 2015
N. Bansal and I. Finocchi (Eds.): ESA 2015, LNCS 9294, pp. 399–410, 2015.
DOI: 10.1007/978-3-662-48350-3_34

expanders.[1] Consequently, a rich body of literature in graph theory deals with deterministic constructions of expanders, of which the best known examples are Margulis's construction [6] (with Gabber and Galil's analysis [7]), algebraic constructions involving Cayley graphs such as that of Lubotzky, Phillips, and Sarnak [8], constructions that utilize the zig-zag product [9], and constructions that rely on the concept of 2-lifts [1, 10].

All of these constructions generate an infinite family of d-regular expanders. However, for important applications of expanders that arise in computer networking, this is not enough. Our primary motivating example are datacenters, which network an unprecedented number of computational nodes and are the subject of much recent attention in the networking research community. Consider a datacenter network represented as a graph, in which each vertex represents a rack of servers, and edges represent communication links between these racks (or, more accurately, between the so-called "top-of-rack switches"). Expanders are natural candidates for datacenter network topologies as they fare well with respect to crucial objectives such as fault-tolerance and throughput [3, 2]. However, the number of racks n in a datacenter grows regularly as new equipment is purchased and old equipment is upgraded, calling for an expander construction that can grow gracefully (see discussion of industry experience in [11], and references therein).

We hence seek expander constructions that satisfy an extra constraint: incremental growth, or *expandability*. When a new rack is added to an existing datacenter, it is impractical to require that the datacenter be entirely rewired and reconfigured. Instead, adding a new rack should entail only a small number of local changes, leaving the vast majority of the network intact. From a theoretical perspective, this boils down to requiring that the construction of expanders not only work for all n, but also involve very few edge insertions and deletions from one expander to the next.

Our aim, then, is to explicitly construct an infinite family of expanders such that (1) every member of the family has good (edge) expansion; and (2) every member of the family can be obtained from the previous member via the addition of a single vertex and only "a few" edge insertions and deletions. Can this be accomplished? What are the inherent tradeoffs (e.g., in terms of edge expansion vs. number of edge insertions/deletions)? We formalize this question and take a first step in this direction. Specifically, we present the first construction of explicit expanding expanders and discuss its strengths and limitations.

1.1   Our Results and Techniques

We formally define edge expansion and expansion cost in Section 2. We now provide an informal exposition. The *edge expansion* of a set of vertices is the number of edges leaving the set divided by the size of the set, and the edge expansion of a graph is the worst-case edge expansion across all sets. The *expansion cost* for a graph G_i on n vertices $\{1, \ldots, n\}$ and graph G_{i+1} on $n+1$ vertices $\{1, \ldots, n+1\}$

---

[1] Throughout this paper we will use "explicit" and "deterministic" interchangeably.

is the number of edge insertions and removals required to transition from G_i to G_{i+1}. The expansion cost of a family of graphs $\{G_i = (V_i, G_i)\}$, where V_{i+1} is the union of V_i and an additional vertex, is the worst-case expansion cost across all consecutive pairs of graphs in the family. Observe that adding a new vertex to a d-regular graph while preserving d-regularity involves inserting d edges between that vertex and the rest of the graph, and removing at least $\frac{d}{2}$ edges to "make room" for the new edges. Hence, $\frac{3d}{2}$ is a lower bound on the expansion cost of any family of d-regular graphs.

Our main result is an explicit construction of an infinite family of d-regular expanders with very good edge expansion and small expansion cost:

Theorem 1. *For any even degree* $d \geq 6$, *there exists an infinite sequence of explicitly constructed d-regular expanders* $\{G_i = (V_i, E_i\}$ *such that*

1. $|V_0| = \frac{d}{2} + 1$, *and for every* $i \geq 0$, $|V_{i+1}| = |V_i| + 1$.
2. *The edge expansion of G_i is at least* $\frac{d}{3} - O(\sqrt{d \log^3 d})$ *for every* $i \geq 0$.
3. *The expansion cost of the family* $\{G_i\}$ *is at most* $\frac{5d}{2}$.

The attentive reader might notice that we claim our graphs are d-regular, yet the number of vertices of the first graph in the sequence, G_0, is only $\frac{d}{2} + 1$. This seeming contradiction is due to our use of multigraphs, i.e., graphs with parallel edges. In particular, G_0 is the complete graph on $\frac{d}{2} + 1$ vertices, but where every two vertices are connected by 2 parallel edges. While expanders are traditionally simple graphs, all nice properties of d-regular expanders, including the relationships between edge and spectral expansion, continue to hold with essentially no change for d-regular "expander multigraphs".

Our construction technique is to first deterministically construct an infinite sequence of "extremely good" expanders by starting at $K_{\frac{d}{2}+1}$ and repeatedly "2-lifting" the graph [1]. This standard and well-studied approach to explicitly constructing an infinite sequence of expanders was introduced in the seminal work of Bilu and Linial [1]. However, as every 2-lift doubles the size of the graph, this construction can only generate expanders on n vertices where $n = 2^i(\frac{d}{2} + 1)$ for some $i \geq 1$. We show how to "interpolate" between these graphs. Intuitively, rather than doubling the number of vertices all at once, we insert new vertices one at a time until reaching the next Bilu-Linial expander in the sequence. Our construction and proof crucially utilize the properties of 2-lifts, as well as the flexibility afforded to us by using multigraphs.

While our main focus is on centralized constructions for use as datacenter networks, the fact that our construction is deterministic also allows for improved expander constructions in some *distributed* models. Most notably, we get improved "self-healing" expanders. In the self-healing model, nodes are either inserted or removed into the graph one at a time, and the algorithm must send logarithmic-size messages between nodes (in synchronous rounds) in order to recover to an expander upon node insertion or removal. Clearly small expansion cost is a useful property in this context. The best-known construction of self-healing expanders [12] gives an expander with edge expansion of at least $d/20000$, $O(1)$ maximum degree, $O(1)$ topology changes, and $O(\log n)$ recovery time and message complexity (where the time and complexity bounds hold with high probability, while the other bounds hold deterministically). Our construction gives a self-healing expander with two improvements: much larger edge expansion (approximately $d/6$ rather than $d/20000$), and deterministic complexity bounds. In particular, we prove the following theorem:

Theorem 2. *For any $d \geq 6$, there is a self-healing expander which is completely deterministic, has edge expansion at least $d/6 - o(d)$, has maximum degree d, has $O(d)$ topology changes, and has recovery time and message complexity of $O(\log n)$.*

1.2 Related Work

The immediate precursor of this paper is a recent paper of Singla et al. [11], which proposes random graphs as datacenter network topologies. [11] presents a simple randomized algorithm for constructing a sequence of random regular graphs with small expansion cost. While using random graphs as datacenter topologies constitutes an important and thought-provoking experiment, the inherent unstructuredness of random graphs poses obstacles to their adoption in practice. Our aim, in contrast, is to *explicitly* construct expanders with *provable* guarantees on edge expansion and expansion cost.

The deterministic/explicit construction of expanders is a prominent research area in both mathematics and computer science. See the survey of Hoory, Linial, and Wigderson [2]. Our approach relies on the seminal paper of Bilu and Linial [1], which proposed and studied the notion of 2-lifting a graph. They proved that when starting with any "good" expander, a random 2-lift results in another good expander and, moreover, that this can be derandomized. Thus [1] provides a means to deterministically construct an infinite sequence of expanders: start with a good expander and repeatedly 2-lift. All expanders in this sequence are proven to be quasi-Ramanujan graphs, and are conjectured to be Ramanujan graphs (i.e., have optimal spectral expansion). Marcus, Spielman, and Srivastava [10] recently showed that this is indeed essentially true for *bipartite* expanders.

There has been significant work on using expanders in peer-to-peer networks and in distributed computing. See, in particular, the continuous-discrete approach of Naor and Wieder [13], and the self-healing expanders of [12]. The main focus of this line of research is on the efficient design of *distributed* systems, and so the goal is to minimize metrics like the number of messages between computational nodes, or the time required for nodes to join/leave the system. Moreover, the actual degree does not matter (since edges are logical rather than physical links), as long as it is constant. Our focus, in contrast, is on centralized constructions that work for any fixed degree d.

2   Preliminaries: Expander Graphs and Expansion Cost

All missing proofs can be found in the full version [14]. We adopt most of our notation from the survey of Hoory, Linial, and Wigderson on expanders [2].

Throughout this paper the graphs considered are multigraphs without self-loops, that is, may have parallel edges between any two vertices. We will commonly treat a multigraph as a weighted simple graph, in which the weight of each edge is an integer that specifies the number of parallel edges between the appropriate two vertices. Given such a weighted graph $G = (V, E, w)$, let $n = |V|$ and say that G is d-regular if every vertex in V has weighted degree d. We let $N(u) = \{v \in V : \{u, v\} \in E\}$ be the neighborhood of vertex u for any vertex $u \in V$. Traditionally, expanders are defined as simple graphs, but it is straightforward to see that all standard results on expanders used here continue to hold for multigraphs.

Expansion: For $S, T \subseteq V$, let $E(S, T)$ denote the multiset of edges with one endpoint in S and one endpoint in T, and let $\bar{S} = V \setminus S$. If $G = (V, E, w)$ is a d-regular multigraph, then for every set $S \subseteq V$ with $1 \leq |S| \leq \frac{n}{2}$ the *edge expansion* (referred to simply as the *expansion*) of S is $h_G(S) = \frac{|E(S, \bar{S})|}{|S|}$. We will sometimes omit the subscript when G is clear from context. The edge expansion of G is $h(G) = \min_{S \subseteq V : 1 \leq |S| \leq \frac{n}{2}} h_G(S)$. We say that G is an expander if $h(G)$ is large. In particular, we want $h(G)$ to be at least d/c for some constant c.

While much of our analysis is combinatorial, we also make extensive use of spectral analysis. Given a multigraph G, the adjacency matrix of G is an $n \times n$ matrix $A(G)$ in which the entry A_{ij} specifies the number of edges between vertex i and vertex j. We let $\lambda_1(G) \geq \lambda_2(G) \geq \cdots \geq \lambda_n(G)$ denote the eigenvalues of $A(G)$, and let $\lambda(G) = \max\{\lambda_2(G), |\lambda_n(G)|\}$.

Cheeger's inequality (the discrete version) enables us relate the eigenvalues of a (multi)graph G to the edge expansion of G:

Theorem 3. $\frac{d - \lambda_2}{2} \leq h(G) \leq \sqrt{2d(d - \lambda_2)}$.

We will also use the *Expander Mixing Lemma*, which, informally, states that the number of edges between any two sets of vertices is very close to the expected number of edges between such sets in a random graph.

Theorem 4 ([15]). $\left| |E(S, T)| - \frac{d|S||T|}{n} \right| \leq \lambda \sqrt{|S||T|}$ *for all* $S, T \subseteq V$.

Bilu-Linial: The construction of d-regular expanders using "lifts", due to Bilu and Linial [1], plays a key role in our construction. Informally, a graph H is called a k-*lift* of a (simple) graph G if every vertex in G is replaced by k vertices in H, and every edge in G is replaced with a perfect matching between the two sets of vertices in H that represent the endpoints of that edge in G. To put this formally: a graph H is called a k-*lift* of graph G if there is a function $\pi : V(H) \to V(G)$ such that the following two properties hold. First, $|\pi^{-1}(u)| = k$ for all $u \in V(G)$. Second, if $\{u, v\} \in E(G)$ then for every $x \in \pi^{-1}(u)$ there is exactly one $y \in \pi^{-1}(v)$ such that $\{x, y\} \in E(H)$.

We call the function π the *assignment function* for H. We follow Bilu and Linial in only being concerned with 2-lifts. Observe that if H is a 2-lift of G then $|V(H)| = 2|V(G)|$ and $|E(H)| = 2|E(G)|$, and furthermore that if G is d-regular then so is H. Bilu and Linial proved that when starting out with a

d-regular expander G that also satisfies a certain sparsity condition (see Corollary 3.1 in [1]), one can deterministically and efficiently find a 2-lift H where $\lambda(H) \leq O(\sqrt{d} \log^3 d)$ and moreover H continues to satisfy the sparsity condition. As K_{d+1} (the d-regular complete graph on $d + 1$ vertices) satisfies the sparsity condition, starting out with K_{d+1} and repeatedly 2-lifting generates a deterministic sequence of d-regular expanders, each of which twice as large as the previous, with edge expansion at least $\frac{d - O(\sqrt{d} \log^3 d)}{2}$ throughout (see also Theorem 6.12 in [2]).

Incremental Expansion: We will also be concerned with the expansion cost of an infinite family of expander (multi)graphs. Given two sets A, B, let $A \triangle B = (A \setminus B) \cup (B \setminus A)$ denote their symmetric difference. Let $\mathcal{G} = G_1, G_2, \ldots$ be an infinite family of d-regular expanders, where $V(G_i) \subset V(G_{i+1})$ for all $i \geq 1$.

Definition 5. *The* expansion cost *of* $\mathcal{G}$ *is* $\alpha(\mathcal{G}) = \max_{i \geq 1} |E(G_i) \triangle E(G_{i+1})|$.

As our focus is on multigraphs, the edge sets are in fact multisets, and so the expansion cost is the change in weight from G_i to G_{i+1}. Slightly more formally, if we let x_e^i denote the number of copies of edge e in $E(G_i)$, we have that $\alpha(\mathcal{G}) = \max_{i \geq 1} \sum_{e \in E(G_{i+1}) \cup E(G_i)} |x_e^i - x_e^{i+1}|$. Observe that the expansion cost is defined for any infinite sequence of graphs, and that a large gap in size from one graph to the next trivially implies a large expansion cost. We restrict our attention henceforth to constructions that generate a d-regular graph on n vertices for every integer n. We observe that the expansion cost of any such sequence is at least $\frac{3d}{2}$, since $E(G_{i+1}) \setminus E(G_i)$ must contain d edges incident to the vertex in $V(G_{i+1}) \setminus V(G_i)$, and in order to maintain d-regularity there must be at least $\frac{d}{2}$ edges in $E(G_i) \setminus E(G_{i+1})$.

3   Construction and Some Observations

We now formally present our construction of the sequence $\mathcal{G}$ of d-regular expanders and prove some simple properties of this construction.

We begin with the complete graph on $\frac{d}{2} + 1$ vertices and assign every edge a weight of 2. This will serve as the first graph in $\mathcal{G}$. To simplify exposition, we will refer to this graph as $G_{\frac{d}{2}+1}$. In general, the subscript i in graph $G_i \in \mathcal{G}$ will henceforth refer to the number of vertices in G_i. Clearly, $G_{\frac{d}{2}+1}$ is d-regular and has edge expansion $\frac{d}{2}$. We now embed the Bilu-Linial sequence of graphs starting from $G_{\frac{d}{2}+1}$ in $\mathcal{G}$: for every $i \geq 0$, let $G_{2^{i+1}(\frac{d}{2}+1)}$ be the 2-lift of $G_{2^i(\frac{d}{2}+1)}$ guaranteed by [1] to have $\lambda(G_{2^i(\frac{d}{2}+1)}) \leq O(\sqrt{d} \log^3 d)$ (recall that the next graph in the sequence can be constructed in polynomial time). Assign weight 2 to every edge in this sequence of expanders. We refer to graphs in this subsequence of $\mathcal{G}$ as *BL expanders*, since they are precisely $d/2$-regular BL expanders in which every edge is doubled. Thus each BL expander is d-regular and by the Cheeger inequality has edge expansion at least $\frac{d}{2} - O(\sqrt{d} \log^3 d)$.

We let G_i^* denote $G_{2^i(\frac{d}{2}+1)}$. We know, from the definition of a 2-lift, that for each i there exists a function $\pi : V(G_{i+1}^*) \to V(G_i^*)$ which is surjective and has

$|\pi^{-1}(u)| = 2$ for all $u \in V(G_i^*)$. As we want that $V(G_i^*) \subset V(G_{i+1}^*)$, we identify one element of $\pi^{-1}(u)$ with u, i.e. for each $u \in V(G_i^*)$ we will assume (without loss of generality) that $u \in V(G_{i+1}^*)$ and $\pi(u) = u$.

To construct the infinite sequence $\mathcal{G}$ it is clearly sufficient to show how to create appropriate expanders for all values of n between $2^i(\frac{d}{2}+1)$ and $2^{i+1}(\frac{d}{2}+1)$ for an arbitrary i. Fix some $i \geq 0$, let $\pi : V(G_{i+1}^*) \to V(G_i^*)$ be the assignment function for the BL expanders, and initialize the sets $S = \emptyset$ (called the *split* vertices) and $U = V(G_i^*)$ (called the *unsplit* vertices). We apply the following algorithm to construct G_{n+1} from G_n, starting with $n = 2^i(\frac{d}{2}+1)$ and iterating until $n = 2^{i+1}(\frac{d}{2}+1) - 1$.

1. **Splitting a vertex u into u and u'.** Let u be an arbitrary unsplit vertex. We let the new vertex in G_{n+1} that is not in G_n be u', the vertex in $\pi^{-1}(u)$ that is not u. Let $S(u) = S \cap N(u)$ be the neighbors of u that have already split, and let $U(u) = U \cap N(u)$ be the neighbors of u that are unsplit. Here the neighborhood $N(u)$ is with respect to G_n.
2. **Inserting edges from u and u' to unsplit neighbors.** For every $v \in U(u)$, replace the edge from u to v (which we prove later always exists) with an edge from u to v of weight 1 and an edge from u' to v of weight 1.
3. **Inserting edges from u and u' to split neighbors.** For every pair of vertices $v, v' \in S(u)$ with $\pi(v) = \pi(v')$, decrease the weight of $\{v, v'\}$ by 1 and do one of the following:
   - if $\{u, v\} \in E(G_{i+1}^*)$, assign $\{u, v\}$ a weight of 2, remove $\{u, v'\}$, and add an edge $\{u', v'\}$ of weight 2;
   - otherwise (that is, $\{u, v'\} \in E(G_{i+1}^*)$), assign $\{u, v'\}$ a weight of 2, remove $\{u, v\}$, and add an edge $\{u', v\}$ of weight 2.
4. **Inserting edges between u and u'.** Add an edge between u and u' of weight $|U(u)|$.
5. **Mark u and u' as split.** Remove u from U, add u and u' to S.

We prove the following simple invariants. We will refer to two vertices u, v as *paired* if $\pi(u) = \pi(v)$. Together, these lemmas imply that the algorithm is well-defined and that we have an infinite sequence of d-regular graphs that interpolates between BL expanders.

Lemma 6. *Let u, u' be paired vertices with $\pi(u) = \pi(u') = u$. Then throughout the execution of the algorithm, edge $\{u, u'\}$ exists if u has already split and if there are neighbors of u which are unsplit. If $\{u, u'\}$ exists then it has weight equal to the number of neighbors of u that are unsplit.*

Lemma 7. *Edges between unpaired split vertices always have weight 2, edges between unsplit vertices always have weight 2, and edges with one endpoint unsplit and one split have weight 1.*

Lemma 8. *Every vertex has weighted degree d throughout the execution of the algorithm.*

Lemma 9. *When all vertices have split, G is precisely G_{i+1}^* in which all edges have weight 2.*

4 Analysis: Expansion and Expansion Cost

We next prove that that the expansion cost of our construction is small, and the edge expansion throughout is good. Specifically, we prove that the expansion cost is at most $\frac{5}{2}d$, and then prove some combinatorial lemmas which will immediately imply that the edge expansion is at least $\frac{d}{4} - O(\sqrt{d \log^3 d})$. We show in Section 5 how this bound on edge expansion can be improved to a tight lower bound of $\frac{d}{3} - O(\sqrt{d \log^3 d})$ via a more delicate, spectral analysis combined with the combinatorial lemmas from this section.

We begin by analyzing the expansion cost.

Theorem 10. $\alpha(\mathcal{G}) \leq \frac{5}{2}d$.

Proof. Suppose G_{n+1} is obtained from G_n by splitting vertex u into u and u'. The transition from G_n to G_{n+1} entails the following changes in edge weights:

- **A change of 2 in edge weights per vertex in $U(u)$.** Each edge from vertex u to a vertex $v \in U(u)$ changes its weight from 2 to 1 and an additional edge of weight 1 is added from u' to v, so there are 2 edge changes per vertex in $U(u)$.
- **A change of 5 in edge weights for every two paired vertices in $S(u)$.** Every pair of edges in G_n (of weight 1) from u to paired vertices v, v' in $S(u)$ is replaced by a pair of edges between u, u' and v, v', each of weight 2, which results in a total change in edge weights of 4: 1 for increasing the weight of one of u's outgoing edges to the pair v, v' from 1 to 2, 1 for decreasing an edge of u's other outgoing edge from 1 to 0, and 2 for the new edge from u' the pair v, v'. In addition, the weight of the edge (v, v') is decreased by 1. So, each pair of vertices in $S(u)$ induces a total change of 5 in edge weights.
- **An additional change of $|U(u)|$ in edge weights.** An edge of weight $|U(u)|$ is added between u and u'.

Hence, $|E(G_n) \triangle E(G_{n+1})| = 2|U(u)| + 5|S(u)|/2 + |U(u)| = 3|U(u)| + (5|S(u)| /2)$. As $2|U(u)| + |S(u)| = d$ by Lemma 8, this concludes the proof of the theorem. $\square$

This analysis is tight for our algorithm. At some point in the execution of the algorithm, some vertex u will be split after all of its neighboring vertices have already been split. As this entails a change in weight of 5 for each of the $\frac{d}{2}$ paired vertices in $S(u)$, the resulting total change in edge weights will be $\frac{5}{2}d$.

4.1 Edge Expansion

We show, via a combinatorial argument, that every member of our sequence of graphs $\mathcal{G}$ has edge expansion at least $\frac{d}{4} - O(\sqrt{d \log^3 d})$. To this end, we show that for every n between $2^i(\frac{d}{2} + 1)$ and $2^{i+1}(\frac{d}{2} + 1)$, the graph $G = G_n = (V, E)$ has edge expansion at least $\frac{d}{4} - O(\sqrt{d \log^3 d})$. We will then show in Section 5 how this lower bound on edge expansion can be tightened to $\frac{d}{3} - O(\sqrt{d \log^3 d})$ via spectral analysis combined with the combinatorial lemmas proved here.

Theorem 11. *For every $G \in \mathcal{G}$, $h(G) \geq \frac{d}{4} - O(\sqrt{d \log^3 d})$.*

We now prove Theorem 11. Let $S \subseteq V$ denote the set of vertices that have already split in G, and let $U \subseteq V$ be the set of vertices that are currently unsplit. Let $H = (V_H, E_H) = G^*_{i+1}$ be the next BL expander in the sequence and let π be its assignment function (note that the range of π is the vertices of the previous BL expander, which includes the vertices U in G). For any subset $A \subseteq V$, let $F(A) \subseteq V_H$ denote the "future" set of A, in which all unsplit vertices in A are split and both vertices appear in $F(A)$. More formally, $F(A) = (A \cap S) \cup (\cup_{u \in A \cap U} \pi^{-1}(u))$. For $X, Y \subseteq V_H$ with $X \cap Y = \emptyset$, let $w_H(X, Y)$ denote the total edge weight between X and Y in H. Lastly, for $A, B \subseteq V$ with $A \cap B = \emptyset$ we define $w_G(A, B)$ similarly, except that we *do not* include edge weights between paired vertices. Our proof proceeds by analyzing $w_G(A, B)$ for all possible different subsets of vertices A, B in G. As $w_G(A, B)$ only reflects the edge weights in G between non-paired vertices, the proof below lower bounds the actual edge expansion (which also includes weights between paired vertices).

Lemma 12. *If $A, B \subseteq S$ with $A \cap B = \emptyset$, then $w_H(F(A), F(B)) = w_G(A, B)$.*

Lemma 13. *If $A, B \subseteq U$ with $A \cap B = \emptyset$, then $w_H(F(A), F(B)) = 2 \cdot w_G(A, B)$.*

Lemma 14. *If $A \subseteq S$ and $B \subseteq U$, then $w_H(F(A), F(B)) = 2 \cdot w_G(A, B)$.*

Combining these lemmas proves that every cut in G has weight at least half of that of the associated "future" cut, since we can divide any cut in G into split and unsplit parts. This implies Theorem 11 as $h(H) \geq \frac{d}{2} - O(\sqrt{d \log^3 d})$.

Lemma 15. *If $(A, \bar{A})$ is a cut in G, then $w_G(A, \bar{A}) \geq \frac{1}{2} w_H(F(A), F(\bar{A}))$.*

5   Improved Edge Expansion Analysis

We proved in Section 4.1 that our sequence of graphs has edge expansion at least $\frac{d}{4} - O(\sqrt{d \log^3 d})$. We next apply spectral analysis to improve this lower bound.

Theorem 16. *For every $G \in \mathcal{G}$, $h(G) \geq \frac{d}{3} - O(\sqrt{d \log^3 d})$.*

Interestingly, while we prove this theorem by using spectral properties of Bilu-Linial expanders, we cannot prove such a theorem through a direct spectral analysis of the expanders that we generate.

Theorem 17. *For any $\epsilon > 0$, there are an infinite number of graphs $G \in \mathcal{G}$ which have $\lambda_2(G) \geq d/2 - \epsilon$.*

This implies that if we want to lower bound $h(G)$ by using Theorem 3 (the Cheeger inequalities), the best bound we could prove would be $d/4$. Thus Theorem 16 beats the eigenvalue bound for this graph.

We now begin our proof of Theorem 16. We use the same terminology and notation as in the proof of Theorem 11. The key to improving our analysis lies in leveraging the fact that $H = G_{i+1}^*$, the next BL expander in the sequence of graphs $\mathcal{G}$, is a strong *spectral* expander (i.e., $\lambda(G_{i+1}^*) \leq O(\sqrt{d \log^3 d})$). We first handle the case of unbalanced cuts, then the more difficult case of nearly-balanced cuts. We then show that the analysis in this section is tight.

Unbalanced Cuts. We first show that in a strong spectral expander, unbalanced cuts give large expansion. This is straightforward from the Mixing Lemma (Theorem 4) if the cut is not *too* unbalanced, i.e. if both sides of the cut are of linear size. However, a straightforward application of the Mixing Lemma fails when the small side is very small. We show that this can be overcome by using the full power of the Mixing Lemma: the two sets in Theorem 4 need not be a cut, but can be any two sets.

Lemma 18. *If $X \subseteq V_H$ with $|X| \leq n/2$, then $w_H(X, \bar{X}) \geq |X| \left(d \left(\frac{n - |X|}{n} \right) - 4\lambda \right)$.*

Lemma 19. *If $X \subseteq V$ with $|X| < \frac{n}{5}$, then $h_G(X) \geq \frac{d}{3} - O \left(\sqrt{d \log^3 d} \right)$.*

Balanced Cuts. We next prove that $h_G(X) \geq \frac{d}{3} - O(\sqrt{d \log^3 d})$ when $\frac{n}{5} \leq |X| \leq \frac{n}{2}$. To accomplish this, we use the Mixing Lemma (again) to show that the expansion does not drop by a factor of 2 from the future cut. Intuitively, if X contains many unsplit vertices, then even though G only gets half of the weight from unsplit vertices than H does, there are only half as many vertices and thus the expansion is basically preserved.[2] On the other hand, if X contains many split vertices, then either $\bar{X}$ also contains many split vertices (and so by Lemma 12 we lose nothing), or $\bar{X}$ contains many unsplit vertices (and so the cut is unbalanced enough for the Mixing Lemma to provide stronger bounds).

Lemma 20. *If $X \subseteq V$ with $\frac{n}{5} \leq |X| \leq \frac{n}{2}$, then $h_G(X) \geq \frac{d}{3} - O \left(\sqrt{d \log^3 d} \right)$.*

Proof. As before, let $S(X) = S \cap X, U(X) = U \cap X, S(\bar{X}) = S \cap \bar{X}$, and $U(\bar{X}) = U \cap \bar{X}$. We first analyze the weight of the future cut using the Mixing Lemma (Theorem 4).

$$w_H(F(X), F(\bar{X})) = w_H(F(S(X)), F(S(\bar{X}))) + w_H(F(S(X)), F(U(\bar{X}))) \quad (1)$$
$$+ w_H(F(U(X)), F(S(\bar{X}))) + w_H(F(U(X)), F(U(\bar{X})))$$

$$\geq \frac{d|F(S(X))| \cdot |F(S(\bar{X}))|}{|F(X)| + |F(\bar{X})|} + \frac{d|F(S(X))| \cdot |F(U(\bar{X}))|}{|F(X)| + |F(\bar{X})|} \quad (2)$$
$$+ \frac{d \cdot |F(U(X))| \cdot |F(S(\bar{X}))|}{|F(X)| + |F(\bar{X})|} + \frac{d \cdot |F(U(X))| \cdot |F(U(\bar{X}))|}{|F(X)| + |F(\bar{X})|} - 4\lambda|V_H|$$

$$\geq d\frac{|S(X)|(|S(\bar{X})| + 2|U(\bar{X})|) + 2|U(X)|(|S(\bar{X})| + 2|U(\bar{X})|)}{|X| + |\bar{X}| + |U(X)| + |U(\bar{X})|} - 4\lambda|V_H|.$$
$$(3)$$

---

[2] We point out that this is not quite accurate, since $F(X)$ could be larger than $F(\bar{X})$.

Equation (1) is simply the partition of the edges crossing the cut into the natural four sets. Equation (2) is the application of the Mixing Lemma to each of the four parts, together with an upper bound of $|V_H|$ on all sets to bound the discrepancy due to the Mixing Lemma to $4\lambda|V_H|$. Equation (3) exploits the fact that unsplit vertices in V split into exactly two vertices in V_H to get that $|V_H| = |F(X)| + |F(\bar{X})| = |X| + |\bar{X}| + |U(X)| + |U(\bar{X})|$, and that $|F(S(X))| = |S(X)|$, $|F(S(\bar{X}))| = |S(\bar{X})|$, $|F(U(X))| = 2|U(X)|$, and $|F(U(\bar{X}))| = 2|U(\bar{X})|$.

We can now apply Lemmas 12, 13, and 14 to relate this to the weight in G. The first term in (3) remains unchanged, whereas the second, third, and fourth terms are reduced by a factor of 2, and the final loss term also remains unchanged. With these adjustments, we get that

$$w_G(X, \bar{X}) \geq \frac{d\left(|S(X)|\left(|S(\bar{X})| + |U(\bar{X})|\right) + |U(X)|\left(|S(\bar{X})| + 2|U(\bar{X})|\right)\right)}{|X| + |\bar{X}| + |U(X)| + |U(\bar{X})|} - 4\lambda|V_H|$$

$$= d \cdot \frac{|S(X)| \cdot |\bar{X}| + |U(X)| \cdot \left(|\bar{X}| + |U(\bar{X})|\right)}{|X| + |\bar{X}| + |U(X)| + |U(\bar{X})|} - 4\lambda|V_H|$$

$$= d \cdot \frac{|X| \cdot |\bar{X}| + |U(X)| \cdot |U(\bar{X})|}{|X| + |\bar{X}| + |U(X)| + |U(\bar{X})|} - 4\lambda|V_H|.$$

Note that λ in this expression is $\lambda(H)$, not $\lambda(G)$. We can now get the expansion simply by dividing by $|X|$, the size of the smaller side: $h_G(X) \geq d \cdot \frac{|X| \cdot |\bar{X}| + |U(X)| \cdot |U(\bar{X})|}{|X|\left(|X| + |\bar{X}| + |U(X)| + |U(\bar{X})|\right)} - 40\lambda$, where for the final term we use the fact that $|V_H| \leq 2n$ and $|X| \geq \frac{n}{5}$ to get that $4\lambda|V_H|/|X| \leq \lambda \cdot 8n/(\frac{n}{5}) = 40\lambda$.

We claim that this expression is at least $\frac{d}{3} - O(\sqrt{d \log^3 d})$. As $\lambda = O(\sqrt{d \log^3 d})$, it needs to be shown that $\frac{|X| \cdot |\bar{X}| + |U(X)| \cdot |U(\bar{X})|}{|X|\left(|X| + |\bar{X}| + |U(X)| + |U(\bar{X})|\right)} \geq \frac{1}{3}$. Suppose for the sake of contradiction that this is false. Then rearranging terms gives us that

$$|U(X)| \cdot (3|U(\bar{X})| - |X|) < |X|^2 - 2|X||\bar{X}| + |X||U(\bar{X})|. \tag{4}$$

If $|U(\bar{X})| > \frac{|X|}{3}$, then (4) implies that $|U(X)| < |X|^2 - 2|X||\bar{X}| + |X||U(\bar{X})| \leq |X|^2 - |X||\bar{X}| \leq 0$, where we used the fact that $|U(\bar{X})| \leq |\bar{X}|$ and $|\bar{X}| \geq |X|$. This is a contradiction, since $|U(X)|$ clearly cannot be negative.

Otherwise, if $|U(\bar{X})| \leq \frac{|X|}{3}$, then (4) implies that

$$|U(X)| > \frac{2|X||\bar{X}| - |X|^2 - |X||U(\bar{X})|}{|X| - 3|U(\bar{X})|} \geq \frac{|X|^2 - |X||U(\bar{X})|}{|X| - 3|U(\bar{X})|} \geq |X|,$$

since $|\bar{X}| \geq |X|$. This is also a contradiction, as $U(X) \subseteq X$, and hence the lemma follows.   □

Combining Lemma 19 and Lemma 20 concludes the proof of Theorem 16.

Tightness of Analysis. We show that the bound on the edge expansion from Theorem 16 is essentially tight and, moreover, is tight infinitely often.

Theorem 21. *There exists a graph in $\mathcal{G}$ with edge expansion at most $\frac{d}{3} + \frac{2}{3}$ and, for every $i \geq 1$, there exists a graph in $\mathcal{G}$ between G_i^* and G_{i+1}^* with edge expansion at most $\frac{d}{3} + O(\sqrt{d \log^3 d})$.*

6   Open Questions

The obvious open question is proving better bounds for expansion and expansion cost, and exploring the space of tradeoffs between them. Our construction interpolates between Bilu-Linial (BL) expanders, which are very good spectral expanders ($\lambda \leq O(\sqrt{d \log^3 d})$). But Theorem 17 implies that some of the expanders that appear between the BL expanders in the sequence are only weak spectral expanders. Can a sequence of strong spectral expanders (say, with $\lambda \leq O(\sqrt{d} \cdot \text{polylog}(d))$) be constructed with low expansion cost?

References

[1] Bilu, Y., Linial, N.: Lifts, discrepancy and nearly optimal spectral gap. Combinatorica 26(5), 495–519 (2006)

[2] Hoory, S., Linial, N., Wigderson, A.: Expander graphs and their applications. Bull. Amer. Math. Soc. (N.S.) 43(4), 439–561 (2006)

[3] Bassalygo, L.A., Pinsker, M.S.: The complexity of an optimal non-blocking commutation scheme without reorganization. Problemy Peredači Informacii 9(1), 84–87 (1973)

[4] Bollobás, B.: The isoperimetric number of random regular graphs. Eur. J. Comb. 9(3), 241–244 (1988)

[5] Friedman, J.: A Proof of Alon's Second Eigenvalue Conjecture and Related Problems. Memoirs of the American Mathematical Society. AMS (2008)

[6] Margulis, G.A.: Explicit constructions of expanders. Problemy Peredači Informacii 9(4), 71–80 (1973)

[7] Gabber, O., Galil, Z.: Explicit constructions of linear-sized superconcentrators. J. Comput. Syst. Sci. 22(3), 407–420 (1981)

[8] Lubotzky, A., Phillips, R., Sarnak, P.: Ramanujan graphs. Combinatorica 8(3), 261–277 (1988)

[9] Reingold, O., Vadhan, S., Wigderson, A.: Entropy waves, the zig-zag graph product, and new constant-degree expanders and extractors. In: FOCS (2000)

[10] Marcus, A., Spielman, D.A., Srivastava, N.: Interlacing families i: Bipartite ramanujan graphs of all degrees. In: FOCS, pp. 529–537 (2013)

[11] Singla, A., Hong, C.Y., Popa, L., Godfrey, P.B.: Jellyfish: Networking data centers randomly. In: 9th USENIX Symposium on Networked Systems Design and Implementation (NSDI) (April 2012)

[12] Pandurangan, G., Robinson, P., Trehan, A.: DEX: self-healing expanders. In: 2014 IEEE 28th International Parallel and Distributed Processing Symposium, Phoenix, AZ, USA, May 19-23, pp. 702–711. IEEE (2014)

[13] Naor, M., Wieder, U.: Novel architectures for p2p applications: The continuous-discrete approach. ACM Trans. Algorithms 3(3) (August 2007)

[14] Dinitz, M., Schapira, M., Valadrsky, A.: Explicit expanding expanders. Preprint (2015). http://arxiv.org/abs/1507.01196

[15] Alon, N., Chung, F.: Explicit construction of linear sized tolerant networks. Discrete Mathematics 72(13), 15–19 (1988)

On the Threshold of Intractability [*],[**]

Pål Grønås Drange[1], Markus Sortland Dregi[1],
Daniel Lokshtanov[1], and Blair D. Sullivan[2]

[1] Dept. Informatics, Univ. Bergen, Norway
{pal.drange,markus.dregi,daniello}@ii.uib.no
[2] Dept. Computer Science, North Carolina State University, Raleigh, NC, USA
blair_sullivan@ncsu.edu

Abstract. We study the computational complexity of the graph modification problems THRESHOLD EDITING and CHAIN EDITING, adding and
deleting as few edges as possible to transform the input into a threshold
(or chain) graph. In this article, we show that both problems are NP-
hard, resolving a conjecture by Natanzon, Shamir, and Sharan (2001).
On the positive side, we show that these problems admit quadratic vertex kernels. Furthermore, we give a subexponential time parameterized
algorithm solving THRESHOLD EDITING in $2^{O(\sqrt{k}\log k)} + \text{poly}(n)$ time,
making it one of relatively few natural problems in this complexity class
on general graphs. These results are of broader interest to the field of
social network analysis, where recent work of Brandes (2014) posits that
the minimum edit distance to a threshold graph gives a good measure of
consistency for node centralities. Finally, we show that all our positive
results extend to CHAIN EDITING, as well as the completion and deletion
variants of both problems.

1 Introduction

In this paper we study the computational complexity of two edge modification
problems, namely editing to threshold graphs and editing to chain graphs. Graph
modification problems ask whether a given graph G can be transformed to have a
certain property using a small number of edits (such as deleting/adding vertices
or edges), and have been the subject of significant previous work [6,19,23]. The
THRESHOLD EDITING problem takes as input an n-vertex graph $G = (V, E)$ and
a non-negative integer k. The objective is to find a set F of at most k pairs of

[*] The research leading to these results has received funding from the Research Council
of Norway, Bergen Research Foundation under the project Beating Hardness by Preprocessing and the European Research Council under the European Union's Seventh
Framework Programme (FP/2007-2013) / ERC Grant Agreement n. 267959.

[**] Blair D. Sullivan supported in part by the Gordon & Betty Moore Foundation as
a DDD Investigator and the DARPA GRAPHS program under SPAWAR Grant
N66001-14-1-4063. Any opinions, findings, and conclusions or recommendations expressed in this publication are those of the author(s) and do not necessarily reflect
the views of DARPA, SSC Pacific, or the Moore Foundation.

© Springer-Verlag Berlin Heidelberg 2015
N. Bansal and I. Finocchi (Eds.): ESA 2015, LNCS 9294, pp. 411–423, 2015.
DOI: 10.1007/978-3-662-48350-3_35

vertices such that G minus any edges in F plus all non-edges in F is a *threshold graph* (can be constructed from the empty graph by repeatedly adding either an isolated vertex or a universal vertex [2]).

THRESHOLD EDITING
Input: A graph G and a non-negative integer k
Q: Is there a set $F \subseteq V^2$ with $|F| \leq k$ so that $G \triangle F$ is a threshold graph?

The computational complexity of THRESHOLD EDITING has repeatedly been stated as open, starting from Natanzon et al. [21], and then more recently by Burzyn et al. [3], and again very recently by Liu, Wang, Guo and Chen [15]. We resolve this by showing that the problem is indeed NP-hard.

Theorem 1. THRESHOLD EDITING *is* NP-*complete, even on split graphs.*

Graph editing problems are well-motivated by problems arising in the applied sciences, where we often have a predicted model from domain knowledge, but observed data fails to fit this model exactly. In this setting, edge modification corresponds to correcting false positives (and/or false negatives) to obtain data that is consistent with the model. THRESHOLD EDITING has specifically been of recent interest in the social sciences, where Brandes et al. are using distance to threshold graphs in work on axiomatization of centrality measures [1, 22]. More generally, editing to threshold graphs and their close relatives *chain graphs* arises in the study of sparse matrix multiplications [25]. Chain graphs are the bipartite analogue of threshold graphs (see Definition 2), and we also establish hardness of CHAIN EDITING.

Theorem 2. CHAIN EDITING *is* NP-*complete, even on bipartite graphs.*

Our final complexity result is for CHORDAL EDITING—a problem whose NP-hardness is well-known and widely used. This result also follows from our techniques, and as the authors were unable to find a proof in the literature, we include this argument for the sake of completeness.

Having settled the complexity of these problems, we turn to studying ways of dealing with their intractability. Cai's theorem [4] shows that THRESHOLD EDITING and CHAIN EDITING are *fixed parameter tractable*, i.e., solvable in $f(k) \cdot \text{poly}(n)$ time where k is the edit distance from the desired model (graph class); However, the lower bounds we prove when showing NP-hardness are on the order of $2^{o(\sqrt{k})}$ under ETH, and thus leave a gap. We show that it is in fact the lower bound which is tight (up to logarithmic factors in the exponent) by giving a subexponential time algorithm for both problems.

Theorem 3. THRESHOLD EDITING *and* CHAIN EDITING *admit* $2^{O(\sqrt{k}\log k)} + \text{poly}(n)$ *subexponential time algorithms.*

Since our results also hold for the *completion* and *deletion* variants of both problems (when F is restricted to be a set of non-edges or edges, respectively), this also answers a question of Liu et al. [16] by giving a subexponential time algorithm for CHAIN EDGE DELETION.

A crucial first step in our algorithms is to preprocess the instance, reducing to a kernel of size polynomial in the parameter. We give quadratic kernels for all three variants (of both THRESHOLD EDITING and CHAIN EDITING).

Theorem 4. THRESHOLD EDITING, THRESHOLD COMPLETION, *and* THRESHOLD DELETION *admit polynomial kernels with* $O(k^2)$ *vertices.*

This answers (affirmatively) a recent question of Liu, Wang and Guo [14]— whether the previously known kernel, which has $O(k^3)$ vertices, for THRESHOLD COMPLETION (equivalently THRESHOLD DELETION) can be improved.

2   Preliminaries

Due to page limits, proofs marked with a $\star$, as well as extended background and definitions are deferred to the full version [9]. For a set A, we use $[A]^2$ to denote the set of all unordered pairs of elements of A. For a graph $G = (V, E)$ and $F \subseteq [V]^2$ we define $G \triangle F$ as the graph $(V, E \triangle F)$, where $\triangle$ denotes the standard symmetric difference operator on sets. For an edge set $F \subseteq [V]^2$ and $v \in V$, we write $F(v)$ to denote the set of edges incident to v in F. For a graph G and a vertex v we define the *true twin class* of v, denoted $\text{ttc}(v)$ as the set $\{u \in V(G) \mid N[u] = N[v]\}$. Similarly, we define the *false twin class* of v, denoted $\text{ftc}(v)$ as the set $\{u \in V(G) \mid N(u) = N(v)\}$. Observe that $\text{ttc}(v) = \{v\}$ or $\text{ftc}(v) = \{v\}$. From this we define the *twin class* of v, denoted $\text{tc}(v)$ as $\text{ttc}(v)$ if $|\text{ttc}(v)| > |\text{ftc}(v)|$ and $\text{ftc}(v)$ otherwise.

Split and threshold graphs. A split graph is a graph $G = (V, E)$ whose vertex set can be partitioned into two sets C and I such that $G[C]$ is a complete graph and $G[I]$ is edgeless, i.e., C is a clique and I an independent set [2]. For a split graph G we say that a partition (C, I) of $V(G)$ forms a *split partition* of G if $G[C]$ induces a clique and $G[I]$ an independent set. A split partition (C, I) is called a *complete* split partition if for every vertex $v \in I$, $N(v) = C$. If G admits a complete split partition, we say that G is a complete split graph. Threshold graphs are closely related to split graphs, as seen in the following proposition.

Proposition 1 ([17]). *A graph G is a threshold graph if and only if G has a split partition (C, I) such that the neighborhoods of the vertices in I are nested, i.e., for every pair of vertices v and u, either $N(v) \subseteq N[u]$ or $N(u) \subseteq N[v]$.*

We also use the following alternative characterization of threshold graphs.

Proposition 2 ([2]). *A graph G is a threshold graph if and only if G does not have a C_4, P_4 nor a $2K_2$ as an induced subgraph.*

Finally, we need a structural characterization of threshold graphs with guaranteed behavior with respect to some optimal solution of THRESHOLD EDITING.

Definition 1 (Threshold partition, $\mathrm{lev}(v)$). *We say that $(\mathcal{C}, \mathcal{I})$ forms a threshold partition of G when $(\mathcal{C}, \mathcal{I}) = (\langle C_1, \ldots, C_t \rangle, \langle I_1, \ldots, I_t \rangle)$ so that: (i) (C, I) is a split partition of G, where $C = \bigcup_{i \leq t} C_i$ and $I = \bigcup_{i \leq t} I_i$; (ii) C_i and I_i are twin classes in G for every i; (iii) $N[C_j] \subset N[C_i]$ and $N(\bar{I}_i) \subset N(I_j)$ for every $i < j$. Finally, (iv) we demand that for every $i \leq t$, $(C_i, I_{\geq i})$ form a complete split partition of the graph induced by $C_i \cup I_{\geq i}$. Here we let $I_{\geq i} = \bigcup_{i \leq j \leq t} I_j$. We furthermore define, for every vertex v in G, $\mathrm{lev}(v)$ as the number i such that $v \in C_i \cup I_i$ and we denote each level $L_i = C_i \cup I_i$.*

Proposition 3 ($\star$). *A graph G is a threshold graph if and only if G admits a threshold partition.*

Lemma 1 ($\star$). *For every instance (G, k) of* Threshold Editing *or* Threshold Completion *it holds that there exists an optimal solution F such that for every pair of vertices $u, v \in V(G)$, if $N_G(u) \subseteq N_G[v]$ then $N_{G \triangle F}(u) \subseteq N_{G \triangle F}[v]$.*

Chain Graphs. Chain graphs are the bipartite graphs whose neighborhoods of the vertices on one of the sides form an inclusion chain (are *nested*). This implies the neighborhoods on the opposite side are also nested. The problem of completing edges to chain graphs was introduced by Golumbic [12] and later studied by Yannakakis [25], Feder, Mannila and Terzi [10] and finally by Fomin and Villanger [11] who showed that Chain Completion is solvable in subexponential time on bipartite graphs whose bipartition must be respected.

Definition 2 (Chain graph). *A bipartite graph $G = (A, B, E)$ is a chain graph if there is an ordering of the vertices of A, $a_1, a_2, \ldots, a_{|A|}$ such that $N(a_1) \subseteq N(a_2) \subseteq \cdots \subseteq N(a_{|A|})$.*

Chain Editing
Input: A graph $G = (V, E)$ and a non-negative integer k
Q: Is there a set F of size at most k such that $G \triangle F$ is a chain graph?

The fixed parameter tractability of Chain Editing, as mentioned above, follows from Cai's theorem [4] and the following forbidden induced subgraph characterization.

Proposition 4 ([2]). *Let G be a graph. The following are equivalent: (i) G is a chain graph, (ii) G is bipartite and $2K_2$-free, (iii) G is $\{2K_2, C_3, C_5\}$-free, and (iv) G is obtained from a threshold graph by removing all edges on the clique side.*

3   Hardness

In this section, we show that Threshold Editing and Chain Editing are NP-complete. We also give a proof that Chordal Editing is NP-complete—although this has been known for a long time (Natanzon [20], Natanzon et al. [21],

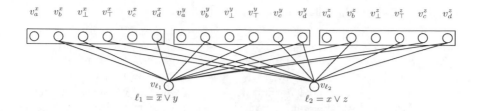

Fig. 1. The connections of a clause and a variable. All the vertices on the top (the variable vertices) belong to the clique, while the vertices on the bottom (the clause vertices) belong to the independent set.

Sharan [24]), the authors were unable to find a proof in the literature, and thus include the observation. A more general version of CHORDAL EDITING was recently shown to be FPT by Cao and Marx [5]. We note their variant is well-known to be NP-complete, as it also generalizes CHORDAL VERTEX DELETION.

NP-completeness of Threshold Editing. Our hardness reduction is from the problem 3SAT, where we are given a 3-CNF-SAT formula ϕ and asked to decide whether ϕ admits a satisfying assignment. We let $\mathcal{C}_\phi$ denote the set of clauses, and $\mathcal{V}_\phi$ the set of variables in a given formula ϕ. An *assignment* for a formula ϕ is a function $\alpha \colon \mathcal{V}_\phi \to \{\texttt{true}, \texttt{false}\}$. Furthermore, we assume we have some natural lexicographical ordering $<_{\text{lex}}$ of the clauses $\ell_1, \ldots, \ell_{|\mathcal{C}_\phi|}$ and the variables $v_1, \ldots, v_{|\mathcal{V}_\phi|}$.

We design a split graph G_ϕ and pick an integer $k_\phi = |\mathcal{C}_\phi| \cdot (3|\mathcal{V}_\phi| - 1)$ so that (G_ϕ, k_ϕ) is a yes-instance of THRESHOLD EDITING if and only if ϕ is satisfiable. Further, we ensure the split partition must be maintained in any threshold graph within distance k_ϕ of G_ϕ. Given ϕ, we first create a clique of size $6|\mathcal{V}_\phi|$; To each variable $x \in \mathcal{V}_\phi$, we associate six vertices $v_a^x, v_b^x, v_\perp^x, v_\top^x, v_c^x, v_d^x$ with a partial order denoted π_ϕ so that $v_a^x <_{\pi_\phi} v_b^x <_{\pi_\phi} v_\top^x, v_\perp^x <_{\pi_\phi} v_c^x <_{\pi_\phi} v_d^x$. and for every two vertex $v_\star^x$ and $v_\star^y$ with $x <_{\text{lex}} y$, we have $v_\star^x <_{\pi_\phi} v_\star^y$. The choice of whether $v_\top^x$ or $v_\perp^x$ comes first will result in the assignment α for ϕ. We enforce the ordering by adding $O(k_\phi^2)$ vertices in the independent set (using the fact that adding $k_\phi + 1$ new vertices incident to exactly the vertices up to and including a vertex v_i prevents swapping it with v_{i+1} in the solution). Now, for every clause $\ell \in \mathcal{C}_\phi$, we add a vertex v_ℓ to the independent set, giving it total size $O(|\mathcal{C}_\phi| + k_\phi^2)$. If the variable x occurs in ℓ, we make v_ℓ incident to v_b^x and v_d^x. If x appears negatively, we also add $(v_\ell, v_\perp^x)$; if positively, we add $(v_\ell, v_\top^x)$ instead. For a variable z which does not occur in a clause ℓ, we make v_ℓ adjacent to v_b^z, v_c^z, and v_d^z. To complete the reduction, we add $k_\phi + 1$ isolated vertices on each end of both the independent set and the clique (total $4k_\phi + 1$). This ensures that no vertex will move from the clique to the independent set partition or vice versa.

Lemma 2 ($\star$). *A 3-CNF-SAT formula ϕ is satisfiable if and only if (G_ϕ, k_ϕ) is a yes-instance to* THRESHOLD EDITING.

Our proof relies on the following observation: when we consider a fixed permutation of the variable gadget vertices (the clique side), the only thing we need to determine for a clause vertex v_ℓ is the *cut-off point*: the point in π_ϕ at which the vertex v_ℓ will no longer have any neighbors. Since no vertex v_i^x swaps places with any other v_j^x for $i, j \in \{a, b, c, d\}$, and no $v_\star^x$ changes with $v_\star^y$ for $x, y \in \mathcal{V}_\phi$, consider a fixed permutation of the variable vertices. We charge the clause vertices with the edits incident to the clause vertex. Since the budget is $k_\phi = |\mathcal{C}| \cdot (3|\mathcal{V}_\phi| - 1)$, and every clause needs at least $3|\mathcal{V}_\phi| - 1$, to obtain a solution, we need to charge every clause vertex with exactly $3|\mathcal{V}_\phi| - 1$ edits.

Lemma 3 ($\star$). *Let (G_ϕ, k_ϕ) be an instance to* THRESHOLD EDITING *constructed from a 3-CNF-SAT formula ϕ. For any clause vertex v_ℓ, at least $3|\mathcal{V}_\phi| - 1$ edges are needed to edit to eliminate all obstructions v_ℓ is a part of.*

Lemma 4 ($\star$). *If there is an editing set F for an instance (G_ϕ, k_ϕ) constructed from a 3-CNF-SAT formula ϕ so that $|F| \leq k_\phi$ and $|F(v_\ell)| = 3|\mathcal{V}_\phi| - 1$, then the $<_{\mathrm{lex}}$-highest vertex connected to v_ℓ corresponds to a variable satisfying the clause ℓ.*

This shows that there is a polynomial time many-one (Karp) reduction from 3SAT to THRESHOLD EDITING implying Theorem 1.

NP-hardness of Chain and Chordal Editing. All formal statements and proofs for editing to chain and chordal graphs are deferred to the full version [9]. To prove CHAIN EDITING is NP-hard, it is useful to define and reduce through two intermediate problems, SPLIT THRESHOLD EDITING and BIPARTITE CHAIN EDITING[1], which each require the editing set to respect the bipartition. We prove that both these problems, as well as CHAIN EDITING, are NP-complete.

To prove that CHORDAL EDITING is NP-hard, we use the observation of Yannakakis that a bipartite graph can be transformed into a chain graph by adding at most k edges if and only if the cobipartite graph formed by completing the two sides can be transformed into a chordal graph by adding at most k edges [25]. We again use an intermediate problem (COBIPARTITE CHORDAL EDITING) which asks for the bipartition to be respected, which we prove is hard via a reduction from BIPARTITE CHAIN EDITING.

Finally, we remark that together with ETH, these results imply that no algorithm solves SPLIT THRESHOLD EDITING, BIPARTITE CHAIN EDITING, CHAIN EDITING, or CHORDAL EDITING in time $2^{o(\sqrt{k})} \cdot \mathrm{poly}(n)$.

4   Quadratic Kernels for Threshold and Chain Editing

We give kernels with a quadratic number of vertices for THRESHOLD COMPLETION, THRESHOLD DELETION, and THRESHOLD EDITING, answering a recent question of Liu, Wang and Guo [14]. Our kernelization algorithms use methods

---

[1] Which has also been referred to as CHAIN EDITING in the literature [13]

similar to those yielding a polynomial kernel for TRIVIALLY PERFECT EDIT-
ING [8]. Further, our techniques extend to give quadratic kernels for all three
analogous problems for editing to chain graphs; these results are deferred to the
full version [9]. Since the class of threshold graphs is closed under taking complements, for every instance (G, k) of THRESHOLD COMPLETION, $(\bar{G}, k)$ is an
equivalent instance of THRESHOLD DELETION (and vice versa), so we restrict
our attention to the completion and editing variants for the remainder of the
section. A *proper kernel* is obtained by a kernelization algorithm which guarantees to never increase the parameter. We may observe that we obtain proper
kernels, since our kernelization algorithms do not modify any edges, and only
change the budget in the case that we discover that we have a no-instance.

Theorem 5 ($\star$). *The following three problems admit kernels with at most $336k^2 +$
$388k + 92$ vertices:* THRESHOLD DELETION, THRESHOLD COMPLETION *and*
THRESHOLD EDITING.

Motivated by the characterization of threshold graphs in Propositions 2 and 3,
our kernels rely on *threshold obstructions*, and consist of a *twin reduction rule*
and an *irrelevant vertex rule*.

The twin reduction rule is based on the following two observations: First, if an
obstruction contains a vertex v one can replace v by a twin not in the obstruction
and obtain a new, isomorphic obstruction. And second, no solution can interact
with all the vertices in a "large" twin class, for some meaning of large, so any
obstruction containing vertices from a large twin class will have to be handled
by edges not incident to the twin class.

An *irrelevant vertex* is one whose removal does not affect the solutions of
the instance. Most of our work consists of proving that we can either find such
an irrelevant vertex or conclude that the graph is small. A key concept of the
irrelevant vertex rule is what will be referred to as a threshold-modulator. It is
a set of vertices X in G of linear size in k, such that for every obstruction H
in G one can add and remove edges in $[X]^2$ and turn H into a non-obstruction.
Our kernelization algorithm will heavily depend on finding a small threshold-
modulator X and the fact that $G - X$ is a threshold graph.

We can in polynomial time either obtain such a set X or conclude correctly
that the instance is a no-instance. The observation that $G - X$ is a threshold
graph will be exploited heavily and we now fix a threshold decomposition $(\mathcal{C}, \mathcal{I})$
of $G - X$. We then prove that the idea of Proposition 1 can be extended to
vertices in $G - X$ when considering their neighborhoods in G. In other words,
the neighborhoods of the vertices in $G - X$ are nested also when considering G.
This immediately yields that the number of subsets Y of X for which there exists
a vertex in $G - X$ having Y as its X-neighborhood is bounded linearly in the
size of X and hence also in k.

We now either conclude that the graph is small or we identify a sequence of
levels in the threshold decomposition containing many vertices, such that all
the clique vertices and all the independent set vertices in the sequence have
identical neighborhoods in X, respectively. The crux is that in the middle of such

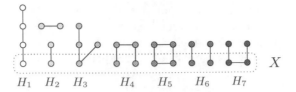

Fig. 2. Some of the intersections of an obstruction with a threshold-modulator X that will not occur by definition. More specifically the ones necessary for the proof of the kernel.

a sequence there will be a vertex that is replaceable by other vertices in every obstruction and hence is irrelevant. Such a sequence is obtained by discarding all levels in the decomposition that are extremal with respect to a subset Y of X, meaning that there either are no levels above or underneath that contain vertices with Y as its X-neighborhood. One can prove that in this process only a quadratic number of vertices are discarded and from this we obtain a kernel.

Since the difference between chain and threshold graphs is in the obstruction set (which is now $\{2K_2, C_3, C_5\}$), we only need to modify the portions of the argument for the threshold kernel that explicitly apply the obstructions—specifically the modulator construction (which will now be of size $5k$), the irrelevant vertex rule, and a lemma giving the nested structure of the neighborhoods in the modulator (which fails in chain graphs, but we prove a weaker version that suffices for our purposes). Combining these, we get quadratic kernels for editing to chain graphs.

Theorem 6 ($\star$). *The following three problems admit kernels with at most $O(k^2)$ vertices:* Chain Deletion, Chain Completion *and* Chain Editing.

5   Subexponential Time Algorithms

Threshold Editing in Subexponential time. In this section we give a subexponential time algorithm for Threshold Editing. We also show that we can modify the algorithm to work with Chain Editing. Combined with the results of Fomin and Villanger [11] and Drange et al. [7], we now have complete information on the subexponentiality of edge modification to threshold and chain graphs. We now aim to prove the following theorem:

Theorem 7. Threshold Editing *admits a* $2^{O(\sqrt{k}\log k)} + \mathrm{poly}(n)$ *subexponential time algorithm.*

The first step of the algorithm is to apply the kernelization algorithm from Theorem 5. Hence, we can assume from this point on that $|V(G)| = O(k^2)$. Recall that the value of the parameter k is not changed during this procedure and hence it is sufficient to aim for a $2^{O(\sqrt{k}\log k)}$ time algorithm also after the kernelization procedure. We fix a solution F of the input instance (G, k) under the assumption that one exists. Observe that $G \triangle F$ is a split graph. The lemma below allows us to iterate over all split partitions of $G \triangle F$ without knowing F in time $2^{O(\sqrt{k} \log k)}$. It follows that we can assume that we have a split partition of $G \triangle F$ at hand and hence we focus on solving instances of SPLIT THRESHOLD EDITING, the problem where the target split partition is fixed.

Lemma 5 ($\star$, Few split partitions). *There is an algorithm that given a graph G and an integer k with $|V(G)| = k^{O(1)}$, can generate a set $\mathcal{P}$ of bipartitions of $V(G)$ such that for every split graph H such that $|E(H) \triangle E(G)| \leq k$ and every split partition (C, I) of H it holds that (C, I) is an element of $\mathcal{P}$. Furthermore, the algorithm terminates in $2^{O(\sqrt{k} \log k)}$ time.*

A crucial concept of our algorithm is the one of *cheap* and *expensive* vertices. A cheap vertex is a vertex that is incident to at most $2\sqrt{k}$ edges of F and every vertex that is not cheap, is expensive. It follows immediately that we can guess the expensive vertices and the neighborhoods of cheap vertices in time $2^{O(\sqrt{k} \log k)}$.

Splitting Pairs and Unbreakable Segments.

Definition 3 (Splitting pair). *Let G be a graph, k an integer, F a solution of (G, k) and $(\mathcal{C}, \mathcal{I})$ a threshold decomposition of $G \triangle F$. We then say that the vertices $u \in I_a$ and $v \in C_b$ form a splitting pair if (i) $a < b$, (ii) u and v are cheap, and (iii) $\cup_{a < i < b} L_i$ consists of only expensive vertices. Recall from Definition 1 that $L_i = C_i \cup I_i$.*

Definition 4 (Unbreakable). *Let G be a graph, k an integer, F a solution of (G, k) and $(\mathcal{C}, \mathcal{I})$ a threshold decomposition of $G \triangle F$. We then say that a sequence of levels $(C_a, I_a), (C_{a+1}, I_{a+1}), \ldots, (C_b, I_b)$ is an unbreakable segment if there is no splitting pair in the vertex set $\cup_{i \in [a,b]} (C_i \cup I_i)$.*

Furthermore, we say that an instance (G, k) is unbreakable if there exists an optimal solution F and a threshold decomposition $(\mathcal{C}, \mathcal{I})$ of $G \triangle F$ such that the entire decomposition is an unbreakable segment. We also say that such a decomposition is a witness of G being unbreakable.

Definition 5. *Let G be a graph and $(\mathcal{C}, \mathcal{I})$ a threshold decomposition of $G \triangle F$ for some solution F. Then we say that i is a transfer level if*

- *for every $j > i$ it holds that C_j contains no cheap vertices and*
- *for every $j < i$ it holds that I_j contains no cheap vertices.*

Lemma 6 ($\star$). *Let (G, k) be a yes instance of SPLIT THRESHOLD EDITING with solution F such that G is unbreakable and $(\mathcal{C}, \mathcal{I})$ a witness. Then there is a transfer level in $(\mathcal{C}, \mathcal{I})$.*

Lemma 7 ($\star$). *Let (G, k) be an instance of SPLIT THRESHOLD EDITING such that G is unbreakable and $(\mathcal{C}, \mathcal{I})$ a witness of this. Then the number of levels in $(\mathcal{C}, \mathcal{I})$ is at most $2\sqrt{k} + 1$.*

Lemma 8 ($\star$). *Let (G, k) be an instance of* SPLIT THRESHOLD EDITING *such that G is unbreakable, $(\mathcal{C}, \mathcal{I})$ is a witness of this and F a corresponding solution. If X is the set of cheap vertices in G then $(G \triangle F)[X]$ forms a complete split graph.*

We will now describe the algorithm `unbreakAlg`. It takes as input an instance $(G, (C, I), k)$ of SPLIT THRESHOLD EDITING, with the assumption that G is unbreakable and has split partition (C, I), and returns either an optimal solution F for (G, k) where $|F| \le k$ or correctly concludes that (G, k) is a no-instance. Assume that (G, k) is a yes-instance. Then there exists an optimal solution F and a threshold decomposition $(\mathcal{C}, \mathcal{I})$ of $G \triangle F$ that is a witness of G being unbreakable. First, we guess the number of levels ℓ in the decomposition, and by Lemma 7, we have that $\ell \in [0, 2\sqrt{k} + 1]$ and the transfer level $t \in [0, \ell]$. Then we guess where the at most $2\sqrt{k}$ vertices that are expensive in G are positioned in $(\mathcal{C}, \mathcal{I})$. Observe that from this information we can obtain all edges between expensive vertices in F. Finally, we put every cheap vertex in the level that minimizes the cost of fixing its adjacencies into the expensive vertices while respecting that t is the transfer level. From this information we can obtain all adjacencies between cheap and expensive vertices in F. Since the cheap vertices induce a complete split graph we have complete information on F and hence may return F.

Lemma 9 ($\star$). *Given an instance (G, k) of* SPLIT THRESHOLD EDITING *with G being unbreakable,* `unbreakAlg` *either gives an optimal solution or correctly concludes that (G, k) is a no-instance in time $2^{O(\sqrt{k} \log k)}$.*

Divide and Conquer. We are now ready to sketch the main algorithm `solveAlg`. It takes as input a split graph G with a fixed split partition (C, I), an integer k and a set of vertices S and either returns an optimal solution of $G[S]$ or correctly concludes that $(G[S], k)$ is a no-instance. Before we continue we fix a threshold decomposition $(\mathcal{C}, \mathcal{I})$ of $G \triangle F$ that respects the split partition of G. Every S will be constructed under the assumption that there exists integers a and b such that $S = \cup_{a \le i \le b} L_i$.

The idea is to use splitting pairs to decompose $G[S]$ into parts that we can solve independently. First, we consider the case that $G[S]$ is unbreakable. We apply `unbreakAlg` and obtain an optimal solution F_1 under the assumption that $G[S]$ is unbreakable. It remains to consider the case when there is a splitting pair. First, we try every pair of cheap vertices u, v with $u \in C$ an $v \in I$ as our splitting pair. We assume that u, v form the upper most splitting pair in $G[S]$ with respect to $(\mathcal{C}, \mathcal{I})$. We also guess the neighborhood of u and v in $G \triangle F$.

The next step is to guess the expensive vertices X in between the levels of u and v and how they are positioned in the decomposition. One can observe that u and v together with their neighborhoods and X is sufficient information to partition $(\mathcal{C}, \mathcal{I})$ into three parts. Namely *(i)* the set of vertices U that are above or at the same level as u in the decomposition, *(ii)* the set of expensive vertices X, and *(iii)* the remaining set of vertices R, the vertices below or at the same level as v. By the selection of u and v it follows that $G[U]$ is unbreakable and hence can be solved by `unbreakAlg`. We then apply `solveAlg` recursively on R. Observe that the edges between the different parts are decided by the definition of a threshold decomposition. Hence, we can construct the resulting F based on our u and v, their neighborhoods and X. We then minimize over all these choices and return the smallest solution we find, also considering F_1.

To obtain the required running time we apply memoization to S, meaning that when we are finished with a call to `solveAlg` we store the solution for this specific S. Then, the next time a similar call is made we can immediately return the solution. Since every splitting pair, together with its neighborhood and the expensive vertices in between uniquely defines a partition of $V(G)$ and there are only $2^{O(\sqrt{k}\log k)}$ such combinations it follows that `solveAlg` is applied to at most $2^{O(\sqrt{k}\log k)}$ many different sets S. Due to the time complexity of the subprocedures of the algorithm, the correctness of Theorem 7 follows.

Chain Editing. By a more involved approach than for threshold graphs, we can with $2^{O(\sqrt{k}\log k)}$ delay assume that we know the bipartition of an optimal solution. We then make one of the sides into a clique and do a query to our threshold algorithm to obtain the following result:

Theorem 8 $(\star)$. CHAIN EDITING *is solvable in time* $2^{O(\sqrt{k}\log k)} + \mathrm{poly}(n)$.

6   Conclusion

In this paper we showed that the problems of editing edges to obtain either a threshold graph or a chain graph are NP-complete. The latter resolves a conjecture of Natanzon et al. [21] and both results answer open questions from Sharan [24], Burzyn et al. [3], and Mancini [18].

On the positive side, we show that both THRESHOLD EDITING and CHAIN EDITING admit quadratic kernels, i.e., given a graph (G, k), we can in polynomial time find an equivalent instance (G', k) where $|V(G')| = O(k^2)$, and furthermore, G' is an induced subgraph of G. We also show that these results hold for the deletion and completion variants as well, and these results answer open questions by Liu et al. in a recent survey on kernelization complexity of graph modification problems [14]. Finally we show that both problems admit subexponential algorithms of time complexity $2^{O(\sqrt{k}\log k)} + \mathrm{poly}(n)$. This answers a recent open question by Liu et al. [16].

In addition, we give a proof for the NP-hardness of CHORDAL EDITING which has been announced several places but which the authors have been unable to find. However, our NP-completeness proof for CHORDAL EDITING suffers a quadratic blow-up from 3SAT, i.e., $k = \Theta(|\phi|^2)$, so we cannot get better than $2^{o(\sqrt{k})} \cdot \mathrm{poly}(n)$ lower bounds from this technique. The current best algorithm for CHORDAL EDITING[2] runs in time $2^{O(k\log k)} \cdot \mathrm{poly}(n)$ [5], and so this leaves a big gap. It would be interesting to see if we can achieve tighter lower bounds,

---

[2] Here, the authors take CHORDAL EDITING to allow vertex deletions.

e.g., $2^{o(k)} \cdot \mathrm{poly}(n)$ time lower bounds for CHORDAL EDITING assuming ETH together with a $2^{O(k)} \cdot \mathrm{poly}(n)$ time algorithm.

References

1. Brandes, U.: Social network algorithmics. ISAAC, Invited talk (2014)
2. Brandstädt, A., Le, V.B., Spinrad, J.P.: Graph Classes. A Survey. SIAM, Philadelphia (1999)
3. Burzyn, P., Bonomo, F., Durán, G.: NP-completeness results for edge modification problems. Discrete Applied Mathematics 154(13), 1824–1844 (2006)
4. Cai, L.: Fixed-parameter tractability of graph modification problems for hereditary properties. Information Processing Letters 58(4), 171–176 (1996)
5. Cao, Y., Marx, D.: Chordal editing is fixed-parameter tractable. In: STACS. LIPIcs, vol. 25, pp. 214–225 (2014)
6. Dehne, F., Langston, M., Luo, X., Pitre, S., Shaw, P., Zhang, Y.: The cluster editing problem: Implementations and experiments. In: IPEC (2006)
7. Drange, P.G., Fomin, F.V., Pilipczuk, M., Villanger, Y.: Exploring subexponential parameterized complexity of completion problems. In: STACS (2014)
8. Drange, P.G., Pilipczuk, M.: A polynomial kernel for trivially perfect editing. In: ESA (to appear, 2015)
9. Drange, P.G., Dregi, M.S., Lokshtanov, D., Sullivan, B.D.: On the threshold of intractability. CoRR, abs/1505.00612 (2015)
10. Feder, T., Mannila, H., Terzi, E.: Approximating the minimum chain completion problem. Information Processing Letters 109(17), 980–985 (2009)
11. Fomin, F.V., Villanger, Y.: Subexponential parameterized algorithm for minimum fill-in. SIAM J. Comput. 42(6), 2197–2216 (2013)
12. Golumbic, M.C.: Algorithmic Graph Theory and Perfect Graphs. Academic Press, New York (1980)
13. Guo, J.: Problem kernels for NP-complete edge deletion problems: Split and related graphs. In: Tokuyama, T. (ed.) ISAAC 2007. LNCS, vol. 4835, pp. 915–926. Springer, Heidelberg (2007)
14. Liu, Y., Wang, J., Guo, J.: An overview of kernelization algorithms for graph modification problems. Tsinghua Science and Technology 19(4), 346–357 (2014)
15. Liu, Y., Wang, J., Guo, J., Chen, J.: Complexity and parameterized algorithms for cograph editing. TCS 461, 45–54 (2012)
16. Liu, Y., Wang, J., You, J., Chen, J., Cao, Y.: Edge deletion problems: Branching facilitated by modular decomposition. Theoretical Computer Science 573, 63–70 (2015)
17. Mahadev, N., Peled, U.: Threshold graphs and related topics, vol. 56. Elsevier (1995)
18. Mancini, F.: Graph modification problems related to graph classes. PhD thesis, University of Bergen (2008)
19. Nastos, J., Gao, Y.: Familial groups in social networks. Social Networks 35(3), 439–450 (2013)
20. Natanzon, A.: Complexity and approximation of some graph modification problems. PhD thesis, Tel Aviv University (1999)
21. Natanzon, A., Shamir, R., Sharan, R.: Complexity classification of some edge modification problems. Discrete Applied Mathematics 113(1), 109–128 (2001)

22. Schoch, D., Brandes, U.: Stars, neighborhood inclusion, and network centrality. In: SIAM Workshop on Network Science (2015)
23. Shamir, R., Sharan, R., Tsur, D.: Cluster graph modification problems. Discrete Applied Mathematics 144(1), 173–182 (2004)
24. Sharan, R.: Graph modification problems and their applications to genomic research. PhD thesis, Tel-Aviv University (2002)
25. Yannakakis, M.: Computing the minimum fill-in is NP-complete. SIAM Journal on Algebraic and Discrete Methods 2(1), 77–79 (1981)

A Polynomial Kernel
for Trivially Perfect Editing*

Pål Grønås Drange[1] and Michał Pilipczuk[2]

[1] Dept. Informatics, Univ. Bergen, Norway
pal.drange@ii.uib.no
[2] Inst. Informatics, Univ. Warsaw, Poland
michal.pilipczuk@mimuw.edu.pl

Abstract. We give a kernel with $O(k^7)$ vertices for TRIVIALLY PERFECT EDITING, the problem of adding or removing at most k edges in order to make a given graph trivially perfect. This answers in affirmative an open question posed by Nastos and Gao (Social Networks, 35(3):439–450, 2013) and by Liu, Wang, and Guo (Tsinghua Science and Technology, 19(4):346–357, 2014). Using our technique one can also obtain kernels of the same size for the related problems, TRIVIALLY PERFECT COMPLETION and TRIVIALLY PERFECT DELETION.

We complement our study of TRIVIALLY PERFECT EDITING by proving that, contrary to TRIVIALLY PERFECT COMPLETION, it cannot be solved in time $2^{o(k)} \cdot n^{O(1)}$ unless the Exponential Time Hypothesis fails. In this manner we complete the picture of the parameterized and kernelization complexity of the classic edge modification problems for the class of trivially perfect graphs.

1 Introduction

Graph modification problems form an important class of problems, where the task is to modify a given graph using a constrained number of modifications in order to make it satisfy some property Π, or equivalently belong to the class $\mathcal{G}$ of graphs satisfying Π. As far as allowed modifications are concerned, probably the most popular variant is to delete vertices only (vertex deletion problems), but there is also much work on modifying the edge set of the graph. Here, there are three natural classes of problems: deletion problems (deleting the least number of edges), completion problems (adding the least number of edges) and editing problems (performing the least number of edge additions or deletions).

In this paper we study edge modification problems from the point of view of parameterized complexity. A parameterized problem is called *fixed-parameter tractable* if it can be solved in time $f(k) \cdot n^{O(1)}$ for some computable function f,

---

* Pilipczuk currently holds a post-doc position at Warsaw Center of Mathematics and Computer Science and is supported by the Polish National Science Centre grant DEC-2013/11/D/ST6/03073. This work has received funding from ERC grant n. 267959 (Drange and Pilipczuk, while the latter was affiliated with Univ. of Bergen).

© Springer-Verlag Berlin Heidelberg 2015
N. Bansal and I. Finocchi (Eds.): ESA 2015, LNCS 9294, pp. 424–436, 2015.
DOI: 10.1007/978-3-662-48350-3_36

where n is the size of the input and k is its parameter. In our case, the natural parameter k is the allowed number of modifications. Cai [3] made a simple observation that for all the aforementioned graph modification problems there is a simple branching algorithm running in time $c^k n^{O(1)}$ for some constant c, as long as $\mathcal{G}$ is *characterized by a finite set of forbidden induced subgraphs*: there is a finite list of graphs $H_1, H_2, \ldots, H_p$ such that any graph G belongs to $\mathcal{G}$ if and only if G does not contain any H_i as an induced subgraph. Although many studied graph classes satisfy this property, there are important examples, like chordal or interval graphs, that are outside this regime.

Hence, the parameterized analysis of modification problems for graph classes characterized by a finite set of forbidden induced subgraphs focused on studying the design of *polynomial kernelization algorithms*[1]; Such an algorithm is required, given an input instance (G, k) of the problem, to preprocess it in polynomial time and obtain an equivalent output instance (G', k'), where $|G'|, k' \le p(k)$ for some polynomial p. That is, the question is the following: can you, using polynomial time preprocessing only, bound the size of the tackled instance by a polynomial function depending only on k?

For vertex deletion problems, as long as $\mathcal{G}$ is characterized by a finite set of forbidden induced subgraphs, the task is to hit all the copies of these subgraphs (so-called *obstacles*) that are originally contained in the graph. Hence, one can construct a simple reduction to the d-HITTING SET problem for a constant d depending on $\mathcal{G}$, and use the classic $O(k^d)$ kernel for the latter that is based on the sunflower lemma [1]. For edge modification problems, however, this approach fails utterly: every edge addition and deletion can create new obstacles, and thus it is not sufficient to hit only the original ones. For this reason, edge modification problems behave counter-intuitively with respect to polynomial kernelization, and up to recently very little was known about their complexity.

On the positive side, kernelization of edge modification problems for well-studied graph classes was explored by Guo [11], who showed that four problems: THRESHOLD COMPLETION, SPLIT COMPLETION, CHAIN COMPLETION, and TRIVIALLY PERFECT COMPLETION, all admit polynomial kernels. However, the study took a turn for the interesting when Kratch and Wahlström [14] showed that there is a graph H on 7 vertices, such that the deletion problem to H-free graphs (the class of graphs not admitting H as an induced subgraph) does not admit a polynomial kernel, unless the polynomial hierarchy collapses. This shows that the subtle differences between edge modification and vertex deletion problems have tremendous impact on the kernelization complexity.

The line of research initiated by Kratch and Wahlström was continued by Guillemot et al. [10], who showed that both for the class of P_ℓ-free graphs (for $\ell \ge 7$) and for the class of C_ℓ-free graphs (for $\ell \ge 4$), the edge deletion problems probably do not have polynomial kernelization algorithms. They simultaneously gave a cubic kernel for the COGRAPH EDITING problem, the problem of editing to

---

[1] It is well-known that a problem is fixed-parameter tractable if and only if the problem admits a kernelization algorithm, hence *polynomial kernelization* is a natural next step.

a graph without induced paths on four vertices. These results were later improved by Cai and Cai [4], who tried to obtain a complete dichotomy of the kernelization complexity of edge modification problems for classes of H-free graphs, for every graph H. The project has been very successful—the question is settled for all 3-connected graphs H as well as all but a finite number of trees. In particular, it turns out that the existence of a polynomial kernel for any of H-FREE EDITING, H-FREE EDGE DELETION, or H-FREE COMPLETION problems is in fact a very rare phenomenon, and basically happens only for some specific small graphs H. For instance, for H being a path or a cycle, the aforementioned three problems admit polynomial kernels if and only if H has at most three edges.

In this paper we study the TRIVIALLY PERFECT EDITING problem, which, for a given graph G and integer k, asks whether G can be transformed into a trivially perfect graph by performing at most k edge additions or removals. Recall that a graph is trivially perfect if it does not contain a P_4 or a C_4 as an induced subgraph. However, there is also an equivalent structural definition that shows trivially perfect graphs model tree-like hierarchical structures; more precisely, they are exactly ancestor-descendant closures of rooted forests. Interest in trivially perfect graphs started with the attempts to prove the strong perfect graph theorem. Recently, a new source of motivation has grown, with the realization that trivially perfect graphs are related to the width parameter *treedepth* (called also vertex ranking number, ordered chromatic number, and minimum elimination tree height). Although it had been known that both the completion and the deletion problem for trivially perfect graphs are NP-hard, it was open for a long time whether the editing version is NP-hard as well [2].

This question was answered very recently by Nastos and Gao [16], who showed that the problem is indeed NP-hard. Actually, the work of Nastos and Gao focuses on exhibiting applications of trivially perfect graphs in social network theory, since this graph class may serve as a model for *familial groups*, that is, communities in social networks showing a hierarchical nature. Specifically, the *editing number* to a trivially perfect graph[2] can be used as a measure of how much a social network resembles a collection of hierarchies. Nastos and Gao also ask whether it is possible to obtain a polynomial kernelization algorithm for this problem. The question about the existence of a polynomial kernel for TRIVIALLY PERFECT EDITING was then restated in a recent survey by Liu, Wang, and Guo [15], which *nota bene* contains a comprehensive overview of the current status of the research on the kernelization complexity of graph modification problems.

Trivially perfect graphs were also studied from the point of view of *subexponential parameterized algorithms*: FPT algorithms with running time $2^{o(k)} \cdot n^{O(1)}$. It has been recently discovered that such algorithms exist for completion problems to multiple subclasses of chordal graphs, and trivially perfect graphs are among them, as proven by a superset of the current authors [7]. On the other hand, TRIVIALLY PERFECT DELETION does not enjoy the existence of such an algorithm unless the Exponential Time Hypothesis (ETH, see [12]) fails [7]. For TRIVIALLY PERFECT EDITING no such analysis was done up to this work.

---

[2] Nastos and Gao use the terminology *quasi-threshold* instead of trivially perfect.

Our Contribution. We answer the question of Nastos and Gao [16] and of Liu, Wang, and Guo [15] in affirmative by proving the following theorem.

Theorem 1. *The problem* TRIVIALLY PERFECT EDITING *admits a proper kernel with $O(k^7)$ vertices.*

Here, we say that a kernel (kernelization algorithm) is *proper* if it can only decrease the parameter, i.e., the output parameter k' satisfies $k' \leq k$.

To prove Theorem 1, we employ an extensive analysis of the tackled instance, based on the structural definition of trivially perfect graphs. The main idea is to construct a small *vertex modulator*, a set of vertices whose removal results in obtaining a trivially perfect graph. However, since we are allowed only edge deletions and additions, this modulator just serves as a tool for exposing the structure of the instance. More specifically, we greedily pack disjoint obstructions into a set X, whose size can be guaranteed to be at most $4k$, with the condition that to get rid of each of these obstructions, at least one edge must be edited inside the modulator per obstruction. Having obtained such a modulator, the rest of the graph, $G - X$, is trivially perfect, and we may apply the structural view on trivially perfect graphs to find irrelevant parts that can be reduced.

While the modulator technique is commonly used in kernelization, the new insight in this work is as follows. Since we work with an edge modification problem, we can be less restrictive about when an obstacle can be packed into the modulator. For example, the obstacle does not need to be completely vertex-disjoint with the so far constructed X; sharing just one vertex is still allowed. This observation allows us to reason about the adjacency structure between X and $V(G) \setminus X$, which is of great help when identifying irrelevant parts.

By modifying our algorithm slightly, we also obtain polynomial kernels for TRIVIALLY PERFECT DELETION and TRIVIALLY PERFECT COMPLETION.

Theorem 2. *The problems* TRIVIALLY PERFECT DELETION *and* TRIVIALLY PERFECT COMPLETION *admit kernels with $O(k^7)$ vertices.*

To the best of our knowledge, no polynomial kernel for TRIVIALLY PERFECT DELETION was known so far. For TRIVIALLY PERFECT COMPLETION, a cubic kernel was shown earlier by Guo [11]. Unfortunately, the work of Guo [11] is published only as a conference extended abstract, where it is only sketched how the approach yielding a quartic kernel for SPLIT DELETION could be used to obtain a cubic kernel for TRIVIALLY PERFECT COMPLETION. The details of this kernelization algorithm are deferred to the full version, which, alas, has not appeared. For this reason, we believe that our proof of Theorem 2 fills an important gap in the literature—the polynomial kernel for TRIVIALLY PERFECT COMPLETION is an important ingredient of the subexponential parameterized algorithm for this problem [7].

Finally, we show that TRIVIALLY PERFECT EDITING, in addition to being NP-complete, cannot admit a subexponential parameterized algorithm, provided that the Exponential Time Hypothesis holds.

Theorem 3. TRIVIALLY PERFECT EDITING *is* NP-*complete, and under ETH cannot be solved in time* $2^{o(k)}n^{O(1)}$ *or* $2^{o(n+m)}$, *even on graphs with maximum degree* 4.

In other words; the familial group measure cannot be computed in time subexponential in terms of the value of the measure. This stands in contrast with TRIVIALLY PERFECT COMPLETION that admits a subexponential parameterized algorithm [7], and shows that TRIVIALLY PERFECT EDITING is more similar to TRIVIALLY PERFECT DELETION, for which a similar lower bound has been proved earlier by Drange et al. [7]. In fact, our reduction can be used as an alternative proof of hardness of TRIVIALLY PERFECT DELETION as well. We remark that our reduction also refutes the existence of a subexponential parameterized algorithm for COGRAPH EDITING, which to the best of our knowledge was not yet known.

Let us note that the NP-hardness reduction for TRIVIALLY PERFECT EDITING presented by Nastos and Gao [16] cannot be used to prove nonexistence of a subexponential parameterized algorithm, since it involves a cubic blow-up of the parameter. To prove Theorem 3, we resort to the technique used for similar hardness results by Komusiewicz and Uhlmann [13] and by Drange et al. [7].

Outline. In this extended abstract we focus on sketching the proof of the main result, i.e., Theorem 1; all the technical proofs have been deferred to the full version of the paper, a preprint of which is available online [8]. In Section 3 we gather some conclusions and open problems. Proofs of Theorems 2 and 3 are also deferred to the full version. We also remark that the full version contains a broader overview of the context of this work in its introductory section.

2   Kernel for Trivially Perfect Editing: Proof of Theorem 1

Trivially Perfect Graphs. A graph G is trivially perfect if and only if it does not contain a C_4 or a P_4 as an induced subgraph. A superset of the current authors [7] proposed the following notion of a structural decomposition for trivially perfect graphs, which exposes their hierarchical nature. In the following, for a rooted tree T and vertex $t \in V(T)$, by T_t we denote the subtree of T rooted at t. The *universal clique* of a graph G is the unique set of universal vertices, and for a set of vertices $X \subseteq V(G)$, we write $G[X]$ to denote the *graph induced* on X.

Definition 1 (Universal clique decomposition, [7]). *A* universal clique decomposition (UCD) *of a connected graph* G *is a pair* $\mathcal{T} = (T = (V_T, E_T), \mathcal{B} = \{B_t\}_{t \in V_T})$, *where* T *is a rooted tree and* $\mathcal{B}$ *is a partition of the vertex set* $V(G)$ *into disjoint nonempty subsets, such that*

- *if* $vw \in E(G)$ *and* $v \in B_t, w \in B_s$, *then either* $t = s$, t *is an ancestor of* s *in* T, *or* s *is an ancestor of* t *in* T, *and*
- *for every node* $t \in V_T$, *the set of vertices* B_t *is the universal clique of* $G[\bigcup_{s \in V(T_t)} B_s]$.

Lemma 1 ([7]). *A connected graph G admits a universal clique decomposition if and only if it is trivially perfect. Moreover, such a decomposition is unique up to isomorphisms.*

We can trivially extend Lemma 1 to disconnected graphs by allowing the decomposition to have a shape of a rooted forest, instead of just a rooted tree.

Overview of the Proof. We first give an overview of the structure of the proof. As usual, the kernelization algorithm will be given as a sequence of *data reduction rules*: simple preprocessing procedures that, if applicable, simplify the instance at hand. For each rule we prove two claims: (a) that applicability of the rule can be recognized in polynomial time, and (b) that the rule is *safe*, i.e., the resulting instance is equivalent to the original one. At the end of the proof we argue that if no rule is applicable, then the size of the instance is bounded by $O(k^7)$. Some rules will decrement the budget k for edits; if this budget drops below zero, we conclude that we are dealing with a no-instance, so we immediately terminate the algorithm and provide a constant-size trivial no-instance as the kernel.

First, we give some preliminary basic rules, which mostly deal with situations where we can find a large number of induced C_4s and P_4s in the graph (henceforth called *obstacles*), which share only one edge or non-edge. We then infer that this edge or non-edge has to be included in any editing set of size at most k, and hence we can perform the necessary edit and decrement the budget.

Then, the idea is to apply a greedy algorithm that iteratively packs "edit-disjoint" induced C_4s and P_4s in the graph. If we are able to pack more than k of them, then this certifies that the considered instance does not have a solution, and we can terminate the algorithm. Hence, if X is the union of vertex sets of the packed obstacles, then $|X| \leq 4k$ and $G - X$ is a trivially perfect graph. Finding such a set X, which we call a TP-modulator, imposes a great deal of structure on the instance, and is the key for further analysis of irrelevant parts of the input.

In this paper we introduce a new twist to the modulator technique; Namely, we observe that since we consider edge editing problems, the packed obstacles do not have to be entirely vertex-disjoint, but the next obstacle can be packed even if it shares one vertex with the union of vertex sets of the previous obstacles; In some limited cases even having two vertices in common is permitted. Thus, the obtained modulator X has the property that not only is there no obstacle in the graph G that is vertex-disjoint with X, but even the existence of obstacles sharing one vertex with X is forbidden. This simple observation enables us to reason about the adjacency structure between X and $V(G) \setminus X$. In Lemma 3 we analyze this structure in order to prove the most important technical result of the proof: The number of subsets of X that are neighborhoods within X of vertices from $V(G) \setminus X$ is bounded polynomially in k.

We then proceed to analyze the trivially perfect graph $G - X$. Having the polynomial bound on the number of neighborhoods within X, we can locate in the UCD of $G - X$ a polynomial (in k) number of *important bags*, where something interesting from the point of view of X-neighborhoods happens. The parts between the important bags have very simple structure. They are either *tassels*: sets of "trees" hanging below some important bag, where each such tree is a module in the whole graph G; or *combs*: long "paths" stretched between two important bags where all the vertices of subtrees attached to the path have exactly the same neighborhood in X. Tassels and combs are treated differently: Large tassels contain large trivially perfect modules in G that can be reduced quite easily, however for combs we need to devise a quite complicated irrelevant vertex rule that locates a vertex that can be safely discarded in a long comb. Since the number of tassels and combs is polynomial in k, reducing the size of each of them to polynomial in k concludes the construction of the kernel.

Simple Reduction Rules. We first apply two "sunflower" reduction rules, which identify single edges or non-edges that have to be edited in any solution of size at most k. In the following, $\overline{G}$ denotes the complement of graph G.

Rule 1. *For an instance (G, k) with $uv \notin E(G)$, if there is a matching of size at least $k + 1$ in $\overline{G}[N(u) \cap N(v)]$, then add edge uv to G and decrease k by one, i.e., return the new instance $(G + uv, k - 1)$.*

Rule 2. *For an instance (G, k) with $uv \in E(G)$ and $N_1 = N(u) \setminus N[v]$ and $N_2 = N(v) \setminus N[u]$, if there is a matching in $\overline{G}$ between N_1 and N_2 of size at least $k + 1$, then delete edge uv from G and decrease k by one, i.e., return the new instance $(G - uv, k - 1)$.*

The safeness of both these rules is easy to argue. We apply these rules exhaustively and then proceed to the further analysis of the instance, which we henceforth call *reduced*. That is, on this instance, Rules 1 and 2 can not be applied. This property will be helpful in the structural analysis.

Modulator Construction. We proceed to the construction of a small modulator whose *raison d'être* is to expose structure in G. We say that a subset $W \subseteq V(G)$ with $|W| = 4$ is an *obstruction* if $G[W]$ is isomorphic to a C_4 or a P_4. Formally, our modulator will be compliant with the following definition.

Definition 2 (TP-modulator). *Let (G, k) be an instance of* TRIVIALLY PERFECT EDITING. *A subset $X \subseteq V(G)$ is a TP-modulator if for every obstruction W the following holds: (i) $|W \cap X| \geq 2$, and (ii) if $|W \cap X| = 2$, then it cannot be that $G[W]$ is a C_4 of the form $x_1 y_1 y_2 x_2 x_1$ or a P_4 of the form $x_1 y_1 y_2 x_2$, where $W \cap X = \{x_1, x_2\}$. We call a TP-modulator X small if $|X| \leq 4k$.*

In particular, observe that for a TP-modulator X there is no obstacle disjoint with X, so $G - X$ is trivially perfect. The following result shows that we can efficiently compute a small TP-modulator for the purpose of further analysis.

Lemma 2. *Given an instance (G, k) for* TRIVIALLY PERFECT EDITING, *we can in polynomial time construct a small TP-modulator $X \subseteq V(G)$, or correctly conclude that (G, k) is a no-instance.*

The proof of Lemma 2 relies on the observation that if we start with $X = \emptyset$ and iteratively pack into X an obstacle W that contradicts the properties asserted by Definition 2, then each packed obstacle needs at least one additional edit to be broken, and hence we cannot pack more than k of them in a yes-instance. By applying Lemma 2, from now on we assume that we are given a small TP-modulator X in G.

Having constructed the modulator, we proceed to the analysis of the adjacency structure between X and $G \setminus X$. For a vertex $v \in V(G) \setminus X$, the X-*neighborhood* of v, denoted $N_G^X(v)$, is the set $N_G(v) \cap X$. The family of X-neighborhoods of G is the set $\{N_G^X(v) : v \in V(G) \setminus X\}$.

Lemma 3. *If (G, k) is a reduced instance for* Trivially Perfect Editing *and X is a small TP-modulator, then the number of different X-neighborhoods is at most $O(k^4)$.*

The proof of Lemma 3 is the cornerstone of our approach. The first step is to understand how, for a pair of vertices of $u, v \in V(G) \setminus X$, the relative positioning of u and v in the UCD of $G - X$ affects the relation between their X-neighborhoods. It appears that this can be very well understood using the maximality properties of the TP-modulator X. Then we move aside $O(k^4)$ "outlier" vertices of $G - X$ that have extraordinary X-neighborhoods, and argue that the remaining X-neighborhoods form within X a set system that has a certain well-defined property; we call such set systems *TP-set systems*. Using a purely combinatorial argumentation we prove that the cardinality of a TP-set system on a universe U cannot exceed $|U| + 1$, and hence the bound of Lemma 3 follows.

Locating Important Bags. Recall that we have just analyzed the structure of neighborhoods that nodes from $V(G) \setminus X$ have in X. Next, our goal is to perform the symmetric analysis: to understand, how the neighborhood of a fixed $x \in X$ in $V(G) \setminus X$ looks like. Let $\mathcal{T} = (T, \mathcal{B})$ be the UCD of $G - X$. First, we aim to locate a family $I \subseteq V(T)$ of $O(k)$ *important bags*, where some non-trivial behavior w.r.t. the neighborhoods of vertices of X happens. Then, we perform the lowest common ancestor-closure on the set I (see Definition 4), thus increasing its size to at most twice. After this step, all the connected components of $T - I$ have very simple structure from the point of view of their neighborhoods in X. As there are only $O(k)$ such components, we will be able to kernelize them separately.

The following definition and lemma explains what are the types of neighborhoods that vertices of X can have in $V(G) \setminus X$. In the following we denote by $\preceq$ the partial order on the vertices of the forest T induced by the ancestor-descendant relation, i.e., $s \preceq t$ if and only if s is an ancestor of t in T (possibly $s = t$).

Definition 3 (Type 0, 1, and 2 neighborhoods). *Let $x \in X$ be any vertex and consider $U_x = N(x) \setminus X$. We say that U_x is a neighborhood of* Type 0 *if U_x is the union of the vertex sets of a collection of connected components of $G - X$;* Type 1 *if there exists a node $t_x \in V(T)$ such that $\bigcup_{s \prec t_x} B_s \subseteq U_x \subseteq \bigcup_{s \preceq t_x} B_s$. In other words, U_x consists of all the vertices contained in bags on the path from t_x*

to the root of its subtree in T, where some vertices of B_{t_x} itself may be excluded; and of Type 2 if there exists a node $t_x \in V(T)$ and a collection $\mathcal{L}_x$ of subtrees of T rooted at children of t_x such that $U_x = \bigcup_{s \preceq t_x} B_s \cup \bigcup_{S \in \mathcal{L}_x} \bigcup_{s \in V(S)} B_s$. In other words, U_x is formed by all the vertices contained in bags on the path from t_x to the root of its subtree in T, plus a selection of subtrees rooted in the children of t_x, where the vertices appearing in the bags of each such subtree are either all included in U_x or all excluded from U_x.

Lemma 4. *Let $x \in X$ be any vertex and consider $U_x = N(x) \setminus X$. Then U_x is of Type 0, 1 or 2.*

The proof of Lemma 4 again exploits the maximality properties of TP-modulator X in order to reason about the structure of neighborhoods of vertices from X in the UCD of $G - X$. Clearly, for every $x \in X$ we can in polynomial time analyze U_x and recognize it as a neighborhood of Type 0, 1, or 2. Let I_0 be the set of nodes t_x for vertices $x \in X$ for which U_x is of Type 1 or 2. To simplify the structure of $T - I_0$, we perform the lowest common ancestor-closure operation on I_0. The following variant of this operation, as well as its two basic properties, are taken verbatim from the work of Fomin et al. [9].

Definition 4 ([9]). *For a rooted tree T and vertex set $M \subseteq V(T)$ the lowest common ancestor-closure (LCA-closure) is obtained by the following process. Initially, set $M' = M$. Then, as long as there are vertices x and y in M' whose least common ancestor w is not in M', add w to M'. When the process terminates, output M' as the LCA-closure of M.*

Lemma 5 ([9]). *Let T be a tree, $M \subseteq V(T)$ and $M' = \text{LCA-closure}(M)$. Then $|M'| \leq 2|M|$ and for every connected component C of $T - M'$, $|N(C)| \leq 2$.*

Construct the set I by taking LCA-closure(I_0) and adding the root of every connected component of T that contains a bag of I_0 (provided it is not already included). The nodes from I will be *marked* as *important nodes*, or *important bags*. From Lemma 5 it follows that $|I| \leq 3|X| \leq 12k$, and by the construction we infer that every connected component C of $T - I$ is of one of the following three forms:

- C is not adjacent to any node of I, and is thus simply a connected component of T that does not contain any important bag.
- C is adjacent to one node a of I, and it is a subtree rooted at a child of a.
- C is adjacent to two nodes a and b of I such that a is an ancestor of b. Then C is formed by the internal nodes of the $a - b$ path in T, plus all the subtrees rooted at the other children of these internal nodes.

Twin and Module Reductions. Two vertices u and v are *true twins* if $N[u] = N[v]$; this relation is an equivalence relation on $V(G)$ and its equivalence classes are called *true twin classes*. A module is a set of vertices M such that for every vertex v in $V(G) \setminus M$, either $M \subseteq N(v)$ or $M \cap N(v) = \emptyset$. The following two reduction rules enable us to reduce large true twin classes and trivially perfect modules.

Rule 3. *If $L \subseteq V(G)$ is a true twin class of size $|L| > 2k + 5$, and $v \in L$ is an arbitrarily picked vertex, then remove v from the graph, i.e., proceed with the instance $(G - v, k)$.*

Rule 4. *Suppose $M \subseteq V(G)$ is a module such that $G[M]$ is trivially perfect and it contains an independent set of size at least $2k + 5$. Then let us take any independent set $I \subseteq M$ of size $2k + 4$, and we delete every vertex of M apart from I, i.e., proceed with the instance $(G - (M \setminus I), k)$.*

The safeness of Rules 3 and 4 follows from a technical check. As for the polynomial time applicability, for Rule 3 it is obvious, whereas we show that we can in polynomial time verify whether Rule 4 can be applied using the *module decomposition* of the graph; we refer to the full version [8] for details. Hence, we can apply Reduction Rules 3 and 4 exhaustively, in addition to the basic rules defined earlier. Rule 3 is helpful in limiting the sizes of non-important bags of the UCD of $G - X$: Provided all the vertices of a bag have the same neighborhood in X, they form a true twin class in G, which has size at most $2k + 5$ if Rule 3 is inapplicable. By the inapplicability of Rule 4, we can bound the sizes of trivially perfect modules in (G, k), as shown in the next two results.

Lemma 6. *A (possibly disconnected) trivially perfect graph with maximum true twin class size t and maximum independent set size α has at most $(2\alpha - 1)t$ vertices in total.*

Corollary 1. *Suppose an instance (G, k) is reduced, and moreover Rules 3 and 4 are not applicable to (G, k). Then for every module $M \subseteq V(G)$ such that $G[M]$ is trivially perfect, we have that $|M| = O(k^2)$.*

From now on we assume that in the considered instance (G, k) we have exhaustively applied all the Rules 1–4. Hence, Corollary 1 can be used. Note that to perform these steps, we do not need to construct the small modulator X at all. However, we hope that the reader already sees that Rules 3 and 4 will be useful for reasoning about too large parts of $G - X$ between the important bags.

Kernelizing Non-important Parts. Recall that we have fixed a small TP-modulator X with $|X| \leq 4k$ such that $G - X$ is a trivially perfect graph with universal clique decomposition $\mathcal{T}$. Moreover, Rules 1–4 are inapplicable to (G, k). By Lemma 3, we have that the number of different X-neighborhoods is $O(k^4)$. By marking the important bags, we have marked a set I of $O(k)$ bags of $\mathcal{T}$ as important, in such a manner that every connected component of $\mathcal{T} - I$ is adjacent to at most two vertices of I, and is in fact of one of the three forms described in the paragraph following Lemma 5.

Thus, the whole vertex set of $G - X$ can be partitioned into four sets:

V_I: vertices contained in bags from I;

V_0: vertices contained in bags of those components of $\mathcal{T} - I$ that are not adjacent to any bag from I;

V_1: vertices contained in bags of those components of $\mathcal{T} - I$ that are adjacent to exactly one bag from I;

V_2: vertices contained in bags of those components of $\mathcal{T} - I$ that are adjacent to exactly two bags from I.

We establish an upper bound on the size of each of these sets separately. Upper bounds for V_I, V_0, and V_1 follow already from the introduced reduction rules.

Lemma 7. $|V_I| \leq O(k^6)$, $|V_0| \leq O(k^6)$, and $|V_1| \leq O(k^7)$.

Let us take a closer look at the bound on $|V_1|$. The vertices of V_1 can be partitioned into $O(k)$ *tassels*, where each tassel comprises all the subtrees rooted at the children of an important node x that do not contain any important node. Each such subtree is a module in G that induces a trivially perfect graph, but different subtrees within one tassel can have different neighborhoods in X. However, from Lemma 3 we know that the number of these neighborhoods is $O(k^4)$. Thus, every tassel can be partitioned into $O(k^4)$ classes of subtrees, each of which is a (possibly disconnected) module that induces a trivially perfect graph. Hence, each of these classes has size $O(k^2)$ by Corollary 1, which together with the number of tassels being $O(k)$ gives us the $O(k^7)$ bound on $|V_1|$.

However, for V_2 we need a new reduction rule. The set V_2 is composed of $O(k)$ *combs*, where each comb is formed by *(i)* vertices appearing in internal bags of a path P in T between an important node $b^{\downarrow}$ and its important ancestor $b^{\uparrow}$, and *(ii)* all the vertices in the subtrees rooted at the children of the internal vertices of P that do not lie on P. The internal nodes of P form the *shaft* of the comb, and the sets of subtrees rooted at the other children of the consecutive nodes of the shaft form its *teeth*. It can be shown that the vertices from shaft have the same neighborhood Y in X, and the vertices from the teeth also have the same neighborhood $Z \subseteq Y$ in X, but it can happen that Z differs from Y. In such a situation, we prove that it is safe to remove the middle tooth of the comb; for a formal statement of the following reduction rule we refer to the full version [8].

Rule 5 (Informal). *Suppose C is a comb of length at least $(4k + 3)^2$, and let R_β be the "middle" tooth of this comb. Then remove R_β from the graph and do not modify the budget. That is, proceed with the instance $(G - R_\beta, k)$.*

The safeness of Rule 5 is proved by an involved technical check, which in particular relies on the exact definition of the "middle" tooth. If we apply Rule 5 on the combs exhaustively, then we know that each of them has length at most $O(k^2)$. Then, using Corollary 1, it can be easily argued that the number of vertices within each comb is $O(k^4)$, which leads to $|V_2| \leq O(k^5)$ since the number of combs is $O(k)$. By combining this with Lemma 7 and the fact that $|X| \leq 4k$, we infer that $|V(G)| \leq O(k^7)$, which concludes the proof of Theorem 1.

3   Conclusions

In this paper we gave the first polynomial kernels for TRIVIALLY PERFECT EDITING and TRIVIALLY PERFECT DELETION, which answers an open problem by

Nastos and Gao [16], and Liu, Wang, and Guo [15]. Also, we showed that TRIV-IALLY PERFECT EDITING, in addition to being NP-complete, is not solvable in subexponential parameterized time unless ETH fails. The same result was known for TRIVIALLY PERFECT DELETION, but contrasts the previous result that the completion variant *does* admit a subexponential parameterized algorithm [7].

The main contribution of the paper is a proof that TRIVIALLY PERFECT EDITING admits a kernel with $O(k^7)$ vertices. We apply the existing technique of constructing a *vertex modulator*, but with a new twist: The fact that we are solving an edge modification problem enables us also to argue about the adjacency structure between the modulator and the rest of the graph, which is helpful in understanding the structure of the instance. We believe that this new insight can be applied to other edge modification problems as well. In fact, the same general approach has been recently used by the first author to achieve a quadratic kernel for THRESHOLD EDITING [6], and by the second author to obtain a polynomial kernel for {CLAW,DIAMOND}-FREE EDGE DELETION [5]. For other questions where the technique might be applicable, we propose the following:

- Improve the $O(k^3)$ kernel for COGRAPH EDITING of [10].
- Do CLAW-FREE EDGE DELETION and LINE GRAPH EDGE DELETION admit polynomial kernels?

References

1. Abu-Khzam, F.N.: Kernelization algorithms for d-hitting set problems. J. Comput. Syst. Sci. 76(7), 524–531 (2010)
2. Burzyn, P., Bonomo, F., Durán, G.: NP-completeness results for edge modification problems. Discrete Applied Mathematics 154(13), 1824–1844 (2006)
3. Cai, L.: Fixed-parameter tractability of graph modification problems for hereditary properties. Information Processing Letters 58(4), 171–176 (1996)
4. Cai, L., Cai, Y.: Incompressibility of H-free edge modification. Algorithmica 71(3), 731–757 (2015)
5. Cygan, M., Pilipczuk, M., Pilipczuk, M., van Leeuwen, E.J., Wrochna, M.: Polynomial kernelization for removing induced claws and diamonds. In: WG (2015)
6. Drange, P.G., Dregi, M.S., Lokshtanov, D., Sullivan, B.D.: On the Intractability of Threshold Editing. In: ESA (2015)
7. Drange, P.G., Fomin, F.V., Pilipczuk, M., Villanger, Y.: Exploring subexponential parameterized complexity of completion problems. In: STACS (2014)
8. Drange, P.G., Pilipczuk, M.: A Polynomial Kernel for Trivially Perfect Editing. CoRR, abs/1412.7558 (2014)
9. Fomin, F.V., Lokshtanov, D., Misra, N., Saurabh, S.: Planar F-deletion: Approximation, kernelization and optimal FPT algorithms. In: FOCS (2012)
10. Guillemot, S., Havet, F., Paul, C., Perez, A.: On the (non-)existence of polynomial kernels for P_ℓ-free edge modification problems. Algorithmica 65(4), 900–926 (2013)
11. Guo, J.: Problem kernels for NP-complete edge deletion problems: Split and related graphs. In: Tokuyama, T. (ed.) ISAAC 2007. LNCS, vol. 4835, pp. 915–926. Springer, Heidelberg (2007)

12. Impagliazzo, R., Paturi, R., Zane, F.: Which problems have strongly exponential complexity? Journal of Computer and System Sciences 63(4), 512–530 (2001)
13. Komusiewicz, C., Uhlmann, J.: Cluster editing with locally bounded modifications. Discrete Applied Mathematics 160(15), 2259–2270 (2012)
14. Kratsch, S., Wahlström, M.: Two edge modification problems without polynomial kernels. Discrete Optimization 10(3), 193–199 (2013)
15. Liu, Y., Wang, J., Guo, J.: An overview of kernelization algorithms for graph modification problems. Tsinghua Science and Technology 19(4), 346–357 (2014)
16. Nastos, J., Gao, Y.: Familial groups in social networks. Social Networks 35(3), 439–450 (2013)

Polymatroid Prophet Inequalities

Paul Dütting[1] and Robert Kleinberg[2]

[1] London School of Economics, London, UK
p.d.duetting@lse.ac.uk
[2] Cornell University, Ithaca, NY, USA
rdk@cs.cornell.edu

Abstract. Prophet inequalities bound the reward of an online algorithm—or gambler—relative to the optimum offline algorithm—the prophet—in settings that involve making selections from a sequence of elements whose order is chosen adversarially but whose weights are random. The goal is to maximize total weight.

We consider the problem of choosing quantities of each element subject to polymatroid constraints when the weights are arbitrary concave functions. We present an online algorithm for this problem that does at least half as well as the optimum offline algorithm. This is best possible, as even in the case where a single number has to be picked no online algorithm can do better.

An important application of our result is in algorithmic mechanism design, where it leads to novel, truthful mechanisms that, under a monotone hazard rate (MHR) assumption on the conditional distributions of marginal weights, achieve a constant-factor approximation to the optimal revenue for this multi-parameter setting. Problems to which this result applies arise, for example, in the context of Video-on-Demand, Sponsored Search, or Bandwidth Markets.

1 Introduction

Prophet inequalities compare the performance of an online algorithm to the optimum offline algorithm in settings that involve making selections from a sequence of random elements. The online algorithm knows the distribution from which the elements will be sampled, while the optimum offline algorithm knows the sequence of sampled elements. Prophet inequalities thus bound the relative power of online and offline algorithms in Bayesian settings. Not surprisingly, they play an important role in the analysis of online and offline algorithms in these settings. Less obviously, but no less importantly, they have a growing number of applications in algorithmic mechanism design. Specifically, they have been used to design simple yet approximately optimal (=revenue maximizing) mechanisms for multi-parameter settings, in which Myerson [17]'s seminal characterization of optimal mechanisms does not apply. Revenue maximization in multi-parameter settings is considered one of the biggest challenges in this field.

A classic result of Krengel and Sucheston [15, 16] shows that when both the online algorithm and the offline algorithm get to pick exactly one element, then

© Springer-Verlag Berlin Heidelberg 2015
N. Bansal and I. Finocchi (Eds.): ESA 2015, LNCS 9294, pp. 437–449, 2015.
DOI: 10.1007/978-3-662-48350-3_37

the online algorithm can do at least half as well as the offline algorithm. More formally, if $w_1, \ldots, w_n$ is a sequence of independent, non-negative, real-valued random variables satisfying $\mathbb{E}[\max_i w_i] < \infty$, then there exists a stopping rule τ such that

$$\mathbb{E}[w_\tau] \geq \frac{1}{2} \cdot \mathbb{E}[\max_i w_i].$$

The bound is achieved, for example, by an elegant algorithm of Samuel-Cahn [18]. This algorithm chooses a threshold T such that $\Pr(\max_i w_i > T) = \frac{1}{2}$, and selects the first element whose weight exceeds this threshold. Alternatively, as described by Kleinberg and Weinberg [14], this bound can be obtained by choosing threshold $T = \mathbb{E}[\max_i X_i]/2$ and picking the first element whose weight exceeds the threshold.

Kleinberg and Weinberg [14] recently extended this result to matroid settings. In a matroid setting we are given a ground set $\mathcal{U}$ and a non-empty downward-closed family of independent sets $\mathcal{I} \subseteq 2^{|\mathcal{U}|}$ satisfying the exchange axiom: for all pairs of sets $I, J \in \mathcal{I}$ and $|I| < |J|$ there exists an element $j \in J$ such that $I \cup \{j\} \in \mathcal{I}$; a maximal element of $\mathcal{I}$ is called a *basis*. For these settings they prove that if both the online and the offline algorithm have to pick an independent set of elements, then the online algorithm again can do at least half as well as the offline algorithm. More formally, if $w_1, \ldots, w_n$ is a sequence of independent, non-negative, real-valued random variables satisfying $\mathbb{E}[\max_i w_i] < \infty$, then there is a way to pick $A \in \mathcal{I}$ in an online fashion such that

$$\mathbb{E}\left[\sum_{i \in A} w_i\right] \geq \frac{1}{2} \cdot \mathbb{E}\left[\max_{B \in \mathcal{I}} \sum_{i \in B} w_i\right].$$

A common restriction of the original result of Krengel and Sucheston and the Kleinberg and Weinberg result is that they only apply to settings with binary decisions (i.e., an element can either be picked or not).

A Prophet Inequality for Polymatroids Our main technical contribution is a prophet inequality for settings in which the gambler and the prophet have to choose quantities of each element subject to polymatroid constraints and the weights are arbitrary concave functions. That is, we consider settings in which we are given a ground set $\mathcal{U}$ and a submodular[1] set function $f : 2^{\mathcal{U}} \to \mathbb{R}$ and a vector of quantities $z \in \mathbb{N}^{|\mathcal{U}|}$ is feasible if $z \in P_f = \{q \in \mathbb{N}^{|\mathcal{U}|} \mid \sum_{u \in S} q(u) \leq f(S) \text{ for all } S \subseteq \mathcal{U}\}$. We will restrict ourselves to integer quantities and integer-valued set functions for ease of exposition; our results trivially extend to rational quantities and rational-valued functions by scaling. For this setting we prove that if the goal of the online and the offline algorithm is to maximize $\sum_{u \in \mathcal{U}} w(u, z(u))$ over feasible z, and if the w's are random concave weights chosen independently for each element, then the online algorithm can do at least half as well as the offline algorithm.

More formally, we show that if $w_1, \ldots, w_n$ is a sequence of independent, non-negative, real-valued concave weight functions for elements $u_1, \ldots, u_n$, then there

---

[1] A set function f is submodular if for all $X \subset Y \subseteq \mathcal{U}$, $f(X \cup Y) + f(X \cap Y) \leq f(X) + f(Y)$.

exists a way to choose a feasible $z = (z_1, \ldots, z_n)$ in an online fashion (i.e.,, choosing z_i when $w_1, \ldots, w_i$ have been revealed but $w_{i+1}, \ldots, w_n$ have not yet been revealed) such that

$$\mathbb{E}\left[\sum_{i=1}^{n} w(u_i, z_i)\right] \geq \frac{1}{2} \cdot \mathbb{E}\left[\max_{q \in P_f} \sum_{i=1}^{n} w(u_i, q_i)\right].$$

Our result contains the previous results as a special case, and is best possible as even in the case where a single element has to be picked no online algorithm can do better.

To prove this result we apply a known reduction from polymatroids to matroids (see, e.g., Section 44.6b of [19]). Applying this reduction, we transform an input sequence to the polymatroid problem to an input sequence of the matroid problem by repeating the (element, weight) pairs in the input sequence to the polymatroid problem. While this construction turns inputs to the polymatroid problem into inputs to the matroid problem, it violates the independence of weights assumption. Different matroid elements corresponding to the same polymatroid element will have identical (and hence dependent) weights.

A second potential difficulty that arises when reducing the polymatroid problem to the matroid problem in this manner is that the canonical way of doing so (by repeating elements of the ground set of the polymatroid and assigning the j-th copy of an element in the resulting matroid problem the marginal weight of the j-th unit of the corresponding element in the polymatroid setting) only leads to a meaningful interpretation if the matroid algorithm always picks contiguous elements from the beginning of each sequence of matroid elements corresponding to the same polymatroid element.

The Kleinberg-Weinberg algorithm does not apply to dependent weights and it also does not necessarily pick consecutive matroid elements. Our main technical workhorse is therefore a novel algorithm for the matroid setting that is capable of handling the dependencies resulting from the reduction, and that ensures that a solution to the matroid problem can be meaningfully translated back to the polymatroid setting. To ensure the latter our algorithm sets increasing thresholds within each block of elements corresponding to the same polymatroid element, and accepts an element precisely if the weight of that element passes the threshold. Once an element fails to pass the threshold it "freezes" the threshold at the current niveau. It thereby ensures that subsequent elements will not be selected as their weight can only be lower. We control for the former, i.e., the potential dependencies across weights of matroid elements corresponding to the same polymatroid element, by introducing the notion of surrogate thresholds and performing large parts of the analysis using these surrogate thresholds as proxies.

Truthful Mechanisms with Near-Optimal Revenue The most important implication of our prophet inequality result are novel, truthful mechanisms that achieve constant-factor approximations to the optimal revenue for a multi-parameter mechanism design problem. The problem to which our mechanisms apply is multi-parameter as each agent can receive multiple units, and can have arbitrary concave valuations. The requirement that the valuations are concave corresponds to the standard economic assumption that valuations have decreasing marginals. Like prior results our mechanisms are posted-price mechanisms; that is, they approach the agents in turn and present them with a price that the agents can either accept or not. However, prior results that have used prophet inequalities to devise posted-price mechanisms were restricted to unit-demand settings (e.g., [5, 1, 14]). To the best of our knowledge, our posted price mechanisms are the first such mechanisms for a multi-unit demand setting, and yield the first constant-factor revenue guarantees for problems with polymatroid structure and valuations with decreasing marginals.

In a Bayesian mechanism design problem with polymatroid structure we are given a set N of n agents. Each agent i has a private, concave valuation function $v_i : \mathbb{N} \to \mathbb{R}_+$, drawn independently from not necessarily identical distributions F_i with support V_i that are common knowledge. A mechanism (x, p) consists of an outcome rule $x : \prod_i V_i \to \mathbb{R}_+^n$, where x_i specifies how much service agent i gets, and a payment rule $p : \prod_i V_i \to \mathbb{R}_+^n$, where p_i specifies the payment of agent i. An outcome is feasible if $\sum_{i \in S} x_i \leq f(S)$ for all $S \subseteq N$, where f is an integer-valued submodular function. Agent i's utility is $u_i(b, v_i) = v_i(x_i(b)) - p_i(b)$, where b denotes the bids of the agents. The welfare of a mechanism is $\sum_{i \in N} v_i(x_i(b))$ and its revenue is $\sum_{i \in N} p_i(b)$. A mechanism is dominant strategy incentive compatible (DSIC) (or truthful) if for every agent i, value v_i, bid b_i and bids $b_{-i} = (b_1, \ldots, b_{i-1}, b_{i+1}, \ldots, b_n)$, $u_i((v_i, b_{-i}), v_i) \geq u_i((b_i, b_{-i}), v_i)$.

Practical mechanism design problems with polymatroid structure include:

(1) **Video-on-Demand** [4]: Consider a collection of spatially dispersed user groups, each of which wants to watch various movies using a streaming service. We can model this via a graph $G = (\bigcup_{i \in N} T_i \cup \{s\}, E)$ in which $T_i \cap T_j = \emptyset$ for all $i, j \in N$ and each edge $e \in E$ has a capacity c_e. The seller is identical with source node s. Each agent $i \in N$ is identified with a number of demand nodes T_i corresponding to the members of user group i. The allocation to each agent $i \in N$ is $\sum_{t \in T_i} x_t$, where x_t is the flow into t. An allocation x is feasible if and only if $\sum_{t \in S} x_t \leq f(S)$ for all $S \subseteq \bigcup_{i \in N} T_i$, where f is the submodular function giving the value of a minimum s-S-cut.

(2) **Local Purchasing Collectives** [3]: Consider a group of buyers which is interested in purchasing a certain good from local providers. We can model this via a bipartite graph with vertices on the left side representing providers and those on the right side representing buyers (elements of N). An edge represents a buyer having access to a particular provider. Suppose that each provider j has a positive supply $s(j)$. A vector of quantities purchased is feasible if each buyer's quantity can be fulfilled by one or more of the adjacent providers without exceeding any provider's supply. More formally, for a set S of buyers let $\Gamma(S)$ denote the set of providers adjacent to at least one element of S, and let $f(S) = \sum_{j \in \Gamma(S)} s(j)$ which is a submodular function. A vector x of quantities purchased is feasible if and only if for every set of buyers S, we have $\sum_{i \in S} x_i \leq f(S)$.

(3) **Sponsored Search** [10]: In sponsored search a set of advertisers seeks to be assigned clicks on ad slots. Denote the set of advertisers by N and the set of ad slots by M. Sort the ad slots $j \in M$ by non-increasing number of clicks $\alpha_j \in \mathbb{N}_{\geq 0}$. An allocation x of clicks to advertisers is feasible if and only if $\sum_{i \in S} x_i \leq f(S)$ for all $S \subseteq N$, where $f(S) = \sum_{j=1}^{|S|} \alpha_j$ is a submodular function.

(4) **Bandwidth Markets** [4]: In wireless communication settings agent $i \in N$ seeks to maximize its transmission rate x_i. In a Gaussian multiple-access channel the set of feasible transmission rates x—the so-called Cover-Wyner region—forms a polymatroid (see [20] for details).

We present two DSIC posted-price mechanisms for these problems. The first combines the thresholds of our algorithm with "eager" reserves, the second combines them with "lazy" reserves [7]. The difference between eager and lazy reserves is that the former are applied during the computation of the allocation, while the latter are applied only after the fact. In our case, however, both can be implemented in an online fashion. We prove that these mechanisms achieve at least a $1/2e^2$ resp. $1/2e$ fraction of the optimal revenue by proving a lower bound in terms of the optimal welfare. For "eager" reserves we use Chebyshev's Integral Inequality and inductively apply a single-sample argument of [7]. For "lazy" reserves we only need the single-sample result.

Related Work. We have already described the result by Krengel and Sucheston [15, 16] for the case in which both the online algorithm and the offline algorithm are allowed to pick one number, showing that the online algorithm can do at least half as well as the offline algorithm. This bound is tight. The result has been extended to the case where both the online algorithm and the offline algorithm can pick k numbers by Alaei [1], showing that the online-to-offline ratio is at least $1 - 1/(\sqrt{k+3})$. This matches the aforementioned tight bound when $k = 1$, and it remains nearly tight for $k > 1$, in the sense that a ratio of $1 - o(1/\sqrt{k})$ is known to be unattainable. Kleinberg and Weinberg [14] have extended the bound of two to settings where the elements picked must be an independent set in a matroid. This bound is also tight, as it subsumes the case where both the online and offline algorithm have to pick one number.

Hajiaghayi et al. [13] observed the following relationship between prophet inequalities and algorithmic mechanism design: algorithms used to prove prophet inequalities can be interpreted as truthful online auction mechanisms, and the prophet inequality in turn can be interpreted as the mechanism's approximation guarantee. Chawla et al. [5] observed an even subtler relationship between the two topics: questions about the approximability of offline Bayesian optimal mechanisms by sequential posted-price mechanisms could be translated into questions about prophet inequalities, via the use of virtual valuation functions. Alaei [1] and Kleinberg and Weinberg [14], armed with stronger prophet inequalities, deepen this relationship even further. More recently, and in parallel to this work, Feldman et al. [9] have designed posted-price mechanisms for combinatorial auctions, which are not based on prophet inequalities.

Another related line of literature is work on secretary problems, which also concerns relations between optimal offline selection rules and suboptimal online stopping rules, but under the assumption of a randomly ordered input rather than independent random numbers in a fixed order. While the polymatroid prophet inequality that we solve here contains the matroid prophet inequality problem as a special case, the matroid secretary problem introduced by Babaioff et al. [2] remains largely unsolved despite recent progress.

A final related direction is work on exponential-sized Markov decision processes (MDP's) [6, 11, 12]. The connection here is that algorithms for prophet inequalities can be formulated as exponential-sized MDP's, whose state reflects the entire set of decisions made prior to a specified point during the algorithm's execution. Most of the algorithms with provable approximation guarantees for exponential-sized MDP's are LP-based, while our algorithm is combinatorial.

2   Preliminaries

In a *Bayesian online selection problem* we are given a ground set $\mathcal{U}$ and for each $x \in \mathcal{U}$ a probability distribution F_x with support $\mathbb{R}^d_{\geq} = \{m \in \mathbb{R}^d_+ : i \leq i' \Rightarrow m_i \geq m_{i'}\}$ of finite dimension $d \in \mathbb{N}_+$. This induces a probability distribution over functions $w : \mathcal{U} \times \{1, \ldots, d\} \to \mathbb{R}_+$ in which the multivariate random variables $\{(w(x, 1), \ldots, w(x, d)) : x \in \mathcal{U}\}$ are independent and the marginals $(w(x, 1) - w(x, 0), \ldots, w(x, d) - w(x, d-1))$ where we set $w(x, 0) = 0$ have distribution F_x. We refer to $w(x, k)$ as the *weight* of k units of x, and to $w(x, k) - w(x, k-1)$ as the *marginal weight* of the k-th unit of x. By our assumption regarding the distributions F_x for $x \in \mathcal{U}$, the marginal weights $w(x, k) - w(x, k-1)$ for all $x \in \mathcal{U}$ and $k \geq 1$ are decreasing and the weight $w(x, k)$ of any given $x \in \mathcal{U}$ is a concave function in k.

The goal is to choose a vector $z \in \mathbb{R}^{|\mathcal{U}|}$ that maximizes $\sum_{x \in \mathcal{U}} w(x, z(x))$. For a given weight function w we use OPT(w), or simply OPT, to denote the optimal value. The vector z will typically be restricted to come from a space of feasible vectors $\mathcal{F} \subseteq \mathbb{R}^{|\mathcal{U}|}$. One common restriction is $\mathcal{F} \subseteq \{0, 1\}^{|\mathcal{U}|}$ in which case $z_i \in \{0, 1\}$ can be thought of as encoding membership in a subset $A \subseteq \mathcal{U}$. Two further restrictions, matroids and polymatroids, were discussed already in Section 1. For matroids the distribution F_x for $x \in \mathcal{U}$ has dimension 1; for polymatroids defined by f taking values in $\{1, \ldots, M\}$ it suffices to consider distributions F_x for $x \in \mathcal{U}$ of dimension M.

An *input sequence* is a sequence σ of ordered pairs $(x_i, w_i)_{i=1,\ldots,|\mathcal{U}|}$ consisting of an element $x_i \in \mathcal{U}$ and marginals $w_i \in \mathbb{R}^d_{\geq}$ such that every $x_i \in \mathcal{U}$ occurs exactly once in the sequence. A *deterministic online selection algorithm* is a function z mapping every input sequence σ to a vector $z(\sigma) \in \mathcal{F}$ such that for any pair of input sequences σ, σ' that match on the first i pairs $(x_1, w_1), \ldots, (x_i, w_i)$ we have $z_j(\sigma) = z_j(\sigma')$ for all $1 \leq j \leq i$. An *online weight-adaptive adversary* that has chosen $x_1 \ldots, x_{i-1}$ and has learned about $w_1, \ldots, w_{i-1}$ chooses x_i without knowing w_i.

Notation. For a real number z, we use z^+ to denote $\max\{z, 0\}$. We use the shorthand $w(S)$ to denote the weight of a feasible (multi-)set S of elements $x \in \mathcal{U}$.

3   Algorithm for Polymatroids

We derive our algorithm for the polymatroid prophet inequality by reducing to the matroid case. We begin by defining *block-structured matroids, block-restricted weight distributions*, and *block-restricted adversaries*. Although a prophet inequality for the resulting matroid problem would translate into a prophet inequality for the polymatroid problem, we cannot simply apply the Kleinberg-Weinberg algorithm for matroids to derive it. The reason is twofold. First, the reduction from polymatroids to matroids leads to weight distributions that are no longer independent. Second, for the weights in the matroid setting to be in one-to-one correspondence to the weights in the polymatroid setting we need to ensure that the matroid algorithm chooses consecutive elements of each block. The crux of our analysis is therefore a novel algorithm for the matroid setting that can handle the dependencies that result from the reduction and that guarantees that weights are consistent.

Due to space limitations, most proofs in this section and the following one are deferred to the full version of the paper [8].

3.1   Block-Structured Matroids

We first define block-structured matroids and show that to every polymatroid defined by an integer-valued submodular function there is an associated block-structured matroid.

Definition 1. *A* block-structured matroid *is one whose ground set is partitioned into blocks $B_1, ..., B_n$ such that the independence relation is preserved under permutations of the ground set that preserve the pieces of the partition.*
    For a set $S \subseteq B_1 \cup \cdots \cup B_n$ we define its cardinality vector $\mathbf{q}(S) = (q_1(S), q_2(S), \ldots, q_n(S))$ by setting $q_i(S) = |S \cap B_i|$ for $i = 1, \ldots, n$.

Lemma 1 (cf. Chapter 44.6b of [19]). *Suppose f is a submodular function on ground set $\mathcal{U} = \{u_1, \ldots, u_n\}$, taking values in $\{0, 1, \ldots, M\}$. There is a block-structured matroid $\mathcal{M}_f$ on ground set $\mathcal{U} \times [M]$ with blocks $B_i = \{u_i\} \times [M]$ ($i = 1, \ldots, n$), whose independent sets are those S satisfying $\mathbf{q}(S) \in P_f$.*

Next we define block-restricted weight distributions and block-restricted adversaries to capture the type of input sequences generated by our reduction. We also define a property of algorithms for block-restricted matroids that ensures that the weights in the matroid setting can be translated back into the polymatroid setting.

Definition 2. *A* block-restricted weight distribution *on a block-structured matroid is a joint distribution of weights for its elements, such that the elements* *of a block receive non-increasing weights, the weights within each block can be arbitrarily correlated, but the weight assignments to different blocks are mutually independent.*

Definition 3. *A block-restricted adversary is one who is restricted to choose an ordering of the input sequence in which the elements of each block appear consecutively, and after any proper subset of the blocks have been presented, the choice of which block is presented next may only depend on the weights of elements that have already been presented.*

Definition 4. *A deterministic algorithm for block-restricted matroids is consistent if whenever it picks the j-th element of a block if it has already picked elements $1, \ldots, j - 1$ from that block. In other words a consistent algorithm for block-restricted matroids may only choose consecutive blocks of items at the beginning of each block in the matroid.*

Note that when all blocks have size 1, a block-structured matroid is simply a matroid, and a block-restricted distribution is simply an independent distribution. Furthermore, a block-restricted adversary is exactly the same as the notion of *online weight-adaptive adversary* defined in [14]. Thus, the special case in which all blocks have size 1 is precisely the setting of the matroid prophet inequality of [14].

3.2 Prophet Inequality for Block-Structured Matroids

Next we describe our algorithm for the matroid problem. The algorithm is similar in spirit to the algorithm of Kleinberg and Weinberg in that it sets a threshold for each element and accepts the element if and only if its weight exceeds the threshold. However, it significantly differs from the Kleinberg-Weinberg algorithm in the way it chooses the thresholds.

Consider a block-restricted matroid $\mathcal{M} = (\mathcal{U}, \mathcal{I})$. Let $w, w' : \mathcal{U} \to \mathbb{R}_+$ denote two assignments of weights to the elements of $\mathcal{U}$ sampled independently from a block-restricted weight distribution. For a given input sequence $\sigma = (x_1, w(x_1)), \ldots, (x_n, w(x_n))$ we compare the set $A = A(\sigma)$ selected by the algorithm to the basis B that maximizes $w'(B)$. The matroid exchange axiom guarantees the existence of a partition of B into disjoint subsets C, R such that $A \cup R$ is also a basis of $\mathcal{M}$. Among all such partitions, let $C(A), R(A)$ denote the one that maximizes $w'(R)$. Let $g(A) = w'(R(A))$.

The selection algorithm when faced with element x_i proceeds as follows: Denote the (possibly empty) set of elements already selected by A_{i-1} and denote the (possibly empty) set of indices of elements belonging to the same block as x_i and that precede x_i in the input sequence by $Pred(x_i)$. Element x_i is accepted if and only if $w(x_i) \geq T_i$ where the threshold T_i is determined as follows. If $A_{i-1} \cup \{x_i\} \notin \mathcal{I}$ then $T_i = \infty$. Otherwise,

$$T_i = \max\{\max_{j \in Pred(x_i)} T_j, \frac{1}{2} \cdot \mathbb{E}[g(A_{i-1}) - g(A_{i-1} \cup \{x_i\})]\}$$

$$= \max\{\max_{j \in Pred(x_i)} T_j, \frac{1}{2} \cdot \mathbb{E}[w'(R(A_{i-1})) - w'(R(A_{i-1} \cup \{x_i\}))]\} \qquad (1)$$

$$= \max\{\max_{j \in Pred(x_i)} T_j, \frac{1}{2} \cdot \mathbb{E}[w'(C(A_{i-1} \cup \{x_i\})) - w'(C(A_{i-1}))]\}. \qquad (2)$$

Note that (1) and (2) define the same quantity: Let B be the maximum weight basis of $\mathcal{M}$ with weights w'. Then, $w'(B) = w'(C(A_{i-1})) + w'(R(A_{i-1}))$ and $w'(B) = w'(C(A_{i-1} \cup \{x_i\})) + w'(R(A_{i-1} \cup \{x_i\}))$. Equalizing and rearranging gives

$$w'(R(A_{i-1})) - w'(R(A_{i-1} \cup \{x_i\})) = w'(C(A_{i-1} \cup \{x_i\})) - w'(C(A_{i-1})).$$

Theorem 1. *For every block-restricted matroid $(\mathcal{U}, \mathcal{I})$ with block-restricted weight distribution there is a deterministic, consistent online selection algorithm that achieves the following performance guarantee against block-restricted adversaries:*

$$\mathbb{E}[w(A)] \geq \frac{1}{2} \cdot OPT.$$

Before we outline how this theorem can be proved, we use it to derive a prophet inequality for polymatroids.

3.3   Prophet Inequality for Polymatroids

The algorithm that achieves the prophet inequality in the polymatroid setting (with integer-valued submodular function f taking values in $\{1, \ldots, M\}$) does so by reducing the problem to the block-structured matroid setting with the matroid $\mathcal{M}_f$ defined in Lemma 1 as follows. If in the polymatroid setting the elements are presented in order $u_1, \ldots, u_n$, then the reduction constructs an input sequence in the matroid setting by presenting the elements in order $(u_1, 1), (u_1, 2), \ldots, (u_2, 1), (u_2, 2), \ldots$ (lexicographic order, $\mathcal{U}$ coordinate first). If in the polymatroid setting the marginal weights of element u_i are $w_i = (w_{i,1}, w_{i,2}, \ldots, w_{i,M})$ then element (u_i, j) is presented in the matroid setting with weight $w_{i,j}$. If the matroid algorithm, while processing elements $(u_i, 1), (u_i, 2), \ldots, (u_i, M)$, selects a subset $\{u_i\} \times S_i$, then the polymatroid algorithm when processing u_i sets $z_i = |S_i|$.

Theorem 2. *For every polymatroid P_f defined by a rational-valued submodular function f and concave weights there exists a deterministic online selection algorithm that satisfies the following performance guarantee against online weight-adaptive adversaries:*

$$\mathbb{E}\left[\sum_{i=1}^{n} w(u_i, z(u_i))\right] \geq \frac{1}{2} \cdot OPT.$$

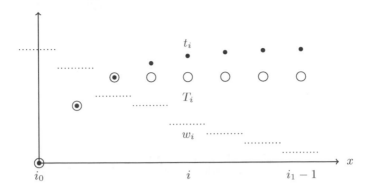

Fig. 1. Visualization of the thresholds set by the algorithm

3.4   Proof of the Block-Restricted Matroid Prophet Inequality

We start with a proposition that provides a lower bound on the sum of the thresholds of the elements that are selected by the algorithm. The proof of this proposition exploits the definition of the thresholds and, in addition, linearity of expectation and a telescoping sum.

Proposition 1. *For every input sequence* σ, *if* $A = A(\sigma)$, *then* $\sum_{x_i \in A} T_i \geq \frac{1}{2} \cdot \mathbb{E}[w'(C(A))]$.

Next we describe our main technical insight. Namely, that the thresholds within a given block have a specific form (see Figure 1 for an illustration). Specifically, consider any block consisting of elements $u_{i_0}, u_{i_0+1}, \ldots, u_{i_1-1}$. For all $i_0 \leq i \leq i_1$ define $A^i = A_{i_0-1} \cup \{x_{i_0}, \ldots, x_i\}$, and

$$t_i = \frac{1}{2} \cdot \mathbb{E}_{w'} \left[g(A^{i-1}) - g(A^i) \right], \tag{3}$$

where for convenience we also set $A^{i_0-1} = A_{i_0-1}$. We will show that the sequence of numbers defined by (3) forms a non-decreasing sequence depending only on the weights associated with previous elements $w_1, w_2, \ldots, w_{i_0-1}$, and that for $i_0 \leq i \leq i_1 - 1$ the algorithm sets threshold $T_i = t_i$ if $t_i \leq w_i$ and $T_i > w_i$ otherwise.

Lemma 2. *Consider a block-structured matroid* $(\mathcal{U}, \mathcal{I})$ *with blocks* $B_1, \ldots, B_n$. *For any input sequence* σ *generated by a block-restricted adversary, and any block* B_j, *let* $i_0, i_0 + 1, \ldots, i_1$ *denote the times when the elements of* B_j *are presented in* σ. *The sequence of numbers* $t_{i_0}, \ldots, t_{i_1}$ *defined by (3) satisfies* $t_{i_0} \leq t_{i_0+1} \leq \cdots \leq t_{i_1}$ *and depends only on the subsequence of* σ *preceding time* i_0. *Moreover, the algorithm is consistent and sets* $T_i = t_i$ *for all* $i_0 \leq i \leq i_1$ *such that* $t_i \leq w_i$, *and* $T_i > w_i$ *otherwise.*

An important corollary of the preceding structural result regarding the thresholds is the following assertion for two weight assignments w, w' drawn independently from a block-restricted weight distribution.

Corollary 1. *Let w, w' be two weight assignments drawn independently from a block-restricted weight distribution. For any input sequence σ generated by a block-restricted adversary, and any block B_j, let $i_0, i_0 + 1, \ldots, i_1$ denote the times when the elements of B_j are presented in σ. Then, for all $i_0 \leq i < i_1$, $(w_i - T_i)^+ = (w_i - t_i)^+$, and $w_i, t_i, w'(x_i)$ are mutually independent, so*

$$\mathbb{E}[(w_i - T_i)^+] = \mathbb{E}[(w_i - t_i)^+] = \mathbb{E}[(w'(x_i) - t_i)^+].$$

The final ingredient is an upper bound on the sum of the surrogate thresholds t_i for the elements x_i in $R(A)$, where A is the set of elements accepted by the algorithm on a given input sequence σ.

Proposition 2. *For every input sequence σ generated by a block-restricted adversary, let $A = A(\sigma)$. Then $\sum_{x_i \in R(A)} t_i \leq \frac{1}{2} \cdot \mathbb{E}[w'(R(A))]$.*

The proof of Theorem 1 uses our structural insight regarding the thresholds to lift the proof from the actual thresholds to the surrogate thresholds. It then uses the upper and lower bounds on the surrogate thresholds from this section to establish the claimed bound.

4   Application to Mechanism Design

We conclude by showing how our prophet inequality algorithm can be used to derive dominant strategy incentive compatible (DSIC), constant factor-approximations to the optimal revenue for a multi-parameter setting in which Myerson's analysis of the revenue-maximizing auction does not apply. Our result applies to concave weights whose distribution satisfies a conditional analog of the monotone hazard rate (MHR) condition. Specifically, we will assume that for each element $u_i \in \mathcal{U}$ the conditional distribution of the marginal weight $w_{i,j}$ of the j-th unit given the marginal weights $w_{i,1}, \ldots, w_{i,j-1}$ of the preceding units is MHR. That is, $\frac{f(w_{i,j}|w_{i,j_0}, \ldots, w_{i,j-1})}{1 - F(w_{i,j}|w_{i,j_0}, \ldots, w_{i,j-1})}$ is non-decreasing in $w_{i,j}$. One example of a distribution satisfying this assumption is obtained by first drawing $w_{i,1} \sim U[0, 1]$, then drawing $w_{i,2} \sim U[0, w_{i,1}]$, and so on.

We obtain posted-price mechanisms by combining the algorithm for polymatroids with "eager" or "lazy" monopoly reserves [7]. The monopoly reserve r^* for a given distribution F over valuations v with density f is $r^* = \phi^{-1}(0)$ where $\phi(v) = v - \frac{1-F(v)}{f(v)}$ is the virtual valuation. In the case of "eager" reserves, we modify the algorithm so that it only awards element x_i if its weight $w(x_i)$ exceeds the threshold T_i and the monopoly reserve r_i^* of the conditional distribution of $w(x_i)$. In the case of "lazy" reserves, we first run the algorithm to determine a tentative allocation, but then we only allocate elements whose weight also exceeds the reserve. Note that this can be done in an online fashion by computing thresholds as if all tentative assignments were made, but only actually awarding an element if it also exceeds the reserve.

Both mechanisms are DSIC as they are posted price. To prove the revenue bounds we need the following single-sample result.

Lemma 3 (Lemma 3.10 of Dhangwatnotai et al. [7]). *Let F be an MHR distribution with monopoly price r^* and revenue function $\hat{R}$. Let $V(t)$ denote the expected welfare of a single-item auction with a posted price of t and a single bidder with valuation drawn from F. For every nonnegative number $t \geq 0$, $\hat{R}(\max\{t, r^*\}) \geq \frac{1}{e} \cdot V(t)$.*

Theorem 3. *For polymatroids P_f defined by rational-valued submodular function f and concave weights that satisfy the conditional analog of the MHR condition, combining the polymatroid prophet inequality algorithm with "eager" or "lazy" reserves yields a DSIC mechanism whose revenue R_{EAGER} or R_{LAZY} on any input sequence σ generated by an online weight-adaptive adversary satisfies*

$$R_{EAGER}(\sigma) \geq \frac{1}{2e^2} \cdot R_{OPT}(\sigma) \text{ or } R_{LAZY} \geq \frac{1}{2e} \cdot R_{OPT}(\sigma),$$

where R_{OPT} denotes the optimal revenue.

Corollary 2. *For MHR valuations with decreasing marginals, there is a truthful $1/2e$ approximation to revenue for video-on-demand, bandwidth markets, sponsored search, and local purchasing collectives.*

Acknowledgments. Part of this work was done while the first author was a Postdoctoral Fellow at Cornell University and Stanford University. His research is supported in part by an SNF Postdoctoral Fellowship. The second author's research is supported in part by NSF award AF-0910940, AFOSR grant FA9550-09-1-0100, a Microsoft Research New Faculty Fellowship, and a Google Research Grant.

References

[1] Alaei, S.: Bayesian combinatorial auctions: Expanding single buyer mechanisms to many buyers. In: Proc. of 52nd FOCS, pp. 512–521 (2011)

[2] Babaioff, M., Immorlica, N., Kleinberg, R.: Matroids, secretary problems, and online mechanisms. In: Proc. of 18th SODA, pp. 434–443 (2007)

[3] Babaioff, M., Lucier, B., Nisan, N.: Bertrand networks. In: Proc. of 14th EC, pp. 33–34 (2013)

[4] Bikhchandani, S., de Vries, S., Schummer, J., Vohra, R.V.: An ascending vickrey auction for selling bases of a matroid. Operations Research 59(2), 400–413 (2011)

[5] Chawla, S., Hartline, J.D., Malec, D.L., Sivan, B.: Multi-parameter mechanism design and sequential posted pricing. In: Proc. of 41th STOC, pp. 311–320 (2010)

[6] Dean, B.C., Goemans, M.X., Vondrák, J.: Approximating the stochastic knapsack problem: The benefit of adaptivity. In: Proc. of 45th FOCS, pp. 208–217 (2004)

[7] Dhangwatnotai, P., Roughgarden, T., Yan, Q.: Revenue maximization with a single sample. Games and Economic Behavior (2014) (in press)

[8] Dütting, P., Kleinberg, R.: Polymatroid prophet inequalities (2013). http://arxiv.org/abs/1307.5299

[9] Feldman, M., Gravin, N., Lucier, B.: Combinatorial auctions via posted prices. In: Proc. of 26th SODA, pp. 123–135 (2015)

[10] Goel, G., Mirrokni, V.S., Paes Leme, R.: Polyhedral clinching auctions and the adwords polytope. In: Proc. of 44th STOC, pp. 107–122 (2012)

[11] Guha, S., Munagala, K.: Multi-armed bandits with metric switching costs. In: Albers, S., Marchetti-Spaccamela, A., Matias, Y., Nikoletseas, S., Thomas, W. (eds.) ICALP 2009, Part II. LNCS, vol. 5556, pp. 496–507. Springer, Heidelberg (2009)

[12] Guha, S., Munagala, K., Shi, P.: Approximation algorithms for restless bandit problems. Journal of the ACM 58, 3:1–3:50 (2010)

[13] Hajiaghayi, M., Kleinberg, R., Sandholm, T.W.: Automated mechanism design and prophet inequalities. In: Proc. of 22nd AAAI, pp. 58–65 (2007)

[14] Kleinberg, R., Weinberg, S.M.: Matroid prophet inequalities. In: Proc. of 44th STOC, pp. 123–136 (2012)

[15] Krengel, U., Sucheston, L.: Semiamarts and finite values. Bulletin of the American Mathematical Society 83, 745–747 (1977)

[16] Krengel, U., Sucheston, L.: On semiamarts, amarts, and processes with finite value. Advances in Probability and Related Topics 4, 197–266 (1978)

[17] Myerson, R.: Optimal auction design. Mathematics of Operations Research 6, 58–73 (1981)

[18] Samuel-Cahn, E.: Comparison of threshold stop rules and maximum for independent nonnegative random variables. Annals of Probability 12, 1213–1216 (1984)

[19] Schrijver, A.: Combinatorial Optimization: Polyhedra and Efficiency. Springer (2002)

[20] Tse, D.N.C., Hanly, S.V.: Multiaccess fading channels-part I: Polymatroid structure, optimal resource allocation and throughput capacities. Transactions on Information Theory 44, 2796–2815 (1998)

Node-Balancing by Edge-Increments

Friedrich Eisenbrand[1], Shay Moran[2,3], Rom Pinchasi[2], and Martin Skutella[4]

[1] EPFL, 1015 Lausanne, Switzerland
[2] Technion, Israel Institute of Technology, Haifa 32000, Israel
[3] Max Planck Institute for Informatics, Saarbrücken, Germany
[4] Technische Universität Berlin, Straße des 17. Juni 136, 10623 Berlin, Germany

Abstract. Suppose you are given a graph $G = (V, E)$ with a weight assignment $w : V \to \mathbb{Z}$ and that your objective is to modify w using legal steps such that all vertices will have the same weight, where in each legal step you are allowed to choose an edge and increment the weights of its end points by 1.

In this paper we study several variants of this problem for graphs and hypergraphs. On the combinatorial side we show connections with fundamental results from matching theory such as Hall's Theorem and Tutte's Theorem. On the algorithmic side we study the computational complexity of associated decision problems.

Our main results are a characterization of the graphs for which any initial assignment can be balanced by edge-increments and a strongly polynomial-time algorithm that computes a balancing sequence of increments if one exists.

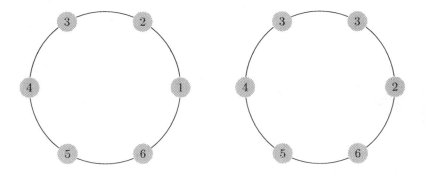

Fig. 1. The node-weights after one incrementing step

1 Introduction

The following puzzle is often used as an introductory puzzle for the method of invariance and potential functions: Six boxes numbered 1 to 6 are arranged in a cycle. For every $1 \le i \le 6$, we start with i oranges in box number i. At each step we are allowed to add one orange to each of two adjacent boxes. Prove that we will never be able to make all boxes contain the same number of oranges.

© Springer-Verlag Berlin Heidelberg 2015
N. Bansal and I. Finocchi (Eds.): ESA 2015, LNCS 9294, pp. 450–458, 2015.
DOI: 10.1007/978-3-662-48350-3_38

One of the simple solutions to this puzzle is to observe that the total number of oranges in boxes $1, 3, 5$ is always smaller than the total number of oranges in boxes $2, 4, 6$ and this never changes through each step of the game.

In this paper we consider the natural generalization of the puzzle above to arbitrary graphs. Let $G = (V, E)$ be a finite graph, let $w : V \rightarrow \mathbb{N}$ be a non-negative integer weight function on its vertices and let $e = \{u, v\} \in E$. A *positive step* on e modifies w by increasing the weights of u, v by 1 unit. We say that w is *equatable in G* if there exists a sequence of only positive steps, $S = s_1, \ldots, s_m$, after which all vertices have the same weight. We also say that the sequence S *positively equates w*.

Our *main results* are the following.

i) We characterize those graphs $G = (V, E)$ for which any initial assigmnet $w : V \rightarrow \mathbb{N}_0$ is equatable. These are the connected graphs with an odd number of nodes for which $G - U$ has less than $|U|$ isolated vertices for any $U \subset V$. Here $G - U$ is the subgraph of G that is induced by $V \setminus U$. (Theorem 1)

ii) We show that the following problem can be solved in strongly polynomial time. Given a graph $G = (V, E)$ and an initial assignment $w : V \rightarrow \mathbb{N}_0$, decide whether w is equatable and compute an equating multiset of edges. (Theorem 3)

iii) An initial assignment w of the nodes of a bipartite graph $G = (L + R, E)$ is not equatable if $w(L) \neq w(R)$, the difference $w(L) - w(R)$ is invariant under edge-increments. However, each balanced assignment with $w(L) = w(R)$ is equatable if and only if the strict Hall condition holds: For any nonempty set of vertices X that is properly contained in L or in R, one has $|X| < |N(X)|$. Here $N(X)$ denotes the neighborhood of X. (Theorem 4)

iv) Finally we show that the analog of the decision problem ii) is NP-hard for hypergraphs. (Theorem 5).

Related Work. The problem of equating the node-weights is closely related to *perfect b-matchings*, [15]. Let $b \in \mathbb{N}_0^{|V|}$ be a vector of non-negative node-weights. A b-matching of a graph $G = (V, E)$ is a vector $x \in \mathbb{N}_0^{|E|}$ that satisfies

$$\sum_{e \in \delta(v)} x_e \leq b_v, \tag{1}$$

where $\delta(v)$ denotes the set of edges of G that are incident to v. A b-matching is *perfect*, if the inequality in (1) can be replaced by equality. Thus b-matchings are a generalization of *matchings*, where b is the all ones vector.

What is the relationship between b-matchings and the process of equating positive weights in graphs by edge-increments? Suppose that the given initial weight assignment $w \in \mathbb{N}_0^{|V|}$ is equatable and that the resulting equated node-weight is $\beta \in \mathbb{N}$. Then, the edge-increments that lead to the balanced node-weight β are a b-matching $x \in \mathbb{N}_0^{|E|}$ with $b_v = \beta - w_v$ for each vertex v. By incrementing the node-weights of each edge e exactly x_e times, one arrives at a balanced assignment with weight β on all the nodes.

Maximum weight b-matchings can be computed in polynomial time [5,4,3]. The currently fastest algorithms for maximum weight matching are by Gabov [6], and Gabov and Tarjan [7]. The fastest algorithm for weighted b-matching is by Anstee [1]. Recent exciting progress for maximum cardinality matching has been given by Madry [13] improving upon the $O(m\sqrt{n})$ running time of Hopcroft and Karp [10] and [12] in the sparse case.

A related notion to equatable graphs is the one of a *regularizable* graph. A graph is regularizable, if there exists a k and a perfect k-matching such that each edge is chosen at least once in this matching. Thus, one obtains a k-regular graph by replacing each edge by as many parallel edges, as its multiplicity in the b-matching. Berge [2] provided the following characterization of regularizable graphs. If G is connected an bipartite, then G is regularizable if and only if $|N(U)| > |U|$ for each non-empty stable set U of G. This is a *strict* Hall condition for stable sets.

2   A Characterization of Equatable Graphs

Which are the graphs $G = (V, E)$ for which any initial assignment $w : V \to \mathbb{N}_0$ is equatable? The following theorem provides the answer to that question.

Theorem 1. *Let $G = (V, E)$ be a finite graph. The following statements are equivalent:*

1. *Every integer assignment $w : V \to \mathbb{N}_0$ is equatable in G.*
2. *G is connected, $|V|$ is odd and for all $U \subseteq V$, the graph $G - U$ has less than $|U|$ isolated vertices, where $G - U$ is the vertex induced subgraph on $V - U$.*

Notice that condition 2) implies that G has at least 3 vertices. We will now provide the proof of this theorem. To do so, we rely on a well known result of Tutte that characterizes the existence of a perfect b-matching in a graph.

Theorem 2 (Tutte [16]). *Let $G = (V, E)$ be a finite graph and let $b : V \to \mathbb{N}_0$ be a weight function on the vertices of G. The following statements are equivalent.*

a) G has a perfect b-matching.
b) For every (possibly empty) subset U of V

$$\sum_{x \in U} b(x) \geq \sum_{x \in I(U)} b(x) + S(G - U), \tag{2}$$

where $I(U)$ is the set of isolated vertices of $G - U$ and $S(G - U)$ is the number of connected components of $G - U$ that are not isolated vertices whose total b-weight is odd.

Proof (Proof of Theorem 1). Suppose that condition 2) holds. In order to show that any $w \in \mathbb{N}_0^{|V|}$ is equatable, it is enough to show that each assignment $w^{(v)}$ with

$$w^{(v)}(u) = \begin{cases} 1, & \text{if } u = v \\ 0, & \text{otherwise} \end{cases}$$

is equatable, since the corresponding steps decrease the weight of v relative to the other vertices by exactly one. We show this by establishing existence of a perfect b-matching with $b(v) = 2 \cdot n$ and $b(u) = 2 \cdot n + 1$ for each other vertex $u \neq v$.

Let $U \subseteq V$. We have to show (2). If $U = \emptyset$, then $G - U = G$. The total b-weight of G is even and $G - \emptyset$ has only one component, since G is connected, thus $S(G - U) = 0$. Also, $G - \emptyset$ does not have isolated vertices. This shows that the right-hand-side of (2) is 0.

If $U \neq \emptyset$ then, by our assumption, $|I(U)| \leq |U| - 1$. We have

$$\sum_{x \in U} b(x) \geq (2n + 1)|U| - 1. \tag{3}$$

Indeed, there is equality in (3) only if $v \in U$.

On the other hand $\sum_{x \in I(U)} b(x) \leq |I(U)|(2n + 1)$. Therefore,

$$\sum_{x \in I(U)} b(x) \leq |I(U)|(2n + 1) \leq (|U| - 1)(2n + 1). \tag{4}$$

Finally, the term $S(G - U)$ is at most the number of components of G that are not isolated vertices. Therefore,

$$S(G - U) \leq \frac{n - 1}{2}. \tag{5}$$

Inequality (2) is, therefore, satisfied because using (3), (4), and (5) inequality (2) reduces to

$$(2n + 1)|U| - 1 \geq (|U| - 1)(2n + 1) + \frac{n - 1}{2},$$

which clearly holds.

Suppose now that every $w \in \mathbb{N}_o^{|V|}$ is equatable. Then clearly, G is connected. Also G has an odd number of vertices since the parity of the sum of weights is invariant under the edge-increment operation. In a graph with an even number of nodes, an equated assignment has even parity which shows that an odd initial assignment is not equatable.

Let U be any non-empty set of vertices of G. Assume to the contrary that $G - U$ has $k \geq |U|$ isolated vertices. Denote by I the set of isolated vertices in $G - U$ and let v be a fixed vertex in U. Consider the weights assignment $w : V \rightarrow \mathbb{N}_0$ such that $w(v) = 1$ and for any other vertex $v' \in V$ we have $w(v') = 0$. We reach a contradiction by showing that w is not equatable. To see this observe that any step that increases by 1 the weight of a vertex in I must increase by 1 the weight of some vertex in U. It follows that at any moment the sum of the weights of the vertices in I is strictly smaller than the sum of the weights of the vertices in U. Because $|I| \geq |U|$, it is not possible to reach a situation where all vertices in $I \cup U$ have the same weight. $\qquad\square$

3   A Polynomial-Time Algorithm to Equate the Weights

In this section, we deal with the computational problem of deciding whether an initial assignment $w : V \to \mathbb{N}_0$ is equatable and, if so, how to compute a multiset of edges that leads to such equated weights. Let us recall the connection to the b-matching problem. If we know a number $\beta \in \mathbb{N}$ such that all node-weights can be brought to β by increment-steps, then the multiset of edges leading to uniform weights β is a perfect b-matching with weights $b(v) = \beta - w(v)$ for each $v \in V$. The primary question is then: Can β be efficiently computed? We will now give a positive answer to this question. The main result of this section is the following theorem. We first provide an upper bound on β.

Theorem 3. *Given a graph $G = (V, E)$ and an integer weights assignment $w : V \to \mathbb{N}_0$, one can determine in strongly polynomial time whether w is positively equatable in G. Moreover, the smallest multiset of edges equating w can be determined efficiently.*

We again make use of Theorem 2. For some target value $\beta \geq \max_{v \in V} w(v)$ let $b_\beta : V \to \mathbb{Z}_{\geq 0}$ with $b_\beta(v) := \beta - w(v)$. By Theorem 2 there is a sequence of positive steps starting from node weights w and yielding uniform node weight β if and only if (2) holds for $b = b_\beta$ and all subsets U of V.

Lemma 1. *If w is equatable, there is such a value β that is bounded from above by $n \max_{v \in V} w(v)$.*

Proof. For the trivial case where w is uniformly zero we can choose $\beta = 0$. Thus, in what follows we might assume that $\max_{v \in V} w(v) \geq 1$. Notice that with respect to (2) the only subsets U of V that might force β to be big are those with $|U| > |I(U)|$ (otherwise, if $|U| \leq |I(U)|$, the left hand side of (2) as a function of β increases at most as fast as the right hand side does). For such subset U, however, and for $\beta = n \max_{v \in V} w(v)$ we get

$$\sum_{x \in U} b_\beta(x) \geq |U|(\beta - \max_{v \in V} w(v))$$

$$\geq |U|\beta - n \max_{v \in V} w(v) + n - |U|$$

$$= (|U| - 1)\beta + n - |U| \geq \sum_{x \in I(U)} b_\beta(x) + S(G - U)$$

which concludes the proof. $\qquad\square$

Proof (Proof of Theorem 3). From now on we fix the parity of β (even or odd) such that $S(G - U)$ no longer depends on the particular value of β (in our algorithm we deal with the two cases sequentially). In particular, for a fixed subset U of V, both the left hand side and the right hand side of (2) are linear functions of β. Therefore, for each $U \subseteq V$, one of three cases holds:

(i) (2) is satisfied for all values of β or for no value of β;
(ii) there is a $\beta_U \in \mathbb{Z}$ such that (2) is satisfied if and only if $\beta \geq \beta_U$;
(iii) there is a $\beta_U \in \mathbb{Z}$ such that (2) is satisfied if and only if $\beta \leq \beta_U$.

This observation finally enables us to find the smallest feasible value of β (with fixed parity) by binary search in polynomial time: Let $\alpha = \max_{v \in V} w(v)$ and $\gamma = n \max_{v \in V} w(v)$. Due to Lemma 1, we can restrict our search for a suitable value β to the interval $[\alpha, \gamma]$. For fixed $\beta' \in [\alpha, \gamma]$, we can test in polynomial time whether (2) is satisfied for all subsets U of V and obtain a violating subset U in the negative case. In fact such such a violating set is found by the algorithm for the perfect $b_{\beta'}$-matching problem that also certifies the non-existence of a perfect $b_{\beta'}$-matching with a set $U \subseteq V$ violating (2).

In the positive case, we can decrease the upper bound γ to β' and continue the search. In the negative case, we distinguish the three cases (i), (ii), and (iii) listed above w.r.t. the violating subset U. In case (i), there is no feasible β and we thus terminate the search. In case (ii) we obtain a new lower bound $\beta_U > \beta'$ and thus continue the search after replacing α with β_U. Finally, in case (iii) we obtain a new upper bound $\beta_U < \beta'$ and thus continue the search after replacing γ with β_U.

Notice that the running time of the resulting binary search algorithm is only weakly polynomial. A strongly polynomial running time can be achieved by replacing binary search with *parametric search* [14]. □

Remark 1. We sketch Megiddo's parametric search technique [14] in our setting. Suppose that we want to find the smallest even β such that there exists a perfect b_β-matching. Consider a fully combinatorial algorithm A for the non-parametric problem, that is, for the perfect b-matching problem. A fully combinatorial algorithm uses only additions, subtractions, and comparisons. In fact, such an algorithm exists for the perfect b-matching problem, if the parities of the $b(v)$ are fixed, as it is for the case of all parametric b_β, if β is even, see [8, p. 186]. More precisely, the algorithm consists then of solving one general network flow problem and one perfect matching problem on graphs that are polynomial in the size of the graph G that is in our input, see, [8]. For both subproblems, there exist fully-combinatorial algorithms, for example the minimum mean-cycle algorithm [9] and Edmond's algorithm [5].

Algorithm A is now modified in order to solve the parametric problem. For this, the modified algorithm has to work with linear functions of the parameter β instead of just constant numbers. Notice that adding or subtracting two linear functions yields a linear function again. Comparing two linear functions, however, imposes a problem. Whenever algorithm A compares two numbers, the modified version first has to determine whether the desired value β_{OPT} is smaller or larger than the unique point β^* at which the two linear functions cross (if at all). This can be decided by calling any algorithm B for the perfect b-matching problem as a subroutine for the fixed parameter value β^*. Depending on the outcome, the corresponding alternative of the comparison is chosen, for example in an `if`-conditional, and one continues to run algorithm A for the parametric problem. The number of calls of B is bounded by the number of comparisons performed by A which is strongly polynomial in the input size. In this way, finding the desired value β_{OPT} is reduced to a series of non-parametric b-matching problems.

4   Bipartite Graphs

Going back to the elementary puzzle presented at the beginning, observe that the corresponding graph G is bipartite where the two parts have the same cardinality. The initial weight w is not equatable, since the sum of the weights of the vertices in one part of the bi-partition is not equal to the sum of the weights.

Let $G = (L, R, E)$ be a bipartite graph. An assignment of weights w to the vertices of G is called *balanced* if $w(L) = w(R)$, where for a subset U of vertices, $w(U)$ is defined as $\sum_{v \in U} w(v)$.

We now characterize those bipartite graphs for which all balanced assignment w are equatable.

Theorem 4. *Let $G = (L + R, E)$ be a bipartite graph. The following statements are equivalent:*

i) Every balanced assignment is positively equatable in G.

ii) For any non empty set of vertices X that is properly contained either in L or in R, $|X| < |N(X)|$.

Here $N(X)$ denotes the *neighborhood* of X, that is, the set of vertices in G that are neighbors of some vertex in X. Notice that condition ii) implies that $|L| = |R|$ holds. Condition ii) is a "strict" version of the well known *Hall's condition* for the existence of a perfect matching. A bipartite graph has a perfect matching if and only if for any non empty set of vertices X that is properly contained either in L or in R, $|X| \leq |N(X)|$.

Proof. Suppose that every balanced assignment to the vertices of G is positively equatable. Assume to the contrary that there exists $X \subset L$, $0 < |X| < |L|$, and $|X| \geq |N(X)|$ (the symmetric case where $X \subset R$ is similar). If $N(X) = \emptyset$, then the vertices in X are isolated and any balanced assignment of weights to the vertices in G where the vertices in X get weight 0 and some other vertex not in X gets a positive weight is not equatable.

If $N(X)$ is not empty, consider the following balanced assignment of weights to the vertices of G. Pick a vertex $v \in L \setminus X$ and a vertex $u \in N(X)$. We define $w(v) = w(u) = 1$ and for every other vertex z of G we define $w(z) = 0$. Clearly, w is balanced. However, the graph G together with the assignment of weights w is not equatable. This is because a positive step increases the total weight of the vertices in $N(X)$ by at least the same amount by which it increases the total weight of the vertices in X. If after a series of positive steps the weights of the vertices in G are the same, then in particular the total weight of the vertices in X is at least as large as the total weight of the vertices in $N(X)$ (because $|X| \geq |N(X)|$), but this is impossible because for the initial assignment of weight the total weight of the vertices in X is 0 while the total weight of the vertices in $N(X)$ is 1.

Now, suppose that ii) holds. As in the proof of Theorem 1, it is enough to show that any assignment with $w(u) = w(v) = 1$ for some $u \in L$ and $v \in R$ and $w(x) = 0$ for any other vertex is equatable, since then, each balanced assignment can be equated. This however follows from the fact that $G - \{u, v\}$ has a perfect matching, since it satisfies Hall's condition.   □

4.1   Hypergraphs

One can naturally generalize the equating problem to hypergraphs. In this setting, one is given a hypergraph $H = (V, E)$ and an integer weights assignment $w : V \to \mathbb{N}_0$. A *a positive step* on $e \in E$ modifies w by increasing the weights of each $u \in E$ by 1 unit. The rest of the definitions are generalized in the obvious way. Not surprisingly, deciding, whether one can equate the weights in a hypergraph is NP-complete. This follows by a reduction from 3-*dimensional matching* [11].

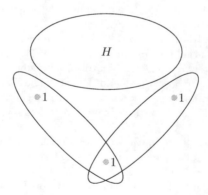

Fig. 2. NP-completeness in the hypergraph setting.

Thus deciding whether a hypergraph has a perfect matching is an NP-complete problem. This can be trivially reduced to the equating problem by adding three new vertices and two new edges, each consisting of two of the three new vertices. The three vertices have weight 1 while all other vertices have weight 0. If these weights can be equated, then they all have weight 1 in an equated assignment. Thus, the weights can be equated if and only if the original hypergraph has a perfect matching. Consequently, the following theorem holds.

Theorem 5. *The decision problem of determining for any hypergraph $H = (V, E)$ and any integer weights assignment $w : V \to \mathbb{Z}$, whether w is positively equatable in H is NP-complete.*

References

1. Anstee, R.P.: A polynomial algorithm for b-matchings: an alternative approach. Information Processing Letters 24(3), 153–157 (1987)
2. Berge, C.: Regularisable graphs i. Discrete Mathematics 23(2), 85–89 (1978)
3. Cunningham, W.H., Marsh, A.B.: A primal algorithm for optimum matching. In: Polyhedral Combinatorics, pp. 50–72. Springer (1978)
4. Edmonds, J.: Maximum matching and a polyhedron with 0,1-vertices. Journal of Research of the National Bureau of Standards 69, 125–130 (1965)
5. Edmonds, J.: Paths, trees and flowers. Canadian Journal of Mathematics 17, 449–467 (1965)
6. Harold, N.: Gabow. Data structures for weighted matching and nearest common ancestors with linking. In: Proceedings of the First Annual ACM-SIAM Symposium on Discrete Algorithms, pp. 434–443. Society for Industrial and Applied Mathematics (1990)
7. Gabow, H.N., Tarjan, R.E.: Faster scaling algorithms for general graph matching problems. Journal of the ACM (JACM) 38(4), 815–853 (1991)
8. Gerards, A.M.H.: Matching. In: Ball, M.O., Magnanti, T.L., Monma, C.L., Nemhauser, G.L. (eds.) Network Models. Handbooks in Operations Research and Management Science, vol. 7, pp. 135–224. North-Holland, Amsterdam (1995)
9. Goldberg, A.V., Tarjan, R.E.: Finding minimum-cost circulations by canceling negative cycles. J. ACM 36(4), 873–886 (1989)
10. Hopcroft, J.E., Karp, R.M.: An n^5/2 algorithm for maximum matchings in bipartite graphs. SIAM Journal on Computing 2(4), 225–231 (1973)
11. Karp, R.M.: Reducibility among combinatorial problems. In: Complexity of Computer Computations, pp. 85–103. Plenum Press, NY (1972)
12. Karzanov, A.V.: On finding a maximum flow in a network with special structure and some applications. Matematicheskie Voprosy Upravleniya Proizvodstvom 5, 81–94 (1973)
13. Madry, A.: Navigating central path with electrical flows: From flows to matchings, and back. In: 2013 IEEE 54th Annual Symposium on Foundations of Computer Science (FOCS), pp. 253–262. IEEE (2013)
14. Megiddo, N.: Combinatorial optimization with rational objective functions. Math. Oper. Res. 4(4), 414–424 (1979)
15. Schrijver, A.: Combinatorial optimization. Polyhedra and efficiency (3 volumes). Algorithms and Combinatorics, vol. 24. Springer, Berlin (2003)
16. Tutte, W.T.: The factors of graphs. Canad. J. Math. 4(3), 314–328 (1952)

The Price of Matching with Metric Preferences

Yuval Emek[1], Tobias Langner[2], and Roger Wattenhofer[2]

[1] Technion, Israel
[2] ETH Zürich, Switzerland

Abstract. We consider a version of the Gale-Shapley stable matching setting, where each pair of nodes is associated with a (symmetric) matching cost and the preferences are determined with respect to these costs. This stable matching version is analyzed through the Price of Anarchy (PoA) and Price of Stability (PoS) lens under the objective of minimizing the total cost of matched nodes (for both the marriage and roommates variants). A simple example demonstrates that in the general case, the PoA and PoS are unbounded, hence we restrict our attention to metric costs. We use the notion of α-stability, where a pair of unmatched nodes defect only if both improve their costs by a factor greater than $\alpha \geq 1$. Our main result is an asymptotically tight trade-off, showing that with respect to α-stable matchings, the Price of Stability is $\Theta\big(n^{\log(1+\frac{1}{2\alpha})}\big)$. The proof is constructive: we present a simple algorithm that outputs an α-stable matching satisfying this bound.

1 Introduction

The aim of this paper is to connect two classic approaches towards *matching*. The first approach tackles matching from a (global) optimization angle à la Edmonds [15, 16]: given $2n$ nodes with pairwise costs $c(x, y) = c(y, x) \in \mathbb{R}_{>0}$, the goal is to construct a perfect matching that minimizes the total cost. The second approach tackles matching from the (local) selfish angle à la Gale and Shapley [18]: each node is equipped with a preference list ranking its potential matches and a matching is *stable* if no two unmatched nodes prefer each other over their current matches.

We consider a restricted case of the stable matching realm, where the nodes preferences are determined based on the aforementioned pairwise costs $c(\cdot, \cdot)$ so that node x prefers node y over node y' if and only if $c(x, y) < c(x, y')$, and focus on the following question: How does the requirement to output a (locally) stable matching affect its (global) total cost? In attempt to provide a quantitative answer to this question, we shall look at matching instances through the *Price of Stability (PoS)* lens that compares the min-cost stable matching to the unrestricted optimum, measuring the ratio of their respective costs. In fact, to provide a deeper understanding of the delicate balance between the global matching cost and its local stability, we generalize the problem by using the notion of an α-*stable* matching for $\alpha \geq 1$, in which no pair of unmatched nodes can defect and thus improve their costs by a factor (strictly) greater than α.

© Springer-Verlag Berlin Heidelberg 2015
N. Bansal and I. Finocchi (Eds.): ESA 2015, LNCS 9294, pp. 459–470, 2015.
DOI: 10.1007/978-3-662-48350-3_39

Unfortunately, in general, the Price of Stability may be unbounded, as the following simple example shows: Let G be a complete graph on four nodes u_1, u_2, v_1, v_2 with edge costs $c(u_1, v_1) = c(u_2, v_2) = 1$, $c(u_1, u_2) = \varepsilon$ for some small $\varepsilon > 0$, and $c(v_1, v_2) = c(u_1, v_2) = c(u_2, v_1) = C$ for some large C. Then, the optimal perfect matching matches u_i to v_i for $i = 1, 2$ with a cost of 2, whereas an α-stable matching for any reasonable value of α must match u_1 to u_2, and hence also v_1 to v_2 which incurs an arbitrarily large cost.

Fortunately, real-world matching instances often exhibit *metric* costs, i.e., costs that satisfy the triangle inequality (or its bipartite counterpart). Metric costs are intuitive for matching instances in which the costs are determined by distances, but we argue that they are present also in matching instances with more complex cost functions, e.g., online dating platforms — refer to the full version of this paper for a comprehensive explanation that also addresses the role that PoS plays in these matching scenarios.

The main result of this paper is an asymptotically tight tradeoff between the parameter α and the PoS considering α-stable matchings: the PoS is roughly $n^{0.58}$ when $\alpha = 1$ and it decreases exponentially as α increases. Since this tradeoff is realized by a simple poly-time algorithm, the designers of a matching system can now efficiently tune the parameter α to balance between the stability and the total cost of their system's output.

1.1   Related Work

Studying the impact of selfish players has been a major theoretical computer science success story in the last decade (see, e.g., the 2012 *Gödel Prize* [28,31,38]). In particular, much effort has been invested in quantifying how the efficiency of a system degrades due to selfishness of its players. The most notable notions in this context are the *Price of Anarchy (PoA)* [28,32] and the *Price of Stability (PoS)* [6,39], comparing the best possible outcome to the outcome of the worst (PoA) or best (PoS) solution with selfish players. Since their introduction, the Price of Anarchy and the Price of Stability have been extensively analyzed in diverse settings such as selfish routing [6,9,12,13,37,38,40], network formation games [3,4,7,11,41], job scheduling [10,14,27,28], and resource allocation [25,36]. While selfish players are traditionally modeled using the *Nash equilibrium* solution concept, where no player can benefit from a unilateral deviation, in matching settings unilateral deviations are not natural. Instead, we want that no *two* unmatched players prefer each other over their current matching partners. This solution concept is generally known as the Gale-Shapley *stable matching* [18].

For the most part, the stable matching realm has been subdivided into two versions: the *marriage* (bipartite) version, where the players are partitioned into *men* and *women* and each man (resp., woman) is equipped with a list of preferences over the set of women (resp., men); and the *roommates* (all-pairs) version, where each player is equipped with a list of preferences over all other players. Gale and Shapley showed that in the bipartite version, a stable matching always exists, and in fact, can be computed by a simple poly-time algorithm. In contrast, the all-pairs version does not necessarily have a solution. Both versions of the stable matching problem and their manifold variants (strictly/weakly ordered preferences, (in-)complete preference lists, (a-)symmetric preferences) admit an abundance of literature; see, e.g., the books of Knuth [26], Gusfield and Irving [19], Roth and Sotomayor [35], and Manlove [29]. The notion of stability studied in this paper has been coined as *weak stability* by Irving [22].

Sometimes, the players' preferences are associated with real costs so that each preference list is sorted in order of non-decreasing costs. This setting gives rise to the *minimum-cost stable matching* problem, where the goal is to construct a stable matching that minimizes the total cost of matched partners. Irving et al. [23] designed a poly-time algorithm for the bipartite (marriage) version of a special case of this problem, referred to as the *egalitarian stable matching* problem, where a cost of j is associated with each player for matching his/her jth preferred partner. Roth et al. [34] gave an LP-based solution to the problem. Irving's work was generalized by Feder [17] who presented a poly-time algorithm for the bipartite version of the general minimum-cost stable matching problem. Moreover, Feder also established the NP-hardness of the all-pairs (roommates) version and showed that it admits a 2-approximation algorithm.

The players' preferences in general stable matching scenarios exhibit no intrinsic correlations. Several approaches have been taken towards introducing consistency in the preference lists [21, 24, 26, 30]. Most relevant to the current paper is the approach of Arkin et al. [8] who studied the *geometric* stable roommate problem, where the players are identified with points in a Euclidean space and the preferences are given by the sorted distances to the other points. They showed that in the geometric setting, a stable matching always exists and that it is unique if the players' preferences exhibit no ties. These results easily generalize to arbitrary metric spaces. Arkin et al. also introduced the notion of an α-*stable* matching, which is central to the current paper.

There is an extensive literature on matching instances whose preferences are determined by the numerical attributes of the edges, interpreted as gains that should be maximized, rather than costs that should be minimized (cf. *correlated two-sided markets*) [1, 2, 5, 20]. Closely related to the goal of the current paper, Anshelevich et al. [5] establish tight tradeoffs between the matching stability parameter α and the PoA and PoS in the bipartite case under this gain maximization variant. In fact, the simple iterative algorithm presented in Sec. 4.1 is equivalent to the algorithm used in the proof of Theorem 2 in [5], but as it turns out, analyzing the quality of the resulting (α-stable) matching under the cost minimization variant studied in the current paper is much more demanding.

Reingold and Tarjan [33] proved that the approximation ratio of some greedy algorithm for minimum-cost perfect matching in metric graphs is $\Theta(n^{\log(3/2)})$ where $\log(3/2) \approx 0.58$.[1] It turns out that this result is equivalent to establishing the same bound for the PoA of minimum-cost perfect matching in such graphs. In the full version of this paper, we give a simpler proof for the PoA-result and extend their result to obtain a lower bound for the PoS for all $\alpha \geq 1$.

---

[1] In this paper, $\log x$ denotes the logarithm of x to the base of 2.

2   Setting and Preliminaries

Consider a graph G with vertex set $V(G)$ and edge set $E(G)$. Each edge $e \in E(G)$ is assigned a positive real *cost* $c(e)$. Unless stated otherwise, our graphs have $2n$ vertices, $n \in \mathbb{Z}_{>0}$, and are either complete ($|E(G)| = \binom{2n}{2}$) or complete bipartite ($V(G) = U_1 \cup U_2$, $|U_1| = |U_2| = n$ and $|E(G)| = n^2$). We say that the complete graph G is *metric* if $c(x, y) \le c(x, z) + c(z, y)$ for every $x, y, z \in V(G)$; we say that the complete bipartite graph G is *metric* if $c(x, y) \le c(x, z) + c(z, z') + c(z', y)$ for every $x, y, z, z' \in V(G)$, where x, z' and y, z are on opposite sides of G. For an arbitrary graph G, the *distance* $\mathrm{dist}_G(x, y)$ of two vertices x and y of G is defined as the weighted length of the shortest path between x and y in G.

A *matching* is a subset $M \subseteq E(G)$ of the edges such that every vertex in $V(G)$ is incident to at most one edge in M. A matching is called *perfect* if every vertex in $V(G)$ is incident to exactly one edge in M, which implies that $|M| = n$ as $|V(G)| = 2n$. For a perfect matching M and a vertex $x \in V(G)$, we denote by $M(x)$ the unique vertex $y \in V(G)$ such that $(x, y) \in M$. Unless stated otherwise, all matchings mentioned hereafter are assumed to be perfect. (Perfect matchings clearly exist in a complete graph with an even number of vertices and in a complete balanced bipartite graph.) Given an edge subset $F \subseteq E(G)$, we define the *cost* of F as the total cost of all edges in F, denoted by $c(F) = \sum_{e \in F} c(e)$; in particular, the cost of a matching is the sum of its edge costs.

Definition (α-Stable Matching). *Consider some (perfect) matching $M \subseteq E(G)$ and some real number $\alpha \ge 1$. An edge $(u, v) \notin M$ is called α-unstable (a.k.a. α-blocking) with respect to M if $\alpha \cdot c(u, v) < \min\{c(u, M(u)), c(v, M(v))\}$. Otherwise, the edge is called α-stable. A matching M is called α-stable if it does not admit any α-unstable edge. We will omit α and call edges as well as matchings just* stable *or* unstable *whenever α is clear from the context or the argumentation holds for every choice of α.*

Let M^* denote a certain (perfect) matching M that minimizes $c(M)$. For simplicity, in what follows, we restrict our attention to complete (rather than complete bipartite) metric graphs, although all our results hold also for the complete bipartite case (following essentially the same lines of arguments).

Definition (α-Price of Stability). *The α-Price of Stability of G, denoted by $PoS_\alpha(G)$, is defined as $PoS_\alpha(G) = \min\{c(M)/c(M^*) : M \text{ is } \alpha\text{-stable matching}\}$. Furthermore, $PoS_\alpha(2n) = \sup\{PoS_\alpha(G) : G \text{ is metric}, |V(G)| = 2n\}$. Unless stated otherwise, when the parameter α is omitted, we refer to the case $\alpha = 1$.*

Definition (Price of Anarchy). *The* Price of Anarchy *of a graph G, denoted by $PoA(G)$, is defined as $PoA(G) = \max\{c(M)/c(M^*) : M \text{ is stable matching}\}$. Furthermore, $PoA(2n) = \sup\{PoA(G) : G \text{ is metric}, |V(G)| = 2n\}$.*

Note that since any stable matching by definition is also α-stable for any $\alpha \ge 1$, the Price of Anarchy does not improve by considering α-stability and hence its definition does not include the parameter α.

3   Price of Anarchy

The following theorem was implicitly proven by Reingold and Tarjan [33] in 1981. They showed that for minimum-cost perfect matching in metric graphs, the approximation ratio of the algorithm that picks edges by ascending costs is $\Theta(n^{\log(3/2)})$. Since the matching returned by this greedy algorithm is stable and since every stable matching can be obtained from the algorithm by an appropriate tie-breaking policy, it follows that the PoA of minimum-cost perfect matching in such graphs is also $\Theta(n^{\log(3/2)})$. A simpler and more intuitive proof for Reingold and Tarjan's 30 years old result is given in the full version of this paper.

Theorem 1. *The PoA of minimum-cost perfect matching in metric graphs with $2n$ vertices is $\Theta(n^{\log(3/2)})$.*

4   Price of Stability

The upper bound established on the PoA in Sec. 3 clearly holds for the PoS too. In the full version of this paper, we show that the proof technique for the $\Omega(n^{\log(3/2)})$-lower bound of Sec. 3 can be easily adapted to establish the same lower bound for the PoS as well. In fact, we generalize this result, showing that $\text{PoS}_\alpha(2n) = \Omega\left(n^{\log(1+1/(2\alpha))}\right)$ for every $\alpha \geq 1$. Consequently, we turn our attention to bounding $\text{PoS}_\alpha(2n)$ from above, establishing the following asymptotically tight upper bound.

Theorem 2. *The PoS_α of minimum-cost perfect matching in metric graphs with $2n$ vertices is at most $3 \cdot n^{\log(1+1/(2\alpha))}$.*

The proof of Theorem 2 is constructive, relying on a simple algorithm presented in Sec. 4.1. Sec. 4.2 provides the analysis of this algorithm, showing that the returned matching indeed satisfies the bound. Full proofs missing from this section can be found in the full version of this paper.

4.1   An Algorithm for α-Stable Matchings

The following algorithm STAB transforms a minimum-cost matching M^* in a metric graph into an α-stable matching M.

ALGORITHM STAB: Start with the minimum-cost matching $M \leftarrow M^*$ and iterate over all edges of G by non-decreasing order of costs. If the edge (u, v) currently considered is α-unstable in the current matching M, replace the edges $(u, M(u))$ and $(v, M(v))$ in M by (u, v) and $(M(u), M(v))$ (this operation is called a *flip* of the edge (u, v)) and continue with the next edge. After having iterated over all edges, return M.

We assume that edge cost ties are resolved in an arbitrary but consistent manner. In the following, we denote by M_i the matching calculated by the above algorithm at the end of iteration i. Moreover, $M_0 = M^*$ is the initial minimum-cost matching and M_S the final matching returned by STAB.

Lemma 3. *For any unstable edge b created by the flip of an edge e, we have $c(b) > c(e)$.*

Corollary 4 follows by induction on i. Lemma 5 then follows by a straightforward analysis of the algorithm's run-time.

Corollary 4. *Let e_i be the edge considered in iteration i. Then for any unstable edge b in M_i it holds that either $c(e_i) < c(b)$ or b will be considered in a later iteration $j > i$.*

Lemma 5. *Algorithm STAB transforms a minimum-cost matching into a valid α-stable matching in time $\mathcal{O}(n^2 \log n)$.*

4.2   Cost Analysis

Our goal in this section is to show that when STAB is invoked with parameter α for any $\alpha \geq 1$, it returns an α-stable matching M_S satisfying $c(M_S) = c(M^*) \cdot \mathcal{O}(n^{\log(1+1/(2\alpha))})$. Since this section makes heavy use of rooted binary trees and their properties, we require a few definitions. In a *full binary tree*, each inner node has exactly two children. The *depth* $d(v)$ of a node v in a tree T is the length of the unique path from the root of T to v and the *height* $h(T)$ of a tree T is defined as the maximal depth of any node in T. The *height* $h(v)$ of a node v of T is defined to be the height of its subtree. The *leaf set* $\mathcal{L}(T)$ or $\mathcal{L}(F)$ of a tree T or a collection F of trees is the set of all leaves in T or F, resp. The *leaf set* $\mathcal{L}(v)$ of a node v in a tree is $\mathcal{L}(T_v)$ where T_v is the subtree rooted at v. Finally, two nodes with the same parent are called *sibling nodes*. We begin with Lemma 6 stating an important property of the edges that are flipped by STAB.

Lemma 6. *If an edge e is flipped in iteration i, then $e \in M_j$ for all $j \geq i$ and, in particular, $e \in M_S$.*

Consider an iteration of STAB where edge (u, v) is flipped because it was unstable at the beginning of the iteration. Then the two edges $(u, M(u))$ and $(v, M(v))$ are replaced by (u, v) and $(M(u), M(v))$. Since the edge (u, v) is selected irrevocably according to Lemma 6, the edges $(u, M(u))$ and $(v, M(v))$ can never be part of M again. The only edge, of the four edges involved, that may be changed again, is the edge $(M(u), M(v))$. Thus, we refer to $(M(u), M(v))$ as an *active* edge. We also refer to all edges in M_0 as active. Using the notion of active edges, we shall now model the changes that STAB applies to the matching during its execution through a logical helper structure called the *flip forest*. To avoid confusion between the basic elements of our graphs and the basic elements of the flip forest, we refer to the former as vertices/edges and to the latter as nodes/links.

Definition (Flip Forest). *The flip forest $F = (U, K)$ for a certain execution of STAB is a collection of disjoint rooted trees and has node set U and link set K. For each edge $e \in V \times V$ that has been active at some stage during the execution, there exists a node $u_e \in U$. This correspondence is denoted by $u_e \sim e$. For each* *flip of an edge (u, v) in G, resulting in the removal of the edges $(u, M(u))$ and $(v, M(v))$ from M, K contains a link connecting the node $y \sim (u, M(u))$ to its parent $x \sim (M(u), M(v))$ and a link connecting the node $z \sim (v, M(v))$ to its parent $x \sim (M(u), M(v))$. (Observe that, by definition, all three edges $(u, M(u))$, $(v, M(v))$, and $(M(u), M(v))$ are active.) Refer to the full version of this paper for an illustration.*[2]

The definition of a flip forest ensures that for each flip of the algorithm, we obtain a binary *flip tree segment*. When we transcribe each flip operation of the complete execution of STAB into a flip tree segment as explained above, we end up with a collection of full binary trees — the *flip forest*. This is because the parent node of a tree segment may appear as a child node of the tree segment corresponding to a later iteration of the algorithm since its corresponding edge is still active and therefore may participate in another flip. Each such tree is called a *flip tree* hereafter.

Observe that all leaves (including isolated nodes) in the flip forest correspond to edges in the minimum-cost matching $M_0 = M^*$. The edges in the matching M_S are implicitly represented by the flip forest: An edge that gets flipped — and is therefore irrevocably selected into M_S — has no corresponding node in F, but we may associate it with the node corresponding to the active edge resulting from the flip. On top of these edges, M_S contains the edges corresponding to the roots of the trees in the flip forest.

We now define a function $\psi : U \mapsto \mathbb{R}$ that maps a real *weight* to each node in the flip forest F as follows. For each leaf ℓ of a flip tree in F, we set $\psi(\ell) := c(e)$, where $\ell \sim e$ and we recall that an edge corresponding to a leaf node in F is part of M^*. The function ψ is extended to an inner node x of a flip tree with child nodes y and z by the recursion

$$\psi(x) := \psi(y) + \psi(z) + (1/\alpha) \cdot \min\{\psi(y), \psi(z)\} \ . \tag{1}$$

For ease of notation, we call the child with smaller (resp., larger) weight as well as the link leading to its parent *light* (resp., *heavy*); ties are resolved arbitrarily. We denote the light child of a node x as x_L and the heavy child as x_H. Then we can rewrite Eq. (1) as $\psi(x) := \psi(x_H) + (1 + 1/\alpha) \cdot \psi(x_L)$.

Lemma 7. *Let x be a node in F and e an edge in G with $x \sim e$. Then $c(e) \leq \psi(x)$.*

At this stage, we would like to relate the weight $\psi(r_T)$ of the roots r_T in F to the cost of the stable matching M_S returned by STAB. To that end, we observe that M_S consists of the edges corresponding to the roots in F and to the edges that have been flipped along the course of the execution; let R and D denote the set of the former and latter edges, respectively. Observe that

$$c(M_S) = \sum_{e \in R} c(e) + \sum_{e \in D} c(e) \ .$$

---

[2] All figures are deferred to the full version of this paper.

Consider the flip of the edge (u, v) resulting in the insertion of the edge $(M(u), M(v)) \sim x$ to M and the removal of the edges $(u, M(u)) \sim x_L$ and $(v, M(v)) \sim x_H$ from M. Since $\psi(x) = \psi(x_H) + (1 + 1/\alpha) \cdot \psi(x_L)$, we have $\psi(x) - (\psi(x_L) + \psi(x_H)) = \psi(x_L)/\alpha$. Lemma 7 then implies that $\psi(x) - (\psi(x_L) + \psi(x_H)) \geq c(u, M(u))/\alpha$, and since edge (u, v) was flipped, we have $\psi(x) - (\psi(x_L) + \psi(x_H)) \geq c(u, v)$. Therefore,

$$\sum_{e \in D} c(e) \leq \sum_{\substack{\text{internal } x \in U}} (\psi(x) - (\psi(x_L) + \psi(x_H)))$$

$$= \sum_{\text{flip trees } T} \left(\psi(r_T) - \sum_{\ell \in \mathcal{L}(T)} \psi(\ell) \right)$$

$$= \sum_{\text{flip trees } T} \psi(r_T) - \sum_{\ell \in \mathcal{L}(F)} \psi(\ell) \ ,$$

where the second equation holds by a telescoping argument. Note further that $\sum_{e \in R} c(e) \leq \sum_{\text{flip trees} T} \psi(r_T)$ and thus

$$c(M_S) \leq 2 \sum_{\text{flip trees } T} \psi(r_T) - \sum_{\ell \in \mathcal{L}(F)} \psi(\ell) \ .$$

Since $c(M^*) = \sum_{\ell \in \mathcal{L}(F)} \psi(\ell)$, Corollary 8 follows.

Corollary 8. *The matching M_S returned by* STAB *satisfies*

$$c(M_S) \leq 2 \sum_{\text{flip trees } T} \psi(r_T) - c(M^*) \ .$$

We will now have a closer look at the properties of our flip trees and their weights. It will be convenient to ignore the relation of the flip trees to the STAB algorithm at this stage; in other words, we consider an abstract full binary tree T with a *leaf weight function* $w : \mathcal{L}(T) \to \mathbb{R}_{\geq 0}$. For any leaf ℓ of T, we set $\psi(\ell) = w(\ell)$ and determine the weight $\psi(x)$ of each inner node x in T following the recursion given by Eq. (1). Note that we allow our tree T to have zero-weight leaves now (this can only make our analysis more general).

Definition (Complete Binary Tree). *A full binary tree T is called* complete *if all leaves are at depth $h(T)$ or $h(T) - 1$. Given some positive integer n that will typically be the number of leaves in some tree, let $h(n) = \lceil \log n \rceil$ and $k(n) = 2^{h(n)} - n$. Note that $0 \leq k(n) < 2^{h(n)-1}$.*

Observe that for a complete full binary T with n leaves, $h(n)$ is the height $h(T)$ of T while $k(n)$ equals the number of missing leaves at the maximum depth $h(T)$.

Definition (ψ-Balanced Binary Tree). *A full binary tree T is called ψ-balanced if for any two sibling nodes x, y in T, we have $\psi(x) = \psi(y)$.*

Consider a full binary tree T. Let $\Lambda(T)$ denote the sum of the weights of the leaves of T, i.e., $\Lambda(T) = \sum_{\ell \in \mathcal{L}(T)} w(\ell) = \sum_{\ell \in \mathcal{L}(T)} \psi(\ell)$, and let $\Psi(T) = \psi(r_T)$ (recall that r_T denotes the root of T). The following observation is established by induction on the node depth.

Observation 1. *For any node v of a ψ-balanced full binary tree T, we have*
$\psi(v) = (2 + 1/\alpha)^{-d(v)} \cdot \Psi(T)$.

Definition (Effect of a Flip Tree). *The* effect $\eta(T)$ *of a full binary tree T is defined as*

$$\eta(T) = \begin{cases} \Psi(T)/\Lambda(T) & \text{if } \Lambda(T) > 0 \\ 1 & \text{if } \Lambda(T) = 0 \end{cases}.$$

An n-leaf full binary tree T is said to be effective *if it maximizes $\eta(T)$, i.e., if there does not exist any n-leaf full binary tree T' such that $\eta(T') > \eta(T)$.*

Intuitively speaking, if we think of T as a flip tree, then its effect is a measure for the factor by which the flips represented by T increase the cost of M^* when applied to it. But, once again, we do not restrict our attention to flip trees at this stage. The effect of a full binary tree is essentially determined by its topology and by the assignment of weights to its leaves. It is important to point out that the effect of a flip tree is invariant to scaling its leaf weights (see full version of this paper). Our upper bound is established by showing that the effect of an effective n-leaf full binary tree is $\mathcal{O}\left(n^{\log(1+1/(2\alpha))}\right)$. We begin by developing a better understanding of the topology of effective ψ-balanced full binary trees.

Lemma 9. *An effective n-leaf ψ-balanced full binary tree must be complete.*

Proof (sketch). Aiming for a contradiction, suppose that T is not complete. Let z be an internal node at depth d with leaf children x, y (whose depth is $d + 1$) and let z' be a leaf at depth $d' < d$. Let T' be the full binary tree obtained from T by deleting x and y and inserting two new leaves x', y' as children of z'. Let w and w' be the leaf weigh functions of T and T', respectively, defined by requiring that T and T' are ψ-balanced and scaled so that $\Psi(T) = \Psi(T') = 1$; this is well defined since by Observation 1, the ψ-values of all nodes in T and T' (and in particular, the leaf weight functions w and w') are fully determined by their topology and the values of $\Psi(T)$ and $\Psi(T')$ (in a top-down fashion).

We establish the proof by arguing that $\Lambda(T') < \Lambda(T)$ which implies $\eta(T') > \eta(T)$, in contradiction to T being effective. To that end, notice that the construction of T' implies $\Lambda(T') = \Lambda(T) + w'(x') + w'(y') + w'(z) - (w(x) + w(y) + w(z'))$. The assertion follows from Observation 1 by a direct calculation. □

Next, we develop a closed-form expression for the effect of complete ψ-balanced full binary trees. We define the function $\varphi : \mathbb{Z}_{>0} \mapsto \mathbb{R}$ as

$$\varphi(n) := \frac{(2 + 1/\alpha)^{h(n)}}{2^{h(n)} + k(n)/\alpha}$$

and recall that $h(n) = \lceil \log n \rceil$ and $k(n) = 2^{h(n)} - n$. Lemma 10 follows from Observation 1 by direct calculation and Lemma 11 follows from φ's definition.

Lemma 10. *The effect of an n-leaf complete ψ-balanced full binary tree T is $\eta(T) = \varphi(n)$.*

Lemma 11. *The function $\varphi(n)$ is strictly increasing.*

Now, we can show that it is sufficient to consider complete ψ-balanced full binary trees.

Lemma 12. *An effective n-leaf full binary tree must be ψ-balanced.*

Proof. We prove the statement by induction on the number of leaves n. The base case of a tree having a single leaf (which is also the root) holds vacuously; the base case of a tree having two leaves is trivial. Assume that the assertion holds for trees with fewer than n leaves and let T be an effective n-leaf full binary tree. Let T_ℓ and T_r be the left and right subtrees of T and let z be the number of leaves in T_ℓ where $1 \leq z \leq n-1$.

Observe that both T_ℓ and T_r have to be effective as otherwise, $\eta(T)$ could be increased. More precisely, if $T_i \in \{T_\ell, T_r\}$ is not effective, then there exists a full binary tree T_i' with the same number of leaves as T_i (either z or $n-z$) such that $\eta(T_i') > \eta(T_i)$; by replacing T_i with T_i' in T and scaling $\Lambda(T_i')$ so that $\Lambda(T_i') = \Lambda(T_i)$, we increase $\Psi(T)$ without affecting $\Lambda(T)$, thus increasing $\eta(T)$, in contradiction to T being effective. By the inductive hypothesis, both T_ℓ and T_r are ψ-balanced, hence Lemma 9 guarantees that both are complete. This allows us to use Lemma 10 to determine the effects of T_ℓ and T_r as $\varphi(z)$ and $\varphi(n-z)$, respectively.

Assume without loss of generality that the leaf weights are scaled such that $\Lambda(T) = \Lambda(T_\ell) + \Lambda(T_r) = 1$ and set $\Lambda(T_\ell) = x$, $\Lambda(T_r) = 1-x$, for some $0 \leq x \leq 1$. We consider a set of $n-1$ functions $f_z : [0,1] \mapsto \mathbb{R}_{>0}$ (parametrized by $1 \leq z \leq n-1$) with

$$
f_z(x) = \begin{cases} \varphi(z) \cdot x + (1+1/\alpha)\varphi(n-z) \cdot (1-x) & \text{if } \varphi(z)x \geq \varphi(n-z)(1-x) \\ (1+1/\alpha)\varphi(z) \cdot x + \varphi(n-z) \cdot (1-x) & \text{if } \varphi(z)x \leq \varphi(n-z)(1-x) \end{cases}
$$

that, by Lemma 10, determine the effect of T given that T_ℓ has $1 \leq z \leq n-1$ leaves and $\Lambda(T_\ell) = x \in [0,1]$. Observe that each f_z is a piecewise linear continuous function, linear in the intervals $[0, b_z]$ and $[b_z, 1]$, where b_z is the *break point* of f_z satisfying $\varphi(z)b_z = \varphi(n-z)(1-b_z)$. Hence, f_z must attain its maximum either at a boundary point 0 or 1, or at the break point b_z, where the latter case corresponds to a ψ-balanced tree.

Consider the function $f(x) = \max_z f_z(x)$ whose maximum corresponds to the effect of an effective n-leaf full binary tree and let $\hat{x} \in \text{argmax}_{x \in [0,1]} f(x)$. We argue that $\hat{x}$ can be neither 0 nor 1. Indeed, if $\hat{x} = 0$, then $\Lambda(T) = \Lambda(T_r)$ and $\Psi(T) = \Psi(T_r)$, hence $\eta(T) = \eta(T_r)$ for the corresponding tree T. But since T_r has fewer leaves than T and is complete and ψ-balanced, Lemmas 10 and 11 dictate that its effect — and thus also the effect of T — must be smaller than the effect of an n-leaf complete ψ-balanced full binary tree, a contradiction to the choice of $\hat{x}$ maximizing $f(x)$. An analogous argument excludes $\hat{x} = 1$. It follows that the maximum of $f(x)$ must be attained at a point $0 < \hat{x} < 1$, which, by the definition of f, is the break point b_z of some function f_z and thus realized by a ψ-balanced tree. $\qquad\square$

Combining Lemmas 9, 10, and 12 and recalling that $h = h(n) = \lceil \log n \rceil \leq \log n + 1$ and $k = k(n) \geq 0$, we conclude that the effect of an n-leaf full binary tree is at most

$$\frac{(2+1/\alpha)^h}{2^h + k/\alpha} \leq \frac{(2+1/\alpha)^h}{2^h} \leq (1 + 1/(2\alpha))^{\log n + 1} \leq 3/2 \cdot n^{\log(1+1/(2\alpha))} \ .$$

Returning to the definition of the flip forest F, we recall that there exists one leaf in F for each of the n edges in the minimum-cost matching M^* and therefore each flip tree has at most n leaves. Furthermore, since

$$c(M^*) = \sum_{\text{flip trees } T} \sum_{\ell \in \mathcal{L}(T)} \psi(\ell) = \sum_{\text{flip trees } T} \Lambda(T) \ ,$$

we can employ Corollary 8 to derive

$$\frac{c(M_S)}{c(M^*)} \leq 2 \cdot \frac{\sum_{\text{flip trees } T} \Psi(T)}{\sum_{\text{flip trees } T} \Lambda(T)} \leq 2 \cdot \max_{\text{flip trees } T} \eta(T) \leq 3 \cdot n^{\log(1+1/(2\alpha))} \ ,$$

thus establishing Theorem 2.

References

1. Abraham, D., Levavi, A., Manlove, D., O'Malley, G.: The stable roommates problem with globally-ranked pairs. In: WINE'07
2. Ackermann, H., Goldberg, P.W., Mirrokni, V.S., Röglin, H., Vöcking, B.: Uncoordinated two-sided matching markets. SICOMP'11
3. Albers, S., Eilts, S., Even-Dar, E., Mansour, Y., Roditty, L.: On nash equilibria for a network creation game. In: SODA'06
4. Andelman, N., Feldman, M., Mansour, Y.: Strong price of anarchy. In: SODA'07
5. Anshelevich, E., Das, S., Naamad, Y.: Anarchy, stability, and utopia: creating better matchings. AAMAS'13
6. Anshelevich, E., Dasgupta, A., Kleinberg, J.M., Tardos, É., Wexler, T., Roughgarden, T.: The price of stability for network design with fair cost allocation. SICOMP'08
7. Anshelevich, E., Dasgupta, A., Tardos, E., Wexler, T.: Near-optimal network design with selfish agents. In: STOC'03
8. Arkin, E.M., Bae, S.W., Efrat, A., Okamoto, K., Mitchell, J.S.B., Polishchuk, V.: Geometric stable roommates. IPL'09
9. Awerbuch, B., Azar, Y., Epstein, A.: The price of routing unsplittable flow. In: STOC'05
10. Awerbuch, B., Azar, Y., Richter, Y., Tsur, D.: Tradeoffs in worst-case equilibria. TCS'06
11. Chen, H.L., Roughgarden, T.: Network design with weighted players. In: SPAA'06
12. Christodoulou, G., Koutsoupias, E.: On the price of anarchy and stability of correlated equilibria of linear congestion games. In: ESA'05
13. Christodoulou, G., Koutsoupias, E.: The price of anarchy of finite congestion games. In: STOC'05
14. Czumaj, A., Vöcking, B.: Tight bounds for worst-case equilibria. In: SODA'02

15. Edmonds, J.: Paths, trees, and flowers. Canadian J. of Math. 1965
16. Edmonds, J.: Maximum matching and a polyhedron with 0, 1 vertices. J. of Research of the National Bureau of Standards 1965
17. Feder, T.: A new fixed point approach for stable networks and stable marriages. JCSS'92
18. Gale, D., Shapley, L.S.: College admissions and the stability of marriage. The American Mathematical Monthly 1962
19. Gusfield, D., Irving, R.W.: The stable marriage problem: structure and algorithms. MIT Press, Cambridge, MA, USA (1989)
20. Hoefer, M., Wagner, L.: Designing profit shares in matching and coalition formation games. In: WINE'13
21. Huang, C.C.: Two's company, three's a crowd: Stable family and threesome roommates problems. In: ESA'07.
22. Irving, R.W.: Stable marriage and indifference. DAM'94
23. Irving, R.W., Leather, P., Gusfield, D.: An efficient algorithm for the "optimal" stable marriage. JACM'87
24. Irving, R.W., Manlove, D.F., Scott, S.: The stable marriage problem with master preference lists. DAM'08
25. Johari, R., Tsitsiklis, J.N.: Efficiency loss in a network resource allocation game. MOR'04
26. Knuth, D.E.: Marriages stables et leurs relations avec d'autres problèmes combinatoires. Les Presses de l'Université de Montréal (1976)
27. Koutsoupias, E., Mavronicolas, M., Spirakis, P.G.: Approximate equilibria and ball fusion. TOCS'03
28. Koutsoupias, E., Papadimitriou, C.: Worst-case equilibria. Computer Science Review 2009
29. Manlove, D.F.: Algorithmics of matching under preferences. World Scientific (2014)
30. Ng, C., Hirschberg, D.S.: Three-dimensional stable matching problems. SIDMA'91
31. Nisan, N., Ronen, A.: Algorithmic mechanism design. GEB'01
32. Papadimitriou, C.: Algorithms, games, and the internet. In: STOC'01
33. Reingold, E.M., Tarjan, R.E.: On a greedy heuristic for complete matching. SICOMP'81
34. Roth, A.E., Rothblum, U.G., Vande Vate, J.H.: Stable matchings, optimal assignments, and linear programming. MOR'93
35. Roth, A.E., Sotomayor, M.A.O.: Two-sided matching: a study in game-theoretic modeling and analysis. Cambridge University Press (1990)
36. Roughgarden, T.: Potential functions and the inefficiency of equilibria. In: ICM'06
37. Roughgarden, T.: The price of anarchy is independent of the network topology. JCSS'03
38. Roughgarden, T., Tardos, E.: How bad is selfish routing? JACM'02
39. Schulz, A.S., Stier Moses, N.E.: On the performance of user equilibria in traffic networks. In: SODA'03
40. Suri, S., Tóth, C.D., Zhou, Y.: Selfish load balancing and atomic congestion games. Algorithmica 2007
41. Vetta, A.: Nash equilibria in competitive societies, with applications to facility location, traffic routing and auctions. In: FOCS'02

Selfish Vector Packing

Leah Epstein[1] and Elena Kleiman[2,*]

[1] Department of Mathematics, University of Haifa, Haifa, Israel
lea@math.haifa.ac.il
[2] Faculty of Industrial Engineering and Management, The Technion, Haifa, Israel
elena.kleiman@gmail.com

Abstract. We study the multidimensional vector packing problem with selfish items. An item is d-dimensional non-zero vector, whose rational components are in $[0, 1]$, and a set of items can be packed into a bin if for any $1 \leq i \leq d$, the sum of the ith components of all items of this set does not exceed 1. Items share costs of bins proportionally to their ℓ_1-norms, and each item corresponds to a selfish player in the sense that it prefers to be packed into a bin minimizing its resulting cost. This defines a class of games called *vector packing games*. We show that any game in this class has a packing that is a strong equilibrium, and that the strong price of anarchy (and the strong price of stability) is logarithmic in d, and provide an algorithm that constructs such a packing. We also show improved and nearly tight lower and upper bounds of $d + 0.657067$ and $d + 0.657143$ respectively, on the price of anarchy, exhibiting a difference between the multidimensional problem and the one dimensional problem, for which that price of anarchy is at most 1.6428.

1 Introduction

Motivation and Framework. Bin Packing is a classical combinatorial optimization problem which has been studied since the early 70's. In addition to its theoretical importance, this problem in its different variants has real-life applications in various areas such as packaging, resource allocation, load balancing and distributed computer system design among many others (see [8] for a survey). In this paper we consider the Multidimensional Bin Packing problem (MBP), also known in the literature as the Vector Packing problem. In this problem, each bin has a d-dimensional capacity of one unit in each of the d dimensions and each item has an associated non-zero multidimensional size, represented by a d-dimensional vector with components in the range $[0, 1]$. The goal is to pack a set of vectors into a minimum number of multidimensional bins without exceeding the capacity of any bin in any dimension. The problem was first studied in one dimension (that is, for $d = 1$) see e.g. [18] and has been extended to multiple dimensions ($d \geq 2$) ([19,15,6]). Such problems naturally arise when packing items

---

* Also affiliated with Department of Software Engineering, ORT Braude College of Engineering, Karmiel, Israel.

© Springer-Verlag Berlin Heidelberg 2015
N. Bansal and I. Finocchi (Eds.): ESA 2015, LNCS 9294, pp. 471–482, 2015.
DOI: 10.1007/978-3-662-48350-3_40

that have multiple, often incomparable, characteristics such as length, weight, and others.

Recently, this problem has received a renewed burst of attention in the context of Cloud Computing, a popular computing service paradigm currently practiced by all major data centers, including those managed by Google and Amazon. In order to reduce maintenance costs and for a more efficient utilization of the physical resources, while considering quality of service requirements, the modern data centers use the technique of *server virtualization*, that consists of abstracting the physical resources and running multiple virtual machines (VMs) on one Physical Machine (PM), that has multiple limited resources, such as memory, CPU, bandwidth etc.. In the heart of this technique lies the assignment of the VMs to the PMs, which is called the *VM Placement problem*. Seeing the VMs as the items and the PMs as the bins, this problem very naturally translates to our MBP problem. It is then not surprising that the MBP problem is one of the main problems encountered in Cloud Computing, and so many works were dedicated to suggesting and applying various MBP algorithms for this purpose, see eg. [29,21,27].

Yet to add to this challenge, in many real-life environments, and Cloud Computing is no exception, the service provider and users can have different, possibly conflicting, interests and behave strategically. Hence, in this paper we take a game-theoretic approach to the MBP, and examine various stability properties of the packings. That demand that the packing is stable is very important, since migration of a VM among the PMs once the assignment was made is highly undesirable, as it interrupts the services running on that VM (until the migration is complete), delaying the service for the user, and on the other hand causing an overhead to the entire system. For this reason, studying scheduling and packing problems under a game-theoretic framework has become a common practice, see e.g. [20,24,1,5,12].

The Model. The multidimensional vector packing problem and the corresponding vector packing game (VPG) are defined as follows. There are n items $I = \{1, 2, \ldots, n\}$, each with $d \geq 2$ components, that are to be partitioned or packed into blocks called *bins*. Let the vector of item $i \in I$ be denoted by $p_i = (p_i^1, \ldots, p_i^d)$, where $0 \leq p_i^j \leq 1$ for $1 \leq j \leq d$, and $\sum_{j=1}^{d} p_k^i > 0$. An infinite supply of bins is given, each with d identical resources. A bin is represented by an all-one vector, and a packed bin B is seen as a set of items it contains, that is, $B \subseteq I$. We let the size of the jth component of bin B be $P_B^j = \sum_{i \in B} p_i^j$, and it is required that $P_B^j = \sum_{i \in B} p_i^j \leq 1$, i.e., for every $1 \leq j \leq d$, it is required that the items do not exceed the capacity of 1. In the optimization problem, the goal is to pack a given set of items into a minimum number of bins. For item i, let $v_i = \sum_{j=1}^{d} p_i^j$ be its ℓ_1-norm, also called its volume. For a bin B, its volume is $v(B) = \sum_{i \in B} v_i = \sum_{j=1}^{d} P_B^j$. A vector packing game is defined by a set of items, such that each item belongs to a different selfish agent. The strategy of an item is the bin into which it is packed, and the outcome is a packing of the items into bins (we assign infinite costs to items packed into invalid bins, and therefore in what follows we assume that all bins are valid). The cost of item i that is packed into bin B in a packing $\mathcal{A}$ is $c_i(\mathcal{A}) = \frac{v_i}{v(B)}$, and thus $\sum_{i \in B} c_i(\mathcal{A}) = 1$ (we use c_i instead of $c_i(\mathcal{A})$ if $\mathcal{A}$ is clear from the context).

We use the ℓ_1-norm in our cost-scheme, as the IaaS market providers offer instances of VMs with fixed amounts of each of the resources, but the users pay a fixed cost for the entire VM, and not per resource.

We will next provide definitions of the game theoretic solution concepts considered in this paper, namely, (pure) Nash equilibria, weakly and strictly Pareto optimal (Nash) equilibria and Strong (Nash) equilibria for VPG.

A packing $\mathcal{A}$ is a (pure) Nash equilibrium (NE) if no item in $\mathcal{A}$ has an incentive to unilaterally move to a different bin (a new bin or a bin where it can be packed legally), given that all other items keep their strategies unchanged, that is, no other item moves at the same time. We measure the inefficiency of equilibria by the (asymptotic) price of anarchy [20] and price of stability [2] of classes of games. For a game G, let $OPT(G)$ denote the minimum number of bins that is required by any valid packing. A solution that uses this number of bins is called *socially optimal* (or optimal). Let J_d denote the class of d-dimensional vector packing games. The price of anarchy is the ratio between the number of (non-empty) bins used by the *worst* packing that is an NE, that is, the largest number of bins in any NE packing for G is ζ, then $\text{POA}(G) = \frac{\zeta}{OPT(G)}$. We let $\text{POA}(d) = \lim\limits_{K \to \infty} \sup\limits_{G \in J_d, OPT(G) \geq K} \text{POA}(G)$, and this is the (asymptotic) price of anarchy for dimension d. The price of stability is defined similarly, but now the *best* packing that is NE is considered for each G.

Strong (Nash) equilibria packings (SNE) [3,25,17,1,14] are packings where there does not exist a subset of items (also called a coalition) that can deviate from their strategies simultaneously (while other items keep their strategies), such that all items participating in the coalition reduce their costs. Note that the items of a coalition can move to existing bins, or they can use a new bin, and they do not necessarily all move to the same bin. Obviously, by this definition, an SNE is an NE.

The grand coalition is defined to be a coalition composed of the entire set of items. A packing is called weakly Pareto optimal if there is no alternative solution to which the grand coalition can deviate simultaneously and every item benefits from it. A packing is called strictly Pareto optimal if there is no alternative solution to which the grand coalition can deviate simultaneously, such that at least one item benefits from it, and no item has a larger cost as a result. The last two concepts are borrowed from welfare economics. The two requirements, that a packing is both (strictly or weakly) Pareto optimal and an NE results in two additional kinds of NE, strictly Pareto optimal NE (SPNE) and weakly Pareto optimal NE (WPNE) [10,7,11,4,9]. By these definitions, every WPNE is an NE, every SPNE is a WPNE, and every SNE is a WPNE. Strictly Pareto optimal points are of particular interest in economics [22]. Even though these concepts are stronger than NE, still for many problems a solution which is an SNE, an SPNE, or a WPNE is not necessarily socially optimal, which is also the case for our game.

The (asymptotic) strong price of anarchy (SPOA) and strong price of stability (SPOA) as well and the strictly and weakly Pareto optimal prices of anarchy (WPPOA and SPPOA) and the strictly and weakly Pareto optimal prices of stability (WPPOS and SPPOS) are defined similarly to POA and POS, but each time the respective stability concepts are considered.

Related Work and Our Contribution. In this section, we survey the known bounds on the game-theoretic measures defined above for bin packing games, for the one dimensional and for the multidimensional cases.

The one dimensional bin packing game (BPG) with proportional cost-sharing scheme (like in this paper) was introduced by Bilò [5], who was the first to study the bin packing problem from this type of game theoretic perspective. He provided the first bounds on the POA, a lower bound of $\frac{8}{5}$ and an upper bound of $\frac{5}{3}$, and showed that POS=1. The quality of NE solutions was further investigated in [12], where nearly tight bounds for the PoA were given; an upper bound of 1.6428 and a lower bound of 1.6416 (see also [28]). Interestingly, the POA is not equal to the approximation ratio of any natural algorithm for bin packing. The SPOA and SPOS were also analyzed in [12], and it was shown that these two measures are equal. Moreover, it was shown the set of SNE and the set of outputs of the Subset Sum algorithm [16] are the same, which gave bounds on the SPOA and SPOS. In the paper [13], the exact SPOA (which is also the approximation ratio of Subset Sum) was determined, and it was shown that its value is approximately 1.6067. In the same article, the parametric problem where the size of every item is upper bounded by a parameter is studied. Some properties of other measures that were not studied in [5,12,13] (the ones related to Pareto optimal solutions) are mentioned in [9]. Specifically, they showed that any optimal packing is Pareto optimal (as we show here, this holds in the multidimensional case, as well), that WPPOA=POA, that SPPOA≥ SPOA and that WPPOS=SPPOS=1. Other variants of this game, with different cost structures, were considered in [9,23].

The multidimensional bin packing (or the vector packing) game (VPG) was first introduced by Ye and Chen in [26], where they call it 'virtual machine placement game'. They considered only NE solutions, showed that any game has a packing that is a Nash equilibrium and provided the first bounds on the POA, a lower bound of d and an upper bound of $d + 16/5$ (which is proved by a reduction to the First Fit algorithm using the result of [15], and holds even for the absolute POA, while the resulting upper bound for the measure studied here is $d + 0.7$, which is the asymptotic approximation ratio of First Fit for multiple dimensions [15]), and showed that POS=1. In this paper we consider this model, which was studied in [26].

In Section 2, we show improved and nearly tight lower and upper bounds of $d + 0.657067$ and $d + 0.657143$ respectively, on the POA for $d \geq 2$, exhibiting a difference between the multidimensional problem and the one dimensional problem, for which that POA is at most 1.6428.

In the same section, we consider Pareto optimal packings, show that they exist for any game, and that any optimal packing exhibits Pareto optimality, hence WPPOS=SPPOS=1 (as for any game there exists a optimal packing that is also

NE), however WPPOA and SPPOA remain linear in d as is the POA, which implies that imposing this stronger demand for stability on the NE solution does not help to reduce the inefficiency of the equilibria. We then proceed to show that restricting the sizes of the coordinates of items in the packing also does not help to reduce this inefficiency, which still remains linear in d.

In Section 3 we consider strong Nash equilibrium packings, show that any game has such packing, and that the SPOA (and the SPOS) is $\ln d + \Theta(1)$. So, only when we make this (very) strong demand for stability it allows us to reduce the inefficiency of the solution. This gap leads us to conclude that this inefficiency is much a result of the fact that the players' actions are not coordinated and not only of their selfishness. In addition, we provide an algorithm called Greedy Set Cover, that constructs (any) SNE packing. It has an exponential running time, but as the problem of computing an SNE packing is NP-hard already for $d = 1$ [12], no polynomial time algorithm can do that, unless P=NP.

Some proofs were omitted from this version due to space constraints.

2   Price of Anarchy

In this section, we prove close bounds on POA(d). We show an upper bound of $d + \frac{23}{35} \approx d + 0.657143$, and a lower bound of approximately $d + 0.657067$. This reveals interesting properties of the problem. The claim that every NE can be obtained by First Fit algorithm (FF in short) is true here as well as in the one dimensional case; sort the bins by non-increasing total size (in all components), and apply FF on the items in this order. Recall that for FF, the approximation ratio for dimension d is equal to the approximation ratio of the one dimensional problem plus $d - 1$. This is not the case here, and the value of the POA is a new value in bin packing problems. In the analysis of FF for multiple dimensions [15], bins for which exactly one component is above $\frac{1}{2}$ and the other components are at most $\frac{1}{2}$ were split into d classes $1, 2, \ldots, d$ (according to which component is the largest one). One property of FF is that the bins of the jth class could have been created by applying FF on one dimensional items (of sizes equal to the jth component). While we use a similar partition in our proof (and partition bins similarly), this property does not hold for NE packings (it can be demonstrated using our lower bound for POA(d), where class 1 is larger by a multiplicative factor of approximately 1.657067 than the number of bins in an optimal solution, while POA(1) ≤ 1.643). The behavior for all values of d such that $d \geq 2$ is more uniform in terms of the difference between the value of the POA and d.

2.1   Upper Bound

Here, we discuss the upper bound. The next theorem shows POA(d) $\leq d + \frac{23}{35}$.

Theorem 1. *Let G' be a d-dimensional game. For every packing $\mathcal{A}'$ that is an NE, $|\mathcal{A}'| \leq (d + \frac{23}{35})OPT(G') + 4$.*

2.2 Lower Bound

We start with defining several parameters. The three parameters A_i, B_i, and C_i are defined for any integer $i \geq 1$. We also let $A_0 = 1$ (while B_0 and C_0 are not defined). Let $A_i = \frac{A_{i-1}}{2^{2i-1}-2^i+1}$ for $i \geq 1$, where $A_1 = A_0$ and $A_i < A_{i-1}$ for $i \geq 2$ (as $2^{2i-1} < 2^i$). Moreover, let $B_i = A_i \cdot (2^i - 2)$ and $C_i = A_i \cdot (2^{2i-1} - 2^{i+1} + 2)$ (and thus $B_1 = C_1 = 0$). For $i \geq 1$, we have $A_i + B_i + C_i = A_i \cdot (1 + (2^i - 2) + (2^{2i-1} - 2^{i+1} + 2)) = A_i \cdot (2^{2i-1} - 2^i + 1) = A_{i-1}$. For any integer $i \geq 1$, let $D_i = \frac{A_i + 2B_i + 2C_i}{2^i - 1} = \frac{2A_{i-1} - A_i}{2^i - 1}$ (thus, $D_1 = 1$), for $j \geq 1$, $\kappa_j = \sum_{i=1}^{j} D_i$, and finally $\kappa_\infty = \lim_{j \to \infty} \kappa_j$ (the approximate value of κ_∞ is 1.657067). Since $0 < A_i \leq 1$ for $i \geq 1$, all values B_i, C_i are positive and smaller than 1.

Theorem 2. *For any $d \geq 2$, POA$(d) \geq \kappa_\infty + d - 1$.*

Proof. We will show that POA$(d) \geq \kappa_j + d - 1$ for any $j \geq 2$. Given $j \geq 2$ and $d \geq 2$, let $N > \max\{d, 2^j\}$ be a large integer divisible by $2^{2i-1} - 2^i + 1$ for any integer i such that $2 \leq i \leq j$. Let $\delta > 0$ be a small constant such that $\delta = \frac{1}{N^{30j+10}}$. For $i \geq 1$, let $\delta_i = N^{10i} \cdot \delta$ and $\rho_i = N^{10i+5} \cdot \delta$. Thus, for $i \geq 1$, $\delta_{i+1} = N^5 \cdot \rho_i = N^{10} \cdot \delta_i$. We have $\delta_i = N^{-30j-10+10i}$ and $\rho_i = N^{-30j-5+10i}$. The values $N \cdot A_i$ are integers for $i \geq 1$, and therefore $N \cdot B_i$ and $N \cdot C_i$ are integers for $i \geq 1$, where $0 < N \cdot B_i < N$ and $0 < N \cdot C_i < N$ (so $1 \leq N \cdot B_i \leq N - 1$ and $1 \leq N \cdot C_i \leq N - 1$).

We have the following types of items.

- For $r = 2, \ldots, d$, items of class $(0, r)$ are $M = N^{30j+1} - N^{30j}$ items whose component r is equal to δ_1, and the other components are equal to δ_1^2. These items are also called items of class 0.
- Items of class 1 have a first component of value $\frac{1}{2} + \delta_1$, and the remaining $d - 1$ components are equal to ρ_1. The number of such items is N.
- Items of class $(i, 1)$ (for $2 \leq i \leq j$) have first components of value $\frac{1}{2^i} + (N \cdot C_i + 1)\delta_i - \rho_{i-1}$. The remaining $d - 1$ components are equal to ρ_i. The number of such items is $N \cdot B_i$.
- Items of class $(i, 2)$ (for $2 \leq i \leq j$) have first components of value $\frac{1}{2^i} - (N \cdot C_i + 1)\delta_i$. The remaining $d - 1$ components are equal to ρ_i. The number of such items is $N \cdot B_i$.
- For $1 \leq t \leq N \cdot C_i$, there is an item of class $(i, 3)$ (for $2 \leq i \leq j$) whose first component has the value $\frac{1}{2^i} + t\delta_i - \rho_{i-1}$. The remaining $d - 1$ components are equal to ρ_i.
- For $1 \leq t \leq N \cdot C_i$, there is an item of class $(i, 4)$ (for $2 \leq i \leq j$) whose first component has the value $\frac{1}{2^i} - t\delta_i$. The remaining $d-1$ components are equal to ρ_i.
- Items of class $(i, 5)$ (for $2 \leq i \leq j$) have first components of value $\frac{1}{2^i} + 2^i(N \cdot C_i + 1)\delta_i - \rho_{i-1}$. The remaining $d - 1$ components are equal to ρ_i. The number of such items is $N \cdot A_i$.

An item of class $(i, 3)$ and an item of class $(i, 4)$ whose sum of first components is $\frac{1}{2^{i-1}} - \rho_{i-1}$ (that is, they are defined for the same value of t) are called a *pair*.

The items of classes $(i, 1)$ and $(i, 2)$ are split into pairs as well (where each pair consists of one item of each of these classes, and the sum of their first components is also $\frac{1}{2^{i-1}} - \rho_{i-1}$). Such a pair (of the first kind or of the second kind) is called *a pair of class i*.

By the definition of δ, every component that is not the first component of an item is at most $0 < \rho_j = N^{-20j-5} \leq N^{-45}$. We show that the first components are in $(0, 1]$. This obviously holds for class 0. For class 1, we have $\delta_1 = N^{-30j} \leq N^{-60} < \frac{1}{2}$. For classes $(i, 1)$, $(i, 3)$, and $(i, 5)$, let the first component be $\frac{1}{2^i} + X\delta_i - \rho_{i-1}$ for $1 \leq X \leq 2^i(N \cdot C_i + 1) \leq 2^j \cdot N \leq N^2$. We have $X\delta_i \geq \delta_i > \rho_{i-1}$, and $X\delta_i \leq N^2\delta_i = N^{2-30j-10+10i} \leq N^{-20j-8} < \frac{1}{2}$. For classes $(i, 2)$ and $(i, 4)$, let the first component be $1 - X\delta_i$ for $1 \leq X \leq N \cdot C_i + 1 \leq N$. We have $X\delta_i \leq N\delta_i \leq N^{-20j-9} < 2^{-20j-9} < \frac{1}{2^{j+1}}$.

Claim 3. *The entire set of these items can be packed into N bins, and this packing is optimal.*

Now, we describe the bins of an alternative packing which uses a much larger number of bins. We will show that it is valid and that that is a NE. We modify the input by removing some items. Obviously, it is still possible to pack the input into N bins. Items can only possibly benefit from moving to non-empty bins, and therefore we will only show that no item can benefit from moving to another packed bin.

For $r = 2, \ldots, d$, there are $N - 1$ bins, each containing N^{30j} items of class $(0, r)$. The rth component of such a bin is 1, and any other component is N^{-30j}. The volume is therefore strictly above 1. As no item has any zero component, no additional item can be packed into such a bin. The other bins will have volumes strictly below 1, and therefore items of class 0 are packed into these $(d-1)(N-1)$ bins can not benefit from moving to another bin.

There are N bins, each containing one item of class 1. The first component for such a bin is $\frac{1}{2} + \delta_1 > \frac{1}{2}$, and the volume is $\frac{1}{2} + \delta_1 + (d-1)\rho_1 < \frac{1}{2} + N\rho_1 = \frac{1}{2} + N^{-30j+6} < 1$.

For $1 \leq i \leq N \cdot C_i - 1$, instead of the pairs, we create modified pairs (only for classes $(i, 3)$ and $(i, 4)$ and not for classes $(i, 1)$ and $(i, 2)$). A *modified pair* for class i is a pair of items whose first components are $\frac{1}{2^i} + (t+1)\delta_i - \rho_{i-1}$ and $\frac{1}{2^i} - t\delta_i$, for some t such that $1 \leq t \leq N \cdot C_i - 1$. The two items of classes $(i, 3)$ and $(i, 4)$ whose first components are $\frac{1}{2^i} + \delta_i - \rho_{i-1}$ and $\frac{1}{2^i} - N \cdot C_i$ are removed from the input as they do not belong to modified pairs (and some modified pairs are removed later). The sum of first components of a modified pair is $\frac{1}{2^{i-1}} + \delta_i - \rho_{i-1}$.

For $2 \leq i \leq j$, there are two kinds of bins of type i. Bins of type $[i, a]$ have one item of class $(i, 5)$ and $2^i - 2$ items of class $(i, 2)$. Bins of type $[i, b]$ have $2^{i-1} - 1$ modified pairs of class i and one item of class $(i, 1)$. Since $B_i = (2^i - 2)A_i$ and $C_i = (2^{i-1} - 1)B_i$, we have $\frac{N \cdot B_i}{2^i - 2} = N \cdot A_i$ and $\frac{N \cdot C_i - 1}{2^{i-1} - 1} = N \cdot B_i - \frac{1}{2^{i-1} - 1} \geq N \cdot B_i - 1$. Thus, by removing one item of class $(i, 1)$ and some items of classes $(i, 3)$ and $(i, 4)$ (that is, by removing some modified pairs), it is possible to create $N \cdot A_i$ bins of type $[i, a]$ and $N \cdot B_i - 1$ bins of type $[i, b]$.

Claim 4. *The packing defined above is a valid NE packing.*

We have proved that the alternative packing is an NE. We find that the total number of bins in this packing is $(d-1)(N-1) + N + \sum_{i=2}^{j}(N(A_i + B_i) - 1)$. We have $A_i + B_i = A_i(2^i - 1) = \frac{2A_{i-1} - A_i}{2^i - 1} = D_i$, as $2A_{i-1} - A_i = (2^{2i} - 2^{i+1} + 1)A_i = (2^i - 1)^2 \cdot A_i$, and $N = N \cdot D_1$. Thus, the number of bins is at least $(d-1)(N-1) - (j-1) + N\kappa_j$. For a sufficiently large value of N, we find that the ratio tends to $\kappa_j + d - 1$. $\square$

2.3   Other Variants

As the POA is fairly high, it is interesting to identify the difficulty. In this section, we show that neither limiting the item components nor requiring that the NE will also be Pareto points affects the property that the price of anarchy is linear. In the next section we will show that the situation is very different for strong equilibria.

The Parametric Case. In the parametric variant, a parameter $0 < \xi < 1$ is given such that all item components do not exceed ξ.

Lemma 5. *The POA is linear in d for any parameter ξ.*

Pareto Price of Anarchy

Proposition 6. *Any optimal packing is strictly Pareto optimal (and hence also weakly Pareto optimal).*

As for any game there exists an optimal packing that is an NE [26], this proves that a SPNE and WPNE always exist, and that SPPOS=WPPOS=1.

Proposition 7. *The WPPOA and SPPOA are linear in d.*

3   Strong Price of Anarchy

We start with showing that any game G has at least one SNE. We will show, however, that SPOS(d) is equal to SPOA(d) and it is logarithmic in the dimension d, and in particular, not every game has a solution that is both an SNE and socially optimal. As for other bin packing games, we consider the following (exponential time) algorithm, called GREEDY SET COVER (GSC). For an input set of items I, in each step, the algorithm selects a subset of items of maximum volume that can be packed into a bin together, and creates this bin.

Proposition 8. *Every run of GSC creates an SNE packing (thus an SNE always exists), and any SNE packing can be created by some run of GSC (with some tie-breaking policy).*

Theorem 9. *For a game $G' \in J_d$ with the set of items I', a packing $\mathcal{A}'$ that is an output of GSC has at most $(\ln 2d + 1)OPT(G') + 1 = (\ln d + \ln 2 + 1)OPT(G') + 1$ bins.*

Proof. First, note that GSC creates at most one bin $\tilde{B}$ such that $v(\tilde{B}) \leq \frac{1}{2}$, as the contents of any two bins of volumes at most $\frac{1}{2}$ can be packed into a single bin. Moreover, if such a bin exists, it must be created last by GSC, as sequence of the volumes of bins created by GSC is monotonically non-increasing. Let G be the game resulting from G' by removing the items of $\tilde{B}$ if it exists, (and otherwise $G = G'$), and let I be its set of items. Obviously $OPT(G) \leq OPT(G')$. Let $\mathcal{A}$ be $\mathcal{A}'$ excluding $\tilde{B}$ if it exists (and otherwise $\mathcal{A} = \mathcal{A}'$). We will show that the number of bins in $\mathcal{A}$, $|\mathcal{A}|$, is at most $(\ln(2d) + 1)OPT(G) \leq (\ln(2d)+1)OPT(G')$. Specifically, since $|\mathcal{A}| = \sum_{i \in I} c_i(\mathcal{A})$, it is sufficient to prove $\sum_{i \in I} c_i(\mathcal{A}) \leq (\ln(2d)+1)OPT(G)$, or alternatively, that given a socially optimal packing $\mathcal{A}''$ for $I' = I \setminus \tilde{B}$, for any bin B' of $\mathcal{A}''$, $\sum_{i \in B'} c_i(\mathcal{A}'') \leq \ln 2d + 1$.

Consider an arbitrary such bin B'', and let $\pi_1, \pi_2, \ldots, \pi_k$ denote its items, ordered in the order that GSC packs them into bins of $\mathcal{A}'$ (items of one bin of $\mathcal{A}'$ are ordered arbitrarily). For the input I, the set of bins of $\mathcal{A}$ is exactly the bins of $\mathcal{A}'$ (as the bin that is possibly removed is packed last by GSC), and therefore we will consider $\mathcal{A}$ in what follows. Let B_ℓ denote the bin of $\mathcal{A}$ that contains item π_ℓ, for $\ell \leq k$. Let $V_\ell = \sum_{j=\ell}^{k} v_{\pi_j}$ (and $V_{k+1} = 0$). For any bin B_ℓ, as this is a bin of $\mathcal{A}$, $v(B_\ell) > \frac{1}{2}$. By the definition of costs, $c_{\pi_\ell}(\mathcal{A}) = \frac{v_{\pi_\ell}}{v(B_\ell)}$.

When item π_ℓ for $\ell < k$ is packed into B_ℓ, items $\pi_{\ell+1}, \ldots, \pi_k$ are also still available for packing (they will be packed into the same bin or a bin opened later). Thus, the bin B_ℓ of $\mathcal{A}$ that contains item π_ℓ has volume of $v(B_\ell) \geq V_\ell$. We find $v(B_\ell) \geq \max\{1/2, V_\ell\}$ and therefore $c_{\pi_\ell}(\mathcal{A}) \leq \frac{v_{\pi_\ell}}{\max\{1/2, V_\ell\}}$. Let $0 \leq k' \leq k$ be the maximum index such that $V_{k'} > \frac{1}{2}$ ($k' = 0$ if $V_1 \leq \frac{1}{2}$). If $k' = 0$, then we have $\sum_{i=1}^{k} c_{\pi_i}(\mathcal{A}) < \sum_{i=1}^{k} 2v_{\pi_i} = 2\sum_{i=1}^{k} v_{\pi_i} = 2V_1 \leq 1$. Otherwise, let $v'_{\pi_{k'}} = V_{k'} - \frac{1}{2} > 0$ and $v''_{\pi_{k'}} = v_{\pi_{k'}} - v'_{\pi_{k'}}$. We have $v''_{\pi_{k'}} = v_{\pi_{k'}} - V_{k'} + \frac{1}{2} = \frac{1}{2} - V_{k'+1} \geq 0$, as $V_{k'+1} \leq \frac{1}{2}$. Thus, $v'_{\pi_k}, v''_{\pi_k} \leq v_{\pi_k}$.

We get

$$c_{\pi_{k'}}(\mathcal{A}) \leq \frac{v_{\pi_{k'}}}{\max\{1/2, V_{k'}\}} = \frac{v'_{\pi_{k'}} + v''_{\pi_{k'}}}{\max\{1/2, V_{k'}\}} \leq \frac{v'_{\pi_{k'}}}{V_{k'}} + 2v''_{\pi_{k'}} .$$

We find that $\sum_{i \in B''} c_i(\mathcal{A}) = \sum_{\ell=1}^{k'-1} c_{\pi_\ell}(\mathcal{A}) + c_{\pi_{k'}}(\mathcal{A}) + \sum_{\ell=k'+1}^{k} c_{\pi_\ell}(\mathcal{A}) \leq \sum_{\ell=1}^{k'-1} \frac{v_{\pi_\ell}}{V_\ell} + \frac{v'_{\pi_{k'}}}{V_{k'}} + 2v''_{\pi_{k'}} + \sum_{\ell=k'+1}^{k} 2v_{\pi_\ell}$.

Using $v''_{\pi_k} = \frac{1}{2} - V_{k'+1}$, we get $2v''_{\pi_{k'}} + \sum_{j=k'+1}^{k} 2v_{\pi_j} = 2(\frac{1}{2} - V_{k'+1}) + 2V_{k'+1} = 1$ (which holds even if $k' = k$).

We use the integral (with $\gamma > 0$ and $\alpha, \beta \geq 0$) $\int_\alpha^\beta \frac{1}{x+\gamma} dx = \ln(\beta+\gamma) - \ln(\alpha+\gamma)$. For $\xi \geq 0$, $v > 0$, we have $\frac{\xi}{\xi+v} \leq \int_0^\xi \frac{1}{x+v} dx = \ln(\xi + v) - \ln(v)$.

For k', taking $\xi = v'_{\pi_{k'}}$, $v = \frac{1}{2}$, we have $\frac{v'_{\pi_{k'}}}{V_{k'}} = \frac{v'_{\pi_{k'}}}{v'_{\pi_{k'}} + \frac{1}{2}} \leq \ln(v'_{\pi_{k'}} + \frac{1}{2}) - \ln(\frac{1}{2}) = \ln(V_{k'}) + \ln 2$, and for $1 \leq \ell \leq k' - 1$, taking $\xi = v_{\pi_\ell}$, $v = V_{\ell+1}$, we get $\frac{v_{\pi_\ell}}{V_\ell} = \frac{v_{\pi_\ell}}{v_{\pi_\ell} + V_{\ell+1}} \leq \ln(V_\ell) - \ln(V_{\ell+1})$.

Thus, $\left(\sum_{j=1}^{k'-1} \frac{v_{\pi_\ell}}{V_\ell}\right) + \frac{v'_{\pi_{k'}}}{V_{k'}} \leq \ln(V_1) + \ln 2 = \ln(V(B'')) + \ln 2 \leq \ln d + \ln 2 = \ln(2d)$, and as $V(B'') \leq d$, we get $\sum_{j=1}^{k} \frac{v_{\pi_\ell}}{v(B_\ell)} \leq \ln(2d) + 1$. $\square$

The next theorem shows that the SPoS is $\Omega(\log d)$, and therefore the SPoA $\Omega(\log d)$ as well.

Theorem 10. *Let $d \geq 2$. For any integer M, there exists a game $G \in J_d$, such that $OPT(G) \geq M$, and any run of GSC uses $H_d \cdot (OPT(G) - 1)$ bins, where $H_d = \sum_{\ell=1}^{d} \frac{1}{\ell}$.*

Proof. Given M, let $N > \max\{M - 1, 4^d\}$ be an integer that is divisible by $d!$. Let $\delta > 0$ such that $\delta = \frac{1}{N^3 d^3}$. For $0 \leq j \leq d+1$, let $\delta_j = \frac{\delta}{4^j}$ (thus $\delta_0 = \delta$, and, $\delta_{j-1} = 4\delta_j$ for $j > 0$), and for $0 \leq j \leq d+1$, $\Delta_j = \sum_{\ell=j}^{d} \delta_\ell$, and therefore for $j \leq d$, $\Delta_j = \sum_{\ell=j}^{d} \delta_d 4^{d-\ell} = \delta_d \cdot \sum_{\ell=0}^{d-j} 4^\ell = \delta_d \cdot (4^{d-j+1} - 1)/3$, and $\Delta_{d+1} = 0$. Note that $\delta < \frac{1}{5000}$ and for $2 \leq j \leq d$, $\Delta_j < \frac{\delta}{4^d} \cdot \frac{4^{d-j+1}}{3} = \frac{\delta}{3 \cdot 4^{j-1}} = \frac{\delta_{j-1}}{3} < \delta_{j-1}$. In particular, $\Delta_2 < \frac{\delta}{12}$, and $\Delta_{j+1} < \delta_j < \frac{1}{2}$.

There is a class of tiny items, all of which are identical items whose d components are all equal to δ_{d+1}. Their total number is $\frac{(4^d-1)N}{3d}$ and the volume of one such item is $d\delta_{d+1}$. There are d classes of large items. Class j (for $1 \leq j \leq d$) has N items in total, partitioned into $\frac{d!}{(d-j)!(j-1)!}$ types. For $2 \leq j \leq d$, every item has $d - j$ components equal to zero, one component equal to $1 - \Delta_{j+1}$ (this component is called the large component of the item, and it is strictly larger than $\frac{1}{2}$), and the remaining $j - 1$ components are equal to δ_j (for $j = d$, there are no components equal to zero). For $j = 1$ there are $\frac{N}{d}$ types of items, every item has one component equal to $1 - \Delta_2 - \delta_{d+1}$ (this is the large component of the item, and it is larger than $\frac{1}{2}$ in this case as well), and the remaining components are equal to zero. The types are determined according to the identity of components (i.e., which components are equal to zero and which component is $1 - \Delta_{j+1}$); there are $\frac{d!}{(d-j)!(j-1)!}$ options for every $1 \leq j \leq d$, and each option gives a different type. There are equal numbers of items of the different types, so there are $\frac{N(d-j)!(j-1)!}{d!}$ items of each type. Let ω_j denote the volume of an item of class j. For $j \geq 2$, $\omega_j = 1 - \Delta_{j+1} + (j-1)\delta_j$, and for $j = 1$, $\omega_1 = 1 - \Delta_2 - \delta_{d+1}$. For $2 \leq j \leq d - 1$, $\omega_j > \omega_{j+1}$ holds as

$$\omega_j - \omega_{j+1} = (1 - \Delta_{j+1} + (j-1)\delta_j) - (1 - \Delta_{j+2} + j\delta_{j+1})$$

$$= \Delta_{j+2} - \Delta_{j+1} + (j-1)\delta_j - j\delta_{j+1} = (j-1)\delta_j - (j+1)\delta_{j+1} = \delta_j((j-1) - (j+1)/4) ,$$

and $((j-1) - (j+1)/4) = (3j-5)/4 > 0$ for $j \geq 2$. Moreover, for $2 \leq j \leq d$, $\omega_j > 1$ as $\omega_d = 1 - \Delta_{d+1} + (d-1)\delta_d = 1 + (d-1)\delta_d > 1$, and $\omega_j > \omega_d$ for $2 \leq j \leq d-1$. However, it holds that $\omega_1 < 1$. On the other hand, $\omega_2 = 1 - \Delta_3 + \delta_2 < 1 + \frac{\delta}{16}$, so for any $1 \leq j \leq d$, $\omega_j < 1 + \frac{\delta}{16}$.

Claim 11. *The entire set of these items can be packed into $N + 1$ bins, and this packing is optimal.*

To analyze the behavior of GSC on this input, we start with several claims.

Claim 12. *If for some j such that $1 \leq j \leq d$, only large items of classes $j, \ldots, d$ are available, while tiny items and large items of classes $1, \ldots, j-1$ have been packed already, then any bin that is packed with available items can contain at most $d - j + 1$ items.*

The next claim discusses the action of GSC as long as not all large items of class 1 are packed.

Claim 13. *Each of the first $\frac{N}{d}$ bins created by GSC contains exactly d items of class 1, all of distinct types, and it contains exactly $\frac{4^d - 1}{3}$ tiny items. After GSC packs the first $\frac{N}{d}$ bins, all remaining items are large, and none of them is of class 1.*

The next claim discusses the action of GSC after all tiny items and all large items of class 1 were packed. For $1 \leq j \leq d+1$, let $\rho_j = \sum_{\ell=1}^{j-1} \frac{N}{\ell}$.

Claim 14. *Each of the bins of indices $\rho_j + 1, \ldots, \rho_{j+1}$ for $2 \leq j \leq d$ contains exactly $d - j + 1$ large items of class j (and no other items).*

Since each bin of GSC contains (excluding tiny items) $d - j + 1$ items of some class j, the total number of bins is $\sum_{j=1}^{d} \frac{N}{d-j+1} = N \sum_{\ell=1}^{d} \frac{1}{\ell} = N \cdot H_d$. Thus, we find that the number of bins created by GSC is exactly $H_d \cdot (OPT(G) - 1)$. □

References

1. Andelman, N., Feldman, M., Mansour, Y.: Strong price of anarchy. Games and Economic Behavior 65(2), 289–317 (2009)
2. Anshelevich, E., Dasgupta, A., Kleinberg, J.M., Tardos, É., Wexler, T., Roughgarden, T.: The price of stability for network design with fair cost allocation. SIAM Journal on Computing 38(4), 1602–1623 (2008)
3. Aumann, R.J.: Acceptable points in general cooperative n-person games. In: Tucker, A.W., Luce, R.D. (eds.) Contributions to the Theory of Games IV, Annals of Mathematics Study 40, pp. 287–324. Princeton University Press (1959)
4. Aumann, Y., Dombb, Y.: Pareto efficiency and approximate Pareto efficiency in routing and load balancing games. In: Kontogiannis, S., Koutsoupias, E., Spirakis, P.G. (eds.) SAGT 2010. LNCS, vol. 6386, pp. 66–77. Springer, Heidelberg (2010)
5. Bilò, V.: On the packing of selfish items. In: Proc. of the 20th International Parallel and Distributed Processing Symposium (IPDPS 2006), IEEE (2006)
6. Chekuri, C., Khanna, S.: On multidimensional packing problems. SIAM Journal on Computing 33(4), 837–851 (2004)
7. Chien, S., Sinclair, A.: Strong and Pareto price of anarchy in congestion games. In: Albers, S., Marchetti-Spaccamela, A., Matias, Y., Nikoletseas, S., Thomas, W. (eds.) ICALP 2009, Part I. LNCS, vol. 5555, pp. 279–291. Springer, Heidelberg (2009)
8. Coffman Jr., E.G., Csirik, J., Galambos, G., Martello, S., Vigo, D.: Bin packing approximation algorithms: Survey and Classification. In: Pardalos, P.M., Du, D.Z., Graham, R.L.L. (eds.) Handbook of Combinatorial Optimization, pp. 455–531. Springer, New York (2013)

9. Dósa, G., Epstein, L.: Generalized selfish bin packing. CoRR, abs/1202.4080 (2012)
10. Dubey, P.: Inefficiency of Nash equilibria. Mathematics of Operations Research 11(1), 1–8 (1986)
11. Epstein, L., Kleiman, E.: On the quality and complexity of Pareto equilibria in the job scheduling game. In: Proc. of the 10th International Conference on Autonomous Agents and Multiagent Systems (AAMAS 2011), pp. 525–532 (2011)
12. Epstein, L., Kleiman, E.: Selfish bin packing. Algorithmica 60(2), 368–394 (2011)
13. Epstein, L., Kleiman, E., Mestre, J.: Parametric packing of selfish items and the subset sum algorithm. In: Algorithmica, pp. 67–78 (2014), doi:10.1007/s00453-014-9942-0
14. Fiat, A., Kaplan, H., Levy, M., Olonetsky, S.: Strong price of anarchy for machine load balancing. In: Arge, L., Cachin, C., Jurdziński, T., Tarlecki, A. (eds.) ICALP 2007. LNCS, vol. 4596, pp. 583–594. Springer, Heidelberg (2007)
15. Garey, M.R., Graham, R.L., Johnson, D.S.: Resource constrained scheduling as generalized bin packing. J. Comb. Theory, Ser. A 21(3), 257–298 (1976)
16. Graham, R.L.: Bounds on multiprocessing anomalies and related packing algorithms. In: Proceedings of the 1972 Spring Joint Computer Conference, pp. 205–217 (1972)
17. Holzman, R., Law-Yone, N.: Strong equilibrium in congestion games. Games and Economic Behavior 21(1–2), 85–101 (1997)
18. Johnson, D.S., Demers, A., Ullman, J.D., Garey, M.R., Graham, R.L.: Worst-case performance bounds for simple one-dimensional packing algorithms. SIAM Journal on Computing 3, 256–278 (1974)
19. Kou, L.T., Markowsky, G.: Multidimensional bin packing algorithms. IBM Journal of Research and Development 21(5), 443–448 (1977)
20. Koutsoupias, E., Papadimitriou, C.H.: Worst-case equilibria. In: Meinel, C., Tison, S. (eds.) STACS 1999. LNCS, vol. 1563, pp. 404–413. Springer, Heidelberg (1999)
21. Lee, S., Panigrahy, R., Prabhakaran, V., Ramasubrahmanian, V., Talwar, K., Uyeda, L., Wieder, U.: Validating heuristics for virtual machine consolidation. Microsoft Research, MSR-TR-2011-9 (2011)
22. Luc, D.T.: Pareto optimality. In: Chinchuluun, A., Pardalos, P.M., Migdalas, A., Pitsoulis, L. (eds.) Pareto Optimality, Game Theory and Equilibria, pp. 481–515. Springer (2008)
23. Ma, R., Dósa, G., Han, X., Ting, H.F., Ye, D., Zhang, Y.: A note on a selfish bin packing problem. J. Global Optimization 56(4), 1457–1462 (2013)
24. Mavronicolas, M., Spirakis, P.G.: The price of selfish routing. Algorithmica 48(1), 91–126 (2007)
25. Rozenfeld, O., Tennenholtz, M.: Strong and correlated strong equilibria in monotone congestion games. In: Spirakis, P.G., Mavronicolas, M., Kontogiannis, S.C. (eds.) WINE 2006. LNCS, vol. 4286, pp. 74–86. Springer, Heidelberg (2006)
26. Ye, D., Chen, J.: Non-cooperative games on multidimensional resource allocation. Future Generation Comp. Syst. 29(6), 1345–1352 (2013)
27. Yin, B., Wang, Y., Meng, L., Qiu, X.: A multi-dimensional resource allocation algorithm in cloud computing. Journal of Information & Computational Science 9(11), 3021–3028 (2012)
28. Yu, G., Zhang, G.: Bin packing of selfish items. In: Papadimitriou, C., Zhang, S. (eds.) WINE 2008. LNCS, vol. 5385, pp. 446–453. Springer, Heidelberg (2008)
29. Zhang, Q., Cheng, L., Boutaba, R.: Cloud computing: state-of-the-art and research challenges. Journal of Internet Services and Applications 1(1), 7–18 (2010)

Approximate Deadline-Scheduling
with Precedence Constraints

Hossein Efsandiari[1,*], MohammadTaghi Hajiaghyi[1,*], Jochen Könemann[2,**],
Hamid Mahini[1,*], David Malec[1,*], Laura Sanità[2,**]

[1] University of Maryland, College Park, MD, USA
[2] University of Waterloo, Waterloo, Ontario N2L 3G1, Canada

Abstract. We consider the classic problem of scheduling a set of n jobs
non-preemptively on a single machine. Each job j has non-negative processing time, weight, and deadline, and a feasible schedule needs to be
consistent with *chain-like* precedence constraints. The goal is to compute a feasible schedule that minimizes the sum of penalties of late
jobs. Lenstra and Rinnoy Kan [Annals of Disc. Math., 1977] in their
seminal work introduced this problem and showed that it is strongly
NP-hard, even when all processing times and weights are 1. We study
the approximability of the problem and our main result is an $O(\log k)$-
approximation algorithm for instances with k distinct job deadlines.

1 Introduction

In an instance of the classic *precedence-constrained single-machine deadline
scheduling* problem we are given a set $[n] := \{1, \ldots, n\}$ of jobs that need to
be scheduled non-preemptively on a single machine. Each job $j \in [n]$ has a non-
negative deadline $d_j \in \mathbb{N}$, a processing time $p_j \in \mathbb{N}$ as well as a non-negative
penalty $w_j \in \mathbb{N}$. A feasible schedule has to be consistent with precedence constraints that are given implicitly by a directed acyclic graph $G = ([n], E)$; i.e.,
job $i \in [n]$ has to be processed before job j if G has a directed i,j-path. A feasible schedule incurs a penalty of w_j if job j is not completed before its deadline
d_j. Our goal is then to find a feasible schedule that minimizes the total penalty
of late jobs. In the standard scheduling notation [10] the problem under consid-
eration is succinctly encoded as $1|\text{prec}|\sum w_j U_j$, where U_j is a binary variable
that takes value 1 if job j is late and 0 otherwise.

Single-machine scheduling with deadline constraints is a practically important
and well-studied subfield of scheduling theory that we cannot adequately survey
here. We refer the reader to Chapter 3 of [22] or Chapter 4 of [3], and focus here
on the literature that directly relates to our problem. The decision version of the
single-machine deadline scheduling problem *without* precedence constraints is part

---

* Supported in part by NSF CAREER award 1053605, NSF grant CCF-1161626, ONR
  YIP award N000141110662, and DARPA/AFOSR grant FA9550-12-1-0423.
** Supported in part by the NSRERC Discovery Grant Program.

© Springer-Verlag Berlin Heidelberg 2015
N. Bansal and I. Finocchi (Eds.): ESA 2015, LNCS 9294, pp. 483–495, 2015.
DOI: 10.1007/978-3-662-48350-3_41

of Karp's list of 21 NP-complete problems [16], and a fully-polynomial-time approximation scheme is known [8,23]. The problem becomes strongly NP-complete in the presence of release dates as was shown by Lenstra et al. [20]. Lenstra and Rinnoy Kan [21] later proved that the above problem is strongly NP-hard even in the special case where each job has unit processing time and penalty, and the precedence digraph G is a collection of vertex-disjoint directed paths.

Despite being classical, and well-motivated, little is known about the approximability of precedence-constrained deadline scheduling. This surprises, given that problems in this class were introduced in the late 70s, and early 80s, and that these are rather natural variants of Karp's original 21 NP-hard problems. The sparsity of results to date suggests that the combination of precedence constraints and deadlines poses significant challenges. We seek to show, however, that these challenges can be overcome to achieve non-trivial approximations for these important scheduling problems. In this paper we focus on the generalization of the problem studied in [21], where jobs are allowed to have arbitrary nonnegative processing times, and where we minimize the *weighted* sum of late jobs. Once more using scheduling notation, this problem is given by $1|\text{chains}| \sum w_j U_j$ (and hereafter referred to as pDLS). Our main result is the following.

Theorem 1. *pDLS has an efficient $O(\log k)$-approximation algorithm, where k is the number of distinct job deadlines in the given instance.*

We note that our algorithm finds a feasible schedule without late jobs if such a schedule exists.

In order to prove this result, we first introduce a novel, and rather subtle *configuration*-type LP. The LP treats each of the directed paths in the given precedence system independently. For each path, the LP has a variable for all nested collections of k suffixes of jobs, and integral solutions set exactly one of these variables per path to 1. This determines which subset of jobs are executed *after* each of the k distinct job deadlines. The LP then has constraints that limit the total processing time of jobs executed before each of the k deadlines. While we can show that integral feasible solutions to our formulation naturally correspond to feasible schedules, the formulation's integrality gap is large (see the full version [7] for details). In order to reduce the gap, we strengthen the formulation using valid inequalities of *Knapsack cover*-type [1,4,14,25] (see also [5,18]).

The resulting formulation has an exponential number of variables and constraints, and it is not clear whether it can be solved efficiently. In the case of chain-like precedences, we are able to provide an alternate formulation that, instead of variables for nested collections of suffixes of jobs, has variables for job-suffixes only. Thereby, we reduce the number of variables to a polynomial of the input size, while increasing the number of constraints slightly. We do not know how to efficiently solve even this alternate LP. However, we are able to provide a *relaxed* separation oracle (in the sense of [4]) for its constraints, and can therefore use the Ellipsoid method [11] to obtain approximate solutions for the alternate LP of sufficient quality.

We are able to provide an efficiently computable map between solutions for the alternate LP, and those of the original exponential-sized formulation. Crucially, we are able to show that the latter solutions are structurally nice; i.e., no two nested families of job suffixes in its support cross! Such *cross-free* solutions to the original LP can then be rounded into high-quality schedules. Because of space limitations, details describing the alternate compact LP, the relaxed separation oracle, and the efficiently computable map are only provided in the full version of the paper [7].

Several comments are in order. First, there is a significant body of research that investigates LP-based techniques for single-machine, precedence-constrained, minimum weighted completion-time problems (e.g., see [13,12,24], and also [6] for a more comprehensive summary of LP-based algorithms for this problem). None of these LPs seem to be useful for the objective of minimizing the total penalty of late jobs. In particular, converting these LPs requires the introduction of so called "big-M"-constraints that invariably yield formulations with large integrality gaps.

Second, using Knapsack-cover inequalities to strengthen an LP formulation for a given covering problem is not new. In the context of approximation algorithms, such inequalities were used by Carr et al. [4] in their work on the Knapsack problem and several generalizations. Subsequently, they also found application in the development of approximation algorithms for *general covering and packing integer programs* [18], in approximating *column-restricted* covering IPs [17,5], as well as in the area of scheduling (without precedence constraints) [2]. Note that our strong formulations for pDLS use variables for (families of) suffixes of jobs in order to encode the chain-like dependencies between jobs. This leads to formulations that are not column-restricted, and they also do not fall into the framework of [18] (as, e.g., their dimension is not polynomial in the input size).

Third, it is not clear how what little work there has been on precedence-constrained deadline scheduling can be applied to the problem we study. The only directly relevant positive result we know of is that of Ibarra and Kim [15], who consider the single-machine scheduling problem in which n jobs need to be scheduled non-preemptively on a single machine while adhering to precedence constraints given by *acyclic directed forests*, with the goal to maximize the total profit of jobs completed before a *common deadline T*. While the allowed constraints are strictly more general than the chain-like ones we study, this is more than outweighed by the fact that all jobs have a common deadline, which significantly reduces the complexity of the problem and renders it similar to the well-studied Knapsack problem. Indeed, we show in the full version [7] of our paper that pDLS with forest precedences and a single deadline admits a pseudo-polynomial time algorithm as well. This implies that the decision version of pDLS is only *weakly* NP-complete in this special case. Given the strong NP-hardness of pDLS (as established in [21]), it is unclear how Ibarra and Kim's results can be leveraged for our problem.

It is natural to ask whether the approximation bound provided in Theorem 1 can be improved. In [7] we provide an example demonstrating that this is unlikely if we use a path-independent rounding scheme (as in the proof of Theorem 1). This example highlights that different paths can play vastly different roles in a solution, and be critical to ensuring that distinct necessary conditions are met.

Thus, rounding paths independently can lead to many independent potential points of failure in the process, and significant boosting of success probabilities must occur if we are to avoid all failures simultaneously. This means, roughly speaking, that our analysis is tight and therefore our approximation factor cannot be improved without significant new techniques. Given the above, it is natural to look for dependent rounding schemes for solutions to our LP. Indeed, such an idea can be made to work for the special case of pDLS with two paths.

Theorem 2. *pDLS with two paths admits a 2-approximation algorithm based on a correlated rounding scheme.*

The proof of Theorem 2 is deferred to [7], and shows that the configurational LP used in the proof of Theorem 1 has an integrality gap of at most 2 for pDLS instances with two paths. This is accomplished using a randomized rounding scheme that samples families of suffix chains from the two paths in a correlated fashion instead of independently. The approach uses the fact that our instances have two paths, and extending it to general instances appears difficult.

We point out that the emphasis in Theorem 2 and its proof is on the techniques used rather than the approximation guarantee obtained. In fact, we provide a dynamic-programming-based exact algorithm for pDLS instances with a fixed number of chains (see [7] for details).

Theorem 3. *pDLS can be solved exactly when the number of chains is fixed.*

1.1   Deadline Scheduling and Technology Diffusion

As we show now, the *precedence-constrained single-machine deadline scheduling* problem is closely related to the *technology diffusion* (TD) problem which was recently introduced by Goldberg and Liu [9] in an effort to model dynamic processes arising in technology adaptation scenarios. In an instance of TD, we are given a graph $G = (V, E)$, and thresholds $\theta(v) \in \{\theta_1, \ldots, \theta_k\}$ for each $v \in V$. We consider dynamic processes in which each vertex $v \in V$ is either *active* or *inactive*, and where an inactive vertex v becomes active if, in the graph induced by it and the active vertices, v lies in a connected component of size at least $\theta(v)$. The goal in TD is now to find a smallest *seed set* S of initially active vertices that eventually lead to the activation of the entire graph. Goldberg and Liu argued that it suffices (albeit at the expense of a constant factor loss in the approximation ratio) to consider the following connected abstraction of the problem: find a permutation $\pi = (v_1, \ldots, v_n)$ of V such that the graph induced by $v_1, \ldots, v_i$ is connected, for all i, and such that

$$S(\pi) = \{v_i : i < \theta(v_i)\}$$

is as small as possible.

As Goldberg and Liu [9] argue, TD has no $o(\log(n))$-approximation algorithm unless NP has quasi-polynomial-time algorithms. The authors also presented an $O(rk \log(n))$-approximation, where r is the diameter of the given graph, and k is the number of distinct thresholds used in the instance. Könemann, Sadeghian, and Sanità [19] recently improved upon this result by presenting a $O(\min\{r, k\} \log(n))$-approximation algorithm. The immediate open question arising from [9] and [19] is whether the dependence of the approximation ratio on r and k is avoidable. As it turns out, our work here provides an affirmative answer for TD instances on *spider* graphs (i.e., trees in which at most one vertex has degree larger than 2).

Theorem 4. *TD is NP-hard on spiders. In these graphs, the problem also admits an $O(\log(k))$-approximation.*

The theorem follows from the fact that TD in spiders and pDLS with unit processing times, and penalties are equivalent. We sketch the proof. Given an instance of TD on spider $G = (V, E)$, we create a job for each vertex $v \in V$, and let $d_v = n - \theta(v) + 1$, and $p_v = w_v = 1$. We also create a dependence chain for each leg of the spider; i.e., the job for vertex v depends on all its descendants in the spider, rooted at its sole vertex of degree larger than 2. It is now an easy exercise to see that the TD instance has a seed set of size s iff the pDLS instance constructed has a schedule that makes s jobs late.

2   Notation

In the rest of the paper we will consider an instance of pDLS given by a collection $[n]$ of jobs. Each job j has non-negative processing time p_j, penalty w_j and deadline d_j. The precedence constraints on $[n]$ are induced by a collection of vertex-disjoint, directed paths $\mathcal{P} = \{P_1, \ldots, P_q\}$. In a feasible schedule job j has to precede job j' if there is a directed j, j'-path in one of the paths in $\mathcal{P}$; we will write $j \preceq j'$ to indicate j has to precede j' from now on for ease of notation, and $j \prec j'$ if we furthermore have $j \neq j'$. We denote the set of distinct deadlines in our instance by $\mathcal{D} = \{D_1, \ldots, D_k\}$, with higher indices corresponding to later deadlines, that is, indexed such that $D_i < D_{i'}$ whenever $i < i'$. We use the notation $i(j) \in [k]$ to denote the index that the deadline of job j has in the set $\mathcal{D}$, so we have that $d_j = D_{i(j)}$ for all $j \in [n]$. We say that a job is *postponed* or *deferred* past a certain deadline D_i if the job is executed after D_i. Our goal is to find a feasible schedule that minimizes the total penalty of late jobs. Given a directed path P, we let

$$P_{\succeq j} := \{j' \in [n] : j \preceq j'\}$$

be the *suffix* induced by job $j \in [n]$. We call a sequence $S = (S_1, S_2, \ldots, S_k)$ of suffixes of a given path $P \in \mathcal{P}$ a *suffix chain* if

$$P \supseteq S_1 \supseteq S_2 \supseteq \cdots \supseteq S_k;$$

while a suffix chain could have arbitrary length, we will only use suffix chains with length $k = |\mathcal{D}|$. Given two suffix chains S and S' with k suffixes each, we say $S \preceq S'$ if $S_i \supseteq S'_i$ for all $i \in [k]$. If we have neither $S \preceq S'$ nor $S' \preceq S$, we say that S and S' *cross*. Given two suffix chains S and S', we obtain their *join* $S \vee S'$ by letting $(S \vee S')_i = S_i \cup S'_i$. Similarly, we let the *meet* of S and S' be obtained by letting $(S \wedge S')_i = S_i \cap S_i$.

3   An Integer Programming Formulation

Our general approach will be to formulate the problem as an integer program, to solve its relaxation, and to randomly round the fractional solution into a feasible schedule of the desired quality. The IP will have a layered structure. For each deadline $D_i \in \mathcal{D}$, we want to decide which jobs in $[n]$ are to be postponed past deadline D_i. We start with the following two easy but crucial observations.

Observation 5. *Consider a path $P \in \mathcal{P}$, and suppose that $j \in P$ is one of the jobs on this path. If j is postponed past D_i then so are all of j's successors on P. Thus, we may assume w.l.o.g. that the collection of jobs of P that are executed after time D_i forms a* suffix *of P.*

Observation 6. *Consider a path $P \in \mathcal{P}$, and suppose that $j \in P$ is one of the jobs on this path. If j is postponed past D_i, then it is also postponed past every earlier deadline $D_{i'} < D_i$. Thus, we may assume w.l.o.g. that the collections $S_1, \ldots, S_k$ of jobs of P that are executed after deadlines $D_1 < \cdots < D_k$, respectively, exhibit a* chain *structure, i.e. $S_1 \supseteq S_2 \supseteq \cdots \supseteq S_k$.*

Combining the above two observations, we see that for each path $P \in \mathcal{P}$, the collections of jobs postponed past each deadline form a suffix chain $S^P = S_1^P \supseteq S_2^P \supseteq \cdots \supseteq S_k^P$. In the following we let $\mathcal{S}^P$ denote the collection of suffix chains for path P; we introduce a binary variable x_S for each suffix chain $S \in \mathcal{S}^P$ and each $P \in \mathcal{P}$. In an IP solution $x_S = 1$ for some $S \in \mathcal{S}^P$ if for each $i \in [k]$ the set of jobs executed past deadline D_i is precisely S_i. We now describe the constraints of the IP in detail.

(C1) **At most one suffix chain of postponed jobs per path.** In a solution, we want to pick at most one suffix chain for each path $P \in \mathcal{P}$, and thus obtain the following natural constraint:

$$\sum_{S \in \mathcal{S}^P} x_S \leq 1 \quad \forall P \in \mathcal{P}. \tag{C1}$$

(C2) **Deferring sufficiently many jobs.** In any feasible schedule, the total processing time of jobs scheduled before time D_i must be at most D_i; conversely, the total processing time of jobs whose execution is deferred past time D_i must be at least $\Gamma - D_i$, where $\Gamma = \sum_{j \in [n]} p_j$ is the total processing time of all jobs. This is captured by the following constraints:

$$\sum_{P \in \mathcal{P}} \sum_{S \in \mathcal{S}^P} p_S^i x_S \geq \Gamma - D_i \quad \forall i \in [k]$$

where p_S^i is the total processing time of the jobs contained in S_i. While the above constraints are certainly valid, in order to reduce the integrality gap of the formulation and successfully apply our rounding scheme we need to *strengthen* them, as we now describe. To this end, suppose that we are given a chain

$$F^P = F_1^P \supseteq F_2^P \supseteq F_3^P \supseteq \ldots \supseteq F_k^P$$

of k suffixes of deferred jobs for each path $P \in \mathcal{P}$, and let $F = \{F^P\}_{P \in \mathcal{P}}$ be the family of these suffix chains. Suppose that we knew that we were looking for a schedule in which the jobs in F_i^P are deferred past deadline D_i for all $P \in \mathcal{P}$. For each $i \in [k]$, a feasible schedule must now defer jobs outside $\bigcup_{P \in \mathcal{P}} F_i^P$ of total processing time at least

$$\Theta^{i,F} := \max \left\{ (\Gamma - D_i) - \sum_{P \in \mathcal{P}} \sum_{j \in F_i^P} p_j, 0 \right\}. \tag{1}$$

We obtain the following valid inequality for any feasible schedule:

$$\sum_{P \in \mathcal{P}} \sum_{S \in \mathcal{S}^P} p_S^{i,F} x_S \geq \Theta^{i,F} \quad \forall i \in [k], \forall F \in \mathcal{S}, \tag{C2}$$

where $\mathcal{S}$ is the collection of all families of suffix chains for $\mathcal{P}$ (including the empty family), and where $p_S^{i,F}$ is the minimum of $\Theta^{i,F}$ and the total processing time of jobs j that are in S_i but not in F_i^P; formally, for $F \in \mathcal{S}$, $i \in [k]$, $P \in \mathcal{P}$, and $S \in \mathcal{S}^P$, we set

$$p_S^{i,F} := \min \left\{ \sum_{j \in S_i \setminus F_i^P} p_j, \Theta^{i,F} \right\}.$$

(C2) falls into the class of *Knapsack Cover* (KC) inequalities [1,4,14,25], and the above *capping* of coefficients is typical for such inequalities.

All that remains to define the IP is to give the objective function. Consider a job j on path $P \in \mathcal{P}$, and suppose that the IP solution x picks suffix chain $S \in \mathcal{S}^P$. Job j is late (i.e., its execution ends after time $d_j = D_{i(j)}$) if j is contained in the suffix $S_{i(j)}$. We can therefore express the penalty of suffix chain S succinctly as

$$w_S := \sum_{j \in P : j \in S_{i(j)}} w_j. \tag{2}$$

We can now state the canonical LP relaxation of the IP as follows

$$\min \left\{ \sum_{P \in \mathcal{P}} \sum_{S \in \mathcal{S}^P} w_S x_S : (C1), (C2), x \geq 0 \right\}. \tag{P}$$

For convenience we introduce auxiliary indicator variables U_j for each job $j \in [n]$. U_j takes value 1 if j's execution ends after time d_j, and hence

$$U_j := \sum_{S \in \mathcal{S}^P : j \in S_{i(j)}} x_S, \tag{3}$$

where P is the chain containing job j.

4   Rounding the Relaxation

Our rounding scheme does not apply only to (suitable) feasible points for (P), but in fact allows us to round a much broader class of (not necessarily feasible) fractional points (U, x) to integral feasible solutions $(\hat{U}, \hat{x})$ of the corresponding IP, while only losing a factor of $O(\log k)$ in the objective value. As we will see later being able to round this broader class of points is crucial for our algorithm. In order to formally describe the class of points we can round, we need to introduce the concept of canonical chain families. Informally, the canonical suffix chain for a path P defers each job $j \in P$ as much as possible, subject to ensuring no job in P is deferred past its deadline. The definition below makes this formal.

Definition 1. *Given an instance of pDLS, we let C_i^P be the longest suffix of path $P \in \mathcal{P}$ that consists only of jobs whose deadline is strictly greater than D_i. Jobs in C_i^P may be scheduled to complete after D_i without incurring a penalty. We call*

$$C^P := C_1^P \supseteq \ldots \supseteq C_k^P$$

the canonical suffix chain for path P, and let $C = \{C^P\}_{P \in \mathcal{P}}$ be the canonical suffix chain family.

Our general approach for rounding a solution (U, x) to program (P) is to split jobs into those with large U_j values and those with small ones. While we can simply think of "rounding up" U_j values when they are already large, we need to utilize the constraints (C1) and (C2) to see how to treat jobs with small U_j values. As it turns out, in order to successfully round (U, x) we need it to satisfy the KC-inequality for a *single* suffix chain family only. Naturally this family will depend on the set of jobs with large U_j value. We can formalize the above as follows.

Consider any instance $\mathcal{I}$ of pDLS, and let (U, x) be a solution to (P). Define the set L of jobs that are late to an extent of at least $1/(\gamma \log k)$ for a parameter $\gamma > 0$ (whose value we will make precise at a later point):

$$L = \{j \,:\, U_j \geq 1/(\gamma \log k)\}.$$

We now obtain a *modified instance of pDLS*, denoted $\mathcal{I}_L$, by increasing the deadline for the jobs in L to Γ. Thus, jobs in L can never be late in the modified instance $\mathcal{I}_L$. Note that since we do not modify the processing time of any job $j \in [n]$, we have that $p_S^{i,F}$ and $\Theta^{i,F}$ remain identical in $\mathcal{I}_L$ and $\mathcal{I}$ for all i, F, and S. Similarly, each job $j \in [n]$ has the same penalty w_j in $\mathcal{I}$ and $\mathcal{I}_L$. Let C be the canonical suffix chain family for $\mathcal{I}_L$. We are able to round a solution (U, x) as long as it satisfies the following conditions:

(a) for each $P \in \mathcal{P}$, the set $\{S \in \mathcal{S}^P : x_S > 0\}$ is cross-free
(b) (U, x) is feasible for a relaxation (P') of (P) that replaces the constraints (C2) by

$$\sum_{P \in \mathcal{P}} \sum_{S \in \mathcal{S}^P} p_S^{i,C} x_S \geq \Theta^{i,C} \quad \forall i \in [k], \tag{C2'}$$

where C is the canonical suffix chain family for the modified pDLS instance $\mathcal{I}_L$.

In the next section, we see how we can find solutions satisfying both of these conditions.

Suppose (U, x) is a solution to (P) that satisfies (a) and (b). Obtain x^0 by letting $x_S^0 = x_S$ if S makes at least one job $j \in [n]$ late in $\mathcal{I}_L$, and let $x_S^0 = 0$ otherwise. Define $U^0 \leq U$ as in (3) (with x^0 in place of x), and note that (U^0, x^0) satisfies (a) and (b). Let us now round (U^0, x^0). We focus on path $P \in \mathcal{P}$, and define the support of (U^0, x^0) induced by P:

$$\mathcal{T}^P := \{S \in \mathcal{S}^P : x_S^0 > 0\}.$$

As this set is cross-free by assumption (a), $\mathcal{T}^P$ has a well-defined maximal element S^* with $S \preceq S^*$ for all $S \in \mathcal{T}^P$ (recall, $S \preceq S^*$ means S defers no less jobs past every deadline D_i than S^* does). By definition, S^* makes at least one job $j \in [n] \setminus L$ late. Since S^* is maximal in $\mathcal{T}^P$ it therefore follows that j is late in all $S \in \mathcal{T}^P$. Using the definition of (U^0, x^0) as well as the fact that $j \notin L$ we obtain

$$\sum_{S \in \mathcal{T}^P} x_S^0 = \sum_{S \in \mathcal{S}^P : j \in S_{i(j)}} x_S^0 = U_j^0 \leq U_j < \frac{1}{\gamma \log k}. \tag{4}$$

We let $(\bar{U}, \bar{x}) = \gamma \log k \cdot (U^0, x^0)$ and obtain the following lemma. The proof of this and several of the following lemmas are deferred to [7] because of space limitations.

Lemma 1. $(\bar{U}, \bar{x})$ *satisfies*

$$\sum_{S \in \mathcal{S}^P} \bar{x}_S \leq 1 \quad \forall i \in [k], \forall P \in \mathcal{P} \tag{$\overline{\text{C1}}$}$$

$$\sum_{P \in \mathcal{P}} \sum_{S \in \mathcal{S}^P} p_S^{i,C} \bar{x}_S \geq \gamma \log k \cdot \Theta^{i,C} \quad \forall i \in [k], \tag{$\overline{\text{C2}}$}$$

where C is the canonical suffix chain family defined for the modified instance $\mathcal{I}_L$ of pDLS.

We now randomly round $(\bar{U}, \bar{x})$ to an integral solution $(\hat{U}, \hat{x})$ as follows. For each $P \in \mathcal{P}$, we independently select a single random suffix chain $S \in \mathcal{S}^P$ using marginals derived from $\bar{x}$, and set the corresponding $\hat{x}_S = 1$. In particular, we set $\hat{x}$ so that for all $P \in \mathcal{P}$ and all $S \in \mathcal{S}^P$ we have

$$\Pr[\hat{x}_S = 1] = \begin{cases} \bar{x}_S & \text{if } S \in \mathcal{T}^P \\ 1 - \sum_{S' \in \mathcal{T}^P} \bar{x}_{S'} & \text{if } S = C^P. \end{cases}$$

Since $(\bar{U}, \bar{x})$ satisfies $(\overline{C1})$, we can see that the above describes a valid randomized process. We run this process independently for each path $P \in \mathcal{P}$ to obtain $\hat{x}$. A job $j \in [n] \setminus L$ is late if it is contained in level $i(j)$ of the suffix chain S chosen for path P by the above process. Thus, we set

$$\hat{U}_j := \sum_{S \in \mathcal{S}^P : j \in S_{i(j)}} \hat{x}_S,$$

and easily obtain the following lemma.

Lemma 2. *For all $j \notin L$, $E[\hat{U}_j] = \bar{U}_j$.*

The preceding lemma shows that the expected penalty of $(\hat{U}, \hat{x})$ in the modified instance $\mathcal{I}_L$ is exactly $\sum_{j \in [n] \setminus L} w_j \bar{U}_j$. The following lemma shows that the schedule induced by $\hat{x}$ postpones at least $\Theta^{i,C}$ jobs past deadline D_i for all $i \in [k]$ with constant probability.

Lemma 3. *With constant probability, we have*

$$\sum_{P \in \mathcal{P}} \sum_{S \in \mathcal{S}^P} p_S^{i,C} \hat{x}_S \geq \Theta^{i,C} \quad \forall i \in [k], \tag{5}$$

where C is the canonical suffix chain family for the modified pDLS instance $\mathcal{I}_L$. In particular, for $\gamma = 4$ the constraint holds with probability at least 0.7.

For each $P \in \mathcal{P}$ let $\hat{S}^P$ be the join of the suffix chain corresponding to solution $\hat{x}$, and the canonical suffix chain C^P; i.e., suppose that $\hat{x}_S = 1$ for $S \in \mathcal{S}^P$. Then

$$\hat{S}^P = S \vee C^P. \tag{6}$$

Clearly, $\hat{S}^P$ is a suffix chain for path P. We use the following greedy algorithm to obtain a schedule.

> **for** $i = 1$ to k **do**
>   **for all** $P \in \mathcal{P}$ **do**
>   Schedule all jobs in $P \setminus \hat{S}_i^P$ not already scheduled respecting the precedence constraints
>   **end for**
> **end for**
> Schedule all remaining jobs respecting the precedence constraints

Theorem 7. *The schedule produced by the above algorithm is feasible. Furthermore, if (5) holds, the schedule has cost at most $\sum_{j \notin L} w_j \hat{U}_j$ in the instance $\mathcal{I}_L$.*

Corollary 1. *The schedule produced by the above algorithm is feasible and incurs penalty at most $8 \log k \cdot \sum_j w_j U_j$ in the original instance of the pDLS with constant probability.*

5   Solving LP (P)

In the preceding sections, we have seen that pDLS can be expressed as an IP, and that solutions to a somewhat weakened LP relaxation of it can be rounded into feasible schedules for the original problem instance. Importantly, we do not know, however, how to obtain solutions for this LP relaxation, and this seems particularly complicated as (P) has an exponential number of variables and constraints. It turns out that, in the special case of chain-like precedence constraints, one can find an equivalent and more compact LP formulation for (P).

The new IP has binary variables x_j^i for each job $j \in [n]$ and for each deadline D_i; in a solution, x_j^i will have value 1 if j and all of its successors are executed after deadline D_i, and $x_j^i = 0$ otherwise. In the revised IP we will add constraints that force a solution to pick at most one suffix of postponed jobs per path and layer (see (D1)). Just like before, we will add a constraint that ensures that the execution of sufficiently many jobs is deferred past each deadline D_i; once more, this involves adding a certain family of Knapsack cover inequalities (see D2). The new IP has a polynomial number of variables. The less expressive nature of these variables forces us to add extra constraints that ensure *chain structure* of the suffixes chosen for each of the paths (see D3). In the following LP relaxation of the new IP, we let

$$U_j = \sum_{j' \preceq j} x_{j'}^{i(j)},$$

which, in an integer solution, has value 1 whenever job j is executed past its deadline.

$$\min \quad \sum_{j \in [n]} w_j U_j \tag{P2}$$

$$\text{s.t.} \quad \sum_{j \in P} x_j^i \leq 1 \quad \forall P \in \mathcal{P}, i \in [k] \tag{D1}$$

$$\sum_{P \in \mathcal{P}} \sum_{j \in P \setminus F_i^P} p_j^{i,F} x_j^i \geq \Theta^{i,F} \quad \forall i \in [k], \forall F \in \mathcal{S} \tag{D2}$$

$$\sum_{j':j' \preceq j} x_{j'}^{i+1} \leq \sum_{j':j' \preceq j} x_{j'}^i, \quad \forall P \in \mathcal{P}, \forall j \in P, \forall i \in [k-1] \tag{D3}$$

$$x \geq 0$$

In the case of chain-like preferences, LPs (P) and (P2) are equivalent, as we can show. More precisely, we are able to obtain an objective-value preserving map between feasible solutions to (P2) and cross-free solutions to (P). Unfortunately, we still do not know how to solve LP (P2): while it has a polynomial number of variables, it has an exponential number of constraints, and we do not know how to employ the Ellipsoid method [11] as we do not know how to separate the Knapsack cover-style constraints in (D2). Instead we design a relaxed separation oracle (in the sense of [4]) that allows us to efficiently find a polynomial-sized subfamily of constraints in (D2). Roughly speaking, replacing (D2) by the constraints in this subfamily yields a compact LP whose solution can be mapped to a solution to a weakened form of (P) that satisfies conditions (a) and (b) in Section 4. Thus, this solution can now be rounded into a feasible schedule (with constant probability). The reader is once more referred to [7] for a detailed description of the solution method sketched in this section.

References

1. Balas, E.: Facets of the knapsack polytope. Math. Programming 8, 146–164 (1975)
2. Bansal, N., Pruhs, K.: The geometry of scheduling. In: Proceedings of the IEEE Symposium on Foundations of Computer Science, pp. 407–414 (2010)
3. Blazewicz, J., Ecker, K.H., Pesch, E., Schmidt, G., Weglarz, J.: Handbook on scheduling: from theory to applications. Springer (2007)
4. Carr, R.D., Fleischer, K.K., Leung, V.J., Phillips, C.A.: Strengthening integrality gaps for capacitated network design and covering problems. In: Proceedings of the ACM-SIAM Symposium on Discrete Algorithms, pp. 106–115 (2000)
5. Chakrabarty, D., Grant, E., Könemann, J.: On column-restricted and priority covering integer programs. In: Eisenbrand, F., Shepherd, F.B. (eds.) IPCO 2010. LNCS, vol. 6080, pp. 355–368. Springer, Heidelberg (2010)
6. Correa, J.R., Schulz, A.S.: Single-machine scheduling with precedence constraints. Math. Oper. Res. 30(4), 1005–1021 (2005)
7. Efsandiari, H., Hajiaghayi, M., Könemann, J., Mahini, H., Malek, D., Sanità, L.: Approximate deadline-scheduling with precedence constraints. Technical report, arXiv (2015)
8. Gens, G.V., Levner, E.V.: Fast approximation algorithm for job sequencing with deadlines. Discrete Appl. Math. 3(4), 313–318 (1981)
9. Goldberg, S., Liu, Z.: Technology diffusion in communication networks. In: Proceedings of the ACM-SIAM Symposium on Discrete Algorithms, pp. 233–240 (2013)
10. Graham, R.L., Lawler, E.L., Lenstra, J.K., Rinnooy Kan, A.H.G.: Optimization and approximation in deterministic sequencing and scheduling: A survey. Annals of Discrete Mathematics 5, 287–326 (1979)
11. Grötschel, M., Lovász, L., Schrijver, A.: The ellipsoid method and its consequences in combinatorial optimization. Combinatorica 1, 169–197 (1981)
12. Hall, L.A., Schulz, A.S., Shmoys, D.B., Wein, J.: Scheduling to minimize average completion time: Off-line and on-line approximation algorithms. Math. Oper. Res. 22(3), 513–544 (1997)
13. Hall, L.A., Shmoys, D.B., Wein, J.: Scheduling to minimize average completion time: Off-line and on-line algorithms. In: Proceedings of the ACM-SIAM Symposium on Discrete Algorithms, pp. 142–151 (1996)
14. Hammer, P.L., Johnson, E.L., Peled, U.N.: Facets of regular 0,1-polytopes. Math. Programming 8, 179–206 (1975)
15. Ibarra, O.H., Kim, C.E.: Approximation algorithms for certain scheduling problems. Math. Oper. Res. 3(3), 197–204 (1978)
16. Karp, R.M.: Reducibility among combinatorial problems. In: Complexity of Computer Computations, pp. 85–103. Plenum Press, NY (1972)
17. Kolliopoulos, S.G.: Approximating covering integer programs with multiplicity constraints. Discrete Appl. Math. 129(2-3), 461–473 (2003)

18. Kolliopoulos, S.G., Young, N.E.: Approximation algorithms for covering/packing integer programs. J. Comput. System Sci. 71(4), 495–505 (2005)
19. Könemann, J., Sadeghian, S., Sanità, L.: Better approximation algorithms for technology diffusion. In: Bodlaender, H.L., Italiano, G.F. (eds.) ESA 2013. LNCS, vol. 8125, pp. 637–646. Springer, Heidelberg (2013)
20. Lenstra, J.K., Rinnooy Kan, A.H.G., Brucker, P.: Complexity of machine scheduling problems. Annals of Discrete Mathematics 1, 343–362 (1977)
21. Lenstra, J.K., Rinnooy Kan, A.H.G.: Complexity results for scheduling chains on a single machine. European J. Operations Research 4(4), 270–275 (1980)
22. Pinedo, M.L.: Scheduling: theory, algorithms, and systems. Springer (2012)
23. Sahni, S.K.: Algorithms for scheduling independent tasks. J. ACM 23(1), 116–127 (1976)
24. Schulz, A.S.: Scheduling to minimize total weighted completion time: Performance guarantees of lp-based heuristics and lower bounds. In: Proceedings of the MPS Conference on Integer Programming and Combinatorial Optimization, pp. 301–315 (1996)
25. Wolsey, L.: Facets for a linear inequality in 0-1 variables. Math. Programming 8, 168–175 (1975)

Prophet Secretary[*]

Hossein Esfandiari[1],[**], MohammadTaghi Hajiaghayi[1],[**],
Vahid Liaghat[1],[**],[***], and Morteza Monemizadeh[2],[†]

[1] University of Maryland, College Park, MD, USA
{hossein,hajiagha,vliaghat}@cs.umd.edu
[2] Computer Science Institute, Charles University, Prague, Czech Republic
monemi@iuuk.mff.cuni.cz

Abstract. Optimal stopping theory is a powerful tool for analyzing scenarios such as online auctions in which we generally require optimizing an objective function over the space of stopping rules for an allocation process under uncertainty. Perhaps the most classic problems of stopping theory are the prophet inequality problem and the secretary problem. The classical prophet inequality states that by choosing the same threshold $OPT/2$ for every step, one can achieve the tight competitive ratio of 0.5. On the other hand, for the basic secretary problem, the optimal strategy achieves the tight competitive ratio of $1/e \approx 0.36$

In this paper, we introduce *prophet secretary*, a natural combination of the prophet inequality and the secretary problems. We show that by using a single uniform threshold one cannot break the 0.5 barrier of the prophet inequality for the prophet secretary problem. However, we show that

- using n distinct non-adaptive thresholds one can obtain a competitive ratio that goes to $(1 - 1/e \approx 0.63)$ as n grows; and
- no online algorithm can achieve a competitive ratio better than 0.73.

Our results improve the (asymptotic) approximation guarantee of single-item sequential posted pricing mechanisms from 0.5 to $(1 - 1/e)$ when the order of agents (customers) is chosen randomly.

We also consider the minimization variants of stopping theory problems and in particular the prophet secretary problem. Interestingly, we show that, even for the simple case in which the input elements are drawn from identical and independent distributions (i.i.d.), there is no constant competitive online algorithm for the minimization variant of the prophet secretary problems. We extend this hardness result to the minimization variants of both the prophet inequality and the secretary problem as well.

---

[*] The full version of this paper is available on http://arxiv.org/abs/1507.01155
[**] Supported in part by NSF CAREER award 1053605, NSF grant CCF-1161626, ONR YIP award N000141110662, DARPA/AFOSR grant FA9550-12-1-0423.
[***] Supported in part by a Google PhD Fellowship.
[†] Partially supported by the project 14-10003S of GA ČR.

© Springer-Verlag Berlin Heidelberg 2015
N. Bansal and I. Finocchi (Eds.): ESA 2015, LNCS 9294, pp. 496–508, 2015.
DOI: 10.1007/978-3-662-48350-3_42

1   Introduction

Optimal stopping theory is a powerful tool for analyzing scenarios in which we generally require optimizing an objective function over the space of stopping rules for an allocation process under uncertainty. One such a scenario is the online auction which is the essence of many modern markets, particularly networked markets where information about goods, agents, and outcomes is revealed over a period of time and the agents must make irrevocable decisions without knowing future information. Combining optimal stopping theory with game theory allows us to model the actions of rational agents applying competing stopping rules in an online market.

Perhaps the most classic problems of stopping theory are the *prophet inequality* and the *secretary problem*. Research investigating the relation between online auction mechanisms and prophet inequalities was initiated by Hajiaghayi, Kleinberg, and Sandholm [11]. They observed that algorithms used in the derivation of prophet inequalities, owing to their monotonicity properties, could be interpreted as truthful online auction mechanisms and that the prophet inequality in turn could be interpreted as the mechanism's approximation guarantee. Later Chawla, Hartline, Malec, and Sivan [7] showed the applications of prophet inequalities in Bayesian optimal mechansim design problems. The connection between the secretary problem and online auction mechanisms has been explored by Hajiaghayi, Kleinberg and Parkes [10] and initiated several follow-up papers (see e.g. [4,5,6,12,14]).
Prophet Inequality. The classical prophet inequality has been studied in the optimal stopping theory since the 1970s when introduced by Krengel and Sucheston [13,16,17] and more recently in computer science Hajiaghayi, Kleinberg and Sandholm [11]. In the prophet inequality setting, given (not necessarily identical) distributions $\{D_1, \ldots, D_n\}$, an online sequence of values $X_1, \cdots, X_n$ where X_i is drawn from D_i, an onlooker has to choose one item from the succession of the values, where X_k is revealed at step k. The onlooker can choose a value only at the time of arrival. The onlooker's goal is to maximize her revenue. The inequality has been interpreted as meaning that a prophet with complete foresight has only a bounded advantage over an onlooker who observes the random variables one by one, and this explains the name *prophet inequality*.

An algorithm for the prophet inequality problem can be described by setting a threshold for every step: we stop at the first step that the arriving value is higher than the threshold of that step. The classical prophet inequality states that by choosing the same threshold $OPT/2$ for every step, one achieves the competitive ratio of $1/2$. Here the optimal solution OPT is defined as $\mathrm{E}\left[\max X_i\right]$. Naturally, the first question is whether one can beat $1/2$. Unfortunately, this is not possible: let $q = \frac{1}{\epsilon}$, and $q' = 0$. The first value X_1 is always 1. The second value X_2 is either q with probability ϵ or q' with probability $1 - \epsilon$. Observe that the expected revenue of any (randomized) online algorithm is $\max(1, \epsilon(\frac{1}{\epsilon})) = 1$. However the prophet, i.e., the optimal offline solution would choose q' if it arrives, and he would choose the first value otherwise. Hence, the optimal offline revenue is

$(1 - \epsilon) \times 1 + \epsilon(\frac{1}{\epsilon}) \approx 2$. Therefore we cannot hope to break the $1/2$ barrier using any online algorithm.

Secretary Problem. Imagine that you manage a company, and you want to hire a secretary from a pool of n applicants. You are very keen on hiring only the best and brightest. Unfortunately, you cannot tell how good a secretary is until you interview him, and you must make an irrevocable decision whether or not to make an offer at the time of the interview. The problem is to design a strategy which maximizes the probability of hiring the most qualified secretary. It is well-known since 1963 by Dynkin in [8] that the optimal policy is to interview the first $t - 1$ applicants, then hire the next one whose quality exceeds that of the first $t - 1$ applicants, where t is defined by $\sum_{j=t+1}^{n} \frac{1}{j-1} \leq 1 < \sum_{j=t}^{n} \frac{1}{j-1}$. As $n \to \infty$, the probability of hiring the best applicant approaches $1/e \approx 0.36$, as does the ratio t/n. Note that a solution to the secretary problem immediately yields an algorithm for a slightly different objective function optimizing the expected value of the chosen element. Subsequent papers have extended the problem by varying the objective function, varying the information available to the decision-maker, and so on, see e.g., [1,9,19,20].

We refer the reader to the full version of this paper for a further survey of previous results.

2   Our Contributions

In this paper, we introduce *prophet secretary* as a natural combination of the prophet inequality problem and the secretary problem with applications to the Bayesian optimal mechanism design. Consider a seller that has an item to sell on the market to a set of arriving customers. The seller knows the types of customers that may be interested in the item and he has a price distribution for each type: the price offered by a customer of a type is anticipated to be drawn from the corresponding distribution. However, the customers arrive in a random order. Upon the arrival of a customer, the seller makes an irrevocable decision to whether sell the item at the offered price. We address the question of maximizing the seller's gain.

More formally, in the prophet secretary problem we are given a set $\{D_1, \ldots, D_n\}$ of (not necessarily identical) distributions. A number X_i is drawn from each distribution D_i and then, after applying a random permutation $\pi_1, \ldots, \pi_n$, the numbers are given to us in an online fashion, i.e., at step k, π_k and X_{π_k} are revealed. We are allowed to choose only one number, which can be done only upon receiving that number. The goal is to maximize the expectation of the chosen value, compared to the expectation of the optimum offline solution that knows the drawn values in advance (i.e., $OPT = \mathrm{E}[\max_i X_i]$). For the ease of notation, in what follows the index i iterates over the distributions while the index k iterates over the arrival steps.

An algorithm for the prophet secretary problem can be described by a sequence of (possibly adaptive) thresholds $\langle \tau_1, \ldots, \tau_n \rangle$: we stop at the first step

k that $X_{\pi_k} \geq \tau_k$. In particular, if the thresholds are non-adaptive, meaning that they are decided in advance, the following is a generic description of an algorithm. The competitive ratio of the following algorithm is defined as $\frac{\mathrm{E}[Y]}{OPT}$.

Algorithm Prophet Secretary
Input: A set of distributions $\{D_1, \ldots, D_n\}$; a randomly permuted stream of numbers $(X_{\pi_1}, \ldots, X_{\pi_n})$ drawn from the corresponding distributions.
Output: A number Y.

1. Let $\langle \tau_1, \ldots, \tau_n \rangle$ be a sequence of thresholds.
2. For $k \leftarrow 1$ to n
   (a) If $X_{\pi_k} \geq \tau_k$ then let $Y = X_{\pi_k}$ and exit the For loop.
3. Output Y as the solution.

Recall that when the arrival order is adversarial, the classical prophet inequality states that by choosing the same threshold $OPT/2$ for every step, one achieves the tight competitive ratio of $1/2$. On the other hand, for the basic secretary problem where the distributions are not known, the optimal strategy is to let $\tau_1 = \cdots = \tau_{\frac{n}{e}} = \infty$ and $\tau_{\frac{n}{e}+1} = \cdots = \tau_n = \max(X_{\pi_1}, \ldots, X_{\pi_{\frac{n}{e}}})$. This leads to the optimal competitive ratio of $\frac{1}{e} \simeq 0.36$. Hence, our goal in the prophet secretary problem is to beat the $1/2$ barrier.

We would like to mention that in an extension of this problem, in which the seller has B identical items to sell, it is indeed easier to track the optimal solution since we have multiple choices. In fact, an algorithm similar to that of [2,3] can guarantee a competitive ratio of $1 - \frac{1}{\sqrt{B+3}}$ which goes to one as B grows.

We first show that unlike the prophet inequality setting, one cannot obtain the optimal competitive ratio by using a single uniform threshold. Indeed, in the full version of this paper we show that $1/2$ is the best competitive ratio one can achieve with uniform thresholds. To beat the $\frac{1}{2}$ barrier, as a warm up we first show that by using two thresholds one can achieve the competitive ratio of $5/9 \simeq 0.55$. This can be achieved by choosing the threshold $\frac{5}{9} \cdot OPT$ for the first half of the steps and then decreasing the threshold to $\frac{OPT}{3}$ for the second half of the steps. Later in Section 4, we show that by setting n distinct thresholds one can obtain the $(1 - 1/e \approx 0.63)$-competitive ratio for the prophet secretary problem.

Theorem 1. *Let $\langle \tau_1, \ldots, \tau_n \rangle$ be a non-increasing sequence of n thresholds, such that (i) $\tau_k = \alpha_k \cdot OPT$ for every $k \in [n]$; (ii) $\alpha_n = \frac{1}{n+1}$; and (iii) $\alpha_k = \frac{n\alpha_{k+1}+1}{n+1}$ for $k \in [n-1]$. The competitive ratio of Algorithm* Prophet Secretary *invoked with thresholds τ_k's is at least α_1. When n goes to infinity, α_1 converges to $1 - 1/e \approx 0.63$.*

Remark 1. We should mention that Yan in [21] establishes a $1 - 1/e$ approximation when the designer is allowed to choose the order of arrival. Thus, Theorem 1 can be viewed as improving that result by showing that a random arrival order is sufficient to obtain the same approximation.

The crux of the analysis of our algorithm is to compute the probability of picking a value x at a step of the algorithm with respect to the threshold factors α_k's. Indeed one source of difficulty arises from the fundamental dependency between the steps: for any step k, the fact that the algorithm has not stopped in the previous steps leads to various restrictions on what we expect to see at the step k. For example, consider the scenario that D_1 is 1 with probability one and D_2 is either 2 or 0 with equal probabilities. Now if the algorithm chooses $\tau_1 = 1$, then it would never happen that the algorithm reaches step two and receives a number drawn from D_2! That would mean we have received a value from D_1 at the first step which is a contradiction since we would have picked that number. In fact, the optimal strategy for this example is to shoot for D_2! We set $\tau_1 = 2$ so that we can ignore the first value in the event that it is drawn from D_1. Then we set $\tau_2 = 1$ so that we can always pick the second value. Therefore in expectation we get $5/4$ which is slightly less than $OPT = 6/4$.

To handle the dependencies between the steps, we first distinguish between the events for $k \in [n]$ that we pick a value between τ_{k+1} and τ_k. We show that the expected value we pick at such events is indeed highly dependent on $\theta(k)$, the probability of passing the first k elements. We then use this observation to analyze competitive ratio with respect to $\theta(k)$'s and the thresholds factors α_k's. We finally show that the competitive ratio is indeed maximized by choosing the threshold factors described in Theorem 1. In Section 3.1, we first prove the theorem for the simple case of $n = 2$. This enables us to demonstrate our techniques without going into the more complicated dependencies for general n. We then present the full proof of Theorem 1 in Section 4. We would like to emphasize that our algorithm only needs to know the value of OPT, thus requiring only a weak access to the distributions themselves.

As mentioned before, Bayesian optimal mechanism design problems provide a compelling application of prophet inequalities in economics. In such a Bayesian market, we have a set of n agents with private types sampled from (not necessary identical) known distributions. Upon receiving the reported types, a seller has to allocate resources and charge prices to the agents. The goal is to maximize the seller's revenue in equilibrium. Chawla et al. [7] pioneered the study the approximability of a special class of such mechanisms, *sequential posted pricing* (SPM): the seller makes a sequence of take-it-or-leave-it offers to agents, offering an item for a specific price. They show although simple, SPMs approximate the optimal revenue in many different settings. Therefore prophet inequalities directly translate to approximation factors for the seller's revenue in these settings through standard machineries. Indeed one can analyze the so-called *virtual values* of winning bids introduced by Roger Myerson [18], to prove via prophet inequalities that the expected virtual value obtained by the SPM mechanism approximates an offline optimum that is with respect to the exact types. Chawla et al. [7] provide a type of prophet inequality in which one can choose the ordering of agents. They show that under matroid feasibility constraints, one can achieve a competitive ratio of 0.5 in this model, and no algorithm can achieve a ratio better 0.8. Kleinberg and Weinberg [15] later improved there result by giving an algorithm with the tight competitive ratio of 0.5 for an adversarial ordering. Our result can be seen as improving their approximation guarantees to 0.63 for the case of single-item SPMs when the order of agents are chosen randomly.

On the other hand, from the negative side the following theorem shows that no online algorithm can achieve a competitive ratio better than 0.73.

Theorem 2. *For any arbitrary small positive number ϵ, there is no online algorithm for the prophet secretary problem with competitive ratio $\frac{11}{15} + \epsilon \approx 0.73 + \epsilon$.*

We also consider the minimization variants of the prophet inequality problem, the prophet secretary problem and the secretary problem. In the minimization variant, we need to select one element of the input and we aim to minimize the expected value of the selected element. In particular, we show that, even for the simple case in which numbers are drawn from *identical and independent distributions (i.i.d.)*, there is no $\frac{(1.11)^n}{6}$ competitive online algorithm for the minimization variants of the prophet inequality and prophet secretary problems.

Theorem 3. *The competitive ratio of any online algorithm for the minimization prophet inequality with n identical and independent distributions is bounded by $\frac{(1.11)^n}{6}$. This bound holds for the minimization prophet secretary problem as well.*

Furthermore, we empower the online algorithm and assume that the online algorithm can withdraw and change its decision once. Indeed, for the minimization variants of all, prophet secretary problem, secretary problem and prophet inequality, we show that there is no C competitive algorithm, where C is an arbitrary large number.

We refer the reader to the full version of this paper for the missing proofs.

3   Preliminaries

We first define some notation. For every $k \in [n]$, let z_k denote the random variable that shows the value we pick at the kth step. Observe that for a fixed sequence of drawn values and a fixed permutation, at most one of z_k's is non-zero since we only pick one number. Let z denote the value chosen by the algorithm. By definition, $z = \sum_{k=1}^{n} z_k$. In fact, since all but one of z_k's are zero, we have the following proposition. We note that since the thresholds are deterministic, the randomness comes from the permutation π and the distributions.

Proposition 1. $\Pr[z \geq x] = \sum_{k \in [n]} \Pr[z_k \geq x]$.

For every $k \in [n]$, let $\theta(k)$ denote the probability that Algorithm Prophet Secretary does not choose a value from the first k steps. For every $i \in [n]$ and $k \in [n-1]$, let $q_{-i}(k)$ denote the probability that the following two events concurrently happen:

(i) Algorithm Prophet Secretary does not choose a value from first k elements.
(ii) None of the first k values are drawn from D_i.

Proposition 2. *If the thresholds of Algorithm Prophet Secretary are non-increasing, then for every $i \in [n]$ and $k \in [n-1]$, we have $\theta(k+1) \leq q_{-i}(k)$.*

3.1   Two Thresholds Breaks $\frac{1}{2}$ Barrier

Since using one threshold is hopeless, we now try using two thresholds. More formally, for the first half of steps, we use a certain threshold, and then we use a different threshold for the rest of steps. We note that similar to the one-threshold algorithm, both thresholds should be proportional to OPT. Furthermore, at the beginning we should be optimistic and try to have a higher threshold, but if we cannot pick a value in the first half, we may need to lower the bar! We show that by using two thresholds one can indeed achieve the competitive ratio of $\frac{5}{9} \simeq 0.55$. In fact, this improvement beyond $1/2$ happens even at $n = 2$. Thus as a warm up before analyzing the main algorithm with n thresholds, we focus on the case of $n = 2$.

Let $\tau_1 = \alpha_1 OPT$ and $\tau_2 = \alpha_2 OPT$ for some $1 \geq \alpha_1 \geq \alpha_2 \geq 0$ to be optimized later. Recall that z_1 and z_2 are the random variables showing the values picked up by the algorithm at step one and two, respectively. We are interested in comparing $E[z]$ with OPT. By Proposition 1 we have

$$E[z] = \int_0^\infty \Pr[z \geq x]\, dx = \int_0^\infty \Pr[z_1 \geq x]\, dx + \int_0^\infty \Pr[z_2 \geq x]\, dx \ .$$

Observe that z_1 (resp. z_2) is either zero or has a value more than τ_1 (resp. τ_2). In fact, since $\tau_1 \geq \tau_2$, z is either zero or has a value more than τ_2. Recall that $\theta(1)$ is the probability of $z_1 = 0$ while $\theta(2)$ is the probability of $z_1 = z_2 = 0$. This observation leads to the following simplification

$$E[z] = \int_0^{\tau_2} \Pr[z_1 \geq x]\, dx + \int_{\tau_2}^{\tau_1} \Pr[z_1 \geq x]\, dx + \int_{\tau_1}^\infty \Pr[z_1 \geq x]\, dx$$

$$+ \int_0^{\tau_2} \Pr[z_2 \geq x]\, dx + \int_{\tau_2}^\infty \Pr[z_2 \geq x]\, dx$$

$$= \int_0^{\tau_2} \Pr[z \geq x]\, dx + \int_{\tau_2}^{\tau_1} \Pr[z_1 \geq x]\, dx + \int_{\tau_1}^\infty \Pr[z_1 \geq x]\, dx + \int_{\tau_2}^\infty \Pr[z_2 \geq x]\, dx$$

$$= \tau_2(1 - \theta(2)) + (\tau_1 - \tau_2)(1 - \theta(1)) + \int_{\tau_1}^\infty \Pr[z_1 \geq x]\, dx + \int_{\tau_2}^\infty \Pr[z_2 \geq x]\, dx \ .$$

Let us first focus on $\Pr[z_1 \geq x]$. The first value may come from any of the two distributions, thus we have

$$\Pr[z_1 \geq x] = \frac{1}{2}\Pr[X_1 \geq x] + \frac{1}{2}\Pr[X_2 \geq x] \ .$$

On the other hand, z_2 is non-zero only if we do not pick anything at the first step. For $i \in \{1, 2\}$, we pick a value of at least x drawn from D_i at step two, if and only if: (i) the value drawn from D_i is at least x; and (ii) our algorithm does not pick a value from the previous step which is drawn from the other distribution. By definitions, the former happens with probability $\Pr[X_i \geq x]$, while the latter happens with probability $q_{-i}(1)$. Since these two events are independent we have

$$\Pr[z_2 \geq x] = \frac{1}{2} \sum_{i \in \{1,2\}} q_{-i}(1)\Pr[X_i \geq x] \geq \frac{\theta(2)}{2} \sum_i \Pr[X_i \geq x] \ ,$$

where the last inequality follows from Proposition 2, although the proposition is trivial for $n = 2$. We can now continue analyzing $E[z]$ from before

$$E[z] = \tau_2(1 - \theta(2)) + (\tau_1 - \tau_2)(1 - \theta(1)) + \int_{\tau_1}^{\infty} \Pr[z_1 \geq x]\, dx + \int_{\tau_2}^{\infty} \Pr[z_2 \geq x]\, dx$$

$$\geq \tau_2(1 - \theta(2)) + (\tau_1 - \tau_2)(1 - \theta(1))$$
$$+ \frac{\theta(1)}{2} \int_{\tau_1}^{\infty} \sum_i \Pr[X_i \geq x]\, dx + \frac{\theta(2)}{2} \int_{\tau_2}^{\infty} \sum_i \Pr[X_i \geq x]\, dx \ .$$

We note that although the $\theta(1)$ factor is not required in the third term of the last inequality, we include it so that the formulas can have the same formation as in the general formula of the next sections. It remains to bound $\int_{\tau_k}^{\infty} \sum_i \Pr[X_i \geq x]$ for $k \in \{1, 2\}$. Recall that $OPT = E[\max_i X_i]$. Hence for every $k \in \{1, 2\}$ we have

$$OPT = \int_0^{\tau_k} \Pr[\max X_i \geq x]\, dx + \int_{\tau_k}^{\infty} \Pr[\max X_i \geq x]\, dx$$

$$\leq \tau_k + \int_{\tau_k}^{\infty} \Pr[\max X_i \geq x]\, dx \qquad\qquad \Pr[\max X_i \geq x] \leq 1$$

$$(1 - \alpha_k)OPT \leq \int_{\tau_k}^{\infty} \Pr[\max X_i \geq x]\, dx \qquad\qquad \tau_k = \alpha_k OPT$$

$$\leq \int_{\tau_k}^{\infty} \sum_i \Pr[X_i \geq x]\, dx \qquad \Pr[\max X_i \geq x] \leq \sum_i \Pr[X_i \geq x]\, dx$$

Therefore we get

$$E[z] \geq \tau_2(1 - \theta(2)) + (\tau_1 - \tau_2)(1 - \theta(1))$$
$$+ \frac{\theta(1)}{2} \int_{\tau_1}^{\infty} \sum_i \Pr[X_i \geq x]\, dx + \frac{\theta(2)}{2} \int_{\tau_2}^{\infty} \sum_i \Pr[X_i \geq x]\, dx$$

$$\geq (\alpha_2 OPT)(1 - \theta(2)) + (\alpha_1 - \alpha_2)OPT(1 - \theta(1)) + \frac{\theta(1)}{2}(1 - \alpha_1)OPT$$
$$+ \frac{\theta(2)}{2}(1 - \alpha_2)OPT = OPT\left(\alpha_1 + \theta(1)(\frac{1 + 2\alpha_2 - 3\alpha_1}{2}) + \theta(2)(\frac{1 - 3\alpha_2}{2})\right)$$

Therefore by choosing $\alpha_2 = 1/3$ and $\alpha_1 = 5/9$, the coefficients of $\theta(1)$ and $\theta(2)$ become zero, leading to the competitive ratio of $5/9 \simeq 0.55$. In the next section, we show how one can generalize the arguments to the case of n thresholds for arbitrary n.

4 $(1 - \frac{1}{e} \approx 0.63)$-Competitive Ratio Using n Thresholds

In this section we prove our main theorem. In particular, we invoke Algorithm Prophet Secretary with n distinct thresholds $\tau_1, \ldots, \tau_n$. The thresholds $\tau_1, \ldots, \tau_n$

that we consider are *non-adaptive* (i.e., Algorithm Prophet Secretary is oblivious to the history) and *non-increasing*. Intuitively, this is because as we move to the end of stream we should be more pessimistic and use lower thresholds to catch remaining higher values.

Formally, for every $k \in [n]$, we consider threshold $\tau_k = \alpha_k \cdot OPT$ where the sequence $\alpha_1, \ldots, \alpha_n$ is non-increasing that is, $\alpha_1 \geq \alpha_1 \geq \ldots \geq \alpha_n$. We invoke Algorithm Prophet Secretary with these thresholds and analyze the competitive ratio of Algorithm Prophet Secretary with respect to coefficients α_k. Theorem 1 shows that there exists a sequence of coefficients α_k that leads to the competitive ratio of $(1 - 1/e) \approx 0.63$.

Proof of Theorem 1. We prove the theorem in two steps: First, we find a lower bound on $\mathrm{E}[z]$ in terms of OPT and coefficients α_i. Second, we set coefficients α_k so that (1) α_1 becomes the competitive ratio of Algorithm Prophet Secretary and (2) α_1 converges to $1 - 1/e$, when n goes to infinity.

We start by proving the following auxiliary lemmas. In the first lemma, we find a lower bound for $\int_{\tau_k}^{\infty} \Pr[\max X_i \geq x] \, dx$ based on $OPT = \mathrm{E}[\max_i X_i]$.

Lemma 1. $\int_{\tau_k}^{\infty} \Pr[\max X_i \geq x] \, dx \geq (1 - \alpha_k)OPT.$

Next, to find a lower bound for $\mathrm{E}[z]$, we first split it into two terms. Later, we find lower bounds for each one of these terms based on $OPT = \mathrm{E}[\max_i X_i]$.

Lemma 2. *Let* $z = \sum_{k=1}^{n} z_k$ *denote the value chosen by Algorithm* Prophet Secretary. *For* z *we have*

$$\mathrm{E}[z] = \sum_{k=1}^{n} \int_{0}^{\tau_k} \Pr[z_k \geq x] \, dx + \sum_{k=1}^{n} \int_{\tau_k}^{\infty} \Pr[z_k \geq x] \, dx.$$

Lemma 3. $\sum_{k=1}^{n} \int_{0}^{\tau_k} \Pr[z_k \geq x] \, dx \geq OPT \sum_{k=1}^{n} (1 - \theta(k))(\alpha_k - \alpha_{k+1}).$

Proof. Suppose $x \leq \tau_k$. Observe that the event $z_k \geq x$ occurs when Algorithm Prophet Secretary chooses a value at step k. In fact, since the thresholds are non-increasing, whatever we pick at the first k steps would be at least x. Recall that for every $k \in [n]$, $\theta(k)$ is the probability that Algorithm Prophet Secretary does not choose a value from the first k steps. Hence, for every $k \in [n]$ and $x \leq \tau_k$ we have

$$\sum_{j \leq k} \Pr[z_j \geq x] = 1 - \theta(k). \tag{1}$$

To simplify the notation, we assume that $\alpha_0 = \infty$ which means $\tau_0 = \infty$ and we let $\alpha_{n+1} = 0$ which means $\tau_{n+1} = 0$. Therefore we have

$$\sum_{k=1}^{n} \int_{0}^{\tau_k} \Pr[z_k \geq x] \, dx = \sum_{k=1}^{n} \int_{\tau_{n+1}}^{\tau_k} \Pr[z_k \geq x] \, dx.$$

Next, we use Equation (1) to prove the lemma as follows.

$$\sum_{k=1}^{n} \int_{0}^{\tau_k} \Pr\left[z_k \geq x\right] dx = \sum_{k=1}^{n} \int_{\tau_{n+1}}^{\tau_k} \Pr\left[z_k \geq x\right] dx$$

$$= \sum_{r=1}^{n} \int_{\tau_{r+1}}^{\tau_r} \sum_{k=1}^{r} \Pr\left[z_k \geq x\right] dx \geq \sum_{r=1}^{n} \int_{\tau_{r+1}}^{\tau_r} (1 - \theta(r)) dx$$

$$= \sum_{r=1}^{n} (1 - \theta(r))(\tau_r - \tau_{r+1}) = OPT \cdot \sum_{k=1}^{n} (1 - \theta(k))(\alpha_k - \alpha_{k+1}) \ . \qquad \square$$

Lemma 4. $\sum_{k=1}^{n} \int_{\tau_k}^{\infty} \Pr\left[z_k \geq x\right] dx \geq OPT \sum_k \frac{\theta(k)}{n}(1 - \alpha_k)$.

Proof. Recall that for every distribution D_i we draw a number X_i. Later, we randomly permute the numbers $X_1, \cdots, X_n$. Let the sequence of indices after the random permutation be $\pi_1, \ldots, \pi_n$, i.e., at step k, number X_{π_k} for $\pi_k \in [n]$ is revealed.

Suppose $x \geq \tau_k$. We break the event $z_k > 0$ to n different scenarios depending on which index of the distributions $D_1, \cdots, D_n$ is mapped to index π_k in the random permutation. Let us consider the scenario in which Algorithm Prophet Secretary chooses the value drawn from a distribution i at step k. Such a scenario happens if (i) Algorithm Prophet Secretary does not choose a value from the first $k - 1$ steps which are not drawn from i, and (ii) $X_i \geq \tau_k$. Observe that the two events are independent. Therefore, we have $\Pr\left[z_k \geq x\right] = \sum_i \Pr\left[\pi_k = i\right] \cdot \Pr\left[X_i \geq x\right] \cdot q_{-i}(k - 1)$, where $q_{-i}(k)$ for every $i \in n$ and $k \in [n - 1]$ is the probability that the following two events concurrently happen: (i) Algorithm Prophet Secretary does not choose a value from the first k elements, and (ii) none of the first k values are drawn from D_i. Since π_k is an index in the random permutation we obtain

$$\Pr\left[z_k \geq x\right] = \sum_i \Pr\left[\pi_k = i\right] \Pr\left[X_i \geq x\right] \cdot q_{-i}(k-1) = \frac{1}{n} \sum_i \Pr\left[X_i \geq x\right] q_{-i}(k-1).$$

Using Proposition 2 and an application of the union bound we then have

$$\Pr\left[z_k \geq x\right] = \sum_i \Pr\left[\pi_k = i\right] \cdot \Pr\left[X_i \geq x\right] \cdot q_{-i}(k - 1)$$

$$= \frac{1}{n} \sum_i \Pr\left[X_i \geq x\right] \cdot q_{-i}(k - 1)$$

$$\geq \frac{\theta(k)}{n} \cdot \sum_i \Pr\left[X_i \geq x\right] \geq \frac{\theta(k)}{n} \cdot \Pr\left[\max_i X_i \geq x\right] \ .$$

Therefore, we obtain the following lower bound on $\sum_{k=1}^{n} \int_{\tau_k}^{\infty} \Pr\left[z_k \geq x\right] dx$.

$$\sum_{k=1}^{n} \int_{\tau_k}^{\infty} \Pr\left[z_k \geq x\right] dx \geq \sum_k \int_{\tau_k}^{\infty} \frac{\theta(k)}{n} \Pr\left[\max X_i \geq x\right] dx$$

$$= \sum_k \frac{\theta(k)}{n} \int_{\tau_k}^{\infty} \Pr\left[\max X_i \geq x\right] dx \ .$$

Finally, we use the lower bound of Lemma 1 for $\int_{\tau_k}^{\infty} \Pr\left[\max X_i \geq x\right] dx$ to prove the lemma.

$$\sum_{k=1}^{n} \int_{\tau_k}^{\infty} \Pr\left[z_k \geq x\right] dx \geq \sum_k \frac{\theta(k)}{n} \int_{\tau_k}^{\infty} \Pr\left[\max X_i \geq x\right] dx$$

$$\geq \sum_k \frac{\theta(k)}{n}(1 - \alpha_k) \cdot OPT = OPT \cdot \sum_k \frac{\theta(k)}{n} \cdot (1 - \alpha_k) \ .$$

$\square$

Now we can plug in the lower bounds of Lemmas 3 and 4 into Lemma 2 to obtain a lower bound for $\mathrm{E}\left[z\right]$.

Corollary 1. *Let $z = \sum_{k=1}^{n} z_k$ denote the value chosen by Algorithm* Prophet Secretary. *For z we have*

$$\mathrm{E}\left[z\right] \geq OPT \cdot (\alpha_1 + \sum_{k=1}^{n} \theta(k)(\frac{1}{n} - \frac{\alpha_k}{n} - \alpha_k + \alpha_{k+1})).$$

We finish the proof of the theorem by proving the following claim.

Lemma 5. *The competitive ratio of Algorithm* Prophet Secretary *is at least α_1 which quickly converges to $1 - 1/e \approx 0.63$ when n goes to infinity.*

Proof. Using Corollary 1, for z we have

$$\mathrm{E}\left[z\right] \geq OPT \left(\alpha_1 + \sum_{k=1}^{n} \theta(k) \left(\frac{1}{n} - \frac{\alpha_k}{n} - \alpha_k + \alpha_{k+1}\right)\right) \ .$$

which means that the competitive ratio depends on the probabilities $\theta(k)$'s. However, we can easily get rid of the probabilities $\theta(k)$'s by choosing α_k's such that for every k, $\left(\frac{1}{n} - \frac{\alpha_k}{n} - \alpha_k + \alpha_{k+1}\right) = 0$.

More formally, by starting from $\alpha_{n+1} = 0$ and choosing $\alpha_k = \frac{1+n\alpha_{k+1}}{1+n}$ for $k \leq n$, the competitive ratio of the algorithm would be α_1. Below, we show that when n goes to infinity, α_1 quickly goes to $1 - 1/e$ which means that the competitive ratio of Algorithm Prophet Secretary converges to $1 - 1/e \approx 0.63$.

First, we show by the induction that $\alpha_k = \sum_{i=0}^{n-k} \frac{n^i}{(1+n)^{i+1}}$. For the base case we have

$$\alpha_n = \frac{1 + n\alpha_{n+1}}{1 + n} = \frac{1 + n \times 0}{1 + n} = \frac{n^0}{(1 + n)^1} \ .$$

Given $\alpha_{k+1} = \sum_{i=0}^{n-(k+1)} \frac{n^i}{(1+n)^{i+1}}$ we show the equality for α_k as follows.

$$\alpha_k = \frac{1 + n\alpha_{k+1}}{1 + n} = \frac{1 + n(\sum_{i=0}^{n-(k+1)} \frac{n^i}{(1+n)^{i+1}})}{1 + n}$$

$$= \frac{n^0}{(1 + n)^1} + \sum_{i=0}^{n-(k+1)} \frac{n^{i+1}}{(1 + n)^{i+2}} = \sum_{i=0}^{n-k} \frac{n^i}{(1 + n)^{i+1}} \ .$$

Now we are ready to show $\alpha_1 \geq 1 - 1/e$ when n goes to infinity.

$$\lim_{n \to \infty} \alpha_1 = \lim_{n \to \infty} \sum_{i=0}^{n-1} \frac{n^i}{(n+1)^{i+1}} = \lim_{n \to \infty} \frac{1}{n+1} \sum_{i=0}^{n-1} (1 - \frac{1}{n+1})^i$$

$$\approx \lim_{n \to \infty} \frac{1}{n+1} \sum_{i=0}^{n-1} e^{-i/n} \approx \int_0^1 e^{-x} dx = 1 - 1/e \ . \qquad \square$$

Acknowledgments. We would like to thank Robert Kleinberg for fruitful discussion on early stages of this project.

References

1. Ajtai, M., Megiddo, N., Waarts, O.: Improved algorithms and analysis for secretary problems and generalizations. SIAM J. Discrete Math. 14(1), 1–27 (2001)
2. Alaei, S., Hajiaghayi, M., Liaghat, V.: Online prophet-inequality matching with applications to ad allocation. In: EC (2012)
3. Alaei, S., Hajiaghayi, M., Liaghat, V.: The online stochastic generalized assignment problem. In: Raghavendra, P., Raskhodnikova, S., Jansen, K., Rolim, J.D.P. (eds.) RANDOM 2013 and APPROX 2013. LNCS, vol. 8096, pp. 11–25. Springer, Heidelberg (2013)
4. Babaioff, M., Immorlica, N., Kempe, D., Kleinberg, R.: A knapsack secretary problem with applications. In: Charikar, M., Jansen, K., Reingold, O., Rolim, J.D.P. (eds.) RANDOM 2007 and APPROX 2007. LNCS, vol. 4627, pp. 16–28. Springer, Heidelberg (2007)
5. Babaioff, M., Immorlica, N., Kempe, D., Kleinberg, R.: Online auctions and generalized secretary problems. SIGecom Exch. 7(2), 1–11 (2008)
6. Babaioff, M., Immorlica, N., Kleinberg, R.: Matroids, secretary problems, and online mechanisms. In: SODA, pp. 434–443 (2007)
7. Chawla, S., Hartline, J., Malec, D., Sivan, B.: Multi-parameter mechanism design and sequential posted pricing (2010)
8. Dynkin, E.B.: The optimum choice of the instant for stopping a markov process. Sov. Math. Dokl. 4, 627–629 (1963)
9. Glasser, K.S., Holzsager, R., Barron, A.: The d choice secretary problem. Comm. Statist. C—Sequential Anal. 2(3), 177–199 (1983)
10. Hajiaghayi, M.T., Kleinberg, R., Parkes, D.C.: Adaptive limited-supply online auctions. In: EC, pp. 71–80 (2004)
11. Hajiaghayi, M.T., Kleinberg, R., Sandholm, T.: Automated online mechanism design and prophet inequalities. In: AAAI, pp. 58–65 (2007)
12. Immorlica, N., Kleinberg, R.D., Mahdian, M.: Secretary problems with competing employers. In: Spirakis, P.G., Mavronicolas, M., Kontogiannis, S.C. (eds.) WINE 2006. LNCS, vol. 4286, pp. 389–400. Springer, Heidelberg (2006)
13. Kennedy, D.P.: Prophet-type inequalities for multi-choice optimal stopping. In: Stoch. Proc. Applic. (1978)
14. Kleinberg, R.: A multiple-choice secretary algorithm with applications to online auctions. In: SODA, pp. 630–631 (2005)
15. Kleinberg, R., Weinberg, S.M.: Matroid prophet inequalities. In: STOC (2012)

16. Krengel, U., Sucheston, L.: Semiamarts and finite values. Bull. Am. Math. Soc. (1977)
17. Krengel, U., Sucheston, L.: On semiamarts, amarts, and processes with finite value. In: Kuelbs, J. (ed.) Probability on Banach Spaces (1978)
18. Myerson, R.B.: Optimal auction design. Mathematics of Operations Research 6(1), 58–73 (1981)
19. Vanderbei, R.J.: The optimal choice of a subset of a population. Math. Oper. Res. 5(4), 481–486 (1980)
20. Wilson, J.G.: Optimal choice and assignment of the best m of n randomly arriving items. Stochastic Process. Appl. 39(2), 325–343 (1991)
21. Yan, Q.: Mechanism design via correlation gap. In: Proceedings of the Twenty-Second Annual ACM-SIAM Symposium on Discrete Algorithms, SODA 2011, San Francisco, California, USA, January 23-25, pp. 710–719 (2011)

Smoothed Analysis of the Squared Euclidean Maximum-Cut Problem[*]

Michael Etscheid and Heiko Röglin

Department of Computer Science, University of Bonn, Germany
{etscheid,roeglin}@cs.uni-bonn.de

Abstract. It is well-known that local search heuristics for the Maximum-Cut problem can take an exponential number of steps to find a local optimum, even though they usually stabilize quickly in experiments. To explain this discrepancy we have recently analyzed the simple local search algorithm FLIP in the framework of smoothed analysis, in which inputs are subject to a small amount of random noise. We have shown that in this framework the number of iterations is quasi-polynomial, i.e., it is polynomially bounded in $n^{\log n}$ and ϕ, where n denotes the number of nodes and ϕ is a parameter of the perturbation.

In this paper we consider the special case in which the nodes are points in a d-dimensional space and the edge weights are given by the squared Euclidean distances between these points. We prove that in this case for any constant dimension d the smoothed number of iterations of FLIP is polynomially bounded in n and $1/\sigma$, where σ denotes the standard deviation of the Gaussian noise. Squared Euclidean distances are often used in clustering problems and our result can also be seen as an upper bound on the smoothed number of iterations of local search for min-sum 2-clustering.

1 Introduction

Clustering is nowadays ubiquitous in computer science. Despite intensive research on sophisticated algorithms, simple local search methods are often the most successful and versatile algorithms in practice. These algorithms are based on a simple principle: start with some feasible clustering and perform local improvements until a local optimum is found. Usually local search methods do not work well in the worst case because in most cases there are rather contrived instances on which they perform poorly.

Motivated by this striking discrepancy between theory and practice, we have recently analyzed the simple local search algorithm FLIP for the Maximum-Cut Problem in the framework of smoothed analysis, which can be considered as a less pessimistic variant of worst-case analysis in which the adversarial input is subject to a small amount of random noise [5]. We continue this line of research and consider the special case of the Maximum-Cut Problem in which the nodes

---

[*] This research was supported by ERC Starting Grant 306465 (BeyondWorstCase).

© Springer-Verlag Berlin Heidelberg 2015
N. Bansal and I. Finocchi (Eds.): ESA 2015, LNCS 9294, pp. 509–520, 2015.
DOI: 10.1007/978-3-662-48350-3_43

are points in a d-dimensional space and the edge weights are given by the squared Euclidean distances between these points. We assume that the input is a finite set $X \subseteq \mathbb{R}^d$ of points that is to be partitioned into two parts X_1 and X_2 such that the weight $\sum_{x \in X_1} \sum_{y \in X_2} \|x - y\|^2$ becomes maximal, where $\|x - y\|$ denotes the Euclidean distance between x and y. The FLIP algorithm starts with an arbitrary cut (X_1, X_2) and iteratively increases the weight of the cut by moving one vertex from X_1 to X_2 or vice versa, as long as such an improvement is possible. Squared Euclidean distances are common in many clustering applications.

In the model we consider, an adversary specifies an arbitrary set $X \subseteq [0, 1]^d$ of $n = |X|$ points. Then each point is randomly perturbed by adding a Gaussian vector of standard deviation σ to it. The parameter σ determines how powerful the adversary is. In the limit for $\sigma \to 0$ the adversary is as powerful as in a classical worst-case analysis, whereas for large σ smoothed analysis almost coincides with average-case analysis. Note that the restriction to $[0, 1]^d$ is merely a scaling issue and entails no loss of generality.

For a given instance of the Maximum-Cut Problem we define the *number of steps of the FLIP algorithm* on that instance to be the largest number of local improvements the FLIP algorithm can make for any choice of the initial cut and any pivot rule determining the local improvement that is chosen if multiple are possible. Formally, this can be described as the longest path in the transition graph of the FLIP algorithm. We are interested in the *smoothed number of steps of the FLIP algorithm*. This quantity depends on the number n of nodes and the standard deviation σ and it is defined as the largest expected number of steps the adversary can achieve by his choice of the point set X.

Theorem 1. *For any constant dimension $d \geq 2$, the smoothed number of steps of the FLIP algorithm for squared Euclidean distances is bounded from above by a polynomial in n and $1/\sigma$. The degree of this polynomial depends linearly on d.*

This result significantly improves upon the exponential worst-case running time of the FLIP algorithm and the quasi-polynomial bound on the smoothed number of steps for general instances. The theorem shows that for squared Euclidean distances worst-case instances are fragile and unlikely to occur in the presence of a small amount of random noise.

We view Theorem 1 as a further step towards understanding the behavior of local search heuristics on semi-random inputs. Its proof is considerably different from our previous analysis for general graphs and also from the smoothed analysis of other local search heuristics in the literature. We believe that the technique used to prove Theorem 1, which we summarize in Section 2, might also be interesting for analyzing local search algorithms for other problems. In that sense, we view Theorem 1 also as a proof of concept of our new technique.

1.1   Related Work

Smoothed analysis has originally been introduced by Spielman and Teng to explain why the simplex method solves linear programs efficiently in practice despite its exponential worst-case running time [13]. Since then it has gained a lot of attention and it has been used to analyze a wide variety of optimization problems and algorithms (see, e.g., the surveys [8,14]).

For many optimization problems, local search heuristics are prime examples of algorithms with exponential worst-case running time that work well and efficiently in practice. Consequently, there has been a considerable amount of research on the smoothed analysis of local search. Englert et al. [4] and Manthey and Veenstra [9] have analyzed the smoothed running time of the popular 2-Opt heuristic for the traveling salesman problem. Arthur and Vassilvitskii initiated the smoothed analysis of the k-means method [2] that culminated in a proof that the smoothed running time of the k-means method is polynomial [1].

Both the worst-case and the smoothed running time of the FLIP algorithm for the Maximum-Cut Problem have been studied. It is known that the problem of computing a locally optimal cut is PLS-complete [11] even for graphs of maximum degree five [3]. This means that, unless PLS $\subseteq$ P, there is no efficient algorithm to compute partitions that are locally optimal and it also implies that there are instances on which there exist initial cuts from which any sequence of local improvements to a local optimum has exponential length. Admittedly, these lower bounds do not carry over immediately to squared Euclidean distances but there is also no sub-exponential worst-case upper bound known for this case.

Elsässer and Tscheuschner [3] were the first who analyzed the smoothed running time of the FLIP algorithm and showed that it is polynomially bounded if the graph G has at most logarithmic degree. Later we [5] analyzed the smoothed running time of the FLIP algorithm for general graphs and we proved a quasi-polynomial bound, i.e., a bound that is polynomial in $n^{\log n}$. While it would also be worthwhile to study the quality of locally optimal cuts in the framework of smoothed analysis, this line of research has not been pursued yet. It is well-known that even in the worst case any locally optimal cut is a 2-approximation of a maximum cut (see, e.g., [7]).

Schulman [12] studied the min-sum 2-clustering problem for squared Euclidean distances. In this problem, the input also consists of a finite set of points $X \subseteq \mathbb{R}^d$ and the goal is to find a partition of X into two classes X_1 and X_2 such that the sum of the edge weights inside the two classes (i.e., $\sum_{x,y \in X_1} \|x - y\|^2 + \sum_{x,y \in X_2} \|x - y\|^2$) becomes minimal. This problem is equivalent to the Maximum-Cut Problem with squared Euclidean distances (not in terms of approximation though) and hence the FLIP algorithm can also be seen as a local search algorithm for min-sum 2-clustering. Schulman gives an algorithm that solves the problem optimally in time $O(n^{d+1})$. The bound proven in Theorem 1 does not improve upon the running time of Schulman's algorithm for computing the optimal cut. However, the worst-case running time of the FLIP algorithm might be much worse than that.

Sankar, Spielman and Teng [10] analyzed the condition number of randomly perturbed matrices and proved that it is unlikely that a matrix whose entries are independent Gaussians has a large condition number. We will use this as one crucial ingredient of our analysis of the FLIP algorithm.

2   Outline of Our Analysis

The analysis of the FLIP algorithm for squared Euclidean distances differs significantly from our previous analysis for general graphs and also from the smoothed analysis of other local search heuristics in the literature. Theorem 1 as well as all results in the literature are based on finding a lower bound for the improvement made by any local improvement or any sequence of consecutive local improvements of a certain length. Since $X \subseteq [0,1]^d$, the value of any cut is bounded polynomially in n with high probability. Hence, proving that in any local improvement or in any sequence of poly(n) consecutive local improvements the value of the cut increases by at least $\varepsilon := 1/\text{poly}(n)$ with high probability suffices for proving that the expected number of local improvements is polynomially bounded. We will call an improvement of at least ε *significant* in the following.

We call a configuration (i.e., a partition of X into X_1 and X_2) *bad* if it admits an insignificant local improvement. Any fixed configuration is bad only with probability at most poly$(n\varepsilon/\sigma)$. With this observation in mind one could try to use a union bound over all possible configurations to bound the probability that there exists a bad configuration. However, since there is an exponential number of configurations, this does not work. In fact, one can even prove that with high probability there do exist bad configurations. We will improve the union bound by not fixing the configuration of all points from X but only some of them, i.e., we will only make a union bound over a small subset of the points. To illustrate this, let us give two examples from the literature.

- An observation that has been exploited by Elsässer and Tscheuschner [3] is that it suffices to fix the flipping vertex and the neighborhood of this vertex in the union bound. For graphs of logarithmic maximum degree, this yields a polynomial bound on the smoothed number of local improvements.
- Another observation that has been used in a much more general form in [5] is the following: Any sequence of constantly many consecutive local improvements that starts and ends with the same vertex flipping yields a significant improvement with high probability. We showed that in order to bound the probability that there exists a sequence of this type in which all improvements are insignificant it suffices to use a union bound over all sequences of this type (there are only polynomially many because they are of constant length). One does, however, not need to specify in the union bound the configuration of the vertices that are not involved in the sequence.

The two examples above have in common that the union bound fixes only the configuration of some of the vertices (which we call *active*). The configuration of the other vertices (which we call *passive*) is not fixed in the union bound. In the examples the active points were chosen such that knowing their configuration suffices to compute the probability that the considered step or sequence of steps is bad. In our analysis we also fix only the configuration of some active vertices. The difference is that the passive vertices are not irrelevant because their configuration has a very essential impact on the improvements made by the flips in the considered sequences.

Let us go into more detail. Remember that we consider complete graphs in which each vertex is a point in $\mathbb{R}^d$ and the weights of the edges are given by the squared Euclidean distance. Our goal is to show that in this setting with high probability there is no sequence in which $9d + 16$ different vertices move making only insignificant local improvements. Observe that the length of such a sequence is at most 2^{9d+16} as otherwise one configuration would repeat. We apply a union bound over all such sequences. We call all vertices that flip in the considered sequence active and apply another union bound over all configurations of the active points. With only the information about the sequence and the configuration of the active points, it is not possible to determine linear combinations of the edge weights that describe the improvements made in the sequence because the configuration of the passive points is unknown.

Assume that the passive points P are partitioned into the sets P_1 and P_2. One crucial observation for our analysis is that in the case of squared Euclidean distances it suffices to know $|P_1|$ and the value $c_P := \sum_{x \in P_1} x - \sum_{x \in P_2} x$ in order to determine the improvements made by the active points. This value is unknown if the configuration of the passive points (i.e., the partition (P_1, P_2)) is unknown and we have to assume that $c_P \in \mathbb{R}^d$ is chosen adversarially. We prove that there is a point $c_P^{\mathrm{apx}} \in \mathbb{R}^d$ such that the first flips of the first $d + 1$ active points can only all be small improvements if c_P is chosen very close to c_P^{apx} (Phase 1). The point c_P^{apx} can be computed as the solution of a system of linear equations whose coefficients are determined by the considered sequence and the active points alone. In particular, c_P^{apx} does not depend on the passive points.

In fact, if c_P is chosen to be c_P^{apx}, then the improvement of each of the steps is exactly equal to zero. The coefficients in this system of linear equations are normally distributed. Hence we can use the result of Sankar et al. [10] to argue that the condition number is not too large with high probability. From this it follows that c_P has to be chosen close to c_P^{apx} in order to guarantee that each step makes only an insignificant improvement. In order to decrease the probability that the condition number is too large, we repeat Phase 1 nine times, i.e., we consider the first flips of the first $9(d + 1)$ active points.

The adversary who determines the position of c_P has no choice but to choose c_P close to c_P^{apx} if he wants to achieve that each of the first 2^{9d+9} steps in the sequence is an insignificant improvement. We then substitute c_P by c_P^{apx} in the formulas describing the improvements of steps. This results in formulas which do not depend on the passive points anymore and by our assumption that c_P is close to c_P^{apx} these formulas are good approximations for the improvements of the last seven active points (Phase 2). We use these formulas to argue that it is unlikely that all of them take values in $(0, \varepsilon)$ without having to use a union bound over the configuration of the passive points. (This approach is remotely inspired by the analysis of the k-means method where approximate centers of clusters are used [1].)

In our analysis we crucially use that the edge weights are given by squared Euclidean distances because for other distance measures the necessary information about the configuration of the passive points is not captured solely by c_P.

3 Preliminaries and Notation

In this section we state some lemmas that we will use later to prove Theorem 1 and we introduce some notation. Throughout the paper, ε denotes the threshold value between an insignificant and a significant step. Due to space limitations, all formal proofs are deferred to the full version. Proof ideas for the most important lemmas are given in this version.

Lemma 2. *Let* $D_{\max} := \sigma\sqrt{2n} + 1$ *and let* X *be a set of* n *Gaussian random vectors in* $\mathbb{R}^d$ *with mean values in* $[0,1]^n$ *and standard deviation* σ. *Then* $\mathbf{Pr}\left[X \nsubseteq [-D_{\max}, D_{\max}]^d\right] \le d/2^n$.

Up to our proof of the main result in Section 7, we assume without further mention that $X \subseteq [-D_{\max}, D_{\max}]^d$ and $\sigma \le 1/\sqrt{2n}$, which implies $D_{\max} \le 2$. Furthermore, we assume $n \ge d$, which is without loss of generality because d is a constant.

Lemma 3. *The weight of any cut is between* 0 *and* $\phi_{\max} := 16dn^2$.

One crucial ingredient of our analysis is the following result.

Lemma 4 (Sankar, Spielman, Teng [10]). *Let* $\bar{A} \in \mathbb{R}^{d \times d}$ *with* $\left\|\bar{A}\right\|_2 \le \sqrt{d}$ *be arbitrary. Let* A *be obtained from* $\bar{A}$ *by adding to each entry an independent Gaussian with mean* 0 *and standard deviation* σ. *Then for all* $\delta \ge 1$, $\mathbf{Pr}\left[\kappa(A) \ge \delta\right] \le \frac{14.1d(1+\sqrt{2\ln(\delta)/9d})}{\delta\sigma}$, *where* $\kappa(A) := \left\|A\right\|_2 \left\|A^{-1}\right\|_2$ *denotes the condition number of* A.

The following lemma follows from elementary probability theory.

Lemma 5. *Let* $k \in \mathbb{N}$ *and* $\lambda_1, \ldots, \lambda_k \in \mathbb{Z}$ *with* $\sum_{i=1}^{k} \lambda_i \ne 0$. *Let* $u, v_1, \ldots, v_k \in \mathbb{R}^d$ *and let* z *denote a* d-*dimensional Gaussian random vector with mean* $\mu \in \mathbb{R}^d$ *and standard deviation* σ. *Then for every* $\tau \in \mathbb{R}$ *and* $\delta > 0$,

$$\mathbf{Pr}\left[u \cdot z + \sum_{i=1}^{k} \lambda_i \cdot \left\|z - v_i\right\|^2 \in [\tau, \tau + \delta]\right] \le \frac{\sqrt{\delta}}{\sigma}.$$

For a point $z \in \mathbb{R}^d$ and a finite set $B \subseteq \mathbb{R}^d$ we write $\Phi(z, B) = \sum_{x \in B} \left\|z - x\right\|^2$. Furthermore we denote by $\mathrm{cm}(B) = \frac{1}{|B|} \sum_{x \in B} x$ the center of mass of B and we use the notation $\Psi(B) = \Phi(\mathrm{cm}(B), B)$.

4 Improvement of a Double Movement

Let us consider the improvement of a single step in which a point $z \in X$ switches sides. If we denote by $X_1^z \subseteq X$ all points on the same side as z before the movement (not including z itself) and by $X_2^z \subseteq X$ all points on the other side then the improvement of the step can be written as

$$\Phi(z, X_1^z) - \Phi(z, X_2^z) \\ = |X_1^z| \cdot \left\|z - \mathrm{cm}(X_1^z)\right\|^2 + \Psi(X_1^z) - \left(|X_2^z| \cdot \left\|z - \mathrm{cm}(X_2^z)\right\|^2 + \Psi(X_2^z)\right), \tag{1}$$

where the equation follows from the following lemma.

Lemma 6 ([6]). *For any $z \in \mathbb{R}^d$ and any finite set $X \subseteq \mathbb{R}^d$ it holds*
$$\Phi(z, X) = |X| \cdot \|z - \operatorname{cm}(X)\|^2 + \Psi(X).$$

Since the occurrence of $\Psi(X_1^z)$ and $\Psi(X_2^z)$ in (1) is problematic for our analysis, we will eliminate these terms by considering two consecutive steps and adding or subtracting their respective improvements. To be more precise consider two consecutive steps in which the points y and z switch sides (in this order) and let X_1^z and X_2^z be defined as above with the only exception that y is contained in neither of them. If y and z are on different sides before they move, then it is easy to see from (1) that the terms $\Psi(X_1^z)$ and $\Psi(X_2^z)$ cancel out if one adds the improvements of the two steps. If y and z are on the same side before they move, then similarly one can see that the terms $\Psi(X_1^z)$ and $\Psi(X_2^z)$ cancel out if one subtracts the improvements of the two steps. In both cases we denote the resulting term $\xi(z)$ (it is only indexed by z and not by y because we define y to be the unique point that moves before z in the considered sequence of steps). If both steps yield an improvement in $(0, \varepsilon]$ then $\xi(z)$ lies in $[-\varepsilon, 2\varepsilon]$.

The following definition makes the reasoning above more formal. For reasons that will become clear later, we assume that the sets X_1^z and X_2^z are both partitioned into two parts, which we call passive and active.

Definition 7. *For a given sequence of steps and an arbitrary point $z \in X$ that moves during this sequence at least once but not in the first step, let $p(z)$ be the point from X that moves last before the first move of z. For any $\varepsilon > 0$, any such point $z \in X$ and the set $P \subseteq X \setminus \{z, p(z)\}$ of passive points that do not move during the considered sequence, we define the following variables and functions, where $y = p(z)$.*

- *$A^z := (X \setminus P) \setminus \{y, z\}$ is the set of active points.*
- *X_1^z is the set of points that are on the same side as z directly before the first movement of z, excluding y and z. Furthermore, let $X_2^z := (X \setminus X_1^z) \setminus \{y, z\}$ be the set of points that are on the other side, excluding y and z.*
- *We partition X_1^z and X_2^z into active and passive points: $A_1^z := A^z \cap X_1^z$, $A_2^z := A^z \cap X_2^z$, $P_1^z := P \cap X_1^z$, $P_2^z := P \cap X_2^z$.*
- *$\pi(z)$ is 1 if y and z jump in different directions; otherwise it is -1.*
- *$\xi(z)$ is defined as the improvement of the z-movement plus $\pi(z)$ times the improvement of the y-movement.*

In the next lemma we break the term $\xi(z)$ into two parts. One part, called $b(z)$, depends only on the active points and $|P_1^z|$ and $|P_2^z|$ but not on the positions of the passive points. All information about the the passive points is subsumed in the other part. It is important for our analysis that all information needed about the passive points is the value $c_P(z)$ as defined in the following lemma.

Lemma 8. *Let $y := p(z)$,*

$$b(z) := (|P_1^z| - |P_2^z|) \cdot (\|z\|^2 - \|y\|^2) + \Phi(z, A_1^z) - \Phi(z, A_2^z)$$

$$- \Phi(y, A_1^z) + \Phi(y, A_2^z) - \begin{cases} 0 & \text{if } \pi(z) = 1, \\ 2\|z - y\|^2 & \text{if } \pi(z) = -1, \end{cases}$$

and $c_P(z) := \sum_{x \in P_1^z} x - \sum_{x \in P_2^z} x$. Then $\xi(z) = 2c_P(z) \cdot (y - z) + b(z)$.

Lemma 9. *If the movements of $p(z)$ and z both yield an improvement of at most $\varepsilon > 0$, then $|\xi(z)| \leq 2\varepsilon$.*

5   Phase 1

In Phase 1 we consider a sequence of length at most 2^{d+1} in which $d+1$ different points $z^0, \ldots, z^d$ move. Then the points $z^0, \ldots, z^d$ are active and all other points are passive. Assume in this section that an arbitrary such sequence is fixed and that the initial configuration of $z^0, \ldots, z^d$ is also fixed. We will later apply a union bound over all choices for such a sequence and the initial configuration of $z^0, \ldots, z^d$. Let $\mathcal{P}_1 \in \{1, 2\}$ be the side on which z^1 is at the beginning of the sequence, and let $\mathcal{P}_2$ be the other side. We define $P := X \setminus \{z^0, \ldots, z^d\}$ and $c_P := c_P(z^1)$. We assume that the cardinalities $|P_1^{z^1}|$ and $|P_2^{z^1}|$ are fixed. We will later also apply a union bound over all choices for these cardinalities.

If any of the movements in the considered sequence yields a significant improvement then we are done. Otherwise we will prove that we have obtained enough information to deduce approximately the position of c_P. In order to see this, observe that by Lemma 8 the first movement of each z^i with $i \geq 1$ determines the following equation:

$$\xi(z^i) = 2c_P(z^i) \cdot (p(z^i) - z^i) + b(z^i).$$

Let σ^i be $+1$ if the first movement of z^i is in the same direction as the first movement of z^1, i.e., from $\mathcal{P}_1$ to $\mathcal{P}_2$, and -1 otherwise. Then $c_P(z^i) = \sigma^i \cdot c_P(z^1)$ holds for every $i \geq 1$. Hence, $\xi(z^i) = 2\sigma^i(p(z^i) - z^i) \cdot c_P + b(z^i)$. This implies that the point c_P satisfies the system $\xi = 2Mc_P + b$ of linear equations where

$$M := \begin{pmatrix} \sigma^1(p(z^1) - z^1) \\ \vdots \\ \sigma^d(p(z^d) - z^d) \end{pmatrix}, \; \xi := \begin{pmatrix} \xi(z^1) \\ \vdots \\ \xi(z^d) \end{pmatrix}, \text{ and } b := \begin{pmatrix} b(z^1) \\ \vdots \\ b(z^d) \end{pmatrix}.$$

If the matrix M is invertible (which it is with probability 1), then $c_P = M^{-1}(\xi - b)/2$. As argued above, we are interested in the case that all movements in Phase 1 yield only a small improvement of at most ε for some $\varepsilon > 0$. In this case each $\xi(z^i)$ satisfies $|\xi(z^i)| \leq 2\varepsilon$ according to Lemma 9. We consider the approximate solution c_P^{apx} of the system of linear equations assuming that each $\xi(z^i)$ is exactly zero: $c_P^{\mathrm{apx}} = -M^{-1}b/2$. If the condition number of M is not too large and each $\xi(z^i)$ is close to zero, then c_P^{apx} is close to c_P. Note that we can calculate c_P^{apx} without uncovering the points in P or knowing their configuration because neither M nor b depends on the positions of the passive points.

Since the sequence of moves is fixed, also the matrix M is fixed. We will first show (using Lemma 4) that it is well-conditioned with high probability.

Lemma 10. *For every $\delta \geq 1$, $\mathbf{Pr}\left[\kappa(M) \geq \delta\right] \leq \frac{72d^3}{\sigma\sqrt{\delta}}$.*

In order for the distance between c_P and c_P^{apx} to be small we do not only need that the matrix M is well-conditioned but also that the norm of the right-hand side of the system of linear equations is not too small.

Lemma 11. *Let $\delta \in [0,1]$. If $\|\xi\|_\infty \leq 2\phi_{\max}\cdot\delta$ then* $\mathbf{Pr}\left[\|b - \xi\|_2 \leq \delta\right] \leq \frac{12d^{3/4}n\sqrt{\delta}}{\sigma}$.

As the quotient $\kappa(M)/\|b - \xi\|$ occurs in our analysis under a condition of the form $\|\xi\|_\infty \leq 2\delta$, we define $q_\delta := \begin{cases} \kappa(M)/\|b - \xi\| & \text{if } \|\xi\|_\infty \leq 2\delta, \\ 0 & \text{otherwise.} \end{cases}$ The second case in this definition is necessary because we will treat the event $\|\xi\|_\infty > 2\delta$ separately and do not want it to have any effect on q_δ.

Lemma 12. *Let $\varepsilon > 0$. If $\|\xi\|_\infty \leq 2\varepsilon$ then $\|c_P^{\mathrm{apx}} - c_P\| \leq 4dn\varepsilon \cdot q_\varepsilon$.*

As we will bound the expected value of the smallest improvement of a sequence later on, we will need a bound for $\int_0^\infty \mathbf{Pr}\left[q_{\phi_{\max}/t} \geq t^c\right] dt$ for some constant $c < 1$. It turns out that the bounds given in Lemma 10 and Lemma 11 are not strong enough to make this integral finite. Therefore, we repeat Phase 1 nine times with $d + 1$ different points each time such that the nine repetitions are mutually independent. We consider active points from a repetition also as active in the other repetitions such that they do not account for c_P and c_P^{apx}. Note that we now need $9(d + 1)$ active points $Z = \{z_i^0, \ldots, z_i^d : i = 1, \ldots, 9\}$ in total. Assume in the following that an arbitrary sequence of length at most $2^{9(d+1)}$ with $9(d + 1)$ active points is fixed and that also the initial configuration of the active points is fixed. We will later apply a union bound over all choices. We get nine approximations c_P^{apx} for the same c_P (possibly negated) and nine different q_δ for the quotient $\frac{\kappa(M)}{\|b-\xi\|}$. Let q_δ^* be the minimum of these q_δ and let i^* be the repetition in which this minimum is obtained.

Lemma 13. $\int_0^\infty \mathbf{Pr}\left[q_{\phi_{\max}/t}^* \geq t^{7/15}\right] dt \leq O\left(\frac{d^{18}n^9}{\sigma^9}\right)$.

6   Phase 2

Assume in the following that an arbitrary sequence of length at most $2^{9(d+1)+7}$ with $9(d + 1) + 7$ active points is fixed and that also the initial configuration of the active points is fixed. We will later apply a union bound over all choices for the sequence and the initial configuration of the active points. The longest prefix of this sequence in which at most $9(d+1)$ points move forms the nine repetitions of Phase 1, which we have analyzed in the previous section. Phase 2, which we analyze in this section, starts with the first move of point number $9(d + 1) + 1$. Hence, Phase 2 contains seven active points that do not move in Phase 1. Let $S = \{s^1, \ldots, s^\ell\}$ denote the set of $\ell := 7$ points that move in Phase 2 in this order (i.e., Phase 2 starts with the first movement of s^1).

We will apply the principle of deferred decisions in the following way: Except from the analysis of the error event $\mathcal{F}_1(\delta)$ (Lemma 16) we assume in this section that the positions of all $9d + 9$ active points Z of Phase 1 are already uncovered. The points from S are passive in Phase 1 and hence, they belong to the set P. This implies, in particular, that c_P (but not c_P^{apx}) depends on these points.

The set of passive points changes now with every new point that gets active. We define the set of passive points during the first move of s^j as $P^j := P \backslash \{s^1, \ldots, s^j\}$ and we assume that in addition to the active points from Phase 1 also the points $s^1, \ldots, s^{j-1}$ are uncovered when s^j moves for the first time.

From now on, we mean with c_P^{apx} the point calculated in repetition i^* of Phase 1 and we denote by ξ the corresponding vector from repetition i^*. Define $\hat{\sigma}^j$ like σ^i in Phase 1 but for the new points moving in Phase 2, i.e., $\hat{\sigma}^j$ is 1 if the first movement of s^j is from $\mathcal{P}_1$ to $\mathcal{P}_2$ and -1 otherwise.

According to Lemma 8, we can write $\xi(s^j) = 2\hat{\sigma}^j c_{P^j} \cdot (p(s^j) - s^j) + b(s^j)$. Due to its definition, the point c_{P^j} is just c_P shifted by s^j and the already uncovered points $s^1, \ldots, s^{j-1}$. Therefore, we get an approximation $c_{P^j}^{\mathrm{apx}}$ for c_{P^j} by shifting c_P^{apx} in the same (now deterministic up to the randomness of s^j) way as c_P. As $c_{P^j}^{\mathrm{apx}}$ and c_{P^j} are near to each other with high probability under the assumption that all steps in Phase 1 yield insignificant improvements, the "approximate improvement" $\xi^{\mathrm{apx}}(s^j) := 2\hat{\sigma}^j c_{P^j}^{\mathrm{apx}} \cdot (p(s^j) - s^j) + b(s^j)$ is nearly the same as $\xi(s^j)$ (Lemma 14). Thus, $|\xi(s^j)|$ can only be small if $|\xi^{\mathrm{apx}}(s^j)|$ is small. But in the definition of $\xi^{\mathrm{apx}}(s^j)$, the only randomness left is the position of the point s^j. Hence, we can derive a bound for the probability of an insignificant improvement by analyzing a term in which only one random point is left. We do this successively for $s^1, \ldots, s^\ell$.

Lemma 14. *Let $1 \le j \le \ell$ and $0 \le \varepsilon \le 1$. If $\|\xi\|_\infty \le 2\varepsilon$ and $|\xi(s^j)| \le 2\varepsilon$, then $|\xi^{\mathrm{apx}}(s^j)| \le 74d^{3/2}nq_\varepsilon^* \cdot \varepsilon$. (Note that $\|\xi\|_\infty \le 2\varepsilon$ does not imply anything for $|\xi(s^j)|$ as ξ is the vector from Phase 1.)*

As already mentioned, the only randomness left in the definition of $\xi^{\mathrm{apx}}(s^j)$ is the point s^j. Hence, we can rewrite $|\xi(s^j)|$ in the following way, where C is the set of all points of the form $-c_P^{\mathrm{apx}} + \sum_{v \in Z \cup S} \alpha_v \cdot v$ with $\alpha_v \in \{-1, 0, 1\}$.

Lemma 15. $|\xi^{\mathrm{apx}}(s^j)| = \nu^j \cdot \|s^j\|^2 + 2s^j \cdot \hat{\sigma}^j \cdot c^j + \tau^j$, *where $\nu^j \in \mathbb{Z}$ and $\tau^j \in \mathbb{R}$ are known constants, and $c^j \in C$ has known coefficients α_v.*

We want to bound the probability that $|\xi^{\mathrm{apx}}(s^j)|$ is close to zero. If $\nu^j = 0$, we have to make sure that $\|c^j\|$ is not too small as otherwise the variance of $2s^j \cdot \hat{\sigma}^j \cdot c^j$ is very small. We cannot guarantee this for every j, but it is unlikely to have three different j with small $\|c^j\|$.

Lemma 16. *For $\delta \ge 0$, let $\mathcal{F}_1(\delta)$ be the event that there are three distinct points $x_1, x_2, x_3 \in C$ with $\|x_i\| \le \sqrt{\delta}/2$ for $i = 1, 2, 3$. Then $\mathbf{Pr}[\mathcal{F}_1(\delta)] \le O(1) \cdot \left(\frac{\sqrt{\delta}}{\sigma}\right)^4$.*

Hence, we know that for at least four different j it is unlikely that $|\xi^{\mathrm{apx}}(s^j)|$ is small. Now if we define $\Delta := \max_j |\xi(s^j)|$ and $\Delta^{\mathrm{apx}} := \max_j |\xi^{\mathrm{apx}}(s^j)|$, we are able to show that it is unlikely that Δ^{apx} and thus Δ is small.

Lemma 17. *For any $\delta \ge 0$, $\mathbf{Pr}[\Delta^{\mathrm{apx}} \le \delta] \le O\left(\frac{\sqrt{\delta}}{\sigma}\right)^4$.*

Corollary 18. $\int_0^\infty \mathbf{Pr}[\Delta^{\mathrm{apx}} \le t^{-8/15}] \, dt \le O(\sigma^{-4})$.

7 Bounding the Expected Number of Steps

With Lemma 13 and Corollary 18 we have all the ingredients that we need for bounding the expected number of steps of the algorithm. We first outsource a calculation which uses the aforementioned lemmata to yield a bound for the probability of a small improvement by a fixed sequence. Then we are able to show our main result.

Lemma 19. $\int_0^\infty \mathbf{Pr}\left[\Delta \leq \frac{2\phi_{\max}}{t}\right] dt \leq O\left(\frac{d^{20.5} \cdot n^{12}}{\sigma^9}\right).$

Proof (Theorem 1). We first stick with our assumption $\sigma \leq 1/\sqrt{2n}$. Let $\mathcal{F}$ be the event that our point set X is not contained in $[-D_{\max}, D_{\max}]^d \subseteq [-2, 2]^d$. Let a block be nine repetitions of Phase 1 followed by a repetition of Phase 2. Let us derive a union bound over all possible choices of blocks: There are $n^{O(d)}$ choices for the active points in Phase 1 and Phase 2. Furthermore, we need another factor n for the choice of $|P_1|$. Instead of fixing the whole sequence of steps, it suffices for our purposes to fix the configuration of the active points before every first move of a point, which results in another factor $2^{O(d^2)}$. Together this results in a factor of $2^{O(d^2)} \cdot n^{O(d)}$.

Let T be the number of blocks that are processed during the FLIP algorithm. Then

$$\mathbf{E}[T] = \int_0^{2^n} \mathbf{Pr}[T \geq t] \, dt \leq \int_0^{2^n} \mathbf{Pr}[\mathcal{F}] + 2^{O(d^2)} \cdot n^{O(d)} \cdot \mathbf{Pr}\left[\Delta \leq \frac{2\phi_{\max}}{t}\right] dt$$

$$\leq 2^n \cdot \frac{d}{2^n} + 2^{O(d^2)} n^{O(d)} \frac{d^{20.5} n^{12}}{\sigma^9} \leq d + \frac{2^{O(d^2)} n^{O(d)}}{\sigma^9} \leq \frac{2^{O(d^2)} n^{O(d)}}{\sigma^9}.$$

As $O(d)$ different points move in a block, the length of a block is at most $2^{O(d)}$. Hence, the total number of steps is bounded by $2^{O(d)} \cdot 2^{O(d^2)} \cdot n^{O(d)} \cdot \sigma^{-9} = 2^{O(d^2)} \cdot n^{O(d)} \cdot \sigma^{-9}$.

If $\sigma > 1/\sqrt{2n}$, we create an equivalent instance by scaling down the mean values by the factor $1/(\sqrt{2n}\sigma)$ (i.e., the mean values remain in $[0, 1]^n$) and setting the standard deviation to $\sigma' = 1/\sqrt{2n} < \sigma$. As these instances are equivalent, we obtain the same expected number of iterations and thus also a bound of $2^{O(d^2)} \cdot n^{O(d)} \cdot (\sqrt{2n})^9 \leq 2^{O(d^2)} \cdot n^{O(d)} \cdot \sigma^{-9}$. $\qquad\square$

8 Concluding Remarks

We proved the first polynomial upper bound on the smoothed number of steps of the FLIP algorithm for the Maximum-Cut problem. Our upper bound applies only to squared Euclidean distances because it uses essentially the identity given in Lemma 6, which is special to squared Euclidean distances. It might be possible to extend our analysis to Bregman divergences because these also satisfy Lemma 6. An immediate extension to general graphs does not seem to be possible and it is still a very interesting open question if the result from [5] for general graphs can be improved.

Our result gives only a polynomial smoothed running time if the dimension is constant because the degree of the polynomial grows linearly with d. We think that it is conceivable to improve the smoothed running time to $2^{O(d)}$ times a polynomial in n and $1/\sigma$ whose degree is independent of d by a more careful analysis of the condition number in Phase 1 that does not use the result by Sankar et al. [10] as a black box. Generally, we hope that our work triggers further improvements like, e.g., the first smoothed analysis of the k-means method by Arthur and Vassilvitskii [2], which only gave a polynomial bound for constant k.

A version of the k-means method that works rather well in experiments is Hartigan's method [15]. Telgarsky and Vattani conjecture that the smoothed running time of this algorithm is polynomial [15]. However, so far this conjecture could not be proven and it seems rather challenging. As Hartigan's method has some similarities with the FLIP algorithm for the Maximum-Cut problem for squared Euclidean distances, we believe that our new proof technique might also be helpful for proving Telgarsky and Vattani's conjecture.

References

1. Arthur, D., Manthey, B., Röglin, H.: Smoothed analysis of the k-means method. JACM 58(5), 19 (2011)
2. Arthur, D., Vassilvitskii, S.: Worst-case and smoothed analysis of the ICP algorithm. SICOMP 39(2), 766–782 (2009)
3. Elsässer, R., Tscheuschner, T.: Settling the complexity of local max-cut (almost) completely. In: Aceto, L., Henzinger, M., Sgall, J. (eds.) ICALP 2011, Part I. LNCS, vol. 6755, pp. 171–182. Springer, Heidelberg (2011)
4. Englert, M., Röglin, H., Vöcking, B.: Worst case and prob. analysis of the 2-Opt algorithm for the TSP. Algorithmica 68(1), 190–264 (2014)
5. Etscheid, M., Röglin, H.: Smoothed analysis of local search for the maximum-cut problem. In: Proc. of 25th SODA, pp. 882–889 (2014)
6. Kanungo, T., Mount, D., Netanyahu, N., Piatko, C., Silverman, R., Wu, A.: A local search appr. algo. for k-means clustering. Comput. Geom. 28, 89–112 (2004)
7. Kleinberg, J., Tardos, É.: Algorithm design. Addison-Wesley (2006)
8. Manthey, B., Röglin, H.: Smoothed analysis: Analysis of algorithms beyond worst case. IT - Information Technology 53(6), 280–286 (2011)
9. Manthey, B., Veenstra, R.: Smoothed analysis of the 2-opt heuristic for the TSP: Polynomial bounds for gaussian noise. In: Cai, L., Cheng, S.-W., Lam, T.-W. (eds.) Algorithms and Computation. LNCS, vol. 8283, pp. 579–589. Springer, Heidelberg (2013)
10. Sankar, A., Spielman, D., Teng, S.-H.: Smoothed analysis of the condition numb. and growth factors of matrices. SIMAX 28(2), 446–476 (2006)
11. Schäffer, A., Yannakakis, M.: Simple local search problems that are hard to solve. SICOMP 20(1), 56–87 (1991)
12. Schulman, L.: Clustering for edge-cost minimization. In: Proc. of 32nd STOC, pp. 547–555 (2000)
13. Spielman, D., Teng, S.-H.: Smoothed analysis of algorithms: Why the simplex algorithm usually takes polynomial time. JACM 51(3), 385–463 (2004)
14. Spielman, D., Teng, S.-H.: Smoothed analysis: An attempt to explain the behavior of algorithms in practice. CACM 52(10), 76–84 (2009)
15. Telgarsky, M., Vattani, A.: Hartigan's method: k-means clustering without voronoi. In: Proc. of 13th AISTATS, pp. 820–827 (2010)

Maximizing Symmetric Submodular Functions

Moran Feldman

School of Computer and Communication Sciences, EPFL,
Lausanne, Switzerland
moran.feldman@epfl.ch

Abstract. Symmetric submodular functions are an important family of submodular functions capturing many interesting cases including cut functions of graphs and hypergraphs. In this work, we identify submodular maximization problems for which one can get a better approximation for symmetric objectives compared to what is known for general submodular functions.

For the problem of maximizing a non-negative symmetric submodular function $f\colon 2^{\mathcal{N}} \to \mathbb{R}^+$ subject to a down-monotone solvable polytope $\mathcal{P} \subseteq [0,1]^{\mathcal{N}}$, we describe an algorithm producing a fractional solution of value at least $0.432 \cdot f(OPT)$, where OPT is the optimal *integral* solution. Our second result is a 0.432-approximation algorithm for the problem $\max\{f(S) : |S| = k\}$ with a non-negative symmetric submodular function $f\colon 2^{\mathcal{N}} \to \mathbb{R}^+$. Our method also applies to non-symmetric functions, in which case it produces $1/e - o(1)$ approximation. Finally, we describe a *deterministic linear-time* $1/2$-approximation algorithm for unconstrained maximization of a non-negative symmetric submodular function.

Keywords: Symmetric submodular functions, Cardinality constraint, Matroid constraint.

1 Introduction

The study of combinatorial problems with submodular objective functions has recently attracted much attention, and is motivated by the principle of economy of scale, prevalent in real world applications. Submodular functions are also commonly used as utility functions in economics and algorithmic game theory. Symmetric submodular functions are an important family of submodular functions capturing, for example, the mutual information function and cut functions of graphs and hypergraphs.

Minimization of symmetric submodular functions subject to various constrains and approximating such functions by other functions received a lot of attention [6,7,13,17,18]. However, maximization of symmetric submodular functions was the subject of only limited research, despite an extensive body of works dealing with maximization of general non-negative submodular functions (see, e.g., [1,3,5,16,19]). In fact, we are only aware of two papers dealing with maximization of symmetric submodular functions. First, Feige et al. [8] show an

© Springer-Verlag Berlin Heidelberg 2015
N. Bansal and I. Finocchi (Eds.): ESA 2015, LNCS 9294, pp. 521–532, 2015.
DOI: 10.1007/978-3-662-48350-3_44

$1/2$-approximation algorithm for the problem of maximizing a symmetric sub-modular function subject to no constraint (which is the best possible). This result was later complemented by an algorithm achieving the same approximation ratio for general submodular functions [1]. Second, Lee et al. [15] show a $1/3$-approximation algorithm for maximizing a symmetric submodular function subject to a general matroid base constraint.

In this work, we identify a few submodular maximization problems for which one can get a better approximation for symmetric objectives than the state of the art approximation for general submodular functions. Our first result is an improved algorithm for maximizing a non-negative symmetric submodular function[1] $f: 2^{\mathcal{N}} \to \mathbb{R}^{+}$ subject to a down-monotone solvable polytope[2] $\mathcal{P} \subseteq [0,1]^{\mathcal{N}}$. More formally, given a set function $f: 2^{\mathcal{N}} \to \mathbb{R}$, its *multilinear* extension is the function $F: [0,1]^{\mathcal{N}} \to \mathbb{R}$ defined by $F(x) = \mathbb{E}[f(\mathcal{R}(x))]$, where $\mathcal{R}(x)$ is a random set containing every element $u \in \mathcal{N}$ with probability x_u, independently. Our result is an approximation algorithm for the problem $\max\{F(x) : x \in \mathcal{P}\}$ whose approximation ratio is about: $1/2 \cdot [1 - (1 - d(\mathcal{P})^{2/d(\mathcal{P})})]$, where $d(\mathcal{P})$ is the density[3] of $\mathcal{P}$. In the following theorem, and throughout the paper, we use n to denote $|\mathcal{N}|$.

Theorem 1. *Given a non-negative symmetric submodular function $f: 2^{\mathcal{N}} \to \mathbb{R}^{+}$, a down-monotone solvable polytope $\mathcal{P} \subseteq [0,1]^{\mathcal{N}}$ and a constant $T \geq 0$, there exists an efficient algorithm that finds a point $x \in [0,1]^{\mathcal{N}}$ such that $F(x) \geq 1/2 \cdot [1 - e^{-2T} - o(1)] \cdot \max\{F(x) : x \in \mathcal{P} \cap \{0,1\}^{\mathcal{N}}\}$. Additionally,*

(a) $x/T \in \mathcal{P}$.
(b) Let $T_{\mathcal{P}} = -\ln(1 - d(\mathcal{P}) + n^{-4})/d(\mathcal{P})$. Then, $T \leq T_{\mathcal{P}}$ implies $x \in \mathcal{P}$.

Theorem 1 improves over the result of [11], who gave an approximation ratio of $e^{-1} - o(1)$ for the case of general submodular functions. More specifically, Theorem 1 provides an approximation ratio of at least $1/2 \cdot [1 - e^{-2}] - o(1) \geq 0.432$ for an arbitrary down-monotone solvable polytope since T can always be set to be at least 1. For many polytopes the result produced by Theorem 1 can be rounded using known rounding methods (see, e.g., pipage rounding [3], swap rounding [4] and contention resolution schemes [5]). For example, matroid polytopes allow rounding without any loss in the approximation ratio. Moreover, due to property (a) of Theorem 1, the combination of our algorithm with the contention resolution schemes rounding described by [5] produces better approximation ratios than those implied by a black box combination (see [11] for details).

---

[1] A set function $f: 2^{\mathcal{N}} \to \mathbb{R}^{+}$ is *symmetric* if $f(S) = f(\mathcal{N} \setminus S)$ for every set $S \in \mathcal{N}$, and *submodular* if $f(A) + f(B) \geq f(A \cup B) + f(A \cap B)$ for every pair of sets $A, B \subseteq \mathcal{N}$.
[2] A polytope $\mathcal{P} \subseteq [0,1]^{\mathcal{N}}$ is *solvable* if one can optimize linear functions over it, and *down-monotone* if for every two vectors $x, y \in [0,1]^{\mathcal{N}}$, $x \leq y$ and $y \in \mathcal{P}$ imply $x \in \mathcal{P}$.
[3] Consider a representation of $\mathcal{P}$ using m inequality constraints, and let us denote the ith inequality constraint by $\sum_{u \in \mathcal{N}} a_{i,u} x_u \leq b_i$. By Sect. 3.A of [9], we may assume all the coefficients are non-negative and each constraint has at least one non-free non-zero coefficient. The *density* $d(\mathcal{P})$ of $\mathcal{P}$ is defined as the maximum value of $\min_{1 \leq i \leq m} \frac{b_i}{\sum_{u \in \mathcal{N}} a_{i,u}}$ for any such representation.

Our next result considers the problem $\max\{f(S) : |S| = k\}$ for a non-negative symmetric submodular function $f : 2^{\mathcal{N}} \to \mathbb{R}^+$. For this problem we prove the following theorem.

Theorem 2. *There exists an efficient algorithm that given a non-negative symmetric submodular function $f : 2^{\mathcal{N}} \to \mathbb{R}^+$ and an integer cardinality parameter $1 \leq k \leq n/2$, achieves an approximation of $1/2[1 - (1 - k/n)^{2n/k}] - o(1)$ for the problem: $\max\{f(S) : |S| = k\}$. If $k > n/2$, then the same result holds with the cardinality parameter replaced by $n - k$.*

Notice that Theorem 2 achieves for the problem $\max\{f(S) : |S| = k\}$ the same approximation ratio achieved by Theorem 1 for the problem $\max\{f(S) : |S| \leq k\}$ (as long as $k \leq n/2$). Using the same technique we get a result also for the more well-studied case of general (non-symmetric) submodular functions.

Theorem 3. *There exists an efficient algorithm that given a non-negative submodular function $f : 2^{\mathcal{N}} \to \mathbb{R}^+$ and an integer cardinality parameter $1 \leq k \leq n$, achieves an approximation of $e^{-1} - o(1)$ for the problem: $\max\{f(S) : |S| = k\}$.*

Theorems 2 and 3 improve over results achieved by [2] when $k/n \leq 0.204$ and $k/n \leq 0.093$, respectively. Most practical applications of maximizing a submodular function subject to a cardinality constraint use instances having relatively small k/n ratios, and thus, can benefit from our improvements (see [2] for a list of such applications). We complement Theorem 2 by showing that one cannot get an approximation ratio better than $1/2$ for any ratio k/n.

Theorem 4. *Consider the problems $\max\{f(S) : |S| = p/q \cdot n\}$ and $\max\{f(S) : |S| \leq p/q \cdot n\}$ where $p < q$ are positive constant integers and f is a non-negative symmetric submodular function $f : 2^{\mathcal{N}} \to \mathbb{R}^+$ obeying $n/q \in \mathbb{Z}$. Then, every algorithm with an approximation ratio of $1/2 + \varepsilon$ for one of the above problems (for any constant $\varepsilon > 0$) uses an exponential number of value oracle queries.*

The result of Theorem 4 follows quite easily from the symmetry gap framework of [19] and is known for the case of general submodular functions as well as for some pairs of p and q (e.g., the case $p/q = 1/2$ follows immediately from the work of [19]). We give the theorem here mainly for completeness reasons, and defer its proof to the full version of this paper.

We also consider the unconstrained submodular maximization problem (i.e., $\max\{f(S) : S \subseteq \mathcal{N}\}$). For symmetric submodular functions, Feige et al. [8] give for this problem a simple linear-time randomized algorithm and a slower deterministic local search, both achieving an optimal approximation ratio of $1/2$ (up to a low order error term in the case of the local search). Buchbinder et al. [1] give a randomized linear-time algorithm for the case of general submodular functions achieving the same approximation ratio. Finding a deterministic algorithm with the same approximation ratio for general submodular functions is still an open problem. Recently, Huang and Borodin [14] showed that a large family of deterministic algorithms resembling the algorithm of [1] fails to achieve $1/2$-approximation. We show that for symmetric submodular functions there exists a *deterministic linear-time* $1/2$-approximation algorithm.

Theorem 5. *There exists a deterministic linear-time* $1/2$-*approximation algorithm for the problem* $\max\{f(S) : S \subseteq \mathcal{N}\}$, *where* $f : 2^{\mathcal{N}} \to \mathbb{R}^+$ *is a non-negative symmetric submodular function.*

Theorem 5 improves over the time complexity of the local search algorithm of [8] and also avoids the low order error term.

1.1   Our Techniques

Some of our results are based on variants of the measured continuous greedy algorithm of [11]. We modify the measured continuous greedy in two main ways.

- The analysis of [11] relies on the observation that $F(\mathbf{1}_{OPT} \vee x) \geq [1 - \max_{u \in \mathcal{N}} x_u] \cdot f(OPT)$ for an arbitrary vector $x \in [0,1]^{\mathcal{N}}$.[4] To get better results for symmetric functions we use an alternative lower bound on $F(\mathbf{1}_{OPT} \vee x)$ given by Lemma 6.

  Lemma 6. *Given a set* S, *a non-negative symmetric submodular function* $f : 2^{\mathcal{N}} \to \mathbb{R}^+$ *and a vector* $x \in [0,1]$ *obeying* $F(y) \leq F(x)$ *for every* $\{y \in [0,1]^{\mathcal{N}} : y \leq x\}$, *then* $F(\mathbf{1}_S \vee x) \geq f(S) - F(x)$.

  Using the bound given by Lemma 6 in the analysis requires a slight modification of the measured continuous greedy algorithm to guarantee that its solution always obeys the requirements of the lemma. We defer the proof of Lemma 6 to Sect. 2.
- The measured continuous greedy algorithm can handle only constraints specified by a down-monotone polytope. Thus, it cannot handle problems of the form $\max\{f(S) : |S| = k\}$. To bypass this difficulty, we use two instances of the measured continuous greedy algorithm applied to the problems $\max\{f(S) : |S| \leq k\}$ and $\max\{f(\mathcal{N} \setminus S) : |S| \leq n - k\}$. Note that the optimal solutions of both problems are at least as good as the optimal solution of $\max\{f(S) : |S| = k\}$. A careful correlation of the two instances preserves their approximation ratios, and allows us to combine their outputs into a solution for $\max\{f(S) : |S| = k\}$ with the same approximation ratio.

Our result for the problem $\max\{f(S) : S \subseteq \mathcal{N}\}$ is based on the linear-time deterministic algorithm suggested by [1] for this problem. Buchbinder et al. [1] showed that this algorithm has an approximation ratio of $1/3$ for general non-negative submodular functions. The algorithm maintains two solutions X and Y that become identical when the algorithm terminates. The analysis of the algorithm is based on a set $OPT(X, Y)$ that starts as OPT and converts gradually to the final value of X (and Y). The key observation of the analysis is showing that in each iteration (of the algorithm) the value of $OPT(X, Y)$ deteriorates by at

---

[4] For every set $S \subseteq \mathcal{N}$, we use $\mathbf{1}_S$ to denote the characteristic vector of S. Given two vectors $x, y \in [0,1]^{\mathcal{N}}$, we use $x \vee y$ to denote the coordinate-wise maximum of x and y. In other words, for every $u \in \mathcal{N}$, $(x \vee y)_u = \max\{x_u, y_u\}$. Similarly, $x \wedge y$ denotes the coordinate-wise minimum of x and y.

most the increase of $f(X)+f(Y)$. In this work we show that the exact same algorithm provides $1/2$-approximation when the objective is also symmetric. To that aim, we consider two sets $OPT(X,Y)$ and $\overline{OPT}(X,Y)$. These sets start as OPT and $\overline{OPT} = \mathcal{N} \setminus OPT$ respectively, and convert gradually into the final value of X (and Y). We prove that the deterioration of $f(OPT(X,Y)) + f(\overline{OPT}(X,Y))$ lower bounds the increase of $f(X) + f(Y)$. Due to space constraints, we omit the development of this idea into a formal proof of Theorem 5.

1.2 Related Work

The literature on submodular maximization problems is very large, and therefore, we mention below only a few of the most relevant works. Feige et al. [8] provided the first constant factor approximation algorithms for $\max\{f(S) : S \subseteq \mathcal{N}\}$. Their best approximation algorithm achieved an approximation ratio of $2/5 - o(1)$. Oveis Gharan and Vondrák [12] used simulated annealing techniques to provide an improved approximation of roughly 0.41. Feldman et al. [10] combined the algorithm of [12] with a new algorithm, yielding an approximation ratio of roughly 0.42. Finally, Buchbinder et al. [1] gave a $1/2$-approximation for this problem, matching a lower bound proved by [8].

The problem of maximizing a (not necessary monotone) submodular function subject to a general matroid constraint was given a 0.309-approximation by [19]. Using simulated annealing techniques this was improved to 0.325 [12], and shortly later was further pushed to $1/e - o(1)$ by [11] via the measured continuous greedy algorithm. Recently, Buchbinder et al. [2] showed that for the problem $\max\{f(S) : |S| \leq k\}$ (which is a special case of a matroid constraint) it is possible to get an approximation ratio in the range $[1/e + \varepsilon, 1/2 - o(1)]$ for some small constant $\varepsilon > 0$ (the exact approximation ratio in this range depends on the ratio k/n). A hardness result of 0.491 was given by [12] for the case $k \ll n$.

The problem of maximizing a (not necessary monotone) submodular function subject to a matroid base constraint was shown to have no constant approximation ratio by [19]. Buchbinder et al. [2] showed that the special case of $\max\{f(S) : |S| = k\}$ admits an approximation ratio from the range $[0.356, 1/2 - o(1)]$ (again, the exact approximation ratio in this range depends on k/n). On the other hand, the hardness of 0.491 by [12] applies also to this problem when $k \ll n$.

2 Preliminaries

For every set $S \subseteq \mathcal{N}$ and an element $u \in \mathcal{N}$, we denote the union $S \cup \{u\}$ by $S + u$, the expression $S \setminus \{u\}$ by $S - u$ and the set $\mathcal{N} \setminus S$ by $\bar{S}$. Additionally, we use $\mathbf{1}_S$ and $\mathbf{1}_u$ to denote the characteristic vectors of S and $\{u\}$, respectively. Given a submodular function $f : 2^{\mathcal{N}} \to \mathbb{R}$ and its corresponding multilinear extension $F : [0,1]^{\mathcal{N}} \to \mathbb{R}$, we denote the partial derivative of F at a point $x \in [0,1]^{\mathcal{N}}$ with respect to an element u by $\partial_u F(x)$. Since F is multilinear, $\partial_u F(x) = F(x \vee \mathbf{1}_u) - F(x \wedge \mathbf{1}_{\mathcal{N}-u})$. Additionally, we use $\bar{f}$ and $\bar{F}$ to denote the functions $\bar{f}(S) = f(\mathcal{N} \setminus S)$ and $\bar{F}(x) = F(\mathbf{1}_{\mathcal{N}} - x)$.

We look for algorithms of polynomial in n (the size of $\mathcal{N}$) time complexity. However, an explicit representation of a submodular function might be exponential in the size of its ground set. The standard way to bypass this difficulty is to assume access to the function via a *value oracle*. For a submodular function $f \colon 2^{\mathcal{N}} \to \mathbb{R}$, given a set $S \subseteq \mathcal{N}$, the value oracle returns the value of $f(S)$. Some of our algorithms assume a more powerful oracle that given a vector $x \in [0,1]^{\mathcal{N}}$, returns the value of $F(x)$. If such an oracle is not available, one can approximate it arbitrarily well using a value oracle to f by averaging enough samples, which results in an $o(1)$ loss in the approximation ratio. This is a standard practice (see, e.g., [3]), and we omit details.

The following lemma gives a few useful properties of submodular functions.

Lemma 7. *If $f \colon 2^{\mathcal{N}} \to \mathbb{R}$ is a submodular function and $F \colon [0,1]^{\mathcal{N}} \to \mathbb{R}$ is its multilinear extension, then:*

- *For every vector $x \in [0,1]^{\mathcal{N}}$, $\bar{F}(x)$ is the multilinear extension of $\bar{f}$.*
- *If f is symmetric, then for every vector $x \in [0,1]^{\mathcal{N}}$, $F(x) = \bar{F}(x)$.*
- *For every three vectors $z \leq y \leq x \in [0,1]^{\mathcal{N}}$, $F(x) - F(y) \leq F(x-z) - F(y-z)$.*

We omit the (standard) proof of Lemma 7 due to space constraints. We are now ready to give the promised proof of Lemma 6.

Proof (Lemma 6). Since f is symmetric, $f(S) - F(x \vee \mathbf{1}_S) = f(\bar{S}) - F((\mathbf{1}_{\mathcal{N}} - x) \wedge \mathbf{1}_{\bar{S}}) \leq F(x \wedge \mathbf{1}_{\bar{S}}) - f(\varnothing) \leq F(x \wedge \mathbf{1}_{\bar{S}}) \leq F(x)$, where the equality and first inequality hold by Lemma 7, the second inequality holds by the non-negativity of f and the last inequality holds since $x \wedge \mathbf{1}_{\bar{S}} \leq x$. $\qquad\square$

The following lemma shows that the multilinear extension behaves like a linear function as long as the change in its input is small. Similar lemmata appear in many works. A proof of this specific lemma can be found in [9] (as Lemma 2.3.7).

Lemma 8. *Consider two vectors $x, x' \in [0,1]^{\mathcal{N}}$ such that for every $u \in \mathcal{N}$, $|x_u - x'_u| \leq \delta$. Then, $F(x') - F(x) \geq \sum_{u \in \mathcal{N}}(x'_u - x_u) \cdot \partial_u F(x) - O(n^3 \delta^2) \cdot \max_{u \in \mathcal{N}} f(\{u\})$.*

We also use the following lemma, which comes handy in proving the feasibility of the solutions produced by some of our algorithms. This lemma is implicitly proved by [11] (some parts of the proof, which are omitted in [11], can be found in [9]).

Lemma 9. *Fix some $\delta \leq n^{-5}$, and let $\{I(i)\}_{i=1}^{\ell}$ be a set of ℓ points in a down-monotone polytope $\mathcal{P} \subseteq [0,1]^{\mathcal{N}}$. Let $\{y(i)\}_{i=0}^{\ell}$ be a a set of $\ell+1$ vectors in $[0,1]^{\mathcal{N}}$ obeying the following constraints. For every element $u \in \mathcal{N}$,*

$$y_u(i) \leq \begin{cases} 0 & \text{if } i = 0 \text{ ,} \\ y_u(i-1) + \delta I_u(i) \cdot (1 - y_u(i-1)) & \text{otherwise .} \end{cases}$$

Then,

- *$y_u(i)/(\delta i) \in \mathcal{P}$.*
- *Let $T_{\mathcal{P}} = -\ln(1 - d(\mathcal{P}) + n^{-4})/d(\mathcal{P})$. Then, $\delta i \leq T_{\mathcal{P}}$ implies $y(i) \in \mathcal{P}$.*

3   Measured Continuous Greedy for Symmetric Functions

In this section we prove Theorem 1. Due to space constraints, many of the proofs of this section are omitted.

Theorem 1. *Given a non-negative symmetric submodular function $f \colon 2^{\mathcal{N}} \to \mathbb{R}^+$, a down-monotone solvable polytope $\mathcal{P} \subseteq [0,1]^{\mathcal{N}}$ and a constant $T \geq 0$, there exists an efficient algorithm that finds a point $x \in [0,1]^{\mathcal{N}}$ such that $F(x) \geq \frac{1}{2} \cdot [1 - e^{-2T} - o(1)] \cdot \max\{F(x) : x \in \mathcal{P} \cap \{0,1\}^{\mathcal{N}}\}$. Additionally,*

(a) $x/T \in \mathcal{P}$.
(b) Let $T_{\mathcal{P}} = -\ln(1 - d(\mathcal{P}) + n^{-4})/d(\mathcal{P})$. Then, $T \leq T_{\mathcal{P}}$ implies $x \in \mathcal{P}$.

To simplify the proof of the theorem, we use the following reduction.

Reduction 10. *When proving Theorem 1, we may assume $\mathbf{1}_u \in \mathcal{P} \ \forall \, u \in \mathcal{N}$.*

Algorithm 1 is a variant of the Measured Continuous Greedy algorithm presented by [11]. Notice that the definition of δ in the algorithm guarantees that $\delta \leq n^{-5}$ and $t = T$ after $\lceil n^5 T \rceil$ iterations. Thus, by Lemma 9, the output of Algorithm 1 obeys properties (a) and (b) guaranteed by Theorem 1. To complete the proof of Theorem 1, it is only necessary to show that the approximation ratio of Algorithm 1 matches the approximation ratio guaranteed by the theorem.

Algorithm 1. Measured Continuous Greedy for Symmetric Functions $(f, \mathcal{P}, T)$

    // Initialization
1 Set: $\delta \leftarrow T(\lceil n^5 T \rceil)^{-1}$.
2 Initialize: $t \leftarrow 0$, $y(0) \leftarrow \mathbf{1}_{\varnothing}$.

    // Main loop
3 **while** $t < T$ **do**
4   **foreach** $u \in \mathcal{N}$ **do** Let $w_u(t) \leftarrow F(y(t) \vee \mathbf{1}_u) - F(y(t))$.
5   Let $I(t)$ be a vector in $\mathcal{P}$ maximizing $I(t) \cdot w(t)$.
6   **foreach** $u \in \mathcal{N}$ **do** Let $y_u(t + \delta) \leftarrow y_u(t) + \delta I_u(t) \cdot (1 - y_u(t))$.
7   **foreach** $u \in \mathcal{N}$ **do** **if** $\partial_u F(y(t + \delta)) < 0$ **then** $y_u(t + \delta) \leftarrow 0$.

8 Return $y(T)$.

First, we need a lower bound on the improvement achieved in each iteration of the algorithm. The following lemma is a counterpart of Corollary III.4 of [11].

Lemma 11. *For every time $0 \leq t < T$, $F(y(t + \delta)) - F(y(t)) \geq \delta \cdot [F(y(t) \vee \mathbf{1}_{OPT}) - F(y(t))] - O(n^3 \delta^2) \cdot f(OPT)$.*

The last lemma gives a lower bound on the improvement achieved in every step of the algorithm in terms of $F(y(t) \vee \mathbf{1}_{OPT})$. To make this lower bound useful, we need to lower bound the term $F(y(t) \vee \mathbf{1}_{OPT})$ using Lemma 6. Observe that Line 7 guarantees that the conditions of Lemma 6 hold.

Corollary 12. *For every time* $0 \leq t < T$, $F(y(T+\delta)) - F(y(T)) \geq \delta \cdot [f(OPT) - 2 \cdot F(y(t))] - O(n^3 \delta^2) \cdot f(OPT)$.

At this point we have a lower bound on the improvement achieved in each iteration in terms of $f(OPT)$ and $F(y(t))$. In order to complete the analysis of the algorithm, we need to derive from it a bound on the value of $F(y(t))$ for every time t. Let $h(t) = 1/2 \cdot [1 - e^{-2t}] \cdot f(OPT)$.

Lemma 13. *For every* $0 \leq t \leq T$, $F(y(t)) \geq h(t) - O(n^3 \delta) \cdot t \cdot f(OPT)$.

Using the last lemma we can prove the approximation ratio of Theorem 1.

Proof (Approximation Ratio of Theorem 1). By Lemma 13,

$$F(y(T)) \geq h(T) - O(n^3 \delta T) \cdot f(OPT) = 1/2 \cdot [1 - 2e^{-T} - O(n^3 \delta T)] \cdot f(OPT) \ .$$

The proof is now complete since T is a constant and $\delta \leq n^{-5}$.   □

4 Equality Cardinality Constraints

In this section we prove Theorem 2.

Theorem 2. *There exists an efficient algorithm that given a non-negative symmetric submodular function* $f \colon 2^{\mathcal{N}} \to \mathbb{R}^+$ *and an integer cardinality parameter* $1 \leq k \leq n/2$, *achieves an approximation of* $1/2[1 - (1 - k/n)^{2n/k}] - o(1)$ *for the problem:* $\max\{f(S) : |S| = k\}$. *If* $k > n/2$, *then the same result holds with the cardinality parameter replaced by* $n - k$.

Due to space constraints, many of the proofs of this section have been omitted. The proof of Theorem 3 is based on similar ideas, and is also omitted. To simplify the proof of Theorem 2, we assume the following reduction was applied.

Reduction 14. *We may assume in the proof of Theorem 2 that* $2k \leq n$.

The algorithm we use to prove Theorem 2 is Algorithm 2. One can think of this algorithm as two synchronized instances of Algorithm 1. One instance starts with the solution $\mathbf{1}_{\varnothing}$ and looks for a solution obeying the constraint $\sum_{u \in \mathcal{N}} x_u \leq k$. The other instance starts with the solution $\mathbf{1}_{\mathcal{N}}$ and looks for a solution obeying the constraint $\sum_{u \in \mathcal{N}} x_u \geq k$ (alternatively, we can think of the second instance as having the objective $\bar{f}$ and the constraint $\sum_{u \in \mathcal{N}} x_u \leq n - k$). The two instances are synchronized in two senses:

- In each iteration, the two instances choose direction vectors I^1 and I^2 obeying $I^1 + I^2 = \mathbf{1}_{\mathcal{N}}$ (i.e., the direction vector of one instance implies the direction vector of the other instance).
- The direction vectors are selected in a way that improves the solutions of both instances.

Algorithm 2. Double Measured Continuous Greedy($f, \mathcal{N}, k$)

```
// Initialization
```
1 Set: $T \leftarrow -n/k \cdot \ln(1 - k/n + n^{-4})$ and $\delta \leftarrow T(\lceil n^5 T \rceil)^{-1}$.
2 Initialize: $t \leftarrow 0$, $y^1(0) \leftarrow \mathbf{1}_\varnothing$ and $y^2(0) \leftarrow \mathbf{1}_\mathcal{N}$.

```
// Main loop
```
3 **while** $t < T$ **do**
4   **foreach** $u \in \mathcal{N}$ **do**
5   Let $w_u^1(t) \leftarrow F(y^1(t) \vee \mathbf{1}_u) - F(y^1(t))$ and
        $w_u^2(t) \leftarrow F(y^2(t) \wedge \mathbf{1}_{\mathcal{N}-u}) - F(y^2(t))$.
6   Let $I^1(t) \in [0,1]^\mathcal{N}$ and $I^2(t) \in [0,1]^\mathcal{N}$ be two vectors maximizing

$$\min\{I^1(t) \cdot w^1(t) + 2 \cdot F(y^1(t)), I^2(t) \cdot w^2(t) + 2 \cdot F(y^2(t))\}$$

     among the vectors obeying $|I^1(t)| = k$, $|I^2(t)| = n - k$ and
     $I^1(t) + I^2(t) = \mathbf{1}_\mathcal{N}$.
7   **foreach** $u \in \mathcal{N}$ **do**
8   Let $y_u^1(t + \delta) \leftarrow y_u^1(t) + \delta I_u^1(t) \cdot (1 - y_u^1(t))$ and
        $y_u^2(t + \delta) \leftarrow y_u^2(t) - \delta I_u^2(t) \cdot y_u^2(t)$.
9   **foreach** $u \in \mathcal{N}$ **do**
10   **if** $\partial_u F(y^1(t + \delta)) < 0$ **then** $y_u^1(t + \delta) \leftarrow 0$.
11   **if** $\partial_u F(y^2(t + \delta)) > 0$ **then** $y_u^2(t + \delta) \leftarrow 1$.
12      $t \leftarrow t + \delta$.

13 **if** $|y^1(T)| = |y^2(T)|$ **then** **return** $y^1(T)$.
14 **else** **return** $y^1(T) \cdot \frac{|y^2(T)| - k}{|y^2(T)| - |y^1(T)|} + y^2(T) \cdot \frac{k - |y^1(T)|}{|y^2(T)| - |y^1(T)|}$.

In Algorithm 2, and throughout this section, given a vector x, we denote $|x| = \sum_{u \in \mathcal{N}} x_u$. The output of Algorithm 2 is a fractional solution. This solution can be rounded into an integral solution using a standard rounding procedure such as pipage rounding [3].

We begin the analysis of Algorithm 2 by observing that its Line 6 can be implemented efficiently using an LP solver. Hence, the algorithm has a polynomial time complexity. The following lemma follows from Lemma 9.

Lemma 15. *For every time $0 \le t \le T$, the vectors $y^1(t)$ and $y^2(t)$ obey:*

- $y^1(t), y^2(t) \in [0,1]^\mathcal{N}$.
- $y^1(t) \le y^2(t)$ *(element-wise).*
- $|y^1(t)| \le k \le |y^2(t)|$.

As a corollary of Lemma 15, we can guarantee feasibility. Let y be the vector produced by Algorithm 2.

Corollary 16. *y is a feasible solution.*

Our next objective is lower bounding the value $F(y)$ in terms of $F(y^1(T))$ and $F(y^2(T))$. Let $r \colon [0,1] \to \mathbb{R}^+$ be the function:

$$r(x) = F(y^1(T) + x(y^2(T) - y^1(T))) \ .$$

Intuitively, $r(x)$ evaluates F on a vector that changes from $y^1(T)$ to $y^2(T)$ as x increases. The following observation follows from the submodularity of f.

Observation 17. *r is a non-negative concave function.*

Corollary 18. $F(y) \geq \min\{F(y^1(T)), F(y^2(T))\}$.

Proof. If $|y^1(T)| = |y^2(T)|$ then $y = y^1(T)$, which makes the corollary trivial. Thus, we may assume from now on: $|y^1(T)| \neq |y^2(T)|$. Observe that in this case:

$$F(y) = F\left(y^1(T) \cdot \frac{|y^2(T)| - k}{|y^2(T)| - |y^1(T)|} + y^2(T) \cdot \frac{k - |y^1(T)|}{|y^2(T)| - |y^1(T)|}\right)$$
$$= r\left(\frac{k - |y^1(T)|}{|y^2(T)| - |y^1(T)|}\right) \ .$$

Notice that $k - |y^1(T)| \in [0, 1]$. Thus, the concavity of r implies:

$$F(y) = r\left(\frac{k - |y^1(T)|}{|y^2(T)| - |y^1(T)|}\right) \geq \min\{r(0), r(1)\} = \min\{F(y^1(T)), F(y^2(T))\} \ .$$

$\square$

The proof of Theorem 2 now boils down to lower bounding the expression $\min\{F(y^1(T)), F(y^2(T))\}$. The following lemma is a counter-part of Lemma III.2 of [11]. Let $\Delta(t) = \min\{F(y^1(t) \vee \mathbf{1}_{OPT}) + F(y^1(t)), F(y^2(t) \wedge \mathbf{1}_{OPT}) + F(y^2(t))\}$.

Lemma 19. *For every time $0 \leq t < T$:*

$$\sum_{u \in \mathcal{N}} (1 - y^1_u(t)) \cdot I^1_u(t) \cdot \partial_u F(y^1(t)) + 2 \cdot F(y^1(t)) \geq \Delta(t) \ ,$$

$$-\sum_{u \in \mathcal{N}} y^2_u(t) \cdot I^2_u(t) \cdot \partial_u F(y^2(t)) + 2 \cdot F(y^2(t)) \geq \Delta(t) \ .$$

We also need the following technical lemma.

Lemma 20. $f(OPT) \geq \max_{u \in \mathcal{N}} f(\{u\})/2$.

Corollary 21. *For every time $0 \leq t < T$,*

$$F(y^1(t + \delta)) - F(y^1(t)) \geq \delta \cdot [\Delta(t) - 2 \cdot F(y^1(t))] - O(n^3 \delta^2) \cdot f(OPT) \ ,$$

$$F(y^2(t + \delta)) - F(y^2(t)) \geq \delta \cdot [\Delta(t) - 2 \cdot F(y^2(t))] - O(n^3 \delta^2) \cdot f(OPT) \ .$$

Proof. Lemmata 8, 19 and 20 imply:

$$F(y^1(t + \delta)) - F(y^1(t)) \geq \delta \cdot [\Delta(t) - 2 \cdot F(y(t))] - O(n^3 \delta^2) \cdot f(OPT) \ ,$$

where $y^1(t + \delta)$ represents its value at the beginning of the loop starting on Line 9. The first part of the corollary now follows by noticing that the last loop can only increase $F(y^1(t + \delta))$. The second part of the corollary is analogous. $\square$

To make the lower bounds given in the above lemma useful, we need a lower bound on $\Delta(t)$. This lower bound is obtained using Lemma 6 and the observation that the loop starting on Line 9 of Algorithm 2 guarantees that the conditions of Lemma 6 hold.

Lemma 22. *For every time $0 \le t < T$, $\Delta(t) \ge f(OPT)$.*

Corollary 21 and Lemma 22 imply together the following counterpart of Corollary 12.

Corollary 23. *For every time $0 \le t < T$,*

$$F(y^1(T + \delta)) - F(y^1(T)) \ge \delta \cdot [f(OPT) - 2 \cdot F(y^1(t))] - O(n^3\delta^2) \cdot f(OPT) \ ,$$

$$F(y^2(T + \delta)) - F(y^2(T)) \ge \delta \cdot [f(OPT) - 2 \cdot F(y^2(t))] - O(n^3\delta^2) \cdot f(OPT) \ .$$

Repeating the same line of arguments used in Sect. 3, the previous corollary implies:

Lemma 24. $F(y^1(T)) \ge \frac{1}{2} \cdot [1 - e^{-2T} - o(1)] \cdot f(OPT)$ and $F(y^2(T)) \ge \frac{1}{2} \cdot [1 - e^{-2T} - o(1)] \cdot f(OPT)$.

We are now ready to prove the approximation ratio guaranteed by Theorem 2.

Proof (Approximation Ratio of Theorem 2). By Corollary 18 and Lemma 24, the approximation ratio of Algorithm 2, up to an error term of $o(1)$, is at least:

$$\frac{1 - e^{-2[-(n/k)\cdot\ln(1-k/n+n^{-4})]}}{2} = \frac{1 - (1 - k/n)^{2n/k} \cdot \left(1 + \frac{n^{-4}}{1-k/n}\right)^{2n/k}}{2}$$

$$\ge \frac{1 - (1 - k/n)^{2n/k} \cdot e^{4n^{-3}k^{-1}}}{2} = \frac{1 - (1 - k/n)^{2n/k}}{2} - o(1) \ . \qquad \square$$

Acknowledgements. This work has been supported in part by ERC Starting Grant 335288-OptApprox.

References

1. Buchbinder, N., Feldman, M., Naor, J.S., Schwartz, R.: A tight linear time (1/2)-approximation for unconstrained submodular maximization. In: 53rd Annual IEEE Symposium on Foundations of Computer Science, pp. 649–658. IEEE Computer Society (2012)
2. Buchbinder, N., Feldman, M., Naor, J.S., Schwartz, R.: Submodular maximization with cardinality constraints. In: The Twenty-Fifth Annual ACM-SIAM Symposium on Discrete Algorithms, pp. 1433–1452. SIAM (2014)
3. Calinescu, G., Chekuri, C., Pal, M., Vondrák, J.: Maximizing a monotone submodular function subject to a matroid constraint. SIAM Journal on Computing 40(6), 1740–1766 (2011)

4. Chekuri, C., Vondrák, J., Zenklusen, R.: Multi-budgeted matchings and matroid intersection via dependent rounding. In: The Twenty-Second Annual ACM-SIAM Symposium on Discrete Algorithms, pp. 1080–1097. SIAM (2011)
5. Chekuri, C., Vondrák, J., Zenklusen, R.: Submodular function maximization via the multilinear relaxation and contention resolution schemes. SIAM J. Comput. 43(6), 1831–1879 (2014)
6. Devanur, N.R., Dughmi, S., Schwartz, R., Sharma, A., Singh, M.: On the approximation of submodular functions. CoRR abs/1304.4948 (2013)
7. Dughmi, S.: Submodular functions: Extensions, distributions, and algorithms. A survey. CoRR abs/0912.0322 (2009)
8. Feige, U., Mirrokni, V.S., Vondrák, J.: Maximizing non-monotone submodular functions. SIAM Journal on Computing 40(4), 1133–1153 (2011)
9. Feldman, M.: Maximization Problems with Submodular Objective Functions. Ph.D. thesis, Computer Science Department, Technion - Israel Institute of Technology (2013)
10. Feldman, M., Naor, J.S., Schwartz, R.: Nonmonotone submodular maximization via a structural continuous greedy algorithm. In: Aceto, L., Henzinger, M., Sgall, J. (eds.) ICALP 2011, Part I. LNCS, vol. 6755, pp. 342–353. Springer, Heidelberg (2011)
11. Feldman, M., Naor, J.S., Schwartz, R.: A unified continuous greedy algorithm for submodular maximization. In: IEEE 52nd Annual Symposium on Foundations of Computer Science, pp. 570–579. IEEE Computer Society (2011)
12. Gharan, S.O., Vondrák, J.: Submodular maximization by simulated annealing. In: The Twenty-Second Annual ACM-SIAM Symposium on Discrete Algorithms, pp. 1098–1117. SIAM (2011)
13. Goemans, M.X., Soto, J.A.: Algorithms for symmetric submodular function minimization under hereditary constraints and generalizations. SIAM J. Discrete Math. 27(2), 1123–1145 (2013)
14. Huang, N., Borodin, A.: Bounds on double-sided myopic algorithms for unconstrained non-monotone submodular maximization. In: Ahn, H.-K., Shin, C.-S. (eds.) ISAAC 2014. LNCS, vol. 8889, pp. 528–539. Springer, Heidelberg (2014)
15. Lee, J., Mirrokni, V.S., Nagarajan, V., Sviridenko, M.: Maximizing non-monotone submodular functions under matroid or knapsack constraints. SIAM Journal on Discrete Mathematics 23(4), 2053–2078 (2010)
16. Lee, J., Sviridenko, M., Vondrák, J.: Submodular maximization over multiple matroids via generalized exchange properties. Math. Oper. Res. 35(4), 795–806 (2010)
17. Nagamochi, H.: Minimum degree orderings. Algorithmica 56(1), 17–34 (2010)
18. Queyranne, M.: Minimizing symmetric submodular functions. Mathematical Programming 82(1–2), 3–12 (1998)
19. Vondrák, J.: Symmetry and approximability of submodular maximization problems. SIAM J. Comput. 42(1), 265–304 (2013)

Approximating LZ77 via Small-Space Multiple-Pattern Matching

Johannes Fischer[1], Travis Gagie[2],*,
Paweł Gawrychowski[3],**, and Tomasz Kociumaka[3],* * *

[1] TU Dortmund, Germany
johannes.fischer@cs.tu-dortmund.de
[2] Helsinki Institute for Information Technology (HIIT),
Department of Computer Science, University of Helsinki, Finland
travis.gagie@cs.helsinki.fi
[3] Institute of Informatics,University of Warsaw, Poland
{gawry,kociumaka}@mimuw.edu.pl

Abstract. We generalize Karp-Rabin string matching to handle multiple patterns in $\mathcal{O}(n \log n + m)$ time and $\mathcal{O}(s)$ space, where n is the length of the text and m is the total length of the s patterns, returning correct answers with high probability. As a prime application of our algorithm, we show how to approximate the LZ77 parse of a string of length n. If the optimal parse consists of z phrases, using only $\mathcal{O}(z)$ working space we can return a parse consisting of at most $2z$ phrases in $\mathcal{O}(n \log n)$ time, and a parse of at most $(1 + \varepsilon)z$ phrases in $\mathcal{O}(n \log^2 n)$ for any constant $\varepsilon > 0$. As previous quasilinear-time algorithms for LZ77 use $\Omega(n/\mathrm{poly} \log n)$ space, but z can be exponentially small in n, these improvements in space consumption are substantial.

1 Introduction

Multiple-pattern matching, the task of locating the occurrences of s patterns of total length m in a single text of length n, is a fundamental problem in the field of string algorithms. The algorithm by Aho and Corasick [2] solves it using $\mathcal{O}(n + m)$ time and $\mathcal{O}(m)$ working space in addition to the space needed for the text. To list all occ occurrences rather than, e.g., the leftmost ones, extra $\mathcal{O}(\mathrm{occ})$ time is necessary. When the space is limited, we can use a compressed Aho-Corasick automaton [12]. In extreme cases, one could apply a linear-time constant-space single-pattern matching algorithm sequentially for each pattern in turn, at the cost of increasing the running time to $\mathcal{O}(n \cdot s + m)$. Well-known examples of such algorithms include those by Galil and Seiferas [9], Crochemore and Perrin [5], and Karp and Rabin [14] (see [3] for a recent survey).

* Supported by Academy of Finland grant 268324.
** Work done while the author held a post-doctoral position at Warsaw Center of Mathematics and Computer Science.
* * * Supported by Polish budget funds for science in 2013-2017 as a research project under the 'Diamond Grant' program.

N. Bansal and I. Finocchi (Eds.): ESA 2015, LNCS 9294, pp. 533–544, 2015.
DOI: 10.1007/978-3-662-48350-3_45

It is easy to generalize Karp-Rabin matching to handle multiple patterns in $\mathcal{O}(n+m)$ expected time and $\mathcal{O}(s)$ working space provided that all patterns are of the same length [11]. To do this, we store the fingerprints of the patterns in a hash table, and then slide a window over the text maintaining the fingerprint of the fragment currently in the window. The hash table lets us check if that fragment is an occurrence of a pattern. If so, we report it and update the hash table so that every pattern is returned at most once. This is a very simple and actually applied idea [1], but it is not clear how to extend it for patterns with many distinct lengths. In this paper we develop a method which works for any set of patterns in $\mathcal{O}(n \log n + m)$ time and $\mathcal{O}(s)$ working space. We assume that read-only random access to the text and the patterns is available throughout the algorithm, and we do not include their sizes while measuring space consumption.

In a very recent independent work Clifford et al. [4] gave a dictionary matching algorithm in the streaming model. In this setting the patterns and later the text are scanned once only (as opposed to read-only random access) and an occurrence needs to be reported immediately after its last character is read. The algorithm presented in [4] uses $\mathcal{O}(s \log \ell)$ space and takes $\mathcal{O}(\log \log(s + \ell))$ time per character where ℓ is the length of the longest pattern ($\frac{m}{s} \leq \ell \leq m$).

As a prime application of our dictionary-matching algorithm, we show how to approximate the Lempel-Ziv 77 (LZ77) parse [19] of a text of length n using working space proportional to the number of phrases (again, we assume read-only random access to the text). Computing the LZ77 parse in small space is an issue of high importance, with space being a frequent bottleneck of today's systems. Moreover, LZ77 is useful not only for data compression, but also as a way to speed up algorithms [16]. We present two solutions to this problem, both of which work in $\mathcal{O}(z)$ space for inputs admitting LZ77 parsing with z phrases. The first one produces a parse consisting of at most $2z$ phrases in $\mathcal{O}(n \log n)$ time, while the other for any positive $\varepsilon < 1$ in $\mathcal{O}(\varepsilon^{-1} n \log^2 n)$ time generates a factorization with no more than $(1 + \varepsilon)z$ phrases.

To the best of our knowledge, approximating LZ77 factorization in small space has not been considered before, and our algorithm is significantly more efficient than methods producing the exact answer. A recent sublinear-space algorithm, due to Kärkkäinen et al. [13], runs in $\mathcal{O}(nd)$ time and uses $\mathcal{O}(n/d)$ space for any parameter d. An earlier online solution by Gasieniec et al. [10] uses $\mathcal{O}(z)$ space and takes $\mathcal{O}(z^2 \log^2 z)$ time for each character appended. Other previous methods use significantly more space when the parse is small relative to n; see [8] for a recent discussion.

Structure of the Paper. Sect. 2 introduces terminology and recalls several known concepts. This is followed by the description of our pattern-matching algorithm. In Sect. 3 we show how to process patterns of length at most s and in Sect. 4 we handle longer patterns, with different procedures for repetitive and non-repetitive ones. Finally, in Sect. 5, we apply the dictionary matching algorithm to construct the approximations of the LZ77 parsing. Selected details and proofs, omitted in this extended abstract due to space restrictions, can be found in the full version of the paper [7].

Model of Computation. Our algorithms are designed for the word-RAM with $\Omega(\log n)$-bit words and assume integer alphabet of polynomial size. The usage of Karp-Rabin fingerprints makes them randomized (Monte Carlo). A correct answer is returned with high probability, i.e., the error probability is inverse polynomial with respect to input size, where the degree of the polynomial can be set arbitrarily large. In the full version of the paper we show how to turn them into Las Vegas algorithms (where the time bounds hold with high probability).

2 Preliminaries

We consider finite words over an integer alphabet $\Sigma = \{0, \ldots, \sigma - 1\}$, where $\sigma = \text{poly}(n + m)$. For a word $w = w[1] \ldots w[n] \in \Sigma^n$, we define the *length* of w as $|w| = n$. For $1 \leq i \leq j \leq n$, a word $u = w[i] \ldots w[j]$ is called a *subword* of w. By $w[i..j]$ we denote the occurrence of u at position i, called a *fragment* of w. A fragment with $i = 1$ is called a *prefix* (and often denoted $w[..j]$) and a fragment with $j = n$ is called a *suffix* (and denoted $w[i..]$).

A positive integer p is called a *period* of w whenever $w[i] = w[i + p]$ for $1 \leq i \leq |w| - p$. In this case, the prefix $w[..p]$ is often also called a period of w. The length of the shortest period of a word w is denoted as $\text{per}(w)$. A word w is called *periodic* if $\text{per}(w) \leq \frac{|w|}{2}$ and *highly periodic* if $\text{per}(w) \leq \frac{|w|}{3}$. The well-known periodicity lemma [6] says that if p and q are both periods of w and $p + q \leq |w|$, then $\gcd(p, q)$ is also a period of w.

2.1 Fingerprints

Our randomized construction is based on Karp-Rabin fingerprints; see [14]. Fix a word $w[1..n]$ over an alphabet $\Sigma = \{0, \ldots, \sigma - 1\}$, a constant $c \geq 1$, a prime number $p > \max(\sigma, n^{c+4})$, and choose $x \in \mathbb{Z}_p$ uniformly at random. We define the fingerprint of a subword $w[i..j]$ as $\Phi(w[i..j]) = w[i] + w[i + 1]x + \ldots + w[j]x^{j-i} \bmod p$. With probability at least $1 - \frac{1}{n^c}$, no two distinct subwords of the same length have equal fingerprints. The situation when this happens for some two subwords is called a *false-positive*. From now on when stating the results we assume that there are no false-positives to avoid repeating that the answers are correct with high probability. For pattern matching, we apply this construction for $w = TP_1 \ldots P_s$, the concatenation of the text with the patterns. Fingerprints let us easily locate many patterns of the same length, as mentioned in the introduction. A straightforward solution described in the introduction builds a hash table mapping fingerprints to patterns. However, the construction of such a table could fail with probability $\Theta(\frac{1}{s^c})$ for some constant c. Since we aim at $O(\frac{1}{(n+m)^c})$ error probability, we provide a solution using a deterministic dictionary [17]. Although it adds $\mathcal{O}(s \log s)$ to the time complexity, this extra term becomes dominated in our main results.

Theorem 1. *Given a text T of length n and patterns $P_1, \ldots, P_s$, each of length exactly ℓ, we can compute the the leftmost occurrence of every pattern P_i in T using $\mathcal{O}(n + s\ell + s \log s)$ total time and $\mathcal{O}(s)$ space.*

2.2   Tries

A trie of a collection of patterns $P_1, \ldots, P_s$ is a rooted tree whose nodes correspond to prefixes of the patterns. The root represents the empty word and the edges are labeled with single characters. The node corresponding to a particular prefix is called its *locus*. In a *compacted* trie unary nodes that do not represent any pattern are *dissolved* and the labels of their incidents edges are concatenated. The dissolved nodes are called *implicit* as opposed to the *explicit* nodes, which remain stored. The locus of a string in a compacted trie might therefore be explicit or implicit. All edges outgoing from the same node are stored on a list sorted according to the first character, which is unique among these edges. The labels of edges of a compacted trie are stored as pointers to the respective fragments of patterns P_i. Consequently, a compacted trie can be stored in space proportional to the number of explicit nodes, which is $O(s)$.

Consider two compacted tries $\mathcal{T}_1$ and $\mathcal{T}_2$. We say that (possibly implicit) nodes $v_1 \in \mathcal{T}_1$ and $v_2 \in \mathcal{T}_2$ are *twins* if they are loci of the same string. Note that every $v_1 \in T_1$ has at most one twin $v_2 \in T_2$. The proof of the following result is omitted due to space constraints.

Lemma 2. *Given two compacted tries $\mathcal{T}_1$ and $\mathcal{T}_2$ constructed for s_1 and s_2 patterns, respectively, in $\mathcal{O}(s_1 + s_2)$ total time and space we can find for each explicit node $v_1 \in \mathcal{T}_1$ a node $v_2 \in \mathcal{T}_2$ such that if v_1 has a twin in $\mathcal{T}_2$, then v_2 is the twin. (If v_1 has no twin in $\mathcal{T}_2$, the algorithm returns an arbitrary node $v_2 \in \mathcal{T}_2$).*

3   Short Patterns

To handle the patterns of length not exceeding a given threshold ℓ, we first build a compacted trie for those patterns. Construction is easy if the patterns are sorted lexicographically: we insert them one by one into the compacted trie first naively traversing the trie from the root, then potentially partitioning one edge into two parts, and finally adding a leaf if necessary. Thus, the following result suffices to efficiently build the tries.

Lemma 3. *One can lexicographically sort strings $P_1, \ldots, P_s$ of total length m in $\mathcal{O}(m + \sigma^\varepsilon)$ time using $\mathcal{O}(s)$ space, for any $\varepsilon > 0$.*

Next, we partition T into $\mathcal{O}(\frac{n}{\ell})$ overlapping blocks $T_1 = T[1..2\ell]$, $T_2 = T[\ell + 1..3\ell]$, $T_3 = T[2\ell + 1..4\ell], \ldots$. Notice that each subword of length at most ℓ is completely contained in some block. Thus, we may seek for occurrences of the patterns in each block separately.

The suffix tree of each block T_i takes $\mathcal{O}(\ell \log \ell)$ time [18] and $\mathcal{O}(\ell)$ space to construct and store (the suffix tree is discarded after processing the block). We apply Lemma 2 to the suffix tree and the compacted trie of patterns; this takes $\mathcal{O}(\ell + s)$ time. For each pattern P_j we obtain a node such that the corresponding subword is equal to P_j provided that P_j occurs in T_i. We compute the leftmost occurrence $T_i[b..e]$ of the subword, which takes constant time if we store additional data at every explicit node of the suffix tree, and then we check whether

$T_i[b..e] = P_j$ using fingerprints. For this, we precompute the fingerprints of all patterns, and for each block T_i we precompute the fingerprints of its prefixes in $\mathcal{O}(\ell)$ time and space, which allows to determine the fingerprint of any of its subwords in constant time.

In total, we spend $\mathcal{O}(m + \sigma^\varepsilon)$ for preprocessing and $\mathcal{O}(\ell \log \ell + s)$ for each block. Since $\sigma = (n+m)^{\mathcal{O}(1)}$, for small enough ε this yields the following result.

Theorem 4. *Given a text T of length n and patterns $P_1, \ldots, P_s$ of total length m, using $\mathcal{O}(n \log \ell + s\frac{n}{\ell} + m)$ total time and $\mathcal{O}(s + \ell)$ space we can compute the leftmost occurrences in T of all patterns P_j of length at most ℓ.*

4   Long Patterns

To handle patterns longer than a certain threshold, we first distribute them into groups according to the value of $\lfloor \log_{4/3} |P_j| \rfloor$. Patterns longer than the text can be ignored, so there are $\mathcal{O}(\log n)$ groups. Each group is handled separately, and from now on we consider only patterns P_j satisfying $\lfloor \log_{4/3} |P_j| \rfloor = i$.

We set $\ell = \lceil (4/3)^{i-1} \rceil$ and define α_j and β_j as, respectively, the prefix and the suffix of length ℓ of P_j. Since $\frac{2}{3}(|\alpha_j| + |\beta_j|) = \frac{4}{3}\ell \geq |P_j|$, the following lemma yields a classification of the patterns into three classes: either P_j is highly periodic, or α_j is not highly periodic, or β_j is not highly periodic.

Lemma 5. *Suppose x and y are, respectively, a prefix and a suffix of a word w such that $|w| \leq \frac{2}{3}(|x| + |y|)$. If w is not highly periodic, then x or y is not highly periodic.*

The intuition behind this classification is as follows. If the prefix or the suffix is not repetitive, then we will not see it many times in a short subword of the text. On the other hand, if both the prefix and suffix are repetitive, then there is some structure that we can take advantage of.

To assign every pattern to the appropriate class, we compute the periods of P_j, α_j and β_j. For this we use the following lemma; a very similar result can be found in [15].

Lemma 6. *Given a read-only string w one can decide in $\mathcal{O}(|w|)$ time and constant space if w is periodic and if so, compute $\mathrm{per}(w)$.*

4.1   Patterns without Long Highly Periodic Prefix

Below we show how to deal with patterns with non-highly periodic prefixes α_j. Patterns with non-highly periodic suffixes β_j can be processed using the same method after reversing the text and the patterns.

Lemma 7. *Let ℓ be an arbitrary integer. Suppose we are given a text T of length n and patterns $P_1, \ldots, P_s$ such that for $1 \leq j \leq s$ we have $\ell \leq |P_j| < \frac{4}{3}\ell$ and $\alpha_j = P_j[1..\ell]$ is not highly periodic. We can compute the leftmost and the rightmost occurrence of each pattern P_j in T in $\mathcal{O}(n + s(1 + \frac{n}{\ell})\log s + s\ell)$ total time using $\mathcal{O}(s)$ space.*

The algorithm scans the text T with a sliding windows of length ℓ. Whenever it encounters a subword equal to the prefix α_j of some P_j, it creates a *request* to verify whether the corresponding suffix β_j of length ℓ occurs at the respective position. The request is processed when the sliding window reaches that position. This way the algorithm detects the occurrences of all the patterns. In particular, we may store the leftmost and rightmost of occurrences of each pattern.

We use the fingerprints to compare the subwords of T with α_j and β_j. To this end, we precompute $\Phi(\alpha_j)$ and $\Phi(\beta_j)$ for each j. We also build a deterministic dictionary $\mathcal{D}$ [17] with an entry mapping $\Phi(\alpha_j)$ to j for every pattern (if there are multiple patterns with the same value of $\Phi(\alpha_j)$, the dictionary maps a fingerprint to a list of indices). These steps take $\mathcal{O}(s\ell)$ and $\mathcal{O}(s \log s)$, respectively. Pending requests are maintained in a priority queue $\mathcal{Q}$, implemented using a binary heap[1] as pairs containing the pattern index (as a value) and the position where the occurrence of β_j is anticipated (as a key).

Algorithm 1. Processing phase for patterns with non-highly periodic α_j.

---

1 **for** $i = 1$ **to** $n - \ell + 1$ **do**
2 $\quad$ $h := \Phi(w[i..i + \ell - 1])$
3 $\quad$ **foreach** $j : \Phi(\alpha_j) = h$ **do**
4 $\quad\quad$ add a request $(i + |P_j| - \ell, j)$ to $\mathcal{Q}$
5 $\quad$ **foreach** *request* $(i, j) \in \mathcal{Q}$ *at position* i **do**
6 $\quad\quad$ **if** $h = \Phi(\beta_j)$ **then**
7 $\quad\quad\quad$ report an occurrence of P_j at $i + \ell - |P_j|$
8 $\quad\quad$ remove (i, j) from $\mathcal{Q}$

Algorithm 1 provides a detailed description of the processing phase. Let us analyze its time and space complexities. Due to the properties of Karp-Rabin fingerprints, line 2 can be implemented in $\mathcal{O}(1)$ time. Also, the loops in lines 3 and 5 takes extra $\mathcal{O}(1)$ time even if the respective collections are empty. Apart from these, every operation can be assigned to a request, each of them taking $\mathcal{O}(1)$ (lines 3 and 5-6) or $\mathcal{O}(\log |\mathcal{Q}|)$ (lines 4 and 8) time. To bound $|\mathcal{Q}|$, we need to look at the maximum number of pending requests.

Fact 8. *For any pattern P_j there are $\mathcal{O}(1 + \frac{n}{\ell})$ requests and at any time at most one of them is pending.*

Proof. Note that there is a one-to-one correspondence between requests concerning P_j and the occurrences of α_j in T. The distance between two such occurrences must be at least $\frac{1}{3}\ell$, because otherwise the period of α_j would be at most $\frac{1}{3}\ell$, thus making α_j highly periodic. This yields the $\mathcal{O}(1 + \frac{n}{\ell})$ upper bound on the total number of requests. Additionally, any request is pending for at most $|P_j| - \ell < \frac{1}{3}\ell$ iterations of the main **for** loop. Hence, the request corresponding to an occurrence of α_j is already processed before the next occurrence appears. $\qquad\square$

---

[1] Hash tables could be used instead of the heap and the deterministic dictionary. Although this would improve the time complexity in Theorem 7, the running time of the algorithm in Thm. 11 would not change and failures with probability inverse polynomial with respect to s would be introduced; see also a discussion before Thm. 1.

Hence, the scanning phase uses $\mathcal{O}(s)$ space and takes $\mathcal{O}(n + s(1 + \frac{n}{\ell}) \log s)$ time. Taking preprocessing into account, we obtain bounds claimed in Lemma 7.

4.2   Highly Periodic Patterns

Lemma 9. *Let ℓ be an arbitrary integer. Given a text T of length n and a collection of highly periodic patterns $P_1, \ldots, P_s$ such that for $1 \leq j \leq s$ we have $\ell \leq |P_j| < \frac{4}{3}\ell$, we can compute the leftmost occurrence of each pattern P_j in T using $\mathcal{O}(n + s(1 + \frac{n}{\ell}) \log s + s\ell)$ total time and $\mathcal{O}(s)$ space.*

Proof (sketch). The solution is basically the same as in the proof of Lemma 7, except that the algorithm ignores certain *shiftable* occurrences. An occurrence of x at position i of T is called *shiftable* if there is another occurrence of x at position $i - \mathrm{per}(x)$. The remaining occurrences are called *non-shiftable*. Notice that the leftmost occurrence is always non-shiftable, so indeed we can safely ignore some of the shiftable occurrences of the patterns. Because $2 \, \mathrm{per}(P_j) \leq \frac{2}{3}|P_j| \leq \frac{8}{9}\ell < \ell$, the following claim implies that if an occurrence of P_j is non-shiftable, then the occurrence of α_j at the same position is also non-shiftable.

Claim. Let y be a prefix of x such that $|y| \geq 2 \, \mathrm{per}(x)$. Suppose x has a non-shiftable occurrence at position i in w. Then the occurrence of y at position i is also non-shiftable.

The proof of the claim, implementation details of shiftable occurrences detection, and the running time analysis are provided in the full version of the paper.   □

4.3   Summary

Theorem 10. *Given a text T of length n and patterns $P_1, \ldots, P_s$ of total length m, using $\mathcal{O}(n \log n + m + s\frac{n}{\ell} \log s)$ total time and $\mathcal{O}(s)$ space we can compute the leftmost occurrences in T of all patterns P_j of length at least ℓ.*

Proof. The algorithm distributes the patterns into $\mathcal{O}(\log n)$ groups according to their lengths, and then into three classes according to their repetitiveness, which takes using $\mathcal{O}(m)$ time and $\mathcal{O}(s)$ space in total. Then, it applies either Lemma 7 or Lemma 9 on every class. It remains to show that the running times of all those calls sum up to the claimed bound. Each of them can be seen as $\mathcal{O}(n)$ plus $\mathcal{O}(|P_j| + (1 + \frac{n}{|P_j|}) \log s)$ per every pattern P_j. Because $\ell \leq |P_j| \leq n$ and there are $\mathcal{O}(\log n)$ groups, this sums up to $\mathcal{O}(n \log n + m + s\frac{n}{\ell} \log s)$.   □

Using Thm. 4 for all patterns of length at most $\min(n, s)$, and (if $s \leq n$) Thm. 10 for patterns of length at least s, we obtain our main theorem.

Theorem 11. *Given a text T of length n and patterns $P_1, \ldots, P_s$ total length m, we can compute the leftmost occurrence of every P_i in T using $\mathcal{O}(n \log n + m)$ total time and $\mathcal{O}(s)$ space.*

5 Approximating LZ77 in Small Space

A non-empty fragment $T[i..j]$ is called a *previous fragment* if the corresponding subword occurs in T at a position $i' < i$. A *phrase* is either a previous fragment or a single letter not occurring before in T. The LZ77-factorization of a text $T[1..n]$ is a greedy factorization of T into z *phrases*, $T = f_1 f_2 \ldots f_z$, such that each f_i is as long as possible. To formalize the concept of LZ77-approximation, we first make the following definition.

Definition 12. *Let $w = g_1 g_2 \ldots g_a$ be a factorization of w into a phrases. We call it c-optimal if the fragment corresponding to the concatenation of any c consecutive phrases $g_i \ldots g_{i+c-1}$ is not a previous fragment.*

A c-optimal factorization approximates the LZ77-factorization in the number of factors, as the following observation states. However, the stronger property of c-optimality is itself useful in certain situations.

Observation 13. *If $w = g_1 g_2 \ldots g_a$ is a c-optimal factorization of w into a phrases, and the LZ77-factorization of w consists of z phrases, then $a \leq c \cdot z$.*

5.1 2-Approximation Algorithm

Outline. Our algorithm is divided into three phases, each of which refines the factorization from the previous phase:

Phase 1. Create a factorization of $T[1..n]$ stored implicitly as up to z *chains* consisting of $\mathcal{O}(\log n)$ phrases each.

Phase 2. Try to merge phrases within the chains to produce an $\mathcal{O}(1)$-optimal factorization.

Phase 3. Try to merge adjacent factors as long as possible to produce the final 2-optimal factorization.

Every phase takes $\mathcal{O}(n \log n)$ time and uses $\mathcal{O}(z)$ working space. In the end, we get a 2-approximation of the LZ77-factorization. Phases 1 and 2 use the very simple multiple pattern matching algorithm for patterns of equal lengths developed in Thm. 1, while Phase 3 requires the general multiple pattern matching algorithm obtained in Thm. 11.

Phase 1. To construct the factorization, we imagine creating a binary tree on top the text T of length $n = 2^k$ – see also Fig. 1 (we implicitly pad w with sufficiently many \$'s to make its length a power of 2). The algorithm works in $\log n$ rounds, and the i-th round works on level i of the tree, starting at $i = 1$ (the children of the root). On level i, the tree divides T into 2^i blocks of size $n/2^i$; the aim is to identify previous fragments among these blocks and declare them as phrases. (In the beginning, no phrases exist, so all blocks are unfactored.) To find out if a block is a previous fragment, we use Thm. 1 and test whether the leftmost occurrence of the corresponding subword is the block itself. The

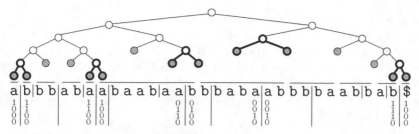

Fig. 1. An illustration of Phase 1 of the algorithm, with the "cherries" depicted in thicker lines. The horizontal lines represent the LZ77-factorization and the vertical lines depict factors induced by the tree. Longer separators are drawn between chains, whose lengths are written in binary with the least significant bits on top.

exploration of the tree is naturally terminated at the nodes corresponding to the previous fragments (or single letters not occurring before), forming the leaves of a (conceptual) binary tree. A pair of leaves sharing the same parent is called a *cherry*. The block corresponding to the common parent is *induced* by the cherry. To analyze the algorithm, we make the following observation:

Fact 14. *A block induced by a cherry is never a previous fragment. Therefore, the number of cherries is at most z.*

Proof. The former part of the claim follows from construction. To prove the latter, observe that the blocks induced by different cherries are disjoint and hence each cherry can be assigned a unique LZ77-factor ending within the block. □

Consequently, while processing level i of the tree, we can afford storing all cherries generated so far on a sorted linked list $\mathcal{L}$. The remaining already generated phrases are not explicitly stored. In addition, we also store a sorted linked list $\mathcal{L}_i$ of all still *unfactored* nodes on the current level i (those for which the corresponding blocks are tested as previous fragments). Their number is bounded by z (because there is a cherry below every node on the list), so the total space is $\mathcal{O}(z)$. Maintaining both lists sorted is easily accomplished by scanning them in parallel with each scan of T, and inserting new cherries/unfactored nodes at their correct places. Furthermore, in the i-th round we apply Thm. 1 to at most 2^i patterns of length $n/2^i$, so the total time is $\sum_{i=1}^{\log n} \mathcal{O}(n + 2^i \log(2^i)) = \mathcal{O}(n \log n)$.

Next, we analyze the structure of the resulting factorization. Let $h_{x-1}h_x$ and $h_y h_{y+1}$ be the two consecutive cherries. The phrases $h_{x+1} \ldots h_{y-1}$ correspond to the right siblings of the ancestors of h_x and to the left siblings of the ancestors of h_y (no further than to the lowest common ancestor of h_x and h_y). This naturally partitions $h_x h_{x+1} \ldots h_{y-1} h_y$ into two parts, called an *increasing chain* and a *decreasing chain* to depict the behaviour of phrase lengths within each part. Observe that these lengths are powers of two, so the structure of a chain of either type is determined by the total length of its phrases, which can be interpreted as a bitvector with bit i' set to 1 if there is a phrase of length $2^{i'}$ in the chain. Those bitvectors can be created while traversing the tree level by level, passing the partially created bitvectors down to the next level $\mathcal{L}_{i+1}$ until finally storing them at the cherries in $\mathcal{L}$.

At the end we obtain a sequence of chains of alternating types, see Fig. 1. Since the structure of each chain follows from its length, we store the sequence of chains rather the actual factorization, which might consist of $\Theta(z \log n) = \omega(z)$ phrases. By Fact 14, our representation uses $\mathcal{O}(z)$ words of space and the last phrase of a decreasing chain concatenated with the first phrase of the consecutive increasing chain never form a previous fragment (these phrases form the block induced by the cherry).

Phase 2. In this phase we merge phrases within the chains. We describe how to process increasing chains; the decreasing are handled, mutatis mutandis, analogously. We partition the phrases $h_\ell \ldots h_r$ within a chain into groups.

For each chain we maintain an *active* group, initially consisting of h_ℓ, and scan the remaining phrases in the left-to-right order. We either append a phrase h_i to the active group g_j, or we output g_j and make $g_{j+1} = h_i$ the new active group. The former action is performed if and only if the fragment of length $2|h_i|$ starting at the same position as g_j is a previous fragment. Having processed the whole chain, we also output the last active group. The proof of the following claim is omitted due to space constraints.

Claim. Within every chain each single group g_j forms a valid phrase, but no concatenation of three adjacent groups $g_j g_{j+1} g_{j+2}$ form a previous fragment.

The procedure described above is executed in parallel for all chains, each of which maintains just the length of its active group. In the i-th round only chains containing a phrase of length 2^i participate (we use bit operations to verify which chains have length containing 2^i in the binary expansion). These chains provide fragments of length 2^{i+1} and Thm. 1 is applied to decide which of them are previous fragments. The chains modify their active groups based on the answers; some of them may output their old active groups. These groups form phrases of the output factorization, so the space required to store them is amortized by the size of that factorization. As far as the running time is concerned, we observe that in the i-th round no more than $\min(z, \frac{n}{2^i})$ chains participate. Thus, the total running time is $\sum_{i=1}^{\log n} \mathcal{O}(n + \frac{n}{2^i} \log \frac{n}{2^i}) = \mathcal{O}(n \log n)$.

Fact 14 combined with the claim above yields 5-optimality of the resulting factorization (see the full version for details).

Phase 3. This phase uses an algorithm which given a c-optimal factorization, computes a 2-optimal factorization using $\mathcal{O}(c \cdot n \log n)$ time and $\mathcal{O}(c \cdot z)$ space.

The procedure consists of c iterations. In every iteration we first detect previous fragments corresponding to concatenations of two adjacent phrases. The total length of the patterns is up to $2n$, so this takes $\mathcal{O}(n \log n + m) = \mathcal{O}(n \log n)$ time and $\mathcal{O}(c \cdot z)$ space using Thm. 11. Next, we scan through the factorization and merge every phrase g_i with the preceding phrase g_{i-1} if $g_{i-1} g_i$ is a previous fragment and g_{i-1} has not been just merged with its predecessor.

We shall prove that the resulting factorization is 2-optimal. Consider a pair of adjacent phrases $g_{i-1} g_i$ in the final factorization and let j be the starting position of g_i. Suppose $g_{i-1}g_i$ is a previous fragment. Our algorithm performs merges only, so the phrase ending at position $j - 1$ concatenated with the phrase starting at position j formed a previous fragment at every iteration. The only reason that these factors were not merged could be another merge of the former factor. Consequently, the factor ending at position $j - 1$ took part in a merge at every iteration, i.e., g_{i-1} is a concatenation of at least c phrases of the input factorization. However, all the phrases created by the algorithm form previous fragments, which contradicts the c-optimality of the input factorization.

5.2   Approximation Scheme

Lemma 15. *Given a text T of length n and its disjoint fragments $T[b_1..e_1], \ldots,$ $T[b_s..e_s]$ (represented by positions), using $\mathcal{O}(n \log^2 n)$ total time and $\mathcal{O}(s)$ space we can find for each j the longest previous fragment being a prefix of $T[b_j..e_j]$.*

Proof. For every j, we use binary search to compute the length ℓ_j of the resulting previous fragment. All binary searches are performed in parallel. That is, we proceed in $\mathcal{O}(\log n)$ phases, and in every phase we check, for every j, if $T[b_j..b_j + \ell_j - 1]$ is a previous fragment using Thm. 11, and then update every ℓ_j accordingly. The space complexity is clearly $\mathcal{O}(s)$ and, because the considered fragments are disjoint, the running time is $\mathcal{O}(n \log n)$ per phase.   □

The starting point is a 2-optimal factorization into a phrases, which can be found in $\mathcal{O}(n \log n)$ time using the previous method. The text is partitioned into $\frac{\varepsilon}{2}a$ blocks consisting of $\frac{2}{\varepsilon}$ consecutive phrases. For each block we construct the greedy factorization: we iteratively partition the block from left to right into maximal phrases. In each step we need to determine the longest previous fragment being a prefix of the part of the block which has not been factorized yet. For this we use Lemma 15 for the unfactored parts, which requires $\mathcal{O}(n \log n)$ time per iteration. The number of iterations is $\mathcal{O}(\frac{1}{\varepsilon})$ since every block can be surely factorized into $\frac{2}{\varepsilon}$ phrases and the greedy factorization is optimal. To bound the approximation guarantee, observe that every phrase inside the block, except possibly for the last one, contains an endpoint of a phrase in the LZ77-factorization. Consequently, the total number of phrases is at most $z + \frac{\varepsilon}{2}a \leq (1 + \varepsilon)z$.

Theorem 16. *Given a text T of length n whose LZ77-factorization consists of z phrases, we can factorize T into at most $2z$ phrases using $\mathcal{O}(n \log n)$ time and $\mathcal{O}(z)$ space. Moreover, for any positive $\varepsilon < 1$ in $\mathcal{O}(\varepsilon^{-1}n \log^2 n)$ time and $\mathcal{O}(z)$ space we can compute a factorization into no more than $(1 + \varepsilon)z$ phrases.*

Epilogue. In the full version of this paper [7] we extend Thm. 11 so that if P_j has no occurrences in T, the algorithm finds (the leftmost occurrence of) the longest prefix of P_j which occurs in T. The same idea lets us obtain an $\mathcal{O}(n \log n)$ time bound in Theorem 15. Hence, in Thm. 16 the running time of the algorithm computing a factorization with $(1 + \varepsilon)z$ phrases becomes $\mathcal{O}(\varepsilon^{-1}n \log n)$.

Acknowledgments. The authors would like to thank the participants of the Stringmasters 2015 workshop in Warsaw, where this work was initiated. We are particularly grateful to Marius Dumitran, Artur Jeż, and Patrick K. Nicholson.

References

1. Rabin-Karp algorithm — Wikipedia, The Free Encyclopedia.
   http://en.wikipedia.org/w/index.php?title=
   Rabin-Karp_algorithm&oldid=665980736
2. Aho, A.V., Corasick, M.J.: Efficient string matching: An aid to bibliographic search.
   Commun. ACM 18(6), 333–340 (1975)
3. Breslauer, D., Grossi, R., Mignosi, F.: Simple real-time constant-space string
   matching. Theor. Comput. Sci. 483, 2–9 (2013)
4. Clifford, R., Fontaine, A., Porat, E., Sach, B., Starikovskaya, T.: Dictionary matching in a stream. In: Bansal, N., Finocchi, I. (eds.) ESA 2015. LNCS, vol. 8737,
   pp. ??–?? Springer, Heidelberg (2015)
5. Crochemore, M., Perrin, D.: Two-way string matching. J. ACM 38(3), 651–675
   (1991)
6. Fine, N.J., Wilf, H.S.: Uniqueness theorems for periodic functions. P. Am. Math.
   Soc. 16(1), 109–114 (1965)
7. Fischer, J., Gagie, T., Gawrychowski, P., Kociumaka, T.: Approximating LZ77 via
   small-space multiple-pattern matching. CoRR abs/1504.06647 (2015)
8. Fischer, J., I, T., Köppl, D.: Lempel Ziv Computation in Small Space (LZ-CISS).
   In: Cicalese, F., Porat, E., Vaccaro, U. (eds.) CPM 2015. LNCS, vol. 9133, pp.
   172–184. Springer, Heidelberg (2015)
9. Galil, Z., Seiferas, J.I.: Time-space-optimal string matching. J. Comput. Syst.
   Sci. 26(3), 280–294 (1983)
10. Gasieniec, L., Karpinski, M., Plandowski, W., Rytter, W.: Efficient algorithms for
    Lempel-Ziv encoding (extended abstract). In: Karlsson, R., Lingas, A. (eds.) SWAT
    1996. LNCS, vol. 1097, pp. 392–403. Springer, Heidelberg (1996)
11. Gum, B., Lipton, R.J.: Cheaper by the dozen: Batched algorithms. In: Kumar, V.,
    Grossman, R.L. (eds.) SDM 2001, pp. 1–11. SIAM, Philadelphia (2001)
12. Hon, W., Ku, T., Shah, R., Thankachan, S.V., Vitter, J.S.: Faster compressed
    dictionary matching. Theor. Comput. Sci. 475, 113–119 (2013)
13. Kärkkäinen, J., Kempa, D., Puglisi, S.J.: Lightweight Lempel-Ziv parsing. In:
    Bonifaci, V., Demetrescu, C., Marchetti-Spaccamela, A. (eds.) SEA 2013. LNCS,
    vol. 7933, pp. 139–150. Springer, Heidelberg (2013)
14. Karp, R.M., Rabin, M.O.: Efficient randomized pattern-matching algorithms. IBM
    J. Res. Dev. 31(2), 249–260 (1987)
15. Kociumaka, T., Starikovskaya, T., Vildhøj, H.W.: Sublinear space algorithms for
    the longest common substring problem. In: Schulz, A.S., Wagner, D. (eds.) ESA
    2014. LNCS, vol. 8737, pp. 605–617. Springer, Heidelberg (2014)
16. Lohrey, M.: Algorithmics on SLP-compressed strings: A survey. Groups Complexity
    Cryptology 4(2), 241–299 (2012)
17. Ružić, M.: Constructing efficient dictionaries in close to sorting time. In: Aceto, L.,
    Damgård, I., Goldberg, L.A., Halldórsson, M.M., Ingólfsdóttir, A., Walukiewicz,
    I. (eds.) ICALP 2008, Part I. LNCS, vol. 5125, pp. 84–95. Springer, Heidelberg
    (2008)
18. Ukkonen, E.: On-line construction of suffix trees. Algorithmica 14(3), 249–260
    (1995)
19. Ziv, J., Lempel, A.: A universal algorithm for sequential data compression. IEEE
    Trans. Inform. Theory 23(3), 337–343 (1977)

Fast Algorithms for Parameterized Problems with Relaxed Disjointness Constraints[*]

Ariel Gabizon[1], Daniel Lokshtanov[2], and Michał Pilipczuk[3]

[1] Department of Computer Science, Technion, Israel
`ariel.gabizon@gmail.com`
[2] Department of Informatics, University of Bergen, Norway
`daniello@ii.uib.no`
[3] Institute of Informatics, University of Warsaw, Poland
`michal.pilipczuk@mimuw.edu.pl`

Abstract. In this paper we consider generalized versions of four well-studied problems in parameterized complexity and exact exponential time algorithms: k-PATH, SET PACKING, MULTILINEAR MONOMIAL TESTING and HAMILTONIAN PATH. The generalization is in every case obtained by introducing a *relaxation parameter*, which relaxes the constraints on feasible solutions. For example, the k-PATH problem is generalized to r-SIMPLE k-PATH where the task is to find a walk of length k that never visits any vertex more than r times. This problem was first considered by Abasi et al. [1]. HAMILTONIAN PATH is generalized to DEGREE BOUNDED SPANNING TREE, where the input is a graph G and integer d, and the task is to find a spanning tree T of G such that every vertex has degree at most d in T.

The generalized problems can easily be shown to be NP-complete for every fixed value of the relaxation parameter. On the other hand, we give algorithms for the generalized problems whose worst-case running time (a) *matches* the running time of the best algorithms for the original problems up to constants in the exponent, and (b) *improves* significantly as the relaxation parameter increases. For example, we give a deterministic algorithm with running time $O^*(2^{O(k\frac{\log r}{r})})$ for r-SIMPLE k-PATH matching up to a constant in the exponent the randomized algorithm of Abasi et al. [1], and a randomized algorithm with running time $O^*(2^{O(n\frac{\log d}{d})})$ for DEGREE BOUNDED SPANNING TREE improving upon an $O(2^{n+o(n)})$ algorithm of Fomin et al. [8].

On the way to obtain our results we generalize the notion of *representative sets* to multisets, and give an efficient algorithm to compute such representative sets. Both the generalization of representative sets to multisets and the algorithm to compute them may be of independent interest.

[*] The research leading to these results has received funding from the European Community's Seventh Framework Programme (FP7/2007-2013) under grant agreement number 257575 and 240258. Mi. Pilipczuk is currently holding a post-doc position at Warsaw Center of Mathematics and Computer Science and is supported by Polish National Science Centre grant DEC-2013/11/D/ST6/03073.

N. Bansal and I. Finocchi (Eds.): ESA 2015, LNCS 9294, pp. 545–556, 2015.
DOI: 10.1007/978-3-662-48350-3_46

1   Introduction

Many of the combinatorial optimization problems studied in theoretical computer science are idealized mathematical models of real-world problems. When the simplest model is well-understood, it can be enriched to better capture the real-world problem one actually wants to solve. Thus it comes as no surprise that many of the well-studied computational problems generalize each other: the CONSTRAINT SATISFACTION PROBLEM generalizes SATISFIABILITY, the problem of finding a spanning tree of maximum degree at most d generalizes HAMILTONIAN PATH, while the problem of packing sets of size 3 generalizes packing sets of size 2, also known as the MAXIMUM MATCHING problem.

By definition, the generalized problem is computationally harder than the original. However it is sometimes the case that the most difficult instances of the generalized problem are actually instances of the original problem. In other words, the "further away" an instance of the generalized problem is from being an instance of the original, the easier the instance is. Abasi et. al [1] initiated the study of this phenomenon in parameterized complexity (we refer the reader to the textbooks [4,5,7,17] for an introduction to parameterized complexity). In particular, they study the r-SIMPLE k-PATH problem. Here the input is a graph G, and integers k and r, and the objective is to determine whether there is an r-simple k-path in G, where an r-simple k-path is a sequence $v_1, v_2, \ldots, v_k$ of vertices such that every pair of consecutive vertices is adjacent and no vertex of G is repeated more than r times in the sequence. Observe that for $r = 1$ the problem is exactly the problem of finding a simple path of length k in G. On the other hand, for $r = k$ the problem is easily solvable in polynomial time, as one just has to look for a walk in G of length k. Thus, gradually increasing r from 1 to k should provide a sequence of computational problems that become easier as r increases. Abasi et al. [1] confirm this intuition by giving a randomized algorithm for r-SIMPLE k-PATH with running time $O(r^{2k/r} n^{O(1)})$ for any $2 \leq r \leq k$.

In this paper we continue the investigation of algorithms for problems with a *relaxation parameter* r that interpolates between an NP-hard and a polynomial time solvable problem. We show that in several interesting cases one can get a sequence of algorithms with better and better running times as the relaxation parameter r increases, essentially providing a smooth transition from the NP hard to the polynomial time solvable case.

Our main technical contribution is a new algorithm for the (r, k)-MONOMIAL DETECTION problem. Here the input is an arithmetic circuit C that computes a polynomial f of degree k in n variables $x_1, \ldots, x_n$. The task is to determine whether the polynomial f has a monomial $\prod_{i=1}^{n} x_i^{a_i}$, such that $0 \leq a_i \leq r$ for every $i \leq n$. The main result of Abasi et al. [1] is a randomized algorithm for (r, k)-MONOMIAL DETECTION with running time $O(r^{2k/r} |C| n^{O(1)})$, and their algorithm for r-SIMPLE k-PATH is obtained using a reduction to (r, k)-MONOMIAL DETECTION. We give a *deterministic* algorithm for the problem with running time $r^{O(k/r)} |C| n^{O(1)}$ in the case when the circuit C is *non-canceling*. Formally, this means that the circuit contains only variables at its leaves (i.e., no constants) and only addition and multiplication gates (i.e, no subtraction gates).

Informally, all monomials of the polynomials computed at intermediate gates of C contribute to the polynomial computed by C. Before stating our theorem, we remark that the formal definitions of problems and notations involved in the algorithmic results presented in the introduction can be found in full version. We use the notation O_k in our theorems to hide $k^{O(1)}$ terms.

Theorem 1 *Given a non-canceling circuit C computing a polynomial $f(X_1, \ldots, X_n) \in \mathbb{Z}[X_1, \ldots, X_n]$, (r, k)-*Monomial Detection *can be solved in deterministic time $O_k(|C| \cdot r^{18k/r} \cdot 2^{O(k/r)} \cdot n \log^3 n)$.*

Comparing our algorithm with the algorithm of Abasi et al. [1], our algorithm is slower by a constant factor in the exponent of r, and only works for non-canceling circuits. However our algorithm is *deterministic* (while the one by Abasi et al. is randomized) and also works for the *weighted* variant of the problem, while the one by Abasi et al. does not. In the weighted variant each variable x_i has a non-negative integer weight w_i, and the weight of a monomial $\Pi_{i=1}^{n} x_i^{a_i}$ is defined as $\sum_{i=1}^{n} w_i a_i$. The task is to determine whether there exists a monomial $\Pi_{i=1}^{n} x_i^{a_i}$, such that $0 \le a_i \le r$ for every $i \le n$, and if so, to return one of minimum weight. As a direct consequence we obtain the first deterministic algorithm for r-Simple k-Path with running time $r^{O(k/r)} |C| n^{O(1)}$, and the first algorithm with such a running time for weighted r-Simple k-Path.

Theorem 2 (Weighted) r-Simple k-Path *can be solved in deterministic time $O_k(r^{12k/r} \cdot 2^{O(k/r)} \cdot n^3 \cdot \log n)$.*

Here, by Weighted r-Simple k-Path we mean the variant where the weights are on vertices, and the weight of an r-simple k-path is the sum of the weights of traversed vertices, including multiplicities. However, we can also solve the edge-weighted variant at a cost of a larger constant in the exponent.

The significance of an in-depth study of (r, k)-Monomial Detection, is that it is the natural "relaxation parameter"-based generalization of the fundamental Multi-linear Monomial Detection problem. The Multi-linear Monomial Detection problem is simply (r, k)-Monomial Detection with $r = 1$. A multitude of parameterized problems reduce to Multi-linear Monomial Detection [2,4,9,14,22]. Thus, obtaining good algorithms (r, k)-Monomial Detection is an important step towards efficient algorithms for relaxation parameter-variants of these problems. For some problems, such as k-Path, efficient algorithms for the relaxation parameter variant (i.e r-Simple k-Path) follow directly from the algorithms for (r, k)-Monomial Detection. In this paper we give two more examples of fundamental problems for which efficient algorithms for (r, k)-Monomial Detection lead to efficient algorithms for their "relaxation parameter"-variant.

Our first example is the (r, p, q)-Packing problem. Here the input is a family $\mathcal{F}$ of sets of size q over a universe of size n, together with integers r and p. The task is to find a subfamily $\mathcal{A} \subseteq \mathcal{F}$ of size at least p such that every element of the universe is contained in at most r sets in $\mathcal{A}$. Observe that (r, p, q)-Packing is the relaxation parameter variant of the classic Set Packing problem ((r, p, q)-Packing with $r = 1$). We give an algorithm for (r, p, q)-Packing with running time $2^{O(pq \cdot \frac{\log r}{r})}|\mathcal{F}|n^{O(1)}$. For $r = 1$ this matches the best known algorithm [2] for SET PACKING, up to constants in the exponent, and when r grows our algorithm is significantly faster than $2^{pq}|\mathcal{F}|n^{O(1)}$. Just as for r-SIMPLE k-PATH, our algorithm also works for weighted variants. We remark that (r, p, q)-PACKING was also studied by Fernau et al. [6] from the perspective of kernelization.

Theorem 3 *Let $k \triangleq p \cdot q$. Then* (WEIGHTED) (r, p, q)-PACKING *can be solved in deterministic time* $O_k(|\mathcal{F}| \cdot r^{12k/r} \cdot 2^{O(k/r)} \cdot n \log^2 n)$.

Again, by WEIGHTED (r, p, q)-PACKING we mean a variant where elements are assigned weights and the weight of a set is equal to the sum of the weights of its elements. However, at a cost of a larger constant in the exponent we can also solve the variant where each set is assigned its own weight.

Our second example is the DEGREE-BOUNDED SPANNING TREE problem. Here, we are given as input a graph G and integer d, and the task is to determine whether G has a spanning tree T whose maximum degree does not exceed d. For $d = 2$ this problem is equivalent to HAMILTONIAN PATH, and hence the problem is NP-complete in general, but for $d = n - 1$ it boils down to checking the connectedness of G. Thus, DEGREE-BOUNDED SPANNING TREE can be thought of as a relaxation parameter variant of HAMILTONIAN PATH. The problem has received significant attention in the field of approximation algorithms: there are classic results of Fürer and Raghavachari [11], Goemans [13], and of Singh and Lau [20] that give additive approximation algorithms for the problem and its weighted variant. From the point of view of exact algorithms, the currently fastest exact algorithm, working for any value of d, is due to Fomin et al. [8] and has running time $O(2^{n+o(n)})$. In this work, we give a randomized algorithm for DEGREE-BOUNDED SPANNING TREE with running time $2^{O(n\frac{\log d}{d})}$, by reducing the problem to an instance of (r, k)-MONOMIAL DETECTION. Thus, our algorithm significantly outperforms the algorithm of Fomin et al. [8] for all super-constant d, and runs in polynomial time for $d = \Omega(n)$. Interestingly, the instance of (r, k)-MONOMIAL DETECTION that we create crucially uses subtraction, since the constructed circuit computes the determinant of some matrix. Thus we are not able to apply our algorithm for non-canceling circuits, and have to resort to the randomized algorithm of Abasi et al. [1] instead. Obtaining a deterministic algorithm for DEGREE-BOUNDED SPANNING TREE that would match the running time of our algorithm, or extending the result to the weighted setting, remains as an interesting open problem.

Our methods. The starting point for our algorithms is the notion of *representative sets*. If $\mathcal{A}$ is a family of sets, with all sets in $\mathcal{A}$ having the same size p, we say that a subfamily $\mathcal{A}' \subseteq \mathcal{A}$ q-*represents* $\mathcal{A}$ if for every set B of size q, whenever there exists a set $A \in \mathcal{A}$ such that A is disjoint from B, then there also exists a set $A' \in \mathcal{A}'$ such that A' is disjoint from B.

Representative sets were defined by Monien [16], and were recently used to give efficient parameterized algorithms for a number of problems [9,10,18,19,23,24], including k-PATH [9,10,19], SET PACKING [19,23,24] and MULTI-LINEAR MONOMIAL DETECTION [9]. It is therefore very tempting to try to use representative sets also for the relaxation parameter variants of these problems. However, it looks very hard to directly use representative sets in this setting. On a superficial level the difficulty lies in that representative sets are useful to guarantee *disjointness*, while the solutions to the relaxation parameter variants of the considered problems may self-intersect up to r times.

We overcome this difficulty by generalizing the notion of representative sets to multisets. When taking the union of two multisets A and B, an element that appears a times in A and b times in B will appear $a+b$ times in the union $A+B$. Thus, if two regular sets A and B are viewed as multisets, they are disjoint if and only if no element appears more than once in $A + B$. We can now relax the notion of disjointedness and require that no element appears more than r times in $A + B$. Specifically, if $\mathcal{A}$ is a family of multisets, with all multisets in $\mathcal{A}$ having the same size p (counting duplicates), we say that a subfamily $\mathcal{A}' \subseteq \mathcal{A}$ q-*represents* $\mathcal{A}$ if the following condition is satisfied. For every multiset B of size q, whenever there exists an $A \in \mathcal{A}$ such that no element appears more than r times in $A + B$, there also exists an $A' \in \mathcal{A}'$ such that no element appears more than r times in $A' + B$. The majority of the technical effort in the paper is spent on proving that every family $\mathcal{A}$ of multisets has a relatively small q-representative family $\mathcal{A}'$ in this new setting, and to give an efficient algorithm to compute $\mathcal{A}'$ from $\mathcal{A}$. The formal statement of this result can be found in Corollary 2.

On the way to develop our algorithm for computing representative sets of multisets, we give a new construction of a pseudo-random object called *lopsided universal sets*. Informally speaking, an (n, p, q)-*lopsided universal set* is a set of strings such that, when focusing on any $k \triangleq p+q$ locations, we see all patterns of hamming weight p. These objects have been of interest for a while in mathematics and in theoretical computer science under the name *Cover Free Families* (Cf. [3]). We give, for the first time, an explicit construction of an (n, p, q)-lopsided universal set whose size is only polynomially larger than optimal for all p and q. See the full version [12] for a formal statement.

Both our algorithm for computing representative sets of multisets, and the new construction of lopsided universal sets may be of independent interest.

Outline of the Paper. In Section 2 we give the necessary definitions and set up notational conventions. In Section 3 we give our construction of representative sets for multisets. This construction requires an auxiliary tool called minimal separating families. The construction of Section 3 assumes that an appropriate construction of minimal separating families is given as a black box, and this construction is deferred to full version. Our new construction of lopsided universal sets is a corollary of the construction of minimal separating families, and is also explained in full version. In Section 4 we briefly discuss the applications of representative sets to (r, k)-Monomial Detection, (r, p, q)-Packing and r-Simple k-Path (details deferred to full version). In Section 5 we present our algorithm for Degree-Bounded Spanning Tree. Finally, we conclude by discussing open problems and directions for future research in Section 6.

The full version of this paper, which contains a natural order of introducing the consecutive concepts and results, is available on arxiv [12].

2   Preliminaries

Notation. Throughout the paper, we use the notation O_k to hide $k^{O(1)}$ terms. We denote $[n] = \{1, 2, \ldots, n\}$. For sets A and B, by $\{A \to B\}$ we denote the set of all functions from A to B. The notation $\triangleq$ is used to introduce new objects defined by formulas on the right hand side.

Separating Families. Recall that, for an integer $t \geq 1$, we say that a family of functions $\mathcal{H} \subseteq \{[n] \to [m]\}$ is a *t-perfect hash family*, if for every $C \subseteq [n]$ of size $|C| = t$ there is $f \in \mathcal{H}$ that is injective on C. We will be interested in constructing families of perfect hash functions that, in addition to being injective on a set C, have the property of sending another large set D to an output *disjoint* from the image of C. We call such a family of functions a *separating family*.

Definition 1 (Separating family). *Fix integers t, k, s, n such that $1 \leq t \leq n$. For disjoint subsets $C, D \subseteq [n]$, we say that a function $h \colon [n] \to [s]$ separates C from D if*

- *h is injective on C; and*
- *there are no collisions between C and D. That is, $h(C) \cap h(D) = \emptyset$.*

   A family of functions $\mathcal{H} \subseteq \{[n] \to [s]\}$ is (t, k, s)-separating if for every disjoint subsets $C, D \subseteq [n]$ with $|C| = t$ and $|D| \leq k - t$, there is a function $h \in \mathcal{H}$ that separates C from D.
   We say that $\mathcal{H}$ is (t, k, s)-separating with probability γ if for any fixed C and D with sizes as above, a function h chosen uniformly at random from $\mathcal{H}$ separates C from D with probability at least γ.

   For us, the most important case of separating families is when the range size is $|C| + 1$. In this case we use the term *minimal separating family*. It will also be convenient to assume in the definition that C is mapped to the first $|C|$ elements in the range.

Definition 2 (Minimal separating family) *A family of functions $\mathcal{H} \subseteq \{[n] \to [t + 1]\}$ is (t, k)-minimal separating if for every disjoint subsets $C, D \subseteq [n]$ with $|C| = t$ and $|D| \leq k - t$, there is a function $h \in \mathcal{H}$ such that*

- *$h(C) = [t]$.*
- *$h(D) \subseteq \{t + 1\}$.*

3   Multiset Separators and Representative Sets

The purpose of this section is to formally define and construct representative sets for multisets. We will use, as an auxiliary result, an efficient construction of a small separating family. This result is actually the most technical part of this paper, and its proof is deferred entirely to the appendix.

Theorem 4 *Fix integers n, t, k such that $1 \leq t \leq \min(n, k)$. Then a (t, k)-minimal separating family of size $O_k((k/t)^{2t} \cdot 2^{O(t)} \cdot \log n)$ can be constructed in time $O_k((k/t)^{2t} \cdot 2^{O(t)} \cdot n \cdot \log n)$.*

Proof. (sketch) We first hash the set C of size t injectively into t 'buckets' using known constructions. At this stage we have 'taken care' of C but the set D of size $\leq k - t$ is scattered somehow among the t buckets. The novel idea is that a combination of guesses as to how the D has been scattered, together with the use of hitting sets for combinatorial rectangles [15] can now be used to separate C from D in a way that is not too costly. □

Our primary tool for the construction of representative sets for multisets is what we call a *multiset separator* (see Definition 3). Informally, a multiset separator is a not too large set of 'witnesses' for the fact that two multisets of bounded size do not jointly contain too many repetitions per element.

Notation for Multisets. Fix integers $n, r, k \geq 1$. We use $[r]_0$ to denote $\{0, \ldots, r\}$. An *r-set* is a multiset A where each element of $[n]$ appears at most r times. It will be convenient to think of A as a vector in $[r]_0^n$, where A_i denotes the number of times i appears in A. We denote by $|A|$ the number of elements in A counting repetitions. That is, $|A| = \sum_{i=1}^{n} A_i$. We refer to $|A|$ as the *size* of A. An (r, k)-set is an *r*-set $A \in [r]_0^n$, where the number of elements with repetitions is at most k. That is, $|A| \leq k$.

Fix *r*-sets $A, B \in [r]_0^n$. We say that $A \leq B$ when $A_i \leq B_i$ for all $i \in [n]$. By $\overline{A} \in [r]_0^n$ we denote the "complement" of *r*-set A, that is, $\overline{A}_i = r - A_i$ for all $i \in [n]$. By $A + B$ we denote the "union" of A and B, that is, $(A + B)_i = A_i + B_i$ for all $i \in [n]$. Suppose now that A and B are (r, k)-sets. We say that A and B are (r, k)-*compatible* if $A + B$ is also an (r, k)-set, and $|A + B| = k$. That is, the total number of elements with repetitions in A and B together is k and any specific element $i \in [n]$ appears in A and B together at most r times. With the notation above at hand, we can define the central object needed for our algorithms.

Definition 3 (Multiset separator) *Let $\mathcal{F}$ be a family of r-sets. We say that $\mathcal{F}$ is an (r, k)-separator if for any (r, k)-sets $A, B \in [r]_0^n$ that are (r, k)-compatible, there exists $F \in \mathcal{F}$ such that $A \leq F \leq \overline{B}$.*

Construction of Multiset Separators. The following theorem shows how an (r, k)-separator can be constructed from a minimal separating family.

Theorem 5 *Fix integers n, r, k such that $1 < r \leq k$, and let $t \triangleq \lfloor 2k/r \rfloor$. Suppose a (t, k)-minimal separating family $\mathcal{H} \subseteq \{[n] \to [t + 1]\}$ can be constructed in time $f(r, k, n)$. Then an (r, k)-separator $\mathcal{F}$ of size $|\mathcal{H}| \cdot (r + 1)^t$ can be constructed in time $O_k(f(r, k, \max(n, t)) \cdot (r + 1)^t)$.*

Proof. (sketch) Given (r, k)-compatible multisets A and B, we wish to 'separate' them by a multiset F such that $A \leq F \leq \overline{B}$. Because of the compatibility, there can only $2k/r$ 'large indices' - where $A_i > r/2$ or $B_i > r/2$. If we knew these indices in advance, we could afford to try all values F_i to find a value such that $A_i \leq F_i \leq \overline{B}_i$. For any other $i \in [n]$, defining $F_i = r/2$ is fine. The problem is that there are many options for the set of large indices. However, a minimal separating family allows us to separate the set of large indices from others while trying a small number of possibilities. □

Combination of Theorems 4 and 5 immediately yields the following.

Corollary 1 $[\star]$[1] *Fix integers n, r, k such that $1 < r \leq k$. Then an (r, k)-separator $\mathcal{F}$ of size $O_k(r^{6k/r} \cdot 2^{O(k/r)} \cdot \log n)$ can be constructed in time $O_k(r^{6k/r} \cdot 2^{O(k/r)} \cdot n \cdot \log n)$*

Multisets Over a Weighted Universe. Before proceeding, we discuss the issue of how the considered multisets will be equipped with *weights*. For simplicity, we assume that the universe $[n]$ is weighted, i.e., each element $i \in [n]$ is assigned an integer weight $\mathbf{w}(i)$. We define the weight of a multiset as the sum of the weights of its elements counting repetitions. Formally, for $A \in [r]_0^n$ we have

$$\mathbf{w}(A) = \sum_{i=1}^{n} A_i \cdot \mathbf{w}(i).$$

Whenever we talk about a *weighted family* of multisets, we mean that the universe $[n]$ is equipped with a weight function and the weights of the multisets are defined as in the formula above.

Representative Sets for Multisets. We are ready to define the notion of a representative set for a family of multisets.

Definition 4 (Representative sets for multisets) *Let $\mathcal{P}$ be a weighted family of (r, k)-sets. We say that a subfamily $\hat{\mathcal{P}} \subseteq \mathcal{P}$ represents $\mathcal{P}$ if for every (r, k)-set Q the following holds. If there exists some $P \in \mathcal{P}$ of weight w that is (r, k)-compatible with Q, then there also exists some $P' \in \hat{\mathcal{P}}$ of weight $w' \leq w$ that is (r, k)-compatible with Q.*

The following definition and lemma show that having an (r, k)-separator is sufficient for constructing representative sets.

Definition 5 *Let $\mathcal{P}$ be a weighted family of r-sets and let $\mathcal{F}$ be a family of (r, k)-sets. The weighted family $\mathrm{Trim}_{\mathcal{F}}(\mathcal{P}) \subseteq \mathcal{P}$ is defined as follows: For each $F \in \mathcal{F}$, and for each $1 \leq i \leq k$, check if there exists some $P \in \mathcal{P}$ with $|P| = i$ and $P \leq F$. If so, insert into $\mathrm{Trim}_{\mathcal{F}}(\mathcal{P})$ some $P \in \mathcal{P}$ that is of minimal weight among those with $|P| = i$ and $P \leq F$.*

Lemma 1. $[\star]$ *Let $\mathcal{F}$ be an (r, k)-separator and let $\mathcal{P}$ be a weighted family of (r, k)-sets. Then $\mathrm{Trim}_{\mathcal{F}}(\mathcal{P})$ represents $\mathcal{P}$.*

---

[1] Proof of statements marked with $\star$ are omitted due to space constraints and may be found in the full version [12].

We can now combine Lemma 1 with the construction of an (r, k)-separator from Corollary 1, and thus obtain a construction of a small representative family for a weighted family of multisets.

Corollary 2 [⋆] *There exists a deterministic algorithm that, given a weighted family $\mathcal{P}$ or (r, k)-sets, runs in time $O_k(|\mathcal{P}| \cdot r^{6k/r} \cdot 2^{O(k/r)} \cdot n \log n)$ and returns returns a family $\hat{P} \subseteq \mathcal{P}$ that represents $\mathcal{P}$ and has size $O_k(r^{6k/r} \cdot 2^{O(k/r)} \cdot \log n)$.*

4 Algorithmic Applications

For lack of space, we defer all details to the full version [12], and now only sketch how representative sets for multisets can be applied to r-SIMPLE k-PATH, (r, p, q)-PACKING, and (r, k)-MONOMIAL DETECTION. In all cases, the algorithm applies the expand-and-shrink strategy that was used by Fomin et al. [10] for k-PATH and by Zehavi [23] for SET PACKING (see also the appropriate chapter of [4]). Basically, we iteratively expand the so far obtained family of partial solutions, e.g. by prolonging all the constructed paths by one vertex, and then trim the obtained expanded family by computing its representative subfamily. For (r, k)-MONOMIAL DETECTION, we compute for each gate a representative set of monomials that appear in the polynomial computed at this gate. This is done by "merging" the representative sets of monomials for the input gates, and computing the representative set of this merge. The assumption that the circuit is non-canceling is necessary to argue that the set of monomials appearing at each gate is indeed a proper merge of the sets appearing on its input gates.

5 Low Degree Monomials and Low Degree Spanning Trees

Let G be a simple, undirected graph with n vertices. Let $\mathcal{T}$ be the family of spanning trees of G; in particular, if G is not connected then $\mathcal{T} = \emptyset$. With every edge $e \in E(G)$ we associate a variable y_e. The *Kirchhoff's polynomial* of G is defined as:

$$K_G((y_e)_{e \in E(G)}) = \sum_{T \in \mathcal{T}} \prod_{e \in E(T)} y_e.$$

Thus, K_G is a polynomial in $\mathbb{Z}[(y_e)_{e \in E(G)}]$. Let $v_1, v_2, \ldots, v_n$ be an arbitrary ordering of $V(G)$. The *Laplacian* of G is an $n \times n$ matrix $L_G = [a_{ij}]$, where

$$a_{ij} = \begin{cases} \sum_{e \text{ incident to } v_i} y_e & \text{if } i = j, \\ -y_{v_i v_j} & \text{if } i \neq j \text{ and } v_i v_j \in E(G), \\ 0 & \text{if } i \neq j \text{ and } v_i v_j \notin E(G). \end{cases}$$

Observe that L_G is symmetric and the entries in every column and in every row of L_G sum up to zero. Then it can be shown that all the *first cofactors* of L_G, i.e., the determinants of matrix L_G after removing a row and a column with the same indices, are equal. Let N_G be this common value; then N_G is again a polynomial over variables $(y_e)_{e \in E(G)}$. The Kirchhoff's Matrix Tree Theorem, in its general form, states that these two polynomials coincide.

Theorem 6 (Matrix Tree Theorem) $K_G = N_G$.

We remark that the Matrix Tree Theorem is usually given in the more specific variant, where all variables y_e are replaced with 1; then the theorem expresses the number of spanning trees of G in terms of the first cofactors of L_G. However, the proof can be easily extended to the above, more general form; cf. [21].

Observe that Theorem 6 provides a polynomial-time algorithm for evaluating K_G over a vector of values of variables $(y_e)_{e \in E(G)}$. Indeed, we just need to construct matrix L_G, remove, say, the first row and the first column, and compute the determinant. We now present how this observation can be used to design a fast exact algorithm for the DEGREE-BOUNDED SPANNING TREE problem.

Theorem 7 *The* DEGREE-BOUNDED SPANNING TREE *problem can be solved in randomized time* $O^*(d^{O(n/d)})$ *with false negatives.*

Proof. Associate every vertex v of the given graph G with a distinct variable x_v. Let $\overline{K}_G \in \mathbb{Z}[(x_v)_{v \in V(G)}]$ be a polynomial defined as K_G with every variable y_{uv}, for $uv \in E(G)$, evaluated to $x_u x_v$. Then it follows that

$$\overline{K}_G((x_v)_{v \in V(G)}) = \sum_{T \in \mathcal{T}} \prod_{v \in V(G)} x_v^{\deg_T(v)}.$$

Observe also that $\overline{K}_G$ is $2(n-1)$-homogeneous, that is, all the monomials of $\overline{K}_G$ have their total degrees equal to $2(n-1)$. Thus, graph G admits a spanning tree with maximum degree at most d if and only if polynomial $\overline{K}_G$ contains a $(d, 2(n-1))$-monomial. Using Theorem 6 we can construct a $n^{O(1)}$-sized circuit evaluating $\overline{K}_G$. Hence, verifying whether $\overline{K}_G$ contains a $(d, 2(n-1))$-monomial boils down to applying the algorithm of Abasi et al. [1] for (r,k)-MONOMIAL DETECTION with $r = d$ and $k = 2(n-1)$. This algorithm runs in randomized time $O^*(d^{O(n/d)})$ and can only produce false negatives. $\square$

Let us repeat that in the proof of Theorem 7 we could not have used Theorem 9 instead of the result of Abasi et al. [1], because the constructed circuit is not non-canceling. Derandomizing the algorithm and extending it to the weighted setting remains hence open.

Interestingly, the running time of the algorithm of Theorem 7 is essentially optimal, up to the $\log d$ factor in the exponent. A similar lower bound for r-SIMPLE k-PATH was given by Abasi et al. [1].

Theorem 8 [⋆] *Unless ETH fails, there exists a constant $s > 0$ such that for no fixed integer $d \geq 2$ the* DEGREE-BOUNDED SPANNING TREE *problem with the degree bound d can be solved in time* $O^*(2^{sn/d})$.

6 Conclusions

In this paper we considered relaxation parameter variants of several well studied problems in parameterized complexity and exact algorithms. We proved, somewhat surprisingly, that instances with moderate values of the relaxation parameter are significantly easier than instances of the original problems. We hope that our work, together with the result of Abasi et al. [1] breaks the ground for a systematic investigation of relaxation parameters in parameterized complexity and exact algorithms. We conclude with mentioning some of the most natural concrete follow up questions to our work.

- We gave a deterministic algorithm for *non-canceling* (r, k)-MONOMIAL DE-TECTION with running time $2^{O(k\frac{\log r}{r})}|C|n^{O(1)}$, while Abasi et al. [1] gave a randomized algorithm with such a running time for (r, k)-MONOMIAL DE-TECTION without the non-cancellation restriction. Is there a deterministic algorithm for (r, k)-MONOMIAL DETECTION with running time $2^{O(k\frac{\log r}{r})}|C|n^{O(1)}$?
- Does there exist a deterministic algorithm with running time $2^{O(n\frac{\log r}{r})}$ for DEGREE-BOUNDED SPANNING TREE? Note that a deterministic algorithm with running time $2^{O(k\frac{\log r}{r})}|C|n^{O(1)}$ for (r, k)-MONOMIAL DETECTION would immediately imply such an algorithm for DEGREE-BOUNDED SPANNING TREE.
- Is there a $2^{O(k\frac{\log r}{r})}n^{O(1)}$ time algorithm for the problem where we are given as input a graph G, integers k and d, and asked whether G contains a subtree T on at least k vertices, such that the maximum degree of T is at most d? Observe that for $k = n$ this is exactly the DEGREE-BOUNDED SPANNING TREE problem.
- Is it possible to obtain kernels for problems with relaxation parameters with smaller and smaller size bounds as the relaxation parameter increases?
- Is there an algorithm for r-SIMPLE k-PATH with running time $2^{O(k/r)}n^{O(1)}$? Or a $2^{O(n/d)}$ time algorithm for DEGREE BOUNDED SPANNING TREE?

Acknowledgements. The first author thanks Fedor V. Fomin for hosting him at a visit in the University of Bergen where this research was initiated. We thank the anonymous reviewers for helpful corrections.

References

1. Abasi, H., Bshouty, N.H., Gabizon, A., Haramaty, E.: On r-Simple k-Path. In: Csuhaj-Varjú, E., Dietzfelbinger, M., Ésik, Z. (eds.) MFCS 2014, Part II. LNCS, vol. 8635, pp. 1–12. Springer, Heidelberg (2014)
2. Björklund, A., Husfeldt, T., Kaski, P., Koivisto, M.: Narrow sieves for parameterized paths and packings. CoRR, abs/1007.1161 (2010)
3. Bshouty, N.H.: Testers and their applications. In: ITCS 2014, pp. 327–352 (2014)
4. Cygan, M., Fomin, F.V., Lokshtanov, D., Kowalik, L., Marx, D., Pilipczuk, M., Pilipczuk, M., Saurabh, S.: Parameterized Algorithms. Springer (in press, 2015)

5. Downey, R.G., Fellows, M.R.: Fundamentals of Parameterized Complexity. Texts in Computer Science. Springer (2013)
6. Fernau, H., López-Ortiz, A., Romero, J.: Kernelization algorithms for packing problems allowing overlaps. In: Jain, R., Jain, S., Stephan, F. (eds.) TAMC 2015. LNCS, vol. 9076, pp. 415–427. Springer, Heidelberg (2015)
7. Flum, J., Grohe, M.: Parameterized Complexity Theory. Texts in Theoretical Computer Science. An EATCS Series. Springer, Berlin (2006)
8. Fomin, F.V., Grandoni, F., Lokshtanov, D., Saurabh, S.: Sharp separation and applications to exact and parameterized algorithms. Algorithmica 63(3), 692–706 (2012)
9. Fomin, F.V., Lokshtanov, D., Panolan, F., Saurabh, S.: Representative sets of product families. In: Schulz, A.S., Wagner, D. (eds.) ESA 2014. LNCS, vol. 8737, pp. 443–454. Springer, Heidelberg (2014)
10. Fomin, F.V., Lokshtanov, D., Saurabh, S.: Efficient computation of representative sets with applications in parameterized and exact algorithms. In: SODA 2014, pp. 142–151 (2014)
11. Fürer, M., Raghavachari, B.: Approximating the minimum-degree Steiner Tree to within one of optimal. J. Algorithms 17(3), 409–423 (1994)
12. Gabizon, A., Lokshtanov, D., Pilipczuk, M.: Fast algorithms for parameterized problems with relaxed disjointness constraints. CoRR, abs/1411.6756 (2014)
13. Goemans, M.X.: Minimum bounded degree spanning trees. In: FOCS 2006, pp. 273–282 (2006)
14. Koutis, I.: Faster algebraic algorithms for path and packing problems. In: Aceto, L., Damgård, I., Goldberg, L.A., Halldórsson, M.M., Ingólfsdóttir, A., Walukiewicz, I. (eds.) ICALP 2008, Part I. LNCS, vol. 5125, pp. 575–586. Springer, Heidelberg (2008)
15. Linial, N., Luby, M., Saks, M.E., Zuckerman, D.: Efficient construction of a small hitting set for combinatorial rectangles in high dimension. Combinatorica 17(2), 215–234 (1997)
16. Monien, B.: How to find long paths efficiently. In: Analysis and Design of Algorithms for Combinatorial Problems, Udine. North-Holland Math. Stud., vol. 109, pp. 239–254. North-Holland, Amsterdam (1982)
17. Niedermeier, R.: Invitation to fixed-parameter algorithms. Oxford Lecture Series in Mathematics and its Applications, vol. 31, p. 300. Oxford University Press, Oxford (2006)
18. Pinter, R.Y., Shachnai, H., Zehavi, M.: Deterministic parameterized algorithms for the graph motif problem. In: Csuhaj-Varjú, E., Dietzfelbinger, M., Ésik, Z. (eds.) MFCS 2014, Part II. LNCS, vol. 8635, pp. 589–600. Springer, Heidelberg (2014)
19. Shachnai, H., Zehavi, M.: Representative families: A unified tradeoff-based approach. In: Schulz, A.S., Wagner, D. (eds.) ESA 2014. LNCS, vol. 8737, pp. 786–797. Springer, Heidelberg (2014)
20. Singh, M., Lau, L.C.: Approximating minimum bounded degree spanning trees to within one of optimal. J. ACM 62(1), 1–1 (2015)
21. Tutte, W.T.: Graph Theory. Cambridge University Press (2001)
22. Williams, R.: Finding paths of length k in $O^*(2^k)$ time. Inf. Process. Lett. 109(6), 315–318 (2009)
23. Zehavi, M.: Deterministic parameterized algorithms for matching and packing problems. CoRR, abs/1311.0484 (2013)
24. Zehavi, M.: Solving parameterized problems by mixing color coding-related techniques. CoRR, abs/1410.5062 (2014)

Medial Axis Based Routing Has Constant Load Balancing Factor

Jie Gao[1] and Mayank Goswami[2]

[1] Department of Computer Science, Stony Brook University, Stony Brook, NY USA 08550
`jgao@cs.stonybrook.edu`
[2] Max-Planck Institute for Informatics, Saarbrücken, Germany 66123
`gmayank@mpi-inf.mpg.de`

Abstract. Load balanced routing is a long standing yet challenging problem. Despite many years' of work there is still a gap between theoretical research and practical algorithms. The main contribution in this paper is to bridge the gap and provide rigorous analysis of a practically interesting algorithm for routing in a large scale sensor network of complex shape – routing by using the medial axis of the network. In this algorithm, a skeleton of the network is extracted such that a virtual coordinate system can be developed for greedy routing achieving good load balance in simulations. We show for the first time a constant approximation factor for this algorithm by a highly technical analysis. The analysis explains the performance observed in previous simulations and is also the first known constant approximation algorithm for load balanced routing in a sensor network with non-trivial geometry.

1 Introduction

Load balanced routing is a fundamental and challenging problem. It is of crucial importance in wireless sensor networks since overloaded nodes may deplete their battery prematurely, severely hampering the utility of the network. It is also more challenging in the sensor network setting as only distributed, lightweight algorithms are useful. Despite many years' of research there is still a separation between algorithms with theoretical guarantees and algorithms that are practically useful. In the theoretical direction, load balanced routing is often formulated as minimizing the maximum traffic load of a given set of routing requests on a network with fixed capacities. Approximation algorithms using global optimizations on graphs [13, 14] do not meet the low resource requirement and do not have constant approximation ratio. In practice, a number of algorithms arrive at a good balance between good performance and low requirement on computation and communication. But nothing provable is known. The results in this paper bridge the gap by providing the first constant approximation ratio for a practically interesting algorithm.

Wireless sensor networks differ from other types of networking scenarios in the rich geometric properties. In this setting, n sensors are embedded (positions chosen uniformly randomly) in a geometric region $\Omega \subset \mathbb{R}^2$ providing dense coverage of Ω. A sensor network is a graph on the vertices. We assume that a unit disk graph is used to model wireless communication: two nodes are connected by an edge if the distance is at most 1. The load of a node is the number of routing paths (out of the total $\binom{n}{2}$

© Springer-Verlag Berlin Heidelberg 2015
N. Bansal and I. Finocchi (Eds.): ESA 2015, LNCS 9294, pp. 557–569, 2015.
DOI: 10.1007/978-3-662-48350-3_47

paths) passing through it under all-pairs communication. One wishes to find paths that minimize the maximum load, and paths that are also easy to store/compute.

Unfortunately, understanding the dependency of load balanced routing on the geometric shape of Ω is still very limited. For sensors uniformly randomly placed inside a disk with all pairs traffic, shortest path routing (i.e., routing paths along straight lines) will create higher traffic load at the center of the disk. But the highest traffic load can be shown to be a constant factor away from the optimal solution (minimizing the maximum traffic) [8]. Nothing is known on the optimal solution in a disk beyond a discrete approximation using simulations [12]. Apart from that, constant approximation solution using greedy routing is known for narrow strips [7] and for simply connected domains admitting a constant stretch area-preserving map to disks [8]. The general question of finding a constant approximation load balanced routing scheme for an *arbitrary* geometric domain is still open.

In this paper we achieve for the first time a constant approximation ratio for load-balanced routing in a network Ω with arbitrarily complex shape. This problem is substantially more challenging than the case of a simply connected domain. In the case of a disk or other simply connected domain, the main challenge is to avoid concentration of routing paths either at the center of the disk or at a reflex vertex (corner) of the domain. But when we move to a domain with holes, a new challenge appears. We need to decide how the routing paths get around the holes and how such traffic is distributed. This decision has to be dependent on the 'width' of the corridors on each side of holes. See Figure 1. Thus the topology of the domain and the topological structure of the routing paths are essential.

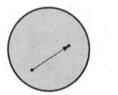 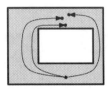

Fig. 1. In simply connected domains, the main challenge is to avoid path concentration (left two figures). In the case of a non-simple domain, we may need to distribute the paths around the holes by how much resource we have available along the 'corridors'.

We show that routing using the medial axis κ of the sensor network field, with some modification, achieves an approximation factor of $O(1)$, when the source and destination are uniformly randomly selected. The medial axis of a 2D domain Ω is the collection of points that have more than one closest points on the boundary of Ω. It is a planar graph homotopic to Ω. Thus as a skeleton of Ω, it has been used in a distributed routing algorithm to guide messages to pass around 'holes' [3]. Basically the medial axis is represented by a compact discrete graph $\hat{\kappa}$ to be disseminated to all nodes in the network. Each node knows its relative position to the medial axis and a greedy routing algorithm directs a message to the destination by first travelling 'horizontally' (i.e., in parallel with the medial axis) and then 'vertically' (i.e., perpendicular to the medial axis) to the destination. In the same paper it has been demonstrated using simulations that this algorithm has guaranteed delivery and good performance in balancing traffic load, compared to other greedy alternatives. In this paper we build on this algorithm. Specifically, our contributions are the following:

- We enhance the original medial axis based routing algorithm by running an LP optimization on the medial axis $\hat{\kappa}$. This LP program helps us decide how to distribute routing paths with respect to the holes in the network. The solution to the LP provides a compact, probabilistic routing guidance. This knowledge is distributed to all nodes in the network. Each node s stores $O(|\hat{\kappa}|)$ information such that for any destination t, a set of paths with probabilities can be extracted to guide how the message should travel with respect to the medial axis κ. The actual routing path is again realized by a local, greedy procedure.
- We show that the above algorithm achieves a constant approximation ratio. The analysis is highly technical. Majority of the proofs are in the full paper and we present the intuition and proof sketch in the main body of the paper.

Related Work. In the field of networking algorithms, load balanced routing on a graph with given source destination pairs is a long standing problem. One way to formulate this problem is to select routes that minimize congestion (the maximum number of messages that any node/link carries), termed the *unsplittable flow problem*. Solving this problem optimally is NP-hard even in very simple networks (such as grid). The best approximation algorithm has an approximation factor of $O(\log n/\log\log n)$ [13, 14] in a network of n vertices. It is also shown that getting an approximation within factor $\Omega(\log\log n)$ is NP-hard [1]. Another popular way to formulate the problem is to consider node disjoint or edge disjoint paths that deliver the largest number of given source destination pairs. This is again NP-hard [9] and the best approximation factor known is $O(\sqrt{n})$ [4]. It is NP-hard to approximate within a factor of $\Omega(\log^{1/2-\varepsilon} n)$ [5]. These approximation algorithms are mostly only of theoretical interest. They require global knowledge and are not suitable for distributed settings.

In the wireless sensor network setting, a number of algorithm make use of the geometric property in designing load balanced routing algorithms. Mei and Stefa [11] suggested to wrap a square network into a torus so as to avoid loading the network center. Yu et. al [16] map the network as the skeleton of a convex polytope using the Thurston's embedding. In [15], a network of multiple holes is converted to the covering space to 'remove' the hole boundaries and prevent them from being heavily loaded. All of them are shown by simulations to have reasonable performance, but no theoretical guarantee is known.

2   Network Model and Medial Axis

In this section we describe the network model for our theoretical analysis. We remark that some of the following assumptions are purely for the proofs and not necessary for the routing algorithm.

Medial Axis in the Continuous Setting: The medial axis κ of a geometric domain Ω is defined as the set of points with more than one closest point to the boundary of Ω. It is known that κ is a planar graph that is homotopic to Ω [2]. κ is composed of branches (called *medial edges*) joined by junction points (called *medial vertices*), that can have degree 1 (being the endpoint of a branch) or degree ≥ 3. Each point a on the medial axis has a maximal empty disk inside Ω that touches at least two points on $\partial\Omega$. The line segment connecting a with a tangent point is called a *chord* of a.

We assume that Ω only has a finite number of arc segments on its boundary $\partial\Omega$, that is, a part of the boundary that coincides with part of a circle. In this case, all but a finite number of points W on the medial axis κ of Ω have only a constant number of chords. In addition, we will punch a point hole at each point of W and recompute the medial axis κ. If there are still points on κ with an infinitely number of chords, we repeat the same procedure until no point on the medial axis has an infinite number of chords. We remark that this procedure will only create a finite number of additional point holes. This way

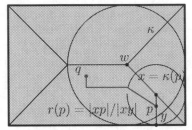

Fig. 2. An example of a medial axis of a domain Ω. Two medial balls centered at x, w are shown. We also show a possible routing path from p to q.

we remove medial vertices of degree 1. We also assume there is no degeneracy, i.e., any maximal empty ball is tangent to at most three points on the boundary. Thus all points on κ have at most three chords. Thus all medial vertices have degree exactly 3. Each canonical piece is bounded by two chords, a medial edge (or a medial vertex) and a piece of $\partial\Omega$.

Medial Axis Based Coordinates: For each point $p \in \Omega$ not on the medial axis, p must be on a unique chord xy, with $x \in \kappa, y \in \partial\Omega$ [3]. Furthermore, y is p's closest point on $\partial\Omega$. By this property we can define the projection of p on κ as the point x and denote it as $\kappa(p)$ and assign coordinates $(x(p), y(p), r(p))$ where $x(p) \in \kappa$ and $y(p) \in \partial\Omega$ are the endpoints of the unique chord p lies on, and $0 < r(p) = |px(p)|/|x(p)y(p)| < 1$ is the normalized distance from $x(p)$.

Discrete Sensors S. We assume that sensors have bounded density ε: inside any disk of radius ε inside Ω, there is at least one sensor inside. Further we assume that any two sensors are at distance at least $\varepsilon/2$ apart. The minimum density requirement is typically guaranteed by sensing coverage requirement. The requirement on minimum distance separation can be obtained by using a greedy algorithm to subsample in a dense region. Basically, select any sensor not yet selected, remove all sensors within distance $\varepsilon/2$, and continue. Any two sensors selected are of distance at least $\varepsilon/2$ apart.

To define a discrete sample of Ω, we also assume a uniform set of sensors along the continuous medial axis κ with density ε, such that any point of κ has a sensor within distance ε along κ.

Discrete Medial Axis. In the sensor network setting the discrete medial axis is used [3]. First, the sample points on κ are connected into a graph as a discrete approximation of κ. In the continuous setting a chord is a line segment connecting a point on the medial axis to a tangent point on the boundary. In the discrete setting, due to the discrete resolution a chord is a tree T_a rooted at a point $a \in S$ on the medial axis. In particular, suppose a is on a medial edge, and its neighbors are a_1 and a_2 to the left and right of a respectively.

And m_1, m_2 are halfway midpoints between a_1 and a, a and a_2 along κ respectively. Suppose the two chords of m_1 and m_2 connect to $p_1, p_2 \in \partial\Omega$ respectively. We build a tree T_a containing all the nodes of S inside the region bounded by four curves: the segment on κ between m_1, m_2, the two chords $m_1 p_1$ and $m_2 p_2$, and the boundary segment between $p_1 p_2$. See Figure 3.

Consider a canonical piece, we define a (continuous) *normalized contour* κ_i of level $i \in [0, 1]$ as the collection of points w such that w lies on a chord xy with $x \in \kappa$, $y \in \partial\Omega$, and $i = |wx|/|xy|$ (the normalized distance to κ). We will use this to redefine chords in the discrete setting. Each node w has a parent in the tree T_a as a node who lies on a lower level contour than w. For each point $p \in S$, if T_a contains p we de-

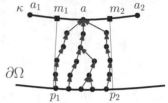

Fig. 3. A discrete chord.

note by $a = \hat{\kappa}(p)$ its projection on $\hat{\kappa}$, to be differentiated from the continuous projection of p on κ. We also denote by $r(a)$ the length of the chord of a. We denote by $\hat{r}(a)$ the depth of the tree T_a, and $\hat{n}(a)$ the total number of nodes in T_a.

For a node a on κ, we first want to understand the relationship between $\hat{r}(a)$ and $\hat{n}(a)$. In the continuous setting, the analog of $\hat{r}(a)$ refers to the length of a chord issued at g, while the analog of $\hat{n}(a)$ refers to the measure of the points on this chord. And these two numbers are the same. In this discrete setting, however, these two measures can be different. In the extreme setting, imagine a perfect disk of radius r. The medial axis is a single point at the center of the disk o. The chords are organized in about $O(1/\varepsilon)$ trees, each one containing roughly $O(r^2/\varepsilon)$ nodes. So in this case $\hat{r}(o) = O(r/\varepsilon)$ and $\hat{n}(o) = O(r^2/\varepsilon)$ – a big discrepancy. However, this can only happen when the point o has infinitely many chords, i.e., the boundary segment is an arc. Under our assumption that medial vertices have degree 3, $\hat{n}(a) = O(\hat{r}(a))$ for any $a \in \kappa$. The proof of the following Lemma is omitted in this abstract.

Lemma 1. *Suppose any point on the medial axis κ of domain Ω has only a finite number of tangents. Then $\hat{n}(a) = O(\hat{r}(a))$ for any $a \in \kappa$.*

By now the continuous medial axis is abstracted by a (discrete) medial axis graph (MAG) $\hat{\kappa}$ in which the (continuous) medial axis are approximated by a discrete sample V and V are connected in the same way as the medial edges. The size of $\hat{\kappa}$ is depends on the geometric complexity of Ω. Typically $|V| = k$ is much smaller than n, the number of sensors. Algorithmically, extracting the medial axis and assigning coordinates proceeds by 1) detecting the boundaries of the sensor domain Ω, 2) flooding to get a discrete medial axis (nodes having equal hop-counts to the boundary) and finally 3) naming each node by a local computation. These steps are described in detail in [3] and we omit the discussion here.

3 Our Medial Axis Based Routing Scheme

Based on [3] we develop the new algorithm as below. The last three steps are new.

1. Extract the medial axis $\hat{\kappa}$ of the discrete sensor network. Assign all the sensors medial axis based coordinates.

2. Run a flow program on the medial axis graph. This returns a routing scheme on the medial axis.
3. Store a compact representation of this routing scheme at the sensors.
4. Extend the routing scheme on the medial axis to get a routing scheme on the entire domain, by routing first "in parallel" and then vertically to the medial axis. By routing in parallel, we move in the direction guided by the routing scheme along the medial axis and keeping the normalized distance to the medial axis to be the same (see Figure 2). By routing vertically, we mean the message travels along a chord towards the destination.

3.1   Step 2: The Flow Program

Let the (discrete) medial axis graph be denoted as $\hat{\kappa} = (V, E)$, $|V| = k$. We denote by $\hat{r}(v)$ the maximum depth of the shortest path tree rooted at $v \in V$, and by $\hat{n}(v)$ the number of sensors in the tree(s) rooted at this node. For each node we assign a capacity $c \cdot \hat{r}(v)$, where $c > 0$ is a parameter.

Our flow program will capture all traffic projected onto the medial axis. Thus we have $k(k-1)$ commodities, one for every ordered pair (i, j) of nodes in V with demand $d_{ij} = d_{ji} = \hat{n}(i)\hat{n}(j)$.

Let the flow of commodity ij along the edge (u, v) be $f_{ij}(u, v)$. Our flow must satisfy the following constraints:

- **Capacity constraints** The load of v is from three kinds of messages: messages with source v, messages with destination v, and messages with neither source nor destination as v.

$$\sum_{j \neq v} \sum_{u} f_{vj}(v, u) + \sum_{j \neq v} \sum_{u} f_{jv}(u, v) + \sum_{ij:i \neq v, j \neq v} \sum_{u} f_{ij}(u, v) \leq c\hat{r}(v)$$

Note that the first two sums each simplify to $m - \hat{n}(v)$, as there are precisely these many messages with v as source and destination.
- **Flow conservation** $\sum_{w \in V} f_{ij}(u, w) = \sum_{w \in V} f_{ij}(w, u)$ $\forall u \notin \{i, j\}$.
- **Demand satisfaction** $\forall i, j \in V$, $\sum_{w \in V} f_{ij}(i, w) = \sum_{w \in V} f_{ij}(w, j) = d_{ij}$.

The program we want to run is to minimize c subject to these conditions. Since the only unknown variables above are the flow quantities, i.e., $f_{ij}(u, v)$ for commodity ij on edge (u, v), we can do the following: we first fix c large enough (say equal to the sum of all the demands, which is $O(n^3)$, where n is the number of sensors in Ω), and find whether the set of three constraints are *feasible*. This would imply that the chosen capacity is sufficient to route the flows according to the specified demands. We then do a binary search on the optimal value of c; each time we halve the current value of c and check for feasibility. Thus we end up with the optimal c, in a runtime that is $O(\log n)$ times the complexity of checking feasibility, which sums up to $O(k^2 \log n)$. This is pretty efficient; note that k is the number of nodes on the medial axis $\hat{\kappa}$, which is much smaller than n (the number of sensors).

Once we get the optimal value of c and the corresponding flows $f_{ij}(u, v)$ on every edge for every commodity, we can find the paths to take from a particular source i to source j using the "path stripping" algorithm in [14]. This returns a set of paths that realize the optimal flow, and these paths constitute the routing scheme on the medial axis graph $\hat{\kappa}$. We call this routing scheme Γ_κ.

3.2   Step 3: Compact Representation of Γ_κ

Once the medial axis is extracted and the coordinates are assigned, we also store a compact representation of the medial axis graph, called CMAG. This graph has the following properties:

1. The number of vertices in the CMAG equals the number of medial vertices.
2. A path (on the medial axis) between two medial vertices that does not go through any other medial vertex, corresponds to an edge between the corresponding pair of vertices on the CMAG. Thus "consecutive" medial vertices have an edge between them.
3. The size of the CMAG is linear in the number of big topological features in Ω; if Ω has h holes (boundaries), then this graph has size $O(h)$.

The proof of the following theorem can be found in the appendix.

Theorem 2. *Let $\hat{\kappa}_c = (V_c, E_c)$ denote the CMAG, with $k = O(h)$ number of vertices and edges. The routing scheme Γ_k can be stored compactly in a way that requires $O(h)$ space for any node on the medial axis, and $O(h2^h)$ space for any medial vertex (a vertex of $\hat{\kappa}_c$).*

3.3   Extending Γ_κ to Γ on Ω

Given a source-destination pair $(s,t) \in \Omega$, we now use Γ_κ to build a path from s to t. Roughly, the path starts from s, follows nodes at the same normalized distance (same r in terms of the medial axis coordinates in Figure 2) to the medial axis as s until it arrives at the chord containing t, and then follows part of the chord to arrive at t. The first part of this path is denoted as routing in parallel to $\hat{\kappa}$. The second part is denoted as routing vertically to $\hat{\kappa}$. The homotopy of this path is the same as the homotopy of the path from $\kappa(s)$ to $\kappa(t)$, as determined by Γ_κ.

Route in Parallel to $\hat{\kappa}$. In the discrete setting, suppose v is on the chord of a_1 and the next node on κ (as suggested by the flow algorithm) is a_2. We choose the next node from the chord of a_2. We may not have a node in T_{a_2} with exactly the same normalized distance to κ as v. Instead, we choose the node u whose normalized distance to κ is the closest to the normalized distance from the source s to κ. By our density constraint, u is within distance ε from the normalized contour κ_i, where i is the normalized distance of s to κ. The same is done for routing in the rotary system if needed.

Route Vertically to $\hat{\kappa}$. When the message gets to a node whose projection to κ is the same as the projection of the destination t, we simply route the message along the tree to the destination.

Route in the Rotary System. At last, we remark that we need to also involve the rotary system as introduced in [3]. Basically the chords of medial vertices partition the domain Ω into *canonical pieces* in which the coordinates we define can be considered as cartesian coordinates. Some canonical pieces share a common chord while some canonical pieces only share a common medial vertex (e.g. the pieces near a medial vertex of degree 3). Now, consider a route P along the medial axis that goes through a medial vertex q with degree 3. If the canonical pieces for the two medial edges on P right before and after q do not share a common chord of q, the extension of P to the domain has to connect the two canonical pieces somehow. To do that, we introduce a rotary system inside the maximal ball centered at each medial vertex of degree 3 using a polar coordinate system. The extension of P will travel along an arc connecting the two canonical pieces. There are two arcs connecting the two pieces, clockwise or counterclockwise. In [3], the direction is arbitrary. Here we always split with probability half among the two choices so each gets half of the total traffic.

Comparison to the Medial Axis Routing Scheme in [3]. The idea of routing in parallel to the medial axis is the same as in [3]; the main innovation here is in the flow program in Step 2. In [3] the routing used on the medial axis was simply the shortest path routing. Our routing scheme given by the flow program helps us prove the approximation guarantee. Theorem 2 guarantees that this routing scheme is also compactly represented and lightweight, much like the shortest path scheme.

4   Proof of Approximation Guarantee

In this section we present the proof of the following theorem:

Theorem 3. *There exists a constant $C \geq 1$ such that the maximum load of the routing scheme Γ obtained above is at most C times the maximum load of the optimal routing scheme Γ^* on Ω.*

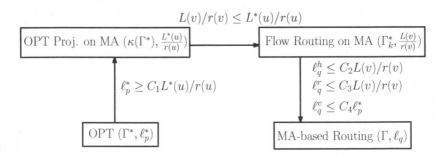

Fig. 4. The outline of the proof for medial-axis based routing.

Proof Sketch. The proof is highly technical and involves a number of new ideas. The outline of the proof is shown in Figure 4. Let the maximum load of Γ^* (Γ) be realized at a point p (or q), and equal ℓ_p^* (ℓ_q). We will show $\ell_q \leq C\ell_p^*$ for some constant C.

There are three main steps to relate ℓ_p^* at node p with ℓ_q at node q. In the first step we consider the projection of optimal routing scheme on medial axis, denoted by $\kappa(\Gamma^*)$.

That is, for each path $\gamma^*(s, t) \in \Gamma^*$, we project each node onto the medial axis, getting a path along the medial axis from $\hat{\kappa}(s)$ to $\hat{\kappa}(t)$. This path might be non-simple, containing duplicated visits to the same node. When such things happen, we simply remove the redundant segments and only keep a simple path along $\hat{\kappa}$. The traffic load caused by $\kappa(\Gamma^*)$ is no greater after this small simplification. It is clear to see that the traffic pattern for $\kappa(\Gamma^*)$ is exactly the same as the traffic pattern for our flow program: node i on the medial axis sends $\hat{n}(i)\hat{n}(j)$ messages to node j.

The three steps in Figure 4 relate the traffic load in the four routing algorithms (the optimal algorithm, the projection of the optimal on κ, the flow program, and the extension of the flow algorithm to Ω), the bottom two on the original domain Ω and top two on the medial axis.

Step One: We relate optimal max load with the projection on $\hat{\kappa}$.

Lemma 4. *Denote by $L^*(i)$ the number of messages passing through node $i \in \hat{\kappa}$ under the projection $\kappa(\Gamma^*)$. Let $u \in \hat{\kappa}$ be the node where the maximum of the quantity $L^*(i)/\hat{r}(i)$ occurs. Then $\ell_p^* \geq C_1 L^*(u)/\hat{r}(u)$, for some constant C_1.*

Proof. Consider the discrete chord at u, which is a tree T_u with $n(u)$ nodes, each has traffic at most ℓ_p^* (the maximum traffic load). Thus the traffic at u by the projection on $\hat{\kappa}$ is $L^*(u) \leq \hat{n}(u)\ell_p^*$. The claim follows from the fact that $\hat{n}(u) = O(\hat{r}(u))$ for any node $u \in \hat{\kappa}$ (Lemma 1).

Step Two: On the medial axis, we run an optimization algorithm to find the routing scheme minimizing $L(x)/\hat{r}(x)$ for any node x on the medial axis, for the given traffic pattern. Minimizing max load is an NP-hard problem when the problem is integral, i.e., only a single path is taken by each node. But in our case, we use a non-integral solution which can be interpreted as a probabilistic solution on a family of paths. Thus we minimize the expected maximum traffic load. This routing paths are denoted by the family Γ_κ. Let $v \in \hat{\kappa}$ be the node where the quantity $L(x)/\hat{r}(x)$ achieves its maximum value ($L(x)$ is the number of messages passing through x under Γ_κ). Optimality of the flow program then implies $L(v)/\hat{r}(v) \leq L^*(u)/\hat{r}(u)$.

Step Three: In the last step, we take the routing paths along the medial axis and convert them to routing paths in the original network. Specifically, for a source s and t in the domain, we first project them to the medial axis along the chord at $\hat{\kappa}(s), \hat{\kappa}(t)$ respectively. Then we use the routing algorithm guided by Γ_κ computed in the previous step to compute a path $\gamma(\hat{\kappa}(s), \hat{\kappa}(t))$ along the medial axis. This path is then converted to a path in the original domain – specifically, the message first travels on a path parallel along the medial axis until it arrives at the chord of t; then it follows the chord to arrive at t. It may also use the rotary system around a medial vertex. Thus the traffic at q (the node with max load) in our routing scheme is divided naturally into horizontal (denoted by ℓ_q^h), rotary (denoted by ℓ_q^r) and vertical (denoted by ℓ_q^v) loads. Thus $\ell_q = \ell_q^h + \ell_q^r + \ell_q^v$. For the horizontal and rotary traffic, we show that there exist constants C_2 and C_3 such that $\ell_q^h \leq C_2 L(v)/\hat{r}(v)$, and $\ell_q^r \leq C_3 L(v)/\hat{r}(v)$. For the vertical traffic, we show that it is bounded by the maximum load of *any* routing scheme on Ω, including that of the optimum Γ^*. Thus we show that $\ell_q^v = O(\ell_p^*)$.

Collecting the inequalities from the three steps, we have that for some constant C, $\ell_q \leq C\ell_p^*$. The complete proof for the third step is in the full paper. In the following we only provide a sketch.

Proof for ℓ_q^h: We first note that by definition of $L(v)/\hat{r}(v)$ as the maximum of the LP program, $L(\hat{\kappa}(q))/\hat{r}(\hat{\kappa}(q)) \leq L(v)/\hat{r}(v)$, where $\hat{\kappa}(q)$ is the projection of q on $\hat{\kappa}$. Hence it suffices to show that there exists C_3 such that $\ell_q^h \leq C_3 L(\hat{\kappa}(q))/\hat{r}(\hat{\kappa}(q))$. For simplicity, set $r := \hat{r}(\hat{\kappa}(q))$. In fact we will prove that for any node v on the chord of $\hat{\kappa}(q)$, the traffic load ℓ_v^h is bounded as required.

The total horizontal traffic load at $\hat{\kappa}(q)$ is actually shared by all the nodes in the chord of $\hat{\kappa}(q)$. If the nodes in the chord share the traffic uniformly, then the claim is trivially true. Note that the flow program, and hence our extension, does not distinguish between nodes on the same chord when it comes to determining homotopy of paths[1].

Therefore, the traffic carried by a node is determined by how many sources stay on the same horizontal level, which can differ. For any node v at normalized distance β, denote by H_β the normalized contour formed by all nodes at normalized distance β. Now, only the nodes with normalized distance within $[\beta - \varepsilon, \beta + \varepsilon]$ may possibly arrive at v, by our definition. By the bounded density condition, the number of sensor nodes with normalized height within $[\beta - \varepsilon, \beta + \varepsilon]$ is proportional to the total area occupied by the points with normalized height in the same range, which is then proportional to the length of H_β. Thus we have that

Observation 1: The ratio of the number of messages passing horizontally through a node v (at height β) and the total number of messages passing through the chord containing v, is proportional to the length of H_β.

Now we need to examine the length of the contours H_β for different normalized distance $\beta \in [0, 1]$. The heaviest traffic load happens at the node whose depth coincides with the longest contour. The worst case is that the contour at one particular normalized height is long, while all the others are very short. The following observation says that this cannot happen (proof omitted).

Observation 2: Suppose H_i is the longest contour, $i \in [0, 1]$. There is a constant δ such that one of the two cases is true: (i) the contours H_j with $j \in [i, i + \delta]$ have length $\Omega(\text{Length}(H_i))$; (ii) the contours H_j with $j \in [i - \delta, i]$ have length $\Omega(\text{Length}(H_i))$.

Within the chord of $\hat{\kappa}(q)$, $O(\delta r)$ nodes have normalized depth within the range $[i, i + \delta]$ or $[i - \delta, i]$. Even if all other contours have length 0 and the rest of the nodes in the chord share no traffic within $L(\hat{\kappa}(q))$, the maximum node will still receive traffic load at most $L(\hat{\kappa}(q))/(\delta r)$. Since δ is a constant, the claim is true.

Proof of ℓ_q^r: Some nodes will carry both horizontal traffic and the rotary traffic. The proof for the rotary traffic being not very high is in fact the same as the argument on horizontal traffic – by analyzing the length of contours at different normalized depth. Thus we omit the discussion here.

Proof of ℓ_q^v: First recall that a message passes through q vertically iff the destination is in the subtree(s) rooted at $\hat{\kappa}(q)$. Thus the number of messages passing vertically is

---

[1] For a given destination t, the distribution over the homotopy types of paths from s_1 to t or s_2 to t is the same, if s_1 and s_2 are on the same chord.

$O(n \cdot w)$ where $w = n(\kappa(q))$, since the source could be any of the n sensors, while the destination must be in the subtree, that has size $w = n(\hat{\kappa}(q))$.

We claim that for any routing scheme Γ on the network, the maximum load is $\Omega(n\sqrt{n})$. This is simple if the network communication graph is assumed to be planar, in which case we can use the planar separator theorem [10]: there exists a cut S of size $O(\sqrt{n})$ such that the removal of this set partitions the graph into two subgraphs A and B, with no edges from A to B, and A and B have at least $n/3$ nodes. This means that during all-pairs communication, there are $\Omega(n^2)$ messages with source in A and destination in B, and all these messages must pass through nodes in S. Thus the average load of a node in S is $\Omega(n^2/\sqrt{n}) = \Omega(n\sqrt{n})$, which is clearly no greater than the maximum load over S. The proof can be generalized to the case when the graph is a *unit-disk-graph*, using arguments similar to those presented in [6]. We omit the technical details here.

Now we show that $\ell^v(q)$ is smaller than the maximum load of any routing scheme. By arguments above, $\ell^v(q) = O(nw)$ and by Lemma 3, $w = n(\kappa(q)) = O(r(\kappa(q))$. Thus if we prove that $r(\kappa(q)) = O(\sqrt{n})$, we are done, as the max load of any routing scheme is $\Omega(n\sqrt{n})$. This is a simple area argument; the maximal ball centered at $\kappa(q)$ and of radius $r(\kappa(q))$ is by definition empty and completely contained inside Ω, since it just touches the boundary at finitely many points. The area of this ball is $\pi r^2(\kappa(q)))$, and because of the uniform density assumption, this area must be $O(n)$. Thus $r = O(\sqrt{n})$ and we are done.

5   Discussions and Open Problems

Randomized to Deterministic: Note that the paths found by the path stripping algorithm are probabilistic; between a given source-destination pair (s, t) the algorithm returns a probability distribution over the set of all possible paths between s and t. One could use randomized rounding as in [14] to get deterministic paths between every pair; however, this will increase our approximation factor from $O(1)$ to a factor that is slightly sublogarithmic in n (the number of sensors on the medial axis).

Other Traffic Distributions: In this paper we showed how to get a constant factor approximation for uniform traffic load. However, our algorithm can easily be generalized to any arbitrary traffic distribution Π on Ω, although we cannot prove the approximation yet. Let $\Pi(s, t)$ denote the probability of communication between source-destination nodes s and t respectively. Let $u = \kappa(s)$ and $v = \kappa(t)$ be their projections on the medial axis. Denote by C_{iu} and C_{jv} denote the set of nodes at depths i and j in the trees rooted at nodes u and v, respectively, and the cardinalities of these sets by n_{iu} and n_{iv}, respectively.

When we set up the multicommodity flow program, instead of assigning a demand of $r_u r_v$ between nodes u and v on the medial axis, we now assign a demand

$$d_{u,v} = \sum_{i=1}^{r(u)} \sum_{j=1}^{r(v)} \sum_{s \in C_{iu}} \sum_{t \in C_{jv}} \frac{\Pi(s,t)}{n_{iu} n_{jv}},$$

and run the flow algorithm. The paths generated by the flow algorithm can be extended to get a routing scheme on Ω that satisfies the traffic distribution Π.

Thus, the two open questions that remain are 1) does the algorithm of this paper for uniformly deployed nodes and arbitrary traffic patterns provide an approximation guarantee?, and 2) can one remove the uniformly deployed nodes condition, even for the uniform traffic distribution?

Acknowledgement. Jie Gao would like to acknowledge NSF support through DMS-1418255, DMS-1221339, CNS-1217823 and Air Force Research AFOSR FA9550-14-1-0193.

References

1. Andrews, M., Zhang, L.: Hardness of the undirected congestion minimization problem. SIAM Journal on Computing 37(1), 112–131 (2007)
2. Attali, D., Boissonnat, J.-D., Edelsbrunner, H.: Stability and computation of the medial axis — a state-of-the-art report. In: Mathematical Foundations of Scientific Visualization, Computer Graphics, and Massive Data Exploration. Springer (2004)
3. Bruck, J., Gao, J., Jiang, A.: MAP: Medial axis based geometric routing in sensor networks. Wireless Networks 13(6), 835–853 (2007)
4. Chekuri, C., Khanna, S., Shepherd, F.B.: An $o(\sqrt{n})$-approximation for EDP in undirected graphs and directed acyclic graphs. Theory of Computing 2, 137–146 (2006)
5. Chuzhoy, J., Khanna, S.: New hardness results for undirected edge-disjoint paths. Manuscript (2005)
6. Gao, J., Zhang, L.: Well-separated pair decomposition for the unit-disk graph metric and its applications. In: Proceedings of the Thirty-Fifth Annual ACM Symposium on Theory of Computing, STOC 2003, pp. 483–492. ACM, New York (2003)
7. Gao, J., Zhang, L.: Load balanced short path routing in wireless networks. In: IEEE InfoCom, vol. 23, pp. 1099–1108 (March 2004)
8. Goswami, M., Ni, C.-C., Ban, X., Gao, J., Gu, D.X., Pingali, V.: Load balanced short path routing in large-scale wireless networks using area-preserving maps. In: Proceedings of the 15th ACM International Symposium on Mobile Ad Hoc Networking and Computing (Mobihoc 2014), pp. 63–72 (August 2014)
9. Karp, R.M.: Reducibility Among Combinatorial Problems. In: Miller, R.E., Thatcher, J.W. (eds.) Complexity of Computer Computations, pp. 85–103. Plenum Press (1972)
10. Lipton, R.J., Tarjan, R.E.: A separator theorem for planar graphs. SIAM Journal on Applied Mathematics 36(2), 177–189 (1979)
11. Mei, A., Stefa, J.: Routing in outer space: fair traffic load in multi-hop wireless networks. In: MobiHoc 2008: Proceedings of the 9th ACM International Symposium on Mobile Ad Hoc Networking and Computing, pp. 23–32. ACM, New York (2008)
12. Popa, L., Rostamizadeh, A., Karp, R., Papadimitriou, C., Stoica, I.: Balancing traffic load in wireless networks with curveball routing. In: MobiHoc 2007: Proceedings of the 8th ACM International Symposium on Mobile Ad Hoc Networking and Computing, pp. 170–179 (2007)
13. Raghavan, P.: Probabilistic construction of deterministic algorithms: approximating packing integer programs. J. Comp. and System Sciences, 130–143 (1988)

14. Raghavan, P., Thompson, C.D.: Provably good routing in graphs: regular arrays. In: Proceedings of the 17th Annual ACM Symposium on Theory of Computing, pp. 79–87 (1985)
15. Sarkar, R., Zeng, W., Gao, J., Gu, X.D.: Covering space for in-network sensor data storage. In: Proc. of the 9th International Symposium on Information Processing in Sensor Networks (IPSN 2010), pp. 232–243 (April 2010)
16. Yu, X., Ban, X., Sarkar, R., Zeng, W., Gu, X.D., Gao, J.: Spherical representation and polyhedron routing for load balancing in wireless sensor networks. In: Proc. of 30th Annual IEEE Conference on Computer Communications (INFOCOM 2011), pp. 612–615 (April 2011)

An Experimental Evaluation
of the Best-of-Many Christofides' Algorithm
for the Traveling Salesman Problem

Kyle Genova[1] and David P. Williamson[2]

[1] Department of Computer Science, Cornell University, Ithaca, NY 14853
kag278@cornell.edu
[2] School of Operations Research and Information Engineering,
Cornell University, Ithaca, NY 14853
dpw@cs.cornell.edu
www.davidpwilliamson.net/work

Abstract. Recent papers on approximation algorithms for the traveling salesman problem (TSP) have given a new variant on the well-known Christofides' algorithm for the TSP, called the *Best-of-Many Christofides' algorithm*. The algorithm involves sampling a spanning tree from the solution to the standard LP relaxation of the TSP, and running Christofides' algorithm on the sampled tree. In this paper we perform an experimental evaluation of the Best-of-Many Christofides' algorithm to see if there are empirical reasons to believe its performance is better than that of Christofides' algorithm. In our experiments, all of the implemented variants of the Best-of-Many Christofides' algorithm perform significantly better than Christofides' algorithm; an algorithm that samples from a maximum entropy distribution over spanning trees seems to be particularly good.

Keywords: traveling salesman problem, Christofides algorithm.

1 Introduction

In the traveling salesman problem (TSP), we are given a complete, undirected graph $G = (V, E)$ as input with costs $c_e \geq 0$ for all $c \in E$, and we must find a tour through all the vertices of minimum cost. In what follows, we will assume that the costs obey the triangle inequality; that is, $c_{(u,w)} \leq c_{(u,v)} + c_{(v,w)}$ for all $u, v, w \in V$. We will sometimes refer to the *asymmetric* traveling salesman problem (ATSP), in which the input is a complete directed graph, and possibly $c_{(u,v)} \neq c_{(v,u)}$.

In 1976, Christofides [6] gave a $\frac{3}{2}$-approximation algorithm for the TSP; an α-approximation algorithm for the TSP is one that runs in polynomial time and returns a solution of cost at most α times the cost of an optimal solution. The value α is sometimes known as the performance guarantee of the algorithm. Christofides' algorithm works as follows: it computes a minimum-cost spanning tree (MST) F of the input graph G, then finds a minimum-cost perfect matching M on all the odd-degree vertices of the tree F. The resulting edge set $F \cup M$ is

© Springer-Verlag Berlin Heidelberg 2015
N. Bansal and I. Finocchi (Eds.): ESA 2015, LNCS 9294, pp. 570–581, 2015.
DOI: 10.1007/978-3-662-48350-3_48

then an Eulerian subgraph of G. By "shortcutting" an Eulerian traversal of the subgraph, we can obtain a tour that visits each vertex exactly once and has cost no greater than the cost of the edges in $F \cup M$.

No approximation algorithm with performance guarantee better than $\frac{3}{2}$ is yet known for the TSP. However, some progress has been made in recent years for special cases and variants of the problem. Asadpour et al. [4] gave an $O(\log n / \log \log n)$-approximation algorithm for the ATSP (where $n = |V|$), improving on a long-standing $O(\log n)$-approximation algorithm of Frieze, Galbiati, and Maffioli [8]. A sequence of improvements has been obtained in the special case of the *graph* TSP, in which the input to the problem is an undirected, not necessarily complete graph G, and the cost $c_{(u,v)}$ for each $u, v \in V$ is the number of edges in the shortest u-v path in G. In this case, Oveis Gharan, Saberi, and Singh [21] were able to improve slightly on the factor of $\frac{3}{2}$. Mömke and Svensson [16] then gave a 1.462-approximation algorithm; Mucha [17] improved the analysis of the Mömke and Svensson algorithm to obtain a $\frac{13}{9}$-approximation algorithm. Sebő and Vygen [24], by adding some additional ideas, gave a 1.4-approximation algorithm for graph TSP.

An idea used in several of these results is to start with a tree that is determined by an LP relaxation, rather than the minimum-cost spanning tree. For the TSP, a well-known relaxation of the problem is as follows:

$$\text{Min} \sum_{e \in E} c_e x_e :$$
$$x(\delta(v)) = 2, \ \forall v \in V; \quad x(\delta(S)) \geq 2, \ \forall S \subset V, S \neq \emptyset; \quad 0 \leq x_e \leq 1, \ \forall e \in E,$$

where $\delta(S)$ is the set of all edges with exactly one endpoint in S and we use the shorthand that $x(F) = \sum_{e \in F} x_e$. This LP relaxation is sometimes called the *Subtour LP*. It is not hard to show that given a feasible solution x to the Subtour LP, $\frac{n-1}{n}x$ is feasible for the spanning tree polytope $\{x \in \Re^{|E|} : x(E) = n-1, x(E(S)) \leq |S| - 1 \ \forall S \subseteq V, |S| \geq 2\}$, where $E(S)$ is the set of all edges with both endpoints in S. Oveis Gharan, Saberi, and Singh [21] propose an algorithm which has since been called (by [2]) *Best-of-Many Christofides*: given the LP solution x^*, we can compute in polynomial time a decomposition of x^* into a convex combination of spanning trees, and we run Christofides' algorithm for each of these trees and output the lowest cost solution found. More precisely, if $\chi_F \in \{0,1\}^{|E|}$ is the characteristic vector of a set of edges F, then given Subtour LP solution x^*, we find spanning trees $F_1, \ldots, F_k$ such that $\frac{n-1}{n}x^* = \sum_{i=1}^{k} \lambda_i \chi_{F_i}$ for $\lambda_i \geq 0$ and $\sum_{i=1}^{k} \lambda_i = 1$. Then for each tree F_i we find a matching M_i of the odd-degree vertices, and we compute a tour by shortcutting $F_i \cup M_i$. We return the cheapest tour found.

An alternative perspective is to consider randomly sampling a spanning tree from a distribution on spanning trees given by the convex combination, then run Christofides' algorithm on the resulting tree found; that is, we sample tree F_i with probability λ_i. This perspective potentially allows us to avoid computing the convex combination explicitly. However, the distribution of trees then depends on the (implicit) convex combination. Asadpour et al. [4] and Oveis

Gharan, Saberi, and Singh [21] use a *maximum entropy* distribution. For the Asadpour et al. ATSP result, the main property used of the maximum entropy distribution is that in some cuts of edges, the appearance of arcs is negatively correlated. Chekuri, Vondrák, and Zenklusen [5] show how to draw a sample with the appropriate negative correlation properties given an explicit convex combination of trees; their distribution over trees is not the same as the maximum entropy distribution.

An exciting possible direction for an improved approximation algorithm for the TSP is to show that some variation of the Best-of-Many Christofides' algorithm gives a performance guarantee strictly better than $3/2$, either by starting with a convex combination of spanning trees, or using some of the stronger properties obtained by sampling a tree from a negatively correlated distribution, or the maximum entropy distribution. In this paper, we experimentally evaluate these different versions of the Best-of-Many Christofides' algorithm in order to see if there is empirical evidence that these algorithmic variants are any better than the standard Christofides' algorithm, and whether any of the variants is more promising than the others.

In particular, we start by implementing Christofides' algorithm. Since most of our instances are geometric, we compute a Delaunay triangulation using the package Triangle [25]; it is known that the edges of an MST for a 2D Euclidean instance are a subset of the edges of the Delaunay triangulation. We use Prim's algorithm to compute the MST from these edges. For non-geometric instances, we use Prim's algorithm to compute the MST. We then use the Blossom V code of Kolmogorov [12] to find a minimum-cost perfect matching on the odd degree vertices of the tree. We compute a tour by shortcutting the resulting Eulerian graph; we perform a simple optimization on the shortcutting. We then use the Concorde TSP solver [3] to compute a solution x^* to the subtour LP. We implement two different ways of finding an explicit convex combination of trees equal to $\frac{n-1}{n}x^*$; in the first, we use a column generation technique suggested by An in his Ph.D. thesis [1] in conjunction with the linear programming solver Gurobi [10]. In the second, we compute a packing of spanning trees via iteratively "splitting off" edges of the LP solution from vertices, then maintaining a convex combination of trees as we "lift back" the split-off edges. For this algorithm, we use a subroutine of Nagamochi and Ibaraki [18] to obtain a complete splitting-off of a vertex. We also implement two methods for obtaining a randomly sampled tree from the support of the LP solution. We first implement the SwapRound procedure of Chekuri et al. [5]; given an explicit convex combination of trees generated by the first two methods, we can sample a spanning tree such that the edges of the tree appearing in any given set are negatively correlated (we define the negative correlation more precisely in Section 2.3). We also implement the method for computing a maximum entropy distribution over spanning trees given the LP solution x^*, and then drawing a sample from this distribution, as given in the ATSP paper of Asadpour et al. [4] and the Ph.D. thesis of Oveis Gharan [19]. Our implementation choices for the maximum entropy routine were influenced by a code shared with us by Oveis Gharan [20].

To test our results, we ran these algorithms on TSPLIB instances of Reinelt [22] (both Euclidean and non-Euclidean instances) and Euclidean VLSI instances from Rohe [23]. We also considered graph TSP instances to see if the performance of the algorithms was better for such instances than for weighted instances. For our graph TSP instances, we used undirected graphs from the Koblenz Network Collection of Kunegis [13].

It is known that the standard Christofides' algorithm typically returns solutions of cost of about 9-10% away from the cost of an optimal solution on average (see, for instance, Johnson and McGeoch [11]); this is better than its worst-case guarantee of at most 50% away from the cost of an optimal solution, but not as good as other heuristics (such as the Lin-Kernighan heuristic [14]) that do not have performance guarantees. We confirm these results for the standard Christofides' algorithm on geometric instances, but Christofides' algorithm appears to do worse on graph TSP instances; we had solutions of about 12% away from optimal. All of the Best-of-Many Christofides' algorithms performed substantially better than the standard Christofides' algorithm, with solutions of cost about 3-7% away from optimal for the Euclidean instances, 2-3% away from optimal for the non-Euclidean instances, and under 1% away from optimal for the graph TSP instances. These results may indicate that the graph TSP instances are easier for LP-based algorithms than geometric instances. The algorithm that used the maximum entropy distribution on average outperformed the other Best-of-Many Christofides' algorithms; however, the algorithm that found a convex combination of spanning trees via splitting off, then used the SwapRound routine, was nearly as good as maximum entropy sampling, and was better in some cases.

Our paper is structured as follows. In Section 2, we give a more detailed description of the algorithms that we implemented. In Section 3, we describe the TSP datasets we used and our machine environment, and in Section 4 we give the results of our experiments, as well as some analysis. We conclude in Section 5. Because of space constraints, we have omitted many details and some figures from this extended abstract; a full version of the paper is available online [9].

2   Algorithms

In this section, we give descriptions of the various algorithms that we implemented. Sections 2.1 and 2.2 describe the two algorithms that generate explicit convex combinations of spanning trees given the Subtour LP solution x^*; as described above, we compute the Subtour LP solution by using the Concorde TSP solver [3]. Section 2.3 describes the SwapRound algorithm of Chekuri, Vondrák, and Zenklusen [5] that generates a randomly sampled tree given an explicit convex combination of spanning trees. Section 2.4 describes the maximum entropy distribution on spanning trees, the algorithm used to compute it, and the algorithm used to draw a sample from this distribution.

2.1   Column Generation

Our first algorithm for decomposing the Subtour LP solution $\frac{n-1}{n}x^*$ into a convex combination of spanning trees follows an algorithm described by An in his Ph.D. thesis [1]; we use column generation to generate the trees in the convex combination. In particular, we would like to solve the following linear program, in which we have a variable y_T for each possible spanning tree T, where we assume that x_e^* is the given solution to the Subtour LP and the graph $G = (V, E)$ is on the edges $E = \{e : x_e^* > 0\}$:

$$\text{Min} \sum_{e \in E} s_e : \sum_{T:e \in T} y_T + s_e = \frac{n-1}{n}x_e^*, \ \forall e \in E; \quad y_T, s_e \geq 0, \ \forall T, \forall e \in E.$$

Since $\frac{n-1}{n}x^*$ can be expressed as a convex combination of spanning trees, the optimal solution to the LP is zero, and for an optimal solution (y^*, s^*), y^* gives a convex combination of spanning trees.

Because of space constraints, we omit details of the implementation of the algorithm. As is typical of column generation algorithms, solving the LP to optimality can take a long time. On instances of 500 cities, it could take up to 10 hours of computation time. Thus in order to make comparisons with other methods, we terminated early. In particular, we store the current objective function value and wait until either the value drops by .1 or 100 iterations have occurred. If the objective has not dropped by .1 in the 100 iterations, we terminate; otherwise, once it has dropped by .1 we restart the iteration count. Our cutoff behavior allows us to avoid the long set of iterations in which the method is near optimal but only makes incremental progress.

2.2   Splitting Off and Tree Packing

Let x^* be a basic feasible solution to the Subtour LP. It is known that x^* is rational, so that there exists a K such that Kx^* is integer. We let $\hat{G}$ be a multigraph with Kx_e^* copies of edge e. Then by the constraints of the Subtour LP, each vertex has $2K$ edges incident on it, and for each $i, j \in V$, there are at least $2K$ edge-disjoint paths between i and j. Lovász [15] showed that given any edge (x, z) incident on z it is possible to find another edge (y, z) incident on z such that if we remove (x, z) and (y, z) from $\hat{G}$ and add edge (x, y), then for all pairs of vertices $i, j \in V$, $i, j \neq z$, there are still $2K$ edge-disjoint paths between i and j. Removing (x, z) and (y, z) and adding (x, y) so as to preserve connectivity in this way is called the *splitting off* operation. A *complete splitting off* at z removes all $2K$ edges incident at z from $\hat{G}$ and adds K edges not incident on z to $\hat{G}$ so that for all pairs of vertices $i, j \in V$, $i, j \neq z$, there are still $2K$ edge-disjoint paths between i and j. Nagamochi and Ibaraki [18] give an $O(nm \log n + n^2 \log^2 n)$ time algorithm for obtaining a complete splitting off at a node. We implemented the Nagamochi-Ibaraki algorithm, and use it to obtain a set of K trees in the original multigraph $\hat{G}$ (following Frank [7, Chapter 10]) by splitting off all edges except those incident to two nodes, then constructing the trees inductively as we "lift back" the split off edges. We omit the details due to space constraints.

2.3   SwapRound and Negatively Correlated Distributions

The algorithms of the previous two sections give a convex combination of spanning trees such that the convex combination of the characteristic vectors of the trees is dominated by the Subtour LP solution x^*. Let $z^* = \sum_{i=1}^{K} \lambda_i \chi_{F_i} \leq x^*$ give the convex combination of the characteristic vectors χ_{F_i} of K trees F_i. One can think about the convex combination as being a distribution on spanning trees. We sample tree F_i with probability λ_i. A nice feature of this sampling scheme is that the expected cost of the sampled tree is $\sum_{e \in E} c_e z_e^* \leq \sum_{e \in E} c_e x_e^*$, at most the value of the Subtour LP. This follows since the probability that a given edge $e \in F$ is in the sampled tree F is

$$\Pr[e \in F] = \sum_{i:e \in F_i} \lambda_i = \sum_{i:e \in F_i} \lambda_i \chi_{F_i}(e) = z_e^* \leq x_e^*.$$

Let X_e be a random variable which is 1 if edge e is in the sampled tree and 0 otherwise; then we have shown that $E[X_e] = z_e^*$.

Asadpour et al. [4] showed that for proving results about the asymmetric TSP, it is useful to think about drawing a sample such that the edges of the spanning tree appearing in a fixed set are *negatively correlated*. We will say that a probability distribution is negatively correlated if $E[X_e] = z_e^*$ and for any set of edges $A \subseteq E$, $E[\Pi_{e \in A} X_e] \leq \prod_{e \in A} z_e^*$, and $E[\Pi_{e \in A}(1 - X_e)] \leq \prod_{e \in A}(1 - z_e^*)$. Negative correlation allows the proof of concentration bounds that get used in the result of Asadpour et al.

Chekuri, Vondrák, and Zenklusen [5] give a sampling scheme they call Swap-Round such that given any convex combination of spanning trees as input, SwapRound gives a sample from a negatively correlated distribution as output (their result applies more generally to matroids). Let $F_1, \ldots, F_k$ be the trees from the convex combination. The algorithm maintains a spanning tree F, which is initially F_1. Then it loops through the other trees $F_2, \ldots, F_k$, and calls a subroutine, MergeBasis, with the two trees F and F_i, and with probability weights $\sum_{j=1}^{i-1} \lambda_j$ and λ_i, and updates F to be the result of MergeBasis. The routine MergeBasis, given two trees F and F' and weights λ and λ', repeatedly interchanges edges between the trees F and F' until the two are the same. While $F \neq F'$, the routine finds edges $e \in F - F'$ and $e' \in F' - F$ such that $F - e + e'$ is a spanning tree and $F' - e' + e$ is a spanning tree (such edges are known to exist if $F \neq F'$). Then with probability $\lambda/(\lambda + \lambda')$, the routine updates F' to $F' - e' + e$, and otherwise the routine sets F to $F - e + e'$. When $F = F'$, the routine returns F.

We implemented the SwapRound and MergeBasis routines in order to see if sampling a spanning tree from a negatively correlated distribution would lead to better overall results. Because both the column generation and splitting off methods give a convex combination of spanning trees, we tried these two methods both with and without the SwapRound routine on the output. Because the sampling can be performed in parallel, we had four threads running to draw the samples. We drew 1000 samples per instance, and output the best tour found.

2.4   The Maximum Entropy Distribution

Asadpour et al. [4] consider sampling spanning trees from the maximum entropy distribution over spanning trees. Given the subtour LP solution x^*, we set $z^* = \frac{n-1}{n}x^*$. If $\mathcal{T}$ is the set of all spanning trees of the graph G, then the maximum entropy distribution is an optimal solution to the following:

$$\text{Inf} \sum_{T \in \mathcal{T}} p(T) \log p(T) : \sum_{T:e \in T} p(T) = z_e^*, \ \forall e \in E; \ \sum_{T \in \mathcal{T}} p(T) = 1; \ p(T) \geq 0, \forall T.$$

Asadpour et al. show that the constraint $\sum_{T \in \mathcal{T}} p(T) = 1$ is redundant. Given that z^* is in the relative interior of the spanning tree polytope, they argue that there must exist γ_e^* for all $e \in E$ such that sampling tree T with probability proportional to $p(T) = e^{\gamma^*(T)}$ (with $\gamma^*(T) \equiv \sum_{e \in T} \gamma_e^*$) results in $\Pr[e \in T] = z_e^*$ and gives the maximum entropy distribution.

Asadpour et al. then give an algorithm for computing values $\tilde{\gamma}_e$ that approximately satisfy the conditions. In particular, the value $\tilde{\gamma}_e$ are such that if we set

$$\tilde{p}(T) \equiv \frac{1}{P} e^{(\sum_{e \in T} \tilde{\gamma}_e)}, \text{ for } P \equiv \sum_{T \in \mathcal{T}} e^{(\sum_{e \in T} \tilde{\gamma}_e)}, \text{ then } \tilde{z}_e \equiv \sum_{T \in \mathcal{T}: T \ni e} \tilde{p}(T) \leq (1 + \epsilon) z_e^*.$$

To compute the $\tilde{\gamma}_e$, we use a combination of the algorithm suggested in Asadpour et al. and one given by code written by Oveis Gharan [20]. We omit the details due to space constraints.

Once the $\tilde{\gamma}_e$ have been computed, we need to be able to sample from the corresponding distribution. Asadpour et al. set $\lambda_e = e^{\tilde{\gamma}_e}$, and then sample a tree with probability proportional to $\prod_{e \in T} \lambda_e$. Asadpour et al. give an algorithm for computing such a sample in polynomial time, which we implemented. We also implemented the following algorithm for sampling λ-random trees used in the code of Oveis Gharan [20]: we pick an arbitrary starting node i as a location, and start with an empty edge set which will become the spanning tree. With probability proportional to the λ_e for each edge $e = (i, j)$ incident on the current node i, pick an incident edge. If the node j has not yet been visited, add edge (i, j) to the tree. The new location becomes node j. Repeat until every node has been added to the tree (and hence the tree is spanning). This algorithm is not guaranteed to find a sample in polynomial time; however, we found that it scaled to larger instances better than the algorithm given in Asadpour et al.

We sampled trees as desired from the graph; as with the SwapRound method, 1000 samples were used for the results presented here. For each sampled tree, we ran Christofides' algorithm, and output the lowest cost result found across all samples. Once the $\tilde{\gamma}$ are computed, we can draw the samples in parallel using four threads.

3   Experiments

The algorithms above were implemented in C++. We ran the algorithms on a machine with a 4.00Ghz Intel i7-875-K processor with 8GB DDR3 memory.

For test data, we used the TSPLIB instances of Reinelt [22]; we considered 59 two-dimensional Euclidean instances with up to 2103 vertices (averaging 524 vertices), and 5 non-Euclidean instances with 120 to 1032 vertices (gr120, si175, si535, pa561, and si1032), averaging 484 vertices. Similarly, we considered 39 two-dimensional Euclidean VLSI instances of Rohe [23] with up to 3694 vertices, averaging 1473 vertices. We also considered instances of the graph TSP: we used 9 instances from the Koblenz Network Collection of Kunegis [13]. Specifically, we considered undirected simple graphs; if the graph had multiple connected components, we used the largest connected component and discarded the rest of the graph. The resulting instances ranged in size from 18 to 1615 vertices, averaging 363.

4   Results

A summary of our results can be found in Table 1. For the two methods that construct an explicit convex combination of spanning trees (column generation and splitting off), the best error is the error of the minimum-cost tour resulting from running Christofides' algorithm over all the trees in the decomposition, while the average error is the average error from running Christofides' algorithm over all the trees in the decomposition. For the methods that sample trees from a distribution (the swap round variants and the maximum entropy distribution), the best error is the smallest error found over all tours resulting from running Christofides' algorithm on all the trees sampled from the distribution, and the average error is the average over all sampled trees. We make several observations based on this summary.

Table 1. Summary of results, giving the percentage in excess of optimal for the algorithms. 'Std' is Christofides algorithm, 'ColGen' is column generation, 'MaxEnt' is maximum entropy sampling, 'Split' is the splitting-off algorithm, and 'SR' is the swap round algorithm. The TSPLIB E instances are two-dimensional Euclidean, and the TSPLIB N instances are non-Euclidean.

	Std	ColGen		ColGen+SR		MaxEnt		Split		Split+SR	
		Best	Ave	Best	Ave	Best	Ave	Best	Ave	Best	Ave
TSPLIB (E)	9.56%	4.03%	6.44%	3.45%	6.24%	3.19%	6.12%	5.23%	6.27%	3.60%	6.02%
VLSI	9.73%	7.00%	8.51%	6.40%	8.33%	5.47%	7.61%	6.60%	7.64%	5.48%	7.52%
TSPLIB (N)	5.40%	2.73%	4.41%	2.22%	4.08%	2.12%	3.99%	2.92%	3.77%	1.99%	3.82%
Graph	12.43%	0.57%	1.37%	0.39%	1.29%	0.31%	1.23%	0.88%	1.77%	0.33%	1.20%

The first is that our results for Christofides' algorithm are very similar to those found by Johnson and McGeoch [11], at least for the Euclidean TSPLIB and VLSI instances, with roughly 9-10% error; the error is less on the non-Euclidean

TSPLIB instances. Somewhat surprisingly, Christofides' algorithm seems to perform significantly worse on the graph TSP instances from the Koblenz Network Collection, with 12% error.

One reason that the performance of Christofides' algorithm on the graph TSP instances is surprising is that for the other algorithms, the graph TSP instances seem to be significantly easier, with error under 1%. The VLSI instances appear to be the hardest overall for the algorithms collectively, but this may be because the average instance size is larger.

Another observation is that using SwapRound to sample trees does improve the overall performance of the output.

Of all the algorithms, drawing from the maximum entropy distribution gives the best overall results, but constructing the convex combination via splitting off and then applying SwapRound was quite close in most cases, and better in some. Column generation is the worst of the variants, but we did not check whether the early termination of the column generation routine contributed to the weak performance of this variant.

Why are the Best-of-Many Christofides' algorithm variants significantly better than Christofides' algorithm? The key is that they trade off significantly higher spanning tree cost against significantly lower matching costs, with the reduction in the matching costs outweighing the increase in the spanning tree cost; these results are summarized in Table 2. The average tree cost for all of the Best-of-Many Christofides' algorithm variants is at most the value of the subtour LP; the subtour LP is known to be very close to the cost of the optimal tour experimentally (about 98%-99% of optimal), and our experiments confirm this, while the minimum-cost spanning trees are 79%-93% of the cost of the optimal tour in our experiments. However, the matching costs are dramatically reduced. For Christofides' algorithm, the cost of the matching is 25%-40% of the cost of the optimal tour, while for the Best-of-Many Christofides' variants, it is 10%-15% in the case of the TSPLIB/VLSI instances, and 4-5% for the graph TSP instances.

Table 2. Costs of trees and matchings for the various methods, all expressed relative to the cost of the optimal tour.

	Tree		Matching					
	Std	BOM	Std	ColGen	ColGen+SR	MaxEnt	Split	Split+SR
TSPLIB (E)	87.47%	98.57%	31.25%	11.43%	11.03%	10.75%	10.65%	10.41%
VLSI	89.85%	98.84%	29.98%	14.30%	14.11%	12.76%	12.78%	12.70%
TSPLIB (N)	92.97%	99.36%	24.15%	9.67%	9.36%	8.75%	8.77%	8.56%
Graph	79.10%	98.23%	39.31%	5.20%	4.84%	4.66%	4.34%	4.49%

One reason the matching costs are so much lower in the Best-of-Many Christofides' algorithm variants is that sampling a spanning tree from the subtour LP solution gives spanning trees such that a very high percentage of the vertices have degree two. See Figure 1 to compare the minimum-cost tree on the VLSI instance XQF131 with a tree produced by the maximum entropy distribution. Thus the number of edges needed in the matching is much smaller. These edges tend to be longer in the Best-of-Many Christofides' variants, because odd-degree vertices are rarer, and thus not as near to each other, but because the number of edges needed is much smaller, there is a significant reduction in the cost of the matching. We summarize information about the matching costs in Table 3, where we give the average fraction of odd-degree nodes for the various algorithms, as well as the average cost of a matching edge (expressed in terms of percentage of the cost of an optimal tour). For the TSPLIB and VLSI instances, 36 to 39% of the min-cost spanning tree vertices are odd, whereas for the Best-of-Many variants, the number is between 8-12%; however, the cost of each matching edge is roughly half to two-thirds more in these instances. The graph instances are quite different; for these instances, the min-cost spanning trees have 66% of vertices having odd degree, while the Best-of-Many variants have about the same percentages as before. For these instances, however, the cost of the matching edges are about the same between the standard Christofides' algorithm and the Best-of-Many variants.

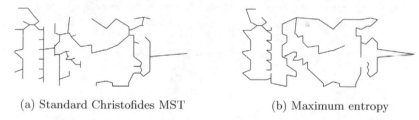

(a) Standard Christofides MST   (b) Maximum entropy

Fig. 1. Sample trees on VLSI instance XQF131 from Rohe [23].

Table 3. Information about matchings for the various algorithms and instances. 'Num' indicates the average percentage of odd-degree vertices in the spanning tree, and 'Cost' is the average cost of a matching edge expressed as a percentage of the cost of the optimal tour.

	Std		ColGen		ColGen+SR		MaxEnt		Split		Split+SR	
	Num	Cost	Num	Cost	Num	Cost	Num	Cost	Num	Cost	Num	Cost
TSPLIB (E)	39%	0.89%	9.3%	1.3%	8.4%	1.4%	8.0%	1.4%	7.6%	1.5%	7.8%	1.4%
VLSI	36%	0.21%	12%	0.34%	11%	0.36%	8.6%	0.38%	8.3%	0.41%	8.3%	0.39%
TSPLIB (N)	38%	0.48%	12%	0.63%	11%	0.67%	10%	0.66%	9.8%	0.69%	9.8%	0.66%
Graph	66%	1.9%	8.9%	1.7%	8.2%	1.7%	7.8%	1.7%	7.6%	1.5%	7.8%	1.5%

5   Conclusions

Our goal in this paper was to determine whether the empirical performance of the Best-of-Many Christofides' algorithms gives any reason to think they might be provably better than the Christofides' algorithm. The answer to this question appears to be yes, with the large caveat that there are many heuristics for the traveling salesman problem (like Lin-Kernighan) with far better performance than Christofides' algorithm which have no provable performance guarantee at all. We also wished to determine which variant might be most promising for further theoretical study. For this question, it seems that the sampling methods have the most promise: that is, maximum entropy sampling or the SwapRound algorithm applied to some initial convex combination of trees. However, because the good performance of these algorithms depends on taking the best result over a large number of samples drawn, one might have to argue that a good tour is produced with reasonable probability after multiple draws; it does not seem that one can argue that a good tour is produced in expectation from a single draw. The average performance of these sampling methods does not seem significantly different from what happens when we construct an explicit convex combination (with column generation, or splitting off).

In our limited experience, it also seems that graph TSP is significantly easier problem for the Best-of-Many variants, and this may bear further investigation both theoretically and empirically. The quality of the solutions found by these algorithms relative to the standard Christofides' algorithm may indicate that these instances really are much easier than the general case of symmetric cost functions with triangle inequality.

Acknowledgments. Both authors were supported in part by NSF grant CCF-1115256.

References

1. An, H.C.: Approximation Algorithms for Traveling Salesman Problems Based on Linear Programming Relaxations. Ph.D. thesis, Department of Computer Science, Cornell University (August 2012)
2. An, H.C., Kleinberg, R., Shmoys, D.B.: Improving Christofides' algorithm for the s-t path TSP. In: Proceedings of the 44th Annual ACM Symposium on Theory of Computing, pp. 875–886 (2012)
3. Applegate, D., Bixby, R., Chvátal, V., Cook, W.: Concorde 03.12.19. http://www.math.uwaterloo.ca/tsp/concorde/index.html
4. Asadpour, A., Goemans, M.X., Madry, A., Oveis Gharan, S., Saberi, A.: An $O(\log n/\log \log n)$-approximation algorithm for the asymmetric traveling salesman problem. In: Proceedings of the 21st Annual ACM-SIAM Symposium on Discrete Algorithms, pp. 379–389 (2010)
5. Chekuri, C., Vondrák, J., Zenklusen, R.: Dependent randomized rounding via exchange properties of combinatorial structures. In: Proceedings of the 51st Annual IEEE Symposium on Foundations of Computer Science, pp. 575–584 (2010), see full version at arxiv.0909:4348
6. Christofides, N.: Worst case analysis of a new heuristic for the traveling salesman problem. Report 388, Graduate School of Industrial Administration, Carnegie-Mellon University, Pittsburgh, PA (1976)
7. Frank, A.: Connections in Combinatorial Optimization. Oxford University Press, Oxford (2011)
8. Frieze, A., Galbiati, G., Maffioli, F.: On the worst-case performance of some algorithms for the asymmetric traveling salesman problem. Networks 12, 23–39 (1982)

9. Genova, K., Williamson, D.P.: An experimental evaluation of the Best-of-Many Christofides' algorithm for the traveling salesman problem, CORR abs/1506.07776 (2015)
10. Gurobi Optimization: Gurobi 5.6.3 (2014). http://www.gurobi.com
11. Johnson, D.S., McGeoch, L.A.: Experimental analysis of heuristics for the STSP. In: Gutin, G., Punnen, A. (eds.) The Traveling Salesman Problem and its Variants, pp. 369–443. Kluwer Academic Publishers (2002)
12. Kolmogorov, V.: Blossom V: a new implementation of a minimum cost perfect matching algorithm. Mathematical Programming Computation 1, 43–67 (2009). http://pub.ist.ac.at/~vnk/software.html
13. Kunegis, J.: KONECT – the Koblenz network collection. In: Proceedings of the International Web Observatory Workshop, pp. 1343–1350 (2013)
14. Lin, S., Kernighan, B.W.: An effective heuristic algorithm for the traveling-salesman problem. Operations Research 21, 498–516 (1973)
15. Lovász, L.: On some connectivity properties of Eulerian graphs. Acta Math. Acad. Sci. Hungar. 28, 129–138 (1976)
16. Mömke, T., Svensson, O.: Approximating graphic TSP by matchings. In: Proceedings of the 52nd Annual IEEE Symposium on Foundations of Computer Science, pp. 560–569 (2011)
17. Mucha, M.: 13/9-approximation for graphic TSP. Theory of Computing Systems 55, 640–657 (2014)
18. Nagamochi, H., Ibaraki, T.: Deterministic $\tilde{O}(mn)$ time edge-splitting in undirected graphs. Journal of Combinatorial Optimization 1, 5–46 (1997)
19. Oveis Gharan, S.: New Rounding Techniques for the Design and Analysis of Approximation Algorithms. Ph.D. thesis, Department of Management Science and Engineering, Stanford University (May 2013)
20. Oveis Gharan, S.: Personal communication (2014)
21. Oveis Gharan, S., Saberi, A., Singh, M.: A randomized rounding approach to the traveling salesman problem. In: Proceedings of the 52nd Annual IEEE Symposium on Foundations of Computer Science, pp. 550–559 (2011)
22. Reinelt, G.: TSPLIB – a traveling salesman problem library. ORSA Journal on Computing, 376–384 (1991)
23. Rohe, A.: Instances found at http://www.math.uwaterloo.ca/tsp/vlsi/index.html (Accessed December 16, 2014)
24. Sebő, A., Vygen, J.: Shorter tours by nicer ears: 7/5-approximation for the graph-TSP, 3/2 for the path version, and 4/3 for two-edge-connected subgraphs. Combinatorica 34, 597–629 (2014)
25. Shewchuk, J.R.: Triangle: Engineering a 2D quality mesh generator and Delaunay triangulator. In: Lin, M.C., Manocha, D. (eds.) FCRC-WS 1996 and WACG 1996. LNCS, vol. 1148, pp. 203–222. Springer, Heidelberg (1996)

Approximating the Smallest Spanning Subgraph for 2-Edge-Connectivity in Directed Graphs

Loukas Georgiadis[1], Giuseppe F. Italiano[2,*], Charis Papadopoulos[1],
and Nikos Parotsidis[1]

[1] University of Ioannina, Greece
{loukas,charis,nparotsi}@cs.uoi.gr
[2] Università di Roma "Tor Vergata", Italy
giuseppe.italiano@uniroma2.it

Abstract. Let G be a strongly connected directed graph. We consider the following three problems, where we wish to compute the smallest strongly connected spanning subgraph of G that maintains respectively: the 2-edge-connected blocks of G (2EC-B); the 2-edge-connected components of G (2EC-C); both the 2-edge-connected blocks and the 2-edge-connected components of G (2EC-B-C). All three problems are NP-hard, and thus we are interested in efficient approximation algorithms. For 2EC-C we can obtain a 3/2-approximation by combining previously known results. For 2EC-B and 2EC-B-C, we present new 4-approximation algorithms that run in linear time. We also propose various heuristics to improve the size of the computed subgraphs in practice, and conduct a thorough experimental study to assess their merits in practical scenarios.

1   Introduction

Let $G = (V, E)$ be a directed graph (digraph), with m edges and n vertices. An edge of G is a *strong bridge* if its removal increases the number of strongly connected components of G. A digraph G is 2-edge-connected if it has no strong bridges. The 2-edge-connected components of G are its maximal 2-edge-connected subgraphs. Let v and w be two distinct vertices: v and w are 2-*edge-connected*, denoted by $v \leftrightarrow_{2e} w$, if there are two edge-disjoint directed paths from v to w and two edge-disjoint directed paths from w to v. (Note that a path from v to w and a path from w to v need not be edge-disjoint.) A 2-*edge-connected block* of $G = (V, E)$ is a maximal subset $B \subseteq V$ such that $u \leftrightarrow_{2e} v$ for all $u, v \in B$. Differently from undirected graphs, in digraphs 2-edge-connected blocks can be different from the 2-edge-connected components, i.e., two vertices may be 2-edge-connected but lie in different 2-edge-connected components. See Figure 1.

Computing a smallest spanning subgraph that maintains the same edge or vertex connectivity properties of the original graph is a fundamental problem in network design, with many practical applications [15]. In this paper we consider

---

* Partially supported by MIUR under Project AMANDA.

N. Bansal and I. Finocchi (Eds.): ESA 2015, LNCS 9294, pp. 582–594, 2015.
DOI: 10.1007/978-3-662-48350-3_49

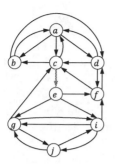 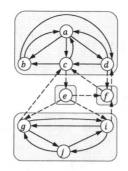 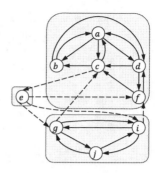

Fig. 1. From left-to-right we show: a strongly connected digraph G with a strong bridge (c, e), the 2-edge-connected components of G, and the 2-edge-connected blocks of G.

the problem of finding the smallest spanning subgraph of G that maintains certain 2-edge-connectivity requirements in addition to strong connectivity. Specifically, we distinguish three problems that we refer to as 2EC-B, 2EC-C and 2EC-B-C. In particular, we wish to compute the smallest strongly connected spanning subgraph of a digraph G that maintains the following properties: the pairwise 2-edge-connectivity of G, i.e., the 2-edge-connected blocks of G (2EC-B); the 2-edge-connected components of G (2EC-C); both the 2-edge-connected blocks and the 2-edge-connected components of G (2EC-B-C). Since all those problems are NP-hard [7], we are interested in designing efficient approximation algorithms.

Related Work. Finding a smallest k-edge-connected (resp. k-vertex-connected) spanning subgraph of a given k-edge-connected (resp. k-vertex-connected) digraph is NP-hard for $k \geq 2$ for undirected graphs, and for $k \geq 1$ for digraphs [7]. Problems of this type, together with more general variants of approximating minimum-cost subgraphs that satisfy certain connectivity requirements, have received a lot of attention, and several important results have been obtained. See, e.g., the survey [13]. Currently, the best approximation ratio for computing the smallest strongly connected spanning subgraph (SCSS) is 3/2 achieved by Vetta [17]. A linear-time algorithm that achieves a 5/3-approximation was given by Zhao et al. [18]. For the smallest k-edge-connected spanning subgraph (kECSS), Laehanukit et al. [14] gave a randomized $(1+1/k)$-approximation algorithm. Regarding hardness of approximation, Gabow et al. [5] showed that there exists an absolute constant $c > 0$ such that for any integer $k \geq 1$, approximating the smallest kECSS on directed multigraphs to within a factor $1 + c/k$ in polynomial time implies P = NP. Jaberi [12] considered various optimization problems related to 2EC-B and proposed corresponding approximation algorithms. The approximation ratio in Jaberi's algorithms, however, is linear in the number of strong bridges, and hence $O(n)$ in the worst case.

Our Results. In this paper we provide both theoretical and experimental contributions to the 2EC-B, 2EC-C and 2EC-B-C problems. A 3/2-approximation for 2EC-C can be obtained by carefully combining the 2ECSS randomized algorithm of Laehanukit et al. [14] and the SCSS algorithm of Vetta [17]. A faster and deterministic 2-approximation algorithm for 2EC-C can be obtained by combining techniques based on edge-disjoint spanning trees [4,16] with the SCSS algorithm of Zhao et al. [18]. We remark that the other two problems considered here, 2EC-B and 2EC-B-C, seem harder to approximate. The only known result is the *sparse certificate* for 2-edge-connected blocks of [8], which implies a linear-time $O(1)$-approximation algorithm for 2EC-B. Unfortunately, no good bound for the approximation constant was previously known, and indeed achieving a small constant seemed to be non-trivial. In this paper, we make a substantial progress in this direction by presenting new 4-approximation algorithms for 2EC-B and 2EC-B-C that run in linear time (the algorithm for 2EC-B-C runs in linear time once the 2-edge-connected components of G are available; if not, they can be computed in $O(n^2)$ time [10]).

From the practical viewpoint, we provide efficient implementations of our algorithms that are very fast in practice. We further propose and implement several heuristics that improve the size (i.e., the number of edges) of the computed spanning subgraphs in practice. Some of our algorithms require $O(mn)$ time in the worst case, so we also present several techniques to achieve significant speedups in their running times. With all these implementations, we conduct a thorough experimental study and report its main findings. We believe that this is crucial to assess the merits of all the algorithms considered in practical scenarios. For lack of space, proofs and some details are omitted and will be given in the full paper.

2   Preliminaries

A *flow graph* is a digraph such that every vertex is reachable from a distinguished start vertex. Let $G = (V, E)$ be a strongly connected digraph. For any vertex $s \in V$, we denote by $G(s) = (V, E, s)$ the corresponding flow graph with start vertex s; all vertices in V are reachable from s since G is strongly connected. The *dominator relation* in $G(s)$ is defined as follows: A vertex u is a *dominator* of a vertex w (u *dominates* w) if every path from s to w contains u; u is a *proper dominator* of w if u dominates w and $u \neq w$. The dominator relation is reflexive and transitive. Its transitive reduction is a rooted tree, the *dominator tree* $D(s)$: u dominates w if and only if u is an ancestor of w in $D(s)$. If $w \neq s$, $d(w)$, the parent of w in $D(s)$, is the *immediate dominator* of w: it is the unique proper dominator of w that is dominated by all proper dominators of w. The dominator tree of a flow graph can be computed in linear time, see, e.g., [1,2]. An edge (u, w) is a *bridge* in $G(s)$ if all paths from s to w include (u, w).[1] Italiano et al. [11] showed that the strong bridges of G can be computed from the bridges of the flow graphs $G(s)$ and $G^R(s)$, where s is an arbitrary start vertex and G^R is the digraph that results from G after reversing edge directions. A spanning tree T of a flow graph $G(s)$ is a tree with root s that contains a path from s to v for all vertices v. Two spanning trees B and R rooted at s are *edge-disjoint* if they have no edge in common. A flow graph $G(s)$ has two such spanning trees if and

---

[1] Throughout, we use consistently the term *bridge* to refer to a bridge of a flow graph $G(s)$ and the term *strong bridge* to refer to a strong bridge in the original graph G.

only if it has no bridges [16]. The two spanning trees are *maximally edge-disjoint* if the only edges they have in common are the bridges of $G(s)$. Two (maximally) edge-disjoint spanning trees can be computed in linear-time by an algorithm of Tarjan [16], using the disjoint set union data structure of Gabow and Tarjan [6]. Two spanning trees B and R rooted at s are *independent* if for all vertices v, the paths from s to v in B and R share only the dominators of v. Every flow graph $G(s)$ has two such spanning trees, computable in linear time [9] which are maximally edge-disjoint.

3   Approximation Algorithms and Heuristics

We describe our main approaches for solving problem 2EC-B. Let $G = (V, E)$ be the input directed graph. The first two algorithms process one edge (x, y) of the current subgraph G' of G at a time, and test if it is safe to remove (x, y). Initially $G' = G$, and the order in which the edges are processed is arbitrary. The third algorithm starts with the empty graph $G' = (V, \emptyset)$, and adds the edges of spanning trees of certain subgraphs of G until the resulting digraph is strongly connected and has the same 2-edge-connected blocks as G.

Two Edge-Disjoint Paths Test. We test if $G' \setminus (x, y)$ contains two edge-disjoint paths from x to y. If this is the case, then we remove edge (x, y). This test takes $O(m)$ time per edge, so the total running time is $O(m^2)$. We refer to this algorithm as Test2EDP-B. Note that Test2EDP-B computes a minimal 2-approximate solution for the 2ECSS problem [3], which is not necessarily minimal for the 2EC-B problem.

2-Edge-Connected Blocks Test. If (x, y) is not a strong bridge in G', we test if $G' \setminus (x, y)$ has the same 2-edge-connected blocks as G'. If this is the case then we remove edge (x, y). We refer to this algorithm as Test2ECB-B. Since the 2-edge-connected blocks of a graph can be computed in linear time [8], Test2ECB-B runs in $O(m^2)$ time. Test2ECB-B computes a minimal solution for 2EC-B and achieves an approximation ratio of 4.

Independent Spanning Trees. We can compute a sparse certificate for 2-edge-connected blocks as in [8], based on a linear-time construction of two independent spanning trees of a flow graph [9]. We refer to this algorithm as IST-B original. We will show later that a suitably modified construction, which we refer to as IST-B, yields a linear-time 4-approximation algorithm.

Test2EDP-B and Test2ECB-B can be combined into a hybrid algorithm (Hybrid-B), as follows: if the tested edge (x, y) connects vertices in the same block (i.e., $x \leftrightarrow_{2e} y$), then apply Test2EDP-B; otherwise, apply Test2ECB-B. One can show that Hybrid-B returns the same sparse subgraph as Test2ECB-B.

In algorithm Hybrid-B we also apply an additional speed-up heuristic for *trivial edges* (x, y): if x belongs to a nontrivial block (i.e., a block of size ≥ 2) and has outdegree two or y belongs to a nontrivial block and has indegree two, then (x, y) must be included in the solution. As we show later in our experiments, such a simple test can yield significant performance gains.

Note that Test2EDP-B, Test2ECB-B and Hybrid-B produce a 4-approximation for 2EC-B in $O(n^2)$ time if they are run on the sparse subgraph computed by IST-B instead of the original digraph. We observed experimentally that this also improves the quality of the computed solutions in practice. Therefore, we applied this idea in all our implementations. See Table 1 in Section 4.

Although all the above algorithms do not maintain the 2-edge-connected components of the original graph, we can still apply them to get an approximation for 2EC-B-C, as follows. First, we compute the 2-edge-connected components of G and solve the 2ECSS problem independently for each such component. Then, we can apply any of the algorithms for 2EC-B (Test2EDP-B, Test2ECB-B, Hybrid-B or IST-B) for the edges that connect different components. To speed them up, we apply them to a *condensed graph* H that is formed from G by contracting each 2-edge-connected component of G into a single supervertex. Note that H is a multigraph since the contractions can create loops and parallel edges. For any vertex v of G, we denote by $h(v)$ the supervertex of H that contains v. Every edge $(h(u), h(v))$ of H is associated with the corresponding original edge (u, v) of G. Algorithms Test2EDP-BC, Test2ECB-BC, Hybrid-BC or IST-BC are obtained by applying to graph H the corresponding algorithm for 2EC-B. Let H' be the obtained subgraph of H, and let G' be the digraph that is obtained after we expand back each supervertex of H with its 2-edge-connected sparse subgraph computed before. Then, G' is a valid solution to the 2EC-B-C problem.

As a special case of applying Test2EDP-B to H, we can immediately remove loops and parallel edges $(h(u), h(v))$ if H has more than two edges directed from $h(u)$ to $h(v)$. To obtain faster implementations, we solve the 2ECSS problems in linear-time using edge-disjoint spanning trees [4,16]. Let C be a 2-edge-connected component of G. We select an arbitrary vertex $v \in C$ as a root and compute two edge-disjoint spanning trees in the flow graph $C(v)$ and two edge-disjoint spanning trees in the reverse flow graph $C^R(v)$. The edges of these spanning trees give a 2-approximate solution C' for 2ECSS on C. Moreover, as in 2EC-B, we can apply algorithms Test2EDP-BC, Test2ECB-BC and Hybrid-BC on the sparse subgraph computed by IST-BC. Then, these algorithms produce a 4-approximation for 2EC-B-C in $O(n^2)$ time. Furthermore, for these $O(n^2)$-time algorithms, we can improve the approximate solution C' for 2ECSS on each 2-edge-connected component C of G, by applying the two edge-disjoint paths test on the edges of C'. We incorporate all these ideas in all our implementations.

We can also use the condensed graph in order to obtain an efficient approximation algorithm for 2EC-C. To that end, we can apply the algorithm of Laehanukit et al. [14] and get a 3/2-approximation of the 2ECSS problem independently for each 2-edge-connected component of G. Then, since we only need to preserve the strong connectivity of H, we can run the algorithm of Vetta [17] on a digraph $\tilde{H}$ that results from H after removing all loops and parallel edges. This computes a spanning subgraph H' of $\tilde{H}$ that is a 3/2-approximation for SCSS in H. The corresponding expanded graph G', where we substitute each supervertex $h(v)$ of H with the approximate smallest 2ECSS, gives a 3/2-approximation for 2EC-C. A faster and deterministic 2-approximation algorithm for 2EC-C can be obtained as follows. For the 2ECSS problems we use the edge-disjoint spanning trees 2-approximation algorithm described above. Then, we solve SCSS on $\tilde{H}$ by applying the linear-time algorithm of Zhao et al. [18]. This yields a 2-approximation algorithm for 2EC-C that runs in linear time once the 2-edge-connected components of G are available (if not, they can be computed in $O(n^2)$ time [10]). We refer to this algorithm as ZNI-C.

Theorem 1. *There is a polynomial-time algorithm for 2EC-C that achieves an approximation ratio of $3/2$. Moreover, if the 2-edge-connected components of G are available, then we can compute a 2-approximate 2EC-C in linear time.*

3.1   Independent Spanning Trees

Here we present our new algorithm IST-B and prove that it gives a linear-time 4-approximation for 2EC-B and 2EC-B-C. Since IST-B is a modified version of the sparse certificate $C(G)$ for the 2-edge-connected blocks of a digraph G [8] (IST-B original), let us review IST-B original first.

Let s be an arbitrarily chosen start vertex of the strongly connected digraph G. The *canonical decomposition* of the dominator tree $D(s)$ is the forest of rooted trees that results from $D(s)$ after the deletion of all the bridges of $G(s)$. Let $T(v)$ denote the tree containing vertex v in this decomposition. We refer to the subtree roots in the canonical decomposition as *marked vertices*. For each marked vertex r we define the *auxiliary graph* $G_r = (V_r, E_r)$ of r as follows. The vertex set V_r of G_r consists of all the vertices in $T(r)$, referred to as *ordinary* vertices, and a set of *auxiliary* vertices, which are obtained by contracting vertices in $V \setminus T(r)$, as follows. Let v be a vertex in $T(r)$. We say that v is a *boundary vertex in $T(r)$* if v has a marked child in $D(s)$. Let w be a marked child of a boundary vertex v: all the vertices that are descendants of w in $D(s)$ are contracted into w. All vertices in $V \setminus T(r)$ that are not descendants of r are contracted into $d(r)$ ($r \neq s$ if any such vertex exists). During those contractions, parallel edges are eliminated. We call an edge in $E_r \setminus E$ *shortcut edge*. Such an edge has an auxiliary vertex as an endpoint. We associate each shortcut edge $(u, v) \in E_r$ with a corresponding original edge $(x, y) \in E$, i.e. x was contracted into u or y was contracted into v (or both). If $G(s)$ has b bridges then all the auxiliary graphs G_r have at most $n + 2b$ vertices and $m + 2b$ edges in total and can be computed in $O(m)$ time. As shown in [8], two ordinary vertices of an auxiliary graph G_r are 2-edge-connected in G if and only if they are 2-edge-connected in G_r. Thus the 2-edge-connected blocks of G are a refinement of the vertex sets in the trees of the canonical decomposition. The sparse certificate of [8] is constructed in three phases. We maintain a list (multiset) L of the edges to be added in $C(G)$; initially $L = \emptyset$. The same edge may be inserted into L multiple times, but the total number of insertions will be $O(n)$. So the edges of $C(G)$ can be obtained from L after we remove duplicates, e.g. by using radix sort. Also, during the construction, the algorithm may choose a shortcut edge or a reverse edge to be inserted into L. In this case we insert the associated original edge instead.

Phase 1. We insert into L the edges of two independent spanning trees, $B(G(s))$ and $R(G(s))$ of $G(s)$.

Phase 2. For each auxiliary graph $H = G_r$ of $G(s)$, that we refer to as the *first-level auxiliary graphs*, we compute two independent spanning trees $B(H^R(r))$ and $R(H^R(r))$ for the corresponding reverse flow graph $H^R(r)$ with start vertex r. We insert into L the edges of these two spanning trees. We note that L induces a strongly connected spanning subgraph of G at the end of this phase.

Phase 3. Finally, in the third phase we process the *second-level auxiliary graphs*, which are the auxiliary graphs of H^R for all first-level auxiliary graphs H. Let (p,q) be a bridge of $H^R(r)$, and let H_q^R be the corresponding second-level auxiliary graph. For every strongly connected component S of $H_q^R \setminus (p,q)$, we choose an arbitrary vertex $v \in S$ and compute a spanning tree of $S(v)$ and a spanning tree of $S^R(v)$, and insert their edges into L.

The above construction inserts $O(n)$ edges into $C(G)$, and therefore achieves a constant approximation ratio for 2EC-B. It is not straightforward, however, to give a good bound for this constant, since the spanning trees that are used in this construction contain auxiliary vertices that are created by applying two levels of the canonical decomposition. In the next section we analyze a modified version of the sparse certificate construction, and show that it achieves a 4-approximation for 2EC-B. Then we show that we also achieve a 4-approximation for 2EC-B-C by applying this sparse certificate on the condensed graph H.

The New Algorithm IST-B. The main idea behind IST-B is to limit the number of edges added to the sparse certificate $C(G)$ because of auxiliary vertices. In particular, we show that in Phase 2 of the construction it suffices to add at most one new edge for each first-level auxiliary vertex, while in Phase 3 at most $2b$ additional edges are necessary for all second-level auxiliary vertices, where b is the number of bridges in $G(s)$.

Consider Phase 2. Let $H = G_r$ be a first-level auxiliary graph. In the sparse certificate we include two independent spanning trees, $B(H^R(r))$ and $R(H^R(r))$, of the reverse flow graph $H^R(r)$ with start vertex r. In our new construction, each auxiliary vertex x in H^R will contribute at most one new edge in $C(G)$. Suppose first that $x = d(r)$, which exists if $r \neq s$. The only edge entering $d(r)$ in H^R is $(r, d(r))$ which is the reverse edge of the bridge $(d(r), r)$ of $G(s)$. So $d(r)$ does not add a new edge in $C(G)$, since all the bridges of $G(s)$ were added in the first phase of the construction. Next we consider an auxiliary vertex $x \neq d(r)$. In H^R there is a unique edge (x, z) leaving x, where $z = d(x)$. This edge is the reverse of the bridge $(d(x), x)$ of $G(s)$. Suppose that x has no children in $B(H^R(r))$ and $R(H^R(r))$. Deleting x and its two entering edges in both spanning trees does not affect the existence of two edge-disjoint paths from v to r in H, for any ordinary vertex v. However, the resulting graph $C(G)$ at the end may not be strongly connected. To fix this, it suffices to include in $C(G)$ the reverse of an edge entering x from only one spanning tree. Finally, suppose that x has children, say in $B(H^R(r))$. Then $z = d(x)$ is the unique child of x in $B(H^R(r))$, and the reverse of the edge (x, z) of $B(H^R(r))$ is already included in $C(G)$ by Phase 1. Therefore, in all cases, we can charge to x at most one new edge.

Now we consider Phase 3. Let H_q^R be a second-level auxiliary graph of H^R. Let e be the strong bridge entering q in H^R, and let S be a strongly connected component in $H_q^R \setminus e$. In our sparse certificate we include the edges of a strongly connected subgraph of S, so we have spanning trees T and T^R of $S(v)$ and $S^R(v)$, respectively, rooted at an arbitrary ordinary vertex v. Let x be an auxiliary vertex of S. If x is a first-level auxiliary vertex in H then it has a unique entering edge (w, x) which is a bridge in $G(s)$ already included in $C(G)$. If x is ordinary in H but a second-level auxiliary vertex in H_q then it has a unique leaving edge (x, z), which is a bridge in $H^R(r)$ and $C(G)$ already contains a corresponding original edge. Consider the first case. If x is a leaf in T^R then we can delete the edge entering x in T^R. Otherwise, w is the unique child of x in T^R, and the corresponding edge (w, x) entering x in H has already been inserted in $C(G)$. The symmetric arguments hold if x is ordinary in H. This analysis implies that we can associate each second-level auxiliary vertex with one edge in each of T and T^R that is either not needed in $C(G)$ or has already been inserted. If all such auxiliary vertices are associated with distinct edges then they do not contribute any new edges in $C(G)$. Suppose now that there are two second-level auxiliary vertices x and y that are associated with a common edge e. This can happen only if one of these vertices, say y, is a first-level auxiliary vertex, and x is ordinary in H. Then y has a unique entering edge in H, which means that $e = (x, y)$ is a strong bridge, and thus already in $C(G)$. Also $e \in T$ and $e^R = (y, x) \in T^R$. In this case, we can treat x and y as a single auxiliary vertex that results from the contraction of e, which contributes at most two new edges in $C(G)$. Since y is a first-level auxiliary vertex, this can happen at most b times in all second-level auxiliary graphs, so the $2b$ bound follows. Using the above construction we obtain the following result (see the full paper for the complete proof):

Theorem 2. *There is a linear-time approximation algorithm for the 2EC-B problem that achieves an approximation ratio of 4. Moreover, if the 2-edge-connected components of the input digraph are known in advance, we can compute a 4-approximation for the 2EC-B-C problem in linear time.*

Heuristics Applied on Auxiliary Graphs. To speed up algorithms from the Test2EDP and Hybrid families, we applied them to the first-level and second-level auxiliary graphs. Our experiments indicated that applying this heuristic to second-level auxiliary graphs yields better results than the ones obtained on first-level auxiliary graphs. We refer to those variants as Test2EDP-B-Aux, Hybrid-B-Aux, Test2EDP-BC-Aux, Hybrid-BC-Aux, depending on the algorithm (Test2EDP or Hybrid) and problem (2EC-B or 2EC-B-C) considered.

4   Experimental Analysis

We implemented the algorithms previously described: 7 for 2EC-B, 6 for 2EC-B-C, and one for 2EC-C, as summarized in Table 1. All implementations were written in C++ and compiled with g++ v.4.4.7 with flag -03. We performed our experiments on a GNU/Linux machine, with Red Hat Enterprise Server v6.6: a

Table 1. The algorithms considered in our experimental study. The worst-case bounds refer to a digraph with n vertices and m edges. [†]These linear running times assume that the 2-edge-connected components of the input digraph are available.

Algorithm	Problem	Technique	Time
ZNI-C	2EC-C	Zhao et al. [18] applied on the condensed graph	$O(m+n)$[†]
IST-B original	2EC-B	Original sparse certificate from [8]	$O(m+n)$
IST-B	2EC-B	Modified sparse certificate	$O(m+n)$
Test2EDP-B	2EC-B	Two edge-disjoint paths test on sparse certificate of input graph	$O(n^2)$
Test2ECB-B	2EC-B	2-edge-connected blocks test on sparse certificate of input graph	$O(n^2)$
Hybrid-B	2EC-B	Hybrid of two edge-disjoint paths and 2-edge-connected blocks test on sparse certificate of input graph	$O(n^2)$
Test2EDP-B-Aux	2EC-B	Test2EDP-B applied on second-level auxiliary graphs	$O(n^2)$
Hybrid-B-Aux	2EC-B	Hybrid-B applied on second-level auxiliary graphs	$O(n^2)$
IST-BC	2EC-B-C	Modified sparse certificate preserving 2-edge-connected components (applied on the condensed graph)	$O(m+n)$[†]
Test2EDP-BC	2EC-B-C	Two edge-disjoint paths test on sparse certificate of condensed graph	$O(n^2)$
Test2ECB-BC	2EC-B-C	2-edge-connected blocks test on sparse certificate of condensed graph	$O(n^2)$
Hybrid-BC	2EC-B-C	Hybrid of two edge-disjoint paths and 2-edge-connected blocks test on sparse certificate of condensed graph	$O(n^2)$
Test2EDP-BC-Aux	2EC-B-C	Test2EDP-BC applied on second-level auxiliary graphs	$O(n^2)$
Hybrid-BC-Aux	2EC-B-C	Hybrid-BC applied on second-level auxiliary graphs	$O(n^2)$

Table 2. Real-world graphs sorted by file size of their largest SCC; n is the number of vertices, m the number of edges, and δ_{avg} is the average vertex indegree; b^* is the number of strong bridges; δ_{avg}^B and δ_{avg}^C are lower bounds on the average vertex indegree of an optimal solution to 2EC-B and 2EC-C, respectively.

Dataset	n	m	file size	δ_{avg}	b^*	δ_{avg}^B	δ_{avg}^C	type
Rome99	3353	8859	100KB	2.64	1474	1.75	1.67	road network
P2p-Gnutella25	5153	17695	203KB	3.43	2181	1.60	1.00	peer2peer
P2p-Gnutella31	14149	50916	621KB	3.59	6673	1.56	1.00	peer2peer
Web-NotreDame	53968	296228	3,9MB	5.48	34879	1.50	1.36	web graph
Soc-Epinions1	32223	443506	5,3MB	13.76	20975	1.56	1.55	social network
USA-road-NY	264346	733846	11MB	2.77	104618	1.80	1.80	road network
USA-road-BAY	321270	800172	12MB	2.49	196474	1.69	1.69	road network
USA-road-COL	435666	1057066	16MB	2.42	276602	1.68	1.68	road network
Amazon0302	241761	1131217	16MB	4.67	73361	1.74	1.64	prod. co-purchase
WikiTalk	111881	1477893	18MB	13.20	85503	1.45	1.44	social network
Web-Stanford	150532	1576314	22MB	10.47	64723	1.62	1.33	web graph
Amazon0601	395234	3301092	49MB	8.35	83995	1.82	1.82	prod. co-purchase
Web-Google	434818	3419124	50MB	7.86	211544	1.59	1.48	web graph
Web-Berkstan	334857	4523232	68MB	13.50	164779	1.56	1.39	web graph

PowerEdge T420 server 64-bit NUMA with two Intel Xeon E5-2430 v2 processors and 16GB of RAM RDIMM memory. Each processor has 6 cores sharing a 15MB L3 cache, and each core has a 2MB private L2 cache and 2.50GHz speed. In our experiments we did not use any parallelization, and each algorithm ran on a single core. We report CPU times measured with the `getrusage` function. All our running times were averaged over ten different runs.

For the experimental evaluation we use the datasets shown in Table 2. We measure the quality of the solution computed by algorithm A on problem $\mathcal{P}$ by a *quality ratio* defined as $q(A, \mathcal{P}) = \delta_{avg}^A / \delta_{avg}^{\mathcal{P}}$, where δ_{avg}^A is the average vertex indegree of the subgraph computed by A and $\delta_{avg}^{\mathcal{P}}$ is a lower bound on the average vertex indegree of the optimal solution for $\mathcal{P}$. Specifically, for 2EC-B and 2EC-B-C we define $\delta_{avg}^B = (n + k)/n$, where n is the total number of vertices of the input digraph and k is the number of vertices that belong in nontrivial 2-edge-connected blocks.[2] We set a similar lower bound δ_{avg}^C for 2EC-C, with the only difference that k is the number of vertices that belong in nontrivial 2-edge-connected components. Note that the quality ratio is an upper bound of the actual approximation ratio. The smaller the values of $q(A, \mathcal{P})$ (i.e., the closer to 1), the better is the approximation obtained by algorithm A for problem $\mathcal{P}$.

We now report the results of our experiments with all the algorithms considered for problems 2EC-B, 2EC-B-C and 2EC-C. As previously mentioned, for the sake of efficiency, all variants of Test2EDP, Test2ECB and Hybrid were run on the sparse certificate computed by either IST-B or IST-BC (depending on the problem at hand) instead of the original digraph. For the 2EC-B problem, the quality ratio of the spanning subgraphs computed by the different algorithms is shown in Table 3, while their running times are plotted in Figure 2 (top). Similarly, for the 2EC-C and 2EC-B-C problems, the quality ratio of the spanning subgraphs computed by the different algorithms is shown in Table 4, while their running times are plotted in Figure 2 (bottom).

There are two peculiarities related to road networks that emerge immediately from the analysis of our experimental data. First, all algorithms achieve consistently better approximations for road networks than for most of the other graphs in our data set. Second, for the 2EC-B problem the Hybrid algorithms (Hybrid-B and Hybrid-B-Aux) seem to achieve substantial speedups on road networks; for the 2EC-B-C problem, this is even true for Test2ECB-BC. The first phenomenon can be explained by taking into account the macroscopic structure of road networks, which is rather different from other networks. Indeed, road networks are very close to be "undirected": i.e., whenever there is an edge (x, y), there is also the reverse edge (y, x) (expect for one-way roads). Roughly speaking, road networks mainly consist of the union of 2-edge-connected components, joined together by strong bridges, and their 2-edge-connected blocks coincide with their 2-edge-connected components. In this setting, a sparse strongly connected subgraph of the condensed graph will preserve both blocks and components. The second phenomenon is mainly due to the *trivial edge* heuristic described in Section 3.

Apart from the peculiarities of road networks, ZNI-C behaves as expected for 2EC-C through its linear-time 2-approximation algorithm. Note that for both problems 2EC-B and 2EC-B-C, all algorithms achieve quality ratio significantly smaller than our theoretical bound of 4. Regarding running times, we observe that the 2EC-B-C algorithms are faster than the 2EC-B algorithms, sometimes

---

[2] This follows from the fact that in the sparse subgraph the k vertices in nontrivial blocks must have indegree at least two, while the remaining $n - k$ vertices must have indegree at least one, since we seek for a strongly connected spanning subgraph.

Table 3. Quality ratio $q(A, \mathcal{P})$ of the solutions computed for 2EC-B.

Dataset	IST-B original	IST-B	Test2EDP-B	Test2ECB-B & Hybrid-B	Test2EDP-B-Aux	Hybrid-B-Aux
Rome99	1.389	1.363	1.171	1.167	1.177	1.174
P2p-Gnutella25	1.656	1.512	1.220	1.143	1.251	1.234
P2p-Gnutella31	1.682	1.541	1.251	1.169	1.291	1.274
Web-NotreDame	1.964	1.807	1.489	1.417	1.500	1.471
Soc-Epinions1	2.047	1.837	1.435	1.379	1.441	1.406
USA-road-NY	1.343	1.245	1.174	1.174	1.175	1.175
USA-road-BAY	1.361	1.307	1.245	1.246	1.246	1.246
USA-road-COL	1.354	1.304	1.251	1.252	1.252	1.252
Amazon0302	1.762	1.570	1.186	1.134	1.206	1.196
WikiTalk	2.181	2.050	1.788	1.588	1.792	1.615
Web-Stanford	1.907	1.688	1.409	1.365	1.418	1.406
Amazon0601	1.866	1.649	1.163	1.146	1.170	1.166
Web-Google	1.921	1.728	1.389	1.322	1.401	1.377
Web-Berkstan	2.048	1.775	1.480	1.427	1.489	1.469

Table 4. Quality ratio $q(A, \mathcal{P})$ of the solutions computed for 2EC-C and 2EC-B-C.

Dataset	ZNI-C	IST-BC	Test2EDP-BC	Test2ECB-BC & Hybrid-BC	Test2EDP-BC-Aux	Hybrid-BC-Aux
Rome99	1.360	1.371	1.197	1.187	1.197	1.195
P2p-Gnutella25	1.276	1.517	1.218	1.141	1.249	1.232
P2p-Gnutella31	1.312	1.537	1.251	1.170	1.290	1.273
Web-NotreDame	1.620	1.747	1.500	1.426	1.510	1.484
Soc-Epinions1	1.790	1.847	1.488	1.435	1.489	1.476
USA-road-NY	1.343	1.341	1.163	1.163	1.163	1.163
USA-road-BAY	1.360	1.357	1.237	1.237	1.237	1.237
USA-road-COL	1.343	1.339	1.242	1.242	1.242	1.242
Amazon0302	1.464	1.580	1.279	1.228	1.292	1.284
WikiTalk	1.891	2.099	1.837	1.630	1.838	1.827
Web-Stanford	1.560	1.679	1.430	1.390	1.436	1.427
Amazon0601	1.709	1.727	1.200	1.186	1.202	1.200
Web-Google	1.637	1.728	1.437	1.381	1.446	1.431
Web-Berkstan	1.637	1.753	1.516	1.472	1.523	1.511

significantly, as they take advantage of the condensed graph that seems to admit small size in real-world applications. In addition, our experiments highlight interesting tradeoffs between practical performance and quality of the obtained solutions. Indeed, the fastest (IST-B and IST-B original for problem 2EC-B; IST-BC for 2EC-B-C) and the slowest algorithms (Test2ECB-B and Hybrid-B for 2EC-B; Test2ECB-BC and Hybrid-BC for 2EC-B-C) tend to produce respectively the worst and the best approximations. Note that IST-B improves the quality of the solution of IST-B original at the price of slightly higher running times, while Hybrid-B (resp., Hybrid-BC) produces the same solutions as Test2ECB-B (resp., Test2ECB-BC) with rather impressive speedups. Running an algorithm on the second-level auxiliary graphs seems to produce substantial performance benefits at the price of a slightly worse approximation (Test2EDP-B-Aux, Hybrid-B-Aux, Test2EDP-BC-Aux and Hybrid-BC-Aux versus Test2EDP-B, Hybrid-B, Test2EDP-BC and Hybrid-BC). Overall, in our experiments Test2EDP-B-Aux and Test2EDP-BC-Aux seem to provide good quality solutions for the problems considered without being penalized too much by a substantial performance degradation.

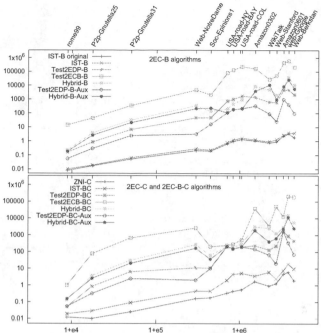

Fig. 2. Running times in seconds with respect to the number of edges (in log-log scale).

References

1. Alstrup, S., Harel, D., Lauridsen, P.W., Thorup, M.: Dominators in linear time. SIAM Journal on Computing 28(6), 2117–2132 (1999)
2. Buchsbaum, A.L., Georgiadis, L., Kaplan, H., Rogers, A., Tarjan, R.E., Westbrook, J.R.: Linear-time algorithms for dominators and other path-evaluation problems. SIAM Journal on Computing 38(4), 1533–1573 (2008)
3. Cheriyan, J., Thurimella, R.: Approximating minimum-size k-connected spanning subgraphs via matching. SIAM J. Comput. 30(2), 528–560 (2000)
4. Edmonds, J.: Edge-disjoint branchings. Combinat. Algorithms, 91–96 (1972)
5. Gabow, H.N., Goemans, M.X., Tardos, E., Williamson, D.P.: Approximating the smallest k-edge connected spanning subgraph by lp-rounding. Networks 53(4), 345–357 (2009)
6. Gabow, H.N., Tarjan, R.E.: A linear-time algorithm for a special case of disjoint set union. Journal of Computer and System Sciences 30(2), 209–221 (1985)
7. Garey, M.R., Johnson, D.S.: Computers and Intractability: A Guide to the Theory of NP-Completeness. W. H. Freeman & Co., New York (1979)
8. Georgiadis, L., Italiano, G.F., Laura, L., Parotsidis, N.: 2-edge connectivity in directed graphs. In: SODA 2015, pp. 1988–2005 (2015)
9. Georgiadis, L., Tarjan, R.E.: Dominator tree verification and vertex-disjoint paths. In: SODA 2005, pp. 433–442 (2005)
10. Henzinger, M., Krinninger, S., Loitzenbauer, V.: Finding 2-edge and 2-vertex strongly connected components in quadratic time. In: ICALP 2015, pp. 713–724 (2015)

11. Italiano, G.F., Laura, L., Santaroni, F.: Finding strong bridges and strong articulation points in linear time. Theor. Comput. Sci. 447, 74–84 (2012)
12. Jaberi, R.: Computing the 2-blocks of directed graphs. RAIRO-Theor. Inf. Appl. 49(2), 93–119 (2015)
13. Kortsarz, G., Nutov, Z.: Approximating minimum cost connectivity problems. Approximation Algorithms and Metaheuristics (2007)
14. Laekhanukit, B., Oveis Gharan, S., Singh, M.: A rounding by sampling approach to the minimum size k-arc connected subgraph problem. In: Czumaj, A., Mehlhorn, K., Pitts, A., Wattenhofer, R. (eds.) ICALP 2012, Part I. LNCS, vol. 7391, pp. 606–616. Springer, Heidelberg (2012)
15. Nagamochi, H., Ibaraki, T.: Algorithmic Aspects of Graph Connectivity, 1st edn. Cambridge University Press (2008)
16. Tarjan, R.E.: Edge-disjoint spanning trees and depth-first search. Acta Informatica 6(2), 171–185 (1976)
17. Vetta, A.: Approximating the minimum strongly connected subgraph via a matching lower bound. In: SODA 2001, pp. 417–426 (2001)
18. Zhao, L., Nagamochi, H., Ibaraki, T.: A linear time 5/3-approximation for the minimum strongly-connected spanning subgraph problem. Information Processing Letters 86(2), 63–70 (2003)

A Probabilistic Approach to Reducing Algebraic Complexity of Delaunay Triangulations

Jean-Daniel Boissonnat[1], Ramsay Dyer[2], and Arijit Ghosh[3]

[1] INRIA Sophia Antipolis, France
Jean-Daniel.Boissonnat@inria.fr
[2] University of Groningen, The Netherlands
r.h.dyer@rug.nl
[3] Max-Planck-Institut für Informatik, Germany
agosh@mpi-inf.mpg.d

Abstract. We propose algorithms to compute the Delaunay triangulation of a point set L using only (squared) distance comparisons (i.e., predicates of degree 2). Our approach is based on the witness complex, a weak form of the Delaunay complex introduced by Carlsson and de Silva. We give conditions that ensure that the witness complex and the Delaunay triangulation coincide and we introduce a new perturbation scheme to compute a perturbed set L' close to L such that the Delaunay triangulation and the witness complex coincide. Our perturbation algorithm is a geometric application of the Moser-Tardos constructive proof of the Lovász local lemma.

1   Introduction

The witness complex was introduced by Carlsson and de Silva [14] as a weak form of the Delaunay complex that is suitable for finite metric spaces and is computed using only distance comparisons. The witness complex $\mathrm{Wit}(L, W)$ is defined from two sets L and W in some metric space X: a finite set of points L on which the complex is built, and a set W of witnesses that serves as an approximation of X. A fundamental result of de Silva [13] states that $\mathrm{Wit}(L, W) = \mathrm{Del}(L)$ if W is the entire Euclidean space $X = \mathbb{R}^d$, and the result extends to spherical, hyperbolic and tree-like geometries. The result has also been extended to the case where $W = X$ is a smoothly embedded curve or surface of $\mathbb{R}^d$ [2]. However, when the set W of witnesses is finite, the Delaunay triangulation and the witness complexes are different and it has been an open question to understand when the two structures are identical. In this paper, we answer this question and present an algorithm to compute a Delaunay triangulation using the witness complex.

We first give conditions on L that ensure that the witness complex and the Delaunay triangulation coincide when $W \subset \mathbb{R}^d$ is a finite set (Section 3). Some of these conditions are purely combinatorial and easy to check. In a second part (Section 4), we show that those conditions can be satisfied by slightly perturbing the input set L. Our perturbation algorithm is a geometric application of the Moser-Tardos constructive proof of the general Lovász local lemma. Its analysis

© Springer-Verlag Berlin Heidelberg 2015
N. Bansal and I. Finocchi (Eds.): ESA 2015, LNCS 9294, pp. 595–606, 2015.
DOI: 10.1007/978-3-662-48350-3_50

uses the notion of protection of a Delaunay triangulation that we have previously introduced to study the stability of Delaunay triangulations [3].

Our algorithm has several interesting properties and we believe that it is a good candidate for implementation in higher dimensions.

1. **Low algebraic degree.** The only numerical operations used by the algorithm are (squared) distance comparisons (i.e., predicates of degree 2). In particular, we do not use orientation or in-sphere predicates, whose degree depends on the dimension d and are difficult to implement robustly in higher dimensions.
2. **Efficiency.** Our algorithm constructs the witness complex $\mathrm{Wit}(L', W) = \mathrm{Del}(L')$ of the perturbed set L' in time sublinear in $|W|$. See Section 5.
3. **Simplex quality.** Differently from all papers on this and related topics, we do not compute the volume or any measure of simplex quality. Nevertheless, through protection, a lower bound on the thickness of the output simplices can be guaranteed (see Theorem 3), and the resulting Delaunay triangulation is stable with respect to small metric or point perturbations [3].
4. **No need for coordinates.** We can construct Delaunay triangulations of points that come from some Euclidean space but whose actual positions are unknown. We simply need to know the interpoint distances.
5. **A thorough analysis.** Almost all papers in Computational Geometry rely on oracles to evaluate predicates exactly and assume that the complexity of those oracles is $O(1)$. Our (probabilistic) analysis is more precise. We only use predicates of degree 2 (i.e. double precision) and the analysis fully covers the case of non generic data.

Previous Work. Millman and Snoeyink [11] developed a degree-2 Voronoi diagram on a $U \times U$ grid in the plane. The diagram of n points can be computed using only double precision by a randomized incremental construction in $O(n \log n \log U)$ expected time and $O(n)$ expected space. The diagram also answers nearest neighbor queries, but it doesn't use sufficient precision to determine a Delaunay triangulation.

Our paper borrows ideas from the controlled perturbation paradigm [9]. The purpose is to actually perturb the input, thereby reducing the required precision of the underlying arithmetic and avoiding explicit treatment of degenerate cases. A specific scheme for Delaunay triangulations in arbitrary dimensions has been proposed by Funke et al. [8]. Their algorithm relies on a careful analysis of the usual predicates of degree $d + 2$ and is much more demanding than ours.

Notation. In order to avoid boundary complications, we work on the flat torus $\mathbb{T}^d = \mathbb{R}^d/\mathbb{Z}^d$. Boundary issues are discussed in the full version [6] of the paper. The *landmarks* form a finite set $L \subset \mathbb{T}^d$, but the set of *witnesses* $W \subseteq \mathbb{T}^d$ is only required to be closed in $\mathbb{T}^d$. If for any $x \in \mathbb{T}^d$ there is a $w \in W$ with $\|w - x\| < \varepsilon$, we say that $W \subset \mathbb{T}^d$ is an ε-*sample*. For *any* finite set $L \subset \mathbb{T}^d$ there is a $\lambda > 0$ such that L is a $\tilde{\lambda}$-sample for $\mathbb{T}^d$ for all $\tilde{\lambda} \geq \lambda$. The parameter λ is called the *sampling radius* of L. Also, there is a $\bar{\mu} > 0$ such that $\|p - q\| \geq \bar{\mu}\lambda$

for all $p, q \in L$. We call $\bar{\mu}$ the *sparsity ratio* of L, and we say that L is $(\lambda, \bar{\mu})$-*net*. Observe that $\bar{\mu} < 2$. Indeed, if p and q belong to a $(\lambda, \bar{\mu})$-net L, and q is the closest point to p in L, then $\bar{\mu}\lambda \leq \|p - q\| < 2\lambda$.

In order to avoid topological complications associated with the periodic boundary conditions, we impose the constraint $\lambda \leq 1/4$. A simplex $\sigma \subset L$ is a finite set. We always assume that L contains a non-degenerate d-simplex (i.e., L is not contained in a lower dimensional flat).

2   Delaunay and Witness Complexes

Definition 1 (Delaunay center and Delaunay complex). *A Delaunay center for a simplex $\sigma \subset L$ is a point $x \in \mathbb{T}^d$ that satisfies $\|p - x\| \leq \|q - x\|$, $\forall p \in \sigma$ and $\forall q \in L$. The Delaunay complex $\mathrm{Del}(L)$ of L is the complex consisting of all simplexes $\sigma \subset L$ that have a Delaunay center.*

Note that x is at equal distance from all the vertices of σ. A Delaunay simplex is *top dimensional* if is not the proper face of any Delaunay simplex. The affine hull of a top dimensional simplex has dimension d. If σ is top dimensional, the Delaunay center is the circumcenter of σ which we denote c_σ. We write R_σ for the circumradius of σ.

Delaunay [7] showed that if the point set L is generic, i.e., if no empty sphere contains $d + 2$ points on its boundary, then $\mathrm{Del}(L)$ is a triangulation of $\mathbb{T}^d$ (see the discussion in Section 3), and any perturbation L' of a finite set L is generic with probability 1. We refer to this as *Delaunay's theorem*.

We introduce now the witness complex that can be considered as a weak variant of the Delaunay complex.

Definition 2 (Witness and witness complex). *Let σ be a simplex with vertices in $L \subset \mathbb{T}^d$, and let w be a point of $W \subseteq \mathbb{T}^d$. We say that w is a witness of σ if $\|w - p\| \leq \|w - q\|$, $\forall p \in \sigma$ and $\forall q \in L \setminus \sigma$. The witness complex $\mathrm{Wit}(L, W)$ is the complex consisting of all simplexes σ such that for any simplex $\tau \subseteq \sigma$, τ has a witness in W.*

Observe that the only predicates involved in the construction of $\mathrm{Wit}(L, W)$ are (squared) distance comparisons, i.e. polynomials of degree 2 in the coordinates of the points. This is to be compared with the predicate that decides whether a point lies inside, on or outside the sphere circumscribing a d-simplex which is a polynomial of degree $d + 2$.

3   Identity of Witness and Delaunay Complexes

In this section, we make the connection between Delaunay and witness complexes more precise. We start with de Silva's result [13]:

Theorem 1. $\mathrm{Wit}(L, W) \subseteq \mathrm{Del}(L)$, *and if $W = \mathbb{T}^d$ then $\mathrm{Wit}(L, W) = \mathrm{Del}(L)$.*

If L is generic, we know that $\mathrm{Del}(L)$ is embedded in $\mathbb{T}^d$ by Delaunay's theorem. It therefore follows from Theorem 1 that the same is true for $\mathrm{Wit}(L, W)$. In particular, the dimension of $\mathrm{Wit}(L, W)$ is at most d.

Identity from protection. When W is not the entire space $\mathbb{T}^d$ but a finite set of points, the equality between $\mathrm{Del}(L)$ and $\mathrm{Wit}(L,W)$ no longer holds. However, by requiring that the d-simplices of $\mathrm{Del}(L)$ be δ-*protected*, a property introduced in [3], we are able to recover the inclusion $\mathrm{Del}(L) \subseteq \mathrm{Wit}(L,W)$, and establish the equality between the Delaunay complex and the witness complex with a discrete set of witnesses.

Definition 3 (δ-protection). *We say that a simplex $\sigma \subset L$ is δ-protected at $x \in \mathbb{T}^d$ if $\|x - q\| > \|x - p\| + \delta$, $\forall p \in \sigma$ and $\forall q \in L \setminus \sigma$.*

We say that $\mathrm{Del}(L)$ is δ-*protected* when each Delaunay d-simplex of $\mathrm{Del}(L)$ has a δ-protected Delaunay center. In this sense, δ-protection is in fact a property of the point set and we also say that L is δ-protected. If $\mathrm{Del}(L)$ is δ-protected for some unspecified $\delta > 0$, we say that L is protected (equivalently L is generic). We always assume $\delta < \lambda$ since it is impossible to have a larger δ if L is a λ-sample. The following lemma is proved in [4]. For simplicity, we use $\mathrm{star}^2(p')$ to denote $\mathrm{star}(\mathrm{star}(p; \mathrm{Del}(L)); \mathrm{Del}(L))$, where $\mathrm{star}(p; K)$ denotes the *star* of p in K, i.e. the smallest subcomplex of K containing the simplices that have p as a vertex. The *link* of p, $\mathrm{link}(p; K)$, is the simplicial complex defined by the simplices in $\mathrm{star}(p)$ that do not contain p.

Lemma 1 (Inheritance of protection). *Let L be a $(\lambda, \bar{\mu})$-net and suppose $p \in L$. If every d-simplex in $\mathrm{star}^2(p)$ is δ-protected, then all simplices in $\mathrm{star}(p; \mathrm{Del}(L))$ are at least δ'-protected where $\delta' = \frac{\bar{\mu}\delta}{4d}$.*

The following lemma is an easy consequence of the previous one (see [6] for a proof).

Lemma 2 (Identity from protection). *Let L be a $(\lambda, \bar{\mu})$-net with $p \in L$. If all the d-simplices in $\mathrm{star}^2(p)$ are δ-protected and W is an ε-sample for $\mathbb{T}^d$ with $\delta \geq \frac{8d\varepsilon}{\bar{\mu}}$, then $\mathrm{star}(p; \mathrm{Wit}(L,W)) = \mathrm{star}(p; \mathrm{Del}(L))$.*

We end this subsection with a result proved in [3, Lemma 3.13] that will be useful in Section 5. For any vertex p of a simplex σ, the *face opposite* p is the face determined by the other vertices of σ, and is denoted by σ_p. The *altitude* of p in σ is the distance $D(p, \sigma) = d(p, \mathrm{aff}(\sigma_p))$ from p to the affine hull of σ_p. The altitude $D(\sigma)$ of σ is the minimum over all vertices p of σ of $D(p, \sigma)$. A poorly-shaped simplex can be characterized by the existence of a relatively small altitude. The *thickness* of a j-simplex σ is the dimensionless quantity $\Theta(\sigma)$ that evaluates to 1 if $j = 0$ and to $\frac{D(\sigma)}{j\Delta(\sigma)}$ otherwise, where $\Delta(\sigma)$ denotes the *diameter* of σ, i.e. the length of its longest edge.

Lemma 3 (Thickness from protection). *Suppose $\sigma \in \mathrm{Del}(L)$ is a d-simplex with circumradius less than λ and shortest edge length greater than or equal to $\bar{\mu}\lambda$. If every $(d-1)$-face of σ is also a face of a δ-protected d-simplex different from σ, then the thickness of σ satisfies $\Theta(\sigma) \geq \frac{\bar{\delta}(\bar{\mu}+\bar{\delta})}{8d}$.*

In particular, suppose $p \in L$, where L is a $(\lambda, \bar{\mu})$-net, and every d-simplex in $\mathrm{star}^2(p)$ is δ-protected, then every d-simplex in $\mathrm{star}(p)$ is $\left(\frac{\bar{\delta}\bar{\mu}}{8d}\right)$-thick.

A combinatorial criterion for identity. The previous result will be useful in our analysis but does not help to compute $\mathrm{Del}(L)$ from $\mathrm{Wit}(L, W)$ since the δ-protection assumption requires knowledge of $\mathrm{Del}(L)$. A more useful result in this context will be given in Lemma 5. Before stating the lemma, we need to introduce some terminology and, in particular, the notion of *good links*.

A complex K is a *k-pseudo-manifold* if it is a pure k-complex and every $(k-1)$-simplex is the face of exactly two k-simplices.

Definition 4 (Good links). *Let K be a complex with vertex set $L \subset \mathbb{T}^d$. We say $p \in L$ has a* good link *if $\mathrm{link}(p; K)$ is a $(d-1)$-pseudo-manifold. If every $p \in L$ has a good link, we say K has* good links.

For our purposes, a simplicial complex K is a *triangulation of $\mathbb{T}^d$* if it is a d-manifold *embedded* in $\mathbb{T}^d$. We observe that a triangulation has good links.

Lemma 4 (Pseudomanifold criterion). *If K is a triangulation of $\mathbb{T}^d$ and $J \subseteq K$ has the same vertex set, then $J = K$ if and only if J has good links.*

A proof is given in [6]. We can now state the lemma that is at the heart of our algorithm. It follows from Theorem 1, Lemma 4, and Delaunay's theorem:

Lemma 5 (Identity from good links). *If L is generic and the vertices of $\mathrm{Wit}(L, W)$ have good links, then $\mathrm{Wit}(L, W) = \mathrm{Del}(L)$.*

4   Turning Witness Complexes into Delaunay Complexes

Let, as before, L be a finite set of landmarks and W a finite set of witnesses. In this section, we intend to use Lemma 5 to construct $\mathrm{Del}(L')$, where L' is close to L, using only comparisons of (squared) distances. The idea is to first construct the witness complex $\mathrm{Wit}(L, W)$ which is a subcomplex of $\mathrm{Del}(L)$ (Theorem 1) that can be computed using only distance comparisons. We then check whether $\mathrm{Wit}(L, W) = \mathrm{Del}(L)$ using the pseudomanifold criterion (Lemma 4). While there is a vertex p of $\mathrm{Wit}(L, W)$ that has a bad link (i.e. a link that is not a pseudomanifold), we perturb p' and the set of vertices $I(p')$, to be exactly defined in Section 4.2, that are responsible for the bad link $L(p') = \mathrm{link}(p', \mathrm{Wit}(L', W))$, and recompute the witness complex for the perturbed points. We write L' for the set of perturbed points at some stage of the algorithm. Each point p' is randomly and independently taken from the so-called *picking ball* $B(p, \rho)$. Upon termination, we have $\mathrm{Wit}(L', W) = \mathrm{Del}(L')$. The parameter ρ, the radius of the picking balls, must satisfy Eq. (2) to be presented later. The steps are described in more detail in [6, Algo. 1]. The analysis of the algorithm relies on the Moser-Tardos constructive proof of Lovász local lemma.

4.1   Lovász Local Lemma

The celebrated Lovász local lemma is a powerful tool to prove the existence of combinatorial objects [1]. Let $\mathcal{A}$ be a finite collection of "bad" events in some probability space. The lemma shows that the probability that none of these events occur is positive provided that the individual events occur with a bounded probability and there is limited dependence among them.

Lemma 6 (Lovász local lemma). *Let $\mathcal{A} = \{A_1, \ldots, A_N\}$ be a finite set of events in some probability space. Suppose that each event A_i is independent of all but at most Γ of the other events A_j, and that $\Pr[A_i] \leq \varpi$ for all $1 \leq i \leq N$. If $\varpi \leq \frac{1}{e(\Gamma+1)}$ (e is the base of the natural logarithm), then $\Pr\left[\bigwedge_{i=1}^{N} \neg A_i\right] > 0$.*

Assume that the events depend on a finite set of mutually independent variables in a probability space. Moser and Tardos [12] gave a constructive proof of Lovász lemma leading to a simple and natural algorithm that checks whether some event $A \in \mathcal{A}$ is violated and randomly picks new values for the random variables on which A depends. We call this a resampling of the event A. Moser and Tardos proved that this simple algorithm quickly terminates, providing an assignment of the random variables that avoids all of the events in $\mathcal{A}$. The expected total number of resampling steps is at most N/Γ.

4.2   Correctness of the Algorithm

We write $\rho = \bar{\rho}\lambda$ and $\mu = \bar{\mu}\lambda$, and we assume $\bar{\rho} < \bar{\mu}/4$. The triangle inequality yields a bound on the sampling radius λ' and the sparsity ratio $\bar{\mu}'$ of any perturbed point set L': $\lambda' = \lambda(1 + \bar{\rho}) < 2\lambda$ and $\bar{\mu}' = \frac{\bar{\mu}-2\bar{\rho}}{1+\bar{\rho}} \geq \frac{\bar{\mu}}{3}$.

We refer to the terminology of the Lovász local lemma. Our *variables* are the points of L' which are randomly and independently taken from the picking balls $B(p, \rho)$, $p \in L$.

The *events* are associated to points of L', the vertices of $\text{Wit}(L', W)$. We say that an event happens at $p' \in L$ when the link $L(p')$ of p' in $\text{Wit}(L', W)$ is not good, i.e., is not a pseudomanifold. We know from Lemma 2 that if p' is a vertex of $\text{Wit}(L', W)$ and $L(p')$ is not good, then there must exist a d-simplex in $\text{star}^2(p')$ that is not δ-protected for $\delta = 8d\varepsilon/\bar{\mu}'$. We will denote by

- $I_1(p')$: the set of points of L' that can be in $\text{star}^2(p')$
- $I_2(p')$: the set of points of L' that can violate the δ-protected zone $Z_\delta(\sigma') = B(c_\sigma, R_\sigma + \delta) \setminus B(c_\sigma, R_\sigma)$ for some d-simplex σ' in $\text{star}^2(p')$
- $I(p') := I_1(p') \cup I_2(p')$
- $S(p')$: the set of d-simplices with vertices in $I_1(p')$ that can belong to $\text{star}^2(p')$

The probability $\varpi_1(p')$ that $L(p')$ is not good is at most the probability $\varpi_2(p')$ that one of the simplices of $S(p')$, say σ', has its δ-protecting zone $Z_\delta(\sigma')$ violated by some point of L'. Write $\varpi_3(q', \sigma')$ for the probability that q' belongs to the δ-protection zone of the d-simplex σ'. We have

$$\varpi_1(p') \leq \varpi_2(p') \leq \sum_{q' \in I_2(p')} \sum_{\sigma' \in S(p')} \varpi_3(q', \sigma') \tag{1}$$

The following lemma, proved in [6] upper bounds $|I(p')|, |S(p')|, \Gamma$ and $\varpi_3(q', \sigma')$. Observe that the events p' and q' are independent if $I(p') \cap I(q') = \emptyset$.

Lemma 7. *(1)* $|I(p')| \leq I = \left(\frac{36}{\bar{\mu}}\right)^d$ *and* $|S(p')| \leq K = \frac{I^{d+1}}{(d+1)!}$. *(2) An event is independent of all but at most* $\Gamma = \left(\frac{66}{\bar{\mu}}\right)^d$ *other events. (3)* $\varpi_3(q',\sigma') \leq 2\pi^{d-1}\frac{\delta}{\rho}$.

Using Eq. (1) and Lemma 7, we conclude that $\varpi_1(p') \leq 2\pi^{d-1} I K \frac{\delta}{\rho}$.

An event depends on at most Γ other events. Hence, to apply the Lovász Local Lemma 6, it remains to ensure that $\varpi_1(p') \leq 1/(e(\Gamma + 1))$. In addition, we also need that $\delta \geq 8d\varepsilon/\bar{\mu}'$ to be able to apply Lemma 2. We thefore need to satisfy $\frac{8d\varepsilon}{\bar{\mu}'} \leq \delta \leq J\rho$ where $J^{-1} \stackrel{\text{def}}{=} 2e\pi^{d-1}IK(\Gamma+1)$. Observe that I, K, Γ and J depend only on $\bar{\mu}$ and d. We conclude that the conditions of the Lovász local lemma hold if the parameter ρ satisfies

$$\frac{\mu}{4} \geq \rho \geq \frac{24d\varepsilon}{\bar{\mu}J} \quad \text{where} \quad J^{-1} \stackrel{\text{def}}{=} 2e\pi^{d-1}IK(\Gamma+1) = \left(\frac{2}{\bar{\mu}}\right)^{O(d^2)} \tag{2}$$

Hence, if ε is sufficiently small, we can fix ρ so that Eq. (2) holds. The algorithm is then guaranteed to terminate. By Lemma 5, the output is Del(L').

It follows from Moser-Tardos theorem that the expected number of times a bad link is encountered is $O\left(\frac{|L|}{\Gamma}\right)$ and since $|I(p')| \leq I$, we get that the number of point perturbations performed by the algorithm is $O\left(\frac{I|L|}{\Gamma}\right)$ on expectation. We sum up the results of this section in

Theorem 2. *Under Eq. (2), the algorithm terminates and outputs the Delaunay triangulation of some set L' whose distance to L is at most ρ. The number of point perturbations performed by the algorithm is* $O\left(\frac{I|L|}{\Gamma}\right)$.

5   Sublinear Algorithm

When the set L' is generic, $K = \text{Wit}(L', W)$ is embedded in $\mathbb{T}^d$ and is therefore d-dimensional. It is well known that the d-skeleton of $\text{Wit}(L', W)$ can be computed in time $O((|W|+|K|)\log|L'|)$ using only distance comparisons [5]. Although easy and general, this construction is not efficient when W is large.

In this section, we show how to implement an algorithm called Algorithm 2 with execution time sublinear in $|W|$. We will assume that the points of W are located at the centers of the cells of a grid, which is no real loss of generality. The idea is to restrict our attention to a subset of W, namely the set of *full-leaf-points* introduced in Section 5.2. These are points that may be close to the circumcenter of some d-simplex. A crucial observation is that if a d-simplex has a bounded thickness, then we can efficiently compute a bound on the number of its full-leaf-points. This observation will also allow us to guarantee some protection (and therefore thickness) on the output simplices, as stated in Theorem 3 below.

The approach is based on the *relaxed Delaunay complex*, which is related to the witness complex, and was also introduced by de Silva [13]. We first introduce this, and the structural observations on which the algorithm is based.

5.1   The Relaxed Delaunay Complex

The basic idea used to get an algorithm sublinear in $|W|$ is to choose witnesses for d-simplices that are close to being circumcenters for these simplices. With this approach, we can in fact avoid looking for witnesses of the lower dimensional simplices. The complex that we will be computing is a subcomplex of a relaxed Delaunay complex:

Definition 5 (Relaxed Delaunay complex). *Let $\sigma \subset L'$ be a simplex. An α-center of σ is any point $x \in \mathbb{T}^d$ such that $\|x - p\| \le \|x - q\| + \alpha \ \forall p, q \in \sigma$. We say that x is an α-Delaunay center of σ if $\|x - p\| \le \|x - q\| + \alpha \ \forall p \in \sigma$ and $\forall q \in L'$. The set of simplices that have an α-Delaunay center in W is a simplicial complex, called the α-relaxed Delaunay complex, and is denoted $\mathrm{Del}^{\alpha}(L', W)$.*

We say $w \in W$ is an α-*witness* for $\sigma \subset L'$ if $\|w - p\| \le \|w - q\| + \alpha$ for all $p \in \sigma$ and all $q \in L' \setminus \sigma$. We observe that $w \in W$ is an α-Delaunay center if and only if it is an α-center and also an α-witness.

Lemma 8. *The distance between an α-Delaunay center for $\sigma \in \mathrm{Del}^{\alpha}(L', W)$ and the farthest vertex in σ is less than $\lambda' + \alpha$. In particular, $\Delta_{\sigma} < 2\lambda' + 2\alpha$.*

If $\tau \in \mathrm{Del}(L')$ and c is a Delaunay center of τ, then any point in $B(c, r)$ is a $2r$-Delaunay center for τ. Thus $\mathrm{Del}(L') \subseteq \mathrm{Del}^{2\varepsilon}(L', W)$.

If, for some $\delta \ge 0$, each of the d-simplices in $\mathrm{Del}^{\alpha}(L', W)$ has a δ-protected circumcenter, then we have that $\mathrm{Del}^{\alpha}(L', W) \subseteq \mathrm{Del}(L')$, and with $\alpha \ge 2\varepsilon$, it follows (Lemma 8) that $\mathrm{Del}^{\alpha}(L', W) = \mathrm{Del}(L')$, and $\mathrm{Del}(L')$ is itself δ-protected and has good links.

Reviewing the analysis of the Moser-Tardos algorithm of Section 4.2, we observe that the exact same estimate of $\varpi_2(p')$ that serves as an upper bound on the probability that one of the simplices in $\mathrm{star}^2(p', \mathrm{Del}(L'))$ is not δ-protected at its circumcenter, also serves as an upper bound on the probability that one of the simplices in $\mathrm{star}^2(p', \mathrm{Del}^{\alpha}(L', \mathbb{T}^d))$ is not δ-protected at its circumcenter, provided that $4\alpha + \delta \le \lambda'$ (using the diameter bound of $2\lambda' + 2\alpha$ from Lemma 8), which we will assume from now on. We can therefore modify the algorithm by replacing $\mathrm{Wit}(L', W)$ by $\mathrm{Del}^{\alpha}(L', W)$.

We now describe how to improve this algorithm to make it efficient. For our purposes it will be sufficient to set $\alpha = 2\varepsilon$. In order to obtain an algorithm sublinear in $|W|$, we will not compute the full $\mathrm{Del}^{2\varepsilon}(L', W)$ but only a subcomplex we call $\mathrm{Del}_0^{2\varepsilon}(L', W)$. The exact definition of $\mathrm{Del}_0^{2\varepsilon}(L', W)$ will be given in Section 5.3, but the idea is to only consider d-simplices that show the properties of being Θ_0-thick for some parameter Θ_0 to be defined later. This will allow us to restrict our attention to points of W that lie near the circumcenter. As explained in Section 5.3, this is done without explicitly computing thickness or circumcenters.

As will be shown in Section 5.4 (Lemma 12), the modification of the witness algorithm ([6, Algo. 1]) that computes $\mathrm{Del}_0^{2\varepsilon}(L', W)$ instead of $\mathrm{Wit}(L', W)$ will terminate and output a complex $\mathrm{Del}_0^{2\varepsilon}(L', W)$ with good links. However, this is not sufficient to guarantee that the output is correct, i.e., that $\mathrm{Del}^{2\varepsilon}(L', W) =$

$\text{Del}(L')$. In order to obtain this guarantee, we insert an extra procedure `check()`, which, without affecting the termination guarantee, will ensure that the simplices of $\text{Del}_0^{2\varepsilon}(L', W)$ have δ^*-protected circumcenters for a positive δ^*. It follows that $\text{Del}_0^{2\varepsilon}(L', W) \subseteq \text{Del}(L')$ and, by Lemma 4, that $\text{Del}_0^{2\varepsilon}(L', W) = \text{Del}(L')$. Pseudocode for this modified perturbation algorithm is presented as Algorithm 2 in [6].

We describe the details of computing $\text{Del}_0^{2\varepsilon}(L', W)$ and of the `check()` procedure in the following subsections.

5.2 Computing Relaxed Delaunay Centers

We observe that the α-Delaunay centers of a d-simplex σ are close to the circumcenter of σ, provided that σ has a bounded thickness:

Lemma 9 (Clustered α-Delaunay centers). *Assume that L' is a $(\lambda', \bar{\mu}')$-sample. Let σ be a non degenerate d-simplex, and let x be an α-center for σ at distance at most $C\lambda'$ from the vertices of σ, for some constant $C > 0$. Then x is at distance at most $\frac{C\alpha}{\Theta(\sigma)\bar{\mu}'}$ from the circumcenter c_σ of σ. In particular, if x is an α-Delaunay center for σ, then $\|c_\sigma - x\| < \frac{2\alpha}{\Theta(\sigma)\bar{\mu}'}$.*

See [6] for a proof. It follows from Lemma 9 that α-Delaunay centers are close to all the bisecting hyperplanes $H_{pq} = \{x \in \mathbb{R}^d \mid \|x - p\| = \|x - q\|, \ p, q \in \sigma\}$. The next simple lemma asserts a kind of qualitative converse:

Lemma 10. *Let σ be a d-simplex and H_{pq} be the bisecting hyperplane of p and q. A point x that satisfies $d(x, H_{pq}) \leq \alpha$, for any $p, q \in \sigma$ is a 2α-center of σ.*

Let σ be a d-simplex of $\text{Del}^\alpha(L', W)$ and let $\bar{\Omega}$ be the smallest box with edges parallel to the coordinate axes that contains σ. Then the edges of $\bar{\Omega}$ have length at most $2\lambda' + 2\alpha$ (Lemma 8). Any α-Delaunay center for σ is at a distance at most $\lambda' + \alpha$ from $\bar{\Omega}$. Therefore all the α-Delaunay centers for σ lie in an axis-aligned hypercube Ω with the same center as $\bar{\Omega}$ and with side length at most $4\lambda' + 4\alpha < 5\lambda'$. Observe that the diameter (diagonal) z of Ω is at most $5\lambda'\sqrt{d}$.

Our strategy is to first compute the α-centers of σ that belong to $\Omega \cap W$ and then to determine which ones are α-witnesses for σ. Deciding if an α-center is an α-witness for σ can be done in constant time since L' is a $(\lambda', \bar{\mu}')$-net and $\Delta_\sigma \leq 2\lambda' + 2\alpha$ (Lemma 8).

We take $\alpha = 2\varepsilon$. To compute the 2ε-Delaunay centers of σ, we will use a pyramid data structure (it is an octree when $d = 3$). The pyramid consists of at most $\log \frac{z}{\varepsilon}$ levels. Each level $h > 0$ is a grid of resolution $2^{-h}z$. The grid at level 0 consists of the single cell, Ω. Each node of the pyramid is associated to a cell of a grid. The children of a node ν correspond to a subdivision into 2^d subcells (of the same size) of the cell associated to ν. The leaves are associated to the cells of the finest grid whose cells have diameter ε.

A node of the pyramid that is intersected by all the bisecting hyperplanes of σ will be called a *full node* or, equivalently, a *full cell*. By our definition of W, a cell of the finest grid contains an element of W at its center. The *full-leaf-points* are the elements of W associated to full cells at the finest level. By Lemma 10, the full-leaf-points are 2ε-centers for σ. In order to identify the full-leaf cells, we traverse the full nodes of the pyramid starting from the root. Note that to decide if a cell is full, we only have to decide if two corners of a cell are on opposite sides of a bisecting hyperplane, which reduces to evaluating a polynomial of degree 2 in the input variables. A simple volume argument leads to the following lemma:

Lemma 11. *The number of full cells is* $\leq n_\sigma(\varepsilon) = \frac{n_0}{(\Theta(\sigma)\bar{\mu}')^d}\log\frac{5\sqrt{d}\lambda'}{\varepsilon}$, *where* n_0 *depends on d. $n_\sigma(\varepsilon)$ is also a bound on the time to compute the full cells.*

5.3   Construction of $\mathrm{Del}_0^{2\varepsilon}(L', W)$

By Lemma 8, all the simplices incident to a vertex p' of $\mathrm{Del}^{2\varepsilon}(L')$ are contained in $N(p') = L' \cap B(p', 2\lambda' + 4\varepsilon)$, and it follows from the fact that L' is a $(\lambda', \bar{\mu}')$-net that $|N(p')| \leq \frac{2^{O(d)}}{(\bar{\mu}')^d}$. In the first step of the algorithm, we compute, for each $p' \in L'$, the set $N(p')$, and the set of d-simplices $C_d(p') = \{\sigma = \{p'\} \cup \tilde{\sigma} : |\tilde{\sigma}| = d$ and $\tilde{\sigma} \subset N(p') \setminus \{p'\}\}$. Observe that $|C_d(p')| = \binom{|N(p')|}{d} = \frac{2^{O(d^2)}}{(\bar{\mu}')^{d^2}}$. We then extract from $C_d(p')$ a subset $WC_d(p')$ of simplices that have a full-leaf-point that is a 2ε-Delaunay center, and have a number of full cells less than or equal to $n_0(\varepsilon) \overset{\text{def}}{=} \frac{n_0}{(\Theta_0\bar{\mu}')^d}\log\frac{5\sqrt{d}\lambda'}{\varepsilon}$. This is done by applying the algorithm of Section 5.2 with a twist. As soon as a d-simplex appears to have more than $n_0(\varepsilon)$ full cells, we stop considering that simplex. The union of the sets $WC_d(p')$ for all $p' \in L'$ is a subcomplex of $\mathrm{Del}^{2\varepsilon}(L')$ called $\mathrm{Del}_0^{2\varepsilon}(L', W)$. It contains every d-simplex σ in $\mathrm{Del}^{2\varepsilon}(L', W)$ that has a 2ε-Delaunay center in W at a distance less than ε from its actual circumcenter, and satisfies the thickness criterion $\Theta(\sigma) \geq \Theta_0$. Note, however, that we do not claim that every simplex that has at most $n_0(\varepsilon)$ full cells is Θ_0-thick. See [6, Algo. 3] for pseudocode describing the algorithm for constructing $\mathrm{Del}_0^{2\varepsilon}(L, W)$. As noted above, $|N(p')| = \frac{2^{O(d)}}{(\bar{\mu}')^d}$ for any $p' \in L'$, and all the $N(p')$ can be computed in $O(|L'|^2)$ time by a brute force method. But assuming we have access to "universal hash functions" then we can use the "grid method" described in [10, Chap. 1] with the sparsity condition of L to get the complexity down to $\frac{2^{O(d)}|L'|}{(\bar{\mu}')^d}$. Using the facts that $\lambda' < 2\lambda$ and $\bar{\mu}' \geq \frac{\bar{\mu}}{3}$ (see beginning of Section 4.2), we conclude that the total complexity of the algorithm is $O\left(\frac{|L|}{\Theta_0^d(\bar{\mu})^{d^2+d}}\log\frac{\lambda}{\varepsilon}\right)$ and is therefore sublinear in $|W|$.

5.4   Correctness of the Algorithm

We will need the following lemma which is an analog of Lemma 2. The lemma also fixes Θ_0. Its proof follows directly from Lemma 3, and the observation that any simplex with a protected circumcenter is a Delaunay simplex.

Lemma 12. *Suppose that the d-simplices in $\mathrm{Del}(L')$ are δ-protected at their circumcenters, with $\delta = \bar{\delta}\lambda'$. If $\Theta_0 = \frac{\bar{\delta}\bar{\mu}'}{8d}$, then $\mathrm{Del}(L') \subseteq \mathrm{Del}_0^{2\varepsilon}(L', W)$ and if* *in addition every d-simplex of* $\mathrm{Del}_0^{2\varepsilon}(L',W)$ *has a protected circumcenter, then* $\mathrm{Del}_0^{2\varepsilon}(L',W) = \mathrm{Del}(L').$

We first show that Algorithm 2 terminates if we desactivate the call to procedure `check()`. As discussed after Lemma 8, the analysis of Section 4.2 implies that the perturbations of Algorithm 2 can be expected to produce a point set L' for which all the d-simplices in $\mathrm{Del}^{2\varepsilon}(L',\mathbb{T}^d)$ have a δ-protected circumcenter. Since this complex includes both $\mathrm{Del}(L')$ and $\mathrm{Del}_0^{2\varepsilon}(L',W)$, Lemma 12 shows that we can expect the algorithm to terminate with the condition that $\mathrm{Del}_0^{2\varepsilon}(L',W)$ has good links.

We now examine procedure `check()` and show that it does not affect the termination guarantee. By Lemma 3, if $\mathrm{Del}(L')$ is δ-protected, then any $\sigma \in \mathrm{Del}(L')$ satisfies $\Theta(\sigma) \geq \Theta_0 = \frac{\bar{\delta}\bar{\mu}'}{8d}$. Consider now $\sigma \in \mathrm{Del}_0^{2\varepsilon}(L',W)$. Since the full leaves of the pyramid data-structure for σ are composed entirely of 2ε-centers at a distance less than $4\sqrt{d}\lambda'$ from any vertex of σ, Lemma 9 implies that, if σ is Θ_0-thick, then $\|x - c_\sigma\| \leq \frac{8\sqrt{d}\varepsilon}{\Theta_0\bar{\mu}'}$. This means that we can restrict our definition of $\mathrm{Del}_0^{2\varepsilon}(L',W)$ to include only simplices for which the set of full leaves has diameter less than $\frac{16\sqrt{d}\varepsilon}{\Theta_0\bar{\mu}'}$. Further, we observe that if σ is δ-protected at its circumcenter, then it will have a $(\delta - 2\varepsilon)$-protected full-leaf-point; this follows from the triangle inequality. The `check()` procedure ensures that all the simplices in $\mathrm{Del}_0^{2\varepsilon}(L',W)$ have these two properties. It follows from the discussion above that activating procedure `check()` does not affect the termination guarantee.

The fact that the algorithm terminates yields $\mathrm{Del}_0^{2\varepsilon}(L',W)$ with good links. In order to apply Lemma 12 to guarantee that $\mathrm{Del}_0^{2\varepsilon}(L',W) = \mathrm{Del}(L')$, we need to guarantee that the simplices of $\mathrm{Del}_0^{2\varepsilon}(L',W)$ are protected. The following lemma, proved in [6], provides a bound on δ to ensure such a protection $\delta^* > 0$.

Lemma 13. *If* $\delta = J\rho$, *with* J *defined in Eq. (2), then the d-simplices in* $\mathrm{Del}_0^{2\varepsilon}(L',W)$ *produced by the modified perturbation algorithm [6, Algo. 2] are* δ^*-*protected, with* $\delta^* = \delta - \left(\frac{34\sqrt{d}}{\Theta_0\bar{\mu}'}\right)\varepsilon.$

In order to have $\delta^* > 0$, we need a lower bound on δ, and hence on the minimal perturbation radius through $\delta = J\rho$. Therefore we require: $\frac{J\mu}{4} \geq \delta > \frac{34\sqrt{d}\varepsilon}{\Theta_0\bar{\mu}'}$ (compare with (2)). Writing $\bar{\delta} = \frac{\delta}{\lambda'}$ and $\Theta_0 = \frac{\bar{\delta}\bar{\mu}'}{8d}$, and using $\lambda' \geq \lambda$, $\bar{\mu}' \geq \bar{\mu}/3$ when $\rho \leq \mu/4$, we obtain the conditions under which Algorithm 2 is guaranteed to produce a δ^*-protected Delaunay triangulation: $\frac{J\bar{\mu}}{4} \geq \bar{\delta} > \frac{2448d^{\frac{3}{2}}}{\bar{\delta}\bar{\mu}^2}\frac{\varepsilon}{\lambda}$. The right-hand inequality is satisfied provided $\bar{\delta} \geq \frac{50d^{\frac{3}{4}}}{\bar{\mu}}\sqrt{\frac{\varepsilon}{\lambda}}$. We have proved

Theorem 3. *If* $\bar{\rho} \leq \bar{\mu}/4$ *and* $\bar{\rho} = \Omega\left(\frac{1}{J\bar{\mu}}\sqrt{\frac{\varepsilon}{\lambda}}\right)$ *(with* J *defined in Eq. (2)),* *Algorithm 2 terminates and outputs the Delaunay triangulation of* L'. *The Delaunay d-simplices are* δ^*-*protected, as defined in Lemma 13, and consequently satisfy a thickness bound of* $\Theta(\sigma) \geq \frac{\bar{\delta}^*(\bar{\mu}/3+\bar{\delta}^*)}{8d}$. *The complexity of the algorithm is* $O\left(\frac{|L|}{\Theta_0^d\bar{\mu}^{d^2+d}}\log\frac{\lambda}{\varepsilon}\right)$. *The constants in* Ω *and* O *depend only on* d *(the dependence being exponential).*

Acknowledgments. Partial support has been provided by the Advanced Grant of the European Research Council GUDHI (Geometric Understanding in Higher Dimensions).

Arijit Ghosh is supported by the Indo-German Max Planck Center for Computer Science (IMPECS).

References

1. Alon, N., Spencer, J.H.: The Probabilistic Method, 3rd edn. Wiley-Interscience, New York (2008)
2. Attali, D., Edelsbrunner, H., Mileyko, Y.: Weak witnesses for Delaunay triangulations of submanifolds. In: Proc. ACM Sympos. Solid and Physical Modeling, pp. 143–150 (2007)
3. Boissonnat, J.D., Dyer, R., Ghosh, A.: The Stability of Delaunay Triangulations. Int. J. on Comp. Geom (IJCGA) 23(4&5), 303–333 (2013)
4. Boissonnat, J.D., Dyer, R., Ghosh, A., Oudot, S.Y.: Only distances are required to reconstruct submanifolds. ArXiv e-prints (October 2014)
5. Boissonnat, J.D., Maria, C.: The Simplex Tree: An Efficient Data Structure for General Simplicial Complexes. Algorithmica 70(3), 406–427 (2014)
6. Boissonnat, J., Dyer, R., Ghosh, A.: A probabilistic approach to reducing the algebraic complexity of computing Delaunay triangulations. CoRR abs/1505.05454 (2015). http://arxiv.org/abs/1505.05454
7. Delaunay, B.: Sur la sphère vide. Izv. Akad. Nauk SSSR, Otdelenie Matematicheskii i Estestvennyka Nauk 7, 793–800 (1934)
8. Funke, S., Klein, C., Mehlhorn, K., Schmitt, S.: Controlled perturbation for Delaunay triangulations. In: Proc. 16th ACM-SIAM Symposium on Discrete Algorithms (SODA), pp. 1047–1056 (2005)
9. Halperin, D.: Controlled perturbation for certified geometric computing with fixed-precision arithmetic. In: Fukuda, K., van der Hoeven, J., Joswig, M., Takayama, N. (eds.) ICMS 2010. LNCS, vol. 6327, pp. 92–95. Springer, Heidelberg (2010)
10. Har-Peled, S.: Geometric Approximation Algorithms. American Mathematical Society (2011)
11. Millman, D.L., Snoeyink, J.: Computing Planar Voronoi Diagrams in Double Precision: A Further Example of Degree-driven Algorithm Design. In: Proc. 26th ACM Symp. on Computational Geometry, pp. 386–392 (2010)
12. Moser, R.A., Tardos, G.: A constructive proof of the generalized Lovász Local Lemma. Journal of the ACM 57(2) (2010)
13. de Silva, V.: A weak characterisation of the Delaunay triangulation. Geometriae Dedicata 135(1), 39–64 (2008)
14. de Silva, V., Carlsson, G.: Topological estimation using witness complexes. In: Proc. Sympos. Point-Based Graphics, pp. 157–166 (2004)

A Characterization of Visibility Graphs
for Pseudo-polygons

Matt Gibson[1], Erik Krohn[2], and Qing Wang[1]

[1] Dept. of Computer Science
University of Texas at San Antonio
San Antonio, TX, USA
[2] Dept. of Computer Science
University of Wisconsin - Oshkosh
Oshkosh, WI, USA

Abstract. In this paper, we give a characterization of the visibility graphs of pseudo-polygons. We first identify some key combinatorial properties of pseudo-polygons, and we then give a set of five necessary conditions based off our identified properties. We then prove that these necessary conditions are also sufficient via a reduction to a characterization of vertex-edge visibility graphs given by O'Rourke and Streinu.

1   Introduction

Geometric covering problems have been a focus of research for decades. Here we are given some set of points P and a set S where each $s \in S$ can cover some subsets of P. The subset of P is generally induced by some geometric object. For example, P might be a set of points in the plane, and s consists of the points contained within some disk in the plane. For most variants, the problem is NP-hard and can easily be reduced to an instance of the combinatorial set cover problem which has a polynomial-time $O(\log n)$-approximation algorithm, which is the best possible approximation under standard complexity assumptions [5]. The main question therefore is to determine for which variants of geometric set cover can we obtain polynomial-time approximation algorithms with approximation ratio $o(\log n)$, as any such algorithm must exploit the geometry of the problem to achieve the result. This area has been studied extensively, see for example [2,13,1], and much progress has been made utilizing algorithms that are based on solving the standard linear programming relaxation.

Unfortunately this technique has severe limitations for some variants of geometric set cover, and new ideas are needed to make progress on these variants. In particular, the techniques are lacking when the points P we wish to cover is a simple polygon, and we wish to place the smallest number of points in P that collectively "see" the polygon. This problem is classically referred to as the *art gallery problem* as an art gallery can be modeled as a polygon and the points placed by an algorithm represent cameras that can "guard" the art gallery. This has been one of the most well-known problems in computational geometry for

© Springer-Verlag Berlin Heidelberg 2015
N. Bansal and I. Finocchi (Eds.): ESA 2015, LNCS 9294, pp. 607–618, 2015.
DOI: 10.1007/978-3-662-48350-3_51

many years, yet still to this date the best polynomial-time approximation algorithm for this problem is a $O(\log n)$-approximation algorithm. The key issue is a fundamental lack of understanding of the combinatorial structure of visibility inside simple polygons. It seems that in order to develop powerful approximation algorithms for this problem, the community first needs to better understand the underlying structure of such visibility.

Visibility Graphs. A very closely related issue which has received a lot of attention in the community is the *visibility graph* of a simple polygon. Given a simple polygon P, the visibility graph $G = (V, E)$ of P has the following structure. For each vertex $p \in P$, there is a vertex in V, and there is an edge connecting two vertices in G if and only if the corresponding vertices in P "see" each other (i.e., the line segment connecting the points does not go outside the polygon). Two major open problems regarding visibility graphs of simple polygons are the visibility graph characterization problem and the visibility graph recognition problem. The *visibility graph characterization* problem seeks to define a set of properties that all visibility graphs satisfy. The *visibility graph recognition* problem is the following. Given a graph G, determine if there exists a simple polygon P such that G is the visibility graph of P in polynomial time.

The problems of characterizing and recognizing the visibility graphs of simple polygons have had partial results given dating back to over 25 years ago [6] and remain open to this day with only a few special cases being solved. Characterization and recognition results have been given in the special cases of "spiral" polygons [4] and "tower polygons" [3]. There have been several results [7,4,11] that collectively have led to four necessary conditions that a simple polygon visibility graph must satisfy. That is, if the graph G does not satisfy all four of the conditions then we know that G is not the visibility graph for *any* simple polygon, and moreover it can be determined if a graph G satisfies all of the necessary conditions in polynomial time. Streinu, however, has given an example of graph that satisfies all of the necessary conditions but is not a visibility graph for any simple polygon [12], implying that the set of conditions is not sufficient and therefore a strengthening of the necessary conditions is needed. Unfortunately it is not even known if simple polygon visibility graph recognition is in NP. See [8] for a nice survey on these problems and other related visibility problems.

Pseudo-polygons. Given the difficulty of understanding simple polygon visibility graphs, O'Rourke and Streinu [9] considered the visibility graphs for a special case of polygons called *pseudo-polygons* which we will now define. An arrangement of *pseudo-lines* $\mathcal{L}$ is a collection of simple curves, each of which separates the plane, such that each pair of pseudo-lines of $\mathcal{L}$ intersects at exactly one point, where they cross. Let $P = \{p_0, p_2, \ldots, p_{n-1}\}$ be a set of points in $\mathbb{R}^2$, and let $\mathcal{L}$ be an arrangement of $\binom{n}{2}$ pseudo-lines such that every pair of points p_i and p_j lie on exactly one pseudo-line in $\mathcal{L}$, and each pseudo-line in $\mathcal{L}$ contains exactly two points of P. The pair $(P, \mathcal{L})$ is called a *pseudo configuration of points* (pcp) in general position.

Intuitively a pseudo-polygon is determined similarly to a standard Euclidean simple polygon except using pseudo-lines instead of straight line segments. Let $L_{i,j}$ denote the pseudo-line through the points p_i and p_j. We view $L_{i,j}$ as having three different components. The subsegment of $L_{i,j}$ connecting p_i and p_j is called the *segment*, and we denote it $p_i p_j$. Removing $p_i p_j$ from $L_{i,j}$ leaves two disjoint *rays*. Let $r_{i,j}$ denote the ray starting from p_i and moving away from p_j, and we let $r_{j,i}$ denote the ray starting at p_j and moving away from p_i. Consider the pseudo line $L_{i,i+1}$ in a pcp (indices taken modulo n and are increasing in counterclockwise order throughout the paper). We let e_i denote the segment of this line. A *pseudo-polygon* is obtained by taking the segments e_i for $i \in \{0, \ldots, n-1\}$ if (1) the intersection of e_i and e_{i+1} is only the point p_{i+1} for all i, and (2) distinct segments e_i and e_j do not intersect for all $j \neq i+1$. We call the segments e_i the *boundary edges*. A pseudo-polygon separates the plane into two regions: "inside" the pseudo-polygon and "outside" the pseudo-polygon, and any two points p_i and p_j see each other if the segment of their pseudo-line does not go outside of the pseudo-polygon. See Fig. 1 for an illustration. Pseudo-polygons can be viewed as a combinatorial abstraction of simple polygons. Note that every simple polygon is a pseudo-polygon (simply allow each $L_{i,j}$ to be the straight line through p_i and p_j), and Streinu showed that there are pseudo-polygons that cannot be "stretched" into a simple polygon [12].

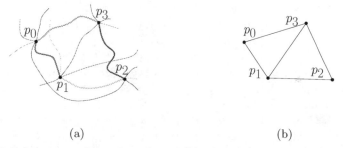

(a)   (b)

Fig. 1. (a) A pcp and pseudo-polygon. (b) The corresponding visibility graph.

O'Rourke and Streinu [9] give a characterization of *vertex-edge* visibility graphs of pseudo-polygons. In this setting, for any vertex v we are told which *edges* v sees rather than which vertices it sees. Unfortunately, O'Rourke and Streinu showed that vertex-edge visibility graphs encode more information about a pseudo-polygon than a regular visibility graph [10], and the regular visibility graph characterization problem has remained open for over fifteen years.

Our Results. In this paper, we give a characterization of the visibility graphs of pseudo-polygons. We first identify some key combinatorial properties of pseudo-polygons, and we then give a set of five necessary conditions based off our identified properties. We then prove that these necessary conditions are also sufficient via a reduction to O'Rourke and Streinu's vertex-edge characterization [9]. That is, for any visibility graph G that satisfies all necessary conditions, we construct a vertex-edge visibility graph G_{VE} that corresponds with G and show that it satisfies the characterization properties. Since all simple polygons are pseudo-polygons, our necessary conditions also apply to simple polygon visibility graphs, and in some cases extend or generalize the previously given necessary conditions given for simple polygon visibility graphs [8]. Each of the four necessary conditions given for simple polygons [8] have been proved using geometric arguments, yet each of them are implied by the necessary conditions we give for pseudo-polygons which are proved without geometric arguments. Given that not all pseudo-polygons are simple polygons [12], additional necessary conditions will be needed to characterize the visibility graphs of simple polygons.

2   Preliminaries

We begin with some preliminaries and definitions that will be relied upon heavily in our proof. Our main focus of this paper is to determine if a graph G is the visibility graph for some pseudo-polygon. Note that the visibility graph G of a pseudo-polygon P must contain a Hamiltonian cycle because each p_i must see p_{i-1} and p_{i+1}. Since determining if a graph contains a Hamiltonian cycle is NP-hard, previous research has assumed that G does have such a cycle C and the vertices are labeled in counterclockwise order according to this cycle. So now suppose we are given an arbitrary graph $G = (V, E)$ with the vertices labeled p_0 to p_{n-1} such that G contains a Hamiltonian cycle $C = (p_0, p_2, \ldots, p_{n-1})$ in order according to their indices. We are interested in determining if G is the visibility graph for some pseudo-polygon P where C corresponds with the boundary of P. For any two vertices p_i and p_j, we let $\partial(p_i, p_j)$ denote the vertices and boundary edges encountered when walking counterclockwise around C from p_i to p_j (inclusive). For any edge $\{p_i, p_j\}$ in G, we say that $\{p_i, p_j\}$ is a *visible pair*, as their points in P must see one another. If $\{p_i, p_j\}$ is not an edge in G, then we call (p_i, p_j) and (p_j, p_i) *invisible pairs*. Note that visible pairs are unordered, and invisible pairs are ordered (for reasons described below).

Consider any invisible pair (p_i, p_j). If G is the visibility graph for a pseudo-polygon P, the segment of $L_{i,j}$ must exit P. For example, suppose we want to construct a polygon P such that the graph in Fig. 2 (a) is the visibility graph of P. Note that p_0 should not see p_2, and thus if there exists such a polygon, it must satisfy that $p_0 p_2$ exits the polygon. In the case of a simple polygon, we view this process as placing the vertices of P in convex position and then contorting the boundary of P to block p_0 from seeing p_2. We can choose p_1 or p_3 to block p_0 from seeing p_2 (see (b) and (c)). Note that as in Fig. 2 (b) when using $p_1 \in \partial(p_0, p_2)$ as the blocker in a simple polygon, the line segment $p_0 p_1$ does not go outside P and the ray $r_{1,0}$ first exits P through a boundary edge in $\partial(p_2, p_0)$. Similarly as in Fig. 2 (c) when using $p_3 \in \partial(p_2, p_0)$ as the blocker, the line segment $p_0 p_3$ does not go outside of the polygon and the ray $r_{3,0}$ first exits the polygon through a boundary edge in $\partial(p_1, p_3)$. The situation is similar in the case of pseudo-polygons, but since we do not have to use straight lines to determine visibility, instead of bending the the boundary of P to block the invisible pair we can instead bend the pseudo-line. See Fig. 2 (d) and (e). Note that the combinatorial structure of the pseudo-line shown in part (d) (resp. part (e)) is the same as the straight line in part (b) (resp. in part (c)). The following definition plays an important role in our characterization. Consider a pseudo-polygon P, and let p_i and p_j be two vertices of P that do not see each other. We say a vertex $p_k \in \partial(p_i, p_j)$ of P is a *designated blocker* for the invisible pair (p_i, p_j) if p_i sees p_k (i.e. the segment $p_i p_k$ is inside the polygon) and the ray $r_{i,k}$ first exits the polygon through an edge in $\partial(p_j, p_i)$. The definition for $p_k \in \partial(p_j, p_i)$ is defined similarly. See Figure 3 (a) for an illustration. Intuitively, a designated blocker is a canonical vertex that prevents the points in an invisible pair from seeing each other. In this section, we will prove a key structural lemma of pseudo-polygons: every invisible pair in any pseudo-polygon P has exactly one designated blocker.

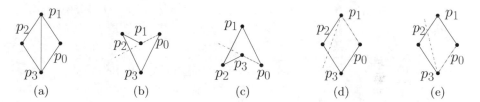

Fig. 2. (a) A visibility graph G. (b) A simple polygon using p_1 to block p_0 and p_2. (c) A simple polygon using p_3 to block p_0 and p_2. (d) A pseudo-polygon using p_1 to block p_0 and p_2. (e) A pseudo-polygon using p_3 to block p_0 and p_2.

We now give several definitions and observations that will be used in the proof of the key lemma. Consider an input graph G with Hamiltonian cycle C, and let (p_i, p_j) be an invisible pair in G. If G is the visibility graph of a pseudo-polygon, then there must be some vertex in G that serves as the designated blocker for (p_i, p_j). The following definition gives a set of at most two candidate vertices for this role. Starting from p_j, walk clockwise towards p_i until we reach the first point p_k such that $\{p_i, p_k\}$ is a visible pair (clearly there must be such a point since $\{p_i, p_{i+1}\}$ is a visible pair). We say that p_k is a *candidate blocker* for (p_i, p_j) if there are no visible pairs $\{p_s, p_t\}$ such that $p_s \in \partial(p_i, p_{k-1})$ and $p_t \in \partial(p_{k+1}, p_j)$. Similarly, walk counterclockwise from p_j to p_i until we reach the first point $p_{k'}$ such that $\{p_i, p_{k'}\}$ is a visible pair. Then $p_{k'}$ is a candidate blocker for (p_i, p_j) if there are no visible pairs $\{p_s, p_t\}$ such that $p_s \in \partial(p_j, p_{k'-1})$ and $p_t \in \partial(p_{k'+1}, p_i)$. Note that a vertex may be a candidate blocker for (p_i, p_j) but not for (p_j, p_i). It clearly follows from the definition that (p_i, p_j) can have at most two candidate blockers: at most one in $\partial(p_i, p_j)$ and at most one in $\partial(p_j, p_i)$. We will see that if a vertex in G is not a candidate blocker for (p_i, p_j), then it cannot serve as a designated blocker for (p_i, p_j) in P.

We utilize some observations regarding the vertex-edge visibility graphs for pseudo-polygons given by O'Rourke and Streinu [9] in the proof of our key lemma as well. We first formally define what it means for a vertex to see a boundary edge in a pseudo-polygon. Vertex p_j is a *witness* for the vertex-edge pair (p_i, e) if and only if either

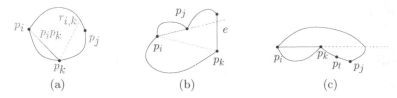

Fig. 3. (a) A designated blocker. (b) The vertex-edge pair (p_i, e) has two witnesses. Therefore p_i sees e. (c) If p_k is the designated blocker for (p_i, p_j) then it also is for (p_i, p_t).

1. p_i and p_j are both endpoints of e (permitting $p_j = p_i$), or
2. p_i is not an endpoint of e, and both of the following occur: (a) p_i sees p_j, and (b) p_j is an endpoint of e, or the first boundary edge intersected by $r_{j,i}$ is e.

Given the definition of a witness, we say vertex p *sees* edge e if and only if there are at least two witnesses for (p, e). See Fig. 3 (b). The definition requires two witnesses as a vertex p_i could see one endpoint of e without seeing any other part of the edge, and in this situation it is defined that p_i does not see e. We now give the following lemma relating edge visibility and vertex visibility. Some similar results for straight-line visibility were given in [10], and we prove them in the context of pseudo-visibility.

Lemma 1. *If a vertex p_i sees edges e_{j-1} and e_j, then it sees vertex p_j. Also if a vertex p_i sees vertex p_j, then it sees at least one of e_{j-1} and e_j.*

The following lemma from [9] is used in the proof of our key lemma. Note that Case A and Case B are symmetric.

Lemma 2. *If $p_k \in \partial(p_{b+1}, p_a)$ sees non-adjacent edges e_a and e_b and no edge $\partial(p_{a+1}, p_b)$, then exactly one of Case A or B holds. **Case A:** (1) p_k sees p_{a+1} but not p_b; and (2) p_{a+1} is a witness for (p_k, e_b); and (3) p_{a+1} sees e_b but p_b does not see e_a. **Case B:** (1) p_k sees p_b but not p_{a+1}; and (2) p_b is a witness for (p_k, e_a); and (3) p_b sees e_a but p_{a+1} does not see e_b.*

We are now ready to present our key structural lemma.

Lemma 3. *For any invisible pair (p_i, p_j) in a pseudo-polygon P, there is exactly one designated blocking vertex p_k. Moreover, p_k is a candidate blocker for the invisible pair (p_i, p_j) in the visibility graph of P.*

Proof. We begin by showing that a designated blocking vertex p_k for an invisible pair (p_i, p_j) is a candidate blocker for the invisible pair (p_i, p_j). Without loss of generality, assume that $p_k \in \partial(p_i, p_j)$. For the sake of contradiction, suppose p_i sees a point $p_t \in \partial(p_{k+1}, p_j)$. The pseudo-lines $L_{i,k}$ and $L_{i,t}$ intersect at p_i, and by the definition of designated blocker, the ray $r_{i,k}$ must intersect $L_{i,t}$ again, a contradiction. Therefore p_k must be the first point that p_i sees when walking clockwise from p_j. It remains to argue that no point $p_s \in \partial(p_{i+1}, p_{k-1})$ sees a point $p_t \in \partial(p_{k+1}, p_j)$. Suppose the contrary. Then the segments $p_i p_k$ and $p_s p_t$ must both be contained inside of the polygon, and therefore they must intersect each other, and we also have $r_{i,k}$ must intersect $p_s p_t$ again following the definition of designated blocker, a contradiction. It follows that the vertex p_k must be a candidate blocker for the invisible pair (p_i, p_j).

It remains to show that there must be exactly one designated blocker for each invisible pair. Since each designated blocker is a candidate blocker, there can clearly be at most two designated blockers. We first show there cannot be two designated blockers for an invisible pair (p_i, p_j). Suppose p_k and $p_{k'}$ are both designated blockers. Since they are both candidate blockers, we can assume without loss of generality that $p_k \in \partial(p_i, p_j)$ and $p_{k'} \in \partial(p_j, p_i)$. It follows from the definition of designated blocker that $L_{i,k}$ and $L_{i,k'}$ intersect twice.

We now show that there must be a designated blocker. Consider an invisible pair (p_i, p_j). Starting from p_j, walk clockwise towards p_i until we reach the first point p_i sees, which we denote p_k. Note that this point must exist since p_i sees p_{i+1}. Similarly walk counter clockwise from p_j until we reach the first point p_i sees, which we denote $p_{k'}$. Clearly it must be that p_i cannot see any point in $\partial(p_{k+1}, p_{k'-1})$. By Lemma 1 we have that p_i must see at least one edge adjacent to p_k and at least one edge adjacent to $p_{k'}$, and we will show that p_i can see exactly one edge in $\partial(p_k, p_{k'})$. First suppose that p_i sees no edges in $\partial(p_k, p_{k'})$. Then it must see e_{k-1} and $e_{k'}$ with no edges in $\partial(p_k, p_{k'})$. Applying Lemma 2, we have that either p_i does not see p_k or it does not see $p_{k'}$, a contradiction. By Lemma 1 we have that p_i cannot see two consecutive edges e_{s-1} and e_s or else p_i would see $p_s \in \partial(p_{k+1}, p_{k'-1})$, a contradiction. So finally suppose p_i sees two non-consecutive edges e_a and e_b in $\partial(p_k, p_{k'})$. Then Lemma 2 implies that either p_i sees p_{a+1} or it sees p_b, a contradiction in either case. It follows that p_i must see exactly one edge in $\partial(p_k, p_{k'})$.

Suppose without loss of generality that the edge $e_a \in \partial(p_k, p_{k'})$ that p_i sees is in $\partial(p_j, p_{k'})$. Then p_i sees e_{k-1} and e_a, and p_i does not see any edge in $\partial(p_k, p_{a-1})$. Applying Lemma 2, we see that we must be in Case A as p_i cannot see p_a. Part (2) from Case A gives us that p_k is a witness for (p_i, e_a), and therefore $r_{i,k}$ first exits the polygon through edge e_a. It follows that p_k is a designated blocker for the invisible pair (p_i, p_j). $\square$

3   Necessary Conditions

In this section, we give a set of five necessary conditions (NCs) that G must satisfy. That is, if G does not satisfy one of the conditions then G is *not* the visibility graph for any pseudo-polygon. Following from Lemma 3, if G is the visibility graph of a pseudo-polygon P then we should be able to assign candidate blockers in G to invisible pairs to serve as the designated blockers in P so that Lemma 3 and other pcp properties hold. The NCs outline a set of properties that this assignment must satisfy if the assignments correspond with a valid set of designated blockers in a pseudo-polygon. The proofs of these conditions use the definition of designated blockers to show that if the assignment of candidate blockers to invisible pairs do not satisfy the condition, then some pseudo-lines intersect twice, intersect but do not cross, etc. We illustrate the conditions with simple polygon examples to develop intuition, but the proofs hold for pseudo-polygons. The proofs are ommitted due to lack of space.

Let (p_i, p_j) be an invisible pair, and let p_k be the candidate blocker assigned to it. The first NC uses the definition of pseudo-lines and designated blockers to provide additional constraints on p_i and p_k. See Fig. 3 (c) for an illustration. Note that while the condition is stated for $p_k \in \partial(p_i, p_j)$, a symmetric condition for when $p_k \in \partial(p_j, p_i)$ clearly holds.

Necessary Condition 1. *If $p_k \in \partial(p_i, p_j)$ is the candidate blocker assigned to invisible pair (p_i, p_j) then both of the following must be satisfied: (1) p_k is assigned to the invisible pair (p_i, p_t) for every $p_t \in \partial(p_{k+1}, p_j)$ and (2) if (p_k, p_j) is an invisible pair then p_i is not the candidate blocker assigned to it.*

Again let p_k be the candidate blocker assigned to an invisible pair (p_i, p_j) such that $p_k \in \partial(p_i, p_j)$. Since p_k is a candidate blocker, we have that (p_s, p_j) is an invisible pair for every $p_s \in \partial(p_i, p_{k-1})$. The next NC is a constraint on the location of designated blockers for (p_s, p_j). In particular, if $\{p_s, p_k\}$ is a visible pair, then p_k must be the designated blocker for (p_s, p_j). See Fig. 4 (a). If (p_s, p_k) is an invisible pair, then it must be assigned a designated blocker p_t. In this case, p_t must also be the designated blocker for (p_s, p_j). See Fig. 4 (b).

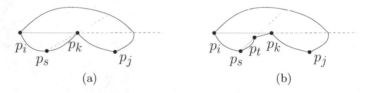

$$(a) \qquad\qquad (b)$$

Fig. 4. (a) If p_k is the designated blocker for (p_i, p_j) and p_s sees p_k then p_k is the designated blocker for (p_s, p_j). (b) If p_s does not see p_k, and p_t is the designated blocker for (p_s, p_k) then p_t is also the designated blocker for (p_s, p_j).

Necessary Condition 2. *Let (p_i, p_j) denote an invisible pair, and suppose p_k is the candidate blocker assigned to this invisible pair. Without loss of generality, suppose $p_k \in \partial(p_i, p_j)$, and let p_s be any vertex in $\partial(p_i, p_{k-1})$. Then exactly one of the following two cases holds: (1) $\{p_s, p_k\}$ is a visible pair, and the candidate blocker assigned to the invisible pair (p_s, p_j) is p_k, or (2) (p_s, p_k) is an invisible pair. If the candidate blocker assigned to (p_s, p_k) is p_t, then (p_s, p_j) is assigned the candidate blocker p_t.*

The next NC is somewhat similar to Necessary Condition 2, except instead of introducing constraints on the designated blockers for (p_s, p_j), it introduces constraints on the designated blockers for (p_j, p_s) (where the order is reversed). Similar to the previous case, if p_j sees p_k then p_k must block p_j from seeing every $p_s \in \partial(p_i, p_{k-1})$, but we can also see that p_k must block p_j from any point p_t such that p_i is the designated blocker for (p_k, p_t). See Fig. 5 (a). If p_j does not see p_k, then there must be a designated blocker p_q for (p_j, p_k). See Fig. 5 (b). We show that in this case, p_q must be the designated blocker for all (p_j, p_s) and (p_j, p_t). Also, (p_i, p_q) must be an invisible pair with designated blocker p_k.

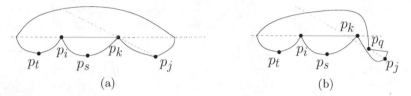

(a)   (b)

Fig. 5. (a) If p_k is the designated blocker for (p_i, p_j) and p_j sees p_k then p_k is the designated blocker for $(p_j, p_s), (p_j, p_i)$, and (p_j, p_t). (b) If p_j does not see p_k, and p_q is the designated blocker for (p_j, p_k) then p_q is the designated blocker for $(p_j, p_s), (p_j, p_i)$, and (p_j, p_t). Moreover, (p_i, p_q) is an invisible pair and p_k is its designated blocker.

Necessary Condition 3. *Let (p_i, p_j) denote an invisible pair, and suppose p_k is the candidate blocker assigned to this invisible pair. Without loss of generality, suppose $p_k \in \partial(p_i, p_j)$. Then exactly one of the following two cases holds:*

1. *(a) $\{p_j, p_k\}$ is a visible pair. (b) For all $p_s \in \partial(p_i, p_{k-1})$, the candidate blocker assigned to the invisible pair (p_j, p_s) is p_k. (c) If p_t is such that p_i is the candidate blocker assigned to the invisible pair (p_k, p_t), then (p_j, p_t) is an invisible pair and is assigned the candidate blocker p_k.*
2. *(a) (p_j, p_k) is an invisible pair. Let p_q denote the candidate blocker assigned to (p_j, p_k). (b) (p_i, p_q) is an invisible pair, and p_k is the candidate blocker assigned to it. (c) For all $p_s \in \partial(p_i, p_k)$, the candidate blocker assigned to the invisible pair (p_j, p_s) is p_q. (d) If p_t is such that p_i is the candidate blocker assigned to the invisible pair (p_k, p_t), then (p_j, p_t) is an invisible pair and is assigned the candidate blocker p_q.*

Suppose p_k is a candidate blocker for an invisible pair (p_i, p_j) (or (p_j, p_i)), and suppose without loss of generality that $p_i \in \partial(p_j, p_k)$. If p_k is also a candidate blocker for an invisible pair (p_s, p_t) such that $p_s, p_t \in \partial(p_k, p_j)$ then we say that the two invisible pairs are a *separable invisible pair*. We have the following condition which is the same as Necessary Condition 3 for simple polygons in [8]. See Fig. 6 (a).

Necessary Condition 4. *Suppose (p_i, p_j) and (p_s, p_t) are a separable invisible pair with respect to a candidate blocker p_k. If p_k is assigned to (p_i, p_j) then it is not assigned to (p_s, p_t).*

We now give the final NC. Let p_i, p_j, p_s, and p_t be four vertices of G in "counter-clockwise order" around the Hamiltonian cycle C. We say that p_i, p_j, p_s,

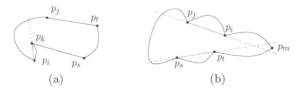

Fig. 6. (a) If p_k blocks one invisible pair of a separable invisible pair then it cannot block the other one as well. (b) p_i, p_j, p_s, and p_t are $\{p_i, p_t\}$-pinched. If p_j blocks p_i from seeing some point, then p_s cannot also block p_t from seeing that point.

and p_t are $\{p_i, p_t\}$-*pinched* if there is a $p_m \in \partial(p_t, p_i)$ such that p_i is the designated blocker for the invisible pair (p_j, p_l) and p_t is the designated blocker for the invisible pair (p_s, p_l). See Fig. 6 (b). The notion of $\{p_j, p_s\}$-pinched is defined symmetrically.

Necessary Condition 5. *Let p_i, p_j, p_s, and p_t be four vertices of G in counterclockwise order around the Hamiltonian cycle C that are $\{p_i, p_t\}$-pinched. Then they are not $\{p_j, p_s\}$-pinched.*

4   Proving the Conditions Are Sufficient

Suppose we are given an assignment of candidate blockers to invisible pairs that satisfies all NCs presented in Section 3. In this section, we prove that G is the visibility graph for some pseudo-polygon. We make use of the characterization of vertex-edge visibility graphs for pseudo-polygons given by O'Rourke and Streinu [9]. That is, we show that the vertex-edge visibility graph associated with G and the assignment of candidate blockers satisfies the necessary and sufficient conditions given in [9].

We begin by giving an important lemma that relates vertex-edge visibility with designated blockers in any pseudo-polygon P.

Lemma 4. *A vertex p_i does not see an edge e_j if and only if one of the two following conditions hold: (1) $p_s \in \partial(p_{i+1}, p_j)$ is the designated blocker for (p_i, p_t) for some $p_t \in \partial(p_{j+1}, p_{i-1})$, or (2) $p_t \in \partial(p_{j+1}, p_{i-1})$ is the designated blocker for (p_i, p_s) for some $p_s \in \partial(p_{i+1}, p_{j-1})$.*

Lemma 4 implies that given any visibility graph G with an assignment of designated blockers to its invisible pairs, there is a unique associated vertex-edge visibility graph. Let us denote this graph G_{VE}. We will show that if the assignment of designated blockers to the invisible pairs satisfies NCs 1-5, then G_{VE} satisfies the following characterization given by O'Rourke and Streinu [9]. This implies that there is a pseudo-polygon P such that G_{VE} is the vertex-edge visibility graph of P *and* G is the visibility graph of P. Note p_j is an *articulation point* of the subgraph of G_{VE} induced by $\partial(p_{i+1}, p_k)$ if and only if p_j is a candidate blocker for the invisible pair (p_s, p_t) for some $p_s \in \partial(p_{i+1}, p_{j-1})$ and some $p_t \in \partial(p_{j+1}, p_k)$.

Theorem 1. *[9] A graph is the vertex-edge visibility graph of a pseudo-polygon P if and only if it satisfies the following. If $p_k \in \partial(p_{j+1}, p_{i-1})$ sees two non-adjacent edges e_i and e_j and no edge in $\partial(p_{i+1}, p_{j-1})$ then it satisfies exactly one of the following two properties: (1) p_{i+1} sees e_j and p_{i+1} is an articulation point of the subgraph induced by $\partial(p_k, p_j)$, or (2) p_j sees e_i and p_j is an articulation point of the subgraph induced by $\partial(p_{i+1}, p_k)$.*

Good Lines and Centers. If $L_{i,j}$ is such that $\{p_i, p_j\}$ is a visible pair, then we say $L_{i,j}$ is a *good line*. Recall $L_{i,j}$ can be decomposed into three portions: the *segment* $p_i p_j$ and two infinite *rays* $r_{i,j}$ and $r_{j,i}$. The ray $r_{i,j}$ starts at p_j and does not include p_i, and $r_{j,i}$ is defined symmetrically. We now define the *center* of $L_{i,j}$ to be the connected subsegment of $L_{i,j}$ consisting of the following: the segment $p_i p_j$, the subsegment of $r_{i,j}$ obtained by starting at p_j and walking along the ray until we first reach exit outside of P (this may or may not be just p_j), and the symmetric subsegment of $r_{j,i}$. Note that the center of $L_{i,j}$ is simply the intersection of $L_{i,j}$ and P if the rays never re-enter P after leaving.

Given the visibility graph G and the assignment of candidate blockers to invisible pairs, we will now describe how to construct a *witness* P' that will be used to show that G is the visibility graph of a pseudo-polygon P. P' has a vertex for each vertex of G, and for every visible pair $\{p_i, p_j\}$ in G, the center of $L_{i,j}$ will appear in P'. The center will behave according to the assignment of candidate blockers to invisible pairs. In other words, if p_j is assigned to the invisible pair (p_i, p_k), then the center will be defined so that it fits the definition of designated blocker for this invisible pair.

For each vertex p_i in G, we add a point p_i to P'. We place these points in $\mathbb{R}^2$ in convex position in "counterclockwise order". That is, indices increase (modulo n) when walking around the convex hull in the counterclockwise direction. Now suppose that p_j is the candidate blocker assigned to an invisible pair (p_i, p_s). We define $r_{i,j}$ to be such that p_j is a designated blocker for (p_i, p_s). First note that if p_j is the candidate blocker assigned to (p_i, p_s) and (p_i, p_t), then it cannot be that one of p_s and p_t is in $\partial(p_i, p_j)$ and the other is in $\partial(p_j, p_i)$ by Necessary Condition 4, so without loss of generality assume that any such point is in $\partial(p_j, p_i)$. Let p_s be such that p_j is assigned to (p_i, p_s) but it is not assigned to (p_i, p_{s+1}). It follows from Necessary Condition 1 that there is exactly one such point p_s that satisfies this condition. We begin the definition of $r_{i,j}$ as a straight line from p_j to the edge e_s. There may be many rays from many different vertices which intersect the edge e_s. If $r_{a,b}$ is another ray intersecting e_s, we "preserve the order" of the rays so that $r_{i,j}$ and $r_{a,b}$ do not intersect. Note that because of property (2) of Necessary Condition 1, these centers do not self-intersect.

Lemma 5. *If G_{VE} does not satisfy the conditions of Theorem 1, then there exists a pair of distinct good line centers that intersect twice in P'.*

Combining Lemma 5 with the following lemma, we get that G_{VE} satisfies Theorem 1 and therefore is the vertex-edge visibility graph for a pseudo-polygon.

Lemma 6. *The centers of any pair of good lines intersects at most once, and if they intersect they cross.*

We now have that G_{VE} is the vertex-edge visibility graph for some pseudo-polygon P. It follows from Lemma 4 that G is the visibility graph of P, giving us the following theorem.

Theorem 2. *A graph G with a given Hamiltonian cycle C is the visibility graph of a pseudo-polygon P if and only if there is an assignment of candidate blockers to the invisible pairs that satisfies Necessary Conditions 1 - 5.*

References

1. Aloupis, G., Cardinal, J., Collette, S., Langerman, S., Orden, D., Ramos, P.: Decomposition of multiple coverings into more parts. In: Proceedings of the Twentieth Annual ACM-SIAM Symposium on Discrete Algorithms, SODA 2009, pp. 302–310. Society for Industrial and Applied Mathematics, Philadelphia (2009)
2. Aronov, B., Ezra, E., Sharir, M.: Small-size epsilon-nets for axis-parallel rectangles and boxes. SIAM J. Comput. 39(7), 3248–3282 (2010)
3. Choi, S.-H., Shin, S.Y., Chwa, K.-Y.: Characterizing and recognizing the visibility graph of a funnel-shaped polygon. Algorithmica 14(1), 27–51 (1995)
4. Everett, H., Corneil, D.G.: Negative results on characterizing visibility graphs. Comput. Geom. 5, 51–63 (1995)
5. Feige, U., Halldórsson, M.M., Kortsarz, G., Srinivasan, A.: Approximating the domatic number. SIAM J. Comput. 32(1), 172–195 (2003)
6. Ghosh, S.K.: On recognizing and characterizing visibility graphs of simple polygons. In: Karlsson, R., Lingas, A. (eds.) SWAT 1988. LNCS, vol. 318, pp. 96–104. Springer, Heidelberg (1988)
7. Ghosh, S.K.: On recognizing and characterizing visibility graphs of simple polygons. Discrete & Computational Geometry 17(2), 143–162 (1997)
8. Ghosh, S.K., Goswami, P.P.: Unsolved problems in visibility graphs of points, segments, and polygons. ACM Comput. Surv. 46(2), 22 (2013)
9. O'Rourke, J., Streinu, I.: Vertex-edge pseudo-visibility graphs: Characterization and recognition. In: Symposium on Computational Geometry, pp. 119–128 (1997)
10. O'Rourke, J., Streinu, I.: The vertex-edge visibility graph of a polygon. Computational Geometry 10(2), 105–120 (1998)
11. Srinivasaraghavan, G., Mukhopadhyay, A.: A new necessary condition for the vertex visibility graphs of simple polygons. Discrete & Computational Geometry 12, 65–82 (1994)
12. Streinu, I.: Non-stretchable pseudo-visibility graphs. Comput. Geom. 31(3), 195–206 (2005)
13. Varadarajan, K.R.: Epsilon nets and union complexity. In: Symposium on Computational Geometry, pp. 11–16 (2009)

Faster and More Dynamic Maximum Flow by Incremental Breadth-First Search

Andrew V. Goldberg[1,*], Sagi Hed[2,*], Haim Kaplan[2,*], Pushmeet Kohli[3],
Robert E. Tarjan[4,*], and Renato F. Werneck[1,*]

[1] Amazon.com Inc.
{andgold,werneck}@amazon.com
[2] School of Computer Science, Tel Aviv University
{sagihed,haimk}@cs.tau.ac.il
[3] Microsoft Research
pkohli@microsoft.com
[4] Department of Computer Science, Princeton University and Intertrust Technologies
ret@cs.princeton.edu

Abstract. We introduce the *Excesses Incremental Breadth-First Search* (Excesses IBFS) algorithm for maximum flow problems. We show that Excesses IBFS has the best overall practical performance on real-world instances, while maintaining the same polynomial running time guarantee of $O(mn^2)$ as IBFS, which it generalizes. Some applications, such as video object segmentation, require solving a series of maximum flow problems, each only slightly different than the previous. Excesses IBFS naturally extends to this dynamic setting and is competitive in practice with other dynamic methods.

1 Introduction

The maximum flow problem and its dual, the minimum s–t cut problem, are fundamental optimization problems with applications in a wide range of areas such as network optimization, computer vision, and signal processing. We present a new robust algorithm for the maximum flow problem that is particularly suitable for dynamic applications. We prove a strongly polynomial running time bound for our algorithm and compare its performance to previous algorithms.

Experimental work has been done on Dinic's blocking flow algorithm [4,16], the Push-Relabel (PR) algorithm [5,12,13], and Hochbaum's Pseudoflow algorithm (HPF) [18,3,9]. All three have a strongly polynomial worst-case time bound. In contrast, the algorithm of Boykov and Kolmogorov (BK) [2] is purely practical: it has no strongly polynomial time bound, but is probably the most widely used algorithm in computer vision. The Incremental Breadth-First Search (IBFS) algorithm [14] shares some features with both BK and PR. It is competitive with BK in practice [14] and has the same strongly polynomial time bounds as PR: $O(mn^2)$ without sophisticated data structures and $O(mn \log(n^2/m))$ with dynamic trees, where m is the number of arcs and n is the number of vertices.

---

* Work partly done while the author was at Microsoft Research Silicon Valley.

© Springer-Verlag Berlin Heidelberg 2015
N. Bansal and I. Finocchi (Eds.): ESA 2015, LNCS 9294, pp. 619–630, 2015.
DOI: 10.1007/978-3-662-48350-3_52

The maximum flow problem has also been studied in a dynamic setting, where one solves a series of maximum flow instances, each obtained from the previous one by relatively few changes in the input. The naive approach is to solve each problem independently, but one can do better. Kohli and Torr [19,20,1] extend the BK algorithm to the dynamic setting for better performance.

Despite improvements over the years, there is still much room for obtaining faster running times in practice. A natural question is whether we can come up with a robust algorithm that is fast for all applications, both static and dynamic, and has a worst-case strongly polynomial time bound.

In this paper, we present the *Excesses IBFS (EIBFS)* algorithm, which generalizes IBFS. We show that EIBFS is the best overall algorithm in practice on real-world data. On most instances it is the fastest compared to all other algorithms (often by orders of magnitude); when it loses, it is by small factors. Unlike IBFS, Excesses IBFS naturally extends to the dynamic setting, where it is competitive in practice with the dynamic extension of BK [19].

Section 2 describes the EIBFS algorithm, proves its correctness, and shows that it has the same worst-case time bounds as IBFS: $O(mn\log(n^2/m))$ with dynamic trees and $O(mn^2)$ without. Section 3 describes improvements that can be implemented in both IBFS and EIBFS. Section 4 has an extensive experimental comparison of all the key players in solving maximum flow in practice. Our benchmark is a superset of previous benchmarks and as comprehensive as we could make it.

IBFS offered a faster, theoretically justified alternative to solving maximum flow. Excesses IBFS offers an even faster, still theoretically justified and dynamic alternative to all existing methods.

Definitions and Notation. The input to the maximum flow problem is a directed graph $G = (V, E)$, a *source* $s \in V$, a *sink* $t \in V$ (with $s \neq t$), and a *capacity function* $u : E \Rightarrow [1, \ldots, U]$.

We assume that every arc a has a reverse arc a^R of capacity 0. A (feasible) flow f is an anti-symmetric function (i.e. $f(a) = -f(a^R)$) on $E \cup E^R$ that satisfies *capacity constraints* on all arcs and *conservation constraints* at all vertices except s and t. The capacity constraint for an arc (v, w) is that $f(v, w) \leq u(v, w)$. The conservation constraint for v is $\sum_{(u,v)\in E} f(u, v) = \sum_{(v,w)\in E} f(v, w)$. The *flow value* is the total flow into the sink: $|f| = \sum_{(v,t)\in E} f(v, t)$. A *cut* is a partitioning of vertices $S \cup T = V$ with $s \in S$ and $t \in T$. The capacity of a cut is defined as $u(S, T) = \sum_{(v,w)\in E|v\in S, w\in T} u(v, w)$. The max-flow/min-cut theorem [10] states that the maximum flow value is equal to the minimum capacity of a cut.

The *residual capacity* of an arc $a \in E \cup E^R$ is defined by $u_f(a) = u(a) - f(a)$. The *residual graph* $G_f = (V, E_f)$ is the graph induced by the arcs in $E \cup E^R$ with strictly positive residual capacity. A *valid distance labeling* from s is an integral function d_s on V that given a flow f satisfies $d_s(s) = 0$ and $d_s(w) \leq d_s(v) + 1$ for every arc $(v, w) \in E_f$. A valid distance labeling to t, d_t, is defined symmetrically. We say that an arc (v, w) is *admissible w.r.t.* d_s if $(v, w) \in E_f$ and $d_s(v) = d_s(w) - 1$, and *admissible w.r.t.* d_t if $(v, w) \in E_f$ and $d_t(w) = d_t(v) - 1$.

2   Excesses IBFS

Unlike IBFS and BK, which always maintain a feasible flow, Excesses IBFS is a generalization of IBFS that maintains a *pseudoflow*, a flow that observes capacity but not conservation constraints. For a vertex v, let $e_f(v) = \sum_{w|(w,v)\in E} f(w,v) - \sum_{w|(v,w)\in E} f(v,w)$. We say v is an *excess* if $e_f(v) > 0$ and a *deficit* if $e_f(v) < 0$. We define s and t to have infinite excess and deficit, respectively.

Pseudoflows often allow to efficiently restart an algorithm after solving a problem to solve related problems. For example, for the global minimum cut problem [17] and the parametric flow problem [11], one gets the same running time bound for a sequence of flow computations as that for a single computation.

EIBFS maintains a pair of vertex-disjoint forests S and T in the admissible subgraph. Each excess is a root of a tree in S, and a root in S must be an excess. Similarly, each deficit is a root of a tree in T, and a root in T must be a deficit. For a non-root vertex v in S or T, we let $p(v)$ be the parent of v in its respective forest. We call a vertex which is not in S nor in T a *free* vertex.

The algorithm maintains distance labels $d_s(v)$ and $d_t(v)$ for every vertex v. The forest arcs in S and T are admissible with respect to d_s and d_t, respectively. Initially, every root r in S or in T has $d_s(r) = 0$ or $d_t(r) = 0$, respectively. New excesses and deficits that form as the algorithm runs may have arbitrary distance labels, so the roots of the forests do not necessarily have zero distance label. Similar forests have been introduced before in an algorithm for finding a global minimum cut [17]. We also maintain $D_s = \max_{v\in S} d_s(v)$ and $D_t = \max_{v\in T} d_t(v)$.

Initially, S contains only s, T contains only t, $d_s(s) = d_t(t) = 0$, $D_s = D_t = 0$ and $p(v)$ is null for every vertex v. The algorithm proceeds in phases. Every phase is either a *forward phase* (where we grow the S forest) or a *reverse phase* (where we grow the T forest). Every phase executes growth steps, which may be interrupted by augmentation steps (when an augmenting path is found) followed by alternating adoption and augmentation steps.

We describe a forward phase; reverse phases are symmetric. The goal of a forward phase is to grow S by one level. If S has vertices at level $D_s + 1$ at the end of the phase, we increment D_s; otherwise we terminate.

We execute growth steps as in IBFS. When the phase starts we make all vertices v in S with $d_s(v) = D_s$ *active*. We then pick an active vertex v and scan v by examining residual arcs (v, w). If w is in S, we do nothing. If w is free, we add w to S, set $p(w) = v$, and set $d_s(w) = D_s + 1$. If w is in T, we perform an augmentation step as described below. We remember (v, w) as the outgoing arc that triggered the augmentation step. If v is still active after the augmentation step, we resume the scan of v from (v, w) to avoid re-scanning the preceding arcs. If (v, w) is still residual and connects the forests, we do more augmentation steps using it. After all arcs out of v have been scanned, v becomes inactive. When all vertices are inactive, the phase ends.

Augmentation steps differ from those of IBFS. When we find a connecting arc (v, w) with v in S and w in T we increase the flow on (v, w) by any feasible amount without violating the capacity constraint of (v, w) (we will discuss the best strategy for choosing the amount later). As a result of adding flow, an excess may be created in T and a deficit may be created in S. We now alternate between augmentation steps and adoption steps as we describe below. Once all excesses have been drained or removed from T and all deficits have been drained or removed from S we continue to perform growth steps.

We describe how we handle excesses created in T. We handle deficits in S symmetrically. We call a vertex $v \in T$ an *orphan* if its parent arc $(v, p(v))$ is not admissible (possibly saturated) and $e_f(v) \geq 0$. We execute an augmentation step by picking an excess $v \in T$ and pushing flow out of v as described below, possibly creating orphans and more excesses in T. If the augmentation step created orphans, we run adoption steps to repair them. After orphans are repaired we execute another augmentation step from another excess. We stop when all excesses are drained or removed from T. The excesses can be picked in any arbitrary order; highest level order seems to work well in practice.

We push flow out of an excess $v \in T$ as follows. We traverse the tree path from v to the root r of its tree in T. For every arc (x, y) along this path, in turn, we increase the flow by $\min\{u_f(x, y), e_f(x)\}$. It follows that we either drain the entire amount of excess from x or saturate the arc (x, y), making x an orphan in T. Root r remains a deficit if we did not drain enough excess into it. Otherwise it has $e_f(r) \geq 0$ and becomes an orphan; it can no longer serve as a root in T.

An adoption step repairs an orphan v in T by either setting a new parent, $p(v)$ in T or removing v from T. There are different methods to performing adoption steps. The simplest one is *round robin* adoption described below. More advanced methods are described in Section 3. In either method, if v is removed from T and v still has excess, then v is added to S as a new root with distance label $d_s(v) = D_s + 1$ in a forward phase or $d_s(v) = D_s$ in a reverse phase.

The original IBFS algorithm can be seen as a restricted version of EIBFS with a specific strategy for choosing the amount of flow to push on a connecting arc (v, w) between S and T. This strategy is to always take the bottleneck residual capacity along the tree path from s to v, the arc (v, w), and the tree path from w to t. Such an augmentation step will never create additional excesses or deficits in its tree. As a result, the S and T forests will simply be BFS trees rooted in s and t, respectively.

Round-Robin Adoption. We describe adoption steps in T. The adoption steps in S are symmetric. For efficiency, we maintain for every vertex a *current arc*, which ensures that each arc incident to a vertex v is scanned at most once following each increase in $d_t(v)$. When a free vertex is added to T or when the distance label of a vertex changes, we set the current arc to the first arc in its adjacency list. We maintain the invariant that the arcs preceding the current arc on the adjacency list of each vertex are not admissible.

The round robin method is based on the *relabel* operation of the push-relabel algorithm [15]. An adoption step on a vertex v works as follows. We first scan v's adjacency list starting from the current arc and stop when we find an admissible outgoing arc or reach the end of the list. If we find an admissible arc (v, u) we set the current arc of v to (v, u) and set $p(v) = u$. If we do not find such an arc, we apply the *orphan relabel* operation to v.

The orphan relabel operation scans v's adjacency list to find a new parent u for v. Vertex u qualifies to be a new parent of v if (1) u is a vertex of minimum $d_t(u)$ such that (v, u) is residual and (2) $d_t(u) < D_t$ in a forward phase and $d_t(u) \leq D_t$ in a reverse phase. If no vertex u qualifies as a new parent of v then we make v a free vertex if $e_f(v) = 0$ or add it to S as a new root if $e_f(v) > 0$.

If there is a vertex u that qualifies to be a parent of v then we choose u to be the first such vertex along v's adjacency list. We set the current arc of v to (v, u), set $p(v) = u$ and set $d_t(v) = d_t(u) + 1$. Every vertex w with $p(w) = v$ now becomes an orphan and needs to be repaired by adoption steps as well.

If v is active and we execute the orphan relabel operation on v, then we make v inactive (v is no longer in T or no longer with distance label $d_t(v) = D_t$).

Pushing Flow on Connecting Arcs. When a growth step finds an arc (v, w) with $v \in S$ and $w \in T$, we must decide how to increase the flow on (v, w). As we show later, the rule below ensures a strongly polynomial time bound.

Let r_v be the root of v's tree in S and let r_w be the root of w's tree in T. Let b_v be the bottleneck capacity along the path from v to r_v and let b_w be the bottleneck capacity along the path from w to r_w. Consider the following cases:

1. If $r_v = s$ and $r_w = t$, we push $u_f(v, w)$.
2. If $r_v = s$ and $r_w \neq t$, we push $\min\{b_v, u_f(v, w)\}$, thus creating no deficits in S (except v temporarily).
3. If $r_v \neq s$ and $r_w = t$, we push $\min\{u_f(v, w), b_w\}$, thus creating no excesses in T (except w temporarily).
4. If $r_v \neq s$ and $r_w \neq t$ we push $\min\{e_f(r_v), b_v, u_f(v, w), b_w, -e_f(r_w)\}$, thus creating no deficits or excesses in S or T (except v or w temporarily).

Correctness and Running Time. Correctness for EIBFS relies on the following lemma, which is the counterpart of Lemma 1 of IBFS [14]. See the full version of this extended abstract for a complete proof of correctness.

Lemma 1. *During a forward phase, if (u, v) is residual: (1) if $u \in S$, $d_s(u) \leq D_s$, and $v \notin S$, then u is an active vertex; (2) if $v \in T$ and $u \notin T$, then $d_t(v) = D_t$; (3) after the increase of D_s, if $u \in S$ and $v \notin S$, then $d_s(u) = D_s$.*

Next we show that the worst-case time complexity of EIBFS is $O(mn^2)$, which can be improved to $O(mn \log(n^2/m))$ using dynamic trees and existing techniques. We first consider the invariants maintained by the algorithm. These are the counterparts of the invariants in Lemma 2 of IBFS [14].

Lemma 2. *The following invariants hold:*

1. *If (v, w) is residual with $v, w \in S$, then $d_s(w) \leq d_s(v) + 1$. If (v, w) is residual with $v, w \in T$, then $d_t(v) \leq d_t(w) + 1$.*
2. *For every vertex u in S, u's current arc precedes the first admissible arc to u or is equal to it. For every vertex u in T, u's current arc precedes the first admissible arc from u or is equal to it.*

3. After an adoption step on u: if u is in S and $e_f(u) \leq 0$ then $(p(u), u)$ is admissible; if u is in T and $e_f(u) \geq 0$ then $(u, p(u))$ is admissible.

4. For every vertex v, $d_s(v)$ and $d_t(v)$ never decrease.

Proof. The proof, by induction on the growth, augmentation and adoption steps, is the same as that of Lemma 2 for IBFS [14]. Although augmentation steps differ a little, the same arguments hold for EIBFS. The only addition is the case of removing an excess from T or a deficit from S during an adoption step.

We consider a vertex v with $e_f(v) > 0$ removed from T during an adoption step; the case of a deficit removed from S is symmetric. Since we reset the current arc of v, (2) holds. Since $e_f(v) > 0$, (3) does not apply. We assign v the highest possible label in S (either $D_s + 1$ for a forward phase or D_s for a reverse phase), so (4) holds. We are left to show that invariant (1) is maintained.

We assume there is a residual arc (u, v) with u in S, otherwise (1) holds vacuously. If u was in S at the beginning of the current phase (v was not) then by Lemma 1 u was on the last level of S and thus $d_s(u) = D_s$ at that time. By induction assumption of (4) we get that now $d_s(u) \geq D_s$ and thus $d_s(v) \leq d_s(u) + 1$. If u was not in S at that time, then by definition of the algorithm it could only have been added to S with a label $D_s + 1$. By induction assumption of (4) we get that now $d_s(u) \geq D_s + 1$ and thus $d_s(v) \leq d_s(u) + 1$. $\square$

Adoption steps on a vertex charge their work to increases in the vertex's distance label. Lemma 3 is the counterpart of Lemma 5 in IBFS [14] and shows why this charging is possible. The proof is the same as in Lemma 5 [14]. Lemma 4 and 5 allow us to bound the maximum label assigned during the algorithm.

Lemma 3. *After an orphan relabel on v in S, $d_s(v)$ increases. After an orphan relabel on v in T, $d_t(v)$ increases.*

Lemma 4. *For every vertex v in S with $e_f(v) > 0$ we have $d_s(v) \leq n$. For every vertex v in T with $e_f(v) < 0$ we have $d_t(v) \leq n$.*

Proof. We prove the lemma for a vertex v in S. The proof for T is symmetric. We put new excesses in S only when we remove an excess from T during an adoption step that follows an augmentation step. Let (x, y) be the connecting arc between S and T that initiated this augmentation step.

Since we created excesses in T and by the definition of the flow increase on (x, y) we get that s is the root of x's tree. By applying Lemma 2 (1) to the path from s to v we get that at the time we initiated the augmentation step we had $d_s(x) \leq n - 1$. By definition of the algorithm we get that at the same time we had $d_s(x) \geq D_s$. It follows that $D_s \leq d_s(x) \leq n - 1$ and therefore $D_s + 1 \leq n$. Since we assign $d_s(v) = D_s + 1$ the lemma follows. $\square$

Lemma 5. *For every vertex v in S we have $d_s(v) < 2n$. For every vertex v in T we have $d_t(v) < 2n$.*

Proof. We prove the lemma for a vertex v in S. The proof for T is symmetric. Let vertex r in S, $e_f(r) > 0$ be the root of v's tree. By applying Lemma 2 (1) to the path from r to v we get that $d_s(v) \leq d_s(r) + n - 1$. By Lemma 4 we get that $d_s(r) \leq n$. It follows that $d_s(v) \leq 2n - 1$. $\square$

By Lemma 5 the maximum d_s or d_t label is $O(n)$. The following theorem follows using the same arguments as in the proof of Lemma 6 for IBFS [14].

Theorem 1. *Excesses IBFS runs in $O(n^2m)$ time.*

Dynamic Setting. We now consider the dynamic setting: after computing a maximum flow in the network, the capacities of some arcs change and we must recompute a maximum flow as fast as possible. IBFS does not seem to provide a robust method for recomputing a maximum flow other than starting the S and T trees from scratch. EIBFS, however, naturally lends itself to this setting. We first restore the invariants that were violated by changing the capacities, then run EIBFS normally continuing with the residual flow and forests from the previous computation.

Consider the network after changing some capacities. There are several types of violations to flow feasibility or to the invariants of the EIBFS that may follow:

1. An arc (v, w) such that now $f(v, w) > u(v, w)$.
2. A new residual arc (v, w) such that v is in S and w is in T.
3. A new residual arc (v, w) such that v and w are in S and $d_s(w) > d_s(v) + 1$, or the symmetric case for T.
4. A new residual arc (v, w) such that v and w are in S, $d_s(w) = d_s(v) + 1$, and (v, w) precedes the current arc of v, or the symmetric case for T.
5. A new residual arc (v, w) such that v is in S, $d_s(v) \leq D_s$ and w not in S, or the symmetric case for T.

Violation (4) can be fixed by reassigning the current arc of v. Violation (1) can be fixed by pushing flow on (w, v); (2), (3) and (5) can be fixed by saturating (v, w). In both cases the end result is the creation of new excesses or deficits in S and T. Excesses in S and deficits in T become new roots and need no further handling. Deficits in S and excesses in T are treated by alternating augmentation and adoption steps as when we find a connecting arc between S and T.

We found that in practice it pays off to reset the forests every $O(m)$ work. After a reset, the S and T forests are composed only of excesses or deficits, respectively, as roots with distance label 0. Note that this scans the nodes array once. This is similar in concept to Push-Relabel's global update operation [5].

3   Improvements to IBFS and Excesses IBFS

Forward or Reverse Phases. IBFS or EIBFS can use different strategies to alternate between forward and reverse phases. The original version of IBFS strictly alternated between them, producing trees with roughly the same height. As observed in [14], IBFS often spends the majority of its time on adoption steps. If arc capacities are distributed independently uniformly at random, balancing by height also tends to balance the amount of adoption work. In practice, however, this strategy is often far from optimal. A more robust alternative is to maintain an operation count that is indicative of the total amount of adoption work in each forest. We run a reverse phase when this counter is higher for S than for T, and a forward phase otherwise. Counting the number of distinct orphans examined (which is fairly oblivious to the choice of adoption method) in every adoption process works well in practice.

Alternative Adoption Strategies. The round-robin adoption strategy tends to be quite fast, but in pathological cases it may process the same vertex a large number of times. We thus propose a *three-pass adoption* strategy, which looks at each arc adjacent to an orphan at most three times during the entire adoption process of one augmentation step. This is more robust, and cannot be outperformed by the round-robin method by more than a constant factor.

This strategy associates a bucket (linked list) with each distance label. We denote by $B(v)$ the distance label associated with the bucket containing v. The method works in two rounds: the first incurs at most one pass of the adjacency list for every orphan; the second round incurs at most two more.

We describe the adoption in T (S is symmetric). The first round examines every orphan v in T in ascending order of distance labels. We scan v's adjacency list starting from the current arc and stop as soon as we find a residual arc (v, u) with $d_t(u) = d_t(v) - 1$. If such a vertex u is found, we set $p(v) = u$ and set the current arc of v to (v, u). If no such u is found, we remove v from T (v becomes a free vertex), put v in bucket $d_t(v) + 1$, and make the children of v in T orphans.

The second round iterates over the buckets in ascending order of distance labels. We examine every orphan v in the bucket. If this is the second pass of v, then we perform an orphan relabel operation as in Section 2. If v finds a potential parent u in T (note that u is not an orphan) we move it to the bucket $d_t(u) + 1$. If vertex v did not find a potential parent, it remains free, but it may be reattached to T later in the round.

If this is the third pass of v, we scan v's adjacency list, performing two operations. The first operation is to find a parent u in T for which $d_t(u) = B(v) - 1$ and (v, u) is residual. At the time of the third pass we are guaranteed to find such a parent. At the end of the scan we set the current arc of v to be (v, u), set $p(v) = u$ and set $d_t(v) = B(v)$. The second operation applies to every neighbor w with (w, v) residual and w either free or in a bucket $B(w) > B(v) + 1$. We put w in the bucket $B(v) + 1$ and remove it from any other bucket.

In practice, we use a *hybrid method*, which works as follows for every adoption process of one augmentation step. It starts with the round-robin method while keeping count of the average number of times each orphan is examined. If this average exceeds 3, it processes all remaining orphans using the three-pass method. We found that this method combines the best of both worlds and outperforms the round-robin and the three-pass methods in practice.

4   Experimental Results

In our experiments we use the implementation of BK version 3.0.1 from `http://www.cs.ucl.ac.uk/staff/V.Kolmogorov/software.html`. That implementation allows for dynamic capacities on the arcs from the source and to the sink. We added the option for dynamic capacities on all arcs. The dynamic version of BK was formulated by Kohli and Torr [19]. We also compare to UBK, an altered version of BK that maintains a consecutive arc structure: the arcs reside in an array grouped by the vertex they originate from (same as in [14]). We denote our implementation of Excesses IBFS with all the optimizations of Section 3 by EIBFS, and use IBFS to denote the implementation from [14]. The implementations of EIBFS and IBFS maintain a consecutive arc structure. We use the implementation of Hochbaum's pseudoflow (HPF) version 2.3 from `http://riot.ieor.berkeley.edu/Applications/Pseudoflow/maxflow.html`. We run the highest label FIFO variant of HPF, as recommended by the download page. We run also an implementation of Two-Level Push-Relabel (P2R) [12]. Our implementation of EIBFS and benchmark data are available at `http://www.cs.tau.ac.il/~sagihed/ibfs/`.

We run our experiments on a 64-bit Windows 8 machine with 8 GB of RAM and an Intel i5-3230M 2.6 GHz processor (two physical cores, 256 KB L1 cache). We used the MinGW g++ compiler with -O3 optimization settings. We compile with 64-bit or 32-bit pointers depending on problem size. We report system times of the maximum flow computation obtained with *ftime*. On the rare occasions in which running times are too small to measure, we round them up to a millisecond. We report absolute times (in seconds) for EIBFS and relative (to EIBFS) times for all algorithms. Factors greater than 1 mean EIBFS was faster.

Batra and Verma [22] noted that initialization times may be significant. Therefore our reported times include any initialization time past the initial reading of arcs. In all implementations, this reading consists of only two operations: writing arcs consecutively to memory and advancing the count of vertex degrees.

Table 1 reports results for static problems. We consider representative instances from a wide variety of families (see the full version of this extended abstract for a complete set of results). Multi-view reconstruction, 3D segmentation, stereo images, surface fitting and the first video segmentation family are provided by University of Western Ontario (`http://vision.csd.uwo.ca/maxflow-data`). Families of deconvolution, decision tree field (DTF), super resolution, texture restoration, automatic labeling environment (ALE) and synthetic segmentation are provided by `http://ttic.uchicago.edu/~dbatra/research/mfcomp/` [22]. Another synthetic family is from the DIMACS maximum flow challenge (`http://dimacs.rutgers.edu/Challenges/`). We run also on families of image and video segmentation with GMM models, multi-label image segmentation [1], lazy-brush image painting [21], road newtwork partitioning (PUNCH) [6], and graph bisection [7,8]. For stereo, ALE, PUNCH, and bisection, we report times summed over similar instances. Capacities are integral for all problems.

On real-world problems, EIBFS is the fastest overall algorithm, sometimes by orders of magnitude. It often improves the performance of IBFS but sometimes slows it down by marginal factors. EIBFS is the overall fastest the stereo, multi-view, 3D segmentation, surface fitting, and video frame families, losing occasionally to other algorithms only by small factors. On DTF problems HPF is the fastest, outperforming EIBFS by 20% on average. On lazy-brush problems EIBFS is fastest

Table 1. Performance on real-world and synthetic static inputs

FAMILY	NAME	$\frac{n}{1024}$	$\frac{m}{n}$	EIBFS TIME[s]	EIBFS	IBFS	BK	UBK	HPF	P2R
stereo	BVZ-tsukuba	108	4.0	0.249	1.00	**0.94**	1.02	1.09	3.50	7.49
	KZ2-venus	294	5.7	1.643	1.00	**0.92**	1.27	1.37	3.23	7.02
multi-view	camel-med	9450	4.0	7.125	**1.00**	1.95	2.67	2.75	1.52	4.44
	gargoyle-med	8640	4.0	8.516	1.00	1.62	11.73	9.35	**0.93**	2.35
3D	adhead6c100(64bit)	12288	6.0	11.250	**1.00**	1.08	2.38	1.95	1.71	1.95
segmentation	babyface6c100	4943	6.0	3.336	**1.00**	1.10	2.22	2.10	3.86	6.22
	bone6c100	7616	6.0	2.180	**1.00**	1.84	1.87	1.54	1.67	2.36
	bone_sx26c100(64bit)	3808	26.0	4.171	**1.00**	1.68	2.90	1.90	1.09	1.25
	bone_sx6c100	3808	6.0	1.492	**1.00**	1.27	2.73	2.28	1.79	1.82
	bone_sxyz26c10	960	26.0	0.742	1.00	1.98	2.04	1.24	**0.71**	1.53
	liver6c100	4064	6.0	3.391	**1.00**	1.16	2.78	2.43	2.15	2.79
surface fitting	bunny-med	6163	6.0	0.687	**1.00**	1.07	1.36	1.15	3.77	22.28
video1	car_32bins	77	9.7	0.015	**1.00**	1.01	9.41	2.92	1.99	5.61
(single frame)	person_16bins	107	9.9	0.015	**1.00**	1.10	154.74	19.74	1.21	5.41
	videoSegA	168	8.0	0.051	**1.00**	1.03	1.46	1.52	5.58	13.62
	videoSegB	225	8.0	0.015	1.00	1.01	**0.70**	1.10	1.58	3.31
	videoSegC	234	8.0	0.042	**1.00**	1.26	1.44	1.48	1.67	3.03
deconvolution	graph3x3	1	21.9	0.001	**1.00**	2.00	1.94	2.00	**1.00**	**1.00**
	graph5x5	1	67.7	0.004	1.00	1.00	7.50	5.19	0.65	**0.25**
DTF	printed_graph1	19	56.7	0.069	1.00	1.09	7.83	4.34	**0.77**	1.31
	printed_graph16	11	55.2	0.034	1.00	1.00	5.69	3.07	**0.80**	1.17
lazy-brush	lbrush-bird	2316	4.0	3.070	**1.00**	3.11	1.28	1.04	1.97	3.81
	lbrush-doctor	2317	4.0	1.140	**1.00**	18.00	1.47	1.15	1.10	9.45
	lbrush-mangagirl	579	4.0	0.273	1.00	4.89	1.23	**0.94**	1.43	8.38
	lbrush-elephant	2314	4.0	2.930	**1.00**	4.95	1.10	1.05	1.13	4.62
texture	texture_graph	42	15.1	0.010	1.00	1.14	**0.57**	1.28	0.86	1.72
resolution	superres_graph	42	15.1	0.007	1.00	1.00	**0.19**	1.19	0.99	1.98
segmentation	butterfly	453	8.0	0.084	**1.00**	1.11	1.52	1.59	5.96	8.22
	comp	236	8.0	0.078	**1.00**	1.40	2.20	2.21	4.01	3.89
	ferro	230	8.0	0.056	**1.00**	1.27	2.11	2.28	6.17	16.57
	flamingo2	468	8.0	0.106	**1.00**	1.06	1.38	1.35	3.94	5.08
PUNCH	punch-eu22p	1825	2.8	29.219	1.00	1.78	3.26	2.65	**0.50**	1.60
	punch-eu22u	1825	2.8	10.517	1.00	2.84	2.37	1.96	**0.89**	3.62
	punch-us22p	1596	2.8	47.189	1.00	1.96	2.98	2.38	**0.32**	0.76
	punch-us22u	1596	2.8	10.859	1.00	6.22	2.40	2.02	**0.85**	2.48
bisection	alue7065	32	3.2	0.110	**1.00**	2.00	1.29	1.42	1.58	4.83
	cal	1761	2.5	7.156	**1.00**	8.45	1.63	1.45	1.64	8.91
	horse	46	6.0	0.313	1.00	1.00	**0.55**	**0.55**	1.94	3.19
	rgg18	254	11.8	5.282	1.00	1.48	0.65	**0.46**	1.75	2.78
ALE	graph_2007_000033	168	27.3	0.915	**1.00**	1.02	175.23	11.32	27.29	24.03
	graph_2007_001288	161	29.0	0.904	**1.00**	1.02	186.73	6.77	69.02	19.55
segmentation	0.000099502487562_1	39	4.0	0.044	1.00	**0.78**	0.99	0.92	**0.78**	1.56
(synthetic)	0.002148473323752_1	39	45.0	0.059	**1.00**	1.06	9.34	6.49	3.22	1.58
	0.004631311287980_1	39	94.5	0.112	**1.00**	1.03	9.56	5.62	2.11	1.47
	0.021520481611665_1	39	432.4	0.584	1.00	1.12	9.52	4.51	**0.69**	**0.69**
	0.100000000000000_1	39	2001.6	3.278	1.00	0.93	9.65	3.93	**0.61**	0.75
DIMACS	ac.n1024	1	1019.0	0.034	1.00	1.07	37.76	4.71	0.91	**0.63**
(synthetic)	ac.n4096	4	4091.0	1.337	1.00	1.18	42.97	7.41	**0.42**	0.50
	rmf-long.n4	264	5.8	11.490	1.00	1.73	7.87	5.18	0.04	**0.02**
	rmf-wide.n4	120	5.8	2.696	1.00	1.21	19.63	13.34	**0.17**	0.30
	wash-line.n16384-64	64	127.7	1.034	1.00	2.68	4.98	3.32	0.51	**0.45**
	wash-line.n8192-45	32	89.8	0.218	1.00	3.15	7.51	4.54	**0.39**	0.53
	wash-rlg-long.n2048	128	6.0	1.621	1.00	6.97	16.42	12.74	**0.07**	**0.07**
	wash-rlg-wide.n2048	128	5.9	1.037	1.00	1.27	213.70	156.68	**0.19**	0.32

Table 2. Performance on real-world dynamic input

FAMILY	INSTANCE NAME	$\frac{n}{1024}$	$\frac{m}{n}$	DYNAMIC EIBFS ITERS	TIME[s]	RELATIVE TIMES EIBFS	BK	UBK	NIBFS
bisection	alue7065	32	3.2	9.0	0.074	1.00	**0.47**	0.50	0.62
	cal	1760	2.5	8.0	5.846	1.00	1.26	1.03	**0.86**
	horse	46	6.0	9.0	0.072	1.00	0.78	**0.65**	0.95
	rgg18	254	11.8	9.0	1.753	1.00	0.52	**0.48**	1.34
video1	car_32bins.inc	77	9.7	4320.0	2.691	**1.00**	2.27	1.10	7.17
	person_16bins.inc	107	9.9	28380.0	21.502	**1.00**	11.78	2.92	9.17
	videoSegA.inc	168	8.0	49.0	0.867	**1.00**	1.96	1.91	1.41
	videoSegB.inc	225	8.0	49.0	0.236	**1.00**	0.61	**0.49**	1.41
video2	gir.inc	405	8.0	4.0	0.078	**1.00**	2.00	1.64	1.36
	highway.inc	75	8.0	40.0	0.177	**1.00**	1.90	1.91	1.31
	office.inc	84	8.0	46.0	0.279	1.00	1.78	1.46	**0.92**
	pedestrians.inc	84	8.0	50.0	0.083	**1.00**	1.13	1.35	1.73
multi-label	cowInc00	405	8.0	16.0	0.001	**1.00**	1.50	**1.00**	29.60
	gardenInc00	20	7.9	28.0	0.001	**1.00**	1.00	**1.00**	4.60

and improves IBFS considerably. BK is the fastest on texture restoration and super resolution problems, since most of the running time is taken up by initialization (as seen by comparing BK to UBK). On image segmentation problems EIBFS is fastest. On PUNCH problems HPF is fastest, outperforming EIBFS by 25% on average. On bisection problems EIBFS and BK/UBK are competitive. On ALE problems EIBFS is faster by orders of magnitude compared to all other algorithms except IBFS. On synthetic problems, EIBFS is faster than IBFS but can still lose by orders of magnitude to HPF and P2R (especially on DIMACS instances). We note that some have very large vertex degrees, with most of the time used for initialization of the arc structure.

Table 2 considers dynamic problems. Dynamic video segmentation aligns maximum flow problems from consecutive video frames as one dynamic maximum flow set. Dynamic multi-label image segmentation aligns maximum flow problems from consecutive alpha expansion iterations over the same label. Dynamic bisection aligns maximum flow problems from nearby branches of a branch-and-bound tree. The table shows that, for dynamic applications, EIBFS is competitive with UBK, which in turn tends to be faster than BK. We also include NIBFS, a more naive implementation of IBFS for the dynamic setting. After every set of incremental changes, it only fixes violations on arcs where the flow is greater than the capacity; it then resets the S and T forests as in the periodic update of dynamic EIBFS. The results show that EIBFS is much more robust: it can outperform NIBFS by large factors but the converse is false.

References

1. Alahari, K., Kohli, P., Torr, P.H.S.: Dynamic hybrid algorithms for MAP inference in discrete mrfs. IEEE PAMI 32(10), 1846–1857 (2010)
2. Boykov, Y., Kolmogorov, V.: An Experimental Comparison of Min-Cut/Max-Flow Algorithms for Energy Minimization in Vision. IEEE PAMI 26(9), 1124–1137 (2004)

3. Chandran, B., Hochbaum, D.: A computational Study of the Pseudoflow and Push-Relabel Algorithms for the Maximum flow Problem. Operations Research 57, 358–376 (2009)
4. Cherkassky, B.V.: A Fast Algorithm for Computing Maximum Flow in a Network. In: Karzanov, A.V. (ed.) Collected Papers, Vol. 3: Combinatorial Methods for Flow Problems, pp. 90–96. The Institute for Systems Studies, Moscow (1979) (in Russian) English translation appears in AMS Trans., 158, 23–30 (1994)
5. Cherkassky, B.V., Goldberg, A.V.: On Implementing Push-Relabel Method for the Maximum Flow Problem. Algorithmica 19, 390–410 (1997)
6. Delling, D., Goldberg, A.V., Razenshteyn, I., Werneck, R.F.: Graph partitioning with natural cuts. In: 25th IEEE IPDPS, pp. 1135–1146 (2011)
7. Delling, D., Goldberg, A.V., Razenshteyn, I., Werneck, R.F.: Exact combinatorial branch-and-bound for graph bisection. In: ALENEX, pp. 30–44 (2012)
8. Delling, D., Werneck, R.F.: Better bounds for graph bisection. In: Epstein, L., Ferragina, P. (eds.) ESA 2012. LNCS, vol. 7501, pp. 407–418. Springer, Heidelberg (2012)
9. Fishbain, B., Hochbaum, D.S., Mueller, S.: Competitive analysis of minimum-cut maximum flow algorithms in vision problems. CoRR, abs/1007.4531 (2010)
10. Ford Jr., L.R., Fulkerson, D.R.: Maximal Flow Through a Network. Canadian Journal of Math. 8, 399–404 (1956)
11. Gallo, G., Grigoriadis, M.D., Tarjan, R.E.: A Fast Parametric Maximum Flow Algorithm and Applications. SIAM J. Comput. 18, 30–55 (1989)
12. Goldberg, A.: Two Level Push-Relabel Algorithm for the Maximum Flow Problem. In: Proc. 5th Alg. Aspects in Info. Management. Springer, New York (2009)
13. Goldberg, A.V.: The partial augment–relabel algorithm for the maximum flow problem. In: Halperin, D., Mehlhorn, K. (eds.) ESA 2008. LNCS, vol. 5193, pp. 466–477. Springer, Heidelberg (2008)
14. Goldberg, A.V., Hed, S., Kaplan, H., Tarjan, R.E., Werneck, R.F.: Maximum flows by incremental breadth-first search. In: Demetrescu, C., Halldórsson, M.M. (eds.) ESA 2011. LNCS, vol. 6942, pp. 457–468. Springer, Heidelberg (2011)
15. Goldberg, A.V., Tarjan, R.E.: A New Approach to the Maximum Flow Problem. J. Assoc. Comput. Mach. 35, 921–940 (1988)
16. Goldfarb, D., Grigoriadis, M.: A Computational Comparison of the Dinic and Network Simplex Methods for Maximum Flow. Ann. Op. Res. 13, 83–123 (1988)
17. Hao, J., Orlin, J.B.: A Faster Algorithm for Finding the Minimum Cut in a Directed Graph. J. Algorithms 17, 424–446 (1994)
18. Hochbaum, D.S.: The pseudoflow algorithm: A new algorithm for the maximum-flow problem. Operations Research 56(4), 992–1009 (2008)
19. Kohli, P., Torr, P.H.S.: Dynamic graph cuts for efficient inference in markov random fields. IEEE Trans. Pattern Anal. Mach. Intell. 29(12), 2079–2088 (2007)
20. Kohli, P., Torr, P.H.S.: Measuring uncertainty in graph cut solutions. Computer Vision and Image Understanding 112(1), 30–38 (2008)
21. Sýkora, D., Dingliana, J., Collins, S.: Lazybrush: Flexible painting tool for hand-drawn cartoons. Comput. Graph. Forum 28(2), 599–608 (2009)
22. Verma, T., Batra, D.: Maxflow revisited: An empirical comparison of maxflow algorithms for dense vision problems. In: BMVC, pp. 1–12 (2012)

The Temp Secretary Problem *

Amos Fiat, Ilia Gorelik, Haim Kaplan, and Slava Novgorodov

Tel Aviv University, Israel
{fiat,haimk}@tau.ac.il, {iliagore,slavanov}@post.tau.ac.il

Abstract. We consider a generalization of the secretary problem where contracts are temporary, and for a fixed duration γ. This models online hiring of temporary employees, or online auctions for re-usable resources. The problem is related to the question of finding a large independent set in a random unit interval graph.

1 Introduction

This paper deals with a variant of the secretary model, where contracts are temporary. *E.g.*, employees are hired for short-term contracts, or re-usable resources are rented out repeatedly, etc. If an item is chosen, it "exists" for a fixed length of time and then disappears.

Motivation for this problem are web sites such as Airbnb and oDesk. Airbnb offers short term rentals in competition with classic hotels. A homeowner posts a rental price and customers either accept it or not. oDesk is a venture capitalizing on freelance employees. A firm seeking short term freelance employees offers a salary and performs interviews of such employees before choosing one of them.

We consider an online setting where items have values determined by an adversary, ("no information" as in the standard model [15]), combined with stochastic arrival times that come from a prior known distribution (in contrast to the random permutation assumption and as done in [21,7,16]). Unlike much of the previous work on online auctions with stochastic arrival/departure timing ([18]), we do not consider the issue of incentive compatibility with respect to timing, and assume that arrival time cannot be misrepresented.

The temp secretary problem can be viewed

1. As a problem related to hiring temporary workers of varying quality subject to workplace capacity constraints. There is some known prior $F(x) = \int_0^x f(z)dz$ on the arrival times of job seekers, some maximal capacity, d, on the number of such workers that can be employed simultaneously, and a bound k on the total number than can be hired over time. If hired, workers cannot be fired before their contract is up.
2. Alternately, one can view the temp secretary problem as dealing with social welfare maximization in the context of rentals. Customers arrive according

---

* Research Supported by The Israeli Centers of Research Excellence (I-CORE) Program (Center No. 4/11), and by ISF Grant no. 822/10.

N. Bansal and I. Finocchi (Eds.): ESA 2015, LNCS 9294, pp. 631–642, 2015.
DOI: 10.1007/978-3-662-48350-3_53

to some distribution. A firm with capacity d can rent out up to d boats simultaneously, possibly constrained to no more than k rentals overall. The firm publishes a rental price, which may change over time *after* a customer is serviced. A customer will choose to rent if her value for the service is at least the current posted price. Such a mechanism is inherently dominant strategy truthful, with the caveat that we make the common assumption that customers reveal their true values in any case.

We give two algorithms, both of which are quite simple and offer posted prices for rental that vary over time. Assuming that the time of arrival cannot be manipulated, this means that our algorithms are dominant strategy incentive compatible.

For rental duration γ, capacity $d = 1$, no budget restrictions, and arrival times from an arbitrary prior, the *time-slice algorithm* gives a $\frac{1}{2e}$ competitive ratio. For arbitrary d the competitive ratio of the time-slice algorithm is at least $(1/2) \cdot (1 - 5/\sqrt{d})$. This can be generalized to more complex settings. The time slice algorithm divides time into slices of length γ. It randomly decides if to work on even or odd slices. Within each slice it uses a variant of some other secretary problem (*E.g.*, [26], [2], [24]) except that it keeps track of the cumulative distribution function rather than the number of secretaries.

The more technically challenging *Charter algorithm* is strongly motivated by the k-secretary algorithm of [24]. For capacity d, employment period γ, and budget $d \leq k \leq d/\gamma$ (the only relevant values), the Charter algorithm does the following:

- Recursively run the algorithm with parameters $\gamma, \lfloor k/2 \rfloor$ on all bids that arrive during the period $[0, 1/2)$.
- Take the bid of rank $\lceil k/2 \rceil$ that appeared during the period $[0, 1/2)$, if such rank exists and set a threshold T to be it's value. If no such rank exists set the threshold T to be zero.
- Greedily accept all items that appear during the period $[1/2, 1)$ that have value at least T — subject to not exceeding capacity (d) or budget (k) constraints.

For $d = 1$ the competitive ratio of the Charter algorithm is at least

$$\frac{1}{1 + k\gamma}\left(1 - \frac{5}{\sqrt{k}} - 7.4\sqrt{\gamma \ln(1/\gamma)}\right).$$

Two special cases of interest are $k = 1/\gamma$ (no budget restriction), in which case the expression above is at least $\frac{1}{2}\left(1 - 12.4\sqrt{\gamma \ln(1/\gamma)}\right)$. We also show an upper bound of $1/2 + \gamma/2$ for $\gamma > 0$. As γ approaches zero the two bounds converge to $1/2$. Another case of interest is when k is fixed and γ approaches zero in which this becomes the guarantee given by Kleinberg's k-secretary algorithm.

For arbitrary d the competitive ratio of the Charter algorithm is at least

$$1 - \Theta\left(\frac{\sqrt{\ln d}}{\sqrt{d}}\right) - \Theta(\gamma \log(1/\gamma)).$$

We remark that neither the time slice algorithm nor the Charter algorithm requires prior knowledge of n, the number of items due to arrive.

At the core of the analysis of the Charter algorithm we prove a bound on the expected size of the maximum independent set of a random unit interval graph. In this random graph model we draw n intervals, each of length γ, by drawing their left endpoints uniformly in the interval $[0, 1)$. We prove that the expected size of a maximum independent set in such a graph is about $n/(1 + n\gamma)$. We say that a set of length γ segments that do not overlap is γ-independent. Similarly, a capacity d γ-independent set allows no more than d segments overlapping at any point.

Note that if $\gamma = 1/n$ then this expected size is about $1/2$. This is intuitively the right bound as each interval in the maximum independent set rules out on average one other interval from being in the maximum independent set.

We show that a random unit interval graph with n vertices has a capacity d γ-independent subset of expected size at least $\min(n, d/\gamma)(1 - \Theta(\sqrt{\ln d}/\sqrt{d}))$. We also show that when $n = d/\gamma$ the expected size of the maximum capacity d γ-independent subset is no more than $n(1 - \Theta(1/\sqrt{d}))$. These results may be of independent interest.

Related Work. Worst case competitive analysis of interval scheduling has a long history, e.g., [30,28]. This is the problem of choosing a set of non-overlapping intervals with various target functions, typically, the sum of values.

[19] introduce the question of auctions for reusable goods. They consider a worst case mechanism design setting. Their main goal is addressing the issue of time incentive compatibility, for some restricted set of misrepresentations.

The secretary problem is arguably due to Johannes Kepler (1571-1630), and has a great many variants, a survey by [15] contains some 70 references. The "permutation" model is that items arrive in some random order, all $n!$ permutations equally likely. Maximizing the probability that the best item is chosen, when the items appear in random order, only comparisons can be made, and the number of items is known in advance, was solved by [27] and by [12]. A great many other variants are described in ([15,11]), differing in the number of items to be chosen, the target function to be maximized, taking discounting into account, etc.

An alternative to the random permutation model is the stochastic arrival model, introduced by Karlin [21] in a "full information" (known distribution on values) setting. Bruss [7] subsequently studied the stochastic arrival model in a no-information model (nothing is known about the distribution of values). Recently, [13] made use of the stochastic arrival model as a tool for the analysis of algorithms in the permutation model.

Much of the recent interest in the secretary problem is due to it's connection to incentive compatible auctions and posted prices [18,24,2,3,1,10].

Most directly relevant to this paper is the k-secretary algorithm by R. Kleinberg [24]. Constrained to picking no more than k secretaries, the total value of the secretaries picked by this algorithm is at least a $(1 - \frac{5}{\sqrt{k}})$ of the value of the best k secretaries.

Babaioff *et al.* [2] introduced the *knapsack secretary problem* in which every secretary has some weight and a value, and one seeks to maximize the sum of values subject to a upper bound on the total weight. They give a $1/(10e)$ competitive algorithm for this problem. (Note that if weights are one then this becomes the k-secretary problem). The Matroid secretary problem, introduced by Babaioff et al. [4], constrains the set of secretaries picked to be an independent set in some underlying Matroid. Subsequent results for arbitrary Matriods are given in [8,26,14].

Another generalization of the secretary problem is the online maximum bipartite matching problem. See [25,22]. Secretary models with full information or partial information (priors on values) appear in [5] and [29]. This was in the context of submodular procurement auctions ([5]) and budget feasible procurement ([29]). Other papers considering a stochastic setting include [23,17].

In our analysis, we give a detailed and quite technical lower bound on the size of the maximum independent set in a random unit interval graph (produced by the greedy algorithm). Independent sets in other random interval graph models were previously studied in [20,9,6].

2   Formal Statement of Problems Considered

Each item x has a value $v(x)$, we assume that for all $x \neq y$, $v(x) \neq v(y)$ by consistent tie breaking, and we say that $x > y$ iff $v(x) > v(y)$. Given a set of items X, define $v(X) = \sum_{x \in X} v(x)$ and $T_k(X) = \max_{T \subseteq X, |T| \leq k} v(T)$.

Given a set X and a density distribution function f defined on $[0, 1)$, let $\theta_f : X \mapsto [0, 1)$ be a random mapping where $\theta_f(x)$ is drawn independently from the distribution f. The function θ_f is called a *stochastic arrival* function, and we interpret $\theta_f(x)$, $x \in X$, to be the time at which item x arrives. For the special case in which f is uniform we refer to θ_f as θ.

In the problems we consider, the items arrive in increasing order of θ_f. If $\theta_f(x) = \theta_f(y)$ the relative order of arrival of x and y is arbitrary. An online algorithm may select an item only upon arrival. If an item x was selected, we say that the online algorithm *holds* x for γ time following $\theta_f(x)$.

An online algorithm A for the temp secretary problem may hold at most one item at any time and may select at most k items in total. We refer to k as the *budget* of A. The goal of the algorithm is to maximize the expected total value of the items that it selects. We denote by $A(X, \theta_f)$ the set of items chosen by algorithm A on items in X appearing according to stochastic arrival function θ_f.

The set of the arrival times of the items selected by an algorithm for the temp secretary problem is said to be γ-*independent*. Formally, a set $S \subset [0, 1)$ is said to be γ-*independent* if for all $t_1, t_2 \in S$, $t_1 \neq t_2$ we have that $|t_1 - t_2| \geq \gamma$.

Given $\gamma > 0$, a budget k, a set X of items, and a mapping $\theta_f : X \mapsto [0, 1)$ we define $\text{Opt}(X, \theta_f)$ to be a γ-independent set S, $|S| \leq k$, that maximizes the sum of values.

Given rental period $\gamma > 0$, distribution f, and budget k, the competitive ratio of an online algorithm A is defined to be

$$\inf_X \frac{E_{\theta_f : X \mapsto [0,1)}[v(A(X, \theta_f))]}{E_{\theta_f : X \mapsto [0,1)}[v(\mathrm{Opt}(X, \theta_f))]}. \tag{1}$$

The competitive ratio of the temp secretary problem is the supremum over all algorithms A of the competitive ratio of A.

Note that when $\gamma \to 0$, the the temp secretary problem reduces to Kleinberg's k-secretary problem.

We extend the γ-temp secretary problem by allowing the algorithm to hold at most d items at any time. Another extension we consider is the *knapsack temp secretary problem* where each item has a weight and we require the set held by the algorithm at any time to be of total weight at most W. Also, we define the *Matroid temp secretary problem* where one restricts the set of items held by the algorithm at any time to be an independent set in some Matroid M.

More generally, one can define a temp secretary problem with respect to some arbitrary predicate P that holds on the set of items held by an online algorithm at all times t. This framework includes all of the variants above. The optimal solution with respect to P is also well defined.

3   The Time-Slice Algorithm

In this section we describe a simple time slicing technique. This gives a reduction from temp secretary problems, with arbitrary known prior distribution on arrival times, to the "usual" continuous setting where secretaries arrive over time, do not depart if hired, and the distribution on arrival times is uniform. The reduction is valid for many variants of the temp secretary problem, including the Matroid secretary problem, and the knapsack secretary problem. We remark that although the Matriod and Knapsack algorithms are stated in the random permutation model, they can be replaced with analogous algorithms in the continuous time model and can therefore be used in our context.

We demonstrate this technique by applying it to the classical secretary problem (hire the best secretary). We obtain an algorithm which we call $Slice_\gamma$ for the temp secretary problem with arbitrary prior distribution on arrival times that is $O(1)$ competitive.

Consider the $1/2\gamma$ time intervals (*i.e.* slices) $I_j = [2\gamma j, \ 2\gamma(j+1)), \ 0 \leq j \leq 1/(2\gamma) - 1$. We split every such interval into two, $I_j^\ell = [2\gamma j, \ 2\gamma j + \gamma), \ I_j^r = [2\gamma j + \gamma, \ 2\gamma(j+1)).$[1]

Initially, we flip a fair coin and with probability $1/2$ decide to pick points only from the left halves (I_j^ℓ's) or only from the right halves (I_j^r's). In each such interval we pick at most one item by running the following modification of the continuous time secretary algorithm.

---

[1] For simplicity we assume that $1/(2\gamma)$ is an integer.

The continuous time secretary algorithm [13] observes the items arriving before time $1/e$, sets the largest value of an observed item as a threshold, and then chooses the first item (that arrives following time $1/e$) of value greater than the threshold. The modified continuous time secretary algorithm observes items as long as the cumulative distribution function of the current time is less than $1/e$, then it sets the largest value of an observed item as a threshold compute a threshold, and choose the next item of value larger than the threshold.

It is clear that any two points picked by this algorithm have arrival times separated by at least γ.

Theorem 1. *The algorithm $Slice_\gamma$ is $1/(2e)$ competitive.*

Proof. The analysis is as follows. Fix the mapping of items to each of the left intervals I_j^ℓ's and to each of the right intervals I_j^r's (leaving free the assignment of items to specific arrival times within their the intervals they are assigned to). Let OPT^ℓ (OPT^r) be the sum of the items of maximum value over all intervals I_j^ℓ (I_j^r). Let OPT be the average optimal value conditioned on this mapping of items to intervals. Clearly,

$$OPT^\ell + OPT^r \geq OPT. \tag{2}$$

For any interval I_j's (I_j^ℓ's) $Slice_\gamma$ gain at least $1/e$ over the top value in the interval conditioned on the event that $Slice_\gamma$ doesn't ignore this interval, this happens with probability $1/2$. Therefore the expected sum of values achieved by $Slice_\gamma$ is at least

$$\frac{1}{2} \cdot \frac{1}{e} OPT^\ell + \frac{1}{2} \cdot \frac{1}{e} OPT^r . \tag{3}$$

Substitution (2) into (3) we get the lemma. □

Appropriately choosing times (rather than number of elements) as a function of the prior distribution allows us to do the same for other variants of the secretary problem, the Knapsack (achieving a competitive ratio of $\frac{1}{2} \cdot \frac{1}{10e}$, see [2]) and Matriod ($O(\ln\ln\rho)$ when ρ is the rank of the Matroid, see [26,14]).

4   Improved Results for the Temp Secretary Problem for the Uniform Arrival Distribution

In this section we give an improved algorithm, referred as the charter algorithm $C_{k,\gamma}$, for the temp secretary problem with uniform arrival times and capacity 1 (at most one secretary can be hired at any time).

As it is never the case that more than $1/\gamma$ items can be selected, setting $k = \lceil 1/\gamma \rceil$ effectively removes the budget constraint. Note that $C_{k,0}$ is Kleinberg's algorithm for the k-secretary problem, with some missing details added to the description.

To analyze the charter algorithm we establish a lower bound on the expected size of the maximum γ-independent subset of a set of uniformly random points in $[0, 1)$. We apply this lower bound to the subset of the items that Kleinberg's algorithm selects.

4.1 The Temp Secretary Algorithm, $C_{k,\gamma}$: A Competitive Ratio of $1/(1 + k\gamma)$

This charter algorithm, $C_{k,\gamma}$ gets parameters k (the maximal number of rentals allowed) and γ (the rental period) as is described in detail in Algorithm 1. As the entire period is normalized to $[0, 1)$, having $k > \lceil 1/\gamma \rceil$ is irrelevant. Thus, we assume that $k \leq \lceil 1/\gamma \rceil$.[2]

We show that $C_{k,\gamma}(X)$ gains in expectation about $1/(1 + k\gamma)$ of the top k values of X, which implies that the competitive ratio (see definition (1)) of $C_{k,\gamma}$ is at least about $1/(1 + k\gamma)$.

Note that for $k = \lceil 1/\gamma \rceil$, $C_{k,\gamma}$ has a competitive ratio close to $1/2$, while for $\gamma = 0$, $C_{k,\gamma}$ has a competitive ratio close to 1.

It is easy to see that $C_{k,\gamma}$, chooses a γ-independent set of size at most k.

The main theorem of this paper is the following generalization of Kleinberg's k-secretary problem:

Theorem 2. *For any set of items $S = \{x_i\}_{i=1}^n$, $0 < \gamma \leq \gamma^* = 0.003176$ and any positive integer $k \leq 1/\gamma$:*

$$E_{\theta:S \mapsto [0,1]}[v(C_{k,\gamma}(S, \theta))] \geq \frac{1}{1 + \gamma k}(1 - \beta(\gamma, k))T_k(S), \qquad (4)$$

where $\beta(\gamma, k) = 7.4\sqrt{\gamma \ln(1/\gamma)} + \frac{5}{\sqrt{k}}$, and the expectation is taken over all uniform mappings of S to the interval $[0, 1)$. (Note that the right hand side of Equation (4) is negative for $\gamma^* < \gamma \leq 0.5$.)

4.2 Outline of the Proof of Theorem 2

We prove Theorem 2 by induction on k. For $k \leq 25$ the theorem holds vacuously.

The profit, $p^{[0,1/2)}$, on those items that arrive during the time interval $[0, 1/2)$ is given by the inductive hypothesis[3]. However, the inductive hypothesis gives this profit, $p^{[0,1/2)}$, in terms of the top $\lfloor k/2 \rfloor$ elements that arrive before time $1/2$, and not in terms of $T_k(X)$, the value of the top k items overall. Thus, we need to relate $p^{[0,1/2)}$ to $T_k(X)$. In the full version of this paper we show that $p^{[0,1/2)}$ is about $1/2$ of $T_k(X)$.

Let $Z_{>T}$ be the set of items that arrive in the time interval $[1/2, 1)$ and have value greater than the threshold T. From $Z_{>T}$ we greedily pick a γ-independent subset[4]. It is easy to see that this set is in fact a maximal γ-independent subset.

To bound the expected profit from the items in $Z_{>T}$ we first bound the size of the maximal γ-independent set amongst these items. To do so we use the following general theorem (see also Section 6 and the full version of this paper).

---

[2] To simplify the presentation we shall assume the in sequel that $k \leq 1/\gamma$.

[3] This profit, $p^{[0,1/2)}$ is $E_{\theta:S \mapsto [0,1)}\left[v(C_{k,\gamma}^{[0,1/2)}(S, \theta))\right]$, where $C_{k,\gamma}^{[0,1/2)}(S, \theta)$ the set of items chosen by the algorithm during the time period $[0, 1/2)$.

[4] Modulo the caveat that the arrival time of the 1st item chosen from the 2nd half must be at least γ after the arrival time of the last item chosen in the 1st half.

ALGORITHM 1. The Charter Algorithm $C_{k,\gamma}$

---

1 **if** $k = 1$ **then**

    /* Use the ''continuous secretary'' algorithm [13]:   */

2   Let x be the largest item to arrive by time $1/e$ (if no item arrives by time $1/e$ — let x be the absolute zero, an item smaller than all other items).

3      $C_{k,\gamma}$ accepts the first item y, $y > x$, that arrives after time $1/e$ (if any)

4 **else**

    /* Process the items scheduled during the time interval $[0, 1/2)$ */

5   Initiate a recursive copy of the algorithm, $C' = C_{\lfloor k/2 \rfloor, 2\gamma}$.

6      $x \leftarrow$ next element   // If no further items arrive $x \leftarrow \emptyset$

7   **while** $x \neq \emptyset$ *AND* $\theta(x) < 1/2$ **do**

8   Simulate C' with input x and modified schedule $\theta'(x) = 2\theta(x)$.

9   **if** C' *accepts* x **then**

10             $C_{k,\gamma}$ accepts x

11         $x \leftarrow$ next element   // If no further items arrive, $x \leftarrow \emptyset$

    /* Determine threshold T */

12   Sort the items that arrived during the time interval $[0, 1/2)$: $y_1 > y_2 > \cdots > y_m$ (with consistent tie breaking).

13   Let $\tau = \lceil k/2 \rceil$.

14   **if** $m < \tau$ **then**

15   set T to be the absolute zero

16   **else**

17   set $T \leftarrow y_\tau$.

    /* Process the items scheduled during the time interval $[1/2, 1)$ */

18   **do**

19   **if** $x > T$ *AND* $(\theta(x) \geq \theta(x') + \gamma$ where x' is the last item accepted by $C_{k,\gamma}$

20   *OR* no items have been previously accepted) **then**

21             $C_{k,\gamma}$ accepts x

22         $x \leftarrow$ next element   // If no further items arrive, $x \leftarrow \emptyset$

23   **until** $x = \emptyset$ *OR* k items have already been accepted

---

Theorem 3. *Let $Z = \{z_1, z_2, \ldots, z_n\}$ be a set of independently uniform samples, z_i, from the real interval $[0, 1)$. For $0 \leq \gamma \leq 1$,*

$$\mathrm{E}_Z[m(Z, \gamma)] \geq \frac{1 - \alpha(\gamma)}{\gamma + 1/n} = \frac{(1 - \alpha(\gamma))n}{1 + n\gamma}, \quad \text{where } \alpha(\gamma) = 3\sqrt{\gamma \ln(1/\gamma)}, \quad (5)$$

where $m(Z, \gamma)$ denotes the size of the largest γ-independent subset of Z.

We apply Theorem 3 to the items in $Z_{>T}$. We can apply this theorem since arrival times of items in $Z_{>T}$ are uniformly distributed in the 2nd half. Specifically, we give a lower bound on the expected profit of the algorithm from the items in the 2nd half as follows:

1. Condition on the size of $Z_{>T}$.
2. Subsequently, condition on the set of arrival times $\{\theta_1, \theta_2, \ldots, \theta_{|Z_{>T}|}\}$ of the items in $Z_{>T}$ but *not* on which item in $Z_{>T}$ arrives when. This conditioning fixes the γ-independent set selected greedily by the algorithm.
3. We take the expectation over all bijections θ whose image on the domain $Z_{>T}$ is the set $\{\theta_1, \theta_2, \ldots, \theta_{|Z_{>T}|}\}$. The expected profit (over the set $Z_{>T}$ and over these bijections) is "approximately"

$$\frac{\text{Size of maximal } \gamma\text{-independent set from } Z_{>T}}{|Z_{>T}|} \cdot \sum_{z \in Z_{>T}} v(z). \tag{6}$$

The "approximately" is because of some technical difficulties:

- We cannot ignore the last item amongst those arriving prior to time $1/2$. If one such item was chosen at some time $1/2 - \gamma < t < 1/2$ then arrivals during the period $[1/2, t + \gamma)$ cannot be chosen.
- We cannot choose more than k items in total, if the algorithm choose λ items from the time interval $[0, 1/2)$, it cannot choose more than $k - \lambda$ items from the time interval $[1/2, 1)$, but $k - \lambda$ may be smaller than the size of the γ-independent set from $Z_{>T}$.

4. To get an unconditional lower bound we average Equation (6) over the possible sizes of the γ-independent set as given by Theorem 3.

5 Upper Bound for the Temp Secretary Problem with Uniform Arrival Times and with No Budget Restriction

Theorem 4. *For the temp secretary problem where item arrival times are taken from the uniform distribution, for any $\gamma \in (0, 1)$, any online algorithm (potentially randomized) has a competitive ratio $\leq 1/2 + \gamma/2$.*

Proof. Let A denote the algorithm. Consider the following two inputs:

1. The set S of n-1 items of value 1.
2. The set $S' = S \cup \{x_n\}$ where $v(x_n) = \infty$.

Note that these inputs are not of the same size (which is ok as the number of items is unknown to the algorithm).

Condition the mapping $\theta : S \mapsto [0, 1)$ (but not the mapping of x_n). If A accepts an item x at time $\theta(x)$ we say that the segment $[x, x + \gamma)$ is *covered*. For a fixed θ let $g(\theta)$ be the expected fraction of $[0, 1)$ which is not covered when running A on the set S with arrival times θ. This expectation is over the coin tosses of A. Let G be $E_{\theta : S \mapsto [0,1)}[g(\theta)]$.

The number of items that A picks on the input S with arrival time θ is at most $\frac{1 - g(\theta)}{\gamma} + 1$. Taking expectation over all mappings $\theta : S \mapsto [0, 1)$ we get that the value gained by A is at most $(1 - G)/\gamma + 1$.

As $n \to \infty$ the optimal solution consists of $\lceil 1/\gamma \rceil$ items of total value $\lceil 1/\gamma \rceil$. Therefore the competitive ratio of A is at most

$$\frac{(1-G)/\gamma + 1}{1/\gamma} = 1 - G + \gamma. \tag{7}$$

Note that $g(\theta)$ is exactly the probability that A picks x_n on the input $S \cup \{x_n\}$ (this probability is over the mapping of x_n to $[0, 1)$ conditioned upon the arrival times of all the items in $S \subset S'$). Therefore the competitive ratio of A on the input S' is

$$\mathrm{E}[g(\theta)] = G . \tag{8}$$

Therefore the competitive ratio of A is no more than the minimum of the two upper bounds (7) and (8)

$$\min(G, 1 - G + \gamma) \le 1/2 + \gamma/2 .$$

$\square$

6   About Theorem 3: A Lower Bound on the Expected Size of the Maximum γ-independent Subset

Recall the definition of Z and $m(Z, \gamma)$ from Theorem 3.

Define the random variable X_i, $1 \le i \le n$ to be the i'th smallest point in Z. Define the random variable C_i to be the number of points from Z that lie in the interval $[X_i, X_i + \gamma)$. Note that at most one of these points can belong to a γ-independent set.

The *greedy algorithm* constructs a maximal γ-independent set by traversing points of Z from the small to large and adding a point whenever possible. Let I_i be a random variable with binary values where $I_i = 1$ iff X_i was chosen by the greedy algorithm. It follows from the definition that $\sum_i I_i$ gives the size of the maximal independent set, $m(Z, \gamma)$, and that $\sum_i I_i C_i = n$.

Note that $\mathrm{E}[C_i] \le 1 + n\gamma$, one for the point X_i itself, and $n\gamma$ as the expected number of uniformly random points that fall into an interval of length γ. If C_i and I_i were independent random variables, it would follow that

$$\mathrm{E}\left[\sum I_i C_i\right] \le (1 + n\gamma) \sum \mathrm{Prob}[I_i = 1],$$

and, thus,

$$m(Z, \gamma) = \sum I_1 \ge n/(1 + n\gamma).$$

Unfortunately, C_i and I_i are not independent, and the full proof of Theorem 3, that deals with such dependencies, appears in the full version of this paper (on the Archive).

7 Discussion and Open Problems

We've introduced online optimization over temporal items under stochastic inputs subject to conditions of two different types:

- "Vertical" constraints: Predicates on the set of items held at all times t. In this class, we've considered conditions such as no more than d simultaneous items held at any time, items held at any time of total weight $\leq W$, items held at any time must be independent in some Matroid.
- "Horizontal" constraints: Predicates on the set of items over all times. Here, we've considered the condition that no more than k employees be hired over time.

One could imagine much more complex settings where the problem is defined by arbitrary constraints of the first type above, and arbitrary constraints of the 2nd type. For example, consider using knapsack constraints in both dimensions. The knapsack constraint for any time t can be viewed as the daily budget for salaries. The knapsack constraint over all times can be viewed as the total budget for salaries. Many other natural constraints suggest themselves.

It seems plausible that the time slice algorithm can be improved, at least in some cases, by making use of information revealed over time, as done by the Charter algorithm.

References

1. Babaioff, M., Dinitz, M., Gupta, A., Immorlica, N., Talwar, K.: Secretary problems: Weights and discounts. In: SODA 2009, pp. 1245–1254 (2009)
2. Babaioff, M., Immorlica, N., Kempe, D., Kleinberg, R.D.: A knapsack secretary problem with applications. In: Charikar, M., Jansen, K., Reingold, O., Rolim, J.D.P. (eds.) RANDOM 2007 and APPROX 2007. LNCS, vol. 4627, pp. 16–28. Springer, Heidelberg (2007)
3. Babaioff, M., Immorlica, N., Kempe, D., Kleinberg, R.: Online auctions and generalized secretary problems. SIGecom Exch. 7(2), 1–11 (2008)
4. Babaioff, M., Immorlica, N., Kleinberg, R.: Matroids, secretary problems, and online mechanisms. In: SODA 2007, pp. 434–443. SIAM (2007)
5. Badanidiyuru, A., Kleinberg, R., Singer, Y.: Learning on a budget: Posted price mechanisms for online procurement. In: EC 2012, pp. 128–145 (2012)
6. Boucheron, S., Fernandez de la Vega, W.: On the independence number of random interval graphs. Combinatorics, Probability and Computing 10, 385–396 (2001)
7. Thomas Bruss, F.: A unified approach to a class of best choice problems with an unknown number of options. Ann. Probab. 12(3), 882–889 (1984)
8. Chakraborty, S., Lachish, O.: Improved competitive ratio for the matroid secretary problem. In: SODA 2012, pp. 1702–1712. SIAM (2012)
9. Coffman Jr., E.G., Poonen, B., Winkler, P.: Packing random intervals. Probability Theory and Related Fields 102(1), 105–121 (1995)
10. Devanur, N.R., Hayes, T.P.: The adwords problem: online keyword matching with budgeted bidders under random permutations. In: EC 2009, pp. 71–78 (2009)

11. Dinitz, M.: Recent advances on the matroid secretary problem. SIGACT News 44(2), 126–142 (2013)
12. Dynkin, E.B.: The optimal choice of the stopping moment for a markov process. Dokl. Akad. Nauk. SSSR, 238–240 (1963)
13. Feldman, M., Naor, J(S.), Schwartz, R.: Improved competitive ratios for submodular secretary problems (Extended abstract). In: Goldberg, L.A., Jansen, K., Ravi, R., Rolim, J.D.P. (eds.) RANDOM 2011 and APPROX 2011. LNCS, vol. 6845, pp. 218–229. Springer, Heidelberg (2011)
14. Feldman, M., Svensson, O., Zenklusen, R.: A simple $O(\log \log(\text{rank}))$-competitive algorithm for the matroid secretary problem. In: SODA 2015, pp. 1189–1201. SIAM (2015)
15. Freeman, P.R.: The secretary problem and its extensions: A review. International Statistical Review / Revue Internationale de Statistique 51(2), 189–206 (1983)
16. Gallego, G., van Ryzin, G.: Optimal dynamic pricing of inventories with stochastic demand over finite horizons. Manage. Sci. 40(8), 999–1020 (1994)
17. Göbel, O., Hoefer, M., Kesselheim, T., Schleiden, T., Vöcking, B.: Online independent set beyond the worst-case: Secretaries, prophets, and periods. In: Esparza, J., Fraigniaud, P., Husfeldt, T., Koutsoupias, E. (eds.) ICALP 2014, Part II. LNCS, vol. 8573, pp. 508–519. Springer, Heidelberg (2014)
18. Hajiaghayi, M.T., Kleinberg, R., Parkes, D.C.: Adaptive limited-supply online auctions. In: EC 2004, pp. 71–80 (2004)
19. Hajiaghayi, M.T., Kleinberg, R.D., Mahdian, M., Parkes, D.C.: Online auctions with re-usable goods. In: EC 2005, pp. 165–174 (2005)
20. Justicz, J., Scheinerman, E.R., Winkler, P.M.: Random intervals. The American Mathematical Monthly 97(10), 881–889 (1990)
21. Karlin, S.: Stochastic models and optimal policy for selling an asset, pp. 148–158 (1962)
22. Kesselheim, T., Radke, K., Tönnis, A., Vöcking, B.: An optimal online algorithm for weighted bipartite matching and extensions to combinatorial auctions. In: Bodlaender, H.L., Italiano, G.F. (eds.) ESA 2013. LNCS, vol. 8125, pp. 589–600. Springer, Heidelberg (2013)
23. Kesselheim, T., Tönnis, A., Radke, K., Vöcking, B.: Primal beats dual on online packing lps in the random-order model. In: STOC 2014, pp. 303–312. ACM (2014)
24. Kleinberg, R.D.: A multiple-choice secretary algorithm with applications to online auctions. In: SODA 2005, pp. 630–631. SIAM (2005)
25. Korula, N., Pál, M.: Algorithms for secretary problems on graphs and hypergraphs. In: Albers, S., Marchetti-Spaccamela, A., Matias, Y., Nikoletseas, S., Thomas, W. (eds.) ICALP 2009, Part II. LNCS, vol. 5556, pp. 508–520. Springer, Heidelberg (2009)
26. Lachish, O.: O(log log rank) competitive-ratio for the matroid secretary problem. CoRR, abs/1403.7343 (2014)
27. Lindley, D.V.: Dynamic programming and decision theory. Appl. Statist., 39–52 (1961)
28. Lipton, R.: Online interval scheduling. In: SODA 1994, pp. 302–311 (1994)
29. Singer, Y., Mittal, M.: Pricing mechanisms for crowdsourcing markets. In: WWW 2013, pp. 1157–1166 (2013)
30. Woeginger, G.J.: On-line scheduling of jobs with fixed start and end times. Theoretical Computer Science 130(1), 5–16 (1994)

How to Sort by Walking on a Tree

Daniel Graf

ETH Zürich, Department of Computer Science, Zurich, Switzerland
grafdan@ethz.ch

Abstract. Consider a graph G with n vertices. On each vertex we place a box. These n vertices and n boxes are both numbered from 1 to n and initially shuffled according to a permutation $\pi \in S_n$. We introduce a sorting problem for a single robot: In every step, the robot can walk along an edge of G and can carry at most one box at a time. At a vertex, it may swap the box placed there with the box it is carrying. How many steps does the robot need to sort all the boxes?

We present an algorithm that produces a shortest possible sorting walk for such a robot if G is a tree. The algorithm runs in time $\mathcal{O}(n^2)$ and can be simplified further if G is a path. We show that for planar graphs the problem of finding a shortest possible sorting walk is $\mathcal{NP}$-complete.

Keywords: Physical Sorting, Shortest Sorting Walk, Warehouse Reorganization, Robot Scheduling, Permutation Properties

1 Introduction

Motivation. Nowadays, many large warehouses are operated by robots. Such automated storage and retrieval systems (abbreviated AS/RS) are used in industrial and retail warehouses, archives and libraries, as well as automated car or bicycle parking systems. When it needs to rearrange the stored goods, such a robot faces a physical sorting task. In contrast to standard sorting algorithms, it does not have constant time access to the stored objects. It might need to travel for a significant amount of time before fetching the object in question, and then moving it to its desired location also takes time. We want to look at the problem of finding the most efficient route for the robot that allows it to permute the stored objects. Our interest in this problem arises from a bike parking system to be built in Basel, for which bike boxes need to be rearranged according to the expected pickup times of the customers.

Problem Description. We consider the following model throughout this paper. Our warehouse holds n boxes. Each box is unique in its content but all the boxes have the same dimensions and can be handled the same way. The storage locations and aisles of the warehouse are represented by a connected graph $G = (V, E)$, where $n = |V|$ and $m = |E|$. Every vertex $v \in V$ represents a location that can hold a single box. Every edge $e = (u, v) \in E$ represents a bidirectional aisle between two locations. We assume that our warehouse is full, meaning

© Springer-Verlag Berlin Heidelberg 2015
N. Bansal and I. Finocchi (Eds.): ESA 2015, LNCS 9294, pp. 643–655, 2015.
DOI: 10.1007/978-3-662-48350-3_54

that at each location there is exactly one box stored initially. The boxes and locations are numbered from 1 to n and are initially shuffled according to some permutation $\pi \in S_n$, representing that the box at vertex i should get moved to vertex $\pi(i)$. The robot is initially placed at a vertex r. In every step the robot can move along a single edge. It can carry at most one box with it at any time. When arriving at a vertex it can either put down the box it was traveling with (if there is no box at this vertex), pick up the box from the current vertex (if it arrived without carrying a box), swap the box it was carrying with the box at this vertex (if there is one) or do nothing.

We refer to each traveled edge of the robot as a *step* of the sorting process. A sequence of steps that lets the robot sort all the boxes according to π and return to r is called a *sorting walk*. We measure the length of a sorting walk as the number of edges that the robot travels along. Therefore, we assume that all aisles are of equal length and that all of the box-handling actions (pickup, swap, putdown) only take a negligible amount of time compared to the time spent traveling along the edges. We are looking for the shortest sorting walk.

Example. Figure 1 shows an example of a warehouse where G is a tree consisting of 8 vertices. It is not obvious how we can find a short walk that allows the robot to sort these 8 boxes. We will see an efficient algorithm that produces such a sorting walk and we will prove that this sorting walk has minimum length.

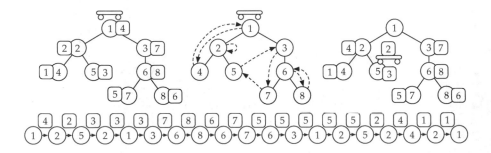

Fig. 1. (left) Initial state of the warehouse with storage locations as circles and boxes as squares. The box at vertex i is labeled with its target vertex $\pi(i)$. (center) The initial state with π drawn as dashed arcs towards their target vertex instead of numbered boxes. (right) This shows the state of the warehouse after two steps have been performed. First the robot brought box 4 to vertex 2. Then it took box 2 to vertex 5. (bottom) A shortest possible sorting walk consisting of 18 steps.

Organization. Section 2 introduces some terminology and shows first lower and upper bounds on the length of a shortest sorting walk on general graphs. We then look for shortest walks for certain classes of graphs. Section 3 shows a way of finding shortest sorting walks on path graphs where the robot starts at one of the ends of the path. Our main result is given in Section 4, where we efficiently construct shortest sorting walks on arbitrary trees with arbitrary starting position. Finally, in Section 5 we show that it is $\mathcal{NP}$-complete to find a shortest sorting walk for planar graphs.

Related Work. Sorting algorithms for physical objects were studied in many different models before. Sorting streams of objects was studied for instance by Knuth [8], where we can use an additional stack to buffer objects for rearrangement. Similar problems were also studied in the context of sorting railway cars, for example by Büsing et al. [1]. Most similar to our solutions is an algorithm called *cycle sort* by Haddon [4] that minimizes the number of writes when sorting an array by looking at the cycles of its permutation. Katsuhisa et al. [10] recently studied the process of sorting n tokens on a graph of n vertices using as few swaps of neighboring tokens as possible. For path graphs the number of swaps is minimized by bubble sort. They give a 2-approximation for tree graphs by simulating cycle sort. Compared to our setting, they do not require that successive actions are applied to nearby vertex positions. Sliding physical objects are also studied in the context of the hardness of many different puzzle games. We refer to Hearn [5] for an overview. An extensive overview of the research on storage yard operation can be found in [2].

2   Notation and General Bounds

Before we look at specific types of graphs, we introduce some notation and show some general lower and upper bounds for the length of a shortest sorting walk.

Formally, we describe the *state* τ of the warehouse by a triple (v, b, σ) where $v \in V$ is the current position of the robot, $b \in \{1, \ldots, n\} \cup \{\square\}$ is the number of the box that the robot is currently traveling with or $\square$ if it is traveling without a box, and σ is the current mapping from vertices to boxes. If there is no box at some vertex i, we will have $\sigma(i) = \square$. At any point there will always be at most one vertex without a box, thus at most one number will not appear in $\{\sigma(i) \mid i \in \{1, \ldots, n\}\}$. In other words: Looking at σ and b together will at all times be a permutation of $\{1, \ldots, n\} \cup \{\square\}$. Given the current state, the next *step* s of the robot can be specified by the pair (p, b), if the robot moves to $p \in V$ with box $b \in \{1, \ldots, n\} \cup \{\square\}$.

We start with $\tau_0 = (r, \square, \pi)$, so the robot is at the starting position and is not carrying a box. Applying a step $s_t = (p, b)$ to a state $\tau_{t-1} = (v_{t-1}, b_{t-1}, \sigma_{t-1})$ transforms it into state $\tau_t = (v_t, b_t, \sigma_t)$ with $v_t = p$, $b_t = b$. σ_t only differs from σ_{t-1} if a swap was performed, so if $b_{t-1} \neq b$, in which case we set $\sigma_t(v_{t-1}) = b_{t-1}$. In order to get $\sigma = id$ in the end, we let the robot put its box down whenever it moves onto an empty location. Thus if $\sigma_{t-1}(p) = \square$, we let $b_t = \square$ and $\sigma_t(p) = b$.

Step s_t is *valid* only if $(v_{t-1}, p) \in E$ and $b \in \{b_{t-1}, \sigma_{t-1}(v_{t-1})\}$, so if the robot moved along an edge of G and carried either the same box as before or the box that was located at the previous vertex. Thus after putting down a box at an empty location, the robot can either immediately pick it up again or continue without carrying a box. A sequence of steps $S = (s_1, \ldots, s_l)$ is a *sorting walk* of length l, if we start with τ_0, all steps are valid, and we end in $\tau_l = (r, \square, id)$. We are looking for the minimum l such that a sorting walk of length l exists.

We denote the set of cycles of the permutation π as $\mathcal{C} = \{C_1, \ldots, C_{|\mathcal{C}|}\}$, where each cycle C_i is an ordered list of vertices $C_i = (v_{i,1}, \ldots, v_{i,|C_i|})$ such that $\pi(v_{i,j}) = v_{i,j+1}$ for all $j < |C_i|$ and $\pi(v_{i,|C_i|}) = v_{i,1}$. In the example shown in Figure 1, we have $\mathcal{C} = \{(1,4), (2), (3,7,5), (6,8)\}$. As cycles of length one represent boxes that are placed correctly from the beginning, we usually ignore such trivial cycles and let $\overline{\mathcal{C}} = \{C \in \mathcal{C} \mid |C| > 1\}$ be the set of non-trivial cycles.

Let $d(u,v)$ denote the distance (length of the shortest path) from u to v in G. So if the robot wants to move a box from vertex u to vertex v, it needs at least $d(u,v)$ steps for that. By $d(C)$ we denote the sum of distances between all pairwise neighbors in the cycle C and by $d(\pi)$ the sum of all such cycle distances for all cycles in π, i.e., $d(\pi) = \sum_{C \in \mathcal{C}} d(C) = \sum_{v \in V} d(v, \pi(v))$.

We distinguish two kinds of steps in a sorting walk: *essential* and *non-essential* steps. A step $s = (p, b)$ is essential if it brings box b one step closer to its target position than it was in any of the previous states, so if $d(p, b)$ is smaller than ever before. We say that such a step is essential for a cycle C if $b \in C$. A single step can be essential for at most one cycle, as at most one box is moved in a step and each box belongs to exactly one cycle. In the example in Figure 1 for instance, the first step was essential for cycle $(1,4)$. Overall, 16 steps (all but s_2 and s_{15}) were essential. This corresponds to the sum of distances of all boxes to their targets $d(\pi)$, which we formalize as follows.

Lemma 1 (Lower bound by counting essential steps). *Every sorting walk for a permutation π on a graph G has length at least $d(\pi) = \sum_{b \in \{1, \ldots, n\}} d(b, \pi(b))$.*

Proof. Throughout any sorting walk, there will be exactly $d(b, \pi(b))$ essential steps that move box b. As the robot cannot move more than one box at a time, the sum of distances between all boxes and their target positions can decrease by at most 1 in each step. Therefore, there will be $d(\pi) = \sum_{b \in \{1, \ldots, n\}} d(b, \pi(b))$ essential steps in every sorting walk and at least as many steps overall.   □

The remaining challenge is to minimize the number of non-essential steps. In case that π consists only of a single cycle, the shortest solution is easy to find. We just pick up the box at r and bring it to its target position $\pi(r)$ in $d(r, \pi(r))$ steps. We continue with the box at $\pi(r)$, bring it to $\pi(\pi(r))$ and so on until we return to r and close the cycle. Therefore, by just following this cycle, the robot can sort these boxes in $d(\pi)$ steps without any non-essential steps. As it brings one box one step closer to its target position in every step, by Lemma 1 no other sorting walk can be shorter.

But what if there is more than one cycle? One idea could be to sort each cycle individually one after the other. This might not give a shortest possible sorting walk, but it might give a reasonable upper bound. So the robot picks up the box at r, brings it to its target, swaps it there, continues with that box and repeats this until it closes the cycle. After that, the robot moves to any box b that is not placed at its correct position yet. These steps will be non-essential as the robot does not carry a box during these steps from r to b. Once it arrives at b, it sorts the cycle in which b is contained. In this way, it sorts cycle after cycle and finally returns to r. The number of non-essential steps in this process depends on the order in which the cycles are processed and which vertices get picked to start the cycles. The following lemma shows that a linear amount of non-essential steps will always suffice.

Lemma 2 (Upper bound from traversal). *There is a sorting walk of length at most $d(\pi) + 2 \cdot (n - 1)$ for a permutation π on a graph G.*

Proof. We let the robot do a depth-first search traversal of G while not carrying a box. Whenever we encounter a box that is not placed correctly yet, we sort its entire cycle. As the robot returns to the same vertex at the end of the cycle we can continue the traversal at the place where we interrupted it. Recall that G is connected, so during the traversal we will visit each vertex at least once and at the end all boxes will be at their target position. The number of non-essential steps is now given by the number of steps in the traversal which is twice the number of edges of the spanning tree produced by the traversal. $\square$

We can see that these sorting walks might not be optimal, for instance in the example shown in Figure 1. Every sorting walk that sorts only one cycle at a time will have length at least 20, while the optimal solution consists of only 18 steps.

As $d(\pi)$ can grow quadratic in n, the linear gap between the upper and lower bound might already be considered negligible. However, for the rest of this paper we want to find sorting walks that are as short as possible.

3   Sorting on Paths

We now look at the case where G is the path graph $P = (V, E)$. Imagine that the vertices v_1 to v_n are ordered on a line from left to right and every vertex is connected to its left and right neighbor, thus $E = \{\{v_i, v_{i+1}\} \mid i \in \{1, \ldots, n-1\}\}$. We further assume that the robot is initially placed at one of the ends of the path, so let $r = v_1$.

By $I(C) = [l(C), r(C)]$, we denote the interval of P covered by the cycle C, where $l(C) = \min_{v_i \in C} i$ and $r(C) = \max_{v_i \in C} i$. We say that two cycles C_1 and C_2 intersect if their intervals intersect. Now let $\mathcal{I} = (\overline{\mathcal{C}}, \mathcal{E})$ be the intersection graph of the non-trivial cycles, so $\mathcal{E} = \{\{C_1, C_2\} \mid C_1, C_2 \in \overline{\mathcal{C}} \text{ s.t. } I(C_1) \cap I(C_2) \neq \emptyset\}$. We then use $\mathcal{D} = \{D_1, \ldots, D_{|\mathcal{D}|}\}$ to represent the partition of $\overline{\mathcal{C}}$ into the connected components of this intersection graph $\mathcal{I}$. Two cycles C_1 and C_2 are in the same connected component $D_i \in \mathcal{D}$, if and only if there exists a sequence of pairwise-intersecting cycles that starts with C_1 and ends with C_2. We let $l(D) = \min_{C \in D} l(C)$ and $r(D) = \max_{C \in D} r(C)$ be the boundary vertices of the connected component D. We index the cycles and components from left to right according to their leftmost vertex, so that $l(C_i) < l(C_j)$ and $l(D_i) < l(D_j)$ whenever $i < j$.

Theorem 1 (Shortest sorting walk on paths). *The shortest sorting walk on a path P with permutation π can be constructed in time $\Theta(n^2)$ and has length*

$$d(\pi) + 2 \cdot \left(l(D_1) - 1 + \sum_{i=1}^{|\mathcal{D}|-1} (l(D_{i+1}) - r(D_i)) \right).$$

Proof. We claim that the number of non-essential steps that are needed is twice the number of edges that are not covered by any cycle interval, and lie between r and the rightmost box that needs to be moved.

We prove the claim by induction on the number of non-trivial cycles of π. We already saw how we can find a minimum sorting walk if π consists of a single cycle only. If there are several cycles but only one of them is non-trivial, so $|\mathcal{C}| > 1$ but $|\bar{\mathcal{C}}| = 1$, the shortest sorting walk is also easy to find: We walk to the right until we encounter the leftmost box of this non-trivial cycle C, then we sort C and return to r. The number of steps is $d(\pi) + 2 \cdot (l(C) - 1)$ and is clearly optimal. Figure 2 (left) gives an example of such a case.

Now let us look at the case where π consists of exactly two non-trivial cycles C_1 and C_2. If C_1 and C_2 intersect, we can interleave the sorting of the two cycles without any non-essential steps. We start sorting C_1 until we first encounter a box that belongs to C_2, so until the first step (p, b) where $p \in C_2$. This will happen eventually, as we assumed that C_1 and C_2 intersect. We then leave box b at position p in order to sort C_2. After sorting C_2, we will be back at position p and can finish sorting C_1, continuing with box b. As we will end in $l(C_1)$ and then return to v_1, we found a minimum walk of length $d(\pi) + 2 \cdot (l(C_1) - 1)$. Figure 2 (center) gives an example of such a case.

Let us assume that C_1 and C_2 do not intersect. This implies that there is no box that has to go from the left of $r(C_1)$ to the right of $l(C_2)$ and vice versa. But the robot still has to visit the vertices of C_2 at some point and then get back to the starting position. So each of the edges between the two cycles will be used for at least two non-essential steps. We construct a sorting walk that achieves this bound of $d(\pi) + 2 \cdot (l(C_1) - 1 + l(C_2) - r(C_1))$. We start by sorting C_1 until we get to $r(C_1)$. We then take the box $\pi(r(C_1))$ from there and walk with it to $l(C_2)$. From there we can sort C_2 starting with box $\pi(l(C_2))$. We again end at $l(C_2)$, where we can pick up box $\pi(r(C_1))$ again and take it back to position $r(C_1)$. From there, we finish sorting C_1 and return back to v_1. Figure 2 (right) gives an example of such a case.

Next, let us assume that we have three or more non-trivial cycles. We look at these cycles from left to right and we assume that by induction we already found a minimum sorting walk S_i for sorting the boxes of the first i cycles C_1 to C_i. For the next cycle C_{i+1} we now distinguish two cases: If C_{i+1} intersects any cycle $C^* \in \{C_1, \ldots, C_i\}$ (which does not necessarily need to be C_i), we can easily insert the essential sorting steps for C_{i+1} into S_i at the point where S_i first walks onto $l(C_{i+1})$ while sorting C^*. As we only add essential steps, this new walk S_{i+1} will still be optimal if S_i was optimal. We have $|S_{i+1}| = |S_i| + d(C_{i+1}) = |S_i| + \sum_{b \in C_{i+1}} d(b, \pi(b))$. In the other case, C_i does not intersect any of the previous cycles. We then know that any sorting walk uses all the edges between $\max_{j \in \{1, \ldots, i\}} r(C_j)$ and $l(C_{i+1})$ for at least two non-essential steps. So

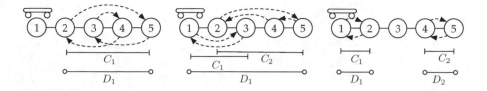

Fig. 2. (left) An example with a single non-trivial cycle. A shortest sorting walk S with $|S| = d(\pi) + 2 \cdot (l(C_1) - 1) = 8 + 2 \cdot (2 - 1) = 10$ is $((2,\square),(3,5),(4,5),(5,5),(4,3),(3,3),(4,4),(3,2),(2,2),(1,\square))$. (center) An example with two intersecting cycles. A shortest sorting walk S with $|S| = d(\pi) = 10$ is $((2,3),(3,5),(4,5),(5,5),(4,4),(3,2),(2,2),(3,3),(2,1),(1,1))$. (right) An example with two non-intersecting cycles. A shortest sorting walk S with $|S| = d(\pi)+2\cdot(l(D_2)-r(D_1)) = 4 + 2 \cdot (4 - 2) = 8$ is $((2,2),(3,1),(4,1),(5,5),(4,4),(3,1),(2,1),(1,1))$.

if we interrupt S_i after the step where it visits $\max_{j\in\{1,\dots,i\}} r(C_j)$ to insert non-essential steps to $l(C_{i+1})$, essential steps to sort C_{i+1} and non-essential steps to get back to $\max_{j\in\{1,\dots,i\}} r(C_j)$ we get a minimum walk S_{i+1}. This case occurs whenever C_{i+1} lies in another connected component than all the previous cycles. So if C_i is the first cycle in some component D_j, we have $|S_{i+1}| = |S_i|+d(C_{i+1})+ 2\cdot(l(D_j)-r(D_{j-1}))$, and so we get exactly the extra steps claimed in the theorem.

$\square$

Algorithmic Construction. The proof of Theorem 1 immediately tells us how we can construct a minimum sorting walk efficiently. Given P and π we first extract the cycles of π and order them according to their leftmost box, which can easily be done in linear time. We then build our sorting walk S in the form of a linked list of steps inductively, starting with an empty walk. While adding cycle after cycle we keep for every vertex v of P a reference to the earliest step of the current walk that arrives at v. We also keep track of the step $s_{\max}$ that reaches the rightmost vertex visited so far.

When adding a new cycle C to the walk, we check whether we stored a step for $l(C)$. If yes, we simply insert the steps to sort C into the walk and update the vertex-references of all the vertices we encounter while sorting C. If $l(C)$ was not visited by the walk so far, we insert the necessary non-essential steps into the walk to get from $s_{\max}$ to $l(C)$ and back after sorting C. In either case we update $s_{\max}$ if necessary. The runtime of adding a new cycle to the walk is linear in the number of steps we add. Overall our construction runs in time $\Theta(n + |S|) \subseteq \Theta(n^2)$, so it is linear in the combined size of the input and output and at most quadratic in the size of the warehouse.

So far, we assumed that the robot works on a path and starts at an endpoint of that path. What if the robot starts at an inner vertex of the path? It is not immediately clear whether its first step should go to the left or to the right then. Instead of going into the details of this scenario, we now study the more general problem of arbitrary trees with arbitrary starting positions.

4   Sorting on Trees

We now want to study the problem of sorting boxes placed on an arbitrary tree. So let $T = (V, E)$ be the underlying tree, let $r \in V$ be the starting vertex and let T be rooted at r. For any cycle C of π we say that it *hits* a vertex v if the box initially placed on v belongs to the cycle C. We denote by $V(C)$ the set of vertices hit by C. We let $T(C)$ denote the minimum subtree of T that contains all vertices hit by C and we say C *covers* v for every $v \in T(C)$. In Figure 1 for example, we have $T((3, 5, 7)) = \{1, 2, 3, 5, 6, 7\}$.

Before describing our solution, we will first derive a lower bound on the length of any sorting walk on T. We describe how we map each sorting walk to an auxiliary structure called *cycle anchor tree* that reflects how the cycles of π are interleaved in the sorting walk. We then bound the length of the sorting walk only knowing its cycle anchor tree. We give an explicit construction of a sorting walk that shows that this bound is tight. In order to find an optimal solution we first find a cycle anchor tree with the minimum possible bound and then apply the tight construction to get a shortest possible sorting walk.

4.1   Cycle Anchor Trees

Definition. A *cycle anchor tree* $\widetilde{T}$ is a directed, rooted tree that contains one vertex $\widetilde{v}_C$ for every non-trivial cycle C of π and an extra root vertex $\widetilde{r}$. Given a sorting walk S we construct from it a cycle anchor tree $\widetilde{T}$ as follows: We start with $\widetilde{T}$ only containing $\widetilde{r}$. We go through the essential steps in S. If step s is the first essential step for some cycle C, we create a vertex $\widetilde{v}_C$ in $\widetilde{T}$. To determine the parent node of $\widetilde{v}_C$ in $\widetilde{T}$ we look for the last essential step s' in S before s and its corresponding cycle C'. We now say that C is *anchored* at C' and add an edge $(\widetilde{v}_{C'}, \widetilde{v}_C)$ to $\widetilde{T}$. If no such step s' exists (which only happens for the very first essential step in S) we use the root $\widetilde{r}$ as the parent of $\widetilde{v}_C$.

Edge Costs. We also assign an integer cost to each edge of a cycle anchor tree. For this we call a sorting step a *down-step* if the robot moves away from the root and an *up-step* otherwise. The cost c for an edge between two nodes of $\widetilde{T}$ is now defined as follows: Let $c((\widetilde{v}_{C_1}, \widetilde{v}_{C_2}))$ be the minimum number of down-steps on the path from any vertex $v \in T(C_1)$ to any vertex $w \in V(C_2)$. Let us fix one such path that minimizes the number of down-steps and let v and w be its endpoints. This path, conceptually, consists of two parts: some up-steps towards the root and then some down-steps away from the root. However, note that we never walk down and then up again, as this would correspond to traversing the same edge twice. Let a be the vertex where this path switches from up-steps to down-steps, also known as the *lowest common ancestor* of v and w. We say that a is an *anchor vertex* for anchoring C_2 at C_1. For the single edge incident to the root, we have $c((\widetilde{r}, \widetilde{v}_C))$ being the minimum number of down-steps on the path from the root to any vertex $v \in V(C)$. The cost $c(\widetilde{T})$ of an entire cycle anchor tree $\widetilde{T}$ is simply the sum of its edge costs. Figure 3 illustrates the definitions and

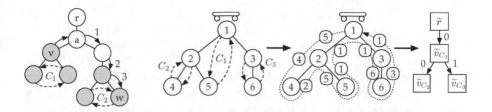

Fig. 3. (first figure on the left) The two pairs of dashed arrows symbolize boxes that need to be swapped. A shortest path from any $v \in T(C_1)$ to any $w \in V(C_2)$ is shown with continuous arrows, three of them being down-steps, so $c((\tilde{v}_{C_1}, \tilde{v}_{C_2})) = 3$. The anchor vertex a is the vertex immediately before the first down step. Note that c is not symmetric as $c((\tilde{v}_{C_2}, \tilde{v}_{C_1})) = 2$. (the three figures on the right) An example of a sorting walk on a tree with three non-trivial cycles. The dashed arrows on the left show the desired shuffling of the boxes. The dotted arrow in the middle shows a minimum sorting walk of ten steps, where each step is labeled with the box it moves. On the right, the corresponding cycle anchor tree is given. The edge from $\tilde{v}_{C_1}$ to $\tilde{v}_{C_3}$ has cost 1 as there is a down-step necessary to get from vertex $1 \in T(C_1)$ to vertex $3 \in V(C_3)$. The edge $(\tilde{v}_{C_1}, \tilde{v}_{C_2})$ is free as vertex 2 is both in $T(C_1)$ and $V(C_2)$.

gives an example of the transformation from a sorting walk to a weighted cycle anchor tree.

Theorem 2 (Lower bound for trees). *Any sorting walk S that sorts a permutation π on a tree T and corresponds to a cycle anchor tree $\tilde{T}$ has length at least $d(\pi) + 2 \cdot c(\tilde{T})$.*

Proof. We partition the steps of S into three sets: essential steps $\mathcal{S}_e$, non-essential down-steps $\mathcal{S}_{n,d}$ and non-essential up-steps $\mathcal{S}_{n,u}$. From Lemma 1 we have $|\mathcal{S}_e| = d(\pi)$. We argue that S contains at least $c(\tilde{T})$ many non-essential down-steps. To do this we look at the segments of S that were relevant when we described how we derive $\tilde{T}$ from S. For an edge $(\tilde{v}_{C_1}, \tilde{v}_{C_2})$ of $\tilde{T}$, we look for the segment S_{C_1,C_2} of S between the first essential step s_2 of C_2 and its most recent preceding essential step s_1 for some other cycle C_1. What do we know about S_{C_1,C_2}? First of all, we know that s_1 is essential for C_1, so s_1 ends at a vertex covered by C_1 and S_{C_1,C_2} starts somewhere in $T(C_1)$. Next, s_2 is the first essential step that moves a box of C_2. Note that some or even all of the boxes of C_2 might have been moved in non-essential steps before s_2, putting them further away from their target position. But as we are on a tree (where there is only a single path between any pair of points), the first time a box gets moved closer to its target position than it was originally is a move away from its initial position, which means that s_2 starts at a vertex hit by C_2. So S_{C_1,C_2} ends somewhere in $V(C_2)$. By definition of $c(\tilde{v}_{C_1}, \tilde{v}_{C_2})$, there are at least $c(\tilde{v}_{C_1}, \tilde{v}_{C_2})$ many down-steps in S_{C_1,C_2}. The same holds for the initial segment $S_{r,C}$. As all these segments of the sorting walk are disjoint, we get that $|\mathcal{S}_{n,d}| \geq c(\tilde{T})$.

Finally we argue that $|\mathcal{S}_{n,d}| = |\mathcal{S}_{n,u}|$ to conclude the proof. Consider any edge e of T and count all steps of S that go along e. Regardless of whether the steps are essential or non-essential, we know that there must be equally many up-steps and down-steps along e, as S is a closed sorting walk and T has no cycles. So for every time we walk down along an edge, we also have to walk up along it once. We see that this equality also holds for the essential up-steps and down-steps along e. Along e there will be as many essential up-steps as there are boxes in the subtree below e whose target is in the tree above e. As π is a permutation, there are equally many boxes that are initially placed above e and have their target in the subtree below e. So as the overall number of steps match and the essential number of steps match, also the number of non-essential up-steps and down-steps must be equal along e. As this holds for any edge e, it also holds for the entire sorting walk.   □

Note that we did not say anything about where these non-essential up-steps are on S, just that there are as many as there are non-essential down-steps.

4.2   Reconstructing a Sorting Walk

We now give a tight construction of a sorting walk of the length of this lower bound.

Theorem 3 (Tight construction). *Given T, π and cycle anchor tree $\widetilde{T}$, we can find a sorting walk of length $d(\pi) + 2 \cdot c(\widetilde{T})$.*

Proof. We perform a depth-first search traversal of $\widetilde{T}$, starting at $\widetilde{r}$ and iteratively insert steps into an initially empty sorting walk S. At any point of the traversal, S is a closed sorting walk that sorts all the visited cycles of the anchor tree. For traversing a new edge of $\widetilde{T}$ from $\widetilde{v}_C$ to $\widetilde{v}_{C'}$, we do the following: Let $v \in T(C)$ and $w \in V(C')$ be the two vertices that have the minimum number of down-steps between them, as in the definition of the edge weights of $\widetilde{T}$. Let a denote the anchor vertex on the path from v to w. Furthermore, let $s = (a, b)$ be the first step of S that ends in a. Note that such a step has to exist, as a either lies in $T(C)$ or on the path from v to the root and all of these vertices already have been visited by S if S sorts C. We now build a sequence $S_{C'}$, which consists of three parts: We first take the box b from a to w, then sort C' starting at w and finally bring b back from w to a. $S_{C'}$ will contain exactly $c(\widetilde{v}_C, \widetilde{v}_{C'})$ down-steps in the first part, then $d(C')$ steps to sort C', and finally $c(\widetilde{v}_C, \widetilde{v}_{C'})$ up-steps. We insert $S_{C'}$ into S immediately after s, making sure that S now also sorts C' and is still a valid sorting walk. After the traversal of all cycles in the anchor tree, S will sort π and be of length $d(\pi) + 2 \cdot c(\widetilde{T})$.   □

Note that the sorting walk S constructed this way does not necessarily map back to $\widetilde{T}$, but its corresponding cycle anchor tree has the same weight as $\widetilde{T}$.

4.3   Finding a Cheapest Cycle Anchor Tree

Let S^* denote a shortest sorting walk for T and π. Using Theorem 3 to find S^* (or another equally long sorting walk), all we need is its corresponding cycle anchor tree $\widetilde{T}^*$. It suffices to find any cycle anchor tree with cost at most $c(\widetilde{T}^*)$. Especially, it suffices to find a cheapest cycle anchor tree $\widetilde{T}_{\min}$ among all possible cycle anchor trees. We then use Theorem 3 to get a sorting walk $S_{\min}$ from $\widetilde{T}_{\min}$. As $c(\widetilde{T}_{\min}) \leq c(\widetilde{T}^*)$ we get

$$|S_{\min}| = d(\pi) + 2 \cdot c(\widetilde{T}_{\min}) \leq d(\pi) + 2 \cdot c(\widetilde{T}^*) \leq |S^*|$$

and therefore $S_{\min}$ is a shortest sorting walk. To find this cheapest cycle anchor tree, we build the complete directed graph $\widetilde{G}$ of potential anchor tree edges. Note that the weights of these edges only depend on T and π but not on a sorting walk.

Optimum Branching. Given this complete weighted directed graph $\widetilde{G}$ we find its minimum directed spanning tree rooted at $\widetilde{r}$ using Edmond's algorithm for optimum branchings [3]. A great introduction to this algorithm, its correctness proof by Karp [7] and its efficient implementation by Dijkstra [9] can be found in the lecture notes of Zwick [11]. Combining these results with Theorem 3 will now allow us to find shortest sorting walks in polynomial time.

Theorem 4 (Efficient solution). *For any sorting problem on a tree T with permutation π, we can find a minimum sorting walk in time $\mathcal{O}(n^2)$.*

Proof. We first extract all the cycles in linear time. We then precompute the weights of all potential cycle anchor tree edges between any pair of cycles or the root. For this we run breadth-first search (BFS) $|\overline{\mathcal{C}}| + 1$ times, starting once with r and once with $T(C)$ for every $C \in \overline{\mathcal{C}}$ and count the number of down-steps along these BFS trees. We also precompute all the anchor points. As we run $\mathcal{O}(n)$ many BFS traversals, this precomputation takes time $\mathcal{O}(n^2)$.

As an efficient implementation of Edmond's algorithm allows us to find $\widetilde{T}_{\min}$ in time $\mathcal{O}(n^2)$, we can find $S_{\min}$ in time $\mathcal{O}(n^2)$ time overall.

In every step of the construction in Theorem 3, we can find step s in constant time, if we keep track of the first step of S visiting each vertex of T. We build S as a linked list of steps in time linear to its length. Thus, as on the path (Theorem 1), we can construct $S_{\min}$ in time $\Theta(n + |S|)$ from $\widetilde{T}_{\min}$.

Combining these three steps gives an algorithm that runs in time $\mathcal{O}(n^2)$. $\square$

5   Sorting on Other Graphs

Our algorithms for G being a path or a tree rely heavily on having unique paths between any pair of vertices. Therefore, these algorithms cannot be applied to graphs with cycles. In this section, we show that no efficient algorithm for general graphs can be found unless $\mathcal{P}$ equals $\mathcal{NP}$.

Theorem 5 ($\mathcal{NP}$-completeness for planar graphs). *Finding a shortest sorting walk for a planar graph $G = (V, E)$ and permutation π is $\mathcal{NP}$-complete.*

Proof. We use a reduction from the problem of finding Hamiltonian circuits in grid graphs [6]. We replace each vertex of the grid by a pair of neighboring vertices with swapped boxes. A formal proof is omitted due to the page limitation.

6   Conclusion

In this paper, we studied a sorting problem on graphs with the simple cost model of counting the number of edges traveled. We presented an efficient algorithm that finds an optimum solution if the graph is a tree, and showed that the problem is hard on general graphs. All our results easily extend to weighted graphs where each edge has an individual travel time. It is open whether there are efficient algorithms for other special kinds of graphs or if there are good approximation algorithms for general graphs.

We provide an implementation of the algorithm for finding shortest sorting walks on paths and trees, as well as an interactive visualization on our website: http://dgraf.ch/treesort

Acknowledgments. I want to thank Kateřina Böhmová and Peter Widmayer for many interesting and helpful discussions as well as the anonymous reviewers for their comments. I acknowledge the support of SNF project 200021L_156620.

References

1. Büsing, C., Maue, J.: Robust algorithms for sorting railway cars. In: de Berg, M., Meyer, U. (eds.) ESA 2010, Part 1. LNCS, vol. 6346, pp. 350–361. Springer, Heidelberg (2010)
2. Carlo, H.J., Vis, I.F., Roodbergen, K.J.: Storage yard operations in container terminals: Literature overview, trends, and research directions. European Journal of Operational Research 235(2), 412–430 (2014)
3. Edmonds, J.: Optimum branchings. Journal of Research of the National Bureau of Standards B 71(4), 233–240 (1967)
4. Haddon, B.K.: Cycle-sort: a linear sorting method. The Computer Journal 33(4), 365–367 (1990)
5. Hearn, R.A.: The complexity of sliding block puzzles and plank puzzles. Tribute to a Mathemagician, pp. 173–183 (2005)
6. Itai, A., Papadimitriou, C.H., Szwarcfiter, J.L.: Hamilton paths in grid graphs. SIAM Journal on Computing 11(4), 676–686 (1982)
7. Karp, R.M.: A simple derivation of edmonds' algorithm for optimum branchings. Networks 1(3), 265–272 (1971)
8. Knuth, D.E.: The art of computer programming: sorting and searching, vol. 3. Pearson Education (1998)
9. Tarjan, R.E.: Finding optimum branchings. Networks 7(1), 25–35 (1977)

10. Yamanaka, K., Demaine, E.D., Ito, T., Kawahara, J., Kiyomi, M., Okamoto, Y., Saitoh, T., Suzuki, A., Uchizawa, K., Uno, T.: Swapping labeled tokens on graphs. In: Ferro, A., Luccio, F., Widmayer, P. (eds.) FUN 2014. LNCS, vol. 8496, pp. 364–375. Springer, Heidelberg (2014)

11. Zwick, U.: Directed minimum spanning trees (April 2013), http://www.cs.tau.ac.il/~zwick/grad-algo-13/directed-mst.pdf

Improved Analysis
of Complete-Linkage Clustering*

Anna Großwendt and Heiko Röglin

Department of Computer Science
University of Bonn, Germany
{grosswen,roeglin}@cs.uni-bonn.de

Abstract. Complete-linkage clustering is a very popular method for computing hierarchical clusterings in practice, which is not fully understood theoretically. Given a finite set $P \subseteq \mathbb{R}^d$ of points, the complete-linkage method starts with each point from P in a cluster of its own and then iteratively merges two clusters from the current clustering that have the smallest diameter when merged into a single cluster.

We study the problem of partitioning P into k clusters such that the largest diameter of the clusters is minimized and we prove that the complete-linkage method computes an $O(1)$-approximation for this problem for any metric that is induced by a norm, assuming that the dimension d is a constant. This improves the best previously known bound of $O(\log k)$ due to Ackermann et al. (Algorithmica, 2014). Our improved bound also carries over to the k-center and the discrete k-center problem.

1 Introduction

In a typical clustering problem, the goal is to partition a given set of objects into clusters such that similar objects belong to the same cluster while dissimilar objects belong to different clusters. Clustering is ubiquitous in computer science with applications ranging from biology to information retrieval and data compression. In applications where the number of clusters is not known a priori, *hierarchical clusterings* are of particular appeal. A hierarchical clustering of a set P of n objects is a sequence $\mathcal{C}_1, \mathcal{C}_2, \ldots, \mathcal{C}_n$, where $\mathcal{C}_i$ is a clustering of P into i non-empty clusters and $\mathcal{C}_{i+1}$ is a refinement of $\mathcal{C}_i$. Besides the advantage that the number of clusters does not have to be specified in advance, hierarchical clusterings are also appealing because they help to understand the hereditary properties of the data and they provide information at different levels of granularity.

In practice, *agglomerative methods* are very popular for computing hierarchical clusterings. An agglomerative clustering method starts with the clustering $\mathcal{C}_n$, in which every object belongs to its own cluster. Then it iteratively merges the two clusters from the current clustering $\mathcal{C}_{i+1}$ with the smallest distance to obtain the next clustering $\mathcal{C}_i$. Depending on how the distance between two clusters is defined, different agglomerative methods can be obtained. A common variant

---

* This research was supported by ERC Starting Grant 306465 (BeyondWorstCase).

N. Bansal and I. Finocchi (Eds.): ESA 2015, LNCS 9294, pp. 656–667, 2015.
DOI: 10.1007/978-3-662-48350-3_55

is the *complete-linkage method* in which the distance between two clusters A and B is defined as the diameter or the (discrete) radius of $A \cup B$, assuming some distance measure on the objects from P is given.

The complete-linkage method is very popular and successful in a wide variety of applications. To name just a few of many recent examples, Rieck et al. [7] have used it for automatic malware detection, Ghaemmaghami et al. [6] have used it to design a speaker attribution system, and Cole et al. [2] use it as part of the Ribosomal Database Project. Yet the complete-linkage method is not fully understood in theory and there is still a considerable gap between the known upper and lower bounds for its approximation guarantee.

1.1   Problem Definitions and Algorithms

Let $P \subseteq \mathbb{R}^d$ denote a finite set of points and let dist: $\mathbb{R}^d \times \mathbb{R}^d \to \mathbb{R}_{\geq 0}$ denote some metric on $\mathbb{R}^d$. A *k-clustering* $\mathcal{C}$ *of* P is a partition of P into k non-empty sets $C_1, \ldots, C_k$. We consider three different ways to measure the quality of the k-clustering $\mathcal{C}$, which lead to three different optimization problems.

- **diameter k-clustering problem**: Find a k-clustering $\mathcal{C}$ with minimum *diameter*. The diameter diam($\mathcal{C}$) of $\mathcal{C}$ is given by the maximal diameter $\max_i$ diam(C_i) of one of its clusters, where the diameter of a set $C \subseteq P$ is defined as diam(C) := $\max_{x,y \in C}$ dist(x, y).
- **k-center problem**: Find a k-clustering $\mathcal{C}$ with minimum *radius*. The radius rad($\mathcal{C}$) of $\mathcal{C}$ is given by the maximal radius $\max_i$ rad(C_i) of one of its clusters, where the radius of a set $C \subseteq P$ is defined as rad(C) := $\min_{y \in \mathbb{R}^d} \max_{x \in C}$ dist(x, y).
- **discrete k-center problem**: Find a k-clustering $\mathcal{C}$ with minimum *discrete radius*. The discrete radius drad($\mathcal{C}$) of $\mathcal{C}$ is given by the maximal discrete radius $\max_i$ drad(C_i) of one of its clusters, where the discrete radius of a set $C \subseteq P$ is defined as drad(C) := $\min_{y \in C} \max_{x \in C}$ dist(x, y).

The *complete-linkage method* **CL** starts with the $|P|$-clustering $\mathcal{C}_{|P|}$ in which every point from P is in its own cluster. Then, for $i = |P| - 1, |P| - 2, \ldots, 1$, it merges two clusters from $\mathcal{C}_{i+1}$ to obtain $\mathcal{C}_i$. Regardless of the choice of which clusters are merged, this yields a hierarchical clustering $\mathcal{C}_1, \ldots, \mathcal{C}_{|P|}$. Which clusters are merged in an iteration depends on the optimization problem we consider. For the diameter k-clustering problem, the complete-linkage method chooses two clusters A and B from $\mathcal{C}_{i+1}$ such that diam($A \cup B$) is minimized. Similarly, for the k-center problem and the discrete k-center problem it chooses two clusters A and B from $\mathcal{C}_{i+1}$ such that rad($A \cup B$) or drad($A \cup B$) is minimized, respectively. Hence, every objective function gives rise to a different variant of the complete-linkage method. When it is not clear from the context which variant is meant, we will use the notation **CL**^{drad}, **CL**^{rad}, and **CL**^{diam} to make the variant clear.

1.2   Related Work

Let $P \subseteq \mathbb{R}^d$ and a metric dist on P be given and let $\mathcal{O}_k^{\text{drad}}$, $\mathcal{O}_k^{\text{rad}}$, and $\mathcal{O}_k^{\text{diam}}$ be optimal k-clusterings of P for the discrete k-center problem, the k-center problem, and the diameter k-clustering problem, respectively. For each of these three problems, it is easy to find examples where no hierarchical clustering $\mathcal{C} = (\mathcal{C}_1, \ldots, \mathcal{C}_{|P|})$ exists such that $\mathcal{C}_k$ is an optimal k-clustering for every k. We say that a hierarchical clustering $\mathcal{C}$ is an α-*approximate hierarchical clustering* for the diameter k-clustering problem if $\operatorname{diam}(\mathcal{C}_k) \leq \alpha \cdot \operatorname{diam}(\mathcal{O}_k^{\mathrm{diam}})$ holds for every k. In general, we allow α to be a function of k and d. We define α-approximate hierarchical clusterings analogously for the (discrete) k-center problem.

Dasgupta and Long [3] gave an efficient algorithm that computes 8-approximate hierarchical clusterings for the diameter k-clustering and the k-center problem, thereby giving a constructive proof of the existence of such hierarchical clusterings. Their result holds true for arbitrary metrics on $\mathbb{R}^d$ and it can even be improved to an expected approximation factor of $2e \approx 5.44$ by a randomized algorithm. They also studied the performance of the complete-linkage method and presented an artificial metric on $\mathbb{R}^2$ for which its approximation factor is only $\Omega(\log k)$ for the diameter k-clustering and the k-center problem. Ackermann et al. [1] showed for the diameter k-clustering and the discrete k-center problem a lower bound of $\Omega(\sqrt[p]{\log k})$ for the ℓ_p-metric for every $p \in \mathbb{N}$, assuming $d = \Omega(k)$.

Ackermann et al. [1] also showed that the complete-linkage method yields an $O(\log k)$-approximation for any metric that is induced by a norm, assuming that d is a constant. Here the constant in the big O notation depends on the dimension d. For the discrete k-center problem the dependence on d is only linear and additive. For the k-center problem the dependence is multiplicative and exponential in d, while for the diameter k-clustering problem it is even multiplicative and doubly exponential in d. The analysis of Ackermann et al. proceeds in two phases. The first phase ends when $2k$ clusters are left and the second phase consists of the last k merge operations. In the first phase a factor depending only on d but not on k is incurred. To make this precise, let $\mathcal{C}_{2k}^{\mathrm{drad}}$, $\mathcal{C}_{2k}^{\mathrm{rad}}$, and $\mathcal{C}_{2k}^{\mathrm{diam}}$ denote the $2k$-clusterings computed by the corresponding variants of **CL**. Ackermann et al. prove that for each objective $X \in \{\mathrm{drad, rad, diam}\}$ there exists a function κ_X such that

$$X(\mathcal{C}_{2k}^X) \leq \kappa_X(d) \cdot X(\mathcal{O}_k^X). \tag{1}$$

The function κ_{drad} is linear in d, the function κ_{rad} is exponential in d, and the function κ_{diam} is doubly exponential in d. The factor $O(\log k)$ is only incurred in the last k merge operations. Let $\mathcal{C}_k^{\mathrm{drad}}$, $\mathcal{C}_k^{\mathrm{rad}}$, and $\mathcal{C}_k^{\mathrm{diam}}$ denote the k-clusterings computed by the corresponding variants of **CL**. Ackermann et al. show that for each objective $X \in \{\mathrm{drad, rad, diam}\}$, it holds

$$X(\mathcal{C}_k^X) \leq O(\log k) \cdot X(\mathcal{C}_{2k}^X),$$

where the constant in the big O notation depends again on the dimension d. Additionally, Ackermann et al. [1] studied the case $d = 1$ separately and proved that the complete-linkage method computes 3-approximate hierarchical clusterings for the diameter k-clustering problem and the k-center problem for $d = 1$.

The approximability of non-hierarchical clustering problems is well understood. Feder and Greene [5] proved that for the Euclidean metric the k-center problem and the diameter k-clustering problem cannot be approximated better than a factor of 1.822 and 1.969, respectively. For the ℓ_1 and the ℓ_∞-metric they prove a lower bound of 2 for the approximability of both problems. On the positive side, they also provide a 2-approximation algorithm for any ℓ_p-metric.

A naive implementation of the complete-linkage method has a running time of $O(|P|^3)$. Defays gave an implementation with running time $O(|P|^2)$ [4].

1.3   Our Results

Our main result is a proof that the complete-linkage method yields an $O(1)$-approximation for the (discrete) k-center problem and the diameter k-clustering problem for any metric on $\mathbb{R}^d$ that is induced by a norm, assuming that d is a constant. This does not contradict the lower bound of Ackermann et al. because this lower bound assumes that the dimension depends linearly on k. In light of our result, the dependence of this lower bound on k is somewhat misleading and it could also be expressed as $\Omega(\sqrt[d]{\log d})$.

In order to obtain our result, we improve the second phase of the analysis of Ackermann et al. [1] and we prove that for each objective $X \in \{\mathrm{drad}, \mathrm{rad}, \mathrm{diam}\}$,

$$X(\mathcal{C}_k^X) \leq O(1) \cdot X(\mathcal{C}_{2k}^X).$$

The constant in the big O notation depends neither on d nor on k. It is 37, 19, and 17 for the discrete k-center problem, the k-center problem, and the diameter k-clustering problem, respectively. Together with (1) this yields the desired bound for the approximation factor.

In our analysis we introduce the concept of *clustering intersection graphs*. Given an ℓ-clustering $C_1, \ldots, C_\ell$ computed by the complete-linkage method and an optimal k-clustering $O_1, \ldots, O_k$, the clustering intersection graph contains a node for each cluster C_j and a hyperedge for every optimal cluster O_i. The hyperedge corresponding to O_i contains all clusters C_j with $O_i \cap C_j \neq \emptyset$. We then observe that merge operations of the complete-linkage method correspond to the contraction of two nodes in the clustering intersection graph. We obtain our results by carefully exploiting the structural properties of clustering intersection graphs.

In Section 2 we introduce formally the concept of clustering intersection graphs and prove some elementary properties. In Section 3 we combine our analysis with the result of Ackermann et al. about the first phase to prove that the complete-linkage method yields an $O(1)$-approximation. Due to space constraints, proofs of statements marked by $(\star)$ are deferred to the full version of this paper.

2   Clustering Intersection Graphs

Our analysis is based on studying the clustering intersection graph induced by **CL** at certain points of time. Before we introduce the concept of clustering intersection graphs formally, we will define these points of time. Let $P \subseteq \mathbb{R}^d$

be arbitrary and let $\mathcal{O}_k$ denote some arbitrary optimal k-clustering of P (w.r.t. the chosen objective function diameter or (discrete) radius). By scaling our point set we may assume that the objective value of $\mathcal{O}_k$ equals 1. We define $t_{\leq x}$ to be the last step before some cluster of size larger than x (w.r.t. the chosen objective function) is obtained and denote the clustering of **CL** at time $t_{\leq x}$ by $\mathcal{A}_x$. The following lemma is crucial for our analysis.

Lemma 1 ($\star$). *Let $x > 0$. In $\mathcal{A}_x$ there do not exist two clusters a_1 and a_2 such that*

$$\text{diam}(a_1) + \text{dist}(a_1, a_2) + \text{diam}(a_2) \leq x, \ \text{for } \boldsymbol{CL}^{\text{diam}},$$

$$\text{rad}(a_1) + \text{dist}(a_1, a_2) + 2\,\text{rad}(a_2) \leq x, \ \text{for } \boldsymbol{CL}^{\text{rad}},$$

$$\text{drad}(a_1) + \text{dist}(a_1, a_2) + 2\,\text{drad}(a_2) \leq x, \ \text{for } \boldsymbol{CL}^{\text{drad}},$$

where $\text{dist}(a_1, a_2)$ is defined as the minimum distance between two points $p_1 \in a_1$ and $p_2 \in a_2$.

This implies that if we have at $t_{\leq x}$ two clusters $a_1, a_2 \in \mathcal{A}_x$ and some cluster $o \in \mathcal{O}_k$ with $a_1 \cap o \neq \emptyset$ and $a_2 \cap o \neq \emptyset$, then depending on the objective function at $t_{\leq 2x+1}$ or $t_{\leq 3x+1}$ either a_1 or a_2 or both were merged.

2.1   Definition and Fundamental Properties

The fact that we can guarantee for certain pairs of clusters that one of it is merged at a certain point of time motivates us to define a clustering intersection graph (which is in general a hypergraph) with the clusters from $\mathcal{A}_x$ as vertices, where two vertices are neighbored if and only if there exists a cluster $o \in \mathcal{O}_k$ with which both have a non-empty intersection.

Definition 2. *Let $\mathcal{O}_k$ be an optimal k-clustering of some finite point set $P \subseteq \mathbb{R}^d$. Let $\mathcal{A}_x$ be the clustering of P computed by $\boldsymbol{CL}$ at time $t_{\leq x}$. We define the clustering intersection graph (CI-graph) $G_x(\mathcal{A}_x, \mathcal{O}_k)$ at point of time $t_{\leq x}$ as a graph with vertex set $\mathcal{A}_x$. A set of vertices $N = \{v_1, \ldots, v_\ell\}$ forms a hyperedge if there exists some cluster $o \in \mathcal{O}_k$ such that for each cluster v_i we have that $v_i \cap o \neq \emptyset$ and furthermore there does not exist a cluster $v \notin N$ with $v \cap o \neq \emptyset$.*

In general, the CI-graph is a hypergraph with exactly k edges and $|\mathcal{A}_x|$ vertices. If a statement holds for arbitrary points of time or the point of time is clear from context we omit the index x and just write G. Note that for each cluster $a \in \mathcal{A}_x$ each point $p \in a$ in the cluster is contained in some optimal cluster o. Thus, the CI-graph does not contain isolated vertices where isolated means that the vertex has no incident edge. We call a vertex ℓ a *leaf* if ℓ is incident to exactly one edge e and moreover ℓ is not the only vertex incident to e. Moreover an edge e is called a *loop* if e is only incident to one vertex. We define the *degree* of a vertex v to be number of non-loop edges that contain v plus twice the number of loops that consist of v. The CI-graph has the crucial property that merging two clusters in $\mathcal{A}_x$ corresponds to contracting the corresponding vertices in the CI-graph.

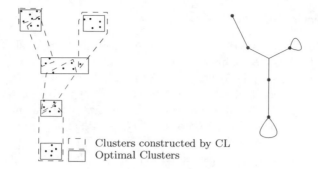

Clusters constructed by CL
Optimal Clusters

Fig. 1. Example of a clustering instance with an optimal clustering and a clustering computed by **CL** (left side) and the corresponding CI-graph (right side). Note that the figure is only schematic and does not depict the actual clustering computed by **CL** on the given instance

Lemma 3. *There is a homomorphism between pairs of clusterings* $(\mathcal{O}, \mathcal{A})$ *where* $\mathcal{O}$ *and* $\mathcal{A}$ *are both clusterings of a finite point set* $P \subseteq \mathbb{R}^d$ *and the set of CI-graphs with respect to the operations* merging two clusters in $\mathcal{A}$ *and* contracting two vertices in the corresponding CI-graph.

Assume that two clusters a_1 and a_2 are merged in a step of **CL**. Then all clusters $o \in \mathcal{O}$ that have a nonempty intersection with a_1 or a_2 clearly have a nonempty intersection with $a_1 \cup a_2$. Let G and G' denote the CI-graph before and after this merge operation, respectively. Then it is easy to see that G' is obtained from G by contracting the two vertices v_1 and v_2 corresponding to a_1 and a_2. The vertex that results from this contraction is incident to each edge that was incident to v_1 or v_2 before.

To prove that the approximation factor of **CL** is at most x, it is sufficient to show that at time $t_{\leq x}$ the CI-graph G_x contains at least as many edges as vertices. Clearly this is equivalent to $|\mathcal{A}_x| \leq k$, which means that **CL** has terminated.

2.2   The One-Dimensional Case

One can prove that **CL** yields a constant approximation factor for all finite point sets $P \subseteq \mathbb{R}$, all metrics dist$: \mathbb{R} \times \mathbb{R} \to \mathbb{R}_{\geq 0}$ and all $k \in \mathbb{N}$ analyzing the structure of the CI-graph after certain time periods showing that at $t_{\leq 3}$ (or $t_{\leq 5}$) the number of vertices is smaller or equal to the number of edges. The result is known for the diameter k-clustering problem and the k-center problem [1]. Our result also holds for the discrete k-center problem. For a detailed proof see the full version of our paper.

Theorem 4 ($\star$). *For* $d = 1$ *and arbitrary* k,
  $\mathbf{CL}^{\mathrm{diam}}$ *computes a 3-approximation for the diameter* k-*clustering problem,*
  $\mathbf{CL}^{\mathrm{rad}}$ *computes a 5-approximation for the* k-*center problem,*
  $\mathbf{CL}^{\mathrm{drad}}$ *computes a 5-approximation for the discrete* k-*center problem.*

2.3   Completion of the CI-Graph

In the one-dimensional case one has the crucial property that all vertices of a CI-graph can be arranged in increasing order on a line such that only neighbored vertices on the line may be contracted. Additionally, it follows from Lemma 1 that at least one vertex of every neighbored pair must be contracted until a certain time step. This implies that each edge is incident to at most 3 vertices at $t_{\leq 1}$, which is essential in the proof of Theorem 4. This property is not true anymore in higher dimensions.

Given a CI-graph G, we construct a weighted multi-graph $\Gamma(G)$, which we call the *completion* of G. The graph $\Gamma(G)$ has the same vertex set as the CI-graph G. For every hyperedge $\{v_1, \ldots, v_\ell\}$ in G, we introduce a clique with edge weights 1 in $\Gamma(G)$. For each pair of vertices v and w from the same connected component that are not adjacent we add an edge (v, w) to $\Gamma(G)$. If p denotes the length of the shortest v-w-path in G then the weight of the edge (v, w) in $\Gamma(G)$ is set to $p + (p - 1)x$ for the objective function diam and $p + (p - 1)2x$ for the objective functions rad and drad. This construction ensures the following important property: the weight of every edge (v, w) in $\Gamma(G)$ is an upper bound for the distance of the corresponding clusters (remember that the distance of two clusters is defined as the smallest distance between any pair of points from these clusters).

Lemma 5 (⋆). *Assume that the shortest v-w-path in a CI-graph G has length p. Then the smallest distance between two points in v and w is at most $p + (p-1)x$ for the objective function* diam *and $p + (p-1)2x$ for the objective functions* rad *and* drad.

For the analysis of **CL** we choose a subgraph H of $\Gamma(G)$. Unfortunately, Lemma 3 cannot be applied to H since H is no CI-graph but we state a weaker version, which is still strong enough for our analysis.

Lemma 6 (⋆). *Let $G_x = G_x(\mathcal{A}_x, \mathcal{O}_k)$ be a CI-graph of a clustering $(\mathcal{A}_x, \mathcal{O}_k)$ at point of time $t_{\leq x}$. Let H_x be a subgraph of $\Gamma(G_x)$ with $V(H_x) = V(G_x)$. Now consider $G_{x'} = G_{x'}(\mathcal{A}_{x'}, \mathcal{O}_k)$ for some point of time $t_{\leq x'}$ with $x' > x$. Let $H_{x'}$ be the graph that arises from H_x by performing the same contractions that are made between G_x and $G_{x'}$. Then $V(G_{x'}) = V(H_{x'})$ and moreover the weight of any edge (v, w) in $H_{x'}$ is an upper bound for the distance of the clusters corresponding to v and w.*

2.4   Analysis of H at Different Time Steps

The analysis of **CL** proceeds as follows. Let G_x be the CI-graph for a fixed point of time $t_{\leq x}$. Assume that there exists a special subgraph H_x of $\Gamma(G_x)$ satisfying the properties

i) $V(H_x) = V(G_x)$,
ii) $|E(H_x)| \leq k$,

iii) and no vertex in H_x is isolated (i.e., every vertex in H_x has at least one incident edge).

We will prove that at a certain point of time t_c, depending on the maximum edge weight in H_x, we have that $|V(H_c)| \leq |E(H_c)|$. Because of property i) and Lemma 6 we conclude $V(H_c) = V(G_c)$. Together with property ii) we obtain $|V(G_c)| = |V(H_c)| \leq |E(H_c)| = |E(H_x)| \leq k$ and thus **CL** terminated. In the following we denote $H_{x'}$ by H if the point of time is clear from context or if a statement holds for all $H_{x'}$ with $x' \geq x$.

First note that H is a multi-graph. Multi-graphs have the crucial property that a connected component has at least as many edges as vertices if and only if a cycle exists (where a loop is considered as a special case of a cycle).

Definition 7. *We call a connected component of H active if the component is a tree. Otherwise we call it inactive.*

Observation 8. *If $H_{x'}$ has no active connected component, then* **CL** *has terminated at* $t_{\leq x'}$.

Leaves of H and their neighbors have a key role in the analysis of the algorithm. We will show that between certain time steps either a leaf or its unique neighbor is merged. Define d_n as an upper bound for the distance between the clusters corresponding to any pair of adjacent vertices v_1 and v_2 in H_x. Because of Lemma 6 we have that d_n is smaller or equal to the maximum edge weight in H at any point of time. We use that fact later when choosing the subgraph H_x. We analyze time steps $t_{\leq x+i(d_n+x)}$ for the diameter k-clustering problem and $t_{\leq x+i(d_n+2x)}$ for the k-center and discrete k-center problem according to Lemma 1 and denote them by t_i. In accordance to that, we define $x_i = x + i(d_n + x)$ for **CL**^{diam} and $x_i = x + i(d_n + 2x)$ for **CL**^{rad} and **CL**^{drad}, respectively.

Definition 9. *We call a vertex $p \in H$ in an active connected component of H a leaf-parent if p is the neighbor of some leaf and has at least degree 2.*

At the beginning of our analysis at $t_{\leq x}$ there does not necessarily exist a leaf-parent in each active component. This follows because the smallest possible active component consists of two connected vertices and is the only possibility of an active component without a leaf-parent (remember that in H there exist no isolated vertices by property iii); any connected component that consists of a single vertex must contain a loop and is hence inactive). Analogous to dimension one we show that at point of time t_1 for each active connected component by **CL** either one vertex was merged with a vertex from another component but thereby some vertex with degree 2 is built or two vertices from one component were merged. The latter means that a cycle was built and the component is no longer active. The following lemma ensures that at a certain point of time there exists a leaf-parent in each active component.

Lemma 10 $(\star)$. *Each active component C of H containing a vertex v of degree 2 contains at least one leaf-parent p. In particular H_{x_1} contains at least one leaf-parent in each active component.*

The proof of Lemma 10 gives a hint that we have in general at least two leaf-parents in each component while components with exactly one leaf-parent are of a special form. We will use this structure later on to prove that if each active component contains at least 2 leaf-parents the algorithm terminates. Therefore we need some statement counting the number of remaining contractions depending on the number of leaf-parents. First, we need some statement how often contraction steps are performed in each component.

Lemma 11 ($\star$). *Let ℓ be some leaf in H_{x_i} at an arbitrary point of time t_i with $i \geq 0$. Then the leaf ℓ is also contained in H_{x_0} and it is not contracted between t_0 and t_i. Moreover between two steps of time t_i and t_{i+1} where $i \in \mathbb{N}_0$ we have that for each leaf ℓ either the leaf ℓ or its corresponding leaf-parent p_ℓ is contracted.*

We denote the number of leaf-parents of H at time t_i for a connected component C by $n_{\ell p}(C)$. Since in each active component the number of leaf-parents is at most the number of leaves, we may conclude that the algorithm performs at least $n_{\ell p}/2$ contractions between t_i and t_{i+1} where $n_{\ell p} = \sum_{i=1}^{r} n_{\ell p}(C_i)$ is the sum over the number of leaf-parents in the active connected components. Now we count the number of leaf-parents contained in one active connected component. The idea is that if each active component contains at least two leaf-parents then we have at least as many contractions as active components and can conclude that the algorithm will terminate. Therefore we show that at a certain point of time every active component must contain at least two leaf-parents. First we will show that if the number of leaf-parents in an active component is at least two, then after contraction the number of leaf-parents does not decrease below two.

Lemma 12 ($\star$). *Assume that two vertices v_1 and v_2 from two different components C_1 and C_2 that contain each at least one leaf-parent are contracted in H. If the resulting component $C = C_1 \cup C_2$ is active then C has at least as many leaf-parents as the maximum of C_1 and C_2, i.e., $n_{\ell p}(C) \geq \max\{n_{\ell p}(C_1), n_{\ell p}(C_2)\}$.*

We may conclude that the only possibility to obtain an active component containing just one leaf-parent is that we contract vertices from two different components which contain only one leaf-parent. In particular for two such components C_1 and C_2 we have to contract the leaf-parents p_1 and p_2. If another vertex and therefore a leaf of C_1 is contracted another component $C_1 \cup C_2$ with at least two leaf-parents is built.

Lemma 13 ($\star$). *For $\mathbf{CL}^{\mathrm{diam}}$ each active component contains at least 2 leaf-parents at point of time t_3. For $\mathbf{CL}^{\mathrm{rad}}$ each active component contains at least 2 leaf-parents at t_2. For $\mathbf{CL}^{\mathrm{drad}}$ each active component contains at least 2 leaf-parents at t_6.*

It remains to prove that $\mathbf{CL}$ terminates if each component contains at least two leaf-parents.

Lemma 14 ($\star$). *If at t_i each active component of H_{x_i} contains at least two leaf-parents then $\mathbf{CL}$ has terminated at t_{i+1}.*

3   Approximation Factor of CL in the Case $d \geq 2$

In this section we combine our analysis with the result of Ackermann et al. [1] for the first phase of **CL** (i.e., the steps until $2k$ clusters are left) in order to prove the main theorem. From the analysis of Ackermann et al. it follows that there is a function κ such that for $x = \kappa(d)$ the CI-graph G_x contains at most $2k$ vertices. We will analyze the last k steps of **CL** more carefully. We consider the completion $\Gamma(G_x)$ of G_x and assume that it is connected. This is not necessarily the case but we will see later that this assumption is without loss of generality because our analysis can be applied to each connected component separately. In fact, the result of Ackermann et al. implies that for each connected component of G_x the number of vertices is at most twice the number of edges.

3.1   CI-Graphs with at most $2k$ Vertices

Let H_x be a subgraph of $\Gamma(G_x)$ with k edges and at most $2k$ vertices such that H_x fulfills the properties i)-iii). The goal is to find such a subgraph H_x whose maximum edge weight is small. Note that properties i), ii), and iii) imply $|V(G_x)| = |V(H_x)| \leq 2|E(H_x)| \leq 2k = 2|E(G_x)|$, which means $|V(G_x)| \leq 2|E(G_x)|$ is a necessary property of G_x to find a subgraph H_x.

We will prove that we can always find a subgraph H_x of G_x that satisfies properties i)-iii) and has the following additional property iv): for each edge $e' = (v, w) \in E(H_x)$ the vertices v and w have distance at most 2 in G_x, i.e., either there is an edge $e \in E(G_x)$ with $\{v, w\} \subseteq e$ or there are two edges $e_v \in E(G_x)$ and $e_w \in E(G_x)$ with $v \in e_v$, $w \in e_w$, and $v \cap w \neq \emptyset$.

Using this we will prove that **CL** terminates at time $t_{\leq O(x)}$ if for each connected component C of the CI-graph G_x we have that $|V(C)| \leq 2|E(C)|$.

In order to find a subgraph H_x of $\Gamma(G_x)$ that satisfies properties i)-iv) we let T be a spanning tree of $\Gamma(G_x)$ that uses only edges of weight 1. Such a spanning tree is guaranteed to exist because we assumed G_x to be connected. Such a spanning tree satisfies all properties except for ii) because the number of edges in T is $|V(G_x)| - 1$ and $|V(G_x)|$ can only be bounded by $2k$.

However, any perfect matching in the spanning tree T is a subgraph H that satisfies the properties i)-iv). If T does not contain a perfect matching, we show how to find a perfect 2-matching (according to the following definition).

Definition 15. *An α-matching in a graph G is a matching M in the complete graph $K_{|V(G)|}$ with $|V(G)|$ vertices such that for each matching edge $(v, w) \in M$ the distance of v and w in G is at most α. Moreover we call an α-matching perfect if M is a perfect matching in $K_{|V(G)|}$.*

Lemma 16 ($\star$). *Each tree T with an even number $|V(T)|$ of vertices has a perfect 2-matching.*

Construction of H_x. We construct a graph H_x that satisfies the properties i), ii), iii), and iv) as follows. First we compute an arbitrary spanning tree T of $\Gamma(G_x)$ that uses only edges of weight 1. If $|V(G_x)| = |V(H_x)|$ is even, then the graph H_x is chosen as a perfect 2-matching of T. Then the properties i), iii), and iv) are satisfied by construction and property ii) is satisfied because of $|E(H_x)| = |V(H_x)|/2 \leq k$. If $|V(G_x)|$ is odd, we choose some leaf v from the spanning tree T. Then we find a perfect 2-matching M in $T \setminus \{v\}$. Since $|V(G_x)| \leq 2|E(G_x)|$ we have that the matching contains at most $|E(G_x)| - 1$ edges. Thus we set H_x to M and may add the edge from T that is incident to v to H_x such that property iii) becomes true.

Now we have a graph H_x fulfilling properties i), ii), iii), and iv). Property iv) and Lemma 6 imply that $d_n \leq 2 + x$ for the objective function diam and $d_n \leq 2 + 2x$ for the objective functions rad and drad. We conclude with the following theorem.

Theorem 17 ($\star$). *Assume that the CI-graph G_x is connected and contains k edges and at most $2k$ vertices at some point of time* $t_{\leq x}$. *Then $\boldsymbol{CL}^{\mathrm{diam}}$ computes a $9x + 8$ approximation for the diameter k-clustering problem. Moreover $\boldsymbol{CL}^{\mathrm{rad}}$ computes a $13x+6$ approximation for the k-center problem and $\boldsymbol{CL}^{\mathrm{drad}}$ computes a $25x + 12$ approximation for the discrete k-center problem.*

3.2   Approximation Factor of CL

Now for each version of the algorithm $\boldsymbol{CL}^{\mathrm{diam}}$, $\boldsymbol{CL}^{\mathrm{rad}}$, and $\boldsymbol{CL}^{\mathrm{drad}}$ we combine our analysis with the special result of [1] corresponding to each of the methods. We state the following lemma from [1] deriving an upper bound for a point of time x where $|V(G_x)| \leq 2k$.

Lemma 18 ([1]). *Let $P \subseteq \mathbb{R}^d$ be finite. Then, for all $k \in \mathbb{N}$ with $2k \leq |P|$, the partition $\mathcal{A}$ of P into $2k$ clusters computed by $\boldsymbol{CL}^{\mathrm{drad}}$ satisfies*

$$\max_{a \in \mathcal{A}} \mathrm{drad}(a) < 20d \cdot \mathrm{drad}(\mathcal{O}_k^{\mathrm{drad}}).$$

Combining this result with Theorem 17 yields the following theorem.

Theorem 19 ($\star$). *For $d \in \mathbb{N}$ and a finite point set $P \subseteq \mathbb{R}^d$ the algorithm $\boldsymbol{CL}^{\mathrm{drad}}$ computes an $O(d)$-approximation for the discrete k-center problem.*

Lemma 20 ([1]). *Let $P \subseteq \mathbb{R}^d$ be finite. Then, for all $k \in \mathbb{N}$ with $2k \leq |P|$, the partition $\mathcal{A}$ of P into $2k$ clusters computed by $\boldsymbol{CL}^{\mathrm{rad}}$ satisfies*

$$\max_{a \in \mathcal{A}} \mathrm{rad}(a) < 24d \cdot e^{24d} \cdot \mathrm{rad}(\mathcal{O}_k^{\mathrm{rad}}).$$

Combining this result with Theorem 17 yields the following theorem.

Theorem 21. *For $d \in \mathbb{N}$ and a finite point set $P \subseteq \mathbb{R}^d$ the algorithm $\boldsymbol{CL}^{\mathrm{rad}}$ computes an $e^{O(d)}$-approximation for the k-center problem.*

Lemma 22 ([1]). *Let $P \subseteq \mathbb{R}^d$ be finite. Then, for all $k \in \mathbb{N}$ with $2k \leq |P|$, the partition $\mathcal{A}$ of P into $2k$ clusters computed by $\boldsymbol{CL}^{\mathrm{diam}}$ satisfies*

$$\max_{a \in \mathcal{A}} \mathrm{diam}(a) < 2^{3(42d)^d}(28d + 6) \cdot \mathrm{diam}(\mathcal{O}_k^{\mathrm{diam}}).$$

Analogously to $\mathbf{CL}^{\mathrm{drad}}$ and $\mathbf{CL}^{\mathrm{rad}}$ we can conclude the following theorem.

Theorem 23. *For $d \in \mathbb{N}$ and a finite point set $P \subseteq \mathbb{R}^d$ the algorithm $\boldsymbol{CL}^{\mathrm{diam}}$ computes a $2^{O(d)^d}$-approximation for the diameter k-clustering problem.*

4   Conclusions

We have shown that the popular complete-linkage method computes $O(1)$-approximate hierarchical clusterings for the diameter k-clustering problem and the (discrete) k-center problem, assuming that d is a constant. For this it was sufficient to improve the second phase of the analysis by Ackermann et al. [1] (i.e., the last k merge operations). We used their results about the first phase to obtain our results. It is a very interesting question if the dependence on the dimension can be improved in the first phase. If we express the known lower bound of Ackermann et al. [1] in terms of d then it becomes $\Omega(\sqrt[d]{\log d})$. Hence, in terms of d, there is still a huge gap between the known upper and lower bounds. Another interesting question is whether the upper bound of $O(\log k)$ holds also for metrics that are not induced by norms.

References

1. Ackermann, M.R., Blömer, J., Kuntze, D., Sohler, C.: Analysis of agglomerative clustering. Algorithmica 69(1), 184–215 (2014)
2. Cole, J.R., Wang, Q., Fish, J.A., Chai, B., McGarrell, D.M., Sun, Y., Brown, C.T., Porras-Alfaro, A., Kuske, C.R., Tiedje, J.M.: Ribosomal database project: data and tools for high throughput rrna analysis. Nucleic Acids Research (2013)
3. Dasgupta, S., Long, P.M.: Performance guarantees for hierarchical clustering. Journal of Computer and System Sciences 70(4), 555–569 (2005)
4. Defays, D.: An efficient algorithm for a complete link method. The Computer Journal 20(4), 364–366 (1977)
5. Feder, T., Greene, D.H.: Optimal algorithms for approximate clustering. In: Proc. of the 20th Annual ACM Symposium on Theory of Computing (STOC), pp. 434–444 (1988)
6. Ghaemmaghami, H., Dean, D., Vogt, R., Sridharan, S.: Speaker attribution of multiple telephone conversations using a complete-linkage clustering approach. In: Proc. of the 2012 IEEE International Conference on Acoustics, Speech and Signal Processing (ICASSP), pp. 4185–4188 (2012)
7. Rieck, K., Trinius, P., Willems, C., Holz, T.: Automatic analysis of malware behavior using machine learning. Journal of Computer Security 19(4), 639–668 (2011)

Structural Parameterizations
of the Mixed Chinese Postman Problem

Gregory Gutin, Mark Jones, and Magnus Wahlström

Royal Holloway, University of London Egham,
Surrey TW20 0EX, UK

Abstract. In the Mixed Chinese Postman Problem (MCPP), given a weighted mixed graph G (G may have both edges and arcs), our aim is to find a minimum weight closed walk traversing each edge and arc at least once. The MCPP parameterized by the number of edges in G or the number of arcs in G is fixed-parameter tractable as proved by van Bevern *et al.* (2014) and Gutin, Jones and Sheng (2014), respectively. In this paper, we consider the unweighted version of MCPP. Solving an open question of van Bevern *et al.* (2014), we show that somewhat unexpectedly MCPP parameterized by the (undirected) treewidth of G is W[1]-hard. In fact, we prove that even the MCPP parameterized by the pathwidth of G is W[1]-hard. On the positive side, we show that MCPP parameterized by treedepth is fixed-parameter tractable. We are unaware of any natural graph parameters between pathwidth and treedepth and so our results provide a dichotomy of the complexity of MCPP.

1 Introduction

A *mixed graph* is a graph that may contain both edges and arcs (i.e., directed edges). A mixed graph G is *strongly connected* if for each ordered pair x, y of vertices in G there is a path from x to y that traverses each arc in its direction. In this paper, we will deal with simple mixed graphs (where for every pair of vertices u, v, at most one of the edge uv, the arc uv and the arc vu exist)[1] and (possibly non-simple) directed multigraphs (with multiple arcs between each pair of vertices). Whenever we refer to the treewidth (pathwidth, treedepth) of a graph, we mean the treewidth (pathwidth, treedepth) of the underlying undirected graph.

In this paper, we study the following well-known problem.

MIXED CHINESE POSTMAN PROBLEM (MCPP)
Instance: A strongly connected mixed graph $G = (V, E \cup A)$, with vertex set V, set E of edges and set A of arcs; a weight function $w : E \cup A \to \mathbb{N}_0$.
Output: A closed walk H of G that traverses each edge and arc at least once, of minimum weight.

---

[1] We can relax our assumption that we deal only with simple mixed graph in the treedepth part of the paper as subdividing parallel arcs/edges increases treedepth by at most one.

© Springer-Verlag Berlin Heidelberg 2015
N. Bansal and I. Finocchi (Eds.): ESA 2015, LNCS 9294, pp. 668–679, 2015.
DOI: 10.1007/978-3-662-48350-3_56

In what follows, we will consider a solution H of MCPP as both a walk in G and a supergraph of G.

There is numerous literature on various algorithms and heuristics for MCPP; for informative surveys, see [2,4,9,18]. When $A = \emptyset$, we call the problem the UNDIRECTED CHINESE POSTMAN PROBLEM (UCPP), and when $E = \emptyset$, we call the problem the DIRECTED CHINESE POSTMAN PROBLEM (DCPP). It is well-known that UCPP is polynomial-time solvable [8] and so is DCPP [1,5,8], but MCPP is NP-complete, even when G is planar with each vertex having total degree 3 and all edges and arcs having weight 1 [17]. It is therefore reasonable to believe that MCPP may become easier the closer it gets to UCPP or DCPP. Indeed, when parameterized by the number of edges in G or the number of arcs in G, MCPP is proved to be fixed-parameter tractable (FPT, defined below) by van Bevern et al. [2] and Gutin, Jones and Sheng [15], respectively.

In this paper, we consider structural parameterizations of MCPP. Van Bevern et al. [2] noted that Fernandes, Lee and Wakabayashi [12] proved that MCPP parameterized by the treewidth of G is in XP (when all edges and arcs have weight 1), and asked whether this parameterization of MCPP is FPT. It is well-known that many graph problems are FPT when parameterized by the treewidth of the input graph (only a few such problems are W[1]-hard; see, e.g., [6,10,14]). In this paper, we show that somewhat unexpectedly the MCPP parameterized by treewidth belongs to a small minority of problems, i.e., it is W[1]-hard. In fact, we prove a stronger result by (i) replacing treewidth with pathwidth, and (ii) assuming that all edges and arcs have weight 1.

To complement this, we obtain a positive result for the parameter treedepth. We prove that if there exists an improvement step for MCPP, then there is also an improvement step where the number of changes is bounded by a function of the treedepth. Thus, to search for a feasible improvement step one can apply Courcelle's theorem on the treedepth decomposition of the graph, because the whole shape of the improvement can be encoded in the formula. Note that the bound on treedepth is used here in two different manners: to prove a structural result about the space of the solutions, and to run the final dynamic programming algorithm.

Following [12], we assume that all weights equal 1, however, we do not foresee any significant difficulty in generalizing our result to the weighted case.

Our paper is organized as follows. In the rest of this section, we provide some basic definitions on parameterized complexity as well as the definitions of treewidth, pathwidth and treedepth. In Section 2 we introduce an intermediate problem PROPERLY BALANCED SUBGRAPH (PBS), and give a W[1]-hardness proof for a restricted variant of it. In Section 3 we reduce this variant of PBS to MCPP parameterized by pathwidth, showing that the latter is also W[1]-hard. In Section 4 we show that PBS is FPT with respect to treedepth, as outlined above, and in Section 5 we reduce MCPP parameterized by treedepth to PBS parameterized by treedepth, showing that this parameterization of MCPP is FPT. We conclude the paper with Section 6, where, in particular, we mention an open question from [2] on another parameterization of MCPP. We assume familiarity with the basic concepts of parameterized algorithms and complexity. For information on these topics, see [7]. For reasons of space, many proofs and figures are omitted, but can be found in the arXiv version [16].

Treewidth, Pathwidth and Treedepth. We use the standard notions of treewidth and pathwidth of an undirected graph [13]. For a directed multigraph H, we will use $pw(H)$ to denote the pathwidth of the underlying undirected graph of H.

The *treedepth* of a connected graph G is defined as follows. Let T be a rooted tree with vertex set $V(G)$, such that if xy is an edge in G then x is either an ancestor or a descendant of y in T. Then we say that G *is embedded in* T. The *depth* of T is the number of vertices in a longest path in T from the root to a leaf. The *treedepth* of G is the minimum t such that G is embedded in a tree of depth t. Thus, for example, a star $K_{1,r}$ has treedepth 2. A path of length n has treedepth $O(\log n)$. A graph of treedepth k has pathwidth at most $k - 1$.

2   Properly Balanced Subgraph Problem

In this section, we introduce the problem PROPERLY BALANCED SUBGRAPH (PBS), and show that it is W[1]-hard parameterized by pathwidth. In Section 4, we will show that a special case of the problem with restricted weights is fixed-parameter tractable with respect to treedepth.

A directed multigraph is called *balanced* if the in-degree of each vertex coincides with its out-degree. A *double arc* is a specified pair of arcs (a, a') such that a and a' have the same heads and tails. We will say that a subgraph D' of D *respects double arcs* if $|A(D') \cap \{a, a'\}| \neq 1$ for every double arc (a, a'). A *forbidden pair* is a specified pair of arcs (b, b') such that b is the reverse of b'. We say that D' *respects forbidden pairs* if $|A(D') \cap \{b, b'\}| \neq 2$ for every forbidden pair (b, b'). We will say that a subgraph D' of D is *properly balanced* if D' is balanced and respects double arcs and forbidden pairs. PBS is then defined as follows.

PROPERLY BALANCED SUBGRAPH (PBS)
Instance: A directed multigraph $D = (V, A)$; a weight function $w : A \to \mathbb{Z}$; a set $X = \{(a_1, a_1'), \ldots, (a_r, a_r')\}$ of *double arcs* with $a_i, a_i' \in A$ for each $i \in [r]$; a set $Y = \{(b_1, b_1'), \ldots, (b_s, b_s')\}$ of *forbidden pairs* with $b_i, b_i' \in A$ for each $i \in [s]$. Each arc occurs at most once in $\bigcup(X \cup Y)$.
Output: A properly balanced subgraph D' of D of negative weight, if one exists.

Throughout the paper, when we talk about a graph in the context of PBS, we implicitly mean a directed multigraph together with a weight function, a set of double arcs and a set of forbidden pairs, as described above.

2.1   Gadgets for PBS

We now describe some simple gadget graphs (for now we do not assign weights; we will do this later). Each gadget will have some number of *input* and *output* arcs. Later, we will combine these gadgets by joining the input and output arcs of different gadgets together using double arcs. Henceforth, for each positive integer n, $[n] = \{1, 2, \ldots, n\}$.

A *Duplication gadget* has one input arc and t output arcs, for some positive integer t. The vertex set consists of vertices x, y, and u_i, v_i for each $i \in [t]$. The arcs form a cycle $xyu_1v_1 \ldots u_tv_tx$. The input arc is the arc xy, and the output arcs are the arcs u_iv_i for each $i \in [t]$.

A *Choice gadget* has one input arc xy and t output arcs $u_iv_i : i \in [t]$, for some positive integer t. The vertex set consists of the vertices x, y, z, w and u_i, v_i for each $i \in [t]$. The arcs consist of a path $wxyz$, and the path zu_iv_iw for each $i \in [t]$.

Finally, a *Checksum gadget* has t_l left input arcs $x_iy_i : i \in [t_l]$ for some positive integer t_l, and t_r right input arcs $u_iv_i : i \in [t_r]$, and no output arcs. The vertex set consists of the vertices w, z together with x_i, y_i for each $i \in [t_l]$ and u_i, v_i for each $i \in [t_r]$. The arc set consists of the path wx_iy_iz for each $i \in [t_l]$, and zu_iv_iw for each $i \in [t_r]$.

Proposition 1 below is easy to prove and thus its proof is omitted.

Proposition 1. *Let D be one of the gadgets described above, let X be a subset of input arcs in D, and let Y be a subset of output arcs in D. Then there exists a properly balanced subgraph D' of D containing all the arcs from X and Y (and no other input or output arcs) if and only if one of the following cases holds: (1) $X = \emptyset$ and $Y = \emptyset$; (2) D is a Duplication gadget, $|X| = 1$ and Y contains all the output arcs of D; (3) D is a Choice gadget, $|X| = 1$ and $|Y| = 1$; or (4) D is a Checksum gadget, and X contains an equal number of left input arcs and right input arcs.*

Observe that in all of our gadgets, the vertices in input or output arcs all have in-degree and out-degree 1. We next describe how to combine these gadgets. For two unjoined arcs uv and xy (possibly in disjoint graphs), the operation of *joining* uv and xy is as follows: Identify u and x, and identify v and y. Keep both uv and xy, and add (uv, xy) as a double arc.

Lemma 1. *Let D_1 and D_2 be disjoint directed multigraphs. Let $u_1v_1, \ldots u_tv_t$ be arcs in D_1, and let $x_1y_1, \ldots x_ty_t$ be arcs in D_2, such that u_i and v_i both have in-degree and out-degree 1 in D_1, and x_i and y_i both have in-degree and out-degree 1 in D_2, for each $i \in [t]$. Let D be the graph formed by joining u_iv_i and x_iy_i, for each $i \in [t]$. Then a subgraph D' of D is a properly balanced graph if and only if (1) $|A(D') \cap \{u_iv_i, x_iy_i\}| \neq 1$ for each $i \in [t]$; and (2) D' restricted to D_1 is a properly balanced subgraph of D_1, and D' restricted to D_2 is a properly balanced subgraph of D_2.*

2.2 W[1]-Hardness of PBS

We now use the gadget behavior, as described in Prop. 1, to construct a W[1]-hardness proof for PBS. By joining an output arc of one gadget to the input arc of another gadget, we have that a solution will only pass through the second gadget if it uses the corresponding arc of the first gadget. Thus for example, if a Duplication gadget has k output arcs, each of which is joined to the input arc of a Choice gadget, then any solution that uses the input arc of the Duplication gadget has to use exactly one output arc from each of the Choice gadgets. By combining gadgets in this way, we can create "circuits" that represent instances of other problems.

We will use this idea to represent the following W[1]-hard problem. In k-MULTICOLORED CLIQUE, we are given a graph $G = (V_1 \cup V_2 \cdots \cup V_k, E)$, such that for each $i \in [k]$, V_i forms an independent set, and asked to decide whether G contains a clique with k vertices, where k is the parameter.

Theorem 1 ([11]). k-MULTICOLORED CLIQUE *is W[1]-hard.*

Theorem 2. PBS *is W[1]-hard parameterized by pathwidth, even when there are no forbidden arcs, there is a single arc a^* of weight -1 and a^* is not part of a double arc, and all other arcs have weight 0.*

We give a sketch of the proof (for a full proof, see the arXiv version [16]):

Let $G = (V_1 \cup \cdots \cup V_k, E)$ be an instance of k-MULTICOLORED CLIQUE. We construct an equivalent instance of PBS as follows.

Initially we have a duplication gadget with k output arcs, whose input arc is the only arc of weight -1. All other arcs will have weight 0. Thus, any solution to the PBS instance will have to use this duplication gadget and all its output arcs. Then for each $i \in [k]$, we choose a vertex $v_i \in V_i$ (represented by a Choice gadget with $|V_i|$ output arcs, whose input arc is joined to the initial Duplication gadget). For each choice of v_i, and for each $j \in [k] \setminus \{i\}$ (enforced by a Duplication gadget with $k-1$ output arcs), we then choose an edge $e_{i \to j}$ that is adjacent to v_i and a vertex in V_j (represented by a choice gadget with $|N(v_i) \cap V_j|$ output arcs).

The graph so far looks like a "tree" of gadgets, and as such has bounded treewidth. It is easy to show that it also has bounded pathwidth (since each gadget has bounded pathwidth and the tree of gadgets has bounded depth). It enforces that we choose a set of vertices $v_1, \ldots, v_k$, and then an edge $e_{i \to j}$ for each ordered pair $(i, j), i \neq j$. (Each possible choice for $e_{i \to j}$ is represented by one output arc on the last layer of Choice gadgets.) Now observe that $v_1, \ldots, v_k$ forms a clique if and only if there are choices of $e_{i \to j}$ such that $e_{i \to j} = e_{j \to i}$ for each (i, j).

We can check for this condition as follows. Firstly, we associate each edge e with a unique number n_e. Then, for each output arc corresponding to the edge e, we join that arc to a Duplication gadget with n_e output arcs. (This increases the pathwidth of the graph by a constant.) Then for each unordered pair $\{i, j\}, i < j$, we create a Checksum gadget CHECKEDGE(i, j). The left input arcs of this gadget are joined to all the output arcs of all Duplication gadgets corresponding to a choice for $e_{i \to j}$, and the right input arcs are joined to all the output arcs of all Duplication gadgets corresponding to a choice for $e_{j \to i}$. This completes the construction of the graph.

It follows that for any solution to the PBS instance, the number of left input arcs of CHECKEDGE(i, j) in the solution is equal to the number associated with the edge chosen for $e_{i \to j}$. Similarly the number of right input arcs in the solution is equal to the number associated with the edge chosen for $e_{j \to}$. As these numbers have to be equal, it follows that there is a solution if and only if the choice for $e_{i \to j}$ is the same as the choice for $e_{j \to i}$ for each $i \neq j$. Thus, our PBS instance has a solution of negative weight if and only if G has a clique. It remains to check the pathwidth of the graph.

Before the addition of the Checksum gadgets, the graph has pathwidth bounded by a constant. As the input arcs of these gadgets are joined to other arcs, adding the Checksum gadgets only requires adding 2 vertices for each (i, j). Thus, the pathwidth of the constructed graph is $O(k^2)$.

3   Reducing PBS to MCPP

We now show how to reduce an instance of PBS, of the structure given in Theorem 2, to MCPP. Let $(D = (V, A), w, X = \{(a_i, a'_i) : i \in [t]\}, Y = \emptyset)$ be an instance of PBS with double arcs X and no forbidden pairs, and where $w(a^*) = -1$ for a single arc a^* and $w(a) = 0$ for every other arc. We may assume that a^* is not in a double arc. We will produce an instance G of MCPP and an integer W, such that G has a solution of weight W, and G has a solution of weight less than W if and only if our instance of PBS has a solution with negative weight. All edges and arcs in our MCPP instance will have weight 1.

We derive G by replacing every double arc and individual arc of D by an appropriate gadget. The gadgets will be such that within each gadget, there are only two behaviors of MCPP solutions of reasonable weight: a solution can be balanced within the gadget (corresponding to not using the original arc/double arc in a solution to D), or a solution can be imbalanced at the vertices by the same amount that the original arc / double arc is (which corresponds to using the original arc / double arc in a solution to D). Thus, every solution for G of reasonable weight corresponds to a properly balanced subgraph of D, and vice versa.

For each gadget, except for the gadget corresponding to the negative weight arc, the weights of the two solutions will be the same. In the case of the negative weight arc, the solution that corresponds to using the arc will be cheaper by 1. Thus, there are two possible weights for a solution to G, and the cheaper weight is only possible if D has a properly balanced subgraph of negative weight.

In what follows, we will construct arcs and edges of two weights, *standard* and *heavy*. Standard arcs and edges have weight 1; heavy arcs and edges have weight M, where M is a large enough (polynomially bounded) integer that we may assume that no solution traverses a heavy arc or edge more than once. This will be useful to impose structure on the possible solutions when constructing gadgets. A heavy arc (edge) is equivalent to a directed (undirected) path of length M, and so we also show W[1]-hardness for the unweighted case.

Given a directed multigraph H (corresponding to part of a solution to an MCPP instance) and a vertex v, the *imbalance* of v is $d_H^+(v) - d_H^-(v)$. The gadgets are constructed as follows. It is straightforward to verify that each gadget has only two solutions that traverse each heavy arc or edge exactly once, and that the imbalances and weights of these solutions are as described above. (For a full proof, see the arXiv version [16].)

For an arc uv of weight 0 that is not part of a double arc: Construct GADGET(u, v) by creating a new vertex z_{uv}, with standard arcs $z_{uv}u$ and $z_{uv}v$, two heavy arcs uz_{uv}, and a heavy arc vz_{uv}.

For an arc uv of weight -1 that is not part of a double arc: Construct GADGET(u, v) by adding two new vertices w_{uv} and z_{uv}, with standard arcs $z_{uv}u, z_{uv}w_{uv}$ and vw_{uv}, two heavy arcs uz_{uv}, one heavy arc $w_{uv}z_{uv}$, and two heavy arcs $w_{uv}v$.

For a double arc from u to v: GADGET(u, v) consists of a heavy arc uv and a heavy edge $\{u, v\}$. Assuming a solution traverses each heavy arc/edge exactly once, the only thing to decide is in which direction to traverse the undirected edge.

We note that each of our gadgets has pathwidth bounded by a constant. It can be shown that replacing the arcs of D with gadgets in this way will only increase the pathwidth by a constant.

We now have that, given an instance (D, w, X, Y) of PBS of the type specified in Theorem 2, we can in polynomial time create an equivalent instance G of MCPP with pathwidth bounded by $O(\text{pw}(D))$. We therefore have a parameterized reduction from this restriction of PBS, parameterized by pathwidth, to MCPP parameterized by pathwidth. As this restriction of PBS is W[1]-hard by Theorem 2, we have the following theorem.

Theorem 3. MCPP *is W[1]-hard parameterized by pathwidth.*

4   PBS Parameterized by Treedepth

In this section we show that a certain restriction of PBS is fixed-parameter tractable with respect to treedepth. The restriction we require is that all arcs in double arcs have weight 0, all arcs in forbidden pairs have weight -1, and all other arcs have weight 1 or -1. We choose this restriction, as this is the version of PBS that we get when we reduce from MCPP.

Aside from some standard dynamic programming techniques, our main technical tool is Lemma 3, which shows that we may assume there exists a solution with size bounded by a function of treedepth. The following simple observation will be useful in the proof of Lemma 3.

Lemma 2. *Let $\{H_i : i \in \mathcal{I}\}$ be a family of pairwise arc-disjoint subgraphs of G, such that each H_i respects double arcs. Then $H = \bigcup_{i \in \mathcal{I}} H_i$ is a properly balanced subgraph of G if and only if H is balanced and H respects forbidden pairs.*

We are now ready to prove that any properly balanced subgraph decomposes into properly balanced subgraphs of size bounded by a function of treedepth. This will allow us to assume, when constructing the algorithm (Theorem 4), that a solution has bounded size.

Lemma 3. *Let G be a directed multigraph (with double arcs and forbidden pairs) of treedepth k, and let H be a properly balanced subgraph of G. Then H is a union of pairwise arc-disjoint graphs H_i, each of which is a properly balanced subgraph of G, with $|A(H_i)| \leq f(k)$ where $f(k) = 2^{2^k}$.*

Proof. We prove the claim by induction on the treedepth k. For the base case, observe that if $k = 1$ then G has no arcs, and the claim is trivially true. So now assume that $k \geq 2$, and that the claim holds for all graphs of treedepth less than k. We also assume that the underlying undirected graph of H is 2-connected, as otherwise a block decomposition of H is a decomposition into properly balanced subgraphs, and we may apply our result to each block of H. Similarly, if the underlying undirected graph of G is not 2-connected but H is, then H lies inside one block of G, and we may restrict our attention to this block. Hence assume that the underlying undirected graph of G is 2-connected as well.

Let G be embedded in a tree T of depth k, and let x be the root of T. Observe that x has only one child in T, as otherwise x is a cut-vertex in G. Let y be this child, and let G' be the multigraph derived from G by identifying x and y and removing loops. Similarly, let H' be the subgraph of G' derived from H by identifying x and y and removing loops. Observe that H' is balanced as H is balanced and so the number of arcs into $\{x, y\}$ equals the number of arcs out of it. Let B be the set of arcs in H between x and y, and observe that there is a one-to-one correspondence between the arcs of H' and the arcs of H not in B. By identifying x and y in T, we get that G' has treedepth at most $k - 1$.

By the induction hypothesis, H' can be partitioned into a family $\{H'_i : i \in \mathcal{I}'\}$ of pairwise arc-disjoint properly balanced subgraphs of G', each having at most $f(k-1)$ arcs. For each $i \in \mathcal{I}'$, let F_i be the subgraph of G corresponding to H'_i. Observe that B can also be partitioned into a family $\{F_i : i \in \mathcal{I}''\}$ of subgraphs with at most 2 arcs, that respect double arcs (we add any double arc from B as a subgraph F_i, and add every other arc as a single-arc subgraph).

Letting $\mathcal{J} = \mathcal{I}' \cup \mathcal{I}''$, we have that $\{F_i : i \in \mathcal{J}\}$ is a partition of H, each F_i has at most $f(k-1)$ arcs, and each F_i respects double arcs and is balanced everywhere except possibly at x and y. We now combine sets of these subgraphs into subgraphs that are balanced everywhere.

For each $i \in \mathcal{J}$, let t_i be the imbalance of F_i at x, i.e. $t_i = d_{F_i}^+(x) - d_{F_i}^-(x)$. Observe that $|t_i| \leq \frac{f(k-1)}{2}$ for each i and, as H is balanced, $\sum_{i \in \mathcal{J}} t_i = 0$. Suppose that there exists a subset $\mathcal{J}' \subseteq \mathcal{J}$ such that $|\mathcal{J}'| \leq f(k-1) - 1$ and $\sum_{i \in \mathcal{J}} t_i = 0$. Then let $H_1 = \bigcup_{i \in \mathcal{J}'} F_i$. By construction, H_1 is balanced at every vertex (as it is balanced for every vertex other than y, and a directed multigraph cannot be imbalanced at a single vertex), and H_1 respects double arcs. As H_1 is a subgraph of H, H_1 also respects forbidden pairs. Therefore H_1 is a properly balanced subgraph, with number of arcs at most $(f(k-1) - 1)f(k-1)$. Observe that $f(k) = 2^{2^k}$ is a solution to the recursion $(f(k-1) - 1)f(k-1) < f(k)$ with $f(1) = 4$. Thus, H_1 has at most 2^{2^k} arcs, as required. By applying a similar argument to $H \setminus H_1$, we get a properly balanced subgraph H_2 with at most $f(k)$ arcs. Repeating this process, we get a partition of H into properly balanced subgraphs each with at most $f(k)$ arcs.

We now show that $\mathcal{J}'$ exists. Let $\mathcal{J}_1$ be a set containing a single t_i, of minimum absolute value, and iteratively construct sets $\mathcal{J}_r$ by adding i such that $t_i < 0$ to $\mathcal{J}_{r-1}$ if $\sum_{p \in \mathcal{J}_{r-1}} t_p > 0$, and adding i such that $t_i > 0$ otherwise. Now note that either $t_i = \pm f(k-1)/2$ for every $i \in \mathcal{J}$, in which case we have a subset $\mathcal{J}'$ with $|\mathcal{J}'| = 2$, or $|\sum_{i \in \mathcal{J}_r} t_i| < \frac{f(k-1)}{2}$ for each r, and therefore there are at most $f(k-1) - 1$ possible values that $\sum_{i \in \mathcal{J}_r} t_i$ can take. Then there exist r, r' such that $r' < r$, $r - r' \le f(k-1) - 1$, and $\sum_{i \in \mathcal{J}_r \setminus \mathcal{J}_{r'}} t_i = 0$. So let $\mathcal{J}' = \mathcal{J}_r \setminus \mathcal{J}_{r'}$. $\qquad\square$

Using Lemma 3, we may now assume that if G has a properly balanced subgraph with negative weight, then it has a properly balanced subgraph of negative weight with at most $f(k)$ arcs (as any negative weight properly balanced subgraph can be partitioned into properly balanced subgraphs of at most $f(k)$ arcs, at least one of which must have negative weight).

4.1   Fixed-Parameter Tractability of PBS

As the treedepth of G is at most k, it follows that it has pathwidth at most $k-1$ [3]. Using this fact, and the fact that we may assume that a solution has at most $f(k)$ arcs, we have the following:

Theorem 4. PBS *is FPT with respect to treedepth, provided that all arcs in double arcs have weight 0, all arcs in forbidden pairs have weight -1, and all other arcs have weight 1 or -1.*

We can prove Theorem 4 using standard dynamic programming techniques. For each node x in a path (or tree) decomposition of D, we may construct a set of partial solutions on the set of vertices covered by x and its descendants. Solutions are indexed by their restriction to $\beta(x)$ and the imbalance they impose on each vertex in $\beta(x)$, where $\beta(x)$ is the set of vertices covered by x. Where multiple solutions share the same index, we store a solution with minimum weight. As the property of being balanced can be decided by checking the imbalance of each vertex, and the other properties of a properly balanced subgraph can be checked by examining the restriction of the subgraph to each pair of adjacent vertices, this indexing gives us enough information to find a minimum weight solution.

The full details of the proof are standard to anyone familiar with dynamic programming techniques, but tedious to write out and verify. Therefore, in the full version of our paper [16] we give a proof using Courcelle's theorem.

5   Positive Result: Reducing MCPP to PBS

In this section, we consider MCPP with all weights equal 1 parameterized by treedepth. In contrast to pathwidth, we will show that MCPP parameterized by treedepth is FPT. Hereinafter, $b_H(v)$ denotes the imbalance of v, i.e. $d_H^+(v) - d_H^-(v)$. In the problem COMP-MCPP, we are given an instance of MCPP together with a solution H, and asked to find a solution H' of weight less than $w(H)$, if one exists. To solve an instance of MCPP, it would be enough to find some (not necessarily optimal) solution of weight M, then repeatedly apply COMP-MCPP to find better solutions, until we find a solution which cannot be improved by COMP-MCPP and is therefore optimal. As COMP-MCPP returns an improved solution if one is available, we would have to apply COMP-MCPP at most M times.

To show that our approach leads to an FPT algorithm for MCPP, we first show that we may assume that M is bounded by an appropriate value.

Lemma 4. *Given an instance (G, w) of MCPP with m arcs and edges, we can, in polynomial time, find a closed walk of of G that traverses each edge and arc at least once, if such a walk exists, and this walk traverses each arc at most $m + 1$ times.*

As with the hardness proof, we will use PBS as an intermediate problem. We now reduce COMP-MCPP to PBS, in the following sense: For any input graph G and initial solution H, we produce a directed multigraph D (with double arcs and forbidden pairs), such that D has a properly balanced subgraph of negative weight if and only if G has a solution of weight less than $w(H)$.

For any adjacent vertices u, v in G, let G_{uv} be the subgraph of G induced by $\{u, v\}$, and similarly let H_{uv} be the subgraph of H induced by $\{u, v\}$. Let $M = w(H)$. Thus, we may assume that any improved solution has weight less than M. By Lemma 4 and the assumption that the weight of every arc and edge is 1, we may assume $M \leq m^2 + m$.

For each edge and arc uv in G, we will produce a gadget D_{uv}, based on G_{uv} and H_{uv} and the value M. The gadget D_{uv} is a directed multigraph, possibly containing double arcs or forbidden pairs, and by combining all the gadgets, we will get an instance D of PBS.

We now construct D_{uv} according to the following cases (roughly speaking, a positive weight arc represents adding an arc in that direction, and a negative weight arc represents removing an arc in the opposite direction):

If G_{uv} is an arc from u to v and H_{uv} traverses uv $t \leq M$ times: Then D_{uv} has $t - 1$ arcs from v to u of weight -1, and $M - t$ arcs from u to v of weight 1.

If G_{uv} is an edge between u and v, and H_{uv} traverses uv from u to v $t \leq M$ times, and from v to u 0 times: Then D_{uv} has a double arc (a, a'), where a and a' are both arcs from v to u of weight 0. In addition, D_{uv} has $t - 1$ arcs from v to u of weight -1, $M - t$ arcs from u to v of weight 1, and $M - 1$ arcs from v to u of weight 1.

If G_{uv} is an edge between u and v and H_{uv} traverses uv from u to v $t > 0$ times, and from v to u $s > 0$ times: Then we may assume $s = t = 1$, as otherwise we may remove a pair of arcs (uv, vu) from H and get a better solution to MCPP. Then D_{uv} has $M - 1$ arcs from u to v of weight 1, $M - 1$ arcs from v to u of weight 1, and a forbidden pair (a, a'), where a is an arc from u to v, a' is an arc from v to u, and both a and a' have weight -1.

Lemma 5. *Let uv be an edge or arc in G, and let B and W be arbitrary integers such that $w(H_{uv}) + W \leq M$. Then the following are equivalent.*

1. *There exists a graph H'_{uv} with vertex set $\{u, v\}$ that covers G_{uv}, such that $w(H'_{uv}) = w(H_{uv}) + W$ and $b_{H'_{uv}}(u) = b_{H_{uv}}(u) + B$;*
2. *D_{uv} has a subgraph D'_{uv} which respects double arcs and forbidden pairs, such that $w(D'_{uv}) = W$ and $b_{D'_{uv}}(u) = B$.*

Note that in a graph H'' with two vertices u and v, $b_{H''}(u) = -b_{H''}(v)$. Thus, in addition to implying that $b_{D'_{uv}}(u) = b_{H'_{uv}}(u) - b_{H_{uv}}(u)$, the claim also implies that $b_{D'_{uv}}(v) = b_{H'_{uv}}(v) - b_{H_{uv}}(v)$.

Lemma 6. *Let D be the directed multigraph derived from G and H by taking the vertex set $V(G)$ and adding the gadget D_{uv} for every arc and edge uv in G. Then there exists a solution H' with weight less than H if and only if D has a properly balanced subgraph of weight less than 0.*

Lemma 6 implies that we have a parameterized reduction from COMP-MCPP parameterized by treedepth to PBS parameterized by treedepth. Then by Theorem 4, we have the following theorem.

Theorem 5. MCPP *with all weights equal 1 is FPT with respect to treedepth.*

6   Discussion

We proved that MCPP parameterized by pathwidth is W[1]-hard, even if all edges and arcs of the input graph G have weight 1. This solves the second open question of van Bevern et al. [2] on parameterizations of MCPP; the first being the parameterization by the number of arcs in G, which was settled in [15].

We also showed that the unweighted version of MCPP is FPT with respect to treedepth. This is the first problem we are aware of that has been shown to be W[1]-hard with respect to treewidth but FPT with respect to treedepth. Note that the pathwidth of a graph lies between its treewidth and treedepth. Open problems include pinning down the tractability border for the problem more precisely, or to find parameterizations that allow for more practical FPT algorithms. Some candidate parameters are *distance to linear forest*, which is weaker than pathwidth, and the *feedback vertex set* number.

Another parameterization of MCPP in [2] is as follows. Call a vertex v of G *even* if the total number of arcs and edges incident to v is even. Motivated by the fact that if each vertex of G is even, then MCPP is polynomial-time solvable [8], van Bevern et al. [2] ask whether MCPP parameterized by the number of non-even vertices is FPT. Here, even membership in XP is open.

Acknowledgment. Research of GG was partially supported by Royal Society Wolfson Research Merit Award.

References

1. Beltrami, E.J., Bodin, L.D.: Networks and vehicle routing for municipal waste collection. Networks 4(1), 65–94 (1974)
2. van Bevern, R., Niedermeier, R., Sorge, M., Weller, M.: Complexity of Arc Routing Problems. In: Arc Routing: Problems, Methods and Applications. SIAM (2014)
3. Bodlaender, H.L., Gilbert, J.R., Hafsteinsson, H., Kloks, T.: Approximating treewidth, pathwidth, frontsize, and shortest elimination tree. Journal of Algorithms 18(2), 238–255 (1995)
4. Brucker, P.: The Chinese postman problem for mixed graphs. In: Noltemeier, H. (ed.) WG 1980. LNCS, vol. 100, pp. 354–366. Springer, Heidelberg (1981)
5. Christofides, N.: The optimum traversal of a graph. Omega 1(6), 719–732 (1973)
6. Dom, M., Lokshtanov, D., Saurabh, S., Villanger, Y.: Capacitated domination and covering: A parameterized perspective. In: Grohe, M., Niedermeier, R. (eds.) IWPEC 2008. LNCS, vol. 5018, pp. 78–91. Springer, Heidelberg (2008)
7. Downey, R.G., Fellows, M.R.: Fundamentals of Parameterized Complexity. Springer (2013)
8. Edmonds, J., Johnson, E.L.: Matching, Euler tours and the Chinese postman. Mathematical Programming 5(1), 88–124 (1973)
9. Eiselt, H.A., Gendreau, M., Laporte, G.: Arc routing problems. I. The Chinese postman problem. Operations Research 43, 231–242 (1995)
10. Fellows, M., Fomin, F., Lokshtanov, D., Rosamond, F., Saurabh, S., Szeider, S., Thomassen, C.: On the complexity of some colorful problems parameterized by treewidth. Inf. Comput. 209(2), 143–153 (2011)
11. Fellows, M.R., Hermelin, D., Rosamond, F., Vialette, S.: On the parameterized complexity of multiple-interval graph problems. Theor. Comput. Sci. 410, 53–61 (2009)
12. Fernandes, C.G., Lee, O., Wakabayashi, Y.: Minimum cycle cover and Chinese postman problems on mixed graphs with bounded tree-width. Discrete Appl. Math. 157(2), 272–279 (2009)
13. Flum, J., Grohe, M.: Parameterized Complexity Theory. Springer (2006)
14. Golovach, P.A., Kratochvil, J., Suchy, O.: Parameterized complexity of generalized domination problems. Discrete Appl. Math. 160(6), 780–792 (2012)
15. Gutin, G., Jones, M., Sheng, B.: Parameterized Complexity of the k-Arc Chinese Postman Problem. In: Schulz, A.S., Wagner, D. (eds.) ESA 2014. LNCS, vol. 8737, pp. 530–541. Springer, Heidelberg (2014)
16. Gutin, G., Jones, M., Wahlström, M.: Structural Parameterizations of the Mixed Chinese Postman Problem. http://arxiv.org/abs/1410.5191
17. Papadimitriou, C.H.: On the complexity of edge traversing. J. ACM 23, 544–554 (1976)
18. Peng, Y.: Approximation algorithms for some postman problems over mixed graphs. Chinese J. Oper. Res. 8, 76–80 (1989)

Online Appointment Scheduling
in the Random Order Model

Oliver Göbel[1], Thomas Kesselheim[2,*], and Andreas Tönnis[1,**]

[1] Department of Computer Science, RWTH Aachen University, Germany
{goebel,toennis}@cs.rwth-aachen.de
[2] Max-Planck-Institut für Informatik and Saarland University,
Saarbrücken, Germany
thomas.kesselheim@mpi-inf.mpg.de

Abstract. We consider the following online appointment scheduling
problem: Jobs of different processing times and weights arrive online
step-by-step. Upon arrival of a job, its (future) starting date has to be
determined immediately and irrevocably before the next job arrives, with
the objective of minimizing the average weighted completion time. In this
type of scheduling problem it is impossible to achieve non-trivial competitive ratios in the classical, adversarial arrival model, even if jobs have
unit processing times. We weaken the adversary and consider random
order of arrival instead. In this model the adversary defines the weight-
processing time pairs for all jobs, but the order in which the jobs arrive
online is a permutation drawn uniformly at random.

For the case of jobs with unit processing time we give a constant-
competitive algorithm. We use this algorithm as a building block for the
general case of variable job processing times and achieve competitive
ratio $O(\log n)$. We complement these algorithms with a lower bound of
$\Omega(n)$ for unit-processing time jobs in the adversarial input model.

1 Introduction

In scheduling problems there is a number of jobs given and the scheduler decides
how, when and where each job is processed. The resulting schedule is evaluated under some objective function, typically minimizing a cost function on the
completion time of the jobs. Often such schedules need to be computed online,
meaning the scheduler has to make decisions without knowing the complete input
for the problem instance. In a standard online variant, once a job is completed,
the scheduler selects which job to process next from the jobs that have arrived
in the meantime. Even worst-case competitive analysis admits surprisingly good
results in this model (see Section 1.1 for an overview).

In stricter settings, worst-case analysis is overly pessimistic and therefore unable to produce meaningful results. Consider the following *appointment scheduling problem*: Jobs arrive online one after the other and directly upon arrival each

---

* Supported in part by the DFG through Cluster of Excellence MMCI.
** Supported by the DFG GRK/1298 "AlgoSyn".

© Springer-Verlag Berlin Heidelberg 2015
N. Bansal and I. Finocchi (Eds.): ESA 2015, LNCS 9294, pp. 680–692, 2015.
DOI: 10.1007/978-3-662-48350-3_57

job has to be assigned a starting and completion time in the future so as to optimize some objective function such as the weighted sum of completion times. As
a motivation, just imagine a doctor's receptionist taking phone calls and making
appointments for the next day. On the one hand he does not want to re-schedule
appointments, on the other hand the most urgent cases should be treated first.
In worst-case analysis, it is impossible to achieve any non-trivial competitive
ratios for this kind of problem as we illustrate by giving a lower bound later on.

An interesting way to bypass these impossibility results is to incorporate a
small stochastic component into the input model. A prime example of this phenomenon from a different domain is the *secretary problem*: A sequence of entities
with different scores show up one after the other. After an entity has arrived,
one has to make an irrevocable decision whether to keep this entity and to stop
the sequence or to discard the entity and continue. Assuming a worst-case input,
it is impossible to achieve a non-trivial competitive ratio. The situation is different if the adversary determines the scores but the arrival order of the entities
is drawn uniformly from all possible permutations. Under these circumstances,
it is possible to pick the highest-scored entity with probability $\frac{1}{e}$. Generalizing
this problem, in many online maximization settings, it is possible to achieve a
constant competitive ratio assuming a random input order whereas worst-case
analyses would be pointless.

These positive results hold for maximization problems, whereas scheduling
problems are typically cost-minimization problems. This can mean a big difference, particularly in a probabilistic setting: In a maximization problem it is
possible to achieve reasonable (expected) competitive ratios, even if the algorithm returns no solution at all with probability $1/2$. Many algorithms indeed
exploit this fact by using a constant fraction of the input only for "statistical"
purposes and dropping it afterwards. In a minimization problem, this is generally impossible as one is usually required to satisfy certain cover contraints such
that dropping input elements is not readily possible. We show that the random-
order assumption makes a significant difference in online minimization problems
nevertheless.

Formal Problem Statement. We assume there are $n \in \mathbb{N}$ jobs and a single processing unit. Each job i has a specific weight $w_i \in \mathbb{R}^+$ as well as a processing time $l_i \in \mathbb{N}$. All jobs are to be processed sequentially without preemption.
Thus a feasible solution is a vector of starting times s such that in no time-step
two jobs are processed simultaneously, i.e. $s_j \notin [s_i, s_i + l_i)$ for all $i, j \in [n]$,
$i \neq j$. The objective is to minimize the weighted sum of completion times, i.e.,
$\sum_{j=1}^{n} w_j C_j = \sum_{j=1}^{n} w_j(s_j + l_j - 1)$. In the scheduling literature, this optimization
problem is also referred to as $1 || \sum_{j=1}^{n} w_j C_j$ in the online list model.

Jobs arrive sequentially and scheduling decisions have to be made immediately
and irrevocably. Upon its arrival, a job's processing time l_i and its weight w_i are
revealed and the algorithm has to assign a position in the schedule to the job[1]. In

---

[1] Note that the *time slots* in this schedule are unrelated to the *rounds* in which the
jobs arrive. Throughout the proofs, we will refer to the schedule slots by s and t,
whereas the rounds will be referred to as r.

a fully adversarial input, even with identical job processing times, no randomized algorithm can be better than $\Omega(n)$-competitive (for a proof see Section 4), which is trivially achieved by any schedule without idle slots. Therefore, we consider the random-order model. Here an adversary constructs the instance and determines the processing times and weights of the jobs, but the arrival order is drawn at random. Technically, a permutation $\pi \in S_n$ is drawn uniformly at random and then the jobs are presented according to this permutation. The algorithm gets to know the processing time of the sequence n before the first round.

We evaluate our online algorithms in terms of the widespread *competitive ratio*. It is defined by $\frac{\mathbf{E}[\text{ALG}]}{\text{OPT}}$, where ALG and OPT denote the cost of the online and the optimal algorithm, respectively. The optimal algorithm, however, works offline and is assumed to know the whole instance in advance. In our algorithms, the expectation of their cost is only with respect to the random input.

Our Contribution. Our main contribution is an algorithm for the case of identical processing times. That is, similarly to the secretary problem, n entities of different weights are revealed online. We have to assign each entity to a slot 1, 2, ... so as to minimize the weighted sum of slot numbers. Our algorithm for this problem is 34-competitive. Specifically, the competitive factor holds for every single job and not only in expectation over all jobs. This means there is no job that suffers a bad position for the sake of the overall solution as each job is guaranteed to lose at most the competitive factor in expectation.

Upon arrival of a job, the algorithm computes the optimal schedule of all jobs seen so far. This solution includes a slot number for the currently considered job. We use this number as a guide to find the permanent slot for the online job. To this end, we scale the locally optimal solution with a factor depending on the fraction of the overall input that we have seen up to this point. This factor decreases as we learn more about the problem instance at hand. Finally we schedule the job to the first free slot after this tentative slot.

As a next step we generalize the setting toward jobs with different processing times. We present an $O(\log n)$-competitive algorithm. It divides the jobs into classes of almost equal processing times. For each processing time class it runs the algorithm for jobs with identical processing time as a subroutine. We devise a labeling scheme that associates slots to processing time classes and guarantees that every instance of the subroutine loses no more than a factor $2 \log n$ on top of its inherent competitive ratio.

We complement this algorithm with the simple lower bound that all online algorithms are $\Omega(n)$-competitive in the worst-case input order.

1.1   Related Work

In offline scheduling minimizing the weighted completion time is well understood. On a single machine the problem can be solved easily with Smith's ratio rule. For more complex versions with identical machines, related machines and release dates there are PTAS known [1,7]. A common online variant of the problem is as follows. The jobs are unknown to the algorithm until their respective release dates. At any point in time, the algorithm decides which of the released jobs to process next. This setting has been studied intensively when only a single machine or identical parallel machines are available as well as when preemption is allowed or not allowed, respectively; see [19,8,27,28,15] for detailed results. More recently a stochastic variant has been considered. Here the scheduler does not learn the true processing time of a job upon its arrival, but he learns a probability distribution over the processing time of the job instead; see [20,21] for the latest results. Stochastic processing times have also been considered in [5] regarding offline appointment scheduling. All these online results are in the adversarial input model, where an oblivious adversary creates the worst-case input sequence online.

Note that this setting (both the deterministic and the stochastic variant) is significantly different from the one studied in this paper. The crucial difference is that in the traditional model scheduling is an ad-hoc decision. It only affects (at most) the time until completing the currently processed job. In contrast, in our problem every job needs to be assigned its starting date irrevocably immediately upon its arrival. To the best of our knowledge, there is only a single previous result in this "online list scheduling model". Fiat and Woeginger [13] show that with worst-case input there is no $O(\log(n))$-competitive randomized online list scheduling algorithm, even on a single machine and with unit weights. We show that in the random-order model we can get constant- or $O(\log n)$-competitive even if jobs have weights.

The random-order model has been studied mostly in the context of packing problems. Motivated by the classical secretary problem, Babaioff et al. [4] introduced the matroid secretary problem and conjectured that it is $O(1)$-competitive. Toward this end several $O(\log \log(\rho))$-competitive algorithms have been proposed recently [18,12]. Here ρ is the rank of the considered matroid. Further variants of the secretary problem are given in [3], among them also one that aims at minimizing the sum of the accepted ranks. Another branch of research focuses on linear constraints. This includes (generalizations of) matching [22,9] and general packing LPs [11,2,26,16]. It is a common assumption in these problems that capacities are large compared to the consumption in a single round. In this case, there are even $1 - \epsilon$-competitive algorithms.

Apart from scheduling, a few other min-sum online optimization problems have been studied. For *facility location* there is a deterministic lower bound of $\Omega(\log(n))$, while Meyerson [23] shows a constant competitive factor in the random order model. For *network design* [25] and the *parking permit problem* [24] only adversarial input has been studied. They admit $O(k)$-competitive algorithms where k is the number of options available. A general framework for linear online covering problems with adversarial input has been presented by Buchbinder and Naor [6]. However, naturally in weighted settings, the competitive ratio is limited by strong impossibility results, such as $\Omega(n)$ in our case.

A number of alternative online input models that combine adversarial and stochastic components have been studied. Devanur et al. [10] use the i.i.d. model and introduce generalization, the adversarial stochastic model. Kleinberg and Weinberg [17] consider the prophet inequality model, in which weights are stochastic but the arrival order is adversarial. In [14], a unifying graph sampling model is introduced that contains the random-order and prophet inequality model as special cases.

2   Jobs with Uniform Processing Time

We start with investigating the online appointment scheduling where all jobs' processing times are uniform, i.e. we consider them to be normalized to 1. The optimal solution is an ordering of the jobs, decreasing in their weight. In this setting the challenge is to order the online incoming jobs according to their weight when they arrive. Given a sufficiently large fraction of the unknown input, a job's relative position in this fraction is a quite good representative for its position in the complete input. This is why Algorithm 1 uses a local solution to guide the online computation. Generally, it works as follows: Upon the arrival of an element i the optimal ordering $\tilde{s}^{(r)}$ on set J containing all jobs that have arrived so far is computed. Afterwards, this local solution is scaled by a factor f_r in order to create sufficiently large gaps between the jobs. This steady distribution is essential for later insertions of jobs as they tend to be ranked between those ones scheduled up to now. The incoming job i is assigned a so-called *tentative slot* $hats_i$ in this scaled solution. As tentative slots are not unique, we solve eventual conflicts by assigning i to the first free slot s_i after its tentative position. At the end of the analysis it will become clear how to choose the parameters c and d.

Algorithm 1. Algorithm for Uniform Jobs

Let J be the set of jobs arrived so far, initially $J = \emptyset$

for each round r with incoming job i **do**

$\quad J := J \cup \{i\}$;

$\quad \tilde{s}^{(r)} :=$ optimal solution in round r on set J;

$\quad \hat{s}_i := \left\lceil f_r \cdot \tilde{s}_i^{(r)} \right\rceil$, where $f_r = c \left(\frac{n}{r} \right)^{1/d}$;

$\quad s_i :=$ earliest free slot after $\hat{s}_i$;

Theorem 1. *The algorithm for jobs with uniform processing time schedules every single job 34-competitive in expectation.*

The proof of this theorem will be split into three parts. First, in Section 2.1, we will bound the expected tentative slot number that is assigned to a job. Next, in Section 2.2, we bound the amount by which a job is shifted due to collisions, i.e., by how much the final slot differs from the tentative slot. Finally, in Section 2.3, we combine these insights to prove the claim.

Note that the job indices are irrelevant for the algorithm. Therefore, in the analysis, we assume that these indices are assigned such that $w_1 > w_2 > \ldots > w_n$. By this assumption, the optimal offline solution is simply $1, 2, \ldots, n$.

2.1   Bound on Tentative Slot Numbers

As a first step, for a fixed job i, we bound its tentative slot number. Note that, by our assumption $w_1 > w_2 > \ldots > w_n$ its slot number in the offline optimum would be i.

Lemma 2. *The expected tentative slot of job i is* $\mathbf{E}\left[\hat{s}_i\right] \leq \frac{cd}{d-1} \cdot i + 1 + O\left(n^{1/d-1}\right)$.

Proof. The algorithm sets $\hat{s}_i = \left\lceil f_{\pi(i)} \tilde{s}_i^{(\pi(i))} \right\rceil \leq f_{\pi(i)} \tilde{s}_i^{(\pi(i))} + 1$. So

$$\mathbf{E}\left[\hat{s}_i - 1\right] \leq \mathbf{E}\left[f_{\pi(i)} \tilde{s}_i^{(\pi(i))}\right] = cn^{1/d} \sum_{r=1}^{n} \mathbf{Pr}\left[r = \pi(i)\right] \frac{1}{r^{1/d}} \mathbf{E}\left[\tilde{s}_i^{(\pi(i))} \,\Big|\, \pi(i) = r\right] \,,$$

where π is the random permutation the jobs are presented in. Observe that $\tilde{s}_i^{(\pi(i))} - 1$ is exactly the number of jobs whose weight is larger than w_i that come in rounds before r. Conditioning on $\pi(i) = r$, the order of the remaining $n-1$ jobs is still uniform. Out of these exactly $i-1$ have a weight larger than w_i. In expectation, a $\frac{r-1}{n-1}$ fraction of these are assigned to rounds $1, \ldots, r-1$. This gives us $\mathbf{E}\left[\tilde{s}_i^{(\pi(i))} \,\Big|\, \pi(i) = r\right] = (i-1)\frac{r-1}{n-1} + 1$. Using that furthermore $\mathbf{Pr}\left[\pi(i) = r\right] = \frac{1}{n}$ for all r, we get

$$\mathbf{E}\left[f_{\pi(i)} \tilde{s}_i^{(\pi(i))}\right] \leq cn^{1/d} \left(\frac{i-1}{n} \sum_{r=1}^{n} \frac{r^{1-1/d}}{n} + \frac{1}{n} \sum_{r=1}^{n} \frac{1}{r^{1/d}}\right) \,.$$

We approximate both sums by the corresponding integrals (see full version). Regarding the bound on $\mathbf{E}\left[\hat{s}_i - 1\right]$, this gives us

$$\mathbf{E}\left[\hat{s}_i - 1\right] \leq cn^{1/d} \left(\frac{i-1}{n^2}\left(\frac{1}{2-\frac{1}{d}} n^{2-1/d} + n^{1-1/d}\right) + \frac{1}{n} + \frac{1}{n^{1/d}}\frac{1}{1-\frac{1}{d}}\right)$$

$$= c\left((i-1)\frac{1}{2-\frac{1}{d}} + \frac{i-1}{n} + \frac{1}{n^{1-1/d}} + \frac{1}{1-\frac{1}{d}}\right) \leq ci\frac{1}{1-\frac{1}{d}} + \frac{c}{n^{1-1/d}}$$

and therefore we get $\mathbf{E}\left[\hat{s}_i\right] \leq \frac{c}{1-\frac{1}{d}} \cdot i + \frac{c}{n^{1-1/d}} + 1$.   □

2.2   From Tentative to Actual Slots

It still remains to bound the number of the actual slot that is assigned to a job i. To this end, we will use the following intuition. Imagine the schedule to be a queue, first all jobs that have tentative slots between slot 1 and $\hat{s}_i$ arrive. While processing these jobs, the arrival continues and more jobs come in. These new jobs are also processed before we start to work off i. We will use the fact that the average expected number of jobs tentative assigned to a slot is bounded by some $q < 1$. This causes the effects of this cascade to be bounded in expectation. To formalize this, we use the following technical lemma for a queueing process.

Lemma 3. *Consider non-negative integer random variables A_t such that there is $q \in (0,1)$ with the property that for any $t \in \mathbb{N}$ and $a_1, \ldots, a_{t-1} \in \mathbb{N}$, we have $\mathbf{E}[A_t \mid A_1 = a_1, \ldots, A_{t-1} = a_{t-1}] \leq q$. Furthermore, let $Q_0 \in \mathbb{N}$ and $Q_{t+1} = \max\{0, Q_t + A_{t+1} - 1\}$ and $T = \min\{t \mid Q_t = 0\}$, then $\mathbf{E}[T] \leq \frac{1}{1-q} Q_0$.*

Proof. We divide the time[2] waiting for $Q_t = 0$ into phases as follows: If phase p ends at time T_{p-1}, then phase p lasts exactly for $Q_{T_{p-1}}$ steps. The intuition is that we are waiting for a FIFO queue to become empty. The initial queue length is Q_0. After Q_0 steps, the initial elements of the queue have been processed. During this processing, additional elements may have arrived that need to be processed. This process continues until $Q_t = 0$ for the first time.

Formally, we set $T_p = T_{p-1} + Q_{T_{p-1}}$, $T_0 = 0$. By this definition we have $Q_{T_p} = Q_{T_{p-1}} + \sum_{t=T_{p-1}+1}^{T_p} A_t + (T_p - T_{p-1}) = \sum_{t=T_{p-1}+1}^{T_p} A_t$. Now induction gives us $\mathbf{E}\left[Q_{T_p} \mid A_1, \ldots, A_{T_{p-1}}\right] = \mathbf{E}\left[\sum_{t=T_{p-1}+1}^{T_p} A_t \mid A_1, \ldots, A_{T_{p-1}}\right] \leq q(T_p - T_{p-1}) = Q_{T_{p-1}}$ using the condition on the expectation. This implies $\mathbf{E}\left[Q_{T_p}\right] \leq q\mathbf{E}\left[Q_{T_{p-1}}\right]$ and by induction $\mathbf{E}\left[Q_{T_p}\right] \leq q^p Q_0$.

We have $T = \max_p T_p$ and therefore $T = \sum_{p=0}^{\infty} Q_{T_p}$. By linearity of expectation, we get $\mathbf{E}[T] = \sum_{p=0}^{\infty} \mathbf{E}\left[Q_{T_p}\right] \leq \sum_{p=0}^{\infty} q^p Q_0 = \frac{1}{1-q} Q_0$.   □

Using this lemma, we will show that the index of the actual slot a job is mapped to is at most a constant factor larger than the tentative slot. This proof is still technically involved because we have to be careful with dependencies. Besides, in each round there are only a few possible options for the respectively assigned tentative slot. So, the arrival is not as balanced as in Lemma 3.

Lemma 4. *Fix a job i and a round r. Conditioned on the event that job i comes in round r and gets tentative slot $\hat{s}_i$, the expected first feasible slot s_i is given by $\mathbf{E}[s_i \mid \pi(i) = r, \hat{s}_i] \leq \frac{1}{1-q} \hat{s}_i$ with $q = \left(\frac{r}{n}\right)^{1/d} \cdot \frac{2d}{c}$.*

Proof. Let A'_t be the random variable counting the number of jobs that are tentatively allocated onto slot t by the end of round $r-1$. Analogously to Lemma 3 we define $Q'_t = \max\{0, Q'_{t-1} + A'_t - 1\}$ and $T' = \min\{t \mid Q'_t = 0\}$. If we set $Q'_0 = \hat{s}_i$ then $T' \geq s_i$ is an upper bound for the first feasible slot after $\hat{s}_i$. Unfortunately, Lemma 3 cannot be applied here because the A'_t are mutually dependent. To apply the lemma nevertheless, we define a set of variables A_t that are coupled to A'_t in such a way, that $\sum_{t' \leq t} A'_{t'} \leq \sum_{t' \leq t} A_{t'}$ holds for all t. Furthermore we choose $Q_0 = Q'_0$. It is easy to see that by this definition $Q'_t \leq Q_t$ for all t and therefore we also have $T' \leq T$. Thus it suffices to consider A_t to prove the lemma.

We choose A_t in such a way that we divert mass away from A'_t onto the $A'_{t'}$ with $t' < t$. As a first step, we balance the load between different slots, which we will exploit later. To this end, let $U_{r'}$ be drawn independently uniformly from $[0, f_{r'}]$, where $f_{r'} = c\left(\frac{n}{r}\right)^{1/d}$ is the scaling factor used in round r'.

---

[2] Note that the notion of time within this proof refers to the queueing perspective, not to the algorithm's input.

Define $Z_{r'} = f_{r'} \tilde{s}^{(r')}_{\pi^{-1}(r')} - U_{r'}$. So, $Z_{r'}$ is a real-valued random variable taking values on $[0, f_{r'} r']$. Indeed it is uniformly distributed on this interval because, conditioned on the set J in a round r', $\pi^{-1}(r')$ can be considered drawn uniformly from J. So, $\tilde{s}^{(r')}_{\pi^{-1}(r')}$ is drawn uniformly from $\{1, 2, \ldots, r'\}$. We will use $\lceil Z_{r'} \rceil$ as a lower bound on the tentative slot used in round r'. Therefore, define $X_{t,r'}$ to be 1 if $\lceil Z_{r'} \rceil = t$ and 0 otherwise. Now, let

$$A_t = \sum_{r' \leq r} \left(\frac{1}{b_{r'}} + \left(1 - \frac{t}{b_{r'}}\right) \cdot X_{t,r'} \right) \quad \text{with } b_{r'} = \lceil f_{r'} r' \rceil = \left\lceil c \left(\frac{n}{r'}\right)^{1/d} r' \right\rceil .$$

To complete the proof of the lemma, it now remains to show the following two claims. (a) $\sum_{t' \leq t} A_{t'} \geq \sum_{t' \leq t} A'_{t'}$ for all t. (b) Given arbitrary numbers $a_1, \ldots, a_{t-1} \in \mathbb{N}$, we have $\mathbf{E}[A_t \mid \bar{A}_1 = a_1, \ldots, A_{t-1} = a_{t-1}] \leq q$. Then Lemma 3 directly gives the desired result. Due to space limitations, the formal proof of Claim (a) can only be found in the full version.

Proof of Claim (b) For an arbitrary matrix $x = (x_{t',r'})_{t',r' \in \mathbb{N}}$, $x_{t',r'} \in \{0, 1\}$, let $\mathcal{E}_x$ be the event that $X_{t',r'} = x_{t',r'}$ for all $t' < t$ and all $r' \leq r$. We now upper-bound the value of $\mathbf{Pr}[X_{t,r'} = 1 \mid \mathcal{E}_x]$. Observe that $\lceil Z_{r'} \rceil \leq \lceil f_{r'} r' \rceil$. Therefore, if $t > f_{r'} r'$, we immediately have $X_{t,r'} = 0$. Furthermore, if $x_{t',r'} = 1$ for some $t' < t$, then also $X_{t,r'} = 0$. So, let us, without loss of generality, assume that $t \leq \lceil f_{r'} r' \rceil$ and $x_{t',r'} = 0$ for all $t' < t$.

We also observe that the algorithm only uses relative ranks to determine the slot allocation: In each round r'' the solution s always allocates the same slots S', regardless of the actual job weights. Therefore, $Z_{r'}$ can considered to be independent of all $Z_{r''}$, $r'' \neq r'$. Consequently, conditioned on $\mathcal{E}_x$, $Z_{r'}$ is uniformly distributed on $(t-1, b_{r'}]$. Therefore, we get

$$\mathbf{Pr}[X_{t,r'} = 1 \mid \mathcal{E}_x] \leq \frac{1}{b_{r'} - (t-1)} .$$

Given arbitrary numbers $a_1, \ldots, a_{t-1} \in \mathbb{N}$, we now bound the conditioned expectation $\mathbf{E}[A_t \mid A_1 = a_1, \ldots, A_{t-1} = a_{t-1}]$. To this end observe, that the event $A_1 = a_1, \ldots, A_{t-1} = a_{t-1}$ can equivalently be expressed by a set $\mathcal{X}$ of 0/1 matrices x with the property that $A_1 = a_1, \ldots, A_{t-1} = a_{t-1}$ if and only if there is $x \in \mathcal{X}$ such that $X_{t',r'} = x_{t',r'}$ for all $t' < t$ and all $r' \leq r$.

Using the above bound on $\mathbf{Pr}[X_{t,r'} = 1 \mid \mathcal{E}_x]$, we get

$$\mathbf{E}[A_t \mid \mathcal{E}_x] = \sum_{r' \leq r} \left(\frac{1}{b_{r'}} + \left(1 - \frac{t}{b_{r'}}\right) \cdot \mathbf{Pr}[X_{t,r'} = 1 \mid \mathcal{E}_x] \right)$$

$$\leq \sum_{r' \leq r} \left(\frac{1}{b_{r'}} + \left(1 - \frac{t-1}{b_{r'}}\right) \cdot \frac{1}{b_{r'} - (t-1)} \right) = \sum_{r' \leq r} 2\frac{1}{b_{r'}} = \frac{2}{cn^{1/d}} \sum_{r' \leq r} (r')^{1/d - 1} .$$

As this bound holds for all $x \in \mathcal{X}$, we also have

$$\mathbf{E}[A_t \mid A_1 = a_1, \ldots, A_{t-1} = a_{t-1}] \leq \frac{2}{cn^{1/d}} \sum_{r' \leq r} (r')^{1/d - 1} \leq \left(\frac{r}{n}\right)^{1/d} \cdot \frac{2d}{c} . \qquad \square$$

2.3 Putting the Pieces Together

Using the insights from the previous sections, the proof of Theorem 1 is relatively straightforward.

Proof (of Theorem 1). Combining Lemmas 2 and 4, we get that for every job i

$$\mathbf{E}\left[s_i\right] \leq \sum_{i=1}^{n} w_i \cdot \frac{1}{1 - \frac{2d}{c}} \mathbf{E}\left[\hat{s}_i\right] \leq \sum_{i=1}^{n} w_i \frac{1}{1 - \frac{2d}{c}} \cdot \left(\frac{cd}{d-1}i + 1 + O\left(n^{1/d-1}\right)\right) .$$

We omit the O-term since it tends towards 0 for large n. Setting $c = 8$ and $d = 2$, we get $\frac{1}{1-\frac{2d}{c}} \cdot \left(\frac{cd}{d-1}i + 1\right) = 2(16i + 1) \leq 34i$. Therefore for every job i the allocated slot s_i in expectation only deviates by a factor of 34 from its optimal slot s_i^* in the offline schedule. $\qquad\square$

3 General Jobs

With the constant competitive algorithm for jobs with uniform processing time as a subroutine we devise an algorithm for jobs with variable processing times. We use several instances of Algorithm 1 to schedule jobs with similar processing time. We sort jobs into processing time classes where every class $\lambda = 2^b$ for $b \in \mathbb{N}$ contains the jobs with processing times between two powers of two $2^{b-1} < l_i \leq 2^b$. To this end we define a labeling scheme that maps the sub-schedules of the different processing time classes onto the overall schedule in such a way, that no job is pushed back by more than a factor of $2\log(n)$ compared to his position in the sub-schedule.

Theorem 5. *The algorithm for jobs with general processing time is $2\alpha\log(n)$-competitive where α is the competitive factor of the algorithm for jobs with uniform jobs used as subroutine.*

We start out with the labeling scheme that allows us to group slots to meta-slots, each associated to a single processing time class. Without loss of generality let the total number of jobs n be a power of two. Let T_k be a complete binary tree of height $\log(n)$. We call the level of the leaves $j = 0$. Now we label a node σ on level j of T_k with $\lambda(\sigma) = 2^{k+j}$. This way, the leaves get label 2^k, their parents get label 2^{k+1} up to the root with label $2^{k+\log n}$.

Now we traverse the tree T_0 in post-order and map its nodes onto the slots in our schedule. In this traversal order we descend left first and map a parent node right after all his children. For each node σ with label $\lambda(\sigma)$ we create a meta-slot of $\lambda(\sigma)$ many neighboring slots. We proceed to map T_1 starting from the first free slot after T_0 and so on.

Observation 6. *The mapping of tree T_k requires $2^k n\log(n)$ slots in the schedule and starts at slot $(2^k - 1)n\log(n) + 1$.*

This follows simply from the fact that every level j in tree T_k contains $\frac{n}{2^j}$ nodes and each node takes 2^{j+k} slots.

Lemma 7. *The γ-th meta slot with label λ ends no later than slot $2\log(n)\gamma\lambda$ in the schedule.*

Proof. The first occurrence of label λ is in tree T_k with $k = 0$ if $\lambda \leq n$ and $k = \log(\lambda) - \log(n)$ otherwise. Therefore tree $T_{k'}$ with $k' = 0$ if $\lambda \leq n$ and $k' = \log(\lambda) + \log(\gamma) - \log(n)$ otherwise is the first tree that contains at least γ meta-slots with label λ.

Now we make a case distinction, if $k' = 0$, then label λ is used on level $j = \log(\lambda)$. In the post-order traversal when reaching the γ-th node with label λ, the subtrees of previous nodes of label λ have been traversed plus the subtree of the current node plus at most a $\frac{\gamma}{n/2^j}$-fraction of all nodes on higher levels. Each subtree takes $\log(\lambda)\lambda$ many slots. There are $\left\lfloor \frac{\gamma}{n/2^j}\frac{n}{2^{-j}} \right\rfloor \leq \gamma 2^{j-j'}$ nodes on higher levels $j' = \log(\lambda) + 1, \ldots, \log n$, each taking $2^{j'}$ slots. So the γ-th node with label λ ends on slot $\gamma\log(\lambda)\lambda + (\log n - \log(\lambda))\gamma 2^j = \log(n)\gamma\lambda$.

In the other case, if $k' \neq 0$ it follows from Observation 6 that tree $T_{k'}$ starts at slot $(2^{k'} - 1)n\log(n) + 1 \leq \lambda + \gamma n\log(n) + 1 = 2^{\log(\gamma)+\log(\lambda)-\log(n)}n\log(n) + 1 = \gamma\lambda\log(n) + 1$. Furthermore label λ is used on level $j = \log(\lambda) - k'$ in tree $T_{k'}$ and the leaves in tree $T_{k'}$ take $2^{k'}$ slots each. Thus the total number of slots used through the post-order traversal on $T_{k'}$ is bounded by $\gamma(\log(\lambda) - k') \cdot 2^{\log(\lambda)-k'} \cdot 2^{k'} + (\log n - (\log(\lambda) - k'))\gamma\lambda = \log(n)\gamma\lambda$. So, the γth slot of label λ ends no later than $2\log(n)\gamma\lambda$. $\qquad\square$

Algorithm 2. Log-Algorithm for Jobs with different processing times

---

for each round r with incoming job i having processing time l_i **do**
  choose b such that $2^{b-1} < l_i \leq 2^b$;
  Let $J_b := \{\text{jobs } j \text{ with } l_j \in (2^{b-1}, 2^b]\}$;
  Let $\Sigma_b := \{\text{meta-slots } \sigma | \lambda(\sigma) = 2^b\}$;
  $s^{(b)} = $ output from uniform algorithm with job i on known J_b;
  schedule i on s_i-th meta-slot in Σ_b;

Proof (of Theorem 5). We run one instance of an α-competitive algorithm for jobs with uniform processing time for every processing time class as a subroutine. These subroutines give for every job i a $\gamma = s_i^{(b)}$ and a $\lambda = 2^b$ such that $2^{b-1} < l_i \leq 2^b$. Now meta-slot γ of the subroutine would end in slot $\gamma\lambda$ if no other labels were interwoven. Therefore the labeling stretches the α-competitive schedule $s^{(b)}$ by an additional factor of $\frac{2\log(n)\gamma\lambda}{\gamma\lambda} = 2\log(n)$. $\qquad\square$

4 Lower Bound for Fully Worst-Case Input

As mentioned before, we motivate the use of the random order model by giving a lower bound when performing classical worst-case analysis in the general setting with different job processing times and weights. There, an adversary is allowed to construct the instance, i.e. to determine the jobs' weights, and then also to present the jobs in any preferred order. This, obviously, is more powerful, but we show that it does not allow designing algorithms that achieve a reasonable competitive ratio. As we use a randomized instance, we can extend our results to even hold for randomized algorithms by applying Yao's principle.

Theorem 8. *For every randomized online algorithm weights can be chosen in a way such that* $\mathbf{E}\,[\text{ALG}] \geq \frac{n}{8} \cdot \text{OPT}$, *even if all jobs have equal processing times.*

Proof. We show this claim by using Yao's principle. We will devise a randomized instance such that for any deterministic algorithm $\mathbf{E}\left[\frac{\text{ALG}}{\text{OPT}}\right] \geq \frac{n}{8}$. To this end, let T be drawn uniformly from $\{1, \ldots, n\}$. We define the weights of a job arriving in round t to be $w_t = M^t$ if $t \leq T$ and $w_t = 0$ otherwise, with $M > n$. First, we compute the cost incurred by an optimal algorithm. According to the construction described above, T many jobs with non-zero weights $M, M^2, \ldots, M^T$ are given. It is obviously the best solution to put the heaviest job first, and then proceed with the jobs decreasing in their weight. Formally, job j is assigned slot $T - j + 1$. This results in cost $\text{OPT} = \sum_{j=1}^{T} M^j (T - j + 1) = \sum_{j=0}^{T-1} M^{T-j}(j+1) \leq 2 \cdot M^T$.

Now we focus on the cost of a deterministic online algorithm ALG. Until (including) round T, the behavior of this algorithm is independent of T. Its behavior after this point is irrelevant for the resulting cost. Therefore, we can express the algorithm's choices as an injective function $\sigma \colon [n] \to \mathbb{N}$, meaning that the ith job is scheduled to slot $\sigma(i)$.

As the function is injective, there is a set $S \subseteq [n]$ of size $\left\lceil \frac{n}{2} \right\rceil$ such that $\sigma(i) \geq \left\lfloor \frac{n}{2} \right\rfloor + 1 \geq \frac{n}{2}$ for all $i \in S$. Note that if $T \in S$, then $\text{ALG} \geq \frac{n}{2} \cdot M^T \geq \frac{n}{2} \cdot \frac{\text{OPT}}{2}$, so $\frac{\text{ALG}}{\text{OPT}} \geq \frac{n}{4}$. As $T \in S$ happens with probability at least $\frac{1}{2}$, we get $\mathbf{E}\left[\frac{\text{ALG}}{\text{OPT}}\right] \geq \frac{n}{8}$. $\qquad\square$

References

1. Afrati, F.N., Bampis, E., Chekuri, C., Karger, D.R., Kenyon, C., Khanna, S., Milis, I., Queyranne, M., Skutella, M., Stein, C., Sviridenko, M.: Approximation schemes for minimizing average weighted completion time with release dates. In: Proc. 40th Symp. Foundations of Computer Science (FOCS), pp. 32–44 (1999)
2. Agrawal, S., Wang, Z., Ye, Y.: A dynamic near-optimal algorithm for online linear programming. Operations Research 62(4), 876–890 (2014)
3. Ajtai, M., Megiddo, N., Waarts, O.: Improved algorithms and analysis for secretary problems and generalizations. SIAM J. Discrete Math. 14(1), 1–27 (2001)
4. Babaioff, M., Immorlica, N., Kleinberg, R.: Matroids, secretary problems, and on-line mechanisms. In: Proc. 18th Symp. Discr. Algorithms (SODA), pp. 434–443 (2007)
5. Begen, M.A., Queyranne, M.: Appointment scheduling with discrete random durations. Math. Oper. Res. 36(2), 240–257 (2011)
6. Buchbinder, N., Naor, J.: Online primal-dual algorithms for covering and packing. Math. Oper. Res. 34(2), 270–286 (2009)

7. Chekuri, C., Khanna, S.: A PTAS for minimizing weighted completion time on uniformly related machines. In: Orejas, F., Spirakis, P.G., van Leeuwen, J. (eds.) ICALP 2001. LNCS, vol. 2076, pp. 848–861. Springer, Heidelberg (2001)
8. Correa, J.R., Wagner, M.R.: Lp-based online scheduling: from single to parallel machines. Math. Program. 119(1), 109–136 (2009)
9. Devanur, N.R., Hayes, T.P.: The adwords problem: online keyword matching with budgeted bidders under random permutations. In: Proc. 10th Conf. Electr. Commerce (EC), pp. 71–78 (2009)
10. Devanur, N.R., Jain, K., Sivan, B., Wilkens, C.A.: Near optimal online algorithms and fast approximation algorithms for resource allocation problems. In: Proc. 12th Conf. Electr. Commerce (EC), pp. 29–38 (2011)
11. Feldman, J., Henzinger, M., Korula, N., Mirrokni, V.S., Stein, C.: Online stochastic packing applied to display ad allocation. In: de Berg, M., Meyer, U. (eds.) ESA 2010, Part I. LNCS, vol. 6346, pp. 182–194. Springer, Heidelberg (2010)
12. Feldman, M., Svensson, O., Zenklusen, R.: A simple O(log log(rank))-competitive algorithm for the matroid secretary problem. In: Proc. 26th Symp. Discr. Algorithms (SODA), pp. 1189–1201 (2015)
13. Fiat, A., Woeginger, G.J.: On-line scheduling on a single machine: Minimizing the total completion time. Acta Inf. 36(4), 287–293 (1999)
14. Göbel, O., Hoefer, M., Kesselheim, T., Schleiden, T., Vöcking, B.: Online independent set beyond the worst-case: Secretaries, prophets, and periods. In: Esparza, J., Fraigniaud, P., Husfeldt, T., Koutsoupias, E. (eds.) ICALP 2014, Part II. LNCS, vol. 8573, pp. 508–519. Springer, Heidelberg (2014)
15. Günther, E., Maurer, O., Megow, N., Wiese, A.: A new approach to online scheduling: Approximating the optimal competitive ratio. In: Proc. 24th Symp. Discr. Algorithms (SODA), pp. 118–128 (2013)
16. Kesselheim, T., Radke, K., Tönnis, A., Vöcking, B.: Primal beats dual on online packing LPs in the random-order model. In: Proc. 46th Symp. Theory of Computing (STOC), pp. 303–312 (2014)
17. Kleinberg, R., Weinberg, S.M.: Matroid prophet inequalities. In: Proc. 44th Symp. Theory of Computing (STOC), pp. 123–136 (2012)
18. Lachish, O.: O(log log rank) competitive ratio for the matroid secretary problem. In: Proc. 55th Symp. Foundations of Computer Science (FOCS), pp. 326–335 (2014)
19. Megow, N., Schulz, A.S.: On-line scheduling to minimize average completion time revisited. Oper. Res. Lett., 32(5):485–490 (2004)
20. Megow, N., Uetz, M., Vredeveld, T.: Models and algorithms for stochastic online scheduling. Math. Oper. Res., 31(3):513–525 (2006)
21. Megow, N., Vredeveld, T.: A tight 2-approximation for preemptive stochastic scheduling. Math. Oper. Res. 39(4), 1297–1310 (2014)
22. Mehta, A., Saberi, A., Vazirani, U.V., Vazirani, V.V.: Adwords and generalized online matching. J. ACM 54(5) (2007)
23. Meyerson, A.: Online facility location. In: Proc. 42nd Symp. Foundations of Computer Science (FOCS), pp. 426–431 (2001)
24. Meyerson, A.: The parking permit problem. In: Proc. 46th Symp. Foundations of Computer Science (FOCS), pp. 274–284 (2005)

25. Meyerson, A., Munagala, K., Plotkin, S.A.: Designing networks incrementally. In: Proc. 42nd Symp. Foundations of Computer Science (FOCS), pp. 406–415 (2001)
26. Molinaro, M., Ravi, R.: The geometry of online packing linear programs. Math. Oper. Res. 39(1), 46–59 (2014)
27. Schulz, A.S., Skutella, M.: The power of α-points in preemptive single machine scheduling. Journal of Scheduling 5, 121–133 (2002)
28. Sitters, R.: Competitive analysis of preemptive single-machine scheduling. Operations Research Letters 38(6), 585–588 (2010)

Approximation Algorithms for Connected Maximum Cut and Related Problems

MohammadTaghi Hajiaghayi[1,*], Guy Kortsarz[2,**],
Robert MacDavid[2], Manish Purohit[1,***], and Kanthi Sarpatwar[3,†]

[1] University of Maryland, College Park, MD 20742, USA
{hajiagha,manishp}@cs.umd.edu
[2] Rutgers University - Camden, Camden, NJ 08102, USA
guyk@camden.rutgers.edu, robertmacdavid@gmail.com
[3] IBM T. J. Watson Research Center, Yorktown Heights, NY 10598
sarpatwa@us.ibm.com

Abstract. An instance of the *Connected Maximum Cut* problem consists of an undirected graph $G = (V, E)$ and the goal is to find a subset of vertices $S \subseteq V$ that maximizes the number of edges in the cut $\delta(S)$ such that the induced graph $G[S]$ is connected. We present the first nontrivial $\Omega(\frac{1}{\log n})$ approximation algorithm for the connected maximum cut problem in general graphs using novel techniques. We then extend our algorithm to an edge weighted case and obtain a poly-logarithmic approximation algorithm. Interestingly, in stark contrast to the classical max-cut problem, we show that the connected maximum cut problem remains NP-hard even on unweighted, planar graphs. On the positive side, we obtain a polynomial time approximation scheme for the connected maximum cut problem on planar graphs and more generally on graphs with bounded genus.

1 Introduction

Submodular optimization problems have, in recent years, received a considerable amount of attention [1, 2, 3, 4, 7, 15, 27] in algorithmic research. In a general *submodular maximization* problem, we are given a non-negative submodular[1] function over the power set of a universe U of elements, $f : 2^U \to \mathbb{R}^+ \cup \{0\}$ and the goal is to find a subset $S \subseteq U$ that maximizes $f(S)$ so that S satisfies certain pre-specified constraints. In addition to their practical relevance, the

---

* Partially supported by NSF CAREER Award 1053605, NSF grant CCF-1161626, DARPA/AFOSR grant FA9550-12-1-0423, and a Google Faculty Research award.
** Partially supported by NSF grant 1218620.
*** Partially supported by NSF grants CCF-1217890 and IIS-1451430.
† Partially supported by NSF grant CCF-1217890. Work done when the author was a student at the University of Maryland, College Park.
[1] A function f is called submodular if $f(S) + f(T) \geq f(S \cup T) + f(S \cap T)$ for all $S, T \subseteq U$.

© Springer-Verlag Berlin Heidelberg 2015
N. Bansal and I. Finocchi (Eds.): ESA 2015, LNCS 9294, pp. 693–704, 2015.
DOI: 10.1007/978-3-662-48350-3_58

study of submodular maximization problems has led to the development of several important theoretical techniques such as the continuous greedy method and multi-linear extensions [4] and the double greedy [2] algorithm, among others.

In this study, we are interested in the problem of maximizing a submodular set function over vertices of a graph, such that the selected vertices induce a connected subgraph. Motivated by applications in coverage over wireless networks, Kuo et al. [25] consider the problem of maximizing a monotone, submodular function f subject to connectivity and cardinality constraints of the form $|S| \leq k$ and provide an $\Omega(\frac{1}{\sqrt{k}})$ approximation algorithm. For a restricted class of monotone, submodular functions that includes the covering function[2], Khuller et al. [24] give a constant factor approximation to the problem of maximizing f subject to connectivity and cardinality constraints.

In the light of these results, it is rather surprising that no non-trivial approximation algorithms are known for the case of general (non-monotone) submodular functions. Formally, we are interested in the following problem, which we refer to as *connected submodular maximization* (CSM): given a simple, undirected graph $G = (V, E)$ and a non-negative submodular set function $f : 2^V \to \mathbb{R}^+ \cup \{0\}$, find a subset of vertices $S \subseteq V$ that maximizes $f(S)$ such that $G[S]$ is connected. We take the first but important step in this direction and study the problem in the case of one of the most important non-monotone submodular functions, namely the *cut* function. Formally, given an undirected graph $G = (V, E)$, the goal is to find a subset $S \subseteq V$, such that $G[S]$ is connected and the number of edges that have exactly one end point in S, referred to as the cut function $\delta(S)$, is maximized. We refer to this as the *connected maximum cut* problem. Further, we also consider an edge weighted variant of this problem, called the *weighted connected maximum cut* problem, where function to be maximized is the total weight of edges in the cut $\delta(S)$.

We now outline an application to the image segmentation problem that seeks to identify "objects" in an image. Graph based approaches for image segmentation [16, 28] represent each pixel as a vertex and weighted edges represent the dissimilarity (or similarity depending on the application) between adjacent pixels. Given such a graph, a connected set of pixels with a large weighted cut naturally corresponds to an object in the image. Vicente et al. [32] show that even for interactive image segmentation, techniques that require connectivity perform significantly better than cut based methods alone.

1.1 Related Work

Max-cut is a fundamental problem in combinatorial optimization that finds applications in diverse areas. A simple randomized algorithm that adds each vertex to S independently with probability $1/2$ gives a 0.5-approximate solution in expectation. In a breakthrough result, Goemans and Williamson [18] gave a 0.878-approximation algorithm using semidefinite programming and randomized

---

[2] In this context, a covering function is defined as $f(S) = \sum_{v \in \mathbb{N}^+(S)} weight(v)$ where $\mathbb{N}^+(S)$ is the closed neighborhood of the set of vertices S

rounding. Further, Khot et al. [23] showed that this factor is optimal assuming the *unique games conjecture*. Interestingly, the max-cut problem can be solved in polynomial time in planar graphs by a curious connection to the matching problem in the dual graph [20]. To the best of our knowledge, the connected maximum cut problem has not been considered before our work. Haglin and Venkatesan [21] showed that a related problem, where we require both sides of the cut, namely $G[S]$ and $G[V \setminus S]$, to be connected, is NP-hard in planar graphs.

We note that the well studied *maximum leaf spanning tree* (MLST) problem (e.g. see [30]) is a special case of the connected submodular maximization problem. We also note that recent work on graph connectivity under vertex sampling leads to a simple constant approximation to the connected submodular maximization for highly connected graphs, i.e., for graphs with $\Omega(\log n)$ vertex connectivity. Proofs of these claims are presented in the full version of this paper [22].

We conclude this section by noting that connected variants of many classical combinatorial problems have been extensively studied in the literature and have been found to be useful. The best example for this is the *connected dominating set* problem. Following the seminal work of Guha and Khuller [19], the problem has found extensive applications (with more than a thousand citations) in the domain of wireless ad hoc networks as a *virtual backbone* (e.g. see [9, 12]). Few other examples of connected variants of classic optimization problems include *group Steiner tree* [17] (which can be seen as a generalization of a connected variant of *set cover*), *connected domatic partition* [5, 6], *connected facility location* [13, 31], and *connected vertex cover* [8].

1.2   Contribution and Techniques

Our key results can be summarized as follows.

1. We obtain the first $\Omega(\frac{1}{\log n})$ approximation algorithm for the connected maximum cut (CMC) problem in general graphs. Often, for basic connectivity problems on graphs, one can obtain simple $O(\log n)$ approximation algorithms using a probabilistic embedding into trees with $O(\log n)$ stretch [14]. Similarly, using the cut-based decompositions given by Räcke [29], one can obtain $O(\log n)$ approximation algorithms for cut problems (e.g. *minimum bisection*). Interestingly, since the CMC problem has the flavors of both cut and connectivity problems simultaneously, neither of these approaches are applicable. Our novel approach is to look for α-*thick trees*, which are basically sub-trees with "high" degree sum on the leaves.

2. For the weighted connected maximum cut problem, we obtain an $\Omega(\frac{1}{\log^2 n})$ approximation algorithm. The basic idea is to group the edges into logarithmic number of weight classes and show that the problem on each weight class boils down to the special case where the weight of every edge is either 0 or 1.

3. We obtain a polynomial time approximation scheme for the CMC problem in planar graphs and more generally in bounded genus graphs. This requires the application of a stronger form of the edge contraction theorem by Demaine, Hajiaghayi and Kawarabayashi [11] that may be of independent interest.

4. We show that the CMC problem remains NP-hard even on unweighted, planar graphs. This is in stark contrast with the regular max-cut problem that can be solved optimally in planar graphs in polynomial time. We obtain a polynomial time reduction from a special case of *3-SAT*, called the *planar monotone 3-SAT* (PM-3SAT), to the CMC problem in planar graphs. This entails a delicate construction, exploiting the so called *rectilinear representation* of a PM-3SAT instance, to maintain planarity of the resulting CMC instance.

2   Approximation Algorithms for General Graphs

In this section, we consider the connected maximum cut problem in general graphs. In fact, we provide an $\Omega(\frac{1}{\log n})$ approximation algorithm for the more general problem in which edges can have weight 0 or 1 and the objective is to maximize the number of edges of weight 1 in the cut. This generalization will be useful later in obtaining a poly-logarithmic approximation algorithm for arbitrary weighted graphs.

We denote the cut of a subset of vertices S in a graph G, i.e., the set of edges in G that are incident on exactly one vertex of S by $\delta_G(S)$ or when G is clear from context, just $\delta(S)$. Further, for two disjoint subsets of vertices S_1 and S_2 in G, we denote the set of edges that have one end point in each of S_1 and S_2, by $\delta_G(S_1, S_2)$ or simply $\delta(S_1, S_2)$. The formal problem definition follows -

Binary Connected Maximum Cut (b-CMC): Given a graph $G = (V, E)$ and a weight function $w : E \rightarrow \{0, 1\}$, find a set $S \subset V$ that maximizes $\sum_{e \in \delta(S)} w(e)$ such that $G[S]$ induces a connected subgraph.

We call an edge of weight 0, a *0-edge* and that of weight 1, a *1-edge*. Further, let $w(\delta(S)) = \sum_{e \in \delta(S)} w(e)$ denote the weight of the cut, i.e., the number of 1-edges in the cut. We first start with a simple reduction rule that ensures that every vertex $v \in V$ has at least one 1-edge incident on it.

Claim 1. *Given a graph $G = (V, E)$, we can construct a graph $G' = (V', E')$ in polynomial time, such that every $v' \in V'$ has at least one 1-edge incident on it and G' has a b-CMC solution S' of weight at least ψ if and only if G has a b-CMC solution S of weight at least ψ.*

Proof Sketch. A vertex v that only has 0-edges incident on it can be deleted from the graph without affecting the value of the optimal cut. To ensure that any solution that contains v remains connected, we add a clique of 0-edges between all neighbors of v. The formal proof is included in the full version [22].   □

From now on, we will assume, without loss of generality, that every vertex of G has at least one 1-edge incident on it. We now introduce some new definitions that would help us to present the main algorithmic ideas. We denote by $W_G(v)$ the total weight of edges incident on a vertex v in G, i.e., $W_G(v) = \sum_{e:v \in e} w(e)$. In other words, $W_G(v)$ is total number of 1-edges incident on v. Further let η be the total number of 1-edges in the graph. The following notion of an α-thick tree is a crucial component of our algorithm.

Definition 1 (α-thick tree). *Let $G = (V, E)$ be a graph with n vertices and η 1-edges. A subtree $T \subseteq G$ (not necessarily spanning), with leaf set L, is said to be α-thick if $\sum_{v \in L} W_G(v) \geq \alpha\eta$.*

The following lemma shows that this notion of an α-thick tree is intimately connected with the b-CMC problem.

Lemma 1. *For any $\alpha > 0$, given a polynomial time algorithm A that computes an α-thick tree T of a graph G, we can obtain an $\frac{\alpha}{4}$-approximation algorithm for the b-CMC problem on G.*

Proof. Given a graph $G = (V, E)$ and weight function $w : E \rightarrow \{0, 1\}$, we use Algorithm A to compute an α-thick tree T, with leaf set L. Let m_L denote the number of 1-edges in $G[L]$, the subgraph induced by L in the graph G. We now partition L into two disjoint sets L_1 and L_2 such that the number of 1-edges in $\delta(L_1, L_2) \geq \frac{m_L}{2}$. This can be done by applying the standard randomized algorithm for max-cut (e.g. see [26]) on $G[L]$ after deleting all the 0-edges. Now, consider the two connected subgraphs $T \setminus L_1$ and $T \setminus L_2$. We first claim that every 1-edge in $\delta(L)$ belongs to either $\delta(T \setminus L_1)$ or $\delta(T \setminus L_2)$. Indeed, any 1-edge e in $\delta(L)$, belongs to one of the four possible sets, namely $\delta(L_2, T \setminus L)$, $\delta(L_1, V \setminus T)$, $\delta(L_1, T \setminus L)$ and $\delta(L_2, V \setminus T)$. In the first two cases, e belongs to $\delta(T \setminus L_2)$ while in the last two cases, c belongs $\delta(T \setminus L_1)$, hence the claim. Further, every 1-edge in $\delta(L_1, L_2)$ belongs to both $\delta(T \setminus L_1)$ and $\delta(T \setminus L_2)$. Hence, we have -

$$\sum_{e \in \delta(T \setminus L_1)} w(e) + \sum_{e \in \delta(T \setminus L_2)} w(e) = \sum_{e \in \delta(L)} w(e) + 2 \sum_{e \in \delta(L_1, L_2)} w(e) \qquad (1)$$

$$\geq \sum_{e \in \delta(L)} w(e) + m_L \geq \frac{1}{2} \sum_{v \in L} W_G(v) \geq \frac{\alpha\eta}{2} \qquad (2)$$

Hence, the better of the two solutions $T \setminus L_1$ or $T \setminus L_2$ is guaranteed to have a cut of weight at least $\frac{\alpha\eta}{4}$, where η is the total number of 1-edges in G. To complete the proof we note that for any optimal solution OPT, $w(\delta(OPT)) \leq \eta$. $\qquad \square$

Thus, if we have an algorithm to compute α-thick trees, Lemma 1 provides an $\Omega(\alpha)$-approximation algorithm for the b-CMC problem. Unfortunately, there exist graphs that do not contain α-thick trees for any non-trivial value of α. For example, let G be a *path graph* with n vertices and $m = n - 1$ 1-edges. It is easy to see that for any subtree T, the sum of degrees of the leaves is at most 4. In spite of this setback, we show that the notion of α-thick trees is still useful in obtaining a good approximation algorithm for the b-CMC problem. In particular, Lemma 3 and Theorem 1 show that path graph is the *only* bad case, i.e., if the graph G does not have a long induced path, then one can find an $\Omega(\frac{1}{\log n})$-thick tree. Lemma 2 shows that we can assume without loss of generality that the b-CMC instance does not have such a long induced path.

Shrinking Thin Paths. A natural idea to handle the above "bad" case is to get rid of such long paths that contain only vertices of degree two by contracting the edges. We refer to a path that only contains vertices of degree two as a d-2 path. Further, we define the length of a d-2 path as the number of *vertices* (of degree two) that it contains. The following lemma shows that we can assume without loss of generality that the graph G contains no "long" d-2 paths.

Lemma 2. *Given a graph G, we can construct, in polynomial time, a graph G' with no d-2 paths of length ≥ 3 such that G' has a b-CMC solution S' of cut weight $(w(\delta(S')))$ at least ψ if and only if G has a b-CMC solution S of cut weight at least ψ. Further, given the solution S' of G', we can recover S in polynomial time.*

Proof Sketch. Suppose the graph G has a d-2 path p of length ≥ 3. We observe that at most 2 edges of p can be cut by any b-CMC solution S since $G[S]$ must be connected. As a result, the long d-2 path p can be replaced by a smaller path p' with at most 2 vertices of degree 2 without affecting the quality of any optimum b-CMC solution. The formal proof follows from a case analysis based on the structure of the optimal solution and its interaction with the path p and is presented in the full version [22]. □

Spanning Tree with Many Leaves. Assuming that the graph has no long d-2 paths, the following lemma shows that we can find a spanning tree T that has $\Omega(n)$ leaves. Note that Claim 1 now guarantees that there are $\Omega(n)$ 1-edges incident on the leaves of T.

Lemma 3. *Given a graph $G = (V, E)$ with no d-2 paths of length ≥ 3, we can obtain, in polynomial time, a spanning tree $T = (V, E_T)$ with at least $\frac{n}{14}$ leaves.*

Proof. Let T be any spanning tree of G. We note that although G does not have d-2 paths of length ≥ 3, such a guarantee does not hold for paths in T. Suppose that there is a d-2 path $\wp$ of length 7 in T. Let the vertices of this path be numbered $v_1, v_2, \ldots, v_7$ and consider the vertices v_3, v_4, v_5. Since G does not have any d-2 path of length 3, there is a vertex $v_i, i \in \{3, 4, 5\}$ such that $deg_G(v_i) \geq 3$. We now add an edge $e = \{v_i, w\}$ in $G \setminus T$ to the tree T. The cycle C that is created as a result must contain either the edge $\{v_1, v_2\}$ or the edge $\{v_6, v_7\}$. We delete this edge to obtain a new spanning tree T'. It is easy to observe that the number of vertices of degree two in T' is strictly less than that in T. This is because, although the new edge $\{v_i, w\}$ can cause w to have degree two in T', we are guaranteed that the vertex v_i will have degree three and vertices v_1 and v_2 (or v_6 and v_7) will have degree one. Hence, as long as there are d-2 paths of length 7 in T, the number of vertices of degree two can be strictly decreased. Thus this process must terminate in at most n steps and the final tree $T^{(1)}$ obtained does not have any d-2 paths of length ≥ 7.

We now show that the tree $T^{(1)}$ contains $\Omega(n)$ leaves by a simple charging argument. Let the tree $T^{(1)}$ be rooted at an arbitrary vertex. We assign each vertex of $T^{(1)}$ a token and redistribute them in the following way : Every vertex v of degree two in $T^{(1)}$ gives its token to its first non degree two descendant, breaking ties arbitrarily. Since there is no d-2 path of length ≥ 7, each non degree two vertex collects at most 7 tokens. Hence, the number of vertices not having degree two in $T^{(1)}$ is at least $\frac{n}{7}$. Further, since the average degree of all vertices in a tree is at most 2, a simple averaging argument shows that $T^{(1)}$ must contain at least $\frac{n}{14}$ vertices of degree one, i.e., $\frac{n}{14}$ leaves. □

2.1 Obtaining an $\Omega(\frac{1}{\log n})$ Approximation

We now have all the ingredients required to obtain the $\Omega(\frac{1}{\log n})$ approximation algorithm. We observe that if the graph G is sparse, i.e. $\eta \leq cn \log n$ (for a suitable constant c), then the tree obtained by using Lemma 3 is an $\Omega(\frac{1}{\log n})$-thick tree and thus we obtain the required approximate solution in this case. On the other hand, if the graph G is sparse, then we use Lemma 3 to obtain a spanning tree, delete the leaves of this tree, and then repeat this procedure until we have no more vertices left. Since, we delete a constant fraction of vertices in each iteration, the total number of iterations is $O(\log n)$. We then choose the "best" tree out of the $O(\log n)$ trees so obtained and show that it must be an α-thick tree, with $\alpha = \Omega(\frac{1}{\log n})$. Finally, using Lemma 1, we obtain an $\Omega(\frac{1}{\log n})$ approximate solution as desired. We refer to Algorithm 1 for the detailed algorithm.

---

1 **Input**: Graph $G = (V, E)$
2 **Output**: A subset $S \subseteq V$, such that $G[S]$ is connected
3 Set $G_1(V_1, E_1) = G$, $n_1 = |V_1|$
4 Let $\eta \leftarrow$ Number of 1-edges in G
5 Use Lemma 3 to obtain a spanning tree T_1 of G_1 with leaf set L_1
6 **if** $\eta \leq cn \log n$ **then**
7 | Use Lemma 1 on T_1 to obtain a set connected S
8 | **return** S
9 **end**
10 $i = 1$
11 **while** $G_i \neq \phi$ **do**
12 | $E_{i+1} \leftarrow E_i \setminus (E[L_i] \cup \delta(L_i))$
13 | $V_{i+1} \leftarrow V_i \setminus L_i$, $n_{i+1} = |V_{i+1}|$
14 | Contract degree-2 vertices in G_{i+1}
15 | Use Lemma 3 to obtain a spanning tree T_{i+1} of G_{i+1} with leaf set L_{i+1}
16 | $i = i + 1$
17 **end**
18 Choose $j = \arg\max_i (\sum_{v \in L_i} deg_G(v))$
19 Use Lemma 1 on T_j to obtain a connected set S
20 **return** S

Algorithm 1. Finding α-thick trees

Theorem 1. *Algorithm 1 gives an $\Omega(\frac{1}{\log n})$ approximate solution for the b-CMC problem.*

Proof. Let us assume that $\eta \leq cn \log n$ (for some constant c). Now, Lemma 3 and Claim 1 together imply that $\sum_{v \in L_1} W_G(v) = \Omega(n)$. Further, since we have $w(\delta(OPT)) \leq \eta \leq cn \log n$, T is an α-thick tree for some $\alpha = \Omega(\frac{1}{\log n})$. Hence, we obtain an $\Omega(\frac{1}{\log n})$ approximate solution using Lemma 1.

On the other hand, if $\eta > cn \log n$, we show that at least one of the trees T_i obtained by the repeated applications of the Lemma 3 is an α-thick tree T of G for $\alpha = \Omega(\frac{1}{\log n})$. We first observe that the While loop in Step 11 runs for at most $O(\log n)$ iterations. This is because we delete $\Omega(n_i)$ leaves in each iteration and hence after $k = O(\log n)$ iterations, we get $G_k = \phi$. We now count the number of 1-edges "lost" in each iteration. We recall that $W_G(v)$ is the total number of 1-edges incident on v in a graph G. In an iteration i, the number of 1-edges lost at Step 12 is at most $\sum_{v \in L_i} W_{G_i}(v)$. In addition, we may lose a total of at most $2n \leq \frac{2\eta}{c \log n}$ edges due to the contraction of degree two vertices in Step 14. Suppose for the sake of contradiction that $\sum_{v \in L_i} W_G(v) < \frac{\eta}{d \log n}, \forall 1 \leq i \leq k$ where d is a suitable constant. Then the total number of 1-edges lost in $k = O(\log n)$ iterations is at most

$$\sum_{i=1}^{k} (\sum_{v \in L_i} W_{G_i}(v)) + \frac{2\eta}{c \log n} < \sum_{i=1}^{k} \frac{\eta}{d \log n} + \frac{2\eta}{c \log n} = \frac{\eta}{\hat{d}} + \frac{\eta}{c \log n} < \eta$$

The equality follows for a suitable constant $\hat{d}$ as $k = O(\log n)$. The final inequality holds for a suitable choice of the constants c and d. But this is a contradiction since we have $G_k = \phi$.

Since we choose j to be the best iteration, we have $\sum_{v \in L_j} W_G(v) \geq \frac{\eta}{d \log n}$ for some constant d. Hence the tree T_j is an α-thick tree of G for $\alpha = \frac{1}{d \log n}$ and the theorem follows by Lemma 1. $\square$

General Weighted Graphs. We now consider the *weighted connected maximum cut* (WCMC) problem. Formally, we are given a graph $G = (V, E)$ and a weight function $w : E \to \mathbb{R}^+ \cup \{0\}$. The goal is to find a subset S of vertices that induces a connected subgraph and maximizes the quantity $\sum_{e \in \delta(S)} w(e)$. We obtain a $\Omega(\frac{1}{\log^2 n})$ approximation algorithm for this problem. Our basic strategy is to group edges having nearly the same weight into a class and thus create $O(\log n)$ classes. We then solve the b-CMC problem for each class independently and return the best solution. Due to space limitations, we defer the details to the full version [22].

3   CMC in Planar and Bounded Genus Graphs

In this section, we consider the CMC problem in planar graphs and more generally, in graphs with genus bounded by a constant. We show that the CMC problem has a PTAS in bounded genus graphs.

3.1   PTAS for Bounded Genus Graphs

We use the following (paraphrased) contraction decomposition theorem by Demaine, Hajiaghayi and Kawarabayashi [11].

Theorem 2. ([11]) *For a bounded-genus graph G and an integer k, the edges of G can be partitioned into k color classes such that contracting all the edges in any color class leads to a graph with treewidth $O(k)$. Further, the color classes are obtained by a radial coloring and have the following property: If edge $e = (u, v)$ is in class i, then every edge e' such that $e' \cap e \neq \phi$ is in class $i - 1$ or i or $i + 1$.*

Given a graph G of constant genus, we use Theorem 2 appropriately to obtain a graph H with constant treewidth. In the full version of the paper [22], we show that one can solve the CMC problem optimally in polynomial time on graphs with constant treewidth by dynamic programming.

Theorem 3 *If the CMC problem can be solved optimally on graphs of constant treewidth, then there exists a polynomial time $(1 - \epsilon)$ approximation algorithm for the CMC problem on bounded genus graphs (and hence on planar graphs).*

Proof. Let $G = (V, E)$ be the graph of genus bounded by a constant and let S denote the optimal CMC of G and $\psi = |\delta(S)|$ be its size. Using Theorem 2 with $k = \frac{3}{\epsilon}$, we obtain a partition of the edges E into $\frac{3}{\epsilon}$ color classes namely $C_1, C_2, \ldots, C_{\frac{3}{\epsilon}}$. We further group three consecutive color classes into $\frac{1}{\epsilon}$ groups $G_1, \ldots, G_{\frac{1}{\epsilon}}$ where $G_j = C_{3j-2} \cup C_{3j-1} \cup C_{3j}$. Let G_{j^*} denote the group that intersects the least with the optimal connected max cut of G, i.e., $j^* = \arg\min_j(|G_j \cap \delta(S)|)^3$. As the $\frac{1}{\epsilon}$ groups partition the edges, we have $|G_{j^*} \cap \delta(S)| \leq \epsilon\psi$. Let $i = 3j^* - 1$, so that $G_{j^*} = C_{i-1} \cup C_i \cup C_{i+1}$. Let $H = (V_H, E_H)$ denote the graph of treewidth $O(\frac{1}{\epsilon})$ obtained by contracting all edges of color C_i.

We first show that H has a CMC of size at least $(1 - \epsilon)\psi$. For a vertex $v \in V_H$, let $\mu(v) \subseteq V$ denote the set of vertices of G that have merged together to form v due to the contraction. We define a subset $S' \subset V_H$ as $S' = \{v \in V_H \mid \mu(v) \cap S \neq \phi\}$. Note that because we contract edges (and not delete them), S' remains connected. We claim that $|\delta(S')| \geq (1 - \epsilon)\psi$. Let $e = (u, v)$ be an edge in $\delta(S)$. Now $e \notin \delta(S')$ implies that at least one edge e' such that $e' \cap e \neq \phi$ has been contracted. By the property guaranteed by Theorem 2, we have that $e \in G_{j^*}$. Hence we have, $|\delta(S')| \geq |\delta(S) \setminus G_{j^*}| = |\delta(S)| - |G_{j^*} \cap \delta(S)| \geq (1 - \epsilon)\psi$.

Finally, given a connected max cut of size ψ in H, we can recover a connected max cut of size at least ψ in G by simply un-contracting all the contracted edges. Hence, by solving the CMC problem on H optimally, we obtain a $(1 - \epsilon)$ approximate solution in G.   □

3.2   NP-Hardness in Planar Graphs

We now describe a non-trivial polynomial time reduction of a *3-SAT* variant known as *planar monotone 3-SAT* (PM-3SAT) to the CMC problem on a planar

---

[3] We "guess" j^* by trying out all the $\frac{1}{\epsilon}$ possibilities.

graph, thereby proving that the latter is NP-hard. The following reduction is interesting as the classical max-cut problem can be solved optimally in polynomial time on planar graphs using duality. In fact, it was earlier claimed erroneously that even CMC can be solved similarly [21].

An instance of PM-3SAT is a 3-CNF boolean formula ϕ such that -

a) A clause contains either all positive literals or all negative literals.

b) The associated bipartite graph $G_\phi{}^4$ is planar.

c) Furthermore, G_ϕ has monotone, rectilinear representation. We refer the reader to Berg and Khosravi [10] for a complete description. Figure 1a illustrates the rectilinear representation by a simple example.

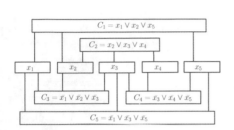

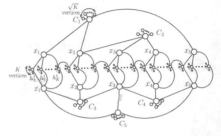

(a) Monotone rectilinear representation

(b) Reduction of PM-3SAT to a *planar* CMC instance

Fig. 1. Example illustrating the rectilinear representation and the reduction to a planar CMC instance of the formula $(x_1 \vee x_2 \vee x_5) \wedge (x_2 \vee x_3 \vee x_4) \wedge (\bar{x}_1 \vee \bar{x}_2 \vee \bar{x}_3) \wedge (\bar{x}_3 \vee \bar{x}_4 \vee \bar{x}_5) \wedge (\bar{x}_1 \vee \bar{x}_3 \vee \bar{x}_5)$.

Given such an instance, the PM-3SAT problem is to decide whether the boolean formula is satisfiable or not. Berg and Khosravi [10] show that the PM-3SAT problem is NP-complete.

The Reduction. Given a PM-3SAT formula ϕ, with a rectilinear representation, we obtain a polynomial time reduction to a planar CMC instance, there by showing that the latter is NP-hard. Let $\{x_i\}_{i=1}^n$ denote the variables of the PM-3SAT instance and $\{C_j\}_{j=1}^m$ denote the clauses. We construct a planar graph H_ϕ as follows. For every variable x_i, we construct the following gadget: We create two vertices $v(x_i)$ and $v(\bar{x}_i)$ corresponding to the literals x_i and $\bar{x}_i$. Additionally, we have $K > m^2$ "helper" vertices, $h_1^i, h_2^i, \ldots, h_K^i$ such that each h_k^i is adjacent to both x_i and $\bar{x}_i$. Further, for every h_k^i we add a set L_k^i of K new vertices that are adjacent only to h_k^i. Now, in the rectilinear representation of the PM-3SAT, we replace each variable rectangle by the above gadget. For two adjacent variable rectangles in the rectilinear representation, say x_i and x_{i+1}, we connect

---

^{4} G_ϕ has a vertex for each clause and each variable and an edge between a clause and the variables that it contains

the helpers h_K^i and h_1^{i+1}. For every clause C_j, H_ϕ has a corresponding vertex $v(C_j)$ with edges to the three literals in the clause. Finally, for each vertex $v(C_j)$, we add a set L_j of $\sqrt{K}$ new vertices adjacent only to $v(C_j)$. It is easy to observe that the reduction maintains the planarity of the graph. Figure 1b illustrates the reduction by an example.

The NP-hardness of the planar connected max cut is a consequence of the following theorem. We defer the proof to the full version [22].

Theorem 4. *Let H_ϕ denote an instance of the planar CMC problem corresponding to an instance ϕ of PM-3SAT obtained as per the reduction above. Then, the formula ϕ is satisfiable if and only if there is a solution S to the CMC problem on H_ϕ with $|\delta_{H_\phi}(S)| \geq m\sqrt{K} + nK + nK^2$.*

References

[1] Badanidiyuru, A., Vondrák, J.: Fast algorithms for maximizing submodular functions. In: SODA, pp. 1497–1514 (2014)

[2] Buchbinder, N., Feldman, M., Naor, J., Schwartz, R.: A tight linear time (1/2)-approximation for unconstrained submodular maximization. In: FOCS, pp. 649–658 (2012)

[3] Buchbinder, N., Feldman, M., Naor, J., Schwartz, R.: Submodular maximization with cardinality constraints. In: SODA, pp. 1433–1452 (2014)

[4] Calinescu, G., Chekuri, C., Pál, M., Vondrák, J.: Maximizing a submodular set function subject to a matroid constraint. In: IPCO, pp. 182–196 (2007)

[5] Censor-Hillel, K., Ghaffari, M., Giakkoupis, G., Haeupler, B., Kuhn, F.: Tight bounds on vertex connectivity under vertex sampling. In: SODA (2015)

[6] Censor-Hillel, K., Ghaffari, M., Kuhn, F.: A new perspective on vertex connectivity. In: SODA, pp. 546–561 (2014)

[7] Chekuri, C., Ene, A.: Submodular Cost Allocation Problem and Applications. In: Aceto, L., Henzinger, M., Sgall, J. (eds.) ICALP 2011, Part I. LNCS, vol. 6755, pp. 354–366. Springer, Heidelberg (2011)

[8] Cygan, M.: Deterministic parameterized connected vertex cover. In: Fomin, F.V., Kaski, P. (eds.) SWAT 2012. LNCS, vol. 7357, pp. 95–106. Springer, Heidelberg (2012)

[9] Das, B., Bharghavan, V.: Routing in ad-hoc networks using minimum connected dominating sets. In: ICC, vol. 1, pp. 376–380 (1997)

[10] de Berg, M., Khosravi, A.: Finding perfect auto-partitions is NP-hard. In: EuroCG 2008, pp. 255–258 (2008)

[11] Demaine, E.D., Hajiaghayi, M., Kawarabayashi, K.-I.: Contraction decomposition in H-minor-free graphs and algorithmic applications. In: STOC, pp. 441–450 (2011)

[12] Du, D.Z., Wan, P.J.: Connected dominating set: theory and applications. Springer optimization and its applications (2013)

[13] Eisenbrand, F., Grandoni, F., Rothvoß, T., Schäfer, G.: Approximating connected facility location problems via random facility sampling and core detouring. In: SODA, pp. 1174–1183 (2008)

[14] Fakcharoenphol, J., Rao, S., Talwar, K.: A tight bound on approximating arbitrary metrics by tree metrics. In: STOC, pp. 448–455 (2003)

[15] Feige, U., Mirrokni, V.S., Vondrak, J.: Maximizing non-monotone submodular functions. SIAM Journal on Computing 40(4), 1133–1153 (2011)

[16] Felzenszwalb, P.F., Huttenlocher, D.P.: Efficient graph-based image segmentation. International Journal of Computer Vision 59(2), 167–181 (2004)

[17] Garg, N., Konjevod, G., Ravi, R.: A polylogarithmic approximation algorithm for the group Steiner tree problem. In: SODA, pp. 253–259 (1998)

[18] Goemans, M.X., Williamson, D.P.: Improved approximation algorithms for maximum cut and satisfiability problems using semidefinite programming. Journal of the ACM 42(6), 1115–1145 (1995)

[19] Guha, S., Khuller, S.: Approximation algorithms for connected dominating sets. Algorithmica 20(4), 374–387 (1998)

[20] Hadlock, F.: Finding a maximum cut of a planar graph in polynomial time. SIAM Journal on Computing 4(3), 221–225 (1975)

[21] Haglin, D.J., Venkatesan, S.M.: Approximation and intractability results for the maximum cut problem and its variants. IEEE Transactions on Computers 40(1), 110–113 (1991)

[22] Hajiaghayi, M.T., Kortsarz, G., MacDavid, R., Purohit, M., Sarpatwar, K.: Approximation algorithms for connected maximum cut and related problems. CoRR (2015), http://arxiv.org/abs/1507.00648

[23] Khot, S., Kindler, G., Mossel, E., O'Donnell, R.: Optimal inapproximability results for MAX-CUT and other 2-variable CSPs? SIAM Journal on Computing 37(1), 319–357 (2007)

[24] Khuller, S., Purohit, M., Sarpatwar, K.K.: Analyzing the optimal neighborhood: Algorithms for budgeted and partial connected dominating set problems. In: SODA, pp. 1702–1713 (2014)

[25] Kuo, T.-W., Lin, K.C.-J., Tsai, M.-J.: Maximizing submodular set function with connectivity constraint: theory and application to networks. In: INFOCOM, pp. 1977–1985 (2013)

[26] Motwani, R., Raghavan, P.: Randomized Algorithms. Cambridge University Press (1995)

[27] Nemhauser, G.L., Wolsey, L.A., Fisher, M.L.: An analysis of approximations for maximizing submodular set functions-I. Mathematical Programming 14(1), 265–294 (1978)

[28] Slav Petrov. Image segmentation with maximum cuts (2005)

[29] Räcke, H.: Optimal hierarchical decompositions for congestion minimization in networks. In: STOC, pp. 255–264 (2008)

[30] Solis-Oba, R.: 2-Approximation algorithm for finding a spanning tree with maximum number of leaves. In: Bilardi, G., Pietracaprina, A., Italiano, G.F., Pucci, G. (eds.) ESA 1998. LNCS, vol. 1461, pp. 441–452. Springer, Heidelberg (1998)

[31] Swamy, C., Kumar, A.: Primal–dual algorithms for connected facility location problems. Algorithmica 40(4), 245–269 (2004)

[32] Vicente, S., Kolmogorov, V., Rother, C.: Graph cut based image segmentation with connectivity priors. In: CVPR, pp. 1–8 (2008)

The Offset Filtration of Convex Objects[*]

Dan Halperin[1], Michael Kerber[2], and Doron Shaharabani[1]

[1] Tel Aviv University, Tel Aviv, Israel
danha@post.tau.ac.il, doron.s@hotmail.com
[2] Max Planck Institute for Informatics, Saarbrücken, Germany
mkerber@mpi-inf.mpg.de

Abstract. We consider offsets of a union of convex objects. We aim for
a filtration, a sequence of nested cell complexes, that captures the topological evolution of the offsets for increasing radii. We describe methods
to compute a filtration based on the Voronoi diagram of the given convex objects. We prove that, in two and three dimensions, the size of
the filtration is proportional to the size of the Voronoi diagram. Our
algorithm runs in $\Theta(n \log n)$ in the 2-dimensional case and in expected
time $O(n^{3+\epsilon})$, for any $\epsilon > 0$, in the 3-dimensional case. Our approach is
inspired by alpha-complexes for point sets, but requires more involved
machinery and analysis primarily since Voronoi regions of general convex objects do not form a good cover. We show by experiments that our
approach results in a similarly fast and topologically more stable method
compared to approximating the input by point samples.

1 Introduction

Motivation. The theory of *persistent homology* has led to a new way of understanding data through its topological properties, commonly referred as *topological data analysis.* The most common setup assumes that the data is given as
a finite set of points and analyzes the sublevel sets of the distance function to
the point set. An equivalent formulation is to take offsets of the point sets with
increasing offset parameter and to study the changes in the hole structure of the
shape obtained by the union of the offset balls (Figure 1).

We pose the question how to generalize the default framework for point sets
to more general input shapes. While there is no theoretical obstacle to consider
distance functions from shapes rather than points (at least for reasonably "nice"
shapes), it raises a computational question: How can the topological information
be encoded in a combinatorial structure of small size?

With the wealth of applications of persistence of point set data, and together
with the challenges raised by the extension from point sets to sets of convex

[*] Work by D.H. and D.S. has been supported in part by the Israel Science Foundation
(grant no. 1102/11), by the German-Israeli Foundation (grant no. 1150-82.6/2011),
and by the Hermann Minkowski–Minerva Center for Geometry at Tel Aviv University. M.K. acknowledges support by the Max Planck Center of Visual Computing
and Communication.

© Springer-Verlag Berlin Heidelberg 2015
N. Bansal and I. Finocchi (Eds.): ESA 2015, LNCS 9294, pp. 705–716, 2015.
DOI: 10.1007/978-3-662-48350-3_59

Fig. 1. From left to right, we see an example shape, three offsets with increasing radii $r_1 < r_2 < r_3$, and the 1-barcode of the shape. While being simply-connected initially, two holes have been formed at radius r_1, one of which disappears for a slightly larger offset value while the other one *persists* for a large range of scales. At r_2, we see the formation of another rather short-lived hole. The barcode summarizes these facts by displaying one bar per hole. The bar spans over the range of offset radii for which the hole is present.

objects, we believe that the latter is a logical next step of investigation. Our motivation originates from the increasingly popular application of 3D printing. A common problem in this context is that often available models of shapes contain features that complicate the printing process, or turn it impossible altogether. A ubiquitous example is the presence of thin features which may easily break, and call for thickening. One work-around is to offset the model by a small value to stabilize it, but the optimal offset parameter is unclear, as it should get rid of many spurious features of the model without introducing too many new ones. Moreover, one would prefer *local thickening* [26], and possibly thickening by different offset size in different parts of the model. A by-product of our work here is a step toward automatically detecting target regions for local thickening that do not incur spurious artifacts. Persistent homology provides a *barcode* which constitutes a summary of the hole structure of the offset shape for any parameter value (Figure 1) which is helpful for the choice of a good offset value.

Problem Definition and Contribution. We design, analyze, implement, and experimentally evaluate algorithms for computing persistence barcodes of convex input objects. More precisely, we concentrate on the problem of computing a *filtration*, a sequence of nested combinatorial cell complexes that undergoes the same topological changes as the offset shapes. Since the input objects are convex, the nerve theorem asserts that the intersection patterns of the offsets (called the *nerve*) reveal the entire topological information. This leads to the generalization of Čech filtrations from point sets to our scenario. The resulting filtration has a size of $O(n^{d+1})$ for n input objects in d-dimensional Euclidean space, where the size of the filtration is defined to be the number of simplex insertions involved over the entire filtration. This size is already problematic for small d and a natural idea to reduce its size is to consider *restricted offsets*, that is, intersecting the offset of an input object with its *Voronoi region*, the portion of the space closest to the object. This approach is again inspired by the analogous case of point sets, where *alpha-complexes* are preferred over the Čech complexes for small dimensions. However, the approach for point sets does *not* directly carry over to arbitrary convex objects: Voronoi regions of convex objects can intersect in non-contractible patterns, which prevents the application of the nerve theorem.

Our first result is that in $\mathbb{R}^2$, the non-contractibility does not really cause problems: the barcode of convex polygons is encoded in the barcode of the nerve of their restricted offsets, despite the non-contractible intersections. As a result we obtain a linear-size filtration, improving the cubic size one obtained from using the unrestricted nerves. The filtration can be computed in time $O(n \log n)$, dominated by the computation time of the Voronoi diagram. While the proof ultimately still relies on the nerve theorem, it requires a deeper investigation of the structure of Voronoi diagrams of convex objects. Moreover, it requires a slight generalization of a result in [9] (see Theorem 1 below), showing that the nerve isomorphism commutes with inclusions, to the case of simplicial complexes connected by certain simplicial maps. The analogous statement in $\mathbb{R}^3$ is not true.

Our second result is a general construction of a cell complex with the desired barcode in three dimensions. Our construction scheme computes the Voronoi diagram of the input sites as a preprocessing step and cuts (subdivides) the edges and faces of the Voronoi diagram into smaller pieces in a controlled way. The resulting refinement of the Voronoi diagram gives rise to a dual cell complex whose size is linear in the size of the Voronoi diagram of the input sites. As the latter is known to be bounded by $O(n^{3+\epsilon})$, our filtration is significantly smaller than $O(n^4)$, as obtained by a Čech-like filtration. The time for computing the filtration is bounded from above by $O(n^{3+\epsilon})$. The correctness proof works by (conceptually) "thicken up" lower-dimensional cells of the Voronoi diagram to obtain a good cover of the space, for which the nerve theorem applies.

We have implemented our algorithm for polygons and report on extensive experimental evaluation. In particular, we compare our approach with the natural alternative to replace the input polygons with sufficiently dense point samples. Although the point sample approach yields very close approximations to the exact barcode in a comparable running time, the approximation error induced by the sampling results in additional noise on a large range of scales and therefore makes the topological analysis of the offset filtration more difficult.

Some proofs throughout the text are omitted due to lack of space. For additional details and proofs see the arxiv version [17].

Related Work. Since its introduction in [14], persistent homology has become an active area of research, including theoretical, algorithmic, and application results; we refer to the textbook [13] and the surveys [8,15] for an overview. The information gathered by persistence is usually displayed either in terms of a barcode (as in this work) or, equivalently, via a *persistence diagram* [13].

The textbook [13] describes the most common approaches for computing filtrations of point sets, including Čech- and alpha-complexes. Another common construction is the *(Vietoris-)Rips* complex: it approximates the Čech complex in the sense that it is nested between two Čech complexes on similar scales. However, this property does not carry over to the case of arbitrary convex objects. A recent research topic is to come up with sparsified versions of Rips [12,24] and Čech complexes [7,22] to lower the filtration size; our work is in the same spirit.

Topological methods for shape analysis have been extensively studied: a commonly used concept are *Reeb graphs* which yield a skeleton representing the connectivity of the shape and can be seen as a special case of persistent homology in dimension 0; see [5] for ample applications. The full theory of persistent homology has also been applied to various tasks in shape analysis, including shape segmentation [25] and partial shape similarity [16]. While these works study the *intrinsic* properties of a shape through descriptor functions independent of the embedding, our problem setup rather asks about *extrinsic* properties, that is, how the shape is embedded in ambient space.

2   Topological Background

We review standard notation in persistent homology and dualizations of set covers through nerves. We assume familiarity with basic topological notions such as simplicial complexes, homology groups and persistent homolgy; the necessary background is covered by the textbook [13] and in more detail by [18].

Filtration. We call a collection of spaces $(Q_\alpha)_{\alpha \geq 0}$ with the property that $Q_\alpha \subseteq Q_{\alpha'}$ whenever $\alpha \leq \alpha'$ a *filtration (induced by inclusion)*. Intuitively, a filtration is merely an increasing sequence of spaces as illustrated in Figure 1 (left). To obtain its *(persistent) barcode*, we apply the *homology functor*: let $H_p(Q)$, the p-th homology group of Q over an arbitrary fixed base field, with $p \geq 0$. Then, the object $(H_p(Q_\alpha))_{\alpha \geq 0}$ together with the induced maps $F_{\alpha,\alpha'} : H_p(Q_\alpha) \to H_p(Q_{\alpha'})$ defines a so-called *persistence module* and encodes the births and deaths of homology classes during the filtration process. This information can be compactly described in terms of a p-barcode. Figure 1 (right) displays the 1-barcode of the example. This construction works generally for filtrations of simplicial complexes (using simplicial homology), of CW-complexes (using cellular homology) and of subsets of $\mathbb{R}^d$ (using singular homology).

Nerves. Let $\mathcal{P} := \{P^1, \ldots, P^n\}$ be a collection of non-empty sets in a common domain. The *underlying space* is defined as $|\mathcal{P}| := \bigcup_{i=1,\ldots,n} P^i$. We call a non-empty subset $\{P^{i_1}, \ldots, P^{i_k}\} \subseteq \mathcal{P}$ *intersecting*, if $\bigcap_{j=1}^{k} P^{i_j} \neq \emptyset$. The *nerve* $\mathrm{Nrv}(\mathcal{P})$ of $\mathcal{P}$ is the collection of all intersecting subsets. It is clear by definition that every singleton set $\{P^i\}$ is in the nerve, and that any non-empty subset of an intersecting set is intersecting. The latter property implies that the nerve is a *simplicial complex*: the singleton sets $\{P_i\}$ are the *vertices* of that complex. We call $\mathcal{P}$ a *good cover* if all sets in the collection are closed and triangulable, and any intersecting subset yields a contractible intersection. For example, any collection of closed convex sets forms a good cover. The *Nerve theorem* states that if $\mathcal{P}$ is a good cover, $|\mathcal{P}|$ is homotopically equivalent to $\mathrm{Nrv}(\mathcal{P})$. In particular, $H_p(|\mathcal{P}|)$ is isomorphic to $H_p(\mathrm{Nrv}(\mathcal{P}))$ for all $p \geq 0$.

Barcodes of Shapes. We let $\mathrm{dist}(\cdot, \cdot)$ denote the Euclidean distance function. For a point set $A \subset \mathbb{R}^d$ and $x \in \mathbb{R}^d$, we set $\mathrm{dist}(x, A) := \inf_{y \in A} \mathrm{dist}(x, y)$. Then, $\mathrm{dist}(\cdot, A) : \mathbb{R}^d \to \mathbb{R}$ is called the *distance function from A*, and $A_\alpha := \{x \in \mathbb{R}^d \mid \mathrm{dist}(x, A) \leq \alpha\}$ is called the *α-offset of A*. With $\mathcal{P}$ as above, we write

$\mathcal{P}_\alpha := \{P_\alpha^1, \ldots, P_\alpha^n\}$ for the collection α-offsets of $\mathcal{P}$. In particular, $\mathcal{P}_0 = \mathcal{P}$. We call $(|\mathcal{P}_\alpha|)_{\alpha \geq 0}$ the *offset-filtration* of $\mathcal{P}$. We pose the question of how to compute the barcode of the offset filtration of convex objects efficiently. See Figure 1 for an illustration of these concepts.

We define the analogue of Čech filtrations: We call $(\mathrm{Nrv}(\mathcal{P}_\alpha))_{\alpha \geq 0}$ the *nerve filtration* of $\mathcal{P}$; it is indeed a filtration because for $\alpha_1 \leq \alpha_2$, $\mathrm{Nrv}(\mathcal{P}_{\alpha_1}) \subseteq \mathrm{Nrv}(\mathcal{P}_{\alpha_2})$. The following result [9] implies immediately that the p-barcodes of the offset filtration and the nerve filtration of $\mathcal{P}$ are equal for all $p \geq 0$.

Theorem 1 (Persistence Nerve Theorem). *Let* $(P_\alpha^1)_{\alpha \geq 0}, \ldots, (P_\alpha^n)_{\alpha \geq 0}$ *be filtrations such that for each* $\alpha \geq 0$, $\mathcal{P}_\alpha := \{P_\alpha^1, \ldots, P_\alpha^n\}$ *is a good cover. Then, the barcodes of* $(|\mathcal{P}_\alpha|)_{\alpha \geq 0}$ *and* $(\mathrm{Nrv}(\mathcal{P}_\alpha))_{\alpha \geq 0}$ *are the same.*

The nerve only changes for values where a collection of offsets of objects becomes intersecting. We call such an offset value *nerve-critical*. Since $|\mathcal{P}| \subset \mathbb{R}^d$, we can restrict to collections of size at most $d + 1$, as the offset filtration has a trivial barcode in dimension d and higher. Sorting the nerve-critical values $0 = \alpha_0 < \alpha_1 < \ldots < \alpha_m$ and setting $K_i := \mathrm{Nrv}(\mathcal{P}_{\alpha_i})$, the nerve filtration simplifies to the finite filtration $K_0 \subset K_1 \subset \ldots \subset K_m$ whose barcode can be computed using standard methods; see [14,28] or [3,6] for an optimized variant. As K_m contains a simplex for any subset of $\mathcal{P}$ of size up to $d + 1$, its size is $\Theta(n^{d+1})$.

3   Barcodes of Restricted Offsets

Let $\mathcal{P} := \{P^1, \ldots, P^n\}$ be convex polyhedra in $\mathbb{R}^d$, that is, each P^i is the intersection of finitely many half-spaces. The major disadvantage of the construction of Section 2 is the sheer size of the resulting filtration, $\Theta(n^{d+1})$. Our goal is to come up with a filtration that yields the same barcode and is substantially smaller in size. Our approach is reminiscent of alpha-complexes for point sets, but it requires additional ideas for being applicable to convex objects.

From now on, we make the following assumptions for simplicity: We refer to the elements of $\mathcal{P}$ as *sites*. We restrict our attention to $d \in \{2, 3\}$, that is, sites are polygons ($d = 2$) or polyhedra ($d = 3$). We assume the sites to be pairwise disjoint and in *general position*, that is, for any pair P^i, P^j of sites, there is a unique pair of points $x^i \in \partial P^i$, $x^j \in \partial P^j$ that realizes the distance between the sites. Moreover, we assume that the number of vertices, edges and faces of each site is bounded by a constant. For a point $p \in \mathbb{R}^d$, the site P^k is *closest* if $\mathrm{dist}(p, P^k) \leq \mathrm{dist}(p, P^\ell)$ for any $1 \leq \ell \leq n$. We assume for simplicity the generic case that no point has more than $d+1$ closest sites. We set $\mathrm{dist}(x) := \mathrm{dist}(x, |\mathcal{P}|)$.

The *Voronoi diagram* [2] $\mathrm{Vor}(\mathcal{P})$ is the partition of the space into maximal connected components with the same set of closest sites. The Voronoi diagram is an *arrangement* in $\mathbb{R}^d$, and its *combinatorial complexity* is the number of cells. The *Voronoi region* of P^k, denoted by V^k, is the (closed) set of points for which P^k is one of its closest sites. For a cell σ of $\mathrm{Vor}(\mathcal{P})$, we call $\mathrm{crit}(\sigma) := \inf_{x \in \sigma} \mathrm{dist}(x)$ the *critical value* of σ and a point x that attains this infimum a *critical point* of σ. Note that critical points of a cell may lie on its boundary.

For any two sites P^i, P^j, the set of all points x with $\text{dist}(x, P^i) = \text{dist}(x, P^j) = \alpha$ is the intersection $\partial(P^i \oplus \mathbb{B}_\alpha) \cap \partial(P^j \oplus \mathbb{B}_\alpha)$, where $\mathbb{B}_\alpha$ is the ball of radius α, and $\oplus$ denoting the Minkowski sum. The *bisector* ρ is the set of points x that satisfy $\text{dist}(x, P^i) = \text{dist}(x, P^j)$. By general position of the sites, there is a unique point on ρ that minimizes $\text{dist}(\cdot, P^i)$. More generally, for $\alpha \geq 0$, let $\rho_\alpha := \{x \in \rho \mid \text{dist}(x, P^i) \leq \alpha\}$. We will frequently use the fact that for any α, ρ_α is empty or contractible. This is implied by the following generalization the well-known *pseudodisk-property* [21], [10, Thm.13.8]. We refer to [17, Appendix B] for the proof of the case $d = 3$.

Theorem 2. *For $d \in \{2, 3\}$, let P_1, P_2 be two convex disjoint polytopes in $\mathbb{R}^d$ in general position and let $\mathbb{B}$ be the unit ball. Then, $\partial(P_1 \oplus \mathbb{B}) \cap \partial(P_2 \oplus \mathbb{B})$ is either empty, a single point, or homeomorphic to a $(d-2)$-sphere.*

The *restricted α-offset* of P^k is defined as $Q_\alpha^k := P_\alpha^k \cap V^k$. We set $\mathcal{Q}_\alpha := \{Q_\alpha^1, \dots, Q_\alpha^n\}$ and $\mathcal{Q} := \mathcal{Q}_0$. In the same way as in Section 2, we define the *restricted nerve filtration* as $(\text{Nrv}(\mathcal{Q}_\alpha))_{\alpha \geq 0}$ and $\mathcal{Q}$-*critical* values as those values where a simplex enters the restricted nerve filtration. The restricted nerve filtration can be expressed by a finite sequence of simplicial complexes that changes precisely at the $\mathcal{Q}$-critical values. The size of the filtration is bounded by the combinatorial complexity of the Voronoi diagram. Moreover, the $\mathcal{Q}$-critical value of a simplex associated with a Voronoi cell σ equals the critical value of σ.

Restricting the offsets to Voronoi regions brings a problem: $\mathcal{Q}_\alpha$ is not necessarily a collection of convex sets, since V^k is not convex in general. Even worse, $\mathcal{Q}_\alpha$ might not be a good cover. For instance, in Figure 2 we see that the Voronoi regions of the two large polygons intersect in two segments. This means that Theorem 1 does not apply to this case.

4   Barcodes of Restricted Offsets in 2D

We first restrict to the case $d = 2$, that means, our input sites are interior-disjoint convex polygons. While the restriction of offsets invalidates the proof strategy of using the persistence nerve theorem, it does not invalidate the statement, at least in dimensions 0 and 1.

Theorem 3. *For convex polygonal sites in $\mathbb{R}^2$, the 0- and 1-barcode of the restricted nerve filtration are equal to the 0- and 1-barcode of the offset filtration, respectively.*

As a consequence of this theorem, we obtain a filtration of size $O(n)$ that has the same barcode as the offset filtration; the size follows from the fact that the complexity of the Voronoi diagram is $O(n)$. This is much smaller than the $O(n^3)$ filtration obtained by the unrestricted nerve. The construction time is dominated by computing the Voronoi diagram and thus bounded by $O(n \log n)$ [27].

The proof of Theorem 3 requires substantially more algebraic machinery than what we have introduced and is omitted due to lack of space. We provide a brief summary of the proof, and refer to [17, Appendix A] for the full proof.

A non-contractible intersection of two restricted offsets of sites is always caused by a bounded connected component in the complement of their union, called a *surrounded region*. Such a region is caused by one or more sites in its inside; we call such sites *inactive* when they become part of a surrounded region, and *active* otherwise. For instance, the two inner sites in Figure 2 become surrounded by the outer sites eventually. The idea of the proof is to ignore inactive sites from the considerations. For any α, the restricted offsets of active sites form a good cover and the nerve theorem applies. Moreover, we can show that the union of restricted offsets does not change when removing the inactive sites, because the surrounding regions "conquer" the entire territory previously occupied by inactive sites. That means, for any $\alpha \geq 0$, the union of restricted offsets has the same homology as the nerve of restricted offset of active sites.

The first major result we show is that the nerves of restricted offsets of active sites and of all sites have the same 0- and 1-homology. The proof (for 1-homology) is based on an explicit construction that transforms any 1-cycle in the nerve of all sites to an homologous cycle that only includes active sites.

To finish the proof, we have to show that the induced isomorphisms commute with inclusions. This case is not covered by Theorem 1 because the offsets do not form a good partition. Instead, we consider the nerves of restricted active sites as an intermediate structure. They do not form a filtration because sites disappear from the nerve when becoming inactive, so the simplicial complexes are not nested. Still, the nerves can be connected by *simplicial maps* instead of inclusions and the concept of barcodes extends to this setup. As second major result, we show that the isomorphism induced by the nerve theorem between offsets sites and nerve of restricted active offsets commutes with these simplicial maps. The proof requires the study of this isomorphism in detail and extends the proof of Theorem 1 in this setting.

Theorem 3 does not generalize to the 2-barcode. For instance, in Figure 2, we see four sites in $\mathbb{R}^2$ where every triple of Voronoi regions intersects, but there is no common intersection of all four of them. Consequently, their nerve consists of the four boundary triangles of a tetrahedron and therefore carries non-trivial 2-homology. We refer to such homology classes as "ghost features". In $\mathbb{R}^2$, the offset filtration can clearly not form any void and we can therefore safely ignore all ghosts. In $\mathbb{R}^3$, however, the 2-barcode carries information about the offset and the ghosts need to be distinguished from real features.

Fig. 2. A ghost sphere

5   Barcodes of Restricted Offsets in 3D

As Theorem 3 does not generalize to higher-dimensions, we now present a refinement of the nerve construction for $\mathbb{R}^3$. Reconsidering the "ghost example" from Section 3, it seems attractive to pass to the multi-nerve [11], that is, introducing a distinct simplex for each lower-dimensional cell of the Voronoi diagram. However, this approach is not sufficient. First, Voronoi cells might be non-simply connected. For example, in Figure 3 (top), the Voronoi cell of the two large polyhedra will contain an unbounded face with a hole in its middle. Furthermore, even if the Voronoi cells form a good cover, this may not be true for the restricted offsets at all scales α. This can be observed at Figure 3 (bottom), where two different connected components are formed when the offsets of the two large polyhedra first intersect, despite the fact that the Voronoi cells form a good cover.

An arrangement $\mathcal{A}$ in $\mathbb{R}^3$ is a *refinement* of $\mathrm{Vor}(\mathcal{P})$ in $\mathbb{R}^3$ if every 0-, 1-, 2-, or 3-dimensional cell of $\mathcal{A}$ is contained in a cell of $\mathrm{Vor}(\mathcal{P})$. For a cell $\sigma \in \mathcal{A}$ and $\alpha \geq 0$, define the *restricted cell* $\sigma_\alpha := \{x \in \sigma \mid \mathrm{dist}(x) \leq \alpha\}$. We call σ *sublevel-contractible* if for all $\alpha \geq 0$, σ_α is empty or contractible. Note that sublevel-contractible cells are contractible. We call an arrangement $\mathcal{A}$ a *sublevel-contractible refinement* of $\mathrm{Vor}(\mathcal{P})$, if $\mathcal{A}$ is a refinement of $\mathrm{Vor}(\mathcal{P})$ and every cell of $\mathcal{A}$ is sublevel-contractible. As before, we define the critical value of a cell $\sigma \in \mathcal{A}$ as $\mathrm{crit}(\sigma) := \inf\{\alpha \in \mathbb{R} \mid \sigma_\alpha \neq \emptyset\}$.

Fig. 3. Problems in 3D

Dualization. An arrangement $\mathcal{A}$ in $\mathbb{R}^3$ gives rise to a dual structure in a natural way: fixing two cells σ, τ of $\mathcal{A}$ with $\dim(\sigma) < \dim(\tau)$, we have that either σ is contained in or completely disjoint from the boundary of τ. In the former case, we say that σ is *incident* to τ. If σ is incident to τ, $\mathrm{crit}(\sigma) \geq \mathrm{crit}(\tau)$. The *dualization* $\mathcal{A}^*$ of $\mathcal{A}$ is defined as follows: for every cell σ of $\mathcal{A}$, $\mathcal{A}^*$ has a *dual cell* σ^* such that their dimensions add up to 3. The *boundary* of a dual

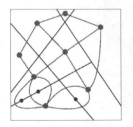

cell σ^* of dimension δ, $\partial(\sigma^*)$, is the set of all dual cells τ^* of dimension $\delta - 1$ such that σ is incident to τ. See the figure to the right for an illustration of a planar arrangement (black) and the dualization (blue).

The critical value of a dual cell is defined as the critical value of its primal counterpart. This turns the dualization into a filtered cell complex, since any dual cell has a critical value not smaller than any dual cell in its boundary. For $\alpha \in [0, \infty]$, we let $\mathcal{A}^*_\alpha$ denote the collection of dual cells with critical value at most α. Since $\partial\partial(\sigma) = 0$ for any σ, there is a well-defined homology group for each $\mathcal{A}^*_\alpha$, and therefore, a barcode of the dualization $\mathcal{A}^*$. We can now state the main result which allows us to express the barcode of the offset of three-dimensional shapes in terms of a combinatorial structure.

Theorem 4. *Let $\mathcal{P}$ be a collection of convex polyhedra in $\mathbb{R}^3$, and let $\mathcal{A}$ be a sublevel-contractible refinement of $\mathrm{Vor}(\mathcal{P})$. Then, the barcode of the offset filtration of $\mathcal{P}$ equals the barcode of $\mathcal{A}^*$.*

Sublevel-contractible Refinements. We are left with the question of how to obtain a sublevel-contractible refinement of a Voronoi diagram of convex polyhedra.

We remark that for $d = 2$, the Voronoi diagram of convex polygons is already sublevel-contractible, so no refinement is needed. The resulting dualization is precisely the multi-nerve [11]. However, we note that in light of Theorem 3, there is no need to consider dualizations at all in the planar case.

We turn to the case $d = 3$. Here, the Voronoi diagram is generally not sublevel-contractible, for the reasons given at the beginning of the section. To make it sublevel-contractible, we will cut every cell into sublevel-contractible pieces. We consider the case of a single connected component of a bisector (that is, a 2-cell of the Voronoi diagram) in isolation: as it will turn out, our method will ensure that all trisectors (i.e., points with the same distance to three sites) on its boundary will become sublevel-contractible as well.

Let us fix a 2-cell σ contained in the bisector ρ of two sites. The boundary of σ consists of 1-cells, where each 1-cell belongs to some trisector. Since ρ is homeomorphic to a plane, we can distinguish between an *outer boundary* of σ (which might be unbounded if σ is unbounded) and an arbitrary number of closed *inner boundaries*. The presence of inner boundaries turns σ non-simply connected, and thus σ is not sublevel-contractible (as in Figure 3 (top)). By Theorem 2 the distance function can only have one local minimum on ρ, and thus on σ. However, it is well possible that dist restricted to σ has local minima on $\partial(\sigma)$, both on inner and outer boundary components. The presence of such local minima turns the restricted cell σ_α disconnected for a certain range of scales and makes σ not sublevel-contractible as well (as in Figure 3 (bottom)).

We now define a sublevel-contractible refinement of σ into 2-cells by introducing cuts in σ. We start by cutting σ along a curve ψ on ρ that goes through the minimum o and that is unimodal for dist$(\cdot)$, that is, has a minimum at o and dist-monotone otherwise. Moreover, we require ψ to intersect every 1-cell in the boundary of σ only a constant number of times. Such a curve indeed exists since we can find two monotone paths from o to the outer boundary that avoid all inner boundaries.

Having cut using ψ, we introduce additional cuts: for any value $\beta > 0$, the β-*isoline* is the curve defined by all points p on the bisector with dist$(p) = \beta$. By Theorem 2, a β-isoline is a closed cycle on the bisector that loops around o. We define a *boundary-critical point* of σ as a local maximum or minimum of dist restricted to $\partial\sigma$. For a boundary-critical point p with $\beta = $ dist(p), an *isoline segment at p* is a maximal connected piece of the β-isoline within $\sigma \setminus \psi$ with p on its boundary. An isoline segment may degenerate into a point. We introduce cuts along all isoline segments for all boundary-critical points. Since ψ is unimodal, unbounded, and goes through o, it intersects any isoline twice. Hence, isoline segments cannot be closed curves. New 0-cells are added at boundary-critical points and at the endpoints of isoline segments. See Figure 4 for an illustration.

Lemma 5. *The cutting scheme from above yields a sublevel-contractible refinement of* Vor$(\mathcal{P})$.

Size and Computation of Sublevel-contractible Refinements. We can compute a sublevel-contractible refinement using a combinatorial algorithm. It traverses the

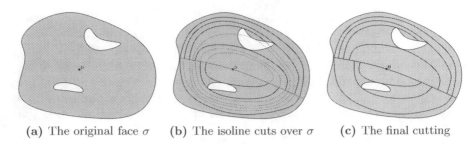

(a) The original face σ (b) The isoline cuts over σ (c) The final cutting

Fig. 4. A sublevel-contractible refinement of a 2-cell. The original 2-cell σ is shown in (a). In (b) we add the cut induced by the unbounded curve ψ (green) and cuts induced by the isoline segments at boundary-critical points (black). The part of an isoline that is not an isoline segment is shown in red. The final cutting is shown in (c).

2-cells σ of Vor($\mathcal{P}$) and cuts them according to the described strategy. For that, it maintains a sweep-line reflecting the intersection of a sweeping isoline with all boundary segments of σ. It performs two sweeps, constructing ψ and adding cuts induced by it in the first sweep, and cutting along isoline segments in the second sweep. It is not hard to argue that the described refinement only increases the complexity of the arrangement by a constant factor. Every cut originates at a 1-cell or a 0-cell of the Voronoi diagram, and we can charge the performed cut to that cell. We can show that every 1- and 0-cell is only charged a constant number of times – the only technical difficulty is to show that the number of local extrema on a trisector is constant. This follows from the fact that a trisector is defined by constantly many algebraic equations, exploiting that every site consists only of constantly many faces. Since every cut introduces only a constant number of new cells to the arrangement, we get the following result:

Theorem 6. *For a set $\mathcal{P}$ of convex, disjoint polyhedra in generic position, let $\pi(n)$ be the complexity of* Vor($\mathcal{P}$). *Then there exists a sublevel-contractible refinement of* Vor($\mathcal{P}$) *with $O(\pi(n))$ cells, which can be computed in $O(\pi(n)\log(\pi(n)))$ time, excluding the computation time of the Voronoi diagram.*

Together with Theorem 4, it follows that we can find a cellular filtration of size $O(\pi(n))$ whose barcode equals the barcode of the original offset filtration. The exact value of $\pi(n)$ is far from being settled but we know that it is bounded from below by $\Omega(n^2)$ (by a straightforward construction even for n points) and from above by $O(n^{3+\epsilon})$ [1,23]. We therefore get a filtration of size $O(n^{3+\epsilon})$, that is significantly smaller than the unrestricted nerve filtration of size $O(n^4)$, and that can be computed in $O(n^{3+\epsilon})$ expected time.

6   Polygons vs. Point Samples: Experimental Comparison

Computing Voronoi diagrams of polyhedra in space is a difficult problem. While efforts to compute it are underway, so far only restricted cases have been completed; see e.g., [19]. We therefore restrict our experiments to the planar case. Still, already in the plane approximating the shapes by point samples introduces noise that is hard to distinguish from real small features.

Comparison Details. We implemented the restricted nerve algorithm for the two-dimensional case as described in Section 4, using CGAL's 2D Segment Delaunay Graphs package [20] for computing the Voronoi diagram of convex polygons. For comparison purposes, we also implemented the unrestricted nerve filtration described in Section 2. In addition, we implemented an approximation of the barcode by approximating the input polygons with a finite point set whose Hausdorff-distance to the input is at most ϵ (by placing a grid of side length $\sqrt{2}\epsilon$ and taking as our sample the centers of all grid cells which intersect the input), and compute the barcode from the alpha complex of the points. The size of the filtration is linear in the number of sampled points. In all variants, after computing the filtration, we obtain the barcode using the PHAT library [4].

Analysis. We compare the three approaches for several inputs of increasing sizes, and report on their running time and filtration size. For brevity, we demonstrate our findings for only one example in Table 1. As expected, the unrestricted nerve yields large filtrations and high running times compared to the restricted nerve. Comparing the running time of the restricted nerve approach with that of the point sampling approach is difficult as it depends on the choice of ϵ. Still, our experiments show that for a comparably fast choice of ϵ, the approximated barcode contains many short bars that can be attributed to noise introduced by the approximation, and that can not be easily distinguished from genuine short bars that exist in the exact barcode. Thus, our approach is comparably fast to that of a "reasonable" approximation of the input, and has the advantage of obtaining an exact barcode. We note that our simple point-sampling approach could be significantly improved, for example, by sampling points only on the boundary of the polygons, however such methods would introduce more noise and would require additional post-processing.

Table 1. Running times (in s) and filtration sizes for an input consisted of 250 randomly generated convex polygons with a total of 1060 vertices

Approach	Time	Size
Restricted nerve	1.285	1473
Unrestricted nerve	9943	2604375
Point sample ($\epsilon = 1$)	0.217	9021
Point sample ($\epsilon = 0.1$)	3.3	505833

References

1. Agarwal, P., Aronov, B., Sharir, M.: Computing envelopes in four dimensions with applications. SIAM J. Comput. 26, 348–358 (1997)
2. Aurenhammer, F., Klein, R.: Chapter 5 - voronoi diagrams. In: Sack, J.-R., Urrutia, J. (eds.) Handbook of Computational Geometry, pp. 201–290. North-Holland (2000)
3. Bauer, U., Kerber, M., Reininghaus, J.: Clear and compress: Computing persistent homology in chunks. In: Topological Methods in Data Analysis and Visualization III. Springer (2014)
4. Bauer, U., Kerber, M., Reininghaus, J., Wagner, H.: Phat - persistent homology algorithms toolbox. In: 4th International Congress of Math. Software (2014)
5. Biasotti, S., Giorgi, D., Spagnuolo, M., Falcidieno, B.: Reeb graphs for shape analysis and applications. Theoretical Computer Science 392, 5–22 (2008)

6. Boissonnat, J.-D., Dey, T.K., Maria, C.: The compressed annotation matrix: An efficient data structure for computing persistent cohomology. In: Eur. Symp. on Alg., pp. 695–706 (2013)
7. Botnan, M., Spreemann, G.: Approximating persistent homology in Euclidean space through collapses. Appl. Algebra Eng. Commun. Comput. 26(1-2), 73–101 (2015)
8. Carlsson, G.: Topology and data. Bull. of the Amer. Math. Soc. 46, 255–308 (2009)
9. Chazal, F., Oudot, S.: Towards persistence-based reconstruction in Euclidean spaces. In: Proc. of the 24th ACM Symp. on Comput. Geom, pp. 232–241 (2008)
10. de Berg, M., van Kreveld, M., Overmars, M., Schwarzkopf, O.: Computational Geometry: Algorithms and Applications, 2nd edn. Springer (2000)
11. Colin de Verdière, É., Ginot, G., Goaoc, X.: Multinerves and Helly numbers of acyclic families. In: Proc. of the 28th ACM Symp. on Comput. Geom., pp. 209–218 (2012)
12. Dey, T., Fan, F., Wang, Y.: Graph induced complex on point data. In: Proc. of the 29th ACM Symp. on Comput. Geom., pp. 107–116 (2013)
13. Edelsbrunner, H., Harer, J.: Computational Topology. An Introduction. Amer. Math. Soc. (2010)
14. Edelsbrunner, H., Letscher, D., Zomorodian, A.: Topological persistence and simplification. Disc. Comput. Geom. 28(4), 511–533 (2002)
15. Edelsbrunner, H., Morozov, D.: Persistent homology: Theory and practice. In: Proc. of the Eur. Congress of Mathematics, pp. 31–50 (2012)
16. Di Fabio, B., Landi, C.: Persistent homology and partial similarity of shapes. Pattern Recognition Letters 33(11), 1445–1450 (2012)
17. Halperin, D., Kerber, M., Shaharabani, D.: The offset filtration of convex objects. CoRR, abs/1407.6132 (2014)
18. Hatcher, A.: Algebraic Topology. Cambridge University Press (2001)
19. Hemmer, M., Setter, O., Halperin, D.: Constructing the exact Voronoi diagram of arbitrary lines in three-dimensional space - with fast point-location. In: Eur. Symp. on Alg., pp. 398–409 (2010)
20. Karavelas, M.: 2D segment Delaunay graphs. In: CGAL User and Reference Manual. CGAL Editorial Board, 4.4 edition (2000)
21. Kedem, K., Livne, R., Pach, J., Sharir, M.: On the union of Jordan regions and collision-free translational motion amidst polygonal obstacles. Disc. Comput. Geom. 1, 59–70 (1986)
22. Kerber, M., Sharathkumar, R.: Approximate čech complex in low and high dimensions. In: 24th Int. Symp. Alg. Comput., pp. 666–676 (2013)
23. Sharir, M.: Almost tight upper bounds for lower envelopes in higher dimensions. Disc. Comput. Geom. 12(1), 327–345 (1994)
24. Sheehy, D.: Linear-size approximation to the Vietoris-Rips filtration. In: Proc. of the 28th ACM Symp. on Comput. Geom., pp. 239–248 (2012)
25. Skraba, P., Ovsjanikov, M., Chazal, F., Guibas, L.: Persistence-based segmentation of deformable shapes. In: Computer Vision and Pattern Recognition Workshops, pp. 2146–2153 (2010)
26. Stava, O., Vanek, J., Benes, B., Carr, N., Mech, R.: Stress relief: improving structural strength of 3d printable objects. ACM Trans. Graph. 31(4), 48 (2012)
27. Yap, C.: An O(n logn) algorithm for the Voronoi diagram of a set of simple curve segments. Disc. Comput. Geom. 2, 365–393 (1987)
28. Zomorodian, A., Carlsson, G.: Computing persistent homology. Disc. Comput. Geom. 33, 249–274 (2005)

Approximation Algorithms for Polynomial-Expansion and Low-Density Graphs*

Sariel Har-Peled and Kent Quanrud

Department of Computer Science, University of Illinois
201 N. Goodwin Avenue, Urbana, IL, 61801, USA
{sariel,quanrud2}@illinois.edu

Abstract. We investigate the family of intersection graphs of low density objects in low dimensional Euclidean space. This family is quite general, and includes planar graphs. This family of graphs has some interesting properties, and in particular, it is a subset of the family of graphs that have polynomial expansion.

We present efficient $(1 + \varepsilon)$-approximation algorithms for polynomial expansion graphs, for Independent Set, Set Cover, and Dominating Set problems, among others, and these results seem to be new. Naturally, PTAS's for these problems are known for subclasses of this graph family.

These results have immediate interesting applications in the geometric domain. For example, the new algorithms yield the only PTAS known for covering points by fat triangles (that are shallow).

We also prove corresponding hardness of approximation for some of these optimization problems, characterizing their intractability with respect to density. For example, we show that there is no PTAS for covering points by fat triangles if they are not shallow, thus matching our PTAS for this problem with respect to depth.

Keywords: Computational geometry, SETH, hardness of approximation.

1 Introduction

Many classical optimization problems are intractable to approximate, let alone solve. Motivated by the discrepancy between the worst-case analysis and real-world success of algorithms, more realistic models of input have been developed, alongside algorithms that take advantage of their properties. We investigate approximation algorithms for two closely-related families of graphs: graphs with *polynomially-bounded expansion*, and intersection graphs of geometric objects with *low-density*.

The first part of our paper investigates a geometric property called *density*. Informally, a set of objects is *low-density* if no ball can intersect too many objects

* Work on this paper was partially supported by a NSF AF awards CCF-1421231, and
  CCF-1217462.

© Springer-Verlag Berlin Heidelberg 2015
N. Bansal and I. Finocchi (Eds.): ESA 2015, LNCS 9294, pp. 717–728, 2015.
DOI: 10.1007/978-3-662-48350-3_60

Objects	Approx. Alg.	Hardness
Disks/pseudo-disks	QPTAS [31]	Exact version is NP-HARD [15]
Fat triangles of same size	$O(1)$ [13]	APX-HARD: [21] I.e., no PTAS possible.
Fat objects in $\mathbb{R}^2$	$O(\log^* \text{opt})$ [5]	APX-HARD: [21]
Objects $\subseteq \mathbb{R}^d$, $O(1)$ density E.g. fat objects, $O(1)$ depth.	PTAS: Theorem 6_{p725}	Exact version is NP-HARD [15]
Objects with polylog density	QPTAS: T6	No PTAS under ETH Lemma 10_{p726}
Objects with density ρ in $\mathbb{R}^d$	PTAS: T6 RT $n^{O(\rho^{(d+1)/d}/\epsilon^d)}$.	No $(1 + \varepsilon)$-approx with RT $n^{\text{polylog}(\rho)}$ assuming ETH: L10

Fig. 1. Known results about the complexity of geometric set-cover. Specifically, the input is a given set of points, and a set of objects, and the task is to find the smallest subset of objects that covers the points. To see that the hardness proof Feder and Greene [15] indeed implies the above, one just has to verify that the input instance their proof generates has bounded depth.

that are larger than it. This notion was introduced by van der Stappen *et al.* [40], although weaker notions involving a single resolution were studied earlier (e.g. in the work by Schwartz and Sharir [38]). A closely related geometric property to density is *fatness*. Informally, an object is fat if it contains a ball, and is contained inside another ball, that up to constant scaling are the same size. Fat objects have low union complexity [3], and in particular, shallow fat objects have low density [39].

We study the intersection graphs arising out of low-density scenes. Specifically, a set $\mathcal{F}$ of objects in $\mathbb{R}^d$ induces an *intersection graph* $\mathsf{G}_{\mathcal{F}}$ having $\mathcal{F}$ as its the set of vertices, and two objects $\mathsf{f}, \mathsf{g} \in \mathcal{F}$ are connected by an edge if and only if $\mathsf{f} \cap \mathsf{g} \neq \emptyset$. Without any restrictions, intersection graphs can represent any graph. There is much work on intersection graphs, from interval graphs, to unit disk graphs, and more. The circle packing theorem [25,4,36] implies that every planar graph can be realized as a coin graph, where the vertices are interior disjoint disks, and there is an edge connecting two vertices if their corresponding disks are touching. This implies that planar graphs are low density. Miller *et al.* [29] studied the intersection graphs of balls (or fat convex object) of bounded depth (i.e., every point is covered by a constant number of balls), and these intersection graphs are readily low density. Some results related to our work include: (i) planar graphs are the intersection graph of segments [9], and (ii) string graphs (i.e., intersection graph of curves in the plane) have small separators [28].

The class of low-density graphs is contained in the class of graphs with polynomial expansion. The class of graphs with polynomial expansion was defined by Nešetřil and Ossona de Mendez as part of a greater investigation on the

Objects	Approx. Alg.	Hardness
Disks/pseudo-disks	PTAS [32]	Exact version is NP-HARD
Fat triangles of similar size.	$O(\log\log \text{opt})$ [6]	APX-HARD Lemma 8p726
Objects with $O(1)$ density.	PTAS: Theorem 6p725	Exact ver. NP-HARD [15]
Objects polylog density.	QPTAS: T6	No PTAS under ETH Lemma 10 / L8
Objects with density ρ in $\mathbb{R}^d$	PTAS: T6 RT $n^{O(\rho^{(d+1)/d}/\epsilon^d)}$	No $(1+\varepsilon)$-approx with RT $n^{\text{polylog}(\rho)}$ assuming ETH: L10

Fig. 2. Known results about the complexity of geometric hitting set. Specifically, the input is a given set of points, and a set of objects, and the task is to find the smallest subset of points such that any object is hit by one of these points.

sparsity of graphs (see the book [35]). Perhaps a motivating observation to their theory is that the sparsity of a graph (the ratio of edges to vertices) is somewhat unstable: a clique (with maximum density) can be disguised as a sparse graph by splitting every edge by a middle vertex. Nešetřil and Ossona de Mendez investigate graph invariants that can sustain small changes to graphs, such as taking (shallow) minors and subdivisions, and their systematic classifications of graphs are complimented by algorithmic results. For the class of *nowhere dense graphs* [35, Section 5.4], Grohe, Kreutzer and Siebertz recently showed that first-order properties are fixed-parameter tractable [18]. In this paper, we study graphs of bounded expansion [35, Section 5.5], which intuitively requires graphs to not only be sparse, but have shallow minors that are sparse as well.

There is a long history of optimization in structured graph classes. Lipton and Tarjan first obtained a PTAS for independent set in planar graphs by using separators [26,27]. Baker [7] developed techniques for covering problems (e.g. dominated set) on planar graphs, which were extended to graphs with bounded local treewidth by Eppstein [14] and graphs excluding minors by Grohe [17]. Separators have also played a key role in geometric optimization algorithms, including a PTAS for independent set and (continuous) piercing set for fat objects [10], a PTAS for piercing half-spaces and pseudo-disks [32], a QPTAS for maximum weighted independent sets of polygons [1,2,20], and a QPTAS for Set Cover by pseudodisks [30], among others.

Lastly, Cabello and Gajser [8] develop PTAS's for some of the problems we study in the specific setting of minor-free graphs.

Our Results. We develop polynomial time approximation schemes for basic independence, packing and covering problems in graphs with polynomial expansion and in low-density graphs. These two graph classes are related as low-density graphs have polynomial expansion, and we leverage properties of polynomial expansion beyond just having small separators.

Specifically, we get PTAS for independent set, dominating set, and subset dominating set for graphs with bounded expansion. These results seems to be new. Naturally, these results immediately extend to low-density graphs. We get faster algorithms for low-density graphs than implied by polynomial expansion by using the underlying geometry of these graphs.

The low-density algorithms are complimented by matching hardness results that suggest our approximations are nearly optimal with respect to depth (under SETH). The context of our results, for geometric settings, is summarized in Figure 1 and Figure 2.

Paper Organization. We describe low-density graphs in Section 2.1 and prove some basic properties. Bounded expansion graphs are surveyed in Section 2.2. Section 3 presents the new approximation algorithms and Section 4 present the hardness results. Proofs, where omitted, can be found in the full version of this paper [21].

2   Preliminaries

2.1   Low-Density Graphs

One of the two main thrusts of this work is to investigate the following family of graphs.

Definition 1. *A set of objects $\mathcal{F}$ in $\mathbb{R}^d$ (not necessarily convex or connected) has* **density** *ρ if any ball b intersects at most ρ objects in $\mathcal{F}$ with diameter larger than the diameter of b, the minimum such quantity is denoted by $\mathrm{density}(\mathcal{F})$. If ρ is a constant, then $\mathcal{F}$ has* **low density**.

*Any graph that is the intersection graph of a set of objects $\mathcal{F}$ in $\mathbb{R}^d$ with density ρ is ρ-***dense***. The class of all graphs that are ρ-dense and are induced by objects in $\mathbb{R}^d$ is denoted by $\mathcal{C}_\rho^d$.*

A set of α-fat convex objects in $\mathbb{R}^d$ with bounded depth had bounded density, where a set has depth k if every point in $\mathbb{R}^d$ is covered at most k times. This fact is well known, see [21].

A graph G is k-*degenerate* if any subgraph of G has a vertex of degree at most k.

Observation 1. *A ρ-dense graph is $(\rho-1)$-degenerate (with degree $\rho-1$ attained by the object with smallest diameter). Therefore, a ρ-dense graph with n vertices has at most $(\rho-1)n$ edges.*

Definition 2. *A metric space $\mathcal{X}$ is a* **doubling space** *if there is a universal constant $c_{\mathrm{dbl}} > 0$, such that any ball b of radius r can be covered by c_{dbl} balls of half the radius. Here c_{dbl} is the* **doubling constant***, and its logarithm is the* **doubling dimension***.*

In $\mathbb{R}^d$ the doubling constant is $c_d = 2^{O(d)}$, and the doubling dimension is $O(d)$ [41], making the doubling dimension a natural abstraction of the notion of dimension in the Euclidean case.

Lemma 1. *Let $\mathcal{F}$ be a set of objects in $\mathbb{R}^d$ with density ρ. Then, for any $\alpha \in (0,1)$, a ball $\mathbb{b} = \mathbb{b}(c, r)$ can intersect at most $\rho c_{\mathrm{dbl}}^{\lceil \lg 1/\alpha \rceil}$ objects of $\mathcal{F}$ with diameter $\geq \alpha \cdot 2r$, where $\lg = \log_2$ and c_{dbl} is the doubling constant of $\mathbb{R}^d$.*

Proof. Cover $\mathbb{b}$ by minimum number balls of radius $\leq \alpha r$. By the definition of the doubling constant, the number of balls needed is $c_{\mathrm{dbl}}^{\lceil \log_2 1/\alpha \rceil}$. Each of these balls, by definition of density, can intersect at most ρ objects of $\mathcal{F}$ of diameter larger than $\alpha \cdot 2r$, which implies the claim.   □

The density definition can be made to be somewhat more flexible, as follows.

Lemma 2. *Let $\alpha > 0$ be a parameter, and let $\mathcal{F}$ be a collection of objects in $\mathbb{R}^d$ such that, for any r, any ball with radius r intersects at most ρ objects with diameter $\geq 2\alpha r$. Then $\mathcal{F}$ has density $c_{\mathrm{dbl}}^{\lceil \log \alpha \rceil} \rho$.*

Proof. Let $\mathbb{b}$ be a ball with radius r. We can cover $\mathbb{b}$ with $c_{\mathrm{dbl}}^{\lceil \log \alpha \rceil}$ balls with radius r/α. Each r/α-radius ball can intersect at most ρ objects with diameter $2\alpha \cdot (r/\alpha) = 2r$, so $\mathbb{b}$ intersects at most $c_{\mathrm{dbl}}^{\lceil \log \alpha \rceil} \rho$ balls with radius r.   ⊔

2.1.1   Minors of Low-Density Objects. Let $\mathcal{F}$ and $\mathcal{G}$ be two collections of objects in $\mathbb{R}^d$. $\mathcal{G}$ is a *minor* of $\mathcal{F}$ if it can be obtained by deleting objects and replacing pairs of overlapping objects f and g (i.e., $\mathsf{f} \cap \mathsf{g} \neq \emptyset$) with their union $\mathsf{f} \cup \mathsf{g}$. A sequence of unions and deletions taking $\mathcal{F}$ to $\mathcal{G}$, where every object of $\mathcal{G}$ corresponds to a *cluster* of objects of $\mathcal{F}$. If the intersection graphs of each of these clusters has radius t, then $\mathcal{G}$ is a *t-shallow minor* of $\mathcal{F}$.

Surprisingly, even for a set $\mathcal{F}$ of fat and convex shapes in the plane with constant density, their intersection graph $\mathsf{G}_{\mathcal{F}}$ can have arbitrarily large cliques as minors (see Figure 3). On the other hand, there is a simple relationship between the depth of a shallow minor of objects and its density.

Lemma 3. *Let $\mathcal{F}$ be a collection of objects with density ρ in $\mathbb{R}^d$, and let $\mathcal{G}$ be t-shallow minor of $\mathcal{F}$. Then $\mathcal{G}$ has density at most $t^{O(1)} \rho$.*

Proof. Every object $\mathsf{g} \in \mathcal{G}$ has a defining subset $\mathcal{F}_{\mathsf{g}} \subseteq \mathcal{F}$. These sets are disjoint, and let $\mathcal{P} = \{\mathcal{F}_{\mathsf{g}} \mid \mathsf{g} \in \mathcal{G}\}$ be the induced partition of $\mathcal{F}$ into clusters. Next, consider any ball $\mathbb{b} = \mathbb{b}(c, r)$, and suppose that $\mathsf{g} \in \mathcal{G}$ intersects $\mathbb{b}$ and it has diameter at least $2r$, and let $\mathcal{F}_{\mathsf{g}} \in \mathcal{P}$ be its defining cluster, and $\mathsf{H} = \mathsf{G}_{\mathcal{F}_{\mathsf{g}}}$ be its associated intersection graph. By assumption H has (graph) diameter $\leq t$.

Now, let h be any object in $\mathcal{F}_{\mathsf{g}}$ that intersect $\mathbb{b}$, let d_{H} denote the shortest path metric of H (under the number of edges), and let h' be the object in $\mathcal{F}_{\mathsf{g}}$ closest to h (under d_{H}), such that $\mathrm{diam}(\mathsf{h}') \geq 2r/t$ (if there is no such object then the diameter of $\mathrm{diam}(\mathsf{g}) < t(2r/t) \leq 2r$, which is a contradiction).

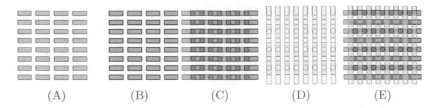

(A)   (B)   (C)   (D)   (E)

Fig. 3. (A) and (B) are two low-density collections of n^2 disjoint horizontal slabs, whose intersection graph (C) contains n rows as minors. (D) is the intersection graph of a low-density collection of vertical slabs that contain n columns as minors. In (E), the intersection graph of all the slabs contain the n rows and n columns as minors that form a $K_{n,n}$ bipartite graph, which in turn contains an $n+1$ vertex clique minor.

Consider the shortest path $\pi \equiv h_1, \ldots, h_\tau$ between $h = h_1$ and $h' = h_\tau$ in H, where $\tau \leq t$. Observe that, for $i = 1, \ldots \tau - 1$, $\mathrm{diam}(h_i) < 2r/t$, and thus the distance between b and h' is bounded by $\sum_{i=1}^{\tau-1} \mathrm{diam}(h_i) \leq (\tau-1)2r/\tau < 2r$. We refer to h' as the *representative* of g, denoted by $\mathrm{rep}(g) \in \mathcal{F}_g$.

Now, let $\mathcal{H} = \Big\{ \mathrm{rep}(g) \in \mathcal{F} \ \Big| \ g \in \mathcal{G}, \mathrm{diam}(g) \geq 2r, \text{ and } g \cap b \neq \emptyset \Big\}$. The representatives in $\mathcal{H}$ are all unique, each is of diameter $\geq 2r/t$, all of them intersect $b(c, 3r)$, and they all belong to $\mathcal{F}$, a set of density ρ. Lemma 1 implies that $|\mathcal{H}| \leq \rho c_{\mathrm{dbl}}^{\lceil \lg t \rceil}$, implying the claim. □

2.2   Graphs with Polynomial Expansion

Let G be an undirected graph. A *minor* of G is a graph H that can be obtained by contracting edges, deleting edges, and deleting vertices from G. If H is a minor of G, then each vertex in H corresponds to a connected subset of vertices in G, based on the contraction of edges. H is a *t-shallow minor* (or a *minor of depth t*) of G, where t is an integer, if each cluster of vertices has radius at most t. Let $\nabla_t(G)$ denote the set of all graphs that are minors[1] of G of depth t.

The *greatest reduced average density of rank r* (denoted d_r) of G is the quantity $d_r(G) = \sup_{H \in \nabla_r(G)} \frac{|E(H)|}{|V(H)|}$ [33]. The *expansion* of a graph class $\mathcal{C}$ is the function $f : \mathbb{N} \to \mathbb{N} \cup \{\infty\}$ defined by $f(r) = \sup_{G \in \mathcal{C}} d_r(G)$. The class $\mathcal{C}$ has *bounded expansion* if $f(r)$ is finite for all r. Specifically, $\mathcal{C}$ has *polynomial expansion* if f is (bounded by) a polynomial, *subexponential expansion* if f is (bounded by) a subexponential function, and so forth. The polynomial expansion is of *order k* if $f(x) = O(x^k)$.

For example, the class of planar graphs has constant expansion because planar graphs are sparse and every minor of a planar graph is planar. More surprisingly, Lemma 3 together with Observation 1 implies that low-density graphs have polynomial expansion.

Lemma 4. *Let $\rho > 0$ be fixed. The graph class $\mathcal{C}_\rho^d$ of ρ-dense graphs in $\mathbb{R}^d$ has polynomial expansion bounded by $f(t) = \rho t^{\lceil \log c_{\mathrm{dbl}} \rceil}$.*

---

[1] I.e., these graphs can not legally drink alcohol.

2.2.1   Separators and Divisions. Nešetřil and Ossona de Mendez showed that graphs with subexponential expansion have subexponential-sized separator. For the simpler case of polynomial expansion, we have the following.

Theorem 2 ([34, Theorem 8.3]). *Let C be a class of graphs with polynomial expansion of order k. For any graph in C with n vertices and m edges, one can compute, in $O\big(mn^{1-\alpha}\log^{1-\alpha} n\big)$ time, a separator of size $O\big(n^{1-\alpha}\log^{1-\alpha} n\big)$, where $\alpha = 1/(2k+2)$.*

For the sake of completeness, a proof is provided in full version of this paper. This result is well known, and we simply retrace the calculations of [34] for polynomial f instead of subexponential f.

Theorem 2 gives us a strongly sublinear separator for low-density graphs of size $O\bigg(\big(\rho^2 n \log n\big)^{1-\frac{1}{O(\log c_{\mathrm{dbl}})}} \bigg)$. Geometric arguments give a slightly stronger separators, of size $O(\rho + \rho^{1/d} n^{1-1/d})$, in $\mathbb{R}^d$. For the sake of completeness, an algorithm to compute such a separator, which also exposes the structure of low-density objects, is provided in the full version.

Consider a set V, and a family W of subsets $C_1, \ldots, C_k \subseteq V$ such that $V = \bigcup_{i=1}^{k} C_i$. A set C_i is a *cluster* and the entire collection $W = \{C_1, \ldots, C_k\}$ is a *cover*. A cover of a graph $G = (V, E)$ is a cover of its vertices. Given a cover W, the *excess* of a vertex $v \in V$ that appears in j clusters is $j - 1$. The *total excess* of the cover W is the sum of excesses of all the vertices in V.

A cover C of G is a λ-*division* if (i) for all $C, C' \in C$, $C \setminus C'$ and $C' \setminus C$ are separated in G (i.e., there is no edge between these sets of vertices in G), and (ii) for all $C \in C$, $|C| \leq \lambda$. A vertex $v \in V$ is an *interior vertex* of a cover W if it appears in exactly one cluster of W (and its excess is zero), and a *boundary vertex* otherwise. By property (i), the entire neighborhood of an interior vertex of a division lies in the same cluster.

The property of a class of graph having λ-divisions is slightly stronger than being weakly hyperfinite. A graph is *weakly hyperfinite* if there is a small subset of vertices whose removal leaves small connected components [35, Section 16.2]. λ-divisions also provide such a set: the set of all boundary vertices. The connected components induced by removing the boundary vertices are not only small, but the neighborhoods of these components are small as well.

As noted by Henzinger *et al.* [23], strongly sublinear separators obtain λ-divisions with total excess εn for $\lambda = \mathrm{poly}(1/\varepsilon)$. Such divisions were first used by Frederickson in planar graphs [16]. A proof of the following well-known result is provided in the full version.

Lemma 5. *Let G be a graph with n vertices, such that any induced subgraph with m vertices has a separator with $O(m^\alpha \log^\beta m)$ vertices, for some $\alpha < 1$ and $\beta \geq 0$. Then, for $\epsilon > 0$, the graph G has λ-divisions with total excess ϵn, where*
$$\lambda O\bigg(\big(\epsilon^{-1} \log^\beta \epsilon^{-1}\big)^{1/(1-\alpha)} \bigg).$$

In this paper, we are concerned specifically with divisions of polynomial expansion graphs and low-density graphs.

Lemma 6. *(a) Let* G *be a graph with* n *vertices and polynomially expansion of degree* k, *and let* $\varepsilon > 0$ *be fixed. Then graph* λ-*divisions with total excess* εn *for* $\lambda = O\big((1/\varepsilon)^{2k+2} \log^{2k+1}(1/\varepsilon)\big)$.

(b) Let G *be a* ρ-*dense graph with* n *vertices. Then* G *has* λ-*divisions with total excess at most* εn, *for* $\lambda = O\big(\rho/\epsilon^d\big)$.

3   Approximation Algorithms

Independent Set. Given an undirected graph $G = (V, E)$, an *independent set* is a set of vertices $S \subseteq V$ such that no two vertices in S are connected by an edge. It is NP-COMPLETE to decide if a graph contains an independent set of size k [24], and one cannot approximate the size of the maximum independent set to within a factor of $n^{1-\varepsilon}$, for any fixed $\varepsilon > 0$, unless $P = NP$ [22].

Chan and Har-Peled [12] gave a PTAS for independent set with planar graphs, and the algorithm and its underlying argument extends to hereditary graph classes with strongly sublinear separators. The algorithm starts with an empty independent set, and repeatedly performs beneficial local exchanges of constant size in the independent set until it reaches a maximal independent set.

Lemma 7. *Let* $\mathcal{C}$ *be a hereditary graph class with* $f(\varepsilon)$-*divisions of total excess* εn. *For any* $\varepsilon > 0$, *and graph* $G = (V, E) \in \mathcal{C}$, *the* $f(\varepsilon)$-*local search algorithm computes, in* $n^{O(f(\varepsilon))}$ *time, a* $(1-2\varepsilon)$-*approximation for the maximum cardinality independent subset of* V.

In particular, we obtain a PTAS for independent set in graph classes with polynomial expansion and low-density.

Theorem 3. *Let* $\mathcal{C}$ *be a graph class with polynomial expansion of order* k, *and let* $\varepsilon > 0$ *be fixed. For* $\lambda = \tilde{O}((1/\varepsilon)^{2k+2})$, *the* λ-*local search algorithm computes a* $(1 - 2\varepsilon)$ *approximation for the maximum size independent set of any graph in* $\mathcal{C}$ *in running time* $n^{O(\lambda)}$.

Theorem 4. *Let* $\varepsilon > 0$ *be given, and* $\mathcal{F}$ *a collection of objects in* $\mathbb{R}^d$. *Then the local search algorithm computes a* $(1 - \varepsilon)$-*approximation for the maximum size independent subset of* $\mathcal{F}$ *in time* $n^{O(\rho/\varepsilon^d)}$.

These bounds are slightly stronger than those obtained by naively applying Theorem 3, because of stronger geometric separators.

The proof of Lemma 7 extends immediately to more general packing problems; namely, to graphs satisfying a hereditary and mergeable property Π, as defined by Har-Peled in [20, Section 5]. Thus, this class of packing problem also has PTAS's on low-density graphs and graph classes with polynomial expansion. These details are deferred to the full version.

Dominating Set. Given an undirected graph $G = (V, E)$, a *dominating set* is a set of vertices $\mathcal{D} \subseteq V$ such that every vertex in G is either in $\mathcal{D}$ or adjacent to a vertex in $\mathcal{D}$ by an edge. It is NP-COMPLETE to decide if a graph contains a dominating set of size k (by a simple reduction from set covering, which is NP-COMPLETE [24]), and one cannot obtain a $c \log n$ approximation (for some constant c) unless $\mathrm{P} = \mathrm{NP}$ [37].

More generally, let $\mathsf{G} = (\mathsf{V}, \mathsf{E})$ be an undirected graph, and let $\mathcal{C}, \mathcal{D} \subseteq \mathsf{V}$ be two subset of vertices. We say $\mathcal{D}$ *dominates* $\mathcal{C}$ if every vertex in $\mathcal{C}$ either is in $\mathcal{D}$ or is adjacent to some vertex in $\mathcal{D}$. In the *subset dominating set problem*, we are given an undirected graph $\mathsf{G} = (\mathsf{V}, \mathsf{E})$ and two subsets of vertices $\mathcal{C}, \mathcal{D} \subseteq \mathsf{V}$ such that $\mathcal{D}$ dominates $\mathcal{C}$, and the goal is to compute the minimum cardinality subset of $\mathcal{D}$ that dominates $\mathcal{C}$.

Theorem 5. *Let $\mathcal{C}$ be a graph class with polynomial expansion of order k, and let $\varepsilon > 0$ be fixed. Given an instance $(\mathsf{G} = (\mathsf{V}, \mathsf{E}), \mathcal{C}, \mathcal{D})$ of the subset dominating set problem with $\mathsf{G} \in \mathcal{C}$ with n vertices, for $\lambda = \widetilde{O}((1/\varepsilon)^{2k+2})$, the λ-local search algorithm computes a $(1 + 3\varepsilon)$-approximation for the smallest cardinality subset of $\mathcal{D}$ that dominates $\mathcal{C}$ in time $n^{O(\lambda)}$.*

We thus immediately obtain PTAS's for dominating set type problems on low-density graphs, and highlight two geometric versions in particular. Let $\mathcal{F}$ be a collection of objects in $\mathbb{R}^d$ and P a collection of points. In the *discrete hitting set problem*, we want to compute the minimum cardinality set $\mathsf{Q} \subseteq \mathsf{P}$ such that for every $f \in \mathcal{F}$, we have $\mathsf{Q} \cap f \neq \emptyset$. In the *discrete geometric set cover problem*, we want to compute the smallest cardinality set $\mathcal{G} \subseteq \mathcal{F}$ such that for every point $\mathsf{p} \in \mathsf{P}$, we have $\mathsf{p} \cap \left(\bigcup_{f \in \mathcal{G}} f \right) \neq \emptyset$.

Theorem 6. *Let $\mathcal{F}$ be a collection of m objects in $\mathbb{R}^d$ with density ρ, and let P be a set of n points in $\mathbb{R}^d$.*
(a) The local search algorithm, for exchanges of size $\lambda = O(\rho/\varepsilon^d)$, computes a subset $\mathsf{Q} \subseteq \mathsf{P}$ that is a $(1 + \varepsilon)$-approximation for the smallest cardinality subset of P that is a hitting set for $\mathcal{F}$. The running time of the algorithm is $mn^{O(\lambda)}$.
(b) The local search algorithm for exchanges of size $\lambda = O(\rho/\varepsilon^d)$, computes a subset $\mathcal{G} \subseteq \mathcal{F}$ that is a $(1 + \varepsilon)$-approximation for the smallest cardinality subset of $\mathcal{F}$ that covers P. The running time of the algorithm is $nm^{O(\lambda)}$.

To our knowledge, these are the first PTAS's for discrete hitting set and discrete set cover with shallow fat triangles and similar fat objects. Previously, such algorithms were known only for disks and points in the plane.

The proof of Theorem 5 for subset dominating set extends to natural variants such as edge cover, vertex cover, distance dominating set, dominating set with multiplicities, and connected dominating set. These reductions, provided in the full version, attest to the robustness of the class of graphs with polynomial expansion.

4   Hardness of Approximation

Some of the results of this section appeared in an unpublished manuscript [19]. Chan and Grant [11] also prove some related hardness results, which were (to some extent) a followup work to the aforementioned manuscript [19]. Proofs of the following hardness results, as well as some other results omitted due to space constraints, are provided in the full version.

In the *fat-triangles discrete hitting set problem*, we are given a set of points in the plane P and a set of fat triangles $\mathcal{T}$, and want to find the smallest subset of P such that each triangle in $\mathcal{T}$ contains at least one point in the set.

Lemma 8. *There is no* PTAS *for the fat-triangle discrete hitting set problem, unless* $\mathrm{P} = \mathrm{NP}$. *One can prespecify an arbitrary constant* $\delta > 0$, *and the claim would hold true even with the following restrictions hold on the given instance* $(\mathsf{P}, \mathcal{T})$: *(A) Every angle of every triangle in* $\mathcal{T}$ *is between* $60 - \delta$ *and* $60 + \delta$ *degrees. (B) No point of* P *is covered by more than three triangles of* $\mathcal{T}$. *(C) The points of* P *are in convex position. (D) All the triangles of* $\mathcal{T}$ *are of similar size. Specifically, each triangle has side length in the range (say)* $(\sqrt{3} - \delta, \sqrt{3} + \delta)$. *(E) The points of* P *are a subset of the vertices of the triangles of* $\mathcal{T}$.

Let P be a set of n points in the plane, and $\mathcal{F}$ be a set of m regions in the plane, such that (I) the shapes of $\mathcal{F}$ are convex, fat, and of similar size, (II) the boundaries of any pair of shapes of $\mathcal{F}$ intersect in at most 6 points, (III) the union complexity of any m shapes of $\mathcal{F}$ is $O(m)$, and (IV) any point of P is covered by a constant number of shapes of $\mathcal{F}$. We are interested in finding the minimum number of shapes of $\mathcal{F}$ that covers all the points of P. This variant is the *friendly geometric set cover* problem.

Lemma 9. *There is no* PTAS *for the friendly geometric set cover problem, unless* $\mathrm{P} = \mathrm{NP}$.

The *exponential time hypothesis* (**ETH**), is that 3SAT can not be solved in time better than $2^{\Omega(n)}$, where n is the number of variables. The *strong exponential time hypothesis* (**SETH**), is that the time to solve kSAT is at least $2^{c_k n}$, where c_k converges to 1 as k increases.

Lemma 10. *Assuming* ETH, *an instance of geometric set cover with n fat triangles, and density at least* $\Omega(\log^c n)$, *can not be* $(1+\varepsilon)$*-approximated in polynomial time, where c is a sufficiently large constant.*

The same holds if the triangles are replaced by objects with low density, and for geometric hitting set.

Acknowledgments. The authors thank Mark de Berg for useful discussions related to the problems studied in this paper. In particular, he pointed out an improved bound of a separator theorem for low-density objects detailed in the full version of this paper [21]. We also thank the anonymous referees. We are particularly grateful to the anonymous referee of a previous version of this paper who pointed out the connection of our work to graphs with bounded expansion.

References

1. Adamaszek, A., Wiese, A.: Approximation schemes for maximum weight independent set of rectangles. In: Proc. 54th Annu. IEEE Sympos. Found. Comput. Sci. (FOCS), pp. 400–409 (2013)
2. Adamaszek, A., Wiese, A.: A QPTAS for maximum weight independent set of polygons with polylogarithmic many vertices. In: Proc. 25th ACM-SIAM Sympos. Discrete Algs. (SODA), pp. 400–409 (2014) **Discrete Algs. (SODA)**
3. Agarwal, P.K., Pach, J., Sharir, M.: State of the union–of geometric objects. In: Goodman, J.E., Pach, J., Pollack, R. (eds.) Surveys in Discrete and Computational Geometry Twenty Years Later. Contemporary Mathematics, vol. 453, pp. 9–48. Amer. Math. Soc. (2008)
4. Andreev, E.M.: On convex polyhedra in lobachevsky spaces. Sbornik: Mathematics 10, 413–440 (1970)
5. Aronov, B., de Berg, M., Ezra, E., Sharir, M.: Improved bounds for the union of locally fat objects in the plane. SIAM J. Comput. 43(2), 543–572 (2014)
6. Aronov, B., Ezra, E., Sharir, M.: Small-size ε-nets for axis-parallel rectangles and boxes. SIAM J. Comput. 39(7), 3248–3282 (2010)
7. Baker, B.S.: Approximation algorithms for NP-complete problems on planar graphs. J. Assoc. Comput. Mach. 41, 153–180 (1994)
8. Cabello, S., Gajser, D.: Simple ptas's for families of graphs excluding a minor. CoRR, abs/1410.5778 (2014)
9. Chalopin, J., Gonçalves, D.: Every planar graph is the intersection graph of segments in the plane: extended abstract. In: Proc. 41st Annu. ACM Sympos. Theory Comput. (STOC), pp. 631–638 (2009)
10. Chan, T.M.: Polynomial-time approximation schemes for packing and piercing fat objects. J. Algorithms 46(2), 178–189 (2003)
11. Chan, T.M., Grant, E.: Exact algorithms and APX-hardness results for geometric set cover. In: Proc. 23rd Canad. Conf. Comput. Geom., CCCG (2011)
12. Chan, T.M., Har-Peled, S.: Approximation algorithms for maximum independent set of pseudo-disks. Discrete Comput. Geom. 48, 373–392 (2012)
13. Clarkson, K.L., Varadarajan, K.R.: Improved approximation algorithms for geometric set cover. Discrete Comput. Geom. 37(1), 43–58 (2007)
14. Eppstein, D.: Diameter and treewidth in minor-closed graph families. Algorithmica 27, 275–291 (2000)
15. Feder, T., Greene, D.H.: Optimal algorithms for approximate clustering. In: Proc. 20th Annu. ACM Sympos. Theory Comput., STOC, pp. 434–444 (1988)
16. Frederickson, G.N.: Fast algorithms for shortest paths in planar graphs, with applications. SIAM J. Comput. 16(6), 1004–1022 (1987)
17. Grohe, M.: Local tree-width, excluded minors, and approximation algorithms. Combinatorica 23(4), 613–632 (2003)
18. Grohe, M., Kreutzer, S., Siebertz, S.: Deciding first-order properties of nowhere dense graphs. In: Proc. 46th Annu. ACM Sympos. Theory Comput., STOC, pp. 89–98 (2014)
19. Har-Peled, S.: Being fat and friendly is not enough. CoRR, abs/0908.2369 (2009)
20. Har-Peled, S.: Quasi-polynomial time approximation scheme for sparse subsets of polygons. In: Proc. 30th Annu. Sympos. Comput. Geom., SoCG, pp. 120–129 (2014)
21. Har-Peled, S., Quanrud, K.: Approximation algorithms for low-density graphs. CoRR, abs/1501.00721 (2015), http://arxiv.org/abs/1501.00721

22. Hastad, J.: Clique is hard to approximate within $n^{1-\epsilon}$. In: Proc. 37th Annu. IEEE Sympos. Found. Comput. Sci., FOCS, pp. 627–636 (1996)
23. Henzinger, M.R., Klein, P., Rao, S., Subramanian, S.: Faster shortest-path algorithms for planar graphs. J. Comput. Sys. Sci. 55, 3–23 (1997)
24. Karp, R.M.: Reducibility among combinatorial problems. In: Complexity of Computer Computations, pp. 85–103 (1972)
25. Koebe, P.: Kontaktprobleme der konformen Abbildung. Ber. Verh. Sächs. Akademie der Wissenschaften Leipzig, Math.-Phys. Klasse 88, 141–164 (1936)
26. Lipton, R.J., Tarjan, R.E.: A separator theorem for planar graphs. SIAM J. Appl. Math. 36, 177–189 (1979)
27. Lipton, R.J., Tarjan, R.E.: Applications of a planar separator theorem. SIAM J. Comput. 9(3), 615–627 (1980)
28. Matoušek, J.: Near-optimal separators in string graphs. Combin., Prob. & Comput. 23(1), 135–139 (2014)
29. Miller, G.L., Teng, S.H., Thurston, W.P., Vavasis, S.A.: Separators for sphere-packings and nearest neighbor graphs. J. Assoc. Comput. Mach. 44(1), 1–29 (1997)
30. Mustafa, N.H., Raman, R., Ray, S.: QPTAS for geometric set-cover problems via optimal separators. ArXiv e-prints (2014)
31. Mustafa, N.H., Raman, R., Ray, S.: Settling the APX-hardness status for geometric set cover. In: Proc. 55th Annu. IEEE Sympos. Found. Comput. Sci., FOCS (2014) (page to appear)
32. Mustafa, N.H., Ray, S.: Improved results on geometric hitting set problems. Discrete Comput. Geom. 44(4), 883–895 (2010)
33. Nesetril, J., Ossona de Mendez, P.: Grad and classes with bounded expansion I. Decompositions. Eur. J. Comb. 29(3), 760–776 (2008)
34. Nesetril, J., Ossona de Mendez, P.: Grad and classes with bounded expansion II. Algorithmic Aspects 29(3), 777–791 (2008)
35. Nesetril, J., Ossona de Mendez, P.: Sparsity. Alg. Combin., vol. 28. Springer (2012)
36. Pach, J., Agarwal, P.K.: Combinatorial Geometry. John Wiley & Sons (1995)
37. Raz, R., Safra, S.: A sub-constant error-probability low-degree test, and a sub-constant error-probability PCP characterization of NP. In: Proc. 29th Annu. ACM Sympos. Theory Comput., STOC, pp. 475–484 (1997)
38. Schwartz, J.T., Sharir, M.: Efficient motion planning algorithms in environments of bounded local complexity. Report 164, Dept, Math. Sci., New York Univ., New York (1985)
39. van der Stappen, A.F.: Motion Planning Amidst Fat Obstacles. PhD thesis, Utrecht University, Netherlands (1992)
40. van der Stappen, A.F., Overmars, M.H., de Berg, M., Vleugels, J.: Motion planning in environments with low obstacle density. Discrete Comput. Geom. 20(4), 561–587 (1998)
41. Verger-Gaugry, J.-L.: Covering a ball with smaller equal balls in $\mathbb{R}^n$. Discrete Comput. Geom. 33(1), 143–155 (2005)

Monotone Drawings of 3-Connected Plane Graphs*

Xin He and Dayu He

Department of Computer Science and Engineering,
State University of New York at Buffalo, Buffalo, NY 14260, USA
{xinhe,dayuhe}@buffalo.edu

Abstract. A monotone drawing of a graph G is a straight-line drawing of G such that, for every pair of vertices u, w in G, there exists a path P_{uw} in G that is monotone on some line l_{uw}. (Namely, the order of the orthogonal projections of the vertices in P_{uw} on l_{uw} is the same as the order they appear in P_{uw}.) In this paper, we show that the classical Schnyder drawing of 3-connected plane graphs is a monotone drawing on a grid of size $f \times f$ ($f \leq 2n - 5$ is the number of internal faces of G), which can be constructed in $O(n)$ time. It also has the advantage that, for any given vertices u, w, the monotone line l_{uw} can be identified in $O(1)$ time.

1 Introduction

A *straight-line drawing* of a plane graph G is a drawing Γ in which each vertex of G is drawn as a distinct point on the plane and each edge of G is drawn as the line segment connecting two end vertices without any edge crossing. A path P in a straight-line drawing Γ is *monotone* if there exists a line l such that the orthogonal projections of the vertices of P on l appear along l in the order they appear in P. We call l a *monotone line* (or *monotone direction*) of P. Γ is called a *monotone drawing* of G if it contains at least one monotone path P_{uw} between every pair of vertices u, w of G. We call the monotone direction l_{uw} of P_{uw} the monotone direction for u, w.

Monotone drawings are introduced by Angelini et. al. as a new visualization paradigm in [1]. Consider the example described in [1]: a traveler uses a road map to find a route from a town u to a town w. He would like to easily spot a path connecting u and w. This task is harder if each path from u to w on the map has legs moving away from u. The traveler rotates the map to better perceive its content. Hence, even if in the original map orientation all the paths from u to w have annoying back and forth legs, the traveler might be happy to find one map orientation where a path from u to w smoothly goes from left to right. This approach is also motivated by human subject experiments: it was shown that the "geodesic tendency" (paths following a given direction) is important in understanding the structure of the underlying graphs [12].

---

* Research supported in part by NSF Grant CCR-1319732.

N. Bansal and I. Finocchi (Eds.): ESA 2015, LNCS 9294, pp. 729–741, 2015.
DOI: 10.1007/978-3-662-48350-3_61

Monotone drawing is also closely related to several other important graph drawing problems. In a monotone drawing, each monotone path is monotone with respect to a different line. In an *upward drawing* [8,10], every directed path is monotone with respect to the positive y direction. Even more related to monotone drawings are *greedy drawings* [2,14,15]. In a greedy drawing, for any two vertices u, v, there exists a path P_{uv} from u to v such that the Euclidean distance from an intermediate vertex of P_{uv} to the destination v decreases at each step. In a monotone drawing, for any two vertices u, v, there exists a path P_{uv} from u to v and a line l_{uv} such that the Euclidean distance from the projection of an intermediate vertex of P_{uv} on l to the projection of the destination v on l decreases at each step.

Related Works. It was shown by Angelini et. al. in [1] that every tree of n vertices has a monotone drawing of size $O(n^2) \times O(n)$, or $O(n^{1.58}) \times O(n^{1.58})$. It was also shown that every biconnected planar graph of n vertices has a monotone drawing in real coordinate space. Several papers have been published after [1]. The focus of the research is to identify the graph classes having monotone drawings and, if so, to find a monotone drawing with size as small as possible. It was shown by Angelini et. al. in [3] that every planar graph has a monotone drawing of size $O(n) \times O(n^2)$. However, the drawing presented in [3] is not straight-line. It may need up to $4n - 10$ bends in the drawing. [13] showed that every tree has a monotone drawing of size $O(n^{1.5}) \times O(n^{1.5})$. Recently it was shown that every planar graph has a monotone drawing of size $O(n^2) \times O(n)$ [11]. The main goal of this paper is to show that every 3-connected plane graph G has a monotone drawing of size $O(n) \times O(n)$.

Let G be a 3-connected plane graph. A straight-line drawing Γ of G is called *convex* if every face of G is drawn as a convex polygon in Γ. Γ is called *strictly convex* if the internal angles of every convex polygon in Γ is strictly less than $180°$. By using a result of [4], Angelini et. al. showed that every strictly convex drawing of G is monotone [1]. The condition *strictly convex* is crucial here. For example, consider a convex drawing Γ of G. Suppose that Γ contains a face F with three consecutive vertices u_1, u_2, u_3 and other three consecutive vertices v_1, v_2, v_3 such that (i) the internal angles of F at u_2 and v_2 are $180°$, and (ii) the line passing through u_1, u_2, u_3 and the line passing through v_1, v_2, v_3 are parallel in Γ. Then, there exists no monotone path in Γ between u_2 and v_2.

Every 3-connected plane graph admits an elegant combinatorial structure called *Schnyder wood* (to be defined later). By using Schnyder wood, one can obtain a convex drawing of G on a grid of size $f \times f$ [7,9]. (f is the number of internal faces of G). Note that $f \leq 2n - 5$ by Euler formula. We call such drawings the *Schnyder drawing*. Bonichon et. al. reduced the drawing size to $(n - 1 - \Delta) \times (n - 1 - \Delta)$ [6], (where Δ is the number of cyclic cycles in the minimal Schnyder wood of G, a parameter with range $0 \leq \Delta \leq (n/2) - 2$). The drawings in [7,6,9] are not strictly convex, hence not known to be monotone.

Rote showed that every 3-connected planar graph has a strictly convex drawing on a grid of size $O(n^{7/3}) \times O(n^{7/3})$ [16]. The size of the drawing grid was reduced to $O(n^2) \times O(n^2)$ in [5]. From the results in [1], this implies a monotone drawing of size $O(n^2) \times O(n^2)$ for 3-connected plane graphs.

Our Results:

- We show that the Schnyder drawing (as described in [9]), although not strictly convex, is actually monotone. Since the size of the drawing is $f \times f$ ($f \leq 2n - 5$), this is the first monotone drawing with area $o(n^3)$ for this subclass of planar graphs.

  In addition, we show that the Schnyder drawing has the following advantages:
- In the monotone drawing in [1], the monotone direction l_{uw} for two vertices u, w depends on u and w. Hence, we may need n^2 different monotone directions l_{uv} for different pairs u and w. In contrast, for the Schnyder drawing, we only need 6 intervals of monotone directions for any vertex pair u, w.
- In the monotone drawing in [1], one needs to search the monotone direction l_{uw} for the given two vertices u, w. The search procedure in [1] is adapted from [4], which takes $O(n \log n)$ time. In contrast, for the Schnyder drawing, we can identify the interval of monotone directions for u, w in $O(1)$ time (after $O(n)$ time pre-processing).

2   Preliminaries

Most definitions in this paper are standard. A *planar graph* is a graph G such that the vertices of G can be drawn in the plane and the edges in G can be drawn as non-intersecting curves. Such a drawing is also called an *embedding*. The embedding of G divides the plane into a number of connected regions called *faces*. The unbounded face is the *external face*. The other faces are *internal faces*. The vertices and edges not on the external face are *internal vertices* and *edges*. A plane graph is a planar graph with a fixed embedding. We abbreviate the words "counterclockwise" and "clockwise" as ccw and cw, respectively.

Let p be a point in the plane and let l be a half-line with p as its starting point. The slope of l, denoted by slope(l), is the angle spanned by a ccw rotation that brings the line in the positive x-axis direction to overlap with l.

In this paper, we only consider *straight-line drawings* (i.e. each edge of G is drawn as a straight-line segment between its end vertices.) Let Γ be such a drawing of G and let $e = (u, w)$ be an edge of G. The *direction* of e, denoted by $d(u, w)$ or $d(e)$, is the half-line starting at u and passing through w. The slope of an edge (u, w), denoted by slope(u, w), is the slope of $d(u, w)$. Observe that slope(u, w) = slope(w, u) − 180°. When comparing directions and their slopes, we assume that they are applied at the origin of the axes.

Let $P(u_1, u_k) = (u_1, \ldots, u_k)$ be a path of G. We also use $P(u_1, u_k)$ to denote the drawing of the path in Γ. $P(u_1, u_k)$ is *monotone with respect to a direction* d if the orthogonal projections of the vertices $u_1, \ldots, u_k$ on d appear in the same order as they appear on the path. $P(u_1, u_k)$ is *monotone* if it is monotone with respect to some direction. A drawing Γ is *monotone* if there exists a monotone path $P(u, w)$ for every pair of vertices u, w in Γ.

Let $e_{\min}$ and $e_{\max}$ be the edges in $P(u_1, u_k)$ with the smallest and the largest slope, respectively. They are called the *extremal edges* of $P(u_1, u_k)$. The closed interval $[\text{slope}(e_{\min}), \text{slope}(e_{\max})]$ is called the *range* of $P(u_1, u_k)$ and denoted by $\text{range}(P(u_1, u_k))$. Note that $\text{slope}(u_i, u_{i+1}) \in \text{range}(P(u_1, u_k))$ for all edges (u_i, u_{i+1}) $(1 \le i \le k-1)$ in $P(u_1, u_k)$. The following property of monotone drawings is proved in [1].

Property 1. A path $P(u_1, u_k)$ with range $[\text{slope}(e_{\min}), \text{slope}(e_{\max})]$ is monotone if and only if $\text{slope}(e_{\max}) - \text{slope}(e_{\min}) < 180°$. (See Fig 1.)

Let $P(u_1, u_k) = (u_1, \ldots, u_k)$ be a monotone path in Γ with the extremal edges $e_{\min}$ and $e_{\max}$. Let $a = \text{slope}(e_{\max}) - 90°$ and $b = \text{slope}(e_{\min}) + 90°$. Note that $a < b$ (because $\text{slope}(e_{\max}) - \text{slope}(e_{\min}) < 180°$). We call the open interval (a, b) the *interval of monotone directions* for $P(u_1, u_k)$.

Property 2. Let $P(u_1, u_k)$ be a monotone path with (a, b) as its interval of monotone directions. For any direction l with $\text{slope}(l) \in (a, b)$, $P(u_1, u_k)$ is monotone with respect to l.

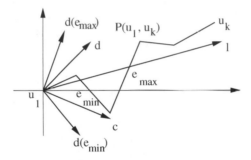

Fig. 1. Illustration of Property 1 and 2; $\text{slope}(c) = a°$; $\text{slope}(d) = b°$.

Proof. Let l be such a direction. For any edge $e \in P(u_1, u_k)$, the angle between e and l is strictly between $-90°$ and $90°$ (see Fig 1). Thus the order of the projections of the vertices in $P(u_1, u_k)$ on l is the same as the order they appear in $P(u_1, u_k)$. □

The following structure of 3-connected plane graphs is defined in [7,18].

Definition 1. *Let G be a 3-connected plane graph with three external vertices v_1, v_2, v_3 in ccw order. A Schnyder wood of G is a triplet of rooted spanning trees $\{T_1, T_2, T_3\}$ of G such that:*

1. *For $i \in \{1, 2, 3\}$, the root of T_i is v_i, and the edges of G are directed from children to parent in T_i.*
2. *Each edge e of G is contained in at least one and at most two spanning trees. If e is contained in two trees, then it has different directions in the two trees.*

3. *For each vertex $v \notin \{v_1, v_2, v_3\}$ of G, v has exactly one edge leaving v in each of T_1, T_2, T_3. The ccw order of the edges incident to v is: leaving in T_1, entering in T_3, leaving in T_2, entering in T_1, leaving in T_3, entering in T_2. Each entering block may be empty.*
4. *An edge with two opposite directions is considered twice. The first and the last incoming edges are possibly coincident with the outgoing edges.*
5. *For $i \in \{1,2,3\}$, all the incoming edges incident to v_i belong to T_i.*

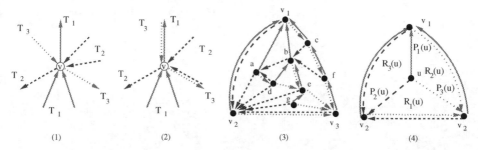

Fig. 2. (1), (2) two examples of edge pattern around an internal vertex v; (3) a 3-connected graph G with its Schnyder wood; (4) the paths $P_i(u)$ and the regions $R_i(u)$.

Figure 2 (1) and (2) show two examples of the edge pattern around an interval vertex v. (The edges in T_1, T_2, T_3 are drawn as solid (red), dashed (blue) and dotted (green) lines, respectively). In the second example, the edge leaving v in T_3 and an edge entering v in T_2 are the same edge. Figure 2 (3) shows an example of the Schnyder wood of a 3-connected plane graph.

In [7], it was shown that every 3-connected plane graph has a Schnyder wood, which can be compute in linear time. For each vertex u of G and $i \in \{1,2,3\}$, $P_i(u)$ denotes the path in T_i from u to the root v_i of T_i. We also use $P_i(u)$ to denote the set of the vertices in $P_i(u)$. The sub-path of the external face of G with end vertices v_1 and v_2 and not containing v_3 is denoted by $\text{ext}(v_1, v_2)$. The sub-paths $\text{ext}(v_2, v_3)$ and $\text{ext}(v_3, v_1)$ are defined similarly.

Let $p_i(u)$ denote the parent of u in T_i. Then, u is an *i-child* of $p_i(u)$. If there is a path in T_i from u to w, then w is an *i-ancestor* of u, and u is an *i-descendant* of w. Schnyder woods have been well studied in [7,9]. They have the following properties:

Property 3. Let G be a 3-connected plane graph with n vertices and m edges. Let $\mathcal{R} = \{T_1, T_2, T_3\}$ be a Schnyder wood of G, where T_i is a spanning tree rooted at the vertex v_i for $i \in \{1,2,3\}$.

1. For each vertex u of G, $P_1(u), P_2(u)$ and $P_3(u)$ have only vertex u in common.
2. For $i, j \in \{1,2,3\}(i \neq j)$ and two vertices u and w, the intersection of $P_i(u)$ and $P_j(w)$ is either empty or a common sub-path.
3. For the vertices v_1, v_2, v_3 the following hold: $P_1(v_2) = P_2(v_1) = \text{ext}(v_1, v_2)$; $P_2(v_3) = P_3(v_2) = \text{ext}(v_2, v_3)$; $P_3(v_1) = P_1(v_3) = \text{ext}(v_3, v_1)$.

For each vertex $u \notin \{v_1, v_2, v_3\}$, the three paths $P_1(u), P_2(u), P_3(u)$ divide G into three regions $R_1(u), R_2(u), R_3(u)$ (see Fig 2 (4)): $R_i(u)$ denotes the region of G bounded by the paths $P_{i-1}(u), P_{i+1}(u)$ and the external path $\text{ext}(v_{i-1}, v_{i+1})$. (We assume a cyclic order on the set $\{1, 2, 3\}$ so that $i - 1$ and $i + 1$ are always defined. Namely if $i = 3$ then $i + 1 = 1$; if $i = 1$ then $i - 1 = 3$). We also use $R_i(u)$ to denote the set of the vertices in the region $R_i(u)$. (For each external vertex v_i ($i \in 1, 2, 3$), $P_i(v_i) = \{v_i\}$. So only $R_i(v_i)$ contains faces in it. $R_{i-1}(v_i)$ and $R_{i+1}(v_i)$ contain no faces.)

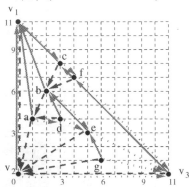

Fig. 3. Schnyder drawing of the graph shown in Fig 2 (3).

Table 1. Coordinates of the vertices in the drawing in Fig 3.

	$x_1(*)$	$x_2(*)$	$x_3(*)$
v_1	11	0	0
v_2	0	11	0
v_3	0	0	11
a	4	6	1
b	6	3	2
c	8	0	3
d	4	4	3
e	3	3	5
f	7	0	4
g	1	4	6

Definition 2. *[7]. For a vertex u in G, $x_i(u)$ ($i \in \{1, 2, 3\}$) denotes the number of faces of G in the region $R_i(u)$. They are called Schnyder coordinates of u.*

The following elegant theorem was proved in [7,9,18].

Theorem 1. *Fix any pair $i, j \in \{1, 2, 3\}$ ($i \neq j$). Take $X(u) = x_i(u)$ and $Y(u) = x_j(u)$ as the x- and y-coordinates for each vertex u in G. Then, we obtain a straight-line convex drawing of G on an $f \times f$ grid (f is the number of internal faces of G).*

By Theorem 1, there are six different Schnyder drawings. In the rest of the paper, we consider a particular Schnyder drawing Γ^* using $x_3(u)$ as the x-coordinate and $x_1(u)$ as the y-coordinate. The results obtained in this paper can be easily adapted to other five Schnyder drawings. From now on, Γ^* always denotes this particular Schnyder drawing. Figure 3 shows the drawing Γ^* of the graph in Figure 2 (3). Table 1 shows the vertex coordinates of Γ^*.

3   Schnyder Drawing of 3-Connected Plane Graphs is Monotone

In this section, we show that the Schnyder drawing Γ^* derived from a Schnyder wood $\mathcal{R} = \{T_1, T_2, T_3\}$ is monotone. First, we show a special property of Γ^* required by our proofs. By Definition 1, any edge $e = (u, w)$ of G belongs to at least one and at most two T_i ($i \in \{1, 2, 3\}$). In the following, the notation

$e = (u, w) \in T_1$ means either e only belongs to T_1, or e belongs to both T_1 and T_2, or e belongs to both T_1 and T_3. In all three cases, u is a child of w in T_1. The meanings of the notations $e \in T_2$ and $e \in T_3$ are similar.

Lemma 1. *Let Γ^* be the Schnyder drawing of G and $e = (u, w)$ an edge of G.*

1. *Suppose $e \in T_1$. If e only belongs to T_1, slope$(u, w) \in (90°, 135°)$. If e belongs to T_1 and T_2, slope$(u, w) = 90°$. If e belongs to T_1 and T_3, slope$(u, w) = 135°$.*
2. *Suppose $e \in T_2$. If e only belongs to T_2, slope$(u, w) \in (180°, 270°)$. If e belongs to T_2 and T_1, slope$(u, w) = 270°$. If e belongs to T_2 and T_3, slope$(u, w) = 180°$.*
3. *Suppose $e \in T_3$. If $e = (u, w)$ only belongs to T_3, slope$(u, w) \in (315°, 360°)$. If e belongs to T_3 and T_1, slope$(u, w) = 315°$. If e belongs to T_3 and T_2, slope$(u, w) = 360°$.*

Proof. Let $X(u) = x_3(u)$ and $Y(u) = x_1(u)$ be the x- and y-coordinate of the vertex u in Γ^*.

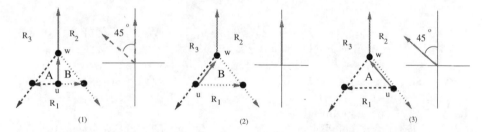

Fig. 4. The proof of Lemma 1. (1) Case 1a; (2) Case 1b; (3) Case 1c.

Case 1: $e = (u, w) \in T_1$. In this case, u is a 1-child of w.

Case 1a: (u, w) belongs to T_1 only (Fig 4 (1)). Then, we have $R_1(w) \supset R_1(u)$, $R_2(w) \subset R_2(u)$ and $R_3(w) \subset R_3(u)$. Let $A = R_3(u) - R_3(w)$ and $B = R_2(u) - R_2(w)$ be the regions shown in Fig 4 (1). Let $|A|$ and $|B|$ denote the number of faces in the region A and B. Note that $|A| > 0$ and $|B| > 0$. Define $\Delta(X) = X(w) - X(u)$ and $\Delta(Y) = Y(w) - Y(u)$. Then, $\Delta(X) = -|A|$ and $\Delta(Y) = |A| + |B|$. This implies slope$(u, w) \in (90°, 135°)$ (see Figure 4 (1)).

Case 1b: (u, v) belongs to T_1 and T_2 (Fig 4 (2)). Then, we have $R_1(w) \supset R_1(u)$, $R_2(w) \subset R_2(u)$ and $R_3(w) = R_3(u)$. Let $B = R_2(u) - R_2(w)$ be the region shown in Fig 4 (2). Then, $\Delta(X) = X(w) - X(u) = 0$ and $\Delta(Y) = Y(w) - Y(v) = |B| > 0$. This implies slope$(u, w) = 90°$ (see Figure 4 (2)).

Case 1c: (u, v) belongs to T_1 and T_3 (Fig 4 (3)). Then, we have $R_1(w) \supset R_1(u)$, $R_2(w) = R_2(u)$ and $R_3(w) \subset R_3(u)$. Let $A = R_3(u) - R_3(w)$ be the region shown in Fig 4 (3). Then, $\Delta(X) = X(w) - X(u) = -|A|$ and $\Delta(Y) = Y(w) - Y(v) = |A|$. This implies slope$(u, w) = 135°$ (see Figure 4 (3)).

Case 2: $e = (u, w) \in T_2$. In this case, u is a 2-child of w (see Figure 5 (1)).

By similar analysis as in Case 1, we have $R_1(w) \subseteq R_1(u)$, $R_2(w) \supseteq R_2(u)$ and $R_3(w) \subseteq R_3(u)$. Let $A = R_1(u) - R_1(w)$ and $B = R_3(u) - R_3(w)$. (Note that either A or B may contain no faces.)

If $e = (u, w)$ only belongs to T_2, then $|A| > 0$ and $|B| > 0$. We have $\Delta(X) = -|B| < 0$ and $\Delta(Y) = -|A| < 0$. This implies slope$(u, w) \in (180°, 270°)$.

If $e = (u, w)$ belongs to T_2 and T_1, then $|B| = 0$. We have $\Delta(X) = 0$ and $\Delta(Y) = -|A| < 0$. This implies slope$(u, w) = 270°$.

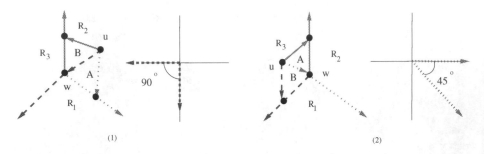

Fig. 5. The proof of Lemma 1. (1) Case 2; (2) Case 3.

If $e = (u, w)$ belongs to T_2 and T_3, then $|A| = 0$. We have $\Delta(Y) = 0$ and $\Delta(X) = -|B| < 0$. This implies slope$(u, w) = 180°$.

Case 3: $e = (u, w) \in T_3$. In this case, u is a 3-child of w (see Figure 5 (2)).

By similar analysis as in Case 1, we have $R_1(w) \subseteq R_1(u)$, $R_2(w) \subseteq R_2(u)$ and $R_3(w) \supseteq R_3(u)$. Let $A = R_2(u) - R_2(w)$ and $B = R_1(u) - R_1(w)$. (Note that either A or B may contain no faces.)

If $e = (u, w)$ only belongs to T_3, then $|A| > 0$ and $|B| > 0$. We have $\Delta(X) = |A| + |B| > 0$ and $\Delta(Y) = -|B| < 0$. This implies slope$(u, w) \in (315°, 360°)$ (see Figure 5 (3)).

If $e - (u, w) \subset T_3$ and T_1, then $|A| = 0$. We have $\Delta(X) = |B|$ and $\Delta(Y) = -|B|$. This implies slope$(u, w) = 315°$.

If edge $e = (u, w) \in T_3$ and T_2, then $|B| = 0$. We have $\Delta(X) = |A| > 0$ and $\Delta(Y) = 0$. This implies slope$(u, w) = 360°$.   □

Definition 3. Let u, w be two vertices of G such that $u \in R_i(w)$ ($i \in \{1, 2, 3\}$).

1. A path $P(u, w)$ in G is called an *i-left path* if there exists a vertex t in $P(u, w)$ (possibly $t = u$ or $t = w$) such that:
   - for any edge e in the sub-path $P(u, t)$, $e \in T_i$.
   - for any edge e in the sub-path $P(t, w)$, $e \in T_{i+1}$.
2. A path $P(u, w)$ in G is called an *i-right path* if there exists a vertex t in $P(u, w)$ ($t \neq u$ and $t \neq w$) such that:
   - for any edge e in the sub-path $P(u, t)$, $e \in T_i$.
   - for any edge e in the sub-path $P(t, w)$, $e \in T_{i-1}$.
   (For $i = 1$, $i - 1$ denotes 3. For $i = 3$, $i + 1$ denotes 1).

Lemma 2. *For any two vertices u and w ($u \neq w$) in G, there always exists an i-left or i-right path $P(u, w)$ for some $i \in \{1, 2, 3\}$.*

Proof. Consider the three regions $R_1(w)$, $R_2(w)$ and $R_3(w)$ defined by the three paths $P_1(w), P_2(w), P_2(w)$. We divide the proof into three cases depending on the location of u.

Case 1: $u \in R_1(w) - P_3(w)$. Follow the path $P_1(u)$ starting from u. Clearly, this path must cross $P_2(w)$ or $P_3(w)$ at a vertex t.

Case 1-left: $t \in P_2(w)$. This case includes the sub-cases:

- $u \in P_2(w)$. In this case $t = u$ (see the vertex u_1 in Fig 6 (1)).
- $u \notin P_2(w)$ and $t = w$ (see the vertex u_2 in Fig 6 (1)).
- $u \notin P_2(w)$ and $t \neq w$ (see the vertex u_3 in Fig 6 (1)).

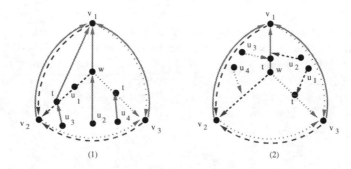

(1)   (2)

Fig. 6. The proof of Lemma 2. (1) The 1-left and 1-right path $P(u, w)$; (2) the 2-left, 2-right, 3-left and 3-right paths $P(u, w)$.

In any case, the concatenation of the sub-path of $P_1(u)$ between u and t and the reverse of the sub-path $P_2(w)$ between t and w is the required 1-left path $P(u, w)$.

Case 1-right: $t \in P_3(w) - \{w\}$ (see the vertex u_4 in Fig 6 (1)).

In this case, the concatenation of the sub-path of $P_1(u)$ between u and t and the reverse of the sub-path $P_3(w)$ between t and w is the required 1-right path $P(u, w)$.

Case 2: $u \in R_2(w) - P_1(w)$. Follow the path $P_2(u)$ starting from u. This path must cross $P_1(w)$ or $P_3(w)$ at a vertex t. If $t \in P_3(w)$, we can obtain a 2-left path $P(u, w)$ (see the vertex u_1 in Fig 6 (2)). If $t \in P_1(w) - \{w\}$, we can obtain a 2-right path $P(u, w)$ (see the vertex u_2 in Fig 6 (2)).

Case 3: $u \in R_3(w) - P_2(w)$. Follow the path $P_3(u)$ starting from u. This path must cross $P_1(w)$ or $P_2(w)$ at a vertex t. If $t \in P_1(w)$, we can obtain a 3-left path $P(u, w)$ (see the vertex u_3 in Fig 6 (2)). If $t \in P_2(w) - \{w\}$, we can obtain a 3-right path $P(u, w)$ (see the vertex u_4 in Fig 6 (2)). □

Lemma 3. *Let $P(u, w)$ be an i-left or i-right path in the Schnyder drawing Γ^*.*

1. *If $P(u, w)$ is a 1-left path, $\text{range}(P(u, w)) = [0°, 135°]$. If $P(u, w)$ is a 1-right path, $\text{range}(P(u, w)) = [90°, 180°]$.*
2. *If $P(u, w)$ is a 2-left path, $\text{range}(P(u, w)) = [180°, 315°]$. If $P(u, w)$ is a 2-right path, $\text{range}(P(u, w)) = [135°, 270°]$.*

3. *If $P(u,w)$ is a 3-left path,* $\operatorname{range}(P(u,w)) = [270°, 360°]$. *If $P(u,w)$ is a 3-right path,* $\operatorname{range}(P(u,w)) = [-45°, 90°]$.

Hence, by Property 1, all i-left and i-right paths $(i \in \{1,2,3\})$ are monotone.

Proof. The proof is divided into six cases depending on the type of $P(u,w)$.

Case 1-left: $P(u,w)$ is a 1-left path. In this case, $P(u,w)$ is divided into two sub-paths $P(u,t)$ and $P(t,w)$. For any edge e_i in $P(u,t)$, e_i is in T_1 and follows the same direction of T_1. By Lemma 1, $\operatorname{slope}(e_i) \in [90°, 135°]$. For any edge $e_j \in P(t,w)$, e_j is in T_2 but follows the opposite direction of T_2. By Lemma 1, $\operatorname{slope}(e_j) \in [0°, 90°]$. Thus, for any edge $e \in P(u,w)$, we have $\operatorname{slope}(e) \in [0°, 135°]$. Fig 7 case 1-left shows the $\operatorname{range}(P(u,w))$. (The left figure shows the path pattern of $P(u,w)$, the right figure shows the corresponding $\operatorname{range}(P(u,w))$).

Case 1-right: $P(u,w)$ is a 1-right path. In this case, $P(u,w)$ is divided into two sub-paths $P(u,t)$ and $P(t,w)$ (see Fig 7 case 1-right). For any edge e_i in $P(u,t)$, e_i is in T_1 and follows the same direction of T_1. By Lemma 1, $\operatorname{slope}(e_i) \in [90°, 135°]$. For any edge $e_j \in P(t,w)$, e_j is in T_3 but follows the opposite direction of T_3. By Lemma 1, $\operatorname{slope}(e_j) \in [135°, 180°]$. Thus, for any edge $e \in P(u,w)$, we have $\operatorname{slope}(e) \in [90°, 180°]$.

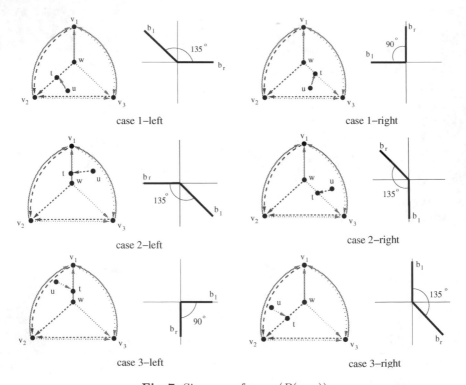

case 1-left   case 1-right

case 2-left   case 2-right

case 3-left   case 3-right

Fig. 7. Six cases of $\operatorname{range}(P(u,w))$

The proof for the other 4 cases are similar.

Case 2-left: $P(u, w)$ is a 2-left path (see Fig 7 case 2-left). In this case, for any edge $e \in P(u, w)$, we have slope$(e) \in [180°, 315°]$.

Case 2-right: $P(u, w)$ is a 2-right path (see Fig 7 case 2-right). In this case, for any edge $e \in P(u, w)$, we have slope$(e) \in [135°, 270°]$.

Case 3-left: $P(u, w)$ is a 3-left path (see Fig 7 case 3-left). In this case, for any edge $e \in P(u, w)$, we have slope$(e) \in [270°, 360°]$.

Case 3-right: $P(u, w)$ is a 3-right path (see Fig 7 case 3-right). In this case, for any edge $e \in P(u, w)$, we have slope$(e) \in [-45°, 90°]$.   □

By Lemmas 2, 3 and Property 1, we immediately have the following:

Theorem 2. *The Schnyder drawing Γ^* of a 3-connected graph G is a monotone drawing.*

Next, we discuss how to find the monotone directions for two given vertices u, w in G. By Lemma 3 and Property 2, we immediately have:

Lemma 4. *Let $P(u, w)$ be a path in Γ^*.*

1. *The interval of monotone directions of a 1-left path $P(u, w)$ is $(45°, 90°)$.*
2. *The interval of monotonc directions of a 1-right palh $P(u, w)$ is $(90°, 180°)$.*
3. *The interval of monotone directions of a 2-left path $P(u, w)$ is $(225°, 270°)$.*
4. *The interval of monotone directions of a 2-right path $P(u, w)$ is $(180°, 225°)$.*
5. *The interval of monotone directions of a 3-left path $P(u, w)$ is $(270°, 360°)$.*
6. *The interval of monotone directions of a 3-right path $P(u, w)$ is $(0°, 45°)$.*

Definition 4. *Let u, w be two vertices of G. If there exists a path $P(u, w)$ that is a 1-left (1-right, 2-left, 2-right, 3-left, 3-right, respectively) path, we say the relative position of u with respect to w is 1-left (1-right, 2-left, 2-right, 3-left, 3-right, respectively).*

Theorem 3. *Let G be a 3-connected plane graph. G can be pre-processed in $O(n)$ time such that, for any two vertices u and w ($u \neq w$) in G, the interval of monotone directions in Γ^* for u, w can be determined in $O(1)$ time.*

Proof. Let $\mathcal{R} = \{T_1, T_2, T_3\}$ be the Schnyder wood of G for the drawing Γ^*. We perform ccw post-order traversal on T_1 and assign post-order number $N_1(u)$ to the vertices u in G. (Namely, we visit and number the vertices in the subtrees rooted at the children of the root v_1 of T_1 recursively in ccw order, then we visit and number the root v_1).

Similarly, let $N_2(u)$ ($N_3(u)$, respectively) be the ccw post-order traversal number of u with respect to T_2 (T_3, respectively). $N_1(u), N_2(u)$ and $N_3(u)$ for all vertices u can be easily computed in $O(n)$ time.

The following facts can be easily verified: the relative position of u with respect to w is:

1-left   if and only if $N_2(u) < N_2(w)$, $N_3(u) > N_3(w)$ and $N_1(u) < N_1(w)$.
1-right   if and only if $N_2(u) < N_2(w)$, $N_3(u) > N_3(w)$ and $N_1(u) > N_1(w)$.
2-left   if and only if $N_3(u) < N_3(w)$, $N_1(u) > N_1(w)$ and $N_2(u) < N_2(w)$.
2-right   if and only if $N_3(u) < N_3(w)$, $N_1(u) > N_1(w)$ and $N_2(u) > N_2(w)$.
3-left   if and only if $N_1(u) < N_1(w)$, $N_2(u) > N_2(w)$ and $N_3(u) < N_3(w)$.
3-right   if and only if $N_1(u) < N_1(w)$, $N_2(u) > N_2(w)$ and $N_3(u) > N_3(w)$.

Knowing $N_1(*), N_2(*)$ and $N_2(*)$, we can determine the relative position of u with respect to w, and hence the interval of the monotone directions for u, w, in $O(1)$ time.   □

4   Conclusion

In this paper, we presented a simple proof that the classical Schnyder drawing of 3-connected plane graphs is a monotone drawing on a $f \times f$ grid, which can be constructed in $O(n)$ time. It also has the advantage that for any two vertices u, w, the interval of monotone directions l_{uw} can be identified in $O(1)$ time. This is the first monotone drawing with $o(n^3)$ area for 3-connected plane graphs.

References

1. Angelini, P., Colasante, E., Di Battista, G., Frati, F., Patrignani, M.: Monotone drawings of graphs. J. of Graph Algorithms and Applications 16(1), 5–35 (2012)
2. Angelini, P., Frati, F., Grilli, L.: An algorithm to construct greedy drawings of triangulations. J. Graph Algorithms and Applications 14(1), 19–51 (2010)
3. Angelini, P., Didimo, W., Kobourov, S., Mchedlidze, T., Roselli, V., Symvonis, A., Wismath, S.: Monotone drawings of graphs with Fixed Embedding. Algorithmica 71(2), 233–257 (2015)
4. Arkin, E.M., Connelly, R., Mitchell, J.S.: On monotone paths among obstacles with applications to planning assemblies. In: SoCG 1989, pp. 334–343 (1989)
5. Bárány, I., Rote, G.: Strictly Convex Drawings of Planar Graphs. Documenta Mathematica 11, 369–391 (2006)
6. Bonichon, N., Felsner, S., Mosbah, M.: Convex Drawings of 3-Connected Plane Graphs. In: Pach, J. (ed.) GD 2004. LNCS, vol. 3383, pp. 60–70. Springer, Heidelberg (2005)
7. Di Battista, G., Tamassia, R., Vismara, L.: Output-sensitive reporting of disjoint paths. Algorithmica 23(4), 302–340 (1999)
8. Di Battista, G., Tamassia, R.: Algorithms for plane representations of acyclic digraphs. Theor. Comput. Sci. 61, 175–198 (1988)
9. Felsner, S.: Convex Drawings of Planar Graphs and the Order Dimension of 3-Polytopes. Orders 18, 19–37 (2001)
10. Garg, A., Tammassia, R.: On the computational complexity of upward and rectilinear planarity testing. SIAM J. Comp. 31(2), 601–625 (2001)
11. Hossain, M. I., Rahman, M. S.: Monotone grid drawings of planar graphs. In: Chen, J., Hopcroft, J.E., Wang, J. (eds.) FAW 2014. LNCS, vol. 8497, pp. 105–116. Springer, Heidelberg (2014)
12. Huang, W., Eades, P., Hong, S.-H.: A Graph Reading Behavior: Geodesic-Path Tendency. In: IEEE Pacific Visualization Symposium, pp. 137–144 (2009)

13. Kindermann, P., Schulz, A., Spoerhase, J., Wolff, A.: On Monotone Drawings of Trees. In: Duncan, C., Symvonis, A. (eds.) GD 2014. LNCS, vol. 8871, pp. 488–500. Springer, Heidelberg (2014)
14. Moitra, A., Leighton, T.: Some results on greedy embeddings in metric spaces. In: FOCS 2008, pp. 337–346 (2008)
15. Papadimitriou, C.H., Ratajczak, D.: On a conjecture related to geometric routing. Theoretical Computer Science 344(1), 3–14 (2005)
16. Rote, G.: Strictly Convex Drawings of Planar Graphs. In: 16th ACM-SIAM SODA, pp. 728–734 (2005)
17. Schnyder, W.: Embedding Planar Graphs on the Grid. In: Proc. 1st Ann. ACM-SIAM Symp. Discrete Algorithms, pp. 138–148 (1990)
18. Schnyder, W., Trotter, W.T.: Convex Embedding of 3-Connected Planar Graphs. Abstracts Amer. Math. Soc. 13(5), 502 (1992)

Faster Fully-Dynamic Minimum Spanning Forest

Jacob Holm, Eva Rotenberg, and Christian Wulff-Nilsen

Department of Computer Science, University of Copenhagen
{jaho,roden,koolooz,}@di.ku.dk
http://www.diku.dk/~koolooz/

Abstract. We give a new data structure for the fully-dynamic minimum spanning forest problem in simple graphs. Edge updates are supported in $O(\log^4 n/\log\log n)$ expected amortized time per operation, improving the $O(\log^4 n)$ amortized bound of Holm et al. (STOC '98, JACM '01). We also provide a deterministic data structure with amortized update time $O(\log^4 n \log\log\log n/\log\log n)$. We assume the Word-RAM model with standard instructions.

1 Introduction

A dynamic graph problem is that of maintaining a dynamic graph on n vertices where edges may be inserted or deleted and possibly where queries regarding properties of the graph are supported. We call the dynamic problem decremental resp. incremental if edges can only be deleted resp. inserted, and fully dynamic if both edge insertions and deletions are supported.

We consider the fully-dynamic minimum spanning forest (MSF) problem which is to maintain a state for each edge of whether it belongs to the current MSF or not. After an edge update, at most one edge becomes a new tree edge in the MSF and at most one edge becomes a non-tree edge, and a data structure needs to output which edge changes state, if any.

Dynamic MSF is one of the most fundamental dynamic graph problems, and the first non-trivial solution was presented by Frederickson [4] in 1983. Frederickson achieved a worst-case update time of $O(\sqrt{m})$ where m is the number of edges at the time of the update. This was later improved by Eppstein et al. [3] to $O(\sqrt{n})$ using the sparsification technique. Henzinger and King made a data structure with amortized update time $O(\sqrt[3]{n}\log n)$. Holm et al. [9] dramatically improved this amortized bound to $O(\log^4 n)$. All these bounds are for simple graphs (no parallel edges), but any MSF structure can be extended to general graphs via a simple reduction that adds $O(\log m)$ to the update time. In the following we will assume all graphs are simple unless otherwise stated.

We show how to support updates in $O(\log^4 n/\log\log n)$ expected amortized time, which is the first improvement in 17 years. To obtain this bound, we assume the Word-RAM model of computation with standard instructions. More generally, our time bound per update can be written as

$$O\left(\frac{\log^4 n}{\log\log n} \cdot \frac{sort(\log^c n, n^2)}{\log^c n}\right),$$

© Springer-Verlag Berlin Heidelberg 2015
N. Bansal and I. Finocchi (Eds.): ESA 2015, LNCS 9294, pp. 742–753, 2015.
DOI: 10.1007/978-3-662-48350-3_62

for some constant $c > 0$, where $sort(k, r)$ is the time for sorting k natural numbers with values in the range from 0 to r. Equivalenty, $sort(k, r)/k$ is the operation time of a priority queue. Thus, the update time of our structure depends on the model of computation, and the choice of the priority queue that our structure uses as a building block. The following table shows both deterministic and randomized variants of the data structure differing only in the choice of priority queue.

Table 1. Update time, depending on choice of priority queue from [2, 5, 11, 13], see Section 1.1

	Deterministic	Randomized
RAM w. AC^0	$O(\log^4 n\sqrt{\log\log\log n}/\sqrt{\log\log n})$	$O(\log^4 n\log\log\log n/\log\log n)$
RAM, AC^0, $O(1)$ time mult.	$O(\log^4 n\log\log\log n/\log\log n)$	$O(\log^4 n/\log\log n)$

1.1 Related Work

Holm et al. [9] gave a deterministic data structure for decremental MSF with $O(\log^2 n)$ amortized update time. Combining this with a slightly modified version of a reduction from fully-dynamic to decremental MSF of Henzinger and King [6], they obtained their $O(\log^4 n)$ bound for fully-dynamic MSF. A somewhat related problem to dynamic MSF is fully-dynamic connectivity. Here a data structure needs to support insertion and deletion of edges as well as connectivity queries between vertex pairs. The problem was first studied by Frederickson [4] who obtained $O(\sqrt{m})$ update time $O(1)$ query time data structure. Update time was improved to $O(\sqrt{n})$ by Eppstein et al. [3]. Henzinger and King [7] obtained expected $O(\log^3 n)$ amortized update time and query time $O(\log n/\log\log n)$. Henzinger and Thorup [8] improved update time to $O(\log^2 n)$ with a clever sampling technique. A deterministic structure with the same bounds was given by Holm et al. [9]. Thorup [12] achieved an expected amortized update-time of $O(\log n(\log\log n)^3)$ and query time $O(\log n/\log\log\log n)$, using randomization. Wulff-Nilsen [14] gave a deterministic, amortized $O(\log^2 n/\log\log n)$ update-time data structure with $O(\log n/\log\log n)$ query time. An $\Omega(\log n)$ lower bound on the operation time for fully-dynamic connectivity and MSF was given by Pătraşcu and Demaine [10].

As indicated above, priority queues are essential to our data structure. Equivalently, we rely on the ability to efficiently sort $l = \log^c n$ elements from $[n^2]$ where c is a constant. Expressed as a function of l, the elements lie in the range $0 \ldots 2^w - 1$, where $w = 2l^{1/c}$. To sort quickly, we rely on $w > l$. In the RAM-model with AC^0 instructions, Raman [11] gave a deterministic bound of $O(l\sqrt{\log l\log\log l})$. Using randomization, Thorup [13] improved this to $O(l\log\log l)$. The same time bounds were achieved without randomization, if assuming constant time multiplication, by Han [5]. Andersson et al. [2] achieve optimal $O(l)$ sorting time, using randomization, and assuming $O(1)$ time multiplication; their algorithm requires $w \gg \log^{2+\varepsilon} l$ for some constant ε, which in our case is satisfied as $w > l^{1/c}$.

1.2 Idea and Paper Outline

Since the data structures of Holm et al. [9] for decremental MSF and fully dynamic connectivity are essentially the same, the question arises of whether the $O(\log^2 n/\log\log n)$

fully-dynamic connectivity structure in [14] can be directly translated to an improved $O(\log^2 n/\log\log n)$ decremental MSF structure. If that were the case, we could immediately use the reduction from fully-dynamic to decremental MSF in [9] to obtain an $O(\log^4 n/\log\log n)$ bound for fully-dynamic MSF. Unfortunately, that is not the case as the data structure in [14] relies on a shortcutting system which can not be easily adapted to decremental MSF. Instead, we make a different analysis of the reduction from fully-dynamic to decremental MSF (Section 2) which surprisingly shows how a slightly slower decremental MSF structure than that in [9] can in fact lead to a slightly faster fully dynamic MSF!

A modified version of the dynamic connectivity structure by Wulff-Nilsen [14] with $O(\log^2 n)$ update time is described in Section 3. It is shown in Section 3.3 how to modify it to a simple decremental MSF structure with the same performance. We then show how to speed up a certain part of this decremental MSF structure in Section 4. The main idea is to extend it with a non-trivial shortcutting system involving fast priority queues in order to speed up the search for replacement edges. This system is the main technical contribution of the paper. We conclude Section 4 by showing that this data structure for decremental MSF speeds up fully-dynamic MSF.

2   Reduction to Decremental MSF

In this section, we present a different analysis of the reduction from fully dynamic MSF to decremental MSF from [9] based on the construction from [6]. The main difference is that in our analysis, we do not insist on all edges being deleted in the decremental MSF problem. The proof of the following Lemma is omitted due to space constraints.

Lemma 1. *Suppose we have a decremental (deletions-only) MSF data structure that for a connected simple graph with n vertices and m edges has a total worst-case running time for the construction and the first d deletions of $O(t_c m + t_r d)$, where t_c and t_r are non-decreasing functions of n. Then there exists a fully dynamic MSF data structure for simple graphs on n vertices with amortized update time $O(\log^3 n + t_c \log^2 n + t_r \log n)$.*

The following corollary is crucial in obtaining our improvement for fully-dynamic MSF. It shows that to obtain a faster data structure for this problem by reduction to decremental MSF, it actually suffices with a decremental MSF structure which is slower than that in [9] in the case where all edges end up being deleted.

Corollary 1. *Given a decremental MSF structure with $t_c = \frac{\log^2 n}{\varepsilon \log\log n}$ and $t_r = \log^{2+\varepsilon} n$ where $\varepsilon < 1$ is a constant, the reduction gives a fully dynamic MSF structure with amortized update time $O(\frac{\log^4}{\log\log n})$.*

3   Simple Data Structures for Dynamic Connectivity and Decremental MSF

In this section, we give a description of the fully-dynamic connectivity data structure in [14] (based on [12]) except that shortcuts are omitted and a spanning forest is maintained. We will modify it in Section 3.3 to support decremental MSF.

Let $G = (V, E)$ denote the dynamic graph. The data structure maintains, for each edge $e \in E$, a *level* $\ell(e)$ which is an integer between 0 and $\ell_{\max} = \lfloor \log_2 n \rfloor$. As we shall see, the level of an edge e starts at 0 and can only increase over time and for the amortization, we can view $\ell_{\max} - \ell(e)$ as the amount of credits left on e.

For $0 \leq i \leq \ell_{\max}$, let E_i denote the set of edges of E with level at least i and let $G_i = (V, E_i)$. The (connected) components of G_i are *level i clusters* or just *clusters*. The following invariant is maintained:

Invariant: For each i, any level i cluster contains at most $\lfloor n/2^i \rfloor$ vertices.

Consider a level i cluster C. By contracting all edges of E_{i+1} in C, we get a connected multigraph of level i-edges where vertices correspond to level $(i + 1)$ clusters contained in C. Our data structure maintains a spanning tree of this multigraph. The union of spanning trees over all clusters is a spanning forest of G.

The data structure maintains a *cluster forest* of G which is a forest $\mathcal{C}$ of rooted trees where a node u at level i is a level i cluster $C(u)$. Roots of $\mathcal{C}$ are components of $G = G_0$ and leaves of $\mathcal{C}$ are vertices of G. A level i-node u which is not a leaf has as children the level $(i + 1)$ nodes v for which $C(v) \subseteq C(u)$. In addition to $\mathcal{C}$, the data structure maintains $n(u)$ for each node $u \in \mathcal{C}$ denoting the number of vertices of G contained in $C(u)$ (equivalently, the number of leaves in the subtree of $\mathcal{C}$ rooted at u).

3.1 Handling Insertions and Deletions

When a new edge $e = (u, v)$ is inserted into G, it is given level 0 and $\mathcal{C}$ is updated by merging the roots r_u and r_v corresponding to the components of G containing u and v, respectively. The new root inherits the children of both r_u and r_v. If $r_u \neq r_v$, e becomes a tree edge in the new level 0 cluster. Otherwise, e becomes a non-tree edge.

Deleting an edge $e = (u, v)$ is more involved. If e is not a tree edge, no structural changes occur in $\mathcal{C}$. Otherwise, let $i = \ell(e)$. The deletion of e splits a spanning tree of a level i cluster C into two subtrees, T_u containing u (inside some level $i + 1$-cluster) and T_v containing v. One of these trees, say T_u, contains at most half the vertices (in V) of C. For each level i edge in T_u, we increase its level to $i + 1$. In C, this amounts to merging all nodes corresponding to level $i + 1$ clusters in T_u into one node, w; see Figure 1(a) and (b). By the choice of T_u, this does not violate the invariant.

Next, we search through (non-tree) level i edges incident to $C(w)$ in some arbitrary order until some edge is found which connects $C(w)$ and T_v (if any). For all visited level i edges which did not reconnect the two trees, their level is increased to $i + 1$, thereby paying for them being visited. If a replacement edge (a, b) was found, no more structural changes occur in $\mathcal{C}$ and (a, b) becomes a new tree edge. Otherwise, w is removed from its parent p (corresponding to C) and a new level i node p' is created having w as its single child and having p as sibling; see Figure 1(b) and (c). This has the effect of splitting $C = C(p)$ into two smaller level i clusters. The same procedure is now repeated recursively at level $i - 1$ where we try to reconnect the two trees of level $i - 1$ edges containing the new level i clusters $C(p)$ and $C(p')$, respectively. If level 0 is reached and no replacement edge was found, a component of G is split in two.

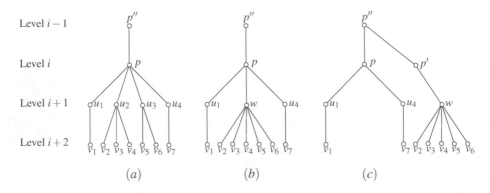

Fig. 1. (a): Part of C before the merge. (b): Level $i+1$ nodes u_2 and u_3 are merged into a new level $i+1$ node w. (c): A replacement level i edge was not found so w is given a new parent p' which becomes the sibling of p.

3.2 Local Trees

To guide the search for level i tree/non-tree edges, we first modify C to a forest C_L of binary trees. This is done by inserting, for each non-leaf node $u \in C$, a binary *local tree* $L(u)$ between u and its children; see Figure 2. To describe the structure of $L(u)$, we first need to define heavy and light children of u. A child v of u in C is *heavy* if $n(v) \geq n(u)/\log^{\varepsilon_h} n$, where $\varepsilon_h > 0$ is a constant that we may pick as small as we like. Otherwise, v is *light*. The root of $L(u)$ has two children, one rooted at *heavy tree* $T_h(u)$ and the other rooted at *light tree* $T_l(u)$. The leaves of $T_h(u)$ resp. $T_l(u)$ are the heavy resp. light children of u. Before describing the structure of these trees, let us associate a *rank* $\mathrm{rank}(v) \leftarrow \lfloor \log n(v) \rfloor$ with each node v in C.

Tree $T_h(u)$ is formed by initially regarding each heavy child of u as a trivial rooted tree with a single node and repeatedly pairing roots r_1 and r_2 of trees with the same rank, creating a new tree with a root r of rank $\mathrm{rank}(r) = \mathrm{rank}(r_1) + 1$ and with children r_1 and r_2. When the process stops, the remaining rooted trees, called *rank trees*, all have distinct ranks and they are attached as children to a rooted *rank path* P such that children with larger rank are closer to the root of P than children of smaller rank. We define the rank of a node on the rank path to be the larger of the ranks of its children.

Tree $T_l(u)$ is more involved. Its leaves are the light children of u and they are divided into groups each having size at most $\log^\alpha n$, where α is a constant that we may pick as

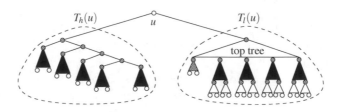

Fig. 2. The structure of local tree $L(u)$ of a node u in C, from [14]. In $T_h(u)$, rank trees are black and the rank path and roots of rank trees are grey. In $T_l(u)$, the buffer tree is grey, top and bottom trees are white, and rank trees are black.

large as we like. The nodes in each group are kept as leaves in a balanced binary search tree (BBST) ordered by $n(v)$-values. One of these trees is the *buffer tree* and the others are *bottom trees*. We define the rank of each bottom tree root as the maximum rank of its leaves and we pair them up into rank trees exactly as we did when forming $T_h(u)$. However, instead of attaching the rank tree roots to a rank path, we instead keep them as leaves of a BBST called the *top tree*[1], where again the ordering is by rank. We also have the buffer tree root as a child of the top tree and we regard it as having smaller rank than all the other leaves. It was shown in [14] that C_L has height $O(\frac{1}{\varepsilon_h} \log n)$. Refer to nodes of C_L belonging to C as *cluster nodes*.

Merging Local Trees. We need to support the merge of local trees $L(u)$ and $L(v)$ in C_L corresponding to a merge of cluster nodes u and v into a new node w. First, we merge the buffer trees of $L(u)$ and $L(v)$ into a new BBST T_b by adding the leaves of the smaller tree to the larger tree. Heavy trees $T_h(u)$ and $T_h(v)$ have their rank paths removed and leaves that should be light in $L(w)$ are removed from $T_h(u)$ and $T_h(v)$ and added as leaves of T_b. For each leaf removed from $T_h(u)$ and $T_h(v)$, we remove their ancestor rank nodes. We end up with subtrees of the original rank trees in $T_h(u)$ and $T_h(v)$ and these subtrees are paired up as before and attached to a new rank path for $T_h(w)$. Tree T_b becomes a buffer tree in $T_l(w)$ if its number of leaves does not exceed $\log^\alpha n$; otherwise, it becomes a bottom tree in $T_l(w)$, leaving an empty buffer tree. Rank trees in $T_l(u)$ and $T_l(v)$ are stripped off from their top trees and paired up into new rank trees as before (here we include T_b if it became a bottom tree) and these are attached as leaves to a new top tree for $T_l(w)$.

In the above merge, let p be the parent of u and v in C. In C_L, we need to delete u and v as leaves of $L(p)$ and to add w as a new leaf of $L(p)$. We shall only describe the deletion of u as v is handled in the same manner. We consider four cases depending on which part of $L(p)$ u belongs to:

- If u is a leaf in the buffer tree of $T_l(p)$, delete it with a standard BBST operation.
- If u is a leaf in a bottom tree B of $T_l(p)$, a similar BBST update happens in B. Additionally the max rank of leaves in B is updated as this rank is associated with the root of B. If the maximum does not decrease, no further updates are needed. Otherwise, remove all ancestor rank nodes of B in $T_l(p)$, pair the resulting rank trees as before and attach them as leaves of the top tree.
- If u is a leaf in $T_h(p)$, remove it and its ancestor rank nodes in $T_h(p)$, pair up the resulting rank trees and attach them to a new rank path for $T_h(p)$.

To add w as a new leaf of $L(p)$, we only have two cases. If w is a heavy node, we regard it as a trivial rank tree, delete the rank path of $T_h(p)$, repeatedly pair up the rank trees (including w) and reattach them with a new rank path to form the updated $T_h(p)$. If instead w is a light node, we add it to the buffer tree of $T_l(p)$ (which may be turned into a bottom tree, as described above).

Handling Cluster Splits. What remains is to describe the updates to local trees after splitting a level i cluster in two. Let w, p, and p' be defined as in the previous subsection

---

[1] Not to be confused with the data structure of that name [1].

and let p'' be the parent of p and p' in C (Figure 1(b) and (c)). Creating $L(p')$ is trivial as p' has only the single child w in C and attaching p' as a leaf of $L(p'')$ is done as above. The removal of w from $L(p)$ decreases $n(p)$ which may cause some light children of p in C to become heavy. In C_L, each corresponding leaf of $T_l(p)$ is removed and added to $T_h(p)$ and $L(p)$ is updated accordingly as described above. Since $n(p)$ decreases, p might change from being a heavy child of p'' to being a light child. If so, we move it from $T_h(p'')$ to the buffer tree of $T_l(p'')$, as described above.

Bitmaps. Having modified C into the forest C_L of binary trees, we add bitmaps to nodes of C_L to guide the search for level i edges. More precisely, each node $u \in C_L$ is associated with two bitmaps $tree(u)$ and $nontree(u)$, where the ith bit of $tree(u)$ ($nontree(u)$) is 1 iff there is at least one level i tree (non-tree) edge incident to a leaf in the subtree of C_L rooted at u. Since C_L is binary, these bitmaps enable us to identify a level i tree/non-tree edge incident to a cluster $C(u)$ by traversing a path down from u in C_L in time proportional to its length by backtracking when bitmaps with ith bit 0 are encountered. When a level i tree edge is removed (which happens if it is deleted from G or has its level increased), then for each of its endpoints u, we set $tree(u)[i] = 0$ and update the bitmaps for all ancestors v of u in C_L bottom-up by taking the logical 'or' of the $tree$-bitmaps of its children. A similar update is done to $nontree$-bitmaps if u is a non-tree edge. When inserting a level i tree/non-tree edge, bitmaps are updated in a similar manner.

3.3 Supporting Decremental MSF

We can convert the above fully dynamical connectivity structure to a decremental MSF structure by using a trick from [9]. For decremental MSF, we can assume that the initial graph is simple and connected and that all weights are distinct and belong to $\{0, 1, \ldots, n^2\}$ by doing an initial comparison sort and then working on ranks of weights instead. All edges start at level 0 and we initialize the spanning forest to the MSF. When searching through the level i non-tree edges incident to $C(w)$ as in Section 3.1, we do so in order of increasing weight. We support this by letting each node of C_L contain the weight of the cheapest level i-edge below it, for each i. To find the cheapest non-tree edge with an endpoint in $C(w)$ we can follow the cheapest level i weight down from w in C_L until we reach a leaf x and then take the cheapest level i-edge incident to x. As shown in [9], this small modification to the connectivity structure suffices to support decremental MSF.

Performance. Finding the initial MSF can be done in $O(m + n \log n)$ time using Prim's algorithm with Fibonacci heaps. We split the time complexity analysis for the rest of the above data structure into three parts: searching for edges down from $C(w)$ in C_L to identify a cheapest level i-edge incident to a leaf x, maintaining the edge weights associated with nodes of C_L, and making structural changes to C_L.

To analyze the time for the first part, note that since C_L has height $O(\log n)$, searching down from $C(w)$ to x takes $O(\log n)$ time. In order to efficiently identify the cheapest level i-edge incident to such a leaf x, we extend the data structure by letting x have an $O(\log n)$ array of doubly-chained lists of edges, so let $E_i(x)$ be the list of level i non-tree edges incident to w in order of increasing weight. The cheapest level i-edge incident to

x is then the first edge of $E_i(x)$ and can thus be found in $O(1)$ time. When increasing the level of an edge $e = (x,y)$ from i to $i+1$, it is not a replacement edge, and is therefore the cheapest level i non-tree edge incident to any vertex in its component. In particular it is the cheapest level i non-tree edge incident to x and y and is therefore at the start of $E_i(x)$ and $E_i(y)$. Furthermore, (as shown in [9]) it is costlier than all other edges that have been moved to level $i+1$ earlier, so when we move it, all we need to do is put it at the end of $E_{i+1}(x)$ and $E_{i+1}(y)$ to keep them sorted. This takes $O(1)$ time.

We have shown how the cheapest non-tree edge incident to $C(w)$ can be found in $O(\log n)$ time. Maintaining edge weights associated with nodes of C_L can also be done in $O(\log n)$ time since for each edge level change (or the deletion of an edge), only the weights along the leaf-to-root paths in C_L from the endpoints of the edge need to be updated. It remains to bound the time for structural changes to C_L. It was shown in [14] that by picking constant ε_h sufficiently small and constant α sufficiently large (see definitions in Section 3.2), this takes amortized $O(\log n/\log\log n)$ time per edge level change plus an additional $O(\log^2 n/\log\log n)$ worst-case time per edge deletion.

We conclude from the above that the total time to build our decremental MSF structure on a simple connected graph with n vertices and m edges, and then deleting d edges is $O(m\log^2 n + d\log^2 n)$. In the next section, we give a variant of this data structure where exactly the same structural changes occur in C_L but where the time to search for edges is sped up using a new shortcutting system together with fast priority queues. Since structural changes take a total of $O(m\log^2 n/\log\log n + d\log^2 n/\log\log n)$ time, these will not be the bottleneck so we ignore them in the time analysis in the next section. Also, the structure in [14] can identify the parent cluster node of a cluster in $O(\log n/\log\log n)$ time, so we shall also ignore this cost.

4   Faster Data Structure for Dynamic MSF

In this section, we present our new data structure for decremental MSF. Assume that the initial graph is connected. If not, we maintain the data structure separately for each component. The total time bound is $O(m\log^2 n/\log\log n + d\log^{2+\varepsilon} n)$ for a constant $\varepsilon < 1$, where the initial graph has m edges and n vertices and where d edges are deleted in total. By Corollary 1, this suffices in order to achieve $O(\log^4 n/\log\log n)$ update time for fully-dynamic MSF.

A bottleneck of the simple data structure for decremental MSF presented in Section 3.3 is moving up and down trees of C_L. The data structure identifies level i-edges incident to a level $(i+1)$-cluster $C(u)$ in order of increasing weight by moving down C_L from node u, always picking the child (or children) with the cheapest level i-edge below it. When a leaf is reached, the cheapest level i-edge e incident to it is traversed. If both endpoints of e were identified in the downward search then we do not need an upwards search. If only one endpoint was identified then we do an upwards search in C_L from the other endpoint until reaching the node for a level $(i+1)$-cluster. Each upwards search can trivially be done in $O(\log n)$ time as this is a bound on the height of trees in C_L. We claim that this is actually fast enough. To see why, note that we only do an upwards search when a reconnecting edge is found. At most one reconnecting edge is found per edge deleted so we can in fact afford to spend $O(\log^{2+\varepsilon} n)$ time on the upwards search.

In the following, we can thus restrict our attention to speeding up downward searches. It suffices to get a search time of $O(\log n / \log \log n)$ since for every two downward searches, we either increase the level of an edge or we find a reconnecting edge.

4.1 A Downwards Shortcutting System

We use a downwards shortcutting system with fast min priority queues to speed up downward searches. Certain nodes of C_L are augmented with min priority queues keyed on edge weights. Since we may assume for our decremental structures that edge weights are in the range $\{0, 1, \ldots, n^2\}$, we can use fast integer priority queues. In the following, we assume constant time for each queue operation. As mentioned in the introduction, a less efficient queue will slow down the performance of our data structure by a factor equal to its operation time.

The nodes of C_L with associated priority queues are referred to as *queue nodes*. The following types of nodes are queue nodes (ε_q is a small constant to be chosen later):

1. cluster nodes whose level is divisible by $\lceil \varepsilon_q \log \log n \rceil$,
2. heavy tree nodes u with a parent v in C_L such that $\operatorname{rank}(u) \le i \lceil \varepsilon_q \log \log n \rceil < \operatorname{rank}(v)$ for an integer i,
3. rank nodes of light trees whose rank is divisible by $\lceil \varepsilon_q \log \log n \rceil$,
4. roots and leaves of buffer, bottom, and top trees.

Each queue node u (excluding leaves of C_L) is associated with an array whose ith entry points to a min-queue $Q_i(u)$, for each level i. If u is a proper ancestor of a level $(i+1)$-node, $Q_i(u)$ is empty. Otherwise, for each nearest descending node v of u in C_L which is either a queue-node or a leaf of C_L, $Q_i(u)$ contains the node v with associated key k denoting the weight of the cheapest level i-edge incident to a leaf of the sub-tree of C_L rooted at v.

The priority queues associated with queue nodes induce our downwards shortcutting system in C_L. To see how, consider a level $(i + 1)$ cluster $C(u)$. To identify the cheapest level i-edge e incident to $C(u)$, assume first that u is a queue node. Then a minimum element in $Q_i(u)$ is a node v below u with e incident to $C(v)$. We refer to (u, v) as a *shortcut*. Whereas our simple data structure would traverse the path from u down to v in C_L, our new data structure can use the shortcut (u, v) to jump directly from u to v within the time it takes to obtain the minimum element in $Q_i(u)$. At v, we identify a minimum element w in $Q_i(v)$ and jump directly to this node along (v, w). This shortcut traversal continues until a leaf of C_L is reached, and e is identified as one of the edges incident to this leaf. If both endpoints of e are below u in C_L, one of the queues contains two distinct minimum elements v and v', corresponding to where the paths down to the endpoints of e branch out. In this case, we search down from both v and v'.

Now assume that u is not a queue node. Then all nearest descending queue nodes v of u are visited and for each of them the minimum element in $Q_i(v)$ is identified and its associated key $k_i(v)$. The search procedure described above is then applied to each of the at most two nodes v with minimum key $k_i(v)$.

Let us analyze the time for the search procedure above. The following lemma (proof omitted) bounds time to identify nearest descending queue nodes.

Lemma 2. *The nearest descending queue nodes of a cluster node can be found in $O(\log^{3\varepsilon_q} n)$ time.*

If our initial node u is not a queue node, we can thus in $O(\log^{3\varepsilon_q} n)$ time find all nearest descending queue nodes of u and among these obtain the at most two nodes v with smallest key in $Q_i(v)$. Now, assume that the initial node u is a queue node and consider the shortcut path P of queue nodes from u to a leaf that the procedure visits. The number of visited queue nodes of type 1 is clearly $O(\log n / (\varepsilon_q \log\log n))$. Since ranks of nodes along P cannot increase and since the difference in rank between two consecutive rank nodes on P is at least $\lceil \varepsilon_q \log\log n \rceil$, the number of queue nodes of type 2 or 3 is also $O(\log n / (\varepsilon_q \log\log n))$. Finally, since the rank difference between a cluster node u and any leaf in $T_l(u)$ is $\Omega(\varepsilon_h \log\log n)$ (see [14]), P contains only $O(\log n / (\varepsilon_h \log\log n))$ queue nodes of type 4. Given our downwards shortcutting system, the cheapest level i-edge incident to $C(u)$ can thus be found in $O(\log^{3\varepsilon_q} n + (\frac{1}{\varepsilon_h} + \frac{1}{\varepsilon_q}) \log n / \log\log n)$ time. Below we show how to maintain this system efficiently under changes to $\mathcal{C}_L$.

4.2 Dealing with Non-Topological Changes

Two types of changes occur in $\mathcal{C}_L$: topological changes when cluster nodes are merged or split and non-topological changes when an edge increases its level or is removed and information about which edges are the cheapest below a cluster node needs to be updated. We start with the non-topological changes.

Suppose a level i-edge e disappears, either because it is deleted or because its level is increased to $i+1$. Then, we need to update priority queues of queue nodes accordingly. If $\ell(e)$ increases, then the two downward paths identified with our shortcutting system contain all the queue nodes whose level i-queues need to be updated. For each endpoint x of e, we traverse each of these paths bottom-up. Let u be the current non-leaf node in one of these traversals and let v be its predecessor wrt. the traversal. If the key of v in $Q_i(u)$ equals the weight $w(e)$ of e, we increase it to the key for the minimum element in $Q_i(v)$, or remove v from $Q_i(u)$ if $Q_i(v)$ is empty. Otherwise, we stop, as the minimum key of $Q_i(v)$ is unchanged, and thus no queue nodes above v need updates. As each queue update takes $O(1)$ time, total time is bounded by the number $O((\frac{1}{\varepsilon_h} + \frac{1}{\varepsilon_q}) \log n / \log\log n)$ of queue nodes considered.

We also need to update priority queues for level $(i+1)$-edges since e has its level increased to $i+1$. Note that all the queue nodes that need to be updated belong to the two downward paths traversed. Again, we traverse each path bottom-up. Let u be the current non-leaf node in one of the traversals and let v be its predecessor. If v is not present in $Q_{i+1}(u)$, we add it with key $w(e)$. Otherwise, if the key of v in $Q_{i+1}(u)$ is greater than $w(e)$, we decrease it to $w(e)$. In both cases, we then proceed upwards. Otherwise, we stop since no queues above u need updates. Total time to update level $(i+1)$-queues is $O((\frac{1}{\varepsilon_h} + \frac{1}{\varepsilon_q}) \log n / \log\log n)$.

It remains to consider the case where e disappears because it was deleted. Then we identify all the queue nodes above e that need to be updated by traversing the leaf-to-root paths in $\mathcal{C}_L$ for the endpoints of e. The queue nodes visited have their queue nodes updated as described above. Since $\mathcal{C}_L$ has height $O(\frac{1}{\varepsilon_h} \log n)$, total time is $O(\frac{1}{\varepsilon_h}\log n + (\frac{1}{\varepsilon_h} + \frac{1}{\varepsilon_q})\log n / \log\log n)$. This completes the description of how to deal with non-topological changes.

4.3 Dealing with Topological Changes

Now, we describe how to maintain queues under topological changes to C_L. We assume that deleting a shortcut is free as it is paid for when the shortcut is formed. In our analysis for bounding the total time to form shortcuts, we use the accounting method; during the course of the algorithm, credits will be associated with certain parts of C_L and each credit can pay for a constant amount of work. Denote by $s_{max} = \log^\alpha n$ the maximum number of leaves of a buffer tree. The following invariants are maintained:

– Each leaf of a heavy tree contains $(2 + \log s_{max})\log n$ credits (*heavy tree invariant*),
– Each leaf of a buffer tree contains $(2 + \log s_{max} - \log s)\log n$ credits where s is the number of leaves in the tree (*buffer tree invariant*).
– Each leaf of a bottom tree contains 1 credit (*bottom tree invariant*).

Lemma 3. *A buffer tree with s_1 leaves has more credits than one with $s_2 < s_1$ leaves.*

Proof. The function $f(x) = x(2 + \log s_{max} - \log x)$ is monotonically increasing on $[1, s_{max}]$ since $f'(x) = 2 + \log s_{max} - \log x - 1/\ln 2 > \log s_{max} - \log x \geq 0$ for all $x \in [1, s_{max}]$.

Recall that ε_h was introduced when defining heavy and light children. We observe that initially, all edges of the decremental MSF structure have level 0 and because of our assumption that the initial graph is connected, C consists of a single root r with each vertex of the graph as a child, implying that C_L is the single local tree $L(r)$. This local tree contains at most $\log^{\varepsilon_h} n$ leaves in the heavy tree and a single buffer tree with at most s_{max} leaves. Furthermore, there are at most n bottom tree leaves. By Lemma 3, the initial amount of credits required is at most $\log^{\varepsilon_h} n(2 + \log s_{max})\log n + 2s_{max}\log n + n$.

The general type of change to C during the deletion of a level i-tree edge was described in Section 3.1 and the corresponding updates to local trees in C_L was described in Section 3.2. The first step is to merge all level $(i+1)$ clusters on the smaller component of the split level i-tree (Figure 1(a) and (b)). We now describe how to update shortcuts accordingly. Assume that only two level $(i+1)$ clusters $C(u)$ and $C(v)$ are merged into a new level $(i+1)$ cluster $C(w)$; the general case is omitted due to space constraints. It may be helpful to consult Figure 2.

We now describe how to update shortcuts through the heavy tree. A shortcut (x,y) *goes through* a node $z \in C_L$ if x is an ancestor of z and y is a descendant of z (where possibly $x = z$ or $y = z$). In the new local tree $L(w)$, we obtain all shortcuts through nodes of $T_h(w)$ bottom-up. Note that queue nodes in the subtrees of C_L rooted at leaves of $T_h(w)$ need not be updated. For each queue node $a \in T_h(w)$, assume all queues of its nearest descending queue nodes have been constructed. Then for each level j, construct $Q_j(a)$ by visiting all nearest descending queue nodes b of a and for each of them adding the cheapest node of $Q_j(b)$ to $Q_j(a)$. By Lemma 2, this takes $O(\log^{3\varepsilon_q} n)$ time for each j, giving a total time of $O(\log^{1+3\varepsilon_q} n)$ to construct the queues associated with a. As $T_h(w)$ has size $O(\log^{\varepsilon_h} n)$, total time to construct all shortcuts through nodes of $T_h(w)$ is $O(\log^{1+3\varepsilon_q+\varepsilon_h} n)$ which over all levels is $O(\log^{2+3\varepsilon_q+\varepsilon_h} n)$. Adding $(2 + \log s_{max})\log n$ credits to each leaf of $T_h(w)$ is dominated by the cost to construct shortcuts.

The procedure for updating shortcuts through the light tree $T_l(w)$ and the local tree for the parent of w in C and handling cluster splits is omitted due to space constraints. It can be shown that with $\varepsilon_h < \frac{1}{2}$ and $\varepsilon_q < \frac{1}{6}$, Corollary 1 implies:

Theorem 1. *There is a data structure for fully-dynamic minimum spanning tree which supports updates in $O(\log^4 n / \log \log n)$ amortized time.*

References

1. Alstrup, S., Holm, J., de Lichtenberg, K., Thorup, M.: Maintaining information in fully dynamic trees with top trees. ACM Trans. Algorithms 1(2), 243–264 (2005)
2. Andersson, A., Hagerup, T., Nilsson, S., Raman, R.: Sorting in linear time? Journal of Computer and System Sciences 57(1), 74–93 (1998), See also STOC 1995
3. Eppstein, D., Galil, Z., Italiano, G.F., Nissenzweig, A.: Sparsification - a technique for speeding up dynamic graph algorithms. J. ACM 44(5), 669–696 (1997)
4. Frederickson, G.N.: Data structures for on-line updating of minimum spanning trees, with applications. SIAM Journal on Computing 14(4), 781–798 (1985), See also STOC 1983
5. Han, Y.: Deterministic sorting in $O(n \log \log n)$ time and linear space. J. Algorithms 50(1), 96–105 (2004), See also STOC 2002
6. Henzinger, M.R., King, V.: Fully dynamic 2-edge connectivity algorithm in polylogarithmic time per operation (1997)
7. Henzinger, M.R., King, V.: Randomized fully dynamic graph algorithms with polylogarithmic time per operation. J. ACM 46(4), 502–516 (1999), See also STOC 1995
8. Henzinger, M.R., Thorup, M.: Sampling to provide or to bound: With applications to fully dynamic graph algorithms. Random Structures and Algorithms 11(4), 369–379 (1997), See also ICALP 1996
9. Holm, J., de Lichtenberg, K., Thorup, M.: Poly-logarithmic deterministic fully-dynamic algorithms for connectivity, minimum spanning tree, 2-edge, and biconnectivity. J. ACM 48(4), 723–760 (2001), See also STOC 1998
10. Pătrașcu, M., Demaine, E.D.: Lower bounds for dynamic connectivity. In: Proc. 36th ACM Symposium on Theory of Computing (STOC), pp. 546–553 (2004)
11. Raman, R.: Fast algorithms for shortest paths and sorting (1996)
12. Thorup, M.: Near-optimal fully-dynamic graph connectivity. In: Proc. 32nd ACM Symposium on Theory of Computing (STOC), pp. 343–350 (2000)
13. Thorup, M.: Randomized sorting in $O(n \log \log n)$ time and linear space using addition, shift, and bit-wise boolean operations. J. Algorithms 42(2), 205–230 (2002), See also SODA 1997
14. Wulff-Nilsen, C.: Faster deterministic fully-dynamic graph connectivity. In: SODA, pp. 1757–1769 (2013)

On the Equivalence among Problems
of Bounded Width

Yoichi Iwata[1,*] and Yuichi Yoshida[2,**]

[1] The University of Tokyo, Tokyo, Japan
y.iwata@is.s.u-tokyo.ac.jp
[2] National Institute of Informatics and Preferred Infrastructure, Inc.
yyoshida@nii.ac.jp

Abstract. In this paper, we introduce a methodology, called decomposition-based reductions, for showing the equivalence among various problems of bounded-width.

First, we show that the following are equivalent for any $\alpha > 0$:
- SAT can be solved in $O^*(2^{\alpha\mathbf{tw}})$ time,
- 3-SAT can be solved in $O^*(2^{\alpha\mathbf{tw}})$ time,
- MAX 2-SAT can be solved in $O^*(2^{\alpha\mathbf{tw}})$ time,
- INDEPENDENT SET can be solved in $O^*(2^{\alpha\mathbf{tw}})$ time, and
- INDEPENDENT SET can be solved in $O^*(2^{\alpha\mathbf{cw}})$ time,

where **tw** and **cw** are the tree-width and clique-width of the instance, respectively.

Then, we introduce a new parameterized complexity class EPNL, which includes SET COVER and TSP, and show that SAT, 3-SAT, MAX 2-SAT, and INDEPENDENT SET parameterized by path-width are EPNL-complete. This implies that if one of these EPNL-complete problems can be solved in $O^*(c^k)$ time, then any problem in EPNL can be solved in $O^*(c^k)$ time.

1 Introduction

SAT is a fundamental problem in complexity theory. Today, it is widely believed that SAT cannot be solved in polynomial time. This is not only because anyone could not find a polynomial-time algorithm for SAT despite many attempts, but also because if SAT can be solved in polynomial time, any problem in NP can be solved in polynomial time (NP-completeness). Actually, even no algorithms faster than the trivial $O^*(2^n)$-time[1] exhaustive search algorithm are known. Impagliazzo and Paturi [12] conjectured that SAT cannot be solved in $O^*((2 - \epsilon)^n)$ time for any $\epsilon > 0$, and this conjecture is called the *Strong Exponential Time Hypothesis (SETH)*. Under the SETH, conditional lower bounds for several

---

* Supported by JSPS Grant-in-Aid for JSPS Fellows (256487).
** Supported by JSPS Grant-in-Aid for Young Scientists (B) (No. 26730009), MEXT Grant-in-Aid for Scientific Research on Innovative Areas (24106001), and JST, ERATO, Kawarabayashi Large Graph Project.
[1] $O^*(\cdot)$ hides a factor polynomial in the input size.

© Springer-Verlag Berlin Heidelberg 2015
N. Bansal and I. Finocchi (Eds.): ESA 2015, LNCS 9294, pp. 754–765, 2015.
DOI: 10.1007/978-3-662-48350-3_63

problems have been obtained, including k-DOMINATING SET [16], problems of bounded tree-width [14,8], and EDIT DISTANCE [3].

When considering polynomial-time tractability, all the NP-complete problems are equivalent, that is, if one of them can be solved in polynomial time, then all of them can be also solved in polynomial time. Similarly, when considering subexponential-time tractability, all the SNP-complete problems are equivalent [13]. However, if we look at the exponential time complexity for solving each NP-complete problem more closely, the situation changes; whereas the current fastest algorithm for SAT is the naive $O^*(2^n)$-time exhaustive search algorithm, faster algorithms have been proposed for many other NP-complete problems such as 3-SAT [11], MAX 2-SAT [18], and INDEPENDENT SET [19]. Although there are many problems, including SET COVER and TSP[2], for which the current fastest algorithms take $O^*(2^n)$ time, we do not know whether a faster algorithm for one of these problems leads to a faster algorithms for SAT and vice versa. Actually, only a few problems, such as HITTING SET and SET SPLITTING, are known to be equivalent to SAT in terms of exponential time complexity [7].

In this paper, we propose a new methodology, called *decomposition-based reductions*. Although the idea of decomposition-based reductions is simple, we can obtain various interesting results. First, we show that when parameterized by *width*, there are many problems that are equivalent to SAT. Second, we show the equivalence among different width; INDEPENDENT SET parameterized by tree-width and INDEPENDENT SET parameterized by clique-width are equivalent. Third, we introduce a new parameterized complexity class EPNL, which includes SET COVER and TSP, and show that many problems parameterized by path-width are EPNL-complete. For these problems, conditional lower-bounds under the SETH are already known [14]. However, our results imply that these problems are at least as hard as not only n-variable SAT but also *any* problem in EPNL. In this sense, our hardness results are more robust.

It has been shown that many NP-hard graph optimization problems can be solved efficiently if the input graph has a nice decomposition. One of the most famous decompositions is *tree-decomposition*, and a graph is parameterized by *tree-width*, the size of the largest bag in the (best) tree-decomposition of the graph. Intuitively speaking, tree-width measures how much a graph looks like a tree. If we are given a graph and its tree-decomposition of width $\mathbf{tw}$[3], many problems can be solved in $O^*(c^{\mathbf{tw}})$ time, where c is a problem-dependent constant. For example, we can solve INDEPENDENT SET and MAX 2-SAT in $O^*(2^{\mathbf{tw}})$ time by standard dynamic programming and DOMINATING SET in $O^*(3^{\mathbf{tw}})$ time by combining with subset convolution [17].[4] Recently, Lokshtanov *et al.* [14] showed

---

[2] For UNDIRECTED HAMILTONICITY, a faster algorithm has been proposed in a recent paper by Björklund [4]. However, for general TSP, the trivial $O^*(2^n)$-time dynamic programming algorithm is still the current fastest.

[3] Obtaining a tree-decomposition of the minimum width is NP-hard. In this paper, we assume that we are given a decomposition as a part of the input, and a problem is parameterized by the width of the given decomposition.

[4] For problems related to SAT, we consider the tree-width of the primal graph of the input. See Section 2 for details.

that many of these algorithms are optimal under the SETH. These results are obtained by reducing an n-variable instance of SAT to an instance of the target problem with tree-width approximately $\frac{n}{\log c}$, where c is a problem dependent constant. However, these reductions are one-way, and thus a faster SAT algorithm may not lead to faster algorithms for these problems. Moreover, there is a possibility that one of these problems has a faster algorithm but the others do not.

The first contribution of this paper is showing the following equivalence among problems of bounded tree-width:

Theorem 1. *For any $\alpha > 0$, the following are equivalent:*

1. SAT *can be solved in $O^*(2^{\alpha \mathbf{tw}})$ time.*
2. 3-SAT *can be solved in $O^*(2^{\alpha \mathbf{tw}})$ time.*
3. MAX 2-SAT *can be solved in $O^*(2^{\alpha \mathbf{tw}})$ time.*
4. INDEPENDENT SET *can be solved in $O^*(2^{\alpha \mathbf{tw}})$ time.*

For all of these problems, the fastest known algorithms run in $O^*(2^{\mathbf{tw}})$ time [15] and Theorem 1 states that this is not a coincidence. Note that an n-variable instance of SAT has tree-width at most $n - 1$. Hence by Theorem 1, for any $\epsilon > 0$, an $O^*((2 - \epsilon)^{\mathbf{tw}})$-time algorithm for INDEPENDENT SET of bounded tree-width implies an $O^*((2 - \epsilon)^n)$-time algorithm for the general SAT. Therefore, our result includes the hardness result by Lokshtanov *et al.* [14]. We believe that the same technique can be applied to many other problems. In practice, SAT solvers are widely used to solve various problems by reductions to SAT. Using our methodology, we can reduce an instance of some problem to an instance of SAT by preserving the tree-width. Since tree-decompositions can be used to speed-up SAT solvers [10], our reductions may be useful in practice.

Clique-width is the number of labels we need to construct the given graph by iteratively performing certain operations. Similarly to the tree-width case, many problems can be solved in $O^*(c^{\mathbf{cw}})$ time if the given graph has a clique-width $\mathbf{cw}$, where c is a problem-dependent constant [6].

The second contribution of this paper is showing the following equivalence between INDEPENDENT SET of bounded tree-width and bounded clique-width:

Theorem 2. *For any $\alpha > 0$, the following are equivalent:*

1. INDEPENDENT SET *can be solved in $O^*(2^{\alpha \mathbf{tw}})$ time.*
2. INDEPENDENT SET *can be solved in $O^*(2^{\alpha \mathbf{cw}})$ time.*

The fastest known algorithms for INDEPENDENT SET parameterized by clique-width runs in $O^*(2^{\mathbf{cw}})$ time [6]. It is surprising that we can obtain such strong connections between problems of bounded tree-width and a problem of bounded clique-width because tree-width and clique-width are very different parameters in nature; a complete graph of n vertices has a clique-width two whereas its tree-width is $n - 1$. Hence, even if there is an efficient algorithm for a problem of bounded tree-width, it does not immediately imply that there is an efficient algorithm for the same problem of bounded clique-width. However, Theorem 2 states that a faster algorithm for INDEPENDENT SET of bounded tree-width implies a faster algorithm for INDEPENDENT SET of bounded clique-width. We note that INDEPENDENT SET is chosen because SAT, 3-SAT, and MAX 2-SAT are still NP-complete when its primal graph is a clique ($\mathbf{cw} = 2$). Hence, these problems parameterized by tree-width and clique-width are not equivalent unless $P = NP$. We believe that we can obtain similar results for many other problems that can be solved efficiently on graphs of bounded clique-width.

The third contribution of this paper is introducing a new parameterized complexity class EPNL (Exactly Parameterized NL) and showing the following complete problems:

Theorem 3. SAT, 3-SAT, MAX 2-SAT, *and* INDEPENDENT SET *parameterized by path-width are* EPNL-*complete.*

Intuitively, EPNL is a class of parameterized problems that can be solved by a non-deterministic Turing machine with the space of $k + O(\log n)$ bits. For the precise definitions of EPNL and EPNL-completeness, see Section 4. Flum and Grohe [9] introduced a similar class, called para-NL, that can be solved in $f(k) + O(\log n)$ space. Although they showed that a trivial parameterization of an NL-complete problem is para-NL-complete under the standard parameterized reduction, this does not hold in our case because we use a different reduction to define the complete problems. If one of the NP-complete problems can be solved in polynomial time, any problem in NP can be solved in polynomial time. Similarly, if one of the EPNL-complete problems can be solved in $O^*(c^k)$ time, any problem in EPNL can be solved in $O^*(c^k)$ time. Since the class EPNL contains many famous problems, such as SET COVER parameterized by the number of elements and TSP parameterized by the number of vertices, for which no $O^*((2 - \epsilon)^n)$-time algorithms are known, our result implies that we can use the hardness of not only SAT but also these problems to establish the hardness of the problems parameterized by path-width.

1.1   Overview of Decomposition-Based Reductions

We explain the basic idea of decomposition-based reductions. Although we deal with three different decompositions in this paper, the basic idea is the same. We believe that the same idea can be used to other decompositions such as branch-decomposition.

A decomposition can be seen as a collection of sets forming a tree. For example, tree-decomposition is a collection of bags forming a tree and clique-decomposition is a collection of labels forming a tree. First, for each node i of a decomposition tree, we create gadgets as follows: (1) for each element x in the corresponding set X_i, create a *path-like* gadget x_i that expresses the *state* of the element (e.g. the value of the variable x for the case of SAT), and (2) create several gadgets to solve *subproblem* corresponding to this node (e.g. simulate clauses inside X_i for the case of SAT). Then, for each node c, its parent p, and each common element $x \in X_c \cap X_p$, by connecting the tail of x_c and the

head of x_p, we establish *local consistency*. From the definition of the decomposition, this leads to *global consistency*. Since the obtained graph has a *locality*, it
has a small width. We may need additional tricks to establish local consistency
without increasing the width.

1.2   Organization

The rest of the paper is organized as follows. In Section 2, we introduce definitions and basic lemmas often used in this paper. In Section 3, we give a tree-width
preserving reduction from MAX 2-SAT to SAT. The reduction is rather simple
but contains an essential idea of tree-decomposition-based reductions. In Section 4, we introduce EPNL and show that SAT parameterized by path-width
is EPNL-complete. Due to space limitations, several reductions and proofs are
omitted. These will be presented in the full version of this paper.

2   Preliminaries

For an integer k, we denote the set $\{1, 2, \ldots, k\}$ by $[k]$ and the set $\{0, 1, \ldots, k-1\}$
by $[k]'$. Let $G = (V, E)$ be an undirected graph. We denote the *degree* of a vertex
v as $d_G(v)$. We denote the *neighborhood* of a vertex u by $N_G(u) = \{v \in V \mid \{u, v\} \in E\}$, and the *closed neighborhood* of u by $N_G[u] = N_G(u) \cup \{u\}$. Similarly,
we denote the neighborhood of a subset $S \subseteq V$ by $N_G(S) = \bigcup_{v \in S} N_G(v) \setminus S$,
and the closed neighborhood by $N_G[S] = N_G(S) \cup S$. We drop the subscript G
when it is clear from the context. For a subset $S \subseteq V$, let $G[S] = (S, \{\{u, v\} \in E \mid u \in S, v \in S\})$ denote the *subgraph induced by* S. For a vertex $v \in V$, let G/v
denote the graph obtained by removing v and making the neighbors of v form
a clique. We call this operation *eliminating* v. Similarly, for a subset $S \subseteq V$, we
denote by G/S the graph obtained by removing S and making the neighbors of
S form a clique.

A *tree-decomposition* of a graph $G = (V, E)$ is a pair (T, χ), where $T = (I, F)$
is a tree and $\chi = \{X_i \subseteq V \mid i \in I\}$ is a collection of subsets of vertices (called
bags), with the following properties:

1. $\bigcup_i X_i = V$.
2. For each edge $uv \in E$, there exists a bag that contains both of u and v.
3. For each vertex $v \in V$, the bags containing v form a connected subtree in T.

In order to avoid confusion between a graph and its decomposition tree T, we
call a vertex of the tree a *node*, and an edge of the tree an *arc*. We identify a node
$i \in I$ of the tree and the corresponding bag X_i. The *width* of a tree-decomposition
is the maximum of $|X_i| - 1$ over all nodes $i \in I$. The *tree-width* of a graph G,
$\mathbf{tw}(G)$, is the minimum width among all the possible tree-decompositions of G.

A *nice tree-decomposition* is a tree decomposition such that the root bag X_r
is an empty set and each node i is one of the following types:

1. Leaf: a leaf node with $X_i = \emptyset$.

2. Introduce(v): a node with one child c such that $X_i = X_c \cup \{v\}$ and $v \notin X_c$.
3. Introduce(uv): a node with one child c such that $u, v \in X_i = X_c$. We require that this node appears exactly once for each edge uv of G.
4. Forget(v): a node with one child c such that $X_i = X_c \setminus \{v\}$ and $v \in X_c$. From the definition of tree-decompositions, this node appears exactly once for each vertex of G.
5. Join: a node with two children l and r with $X_i = X_l = X_r$.

Any tree-decomposition can be easily converted into a nice tree-decomposition of the same width in polynomial time by inserting intermediate bags between each adjacent bags. Thus, in this paper, we use nice tree-decompositions to make discussions simple.

A (nice) *path-decomposition* is a (nice) tree-decomposition (T, χ) such that the decomposition tree $T = (I, F)$ is a path. The *path-width* of a graph G, $\mathbf{pw}(G)$, is the minimum width among all the possible path-decompositions of G.

In order to prove the upper bound on tree-width, we will often use the following lemmas. Due to space limitations, proofs of these lemmas are omitted.

Lemma 1 (Arnborg [1]). *For a graph $G = (V, E)$ and a vertex $v \in V$, $\mathbf{tw}(G) \leq \max(d(v), \mathbf{tw}(G/v))$. Moreover, if we are given a tree-decomposition of G/v of width w, we can construct a tree-decomposition of G of width $\max(d(v), w)$ in linear time.*

Lemma 2. *For a graph $G = (V, E)$ and a vertex subset $S \subseteq V$, $\mathbf{tw}(G) \leq \max(|N[S]| - 1, \mathbf{tw}(G/S))$.*

Lemma 3. *Let X and Y be disjoint vertex sets of a graph G such that for each vertex $x \in X$, $|N(x) \cap Y| \leq 1$. Then, $\mathbf{tw}(G) \leq \max(|N[X] \setminus Y|, \mathbf{tw}(G/X))$.*

Lemma 4. *Let $\{S_i \mid i \in [d]\}$ be a family of disjoint vertex sets of a graph G such that each set has size at most k and there are no edges between S_i and S_j for any $|i - j| > 1$. Then, $\mathbf{tw}(G) \leq \max(2k + |N(S)| - 1, \mathbf{tw}(G/S))$, where $S = \bigcup_{i \in [d]} S_i$.*

For a vertex set S, if we can obtain $\mathbf{tw}(G) \leq \max(d, \mathbf{tw}(G/S))$ by applying one of these lemmas, we say that the elimination has *degree* d. If we can reduce a graph G into a graph G' by a series of eliminations of degree at most d, we can obtain $\mathbf{tw}(G) \leq \max(d, \mathbf{tw}(G'))$.

Let x be a Boolean variable. We denote the negation of x by $\bar{x}$. A *literal* is either a variable or its negation, and a *clause* is a disjunction of several literals $l_1, \ldots, l_k$, where k is called the *length* of the clause. We call a clause of length k a *k-clause*. A *CNF* is a conjunction of clauses. If all the clauses have length at most k, it is called a *k-CNF*. We say that a CNF on a variable set X is *satisfiable* if there is an assignment to X that makes the CNF true. (*k-*)SAT is a problem in which, given a variable set X and a (k-)CNF $\mathcal{C}$, the objective is to determine whether $\mathcal{C}$ is satisfiable or not. MAX 2-SAT is a problem in which, given a variable set X, a 2-CNF $\mathcal{C}$, and an integer k, the objective is to determine whether there exists an assignment that satisfies at least k clauses in $\mathcal{C}$.

Let $\mathcal{C}$ be a CNF on variables X. The *primal graph* of $\mathcal{C}$ is the graph $G = (X, E)$ such that there exists an edge between two vertices if and only if their corresponding variables appear in the same clause. For readability, we identify a variable or a literal as the corresponding vertex in the primal graph. That is, we may use the same symbol x to indicate both a variable in a CNF and the corresponding vertex in the primal graph, and both literals x and $\bar{x}$ correspond to the identical vertex in the primal graph. For a CNF $\mathcal{C}$, we slightly change the definition of the nice tree-decomposition as follows:

3'. Introduce(C): an internal node with one child c such that $X_i = X_c$ and all the variables in C are in X_i. We require that this node appears exactly once for each clause $C \in \mathcal{C}$.

Note that because the variables in the same clause form a clique in the primal graph, there always exists a bag that contains all of them.

In our reductions, we will use a binary representation of an integer. Let $\{a_1, a_2, \ldots, a_M\}$ be Boolean variables. We denote the integer $\sum_{i \in [M], a_i = \text{true}} 2^{i-1}$ by $(a_1 a_2 \ldots a_M)_2$, or $(a_*)_2$ for short. For readability, we will frequently use (arithmetic) constraints such as $(a_*)_2 = (b_*)_2 + (c_*)_2$. Note that any arithmetic constraint on M variables can be trivially simulated by at most 2^M M-clauses. Thus, if M is logarithmic in the input size, the number of required clauses is polynomial in the input size.

3   Tree-width Preserving Reduction from Max 2-SAT to SAT

Let $(X, \mathcal{C} = \{C_1, \ldots, C_m\}, k)$ be an instance of MAX 2-SAT. We want to construct an instance $(X', \mathcal{C}')$ of SAT such that $\mathcal{C}'$ is satisfiable if and only if at least k clauses of $\mathcal{C}$ can be satisfied. Let $M = \lceil \log(m+1) \rceil$. In the following reductions, we will use arithmetic constraints on $O(M)$ variables, which can be simulated by poly(m) clauses.

Let $T = (I, F)$ be a given nice tree-decomposition of width $\mathbf{tw}$. We will create an instance of SAT whose tree-width is at most $\mathbf{tw} + O(\log m)$. We note that the additive $O(\log m)$ factor is allowed because $O^*(2^{\alpha(\mathbf{tw} + O(\log m))}) = O^*(2^{\alpha \mathbf{tw}} \text{poly}(m)) = O^*(2^{\alpha \mathbf{tw}})$. For each node $i \in I$, we create variables $\{x_i \mid x \in X_i\} \cup \{s_{i,j} \mid j \in [M]\} \cup \{w_i\}$. The value $(s_{i,*})_2$ will represent the number of satisfied clauses in the subtree rooted at i. For each node i and its parent p, we create a constraint $x_i = x_p$ for each variable $x \in X_i \cap X_p$. Because the nodes containing the same variable form a connected subtree in T, these constraints ensure that for any variable $x \in X$, all the variables $\{x_i \mid x \in X_i\}$ take the same value. For each node i, according to its type, we do as follows:

1. Leaf: create a clause $(\overline{s_{i,j}})$ for each $j \in [M]$.
2. Introduce(v): create a constraint $s_{i,j} = s_{c,j}$ for each $j \in [M]$.

3. Introduce($x \vee y$): create a constraint $w_i \Leftrightarrow (x_i \vee y_i)$ and a constraint $(s_{i,*})_2 = (s_{c,*})_2 + (w_i)_2$.

4. Forget(v): create a constraint $s_{i,j} = s_{c,j}$ for each $j \in [M]$.

5. Join: create a constraint $(s_{i,*})_2 = (s_{l,*})_2 + (s_{r,*})_2$.

Finally for the root node r, we create a constraint $(s_{r,*})_2 \geq k$. Now, we have obtained an instance $(X', \mathcal{C}')$ of polynomial size. We note that, from the definition of a nice tree-decomposition, there exists exactly one Introduce(C) node for each clause $C \in \mathcal{C}$. Thus, the sum $\sum_{i \in I}(w_i)_2$, which is equal to $(s_{r,*})_2$, represents the number of satisfied clauses. Therefore, $\mathcal{C}'$ is satisfiable if and only if at least k clauses of $\mathcal{C}$ can be satisfied. Finally, we show that the reduction preserves the tree-width.

Lemma 5. *$\mathcal{C}'$ has tree-width at most* **tw** $+ O(\log m)$.

Proof. We will prove the bound by reducing the primal graph of $\mathcal{C}'$ into an empty graph by a series of eliminations of degree at most **tw** $+ O(\log m)$. For a node i, let Y_i denote the vertex set $\{x_i \mid x \in X_i\}$ and V_i denote the vertex set $Y_i \cup \{w_i\} \cup \{s_{i,j} \mid j \in [M]\}$. Starting from the primal graph of $\mathcal{C}'$ and the given tree-decomposition T of $\mathcal{C}$, we eliminate the vertices as follows. First, we choose an arbitrary leaf i of T. Then, we eliminate all the vertices of V_i in a certain order, which will be described later. Finally, we remove i from T and repeat the process until T becomes empty.

Let i be a leaf and p be its parent. If i is the only child of p, we have $N(V_i) \subseteq V_p$. Thus, the eliminations of V_i can create edges only inside V_p. If p has another child q, we have $N(V_i) \subseteq V_p \cup \{s_{q,j} \mid j \in [M]\}$. Thus, the eliminations of V_i can create edges only inside $V_p \cup \{s_{q,j} \mid j \in [M]\}$. Therefore, after processing each node, we can ensure that the edges created by previous eliminations are only inside $V_i \cup \{s_{c,j} \mid c$ is a child of i and $j \in [M]\}$ for each node i.

Now, we describe the details of the eliminations. Let i be the current node to process. If i is the root, the number of remaining vertices is $O(\log m)$. Thus, the elimination of these vertices has degree $O(\log m)$. Otherwise, let p be the parent of i. First, we eliminate the vertices Y_i. Because each vertex of Y_i is adjacent to at most one vertex of Y_p, Lemma 3 gives the elimination of degree $|N[Y_i] \setminus Y_p| \leq |V_i| \leq$ **tw** $+ O(\log m)$. Then, we eliminate the remaining vertices $V_i \setminus Y_i$. If i is the only child of p, let $V_q = Y_q = \emptyset$, and otherwise, let q be the another child of p. By applying Lemma 2, we obtain the elimination of degree $|N[V_i \setminus Y_i]| - 1 \leq |V_i \setminus Y_i| + |V_p| + |V_q \setminus Y_q| \leq$ **tw** $+ O(\log m)$. □

4   Exactly Parameterized NL

By extending the classical complexity class NL (Non-deterministic Logspace), we define a class of parameterized problems EPNL (Exactly Parameterized NL) which can be solved by a non-deterministic Turing machine with the space of $k + O(\log n)$ bits.

Definition 1 (EPNL). *A parameterized problem (L, κ), where, $L \subseteq \{0,1\}^*$ is a language and $\kappa : \{0,1\}^* \to \mathbb{N}$ is a parameterization, is in EPNL if there exists a polynomial $p : \mathbb{N} \to \mathbb{N}$ and a verifying polynomial-time deterministic Turing machine $M : \{0,1\}^* \times \{0,1\}^* \to \{0,1\}$ with four binary tapes, a read-only input tape, a read-only read-once certificate tape, and two read/write working tapes called the k-bit tape and the logspace tape with the following properties.*

- *For any input $x \in \{0,1\}^*$, it holds that $x \in L$ if and only if there exists a certificate $y \in \{0,1\}^{p(|x|)}$ such that $M(x, y) = 1$.*
- *For any $x \in \{0,1\}^*$ and $y \in \{0,1\}^{p(|x|)}$, the machine M uses at most $\kappa(x)$ space from the k-bit tape and $O(\log |x|)$ space from the logspace tape.*

Note that the machine M is not allowed to use $O(\kappa(x))$ bits from the k-bit tape but at most $\kappa(x)$ bits. This is why we use two separated working tapes instead of one long working tape of length $\kappa(x) + O(\log |x|)$; in the latter case, because there is only one head, it may be difficult to simulate a random-access $\kappa(x)$-bit array.

We give several examples of problems in EPNL. Due to the space constraints, the proofs are omitted. For all the problems in Lemma 7, the current fastest algorithms take $O^*(2^n)$ time [5].

Lemma 6. SAT, 3-SAT, MAX 2-SAT, *and* INDEPENDENT SET *parameterized by path-width are in EPNL.*

Lemma 7. TRAVELLING SALESMAN PROBLEM(TSP), OPTIMAL LINEAR ARRANGEMENT, DIRECTED FEEDBACK ARC SET *parameterized by the number of vertices of the input graph, and* SET COVER *parameterized by the number of elements are in EPNL.*

Now, we define *logspace parameter-preserving reduction* and introduce EPNL-complete problems.

Definition 2 (Reducibility). *A parameterized problem $A = (L, \kappa)$ is logspace parameter-preserving reducible to a parameterized problem $B = (L', \kappa')$, denoted by $A \leq_L^{pp} B$, if there exists a logspace computable function $\phi : \{0,1\}^* \to \{0,1\}^*$ such that*

- $x \in L \iff \phi(x) \in L'$, *and*
- $\kappa'(\phi(x)) \leq \kappa(x) + O(\log |x|)$.

Note that in the standard parameterized reduction, the computation can take $f(\kappa(x))\text{poly}(|x|)$ time and the parameter $\kappa'(\phi(x))$ of the reduced instance can be increased to any function of the original parameter $\kappa(x)$. However, in our reduction, we allow only a logspace computation and an additive increase by $O(\log |x|)$ of the parameter.

Proposition 1. *If $A \leq_L^{pp} B$ and $B \in$ EPNL, then $A \in$ EPNL.*

The proof of the proposition is an easy extension of the case for NL (see the textbook by Arora and Barak [2, Chap.4.3.]), so we omit it here.

Definition 3 (EPNL-complete). *A parameterized problem A is called* EPNL-*hard if for any $B \in$ EPNL, we have $B \leq_L^{pp} A$. Moreover, if $A \in$ EPNL, A is called* EPNL-complete.

Since there are at most $2^{k+O(\log |x|)}\mathrm{poly}(|x|) = O^*(2^k)$ states, any problem in EPNL can be solved in $O^*(2^k)$ time by dynamic programming. The following proposition follows from the definitions.

Proposition 2. *Any problem in EPNL can be solved in $O^*(2^k)$ time. If one of the EPNL-hard problem can be solved in $O^*(c^k)$ time, then any problem in EPNL can also be solved in $O^*(c^k)$ time.*

Now, we show that the problems in Lemma 6 are EPNL-complete. Due to the space constraints, we give a proof only for SAT here. The proofs for the other problems can be obtained by series of (decomposition-based) reductions and will be presented in the full version of this paper.

Theorem 4. *SAT parameterized by path-width is EPNL-complete.*

Proof. SAT parameterized by path-width is in EPNL. So it is sufficient to show that any parameterized problem $A = (L, \kappa)$ in EPNL can be reduced to SAT parameterized by path-width. Let M be a Turing machine that accepts L, Q be the set of (internal) states of M, and $t, s : \mathbb{N} \to \mathbb{N}$ be the polynomial time bound and logarithmic space bound of M, respectively. We reduce an instance x of A with a parameter $k = \kappa(x)$ to SAT as follows.

For each step $i \in [t(|x|)]$, we create the following variables:

- $Q_{i,q}$ for each $q \in Q$, which indicates that M is in state q,
- $H_{i,j}^I$ for each $j \in [\lceil \log |x| \rceil]$, which indicates the position of the input tape head in binary,
- $H_{i,j}^K$ for each $j \in [\lceil \log k \rceil]$, which indicates the position of the k-bit tape head in binary,
- $H_{i,j}^L$ for each $j \in [\lceil \log r(|x|) \rceil]$, which indicates the position of the logspace tape head in binary,
- $T_{i,h}^K$ for each $h \in [k]'$, which indicates the symbol written in the h-th cell of the k-bit tape,
- $T_{i,h}^L$ for each $h \in [s(|x|)]'$, which indicates the symbol written in the h-th cell of the logspace tape, and
- T_i^C, which represents the symbol in the cell of the certificate tape.

Now, we create clauses. Let $q_s \in Q$ be the initial state and $q_t \in Q$ be the accepting state. First, we create the following clauses (consisting of single literals) to express the initial and the final configuration:

- Q_{1,q_s} (the machine is in the state q_s),
- $\overline{H_{1,j}^I}$ for each $j \in [\lceil \log |x| \rceil]$ (the input tape head is at the position 0),
- $\overline{H_{1,j}^K}$ for each $j \in [\lceil \log k \rceil]$ (the k-bit tape head is at the position 0),
- $\overline{H_{1,j}^L}$ for each $j \in [\lceil \log s(|x|) \rceil]$ (the logspace tape head is at the position 0),

- $\overline{T_{1,h}^K}$ for each $h \in [k]'$ (each cell of the k-bit tape has symbol 0),
- $\overline{T_{1,h}^L}$ for each $h \in [r(|x|)]'$ (each cell of the logspace tape has symbol 0), and
- $Q_{t(|x|),q_t}$ (the machine must finish in the state q_t).

Then, for each step $i \in [t(|x|)]$, we create clauses to express transitions. The machine can take only one state, so we create a clause $\overline{Q_{i,q}} \vee \overline{Q_{i,q'}}$ for each $q \neq q'$. If a cell changes, the head must be there (or equivalently, cells not pointed by the head must remain unchanged), so we create the following clauses:

- $T_{i,h^K}^K \neq T_{i+1,h^K}^K \to (H_{i,*}^K)_2 = h^K$ for each $h^K \in [k]'$, and
- $T_{i,h^L}^L \neq T_{i+1,h^L}^L \to (H_{i,*}^L)_2 = h^L$ for each $h^L \in [s(|x|)]'$.

Let $\delta : (q, c^I, c^K, c^L, c^C) \mapsto (q', c'^K, c'^L, d^I, d^K, d^L, d^C)$ be the transition function, which indicates that if the machine is in the state q, the symbol in the input tape is c^I, the symbol in the k-bit tape is c^K, the symbol in the logspace tape is c^L, and the symbol in the certificate tape is c^C, then the machine changes the state to q', write c'^K to the cell of the k-bit tape, write c'^L to the cell of the logspace tape, move the input tape head by d^I, move the k-bit tape head by d^K, move the logspace tape head by d^L, and move the certificate tape head by d^C. Note that since the certificate tape is read-once, $d^C \geq 0$. For each $h^I \in [|x|]'$, $h^K \in [k]'$, $h^L \in [s(|x|)]'$, and transition $(q, c^I, c^K, c^L, c^C) \mapsto (q', c'^K, c'^L, d^I, d^K, d^L, d^C)$, we create clauses as follows. If a symbol in the h^I-th position of the input tape is not c^I, this transition never occurs. Otherwise, let C be the constraint $Q_{i,q} \wedge (H_{i,*}^I)_2 = h^I \wedge (H_{i,*}^K)_2 = h^K \wedge (H_{i,*}^L)_2 = h^L \wedge T_{i,h^K}^K = c^K \wedge T_{i,h^L}^L = c^L \wedge T_i^C = c^C$. Then, we create the following clauses:

- $C \to Q_{i+1,q'}$ (the machine changes the state to q'),
- $C \to T_{i+1,h^K}^K = c'^K$ (c'^K is written in the cell of the k-bit tape),
- $C \to T_{i+1,h^L}^L = c'^L$ (c'^L is written in the cell of the the logspace tape),
- $C \to (H_{i+1,*}^I)_2 = h^I + d^I$ (the input tape head moves by d^I),
- $C \to (H_{i+1,*}^K)_2 = h^K + d^K$ (the k-bit tape head moves by d^K),
- $C \to (H_{i+1,*}^L)_2 = h^L + d^L$ (the logspace tape head moves by d^L), and
- $C \to T_i^C = T_{i+1}^C$ if $d^C = 0$ (if the certificate tape head does not move, then the symbol in the certificate tape does not change).

It is not difficult to check that the reduction can be done in logspace and the obtained CNF is satisfiable if and only if there is a certificate such that the machine finishes in the accepting state. Finally, we show that the obtained CNF has path-width $k + O(\log |x|)$.

For a step i, let $T_i^K = \{T_{i,h}^K \mid h \in [k]'\}$ and X_i be the set of other variables. The primal graph of the obtained CNF has the following properties:

- $N[X_i] \subseteq T_{i-1}^K \cup X_{i-1} \cup T_i^K \cup X_i \cup T_{i+1}^K \cup X_{i+1}$,
- $N(T_{i,j}^K) \subseteq \{T_{i-1,j}^K, T_{i+1,j}^K\} \cup X_{i-1} \cup X_i \cup X_{i+1}$.

We can construct a path-decomposition as follows: starting from a bag $T_1^K \cup X_1$ and $i = 1$, introduce X_{i+1}, introduce $T_{i+1,1}^K$, forget $T_{i,1}^K$, ..., introduce $T_{i+1,k}^K$, forget

$T_{i,k}^K$, forget X_i (the current bag consists of $T_{i+1}^K \cup X_{i+1}$), and then increase i. Since the size of X_i is $O(\log |x|)$ and the size of T_i^K is exactly k, the width of the obtained path-decomposition is $k + O(\log |x|)$. □

References

1. Arnborg, S.: Efficient algorithms for combinatorial problems with bounded decomposability - a survey. BIT Numerical Mathematics 25(1), 2–23 (1985)
2. Arora, S., Barak, B.: Computational Complexity - A Modern Approach. Cambridge University Press (2009)
3. Backurs, A., Indyk, P.: Edit distance cannot be computed in strongly subquadratic time (unless SETH is false). In: STOC, pp. 51–58 (2015)
4. Björklund, A.: Determinant sums for undirected hamiltonicity. SIAM J. Comput. 43(1), 280–299 (2014)
5. Bodlaender, H.L., Fomin, F.V., Koster, A.M.C.A., Kratsch, D., Thilikos, D.M.: A note on exact algorithms for vertex ordering problems on graphs. Theory Comput. Syst. 50(3), 420–432 (2012)
6. Courcelle, B., Makowsky, J.A., Rotics, U.: Linear time solvable optimization problems on graphs of bounded clique-width. Theor. Comput. Syst. 33(2), 125–150 (2000)
7. Cygan, M., Dell, H., Lokshtanov, D., Marx, D., Nederlof, J., Okamoto, Y., Paturi, R., Saurabh, S., Wahlström, M · On problems as hard as CNF-SAT. In: CCC, pp. 74–84 (2012)
8. Cygan, M., Nederlof, J., Pilipczuk, M., Pilipczuk, M., van Rooij, J.M.M., Wojtaszczyk, J.O.: Solving connectivity problems parameterized by treewidth in single exponential time. In: FOCS, pp. 150–159 (2011)
9. Flum, J., Grohe, M.: Describing parameterized complexity classes. Inf. Comput. 187(2), 291–319 (2003)
10. Habet, D., Paris, L., Terrioux, C.: A tree decomposition based approach to solve structured SAT instances. In: ICTAI, pp. 115–122 (2009)
11. Hertli, T.: 3-SAT faster and simpler - unique-SAT bounds for PPSZ hold in general. SIAM J. Comput. 43(2), 718–729 (2014)
12. Impagliazzo, R., Paturi, R.: On the complexity of k-SAT. J. Comput. System Sci. 62(2), 367–375 (2001)
13. Impagliazzo, R., Paturi, R., Zane, F.: Which problems have strongly exponential complexity? J. Comput. Syst. Sci. 63(4), 512–530 (2001)
14. Lokshtanov, D., Marx, D., Saurabh, S.: Known algorithms on graphs on bounded treewidth are probably optimal. In: SODA, pp. 777–789 (2011)
15. Niedermeier, R.: Invitation to Fixed-Parameter Algorithms. Oxford University Press (2006)
16. Patrascu, M., Williams, R.: On the possibility of faster sat algorithms. In: SODA, pp. 1065–1075 (2010)
17. van Rooij, J.M.M., Bodlaender, H.L., Rossmanith, P.: Dynamic programming on tree decompositions using generalised fast subset convolution. In: Fiat, A., Sanders, P. (eds.) ESA 2009. LNCS, vol. 5757, pp. 566–577. Springer, Heidelberg (2009)
18. Williams, R.: A new algorithm for optimal 2-constraint satisfaction and its implications. Theor. Comput. Sci. 348(2-3), 357–365 (2005)
19. Xiao, M., Nagamochi, H.: Exact algorithms for maximum independent set. In: Cai, L., Cheng, S.-W., Lam, T.-W. (eds.) Algorithms and Computation. LNCS, vol. 8283, pp. 328–338. Springer, Heidelberg (2013)

Fast Output-Sensitive Matrix Multiplication

Riko Jacob[1] and Morten Stöckel[2, *]

[1] IT University of Copenhagen
rikj@itu.dk
[2] University of Copenhagen
most@di.ku.dk

Abstract. We consider the problem of multiplying two $U \times U$ matrices A and C of elements from a field $\mathbb{F}$. We present a new randomized algorithm that can use the known fast square matrix multiplication algorithms to perform fewer arithmetic operations than the current state of the art for output matrices that are sparse.

In particular, let ω be the best known constant such that two dense $U \times U$ matrices can be multiplied with $\mathcal{O}(U^\omega)$ arithmetic operations. Further denote by N the number of nonzero entries in the input matrices while Z is the number of nonzero entries of matrix product AC. We present a new Monte Carlo algorithm that uses $\tilde{\mathcal{O}}\left(U^2 \left(\frac{Z}{U}\right)^{\omega-2} + N\right)$ arithmetic operations and outputs the nonzero entries of AC with high probability. For dense input, i.e., $N = U^2$, if Z is asymptotically larger than U, this improves over state of the art methods, and it is always at most $\mathcal{O}(U^\omega)$. For general input density we improve upon state of the art when N is asymptotically larger than $U^{4-\omega}Z^{\omega-5/2}$.

The algorithm is based on dividing the input into "balanced" subproblems which are then compressed and computed. The new subroutine that computes a matrix product with balanced rows and columns in its output uses time $\tilde{\mathcal{O}}\left(UZ^{(\omega-1)/2} + N\right)$ which is better than the current state of the art for balanced matrices when N is asymptotically larger than $UZ^{\omega/2-1}$, which always holds when $N = U^2$.

In the I/O model — where M is the memory size and B is the block size — our algorithm is the first nontrivial result that exploits cancellations and sparsity of the output. The I/O complexity of our algorithm is $\tilde{\mathcal{O}}\left(U^2(Z/U)^{\omega-2}/(M^{\omega/2-1}B) + Z/B + N/B\right)$, which is asymptotically faster than the state of the art unless M is large.

1   Introduction

In this paper we consider computing the matrix product AC of two matrices A and C in the case where the number of nonzero entries of the output AC is sparse. In particular we consider the case where matrix elements are from an

---

* This work was done while at IT University of Copenhagen. Supported by the Danish National Research Foundation / Sapere Aude program and VILLUM FONDEN.

© Springer-Verlag Berlin Heidelberg 2015
N. Bansal and I. Finocchi (Eds.): ESA 2015, LNCS 9294, pp. 766–778, 2015.
DOI: 10.1007/978-3-662-48350-3_64

arbitrary field and cancellations can be exploited, i.e., "Strassen-like" methods are allowed.

The case of sparse output is well-motivated by real-world applications such as computation of covariance matrices in statistical analysis. In general, matrix multiplication is a fundamental operation in computer science and mathematics, due to the wide range of applications and reductions to it — e.g. computing the determinant and inverse of a matrix, or Gaussian elimination. Matrix multiplication has also seen lots of use in non-obvious applications such as bioinformatics [18], computing matchings [15,12] and algebraic reasoning about graphs, e.g. cycle counting [2,3].

Our main result is a new output-sensitive Monte Carlo algorithm, that given as input the $U \times U$ matrices A and C computes the nonzero entries of AC with high probability. For Z the number of nonzero output entries of AC, and N the number of nonzero entries of A and C, our algorithm uses $\tilde{O}\left(U^2 \left(\frac{Z}{U}\right)^{\omega-2} + N\right)$ arithmetic operations and its I/O complexity, where memory size M and block size B are parameters in the model, is given by $\tilde{O}\left(U^2(Z/U)^{\omega-2}/(M^{\omega/2-1}B) + Z/B + N/B\right)$.

The algorithm exploits cancellations, both to avoid computing zero entries of the output and by calling a Strassen-like algorithms.

When the input is dense, i.e., $N = U^2$, the RAM bound is strictly better than all state of the art methods when $Z \gg U$ and is never worse than $O(U^\omega)$. The I/O bound is strictly better than state of the art unless M is large — see Section 1.3. We note that the algorithm works over any field, but not over any semiring.

1.1   Preliminaries

We analyze our algorithms in the standard word-RAM model we assume that the word size is large enough to fit a field element. We further analyze our algorithms in the external memory model [1], where we have a disk containing an infinite number of words and a internal memory that can hold M words. A word can only be manipulated if it is residing in internal memory and words are transferred to internal memory in blocks of B words at a time and the number of such block transfers is called the I/O performance of the algorithm. Here we assume that a word can hold a field element as well as its position in the matrix.

Let A be a $U_1 \times U_3$ matrix and C be a $U_3 \times U_2$ matrix over any field $\mathbb{F}$, then $A_{i,j}$ is the entry of A located in the i'th row and j'th column and $A_{i,*}$ will be used as shorthand for the entire i'th row ($A_{*,i}$ for column i). The matrix product is given as $(AC)_{i,j} = \sum_{k=1}^{U_3} A_{i,k} C_{k,j}$. We say that a sum of elementary products *cancel* if the sum over them equals zero even though there are nonzero summands. We allow ourselves to use cancellations, i.e., "Strassen-like" methods, and our algorithms does not produce output results that are zero. We use log for the logarithm with base 2 and ln for the natural logarithm and when used in a context that requires integers, we let log stand for $\lceil \log \rceil$ and ln stand for $\lceil \ln \rceil$. Throughout this paper we will use $f(n) = \tilde{O}(g(n))$ as shorthand for $f(n) = O\left(g(n) \log^c g(n)\right)$ for any constant c. Here n stands for the input size, in our context it can always be taken as U. We let $\mathrm{sort}(n) = \mathcal{O}((n/B)\log_{M/B}(n/B))$ be shorthand for the I/O complexity [1] of sorting n elements. Note that input layout is irrelevant for our complexities, as sorting the input to a different layout is hidden in $\tilde{\mathcal{O}}(\cdot)$, i.e., $O(sort(n)) = \tilde{\mathcal{O}}(n/B)$. When we use this notation there is usually at least a binary logarithm hidden in the $\tilde{\mathcal{O}}()$-notation and not only the $\log_{M/B}(n/B)$-factor of sorting.

We will let $f \gg g$ denote that f is asymptotically larger than g. Let ω denote the real number for which the matrix product of two $U \times U$ matrices can be computed in time $\mathcal{O}(U^\omega)$. For a hash function $h\colon [U] \to [d]$ we define the binary projection matrix $P_h \in \{0,1\}^{U \times d}$ by $(P_h)_{i,j} = 1 \iff j = h(i)$. Finally we say that an algorithm has a *silent failure* to mean that the algorithm finished without an error, but the output is not correct.

We will use the following easy fact about the number of arithmetic operations $F_{\mathrm{RAM}}(U,V,W)$, and I/Os $F_{\mathrm{I/O}}(U,V,W)$ needed to multiply a $U \times V$ matrix by a $V \times W$ matrix.

Fact 1 (Folklore). *Let ω be the smallest constant such that an algorithm to multiply two $n \times n$ matrices that runs in time $\mathcal{O}(n^\omega)$ is known. Let $\beta = \min\{U, V, W\}$.*

Fast matrix multiplication has $F_{\mathrm{RAM}}(U,V,W) = \mathcal{O}(UVW \cdot \beta^{\omega-3})$ running time on a RAM, and uses
$$F_{\mathrm{I/O}}(U,V,W) = \mathcal{O}\left(F_{\mathrm{RAM}}(U,V,W)/(M^{\omega/2-1}B) + (UV + VW + UW)/B\right) \text{ I/Os.}$$

Proof. Assume wlog that β divides $\alpha = UVW/\beta$. Since β is the smallest dimension we can divide the matrices into α/β^2 submatrices of size $\beta \times \beta$, which can each be solved in $\mathcal{O}(\beta^\omega)$ operations. The I/O complexity follows from an equivalent blocking argument [10]. ■

For U, V, W and U', V', W' with $UVW = U'V'W'$ we have $F(U,V,W) > F(U',V',W')$ if $\min\{U,V,W\} < \min\{U',V',W'\}$, i.e., the "closer to square" the situation is, the faster the fast matrix multiplication algorithm runs, both in terms of RAM and I/O complexity.

1.2 Our Results

We show the following theorem, that provides an output sensitive fast matrix multiplication algorithm granted that the output is balanced.

Theorem 2. *Let A and C be $U \times U$ matrices over the field $\mathbb{F}$ that contain at most N nonzero entries and the product AC contains at most $Z \geq U$ nonzero entries in total and at most $\Theta(Z/U)$ per row and per column. Then there exists an algorithm for which it holds:*

(a) *The algorithm uses $\tilde{\mathcal{O}}\left(UZ^{\frac{\omega-1}{2}} + N\right)$ time in the RAM model.*

(b) *The algorithm uses $\tilde{\mathcal{O}}\left(UZ^{\frac{\omega-1}{2}}/(M^{\omega/2-1}B) + Z/B + N/B\right)$ I/Os*

(c) *With probability at least $1 - 1/U^2$ the algorithm outputs the nonzero entries of AC.*

We then show the main theorem, a fast matrix multiplication algorithm that works on any input and is sensitive to the average number of nonzero entries in the rows and columns of the output.

Theorem 3. *Let A and C be $U \times U$ matrices over field $\mathbb{F}$ that contain at most N nonzero entries and the product AC contains at most Z nonzero entries in total. Then there exists an algorithm for which it holds:*

(a) The algorithm uses time $\tilde{O}\left(U^2(Z/U)^{\omega-2} + N\right)$ time in the RAM model.
(b) The algorithm uses $\tilde{O}\left(U^2(Z/U)^{\omega-2}/(M^{\omega/2-1}B) + U^2/B\right)$ I/Os
(c) With probability at least $1 - 1/U^2$ the algorithm outputs the nonzero entries of AC.

The algorithm of Theorem 2 has asymptotically lower running time compared to that of Theorem 3 for $1 < Z < U^2$ (for current ω) and for $Z = U^2$ they both match $\tilde{O}(U^\omega)$. However, the algorithm from Theorem 2 requires balanced rows and columns and in fact the algorithm from Theorem 3, which works in the general case, is based on calling it on balanced partitions. For the sake of simplicity we state and proof Theorem 3 for square inputs only and note that there is nothing in our arguments prevents the generalization to the rectangular case. To the knowledge of the authors there are no previously known output-sensitive I/O-efficient algorithms that exploits cancellations and we outperform the general dense as well as the optimal sparse algorithm by Pagh-Stöckel unless M is large. We summarize the results of this section and the closest related results in Table 1.

Table 1. Comparison of matrix multiplication algorithms of two $U \times U$ in the RAM and I/O model. N denotes the number of nonzeros in the input matrices, Z the number of nonzeros in the output matrix and ω is the currently lowest matrix multiplication exponent.

Method	word-RAM complexity	Notes
GENERAL DENSE	$\mathcal{O}\left(U^\omega\right)$	
LINGAS	$\tilde{O}\left(U^2 Z^{\omega/2-1}\right)$	Requires boolean matrices.
IWEN-SPENCER, LE GALL	$\mathcal{O}\left(U^{2+\varepsilon}\right)$	Requires $\mathcal{O}\left(n^{0.3}\right)$ nonzeros per column.
WILLIAMS-YU, PAGH	$\tilde{O}\left(U^2 + UZ\right)$	
VAN GUCHT ET AL.	$\tilde{O}\left(N\sqrt{Z} + Z + N\right)$	
THIS PAPER, THM. 2	$\tilde{O}\left(UZ^{(\omega-1)/2} + N\right)$	Requires balanced rows and columns.
THIS PAPER, THM. 3	$\tilde{O}\left(U^2(Z/U)^{\omega-2} + N\right)$	
Method	I/O complexity	Notes
GENERAL DENSE	$\tilde{O}\left(U^\omega/(M^{\omega/2-1}B)\right)$	
PAGH-STÖCKEL	$\tilde{O}\left(N\sqrt{Z}/(B\sqrt{M})\right)$	Elements from semirings.
THIS PAPER, THM. 2	$\tilde{O}\left(UZ^{\frac{\omega-1}{2}}/(M^{\omega/2-1}B) + Z/B + N/B\right)$	Requires balanced rows and columns.
THIS PAPER, THM. 3	$\tilde{O}\left(U^2(Z/U)^{\omega-2}/(M^{\omega/2-1}B) + U^2/B\right)$	

Result Structure. The paper is split into three parts: the row balanced case, subdivision into balanced parts, and the balanced case. After a discussion of the related work in Section 1.3, we consider in Section 2 the case where we have an upper bound on the number of output entries in each row of the output. In this case we can compress the computation by reducing the number of columns, where the magnitude of the reduction is based on the upper bound on the output entries. Then in Section 3 we make use of the row balanced compression, by showing that a potentially unbalanced output can be divided into such balanced cases, which gives Theorem 3. In Section 4 we show that if there is an upper bound on the number of output entries in *both* rows and columns, then we can compress in both directions which yields Theorem 2.

1.3   Related Work

Two $U \times U$ matrices can trivially be multiplied in $\mathcal{O}(U^3)$ arithmetic operations by U^2 inner products of length U vectors. The first one to improve upon the $\mathcal{O}(U^3)$ barrier was Strassen [17] who for $\omega = \log_2 7$ showed an $\mathcal{O}(U^\omega)$ algorithm by exploiting clever cancellations. Since then there has been numerous advances on ω, e.g. [16,6,20] and most recently $\omega < 2.3728639$ was shown due to Le Gall [8].

The closest algorithm in spirit to our general algorithm of Theorem 3 is due to Williams and Yu [22]. They recursively, using time $\tilde{\mathcal{O}}\left(U^2\right)$, with high probability compute all positions of nonzero output entries. After this they compute each output value in time $\mathcal{O}(U)$ for a total number of $\tilde{\mathcal{O}}\left(U^2 + UZ\right)$ operations. This matches the exact case of Pagh's compressed matrix multiplication result [13], which is significantly more involved but also gives a stronger guarantee: given a parameter b, using time $\tilde{\mathcal{O}}\left(U^2 + Ub\right)$, it gives the exact answer with high probability if $Z \leq b$, otherwise it outputs a matrix where each entry is close to AC in terms of the Frobenius norm. Our general algorithm of Theorem 3 improves upon both when $Z \gg U$.

In the case of sparse *input* matrices, Yuster and Zwick [23] showed how to exploit sparseness of input using an elegant and simple partitioning method. Their result was extended to be both input and output sensitive by Amossen and Pagh [4], leading to a time bound of $\tilde{\mathcal{O}}\left(N^{2/3}Z^{2/3} + N^{0.862}Z^{0.546}\right)$ based on current ω. In our (non-input sensitive) case where $N = U^2$ we are strictly better for all $U > 1$ and $Z > U$. We note that the algorithm of Amossen and Pagh is presented for boolean input and claimed to work over any ring, however both the result of Yuster-Zwick and Amossen-Pagh do not support cancellations, i.e., their bounds are in terms of the number of vector pairs of the input that have nonzero elementary products. The above, as well as Pagh [13] and Williams-Yu [22] were improved recently by Van Gucht et al. [19] to be $\tilde{\mathcal{O}}\left(Z + N\sqrt{Z} + N\right)$ operations, stated in the binary case but claimed to work over any field. Compared to this we use $\tilde{\mathcal{O}}\left(U^2(Z/U)^{\omega-2}\right)$ operations by Theorem 3, which for $N > U^{4-\omega}Z^{\omega-5/2}$ improves Van Gucht et al. in the general case. For dense inputs, $N = U^2$, this threshold on N simplifies to $Z > U^{(\omega-2)/(\omega-5/2)}$ which holds for all positive integers $Z \geq 1$ and $U > 1$.

In the balanced case, Iwen and Spencer [9] showed that if every column of the output matrix has $\mathcal{O}(U^{0.29462})$ nonzero entries, then the matrix product can be computed using time $\mathcal{O}(U^{2+\varepsilon})$ for constant $\varepsilon > 0$. Recently due to Le Gall [8] this result now holds for output matrices with columns with at most $\mathcal{O}(U^{0.3})$ nonzeros. In this case our balanced matrix multiplication algorithm of Theorem 2 uses time $\tilde{\mathcal{O}}\left(U^{2.19}\right)$ (for $\omega = 2.3728639$), which is asymptotically worse, but our method applies to general balanced matrices. Compared to Van Gucht et al., in the balanced case, we improve upon their operation count when $N > UZ^{\omega/2-1}$. When $N = U^2$ this always holds since $Z \leq U^2$.

For boolean input matrices, the output sensitive algorithm of Lingas [11] runs in time $\tilde{\mathcal{O}}\left(U^2 Z^{\omega/2-1}\right)$, which we improve on for $1 \leq Z < U^2$ by a relative factor of $Z^{1-\omega/2}$ and match when $Z = U^2$. Additionally our algorithm works over any field. Lingas however shows a partial derandomization that achieves the same bound using only $\mathcal{O}(\log^2 U)$ random bits, which is a direction not pursued in this paper. The general case, i.e., dense input and output without fast matrix multiplication allowed, multiplying two boolean matrices has time complexity $\mathcal{O}\left(U^3 \mathrm{poly}(\log\log)/\log^4 U\right)$ due to Yu [21], which is an improvement to Chans time $\mathcal{O}\left(U^3 (\log\log U)^3 / \log^3 U\right)$ algorithm [5]. Both are essentially improvements on the four russians trick: 1) divide the matrix into small $t \times t$ blocks 2) pre-compute results of all $t \times t$ blocks and store them in a dictionary. Typically $t - \mathcal{O}(\log U)$ and the gain is that there are $(U/t)^2 = U^2/\log^2 U$ blocks instead of U^2 cells.

In terms of I/O complexity, an optimal general algorithm was presented by Hong and Kung [10] that uses $\mathcal{O}(U^3/(B\sqrt{M})$ I/Os using a blocking argument. An equivalent blocking argument gives in our case where we are allowed to exploit cancellations an I/O complexity of $\mathcal{O}(U^\omega/(M^{\omega/2-1}B))$. This is the complexity of the black box we will invoke on dense subproblems (also stated in Fact 1). For sparse matrix multiplication in the I/O model over semirings, Pagh and Stöckel [14] showed a $\tilde{\mathcal{O}}(N\sqrt{Z}/(B\sqrt{M})$ algorithm and a matching lower bound. To the knowledge of the authors the algorithm presented in this paper is the first algorithm that exploits cancellations and is output sensitive. Our algorithm is asymptotically better than the general dense for all $1 \leq Z < U^2$ and we match their complexity for $Z = U^2$. Our new algorithm is asymptotically faster than Pagh-Stöckel precisely when $Z > \left(U^{4-\omega}M^{3/2-\omega/2}/N\right)^{1/(5/2-\omega)}$. To simplify, consider the case of dense input, i.e., $N = U^2$, then our new algorithm is better unless M much is larger than Z, which is typically assumed to not be the case.

Comparison summary. The general algorithm of Theorem 3 works over any field and can exploit cancellations (and supports cancellation of terms). In the RAM model it is never worse than $\mathcal{O}(U^\omega)$ and in the case of dense input, which is the case that suits our algorithms best, we improve upon state of the art (Pagh [13] and Williams-Yu [22]) when $Z \gg U$. For arbitrary input density we improve upon state of the art (Van Gucht et al [19] when $N \gg U^{4-\omega}Z^{\omega-5/2}$. In the I/O model our algorithm is the first output sensitive algorithm that exploits cancellations and it outperforms the known semiring algorithms for almost the entire parameter space.

2   The Row Balanced Case

Lemma 1. *Let $A \in \mathbb{F}^{U' \times U}$ and $C \in \mathbb{F}^{U \times U}$ be two matrices with $U' \leq U$. Assume each row of the $U' \times U$ product matrix AC has at most $d/5$ non-zero entries. For any natural constant c there is a data structure where*

(a) *initializing the data structure takes time $\tilde{O}\left(\text{nnz}(C) + F_{\text{RAM}}(U', U, d)\right)$ on the RAM and $\tilde{O}\left(\text{sort}_{M,B}(\text{nnz}(C)) + F_{\text{I/O}}(U', U, d)\right)$ I/Os*

(b) *the silent failure probability of initialization is at most U^{-c}*

(c) *the data structure can answer queries for entries of (AC) in time $O(\log(U^{2+c}))$*

(d) *the data structure can report all (non-zero) entries with $O(U' \text{sort}(kU))$ I/Os*

Proof. We chose a parameter k (number of repetitions) by $k = 6 \ln(U^{2+c})$. The data structure consists of k independent compressed versions of AC, $G_i \in \mathbb{F}^{U' \times d}$. We choose k independent random (hash) functions $h_l : [U] \to [d]$ where each $h_l(j)$ is chosen independently and uniformly from $[d]$. Each h_l is stored as a table in the data structure. We compute k compressed matrices $G_l = AC P_{h_l}$. See also Algorithm 1. This computation can be achieved by scanning C and changing all column indexes by applying the hash function. The resulting matrix C' with d columns is multiplied by A using a fast dense method described in Fact 1.

There are $U'U \leq U^2$ different queries that are answered using Algorithm 2. A particular $z_l = (G_l)_{i,j}$ is correct if no (other) non-zero entry of AC has been hashed to the same position of G_l. Because we are hashing only within rows, there are at most $d/5$ such non-zero entries. For a random hash function the probability that a particular one does so is $1/d$, so the probability that z_l is correct is at least

$$(1 - 1/d)^{d/5} \geq (1/4)^{1/5} \geq 3/4 \,.$$

Here, the first inequality stems from $(1 - 1/d)^d$ approaching e^{-1} from below and being $\geq 1/4$ for $d \geq 2$, and the second by $4^4 > 3^5$. Now, the probability of the median (or a majority vote) being false is at most that of the following Bernoulli experiment: If heads (errors) turn up with probability $1/4$ and k trials are performed, at least $k/2$ heads show up. By a Chernoff-Hoeffding inequality [7, Theorem 1.1, page 6] (equation (1.6) with $\epsilon = 1$) this failure probability is at most $\exp(-\frac{1}{3}\frac{k}{2}) = \exp(-k/6) \leq U^{-2-c}$. Hence a union bound over all U^2 possible queries shows that a randomly chosen set of compression matrices leads to a correct data structure with probability at least $1 - U^{-c}$.

To report all entries of AC we proceed row by row. We can assume that all G_l are stored additionally row wise, i.e., first the first row of G_1, then the first row of G_2 and so on, after that the second row of each G_l and so on. This reordering of the data structure does not change the asymptotic running time of the initialization phase. For row i we copy each entry $g_{i,j'} \in G_l$ as many times as there is an $j \in U$ with $h_l(j) = j'$ and annotate the copy with the index j.

This can be achieved by scanning if the graph of the hash function is sorted by target element. Then all these copies are sorted mainly by the annotated index and secondarily by their value. With this, for each j, the entries of the G_l corresponding to (i,j) are sorted together and a median or majority vote can easily be computed. If desired, the zero entries can be filtered out. The I/O complexity is obviously as claimed. ∎

Input: Matrices $A \in \mathbb{F}^{U' \times U}$ and $C \in \mathbb{F}^{U \times U}$
Output: k, hash functions $h_1, \ldots, h_k : [U] \to [d]$,
        compressed matrices $G_i \in \mathbb{F}^{U' \times d}$
1   Set $k = k = 6 \ln(U^2/\delta)$
2   **for** $l \leftarrow 1$ **to** k **do**
3   Create random function $h_l : [U] \to [d]$
4   Create $C' \in \mathbb{F}^{U \times d}$, initialized to 0
5   **foreach** (i,j) *with* $C_{i,j} \neq 0$ **do**
6         $C'_{i,h_l(j)} += C_{i,j}$;   /* $C' = CP_{h_l}$ */
7   Fast Multiply $G_i = AC'$

Algorithm 1: Initialization Balanced Rows

Input: k, Hash Functions $h_1, \ldots, h_k : U \to d$, i, j
Output: $(AC)_{i,j}$
1   **for** $l \leftarrow 1$ **to** k **do**
2   Set $z_l = (G_l)_{i,h_l(j)}$
3   Report Median of z_l

Algorithm 2: Query Balanced Rows

---

3   Subdividing into Balanced Parts

To make use of Lemma 1 for general matrices, it is important that we can estimate the number of nonzero elements in the rows of the output:

Fact 4 ([14], Lemma 1). *Let A and C be $U \times U$ matrices with entries of field $\mathbb{F}$, $N = \mathrm{nnz}(A) + \mathrm{nnz}(C)$ and let $0 < \varepsilon, \delta \leq 1$.*
We can compute estimates $z_1, \ldots, z_U$ using $\mathcal{O}(\varepsilon^{-3} N \log(U/\delta) \log U)$ RAM operations and $\tilde{\mathcal{O}}(\varepsilon^{-3} N/B)$ I/Os, such that with probability at least $1 - \delta$ it holds that $(1 - \varepsilon) \mathrm{nnz}([AC]_{k*}) \leq z_k \leq (1 + \varepsilon) \mathrm{nnz}([AC]_{k*})$ for all $1 \leq k \leq U$.

With this we can get the following data structure for potentially unbalanced output matrices.

Lemma 2. *Let* $A, C \in \mathbb{F}^{U \times U}$ *be two square matrices, and let* $N = \mathrm{nnz}(A) + \mathrm{nnz}(C)$, $Z = \mathrm{nnz}(AC)$. *For any natural constant* c *there is a data structure where*

(a) *initializing the data structure takes*
    time $\tilde{\mathcal{O}}\left(F_{\mathrm{RAM}}(U, Z/U, U) + N\right) = \tilde{\mathcal{O}}\left(U^2(Z/U)^{\omega-2} + N\right)$ *on the RAM*
    and $\tilde{\mathcal{O}}\left(F_{\mathrm{I/O}}(U, Z/U, U)\right)$ *I/Os*
(b) *the silent failure probability of initialization is at most* U^{-c}
(c) *the data structure can answer queries for entries of* (AC) *in time* $O(\log(U^2/\delta))$
(d) *the data structure can report all (non-zero) entries with* $O(U\,\mathrm{sort}(kU) + \mathrm{sort}(U^2))$ *I/Os*

Proof. We use Fact 4 to partition the output matrix into blocks of rows, for each of which we use the data structure of Lemma 1.

Observe that permuting the rows of AC is the same as permuting the rows of A. We use Fact 4 with $\varepsilon = .3$ and $\delta = U^{1+c}/2$ to estimate the number of non-zeros in row i of AC as z_i. We group the rows accordingly, row i belongs to group l if $U \cdot 2^{-l-1} < 1.3 z_i \leq U \cdot 2^{-l}$. We create a table that states for each original row its group and position in the group (row in the smaller matrix) Hence with overall probability at least $1 - U^{-c}/2$, each group l contains x_l rows where the number of non-zeros is between $U \cdot 2^{-l-2}$ and $U \cdot 2^{-l}$. At the cost of sorting A we make these at most $\log U$ matrices $A_l \in \mathbb{F}^{x_l \times U}$ explicit in sparse format. We create the overall data structure from smaller ones, one for each $R_l = A_l C$ using Lemma 1 with $c' = c+1$. These data structures have an individual success probability of at least $1 - U^{-c'}$, and because there are at most $\log U$ such data structures, by a union bound, an overall success probability of $1 - U^{-c}$.

The overall creation cost hinges on the cost for multiplying the smaller matrices, i.e., $\sum_l F_{\mathrm{RAM}}(U, x_l, 5U2^{-l})$. To bound this, we estimate x_l by the upper bound $x_l \leq 4 \cdot Z \cdot 2^l / U$ which stems from $U \cdot 2^{-l-2} \cdot x_l \leq Z$. For $l > \log(Z/U)$ this bound is bigger than U and we should estimate such x_l as U. This implies that this part of the sum forms a geometric series, and asymptotically it is sufficient to consider

$$\sum_{l=1}^{\log(Z/U)} F_{\mathrm{RAM}}(U, 4Z \cdot 2^l/U, 5U2^{-l}).$$

For this sum of fast matrix multiply running times we realize that the product of the dimensions is always $20UZ$, and hence each term is at most that of the most unbalanced case. Hence we can estimate the overall running time on the RAM as $F_{\mathrm{RAM}}(U, 4Z/U, 5U) \log(U)$. The same argument works for the overall number of I/Os. By observing that the required work on A takes $\tilde{\mathcal{O}}\left(\mathrm{sort}(U^2)\right) = \tilde{\mathcal{O}}\left(F_{\mathrm{I/O}}(U, Z/U, U)\right)$ we get that the I/O performance of initializing is as stated.

To report all output entries, we use the reporting possibility of the smaller data structures. That output has to be annotated with the original row number and sorted accordingly. This yields the claimed I/O-bound.   ∎

Proof (of Theorem 3). We use the data structure of Lemma 2 with $c = 2$. To produce the output RAM efficiently, we perform U^2 queries to the data structure which is $\tilde{\mathcal{O}}\left(U^2\right) = \tilde{\mathcal{O}}\left(U^2(Z/U)^{\omega-2}\right)$. To produce the output I/O efficiently, we use the procedure of item (d). Because $O(U\mathrm{sort}(kU) + sort(U^2)) = \tilde{\mathcal{O}}\left(U^2/B\right) = \tilde{\mathcal{O}}\left(F_{\mathrm{I/O}}(U, Z/U, U)\right)$ by Fact 1 we also get the claimed I/O performance.   ■

4   The Balanced Case

If the matrices are square and have d as an upper bounds on the number of nonzero entries both in columns and row we can hash in both columns and rows. This is beneficial because we achieve a more balanced setting of the dimensions when calling the fast matrix multiplication algorithm.

4.1   A Data Structure for Balanced Output

Lemma 3. *Let* $A \in \mathbb{F}^{U' \times U}$ *and* $C \in \mathbb{F}^{U \times U}$ *be two matrices with* $U' \leq U$. *Let* $Z = \mathrm{nnz}(AC)$ *and* $N = \mathrm{nnz}(A) + \mathrm{nnz}(C)$. *Assume each row and each column of the* $U' \times U$ *product matrix* AC *has at most* $d/5$ *entries. Assume further that* $Z > Ud/O(1)$. *For any constant* c *there is a data structure where*

(a) *initializing the data structure takes time* $\tilde{\mathcal{O}}\left(N + F_{\mathrm{RAM}}(\sqrt{Ud}, U, \sqrt{Ud})\right) = \tilde{\mathcal{O}}_{\delta}\left(UZ^{\frac{\omega-1}{2}} + N\right)$ *on the RAM and* $\tilde{\mathcal{O}}\left(N/B + F_{\mathrm{I/O}}(\sqrt{Ud}, U, \sqrt{Ud})\right)$ *I/Os*

(b) *the silent failure probability of initialization is at most* $3U^{-c}$

(c) *the data structure can answer queries for entries of* (AC) *in time* $O(\log(U^{2+c}))$

(d) *A batched query for up to* $2Z$ *entries can be performed with* $\tilde{\mathcal{O}}\left((Z)/B\right)$ *I/Os.*

Proof. (Sketch) The construction is as in Lemma 1 (including $k = 6\ln(U^{2+c})$), only that both A and C are compressed using hash functions. The compression is only down to $\sqrt{Ud} > d$ columns and rows. Additionally to collisions within a row or within a column, now also collision of arbitrary elements are possible. The failure probability of all three cases can be derived just as in Lemma 1. For the chosen parameters the sum of the three failure probabilities yields the claim.

To perform a batched query for $2Z$ entries, the following I/O-efficient algorithm can be used. Observe that all described sorting steps are on the queries or on compressed matrices and hence operate on $O(Z)$ elements. We first consider the compressed matrices individually. With a first sorting step (by row), the queries are annotated by the hashed row. In a second sorting step they are annotated by column. Now all annotated queries are sorted into the compressed matrix. For each query, the corresponding entry of the compressed matrix is extracted and annotated with the query. When this is done for all compressed matrices, the annotated entries are sorted by the index of query and a median or majority vote is performed.   ■

4.2   Creating the Output as Sparse Matrix

Unlike in the other settings, here the time to query for all possible entries of the output matrix might dominate the overall running time. Hence, in this section we propose an additional algorithm to efficiently extract the entries of AC that are nonzero, relying on the bulk query possibility of the data structure.

Lemma 4. *Let $A, C \in \mathbb{F}^{U \times U}$ be two matrices. Let $Z = \mathbf{nnz}(AC)$ and $N = \mathbf{nnz}(A) + \mathbf{nnz}(C)$. Assume each row and each column of the $U \times U$ product matrix AC has at most $d/5$ entries. Assume further that $Z > Ud/O(1)$. For any natural constant c there is an algorithm that*

(a) takes time $\tilde{O}\left(Z + N + F_{\mathrm{RAM}}(\sqrt{Ud}, U, \sqrt{Ud})\right)$ on the RAM

      and $\tilde{O}\left(\mathrm{sort}_{M,B}(N + Z) + F_{\mathrm{I/O}}(\sqrt{Ud}, U, \sqrt{Ud})\right)$ I/Os

(b) the silent failure probability is at most $5U^{-c}$

Proof. For each column of AC, we create a perfectly balanced binary tree where the leafs are the individual entries of the column. An internal node of the tree is annotated with the information whether there is a nonzero leaf in the subtree. A simultaneous BFS traversal of all these trees (one for each column) allows to identify all positions of nonzero entries. The number of elements on the "wavefront" of the BFS algorithm is at most Z. To advance the wavefront by one level, a batched query to at most $2Z$ positions of the tree is sufficient. Finally, the matrix AC itself is stored in a data structure as in Lemma 3 with failure probability $3U^{-c}$.

Instead of annotating the tree nodes directly, we compute several random dot products with the leaf-values. More precisely, for each leaf we choose a coefficient uniformly from $\{0, 1\} \subset \mathbb{F}$. Now, if one leaf has value $r \neq 0$, then with probability at least $1/2$ the dot product is non-zero: Assume all other coefficients are fixed, leading to a certain partial sum s; now it is impossible that both s and $s + 1 \cdot r$ are zero. If we have $(c + 3) \log U$ many such coefficients, the failure probability is at most $1/U^{c+3}$.

Observe that for one level of the tree and one random choice, the dot products form a matrix that can be computed as HAC for a $\{0, 1\}$-matrix H, and HA can be computed by sorting A. Observe further that the number of nonzero entries of the columns and rows of HAC are such that each of these matrices can be encoded using the data structure of Lemma 3 with failure probability $3U^{-c-1}$.

If there is no failure, the algorithm, using the batch queries, achieves the claimed running times and I/Os.

The probability that there is an encoding error in any of the matrices encoding the trees is, for sufficiently big U, at most U^{-c} because there are only $\log(U)(c + 3)\log(U) < U/3$ such matrices. The probability of a single tree node being encoded incorrectly is at most $3/U^{c+3}$. Because there are $U^2 \log U < U^3/3$ tree nodes in total, all trees are correct with probability at least $1 - U^{-c}$. Hence the overall failure probability, including that the data structure for the entries of AC failed, is hence as claimed. ∎

Proof (of Theorem 2). We invoke Lemma 4 with $c = 2$. Combining the statements about the RAM running time with Fact 1 and the calculation $ZU\sqrt{\tilde{Z}}^{\omega-3} = UZ^{\frac{\omega-1}{2}}$ gives the claimed RAM running time. Doing the same for the I/O bound and using that $\tilde{\mathcal{O}}\left(\text{sort}_{M,B}(N + Z)\right) = \tilde{\mathcal{O}}\left(N/B + Z/B\right)$ gives the claimed I/O performance. ∎

References

1. Aggarwal, A., Vitter, S., Jeffrey: The input/output complexity of sorting and related problems. Commun. ACM 31(9), 1116–1127 (1988)
2. Alon, N., Yuster, R., Zwick, U.: Color-coding. J. ACM 42(4), 844–856 (1995)
3. Alon, N., Yuster, R., Zwick, U.: Finding and counting given length cycles. Algorithmica 17(3), 209–223 (1997)
4. Amossen, R.R., Pagh, R.: Faster join-projects and sparse matrix multiplications. In: International Conference on Database Theory, ICDT 2009 (2009)
5. Chan, T.M.: Speeding up the four russians algorithm by about one more logarithmic factor. In: SODA 2015, San Diego, CA, USA, 2015, January 4-6, pp. 212–217 (2015)
6. Coppersmith, D., Winograd, S.: Matrix multiplication via arithmetic progressions. Journal of Symbolic Computation 9(3), 251–280 (1990)
7. Dubhashi, D.P., Panconesi, A.: Concentration of Measure for the Analysis of Randomized Algorithms. Cambridge University Press (2009)
8. Gall, F.L.: Powers of tensors and fast matrix multiplication. arXiv preprint arXiv:1401.7714 (2014)
9. Iwen, M., Spencer, C.: A note on compressed sensing and the complexity of matrix multiplication. Information Processing Letters 109(10), 468–471 (2009)
10. Jia-Wei, H., Kung, H.T.: I/O complexity: The red-blue pebble game. In: ACM Symposium on Theory of Computing, STOC 1981 (1981)
11. Lingas, A.: A fast output-sensitive algorithm for boolean matrix multiplication. In: Fiat, A., Sanders, P. (eds.) ESA 2009. LNCS, vol. 5757, pp. 408–419. Springer, Heidelberg (2009)
12. Mulmuley, K., Vazirani, U., Vazirani, V.: Matching is as easy as matrix inversion. Combinatorica 7(1), 105–113 (1987)
13. Pagh, R.: Compressed matrix multiplication. ACM Trans. Comput. Theory 5(3), 9:1–9:17 (2013)
14. Pagh, R., Stöckel, M.: The input/output complexity of sparse matrix multiplication. In: Schulz, A.S., Wagner, D. (eds.) ESA 2014. LNCS, vol. 8737, Springer, Heidelberg (2014)
15. Rabin, M.O., Vazirani, V.V.: Maximum matchings in general graphs through randomization. J. Algorithms 10(4), 557–567 (1989)
16. Stothers, A.J.: On the complexity of matrix multiplication. PhD thesis, The University of Edinburgh (2010)
17. Strassen, V.: Gaussian elimination is not optimal. Numerische Mathematik 13(4), 354–356 (1969)
18. van Dongen, S.: Graph Clustering by Flow Simulation. PhD thesis, University of Utrecht (2000)
19. Van Gucht, D., Williams, R., Woodruff, D.P., Zhang, Q.: The communication complexity of distributed set-joins with applications to matrix multiplication. In: Proceedings of the 34th ACM Symposium on Principles of Database Systems, PODS 2015, pp. 199–212. ACM, New York (2015)

20. Williams, V.V.: Multiplying matrices faster than coppersmith-winograd. In: Proceedings of the Forty-Fourth Annual ACM Symposium on Theory of Computing (2012)
21. Yu, H.: An improved combinatorial algorithm for boolean matrix multiplication. In: ICALP (2015)
22. Yu, H., Williams, R.: Finding orthogonal vectors in discrete structures, ch. 135, pp. 1867–1877. SIAM (2014)
23. Yuster, R., Zwick, U.: Fast sparse matrix multiplication. ACM Trans. Algorithms 1(1), 2–13 (2005)

A Structural Approach to Kernels for ILPs: Treewidth and Total Unimodularity

Bart M.P. Jansen[1,*] and Stefan Kratsch[2,**]

[1] Eindhoven University of Technology, The Netherlands
b.m.p.jansen@tue.nl
[2] University of Bonn, Germany
kratsch@cs.uni-bonn.de

Abstract. Kernelization is a theoretical formalization of efficient pre-processing for NP-hard problems. Empirically, preprocessing is highly successful in practice, for example in state-of-the-art ILP-solvers like CPLEX. Motivated by this, previous work studied the existence of kernelizations for ILP related problems, e.g., for testing feasibility of $Ax \leq b$. In contrast to the observed success of CPLEX, however, the results were largely negative. Intuitively, practical instances have far more useful structure than the worst-case instances used to prove these lower bounds.

In the present paper, we study the effect that subsystems that have (a Gaifman graph of) bounded treewidth or that are totally unimodular have on the kernelizability of the ILP feasibility problem. We show that, on the positive side, if these subsystems have a small number of variables on which they interact with the remaining instance, then we can efficiently replace them by smaller subsystems of size polynomial in the domain without changing feasibility. Thus, if large parts of an instance consist of such subsystems, then this yields a substantial size reduction. Complementing this we prove that relaxations to the considered structures, e.g., larger boundaries of the subsystems, allow worst-case lower bounds against kernelization. Thus, these relaxed structures give rise to instance families that cannot be efficiently reduced, by any approach.

1 Introduction

The notion of kernelization from parameterized complexity is a theoretical formalization of preprocessing (i.e., data reduction) for NP-hard combinatorial problems. Within this framework it is possible to prove worst-case upper and lower bounds for preprocessing; see, e.g., recent surveys on kernelization [13,12]. Arguably one of the most successful examples of preprocessing in practice are the simplification routines within modern integer linear program (ILP) solvers like CPLEX (see also [8,1,14]). Since ILPs have high expressive power, already the problem of testing feasibility of an ILP is NP-hard; there are immediate reductions from a variety of well-known NP-hard problems. Thus, the problem also inherits many lower bounds, in particular, lower bounds against kernelization.

* Supported by ERC Starting Grant 306992 "Parameterized Approximation".
** Supported by the German Research Foundation (DFG), KR 4286/1.

© Springer-Verlag Berlin Heidelberg 2015
N. Bansal and I. Finocchi (Eds.): ESA 2015, LNCS 9294, pp. 779–791, 2015.
DOI: 10.1007/978-3-662-48350-3_65

INTEGER LINEAR PROGRAM FEASIBILITY – ILPF
Input: A matrix $A \in \mathbb{Z}^{m \times n}$ and a vector $b \in \mathbb{Z}^m$.
Question: Is there an integer vector $x \in \mathbb{Z}^n$ with $Ax \leq b$?

Despite this negative outlook, a formal theory of preprocessing, such as kernelization aims to be, needs to provide a more detailed view on one of the most successful practical examples of preprocessing, even if worst-case bounds will rarely match empirical results. With this premise we take a structural approach to studying kernelization for ILPF. We pursue two main structural aspects of ILPs. The first one is the treewidth of the so-called *Gaifman graph* underlying the constraint matrix A. As a second aspect we consider ILPs whose constraint matrix has large parts that are totally unimodular. Both bounded treewidth and total unimodularity of the whole system $Ax \leq b$ imply that feasibility (and optimization) are tractable. We study the effect of having *subsystems* that have bounded treewidth or that are totally unimodular. We determine when such subsystems allow for a substantial reduction in instance size. Our approach differs from previous work [10,11] in that we study structural parameters related to treewidth and total unimodularity rather than considering parameters such as the dimensions n and m of the constrain matrix A or the sparsity thereof.

Treewidth and ILPs. The Gaifman graph $G(A)$ of a matrix $A \in \mathbb{Z}^{m \times n}$ is a graph with one vertex per column of A, i.e., one vertex per variable, such that variables that occur in a common constraint form a clique in $G(A)$ (see Section 3.1). This perspective allows us to consider the structure of an ILP by graph-theoretical means. In the context of graph problems, a frequently employed preprocessing strategy is to replace a simple (i.e., constant-treewidth) part of the graph that attaches to the remainder through a constant-size boundary, by a smaller gadget that enforces the same restrictions on potential solutions. There are several meta-kernelization theorems (cf. [9]) stating that large classes of graph problems can be effectively preprocessed by repeatedly replacing such *protrusions* by smaller structures. It is therefore natural to consider whether large protrusions in the Gaifman graph $G(A)$, corresponding to subsystems of the ILP, can safely be replaced by smaller subsystems.

We give an explicit dynamic programming algorithm to determine which assignments to the boundary variables (see Section 3.3) of the protrusions can be extended to feasible assignments to the remaining variables in the protrusion. Then we show that, given a list of feasible assignments to the boundary of the protrusion, the corresponding subsystem of the ILP can be replaced by new constraints. If there are r variables in the boundary and their domain is bounded by d, we find a replacement system with $\mathcal{O}(r \cdot d^r)$ variables and constraints that can be described in $\tilde{\mathcal{O}}(d^{2r})$ bits. By an information-theoretic argument we prove that equivalent replacement systems require $\Omega(d^r)$ bits to encode. Moreover, we prove that large-domain structures are indeed an obstruction for effective kernelization by proving that a family of instances with a single variable of large domain (all others have $\{0, 1\}$), and with given Gaifman decompositions into protrusions and a small shared part of encoding size N, admit no kernelization or compression to size polynomial in N.

On the positive side, we apply the replacement algorithm to protrusion decompositions of the Gaifman graph to shrink ILPF instances. When an ILPF instance can be decomposed into a small number of protrusions with small boundary domains, replacing each protrusion by a small equivalent gadget yields an equivalent instance whose overall size is bounded. The recent work of Kim et al. [9] on meta-kernelization has identified a structural graph parameter such that graphs from an appropriately chosen family with parameter value k can be decomposed into $\mathcal{O}(k)$ protrusions. If the Gaifman graph of an ILPF instance satisfies these requirements, the ILPF problem has kernels of size polynomial in k. Concretely, one can show that bounded-domain ILPF has polynomial kernels when the Gaifman graph excludes a fixed graph H as a topological minor and the parameter k is the size of a modulator of the graph to constant treewidth. We do not pursue this application further in the paper, as it follows from our reduction algorithms in a straight-forward manner.

Total Unimodularity. Recall that a matrix is totally unimodular (TU) if every square submatrix has determinant 1, -1, or 0. If A is TU then feasibility of $Ax \leq b$, for any integral vector b, can be tested in polynomial time. (Similarly, one can efficiently optimize any function $c^T x$ subject to $Ax \leq b$.) We say that a matrix A is *totally unimodular plus p columns* if it can be obtained from a TU matrix by changing entries in at most p columns. Clearly, changing a single entry may break total unimodularity, but changing only few entries should still give a system of constraints $Ax \leq b$ that is much simpler than the worst-case. Indeed, if, e.g., all variables are binary (domain $\{0, 1\}$) then one may check feasibility by simply trying all 2^p assignments to variables with modified column in A. The system on the remaining variables will be TU and can be tested efficiently.

From the perspective of kernelization it is interesting whether a small value of p allows a reduction in size for $Ax \leq b$ or, in other words, whether one can efficiently find an equivalent system of size polynomial in p. We prove that this depends on the structure of the system on variables with unmodified columns. If this remaining system decomposes into separate subsystems, each of which depends only on a *bounded number of variables in non-TU columns*, then by a similar reduction rule as for the treewidth case we get a reduced instance of size polynomial in p and the domain size d. Complementing this we prove that without this bounded dependence there is no kernelization to size polynomial in $p + d$; this also holds even if p counts the *number of entry changes* to obtain A from a TU matrix, rather than the (usually smaller) number of modified columns.

Related Work. Several lower bounds for kernelization for ILPF and other ILP-related problems follow already from lower bounds for other (less general) problems. For example, unless $\mathsf{NP} \subseteq \mathsf{coNP/poly}$ and the polynomial hierarchy collapses, there is no efficient algorithm that reduces every instance (A, b) of ILPF to an equivalent instance of size polynomial in n (here n refers to the number of columns in A); this follows from lower bounds for HITTING SET [6] or for SATISFIABILITY [5] and, thus, holds already for binary variables (0/1-ILPF). The direct study of kernelization properties of ILPs was initiated in [10,11] and focused on the influence of row- and column-sparsity of A on having kernelization results in terms of the dimensions n and m of A. At high level, the outcome is that unbounded domain variables rule out essentially all nontrivial attempts at polynomial kernelizations. In particular, ILPF admits no kernelization to size polynomial in $n + m$ when variable domains are unbounded, unless $\mathsf{NP} \subseteq \mathsf{coNP/poly}$; this remains true under strict bounds on sparsity [10]. For bounded domain variables the situation is a bit more positive: there are generalizations of positive results for d-HITTING SET and d-SATISFIABILITY (when sets/clauses have size at most d). One can reduce to size polynomial in n in general [11], and to size polynomial in k when seeking a feasible $x \geq 0$ with $|x|_1 \leq k$ for a sparse covering ILP [10].

Organization. Section 2 contains preliminaries about parameterized complexity, graphs, and treewidth. In Section 3 we analyze the effect of treewidth on preprocessing ILPs, while we consider the effect of large totally unimodular submatrices in Section 4. We conclude in Section 5. Due to space constraints, most proofs are deferred to the full version.

2   Preliminaries

Parameterized Complexity and Kernelization. A *parameterized problem* is a set $Q \subseteq \Sigma^* \times \mathbb{N}$ where Σ is any finite alphabet and $\mathbb{N}$ denotes the non-negative integers. In an instance $(x, k) \in \Sigma^* \times \mathbb{N}$ the second component is called the *parameter*. A parameterized problem Q is *fixed-parameter tractable* (FPT) if there is an algorithm that, given any instance $(x, k) \in \Sigma^* \times \mathbb{N}$, takes time $f(k)|x|^{O(1)}$ and correctly determines whether $(x, k) \in Q$; here f is any computable function. A *kernelization* for Q is an algorithm K that, given $(x, k) \in \Sigma^* \times \mathbb{N}$, takes time polynomial in $|x| + k$ and returns an instance $(x', k') \in \Sigma^* \times \mathbb{N}$ such that $(x, k) \in Q$ if and only if $(x', k') \in Q$ (i.e., the two instances are equivalent) and $|x'| + k' \leq h(k)$; here h is any computable function, and we also call it the *size of the kernel*. If $h(k)$ is polynomially bounded in k, then K is a *polynomial kernelization*. We also define *(polynomial) compression*; the only difference with kernelization is that the output is any instance $x' \in \Sigma'^*$ with respect to any fixed language $\mathcal{L}$, i.e., we demand that $(x, k) \in Q$ if and only if $x' \in \mathcal{L}$ and that $|x'| \leq h(k)$. A polynomial-parameter transformation from a parameterized problem P to a parameterized problem Q is a polynomial-time mapping that transforms each instance (x, k) of P into an equivalent instance (x', k') of Q, with the guarantee that $(x, k) \in P$ if and only if $(x', k') \in Q$ and $k' \leq p(k)$ for some polynomial p.

Graphs. All graphs in this work are simple, undirected, and finite. For a finite set X and positive integer n, we denote by $\binom{X}{n}$ the family of size-n subsets of X. The set $\{1, \ldots, n\}$ is abbreviated as $[n]$. An undirected graph G consists of a vertex set $V(G)$ and edge set $E(G) \subseteq \binom{V(G)}{2}$. For a set $X \subseteq V(G)$ we use $G[X]$ to denote the subgraph of G induced by X. We use $G - X$ as a shorthand for $G[V(G) \setminus X]$. For $v \in V(G)$ we use $N_G(v)$ to denote the open neighborhood of v. For $X \subseteq V(G)$ we define $N_G(X) := \bigcup_{v \in X} N_G(v) \setminus X$. The *boundary* of X in G, denoted $\partial_G(X)$, is the set of vertices in X that have a neighbor in $V(G) \setminus X$.

Protrusion Decompositions. For a positive integer r, an r-*protrusion* in a graph G is a vertex set $X \subseteq V(G)$ such that $\mathrm{tw}(G[X]) \leq r - 1$ and $\partial_G(X) \leq r$. An (α, r)-protrusion decomposition of a graph G is a partition $\mathcal{P} = Y_0 \uplus Y_1 \uplus \ldots \uplus Y_\ell$ of $V(G)$ such that (1) for every $1 \leq i \leq \ell$ we have $N_G(Y_i) \subseteq Y_0$, (2) $\max(\ell, |Y_0|) \leq \alpha$, and (3) for every $1 \leq i \leq \ell$ the set $Y_i \cup N_G(Y_i)$ is an r-protrusion in G. We sometimes refer to Y_0 as the *shared part*.

3   ILPs of Bounded Treewidth

We analyze the influence of treewidth for preprocessing ILPF. In Section 3.1 we give formal definitions to capture the treewidth of an ILP, and introduce a special type of tree decompositions to solve ILPs efficiently. In Section 3.2 we study the parameterized complexity of ILPF parameterized by treewidth. Tractability turns out to depend on the domain of the variables. An instance (A, b) of ILPF has *domain size* d if, for every variable x_i, there are constraints $-x_i \leq d'$ and $x_i \leq d''$ for some $d' \geq 0$ and $d'' \leq d-1$. (All positive results work also under more relaxed definitions of domain size d, e.g., any choice of d integers for each variable, at the cost of technical complication.) The feasibility of bounded-treewidth, bounded-domain ILPs is used in Section 3.3 to formulate a protrusion replacement rule. It allows the number of variables in an ILP of domain size d that is decomposed by a (k, r)-protrusion decomposition to be reduced to $\mathcal{O}(k \cdot r \cdot d^r)$. In Section 3.4 we discuss limitations of the protrusion-replacement approach.

3.1   Tree Decompositions of Linear Programs

Given a constraint matrix $A \in \mathbb{Z}^{m \times n}$ we define the corresponding *Gaifman graph* $G = G(A)$ as follows [7, Chapter 11]. We let $V(G) = \{x_1, \ldots, x_n\}$, i.e., the variables in $Ax \leq b$ for $b \in \mathbb{Z}^m$. We let $\{x_i, x_j\} \in E(G)$ if and only if there is an $r \in [m]$ with $A[r, i] \neq 0$ and $A[r, j] \neq 0$. Intuitively, two vertices are adjacent if the corresponding variables occur together in some constraint.

Observation 1. *For every row r of $A \in \mathbb{Z}^{m \times n}$, the variables Y_r with nonzero coefficients in row r form a clique in $G(A)$. Consequently (cf. [3]), any tree decomposition $(T, \mathcal{X})$ of $G(A)$ has a node i with $Y_r \subseteq X_i$.*

To simplify the description of our dynamic programming procedure, we will restrict the form of the tree decompositions that the algorithm is applied to. This is common practice when dealing with graphs of bounded treewidth: one works with *nice tree decompositions* consisting of *leaf, join, forget*, and *introduce* nodes. When using dynamic programming to solve ILPs it will be convenient to have another type of node, the *constraint node*, to connect the structure of the Gaifman graph to the constraints in the ILP. In the full version we define the notion of a *nice Gaifman decomposition* including constraint nodes. It can be derived efficiently from a nice tree decomposition of the Gaifman graph.

Proposition 1. *There is an algorithm that, given $A \in \mathbb{Z}^{m \times n}$ and a width-w tree decomposition $(T, \mathcal{X})$ of the Gaifman graph of A, computes a nice Gaifman decomposition $(T', \mathcal{X}', \mathcal{Z}')$ of A having width w in $\mathcal{O}(w^2 \cdot |V(T)| + n \cdot m \cdot w)$ time.*

3.2   Feasibility on Gaifman Graphs of Bounded Treewidth

We discuss the influence of treewidth on the complexity of ILPF. It turns out that for unbounded domain variables the problem remains weakly NP-hard on instances with Gaifman graphs of treewidth at most two (Theorem 1). On the other hand, the problem can be solved by a simple dynamic programming algorithm with runtime $\mathcal{O}^*(d^{w+1})$, where d is the domain size and w denotes the width of a given tree decomposition of the Gaifman graph (Theorem 2). In other words, the problem is fixed-parameter tractable in terms of $d+w$, and efficiently solvable for bounded treewidth and d polynomially bounded in the input size.

Both results are not hard to prove and fixed-parameter tractability of ILPF($d+w$) can also be derived from Courcelle's theorem (cf. [7, Corollary 11.43]). Explicit proofs are provided in the full version of this work. Theorem 2 is a subroutine of our protrusion reduction algorithm.

Theorem 1. ILP FEASIBILITY *remains weakly* NP-*hard when restricted to instances (A, b) whose Gaifman graph $G(A)$ has treewidth two.*

Theorem 2. *Instances $(A \in \mathbb{Z}^{m \times n}, b)$ of* ILPF *of domain size d with a given nice Gaifman decomposition of width w can be solved in time $\mathcal{O}(d^{w+1} \cdot w \cdot (n+m))$.*

If a nice Gaifman decomposition is not given, one can be computed by combining an algorithm for computing tree decompositions [2,4] with Proposition 1.

3.3   Protrusion Reductions

To formulate the protrusion replacement rule, which is the main algorithmic asset used in this section, we need some terminology. For a non-negative integer r, an r-*boundaried ILP* is an instance (A, b) of ILPF in which r distinct *boundary variables* $x_{t_1}, \ldots, x_{t_r}$ are distinguished among the total variable set $\{x_1, \ldots, x_n\}$. If $Y = (x_{i_1}, \ldots, x_{i_r})$ is a sequence of variables of $Ax \leq b$, we will also use (A, b, Y) to denote the corresponding r-boundaried ILP. The *feasible boundary assignments* of a boundaried ILP are those assignments to the boundary variables that can be completed into an assignment that is feasible for the entire system.

Definition 1. *Two r-boundaried ILPs $(A, b, x_{t_1}, \ldots, x_{t_r})$ and $(A', b', x'_{t'_1}, \ldots, x'_{t'_r})$ are equivalent if they have the same feasible boundary assignments.*

The following lemma shows how to compute equivalent boundaried ILPs for any boundaried input ILP. The replacement system is built by adding, for each infeasible boundary assignment, a set of constraints on auxiliary variables that explicitly blocks that assignment.

Lemma 1. *There is an algorithm with the following specifications: (1) It gets as input an r-boundaried ILP $(A, b, x_{t_1}, \ldots, x_{t_r})$ with domain size d, with $A \in \mathbb{Z}^{m \times n}$, $b \in \mathbb{Z}^m$, and a width-w nice Gaifman decomposition $(T, \mathcal{X}, \mathcal{Z})$ of A. (2) Given such an input it takes time $\mathcal{O}(d^r \cdot (d^{w+1} w(n+m)) + r^2 \cdot d^{2r})$. (3) Its output is an equivalent r-boundaried ILP $(A', b', x'_{t'_1}, \ldots, x'_{t'_r})$ of domain size d containing $\mathcal{O}(r \cdot d^r)$ variables and constraints, and all entries of (A', b') in $\{-d, \ldots, d\}$.*

Intuitively, we can simplify an ILPF instance (A, b) with a given protrusion decomposition by replacing all protrusions with equivalent boundaried ILPs of small size via Lemma 1. We get a new instance containing all replacement constraints plus all original constraints that are fully contained in the shared part.

Theorem 3. *For each constant r there is an algorithm that, given an instance (A, b) of ILPF with domain size d, along with a (k, r)-protrusion decomposition $Y_0 \uplus Y_1 \uplus \ldots \uplus Y_\ell$ of the given Gaifman graph $G(A)$, outputs an equivalent instance (A', b') of ILPF with domain size d on $\mathcal{O}(k \cdot d^r)$ variables in time $\mathcal{O}(n \cdot m + d^{2r}(n + m) + k \cdot m \cdot d^r + k^2 \cdot d^{2r})$. Each constraint of (A', b') is either a constraint in (A, b) involving only variables from Y_0, or one of $\mathcal{O}(k \cdot d^r)$ new constraints with coefficients and right-hand side among $\{-d, \ldots, d\}$.*

3.4   Limitations for Replacing Protrusions

In this section, we discuss limitations regarding the replacement of protrusions in an ILP. First of all, there is an information-theoretic limitation for the worst-case size replacement of any r-boundaried ILP with variables $x_{t_1}, \ldots, x_{t_r}$ each with domain size d. Clearly, there are d^r different assignments to the boundary. For any set A of assignments to the boundary variables, using auxiliary variables and constraints one can construct an r-boundaried ILP whose feasible boundary assignments are exactly A. This gives a lower bound of d^r bits for the encoding size of a general r-boundaried ILP, since we have 2^{d^r} subsets. Our first result regarding limitations for replacing protrusions is that this lower bound even holds for boundaried ILPs of bounded treewidth.

Proposition 2. *For any $d, r \in \mathbb{N}$ and $A \subseteq \{0, \ldots, d - 1\}^r$ there is an r-boundaried ILP of treewidth $3r$ with domain size d, whose feasible boundary assignments are A.*

The proposition follows from the fact that the encoding in Lemma 1 produces a boundaried ILP of treewidth at most $3r$. To find an r-boundaried ILP of small treewidth whose feasible assignments are A, we may therefore first construct an arbitrary r-boundaried ILP whose feasible boundary assignments are A, and then invoke Lemma 1. Our used encoding in Lemma 1 uses size $\tilde{\mathcal{O}}(d^{2r})$. Note that, when using an encoding for sparse matrices, our replacement size comes fairly close to the information-theoretic lower bound.

Second, the lower bounds for $0/1$-ILPF(n), which follow, e.g., from lower bounds for HITTING SET parameterized by ground set size, imply that there is no hope for a kernelization just in terms of deletion distance to a system of bounded treewidth. (This distance is upper bounded by n.) Note that the bound relies on a fairly direct formulation of HITTING SET instances as ILPs, which creates huge cliques in the Gaifman graph when expressing sets as large inequalities (over indicator variables). The lower bound can be strengthened somewhat by instead representing sets less directly: For each set, "compute" the sum of its indicator variables using auxiliary variables for partial sums. Similarly to the example of SUBSET SUM, this creates a structure of bounded treewidth. Note, however, that this structure is not a (useful) protrusion because its boundary can be as large as n; this is indeed the crux of having only a modulator to bounded treewidth but no guarantee for (or means of proving of) the existence of protrusions with small boundaries.

Finally, we prove in the following theorem that the mentioned information-theoretic limitation also affects the possibility of strong preprocessing, rather than being an artifact of the definition of equivalent boundaried ILPs. In other words, there is a family of ILPF instances that already come with a protrusion decomposition, and with a single variable of large domain, but that cannot be reduced to size polynomial in the parameters of this decomposition. Note that this includes all other ways of handling these instances, which establishes that protrusions with even a single large domain boundary variable can be the crucial obstruction from achieving a polynomial kernelization.

Theorem 4. *Assuming* NP $\not\subseteq$ coNP/poly, *there is no polynomial-time algorithm that compresses instances* $(A \in \mathbb{Z}^{m \times n}, b \in \mathbb{Z}^m)$ *of* ILPF *with entries in* $\{-n, \ldots, n\}$ *that consist of* $\{0,1\}$*-variables except for a single variable of domain* $d \leq n$, *which are given together with a* $(k,5)$*-protrusion decomposition* $Y_0 \uplus Y_1 \uplus \ldots \uplus Y_\ell$ *of* $V(G(A))$, *to size polynomial in* $k + \hat{m} + \log d$, *where* $\hat{m}$ *is the number of constraints that affect only variables of* Y_0.

Intuitively, the parameterization chosen in the theorem implies that everything can be bounded to size polynomial in the parameters except for the variables in $Y_1, \ldots, Y_\ell$ and the (encoding size of the) constraints that are fully contained in protrusions (recall that constraints give cliques in $G(A)$). To put this lower bound into context, we prove that a more general (and less technical) variant is fixed-parameter tractable.

Theorem 5. *The following variant of* ILPF *is* FPT*: We allow a constant number* c *of variables with polynomially bounded domain; all other variables have domain size* d. *Furthermore, there is a specified set of variables* $S \subseteq \{x_1, \ldots, x_n\}$ *such that the graph* $G(A) - S$ *has bounded treewidth. The parameter is* $d + |S|$.

4   Totally Unimodular Subproblems

We say that a matrix A is *totally unimodular plus p entries* if A can be obtained from a totally unimodular matrix by replacing any p entries by new values. (This is more restrictive than the equally natural definition of adding p arbitrary rows or columns.) We note that ILPF is FPT with respect to parameter $p + d$, where d bounds the domain: It suffices to try all d^p assignments for variables whose column in A has at least one modified entry. After simplification the obtained system is TU and, thus, existence of a feasible assignment for the remaining entries can be tested in polynomial time.

Our following result shows that, despite fixed-parameter tractability for parameter $p + d$, the existence of a polynomial kernelization is unlikely; this holds already when $d = 2$.

Theorem 6. ILP FEASIBILITY *restricted to instances* (A, b, p) *where* A *is totally unimodular plus* p *entries does not admit a kernel or compression to size polynomial in* p *unless* $\mathsf{NP} \subseteq \mathsf{coNP/poly}$, *even if all domains are* $\{0, 1\}$.

Proof. We reduce from the HITTING SET(n) problem, in which we are given a set U, a set $\mathcal{F} \subseteq 2^U$ of subsets of U, and an integer $k \in \mathbb{N}$, and we have to decide whether there is a choice of at most k elements of U that intersects all sets in $\mathcal{F}$; the parameter is $n := |U|$. Dom et al. [6] proved that HITTING SET(n) admits no polynomial kernelization or compression (in terms of n) unless $\mathsf{NP} \subseteq \mathsf{coNP/poly}$. We present a polynomial-parameter transformation from HITTING SET(n) to ILPF(p) with domain size 2. The ILP produced by the reduction will be of the form $Ax \leq b$ where A is totally unimodular plus $p = n$ entries. Thus, any polynomial kernelization or compression in terms of p would give a polynomial compression for HITTING SET(n) and, thus, imply $\mathsf{NP} \subseteq \mathsf{coNP/poly}$ as claimed.

Construction. Let an instance $(U, \mathcal{F}, k)$ be given. We construct an ILP with 0/1-variables that is feasible if and only if $(U, \mathcal{F}, k)$ is *yes* for HITTING SET(n).

- Our ILP has two types of variables: $x_{u,F}$ for all $u \in U, F \in \mathcal{F}$ and x_u for all $u \in U$. For all variables we enforce domain $\{0, 1\}$ by $x_{u,F} \geq 0$, $x_{u,F} \leq 1$, $x_u \geq 0$, and $x_u \leq 1$ for $u \in U$ and $F \in \mathcal{F}$.
- The variables $x_{u,F}$ are intended to encode what elements of u "hit" which sets $F \in \mathcal{F}$. We enforce that each set $F \in \mathcal{F}$ is "hit" by adding the following constraint.

$$1 \leq \sum_{u \in F} x_{u,F} \qquad\qquad \forall F \in \mathcal{F} \qquad\qquad (1)$$

- The variables x_u are used to control which variables $x_{u,F}$ may be assigned values greater than zero; effectively, they correspond to the choice of a hitting set from U. Control of the $x_{u,F}$ variables comes from the following constraints.

$$\sum_{F \in \mathcal{F}} x_{u,F} \leq |\mathcal{F}| \cdot x_u \qquad\qquad \forall u \in U \qquad\qquad (2)$$

Additionally, we constrain the sum over all x_u to be at most k, in line with the concept of having x_u select a hitting set of size at most k.

$$\sum_{u \in U} x_u \leq k \qquad\qquad (3)$$

Clearly, the construction can be performed in polynomial time. The proof that the obtained ILP is feasible if and only $(U, \mathcal{F}, k)$ is *yes* for HITTING SET(n) is deferred to the full version. In the remainder of the proof, we show that the constraints can be written as $Ax \leq b$ where A is totally unimodular plus n entries (here **x** stands for the vector of all variables $x_{u,F}$ and x_u over all $u \in U$ and $F \in \mathcal{F}$). First, we need to write our constraints in the form

$A \begin{pmatrix} x_{U,\mathcal{F}} \\ x_U \end{pmatrix} \leq b$, where, e.g., x_U stands for the column vector of all variables x_u with $u \in U$. For now, we translate constraints (1), (2), and (3) into this form; domain-enforcing constraints will be discussed later. We obtain the following, where $(\mathbf{1/0})$ and $(\mathbf{-1/0})$ are shorthand for submatrices that are entirely 0 except for exactly one 1 or one -1 per column, respectively.

$$
A'\mathbf{x} = \left(\left(\begin{array}{c} \overbrace{\begin{pmatrix} \mathbf{-1/0} \end{pmatrix}}^{|\mathcal{F}| \times |U|} \\[2ex] \begin{pmatrix} \mathbf{1/0} \end{pmatrix} \\[2ex] 0 \cdots 0 \end{array} \right. \left. \begin{array}{c} \overbrace{\begin{pmatrix} 0 & \cdots & 0 \\ \vdots & \ddots & \vdots \\ 0 & \cdots & 0 \\ -|\mathcal{F}| & 0 & 0 \\ 0 & -|\mathcal{F}| & 0 \\ 0 & 0 & -|\mathcal{F}| \\ 1 & \cdots & 1 \end{pmatrix}}^{|U|} \end{array} \right) \right) \begin{pmatrix} x_{U,\mathcal{F}} \\ x_U \end{pmatrix} \leq \begin{pmatrix} -1 \\ \vdots \\ -1 \\ 0 \\ \vdots \\ 0 \\ k \end{pmatrix} \begin{array}{l} (1) \\[6ex] (2) \\[3ex] (3) \end{array}
$$

Let us denote by $\hat{A}'$ the matrix obtained from A' by replacing all n entries $-|\mathcal{F}|$ by zero.

$$
\hat{A}' = \begin{pmatrix} (\mathbf{-1/0}) & \mathbf{0} \\ (\mathbf{1/0}) & \mathbf{0} \\ 0 \cdots 0 & 1 \cdots 1 \end{pmatrix}
$$

It is known that any matrix over $\{-1, 0, 1\}$ in which every column has at most one entry 1 and at most one entry -1, is totally unimodular (cf. [15, Theorem 13.9]). Since $\hat{A}'$ is of this form, it is clear that $\hat{A}'$ is totally unimodular. To obtain the whole constraint matrix A we need to add rows corresponding to domain-enforcing constraints for all variables, and reset the $-|\mathcal{F}|$ values that we replaced by zero. Clearly, putting back the latter breaks total unimodularity (and this is why A is only almost TU), but let us add everything else and see that the obtained matrix $\hat{A}$ is totally unimodular. The domain-enforcing constraints affect only one variable each and, thus, each of them corresponds to a row in $\hat{A}$ that contains only a single nonzero entry of value 1 or -1. It is well known that adding such rows (or columns) preserves total unimodularity. (The determinant of any square submatrix containing such a row can be reduced to that of a smaller submatrix by expanding along a row that has only one nonzero of 1 or -1.)

Finally, $\hat{A}$ and A are only distinguished by the n entries of value $-|\mathcal{F}|$ that are present in A but which are 0 in $\hat{A}$. Since $\hat{A}$ is totally unimodular, it follows that A is totally unimodular plus $p = n$ entries, as claimed. □

Complementing Theorem 6, we prove that TU subsystems of an ILP can be reduced to a size that is polynomial in the domain, with degree depending on the number of variables that occur also in the rest of the ILP. We again phrase this in terms of replacing boundaried ILPs and prove that any r-boundaried TU subsystem can be replaced by a small equivalent system of size polynomial in the domain d with degree depending on r.

Lemma 2. *There is an algorithm with the following specifications: (1) It gets as input an r-boundaried ILP $(A, b, x_{t_1}, \ldots, x_{t_r})$ with domain size d, with $A \in \mathbb{Z}^{m \times n}$, $b \in \mathbb{Z}^m$, and such that the restriction of A to columns $[m] \setminus \{t_1, \ldots, t_r\}$ is totally unimodular. (2) Given such an input it takes time $\mathcal{O}(d^r \cdot g(n, m) + d^{2r})$ where $g(n, m)$ is the runtime for an LP solver for determining feasibility of a linear program with n variables and m constraints. (3) Its output is an equivalent r-boundaried ILP $(A', b', x'_{t'_1}, \ldots, x'_{t'_r})$ of domain size d containing $\mathcal{O}(r \cdot d^r)$ variables and constraints, and all entries of (A', b') in $\{-d, \ldots, d\}$.*

Lemma 2 implies that if Y_0 is a set of (at most) p variables whose removal makes the remaining system TU, then the number of variables in the system can efficiently be reduced to a polynomial in $d + p$ with degree depending on r, if each TU subsystem depends on at most r variables in Y_0. To get this, it suffices to apply Lemma 2 once for each choice of at most r boundary variables in Y_0. (Note that without assuming a bounded value of r we only know $r \leq p$, so the worst-case bound obtained is not polynomial, but exponential, in $p + d$.)

5   Discussion and Future Work

We have studied the effect that subsystems with bounded treewidth or total unimodularity have regarding kernelization of the ILP FEASIBILITY problem. We show that if such subsystems have a constant-size boundary to the rest of the system, then they can be replaced by an equivalent subsystem of size polynomial in the domain size (with degree depending on the boundary size). Thus, if an ILPF instance can be decomposed by specifying a set of p shared variables whose deletion (or replacing with concrete values) creates subsystems that are all TU or bounded treewidth and have bounded dependence on the p variables, then this can be replaced by an equivalent system whose number of variables is polynomial in p and the domain size d. We point out that for the case of binary variables (at least in the boundary) the replacement structures get much simpler, using no additional variables and with only a single constraint per forbidden assignment. Using a similar approach and binary encoding for boundary variables should reduce the number of additional variables to $\mathcal{O}(\log d)$ per boundary variable.

Complementing this, we established several lower bounds regarding limitations of replacing such subsystems. Inherently, the replacement rules rely on having subsystems with small boundary size for giving polynomial bounds. We showed that this is indeed necessary by giving lower bounds for fairly restricted settings where we do not have the guarantee of constant boundary size, independent of the means of data reduction. In the case of treewidth we could also show that boundaries with only one large-domain variable can be a provable obstacle. For the case of totally unimodular subsystems a discussion in the full version shows that these behave in a slightly simpler way than bounded-treewidth subsystems: By ad hoc arguments we can save a factor of d in the encoding size by essentially dropping the contribution of any one boundary variable; thus we rule out a lower bound proof for the case of one boundary variable having large domain. Asymptotically, we can save a factor of almost d^2, which is tight. It would be interesting whether having two or more large domain variables (in a boundary of constant size) would again allow a lower bound against kernelization.

A natural extension of our work is to consider the optimization setting where we have to minimize or maximize a linear function over the variables, and may or may not already know that the system is feasible. In part, our techniques are already consistent with this since the reduction routine based on treewidth dynamic programming or optimization over a TU subsystem can be easily augmented to also optimize a target function over the variables. A technical caveat, however, is the following: If we simplify a protrusion, then along with each feasible assignment to the boundary, we have to store the best target function contribution that could be obtained with the variables that are removed; this value can, theoretically, be unbounded in all other parameters. If a binary encoding of such values is sufficiently small (or if the needed space is allowed through an additional parameter), then our results also carry over to optimization. Apart from that, a rigorous analysis of both weight reduction techniques and possible lower bounds is left as future work.

References

1. Atamtürk, A., Savelsbergh, M.W.P.: Integer-programming software systems. Annals OR 140(1), 67–124 (2005)
2. Bodlaender, H.L.: A linear-time algorithm for finding tree-decompositions of small treewidth. SIAM Journal on Computing 25(6), 1305–1317 (1996)
3. Bodlaender, H.L.: A partial k-arboretum of graphs with bounded treewidth. Theoretical Computer Science 209(1-2), 1–45 (1998)
4. Bodlaender, H.L., Drange, P.G., Dregi, M.S., Fomin, F.V., Lokshtanov, D., Pilipczuk, M.: An $O(c^k n)$ 5-approximation algorithm for treewidth. In: FOCS, pp. 499–508. IEEE Computer Society (2013)
5. Dell, H., van Melkebeek, D.: Satisfiability allows no nontrivial sparsification unless the polynomial-time hierarchy collapses. J. ACM 61(4), 23 (2014)
6. Dom, M., Lokshtanov, D., Saurabh, S.: Incompressibility through colors and IDs. In: Albers, S., Marchetti-Spaccamela, A., Matias, Y., Nikoletseas, S., Thomas, W. (eds.) ICALP 2009, Part I. LNCS, vol. 5555, pp. 378–389. Springer, Heidelberg (2009)
7. Flum, J., Grohe, M.: Parameterized Complexity Theory. Springer (2006)
8. Genova, K., Guliashki, V.: Linear integer programming methods and approaches – a survey. Cybernetics and Information Technologies 11(1) (2011)
9. Kim, E.J., Langer, A., Paul, C., Reidl, F., Rossmanith, P., Sau, I., Sikdar, S.: Linear kernels and single-exponential algorithms via protrusion decompositions. In: Fomin, F.V., Freivalds, R., Kwiatkowska, M., Peleg, D. (eds.) ICALP 2013, Part I. LNCS, vol. 7965, pp. 613–624. Springer, Heidelberg (2013)
10. Kratsch, S.: On polynomial kernels for integer linear programs: Covering, packing and feasibility. In: Bodlaender, H.L., Italiano, G.F. (eds.) ESA 2013. LNCS, vol. 8125, pp. 647–658. Springer, Heidelberg (2013)
11. Kratsch, S.: On polynomial kernels for sparse integer linear programs. In: STACS. LIPIcs, vol. 20, pp. 80–91 (2013)

12. Kratsch, S.: Recent developments in kernelization: A survey. Bulletin of the EATCS, 113 (2014)
13. Lokshtanov, D., Misra, N., Saurabh, S.: Kernelization - preprocessing with a guarantee. In: Bodlaender, H.L., Downey, R., Fomin, F.V., Marx, D. (eds.) The Multivariate Algorithmic Revolution and Beyond. LNCS, vol. 7370, pp. 129–161. Springer, Heidelberg (2012)
14. Savelsbergh, M.W.P.: Preprocessing and probing techniques for mixed integer programming problems. INFORMS Journal on Computing 6(4), 445–454 (1994)
15. Schrijver, A.: Combinatorial Optimization: Polyhedra and Efficiency. Springer (2003)

On the Approximability of Digraph Ordering

Sreyash Kenkre[1], Vinayaka Pandit[1], Manish Purohit[2*], and Rishi Saket[1]

[1] IBM Research, Bangalore, Karnataka 560045, India
{srekenkr,pvinayak,rissaket}@in.ibm.com
[2] University of Maryland, College Park, MD 20742, USA
manishp@cs.umd.edu

Abstract. Given an n-vertex digraph $D = (V, A)$ the MAX-k-ORDERING problem is to compute a labeling $\ell : V \to [k]$ maximizing the number of forward edges, i.e. edges (u, v) such that $\ell(u) < \ell(v)$. For different values of k, this reduces to *Maximum Acyclic Subgraph* $(k = n)$, and MAX-DICUT $(k = 2)$. This work studies the approximability of MAX-k-ORDERING and its generalizations, motivated by their applications to job scheduling with *soft* precedence constraints. We give an LP rounding based 2-approximation algorithm for MAX-k-ORDERING for any $k = \{2, \ldots, n\}$, improving on the known $2k/(k-1)$-approximation obtained via random assignment. The tightness of this rounding is shown by proving that for any $k = \{2, \ldots, n\}$ and constant $\varepsilon > 0$, MAX-k-ORDERING has an LP integrality gap of $2 - \varepsilon$ for $n^{\Omega(1/\log\log k)}$ rounds of the Sherali-Adams hierarchy.

A further generalization of MAX-k-ORDERING is the *restricted maximum acyclic subgraph* problem or RMAS, where each vertex v has a finite set of allowable labels $S_v \subseteq \mathbb{Z}^+$. We prove an LP rounding based $4\sqrt{2}/(\sqrt{2}+1) \approx 2.344$ approximation for it, improving on the $2\sqrt{2} \approx 2.828$ approximation recently given by Grandoni et al. [5]. In fact, our approximation algorithm also works for a general version where the objective counts the edges which go forward by at least a positive *offset* specific to each edge.

The minimization formulation of digraph ordering is *DAG edge deletion* or DED(k), which requires deleting the minimum number of edges from an n-vertex directed acyclic graph (DAG) to remove all paths of length k. We show that both, the LP relaxation and a local ratio approach for DED(k) yield k-approximation for any $k \in [n]$. A vertex deletion version was studied earlier by Paik et al. [16], and Svensson [17].

1 Introduction

One of the most well studied combinatorial problems on directed graphs (digraphs) is the *Maximum Acyclic Subgraph* problem (MAS): given an n-vertex digraph, find a subgraph[1] of maximum number of edges containing no directed

---

* Partially supported by NSF grants CCF-1217890 and IIS-1451430.
[1] Unless specified, throughout this paper a *subgraph* is not necessarily induced.

© Springer-Verlag Berlin Heidelberg 2015
N. Bansal and I. Finocchi (Eds.): ESA 2015, LNCS 9294, pp. 792–803, 2015.
DOI: 10.1007/978-3-662-48350-3_66

cycles. An equivalent formulation of MAS is to obtain a linear ordering of the vertices which maximizes the number of directed edges going forward. A natural generalization is MAX-k-ORDERING where the goal is to compute the best k-*ordering*, i.e. a labeling of the vertices from $[k] = \{1, \ldots, k\}$ ($2 \le k \le n$), which maximizes the number of directed edges going forward in this ordering. It can be seen – and we show this formally – that MAX-k-ORDERING is equivalent to finding the maximum subgraph which has no directed cycles, and no directed paths[2] of length k. Note that MAS is the special case of MAX-k-ORDERING when $k = n$, and for $k = 2$ MAX-k-ORDERING reduces to the well known MAX-DICUT problem.

A related problem is the *Restricted Maximum Acyclic Subgraph* problem or RMAS, in which each vertex v of the digraph has to be assigned a label from a finite set $S_v \subseteq \mathbb{Z}^+$ to maximize the number of edges going forward in this assignment. Khandekar et al. [9] introduced RMAS in the context of *graph pricing* problems and its approximability has recently been studied by Grandoni et al. [5]. A further generalization is OFFSETRMAS where each edge (u, v) has an offset $o_e \in \mathbb{Z}^+$ and is satisfied by a labeling ℓ if $\ell(u) + o_e \le \ell(v)$. Note that when all offsets are unit OFFSETRMAS reduces to RMAS, which in turn reduces to MAX-k-ORDERING when $S_v = [k]$ for all vertices v.

This study focuses on the approximability of MAX-k-ORDERING and its generalizations and is motivated by their applicability in scheduling jobs with *soft* precedences under a hard deadline. Consider the following simple case of discrete time scheduling: given n unit length jobs with precedence constraints and an infinite capacity machine, find a schedule so that all the jobs are completed by timestep k. Since it may not be feasible to satisfy all the precedence constraints, the goal is to satisfy the maximum number. This is equivalent to MAX-k-ORDERING on the corresponding precedence digraph. One can generalize this setting to each job having a set of allowable timesteps when it can be scheduled. This can be abstracted as RMAS and a further generalization to each precedence having an associated lag between the start-times yields OFFSETRMAS as the underlying optimization problem.

Also of interest is the minimization version of MAX-k-ORDERING on directed *acyclic* graphs (DAGs). We refer to it as *DAG edge deletion* or DED(k) where the goal is to delete the minimum number of directed edges from a DAG so that the remaining digraph does not contain any path of length k. Note that the problem for arbitrary k does not admit any approximation factor on general digraphs since even detecting whether a digraph has a path of length k is the well studied NP-hard longest path problem. A vertex deletion formulation of DED(k) was introduced as an abstraction of certain VLSI design and communication problems by Paik et al. [16] who gave efficient algorithms for it on special cases of DAGs, and proved it to be NP-complete in general. More recently, its connection to project scheduling was noted by Svensson [17] who proved inapproximability results for the vertex deletion version.

---

[2] The length of a directed path is the number of directed edges it contains.

The rest of this section gives a background of previous related work, describes our results, and provides an overview of the techniques used.

1.1   Related Work

It is easy to see that MAS admits a trivial 2-approximation, by taking any linear ordering or its reverse, and this is also obtained by a random ordering. For MAX-k-ORDERING the random k-ordering yields a $2k/(k-1)$-approximation for any $k \in \{2, \ldots, n\}$. For $k = 2$, which is MAX-DICUT, the semidefinite programming (SDP) relaxation is shown to yield a ≈ 1.144-approximation in [13], improving upon previous analyses of [14], [19], and [3]. As mentioned above, RMAS is a generalization of MAX-k-ORDERING, and a $2\sqrt{2}$-approximation for it based on linear programming (LP) rounding was shown recently by Grandoni et al. [5] which is also the best approximation for MAX-k-ORDERING for $k = 3$. For $4 \leq k \leq n-1$, to the best of our knowledge the proven approximation factor for MAX-k-ORDERING remains $2k/(k-1)$.

On the hardness side, Newman [15] showed that MAS is NP-hard to approximate within a factor of $66/65$. Assuming Khot's [10] Unique Games Conjecture (UGC), Guruswami et al. [6] gave a $(2 - \varepsilon)$-inapproximability for any $\varepsilon > 0$. Note that MAX-DICUT is at least as hard as MAX-CUT. Thus, for $k = 2$, MAX-k-ORDERING is NP-hard to approximate within factor $(13/12 - \varepsilon)$ [7], and within factor 1.1382 assuming the UGC [11]. For larger constants k, the result of Guruswami et al. [6] implicitly shows a $(2 - o_k(1))$-inapproximability for MAX-k-ORDERING, assuming the UGC.

For the vertex deletion version of $DED(k)$, Paik et al. [16] gave linear time and quadratic time algorithms for rooted trees and series-parallel graphs respectively. The problem reduces to vertex cover on k-uniform hypergraphs for any constant k thereby admitting a k-approximation, and a matching $(k-\varepsilon)$-inapproximability assuming the UGC was obtained by Svensson [17].

1.2   Our Results

The main algorithmic result of this paper is the following improved approximation guarantee for MAX-k-ORDERING.

Theorem 1. *There exists a polynomial time 2-approximation algorithm for* MAX-k-ORDERING *on n-vertex weighted digraphs for any* $k \in \{2, \ldots, n\}$.

The above approximation is obtained by appropriately rounding the standard LP relaxation of the CSP formulation of MAX-k-ORDERING. For small values of k this yields significant improvement on the previously known approximation factors: $2\sqrt{2}$ for $k = 3$ (implicit in [5]), $8/3$ for $k = 4$, and 2.5 for $k = 5$. The latter two factors follow from the previous best $2k/(k-1)$-approximation given by a random k-ordering for $4 \leq k \leq n-1$. The detailed proof of Theorem 1 is given in Section 3.

Using an LP rounding approach similar to Theorem 1, in Section 4 we show the following improved approximation for OFFSETRMAS which implies the same for RMAS. Our result improves the previous $2\sqrt{2} \approx 2.828$-approximation for RMAS obtained by Grandoni et al. [5].

Theorem 2. *There exists a polynomial time $4\sqrt{2}/(\sqrt{2}+1) \approx 2.344$ approximation algorithm for OFFSETRMAS on weighted digraphs.*

Our next result gives a lower bound that matches the approximation obtained in Theorem 1. We show that even after strengthening the LP relaxation of MAX-k-ORDERING with a large number of rounds of the Sherali-Adams hierarchy, its integrality gap remains close to 2, and hence Theorem 1 is tight.

Theorem 3. *For any small enough constant $\varepsilon > 0$, there exists $\gamma > 0$ such that for MAX-k-ORDERING on n-vertex weighted digraphs and any $k \in \{2,\ldots,n\}$, the LP relaxation with $n^{(\gamma/\log\log k)}$ rounds of Sherali-Adams constraints has a $(2 - \varepsilon)$ integrality gap.*

For DED(k) on DAGs we prove the following approximation for any k, not necessarily a constant.

Theorem 4. *The standard LP relaxation for DED(k) on n-vertex DAGs can be solved in polynomial time for $k = \{2,\ldots,n-1\}$ and yields a k-approximation. The same approximation factor is also obtained by a combinatorial algorithm.*

Due to lack of space the proofs of Theorem 3 and Theorem 4 are deferred to the full version [8]. In the full version, we also complement the above theorem by showing a $(\lfloor k/2 \rfloor - \varepsilon)$ hardness factor for DED(k) via a simple gadget reduction from Svensson's [17] $(k-\varepsilon)$-inapproximability for the vertex deletion version for constant k, assuming the UGC.

1.3   Overview of Techniques

The approximation algorithms we obtain for MAX-k-ORDERING and its generalizations are based on rounding the standard LP relaxation for the instance. MAX-k-ORDERING is viewed as a constraint satisfaction problem (CSP) over alphabet $[k]$, and the corresponding LP relaxation has $[0,1]$-valued variables x_i^v for each vertex v and label $i \in [k]$, and y_{ij}^e for each edge (u,v) and pairs of labels i and j to u and v respectively. We show that a generalization of the rounding algorithm used by Trevisan [18] for approximating q-ary boolean CSPs yields a 2-approximation in our setting. The key ingredient in the analysis is a lower bound on a certain product of the $\{x_i^u\}, \{x_i^v\}$ variables corresponding to the end points of an edge $e = (u,v)$ in terms of the $\{y_{ij}^e\}$ variables for that edge. This improves a weaker bound shown by Grandoni et al. [5]. For OFFSETRMAS, a modification of this rounding algorithm yields the improved approximation.

The construction of the integrality gap for the LP augmented with Sherali-Adams constraints for MAX-k-ORDERING begins with a simple integrality gap instance for the basic LP relaxation. This instance is appropriately sparsified to ensure that subgraphs of polynomially large (but bounded) size are *tree-like*. On trees, it is easy to construct a distribution over labelings from $[k]$ to the vertices (thought of as k-orderings), such that the marginal distribution on each vertex is uniform over $[k]$ and a large fraction of edges are satisfied in expectation. Using this along with the sparsification allows us to construct distributions for each bounded subgraph, i.e. good local distributions. Finally a geometric *embedding* of the marginals of these distributions followed by Gaussian rounding yields modified local distributions which are *consistent* on the common vertex sets. These distributions correspond to an LP solution with a high objective value, for large number of rounds of Sherali-Adams constraints. Our construction follows the approach in a recent work of Lee [12] which is based on earlier works of Arora et al. [1] and Charikar et al. [2].

For the DED(k) problem, the approximation algorithms stated in Theorem 4 are obtained using the acyclicity of the input DAG. In particular, we show that both, the LP rounding and the local ratio approach, can be implemented in polynomial time on DAGs yielding k-approximate solutions.

2   Preliminaries

This section formally defines the problems studied in this paper. We begin with MAX-k-ORDERING.

Definition 1. MAX-k-ORDERING: *Given an n-vertex digraph $D = (V, A)$ with a non-negative weight function $w : A \to \mathbb{R}^+$, and an integer $2 \le k \le n$, find a labeling to the vertices $\ell : V \to [k]$ that maximizes the weighted fraction of edges $e = (u, v) \in A$ such that $\ell(u) < \ell(v)$, i.e. forward edges.*

It can be seen that MAX-k-ORDERING is equivalent to the problem of computing the maximum weighted subgraph of D which is acyclic and does not contain any directed path of length k. The following lemma implies this equivalence and its proof is included in the full version [8].

Lemma 1. *Given a digraph $D = (V, A)$, there exists a labeling $\ell : V \to [k]$ with each edge $e = (u, v) \in A$ satisfying $\ell(u) < \ell(v)$, if and only if D is acyclic and does not contain any directed path of length k.*

The generalizations of MAX-k-ORDERING studied in this work, viz. RMAS and OFFSETRMAS, are defined as follows.

Definition 2. OFFSETRMAS: *The input is a digraph $D = (V, A)$ with a finite subset $S_v \subseteq \mathbb{Z}^+$ of labels for each vertex $v \in V$, a non-negative weight function $w : A \to \mathbb{R}^+$, and offsets $o_e \in \mathbb{Z}^+$ for each edge $e \in A$. A labeling ℓ to V s.t. $\ell(v) \in S_v, \forall v \in V$ satisfies an edge $e = (u, v)$ if $\ell(u) + o_e \le \ell(v)$. The goal is to compute a labeling that maximizes the weighted fraction of satisfied edges. RMAS is the special case when each offset is unit.*

As mentioned earlier, DED(k) is not approximable on general digraphs. Therefore, we define it only on DAGs.

Definition 3. DED(k): *Given a DAG $D = (V, A)$ with a non-negative weight function $w : A \to \mathbb{R}^+$, and an integer $2 \le k \le n - 1$, find a minimum weight set of edges $F \subseteq A$ such that $(V, A \setminus F)$ does not contain any path of length k.*

The rest of this section describes the LP relaxations for MAX-k-ORDERING and OFFSETRMAS studied in this paper.

2.1 LP Relaxation for MAX-k-ORDERING

From Definition 1, an instance $\mathcal{I}$ of MAX-k-ORDERING is given by $D = (V, A)$, k, and w. Viewing it as a CSP over label set $[k]$, the standard LP relaxation given in Figure 1 is defined over variables x_i^v for each vertex v and label i, and y_{ij}^e for each edge $e = (u, v)$ and labels i to u and j to v.

$$\max \sum_{e \in A} w(e) \cdot \sum_{\substack{i,j \in [k] \\ i < j}} y_{ij}^e \tag{1}$$

subject to,

$$\forall v \in V, \qquad\qquad\qquad \sum_{i \in [k]} x_i^v = 1. \tag{2}$$

$$\forall e = (u, v) \in A, \text{ and } i, j \in [k], \qquad \sum_{\ell \in [k]} y_{i\ell}^e = x_i^u, \tag{3}$$

$$\text{and,} \qquad\qquad\qquad \sum_{\ell \in [k]} y_{\ell j}^e = x_j^v. \tag{4}$$

$$\forall v \in V, \text{ and } i \in [k], \qquad\qquad x_i^v \geq 0. \tag{5}$$
$$\forall e \in A, \text{ and } i, j \in [k], \qquad\qquad y_{ij}^e \geq 0. \tag{6}$$

Fig. 1. LP Relaxation for instance $\mathcal{I}$ of MAX-k-ORDERING.

Sherali-Adams Constraints. Let $z_\sigma^S \in [0, 1]$ be a variable corresponding to a subset S of vertices, and a labeling $\sigma : S \to [k]$. The LP relaxation in Figure 1 can augmented with r rounds of Sherali-Adams constraints which are defined over the variables $\{z_\sigma^S \mid 1 \leq |S| \leq r+1\}$. The additional constraints are given in Figure 2. The Sherali-Adams variables define, for each subset S of at most $(r+1)$ vertices, a distribution over the possible labelings from $[k]$ to the vertices in S. The constraints given by Equation (7) ensure that these distributions are consistent across subsets. Additionally, Equations (9) and (10) ensure the consistency of these distributions with the variables of the standard LP relaxation given in Figure 1.

LP Relaxation for RMAS **and** OFFSETRMAS. The LP relaxation for RMAS is a generalization of the one in Figure 1 for MAX-k-ORDERING and we omit a

$\forall S \subseteq T \subseteq V,$
$1 \leq |S|, |T| \leq r + 1,$
and $\sigma : S \to [k],$
$$z_\sigma^S = \sum_{\substack{\rho : T \to [k] \\ \rho|_S = \sigma}} z_\rho^T. \qquad (7)$$

$\forall S \subseteq V, 1 \leq |S| \leq r + 1,$
and $\sigma : S \to [k],$
$$0 \leq z_\sigma^S \leq 1. \qquad (8)$$

$\forall v \in V,$ and $\sigma : \{v\} \to [k],$
s.t. $\sigma(v) = i,$
$$x_i^v = z_\sigma^{\{v\}}. \qquad (9)$$

$\forall e = (u, v) \in A,$ and,
$\sigma : \{u, v\} \to [k],$
s.t. $(\sigma(u), \sigma(v)) = (i, j),$
$$y_{ij}^e = z_\sigma^{\{u,v\}}. \qquad (10)$$

Fig. 2. r-round Sherali-Adams constraints for LP relaxation in Figure 1.

detailed definition. Let $\mathcal{S} = \cup_{v \in V} S_v$ denote the set of all labels. For convenience, we define variables $\{x_i^v \mid v \in V, i \in \mathcal{S}\}$ and $\{y_{ij}^e \mid e = (u, v) \in A, i, j \in \mathcal{S}\}$ and force the infeasible assignments to be zero, i.e. $x_i^v = 0$ for $i \notin S_v$. The other constraints are modified accordingly. For OFFSETRMAS, an additional change is that the contribution to the objective from each edge $e = (u, v)$ is $\sum_{i \in S_u, j \in S_v, i + o_e \leq j} y_{ij}^e$.

3   A 2-Approximation for MAX-k-ORDERING

This section proves the following theorem that implies Theorem 1.

Theorem 5. *Let $\{x_i^v\}, \{y_{ij}^e\}$ denote an optimal solution to the LP in Figure 1. Let $\ell : V \to [k]$ be a randomized labeling obtained by independently assigning to each vertex v label i with probability $1/2k + x_i^v/2$. Then, for any edge $e = (u, v)$,*

$$\Pr[\ell(u) < \ell(v)] \geq \frac{1}{2} \left(\sum_{\substack{i, j \in [k] \\ i < j}} y_{ij}^e \right).$$

To analyze the rounding given above, we need the following key lemma that bounds the sum of products of row and column sums of a matrix in terms of the matrix entries. It improves a weaker bound shown by Grandoni et al. [5] and also generalizes to arbitrary offsets.

Lemma 2. *Let* $\mathbb{A} = [a_{ij}]$ *be a* $k \times k$ *matrix with non-negative entries. Let* $r_i = \sum_j a_{ij}$ *and* $c_j = \sum_i a_{ij}$ *denote the sum of entries in the* i^{th} *row and* j^{th} *column respectively, and let* $1 \leq \theta \leq k - 1$ *be an integer offset . Then,*

$$\sum_{\substack{i+\theta \leq j \\ i,j \in [k]}} r_i c_j \geq \frac{k - \theta + 1}{2(k - \theta)} \left(\sum_{\substack{i+\theta \leq j \\ i,j \in [k]}} a_{ij} \right)^2 . \tag{11}$$

Proof. The LHS of the above is simplified as,

$$\sum_{i+\theta \leq j} r_i c_j = \sum_{i+\theta \leq j} \left[\left(\sum_{j'} a_{ij'} \right) \left(\sum_{i'} a_{i'j} \right) \right] \tag{12}$$

$$\geq \sum_{x+\theta \leq y} a_{xy}^2 + 2 \cdot \sum_{\substack{x+\theta \leq y \\ x+\theta \leq y' \\ y<y'}} a_{xy}a_{xy'} + \sum_{\substack{x+\theta \leq y \\ x'+\theta \leq y' \\ x<x'}} a_{xy}a_{x'y'}, \tag{13}$$

where all the indices above are in $[k]$. Note that (13) follows from (12) because:
(i) For any $x + \theta \leq y$, a_{xy}^2 appears in the RHS of (12) when $i = x$ and $j = y$.
(ii) For $x+\theta \leq y$ and $x+\theta \leq y'$, $a_{xy}a_{xy'}$ appears in the RHS of (12) both, when $i = x, j = y$, and when $i = x, j = y'$.
(iii) For any $x + \theta \leq y$ and $x' + \theta \leq y'$ (say $x < x'$), it must be that $x+\theta \leq y'$, and hence $a_{xy}a_{x'y'}$ appears in the RHS of (12) when $i = x$ and $j = y'$.
Thus, we obtain,

$$\sum_{i+\theta \leq j} r_i c_j \geq \left(\sum_{x+\theta \leq y} a_{xy} \right)^2 - \left(\sum_{\substack{x+\theta \leq y \\ x'+\theta \leq y' \\ x<x'}} a_{xy}a_{x'y'} \right) . \tag{14}$$

Therefore, it is sufficient to show that

$$\sum_{\substack{x+\theta \leq y \\ x'+\theta \leq y' \\ x<x'}} a_{xy}a_{x'y'} \leq \frac{k - \theta - 1}{2(k - \theta)} \left(\sum_{x+\theta \leq y} a_{xy} \right)^2 . \tag{15}$$

Substituting,

$$\left(\sum_{x+\theta \leq y} a_{xy} \right)^2 = \sum_{x+\theta \leq y} a_{xy}^2 + 2 \cdot \sum_{\substack{x+\theta \leq y \\ x+\theta \leq y' \\ y<y'}} a_{xy}a_{xy'} + 2 \cdot \sum_{\substack{x+\theta \leq y \\ x'+\theta \leq y' \\ x<x'}} a_{xy}a_{x'y'},$$

and simplifying, inequality (15) can be rewritten as,

$$\sum_{x+\theta\leq y} a_{xy}^2 + 2 \sum_{\substack{x+\theta\leq y \\ x+\theta\leq y' \\ y<y'}} a_{xy}a_{xy'} - \left(\frac{2}{k-\theta-1}\right)\cdot\sum_{\substack{x+\theta\leq y \\ x'+\theta\leq y' \\ x<x'}} a_{xy}a_{x'y'} \geq 0, \qquad (16)$$

$$\Leftrightarrow \bar{a}^\mathsf{T} M\bar{a} \geq 0, \qquad (17)$$

where $\bar{a} \in \mathbb{R}^{\mathcal{Z}}$, $\mathcal{Z} := \{(x,y) \mid x+\theta \leq y \text{ and } x,y \in [k]\}$ with $\bar{a}_{(x,y)} := a_{xy}$, and $M \in \mathbb{R}^{\mathcal{Z}\times\mathcal{Z}}$ is a symmetric matrix defined as follows:

$$M_{(x,y)(x',y')} = \begin{cases} 1 & \text{if } (x,y) = (x',y'), \\ 1 & \text{if } x' = x, \text{ and } y \neq y', \\ -1/(k-\theta-1) & \text{if } x \neq x'. \end{cases} \qquad (18)$$

To complete the proof of the lemma we show that M is positive semidefinite. Consider the set of unit vectors $\{v_x \mid 1 \leq x \leq k-\theta\}$ given by the normalized corner points of the $(k-\theta-1)$-dimensional simplex centered at the origin. It is easy to see (for e.g. in Lemma 3 of [4]) that, $\langle v_x, v_{x'}\rangle = -1/(k-\theta-1)$ if $x \neq x'$. Thus, $M = L^\mathsf{T}L$, where L is a matrix whose columns are indexed by $\mathcal{Z}$ such that the (x,y) column is v_x. Therefore, M is positive semidefinite. □

Proof (of Theorem 5). For brevity, let $z_e = \sum_{i<j} y_{ij}^e$ denote the contribution of the edge e to the LP objective. From the definition of the rounding procedure we have,

$$\Pr[\ell(u) < \ell(v)] = \sum_{i<j} \Pr[\ell(u) = i]\Pr[\ell(v) = j]$$

$$= \sum_{i<j}\left(\frac{1}{2k} + \frac{x_i^u}{2}\right)\left(\frac{1}{2k} + \frac{x_j^v}{2}\right)$$

$$= \frac{1}{4}\left(\frac{(k-1)}{2k} + \frac{1}{k}\sum_{i<j}(x_i^u + x_j^v) + \sum_{i<j}x_i^u x_j^v\right)$$

We can now apply Lemma 2 to the $k \times k$ matrix $[y_{ij}^e]$. The LP constraints guarantee that $r_i = x_i^u$ and $c_j = x_j^v$ are equal to the row and column sums respectively. Further, substituting offset $\theta = 1$, we obtain

$$\Pr[\ell(u) < \ell(v)] \geq \frac{1}{4}\left(\frac{(k-1)}{2k} + \frac{1}{k}\sum_{i<j}(x_i^u + x_j^v) + \frac{k}{2(k-1)}z_e^2\right). \qquad (19)$$

On the other hand,

$$\sum_{i<j}(x_i^u + x_j^v) = \sum_{i=1}^{k-1}(k-i)x_i^u + \sum_{j=2}^{k}(j-1)x_j^v$$

$$\geq \sum_{i=1}^{k-1}\left[(k-i)\sum_{j'>i}y_{ij'}^e\right] + \sum_{j=2}^{k}\left[(j-1)\sum_{i'<j}y_{i'j}^e\right]. \tag{20}$$

For $a < b$, y_{ab}^e appears $(k-a)$ times in the RHS of the above inequality when $i = a$, and $(b-1)$ times when $j = b$. Since $k - a + b - 1 \geq k$, we obtain that RHS of Equation (20) is lower bounded by $k\sum_{a<b}y_{ab}^e = kz_e$. Substituting back into Equation (19) and simplifying gives us that $\Pr[\ell(u) < \ell(v)]$ is at least,

$$\frac{z_e}{4}\left[1 + \frac{1}{2}\left(\frac{(k-1)}{kz_e} + \frac{kz_e}{(k-1)}\right)\right] \geq \frac{z_e}{4}(1+1) = \frac{z_e}{2}, \tag{21}$$

where we use $t + 1/t \geq 2$ for $t > 0$. $\qquad\square$

4   Approximation for OFFSETRMAS

Let $D = (V, A)$, $\{S_v\}_{v\in V}$, w, and $\{o_e\}_{e\in A}$ constitute an instance of OFFSETR-MAS as given in Definition 2. Without loss of generality, one can assume that for each edge $e = (u, v) \in A$, $\min(S_u) + o_e \leq \max(S_v)$, otherwise no feasible solution can satisfy e and that edge can be removed. A simple randomized strategy that independently assigns each vertex v either $\ell_{min}^v := \min(S_v)$ or $\ell_{max}^v := \max(S_v)$ with equal probability is a 4-approximation. The recent work of Grandoni et al. [5] show that combining this randomized scheme with an appropriate LP-rounding yields a $2\sqrt{2} \approx 2.828$ approximation algorithm for RMAS.

We show that a variant of the rounding scheme developed in Section 3 yields an improved approximation factor for OFFSETRMAS. In particular, we prove the following theorem which implies Theorem 2.

Theorem 6. *Let* $\{x_i^v\}$, $\{y_{ij}^e\}$ *denote an optimal solution to the linear programming relaxation of* OFFSETRMAS *described in Section 2. Let* ℓ *be a randomized labeling obtained by independently assigning labels to each vertex* v *with the following probabilities:*

$$\Pr[\ell(v) = i] = \begin{cases} \frac{1}{4} + \frac{x_i^v}{2} & \text{if } i \in \{\ell_{min}^v, \ell_{max}^v\} \\ \frac{x_i^v}{2} & \text{if } i \in S_v \setminus \{\ell_{min}^v, \ell_{max}^v\} \end{cases} \tag{22}$$

Then, for any edge $e = (u, v)$ *we have*

$$\Pr[\ell(u) + o_e \leq \ell(v)] \geq \frac{1}{4}\left(1 + \frac{1}{\sqrt{2}}\right)\left(\sum_{\substack{i\in S_u, j\in S_v \\ i+o_e \leq j}} y_{ij}^e\right).$$

Proof. Let $\mathcal{S} = \cup_{v \in V} S_v$ denote the set of all labels and let $z_e = \left(\sum_{i+o_e \leq j} y_{ij}^e\right)$ denote the contribution of the edge e to the LP objective. We have,

$$\Pr[\ell(u) + o_e \leq \ell(v)] = \sum_{\substack{i+o_e \leq j \\ i \in S_u, j \in S_v}} \Pr[\ell(u) = i] \Pr[\ell(v) = j]$$

Substituting the assignment probabilities from (22) into the above and simplifying we obtain,

$$\Pr[\ell(u) + o_e \leq \ell(v)]$$

$$= \frac{1}{16} + \frac{1}{8} \left(\sum_{\substack{i \leq \ell_{max}^v - o_e \\ i \in \mathcal{S}}} x_i^u + \sum_{\substack{j \geq \ell_{min}^u + o_e \\ j \in \mathcal{S}}} x_j^v \right) + \frac{1}{4} \left(\sum_{\substack{i+o_e \leq j \\ i,j \in \mathcal{S}}} x_i^u x_j^v \right) \qquad (23)$$

Note that we allow $i, j \in \mathcal{S}$ in the above sums instead of S_u and S_v. This does not affect the analysis as the LP forces $x_i^u = 0$ for $i \notin S_u$ and similarly for v. Now, consider the $|\mathcal{S}| \times |\mathcal{S}|$ matrix $[y_{ij}^e]$. Since x_i^u and x_j^v are equal to the row sums and column sums of this matrix respectively, Lemma 2 guarantees that,

$$\sum_{\substack{i+o_e \leq j \\ i,j \in \mathcal{S}}} x_i^u x_j^v \geq \frac{|\mathcal{S}| - o_e + 1}{2(|\mathcal{S}| - o_e)} \left(\sum_{\substack{i \in S_u, j \in S_v \\ i+o_e \leq j}} y_{ij}^e \right)^2 \geq \frac{(|\mathcal{S}| - o_e + 1)}{2(|\mathcal{S}| - o_e)} z_e^2 \geq \frac{z_e^2}{2}.$$

We thus have,

$$\Pr[\ell(u) + o_e \leq \ell(v)]$$

$$\geq \frac{1}{16} + \frac{1}{8} \left(\sum_{i \leq \ell_{max}^v - o_e} x_i^u + \sum_{j \geq \ell_{min}^u + o_e} x_j^v \right) + \frac{z_e^2}{8}$$

$$\geq \frac{1}{16} + \frac{1}{8} \left(\sum_{i \leq \ell_{max}^v - o_e} \left(\sum_{j \geq i+o_e} y_{i,j}^e \right) + \sum_{j \geq \ell_{min}^u + o_e} \left(\sum_{i+o_e \leq j} y_{i,j}^e \right) \right) + \frac{z_e^2}{8}$$

$$= \frac{1}{16} + \frac{1}{8} \left(2 \sum_{\substack{i+o_e \leq j \\ i \in S_u, j \in S_v}} y_{i,j}^e \right) + \frac{z_e^2}{8}$$

$$\geq \frac{1}{16} + \frac{z_e}{4} + \frac{z_e^2}{8}$$

$$= \frac{z_e}{4} \left(1 + \frac{1}{2} \left(\frac{1}{2z_e} + z_e \right) \right) \geq \frac{z_e}{4} \left(1 + \frac{1}{\sqrt{2}} \right), \qquad (24)$$

where the last inequality uses $t + 1/at \geq 2/\sqrt{a}$ for $a, t > 0$. $\qquad \square$

References

[1] Arora, S., Bollobás, B., Lovász, L.: Proving integrality gaps without knowing the linear program. In: Proc. FOCS, pp. 313–313 (2002)

[2] Charikar, M., Makarychev, K., Makarychev, Y.: Integrality gaps for Sherali-Adams relaxations. In: Proc. STOC, pp. 283–292 (2009)

[3] Feige, U., Goemans, M.X.: Approximating the value of two prover proof systems, with applications to MAX 2SAT and MAX DICUT. In: Proc. ISTCS, pp. 182–189 (1995)

[4] Frieze, A., Jerrum, M.: Improved approximation algorithms for MAX k-CUT and MAX BISECTION. Algorithmica 18(1), 67–81 (1997)

[5] Grandoni, F., Kociumaka, T., Włodarczyk, M.: An LP-rounding-approximation for restricted maximum acyclic subgraph. Information Processing Letters 115(2), 182–185 (2015)

[6] Guruswami, V., Manokaran, R., Raghavendra, P.: Beating the random ordering is hard: Inapproximability of maximum acyclic subgraph. In: Proc. FOCS, pp. 573–582 (2008)

[7] Håstad, J.: Some optimal inapproximability results. JACM 48(4), 798–859 (2001)

[8] Kenkre, S., Pandit, V., Purohit, M.: Saket R. On the approximability of digraph ordering. CoRR (2015). http://arxiv.org/abs/1507.00662

[9] Khandekar, R., Kimbrel, T., Makarychev, K., Sviridenko, M.: On hardness of pricing items for single-minded bidders. In: Dinur, I., Jansen, K., Naor, J., Rolim, J. (eds.) APPROX and RANDOM 2009. LNCS, vol. 5687, pp. 202–216. Springer, Heidelberg (2009)

[10] Khot, S.: On the power of unique 2-prover 1-round games. In: Proc. STOC, pp. 767–775 (2002)

[11] Khot, S., Kindler, G., Mossel, E., O'Donnell, R.: Optimal inapproximability results for MAX-CUT and other 2-variable CSPs? SIAM Journal on Computing 37(1), 319–357 (2007)

[12] Lee, E.: Hardness of graph pricing through generalized max-dicut. In: Proc. STOC, pp. 391–399 (2015)

[13] Lewin, M., Livnat, D., Zwick, U.: Improved rounding techniques for the MAX 2-SAT and MAX DI-CUT problems. In: Proc. IPCO, pp. 67–82 (2002)

[14] Matuura, S., Matsui, T.: 0.863-approximation algorithm for MAX DICUT. In: Goemans, M.X., Jansen, K., Rolim, J.D.P., Trevisan, L. (eds.) RANDOM 2001 and APPROX 2001. LNCS, vol. 2129, pp. 138–146. Springer, Heidelberg (2001)

[15] Newman, A.: Approximating the maximum acyclic subgraph. PhD thesis, Massachusetts Institute of Technology (2000)

[16] Paik, D., Reddy, S., Sahni, S.: Deleting vertices to bound path length. IEEE Transactions on Computers 43(9), 1091–1096 (1994)

[17] Svensson, O.: Hardness of vertex deletion and project scheduling. In: Gupta, A., Jansen, K., Rolim, J., Servedio, R. (eds.) APPROX 2012 and RANDOM 2012. LNCS, vol. 7408, pp. 301–312. Springer, Heidelberg (2012)

[18] Trevisan, L.: Parallel approximation algorithms by positive linear programming. Algorithmica 21(1), 72–88 (1998)

[19] Zwick, U.: Analyzing the MAX 2-SAT and MAX DI-CUT approximation algorithms of Feige and Goemans. Manuscript (2000)

Welfare Maximization with Deferred Acceptance Auctions in Reallocation Problems

Anthony Kim*

Department of Computer Science, Stanford University, USA
tonyekim@stanford.edu

Abstract. We design approximate weakly group strategy-proof mechanisms for resource reallocation problems using Milgrom and Segal's deferred acceptance auction framework: the radio spectrum and network bandwidth reallocation problems in the procurement auction setting and the cost minimization problem with set cover constraints in the selling auction setting. Our deferred acceptance auctions are derived from simple greedy algorithms for the underlying optimization problems and guarantee approximately optimal social welfare (cost) of the agents retaining their rights (contracts). In the reallocation problems, we design procurement auctions to purchase agents' broadcast/access rights to free up some of the resources such that the unpurchased rights can still be exercised with respect to the remaining resources. In the cost minimization problem, we design a selling auction to sell early termination rights to agents with existing contracts such that some minimal constraints are still satisfied with remaining contracts. In these problems, while the "allocated" agents transact, exchanging rights and payments, the objective and feasibility constraints are on the "rejected" agents.

1 Introduction

Motivated by the US government's effort to reallocate channels currently allocated for television broadcasting for wireless broadband services, Milgrom and Segal [9] introduced a class of mechanisms called deferred acceptance (DA) auctions for resource reallocation problems. DA auctions greedily choose an allocation by iteratively rejecting the least attractive bid determined by some scoring functions and can be implemented with adaptive reverse greedy algorithms. Milgrom and Segal showed that DA auctions satisfy several important properties: they are strategyproof, weakly group (WG) strategy-proof, can be implemented using ascending clock auctions, and lead to the same outcomes as the complete-information Nash equilibria of corresponding paid-as-bid auctions.

Subsequently, Dütting et al. [3] studied the strengths and limitations of DA auctions with respect to achievable approximation guarantees on social welfare in two selling auction design problems. For the knapsack auction problem, they showed a separation between approximation guarantees by DA auctions and by

---

* Supported in part by an NSF Graduate Research Fellowship.

© Springer-Verlag Berlin Heidelberg 2015
N. Bansal and I. Finocchi (Eds.): ESA 2015, LNCS 9294, pp. 804–815, 2015.
DOI: 10.1007/978-3-662-48350-3_67

more general strategyproof mechanisms. For the combinatorial auction problem with single-minded bidders, they designed an $O(d)$-approximate DA auction when bidders' desired bundles' sizes are at most d and an $O(\sqrt{m \log m})$-approximate DA auction where m is the number of items. In a different work, Dütting et al. [4] studied double auctions for settings in which unit-demand buyers and unit-supply sellers must be matched one-to-one subject to certain constraints. In particular, they showed WG-strategy-proof DA double auctions can be designed by composing two greedy algorithms, one for each side, that use DA rules.

In this paper, we further develop connections between DA auctions and greedy algorithms in the context of the resource reallocation problems that motivated Milgrom and Segal [9]'s DA auction framework. More specifically, we show that several simple greedy approximation algorithms lead to DA auctions with the same approximation guarantees. We consider welfare maximization in the radio spectrum and network bandwidth reallocation problems and the cost minimization problem with set cover constraints.

In the radio spectrum reallocation problem, we (the government) want to reallocate channels currently allocated for television broadcasting for wireless broadband services, effectively reducing the number of channels available for broadcasting, by buying the television stations' broadcasting rights. The reallocation process involves purchasing some of the rights and reassigning the remaining stations with rights into a smaller set of channels; the cleared spectrum will, then, be used for wireless broadband services. The reassignment should be accomplished in a way that respects constraints stipulating that two interfering neighboring stations are not assigned to the same channel. Assuming the stations bid their values of keeping broadcast rights, we want to maximize the social welfare of the stations to keep their rights.

Similarly, in the network bandwidth reallocation problem, we (the network operator) want to reallocate some network connections currently in use for other purposes. We want to buy access rights of some firms using the network such that the demands of firms still with their rights can be served in a smaller network, with the goal of maximizing the social welfare of these latter firms.

In addition, we consider the cost minimization problem with set cover constraints in the selling auction setting. The bidders are looking to terminate their contracts, by paying penalties if necessary, and we (the government) want to agree to such requests while ensuring that some minimal constraints, modeled by the well-known set cover problem, are satisfied. We sell early termination rights to these bidders with the goal of minimizing the social cost, i.e., the total bid value, of those bidders whose requests are not honored.

Our Contributions Using Milgrom and Segal [9]'s DA auction framework, we design approximate DA auctions for reallocation problems, in which the "allocated" agents transact, exchanging rights and payments, while the objective and feasibility constraints are on the "rejected" agents. We show simple forward greedy algorithms, not the reverse kind, are sufficient to derive DA auctions that guarantee approximately optimal social welfare (cost) of agents to retain their rights (contracts). Our DA auctions are computationally efficient and can be computable in polynomial time. More interestingly, the scoring functions of the auctions are algorithmic in nature and might not be expressible in a closed form; they use helper variables to track the progress of allocation.

In the radio spectrum reallocation problem, we design DA auctions that are approximately optimal for certain interference graphs: interval graphs, disk graphs, and bounded degree-d graphs. Unlike the near-optimality result in [9] that relies on the existence of a non-trivial partitioning scheme, our approximation ratios depend solely on simple graph parameters. Disk graphs, in particular, are a natural modeling representation of the interference graphs of the stations' circular broadcast ranges. For the network bandwidth reallocation problem, we design an approximate DA auction with scoring functions derived from a result due to Briest et al. [2]. For the cost minimization problem with set cover constraints, we show that a well-known primal-dual greedy approximation algorithm can be implemented by a DA auction.

Future Work. It would be interesting to investigate whether or not we can further improve the results in this paper. For instance, in the radio spectrum reallocation problem, a better performing DA auction with more complex scoring functions might be possible on different kinds of interference graphs. More generally, it would be interesting to have black box results formalizing the conditions under which greedy algorithms can be implemented as DA auctions and comparison-type results comparing DA auctions to other kinds of auctions with different incentive properties in terms of social welfare/cost.

Other Related Work. Mechanisms derived from greedy algorithms have been studied previously for other various mechanism design problems, but not within the DA auction framework and without consideration of WG strategy-proofness. We refer to these work and the references therein: Lehmann et al. [7], Borodin and Lucier [8, 1], and Briest at al. [2].

Due to space constraints, we refer to the long version of this paper [6] for more details.

2   Preliminaries

We consider procurement auctions in the resource reallocation problems; the setting for selling auctions in the cost minimization problem is equivalent but inverted.

Let N be the set of bidders. Each bidder submits a bid and the auction decides on the allocation and payments such that the bidders either "win" or "lose", i.e., his bid to supply an item is accepted or not, and only the winning bidders receive payments. We assume that bids $b = (b_1, \ldots, b_{|N|})$ are from the bid profile space $B = B_1 \times \cdots \times B_{|N|}$, and that each bidder i's value v_i is in the range $[0, \bar{v}_i]$ and

Algorithm 1. DA Auction with Scoring Functions $\{s_i^A\}_{A \subseteq N, i \in A}$

1: Accept bids $b_1, \ldots, b_{|N|}$
2: $A = N$
3: **while** $\exists i \in A : s_i^A(b_i, b_{N \setminus A}) > 0$ **do**
4:      $i = \mathrm{argmax}_{i \in A}\, s_i^A(b_i, b_{N \setminus A})$
5:      $A = A \setminus \{i\}$
6: **end while**
7: **return** A

his bid b_i is restricted to a finite set B_i such that $\max B_i > \bar{v}_i$.[1] A procurement auction has allocation rule $a : B \to 2^N$ and payment rule $p : B \to \mathbb{R}^{|N|}$ such that $p_i(b) = 0$ for $i \in N \setminus a(b)$. In settings we consider, there will be constraints on the possible allocations allowed. Note each bidder is a strategic agent that seeks to maximize his utility (or payoff) which is $p_i(b) - v_i$ if he wins, and 0 otherwise.

We define important properties of auctions: An auction is *strategy-proof* if for every bidder i, $v_i \in [0, \bar{v}_i]$, and other bids $b_{-i} \in B_{-i}$, it is optimal for the bidder to truthfully bid his value $v_i^+ := \min\{b_i \in B_i : b_i > v_i\}$. An auction is *weakly group (WG) strategy-proof* if for every profile of values v, a set of bidders $S \subseteq N$, and coordinated false bids by these deviating bidders, there exists at least one bidder in S who does not get a strictly better payoff than under truthful reporting. An auction is α-*approximate* if it achieves at least the α fraction of the optimal welfare in all problem instances.

In reallocation problems, we want to design a procurement auction to purchase agents' broadcast/access rights to free up some of the resources such that the unpurchased rights can still be exercised with respect to the remaining resources. Agents report their private values of rights and receive payments when their rights are purchased. Our objective is to maximize the social welfare of agents retaining their rights. Note that the "allocated" agents transact, exchanging rights and payments, but the objective and feasibility constraints are on the "rejected" agents in these auctions.[2]

Deferred Acceptance Auctions Deferred acceptance (DA) auctions are auctions in which the allocation is determined by an iterative process of rejecting bidders one by one from the whole set. Milgrom and Segal [9] showed they satisfy strong incentive guarantees such as strategy-proofness, WG-strategy-proofness, and other useful implementation properties. Note the well-studied Vickrey auction is known to be not WG-strategy-proof.

DA auctions can be described as reverse greedy algorithms that use certain scoring functions $\{s_i^A\}_{A \subseteq N, i \in A}$ as shown in Algorithm 1. Bidders are removed from A one at a time until termination of the while-loop and the resulting A is the allocation. Note the scoring functions $s_i^A : B_i \times B_{N \setminus A} \to \mathbb{R}^+$ are nondecreasing in the first argument. The bidder i's score during an iteration with the current

---

[1] The restriction of finite bid spaces can be removed.
[2] In more common auction settings, the objective and feasibility constraints are on the "allocated" agents.

set of active bidders A is dependent on its own bid b_i and bids of those inactive bidders in $N \setminus A$, but not on the bids of other active bidders. For the allocation rule a determined by Algorithm 1, the corresponding payment rule p is defined:

$$p_i(b_{-i}) = \max\{b_i' \in B_i : i \in a(b_i', b_{-i})\} \ , \qquad (1)$$

i.e., the winning bidder i's payment is the maximum bid value with which he remains allocated. For more details, see [9].

We study greedy algorithms with allocation rules implementable by DA auctions. These algorithms include the standard greedy-by-weight algorithms that process elements in order of decreasing weight and several primal-dual greedy algorithms that utilize variables from the linear programming formulations of the underlying problems. In this paper, we are primarily interested in the greedy algorithms of the "single pass" nature, i.e., those that start with an empty (and feasible) solution and iteratively augment it without any post-processing steps that might undo some part of the solution.

We formalize a notion to capture which greedy algorithms can be implemented as DA auctions. Note we have an analogous definition for reverse greedy algorithms that instead return A as the final solution[3], but the following version for forward greedy algorithms is sufficient for the reallocation problems:

Definition 1. *Let N be the element set and $w : N \to \mathbb{R}^+$ be a weight function. A greedy algorithm* ALG *is DA-implementable if it can be implemented with active set A which is initialized to N and scoring functions $s_i^A : B_i \times B_{N \setminus A} \to \mathbb{R}^+$ for each $A \subseteq N, i \in A$ such that:*

1. *In each iteration, element i to be selected and removed from A is the highest scoring element $\mathrm{argmax}_{i \in A} \, s_i^A(w_i, w_{N \setminus A})$;*
2. *When $s_i^A(w_i, w_{N \setminus A}) = 0$ for all $i \in A$,* ALG *terminates and returns $N \setminus A$.*

Notations We use the common notation $-i$ to denote the bidders other than bidder i. We use $x(i)$ and x_i interchangeably to indicate the i-th component value (representing the i-th bidder, the i-th edge, etc.) for any vector variable x. Without loss of generality, we assume argmax and argmin operators return the lowest indexed argument according to a consistent global order in the case of a tie.

3   Radio Spectrum Reallocation

We design a simple DA auction that achieves near-optimal social welfare for certain classes of interference graphs in the spectrum reallocation problem. Without relying on the assumption of a partitioning scheme in Milgrom and Segal [9], we show that our DA auction achieves an approximation ratio dependent only on

---

[3] This is for the more common auction setting described in Footnote 2.

simple structural parameters of the interference graphs.[4] For our analysis, we proceed with a series of reductions from the spectrum reallocation problem to the maximum weight k-colorable subgraph problem to the submodular function maximization and maximum weight independent set problems.

The spectrum reallocation problem is as follows: Let $G = (V, E)$ be the interference graph where V is the vertex set representing the television stations and E is the edge set where there is an edge between two vertices if the corresponding stations interfere were they to be assigned to the same channel; we use N and V interchangeably. Let k be the number of channels available for reassignment. Given stations' bids $b_1, b_2, \ldots$, we want to find a set A of stations to allocate, or buy their broadcast rights, to maximize the welfare of those retaining their rights such that the subgraph $G(V \setminus A, E)$ induced by them is k-colorable, i.e., colorable with k colors. We interpret the bids to be the stations' reported values of retaining the broadcast rights.

We consider the following classes of G: interval graphs, disk graphs, and bounded degree-d graphs. The interval graphs and disk graphs are intersection graphs with a geometric representation; in interval (disk) graphs, the vertices represent line-segments (disks) in a 1D-space (2D-space) and an edge exists between two vertices if their corresponding representations intersect or even just touch. In bounded degree-d graphs, the degree of every vertex is at most d.

3.1   Main Results

We reduce the spectrum reallocation problem to the NP-hard problem of finding the maximum weight k-colorable subgraph of G: Given $G(V, E)$, k and a weight function $w : V \to \mathbb{R}^+$, we want to find the maximum weight subgraph $V' \subset V$ such that $G(V', E)$ is k-colorable. As the weight function is derived from the bids $b_1, b_2, \ldots$, we have the same objective in both problems. In particular, if there is a DA-implementable approximate greedy algorithm for the maximum weight k-colorable subgraph problem, we get a DA auction with the same performance guarantee on social welfare for the spectrum reallocation problem.

In fact, Algorithm 2 is a DA-implementable greedy algorithm with good approximation guarantees:

Theorem 1. *Algorithm 2 is a DA-implementable $(1 - e^{-1/\alpha})$-approximation algorithm for the maximum weight k-colorable subgraph problem where α is $2 + \gamma$ for interval graphs, $(2 + \gamma)^2$ for disk graphs, and d for bounded degree-d graphs, for $\gamma = l_{\max}/l_{\min}$, i.e., the ratio between the maximum and minimum lengths (radii) for interval (disk) graphs.*

Proof. (DA-implementability) Note that Algorithm 2 upon termination returns a subgraph along with a valid k-coloring given by the independent sets $I_1, \ldots, I_k$.

---

[4] Milgrom and Segal's result assumes the existence of a ordered partition of N into m disjoint sets $N_1, \ldots, N_m$ such that: (1) the edge (i, j) exists for each $i, j \in N_k, 1 \leq k \leq m$; and (2) there exists some $d < n$ such that $|S| + |\cap_{i \in S} \cup_{l < k} \{j \in N_l : (i, j) \in E\}| \leq n$ for each $1 \leq k \leq m$ and $S \subseteq N_k$ with $|S| \leq n - d$.

Algorithm 2. Greedy Algorithm for the Max Weight k-colorable Subgraph Prob.

1: $I_1 = \cdots = I_k = \emptyset$
2: **for** $v \in V$ in decreasing order of weight **do**
3:    **if** $\exists\, i : I_i \cup \{v\}$ remains independent **then**
4:       $i = \min\{i : I_i \cup \{v\}$ remains independent$\}$
5:       $I_i = I_i \cup \{v\}$
6:    **end if**
7: **end for**
8: **return** $\bigcup_i I_i$

We show its implementation with the active set and scoring functions as follows: Let the active set A be the vertices not yet selected and $I_1^{N \setminus A}, \ldots, I_k^{N \setminus A}$ be the associated independent sets consisting of vertices in $N \setminus A$. We define the scoring functions:

$$s_v^A(w_v, w_{N \setminus A}) = \begin{cases} w(v), & \text{if } \exists\, i : I_i^{N \setminus A} \cup \{v\} \text{ is independent} \\ 0, & \text{otherwise} \end{cases} \tag{2}$$

Note that each scoring function s_v^A is nondecreasing in the first argument, and that the next element v to be greedily selected into $N \setminus A$ (and out of A) is the highest scoring element $\operatorname{argmax}_v s_v^A$.

(Approximation) We defer this part of the proof to Section 3.2.

Hence, our main result follows. In the special cases of the unit-interval and unit-disk graphs (i.e., $\gamma = 1$), we get constant approximations:

Corollary 1. *The DA auction with scoring functions* (2) *is a* $(1 - e^{-1/\alpha})$-*approximate WG-strategy-proof mechanism for the spectrum reallocation problem, where* α *is* $2 + \gamma$ *for interval graphs,* $(2 + \gamma)^2$ *for disk graphs, and* d *for bounded degree-d graphs; and* $\gamma = l_{\max}/l_{\min}$.

3.2 Proof of Theorem 1 (Approximation)

We show that Algorithm 2 achieves the stated approximation ratios for the interval graphs, disk graphs, and bounded degree-d graphs by analyzing a related algorithm, Algorithm 3, which will be shown to be equivalent for a choice of ALG. We further reduce the maximum weight k-colorable subgraph problem to a monotone submodular function maximization problem. For a short review of submodular functions, we refer to Appendix A in [6].

Let M be the set of all independent sets of G and $f : 2^M \to \mathbb{R}$ be a set function defined as $f(S) = \sum_{v \in \cup_{I \in S} I} w(v)$. Note that $f(\emptyset) = 0$ and f is a monotone submodular function. Since we want to find k independent sets $I_1, \ldots, I_k$ (for k colors) such that the total weight of vertices covered by them is maximized, the maximum weight k-colorable subgraph problem is equivalent to the

---

Algorithm 3. A Greedy Algorithm with Subroutine ALG

1: $V' = V$
2: **for** $i = 1, \ldots, k$ **do**
3:     Run ALG on $G(V', E)$ and get an (approx.) max weight independent set I_i
4:     $V' = V' \setminus I_i$
5: **end for**
6: **return** $\bigcup_i I_i$

---

maximization problem of

$$\max_{S \subseteq M : |S| \leq k} f(S) \ . \tag{3}$$

Note the k independent sets should be disjoint to be a valid coloring, but as any k independent sets can be modified to be disjoint, the above maximization problem still gives the same optimal value.

The well-known greedy algorithm due to Nemhauser et al. [10], that is Algorithm 3 with the optimal ALG that returns the maximum weight independent set, is a $(1 - e^{-1})$-approximation algorithm. However, it is not computationally efficient as the maximum weight independent set problem is difficult for general graphs; even the unweighted version is known to be NP-hard and cannot be approximated in polynomial time within a factor of $|V|^{1-\epsilon}$ for any fixed $\epsilon > 0$, unless P = NP [5].

Instead, we use the following lemma (with its proof in Appendix A in [6]) to show that Algorithm 3 with an approximation algorithm ALG has a similar approximation guarantee, and design a computationally efficient ALG:

Lemma 1. *Algorithm 3 with an α-approximation algorithm ALG is a $(1 - e^{-1/\alpha})$-approximation algorithm for the maximization problem* (3).

For the classes of interference graphs in consideration, the following ALG is a polynomial time approximation algorithm:

ALG := Given graph G, select vertices in decreasing order of weight

as long as those selected form an independent set.

Lemma 2. *Algorithm ALG is an α-approximation algorithm for the maximum weight independent set problem, where α is $2 + \gamma$ for interval graphs, $(2 + \gamma)^2$ for disk graphs, and d for bounded degree-d graphs; and $\gamma = l_{max}/l_{min}$.*

Proof. (Sketch) Note the graphs are bounded claw-free graphs; a graph is $(\tau+1)$-*claw free* if each vertex has at most τ mutually non-adjacent vertices. More generally, ALG is a τ-approximation algorithm on $(\tau + 1)$-claw free graphs. For each vertex selected by the optimal algorithm, ALG either selects it or does not select it in favor of another. For each vertex selected by ALG, it can be involved in at most τ such instances of the latter case. For details, see Appendix B in [6].

Finally, note that Algorithm 3 is equivalent to Algorithm 2; we construct the independent sets one by one in the former and all at once in parallel in the latter. By Lemmas 1 and 2, Theorem 1 follows.

4   Network Bandwidth Reallocation

In this section, we show an approximate DA auction for the network bandwidth reallocation problem. We reduce the problem to the optimization problem of network unicast/multicast routing and show a primal-dual greedy algorithm due to Briest et al. [2] is DA-implementable.

The network bandwidth reallocation problem for unicast routing is defined as follows: Let $G = (V, E)$ be the network graph with $|V| = n$, $|E| = m$, and edge capacities $c(e), \forall e \in E$. Let N be the firms (the bidders) with access rights such that each firm i has a terminal pair (s_i, t_i) and demand $d(i)$ and private value $v(i)$ for his right. Without loss of generality, we assume that $d(i) \in [0, 1], \forall i$; $c(e) \geq 1, \forall e$; and $C := \min_e c(e) > 1$. Given the reports of firms' values for their rights $b_1, b_2, \ldots$, we buy access rights of some firms such that the demands of those still retaining rights can be satisfied in the network, i.e., there is a feasible solution that routes each unsplittable flow of the demanded amount between terminals subject to the edge capacity constraints. We want to maximize the social welfare of those still holding rights. Note G is the reduced smaller network after removing reallocated edges.

In the multicast routing version of the problem, each firm has a set of terminal vertices, with one being the source, and demands a (unsplittable) multicast tree (a.k.a., Steiner tree) spanning the terminals.

For the corresponding optimization problem of network routing, we know the values $v(i), \forall i \in N$ and want to compute a subset of firms with the maximum total value such that their demands can be satisfied. Note that the objectives of both mechanism design and algorithmic problems are on the same set of firms.

Algorithm 4, due to Briest et al. [2], is a polynomial time greedy algorithm based on the primal-dual scheme (see Appendix C in [6] for the primal-dual linear programming relaxations). Let $\mathcal{S}_i$ be the set of all paths from s_i to t_i in G and $\mathcal{S} = \bigcup_i \mathcal{S}_i$; in the multicast routing case, $\mathcal{S}_i$ is the set of all Steiner trees spanning the firm i's terminals. Given $S \in \mathcal{S}_i$, let $q_S(e) - d(i)$ if $e \in S$, and $q_S(e) = 0$ otherwise. Note $\bar{e} \approx 2.718$ is the Euler number. In Line 3, we need to compute the shortest path with respect to the dual variables y in the unicast routing case and the minimum weight Steiner tree in the multicast routing case. We can compute the shortest path exactly using any shortest path algorithm. For the NP-hard Steiner tree problem, we use the polynomial time 1.55-approximation algorithm due to Robins and Zelikovsky [11].

We show that Algorithm 4 is DA-implementable and obtain a DA auction for the network bandwidth reallocation problem:

Theorem 2. *Algorithm 4 is a DA-implementable $\left(\frac{\bar{e}\gamma C}{C-1} m^{1/(C-1)} \right)$-approximation algorithm for the network bandwidth reallocation problem where $\gamma = 1$ for unicast routing and $\gamma = 1.55$ for multicast routing.*

Proof. (DA-implementability) We show an implementation with an active set and scoring functions as follows: Let the active set A be the set of firms not yet selected, so N' in Algorithm 4, and $\{y_e^{N \setminus A}\}_{e \in E}$ be the associated dual variables.

Algorithm 4. Greedy Algorithm for the Network Routing Problem

---
1: $\mathcal{T} = \emptyset$; $N' = N$; $y(e) = 1/c(e), \forall e \in E$
2: **while** $N' \neq \emptyset$ and $\sum_{e \in E} c(e)y(e) < \bar{e}^{C-1}m$ **do**
3:     $S_i = \text{argmin}_{S \in \mathcal{S}_i} \sum_{e \in S} y(e), \forall i \in N'$
4:     $i = \text{argmax}_{i \in N'} \left\{ \frac{v(i)}{d(i) \cdot \sum_{e \in S_i} y(e)} \right\}$
5:     $\mathcal{T} = \mathcal{T} \cup \{S_i\}$; $N' = N' \setminus \{i\}$
6:     Update $y(e) = y(e) \cdot \left(\bar{e}^{C-1}m \right)^{q_{S_i}(e)/(c(e)-1)}, \forall e \in S_i$
7: **end while**
8: **return** $\mathcal{T}$

We define the scoring functions:

$$s_i^A(v_i, v_{N \setminus A}) = \frac{v(i)}{d(i) \sum_{e \in S_i} y_e^{N \setminus A}}, \text{ where } S_i = \text{argmin}_{S \in \mathcal{S}_i} \sum_{e \in S} y_e^{N \setminus A}. \quad (4)$$

Note A changes when a firm is selected in Lines 4-5 and the scoring functions change correspondingly when the dual variables y_e change. The scoring functions are nondecreasing in the first argument and functions of attributes of firms in $N \setminus A$. Also, the next firm to be added to $N \setminus A$ is the highest scoring firm.

(Approximation) We refer this part of the proof to Briest et al. [2].

Corollary 2. *The DA auction with scoring functions (4) is a $\left(\frac{\bar{e}\gamma C}{C-1} m^{1/(C-1)} \right)$- approximate WG-strategy-proof mechanism for the network bandwidth realloca- tion problem, where $\gamma = 1$ for unicast routing and $\gamma = 1.55$ for multicast routing.*

5   Cost Minimization with Set Cover Constraints

We apply our approach to the cost minimization problem with set cover con- straints in the selling auction setting.

In the selling auction setting, the DA auction (Algorithm 1) is "inverted": the while-loop's stopping condition becomes $\exists i \in A : s_i^A(b_i, b_{N \setminus A}) < \infty$ (so, it terminates when all the scores are ∞); the next agent to be rejected in Lines 4-5 becomes the lowest scoring agent $\text{argmin}_{i \in A} s_i^A(b_i, b_{N \setminus A})$; and the payment rule (1) becomes $p_i(b_{-i}) = \min\{b_i' \in B_i : i \in a(b_i', b_{-i})\}$. Similarly, each bidder wants to maximize his utility which is $v_i - p_i(b)$ if he wins, and 0 otherwise.

Let N be the set of bidders with cost function $c : N \to \mathbb{R}^+$ and E be a universe of elements. For each $i \in N$, there is an associated set $S_i \subseteq E$; we assume each $e \in E$ is covered by, i.e., contained in, at least one S_i. For example, N is a set of firms with private costs c that are bidding $b_1, b_2, \ldots$ to prematurely terminate their contracts and E is a representation of their responsibilities. We want to honor requests while rejecting some to ensure that all the responsibilities are covered by at least one rejected firm. We interpret $c(i)$ to be the value of early termination for firm i and want to minimize the social cost, the total bid value, of those still with contracts.

Algorithm 5. Greedy Algorithm for the Set Cover Problem

1: $I = \emptyset$; $y(e) = 0, \forall e \in E$
2: **while** $\exists e \notin \bigcup_{i \in I} S_i$ **do**
3:     Increase $y(e)$ until there is some i such that $\sum_{e' \in S_i} y(e') = c(i)$
4:     $I = I \cup \{i\}$
5: **end while**
6: **return** I

The cost minimization problem with set cover constraints reduces to the well-known set cover problem with the same objective: given the cost function c, we want to select a subset of N with minimum cost whose sets cover E. The set cover problem includes many other algorithm design problems such as the minimum spanning tree problem, the Steiner tree problem, etc.

The set cover problem has primal-dual linear programming relaxations (see Appendix D in [6]) and a polynomial time greedy approximation algorithm, Algorithm 5. We show that Algorithm 5 is DA-implementable and obtain a DA auction:

Theorem 3. *Algorithm 5 is a DA-implementable f-approximation algorithm for the set cover problem where $f = \max_e |\{i : e \in S_i\}|$.*

Proof. (DA-implementability) We show an implementation with an active set and scoring functions as follows: Let A be the set of bidders not yet selected and $y^{N \setminus A}(e), \forall e \in E$ be the associated dual variables, dependent on $N \setminus A$, in Algorithm 5. We define the scoring functions as follows:

$$
s_i^A(c_i, c_{N \setminus A}) = \begin{cases} c(i) - \sum_{e \in S_i} y^{N \setminus A}(e), & \text{if } T_i^{N \setminus A} \text{ is not empty} \\ \infty, & \text{otherwise} \end{cases} , \qquad (5)
$$

where $T_i^{N \setminus A} = \{e : c \in S_i, e \notin \bigcup_{j \in N \setminus A} S_j\}$, i.e., those elements in S_i not covered by the selected sets in $N \setminus A$. When a bidder i is selected from A, there is an element $e \in T_i^{N \setminus A}$, the lowest indexed one if many exist, that we can associate to the bidder and increase the corresponding dual variable $y(e)$ by the amount $c(i) - \sum_{e' \in S_i} y^{N \setminus A}(e')$.

Note that the scoring functions are nondecreasing in the first argument. Assume the lowest scoring bidder i_t is associated with the element $e_t \in T_i^{N \setminus A}$ at each iteration t. Then, the steps of the DA auction with the above scoring functions can be realized as the steps of Algorithm 5 when the elements to be used in Line 3 are exactly the associated elements $e_1, e_2, \ldots$, and the next bidder to be selected by Algorithm 5 is the lowest scoring bidder at each iteration.

(Approximation) The proof that Algorithm 5 has the approximation ratio of f can be found, for instance, in [12]. $\blacksquare$

Corollary 3. *The ("inverted") DA auction with scoring functions (5) is a f-approximate WG-strategy-proof mechanism for the cost minimization problem with set cover constraints where $f = \max_e |\{i : e \in S_i\}|$.*

Acknowledgments. We would like to thank Vasilis Gkatzelis, Afshin Nikzad, and Amin Saberi for their helpful discussions.

References

[1] Borodin, A., Lucier, B.: On the limitations of greedy mechanism design for truthful combinatorial auctions. In: Abramsky, S., Gavoille, C., Kirchner, C., Meyer auf der Heide, F., Spirakis, P.G. (eds.) ICALP 2010. LNCS, vol. 6198, pp. 90–101. Springer, Heidelberg (2010)

[2] Briest, P., Krysta, P., Vöcking, B.: Approximation techniques for utilitarian mechanism design. SIAM Journal on Computing 40(6), 1587–1622 (2011)

[3] Dütting, P., Gkatzelis, V., Roughgarden, T.: The performance of deferred-acceptance auctions. In: Proceedings of the Fifteenth ACM Conference on Economics and Computation, EC 2014, pp. 187–204. ACM, New York (2014)

[4] Dütting, P., Roughgarden, T., Talgam-Cohen, I.: Modularity and greed in double auctions. In: Proceedings of the Fifteenth ACM Conference on Economics and Computation, EC 2014, pp. 241–258. ACM, New York (2014)

[5] Håstad, J.: Clique is hard to approximate within $n^{1-\epsilon}$. Acta Mathematica 182(1), 105–142 (1999)

[6] Kim, A.: Welfare maximization with deferred acceptance auctions in reallocation problems. arXiv:1507.01353 (2015)

[7] Lehmann, D., Oćallaghan, L.I., Shoham, Y.: Truth revelation in approximately efficient combinatorial auctions. J. ACM 49(5), 577–602 (2002)

[8] Lucier, B., Borodin, A.: Price of anarchy for greedy auctions. In: Proceedings of the Twenty-First Annual ACM-SIAM Symposium on Discrete Algorithms, SODA 2010, pp. 537–553. Society for Industrial and Applied Mathematics, Philadelphia (2010)

[9] Milgrom, P., Segal, I.: Deferred-acceptance auctions and radio spectrum reallocation. In: Proceedings of the Fifteenth ACM Conference on Economics and Computation, EC 2014, pp. 185–186. ACM, New York (2014)

[10] Nemhauser, G., Wolsey, L., Fisher, M.: An analysis of approximations for maximizing submodular set functions - i. Mathematical Programming 14(1), 265–294 (1978)

[11] Robins, G., Zelikovsky, A.: Improved steiner tree approximation in graphs. In: Proceedings of the Eleventh Annual ACM-SIAM Symposium on Discrete Algorithms, SODA 2000, pp. 770–779. Society for Industrial and Applied Mathematics, Philadelphia (2000)

[12] Williamson, D.P., Shmoys, D.B.: The Design of Approximation Algorithms, 1st edn. Cambridge University Press, New York (2011)

On the Pathwidth
of Almost Semicomplete Digraphs

Kenta Kitsunai[1], Yasuaki Kobayashi[2], and Hisao Tamaki[3]

[1] NTT DATA Corporation
mizuna0719@gmail.com
[2] Computer Center, Gakushuin University
yasuaki.kobayashi@gakushuin.ac.jp
[3] Department of Computer Science, Meiji University
tamaki@cs.meiji.ac.jp

Abstract. We call a digraph *h-semicomplete* if each vertex of the digraph has at most h non-neighbors, where a non-neighbor of a vertex v is a vertex $u \neq v$ such that there is no edge between u and v in either direction. This notion generalizes that of semicomplete digraphs which are 0-semicomplete and tournaments which are semicomplete and have no anti-parallel pairs of edges. Our results in this paper are as follows. (1) We give an algorithm which, given an h-semicomplete digraph G on n vertices and a positive integer k, in $(h + 2k + 1)^{2k}n^{O(1)}$ time either constructs a path-decomposition of G of width at most k or concludes correctly that the pathwidth of G is larger than k. (2) We show that there is a function $f(k, h)$ such that every h-semicomplete digraph of pathwidth at least $f(k, h)$ has a semicomplete subgraph of pathwidth at least k.

One consequence of these results is that the problem of deciding if a fixed digraph H is topologically contained in a given h-semicomplete digraph G admits a polynomial-time algorithm for fixed h.

1 Introduction

A *tournament* is a digraph obtained from a complete graph by orienting each edge. A *semicomplete digraph* generalizes a tournament, allowing each pair of distinct vertices to optionally have two edges in both directions between them. Tournaments and semicomplete digraphs are well-studied (see [1], for example) and have recently been attracting renewed interests in the following context.

There are many problems on undirected graphs that admit polynomial time algorithms but have digraph counterparts that are NP-complete. For example, Robertson and Seymour [14], in their Graph Minors project, proved that the k disjoint paths problem (and the k edge-disjoint paths problem) can be solved in polynomial for fixed k. On the other hand, digraph versions of these problems are NP-complete even for $k = 2$ due to Fortune, Hopcroft, and Wyllie [5]. Recently, Chudnovsky, Scot, and Seymour [3] showed that the k directed disjoint paths problem can be solved in polynomial time for fixed k if the digraph is restricted

© Springer-Verlag Berlin Heidelberg 2015
N. Bansal and I. Finocchi (Eds.): ESA 2015, LNCS 9294, pp. 816–827, 2015.
DOI: 10.1007/978-3-662-48350-3_68

to be semicomplete. The edge-disjoint version of the problem is also polynomial time solvable on semicomplete digraphs, due to Fradkin and Seymour [8]. The situation is similar for the topological containment problem, which asks if a given graph (digraph) contains a subgraph isomorphic to a subdivision of a fixed graph (digraph) H: the undirected version is polynomial time solvable due to the disjoint paths result and the directed version is NP-complete on general digraphs [5], while the question on semicomplete digraphs is polynomial time solvable due to Fradkin and Seymour [7] and moreover is fixed-parameter tractable due to Fomin and Pilipczuk [6,13]. In addition to these algorithmic results, some well-quasi-order results that are similar to the celebrated Graph Minors theorem of Robertson and Seymour [15] have been proved on the class of semicomplete digraphs [4,11]. These developments seem to suggest that the class of semicomplete digraphs is a promising stage for pursuing digraph analogues of the splendid outcomes, direct and indirect, from the Graph Minors project.

Given this progress on semicomplete digraphs, it is natural to look for more general classes of digraphs on which similar results hold. Indeed, the results on disjoint paths problems cited above are proved for some generalizations of semicomplete digraphs. The vertex-disjoint path algorithm given in [3] works for a digraph class called d-path dominant digraphs, which contains semicomplete digraphs ($d = 1$) and digraphs with multipartite underlying graphs ($d - 2$). The edge-disjoint path algorithm given in [8] works for digraphs with independence number (of the underlying graph) bounded by some fixed integer. On the other hand, the results for topological containment in [7,6,13] are strictly for the class of semicomplete graphs.

The *pathwidth* of digraphs, which plays an essential role in some of the above results, is defined as follows. Let G be a digraph. A *path-decomposition* of G is a sequence $(X_1, \ldots, X_m)$ of vertex sets $X_i \subseteq V(G)$, called *bags*, such that the following three conditions are satisfied:

1. $\bigcup_{1 \leq i \leq m} X_i = V(G)$,
2. for each edge (u, v) of G, $u \in X_i$ and $v \in X_j$ for some $i \geq j$, and
3. for every $v \in V(G)$, the set $\{i \mid v \in X_i\}$ of indices of the bags containing v forms a single integer interval.

The first and the third conditions are the same as in the definition of the path-width of undirected graphs; the second condition, on each edge, is different and depends on the direction of the edge. Note that some authors, including the present authors in previous work in different contexts, reverse the direction of edges in this condition. We follow the convention of the papers cited above. As in the case of undirected graphs, the *width* of a path-decomposition $(X_1, \ldots, X_m)$ is $\max_{1 \leq i \leq m} |X_i| - 1$ and the *pathwidth* of G, denoted by $\mathrm{pw}(G)$, is the smallest integer k such that there is a path-decomposition of G of width k.

Unlike for the pathwidth of undirected graphs, which is linear-time fixed-parameter tractable [2], no FPT-time algorithm is known for computing the pathwidth of general digraphs: only XP-time algorithms (of running time $n^{O(k)}$) are known. The third author of the current paper proposed one in [16], which was unfortunately flawed and has recently been corrected in [9] by the current and two more authors. Another XP algorithm is due to Nagamochi [12], which is formulated for a more general problem of optimizing linear layouts in submodular systems.

In this paper, we consider another direction of generalizing semicomplete digraphs and study the pathwidth of digraphs in the generalized class. For nonnegative integer h, we say that a simple digraph G is h-semicomplete if each vertex of G has at most h non-neighbors, where a non-neighbor of vertex v is a vertex u distinct from v such that there is no edge of G between u and v in either direction. Thus, semicomplete digraphs are 0-semicomplete. Our main results are as follows.

Theorem 1. *There is an algorithm which, given an h-semicomplete digraph G on n vertices and a positive integer k, in $(h + 2k + 1)^{2k}n^{O(1)}$ time either constructs a path-decomposition of G of width at most k or concludes correctly that the pathwidth is larger than k.*

This theorem generalizes the $k^{O(k)}n^2$ time result of Pilipczuk [13] on semicomplete digraphs. Compared on semicomplete digraphs, his algorithm has smaller dependence on n (our $O(1)$ exponent on n is naively 4), while the hidden constant in the exponent on k can be large.

Theorem 2. *There is a function $f(h, k)$ on positive integers h and k such that each h-semicomplete digraph with pathwidth at least $f(h, k)$ has a semicomplete subgraph of pathwidth at least k.*

The topological containment result in [7] is based on two components. One is a combinatorial result that, for each fixed digraph H, there is a positive integer k such that every semicomplete digraph G of pathwidth larger than k topologically contains H. The second component is a dynamic programming algorithm that, given a digraph G on n vertices together with a path-decomposition of width k and a digraph H on r vertices with s edges, decides if G topologically contains H in $O(n^{3(k+rs)+4})$ time. Note that this algorithm does not require G to be semicomplete. Theorem 2 enables us to generalize the first component to h-semicomplete digraphs and Theorem 1 gives us the path-decomposition to be used in the dynamic programming. Thus, we have the following theorem.

Theorem 3. *For fixed positive integer h and fixed digraph H, the problem of deciding if a given h-semicomplete digraph topologically contains H can be solved in polynomial time.*

We should remark that extending the FPT result of [6,13] in this direction using the approach of this paper appears difficult, as the FPT-time dynamic programming algorithm therein heavily relies on the strict semicompleteness of the input digraph.

Techniques. Our algorithm in Theorem 1 borrows the notion of separation chains from [13] but the algorithm itself is completely different from the one in [13]. The advantage of our algorithm is that it works correctly on general digraphs, in contrast to the one in [13] which is highly specialized for semicomplete digraphs. We need a property of h-semicomplete digraphs only in the analysis of the running time.

Our algorithm is based on the one due to Nagamochi [12] for more general problem of finding an optimal linear layout for submodular systems. Informally, his algorithm applied to the pathwidth computation works as follows. Fix digraph G and let $d^+(U)$ for each $U \subseteq V(G)$ denote the number of out-neighbors of U. The *width* of permutation π of $V(G)$ is defined to be the maximum of $d^+(V(\pi'))$ where π' ranges over all the prefixes of π and $V(\pi')$ denotes the set of vertices in π'. The smallest integer k such that there is a permutation of width k is called the *vertex separation number* of G and is equal to the pathwidth of G [17]. Thus, our goal is to decide, given k, if there is a permutation of $V(G)$ of width at most k.

Nagamochi's algorithm is a combination of divide-and-conquer and branching from both sides of the permutation. For disjoint subsets S and T of $V(G)$, call a permutation π of $V(G)$ an (S,T)-*permutation*, if it has a prefix π' with $V(\pi') = S$ and a suffix π'' with $V(\pi'') = T$. A vertex set X that minimize $d^+(X)$ subject to $S \subseteq X \subseteq V(G) \setminus T$ is called a minimum (S,T) separator. A crucial observation, based on the submodularity of set function d^+ is the following. Let X be a minimum (S,T)-separator. Then, if there is an (S,T)-permutation of width at most k then there is such a permutation that is an $(S, V(G) \setminus X)$-permutation and an (X,T)-permutation at the same time. Thus if there is a minimum (S,T)-separator distinct from both S and $V(G) \setminus T$, then we can divide the problem into two smaller subproblems. When there is no minimum (S,T)-separator other than S or $V(G) \setminus T$, we need to branch on vertices to add to S or T. For general digraphs, the running time is $n^{2k+O(1)}$: we need to branch on $O(n)$ vertices from both sides, and the depth of branching is bounded by k, as the value $d^+(X)$ of the minimum separator X increases at least by one after we branch from both sides.

For h-semicomplete digraphs, we observe that the number of vertices v such that $d^+(S \cup \{v\}) \leq k$ is at most $h + 2k + 1$ (see Proposition 1) and therefore, we need to branch on at most $h + 2k + 1$ vertices when extending from S. Unfortunately, we do not have a similar bound on the number of vertices to branch on from the side of T. For example, if $|T| < k$, then $d^+(V(G) \setminus (T \cup \{v\})) \leq k$ for every $v \notin T$ and therefore we need to branch on every vertex not in $T \cup S \cup N^+(S)$, where $N^+(S)$ denotes the set of out-neighbors of S.

This asymmetry comes from the asymmetry inherent in the vertex separation number characterization: the width of a permutation π in G is not equal in general to the width of a reversal of π in G^{-1}, the digraph obtained from G by reversing all of its edges. We use separation chains [13] to give a symmetric characterization of pathwidth and formulate a variant of Nagamochi's algorithm which branches from each side on at most $(h + 2k + 1)$ vertices. This is how we get the running time stated in Theorem 1. We remark that a similar result on cutwidth is an immediate corollary of the Nagamochi's result, since we have the desired symmetry in the definition of cutwidth: the cutwidth of a permutation π in G equals the cutwidth of the reversal of π in G^{-1}.

The scenario for the combinatorial result in Theorem 2 is rather straightforward. Given an h-semicomplete graph G of pathwidth at least $f(h, k)$, we complete it into a semicomplete graph G' on $V(G)$, which must have pathwidth at least $f(h, k)$. We then find an obstacle $T \subseteq V(G)$ in G' for small pathwidth, of one of the types defined in [13]. Then we consider a random semicomplete subgraph G'' of G and show that G'' inherits an obstacle T' from T with high probability such that the existence of T' in G'' implies $\mathrm{pw}(G'') \geq k$. We need to overcome, however, some difficulties in carrying out this scenario. To be more specific, consider one type of obstacles, namely *degree tangles* [13]. An (l, k)-degree tangle of G is a vertex set T with $|T| = l$ such that $\max_{v \in T} d^+(v) - \min_{v \in T} d^+(v) \leq k$. In order for a degree tangle T in G' to give rise to a degree-tangle T' of the random subgraph G'', we need the out-degrees of vertices in T' to "shrink" almost uniformly. To this end, we wish our sampling to be such that (1) each vertex $v \in V(G)$ is in $V(G'')$ with a fixed probability p and (2) for each vertex set $S \subseteq V(G)$, the intersection $S \cap V(G'')$ has cardinality sharply concentrated around its expectation $p|S|$. The following theorem, which may be of independent interest, makes this possible: we apply this theorem to the complement of the underlying graph of G with $d = h$.

Theorem 4. *Let G be an undirected graph on n vertices with maximum degree d or smaller. Let $p = \frac{1}{2d+1}$. Then, it is possible to sample a set I of independent vertices of G so that $\mathbf{Pr}(v \in I) = p$ for each $v \in V(G)$ and, for each $S \subseteq V(G)$, we have*

$$\mathbf{Pr}(|S \cap I| > p|S| + t) < \exp\left(-\frac{t^2}{9|S|}\right)$$

and

$$\mathbf{Pr}(|S \cap I| < p|S| - t) < \exp\left(-\frac{t^2}{9|S|}\right).$$

Even with this sampling method, it is still not clear if we can have the desired "uniform shrinking" of out-degrees of the vertices in the degree tangle, since if the set S of out-neighbors of a vertex has cardinality $\Omega(n)$, then the deviation of $|S \cap V(G'')|$ from its expectation $p|S|$ is necessarily $\Omega(\sqrt{n})$. To overcome this difficulty, we introduce several types of obstacles that are robust against random sampling and show that (1) if G' has an obstacle of a type in [13] then it has a robust obstacle and (2) each robust obstacle in G' indeed gives rise to a strong enough obstacle in $G(V'')$ with high probability.

The rest of this paper is organized as follows. In Section 2 we define some notation. In Section 3, we describe our algorithm and prove Theorem 1, omitting the proofs of some lemmas. In Section 4, we sketch our proof of Theorem 2. All the omitted proofs and details, including the proof of Theorem 4, can be found in the full version of this paper [10].

2   Notation

Digraphs in this paper are simple: there are no self-loops and, between each pair of distinct vertices, there is at most one edge in each direction. For digraph G, $V(G)$ denotes the set of vertices of G and $E(G) \subseteq V(G) \times V(G)$ the set of edges of G. If $(u, v) \in E(G)$, then v is an *out-neighbor* of u and u is an *in-neighbor* of v. For each $v \in V(G)$, we denote the set of in-neighbors of v by $N_G^-(v) = \{u \mid (u, v) \in E(G)\}$ and write $N_G^-[v]$ for $N_G^-(v) \cup \{v\}$. For $U \subseteq V(G)$, we define $N_G^-[U] = \bigcup_{v \in U} N_G^-[v]$ and $N_G^-(U) = N_G^-[U] \setminus U$. We define the notation for out-neighbors N^+ similarly. In this paper, the *in-degree* and *out-degree* of vertex v in G, denoted by $d_G^-(v)$ and $d_G^+(v)$, respectively, counts the in-neighbors and out-neighbors rather than the incoming and outgoing edges: $d_G^-(v) = |N_G^-(v)|$ and $d_G^+(v) = |N_G^+(v)|$; we also define $d_G^-(U) = |N_G^-(U)|$ and $d_G^+(U) = |N_G^+(U)|$ for $U \subseteq V(G)$. We omit the reference to G from the above notation when it is clear from the context which digraph is meant.

3   Algorithm

In this section, we describe the algorithm claimed in Theorem 1, prove its correctness, and analyze its running time. As suggested in the introduction, our first task is to give a symmetric characterization of pathwidth to which the Nagamochi's algorithm is adaptable.

Let G be a digraph. A pair (A, B) of vertex sets of G is a *separation* of G if $A \cup B = V$ and there is no edge from $A \setminus B$ to $B \setminus A$. The *order* of separation (A, B) is $|A \cap B|$. For $S, T \subseteq V$ such that $S \cap T = \emptyset$, separation (A, B) is an *S–T separation* if $S \cap B = \emptyset$ and $T \cap A = \emptyset$. We call an S–T separation (A, B) *trivial* if $B = V(G) \setminus S$ or $A = V(G) \setminus T$.

An important role in our algorithm is played by a *minimum S-T separation*, which is defined to be an S–T separation of the smallest order. Note that if a minimum S-T separation is trivial, then it must be either $(N^+[S], V(G) \setminus S)$ or $(V(G) \setminus T, N^-[T])$. As will be seen later, we may use non-trivial minimum S-T separations to divide-and-conquer subproblems in our pathwidth computation.

A sequence of separations $((A_0, B_0), (A_1, B_1), \ldots, (A_r, B_r))$ is a *separation chain* if $A_0 \subseteq A_1 \subseteq \ldots \subseteq A_r$ and $B_r \subseteq B_{r-1} \subseteq \ldots \subseteq B_0$. The *order* of this separation chain is the maximum order of its member separations. We use operator $+$ for concatenating sequences of separations and for appending a separation to a sequence of separations: for sequences C and C' of separations and a separation (A, B), $C + C'$ is the concatenation of C and C', $(A, B) + C$ is the sequence C preceded by (A, B), and $C + (A, B)$ is the sequence C followed by (A, B).

Let $C = ((A_0, B_0), (A_1, B_1), \ldots, (A_r, B_r))$ be a separation chain. We say that C is *gapless* if, for every $0 < i \leq r$, either $|A_i \setminus A_{i-1}| \leq 1$ or $|B_{i-1} \setminus B_i| \leq 1$ holds. Note that this definition allows a repetition of an identical separation. We say that C is an *S–T chain*, if $B_0 = V(G) \setminus S$ and $A_r = V(G) \setminus T$, that is, both ends of C are trivial S–T separations. Note that every separation in an S–T chain is an S–T separation.

As observed in [13],
(1) if $(X_1, X_2, \ldots, X_r)$ is a path-decomposition of G then $((A_0, B_0), (A_1, B_1),$ $\ldots, (A_r, B_r))$, where $A_i = \bigcup_{j \leq i} X_j$ and $B_i = \bigcup_{i < j} X_j$, is an $\emptyset$–$\emptyset$ chain in G, and
(2) if $((A_0, B_0), (A_1, B_1), \ldots, (A_r, B_r))$ is an $\emptyset$–$\emptyset$ chain in G, then $(W_1, W_2, \ldots, W_r)$, where $W_i = A_i \cap B_{i-1}$ for $1 \leq i \leq r$, is a path-decomposition of G.

These observations lead to the following characterization of pathwidth by means of gapless separation chains.

Lemma 1. *Digraph G has a path-decomposition of width k if and only if it has a gapless $\emptyset$–$\emptyset$ chain of order k.*

Proof. Suppose G has a path-decomposition $(X_1, X_2, \ldots, X_r)$ of width k. We may assume that this path-decomposition is nice: $X_1 = X_r = \emptyset$ and, for $1 \leq i < r$, either $X_{i+1} = X_i \cup \{v\}$ for some $v \in V(G) \setminus X_i$ or $X_{i+1} = X_i \setminus \{v\}$ for some $v \in X_i$. If we set $A_i = \bigcup_{j \leq i} X_j$ and $B_i = \bigcup_{j > i} X_j$ for $0 \leq i \leq r$ as in observation (1), then $((A_0, B_0), (A_1, B_1), \ldots, (A_r, B_r))$ is a gapless $\emptyset$–$\emptyset$ chain. The order of this separation chain is $\max_{0 \leq i \leq r} |A_i \cap B_i| = \max_{1 \leq i \leq r-1} |X_i \cap X_{i+1}| = k$. Conversely, suppose a gapless separation chain $((A_0, B_0), (A_1, B_1), \ldots, (A_r, B_r))$ of order k is given. We set $X_i = A_i \cap B_{i-1}$ for $1 \leq i \leq r$. Then, $(X_1, X_2, \ldots, X_r)$ is a path-decomposition by observation (2). Since our separation chain is gapless, we have either $|A_i \setminus A_{i-1}| \leq 1$ or $|B_{i-1} \setminus B_i| \leq 1$ for $1 \leq i \leq r$. In the former case, we have $|A_i \cap B_{i-1}| \leq |A_{i-1} \cap B_{i-1}| + 1 = k + 1$ and, in the latter case, we have $|A_i \cap B_{i-1}| \leq |A_i \cap B_i| + 1 = k + 1$. Therefore, the width of path-decomposition $(X_1, X_2, \ldots, X_r)$ is at most k and hence G has a path-decomposition of width k. $\square$

We say that a pair (S, T) of vertex sets of G is k-*admissible* if $N^+[S] \cap T = \emptyset$ (and hence $S \cap N^-[T] = \emptyset$), $d^+(S) \leq k$, and $d^-(T) \leq k$. It is clear that (S, T) must be k-admissible in order for G to have a gapless S–T chain of order at most k. Our algorithm solves the following problem with parameter k: given digraph G and a k-admissible pair (S, T), compute a gapless S–T chain of order at most k if one exists and otherwise report the non-existence. The algorithm in Theorem 1 applies this algorithm to $(S, T) = (\emptyset, \emptyset)$ and, if it returns an $\emptyset$–$\emptyset$ chain of order k, converts it to a path-decomposition of width at most k, using the proof of Lemma 1.

The following lemma provides the base case for our algorithm.

Lemma 2. *If pair (S, T) is k-admissible and satisfies $|V(G) \setminus (S \cup T)| \leq k + 1$ then G has a gapless S–T chain of order at most k.*

The proof is by induction on $|V(G) \setminus (S \cup T)|$ and is constructive.

We have two types of recurrences: divide-and-conquer and branching. The following lemma, which corresponds to the main lemma in [12] underlying the algorithm for submodular systems, provides a recurrence of the first type.

Lemma 3. *Suppose G has a gapless S–T chain of order k and let (X, Y) be a minimum S–T separation of G. Then G has a gapless S–T chain of order at most k of the form $C_1 + (X, Y) + C_2$, where C_1 is a gapless S–$(Y \setminus X)$ chain and C_2 is a gapless $(X \setminus Y)$–T chain.*

The recurrence of the second type is provided by the following lemma.

Lemma 4. *Suppose G has a gapless S–T chain of order at most k and suppose that $|V(G) \setminus (S \cup T)| \geq k + 2$ holds. Then, there are a gapless S–T chain $((A_0, B_0), \ldots, (A_r, B_r))$ of order at most k and a pair of distinct vertices $u \in V(G) \setminus (S \cup N^-[T])$ and $v \in V(G) \setminus (T \cup N^+[S])$ such that the following holds:*

1. *$((A_1, B_1), \ldots, (A_r, B_r))$ is an $(S \cup \{u\})$–T chain,*
2. *$((A_0, B_0), \ldots, (A_{r-1}, B_{r-1}))$ is an S–$(T \cup \{v\})$ chain, and*
3. *$((A_1, B_1), \ldots, (A_{r-1}, B_{r-1}))$ is an $(S \cup \{u\})$–$(T \cup \{v\})$ chain.*

Given these recurrences and the base case above, our algorithm is straightforward. Suppose we are given a k-admissible pair (S, T). If $|V(G) \setminus (S \cup T)| \leq k+1$ holds then we apply Lemma 2 and return the gapless S–T chain it provides. Suppose otherwise. We test if there is a minimum S–T separation that is non-trivial: a minimum S–T separation (X, Y) that is not equal to either $(N^+[S], V(G) \setminus S)$ or $(V(G) \setminus T, N^-[T])$. If we find one, we apply Lemma 3 and recurse on subproblems $(S, Y \setminus X)$ and $(X \setminus Y, T)$. If either of the recursive calls returns a negative answer, we return a negative answer. Otherwise, we concatenate the solutions from the subproblems as prescribed in Lemma 3 and return the result. Finally suppose that there is no minimum S–T separation that is non-trivial. If $(N^+[S], V(G) \setminus S)$ is the only minimum S–T separation, then we recurse on $(S \cup \{v\}, T)$ for every $v \in V(G) \setminus (S \cup T)$ such that $(S \cup \{v\}, T)$ is k-admissible. If $(V(G) \setminus T, N^-[T])$ is the only minimum S–T-separation, then we similarly branch from T. If both $(N^+[S], V(G) \setminus S)$ and $(V(G) \setminus T, N^-[T])$ are the minimum S–T separations, then we branch from both sides. In either case, if any of the recursive call returns a gapless separation chain of order at most k, we trivially extend the chain into a gapless S–T separation of order at most k and return this chain. Otherwise, that is, if all the recursive calls return negative answers, we return a negative answer.

The correctness of this algorithm is proved by a straightforward induction for which the above Lemmas provide the base case and the induction steps.

We analyze the running time of the algorithm. The following observation extends the one in [13] that the number of vertices of out-degree at most k in a semicomplete digraph is at most $2k + 1$.

Proposition 1. *Let G be an h-semicomplete digraph and let $U \subseteq V(G)$. Then the number of vertices $v \in V(G) \setminus U$ such that $d^+(U \cup \{v\}) \leq k$ is at most $h + 2k + 1$ for every $k > 0$. The similar statement with the out-degree replaced by the in-degree also holds.*

Proof. Fix U, let $X \subset V(G) \setminus U$ be arbitrary, and set $|X| = b$. By the definition of h-semicomplete digraphs, $G[X]$ contains at least $b(b - h - 1)/2$ edges and hence the average out-degree of vertices in $G[X]$ is at least $(b - h - 1)/2$. For each $v \in X$, $N_G^+(U \cup \{v\})$ contains $N_{G[X]}^+(v)$ and hence if $b > h + 2k + 1$ then there is at least one $v \in X$ such that $|N_G^+(U \cup \{v\})| > k$. This proves the first statement. The second statement is immediate by symmetry. □

Thus, the number of vertices to branch on from each side in the above algorithm is bounded by $h + 2k + 1$.

To measure the "size" of the problem instance (S, T), we introduce the following two functions. Let $\gamma(S, T)$ denote the order of the minimum S–T separation. Let $\mu(S, T)$ be defined by

$$\mu(S, T) = 2|V(G) \setminus (N^+[S] \cup N^-[T])| + |N^+(S) \Delta N^-(T)|,$$

where $X \Delta Y$ is the symmetric difference between X and Y.

Lemma 5. *Let (X, Y) be a minimum S–T separation. Then, we have*

$$\mu(S, \ Y \setminus X) + \mu(X \setminus Y, \ T) = \mu(S, T).$$

Lemma 6. *Let (X, Y) be a non-trivial S–T separation: $X \setminus Y \neq S$ and $Y \setminus X \neq T$. Then, we have $\mu(S, \ Y \setminus X) \geq 1$ and $\mu(X \setminus Y, \ T) \geq 1$.*

Let $R(S, T)$ denote the number of problem instances recursively considered when we solve the instance (S, T), not counting the instances in the base case, but counting the instance (S, T) itself unless it is in the base case. Let $\mu'(S, T) = \max\{0, 2\mu(S, T) - 1\}$.

Lemma 7. *Let G be an h-semicomplete digraph and k a positive integer. Then, for each k-admissible pair (S, T), we have*

$$R(S, T) \leq \mu'(S, T) \cdot (h + 2k + 1)^{2(k - \gamma(S, T))}$$

The proof of this lemma is by a straightforward induction on the structure of the computation.

The time for processing each pair (S, T) excluding the time consumed by subsequent recursive calls is dominated by the time for finding minimum S–T separation and for deciding if there is a minimum S–T separation that is not trivial. This can be done in $n^{O(1)}$ time by the repeated use of a standard augmenting path algorithm for minimum S-T cut. Since $\mu'(\emptyset, \emptyset) = O(n)$, we have the running time claimed in Theorem 1.

4   Tame Obstacles Survive Random Sampling: Proof Sketch of Theorem 2

We sketch the proof of Theorem 2 in this section.

Let G be a semicomplete digraph with n vertices. For $0 \leq d \leq n$, let $V_{\leq d}^-(G)$, $V_{\geq d}^-(G)$, $V_{\leq d}^+(G)$, and $V_{\geq d}^+(G)$ denote the set of vertices v with $d_G^-(v) \leq d$, $d_G^-(v) \geq d$, $d_G^+(v) \leq d$, and $d_G^+(v) \geq d$, respectively. We omit the reference to G and write $V_{\leq d}^-$ etc. when G is clear from the context.

We first define several types of obstacles against small pathwidth in semicomplete digraphs.

Definition 1. *[13] Let G be a semicomplete digraph and let $d \geq 0$, $l > 0$ and $k > 0$ be integers. A (d, l, k)-degree tangle of G is a vertex set $T \subseteq V_{\geq d}^+ \cap V_{\leq d+k}^+$ with $|T| = l$. A (d, l, k)-matching tangle of G is a pair of vertex sets (T_1, T_2) with $|T_1| = |T_2| = l$ such that:*

1. *$T_1 \subseteq V_{\leq d}^+$, $T_2 \subseteq V_{\geq d+k+1}^+$, and*
2. *there is some bijection $\phi : T_1 \to T_2$ such that $(v, \phi(v)) \in E(G)$ for every $v \in T_1$.*

We will often refer to a (d, l, k)-degree (-matching) tangle as an (l, k)-degree (-matching) tangle without specifying d.

Pilipczuk [13] showed that a $(5k+2, k)$-degree tangle in G implies $\mathrm{pw}(G) \geq k+1$ and an (l, k)-matching tangle implies $\mathrm{pw}(G) \geq \min\{l, k+1\}$. We prove and use a slightly stronger statement on degree tangles: an (l, k)-degree tangle in G implies $\mathrm{pw}(G) \geq (l - k - 1)/2$.

We follow the scenario described in the introduction. Given an h-semicomplete digraph G of pathwidth at least $f(h, k)$, we complete it into a semicomplete digraph G' on $V(G)$, in which we find a large obstacle, say a degree tangle T. Then, we apply Theorem 4 to obtain a random independent set I of the complement of the underlying graph of G. We hope that $T \cap I$ is a tangle of $G[I]$ that is strong enough to conclude $\mathrm{pw}(G[I]) \geq k$. For this to happen, we need to have the out-degrees $|N_{G'}^+(v) \cap I|$ of v, for $v \in T \cap I$, to be close to each other.

As observed in [13], the optimal vertex separation sequence lists the vertices roughly in the order of increasing out-degrees and therefore each vertex has most vertices of smaller degree as its out-neighbors, with some exceptions. The following notion of the wildness of vertices measures how exceptional a vertex is.

Definition 2. *For each vertex $v \in G$, we define the wildness $\mathrm{wld}(v)$ of v by*

$$\mathrm{wld}(v) = |V_{\leq d^+(v)}^+ \setminus N^+(v)|.$$

If the vertices of a degree-tangle T have small wildness, then most of their out-neighbors are shared and we may expect that their degrees in the sampled subgraph $G[I]$ will be close together. We call such a degree-tangle *tame*.

Definition 3. *We say that an (l, w)-degree tangle T of G is tame (relative to the parameters l and w), if $\mathrm{wld}(v) \leq 3l + w + 2\mathrm{pw}(G)$ for each $v \in T$.*

A degree-tangle is not necessarily tame, but a large number of wild vertices in a degree-tangle are themselves an evidence of large pathwidth. We capture this fact by another type of obstacles we call spiders.

Definition 4. *Let G be a semicomplete digraph and let d, l, and w be positive integers. A (d, l, w)-spider is a triple (T, L, R), where T is a vertex set with $|T| \geq l$, L is a family $\{L_v \mid v \in T\}$ of vertex sets, and R is a family $\{R_v \mid v \in T\}$ of vertex sets, such that the following holds for each $v \in T$:*

1. $L_v \subseteq N^-(v)$, $|L_v| \geq 3l$, and $d^+(u) \leq d$ for each $u \in L_v$, and
2. $R_v \subseteq N^+(v)$, $|R_v| \geq 3l$, and $d^+(u) \geq d + w$ for each $u \in R_v$.

We will sometimes refer to a (d, l, w)-spider as an (l, w)-spider, without specifying d.

Using an argument similar to the one in [13] used to show that a matching-tangle is indeed an obstacle, we have the following lemma.

Lemma 8. *If a semicomplete digraph G has an (l, w)-spider then $\mathrm{pw}(G) > \min\{l, w\}$.*

The following lemma shows that spiders capture what we intended them to capture.

Lemma 9. *Suppose G has a $(2l, w)$-degree tangle T. Then, G has either a tame (l, w)-degree tangle or an (l, w)-spider.*

We similarly define the tameness of matching tangles and of spiders. We then show that if we have a matching-tangle then we have either a tame matching-tangle or a spider. We also show that if we have a spider then we have a tame spider. This gets us ready for carrying out our scenario.

Fix positive integer h. Let k_h be a constant large enough as required in technical Lemmas used below. We set $f(k, h) = 128(h + 1)k$ for $k \geq k_h$ and $f(k, h) = f(k_h, h)$ for $k < k_h$.

Let G be an h-semicomplete digraph of pathwidth at least $f(k, h)$. In the following proof that G contains a semicomplete subgraph of pathwidth at least k, we assume $k \geq k_h$; otherwise we would prove that G contains a semicomplete subgraph of pathwidth at least $k_h \geq k$. We set $K = (h + 1)k$ for readability.

List the vertices of G as $v_1, \ldots, v_n$, in the non-decreasing order of out-degrees. Let G' be the semicomplete digraph obtained from G by adding edge (v_i, v_j) for each pair $i > j$ such that neither (v_i, v_j) nor (v_j, v_i) is an edge of G. By our assumption, $\mathrm{pw}(G') \geq \mathrm{pw}(G)$ is at least $128K$. We assume that $\mathrm{pw}(G') \leq 140K$ in our construction; if this assumption does not hold, we choose $k' \geq k$ such that $128(h + 1)k' \leq \mathrm{pw}(G') \leq 140(h + 1)k'$ and prove that G has a semicomplete subgraph of pathwidth at least k'.

We first obtain a tame $(46K, 18K)$-degree tangle, a tame $(6K, w)$-spider for some $w \geq 18K$, or a tame $(6K, 18K)$-matching tangle of G'. The crucial part of the proof that this is possible is the algorithm due to Pilipczuk (Theorem 32 in [13]) that finds in the given semicomplete digraph either a large degree-tangle, a large matching-tangle, or a path-decomposition of some width if not obstructed by those tangles.

If G' has a tame $(46K, 18K)$-degree tangle, then we may show that the random subgraph $G[I]$ has a $(21k, 10k)$-degree tangle with high probability. If G' has a tame $(6K, w)$-spider for $w \geq 18K$, we may show that $G[I]$ contains a (k, k)-spider with high probability. For the third case where we have a tame $(6K, 18K)$-matching tangle, we slightly modify the sampling method: we contract each edge representing the matching bijection of the matching tangle and sample on this contracted graph. With this care taken, we may show that the resulting random subgraph contains with a (k, k)-matching tangle. In either case, we conclude that G contains a semicomplete subgraph of pathwidth at least k. This completes a sketch of the proof of Theorem 2.

References

1. Bang-Jensen, J., Gutin, G.Z.: Digraphs: theory, algorithms and applications. Springer Science & Business Media (2008)
2. Bodlaender, H.L.: A linear-time algorithm for finding tree-decompositions of small treewidth. SIAM Journal on Computing 25, 1305–1317 (1996)
3. Chudnovsky, M., Scot, A., Seymour, P.: Disjoint paths in tournaments. Advances in Mathematics 270, 582–597 (2015)
4. Chudnovsky, M., Seymour, P.: A well-quasi-order for tournaments. Journal of Combinatorial Theory, Series B 101(1), 47–53 (2011)
5. Fortune, S., Hopcroft, J., Wyllie, J.: The directed subgraph homeomorphism problem. Theoretical Computer Science 10(2), 111–121 (1980)
6. Fomin, F.V., Pilipczuk, M.: Jungles, bundles, and fixed-parameter tractability. In: Proceedings of the Twenty-Fourth Annual ACM-SIAM Symposium on Discrete Algorithms, pp. 396–413 (2013)
7. Fradkin, A.O., Seymour, P.D.: Tournament pathwidth and topological containment. Journal of Combinatorial Theory, Series B 103(3), 374–384 (2013)
8. Fradkin, A., Seymour, P.: Edge-disjoint paths in digraphs with bounded independence number. Journal of Combinatorial Theory, Series B 110, 19–46 (2015)
9. Kistunai, K., Kobayashi, Y., Komuro, K., Tamaki, H., Tano, T.: Computing directed pathwidth in $O(1.89^n)$ time. Algorithmica (2015) (accepted for publication)
10. Kistunai, K., Kobayashi, Y., Tamaki, H.: On the pathwidth of almost semicomplete digraphs. arXiv preprint arXiv:1507.01934 (2015)
11. Kim, I., Seymour, P.: Tournament minors. Journal of Combinatorial Theory, Series B 112, 138–153 (2015)
12. Nagamochi, H.: Linear layouts in submodular systems. In: Proceedings of the 23rd International Symposium on Algorithms and Computation, pp. 475–484 (2012)
13. Pilipczuk, M.: Computing cutwidth and pathwidth of semi-complete digraphs via degree orderings. arXiv preprint arXiv:1210.5363 (2012) Conference version in Proceedings of the 30th International Symposium on Theoretical Aspects of Computer Science, pp. 197–208 (2013)
14. Robertson, N., Seymour, P.D.: Graph minors. XIII. The disjoint paths problem. Journal of Combinatorial Theory, Series B 63(1), 65–110 (1995)
15. Robertson, N., Seymour, P.D.: Graph minors. XX. Wagner's conjecture. Journal of Combinatorial Theory, Series B 92(2), 325–357 (2004)
16. Tamaki, H.: A Polynomial Time Algorithm for Bounded Directed Pathwidth. In: Proceedings of the 37th International Workshop on Graph-Theoretic Concepts in Computer Science, WG 2011, pp. 331–342 (2011)
17. Yang, B., Cao, Y.: Digraph searching, directed vertex separation and directed pathwidth. Discrete Applied Mathematics 156(10), 1822–1837 (2008)

Quicksort, Largest Bucket, and Min-Wise Hashing with Limited Independence

Mathias Bæk Tejs Knudsen* and Morten Stöckel**

University of Copenhagen
{knudsen,most}@di.ku.dk

Abstract. Randomized algorithms and data structures are often analyzed under the assumption of access to a perfect source of randomness. The most fundamental metric used to measure how "random" a hash function or a random number generator is, is its *independence*: a sequence of random variables is said to be k-independent if every variable is uniform and every size k subset is independent.

In this paper we consider three classic algorithms under limited independence. Besides the theoretical interest in removing the unrealistic assumption of full independence, the work is motivated by lower independence being more practical. We provide new bounds for randomized quicksort, min-wise hashing and largest bucket size under limited independence. Our results can be summarized as follows.

- *Randomized Quicksort.* When pivot elements are computed using a 5-independent hash function, Karloff and Raghavan, J.ACM'93 showed $\mathcal{O}(n \log n)$ expected worst-case running time for a special version of quicksort. We improve upon this, showing that the same running time is achieved with only 4-independence.
- *Min-Wise Hashing.* For a set A, consider the probability of a particular element being mapped to the smallest hash value. It is known that 5-independence implies the optimal probability $\mathcal{O}(1/n)$. Broder et al., STOC'98 showed that 2-independence implies it is $\mathcal{O}(1/\sqrt{|A|})$. We show a matching lower bound as well as new tight bounds for 3- and 4-independent hash functions.
- *Largest Bucket.* We consider the case where n balls are distributed to n buckets using a k-independent hash function and analyze the largest bucket size. Alon et. al, STOC'97 showed that there exists a 2-independent hash function implying a bucket of size $\Omega(n^{1/2})$. We generalize the bound, providing a k-independent family of functions that imply size $\Omega(n^{1/k})$.

* Research partly supported by Mikkel Thorup's Advanced Grant from the Danish Council for Independent Research under the Sapere Aude programme and the FNU project AlgoDisc - Discrete Mathematics, Algorithms, and Data Structures.
** This work was done while at IT University of Copenhagen. Supported by the Danish National Research Foundation / Sapere Aude program and VILLUM FONDEN.

© Springer-Verlag Berlin Heidelberg 2015
N. Bansal and I. Finocchi (Eds.): ESA 2015, LNCS 9294, pp. 828–839, 2015.
DOI: 10.1007/978-3-662-48350-3_69

1 Introduction

A unifying metric of strength of hash functions and pseudorandom number generators is the *independence* of the function. We say that a sequence of random variables is k-independent if every random variable is uniform and every size k subset is independent. A question of theoretical interest is, regarding each algorithmic application, *how much independence is required?*. With the standard implementation of a random generator or hash function via a k-degree polynomial k determines both the space used and the amount of randomness provided. A typical assumption when performing algorithmic analysis is to just assume full independence, i.e., that for input size n then the hash function is n-independent. Besides the interest from a theoretic perspective, the question of how much independence is required is in fact interesting from a practical perspective: hash functions and generators with lower independence are as a rule of thumb faster in practice than those with higher independence, hence if it is proven that the algorithmic application needs only k-independence to work, then it can provide a speedup for an implementation to specifically pick a fast construction that provides the required k-independence. In this paper we consider three fundamental applications of random hashing, where we provide new bounds for limited independence. We note that due to space constraints, this version should be considered an extended abstract and we refer to the full version for full proofs and all technical details.

Min-Wise Hashing. We consider the commonly used scheme *min-wise hashing*, which was first introduced by Broder [2] and has several practical applications (see Section 2). Here we study families of hash functions, where a function h is picked uniformly at random from the family and applied to all elements of a set A of size n. We say that h is min-wise independent if for any element $x \in A$ then $\mathbf{Pr}(\min h(A) = h(x)) = 1/n$ and ε-min-wise independent if $\mathbf{Pr}(\min h(A) = h(x)) \le (1 + \varepsilon)/n$. For families of k-independent hash functions we show new tight bounds for $k = 2, 3, 4$ of $\varepsilon = \Theta(\sqrt{n}), \Theta(\log n), \Theta(\log n)$ respectively and for $k = 5$ it is folklore that $O(1)$-min-wise ($\varepsilon = \mathcal{O}(1)$) can be achieved. Since tight bounds for $k \ge 5$ exist (see Section 2), our contribution closes the problem.

Randomized Quicksort. Next we consider a classic sorting algorithm presented in many randomized algorithms books, e.g. already on page three of Motwani-Raghavan [11]. In the setting where we assign a hash value to each element, and the pivot element is chosen to be the one with the smallest hash value, the classic analysis of quicksort in Motwani-Raghavan gives that the expected worst-case running time is $\mathcal{O}(n(\log n) \cdot (1 + \varepsilon))$ if n elements are sorted using an ε-min-wise hash function. The new tight bounds for min-wise hashing show the limitations of this classical analysis, and for $k = 2, 4$ we get stronger bounds using a new approach. A special version of randomized quicksort was shown by Karloff and Raghavan to use expected worst-case time $\mathcal{O}(n \log n)$ when the pivot elements are chosen using a 5-independent hash function [10]. Our main result is a new general bound for the number of comparisons performed under limited independence, which applies to several settings of quicksort, including the setting of Karloff-Raghavan where we show the same running time using only 4-independence. Furthermore, we show that $k = 2$ and $k = 3$ can imply expected worst-case time $\Omega\left(n \log^2 n\right)$. An interesting observation is that our new bounds for $k = 4$ and $k = 2$ shows that the classic analysis using min-wise hashing is not tight, as we go below those bounds by a factor $\log n$ for $k = 4$ and a factor $\sqrt{n}/\log n$ for $k = 2$. Our findings imply that a faster 4-independent hash function can be used to guarantee the optimal running time for randomized quicksort, which could potentially be of practical interest. Interestingly, our new bounds on the number of performed comparisons under limited independence has implications on classic algorithms for binary planar partitions and treaps. For binary planar partitions our results imply expected partition size $\mathcal{O}(n \log n)$ for the classic randomized algorithm for computing binary planar partitions [11, Page 10] under 4-independence. For randomized treaps [11, Page 201] our new results imply $\mathcal{O}(\log n)$ worst-case depth for 4-independence.

Largest Bucket Size. The last setting we consider is throwing n balls into n buckets using a k-independent hash function and analyzing the size of the largest bucket. This can be regarded as a load balancing as the balls can represent "tasks" and the buckets can represent processing units. Our main result is a family of k-independent hash functions, which when used in this setting implies largest bucket size $\Omega(n^{1/k})$ with constant probability. This result was previously known only for $k = 2$ due to Alon et al. [1] and our result is a generalization of their bound. As an example of the usefulness of such bucket size bounds, consider the fundamental data structure; the dictionary. Widely used algorithms books such as Cormen et al. [7] teaches as the standard method to implement a dictionary to use an array with *chaining*. Chaining here simply means that for each key, corresponding to an entry in the array, we have a linked list (chain) and when a new key-value pair is inserted, it is inserted at the end of the linked list. Clearly then, searching for a particular key-value pair takes worst-case time proportional to the size of the largest chain. Hence, if one is interested in worst-case lookup time guarantees then the expected largest bucket size formed by the keys in the dictionary is of great importance.

2   Relation to Previous Work

We will briefly review related work on the topic of bounding the independence used as well as mention some of the popular hash function constructions.

The line of research that considers the amount of independence required is substantial. As examples, Pagh et al. [12] showed that linear probing works with 5-independence. For the case of ε-min-wise hashing Indyk [9] showed that $\mathcal{O}(\log \frac{1}{\varepsilon})$-independence is sufficient. For both of the above problems Thorup and Pătraşcu [14] showed optimality: They show existence of explicit families of hash functions that for linear probing is 4-independent leading to $\Omega(\log n)$ probes and for ε-min-wise hashing is $\Omega(\log \frac{1}{\varepsilon})$-independent implying (2ε)-min-wise hashing. Additionally, they show that the popular multiply-shift hashing scheme by Dietzfelbinger et al. [8] is not sufficient for linear probing and ε-min-wise hashing.

In terms of lower bounds, it was shown by Broder et al. [3] that $k = 2$ implies $\mathbf{Pr}(\min h(A) = h(x)) = 1/\sqrt{|A|}$. We provide a matching lower bound and new tight bounds for $k = 3, 4$. Additionally, we review a folklore $\mathcal{O}(1/n)$ upper bound for $k = 5$. Our lower bound proofs for min-wise hashing (see Table 1) for $k = 3, 4$ are similar to those of Thorup and Pătrașcu for linear probing, in fact we use the same bad families of hash functions but with a different analysis. Further the same families imply the same multiplicative factors relative to the optimal. Our new tight bounds together with the bounds for $k \geq 5$ due to [9,14] provide the full picture of how min-wise hashing behaves under limited independence.

Randomized quicksort [11] is well known to sort n elements in expected time $\mathcal{O}(n \log n)$ under full independence. Given that pivot elements are picked by having n random variables with outcomes $0, \ldots, n - 1$ and the outcome of variable i in the sequence determines the ith pivot element, then running time $\mathcal{O}(n \log n)$ has been shown [10] for $k = 5$. We improve this and show $\mathcal{O}(n \log n)$ time for $k = 4$ in the same setting. To the knowledge of the authors, it is still an open problem to analyze the version of randomized quicksort under limited independence as presented by e.g. Motwani-Raghavan. The analysis of both the randomized binary planar partition algorithm and the randomized treap in Motwani-Raghavan is done using the exact same argument as for quicksort, namely using min-wise hashing which we show cannot be improved further and is not tight. Our new quicksort bounds directly translates to improvements for these two applications. The randomized binary planar partition algorithm is hence improved to be of expected size $\mathcal{O}(n \log^2 n)$ for $k = 2$ and $\mathcal{O}(n \log n)$ for $k = 4$, and the expected worst case depth of any node in a randomized treap is improved to be $\mathcal{O}(\log^2 n)$ for $k = 2$ and $\mathcal{O}(\log n)$ for $k = 4$.

As briefly mentioned earlier, our largest bucket size result is related to the generalization of Alon et al., STOC'97, specifically [1, Theorem 2]. They show that for a (perfect square) field $\mathbb{F}$ then the class $\mathcal{H}$ of all linear transformations between $\mathbb{F}^2$ and $\mathbb{F}$ has the property that when a hash function is picked uniformly at random from $h \in \mathcal{H}$ then an input set of size n exists so that the largest bucket has size at least $\sqrt{n}$. In terms of upper bounds for largest bucket size, remember that a family $\mathcal{H}_u$ of hash functions that map from $\mathcal{U}$ to $[n]$ is *universal* [4] if for a h picked uniformly from $\mathcal{H}_u$ it holds

$$\forall x \neq y \in \mathcal{U} : \mathbf{Pr}(h(x) = h(y)) \leq 1/n.$$

Universal hash functions are known to have expected largest bucket size at most $\sqrt{n} + 1/2$, hence essentially tight compared to the bound $\sqrt{n}$ lower bound of Alon et al. On the other end of the spectrum, full independence is known to give expected largest bucket size $\Theta(\log n/\log\log n)$ due to a standard application of Chernoff bounds. This bound was proven to hold for $\Theta(\log n/\log\log n)$-independence as well [15]. In Section 7.1 we additionally review a folklore upper bound coinciding with our new $\Omega(n^{1/k})$ lower bound.

Since the question of how much independence is needed from a practical perspective can often be represented as "how fast a hash function can I use and maintain algorithmic guarantees?" we will briefly recap some commonly used hash functions and pseudorandom generators. Functions with lower independence are typically faster in practice than functions with higher independence. The formalization of this is due to Siegel's lower bound [16] in which it is shown that in the cell probe model, to achieve k-independence and number of probes $t < k$ then you need space $k(n/k)^{1/t}$. Since space usage scales with the independence k then for high k the effects of the memory hierarchy will mean that even if the time is held constant the practical time will scale with k as cache effects impact the running time.

The most used hashing scheme in practice is, as mentioned, the 2-independent multiply-shift by Dietzfelbinger et al. [8], which can be twice as fast [18] compared to even the simplest linear transformation $x \mapsto (ax + b) \mod p$. For 3-independence we have due to (analysis by) Thorup and Pătraşcu the simple tabulation scheme [13], which can be altered to give 5-universality [19]. For general k-independent hash functions the standard solution is degree $k - 1$ polynomials, however especially for low k these are known to run slowly, e.g. for $k = 5$ then polynomial hashing is 5 times slower than the tabulation based solution of [19]. Alternatively for high independence the double tabulation scheme by Thorup [17], which builds on Siegels result [16], can potentially be practical. On smaller universes Thorup gives explicit and practical parameters for 100-independence. Also for high independence, the nearly optimal hash function of Christiani et al. [6] should be practical. For generating k-independent variables then Christiani and Pagh's constant time generator [5] performs well — their method is at an order of magnitude faster than evaluating a polynomial using fast fourier transform. We note that even though constant time generators as the above exist, the practical evaluation time will scale with the independence. This comes from the space usage of the generators scaling with the independence of the generated variables, and increasing the working set incurs more cache misses and hence increases the number of block transfers performed.

Finally, we would like to note that the paradigm of independence has its limitations in the sense that even though one can prove that k-independence by itself does not imply certain algorithmic guarantees, it can not be ruled out that k-independent hash functions exist that do. That is, lower bound proofs typically construct artificial families to provide counter examples, which in practice would not come into play. As an example, consider that linear probing needs 5-independence to work as mentioned above but it has been proven to work with simple tabulation hashing [13], which only has 3-independence.

3   Our Results

With regard to min-wise hashing, we close this version of the problem by providing new and tight bounds for $k = 2, 3, 4$. We consider the following setting: let A be a set of size n and let $\mathcal{H}$ be a k-independent family of hash functions. We examine the probability of any element $x \in A$ receiving the smallest hash value $h(x)$ out of all elements in A when $h \in \mathcal{H}$ is picked uniformly at random. For the case of $k = 2, 3, 4$-independent families we provide the new bounds

Table 1. Result overview for min-wise hashing. Results in this paper are marked with *. For a set A of size n and an element $x \in A$, the cells correspond the probability $\mathbf{Pr}(\min h(A) = h(x))$ for a hash function h picked uniformly at random from a k-independent family $\mathcal{H}$.

	$k = 2$	$k = 3$	$k = 4$	$k \geq 5$
UPPER BOUND	$\mathcal{O}(\sqrt{n}/n)$	$\mathcal{O}((\log n)/n)^*$	$\mathcal{O}((\log n)/n)^*$	$\mathcal{O}(1/n)$
LOWER BOUND	$\Omega(\sqrt{n}/n)^*$	$\Omega((\log n)/n)^*$	$\Omega((\log n)/n)^*$	$\Omega(1/n)$

shown in Table 1, which provides a full understanding of the parameter space as a tight bound of $\mathbf{Pr}(\min h(A) = h(x)) = \mathcal{O}(1/n)$ is known for $k \geq 5$ due to Indyk [9]. We note that our lower bound proofs on min-wise hashing, which work by providing explicit bad families of functions, share similarity with Thorup and Pătraşcu's [14, Table 1] lower bounds on linear probing. In fact, our bad families of functions used are exactly the same, while the analysis is different. Surprisingly, the constructions imply the same factor relative to optimal as in linear probing, for every examined value of k.

Next, we consider randomized quicksort under limited independence. In the same setting as Karloff and Raghavan [10], our main result is that 4-independence is sufficient for the optimal $\mathcal{O}(n \log n)$ expected worst-case running time. The setting is essentially that pivot elements are picked from a sequence of k-independent random variables that are pre-computed. Our results apply to a related setting of quicksort as well as to the analysis of binary planar partitions and randomized treaps. Our results are summarized in Table 2.

Table 2. Result overview for randomized quicksort. Results in this paper are marked with *. When our hash function h is picked uniformly from k-independent family $\mathcal{H}$ then the cells in the table denote the expected running time to sort n distinct elements. The 5-independent upper bound is from Karloff-Raghavan [10].

	$k = 2$	$k = 3$	$k = 4$	$k \geq 5$
UPPER BOUND	$\mathcal{O}(n \log^2 n)^*$	$\mathcal{O}(n \log^2 n)^*$	$\mathcal{O}(n \log n)^*$	$\mathcal{O}(n \log n)$
LOWER BOUND	$\Omega(n \log n)$	$\Omega(n \log n)$	$\Omega(n \log n)$	$\Omega(n \log n)$

Finally, we consider the fundamental problem of throwing n balls into n buckets. The main result is a simple k-independent family of functions which when used to throw the balls imply that with constant probability the largest bucket has $\Omega(n^{1/k})$ balls. We show the theorem below.

Theorem 1. *Consider the setting where n balls are distributed among n buckets using a random hash function h. For $m \leq n$ and any $k \in \mathbb{N}$ such that $k < n^{1/k}$ and $m^k \geq n$ a distribution over k-independent hash functions exists such that the largest bucket size is $\Omega(m)$ with probability $\Omega\left(\frac{n}{m^k}\right)$ when h is chosen according to this distribution.*

An implication of Theorem 1 is that we now have the full understanding of the parameter space for this problem, as it was well known that independence

$k = \mathcal{O}(\log n/\log\log n)$ implied $\Theta(\log n/\log\log n)$ balls in the largest bucket. We summarize with the corollary below.

Corollary 1. *Consider the setting where n balls are distributed among n buckets using a random hash function h. Given an integer k a distribution over hash functions exists such that if h is chosen according to this distribution then with L being the size of the largest bucket*

(a) if $k \leq n^{1/k}$ then $L = \Omega\left(n^{1/k}\right)$ with probability $\Omega(1)$.
(b) if $k > n^{1/k}$ then $L = \Omega\left(\log n/\log\log n\right)$ with probability $\Omega(1)$.

We note that the result of Theorem 1 is not quite the generalization of the lower bound of Alon et al. since they show $\Omega(n^{1/2})$ largest bucket size for a special class of linear transformations while our result provides an explicit worst-case k-independent scheme to achieve largest bucket size $\Omega(n^{1/k})$. However, as is evident from the proof of Theorem 1, our scheme is not that artificial: In fact it is "nearly" standard polynomial hashing.

4   Preliminaries

We will introduce some notation and fundamentals used in the paper. For an integer n we let $[n]$ denote $\{0, \ldots, n-1\}$. For an event E we let $[E]$ be the variable that is 1 if E occurs and 0 otherwise. Unless explicitly stated otherwise, $\log n$ refers to the base 2 logarithm of n. For a real number x and a non-negative integer k we define $x^{\underline{k}}$ as $x(x-1)\ldots(x-(k-1))$.

The paper is about application bounds when the independence of the random variables used is limited. We define independence of a hash function formally below.

Definition 1. *Let $h\colon \mathcal{U} \to V$ be picked uniformly at random from a family $\mathcal{H}$ of functions, $k \in \mathbb{N}$ and let $u_1, \ldots, u_k$ be any distinct k elements from $\mathcal{U}$ and $v_1, \ldots, v_k$ be any k elements from V.*
Then the family $\mathcal{H}$ is k-independent if it holds that

$$\mathbf{Pr}_{h\in\mathcal{H}}\left(h(u_1) = v_1 \wedge \ldots \wedge h(u_k) = v_k\right) = \frac{1}{|V|^k}.$$

Note that an equivalent definition for a sequence of random variables hold: they are k-independent if any element is uniformly distributed and every k-tuple of them is independent.

5   Min-Wise Hashing

In this section we show the bounds that can be seen in Table 1. As mentioned earlier, there is a close relationship between the worst case query time of an element in linear probing and min-wise hashing when analyzed under the assumption of hash functions with limited independence. Intuitively, long query time for linear probing is caused by many hash values being "close" to the hash value of the query element. On the other hand, a hash value is likely to be the minimum if it is "far away" from the other hash values. So intuitively, min-wise hashing and linear probing are related by the fact that good guarantees require a "sharp" concentration on how close to the hash value of the query element the other hash values are.

We refer to the full version for the proof details.

5.1   Upper Bounds

We show the following theorem which results in the upper bounds shown in Table 1. Note that the bound for 4-independence follows trivially from the bound for 3-independence and that the 5-independence bound is folklore but included for completeness.

Theorem 2. *Let* $X = \{x_0, x_1, \ldots, x_n\}$ *and* $h : X \to (0, 1)$ *be a hash function. If* h *is 3-independent then*

$$\mathbf{Pr}\left(h(x_0) < \min_{i \in \{1,\ldots,n\}} h(x_i)\right) = \mathcal{O}\left(\frac{\log(n+1)}{n+1}\right)$$

If h *is 5-independent then*

$$\mathbf{Pr}\left(h(x_0) < \min_{i \in \{1,\ldots,n\}} h(x_i)\right) = \mathcal{O}\left(\frac{1}{n+1}\right)$$

5.2   Lower Bounds

We first show the $k = 4$ lower bound seen in Table 1. As mentioned earlier, the argument follows from the same "bad" distribution as Thorup and Pătraşc [14], but with a different analysis.

Theorem 3. *For any key set* $X = \{x_0, x_1, \ldots, x_n\}$ *there exists a random hash function* $h : X \to (0, 1)$ *that is 4-independent such that*

$$\mathbf{Pr}\left(h(x_0) < \min \{h(x_1), \ldots, h(x_n)\}\right) = \Omega\left(\frac{\log(n+1)}{n+1}\right) \qquad (1)$$

The lower bound for $k = 2$ is shown in the following theorem.

Theorem 4. *For any key set* $X = \{x_0, x_1, \ldots, x_n\}$ *there exists a random hash function* $h : X \to [0, 1)$ *that is 2-independent such that*

$$\mathbf{Pr}\left(h(x_0) < \min_{i \in \{1,\ldots,n\}} h(x_i)\right) = \Omega\left(\frac{1}{\sqrt{n}}\right)$$

6   Quicksort

The textbook version of the quicksort algorithm, as explained in [11], is the following. As input, we are given a set of n numbers $S = \{x_0, \dots, x_{n-1}\}$ and we choose a pivot element x_i uniformly at random . We then compare each element in S with x_i and determine the sets S_1 and S_2 which consist of the elements that are smaller and greater than x_i respectively. Then we recursively call the procedure on S_1 and S_2 and output the sorted sequence S_1 followed by x_i and S_2. For this setting, to the knowledge of the authors, there are no known bounds under limited independence.

We consider two different settings where our results seen in Table 2 apply.

Setting 1. Firstly, we consider the same setting as in [10]. Let the input again be $S = \{x_0, \dots, x_{n-1}\}$. The pivot elements are pre-computed the following way: let random variables $Y_1, \dots, Y_n$ be k-independent and each Y_i is uniform over $[n]$. The ith pivot element is chosen to be x_{Y_i}. Note that the sequence of Y_i's is not always a permutation, hence a cleanup phase is necessary afterwards in order to ensure pivots have been performed on all elements.

Setting 2. The second setting we consider is the following. Let $Z = Z_1, \dots, Z_n$ be a sequence of k-independent random variables that are uniform over the interval $(0, 1)$. The first pivot element is x_i where i is the index of the smallest Z_i. Then, recursively, the pivot elements are found in the same manner in the subproblems. We note that finding the smallest Z_i in each interval incurs an additional cost that is of the same order as sorting the sequence of Z_i's.

In this section we show the results of Table 2 in Setting 1. We refer to the full version of this article for proofs for Setting 2 and note that the same bounds apply to both settings.

Recall that we can use the results on min-wise hashing to show upper bounds on the running time. The key to sharpening this analysis is to consider a problem related to that of min-wise hashing. In Lemma 1, we show that for two sets A, B satisfying $|A| \leq |B|$ there are only $O(1)$ pivot elements chosen from A before the first element is chosen from B. We could use a min-wise type of argument to show that a single element $a \in A$ is chosen as a pivot element before the first pivot element is chosen from B with probability at most $\mathcal{O}\left(\frac{\log n}{|B|}\right)$. However, this would only give us an upper bound of $\mathcal{O}(\log n)$ and not $\mathcal{O}(1)$. We refer to the full version for the proof details.

Lemma 1. *Let* $h : [n] \to [n]$ *be a 4-independent hash function and let* $A, B \subseteq [n]$ *be disjoint sets such that* $|A| \leq |B|$. *Let* $j \in [n]$ *be the smallest value such that* $h(j) \in B$, *and* $j = n$ *if no such* j *exist. Then let* C *be the* $i \in [j]$ *such that* $h(i) \in A$, *i.e.*

$$C = \{i \in [n] \mid h(i) \in A, \ h(0), \dots, h(i-1) \notin B\}$$

Then $\mathbb{E}(|C|) = O(1)$.

We show how we apply Lemma 1 to guarantee that quicksort only makes $\mathcal{O}(n \log n)$ comparisons in expectation.

Theorem 5. *Consider quicksort in Setting 1 where we sort a set $S = \{x_0, \ldots, x_{n-1}\}$ and pivot elements are chosen using a 4-independent hash function. For any i the expected number of times x_i is compared with another element $x_j \in S \setminus \{x_i\}$ when x_j is chosen as a pivot element is $\mathcal{O}(\log n)$. In particular the expected running time is $\mathcal{O}(n \log n)$.*

Next we show that the cleanup phase as described by Setting 1 takes $\mathcal{O}(n \log n)$ for $k = 2$, which means it makes no difference to asymptotic running time of quicksort.

Lemma 2. *Consider quicksort in Setting 1 where we sort a set $S = \{x_0, \ldots, x_{n-1}\}$ with a 2-independent hash function. The cleanup phase takes $\mathcal{O}(n \log n)$ time.*

Finally we show the new 2-independent bound. The argument follows as the 4-independent argument, except with 2nd moment bounds instead of 4th moment bounds.

Theorem 6. *Consider quicksort in Setting 1 where we sort a set $S = \{x_0, \ldots, x_{n-1}\}$ and pivot elements are chosen using a 2-independent hash function. For any i the expected number of times x_i is compared with another element $x_j \in S \setminus \{x_i\}$ when x_j is chosen as a pivot element is $\mathcal{O}(\log^2 n)$. In particular the expected running time is $\mathcal{O}(n \log^2 n)$.*

6.1 Binary Planar Partitions and Randomized Treaps

The result for quicksort shown in Theorem 5 has direct implications for two classic randomized algorithms. Both algorithms are explained in common text books, e.g. Motwani-Raghavan.

A straightforward analysis of the randomized algorithm [11, Page 12] for constructing binary planar bipartitions simply uses min-wise hashing to analyze the expected size of the partition. In the analysis, the size of the constructed partition depends on the probability of the event happening that a line segment u comes before a line segment v in the random permutation $u, \ldots, u_i, v$. Using the the min-wise probabilities of Table 1 directly we get the same bounds on the partition size as running times on quicksort using the min-wise analysis. This analysis is tightened through Theorem 5 for both $k = 2$ and $k = 4$.

By an analogous argument, the randomized treap data structure of [11, Page 201] achieves expected node depth $\mathcal{O}(\log n)$ when a treap is built over a size n set using the min-wise bounds. Under limited independence using the min-wise analysis, the bounds achieved are then $\mathcal{O}(\sqrt{n}), \mathcal{O}(\log^2 n), \mathcal{O}(\log^2 n), \mathcal{O}(\log n)$ for $k = 2, 3, 4, 5$ respectively. By Theorem 5 we get $\mathcal{O}(\log^2 n)$ for $k = 2$ and $\mathcal{O}(\log n)$ for $k = 4$.

7   Largest Bucket Size

We explore the standard case of throwing n balls into n buckets using a random hash function. We are interested in analyzing the bucket that has the largest number of balls mapped to it. Particularly, for this problem our main contribution is an explicit family of hash functions that are k-independent (remember Definition 1) and where the largest bucket size is $\Omega\left(n^{1/k}\right)$. However we start by stating the matching upper bound. For the proof details of both bounds we refer to the full version of the article.

7.1   Upper Bound

We will briefly show the upper bound that matches our lower bound presented in the next section. We are unaware of literature that includes the upper bound, but note that it follows from a standard argument and is included for the sake of completeness.

Lemma 3. *Consider the setting where n balls are distributed among n buckets using a random hash function h. For $m = \Omega\left(\frac{\log n}{\log \log n}\right)$ and any $k \in \mathbb{N}$ such that $k < n^{1/k}$ then if h is k-independent the largest bucket size is $\mathcal{O}(m)$ with probability at least $1 - \frac{n}{m^k}$.*

7.2   Lower Bound

At a high level, our hashing scheme is to divide the buckets into sets of size p, for prime a $p = \Theta(m)$, and in each set polynomial hashing is used on the keys that do not "fill" the set. The crucial point is then to see that for polynomial hashing, the probability that a particular polynomial hashes a set of keys to the same value can be bounded by the probability of all coefficients of the polynomial being zero. Having a bound on this probability, the set size can be picked such that with constant probability the coefficients of one of the polynomials is zero, resulting in a large bucket.

Since it is well known that using $\mathcal{O}(\log n / \log \log n)$-independent hash function to distribute the balls will imply largest bucket size $\Omega(\log n / \log \log n)$, Corollary 1 provides the full understanding of the largest bucket size.

Proof. (of Corollary 1) Part (a) follows directly from Theorem 1. Part (b) follows since $k > n^{1/k}$ implies $k > \log n / \log \log n$ and so we apply the $\Omega(\log n / \log \log n)$ bound from [15]. ∎

References

1. Alon, N., Dietzfelbinger, M., Miltersen, P.B., Petrank, E., Tardos, G.: Is linear hashing good? In: Symposium on Theory of Computing, STOC 1997 (1997)
2. Broder, A.Z.: On the resemblance and containment of documents. In: Compression and Complexity of Sequences (SEQUENCES), pp. 21–29 (1997)

3. Broder, A.Z., Charikar, M., Frieze, A.M., Mitzenmacher, M.: Min-wise independent permutations. Journal of Computer and System Sciences 60
4. Carter, J.L., Wegman, M.N.: Universal classes of hash functions. Journal of Computer and System Sciences 18(2), 143–154 (1979)
5. Christiani, T., Pagh, R.: Generating k-independent variables in constant time. In: Foundations of Computer Science (FOCS) (2014)
6. Christiani, T., Pagh, R., Thorup, M.: From independence to expansion and back again. In: Proceedings of the Forty-Seventh Annual ACM on Symposium on Theory of Computing, STOC 2015, pp. 813–820. ACM, New York (2015)
7. Cormen, T.H., Stein, C., Rivest, R.L., Leiserson, C.E.: Introduction to algorithms, 2nd edn. McGraw-Hill Higher Education (2001)
8. Dietzfelbinger, M., Hagerup, T., Katajainen, J., Penttonen, M.: A reliable randomized algorithm for the closest-pair problem. Journal of Algorithms 25(1), 19–51 (1997)
9. Indyk, P.: A small approximately min-wise independent family of hash functions. In: ACM-SIAM Symposium on Discrete Algorithms, SODA 1999 (1999)
10. Karloff, H., Raghavan, P.: Randomized algorithms and pseudorandom numbers. In: Proceedings of the Twentieth Annual ACM Symposium on Theory of Computing, STOC 1988, pp. 310–321. ACM, New York (1988)
11. Motwani, R., Raghavan, P.: Randomized algorithms. Cambridge University Press, New York (1995)
12. Pagh, A., Pagh, R., Ruzic, M.: Linear probing with constant independence. In: Proceedings of the Thirty-ninth Annual ACM Symposium on Theory of Computing, STOC 2007, pp. 318–327. ACM (2007)
13. Pătraşcu, M., Thorup, M.: The power of simple tabulation hashing. In: Proceedings of the Forty-third Annual ACM Symposium on Theory of Computing, STOC 2011, pp. 1–10. ACM, New York (2011)
14. Pătraşcu, M., Thorup, M.: On the k-independence required by linear probing and minwise independence, CoRR abs/1302.5127 (2013)
15. Schmidt, J.P., Siegel, A., Srinivasan, A.: Chernoff-Hoeffding bounds for applications with limited independence. SIAM J. Discret. Math. 8(2), 223–250 (1995)
16. Siegel, A.: On universal classes of extremely random constant-time hash functions. SIAM J. Comput. 33(3), 505–543 (2004)
17. Thorup, M.: Simple tabulation, fast expanders, double tabulation, and high independence. In: Proc. FOCS 2013, pp. 90–99 (2013)
18. Thorup, M.: Even strongly universal hashing is pretty fast. In: Proceedings of the Eleventh Annual ACM-SIAM Symposium on Discrete Algorithms, SODA 2000, Philadelphia, PA, USA, pp. 496–497. Society for Industrial and Applied Mathematics (2000)
19. Thorup, M., Zhang, Y.: Tabulation based 5-universal hashing and linear probing. In: ALENEX 2010, pp. 62–76 (2010)

Maximum Matching in Turnstile Streams*

Christian Konrad

Reykjavik University, Reykjavik, Iceland
christiank@ru.is

Abstract. We consider the unweighted bipartite maximum matching problem in the one-pass turnstile streaming model where the input stream consists of edge insertions and deletions. In the insertion-only model, a one-pass 2-approximation streaming algorithm can be easily obtained with space $O(n \log n)$, where n denotes the number of vertices of the input graph. We show that no such result is possible if edge deletions are allowed, even if space $O(n^{3/2-\delta})$ is granted, for every $\delta > 0$. Specifically, for every $0 \le \epsilon \le 1$, we show that in the one-pass turnstile streaming model, in order to compute a $O(n^\epsilon)$-approximation, space $\Omega(n^{3/2-4\epsilon})$ is required for constant error randomized algorithms, and, up to logarithmic factors, space $\tilde{O}(n^{2-2\epsilon})$ is sufficient.

Our lower bound result is proved in the simultaneous message model of communication and may be of independent interest.

1 Introduction

Massive graphs are usually dynamic objects that evolve over time in structure and size. For example, the Internet graph changes as webpages are created or deleted, the structure of social network graphs changes as friendships are established or ended, and graph databases change in size when data items are inserted or deleted. Dynamic graph algorithms can cope with evolving graphs of moderate sizes. They receive a sequence of updates, such as edge insertions or deletions, and maintain valid solutions at any moment. However, when considering massive graphs, these algorithms are often less suited as they assume random access to the input graph, an assumption that can hardly be guaranteed in this context. Consequently, research has been carried out on dynamic graph streaming algorithms that can handle both edge insertions and deletions.

Dynamic Graph Streams. A data streaming algorithm processes an input stream $X = X_1, \ldots, X_n$ sequentially item by item from left to right in passes while using a memory whose size is sublinear in the size of the input [24]. Graph streams have been studied for almost two decades. However, until recently, all graph streams considered in the literature were *insertion-only*, i.e., they process streams consisting of sequences of edge insertions. In 2012, Ahn, Guha and McGregor [2] initiated the study of *dynamic graph streaming algorithms* that process streams consisting of both edge insertions and deletions. Since then, it has been shown that a variety of problems for which space-efficient streaming

---

* Supported by Icelandic Research Fund grant-of-excellence no. 120032011.

N. Bansal and I. Finocchi (Eds.): ESA 2015, LNCS 9294, pp. 840–852, 2015.
DOI: 10.1007/978-3-662-48350-3_70

algorithms in the insertion-only model are known, such as testing connectivity and bipartiteness, computing spanning trees, computing cut-preserving sparsifiers and spectral sparsifiers, can similarly be solved well in small space in the dynamic model [2,3,19,18]. An exception is the maximum matching problem which, as we will detail later, is probably the most studied graph problem in streaming settings. In the insertion-only model, a 2-approximation algorithm for this problem can easily be obtained in one pass with $O(n \log n)$ space, where n is the number of vertices in the input graph. Even in the sliding-window model [1], which can be seen as a model located between the insertion-only model and the dynamic model, the problem can be solved well [7]. The status of the problem in the dynamic model has been open so far, and, in fact, it is one of the open problems collected at the Bertinoro 2014 workshop on sublinear algorithms [2].

Results on dynamic matching algorithms [5,4] show that even when the sequence of graph updates contains deletions, then large matchings can be maintained without too many reconfigurations. These results may give reasons for hope that constant or poly-logarithmic approximations could be achieved in the one-pass dynamic streaming model. We, however, show that if there is such an algorithm, then it uses a huge amount of space.

Summary of Our Results. In this paper, we present a one-pass dynamic streaming algorithm for maximum bipartite matching and a space lower bound for streaming algorithms in the turnstile model, a slightly more general model than the dynamic model (see Section 2 for a discussion), the latter constituting the main contribution of this paper. We show that in one pass, an $O(n^\epsilon)$-approximation can be computed in space $\tilde{O}(n^{2-2\epsilon})$ (**Theorem 4**), and space $\Omega(n^{3/2-4\epsilon})$ is necessary for such an approximation (**Corollary 1**).

Lower Bound via Communication Complexity. Many lower bounds on the space requirements of algorithms in the insertion-only model are proved in the one-way communication model. In the one-way model, party one sends a message to party two who, upon reception, sends a message to party three. This process continues until the last party receives a message and outputs the result. A recent result by Li, Nguyên and Woodruff [21] shows that space lower bounds for turnstile streaming algorithms can be proved in the more restrictive *simultaneous model of communication* (SIM model). In this model, the participating parties simultaneously each send a single message to a third party, denoted the referee, who computes the output of the protocol as a function of the received messages. A lower bound on the size of the largest message of the protocol is then a lower bound on the space requirements of a turnstile one-pass streaming algorithm. Our paper is the first that uses this connection in the context of graph problems.

A starting point for our lower bound result is a work of Goel, Kapralov and Khanna [12], and a follow-up work by Kapralov [16]. In [12], via a one-way two-

---

[1] In the sliding-window model, an algorithm receives a potentially infinite insertion-only stream, however, only a fixed number of most recent edges are considered by the algorithm. Edges outside the most recent window of time are seen as deleted.

[2] See also http://sublinear.info/64

party communication lower bound, it is shown that in the insertion-only model, every algorithm that computes a $(3/2 - \epsilon)$-approximation, for $\epsilon > 0$, requires $\Omega(n^{1+\frac{1}{\log\log n}})$ space. This lower bound has then been strengthened in [16] to hold for $(e/(e-1)-\epsilon)$-approximation algorithms. Both lower bound constructions heavily rely on *Ruzsa-Szemerédi graphs*. A graph G is an (r, s)-Ruzsa-Szemerédi graph (in short: RS-graph), if its edge set can be partitioned into r disjoint induced matchings each of size at least s. The main argument of [12] can be summarized as follows: Suppose that the first party holds a relatively dense Ruzsa-Szemerédi graph G_1. The second party holds a graph G_2 whose edges render one particular induced matching $M \subseteq E(G_1)$ of the first party indispensable for every large matching in the graph $G_1 \cup G_2$, while all other induced matchings are rendered redundant. Note that as M is an induced matching, there are no alternative edges in G_1 different from M that interconnect the vertices that are matched by M. As the first party is not aware which of its induced matchings is required, and as the communication budget is restricted, only few edges of M on average will be sent to the second party. Hence, the expected size of the output matching is bounded.

When implementing the previous idea in the SIM setting, the following issues have to be addressed:

Firstly, the number of parties in the simultaneous message protocol needs to be at least as large as the desired bound on the approximation factor. The trivial protocol where every party sends a maximum matching of its subgraph, and the referee outputs the largest received matching, shows that the approximation factor cannot be larger than the number of parties, even when message sizes are as small as $\tilde{O}(n)$. Hence, proving hardness for polynomial approximation factors requires a polynomial number of participating parties. On the other hand, the number of parties can neither be chosen too large: If the input graph is equally split among p parties, for a large p, then the subgraphs of the parties are of size $O(n^2/p)$. Thus, with messages of size $\tilde{O}(n^2/p)$, all subgraphs can be sent to the referee who then computes and outputs an optimal solution. Hence, the larger the number of parties, the weaker a bound on the message sizes can be achieved.

Secondly, there is no "second party" as in the one-way setting whose edges could render one particular matching of every other party indispensable. Instead, a construction is required so that every party both has the function of party one (one of its induced matchings is indispensable for every large matching) and of party two (some of its edges render many of the induced matchings of other parties redundant). This suggests that the RS-graphs of the parties have to overlap in many vertices. While arbitrary RS-graphs with good properties can be employed for the lower bounds of [12] and [16], we need RS-graphs with simple structure in order to coordinate the overlaps between the parties.

We show that both concerns can be handled. In Section 3, we present a carefully designed input distribution where each party holds a highly symmetrical RS-graph. The RS-graph of a party overlaps almost everywhere with the RS-graphs of other parties, except in one small induced matching. This matching, however, cannot be distinguished by the party, and hence, as in the one-way setting, the referee will not receive many edges of this matching.

Upper Bound. Our upper bound result is achieved by an implementation of a simple matching algorithm in the dynamic streaming model: For an integer k, pick a random subset $A' \subseteq A$ of size k of one bipartition of the bipartite input graph $G = (A, B, E)$; for each $a \in A'$, store arbitrary $\min\{k, \deg(a)\}$ incident edges, where $\deg(a)$ denotes the degree of a in the input graph; output a maximum matching in the graph induced by the stored edges. We prove that this algorithm has an approximation factor of n/k. In order to collect k incident edges of a given vertex in the dynamic streaming model, we employ the l_0-samplers of Jowhari, Sağlam, Tardos [15], which have previously been used for dynamic graph streaming algorithms [2]. By chosing $k = \Theta(n^{1-\epsilon})$, this construction leads to a $O(n^\epsilon)$-approximation algorithm with space $\tilde{O}(n^{2-2\epsilon})$. While this algorithm in itself is rather simple and standard, it shows that non-trivial approximation ratios for maximum bipartite matching in the dynamic streaming model are possible with sublinear space. Our upper and lower bounds show that in order to compute a n^ϵ-approximation, space $\tilde{O}(n^{2-2\epsilon})$ is sufficient and space $\Omega(n^{3/2-4\epsilon})$ is required. Improving on either side is left as an open problem.

Further Related Work. Matching problems are probably the most studied graph problem in the streaming model [11,22,8,9,1,2,20,25,12,16,13,7,6,18,23,17,10]. Closest to our work are the already mentioned lower bounds [12] and [16]. Their arguments are combinatorial and so are the arguments in this paper. Note that lower bounds for matching problems in communication settings have also been obtained via information complexity in [13,14].

In the dynamic streaming model, Ahn et al. [2] provide a multi-pass algorithm with $O(n^{1+1/p} \operatorname{poly} \epsilon^{-1})$ space, $O(p \cdot \epsilon^{-2} \cdot \log \epsilon^{-1})$ passes, and approximation factor $1 + \epsilon$ for the weighted maximum matching problem, for a parameter p. This is the only result on matchings known in the dynamic streaming setting.

Outline. After a section on preliminaries, we present our hard input distribution in Section 3 which is then used in Section 4 in order to prove our lower bound in the SIM model. Finally, our upper bound is presented in Section 5.

Due to space restrictions, lemmas and theorems marked with (*) are postponed to the full version of this article.

2   Preliminaries

For an integer $a \geq 1$, we write $[a]$ for $\{1, \ldots, a\}$. We use the notation $\tilde{O}()$, which equals the standard $O()$ notation where all poly-logarithmic factors are ignored.

Simultaneous Communication Complexity. Let $G = (A, B, E)$ denote a simple bipartite graph, and, for an integer $P \geq 2$, let $G_1, \ldots, G_P$ be edge-disjoint subgraphs of G. In the simultaneous message complexity setting, for $p \in [P]$, party p is given G_p, and sends a single message μ_p of limited size to a third party denoted the referee. Upon reception of all messages, the referee outputs a matching M in G. Note that the participating parties cannot communicate with each other, but they have access to an infinite number of shared random coin flips which can be used to synchronize their messages.

We say that an algorithm/protocol is a constant error algorithm/protocol if it errs with probability at most ϵ, for $0 \leq \epsilon < 1/2$. We also assume that a algorithm/protocol never outputs edges that do not exist in the input graph.

Turnstile Streams. For a bipartite graph $G = (A, B, E)$, let $X = X_1, X_2, \ldots$ be the input stream with $X_i \in E \times \{+1, -1\}$, where $+1$ indicates that an edge is inserted, and -1 indicates that an edge is deleted. Edges could potentially be inserted multiple times, or be removed before they have been inserted, as long as once the stream has been fully processed, the multiplicity of an edge is in $\{-c, -c+1, \ldots, c-1, c\}$, for some integer c. The reduction of [21] and hence our lower bound holds for algorithms that can handle this type of dynamic streams, also known as *turnstile streams*. Such algorithms may for instance abort if negative edge multiplicities are encountered, or they output a solution among the edges with non-zero multiplicity.

In [21] it is shown that every turnstile algorithm can be seen as an algorithm that solely computes a linear sketch of the input stream. As linear sketches can be implemented in the SIM model, lower bounds in the SIM model are lower bounds on the sketching complexity of problems, which in turn imply lower bounds for turnstile algorithms. We stress that our lower bound holds for linear sketches. Note that *all* known dynamic graph algorithms[3] solely compute linear sketches (e.g. [2,3,19,18]). This gives reasons to conjecture that also all dynamic algorithms can be seen as linear sketches, and, as a consequence, our lower bound not only holds for turnstile algorithms but for all dynamic algorithms.

3   Hard Input Distribution

In this section, we construct our hard input distribution. First, we describe the construction of the distribution from a global point of view in Subsection 3.1. Restricted to the input graph G_p of any party $p \in [P]$, the distribution of G_p can be described by a different construction which is simpler and more suitable for our purposes. This will be discussed in Subsection 3.2.

3.1   Hard Input Distribution: Global View

Denote by P the number of parties of the simultaneous message protocol. Let k, Q be integers so that $P \leq k \leq \frac{n}{P}$, and $Q = o(P)$. The precise values of k and Q will be determined later. First, we define a bipartite graph $G' = (A, B, E)$ on $O(n)$ vertices with $A = B = [(Q + P)k]$ from which we obtain our hard input distribution. For $1 \leq i \leq Q + P$, let $A_i = \{1 + (i - 1)k, ik\}$ and let $B_i = \{1 + (i - 1)k, ik\}$. The edge set E is a collection of matchings as follows:

$$E = \bigcup_{i,j \in [Q], p \in [P]} M_{i,j}^p \cup \bigcup_{i \in \{Q+1, \ldots, Q+P\}, j \in [Q]} (M_{i,j} \cup M_{j,i}) \cup \bigcup_{i \in \{Q+1, \ldots, Q+P\}} M_{i,i},$$

---

[3] Some of those algorithms couldn't handle arbitrary turnstile streams as they rely on the fact that all edge multiplicities are in $\{0, 1\}$.

where $M_{i,j}$ is a perfect matching between A_i and B_j, and $M_{i,j}^1, \ldots, M_{i,j}^P$ are P edge-disjoint perfect matchings between A_i and B_j. Note that as we required that $k \geq P$, the edge-disjoint matchings $M_{i,j}^1, \ldots, M_{i,j}^P$ can be constructed[4].

From G', we construct the input graphs of the different parties as follows:

1. For every $p \in [P]$, let $G'_p = (A, B, E'_p)$ where E'_p consists of the matchings $M_{i,j}^p$ for $i, j \in [Q]$, the matching $M_{Q+p,Q+p}$ and the matchings $M_{Q+p,j}$ and $M_{j,Q+p}$ for $j \in [Q]$.

2. For every $p \in [P]$, for every matching M of G'_p, pick a subset of edges of size $k/2$ from M uniformly at random and replace M by this subset.

3. Pick random permutations $\pi_A, \pi_B : [Q + P] \to [Q + P]$. Permute the vertex IDs of the graphs G'_p, for $1 \leq p \leq P$, so that if $\pi_A(i) = j$ then A_i receives the IDs of A_j as follows: The vertices $a_1 = 1 + k(i - 1), a_2 = 2 + k(i - 1), \ldots, a_k = ki$ receive new IDs so that after the change of IDs, we have $a_1 = 1 + k(j - 1), a_2 = 2 + k(j - 1), \ldots, a_k = kj$. The same procedure is carried out with vertices B_i and permutation π_B. Denote by G_p the graph G'_p once half of the edges have been removed and the vertex IDs have been permuted. Let G be the union of the graphs G_p.

The structure of G' and a subgraph G'_p is illustrated in Figure 1.

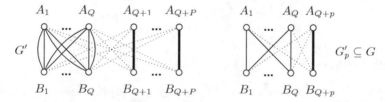

$$A_1 \quad A_Q \quad A_{Q+1} \quad A_{Q+P} \qquad\qquad A_1 \quad A_Q \ A_{Q+p}$$

G' $\qquad\qquad\qquad\qquad\qquad\qquad\qquad\qquad\qquad\qquad$ $G'_p \subseteq G$

$$B_1 \quad B_Q \quad B_{Q+1} \quad B_{Q+P} \qquad\qquad B_1 \quad B_Q \ B_{Q+p}$$

Fig. 1. Left: Graph G'. A vertex corresponds to a group of k vertices. Each edge indicates a perfect matching between the respective vertex groups. The bold edges correspond to the matchings $M_{Q+p,Q+p}$, for $1 \leq p \leq P$, the solid edges correspond to matchings $M_{i,j}^p$, for $1 \leq i, j \leq Q$, $1 \leq p \leq P$, and the dotted edges correspond to matchings $M_{Q+p,i}, M_{i,Q+p}$, for $1 \leq i \leq Q$ and $1 \leq p \leq P$. Right: Subgraph $G'_p \subseteq G$.

Properties of the input graphs. Graph G' has a perfect matching of size $(Q + P)k$ which consists of a perfect matching between vertices $A_1, \ldots, A_Q$ and $B_1, \ldots, B_Q$, and the matchings $M_{Q+p,Q+p}$ for $1 \leq p \leq P$. As by Step 2 of the construction of the hard instances, we remove half of the edges of every matching, a maximum matching in graph G is of size at least $\frac{(Q+P)k}{2}$. Note that while there are many possibilities to match the vertex groups $A_1, \ldots, A_Q$ and $B_1, \ldots, B_Q$, in every large matching, many vertices of A_{Q+i} are matched to vertices of B_{Q+i} using edges from the matching $M_{Q+i,Q+i}$. For some $p \in [P]$, consider now the

---

[4] For instance, define G' so that $G'|_{A_i \cup B_i}$ is a P-regular bipartite graph. It is well-known (and easy to see via Hall's theorem) that any P-regular bipartite graph is the union of P edge-disjoint perfect matchings.

graph G'_p from which the graph G_p is constructed. G'_p consists of perfect matchings between the vertex groups A_i and B_j for every $i, j \in [Q] \cup \{Q+p\}$. In graph G_p, besides the fact that only half of the edges of every matching are kept, the vertex IDs are permuted. We will argue that due to the permuted vertices, given G_p, it is difficult to determine which of the matchings corresponds to the matching $M_{Q+p,Q+p}$ in G'. Therefore, if the referee is able to output edges from the matching $M_{Q+p,Q+p}$, then many edges from every matching have to be included into the message μ_p sent by party p.

3.2   Hard Input Distribution: Local View

From the perspective of an individual party, by symmetry of the previous construction, the distribution from which the graph G_p is chosen can also be described as follows:

1. Pick $I_A, I_B \subseteq [Q + P]$ so that $|I_A| = |I_B| = Q + 1$ uniformly at random.
2. For every $i \in I_A$ and $j \in I_B$, introduce a matching of size $k/2$ between A_i and B_j chosen uniformly at random from all possible matching between A_i and B_j of size $k/2$.

G_p can be seen as a $((Q+1)^2, k/2)$-Ruzsa-Szemerédi graph or as a $(Q+1, k(Q+1)/2)$-Ruzsa-Szemerédi graph. Let $\mathcal{G}_p$ denote the possible input graphs of party p. We prove now a lower bound on $|\mathcal{G}_p|$.

Lemma 1. *There are at least* $|\mathcal{G}_p| > \binom{Q+P}{Q+1} \frac{(Q+P)!}{(P-1)!} \left(\frac{2^k}{k+1} \right)^{(Q+1)^2}$ *possible input graphs for every party p. Moreover, the input distribution is uniform.*

Proof. The vertex groups I_A and I_B are each of cardinality $Q+1$ and chosen from the set $[Q + P]$. There are $\binom{Q+P}{Q+1}$ choices for I_A. Consider one particular choice of I_A. Then, there are $\frac{(Q+P)!}{(P-1)!}$ possibilities to pair those with $Q+1$ vertex groups of the B nodes. Each matching is a subset of $k/2$ edges from k potential edges. Hence, there are $\binom{Q+P}{Q+1} \frac{(Q+P)!}{(P-1)!} \left(\frac{k}{\frac{1}{2}k} \right)^{(Q+1)^2}$ input graphs for each party. Using a bound on the central binomial coefficient, this term can be bounded from below by $\binom{Q+P}{Q+1} \frac{(Q+P)!}{(P-1)!} \left(\frac{2^k}{k+1} \right)^{(Q+1)^2}$. $\square$

The matching in G_p that corresponds to the matching between A_{Q+p} and B_{Q+p} in G'_p will play an important role in our argument. In the previous construction, every introduced matching in G_p plays the role of matching $M_{Q+p,Q+p}$ in G'_p with equal probability. In the following, we will denote by M_p the matching in G_p that corresponds to the matching $M_{Q+p,Q+p}$ in G'_p.

4   Simultaneous Message Complexity Lower Bound

We prove now that no communication protocol with limited maximal message size performs well on the input distribution described in Section 3. First, we focus on deterministic protocols, and we prove a lower bound on the expected approximation ratio (over all possible input graphs) of any deterministic protocol (Theorem 1). Then, via an application of Yao's lemma, we obtain our result for randomized constant error protocols (Theorem 2). Our lower bound for dynamic one-pass streaming algorithms, Corollary 1, is then obtained as a corollary of Theorem 2 and the reduction of [21].

Lower Bound For Deterministic Protocols. Consider a deterministic protocol that runs on a hard instance graph G and uses messages of length at most s. As the protocol is deterministic, for every party $p \in [P]$, there exists a function m_p that maps the input graph G_p of party p to a message μ_p. As the maximum message length is limited by s, there are 2^s different possible messages. Our parameters Q, k will be chosen so that s is much smaller than the number of input graphs G_p for party p, as stated in Lemma 1. Consequently, many input graphs are mapped to the same message.

Consider now a message μ_p and denote by μ_p^{-1} the set of graphs G_p that are mapped by m_p to message μ_p. Upon reception of μ_p, the referee can only output edges that are contained in *every* graph of μ_p^{-1}, since all outputted edges have to be contained in the input graph.

Let N denote the matching outputted by the referee, and let $N_p = N \cap M_p$ denote the outputted edges from matching M_p. Furthermore, for a given message μ_p, denote by $G_{\mu_p} := M_p \cap \bigcap_{G_p \in \mu_p^{-1}} G_p$.

In the following, we will bound the quantity $\mathbb{E}|N_p|$ from above (Lemma 2). By linearity of expectation, this allows us to argue about the expected number of edges of the matchings $\cup_p N_p$ outputted by the referee. We can hence argue about the expected size of the outputted matching, which in turn implies a lower bound on the approximation guarantee of the protocol (Theorem 1).

Lemma 2. *For every party $p \in [P]$, we have $\mathbb{E}|N_p| = \mathrm{O}\left(\frac{\sqrt{sk}}{Q}\right)$.*

Proof. Let Γ denote the set of potential messages from party p to the referee. As the maximum message length is bounded by s, we have $|\Gamma| \leq 2^s$. Let $V = \frac{|\mathcal{G}_p|}{k2^s}$ be a parameter which splits the set Γ into two parts as follows. Denote by $\Gamma_{\geq} \subseteq \Gamma$ the set of messages μ_p so that $|\mu_p^{-1}| \geq V$, and let $\Gamma_{<} = \Gamma \setminus \Gamma_{\geq}$. In the following, for a message $\mu_p \in \Gamma$, we denote by $\mathbb{P}[\mu_p]$ the probability that message μ_p is sent by party p. Note that $\sum_{\mu_p \in \Gamma_{<}} \mathbb{P}[\mu_p] < \frac{2^s V}{|\mathcal{G}_p|}$, since there are at most $2^s V$ input graphs that are mapped to messages in $\Gamma_{<}$. We hence obtain:

$$\mathbb{E}|N_p| \leq \sum_{\mu_p \in \Gamma} \mathbb{P}[\mu_p]\,\mathbb{E}|G_{\mu_p}| = \sum_{\mu_p \in \Gamma_{\geq}} \left(\mathbb{P}[\mu_p]\,\mathbb{E}|G_{\mu_p}|\right) + \sum_{\mu_p \in \Gamma_{<}} \left(\mathbb{P}[\mu_p]\,\mathbb{E}|G_{\mu_p}|\right)$$

$$\leq \sum_{\mu_p \in \Gamma_{\geq}} \left(\frac{|\mu_p^{-1}|}{2^s}\mathbb{E}|G_{\mu_p}|\right) + \sum_{\mu_p \in \Gamma_{<}} \left(\mathbb{P}[\mu_p]\right) k$$

$$< \max\{\mathbb{E}|G_{\mu_p}| : \mu_p \in \Gamma_{\geq}\} + \frac{2^s V}{|\mathcal{G}_p|}k = \max\{\mathbb{E}|G_{\mu_p}| : \mu_p \in \Gamma_{\geq}\} + 1,$$

where we used the definition of V for the last equality. In Lemma 3, we prove that $\forall \mu_p \in \Gamma_\geq : \mathbb{E}|G_{\mu_p}| = O\left(\frac{\sqrt{sk}}{Q}\right)$. This then implies the result.   □

Lemma 3. *Suppose* μ_p *is so that* $|\mu_p^{-1}| \geq V = \frac{|\mathcal{G}_p|}{k2^s}$. *Then,* $\mathbb{E}|G_{\mu_p}| = O\left(\frac{\sqrt{sk}}{Q}\right)$.

Proof. Remember that every graph $G_p \in \mu_p^{-1}$ consists of $(Q+1)^2$ edge-disjoint matchings, and M_p is a randomly chosen one of those. We define

$$I_l = \{(i,j) \in [Q+P] \times [Q+P] : G_{\mu_p}|_{A_i \cup B_j} \text{ contains a matching of size } l\}.$$

We prove first that if $|I_l|$ is large, then μ_p^{-1} is small.
Claim (*). Let $l = o(k)$. Then, $|I_l| \geq x \Rightarrow |\mu_p^{-1}| < \binom{Q+P}{Q+1} \frac{(Q+P)!}{(P-1)!} (\frac{3}{4})^{lx} \left(\frac{2^k}{\sqrt{k}}\right)^{(Q+1)^2}$.

Then, we can bound:

$$\mathbb{E}|G_{\mu_p}| \leq \frac{|I_l|}{\binom{Q+1}{2}} \cdot k + \left(1 - \frac{|I_l|}{\binom{Q+1}{2}}\right)l < \frac{|I_l|}{\binom{Q+1}{2}} \cdot k + l. \tag{1}$$

Note that by assumption, we have $\mu_p^{-1} \geq V$. Let l, x be two integers so that:

$$\binom{Q+P}{Q+1} \frac{(Q+P)!}{(P-1)!} (\frac{3}{4})^{lx} \left(\frac{2^k}{\sqrt{k}}\right)^{(Q+1)^2} = V. \tag{2}$$

Then, by the previous claim, we obtain $|I_l| < x$. Solving Equality 2 for variable x, and further bounding it yields:

$$x \leq \frac{1}{l}\left((Q+1)^2(k - \frac{1}{2}\log k) + \log\left(\binom{Q+P}{Q+1}\frac{(Q+P)!}{(P-1)!}\right) - \log V\right). \tag{3}$$

Remember that V was chosen as $V = \frac{|\mathcal{G}_p|}{k2^s}$, and hence $\log V \geq (Q+1)^2(k - \log(k+1)) + \log\left(\binom{Q+P}{Q+1}\frac{(Q+P)!}{(P-1)!}\right) - s - \log(k)$. Using this bound in Inequality 3 yields

$$x \leq \frac{1}{l}\left((Q+1)^2(\log(k+1) - \frac{1}{2}\log k) + s - \log k\right)$$

$$\leq \frac{1}{l}\left((Q+1)^2(\log(k+1)) + s\right).$$

Now, using $|I_l| \leq x$ and the previous inequality on x, we continue simplifying Inequality 1 as follows:

$$\mathbb{E}|G_\mu| \leq \cdots < \frac{|I_l|}{(Q+1)^2} \cdot k + l \leq \frac{(Q+1)^2(\log(k+1)) + s}{l(Q+1)^2} \cdot k + l$$

$$\leq \frac{\log(k+1)k}{l} + \frac{sk}{l(Q+1)^2} + l = O\left(\frac{sk}{l(Q+1)^2} + l\right),$$

since $s = \omega((Q+1)^2 \log(k+1))$. We optimize by choosing $l = \frac{\sqrt{sk}}{Q}$, and we conclude $\mathbb{E}|G_\mu| = O(\frac{\sqrt{sk}}{Q})$.   □

Theorem 1. *For any $P \leq \sqrt{n}$, let $\mathcal{P}_{det}$ be a P-party deterministic simultaneous message protocol for maximum matching where all messages are of size at most s. Then, $\mathcal{P}_{det}$ has an expected approximation factor of $\Omega\left(\left(\frac{Pn}{s}\right)^{\frac{1}{4}}\right)$.*

Proof. For every matching M' in the input graph G, the size of M' can be bounded by $|M'| \leq 2Qk + \sum_{p=1}^{P} |M' \cap M_p|$, since at most $2Qk$ edges can be matched to the vertices of the vertex groups $\bigcup_{i \in [Q]} A_i \cup B_i$, and the edges of matchings M_p are the only ones not incident to any vertex in $\bigcup_{i \in [Q]} A_i \cup B_i$. Hence, by linearity of expectation, and the application of Lemma 2, we obtain:

$$\mathbb{E}|N| \leq 2Qk + \sum_{p=1}^{P} \mathbb{E}|N_p| \leq 2Qk + P \cdot \mathrm{O}\left(\frac{\sqrt{sk}}{Q}\right). \tag{4}$$

A maximum matching in G is of size at least $\frac{k(Q+P)}{2}$. We hence obtain the expected approximation factor:

$$\mathbb{E}\frac{\frac{1}{2}k(Q+P)}{|N|} \geq \frac{\frac{1}{2}k(Q+P)}{\mathbb{E}|N|} = \Omega\left(\frac{k(Q+P)}{\left(Qk + P \cdot \frac{\sqrt{sk}}{Q}\right)}\right) = \Omega\left(\frac{(Q+P)Q\sqrt{k}}{Q^2\sqrt{k} + P\sqrt{s}}\right)$$

$$= \Omega\left(\frac{PQ\sqrt{k}}{Q^2\sqrt{k} + P\sqrt{s}}\right) = \Omega\left(\min\{\frac{P}{Q}, \frac{Q\sqrt{k}}{\sqrt{s}}\}\right), \tag{5}$$

where the first inequality follows from Jensen's inequality, and the third equality uses $Q = o(P)$. The previous expression is maximized for $Q = \left(\frac{P\sqrt{s}}{\sqrt{k}}\right)^{1/2}$, and we obtain an approximation factor of $\Omega\left(\frac{P^{\frac{1}{2}}k^{\frac{1}{4}}}{s^{\frac{1}{4}}}\right)$. In turn, this expression is maximized when k is as large as possible, that is, $k = n/P$ (remember that the possible range for k is $P \leq k \leq n/P$). We hence conclude that the approximation factor is $\Omega((\frac{Pn}{s})^{\frac{1}{4}})$. $\square$

Lower Bound for Randomized Protocols. Last, in Theorem 2, we extend our determinstic lower bound to randomized ones.

Theorem 2 (*). *For any $P \leq \sqrt{n}$, let $\mathcal{P}_{rand}$ be a P-party randomized simultaneous message protocol for maximum matching with error at most $\epsilon < 1/2$, and all messages are of size at most s. Then, $\mathcal{P}_{rand}$ has an approximation factor of $\Omega\left(\left(\frac{Pn}{s}\right)^{\frac{1}{4}}\right)$.*

Our lower bound for one-pass turnstile algorithms now follows from the reduction given in [21] and the application of Theorem 2 for $P = \sqrt{n}$.

Corollary 1. *For every $0 \leq \epsilon \leq 1$, every randomized constant error turnstile one-pass streaming algorithm for maximum bipartite matching with approximation ratio n^ϵ uses space $\Omega\left(n^{\frac{3}{2}-4\epsilon}\right)$.*

5   Upper Bound

In this section, we first present a simple randomized algorithm for bipartite matching. Then, we will discuss implementations of this algorithm as a simultaneous message protocol and as a dynamic one-pass streaming algorithm.

Algorithm 1. Bipartite Matching algorithm

Require: $G = (A, B, E)$ {Bipartite input graph}
 1: $A' \leftarrow$ subset of A of size k chosen uniformly at random
 2: $\forall a \in A' : E'[a] \leftarrow$ arbitrary subset of incident edges of a of size $\min\{k, \deg_G(a)\}$
 3: **return** maximum matching in $\bigcup_{a \in A'} E'[a]$

Bipartite Matching Algorithm. Consider Algorithm 1. First, a subset $A' \subseteq A$ consisting of k vertices is chosen uniformly at random. Then, for each vertex $a \in A'$, the algorithm picks arbitrary k incident edges. Finally, a maximum matching among the retained edges is computed and returned.

Clearly, the algorithm stores at most k^2 edges. In the next lemma, we prove that Algorithm 1 has an expected approximation ratio of $\frac{n}{k}$.

Lemma 4 (*). *Let $G = (A, B, E)$ be a bipartite graph with $|A| + |B| = n$. Then, Algorithm 1 has an expected approximation ratio of $\frac{n}{k}$.*

Implementation of Algorithm 1 as a Simultaneous Message Protocol. Algorithm 1 can be implemented in the simultaneous message model as follows. Using shared random coins, the P parties agree on the subset $A' \subseteq A$. Then, for every $a \in A'$, every party chooses arbitrary $\min\{\deg_{G_i}(a), k\}$ edges incident to a and sends them to the referee. The referee computes a maximum matching in the graph induced by all received edges. As the referee receives a superset of the edges as described in Algorithm 1, the same approximation factor as in Lemma 4 holds. We hence obtain the following theorem:

Theorem 3. *For every $P \geq 1$, there is a randomized P-party simultaneous message protocol for maximum matching with expected approximation factor n^α and all messages are of size $\tilde{O}(n^{2-2\alpha})$.*

Implementation of Algorithm 1 as a Dynamic Streaming Algorithm. We employ the technique of l_0 sampling in our algorithm [15]. For a turnstile stream that describes a vector x, a l_0-sampler samples uniformly at random from the non-zero coordinates of x. Similar to Ahn, Guha, and McGregor [2], we employ the l_0-sampler by Jowhari et al. [15]. Their result can be summarized as follows:

Lemma 5 ([15]). *There exists a turnstile streaming algorithm that performs l_0-sampling using space $O(\log^2 n \log \delta^{-1})$ with error probability at most δ.*

In order to implement Algorithm 1 in the dynamic streaming setting, for every $a \in A'$, we use enough l_0-samplers on the sub-stream of incident edges of a in order to guarantee that with large enough probability, at least $\min\{k, \deg_G(a)\}$ different incident edges of a are sampled. It can be seen that, for a large enough constant c, $c \cdot k \log n$ samplers are enough, with probability $1 - \frac{1}{n^{\Theta(c)}}$. We make use of the following technical lemma.

Lemma 6 (*). *Let S be a finite set, k an integer, and c a large enough constant. When sampling $c \cdot k \log n$ times from S, then with probability $1 - \frac{1}{n^{\Theta(c)}}$, at least $\min\{k, |S|\}$ different elements of S have been sampled.*

This allows us to conclude with the main theorem of this section.

Theorem 4. *There exists a one-pass randomized dynamic streaming algorithm for maximum bipartite matching with expected approximation ratio n^α using space $\tilde{O}(n^{2-2\alpha})$.*

References

1. Ahn, K.J., Guha, S.: Linear programming in the semi-streaming model with application to the maximum matching problem. pp. 526–538. ICALP (2011)
2. Ahn, K.J., Guha, S., McGregor, A.: Analyzing graph structure via linear measurements. pp. 459–467. SODA (2012)
3. Ahn, K.J., Guha, S., McGregor, A.: Spectral sparsification in dynamic graph streams. In: APPROX/RANDOM. pp. 1–10 (2013)
4. Bosek, B., Leniowski, D., Sankowski, P., Zych, A.: Online bipartite matching in offline time. pp. 384–393. FOCS (2014)
5. Chaudhuri, K., Daskalakis, C., Kleinberg, R.D., Lin, H.: Online bipartite perfect matching with augmentations. pp. 1044–1052. INFOCOM (2009)
6. Crouch, M., Stubbs, D.S.: Improved streaming algorithms for weighted matching, via unweighted matching. In: APPROX/RANDOM. pp. 96–104 (2014)
7. Crouch, M.S., McGregor, A., Stubbs, D.: Dynamic graphs in the sliding-window model. In: ESA. pp. 337–348 (2013)
8. Eggert, S., Kliemann, L., Munstermann, P., Srivastav, A.: Bipartite matching in the semi-streaming model. Algorithmica 63(1-2), 490–508 (2012)
9. Epstein, L., Levin, A., Mestre, J., Segev, D.: Improved approximation guarantees for weighted matching in the semi-streaming model. pp. 347–358. STACS (2010)
10. Esfandiari, H., Hajiaghayi, M.T., Liaghat, V., Monemizadeh, M., Onak, K.: Streaming algorithms for estimating the matching size in planar graphs and beyond. SODA (2015)
11. Feigenbaum, J., Kannan, S., McGregor, A., Suri, S., Zhang, J.: On graph problems in a semi-streaming model. Theor. Comput. Sci. 348(2), 207–216 (2005)
12. Goel, A., Kapralov, M., Khanna, S.: On the communication and streaming complexity of maximum bipartite matching. pp. 468–485. SODA (2012)
13. Guruswami, V., Onak, K.: Superlinear lower bounds for multipass graph processing. In: CCC. pp. 287–298 (2013)
14. Huang, Z., Radunovic, B., Vojnovic, M., Zhang, Q.: Communication complexity of approximate matching in distributed graphs. STACS (2015)
15. Jowhari, H., Sağlam, M., Tardos, G.: Tight bounds for Lp samplers, finding duplicates in streams, and related problems. PODS, New York, NY, USA (2011)
16. Kapralov, M.: Better bounds for matchings in the streaming model. SODA (2013)

17. Kapralov, M., Khanna, S., Sudan, M.: Approximating matching size from random streams. pp. 734–751. SODA (2014)
18. Kapralov, M., Lee, Y.T., Musco, C., Musco, C., Sidford, A.: Single pass spectral sparsification in dynamic streams. In: FOCS. pp. 561–570 (2014)
19. Kapralov, M., Woodruff, D.: Spanners and sparsifiers in dynamic streams. In: PODC. pp. 272–281. PODC '14, ACM, New York, NY, USA (2014)
20. Konrad, C., Magniez, F., Mathieu, C.: Maximum matching in semi-streaming with few passes. APPROX/RANDOM (2012)
21. Li, Y., Nguyen, H.L., Woodruff, D.P.: Turnstile streaming algorithms might as well be linear sketches. pp. 174–183. STOC, New York, NY, USA (2014)
22. McGregor, A.: Finding graph matchings in data streams. pp. 170–181. APPROX/RANDOM (2005)
23. McGregor, A.: Graph stream algorithms: A survey. SIGMOD Rec. 43(1) (2014)
24. Muthukrishnan, S.: Data streams: Algorithms and applications. Now Publishers Inc (2005)
25. Zelke, M.: Weighted matching in the semi-streaming model. Algorithmica 62(1-2), 1–20 (2012)

A Lasserre Lower Bound for the Min-Sum Single Machine Scheduling Problem[*]

Adam Kurpisz, Samuli Leppänen, and Monaldo Mastrolilli

IDSIA, 6928 Manno, Switzerland
{adam,samuli,monaldo}@idsia.ch

Abstract. The Min-sum single machine scheduling problem (denoted $1||\sum f_j$) generalizes a large number of sequencing problems. The first constant approximation guarantees have been obtained only recently and are based on natural time-indexed LP relaxations strengthened with the so called *Knapsack-Cover* inequalities (see Bansal and Pruhs, Cheung and Shmoys and the recent $(4 + \epsilon)$-approximation by Mestre and Verschae). These relaxations have an integrality gap of 2, since the Min-knapsack problem is a special case. No APX-hardness result is known and it is still conceivable that there exists a PTAS. Interestingly, the Lasserre hierarchy relaxation, when the objective function is incorporated as a constraint, reduces the integrality gap for the Min-knapsack problem to $1 + \epsilon$.

In this paper we study the complexity of the Min-sum single machine scheduling problem under algorithms from the Lasserre hierarchy. We prove the first lower bound for this model by showing that the integrality gap is unbounded at level $\Omega(\sqrt{n})$ even for a variant of the problem that is solvable in $O(n \log n)$ time, namely Min-number of tardy jobs. We consider a natural formulation that incorporates the objective function as a constraint and prove the result by partially diagonalizing the matrix associated with the relaxation and exploiting this characterization.

1 Introduction

The MIN-SUM SINGLE MACHINE scheduling problem (often denoted $1||\sum f_j$) is defined by a set of n jobs to be scheduled on a single machine. Each job has an integral processing time, and there is a monotone function $f_j(C_j)$ specifying the cost incurred when the job j is completed at a particular time C_j; the goal is to minimize $\sum f_j(C_j)$. A natural special case of this problem is given by the MIN-NUMBER OF TARDY JOBS (denoted $1||\sum w_j U_j$), with $f_j(C_j) = w_j$ if $C_j > d_j$, and 0 otherwise, where $w_j \geq 0$, $d_j > 0$ are the specific cost and due date of the job j respectively. This problem is known to be NP-complete [11]. However, restricting to unit weights, the problem can be solved in $O(n \log n)$ time [17].

---

[*] Supported by the Swiss National Science Foundation project 200020-144491/1 "Approximation Algorithms for Machine Scheduling Through Theory and Experiments".

© Springer-Verlag Berlin Heidelberg 2015
N. Bansal and I. Finocchi (Eds.): ESA 2015, LNCS 9294, pp. 853–864, 2015.
DOI: 10.1007/978-3-662-48350-3_71

The first constant approximation algorithm for $1||\sum f_j$ was obtained by Bansal and Pruhs [1], who considered an even more general scheduling problem. Their 16-approximation has been recently improved to $4 + \epsilon$: Cheung and Shmoys [5] gave a primal-dual algorithm and claimed that is a $(2 + \epsilon)$-approximation; recently, Mestre and Verschae [16] showed that the analysis in [5] cannot yield an approximation better than 4 and provided a proof that the algorithm in [5] has an approximation ratio of $4 + \epsilon$.

A particular difficulty in approximating this problem lies in the fact that the ratio (*integrality gap*) between the optimal IP solution to the optimal solution of "natural" LPs can be arbitrarily large, since the MIN-KNAPSACK LP is a common special case. Thus, in [1,5] the authors strengthen natural time-indexed LP relaxations by adding (exponentially many) *Knapsack-Cover* (KC) inequalities introduced by Wolsey [24] (see also [4]) that have proved to be a useful tool to address capacitated covering problems.

One source of improvements could be the use of semidefinite relaxations such as the powerful Lasserre/Sum-of-Squares hierarchy [13,19,22] (we defer the definition and related results to Section 2). Indeed, it is known [10] that for MIN-KNAPSACK the Lasserre hierarchy relaxation, when the objective function is incorporated as a constraint in the natural LP, reduces the gap to $(1 + \varepsilon)$ at level $O(1/\varepsilon)$, for any $\varepsilon > 0$.[1] In light of this observation, it is therefore tempting to understand whether the Lasserre hierarchy relaxation can replace the use of exponentially many KC inequalities to get a better approximation for the problem $1||\sum f_j$.[2]

In this paper we study the complexity of the MIN-SUM SINGLE MACHINE scheduling problem under algorithms from the Lasserre hierarchy. Our contribution is two-fold. We provide a novel technique that is interesting in its own for analyzing integrality gaps for the Lasserre hierarchy. We then use this technique to prove the first lower bound for this model by showing that the integrality gap is unbounded at level $\Omega(\sqrt{n})$ even for the unweighted MIN-NUMBER OF TARDY JOBS problem, a variant of the problem that admits an $O(n \log n)$ time algorithm [17]. This is obtained by formulating the hierarchy as a sum of (exponentially many) rank-one matrices (Section 2) and, for every constraint, by choosing a dedicated collection (Section 3) of rank-one matrices whose sum can be shown to be positive definite by diagonalizing it; it is then sufficient to compare its smallest eigenvalue to the smallest eigenvalue of the remaining part of the sum of the rank-one matrices (Theorem 1). Furthermore, we complement the result by proving a tight characterization of the considered instance by analyzing the sign of the Rayleigh quotient (Theorem 2).

Finally, we show a different use of the above technique to prove that the class of unconstrained k $(\leq n)$ degree 0/1 n-variate polynomial optimization problems cannot be solved exactly within $k - 1$ levels of the Lasserre hierarchy relaxation.

---

[1] The same holds even for the weaker Sherali-Adams hierarchy relaxations.

[2] Note that in order to claim that one can optimize over the Lasserre hierarchy in polynomial time, one needs to assume that the number of constraint of the starting LP is polynomial in the number of variables (see the discussion in [14]).

We do this by exhibiting for each k a 0/1 polynomial optimization problem of degree k with an integrality gap. This complements the recent results in [7,12]: in [7] it is shown that the Lasserre relaxation does not have any gap at level $\lceil \frac{n}{2} \rceil$ when optimizing n-variate 0/1 polynomials of degree 2; in [12] the authors of this paper prove that the only polynomials that can have a gap at level $n-1$ must have degree n.

Duo to space limitation the some of the proofs are omitted.

2   The Lasserre Hierarchy

In this section we provide a formal definition of the Lasserre hierarchy [13] together with a brief overview of the literature.

Related work. The Lasserre/Sum-of-Squares hierarchy [13,19,22] is a systematic procedure to strengthen a relaxation for an optimization problem by constructing a sequence of increasingly tight formulations, obtained by adding additional variables and SDP constraints. The hierarchy is parameterized by its level t, such that the formulation gets tighter as t increases, and a solution can be found in time $n^{O(t)}$. This approach captures the convex relaxations used in the best available approximation algorithms for a wide variety of optimization problems. Due to space restrictions, we refer the reader to [6,14,18,20] and the references therein.

The limitations of the Lasserre hierarchy have also been studied, but not many techniques for proving lower bounds are known. Most of the known lower bounds for the hierarchy originated in the works of Grigoriev [8,9] (also independently rediscovered later by Schoenebeck [21]). In [9] it is shown that random 3XOR or 3SAT instances cannot be solved by even $\Omega(n)$ rounds of Lasserre hierarchy. Lower bounds, such as those of [3,23] rely on [9,21] plus gadget reductions. For different techniques to obtain lower bounds see [2,12,15].

Notation and the formal definition. In the context of this paper, it is convenient to define the hierarchy in an equivalent form that follows easily from "standard" definitions (see e.g. [14]) after a change of variables.[3]

For the applications that we have in mind, we restrict our discussion to optimization problems with 0/1-variables and m linear constraints. We denote $K = \{x \in \mathbb{R}^n \mid g_\ell(x) \geq 0, \forall \ell \in [m]\}$ to be the feasible set of the linear relaxation. We are interested in approximating the convex hull of the integral points in K. We refer to the ℓ-th linear constraint evaluated at the set $I \subseteq [n]$ ($x_i = 1$ for $i \in I$, and $x_i = 0$ for $i \notin I$) as $g_\ell(x_I)$. For each integral solution x_I, where $I \subseteq N$, in the Lasserre hierarchy defined below there is a variable y_I^n that can be interpreted as the "relaxed" indicator variable for the solution x_I.

---

[3] Notice that the used formulation of the Lasserre hierarchy given in Definition 1 has exponentially many variables y_I^n, due to the change of variables. This is not a problem for our purposes, since we are interested in showing an integrality gap rather than solving an optimization problem.

For a set $I \subseteq [n]$ and fixed integer t, let $\mathcal{P}_t(I)$ denote the set of the subsets of I of size at most t. For simplicity we write $\mathcal{P}_t([n]) = \mathcal{P}_t(n)$. Define d-zeta vectors: $Z_I \in \mathbb{R}^{\mathcal{P}_d(n)}$ for every $I \subseteq [n]$, such that for each $|J| \le d$, $[Z_I]_J = \begin{cases} 1, & \text{if } J \subseteq I \\ 0, & \text{otherwise} \end{cases}$. In order to keep the notation simple, we do not emphasize the parameter d as the dimension of the vectors should be clear from the context (we can think of the parameter d as either t or $t+1$).

Definition 1. *The Lasserre hierarchy relaxation at level t for the set K, denoted by $\mathrm{LAS}_t(K)$, is given by the set of values $y_I^n \in \mathbb{R}$ for $I \subseteq [n]$ that satisfy*

$$\sum_{I \subseteq [n]} y_I^n = 1, \tag{1}$$

$$\sum_{I \subseteq [n]} y_I^n Z_I Z_I^\top \succeq 0, \text{ where } Z_I \in \mathbb{R}^{\mathcal{P}_{t+1}(n)} \tag{2}$$

$$\sum_{I \subseteq [n]} g_\ell(x_I) y_I^n Z_I Z_I^\top \succeq 0, \ \forall \ell \in [m], \text{ where } Z_I \in \mathbb{R}^{\mathcal{P}_t(n)} \tag{3}$$

It is straightforward to see that the Lasserre hierarchy formulation given in Definition 1 is a relaxation of the integral polytope. Indeed consider any feasible integral solution $x_I \in K$ and set $y_I^n = 1$ and the other variables to zero. This solution clearly satisfies Condition (1), Condition (2) because the rank one matrix $Z_I Z_I^\top$ is positive semidefinite (PSD), and Condition (3) since $x_I \in K$.

3   Partial Diagonalization

In this section we describe how to partially diagonalize the matrices associated to Lasserre hierarchy. This will be used in the proofs of Theorem 1 and Theorem 2.

Below we denote by w_I^n either y_I^n or $y_I^n g_\ell(x_I)$. The following simple observation describes a congruent transformation ($\cong$) to obtain a partial diagonalization of the matrices used in Definition 1. We will use this partial diagonalization in our bound derivation.

Lemma 1. *Let $\mathcal{C} \subseteq \mathcal{P}_n(n)$ be a collection of size $|\mathcal{P}_d(n)|$ (where d is either t or $t+1$). If $\mathcal{C}$ is such that the matrix Z with columns Z_I for every $I \in \mathcal{C}$ is invertible, then*

$$\sum_{I \subseteq [n]} w_I^n Z_I Z_I^\top \cong D + \sum_{I \in \mathcal{P}_n(n) \setminus \mathcal{C}} w_I^n Z^{-1} Z_I (Z^{-1} Z_I)^\top$$

where D is a diagonal matrix with entries w_I^n, for $I \in \mathcal{C}$.

Proof. It is sufficient to note that $\sum_{I \in \mathcal{C}} w_I^n Z_I Z_I^\top = ZDZ^\top$.   □

Since congruent transformations are known to preserve the sign of the eigenvalues, the above lemma in principle gives us a technique to check whether or not (2) and (3) are satisfied: show that the sum of the smallest diagonal element of D and the smallest eigenvalue of the matrix $\sum_{I \in [n] \setminus C} w_I^n Z^{-1} Z_I (Z^{-1} Z_I)^\top$ is non-negative. In what follows we introduce a method to select the collection C such that the matrix Z is invertible.

Let Z_d denote the matrix with columns $[Z_d]_I = Z_I$ indexed by sets $I \subseteq [n]$ of size at most d. The matrix Z_d is invertible as it is upper triangular with ones on the diagonal. It is straightforward to check that the inverse Z_d^{-1} is given by $\left[Z_d^{-1}\right]_{I,J} = (-1)^{|J \setminus I|}$ if $I \subseteq J$ and 0 otherwise (see e.g. [14]). In Lemma 1 we require a collection C such that the matrix, whose columns are the zeta vectors corresponding to elements in C, is invertible. The above indicates that if we take C to be the set of subsets of $[n]$ with size less or equal to d, then this requirement is satisfied. We can think that the matrix Z_d contains as columns the zeta vectors corresponding to the set $\emptyset$ and all the symmetric differences of the set $\emptyset$ with sets of size at most d. The observation allows us to generalize this notion: fix a set $S \subseteq [n]$, and define C to contain all the sets $S \oplus I$ for $|I| \leq d$ (here $\oplus$ denotes the symmetric difference). More formally, consider the following $|\mathcal{P}_d(n)| \times |\mathcal{P}_d(n)|$ matrix $Z_{d(S)}$, whose generic entry $I, J \subseteq \mathcal{P}_d(n)$ is

$$\left[Z_{d(S)}\right]_{I,J} = \begin{cases} 1 \text{ if } I \subseteq J \oplus S, \\ 0 \text{ otherwise.} \end{cases} \tag{4}$$

Note that $Z_{d(\emptyset)} = Z_d$. In order to apply Lemma 1, we show that $Z_{d(S)}$ is invertible.

Lemma 2. *Let $A_{d(S)}$ be a $|\mathcal{P}_d(n)| \times |\mathcal{P}_d(n)|$ matrix defined as*

$$\left[A_{d(S)}\right]_{I,K} = \begin{cases} (-1)^{|K \cap S|} & \text{if } (I \setminus S) \subseteq K \subseteq I \\ 0 & \text{otherwise.} \end{cases} \tag{5}$$

Then $Z_{d(S)}^{-1} = Z_d^{-1} A_{d(S)}$.

We also give a closed form of the elements of the matrix $Z_{d(S)}^{-1}$.

Lemma 3. *For each $I, J \subseteq \mathcal{P}_d(N)$ the generic entry (I, J) of $Z_{d(S)}^{-1}$ is*

$$\left[Z_{d(S)}^{-1}\right]_{I,J} = (-1)^{|J \cap S| + |J \setminus I|} \begin{cases} (-1)^{d-|I \cup J|} \binom{|S \setminus (I \cup J)|-1}{d-|I \cap J|}, & \text{if } I \setminus S \subseteq J \\ 0, & \text{otherwise.} \end{cases} \tag{6}$$

4 A Lower Bound for Min-Number of Tardy Jobs

We consider the single machine scheduling problem to minimize the number of tardy jobs: we are given a set of n jobs, each with a processing time $p_j > 0$, and a due date $d_j > 0$. We have to sequence the jobs on a single machine such that no two jobs overlap. For each job j that is not completed by its due date, we pay the cost w_j.

4.1   The Starting Linear Program

Our result is based on the following "natural" linear programming (LP) relaxation that is a special case of the LPs used in [1,5] (therefore our gap result also holds if we apply those LP formulations). For each job we introduce a variable $x_j \in [0,1]$ with the intended (integral) meaning that $x_j = 1$ if and only if the job j completes after its deadline. Then, for any time $s \in \{d_1, \ldots, d_n\}$, the sum of the processing times of the jobs with deadlines less than s, and that complete before s, must satisfy $\sum_{j:d_j \leq s}(1 - x_j)p_j \leq s$. The latter constraint can be rewritten as a capacitated covering constraint, $\sum_{j:d_j \leq t} x_j p_j \geq D_t$, where $D_s := \sum_{j:d_j \leq s} p_j - s$ represents the *demand* at time s. The goal is to minimize $\sum_j w_j x_j$.

4.2   The Integrality Gap Instance

Consider the following instance with $n = m^2$ jobs of unit costs. The jobs are partitioned into m blocks $N_1, N_2, \ldots, N_m$, each with m jobs. For $i \in [m]$, the jobs belonging to block N_i have the same processing time P^i, for $P > 1$, and the same deadline $d_i = m \sum_{j=1}^{i} P^j - \sum_{j=1}^{i} P^{j-1}$. Then the demand at time d_i is $D_i = \sum_{j=1}^{i} P^{j-1}$. For any $t \geq 0$, let T be the smallest value that makes $\text{LAS}_t(LP(T))$ feasible, where $LP(T)$ is defined as follows for $x_{ij} \in [0,1]$, for $i, j \in [m]$:

$$LP(T) \qquad \sum_{i=1}^{m}\sum_{j=1}^{m} x_{ij} \leq T, \qquad\qquad\qquad (7a)$$

$$\sum_{i=1}^{\ell}\sum_{j=1}^{m} x_{ij} \cdot P^i \geq D_\ell, \qquad\qquad \text{for } \ell \in [m] \qquad (7b)$$

Note that, for any feasible *integral* solution for $LP(T)$, the smallest T (i.e. the optimal integral value) can be obtained by selecting one job for each block, so the smallest T for integral solutions is $m = \sqrt{n}$. The *integrality gap* of $\text{LAS}_t(LP(T))$ (or $LP(T)$) is defined as the ratio between $\sqrt{n}$ (i.e. the optimal integral value) and the smallest T that makes $\text{LAS}_t(LP(T))$ (or $LP(T)$) feasible. It is easy to check that $LP(T)$ has an integrality gap P for any $P \geq 1$: for $T = \sqrt{n}/P$, a feasible fractional solution for $LP(T)$ exists by setting $x_{ij} = \frac{1}{\sqrt{n}P}$.

4.3   Proof of Integrality Gap for $\text{LAS}_t(LP(T))$

Theorem 1. *For any $k \geq 1$ and n such that $t = \frac{\sqrt{n}}{2k} - \frac{1}{2} \in \mathbb{N}$, the following solution is feasible for $\text{LAS}_t(LP(\sqrt{n}/k))$*

$$y_I^n = \begin{cases} \alpha, \ \forall I \in \mathcal{P}_{2t+1}(n) \\ 0, \ \textit{otherwise} \end{cases} \qquad (8)$$

where $\alpha > 0$ is such that $\sum_{I \subseteq [n]} y_I^n = 1$ and the parameter P is large enough.

Proof. We need to show that the solution (8) satisifies the feasibility conditions (1)–(2) for the variables and the condition (3) for every constraint. The condition (1) is satisfied by definition of the solution, and (2) becomes a sum of positive semidefinite matrices $Z_I Z_I^\top$ with non-negative weights y_I^n, so it is satisfied as well.

It remains to show that (3) is satisfied for both (7a) and (7b). Consider the equation (7a) first, and let $g(x_I) = T - \sum_{i,j} x_{ij}$ be the value of the constraint when the decision variables are $x_{ij} = 1$ whenever $(i,j) \in I$, and 0 otherwise.[4] Now for every $I \subseteq [n]$, it holds $g(x_I)y_I^n \geq 0$, as we have $y_I^n = 0$ for every I containing more than $2t + 1 = \frac{\sqrt{n}}{k} = T$ elements. Hence the sum in (3) is again a sum of positive semidefinite matrices with non-negative weights, and the condition is satisfied.

Finally, consider the ℓ-th constraint of the form (7b), and let $g_\ell(x_I) = \sum_{i=1}^{l} \sum_{j=1}^{m} x_{ij} \cdot P^i - D_\ell$. In order to prove that (3) is satisfied, we apply Lemma 1 with the following collection of subsets of $[n]$: $\mathcal{C} = \{I \oplus S \mid I \subseteq [m], |I| \leq t\}$, where we take $S = \{(\ell, j) \mid j \in [t + 1]\}$. Now, any solution given by the elements of $\mathcal{C}$ contains at least one job from the block ℓ, meaning that the corresponding allocation x_I satisfies the constraint.

By Lemma 2, the matrix $Z_{t(S)}$ is invertible and by Lemma 1 we have for (3) that $\sum_{I \subseteq [n]} g_\ell(x_I)y_I^n Z_I Z_I^\top \cong D + \sum_{I \in [n] \setminus \mathcal{C}} g_\ell(x_I)y_I^n Z_{t(S)}^{-1} Z_I (Z_{t(S)}^{-1} Z_I)^\top$, where D is a diagonal matrix with elements $g_\ell(x_I)y_I^n$ for each $I \in \mathcal{C}$. We prove that the latter is positive semidefinite by analysing its smallest eigenvalue $\lambda_{\min}$. Writing $R_I = Z_{t(S)}^{-1} Z_I (Z_{t(S)}^{-1} Z_I)^\top$, we have by Weyl's inequality

$$\lambda_{\min}\left(D + \sum_{I \in [n] \setminus \mathcal{C}} g_\ell(x_I)y_I^n R_I\right) \geq \lambda_{\min}(D) + \lambda_{\min}\left(\sum_{I \in [n] \setminus \mathcal{C}} g_\ell(x_I)y_I^n R_I\right)$$

Since D is a diagonal matrix with entries $g_\ell(x_I)y_I^n$ for $I \in \mathcal{C}$, and for every $I \in \mathcal{C}$ the constraint $g_\ell(x_I)$ is satisfied, we have $\lambda_{\min}(D) \geq \alpha\left(P^\ell - D_\ell\right) = \alpha\left(P^\ell - \frac{P^\ell - 1}{P - 1}\right)$.

On the other hand for every $I \subseteq [n]$, $g_\ell(x_I) \geq -\sum_{j=1}^{\ell} P^{j-1} = -\frac{P^\ell - 1}{P - 1}$. The nonzero eigenvalue of the rank one matrix R_I is $\left(Z_{t(S)}^{-1} Z_I\right)^\top Z_{t(S)}^{-1} Z_I \leq |\mathcal{P}_t(n)|^3 t^{O(t)} = n^{O(\sqrt{n})}$. This is because by Lemma 3, for every $I, J \in \mathcal{P}_t(n)$, $|[Z_{t(S)}^{-1}]_{I,J}| \leq t^{O(t)}$, for $|S| = t + 1$, and $[Z_I]_J \in \{0, 1\}$. Thus

$$\lambda_{\min}\left(D + \sum_{I \in [n] \setminus \mathcal{C}} g_\ell(x_I)y_I^n R_I\right) \geq \alpha\left(P^k - \frac{P^k - 1}{P - 1}\right) - \alpha\frac{P^k - 1}{P - 1}2^n n^{O(\sqrt{n})} \geq 0$$

for $P = n^{O(\sqrt{n})}$. $\qquad\square$

---

[4] Strictly speaking $I \subseteq [n]$ is a set of numbers, so we associate to each pair i, j a number via the one-to-one mapping $(i - 1)m + j$. Hence, to keep the notation simple, we here understand $(i, j) \in I$ to mean $(i - 1)m + j \in I$.

The above theorem states that the Lasserre hierarchy has an arbitrarily large integrality gap k even at level $t = \frac{\sqrt{n}}{2k} - \frac{1}{2}$. In the following we provide a tight analysis characterization for this instance, namely we prove that the Lasserre hierarchy admits an arbitrarily large gap k even at level $t = \frac{\sqrt{n}}{k} - 1$. Note that at the next level, namely $t + 1 = \sqrt{n}/k$, $\mathrm{LAS}_{t+1}(LP(\sqrt{n}/k))$ has no feasible solution for $k > 1$,[5] which gives a tight characterization of the integrality gap threshold phenomenon. The claimed tight bound is obtained by utilizing a more involved analysis of the sign of the Rayleigh quotient for the almost diagonal matrix characterization of the Lasserre hierarchy.

Theorem 2. *For any $k \geq 1$ and n such that $t = \frac{\sqrt{n}}{k} - 1 \in \mathbb{N}$, the following solution is feasible for $\mathrm{LAS}_t(LP(\sqrt{n}/k))$*

$$y_I^n = \begin{cases} \alpha, & \forall I \in \mathcal{P}_{t+1}(n) \\ 0, & otherwise \end{cases} \tag{9}$$

where $\alpha > 0$ is such that $\sum_{I \subseteq [n]} y_I^n = 1$ and the parameter P is large enough.

Proof. The solution satisfies the conditions (1), (2) and (3) for (7a) by the same argument as in the proof of Theorem 1.

We prove that the solution satisfies the condition (3) for any constraint ℓ of the form (7b). Since $M \succeq 0$ if and only if $v^\top M v \geq 0$, for every unit vector v of appropriate size, by Lemma 1 (for the collection $\mathcal{C} = \mathcal{P}_t(n)$) and using the solution (9) we can transform (3) to the following semi-infinite system of linear inequalities

$$\sum_{I \in \mathcal{P}_t(n)} g_\ell(x_I) v_I^2 + \sum_{J \subseteq [n] : |J| = t+1} \left(\sum_{\substack{I \in \mathcal{P}_t(n) \\ I \subset J}} v_I (-1)^{|I|} \right)^2 g_\ell(x_J) \geq 0, \quad \forall v \in \mathbb{S}^{|\mathcal{P}_t(n)|-1} \tag{10}$$

Consider the ℓ-th covering constraint $g_\ell(x) \geq 0$ of the form (7b) and the corresponding semi-infinite set of linear inequalities (10). Then consider the following partition of $\mathcal{P}_{t+1}(n)$: $A = \{I \in \mathcal{P}_{t+1}(n) : I \cap N_\ell \neq \emptyset\}$ and $B = \{I \in \mathcal{P}_{t+1}(n) : I \cap N_\ell = \emptyset\}$.

Note that A corresponds to the assignments that are guaranteed to satisfy the constraint ℓ. More precisely, for $S \in A$ we have $g_\ell(x_S) \geq \left(P^\ell - \sum_{j=1}^{\ell} P^{j-1}\right) = P^\ell \left(1 - \frac{P^\ell - 1}{P^\ell(P-1)}\right) \geq P^\ell \left(1 - \frac{1}{P-1}\right)$, and for $S \in B$ we have $g_\ell(x_S) \geq -\sum_{j=1}^{\ell} P^{j-1} \geq P^\ell \left(-\frac{1}{P-1}\right)$. Since $P > 0$, by scaling $g_\ell(x) \geq 0$ (see (7b)) by P^ℓ, we will assume, w.l.o.g., that

---

[5] The constraint (7b) implies that any feasible solution for $\mathrm{LAS}_{t+1}(LP(\sqrt{n}/k))$ has $y_I^n = 0$ for all $|I| > \sqrt{n}/k$. This in turn implies, with Lemma 1 for $\mathcal{C} = \mathcal{P}_t(n)$, that $\sum_{I \subseteq [n]} g_\ell(x_I) y_I^n Z_I Z_I^\top \cong D_\ell$, where D_ℓ is a diagonal matrix with entries $g_\ell(x_I) y_I^n$, for every $|I| \leq t$, there exists ℓ such that $g_\ell(x_I) < 0$ which, in any feasible solution implies $y_I^n = 0$, contradicting (1).

$$g_\ell(x_S) \geq \begin{cases} 1 - \frac{1}{P-1}, & \text{if } S \in A \\ -\frac{1}{P-1}, & \text{if } S \in B \end{cases}$$

Note that, since v is a unit vector, we have $v_I^2 \leq 1$, and for any $J \subseteq [n]$ such that $|J| = t+1$, the coefficient of $g_\ell(x_J)$ is bounded by $\left(\sum_{\substack{I \in \mathcal{P}_t(n) \\ I \subset J}} v_I(-1)^{|I|} \right)^2 \leq 2^{O(t)}$. For all unit vectors v, let β denote the smallest possible total sum of the negative terms in (10) (these are those related to $g_\ell(x_I)$ for $I \in B$). Note that $\beta \geq -\frac{|B|2^{O(t)}}{P} = -\frac{n^{O(t)}}{P}$.

In the following, we show that, for sufficiently large P, the claimed solution satisfies (10). We prove this by contradiction.

Assume that there exists a unit vector v such that (10) is not satisfied by the solution. We start by observing that under the previous assumption the following holds $v_I^2 = \frac{n^{O(t)}}{P}$ for all $I \in A \cap \mathcal{P}_t(n)$. If not, we would have an $I \in A \cap \mathcal{P}_t(n)$ such that $v_I^2 g_\ell(x_I) \geq -\beta$ contradicting the assumption that (10) is not satisfied. We claim that under the contradiction assumption, the previous bound on v_I^2 can be generalized to $v_I^2 = \frac{n^{O(t^2)}}{P}$ for any $I \in \mathcal{P}_t(n)$. Then, by choosing P such that $v_I^2 < 1/n^{2t}$, for $I \in \mathcal{P}_t(n)$, we have $\sum_{I \in \mathcal{P}_t(n)} v_I^2 < 1$, which contradicts the assumption that v is a unit vector.

The claim follows by showing that $\forall I \in B \cap \mathcal{P}_t(n)$ it holds $v_I^2 \leq n^{O(t^2)}/P$. The proof is by induction on the size of I for any $I \in B \cap \mathcal{P}_t(n)$.

Consider the empty set, since $\emptyset \in B \cap \mathcal{P}_t(n)$. We show that $v_\emptyset^2 = n^{O(t)}/P$. With this aim, consider any $J \subseteq N_\ell$ with $|J| = t + 1$. Note that $J \in A$, so $g_\ell(x_J) \geq t+1-1/(P-1)$ and its coefficient $u_J^2 = \left(\sum_{\substack{I \in \mathcal{P}_t(n) \\ I \subset J}} v_I(-1)^{|I|} \right)^2$ is the square of the sum of $v_\emptyset$ and other terms v_I, all with $I \in A \cap \mathcal{P}_t(n)$. Ignoring all the other positive terms apart from the one corresponding to J in (10), evaluating the sum of all the negative terms as β and using a loose bound $g_\ell(x_J) \geq 1/2$ for large P, we obtain the following bound b_0

$$|v_\emptyset| \leq \sqrt{-2\beta} + \sum_{\emptyset \neq I \subset J} |v_I| \leq b_0 = O\left(\sqrt{-\beta} + 2^{O(t)} \frac{n^{O(t)}}{\sqrt{P}} \right) = \frac{n^{O(t)}}{\sqrt{P}} \qquad (11)$$

which implies that $v_\emptyset^2 = n^{O(t)}/P$.

Similarly as before, consider any singleton set $\{i\}$ with $\{i\} \in B \cap \mathcal{P}_t(n)$ and any $J \subseteq N_\ell$ with $|J| = t$. Note that $J \in A$, $g_\ell(x_J) \geq t - 1/(P-1)$ and its coefficient $u_J^2 = \left(\sum_{\substack{I \in \mathcal{P}_t(n) \\ I \subset J \cup \{i\}}} v_I(-1)^{|I|} \right)^2$ is the square of the sum of $v_{\{i\}}$, $v_\emptyset$ and other terms v_I, with $I \subseteq J$ and therefore $v_I^2 = \frac{n^{O(t)}}{P}$. Moreover, again note that u_J^2 is smaller than $-\beta$ (otherwise (10) is satisfied). Therefore, for any singleton set $\{i\} \in B \cap \mathcal{P}_t(n)$, we have that $|v_{\{i\}}| \leq |v_\emptyset| + \sqrt{-2\beta} + \sum_{\emptyset \neq I \subset J} |v_I| \leq 2b_0$.

Generalizing by induction, consider any set $S \in B \cap \mathcal{P}_t(n)$ and any $J \subseteq N_\ell$ with $|J| = t + 1 - |S|$. We claim that $|v_{|S|}| \leq b_{|S|}$ where

$$b_{|S|} = b_0 + \sum_{i=0}^{|S|-1} n^i b_i \tag{12}$$

This follows by induction hypothesis and by because again $g_\ell(J \cup S) u_{J \cup S} \leq -\beta$ and therefore, $|v_S| \leq \sum_{i=0}^{|S|-1} \left(\sum_{\substack{I \in B \\ |I|=i}} |v_I| \right) + \sqrt{-2\beta} + \sum_{\emptyset \neq I \subset J} |v_I|$.

From (12), for any $S \in B \cap \mathcal{P}_t(n)$, we have that $|v_S|$ is bounded by $b_t = (n^{t-1} + 1)b_{t-1} = n^{O(t^2)} b_0 = \frac{n^{O(t^2)}}{\sqrt{P}}$. □

5   Application in 0/1 Polynomial Optimization

In this section we use the developed technique to prove an integrality gap result for the unconstrained 0/1 n-variate polynomial optimization problem. We start with the following definition of Lasserre hierarchy.

Definition 2. *The Lasserre hierarchy at level t for the unconstrained 0/1 optimization problem with the objective function $f(x) : \{0,1\}^n \to \mathbb{R}$, denoted by $\mathrm{LAS}_t(f(x))$, is given by the feasible points y_I^n for each $I \subseteq [n]$ of the following semidefinite program*

$$\sum_{I \subseteq [n]} y_I^n = 1, \tag{13}$$

$$\sum_{I \subseteq [n]} y_I^n Z_I Z_I^\top \succeq 0, \text{ where } Z_I \in \mathbb{R}^{\mathcal{P}_t(n)} \tag{14}$$

The main result of this section is the following theorem.

Theorem 3. *The class of unconstrained k–degree 0/1 n-variate polynomial optimization problems cannot be solved exactly with a $k-1$ level of Lasserre hierarchy.*

Proof. For every $k \leq n$ we give an unconstrained n-variate polynomial optimization problem with an objective function $f(x)$ of degree k such that $\mathrm{LAS}_{k-1}(f(x))$ has an integrality gap. Consider a maximization problem with the following objective function over $\{0,1\}^n$: $f(x) = \sum_{\substack{I \subseteq [n] \\ |I| \leq k}} \binom{n-|I|}{k-|I|}(-1)^{|I|+1} \prod_{i \in I} x_i$. We prove that the following solution is super-optimal and feasible for $\mathrm{LAS}_{k-1}(f(x))$

$$y_I^n = \begin{cases} \alpha, & \forall I \in [n], \ |I| \geq n-k+1 \\ -\epsilon, & I = \emptyset \\ 0, & \text{otherwise} \end{cases} \tag{15}$$

where $\alpha > 0$ is such that $\sum_{\emptyset \neq I \subseteq [n]} y_I^n = 1 + \epsilon$ and the ϵ is small enough.

It is easy to check that the objective function is equivalent to

$$f(x) = \sum_{\substack{K \subseteq [n] \\ |K|=k}} \sum_{\substack{J \subseteq K \\ J \neq \emptyset}} (-1)^{|J|+1} \prod_{j \in J} x_j$$

Now, consider any integral $0/1$ solution, for every $K \subseteq [n]$ of size $|K| = k$, a partial summation $\sum_{\emptyset \neq J \subseteq K} (-1)^{|J|+1} \prod_{j \in J} x_j$ takes value one, if for at least one $j \in K$, $x_j = 1$, and zero otherwise. Thus the integral optimum is $\binom{n}{k}$ for any solution $x \in \{0,1\}^n$ such that at least $n - k + 1$ coordinates are set to 1.

On the other hand the objective value for the Lasserre solution (15) is given by the formula[6]

$$\sum_{I \in [n]} f(x_I) y_I^n = \sum_{\substack{I \in [n] \\ |I| \geq n-k+1}} f(x_I) y_I^n = \binom{n}{k} \sum_{\substack{I \in [n] \\ |I| \geq n-k+1}} y_I^n = (1+\epsilon) \binom{n}{k}$$

where the first equality comes from the fact that $f(x_\emptyset) = 0$ and the second from the fact that $f(x_I) = \binom{n}{k}$ for any $I \subseteq [n]$, $|I| \geq n - k + 1$.

Finally, we prove that the solution (15) is feasible for $\mathrm{LAS}_{k-1}(f(x))$. The constraint (13) is satisfied by definition. In order to prove that the constraint (14) is satisfied, we apply Lemma 1 with the collection $\mathcal{C} = \{I \oplus S \mid I \subseteq [n], |I| \leq k-1\}$ of subsets of $[n]$, for $S = [n]$, and get that

$$D + \sum_{I \in [n] \setminus \mathcal{C}} y_I^n Z_{t(S)}^{-1} Z_I (Z_{t(S)}^{-1} Z_I)^\top = D - \epsilon Z_{t(S)}^{-1} Z_\emptyset (Z_{t(S)}^{-1} Z_\emptyset)^\top \qquad (16)$$

where D is a diagonal matrix with diagonal entires equal to $\alpha \geq 1/2^n$. Since the nonzero eigenvalue of the rank one matrix $Z_{t(S)}^{-1} Z_\emptyset (Z_{t(S)}^{-1} Z_\emptyset)^\top$ is equal to $\left(Z_{t(S)}^{-1} Z_\emptyset \right)^\top Z_{t(S)}^{-1} Z_\emptyset \leq |\mathcal{P}_t(n)| t^{2t} = n^{O(t)}$, one can choose $\epsilon = 1/n^{O(t)}$ such that by the Weyl's inequality we have that the matrix in (16) is PSD. $\qquad\square$

References

1. Bansal, N., Pruhs, K.: The geometry of scheduling. In: FOCS, pp. 407–414 (2010)
2. Barak, B., Chan, S.O., Kothari, P.: Sum of squares lower bounds from pairwise independence. In: STOC (2015)
3. Bhaskara, A., Charikar, M., Vijayaraghavan, A., Guruswami, V., Zhou, Y.: Polynomial integrality gaps for strong sdp relaxations of densest k-subgraph. In: SODA, pp. 388–405 (2012)
4. Carr, R.D., Fleischer, L., Leung, V.J., Phillips, C.A.: Strengthening integrality gaps for capacitated network design and covering problems. In: SODA, pp. 106–115 (2000)

---

[6] The objective function for Lasserre hierarchy at level t is $\sum_{I \in \mathcal{P}_{2t}(n)} f_I y_I$, for f being the vector of coefficients in $f(x)$. This, with Definition 2, implies the objective function of the form $\sum_{I \subseteq [n]} f(x_I) y_I^n$.

5. Cheung, M., Shmoys, D.B.: A primal-dual approximation algorithm for min-sum single-machine scheduling problems. In: Goldberg, L.A., Jansen, K., Ravi, R., Rolim, J.D.P. (eds.) RANDOM 2011 and APPROX 2011. LNCS, vol. 6845, pp. 135–146. Springer, Heidelberg (2011)
6. Chlamtac, E., Tulsiani, M.: Convex relaxations and integrality gaps. In: Handbook on Semidefinite, Conic and Polynomial Optimization. Springer (to appear)
7. Fawzi, H., Saunderson, J., Parrilo, P.: Sparse sum-of-squares certificates on finite abelian groups. CoRR, abs/1503.01207 (2015)
8. Grigoriev, D.: Complexity of positivstellensatz proofs for the knapsack. Computational Complexity 10(2), 139–154 (2001)
9. Grigoriev, D.: Linear lower bound on degrees of positivstellensatz calculus proofs for the parity. Theoretical Computer Science 259(1-2), 613–622 (2001)
10. Karlin, A.R., Mathieu, C., Nguyen, C.T.: Integrality gaps of linear and semidefinite programming relaxations for knapsack. In: Günlük, O., Woeginger, G.J. (eds.) IPCO 2011. LNCS, vol. 6655, pp. 301–314. Springer, Heidelberg (2011)
11. Karp, R.M.: Reducibility among combinatorial problems. In: Proceedings of a symposium on the Complexity of Computer Computations, March 20-22, pp. 85–103. Thomas J. Watson Research Center, Yorktown Heights (1972)
12. Kurpisz, A., Leppänen, S., Mastrolilli, M.: On the hardest problem formulations for the 0/1 Lasserre hierarchy. In: Halldórsson, M M., Iwama, K., Kobayashi, N., Speckmann, B. (eds.) ICALP 2015. LNCS, vol. 9134, pp. 872–885. Springer, Heidelberg (2015)
13. Lasserre, J.B.: Global optimization with polynomials and the problem of moments. SIAM Journal on Optimization 11(3), 796–817 (2001)
14. Laurent, M.: A comparison of the Sherali-Adams, Lovász-Schrijver, and Lasserre relaxations for 0-1 programming. Mathematics of Operations Research 28(3), 470–496 (2003)
15. Meka, R., Potechin, A., Wigderson, A.: Sum-of-squares lower bounds for planted clique. CoRR, abs/1503.06447. In: STOC 2015 (2015) (to appear)
16. Mestre, J., Verschae, J.: A 4-approximation for scheduling on a single machine with general cost function. CoRR, abs/1403.0298 (2014)
17. Moore, M.J.: An n job, one machine sequencing algorithm for minimizing the number of late jobs. Management Science 15, 102–109 (1968)
18. O'Donnell, R., Zhou, Y.: Approximability and proof complexity. In: Khanna, S. (ed.) SODA, pp. 1537–1556. SIAM (2013)
19. Parrilo, P.: Structured Semidefinite Programs and Semialgebraic Geometry Methods in Robustness and Optimization. PhD thesis, California Institute of Technology (2000)
20. Rothvoß, T.: The lasserre hierarchy in approximation algorithms. Lecture Notes for the MAPSP 2013 - Tutorial (June 2013)
21. Schoenebeck, G.: Linear level Lasserre lower bounds for certain k-csps. In: FOCS, pp. 593–602 (2008)
22. Shor, N.: Class of global minimum bounds of polynomial functions. Cybernetics 23(6), 731–734 (1987)
23. Tulsiani, M.: Csp gaps and reductions in the Lasserre hierarchy. In: STOC, pp. 303–312 (2009)
24. Wolsey, L.A.: Facets for a linear inequality in 0–1 variables. Mathematical Programming 8, 168–175 (1975)

Optimal Parameterized Algorithms for Planar Facility Location Problems Using Voronoi Diagrams*

Dániel Marx[1] and Michał Pilipczuk[2]

[1] Institute for Computer Science and Control,
Hungarian Academy of Sciences (MTA SZTAKI), Budapest, Hungary
dmarx@cs.bme.hu
[2] Institute of Informatics, University of Warsaw, Poland
michal.pilipczuk@mimuw.edu.pl

Abstract. We study a general family of facility location problems defined on planar graphs and on the 2-dimensional plane. In these problems, a subset of k objects has to be selected, satisfying certain packing (disjointness) and covering constraints. Our main result is showing that, for each of these problems, the $n^{\mathcal{O}(k)}$ time brute force algorithm of selecting k objects can be improved to $n^{\mathcal{O}(\sqrt{k})}$ time. The algorithm is based on focusing on the Voronoi diagram of a hypothetical solution of k objects; this idea was introduced recently in the design of geometric QPTASs, but was not yet used for exact algorithms and for planar graphs. As concrete consequences of our main result, we obtain $n^{\mathcal{O}(\sqrt{k})}$ time algorithms for the following problems: d-SCATTERED SET in planar graphs (find k vertices at pairwise distance d); d-DOMINATING SET/(k, d)-CENTER in planar graphs (find k vertices such that every vertex is at distance at most d from these vertices); select k pairwise disjoint connected vertex sets from a given collection; select k pairwise disjoint disks in the plane (of possibly different radii) from a given collection; cover a set of points in the plane by selecting k disks/axis-parallel squares from a given collection. We complement these positive results with lower bounds suggesting that some similar, but slightly more general problems (such as covering points with axis-parallel rectangles) do not admit $n^{\mathcal{O}(\sqrt{k})}$ time algorithms.

1 Introduction

Parameterized problems often become easier when restricted to planar graphs: usually significantly better running times can be achieved and sometimes problems that are W[1]-hard on general graphs become fixed-parameter tractable

* M. Pilipczuk was partially supported by ERC Grant no. 267959, by the Polish National Science Centre grant DEC-2013/11/D/ST6/03073 and by the Foundation for Polish Science via the START stipend programme. M. Pilipczuk also holds a post-doc position at Warsaw Center of Mathematics and Computer Science. D. Marx was supported by ERC Grant PARAMTIGHT (No. 280152) and OTKA grant NK105645.

N. Bansal and I. Finocchi (Eds.): ESA 2015, LNCS 9294, pp. 865–877, 2015.
DOI: 10.1007/978-3-662-48350-3_72

on planar graphs. In most cases, the improved running time involves a square root dependence on the parameter: it is often of the form $2^{\mathcal{O}(\sqrt{k})} \cdot n^{\mathcal{O}(1)}$ or $n^{\mathcal{O}(\sqrt{k})}$. The appearance of the square root can be usually traced back to the fact that a planar graph with n vertices has treewidth $\mathcal{O}(\sqrt{n})$. Indeed, the theory of bidimensionality gives a quick explanation why problems such as INDEPENDENT SET, LONGEST PATH, FEEDBACK VERTEX SET, DOMINATING SET, or even distance-r versions of INDEPENDENT SET and DOMINATING SET (for fixed r) have algorithms with running time $2^{\mathcal{O}(\sqrt{k})} \cdot n^{\mathcal{O}(1)}$ (cf. survey [6]). In all these problems, there is a relation between the size of the largest grid minor and the size of the optimum solution, which allows us to bound the treewidth of the graph in terms of the parameter of the problem. More recently, subexponential parameterized algorithms have been explored also for problems where there is no such straightforward parameter-treewidth bound: for examples, see [5,9,10,14].

A similar "square root phenomenon" has been observed in the case of geometric problems: it is usual to see a square root in the exponent of the running time of algorithms for NP-hard problems defined in the 2-dimensional Euclidean plane. Most relevant to our paper is the fact that INDEPENDENT SET for unit disks (given a set of n unit disks, select k of them that are pairwise disjoint) and the discrete k-center problem (given a set of n points and a set of n unit disks, select k disks whose union covers every point) can be solved in time $n^{\mathcal{O}(\sqrt{k})}$ by geometric separation theorems and shifting arguments [3,4,8,12], improving on the trivial $n^{\mathcal{O}(k)}$ time brute force algorithm. However, all of these algorithms are crucially based on a notion of area and rely on the property that all the disks have the same size (at least approximately). Therefore, it seems unlikely that these techniques can be generalized to the case when the disks can have very different radii or to planar-graph versions of the problem, where the notion of area is meaningless. Using similar techniques, one can obtain approximation schemes for these and related geometric problems, again with the limitation that the objects need to have (roughly) the same area. Very recently, a new and powerful technique emerged from a line of quasi-polynomial time approximation schemes (QPTAS) for geometric problems [1,2,7,13]. As described explicitly by Har-Peled [7], the main idea is to reason about the Voronoi diagram of the k objects in the solution. In particular, we are trying to guess a separator consisting of $\mathcal{O}(\sqrt{k})$ segments that corresponds to a balanced separator of the Voronoi diagram. In this paper, we show how this basic idea and its extensions can be implemented to obtain $n^{\mathcal{O}(\sqrt{k})}$ time exact algorithms for a wide family of geometric packing and covering problems in a uniform way. In fact, we show that the algorithms can be made to work in the much more general context of planar graph problems.

Algorithmic Results. We study a general family of facility location problems for planar graphs, where a set of k objects has to be selected, subject to certain independence and covering constraints. Two archetypal problems from this family are (1) selecting k vertices of an edge-weighted planar graph that are at distance at least d from each other (d-SCATTERED SET) and (2) selecting k vertices of an edge-weighted planar graph such that every vertex of the graph is at distance at most d from a selected vertex (d-DOMINATING SET); for both problems, d is a real value being part of the input. We show that, under very general conditions, the trivial $n^{\mathcal{O}(k)}$ time brute force algorithm can be improved to $n^{\mathcal{O}(\sqrt{k})}$ time for problems in this family. Our result is not just a simple consequence of bidimensionality and bounding the treewidth of the input graph. Instead, we focus on the Voronoi diagram of a hypothetical solution, which can be considered as a planar graph with $\mathcal{O}(k)$ vertices. It is known that such a planar graph has a balanced separator cycle of length $\mathcal{O}(\sqrt{k})$, which can be translated into a separator that breaks the instance in way suitable for using recursion on the resulting subproblems. Of course, we do not know the Voronoi diagram of the solution and its balanced separator cycle, but we argue that only $n^{\mathcal{O}(\sqrt{k})}$ separator cycles can be potential candidates. Hence, by guessing one of these cycles, we define and solve $n^{\mathcal{O}(\sqrt{k})}$ subproblems. The running time of the algorithm is thus governed by a recurrence relation of the form $f(k) = n^{\mathcal{O}(\sqrt{k})} f(k/2)$, which resolves to $f(k) = n^{\mathcal{O}(\sqrt{k})}$.

In Section 3, we define a general facility location problem DISJOINT NETWORK COVERAGE that contains numerous concrete problems of interest as special cases. Now, we discuss specific algorithmic results following from the general result.

Informally, the input of DISJOINT NETWORK COVERAGE consists of an edge-weighted planar graph G, a set $\mathcal{D}$ of objects (which are connected sets of vertices in G) and a set $\mathcal{C}$ of clients (which are vertices of G). The task is to select a set of exactly k pairwise-disjoint[1] objects that maximizes the total number of the covered clients. We define covering as follows: the input contains a radius for each object in $\mathcal{D}$ and a sensitivity for each client in $\mathcal{C}$, and a client is considered covered by an object if the sum of the radius and the sensitivity is at least the distance between the object and the client. When both the radius and the sensitivity are 0, then this means that the client is inside the object; when the radius is r and the sensitivity is 0, then this means that the client is at distance at most r from the object. The objects and the clients may be equipped with costs and prizes, and we may want to maximize/minimize the total revenue of the solution.

The first special case of the problem is when there are no clients at all: then the task is to select k objects that are pairwise disjoint. Our algorithm solves this problem in complete generality: the only condition is that each object is a connected vertex set (i.e. it induces a connected subgraph of G).

Theorem 1.1 (packing connected sets). *Let G be a planar graph, $\mathcal{D}$ be a family of connected vertex sets of G, and k be an integer. In time $|\mathcal{D}|^{\mathcal{O}(\sqrt{k})} \cdot n^{\mathcal{O}(1)}$, we can find a set of k pairwise disjoint objects in $\mathcal{D}$, if such a set exists.*

We can also solve the weighted version, where we want to select k members of $\mathcal{D}$ maximizing the total weight. As a special case, Theorem 1.1 gives us an $n^{\mathcal{O}(\sqrt{k})}$ time algorithm for d-SCATTERED SET, which asks for k vertices that are at distance at least d from each other (with d being part of the input).

---

[1] More precisely, if objects have different radii, then instead of requiring disjointness, we set up a technical condition called "normality," which we define in Section 3.

If each object in $\mathcal{D}$ is a single vertex and $r(\cdot)$ assigns a radius to each object (potentially different radii for different objects), then we get a natural covering problem. Thus, the following theorem is also a corollary of our general result.

Theorem 1.2 (covering vertices with centers of different radii). *Let G be a planar graph, let $D, C \subseteq V(G)$ be two subsets of vertices, let $r \colon D \to \mathbb{Z}^+$ be a function, and k be an integer. In time $|D|^{\mathcal{O}(\sqrt{k})} \cdot n^{\mathcal{O}(1)}$, we can find a set $S \subseteq D$ of k vertices that maximizes the number of vertices covered in C, where a vertex $u \in C$ is covered by $v \in S$ if the distance between u and v is at most $r(v)$.*

If $D = C = V(G)$, $r(v) = d$ for every $v \in V(G)$, and we are looking for a solution fully covering C, then we obtain as a special case d-DOMINATING SET (also called (k, d)-CENTER). Theorem 1.2 gives an $n^{\mathcal{O}(\sqrt{k})}$ time algorithm for this problem (with d being part of the input). Theorem 1.2 can be also interpreted as covering the vertices in C by very specific objects: balls of radius $r(v)$ around a center v. If we require that the selected objects of the solution are pairwise disjoint, then we can generalize this problem to arbitrary objects.

Theorem 1.3 (covering vertices with independent objects). *Let G be a planar graph, let $\mathcal{D}$ be a set of connected vertex sets in G, let $C \subseteq V(G)$ be a set of vertices, and let k be an integer. In time $|\mathcal{D}|^{\mathcal{O}(\sqrt{k})} \cdot n^{\mathcal{O}(1)}$, we can find a set S of at most k pairwise disjoint objects in $\mathcal{D}$ that maximizes the number of vertices of C in the union of the vertex sets in S.*

By simple reductions, geometric packing/covering problems can be reduced to problems on planar graphs. In particular, given a set of disks (of possibly different radii), the problem of selecting k disjoint disks can be reduced to selecting disjoint connected vertex sets in a planar graph, and Theorem 1.1 can be applied.

Theorem 1.4 (packing disks). *Given a set $\mathcal{D}$ of disks (of possibly different radii) in the plane, in time $|\mathcal{D}|^{\mathcal{O}(\sqrt{k})}$ we can find a set of k pairwise disjoint disks, if such a set exists.*

This is a strong generalization of the results of Alber and Fiala [4], which gives an $|\mathcal{D}|^{\mathcal{O}(\sqrt{k})}$ time algorithm only if the ratio of the radii of the smallest and largest disks can bounded by a constant (in particular, if all the disks are unit disks). As Theorem 1.1 works for arbitrary connected sets of vertices, we can prove the analog of Theorem 1.4 for most reasonable sets of connected geometric objects.

Theorem 1.5 (packing simple polygons). *Given a set $\mathcal{D}$ of simple polygons in the plane, in time $|\mathcal{D}|^{\mathcal{O}(\sqrt{k})} \cdot n^{\mathcal{O}(1)}$ we can find a set of k polygons in $\mathcal{D}$ with pairwise disjoint closed interiors, if such a set exists. Here n is the total number of vertices of the polygons in $\mathcal{D}$.*

Similarly, the problem of covering the maximum number of points by selecting k disks from a given set $\mathcal{D}$ of disks can be reduced to a problem on planar graphs and then Theorem 1.2 can be invoked.

Theorem 1.6 (covering with disks). *Given a set $\mathcal{C}$ of points and a set $\mathcal{D}$ of disks (of possibly different radii) in the plane, in time $|\mathcal{D}|^{\mathcal{O}(\sqrt{k})} \cdot |\mathcal{C}|^{\mathcal{O}(1)}$ we can find a set of k disks in $\mathcal{D}$ maximizing the total number of points they cover in $\mathcal{C}$.*

Covering points with axis-parallel squares (of different sizes) can be handled similarly, by treating axis-parallel squares as balls in the *in the ℓ_∞ metric*.

Theorem 1.7 (covering with squares). *Given a set $\mathcal{C}$ of points and a set $\mathcal{D}$ of axis-parallel squares (of possibly different size) in the plane, in time $|\mathcal{D}|^{\mathcal{O}(\sqrt{k})} \cdot |\mathcal{C}|^{\mathcal{O}(1)}$ we can find a set of k squares in $\mathcal{D}$ maximizing the total number of points they cover in $\mathcal{C}$.*

Hardness Results. Comparing packing results Theorems 1.1 and 1.5 with covering results Theorems 1.2, 1.6, and 1.7, one can observe that our algorithm solves packing problems in much wider generality than covering problems. It seems that we can handle arbitrary objects in packing problems, while it is essential for covering problems that each object is a ball in some metric. We present a set of hardness results suggesting that this apparent difference is not a shortcoming of our algorithm, but it is inherent to the problem: there are natural geometric covering problems where the square root phenomenon does not occur.

Using a result of Pătraşcu and Williams [15] and a simple reduction from Dominating Set, we show that if the task is to cover points with convex polygons, then improving upon a brute-force algorithm is unlikely.

Theorem 1.8 (covering with convex polygons, lower bound). *Let $\mathcal{D}$ be a set of convex polygons and let $\mathcal{P}$ be a set of points in the plane. Assuming SETH, there is no $f(k) \cdot (|\mathcal{D}| + |\mathcal{P}|)^{k-\epsilon}$ time algorithm for any computable function f and $\epsilon > 0$ that decides if there are k polygons in $\mathcal{D}$ that together cover $\mathcal{P}$.*

Theorem 1.8 gives a lower bound only if the covering problem allows arbitrary convex polygons. We present also two lower bounds in the much more restricted setting of covering with axis-parallel rectangles.

Theorem 1.9 (covering with rectangles, lower bound). *Consider the problem of covering a point set $\mathcal{P}$ by selecting k axis-parallel rectangles from a set $\mathcal{D}$.*

1. *Assuming ETH, there is no algorithm for this problem with running time $f(k) \cdot (|\mathcal{P}| + |\mathcal{D}|)^{o(k)}$ for any computable function f, even if each rectangle in $\mathcal{D}$ is of size $1 \times k$ or $k \times 1$.*
2. *Assuming ETH, for every $\epsilon_0 > 0$, there is no algorithm for this problem with running time $f(k) \cdot (|\mathcal{P}| + |\mathcal{D}|)^{o(k/\log k)}$ for any computable function f, even if each rectangle in $\mathcal{D}$ has both width and height in the range $[1 - \epsilon_0, 1 + \epsilon_0]$.*

This shows that even a minor deviation from the setting of Theorem 1.7 makes the existence of $n^{\mathcal{O}(\sqrt{k})}$ algorithms implausible. It seems that for covering problems, the square root phenomenon depends not on the objects being simple, or fat, or similar in size, but really on the fact that the objects are balls in a metric.

2   Geometric Problems

Our main algorithmic result is a technique for solving a general facility location problem on planar graphs in time $n^{\mathcal{O}(\sqrt{k})}$. With simple reductions, we can use this algorithm to solve 2-dimensional geometric problems. However, our main algorithmic ideas can be implemented also directly in the geometric setting, giving self-contained geometric algorithms. These algorithms avoid some of the technical complications that arise in the planar graph counterparts, such as the Voronoi diagram having bridges or shortest paths sharing subpaths. The full algorithm appears in the full version [11], here we focus on these simpler cases.

Packing Unit Disks. We start with INDEPENDENT SET for unit disks: given a set $\mathcal{D}$ of closed disks of unit radius in the plane, the task is to select k disjoint disks. This problem is known to be solvable in time $n^{\mathcal{O}(\sqrt{k})}$ [4,12]. We present another $n^{\mathcal{O}(\sqrt{k})}$ algorithm for the problem, demonstrating how we can solve it recursively by focusing on the Voronoi diagram of a hypothetical solution.

The main combinatorial idea behind the algorithm is the following. Let $\mathcal{P}$ be a set of points in the plane. The *Voronoi region* of $p \in \mathcal{P}$ is the set of those points x in the plane that are "closest" to p in the sense that the distance of x and $\mathcal{P}$ is exactly the distance of x and p. Consider a hypothetical solution consisting of k independent disks and let us consider the Voronoi diagram of the centers of these k disks (see Figure 1(a)). To emphasize that we consider the Voronoi diagram of the centers of the k disks in the solution and *not* the centers of the n disks in the input, we call this diagram the *solution Voronoi diagram*. For simplicity, let us assume that the solution Voronoi diagram is a 2-connected 3-regular graph embedded on a sphere. We need a balanced separator theorem of the following form. A *noose* of a plane graph G is a closed curve δ on the sphere such that δ alternately travels through faces and vertices of G, and every vertex and face of G is visited at most once. It is possible to show that every 3-regular planar graph G with k faces has a noose δ of length $\mathcal{O}(\sqrt{k})$ (that is, going through $\mathcal{O}(\sqrt{k})$ faces and vertices) that is *face balanced*, in the sense that there are at most $\frac{2}{3}k$ faces of G strictly inside δ and at most $\frac{2}{3}k$ faces of G strictly outside δ.

Consider a face-balanced noose δ of length $\mathcal{O}(\sqrt{k})$ as above (see Figure 1(b)). Noose δ goes through $\mathcal{O}(\sqrt{k})$ faces of the solution Voronoi diagram, which correspond to a set Q of $\mathcal{O}(\sqrt{k})$ disks of the solution. The noose can be turned into a polygon Γ with $\mathcal{O}(\sqrt{k})$ vertices the following way (see Figure 1(c)). Consider a subcurve of δ that is contained in the face corresponding to disk $p \in Q$ and its endpoints are vertices x and y of the solution Voronoi diagram. Then we can "straighten" this subcurve by replacing it with straight line segments connecting the center of p with x and y. Thus, the vertices of polygon Γ are center points of disks in Q and vertices of the solution Voronoi diagram. Observe that Γ intersects the Voronoi regions of the points in Q only; this follows from the convexity of the Voronoi regions. In particular, among the disks in the solution, Γ does not intersect any disk outside Q.

The main idea is to use this polygon Γ to separate the problem into two subproblems. Of course, we do not know the solution Voronoi diagram and hence

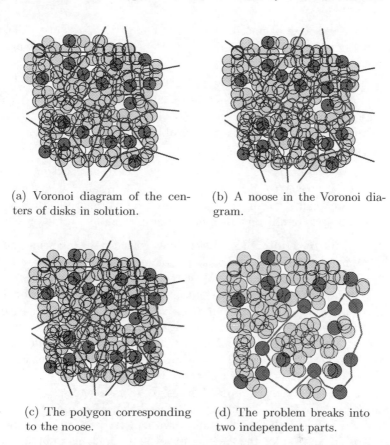

(a) Voronoi diagram of the centers of disks in solution.

(b) A noose in the Voronoi diagram.

(c) The polygon corresponding to the noose.

(d) The problem breaks into two independent parts.

Fig. 1. Using a noose in the Voronoi diagram for divide and conquer.

we have no way of computing from it the balanced noose δ and the polygon Γ. However, we can efficiently list $n^{\mathcal{O}(\sqrt{k})}$ candidate polygons. By definition, every vertex of the polygon Γ is either the center of a disk in $\mathcal{D}$ or a vertex of the solution Voronoi diagram. Every vertex of the solution Voronoi diagram is equidistant from the centers of three disks in $\mathcal{D}$ and for any three such centers (in general position) there is a unique point in the plane equidistant from them. Thus every vertex of the polygon Γ is either a center of a disk in $\mathcal{D}$ or can be described by a triple of disks in $\mathcal{D}$, and hence Γ can be described by an $\mathcal{O}(\sqrt{k})$-tuple of disks from $\mathcal{D}$. By branching into $n^{\mathcal{O}(\sqrt{k})}$ directions, we may assume that we have correctly guessed the subset Q of the solution and the polygon Γ.

Provided Q is indeed part of the solution (which we assume), we may remove these disks from $\mathcal{D}$ and decrease the target number of disks by $|Q|$. We can also perform the following cleaning steps: (1) Remove any disk that intersects a disk in Q. (2) Remove any disk that intersects Γ. The correctness of the cleaning steps above follows directly from our observations on the properties of Γ.

After these cleaning steps, the instance falls apart into two independent parts: each remaining disk is either strictly inside Γ or strictly outside Γ (see Figure 1(d)). As δ was face balanced, there are at most $\frac{2}{3}k$ faces of the solution Voronoi diagram inside/outside δ, and hence the solution contains at most $\frac{2}{3}k$ disks inside/outside Γ. Therefore, for $k' := 1, \ldots, \lfloor \frac{2}{3}k \rfloor$, we recursively try to find exactly k' independent disks from the input restricted to the inside/outside Γ, resulting in $2 \cdot \frac{2}{3}k$ recursive calls. Taking into account the $n^{\mathcal{O}(\sqrt{k})}$ guesses for Q and Γ, the number of subproblems we need to solve is $2 \cdot \frac{2}{3}k \cdot n^{\mathcal{O}(\sqrt{k})} = n^{\mathcal{O}(\sqrt{k})}$ and the parameter value is at most $\frac{2}{3}k$ in each subproblem. Therefore, the running time of the algorithm is governed by the recursion $T(n, k) = n^{\mathcal{O}(\sqrt{k})} \cdot T(n, (2/3)k)$, which solves to $T(n, k) = n^{\mathcal{O}(\sqrt{k})}$. This proves the first result: packing unit disks in the plane in time $n^{\mathcal{O}(\sqrt{k})}$. Let us repeat that this result was known before [4,12], but as we shall see, our algorithm based on Voronoi diagrams can be generalized to objects of different size, planar graphs, and covering problems.

Covering Points by Unit Disks. Let us now consider the following problem: given a set $\mathcal{D}$ of unit disks and a set $\mathcal{C}$ of client points, we need to select k disks from $\mathcal{D}$ that together cover every point in $\mathcal{C}$. We show that this problem can be solved in time $n^{\mathcal{O}(\sqrt{k})}$ using a similar approach. Note that this time the disks in the solution are not necessarily disjoint, but this does not change the fact that their centers (which can be assumed to be distinct) define a Voronoi diagram. Therefore, it will be convenient to switch to an equivalent formulation of the problem described in terms of the centers of the disks: $\mathcal{D}$ is a set of points and we say that a selected point in $\mathcal{D}$ covers a point in $\mathcal{C}$ if their distance is at most 1.

As before, we can try $n^{\mathcal{O}(\sqrt{k})}$ possibilities to guess a set $Q \subseteq \mathcal{D}$ of center points and a polygon Γ corresponding to a face-balanced noose. The question is how to use Γ to split the problem into two independent subproblems. The cleaning steps (1) and (2) for the packing problem are no longer applicable: the solution may contain disks intersecting the disks with centers in Q as well as further disks intersecting Γ. Instead we do as follows. First, if we assume that Q is part of the solution, then any point in $\mathcal{C}$ covered by some point in Q can be removed. Second, we know that in the solution Voronoi diagram every point of Γ belongs to the Voronoi region of some point in Q. Hence we can remove any point from $\mathcal{D}$ that contradicts this assumption. That is, if there is $p \in \mathcal{D}$ and $v \in \Gamma$ such that v is closer to p than to every point in Q, then we can safely remove p from $\mathcal{D}$. Thus we have the following cleaning steps: (1) Remove every point of $\mathcal{C}$ covered by Q. (2) Remove every point of $\mathcal{D}$ that is closer to a point of Γ than every point in Q. Let $\mathcal{D}_{\mathrm{in}}, \mathcal{D}_{\mathrm{out}}$ ($\mathcal{C}_{\mathrm{in}}, \mathcal{C}_{\mathrm{out}}$) be the remaining points in $\mathcal{D}$ ($\mathcal{C}$) strictly inside and outside Γ, respectively. We know that the solution contains at most $\frac{2}{3}k$ center points inside/outside Γ. Hence, for $1 \leq k' \leq \lfloor \frac{2}{3}k \rfloor$, we solve two subproblems, with point sets $(\mathcal{D}_{\mathrm{in}}, \mathcal{C}_{\mathrm{in}})$ and $(\mathcal{D}_{\mathrm{out}}, \mathcal{C}_{\mathrm{out}})$.

If there is a set of k_{in} center points in $\mathcal{D}_{\mathrm{in}}$ covering $\mathcal{C}_{\mathrm{in}}$ and there is a set of k_{out} center points in $\mathcal{D}_{\mathrm{out}}$ covering $\mathcal{C}_{\mathrm{out}}$, then, together with Q, they form a solution of $|Q| + k_{\mathrm{in}} + k_{\mathrm{out}}$ center points. By solving the defined subproblems optimally, we know the minimum value of k_{in} and k_{out} required to cover $\mathcal{C}_{\mathrm{in}}$ and $\mathcal{C}_{\mathrm{out}}$, and hence we can determine the smallest solution that can be put together this way. But is it true that we can always put together an *optimum* solution this way? The problem is that, in principle, the solution may contain a center point $p \in \mathcal{D}_{\text{out}}$ that covers some point $q \in \mathcal{C}_{\text{in}}$ that is not covered by any center point in $\mathcal{D}_{\text{in}}$. In this case, in the optimum solution the number of center points selected from $\mathcal{D}_{\text{in}}$ can be strictly less than what is needed to cover $\mathcal{C}_{\text{in}}$.

Fortunately, we can show that this problem never arises. Suppose that there is such a $p \in \mathcal{D}_{\text{out}}$ and $q \in \mathcal{C}_{\text{in}}$. As p is outside Γ and q is inside Γ, the segment connecting p and q intersects Γ at some point $v \in \Gamma$, which means $\text{dist}(p,q) = \text{dist}(p,v) + \text{dist}(v,q)$. By cleaning step (2), there has to be a $p' \in Q$ such that $\text{dist}(p',v) \leq \text{dist}(p,v)$, otherwise p would be removed from $\mathcal{D}$. This means that $\text{dist}(p,q) = \text{dist}(p,v) + \text{dist}(v,q) \geq \text{dist}(p',v) + \text{dist}(v,q) \geq \text{dist}(p',q)$. Therefore, if p covers q, then so does $p' \in Q$. But in this case we would have removed q from $\mathcal{C}$ in the first cleaning step. Thus we can indeed obtain an optimum solution the way we proposed, by solving optimally the defined subproblems. Again we have $n^{\mathcal{O}(\sqrt{k})}$ subproblems, with parameter value at most $\frac{2}{3}k$. Hence, the same recursion applies to the running time, resulting in an $n^{\mathcal{O}(\sqrt{k})}$ time algorithm.

Packing in Planar Graphs. How can we translate the geometric ideas explained above to the context of planar graphs? Let G be an edge-weighted planar graph and let $\mathcal{F}$ be a set of disjoint "objects" — connected sets of vertices in G. Then we can define the analog of the Voronoi regions: for every $p \in \mathcal{F}$, let M_p contain every vertex v to which p is the closest object in $\mathcal{F}$, that is, $\text{dist}(v, \mathcal{F}) = \text{dist}(v, p)$. It is easy to verify that region M_p has the following convexity property: if $v \in M_p$ and P is a shortest path between v and p, then every vertex of P is in M_p.

While Voronoi regions are easy to define in graphs, the proper definition of Voronoi diagrams and the construction of polygon Γ are far from obvious. We omit the discussion of these technical details, and we only state in Lemma 2.1 below (a simplified version of) the main technical tool that is at the core of the algorithm. Note that the statement of Lemma 2.1 involves only the notion of Voronoi regions, hence there are no technical issues in interpreting and using it. However, in the proof we have to define the analog of the Voronoi diagram for planar graphs and address issues such that this diagram is not 2-connected etc.

Let us consider first the packing problem: given an edge-weighted graph G, a set $\mathcal{D}$ of d objects (connected subsets of vertices), and an integer k, find a subset $\mathcal{F} \subseteq \mathcal{D}$ of k disjoint objects. Looking at the algorithm for packing unit disks, what we need is a suitable *guarded separator*: a pair (Q, Γ) consisting of a set $Q \subseteq \mathcal{D}$ of $\mathcal{O}(\sqrt{k})$ objects and a subset $\Gamma \subseteq V(G)$ of vertices. If there is a hypothetical solution $\mathcal{F} \subseteq \mathcal{D}$ consisting of k disjoint objects, then a suitable guarded separator (Q, Γ) should satisfy the following three properties: (1) $Q \subseteq \mathcal{F}$, (2) Γ is fully contained in the Voronoi regions of the objects in Q, and (3) Γ separates the objects in $\mathcal{F}$ in a balanced way. Our main technical result is that it is possible to enumerate a set of $d^{\mathcal{O}(\sqrt{k})}$ guarded separators such that for every solution $\mathcal{F}$, one of the enumerated guarded separators satisfies these three properties. We state here a simplified version that is suitable for packing problems.

Lemma 2.1. *Let G be an n-vertex edge-weighted planar graph, $\mathcal{D}$ a set of d connected subsets of $V(G)$, and k an integer. We can enumerate (in time polynomial in the size of the output and n) a set $\mathcal{N}$ of $d^{\mathcal{O}(\sqrt{k})}$ pairs (Q, Γ) with $Q \subseteq \mathcal{D}$, $|Q| = \mathcal{O}(\sqrt{k})$, $\Gamma \subseteq V(G)$ such that the following holds. If $\mathcal{F} \subseteq \mathcal{D}$ is a set of k pairwise disjoint objects, then there is a pair $(Q, \Gamma) \in \mathcal{N}$ such that*

1. *$Q \subseteq \mathcal{F}$,*
2. *if $(M_p)_{p \in \mathcal{F}}$ are the Voronoi regions of $\mathcal{F}$, then $\Gamma \subseteq \bigcup_{p \in Q} M_p$,*
3. *for every connected component C of $G - \Gamma$, there are at most $\frac{2}{3}k$ objects of $\mathcal{F}$ that are fully contained in C.*

The proof goes along the same lines as the argument for the geometric setting. After carefully defining the analog of the Voronoi diagram, we can use the planar separator result to obtain a noose δ. As before, we "straighten" the noose into a closed walk in the graph using shortest paths connecting $\mathcal{O}(\sqrt{k})$ objects and $\mathcal{O}(\sqrt{k})$ vertices of the Voronoi diagram. The vertices of this walk separate the objects that are inside/outside the noose, hence it has the required properties. Thus by trying all sets of $\mathcal{O}(\sqrt{k})$ objects and $\mathcal{O}(\sqrt{k})$ vertices of the Voronoi diagram, we can enumerate a suitable set $\mathcal{N}$. A technical difficulty in the proof is that the definition of the vertices of the Voronoi diagram is nontrivial. Moreover, to achieve the bound $d^{\mathcal{O}(\sqrt{k})}$ instead of $n^{\mathcal{O}(\sqrt{k})}$, we need a more involved way of finding a set of $d^{\mathcal{O}(1)}$ candidate vertices; unlike in the geometric setting, enumerating vertices equidistant from three objects is not sufficient.

Armed with set $\mathcal{N}$ from Lemma 2.1, the packing problem can be solved in a way analogous to the case of unit disks. We guess a pair $(Q, \Gamma) \in \mathcal{N}$ that satisfies the properties of Lemma 2.1. Then objects of Q as well as those intersecting Γ can be removed from $\mathcal{D}$. In other words, we have to solve the problem on graph $G - \Gamma$, so we can focus on each connected component separately. However, we know that each such component contains at most $\frac{2}{3}k$ objects of the solution. Hence, for each component C of $G - \Gamma$ containing at least one object of $\mathcal{D}$ and for $k' = 1, \ldots, \lfloor \frac{2}{3}k \rfloor$, we recursively solve the problem on C with parameter k'. A similar reasoning as before shows that we can put together an optimum solution for the original problem from optimum solutions for the subproblems. As at most d components of $G - \Gamma$ contain objects from $\mathcal{D}$, we recursively solve at most $d \cdot \frac{2}{3}k$ subproblems for a given (Q, Γ). Hence, the total number of subproblems we solve is at most $d \cdot \frac{2}{3}k \cdot |\mathcal{N}| = d \cdot \frac{2}{3}k \cdot d^{\mathcal{O}(\sqrt{k})} = d^{\mathcal{O}(\sqrt{k})}$. The same analysis of the recurrence shows that the running time of the algorithm is $d^{\mathcal{O}(\sqrt{k})} \cdot n^{\mathcal{O}(1)}$.

Covering in Planar Graphs. Let us consider now the following analog of covering points by unit disks: given an edge-weighted planar graph G, two sets of vertices $\mathcal{D}$ and $\mathcal{C}$, and integers k and r, the task is to find a set $\mathcal{F} \subseteq \mathcal{D}$ of k vertices that covers every vertex in $\mathcal{C}$. Here $p \in \mathcal{D}$ covers $q \in \mathcal{C}$ if $\text{dist}(p, q) \leq r$, i.e., we can imagine that p represents a ball of radius r in the graph with center at p. Unlike in the case of packing, $\mathcal{D}$ is a set of vertices, not a set of connected sets.

Let $\mathcal{F}$ be a hypothetical solution. We can construct the set $\mathcal{N}$ given by Lemma 2.1 and guess a guarded separator (Q, Γ) satisfying the three properties. As we assume that Q is part of the solution, we remove from $\mathcal{C}$ every vertex that is covered by some vertex in Q; let $\mathcal{C}'$ be the remaining vertices. By the third property of Lemma 2.1, we can assume that in the solution $\mathcal{F}$, the set Γ is fully contained in the Voronoi regions of the vertices in Q. This means that if there is a $p \in \mathcal{D} \setminus Q$ and $v \in \Gamma$ such that $\mathrm{dist}(p, v) < \mathrm{dist}(p, Q)$, then p can be removed from $\mathcal{D}$. Let $\mathcal{D}'$ be the remaining set of vertices. For every component C of $G - S$ and $k' = 1, \ldots, \lfloor \frac{2}{3} k \rfloor$, we recursively solve the problem restricted to C, that is, with the restrictions $\mathcal{D}'[C]$ and $\mathcal{C}'[C]$ of the object and client sets. It is very important to point out that now (unlike how we did the packing problem) we *do not* change the graph G in each call: we use the same graph G, only with the restricted sets $\mathcal{D}'[C]$ and $\mathcal{C}'[C]$. The reason is that restricting to the graph $G[C]$ can change the distances between vertices in C.

If k_C is the minimum number of vertices in $\mathcal{D}'[C]$ that can cover $\mathcal{C}'[C]$, then we know that there are $|Q| + \sum k_C$ vertices in $\mathcal{D}$ that cover every vertex in $\mathcal{C}$. As in the case of covering with disks, we argue that if there is a solution, then we can obtain a solution this way. The reasoning is basically the same: we just replace the Euclidean metric with the graph metric, and use the fact that the shortest path connecting any two points of $\mathcal{D}' \cup \mathcal{C}'$ lying in different components of $G - \Gamma$ must intersect Γ. As in the case of packing, we have at most $d \cdot \frac{2}{3} k \cdot d^{\mathcal{O}(\sqrt{k})}$ subproblems and the running time $d^{\mathcal{O}(\sqrt{k})} \cdot n^{\mathcal{O}(1)}$ follows the same way.

Nonuniform Radius. A natural generalization of the covering problem is when every vertex $p \in \mathcal{D}$ is given a radius $r(p) \geq 0$ and p covers a $q \in \mathcal{C}$ if $\mathrm{dist}(p, q) \leq r(p)$. That is, now the vertices in $\mathcal{D}$ represent balls with possibly different radii.

There are two ways in which we can handle this more general problem. The first is a simple graph-theoretic trick. For every $p \in \mathcal{D}$, attach a path of length $R - r(p)$ to p, and replace p in $\mathcal{D}$ with the other end p' of this path, where $R = \max_{p' \in \mathcal{D}} r(p')$. Now a vertex $q \in \mathcal{C}$ is at distance at most $r(p)$ from p iff it is at distance at most R from p', so we can solve the problem by applying the algorithm for uniform radius R. The second way is somewhat more complicated, but it seems to be the robust solution of the issue. We can namely work with the *additively weighted* Voronoi diagram, that is, instead of defining the Voronoi regions of $\mathcal{F}$ by comparing distances $\mathrm{dist}(p, v)$ for $p \in \mathcal{F}$, we compare the weighted distances $\mathrm{dist}(p, v) - r(p)$. It can be verified that the main arguments of the algorithm, like convexity of the Voronoi regions or the separability of subproblems in the covering setting, all go through after redoing the same calculations.

3 The General Problem

Suppose we are given an edge-weighted undirected graph G, a family of *objects* $\mathcal{D}$, and a family of *clients* $\mathcal{C}$. Every object $p \in \mathcal{D}$ has three attributes. It has its *location* $\mathbf{loc}(p)$, which is a nonempty subset of vertices of G such that $G[\mathbf{loc}(p)]$ is connected. It has its *cost* $\lambda(p)$, which is a real number (possibly negative). Finally, it has its *radius* $r(p)$, which is a nonnegative real value denoting the strength of domination imposed by p. Every client $q \in \mathcal{C}$ has three attributes. It has its *placement* $\mathbf{pla}(q)$, which is a vertex of G where the client resides. It has also its *sensitivity* $s(q)$, which is a real value denoting how sensitive the client is to domination from objects. Finally, it has *prize* $\pi(q)$, which is a real value denoting the prize for dominating the client. Note that there can be multiple clients placed in the same vertex and the prizes may be negative.

We say that a subfamily $\mathcal{F} \subseteq \mathcal{D}$ is *normal* if locations of objects from $\mathcal{F}$ are disjoint, and moreover $\mathrm{dist}(\mathbf{loc}(p_1), \mathbf{loc}(p_2)) > |r(p_1) - r(p_2)|$ for all pairs (p_1, p_2) of different objects in $\mathcal{F}$. In particular, normality implies disjointness of locations of objects from $\mathcal{F}$, but if all the radii are equal, then the two notions coincide. We say that a client q is *covered* by an object p if $\mathrm{dist}(\mathbf{pla}(q), \mathbf{loc}(p)) \leq s(q) + r(p)$.

We are finally ready to define DISJOINT NETWORK COVERAGE. As input we get an edge-weighted graph G embedded on a sphere, families of objects $\mathcal{D}$ and clients $\mathcal{C}$ (described using their attributes), and an integer k. For a subfamily $\mathcal{Z} \subseteq \mathcal{D}$, we define its *revenue* $\Pi(\mathcal{Z})$ as the total sum of prizes of clients covered by at least one object from $\mathcal{Z}$ minus the total sum of costs of objects from $\mathcal{Z}$. In the DISJOINT NETWORK COVERAGE problem, the task is to find a subfamily $\mathcal{Z} \subseteq \mathcal{D}$ such that the following holds: (1) family $\mathcal{Z}$ is normal and has cardinality *exactly* k and (2) subject to the previous constraint, family $\mathcal{Z}$ maximizes the revenue $\Pi(\mathcal{Z})$. The main result of this paper is the following theorem.

Theorem 3.1 (Main result). DISJOINT NETWORK COVERAGE *can be solved in time* $|\mathcal{D}|^{\mathcal{O}(\sqrt{k})} \cdot (|\mathcal{C}| \cdot |V(G)|)^{\mathcal{O}(1)}$.

References

1. Adamaszek, A., Wiese, A.: Approximation schemes for maximum weight independent set of rectangles. In: FOCS, pp. 400–409 (2013)
2. Adamaszek, A., Wiese, A.: A QPTAS for maximum weight independent set of polygons with polylogarithmically many vertices. In: SODA, pp. 645–656 (2014)
3. Agarwal, P.K., Overmars, M.H., Sharir, M.: Computing maximally separated sets in the plane. SIAM J. Comput. 36(3), 815–834 (2006)
4. Alber, J., Fiala, J.: Geometric separation and exact solutions for the parameterized independent set problem on disk graphs. J. Algorithms 52(2), 134–151 (2004)
5. Chitnis, R.H., Hajiaghayi, M., Marx, D.: Tight bounds for Planar Strongly Connected Steiner Subgraph with fixed number of terminals (and extensions). In: SODA, pp. 1782–1801 (2014)
6. Demaine, E.D., Hajiaghayi, M.: The bidimensionality theory and its algorithmic applications. Comput. J. 51(3), 292–302 (2008)
7. Har-Peled, S.: Quasi-polynomial time approximation scheme for sparse subsets of polygons. In: SoCG, pp. 120 (2014)
8. Hwang, R.Z., Lee, R.C.T., Chang, R.C.: The slab dividing approach to solve the Euclidean p-center problem. Algorithmica 9(1), 1–22 (1993)
9. Klein, P.N., Marx, D.: Solving PLANAR k-TERMINAL CUT in $O(n^{c\sqrt{k}})$ Time. In: Czumaj, A., Mehlhorn, K., Pitts, A., Wattenhofer, R. (eds.) ICALP 2012, Part I. LNCS, vol. 7391, pp. 569–580. Springer, Heidelberg (2012)
10. Klein, P.N., Marx, D.: A subexponential parameterized algorithm for Subset TSP on planar graphs. In: SODA, pp. 1812–1830 (2014)

11. Marx, D., Pilipczuk, M.: Optimal parameterized algorithms for planar facility location problems using Voronoi diagrams. CoRR, abs/1504.05476 (2015)
12. Marx, D., Sidiropoulos, A.: The limited blessing of low dimensionality: when $1-1/d$ is the best possible exponent for d-dimensional geometric problems. In: SoCG, p. 67 (2014)
13. Mustafa, N.H., Raman, R., Ray, S.: QPTAS for geometric set-cover problems via optimal separators. CoRR, abs/1403.0835 (2014)
14. Pilipczuk, M., Pilipczuk, M., Sankowski, P., van Leeuwen, E.J.: Network sparsification for Steiner problems on planar and bounded-genus graphs. In: FOCS, pp. 276–285. IEEE Computer Society (2014)
15. Pătraşcu, M., Williams, R.: On the possibility of faster SAT algorithms. In: SODA, pp. 1065–1075 (2010)

Randomization Helps Computing a Minimum Spanning Tree under Uncertainty[*]

Nicole Megow[1], Julie Meißner[2], and Martin Skutella[2]

[1] Center for Mathematics, Technische Universität München, Germany
nicole.megow@tum.de
[2] Department of Mathematics, Technische Universität Berlin, Germany
{jmeiss,skutella}@math.tu-berlin.de

Abstract. We consider the problem of finding a minimum spanning tree (MST) in a graph with uncertain edge weights given by open intervals on the edges. The exact weight of an edge in the corresponding uncertainty interval can be queried at a given cost. The task is to determine a possibly adaptive query sequence of minimum total cost for finding an MST. For uniform query cost, a deterministic algorithm with best possible competitive ratio 2 is known [7].

We solve a long-standing open problem by showing that randomized query strategies can beat the best possible competitive ratio 2 of deterministic algorithms. Our randomized algorithm achieves expected competitive ratio $1 + 1/\sqrt{2} \approx 1.707$. This result is based on novel structural insights to the problem enabling an interpretation as a generalized online bipartite vertex cover problem. We also consider arbitrary, edge-individual query costs and show how to obtain algorithms matching the best known competitive ratios for uniform query cost. Moreover, we give an optimal algorithm for the related problem of computing the exact weight of an MST at minimum query cost. This algorithm is based on an interesting relation between different algorithmic approaches using the cycle-property and the cut-property characterizing MSTs. Finally, we argue that all our results also hold for the more general setting of matroids. All our algorithms run in polynomial time.

1 Introduction

Uncertainty in the input data is an omnipresent issue in most real world planning processes. The quality of solutions for optimization problems with uncertain input data crucially depends on the amount of uncertainty. More information, or even knowing the exact data, allows for significantly improved solutions (see, e. g., [16]). It is impossible to fully avoid uncertainty. Nevertheless, it is sometimes possible to obtain exact data, but it may involve certain exploration cost in time, money, energy, bandwidth, etc. A classical application are estimated user

---

[*] This research was carried out in the framework of MATHEON supported by Einstein Foundation Berlin. The first two authors were additionally supported by the German Science Foundation (DFG) under contract ME 3825/1.

N. Bansal and I. Finocchi (Eds.): ESA 2015, LNCS 9294, pp. 878–890, 2015.
DOI: 10.1007/978-3-662-48350-3_73

demands that can be specified by undertaking a user survey, but this is an investment in terms of time and/or cost.

In this paper we are concerned with fundamental combinatorial optimization problems with uncertain input data which can be explored at certain cost. We mainly focus on the minimum spanning tree (MST) problem with uncertain edge weights. In a given graph, we know initially for each edge only an interval containing the edge weight. The true value is revealed upon request (we say query) at a given cost. The task is to determine a minimum-cost adaptive sequence of queries to find a minimum spanning tree. In the basic setting, we only need to guarantee that the obtained spanning tree is minimal and do not need to compute its actual weight, i.e., there might be tree edges whose weight we never query, as they appear in an MST independent of their exact weight. We measure the performance of an algorithm by competitive analysis. For any realization of edge weights, we compare the query cost of an algorithm with the optimal query cost. This is the cost for verifying an MST for a given fixed realization.

As our main result we develop a randomized algorithm that improves upon the competitive ratio of any deterministic algorithm. This solves an important open problem in this area (cf. [4]). We also present the first algorithms for non-uniform query costs and generalize the results to uncertainty matroids, in both settings matching the best known competitive ratios for MST with uniform query cost.

Related Work. The huge variety of research streams dealing with optimization under uncertainty reflects its importance for theory and practice. The major fields are online optimization [3], stochastic optimization [2], and robust optimization [1], each modeling uncertain information in a different way. Typically these models do not provide the possibility to influence when and how uncertain data is revealed. Kahan [12] was probably the first to study algorithms for explicitly exploring uncertain information in the context of finding the maximum and median of a set of values known to lie in given uncertainty intervals.

The MST problem with uncertain edge weights was introduced by Erlebach et al. [7]. Their deterministic Algorithm U-RED achieves competitive ratio 2 for uniform query cost when all uncertainty intervals are open intervals or trivial (i.e., containing one point only). They also show that this ratio is optimal and can be generalized to the problem of finding a minimum weight basis of a matroid with uncertain weights [6]. According to Erlebach [4] it remained a major open problem whether randomized algorithms can beat competitive ratio 2. The offline problem of finding the optimal query set for given exact edge weights can be solved optimally in polynomial time [5].

Further problems studied in this uncertainty model include finding the k-th smallest value in a set of uncertainty intervals [9, 11, 12] (also with non-uniform query cost [9]), caching problems in distributed databases [15], computing a function value [13], and classical combinatorial optimization problems, such as shortest path [8], finding the median [9], and the knapsack problem [10].

A generalized exploration model was proposed in [11]. The OP-OP model reveals, upon an edge query, a refined open or trivial subinterval and might, thus, require multiple queries per edge. They show that Algorithm U-RED [7] can be adopted and still achieves competitive ratio 2. The restriction to open intervals is not crucial as slight model adaptions allow to deal with closed intervals [11, 12].

While most works aim for minimal query sets to guarantee exact optimal solutions, Olsten and Widom [15] initiate the study of trade-offs between the number of queries and the precision of the found solution. They are concerned with caching problems. Further work in this vein can be found in [8, 9, 13].

Our Contribution. After presenting some structural insights in Section 2, we affirmatively answer the question if randomization helps to minimize query cost in order to find an MST. In Section 3 we present a randomized algorithm with tight competitive ratio 1.707, thus beating the best possible competitive ratio 2 of any deterministic algorithm. On the other hand, one can easily achieve a lower bound of 1.5 for any randomized algorithm by considering a graph consisting of two parallel edges with crossing uncertainty intervals, e.g., $(1, 3)$ and $(2, 4)$.

One key observation is that the minimum spanning tree problem under uncertainty can be interpreted as a generalized online bipartite vertex cover problem. A similar connection for a *given* realization of edge weights was established in [5] for the related MST verification problem. In our case, new structural insights allow for a preprocessing which suggests a unique bipartition of the edges for *all* realizations simultaneously. Our algorithm borrows and refines ideas from a recent water-filling algorithm for the online bipartite vertex cover problem [17].

In Section 4 we consider the more general non-uniform query cost model in which each edge has an individual query cost. We observe that this problem can be reformulated within a different uncertainty model, called OP-OP, presented in [11]. The 2-competitive algorithm in [11] is a pseudo-polynomial 2-competitive algorithm for our problem with non-uniform query cost. We design new direct and polynomial-time algorithms that are 2-competitive and 1.707-competitive in expectation. To that end, we employ a new strategy carefully balancing the query cost of an edge and the number of cycles it occurs in.

In Section 5 we consider the problem of computing the exact value of an MST under uncertain edge weights. While previous algorithms (U-RED [7], our algorithms) aim for removing the largest-weight edge from a cycle, we now attempt to detect minimum-weight edges separating the graph into two components. Interestingly, the latter cut-based algorithm can be shown to solve the original problem (not computing the exact MST value) achieving the same best possible competitive ratio of 2 as the cycle-based algorithm presented in [7].

Finally, in Section 6 we observe in a broader context that these two algorithms can be interpreted as the best-in and worst-out greedy algorithm on matroids.

Due to space constraints some details are omitted and will be presented in a full version.

2   Problem Definition, Notation and Structural Insights

Problem Description. Initially we are given a weighted, undirected, connected graph $G = (V, E)$, with $|V| = n$ and $|E| = m$. Each edge $e \in E$ comes with an uncertainty interval A_e and possibly a query cost c_e. The uncertainty interval A_e

gives the only information about e's unknown weight $w_e \in A_e$. We assume that an interval is either trivial, i.e., $A_e = [w_e, w_e]$, or, it is open $A_e = (L_e, U_e)$ with lower limit L_e and upper limit $U_e > L_e$. A realization of edge weights $(w_e)_{e \in E}$ for an uncertainty graph G is *feasible*, if all edge weights w_e, $e \in E$, lie in their corresponding uncertainty intervals, i.e., $w_e \in A_e$.

The task is to find a minimum spanning tree (MST) in the uncertainty graph G for an unknown, feasible realization of edge weights. To that end, we may query any edge $e \in E$ at cost c_e and obtain the exact weight w_e. The goal is to design an algorithm that constructs a sequence of queries that determines an MST at minimum total query cost. A set of queries $Q \subseteq E$ is *feasible*, if an MST can be determined given the exact edge weights for edges in Q; that is, given w_e for $e \in Q$, there is a spanning tree which is minimal for any realization of edge weights $w_e \in A_e$ for $e \in E \setminus Q$. We denote this problem as *MST with edge uncertainty* and say *MST under uncertainty* for short.

Note that this problem does not necessarily involve computing the actual MST weight. We refer to the problem variant in which the actual MST weight must be computed as *computing the MST weight under uncertainty*.

We evaluate our algorithms by standard competitive analysis. An algorithm is α-*competitive* if, for any realization $(w_e)_{e \in E}$, the solution query cost is at most α times the optimal query cost for this realization. The optimal query cost is the minimum query cost that an offline algorithm must pay when it is given the realization (and thus an MST) and has to verify an MST. The *competitive ratio* of an algorithm ALG is the infimum over all α such that ALG is α-competitive. For randomized algorithms we consider the expected query cost. Competitive analysis addresses the problem complexity evolving from the uncertainty in the input, possibly neglecting any computational complexity. However, we note that all our algorithms run in polynomial time unless explicitly stated otherwise.

Structural Insight. We derive a structural property that allows to reduce MST under uncertainty to a set of crucial instances. Given an uncertainty graph $G = (V, E)$, consider the following two MSTs for extreme realizations. Let $T_L \subseteq E$ be an MST for the realization w^L, in which all edge weights of edges with non-trivial uncertainty interval are close to their lower limit, more precisely $w_e = L_e + \varepsilon$ for infinitesimally small $\varepsilon > 0$. Symmetrically, let $T_U \subseteq E$ be an MST when the same edges have weight $w_e = U_e - \varepsilon$.

Theorem 1. *Given an uncertainty graph with trees T_L and T_U, any edge $e \in T_L \setminus T_U$ with $L_e \neq U_e$ is in every feasible query set for any feasible realization.*

Proof. Given an uncertainty graph, let h be an edge in $T_L \setminus T_U$ with non-trivial uncertainty interval. Assume all edges apart from h have been queried and thus have fixed weight w_e. As edge h is in T_L, we can choose its edge weight such that edge h is in any MST. We set $w_h = L_h + \varepsilon$ and choose ε so small, that all edges with at least the same weight in w^L now have a strictly larger edge weight. Symmetrically, if we choose the edge weight w_h sufficiently close to the upper limit U_h, no MST contains edge h. Consequently we cannot decide whether edge h is in an MST without querying it. $\square$

Any edge in the set $T_L \backslash T_U$ with non-trivial uncertainty area is in every feasible query set and thus can be queried in a preprocessing step. Its existence increases the size of every feasible query set and hence decreases the competitive ratio of an instance. Thus we restrict our analysis to instances for which the set $T_L \backslash T_U$ contains only edges with $L_e = U_e$, to find the worst-case competitive ratio of an algorithm.

Assumption 1 *We restrict to uncertainty graphs for which all edges $e \in T_L \backslash T_U$ have trivial uncertainty area, i.e. $L_e = U_e$.*

We call an edge in a cycle *maximal*, if any realization has an MST that does not contain this edge. Symmetrically, an edge in a cut is *minimal* if every realization has an MST containing it. Whenever we sort by increasing lower (decreasing upper) limit, we break ties by preferring the smaller upper (greater lower) limit.

3   A Randomized Algorithm for MST under Uncertainty

We give an algorithm that solves MST under uncertainty with competitive ratio $1 + 1/\sqrt{2} \approx 1.707$ in expectation. As U-RED in [7], our algorithm is based on Kruskal's algorithm [14], that iteratively deletes maximal edges from cycles. We decide how to resolve cycles, by maintaining an edge potential for each edge $e \in T_L$ describing the probability to query it. The edge potentials are increased in every cycle we consider throughout the algorithm. To determine the increase, we carefully adapt a water-filling scheme presented in [17] for online bipartite vertex cover. In this section we assume uniform query cost $c_e = 1$, $e \in E$, and explain the generalization to non-uniform query costs in Section 4.

Our algorithm RANDOM is structured as follows: We maintain a tree Γ which initially is set to T_L and sort all remaining edges in $R = E \setminus T_L$ by increasing lower limit. We choose a query bound $b \in [0,1]$ uniformly at random. Then we iteratively add edges $f_i \in R$ to Γ, closing a unique cycle C_i in each iteration i. Edges in $C_i \cap T_L$ with uncertainty interval overlapping that of edge f_i compose the *neighbor set* $X(f_i)$. In each cycle we query edges until we identify a maximal edge.

To decide which edge to query in cycle C_i, we consider the potentials y_e of edges $e \in C_i \cap T_L$. In each iteration we evenly increase the potential of all neighbors $X(f_i)$ of edge f_i. We choose a threshold $t(f_i)$ as large as possible, such that when we increase y_e to $\max\{t(f_i), y_e\}$ for all neighbors $e \in X(f_i)$, the total increase in the potential sums up to no more than a fixed parameter α whose optimal value is determined later in the analysis.

Now we compare the edge potentials to the query bound b and decide if we query the edges in $X(f_i)$ or edge f_i. If these queries do not suffice to identify a maximal edge, we repeatedly query the edge with the largest upper limit in the cycle. A formal description of our algorithm is given in Algorithm 1.

RANDOM computes a feasible query set, since it deletes in each cycle a maximal edge. It terminates, as in each iteration of the while loop one edge is queried. When all edges in a cycle have been queried, we always find a maximal edge.

For the analysis of RANDOM we combine the Kruskal-MST structure of our algorithm with Assumption 1. Similar to the analysis in [7] we derive two lemmas.

Algorithm 1. RANDOM

Input: An uncertainty graph $G = (V, E)$ and a parameter $\alpha \in [0, 1]$.
Output: A feasible query set Q.
1. Determine tree T_L, set the temporary graph Γ to T_L, and initialize $Q = \emptyset$.
2. Index the edges in $R = E \backslash T_L$ by increasing lower limit $f_1, \ldots, f_{m-n+1}$.
3. For all edges $e \in T_L$ set the potential y_e to 0.
4. Choose the query bound $b \in [0, 1]$ uniformly at random.
5. **for** $i = 1$ to $m - n + 1$ **do**
6.   Add edge f_i to the temporary graph Γ and let C_i be the unique cycle closed.
7.   Let $X(f_i)$ be the set of edges $g \in T_L \cap C_i$ with $U_g > L_{f_i}$.
8.   **if** no edge in the cycle C_i is maximal **then**
9.   Maximize the threshold $t(f_i) \leq 1$ s.t. $\sum_{e \in X(f_i)} \max\{0, t(f_i) - y_e\} \leq \alpha$.
10.   Increase the edge potential $y_e = \max\{t(f_i), y_e\}$ for all edges $e \in X(f_i)$.
11.   **if** $t(f_i) \leq b$ **then**
12.   Add edge f_i to the query set Q and query it.
13.   **else**
14.   Add all edges in $X(f_i)$ to query set Q and query them.
15.   **while** no edge in the cycle C_i is maximal **do**
16.   Query the edge $e \in C_i$ with maximum U_e and add it to the query set Q.
17.   Delete a maximal edge from Γ.
18. **return** The query set Q.

Lemma 1. *Any feasible query set contains for every cycle-closing edge f_i either edge f_i or its neighborhood $X(f_i)$.*

Lemma 2. *Any edge queried in Line 16 of* RANDOM *is contained in any feasible query set.*

This concludes the preliminaries to prove the algorithm's competitive ratio.

Theorem 2. *For $\alpha = \frac{1}{\sqrt{2}}$,* RANDOM *has competitive ratio $1 + \frac{1}{\sqrt{2}} (\approx 1.707)$.*

Proof. Consider a fixed realization and an optimal query set Q^*. We denote the potential of an edge $e \in T_L$ at the start of iteration i by y_e^i and use y_e to denote the edge potential after the last iteration of the algorithm. The increase of potentials in the algorithm depends on the cycles that are closed and thus on the realization, but not the queried edges. This means the edge potentials are chosen independently of the query bound b in the algorithm. Edges queried in Line 16 are in Q^* by Lemma 2, therefore an edge $e \in T_L \backslash Q^*$ is queried with probability $P(y_e > b) = y_e$ and an edge $f_i \in R \backslash Q^*$ is queried with probability $P(t(f_i) \leq b) = 1 - t(f_i)$. Hence, we can bound the total expected query cost by

$$\mathbb{E}\left[|Q|\right] \leq |Q^*| + \sum_{e \in T_L \backslash Q^*} y_e + \sum_{i: f_i \in R \backslash Q^*} (1 - t(f_i)). \tag{1}$$

For any edge $e \in T_L \backslash Q^*$, Lemma 1 states that all edges $f \in R$ with $e \in X(f)$ must be in the optimal query set Q^*. The potential y_e is the sum of the potential increases caused by edges $f \in R$ with $e \in X(f)$. As in each iteration of the algorithm the total increase of potential is bounded by α, we have

$$
\sum_{e \in T_L \setminus Q^*} y_e = \sum_{e \in T_L \setminus Q^*} \sum_{\substack{i : f_i \in R \cap Q^*, \\ e \in X(f_i)}} \max\left\{ t(f_i) - y_e^i, 0 \right\}
$$

$$
\leq \sum_{i : f_i \in R \cap Q^*} \sum_{e \in X(f_i)} \max\left\{ t(f_i) - y_e^i, 0 \right\} \leq \sum_{i : f_i \in R \cap Q^*} \alpha = \alpha \cdot |R \cap Q^*|. \quad (2)
$$

For an edge $f_i \in R \setminus Q^*$ with $t(f_i) < 1$ we distribute exactly potential α among its neighbors $X(f_i)$ in Lines 9 and 10 of the algorithm. By Lemma 1, the neighbor set $X(f_i)$ is part of the optimal query set Q^*. We consider the share of the total potential increase each neighbor receives and distribute the term $1 - t(f_i)$ (see (1)) according to these shares. Hence,

$$
\sum_{i : f_i \in R \setminus Q^*} (1 - t(f_i)) = \sum_{i : f_i \in R \setminus Q^*} \frac{1 - t(f_i)}{\alpha} \sum_{e \in X(f_i)} \max\{ t(f_i) - y_e^i, 0 \}
$$

$$
= \sum_{e \in T_L \cap Q^*} \sum_{\substack{i : f_i \in R \setminus Q^*, \\ e \in X(f_i)}} \frac{1 - t(f_i)}{\alpha} \left(y_e^{i+1} - y_e^i \right). \quad (3)
$$

In the last equation we have used $y_e^{i+1} = \max\{ t(f_i), y_e^i \}$. We consider the inner sum in (3) and bound the summation term from above by an integral from y_e^i to y_e^{i+1} of the function $\frac{1-z}{\alpha}$. This yields a valid upper bound as the function is decreasing in z and $t(f_i) = y_e^{i+1}$, unless $y_e^{i+1} - y_e^i = 0$. This yields

$$
\sum_{\substack{i : f_i \in R \setminus Q^*, \\ e \in X(f_i)}} \frac{1 - t(f_i)}{\alpha} \left(y_e^{i+1} - y_e^i \right) \leq \sum_{\substack{i : f_i \in R \setminus Q^*, \\ e \in X(f_i)}} \int_{y_e^i}^{y_e^{i+1}} \frac{1 - z}{\alpha} dz \leq \int_0^1 \frac{1 - z}{\alpha} dz = \frac{1}{2\alpha}.
$$

Now we use this bound in Equation (3) and conclude

$$
\sum_{i : f_i \in R \setminus Q^*} (1 - t(f_i)) \leq \frac{1}{2\alpha} \cdot |T_L \cap Q^*|.
$$

Plugging this bound and (2) into (1) yields total query cost

$$
\mathbb{E}[|Q|] \leq |Q^*| + \alpha \cdot |R \cap Q^*| + \frac{1}{2\alpha} \cdot |T_L \cap Q^*|.
$$

Choosing $\alpha = 1/\sqrt{2}$ gives the desired competitive ratio $1 + 1/\sqrt{2}$ for RANDOM.

A simple example shows that this analysis is tight. Consider two parallel edges f and g with overlapping uncertainty intervals. Let f be the edge with larger upper limit. In RANDOM we distribute potential α to g and potential $1 - \alpha$ to edge f. However, the realization with $L_f < w_g < U_g < w_f$ has optimal query set $\{f\}$, while $\{g\}$ is not a feasible query set. The algorithm queries edge g first with probability α and has query cost 2 in this case. Thus the algorithm has expected competitive ratio $1 + \alpha$ for this instance. □

4   Non-uniform Query Cost

Consider the problem MST under uncertainty in which each edge $e \in E$ has associated an individual query cost c_e. W.l.o.g. we assume $c_e > 0$, for all $e \in E$, since querying all other edges only decreases the total query cost. We give a new polynomial-time 2-competitive algorithm, which is deterministically best possible. Furthermore, we adapt our algorithm RANDOM (Sec. 3) to handle non-uniform query costs achieving the same competitive ratio $1 + 1/\sqrt{2}$.

Before showing the main results, we remark that the problem can be transformed into the OP-OP model [11]. This model allows multiple queries per edge and each query returns an open or trivial subinterval. Given an uncertainty graph, we model the query cost $c_e, e \in E$, in the OP-OP model as follows: querying an edge e returns the same interval for $c_e - 1$ queries and then the exact edge weight. The 2-competitive algorithm for the OP-OP model [11] has a running time depending on the query cost of our original problem.

Theorem 3. *There is a pseudo-polynomial 2-competitive algorithm for MST under uncertainty and non-uniform query cost.*

4.1   Balancing Algorithm

Our polynomial-time algorithm BALANCE relies on the property that an MST is cycle-free, similar to previous algorithms for uniform query cost. The key idea is as follows: To decide which edge to query in a cycle, we use a value function $v : E \rightarrow \mathbb{R}_{\geq 0}$ that represents for an edge $e \in E$ the cost difference between a local solution containing e and one that does not. Initially we are locally only aware of each edge individually and thus initialize its value at c_e. We design a balancing scheme that queries among two well-chosen edges the one with smaller value and charges the value of the queried edge to the non-queried alternative.

More formally, in BALANCE (cf. Algorithm 2) we choose a tree T_L and iteratively add the other edges in increasing order of lower limit to T_L. In an occurring cycle, we consider an edge f with maximal upper limit and an edge g with overlapping uncertainty interval. We query the edge $e \in \{f, g\}$ with the smaller value $v(e)$ and decrease the value of the non-queried edge by $v(e)$. We repeat this until we identify a maximal edge in the cycle.

BALANCE computes a feasible query set, since it deletes in each cycle a maximal edge. It terminates, as in each iteration of the while loop one edge is deleted or queried. When all edges in a cycle are queried, we always find a maximal edge.

Now we consider a query set computed by BALANCE. For each edge $e \in E$, let the set of its *children* $C(e) \subseteq E$ be the set of edges that decreased $v(e)$ in the algorithm (cf. Line 14 and 16). Furthermore we define recursively the set of *related edges* $S_e \subseteq E$ to be the union of edge e and the sets S_h of all children $h \in C(e)$.

Every edge is the child of at most one edge, because when it contributes to some value in Line 14 or 16, it is queried. Thus we can interpret a set of related edges S_e and its children-relation as a tree. Slightly abusing notation we speak of a *vertex cover* $VC(S_e)$ of the set of related edges S_e and mean a minimum weight

---

Algorithm 2. BALANCE

Input: An uncertainty Graph $U = (V, E)$ and a query cost function $c : E \to \mathbb{R}_{\geq 0}$.
Output: A feasible query set Q.
1. Choose a tree T_L and let the temporary graph $\Gamma = T_L$.
2. Index the edges in $E \backslash T_L$ by increasing lower limit $e_1, e_2, \ldots, e_{m-n+1}$.
3. Set a value function $v : E \to \mathbb{R}_{\geq 0}$ to c_e for all edges.
4. **for** $i = 1$ to $m - n + 1$ **do**
5.   Add e_i to Γ.
6.   **while** Γ has a cycle C **do**
7.   **if** C contains a maximal edge e **then**
8.   Delete e from Γ.
9.   **else**
10.   Choose $f \in C$ such that $U_f = \max\{U_e | e \in C\}$ and $g \in C \backslash f$ with $U_g > L_f$.
11.   **if** A_g is trivial **then**
12.   Query edge f and add f to Q.
13.   **else if** $v(f) \geq v(g)$ **then**
14.   Query edge g, add g to Q, and subtract $v(g)$ from $v(f)$.
15.   **else**
16.   Query edge f, add f to Q, and subtract $v(f)$ from $v(g)$.
17. **return** The query set Q.

---

vertex cover in the corresponding tree. We use VC_e for a vertex cover containing edge e and $VC_{\backslash e}$ for one not containing e. Similar to Lemmas 1 and 2 we have:

Lemma 3. *For every feasible query set Q and every set of related edges S_e in* BALANCE, *the set Q contains a vertex cover of S_e.*

Lemma 4. *Every edge queried in Line 12 is in every feasible query set.*

The following two lemmas establish a relation between the value $v(e)$ of an edge $e \in E$ and the cost of its related edges S_e. The proof is by induction.

Lemma 5. *The value function after an execution of* BALANCE *fulfills for every edge $e \in E$ and its set of related edges S_e:*

$$\sum_{f \in VC_e(S_e)} c_f = v(e) + \sum_{f \in VC_{\backslash e}(S_e)} c_f.$$

Lemma 6. *The value function after an execution of* BALANCE *fulfills for every edge $e \in E$ and its set of related edges S_e:*

$$2 \cdot \sum_{f \in VC_{\backslash e}(S_e)} c_f = -v(e) + \sum_{f \in S_e} c_f.$$

This concludes all preliminaries we need to prove the main theorem.

Theorem 4. BALANCE *has a competitive ratio of 2 and this is best possible.*

Proof. For some realization, let Q^* denote an optimal query set and Q a query set computed by BALANCE. Let A be the set of all edges not in Q and let B be the set of all edges queried in Line 12 of BALANCE. As any edge in Q is either queried in Line 12, or the child of some other edge, Q is the disjoint union of the sets $S_a \setminus \{a\}, a \in A$, and the sets $S_b, b \in B$.

We bound the cost of a set $S_a \setminus \{a\}, a \in A$, by applying Lemmas 6 and then 5 and concluding from Lemma 3 that Q^* contains a vertex cover of S_a. Hence,

$$\sum_{e \in S_a \setminus \{a\}} c_e \leq 2 \cdot \sum_{e \in VC_{\setminus a}(S_a)} c_e = 2 \cdot \sum_{e \in VC(S_a)} c_e \leq 2 \cdot \sum_{e \in Q^* \cap S_a} c_e.$$

By definition, every edge $b \in B$ is queried in Line 12 and is thus by Lemma 4 an element of Q^*. Applying Lemma 3 this means the cost of $Q^* \cap S_b$ is at least the of cost $VC_b(S_b)$. We use this fact after applying Lemmas 5 and 6 and deduce

$$\sum_{e \in S_b} c_e = v(b) + 2 \cdot \sum_{e \in VC_{\setminus b}(S_b)} c_e \leq 2 \cdot \sum_{e \in VC_b(S_b)} c_e \leq 2 \cdot \sum_{e \in Q^* \cap S_b} c_e.$$

As the set Q is a disjoint union of all sets $S_a \setminus \{a\}, a \in A$, and $S_b, b \in B$, this yields the desired competitive ratio of 2. This factor is best possible for deterministic algorithms, even in the special case of uniform query costs [7].   □

4.2   Randomization for Non-uniform Query Cost

We generalize the algorithm RANDOM (Sec. 3) to the non-uniform query cost model. The adaptation is similar to one for the weighted online bipartite vertex cover problem in [17]. For each edge $f_i \in R$ we distribute at most $1/\alpha \cdot c_{f_i}$ new potential to its neighborhood $X(f_i)$. We replace Line 9 of RANDOM by:

$$\text{Maximize } t(f_i) \leq 1 \text{ s.t.} \sum_{e \in X(f_i)} c_e \max\{t(f_i) - y_e), 0\} \leq \frac{c_{f_i}}{\alpha} \text{ holds.} \quad (4)$$

Using exactly the same analysis as presented in Section 3 this yields:

Theorem 5. *For the non-uniform query cost setting* RANDOM *adapted by* (4) *achieves expected competitive ratio* $1 + \frac{1}{\sqrt{2}}$.

5   Computing the MST Weight under Uncertainty

In this section we give an optimal polynomial-time algorithm for computing the exact MST weight in an uncertainty graph. As a key to our result, we algorithmically utilize the well-known characterization of MSTs through the *cut property* - in contrast to previous algorithms for the MST under uncertainty problem which relied on the *cycle property* (cf. RANDOM, BALANCE, and U-RED [7]).

In CUT-WEIGHT, we consider a tree T_U and iteratively delete its edges in decreasing order of upper limits. In each iteration, we consider the cut which

---

Algorithm 3. CUT-WEIGHT

Input: An uncertainty graph $G = (V, E)$.
Output: A feasible query set Q.
1. Choose a tree T_U and let the temporary graph $\Gamma = T_U$. Initialize $Q = \emptyset$.
2. Index all edges of T_U by decreasing upper limit $e_1, e_2, ..., e_{n-1}$.
3. **for** $i = 1$ to $n - 1$ **do**
4.   Delete e_i from Γ.
5.   **while** Γ has two components **do**
6.   Let S be the cut containing all edges in G between the two components of Γ.
7.   **if** S contains a minimal edge e **then**
8.   Query edge e and add it to Q.
9.   Replace edge e_i in Γ with e and contract edge e.
10.   **else**
11.   Choose $g \in S$ such that $L_g = \min\{L_e | e \in S\}$, query it and add it to Q.
12. **return** The query set Q.

---

is defined in the original graph and query edges in increasing order of lower limits until we identify a minimal edge. Then we exchange the tree edge with the minimal edge and contract it. Applying this procedure, we only query edges that are in any feasible query set.

Theorem 6. CUT-WEIGHT *determines the optimal query set and the exact MST weight in polynomial time.*

It may seem surprising that CUT-WEIGHT solves the problem optimally whereas cycle-based algorithms do not. However, there is an intuition. CUT-WEIGHT identifies a *minimum weight* edge in each cut which characterizes an MST. Informally speaking, it has a bias to query edges of the MST. In contrast, cycle-based algorithms identify *maximum weight* edges in cycles, which are not in the tree.

6   Matroids under Uncertainty

We briefly consider a natural generalization of MST under uncertainty: given an *uncertainty matroid*, i.e., a matroid with a ground set of elements with unknown weights, find a minimum weight matroid base. Erlebach et al. [6] show that the algorithm U-RED [7] can be applied to uncertainty matroids with uniform query cost and yields again a competitive ratio of 2. Similarly, our algorithms RANDOM and BALANCE can be generalized to matroids with non-uniform cost, and CUT-WEIGHT can determine the total weight of a minimum weight matroid base.

Theorem 7. *There are deterministic resp. randomized online algorithms with competitive ratio 2 resp. 1.707 for finding a minimum weight matroid base in an uncertainty matroid with non-uniform query cost.*

Theorem 8. *There is an algorithm that determines an optimal query set and the exact weight of a min-weight matroid base in an uncertainty matroid.*

In a matroid with known weights we can find a minimum weight base greedily; we distinguish between *best-in* greedy and *worst-out* greedy algorithms (cf. [14]). They are dual in the sense that both solve the problem on a matroid and take the role of the other on the corresponding dual matroid. The best-in greedy algorithm adds elements in increasing order of weights as long as the system stays independent. Merging ideas from our algorithm RANDOM and U-RED2 in [6] yields a best-in greedy algorithm, CYCLE-ALG, for uncertainty matroids. A worst-out greedy algorithm deletes elements in decreasing order of weights as long as a basis is contained. We can adapt our algorithm CUT-WEIGHT in Section 5 to a worst-out greedy algorithm, CUT-ALG, for uncertainty matroids.

Proposition 1. *The algorithms* CYCLE-ALG *and* CUT-ALG *are dual to each other in the sense that they solve the same problem on a matroid and its dual.*

Acknowledgments. We thank the anonymous referees for numerous helpful comments that improved the presentation of the paper.

References

1. Ben-Tal, A., El Ghaoui, L., Nemirovski, A.S.: Robust Optimization. Princeton Series in Applied Mathematics. Princeton University Press (2009)
2. Birge, J.R., Louveaux, F.: Introduction to Stochastic Programming. Springer Series in Operations Research. Springer, Heidelberg (1997)
3. Borodin, A., El-Yaniv, R.: Online Computation and Competitive Analysis. Cambridge University Press (1998)
4. Erlebach, T.: Computing with uncertainty. Invited lecture, Graduate Program MDS, Berlin (2013)
5. Erlebach, T., Hoffmann, M.: Minimum spanning tree verification under uncertainty. In: Kratsch, D., Todinca, I. (eds.) WG 2014. LNCS, vol. 8747, pp. 164–175. Springer, Heidelberg (2014)
6. Erlebach, T., Hoffmann, M., Kammer, F.: Query-competitive algorithms for cheapest set problems under uncertainty. In: Csuhaj-Varjú, E., Dietzfelbinger, M., Ésik, Z. (eds.) MFCS 2014, LNCS, vol. 8635, pp. 263–274. Springer, Heidelberg (2014)
7. Erlebach, T., Hoffmann, M., Krizanc, D., Mihalák, M., Raman, R.: Computing minimum spanning trees with uncertainty. In: Proc. STACS, pp. 277–288 (2008)
8. Feder, T., Motwani, R., O'Callaghan, L., Olston, C., Panigrahy, R.: Computing shortest paths with uncertainty. Journal of Algorithms 62, 1–18 (2007)
9. Feder, T., Motwani, R., Panigrahy, R., Olston, C., Widom, J.: Computing the median with uncertainty. SIAM Journal on Computing 32, 538–547 (2003)
10. Goerigk, M., Gupta, M., Ide, J., Schöbel, A., Sen, S.: The robust knapsack problem with queries. Computers & OR 55, 12–22 (2015)
11. Gupta, M., Sabharwal, Y., Sen, S.: The update complexity of selection and related problems. In: Proc. of FSTTCS. LIPIcs, vol. 13, pp. 325–338 (2011)
12. Kahan, S.: A model for data in motion. In: Proc. of STOC, pp. 267–277 (1991)
13. Khanna, S., Tan, W.C.: On computing functions with uncertainty. In: Proceedings of PODS, pp. 171–182 (2001)
14. Korte, B., Vygen, J.: Combinatorial optimization. Springer (2012)
15. Olston, C., Widom, J.: Offering a precision-performance tradeoff for aggregation queries over replicated data. In: Proceedings of VLDB, pp. 144–155 (2000)

16. Patil, P., Shrotri, A.P., Dandekar, A.R.: Management of uncertainty in supply chain. Int. J. of Emerging Technology and Advanced Engineering 2, 303–308 (2012)
17. Wang, Y., Wong, S.C.-w.: Two-sided online bipartite matching and vertex cover: Beating the greedy algorithm. In: Halldórsson, M.M., Iwama, K., Kobayashi, N., Speckmann, B. (eds.) ICALP 2015. LNCS, vol. 9134, pp. 1070–1081. Springer, Heidelberg (2015)

Compressed Data Structures
for Dynamic Sequences

J. Ian Munro and Yakov Nekrich

David R. Cheriton School of Computer Science, University of Waterloo

Abstract. We consider the problem of storing a dynamic string S over an alphabet $\Sigma = \{1, \ldots, \sigma\}$ in compressed form. Our representation supports insertions and deletions of symbols and answers three fundamental queries: $\mathrm{access}(i, S)$ returns the i-th symbol in S, $\mathrm{rank}_a(i, S)$ counts how many times a symbol a occurs among the first i positions in S, and $\mathrm{select}_a(i, S)$ finds the position where a symbol a occurs for the i-th time. We present the first fully-dynamic data structure for arbitrarily large alphabets that achieves optimal query times for all three operations and supports updates with worst-case time guarantees. Ours is also the first fully-dynamic data structure that needs only $nH_k + o(n \log \sigma)$ bits, where H_k is the k-th order entropy and n is the string length. Moreover our representation supports extraction of a substring $S[i..i + \ell]$ in optimal $O(\log n / \log \log n + \ell / \log_\sigma n)$ time.

1 Introduction

In this paper we consider the problem of storing a sequence S of length n over an alphabet $\Sigma = \{1, \ldots, \sigma\}$ so that the following operations are supported:
- $\mathrm{access}(i, S)$ returns the i-th symbol, $S[i]$, in S
- $\mathrm{rank}_a(i, S)$ counts how many times a occurs among the first i symbols in S, $\mathrm{rank}_a(i, S) = |\{\, j \mid S[j] = a \text{ and } 1 \leq j \leq i \,\}|$
- $\mathrm{select}_a(i, S)$ finds the position in S where a occurs for the i-th time, $\mathrm{select}_a(i, S) = j$ where j is such that $S[j] = a$ and $\mathrm{rank}_a(j, S) = i$.
This problem, also known as the rank-select problem, is one of the most fundamental problems in compressed data structures. There are many data structures that store a string in compressed form and support three above defined operations efficiently. There are static data structures that use $nH_0 + o(n \log \sigma)$ bits or even $nH_k + o(n \log \sigma)$ bits for any $k \leq \alpha \log_\sigma n - 1$ and a positive constant $\alpha < 1$[1]. Efficient static rank-select data structures are described in [11,10,8,18,19,2,14,26,4]. We refer to [4] for most recent results and a discussion of previous static solutions.

---

[1] Henceforth $H_0(S) = \sum_{a \in \Sigma} \frac{n_a}{n} \log \frac{n}{n_a}$, where n_a is the number of times a occurs in S, is the 0-th order entropy and $H_k(S)$ for $k \geq 0$ is the k-th order empirical entropy. $H_k(S)$ can be defined as $H_k(S) = \sum_{A \in \Sigma^k} |S_A| H_0(S_A)$, where S_A is the subsequence of S generated by symbols that follow the k-tuple A; $H_k(S)$ is the lower bound on the average space usage of any statistical compression method that encodes each symbol using the context of k previous symbols [22].

© Springer-Verlag Berlin Heidelberg 2015
N. Bansal and I. Finocchi (Eds.): ESA 2015, LNCS 9294, pp. 891–902, 2015.
DOI: 10.1007/978-3-662-48350-3_74

In many situations we must work with dynamic sequences. We must be able to insert a new symbol at an arbitrary position i in the sequence or delete an arbitrary symbol $S[i]$. The design of dynamic solutions, that support insertions and deletions of symbols, is an important problem. Fully-dynamic data structures for rank-select problem were considered in [15,7,5,20,6,13,21,16]. Recently Navarro and Nekrich [24,25] obtained a fully-dynamic solution with $O(\log n/\log\log n)$ times for rank, access, and select operations. By the lower bound of Fredman and Saks [9], these query times are optimal. The data structure described in [24] uses $nH_0(S) + o(n\log\sigma)$ bits and supports updates in $O(\log n/\log\log n)$ amortized time. It is also possible to support updates in $O(\log n)$ worst-case time, but then the time for answering a rank query grows to $O(\log n)$ [25]. All previously known fully-dynamic data structures need at least $nH_0(S) + o(n\log\sigma)$ bits. Two only exceptions are data structures of Jansson et al. [17] and Grossi et al. [12] that keep S in $nH_k(S) + o(n\log\sigma)$ bits, but do not support rank and select queries. A more restrictive dynamic scenario was considered by Grossi et al. [12] and Jansson et al. [17]: an update operation *replaces* a symbol $S[i]$ with another symbol so that the total length of S does not change, but insertions of new symbols or deletions of symbols of S are not supported. Their data structures need $nH_k(S) + o(n\log\sigma)$ bits and answer access queries in $O(1)$ time; the data structure of Grossi et al. [12] also supports rank and select queries in $O(\log n/\log\log n)$ time.

In this paper we describe the first fully-dynamic data structure that keeps the input sequence in $nH_k(S) + o(n\log\sigma)$ bits; our representation supports rank, select, and access queries in optimal $O(\log n/\log\log n)$ time. Symbol insertions and deletions at any position in S are supported in $O(\log n/\log\log n)$ worst-case time. We list our and previous results for fully-dynamic sequences in Table 1. Our representation of dynamic sequences also supports the operation of extracting a substring. Previous dynamic data structures require $O(\ell)$ calls of access operation in order to extract the substring of length ℓ. Thus the previous best fully-dynamic representation, described in [24] needs $O(\ell(\log n/\log\log n))$ time to extract a substring $S[i..i + \ell - 1]$ of S. Data structures described in [12] and [17] support substring extraction in $O(\log n/\log\log n + \ell/\log_\sigma n)$ time but they either do not support rank and select queries or they support only updates that replace a symbol with another symbol. Our dynamic data structure can extract a substring in optimal $O(\log n/\log\log n + \ell/\log_\sigma n)$ time without any restrictions on updates or queries.

In Section 2 we describe a data structure that uses $O(\log n)$ bits per symbol and supports rank, select, and access in optimal $O(\log n/\log\log n)$ time. This data structure essentially maintains a linked list L containing all symbols of S; using some auxiliary data structures on L, we can answer rank, select, and access queries on S. In Section 3 we show how the space usage can be reduced to $O(\log\sigma)$ bits per symbol. A compressed data structure that needs $H_0(S)$ bits per symbol is presented in Section 4. The approach of Section 4 is based on dividing S into a number of subsequences. We store a fully-dynamic data structure for only one such subsequence of appropriately small size. Updates on

Table 1. Previous and New Results for Fully-Dynamic Sequences. The rightmost column indicates whether updates are amortized (A) or worst-case (W). We use notation $\lambda = \log n / \log \log n$ in this table.

Ref.	Space	Rank	Select	Access	Insert/ Delete	
[14]	$nH_0(S) + o(n \log \sigma)$	$O((1 + \log \sigma / \log \log n)\lambda)$			$O((1 + \log \sigma / \log \log n)\lambda)$	W
[26]	$nH_0(S) + o(n \log \sigma)$	$O((\log \sigma / \log \log n)\lambda)$			$O((\log \sigma / \log \log n)\lambda)$	W
[24]	$nH_0(S) + o(n \log \sigma)$	$O(\lambda)$	$O(\lambda)$	$O(\lambda)$	$O(\lambda)$	A
[24]	$nH_0(S) + o(n \log \sigma)$	$O(\log n)$	$O(\lambda)$	$O(\lambda)$	$O(\log n)$	W
[17]	$nH_k + o(n \log \sigma)$	-	-	$O(\lambda)$	$O(\lambda)$	W
[12]	$nH_k + o(n \log \sigma)$	-	-	$O(\lambda)$	$O(\lambda)$	W
New	$nH_k + o(n \log \sigma)$	$O(\lambda)$	$O(\lambda)$	$O(\lambda)$	$O(\lambda)$	W

other subsequences are supported by periodic re-building. In Section 5 we show that the space usage can be reduced to $nH_k(S) + o(n \log \sigma)$.

2 $O(n \log n)$-Bit Data Structure

We start by describing a data structure that uses $O(\log n)$ bits per symbol.

Lemma 1. *A dynamic string $S[1, m]$ for $m \leq n$ over alphabet $\Sigma = \{1, \ldots, \sigma\}$ can be stored in a data structure that needs $O(m \log m)$ bits, and answers queries access, rank and select in time $O(\log m / \log \log n)$. Insertions and deletions of symbols are supported in $O(\log m / \log \log n)$ time. The data structure uses a universal look-up table of size $o(n^\varepsilon)$ for an arbitrarily small $\varepsilon > 0$.*

Proof: We keep elements of S in a list L. Each entry of L contains a symbol $a \in \Sigma$. For every $a \in \Sigma$, we also maintain the list L_a. Entries of L_a correspond to those entries of L that contain the symbol a. We maintain data structures $D(L)$ and $D(L_a)$ that enable us to find the number of entries in L (or in some list L_a) that precede an entry $e \in L$ (resp. $e \in L_a$); we can also find the i-th entry e in L_a or L using $D(L.)$. We will prove in Lemma 4 that $D(L)$ needs $O(m \log m)$ bits and supports queries and updates on L in $O(\log m / \log \log n)$ time.

We can answer a query $\text{select}_a(i, S)$ by finding the i-th entry e_i in L_a, following the pointer from e_i to the corresponding entry $e' \in L$, and counting the number v of entries preceding e' in L. Clearly[2], $\text{select}_a(i, S) = v$. To answer a query $\text{rank}_a(i, S)$, we first find the i-th entry e in L. Then we find the last entry e_a that precedes e and contains a. Such queries can be answered in $O((\log \log \sigma)^2 \log \log m)$ time as will be shown in the full version of this paper [23]. If e'_a is the entry that corresponds to e_a in L_a, then $\text{rank}_a(i, S) = v$, where v is the number of entries that precede e'_a in L_a.   □

---

[2] To simplify the description, we assume that a list entry precedes itself.

3 $O(n \log \sigma)$-Bit Data Structure

Lemma 2. *A dynamic string $S[1, n]$ over alphabet $\Sigma = \{\, 1, \ldots, \sigma \,\}$ can be stored in a data structure using $O(n \log \sigma)$ bits, and supporting queries access, rank and select in time $O(\log n / \log \log n)$. Insertions and deletions of symbols are supported in $O(\log n / \log \log n)$ time.*

Proof: If $\sigma = \log^{O(1)} n$, then the data structures described in [26] and [14] provide desired query and update times. The case $\sigma = \log^{\Omega(1)} n$ is considered below.

We show how the problem on a sequence of size n can be reduced to the same problem on a sequence of size $O(\sigma \log n)$. The sequence S is divided into chunks. We can maintain the size n_i of each chunk C_i, so that $n_i = O(\sigma \log n)$ and the total number of chunks is bounded by $O(n/\sigma)$. We will show how to maintain chunks in the full version of this paper [23]. For each $a \in \Sigma$, we keep a global bit sequence B_a. $B_a = 1^{d_1} 0 1^{d_2} 0 \ldots 1^{d_i} 0 \ldots$ where d_i is the number of times a occurs in the chunk C_i. We also keep a bit sequence $B_t = 1^{n_1} 0 1^{n_2} 0 \ldots 1^{n_i} 0 \ldots$. We can compute $\text{rank}_a(i, S) = v_1 + v_2$ where $v_1 = \text{rank}_1(\text{select}_0(j_1, B_a), B_a)$, $j_1 = \text{rank}_0(\text{select}_1(i, B_t), B_t)$, $v_2 = \text{rank}_a(i_1, C_{i_2})$, $i_2 = j_1 + 1$ and $i_1 = i - \text{rank}_1(\text{select}_0(j_1, B_t), B_t)$. To answer a query $\text{select}_a(i, S)$, we first find the index i_2 of the chunk C_{i_2} that contains the i-th occurrence of i, $i_2 = \text{rank}_0(\text{select}_1(i, B_a), B_a) + 1$. Then we find $v_a = \text{select}_a(C_{i_2}, i - i_1)$ for $i_1 = \text{rank}_1(\text{select}_0(i_2 - 1, B_a), B_a)$; v_a identifies the position of the $(i - i_1)$-th occurrence of a in the chunk C_{i_2}, where i_1 denotes the number of a's in the first $i_2 - 1$ chunks. Finally we compute $\text{select}_a(i, S) = v_a + s_p$ where $s_p = \text{rank}_1(\text{select}_0(i_2 - 1, B_t), B_t)$ is the total number of symbols in the first $i_2 - 1$ chunks. We can support queries and updates on B_t and on each B_a in $O(\log n / \log \log n)$ time [26]. By Lemma 1, queries and updates on C_i are supported in $O(\log \sigma / \log \log n)$ time. Hence, the query and update times of our data structure are $O(\log n / \log \log n)$.

B_t can be kept in $O((n/\sigma) \log \sigma)$ bits [26]. The array B_a uses $O(n_a \log \frac{n}{n_a})$ bits, where n_a is the number of times a occurs in S. Hence, all B_a and B_t use $O((n/\sigma) \log \sigma + \sum_a n_a \log \frac{n}{n_a}) = O(n \log \sigma)$ bits. By Lemma 1, we can also keep the data structure for each chunk in $O(\log \sigma + \log \log n) = O(\log \sigma)$ bits per symbol. $\qquad\square$

4   Compressed Data Structure

In this Section we describe a data structure that uses $H_0(S)$ bits per symbol. We start by considering the case when the alphabet size is not too large, $\sigma \leq n / \log^3 n$. The sequence S is split into subsequences $S_0, S_1, \ldots S_r$ for $r = O(\log n / (\log \log n))$. The subsequence S_0 is stored in $O(\log \sigma)$ bits per element as described in Lemma 2. Subsequences $S_1, \ldots S_r$ are substrings of $S \setminus S_0$. $S_1, \ldots S_r$ are stored in compressed static data structures. New elements are always inserted into the subsequence S_0. Deletions from S_i, $i \geq 1$, are implemented as lazy deletions: an element in S_i is marked as deleted. We guarantee that the number of elements that are marked as deleted is bounded by $O(n/r)$. If a subsequence S_i contains many elements marked as deleted, it is re-built: we create a new instance of S_i that does not contain deleted symbols. If a symbol sequence S_0 contains too many elements, we insert the elements of S_0 into S_i and re-build S_i for $i \geq 1$. Processes of constructing a new subsequence and re-building a subsequence with too many obsolete elements are run in the background.

The bit sequence M identifies elements in S that are marked as deleted: $M[j] = 0$ if and only if $S[j]$ is marked as deleted. The bit sequence R distinguishes between the elements of S_0 and elements of S_i, $i \geq 1$: $R[j] = 0$ if the j-th element of S is kept in S_0 and $R[j] = 1$ otherwise.

We further need auxiliary data structures for answering select queries. We start by defining an auxiliary subsequence $\tilde{S}$ that contains copies of elements already stored in other subsequences. Consider a subsequence $\overline{S}$ obtained by merging subsequences $S_1, \ldots, S_r$ (in other words, $\overline{S}$ is obtained from S by removing elements of S_0). Let S'_a be the subsequence obtained by selecting (roughly) every r-th occurrence of a symbol a in $\overline{S}$. The subsequence S' is obtained by merging subsequences S'_a for all $a \in \Sigma$. Finally $\tilde{S}$ is obtained by merging S' and S_0. We support queries $\text{select}'_a(i, \tilde{S})$ on $\tilde{S}$, defined as follows: $\text{select}'_a(i, \tilde{S}) = j$ such that (i) a copy of $S[j]$ is stored in $\tilde{S}$ and (ii) if $\text{select}_a(i, S) = j_1$, then $j \leq j_1$ and copies of elements $S[j+1], S[j+2], \ldots, S[j_1]$ are not stored in $\tilde{S}$. That is, $\text{select}'_a(i, \tilde{S})$ returns the largest index j, such that $S[j]$ precedes $S[\text{select}_a(i, S)]$ and $S[j]$ is also stored in $\tilde{S}$. The data structure for $\tilde{S}$ delivers approximate answers for select queries; we will show later how the answer to a query $\text{select}_a(i, S)$ can be found quickly if the answer to $\text{select}'_a(i, \tilde{S})$ is known. Queries $\text{select}'(i, \tilde{S})$ can be implemented using standard operations on a bit sequence of size $O((n/r) \log \log n)$ bits; for completeness, we provide a description in the full version of this paper [23]. We remark that $\overline{S}$ and S' are introduced to define $\tilde{S}$; these two subsequences are not stored in our data structure. The bit sequence $\tilde{E}$ indicates what symbols of S are also stored in $\tilde{S}$: $\tilde{E}[i] = 1$ if a copy of $S[i]$ is stored in $\tilde{S}$ and $\tilde{E}[i] = 0$ otherwise. The bit sequence $\tilde{B}$ indicates what symbols in $\tilde{S}$ are actually from S_0: $\tilde{B}[i] = 0$ iff $\tilde{S}[i]$ is stored in the subsequence S_0. Besides, we keep bit sequences D_a for each $a \in \Sigma$. Bits of D_a correspond to occurrences of a in S. If the l-th occurrence of a in S is marked as deleted, then $D_a[l] = 0$. All other bits in D_a are set to 1.

We provide the list of subsequences in Table 2. Each subsequence is augmented with a data structure that supports rank and select queries. For simplicity we will not distinguish between a subsequence and a data structure on its elements. If a subsequence supports updates, then either (i) this is a subsequence over a small alphabet or (ii) this subsequence contains a small number of elements. In case (i), the subsequence is over an alphabet of constant size; by [26,14] queries on such subsequences are answered in $O(\log n / \log \log n)$ time. In case (ii) the subsequence contains $O(n/r)$ elements; data structures on such subsequences are implemented as in Lemma 2. All auxiliary subsequences, except for $\tilde{S}$, are of type (i). Subsequence S_0 and an auxiliary subsequence $\tilde{S}$ are of type (ii). Subsequences S_i for $i \geq 1$ are static, i.e. they are stored in data structures that do not support

Table 2. Auxiliary subsequences for answering rank and select queries. A subsequence is dynamic if both insertions and deletions are supported. If a subsequence is static, then updates are not supported. Static subsequences are re-built when they contain too many obsolete elements.

Name	Purpose	Alph. Size	Dynamic/ Static
S_0	Subsequence of S	-	Dynamic
$S_i,\ 1 \le i \le r$	Subsequence of S	-	Static
M	Positions of symbols in S_i, $i \ge 1$, that are marked as deleted	const	Dynamic
R	Positions of symbols from S_0 in S	const	Dynamic
$\tilde{S}$	Delivers an approximate answer to select queries	-	Dynamic
$S'_a,\ a \in \Sigma$	Auxiliary sequences for $\tilde{S}$	-	Dynamic
$\tilde{E}$	Positions of symbols from $\tilde{S}$ in S	const	Dynamic
$\tilde{B}$	Positions of symbols from S_0 in $\tilde{S}$	const	Dynamic
D_a	Positions of symbols marked as deleted among all a's	const	Dynamic

updates. We re-build these subsequences when they contain too many obsolete elements. Thus dynamic subsequences support rank, select, access, and updates in $O(\log n / \log \log n)$ time. It is known that we can implement all basic operations on a static sequence in $O(\log n / \log \log n)$ time[3]. Our data structures on static subsequences are based on the approach of Barbay et al. [3]; however, our data structure can be constructed faster when the alphabet size is small and supports a substring extraction operation. A full description will be given in the full version of this paper [23]. We will show below that queries on S are answered by $O(1)$ queries on dynamic subsequences and $O(1)$ queries on static subsequences.

We also maintain arrays $Size[]$ and $Count_a[]$ for every $a \in \Sigma$. For any $1 \le i \le r$, $Size[i]$ is the number of symbols in S_i and $Count_a[i]$ specifies how many times a occurs in S_i. We keep a data structure that computes the sum of the first $i \le r$ entries in $Size[i]$ and find the largest j such that $\sum_{t=1}^{j} Size[t] \le q$ for any integer q. The same kinds of queries are also supported on $Count_a[]$. Arrays $Size[]$ and $Count_a[]$ use $O(\sigma \cdot r \cdot \log n) = O(n / \log n)$ bits.

Queries. To answer a query $\text{rank}_a(i, S)$, we start by computing $i' = \text{select}_1(i, M)$; i' is the position of the i-th element that is not marked as deleted. Then we find $i_0 = \text{rank}_0(i', R)$ and $i_1 = \text{rank}_1(i', R)$. By definition of R, i_0 is the number of elements of $S[1..i]$ that are stored in the subsequence S_0. The number of a's in $S_0[1..i_0]$ is computed as $c_1 = \text{rank}_a(i_0, S_0)$. The number of a's in $S_1, \ldots, S_r$ before the position i' is found as follows. We identify the index t, such that $\sum_{j=1}^{t} Size[j] < i_1 \le \sum_{j=1}^{t+1} Size[j]$. Then we compute how many times a occurred in $S_1, \ldots, S_t$, $c_{2,1} = \sum_{j=1}^{t} Count_a[j]$, and in the relevant prefix of S_{t+1}, $c_{2,2} = \text{rank}_a(i_1 - \sum_{j=1}^{t} Size[j], S_{t+1})$. Let $c_2 = \text{rank}_1(c_{2,1} + c_{2,2}, D_a)$. Thus c_2 is the number of symbols 'a' that are not marked as deleted among the first $c_{2,1} + c_{2,2}$ occurrences of a in $S \setminus S_0$. Hence $\text{rank}_a(i, S) = c_1 + c_2$.

---

[3] Static data structures also achieve significantly faster query times, but this is not necessary for our implementation.

To answer a query $select_a(i, S)$, we first obtain an approximate answer by asking a query $select'_a(i, \tilde{S})$. Let $i' = select_1(i, D_a)$ be the rank of the i-th symbol a that is not marked as deleted. Let $l_0 = select'_a(i', \tilde{S})$. We find $l_1 = rank_1(l_0, \tilde{E})$ and $l_2 = select_a(rank_a(l_1, \tilde{S}) + 1, \tilde{S})$. Let $first = select_1(l_1, \tilde{E})$ and $last = select_1(l_2, \tilde{E})$ be the positions of $\tilde{S}[l_1]$ and $\tilde{S}[l_2]$ in S. By definition of $select'$, $rank_a(first, S) \leq i$ and $rank_a(last, S) > i$. If $rank_a(first, S) = i$, then obviously $select_a(i, S) = first$. Otherwise the answer to $select_a(i, S)$ is an integer between $first$ and $last$. By definition of $\tilde{S}$, the substring $S[first]$, $S[first + 1]$, $\ldots$, $S[last]$ contains at most r occurrences of a. All these occurrences are stored in subsequences S_j for $j \geq 1$. We compute $i_0 = rank_a(rank_0(first, R), S_0)$ and $i_1 = i' - i_0$. We find the index t such that $\sum_{j=1}^{t-1} Count_a[j] < i_1 \leq \sum_{j=1}^{t} Count_a[j]$. Then $v_1 = select_a(i_1 - \sum_{j=1}^{t-1} Count_a[j], S_t)$ is the position of $S[select_a(i, S)]$ in S_t. We find its index in S by computing $v_2 = v_1 + \sum_{j=1}^{t-1} Size[j]$ and $v_3 = select_1(v_2, R)$. Finally $select_a(i, S) = rank_1(v_3, M)$.

Answering an access query is straightforward. We determine whether $S[i]$ is stored in S_0 or in some S_j for $j \geq 1$ using R. Let $i' = select_1(i, M)$. If $R[i'] = 0$ and $S[i]$ is stored in S_0, then $S[i] = S_0[rank_0(i', R)]$. If $R[i'] = 1$, we compute $i_1 = rank_1(i', R)$ and find the index j such that $\sum_{t=1}^{j-1} Size[t] < i_1 \leq \sum_{t=1}^{j} Size[t]$. The answer to $access(i, S)$ is $S[i] = S_j[i_2]$ for $i_2 = i_1 - \sum_{t=1}^{j-1} Size[t]$.

Space Usage. The redundancy of our data structure can be estimated as follows. The space needed to keep the symbols that are marked as deleted in subsequences S_j is bounded by $O((n/r) \log \sigma)$. S_0 also takes $O((n/r) \log \sigma)$ bits. The bit sequences R and M need $O((n/r) \log r) = o(n)$ bits; $\tilde{B}, \tilde{E}$ also use $O((n/r) \log r)$ bits. Each bit sequence D_a can be maintained in $O(n'_a \log(n_a/n'_a))$ bits where n_a is the total number of symbols a in S and n'_a is the number of symbols a that are marked as deleted. All D_a take $O(\sum n'_a \log \frac{n_a}{n'_a})$; the last expression can be bounded by $O((n/r)(\log r + \log \sigma))$. The subsequence $\tilde{S}$ can be stored in $O((n/r) \log \sigma)$ bits. Thus all auxiliary subsequences use $O((n/r)(\log \sigma + \log r)) = o(n \log \sigma)$ bits. Data structures for subsequences S_i, $r \geq i \geq 1$, use $\sum_{i=1}^{r} (n_i H_k(S_i) + o(n_i \log \sigma)) = n H_k(S \setminus S_0) + o(n \log \sigma)$ bits for any $k = o(\log_\sigma n)$, where n_i is the number of symbols in S_i. Since $H_k(S) \leq H_0(S)$ for $k \geq 0$, all subsequences S_i are stored in $n H_0(S) + o(n \log \sigma)$ bits.

Updates. When a new symbol is inserted, we insert it into the subsequence S_0 and update the sequence R. The data structure for $\tilde{S}$ is also updated accordingly. We also insert a 1-bit at the appropriate position of bit sequences M and D_a where a is the inserted symbol. Deletions from S_0 are symmetric. When an element is deleted from S_i, $i \geq 1$, we replace the 1-bit corresponding to this element in M with a 0-bit. We also change the appropriate bit in D_a to 0, where a is the symbol that was deleted from S_i.

We must guarantee that the number of elements in S_0 is bounded by $O(n/r)$; the number of elements marked as deleted must be also bounded by $O(n/r)$. Hence we must re-build the data structure when the number of symbols in S_0 or the number of deleted symbols is too big. Since we aim for updates with worst-case bounds, the cost of re-building is distributed among $O(n/r)$ updates. We run two processes in the background. The first background process moves elements of S_0 into subsequences S_i. The second process purges sequences S_1, ..., S_r and removes all symbols marked as deleted from these sequences. Details are given in the full version of this paper.

We assumed in the description of updates that $\log n$ is fixed. In the general case we need additional background processes that increase or decrease sizes of subsequences when n becomes too large or too small. These processes are organized in a standard way. Thus we obtain the following result

Lemma 3. *A dynamic string $S[1,n]$ over alphabet $\Sigma = \{1,\ldots,\sigma\}$ for $\sigma < n/\log^3 n$ can be stored in a data structure using $nH_0 + o(n\log\sigma) + O(n\log\log n)$ bits, and supporting queries* access, rank *and* select *in time $O(\log n/\log\log n)$. Insertions and deletions of symbols are supported in $O(\log n/\log\log n)$ time.*

In the full version of this paper [23] we show that the space usage of the above described data structure can be reduced to $nH_k + o(n\log\sigma)$ bits. We also show how the result of Lemma 3 can be extended to the case when $\sigma \geq n/\log^3 n$. The full version also contains the description of the static data structure and presents the procedure for extracting a substring $S[i..i+\ell]$ of S in $O(\log n/\log\log n + \ell)$ time.

4.1 Compressed Data Structure for $\sigma > n/\log^3 n$

If the alphabet size σ is almost linear, we cannot afford storing the arrays $Count_a[]$. Instead, we keep a bit sequence $BCount_a$ for each alphabet symbol a. Let $s_{a,i}$ denote the number of a's occurrences in the subsequence S_i and $s_a = \sum_{i=1}^{r} s_{a,i}$. Then $BCount_a = 1^{s_{a,1}}01^{s_{a,2}}0\ldots1^{s_{a,r}}$. If $s_a < r\log^2 n$, we can keep $BCount_a$ in $O(s_a \log\frac{r+s_a}{s_a}) = O(s_a \log\log n)$ bits. If $s_a > r\log^2 n$, we can keep $BCount_a$ in $O(r\log\frac{r+s_a}{s_a}) = O((s_a/\log^2 n)\log n) = O(s_a/\log n)$ bits. Using $BCount_a$, we can find for any q the subsequence S_j, such that $Count_a[j] < q \leq Count_a[j+1]$ in $O(\log n/\log\log n)$ time.

We also keep an effective alphabet[4] for each S_j. We keep a bit vector $Map_j[]$ of size σ, such that $Map_j[a] = 1$ if and only if a occurs in S_j. Using $Map_j[]$, we can map a symbol $a \in [1,n]$ to a symbol $map_j(a) = rank_1(a, Map_j)$ so that $map_j(a) \in [1,|S_j|]$ for any a that occurs in S_j. Let $\Sigma_j = \{map_j(a) \mid a$ occurs in $S_j\}$. For every $map_j(a)$ we can find the corresponding symbol a using a select query on Map_j. We keep a static data structure for each sequence S_j over Σ_j. Queries and updates are supported in the same way as in Lemma 3. Combining the result of this subsection and Lemma 3, we obtain the data structure for an arbitrary alphabet size.

Theorem 1. *A dynamic string $S[1,n]$ over alphabet $\Sigma = \{1,\ldots,\sigma\}$ can be stored in a data structure using $nH_0 + o(n\log\sigma)$ bits, and supporting queries*

---

[4] An alphabet for S_j is effective if it contains only symbols that actually occurred in S_j.

access, rank *and* select *in time* $O(\log n/ \log\log n)$. *Insertions and deletions of symbols are supported in* $O(\log n/ \log\log n)$ *time.*

5   Compressed Data Structure II

By slightly modifying the data structure of Theorem 1 we can reduce the space usage to essentially $H_k(S)$ bit per symbol for any $k = o(\log_\sigma n)$ simultaneously. First, we observe that any sub-sequence S_i for $i \geq 1$ is kept in a data structures that consumes $H_k(S_i) + o(|S_i| \log \sigma)$ bits of space. Thus all S_i use $\sum_{i=1}^{r}(n_i H_k(S_i)+o(n_i \log \sigma)) = nH_k(S\backslash S_0)+o(n\log\sigma)$ bits. It can be shown that $nH_k(S\backslash S_0)+o(n_i\log\sigma)) = nH_k(S\backslash S_0)+O(n\frac{\log n}{r})+o(n\log\sigma)$ bits; for completeness, we prove this bound in the full version [23]. Since $r = O(\log n/\log\log n)$, the data structure of Theorem 1 uses $nH_k + o(n\log\sigma) + O(n\log\log n)$ bits.

In order to get rid of the $O(n\log\log n)$ additive term, we use a different static data structure; our static data structure is described in the full version. As before, the data structure for a sequence S_i uses $|S_i|H_k + o(|S_i|\log\sigma)$ bits. But we also show in the full version that our static data structure can be constructed in $O(|S_i|/\log^{1/6} n)$ time if the alphabet size σ is sufficiently small, $\sigma \leq 2^{\log^{1/3} n}$. The space usage $nH_k(S) + o(n\log\sigma)$ can be achieved by appropriate change of the parameter r. If $\sigma > 2^{\log^{1/3} n}$, we use the data structure of Theorem 1. As explained above, the space usage is $nH_k + o(n\log\sigma) + O(n\log\log n) = nH_k + o(n\log\sigma)$. If $\sigma \leq 2^{\log^{1/3} n}$ we also use the data structure of Theorem 1, but we set $r = O(\log n\log\log n)$. The data structure needs $nH_k(S) + O(n/\log\log n) + o(n\log\sigma) = nH_k(S) + o(n\log\sigma)$ bits. Since we can re-build a static data structure for a sequence S_i in $O(|S_i|\log^{1/6} n)$ time, background processes incur an additional cost of $O(\log n/\log\log n)$. Hence the cost of updates does not increase.

6   Substring Extraction

Our representation of compressed sequences also enables us to retrieve a substring $S[i..i + \ell - 1]$ of S. We can retrieve a substring of S by extracting a substring of S_0 and a substring of some S_i for $i \geq 1$ and merging the result. A detailed description is provided in the full version of this paper [23]. Our result can be summed up as follows.

Theorem 2. *A dynamic string* $S[1,n]$ *over alphabet* $\Sigma = \{1,\ldots,\sigma\}$ *can be stored in a data structure using* $nH_k + o(n\log\sigma)$ *bits, and supporting queries* access, rank *and* select *in time* $O(\log n/\log\log n)$. *Insertions and deletions of symbols are supported in* $O(\log n/\log\log n)$ *time. A substring of* S *can be extracted in* $O(\log n/\log\log n + \ell/\log_\sigma n)$ *time, where* ℓ *denotes the length of the substring.*

References

1. Arge, L., Vitter, J.S.: Optimal external memory interval management. SIAM J. Comput. 32(6), 1488–1508 (2003)
2. Barbay, J., Gagie, T., Navarro, G., Nekrich, Y.: Alphabet partitioning for compressed rank/select and applications. In: Cheong, O., Chwa, K.-Y., Park, K. (eds.) ISAAC 2010, Part II. LNCS, vol. 6507, pp. 315–326. Springer, Heidelberg (2010)
3. Barbay, J., He, M., Munro, J.I., Rao, S.S.: Succinct indexes for strings, binary relations and multi-labeled trees. ACM Transactions on Algorithms 7(4), article 52 (2011)
4. Belazzougui, D., Navarro, G.: New lower and upper bounds for representing sequences. In: Epstein, L., Ferragina, P. (eds.) ESA 2012. LNCS, vol. 7501, pp. 181–192. Springer, Heidelberg (2012)
5. Blandford, D., Blelloch, G.: Compact representations of ordered sets. In: Proc. 15th SODA, pp. 11–19 (2004)
6. Chan, H., Hon, W.-K., Lam, T.-H., Sadakane, K.: Compressed indexes for dynamic text collections. ACM Transactions on Algorithms 3(2), article 21 (2007)
7. Chan, H.-L., Hon, W.-K., Lam, T.-W.: Compressed index for a dynamic collection of texts. In: Sahinalp, S.C., Muthukrishnan, S.M., Dogrusoz, U. (eds.) CPM 2004. LNCS, vol. 3109, pp. 445–456. Springer, Heidelberg (2004)
8. Ferragina, P., Manzini, G., Mäkinen, V., Navarro, G.: Compressed representations of sequences and full-text indexes. ACM Transactions on Algorithms 3(2), article 20 (2007)
9. Fredman, M., Saks, M.: The cell probe complexity of dynamic data structures. In: Proc. 21st STOC, pp. 345–354 (1989)
10. Golynski, A., Munro, J.I., Rao, S.S.: Rank/select operations on large alphabets: a tool for text indexing. In: Proc. 17th SODA, pp. 368–373 (2006)
11. Grossi, R., Gupta, A., Vitter, J.S.: High-order entropy-compressed text indexes. In: Proc. 14th SODA, pp. 841–850 (2003)
12. Grossi, R., Raman, R., Satti, S.R., Venturini, R.: Dynamic compressed strings with random access. In: Fomin, F.V., Freivalds, R., Kwiatkowska, M., Peleg, D. (eds.) ICALP 2013, Part I. LNCS, vol. 7965, pp. 504–515. Springer, Heidelberg (2013)
13. Gupta, A., Hon, W.-K., Shah, R., Vitter, J.S.: A framework for dynamizing succinct data structures. In: Arge, L., Cachin, C., Jurdziński, T., Tarlecki, A. (eds.) ICALP 2007. LNCS, vol. 4596, pp. 521–532. Springer, Heidelberg (2007)
14. He, M., Munro, J.I.: Succinct representations of dynamic strings. In: Chavez, E., Lonardi, S. (eds.) SPIRE 2010. LNCS, vol. 6393, pp. 334–346. Springer, Heidelberg (2010)
15. Hon, W.-K., Sadakane, K., Sung, W.-K.: Succinct data structures for searchable partial sums. In: Ibaraki, T., Katoh, N., Ono, H. (eds.) ISAAC 2003. LNCS, vol. 2906, pp. 505–516. Springer, Heidelberg (2003)
16. Hon, W.-K., Sadakane, K., Sung, W.-K.: Succinct data structures for searchable partial sums with optimal worst-case performance. Theoretical Computer Science 412(39), 5176–5186 (2011)
17. Jansson, J., Sadakane, K., Sung, W.-K.: CRAM: Compressed random access memory. In: Czumaj, A., Mehlhorn, K., Pitts, A., Wattenhofer, R. (eds.) ICALP 2012, Part I. LNCS, vol. 7391, pp. 510–521. Springer, Heidelberg (2012)
18. Lee, S., Park, K.: Dynamic rank-select structures with applications to run-length encoded texts. In: Ma, B., Zhang, K. (eds.) CPM 2007. LNCS, vol. 4580, pp. 95–106. Springer, Heidelberg (2007)

19. Lee, S., Park, K.: Dynamic rank/select structures with applications to run-length encoded texts. Theoretical Computer Science 410(43), 4402–4413 (2009)
20. Mäkinen, V., Navarro, G.: Dynamic entropy-compressed sequences and full-text indexes. In: Lewenstein, M., Valiente, G. (eds.) CPM 2006. LNCS, vol. 4009, pp. 307–318. Springer, Heidelberg (2006)
21. Mäkinen, V., Navarro, G.: Dynamic entropy-compressed sequences and full-text indexes. ACM Transactions on Algorithms 4(3), article 32 (2008)
22. Manzini, G.: An analysis of the burrows-wheeler transform. J. ACM 48(3), 407–430 (2001)
23. Munro, J.I., Nekrich, Y.: Compressed data structures for dynamic sequences. ArXiv e-prints 1507.06866 (2015)
24. Navarro, G., Nekrich, Y.: Optimal dynamic sequence representations. In: Proc. 24th Annual ACM-SIAM Symposium on Discrete Algorithms (SODA 2013), pp. 865–876 (2013)
25. Navarro, G., Nekrich, Y.: Optimal dynamic sequence representations (full version). submitted for publication (2013)
26. Navarro, G., Sadakane, K.: Fully functional static and dynamic succinct trees. ACM Transactions on Algorithms 10(3), 16 (2014)
27. Patrascu, M., Demaine, E.D.: Tight bounds for the partial-sums problem. In: Proc. 15th Annual ACM-SIAM Symposium on Discrete Algorithms (SODA 2004), pp. 20–29 (2004)

A.1   Prefix Sum Queries on a List

In this section we describe a data structure on a list L that is used in the proof of Lemma 1 in Section 2.

Lemma 4. *We can keep a dynamic list L in an $O(m \log m)$-bit data structure $D(L)$, where m is the number of entries in L. $D(L)$ can find the i-th entry in L for $1 \leq i \leq m$ in $O(\log m / \log \log n)$ time. $D(L)$ can also compute the number of entries before a given element $e \in L$ in $O(\log m / \log \log n)$ time. Insertions and deletions are also supported in $O(\log m / \log \log n)$ time.*

Proof: $D(L)$ is implemented as a balanced tree with node degree $\Theta(\log^\varepsilon n)$. In every internal node we keep a data structure $Pref(u)$; $Pref(u)$ contains the total number $n(u_i)$ of elements stored below every child u_i of u. $Pref(u)$ supports prefix sum queries (i.e., computes $\sum_{i=1}^{t} n(u_i)$ for any t) and finds the largest j, such that $\sum_{i=1}^{j} n(u_i) \leq q$ for any integer q. We implement $Pref(u)$ as in Lemma 2.2 in [27] so that both types of queries are supported in $O(1)$ time. $Pref(u)$ uses linear space (in the number of its elements) and can be updated in $O(1)$ time. $Pref(u)$ needs a look-up table of size $o(n^\varepsilon)$. To find the i-th entry in a list, we traverse the root-to-leaf path; in each visited node u we find the child that contains the i-th entry using $Pref(u)$. To find the number of entries preceding a given entry e in a list, we traverse the leaf-to-root path π that starts in the leaf containing e. In each visited node u we answer a query to $Pref(u)$: if the j-th child u_j of u is on π, then we compute $s(u) = \sum_{i=1}^{j-1} n(u_i)$ using $Pref(u)$. The total number of entries to the left of e is the sum of $s(u)$ for all nodes u on π. Since we spend $O(1)$ time in each visited node, both types of queries are answered in $O(1)$ time. An update operation leads to $O(\log m / \log \log n)$ updates of data structures $Pref(u)$. The tree can be re-balanced using the weight-balanced B-tree [1], so that its height is always bounded by $O(\log m / \log \log n)$.   □

Geometric Hitting Sets for Disks: Theory and Practice

Norbert Bus[1], Nabil H. Mustafa[1,*], and Saurabh Ray[2]

[1] Université Paris-Est, LIGM, Equipe A3SI, ESIEE Paris, France
{busn,mustafan}@esiee.fr
[2] Computer Science Department, New York University, Abu Dhabi, United Arab Emirates
saurabh.ray@nyu.edu

Abstract. The geometric hitting set problem is one of the basic geometric combinatorial optimization problems: given a set P of points, and a set $\mathcal{D}$ of geometric objects in the plane, the goal is to compute a small-sized subset of P that hits all objects in $\mathcal{D}$. In 1994, Bronniman and Goodrich [6] made an important connection of this problem to the size of fundamental combinatorial structures called ϵ-nets, showing that small-sized ϵ-nets imply approximation algorithms with correspondingly small approximation ratios. Finally, recently Agarwal-Pan [5] showed that their scheme can be implemented in near-linear time for disks in the plane.

This current state-of-the-art is lacking in three ways. First, the constants in current ϵ-net constructions are large, so the approximation factor ends up being more than 40. Second, the algorithm uses sophisticated geometric tools and data structures with large resulting constants. Third, these have resulted in a lack of available software for fast computation of small hitting-sets. In this paper, we make progress on all three of these barriers: i) we prove improved bounds on sizes of ϵ-nets, ii) design hitting-set algorithms without the use of these data-structures and finally, iii) present dnet, a public source-code module that incorporates both of these improvements to compute small-sized hitting sets and ϵ-nets efficiently in practice.

Keywords: Geometric Hitting Sets, Approximation Algorithms, Computational Geometry.

1 Introduction

The minimum hitting set problem is one of the most fundamental combinatorial optimization problems: given a range space $(P, \mathcal{D})$ consisting of a set P and a set $\mathcal{D}$ of subsets of P called the *ranges*, the task is to compute the smallest subset $Q \subseteq P$ that has a non-empty intersection with each of the ranges in $\mathcal{D}$. This problem is strongly NP-hard. If there are no restrictions on the set system $\mathcal{D}$, then it is known that it is NP-hard to approximate the minimum hitting set within a logarithmic factor of the optimal. The problem is NP-complete even for the case where each range has exactly two points since this problem is equivalent to the vertex cover problem which is known to be NP-complete. A natural occurrence of the hitting set problem occurs when the range space

* The work of Nabil H. Mustafa in this paper has been supported by the grant ANR SAGA (JCJC-14-CE25-0016-01).

N. Bansal and I. Finocchi (Eds.): ESA 2015, LNCS 9294, pp. 903–914, 2015.
DOI: 10.1007/978-3-662-48350-3_75

$\mathcal{D}$ is derived from geometry – e.g., given a set P of n points in $\mathbb{R}^2$, and a set $\mathcal{D}$ of m triangles containing points of P, compute the minimum-sized subset of P that hits all the triangles in $\mathcal{D}$. Unfortunately, for most natural geometric range spaces, computing the minimum-sized hitting set remains NP-hard. For example, even the (relatively) simple case where $\mathcal{D}$ is a set of unit disks in the plane is strongly NP-hard [10].

Given a range space $(P, \mathcal{D})$, a positive measure μ on P (e.g., the counting measure), and a parameter $\epsilon > 0$, an ϵ-net is a subset $S \subseteq P$ such that $D \cap S \neq \emptyset$ for all $D \in \mathcal{D}$ with $\mu(D \cap P) \geq \epsilon \cdot \mu(P)$. The ϵ-net theorem [9] implies that for a large family of geometric hitting set systems (e.g., disks, half-spaces, k-sided polytopes, r-admissible set of regions in $\mathbb{R}^d$) there exists an ϵ-net of size $O(d/\epsilon \log d/\epsilon)$. For certain range spaces, one can even show the existence of ϵ-nets of size $O(1/\epsilon)$ – an important case being for disks in $\mathbb{R}^2$ [12]. In 1994, Bronnimann and Goodrich [6] proved the following interesting connection between the hitting-set problem, and ϵ-nets: if one can compute an ϵ-net of size c/ϵ for the ϵ-net problem for $(P, \mathcal{D})$ in polynomial time, then one can compute a hitting set of size at most $c \cdot \text{OPT}$ for $(P, \mathcal{D})$, where OPT is the size of the optimal (smallest) hitting set, in polynomial time. Until very recently, the best algorithms based on this observation, referred to as rounding techniques, had running times of $\Omega(n^2)$, and it had been a long-standing open problem to compute a $O(1)$-approximation to the hitting-set problem for disks in the plane in near-linear time. In a recent break-through, Agarwal-Pan [5] presented the first near-linear algorithm for computing $O(1)$-approximations for hitting sets for disks.

The limitation of the rounding technique – that it cannot give a PTAS – was overcome using an entirely different technique: local search [11,4]. It has been shown that the local search algorithm for the hitting set problem for disks in the plane gives a PTAS. Unfortunately the running time of the naive algorithm to compute a $(1 + \epsilon)$-approximation is $O(n^{O(1/\epsilon^2)})$. Based on local search, an $\tilde{O}(n^{2.34})$ time algorithm was proposed [7] yielding an $(8 + \epsilon)$-approximation.

Our Contributions

All approaches towards approximating geometric hitting sets for disks have to be evaluated on the questions of computational efficiency as well as approximation quality. In spite of all the progress, there remains a large gap – mainly due to the ugly trade-offs between running times and approximation factors. The breakthrough algorithm of Agarwal-Pan [5] suffers from two main problems:

- It rounds via ϵ-nets to design a $\tilde{O}(n)$-time algorithm, but the constant in the approximation depends on the constant in the size of ϵ-nets, which are large. For disks in the plane, the current best size of ϵ-net is at least $40/\epsilon$ [12], yielding at best a 40-approximation algorithm. Furthermore, there is no implementation or software solution available that can even compute such ϵ-nets efficiently.
- It uses sophisticated data-structures that have large constants in the running time. In particular, it uses the $O(\log n + k)$-time algorithm for range reporting for disk ranges in the plane (alternatively, for halfspaces in $\mathbb{R}^3$) as well as a dynamic data-structure for maintaining approximate weighted range-counting under disk ranges in polylogarithmic time. We have not been able to find efficient implementations of any of these data-structures.

It will turn out that all ideas for an efficient practical solution for the geometric hitting
set problem for disks are unified by one of the basic structures in the study of planar
geometry: Delaunay triangulations. Delaunay triangulations will be the key structure
for computing these improved ϵ-nets, and the Delaunay structures *already computed*
for constructing these nets will turn out to be crucial in computing small-sized hitting
sets. More precisely, our contributions are:

1. Constructing small ϵ-nets (Section 2). We show that the sample-and-refine approach
of Chazelle-Friedman [8] together with additional structural properties of Delaunay
triangulation results in ϵ-nets of surprisingly low size:

Theorem 1. *Given a set P of n points in $\mathbb{R}^2$, there exists an ϵ-net under disk ranges of
size at most $13.4/\epsilon$. Furthermore it can be computed in expected time $O(n \log n)$.*

The algorithm is simple to implement. We have implemented it, and present the sizes of
ϵ-nets for various real-world data-sets; the results indicate that our theoretical analysis
closely tracks the actual size of the nets.

2. Engineering a hitting-set algorithm (Section 3). Together with the result of Agarwal-
Pan, this immediately implies:

Corollary 1. *For any $\delta > 0$, one can compute a $(13.4 + \delta)$-approximation to the minimum hitting set for $(P, \mathcal{D})$ in time $\tilde{O}(n)$.*

We then present a modification of the algorithm of Agarwal-Pan that does not use any
complicated data-structures – just Delaunay triangulations, ϵ-nets and binary search
(e.g., it turns out that output sensitive range reporting is not required). This comes with
a price: although experimental results indicate a near-linear running time, we have been
unable to theoretically prove that the algorithm runs in expected near-linear time.

3. Implementation and experimental evaluation (Section 4). We present dnet, a public
source-code module that incorporates all these ideas to efficiently compute small-sized
hitting sets in practice. We give detailed experimental results on both synthetic and
real-world data sets, which indicates that the algorithm computes, on average, a 1.3-
approximation in near-linear time. This surprisingly low approximation factor compared to the proven worst case bound is the result of fine tuning the parameters of the
algorithm.

Due to lack of space, most of the proofs are left for the full paper.

2 A Near Linear Time Algorithm for Computing ϵ-nets for Disks in the Plane

Through a more careful analysis, we present an algorithm for computing an ϵ-net of
size $\frac{13.4}{\epsilon}$, running in expected near linear time. The method, shown in Algorithm 1,
computes a random sample and then solves subproblems involving subsets of the points
located in pairs of Delaunay disks circumscribing adjacent triangles in the Delaunay
triangulation of the random sample. The key to improved bounds is i) using additional

structural properties of Delaunay triangulations, and ii) new improved constructions of ϵ-nets for large values of ϵ. The presented algorithm can be extended to handle the case when the ϵ-net is with respect to a measure on the point set taking only rational values.

Let $\Delta(abc)$ denote the triangle defined by the three points a, b and c. D_{abc} denotes the disk through a, b and c, while $D_{ab\bar{c}}$ denotes the halfspace defined by a and b not containing the point c. Let $c(D)$ denote the center of the disk D.

Let $\Xi(R)$ be the Delaunay triangulation of a set of points $R \subseteq P$ in the plane. We will use Ξ when R is clear from the context. For any triangle $\Delta \in \Xi$, let D_Δ be the Delaunay disk of Δ, and let P_Δ be the set of points of P contained in D_Δ. Similarly, for any edge $e \in \Xi$, let Δ_e^1 and Δ_e^2 be the two triangles in Ξ adjacent to e, and $P_e = P_{\Delta_e^1} \bigcup P_{\Delta_e^2}$. If e is on the convex-hull, then one of the triangles is taken to be the halfspace whose boundary line is supported by e and not containing R.

Algorithm 1. Compute ϵ-nets

Data: Compute ϵ-net, given P: set of n points in $\mathbb{R}^2$, $\epsilon > 0$ and c_0.

1 **if** $\epsilon n < 13$ **then**
2 $\quad \lfloor$ Return P

3 Pick each point $p \in P$ into R independently with probability $\frac{c_0}{\epsilon n}$.
4 **if** $|R| \le c_0/2\epsilon$ **then**
5 $\quad \lfloor$ restart algorithm.

6 Compute the Delaunay triangulation Ξ of R.
7 **for** *triangles* $\Delta \in \Xi$ **do**
8 $\quad \lfloor$ Compute the set of points $P_\Delta \subseteq P$ in Delaunay disk D_Δ of Δ.

9 **for** *edges* $e \in \Xi$ **do**
10 $\quad$ Let Δ_e^1 and Δ_e^2 be the two triangles adjacent to e, $P_e = P_{\Delta_e^1} \cup P_{\Delta_e^2}$.
11 $\quad$ Let $\epsilon' = (\frac{\epsilon n}{|P_e|})$ and compute an ϵ'-net R_e for P_e depending on the cases below:
12 $\quad$ **if** $\frac{1}{2} < \epsilon' < 1$ **then**
13 $\quad\quad \lfloor$ compute using Lemma 1.
14 $\quad$ **if** $\epsilon' \le \frac{1}{2}$ **then**
15 $\quad\quad \lfloor$ compute recursively.

16 **Return** $\left(\bigcup_e R_e\right) \cup R$.

In order to prove that the algorithm gives the desired result, the following lemma regarding the size of an ϵ-net will be useful. Let $f(\epsilon)$ be the upper bound of the size of the smallest ϵ-net for any set P of points in $\mathbb{R}^2$ under disk ranges.

Lemma 1. *For $\frac{2}{3} < \epsilon < 1$, $f(\epsilon) \le 2$, and for $\frac{1}{2} < \epsilon \le \frac{2}{3}$, $f(\epsilon) \le 10$. In both cases the ϵ-net can be computed in $O(n \log n)$ time.*

Call a tuple $(\{p,q\},\{r,s\})$, where $p,q,r,s \in P$, a *Delaunay quadruple* if $int(\Delta(pqr)) \cap int(\Delta(pqs)) = \emptyset$ where $int(\cdot)$ denotes the interior of a set. Define its *weight*, denoted $W_{(\{p,q\},\{r,s\})}$, to be the number of points of P in $D_{pqr} \cup D_{pqs}$. Let $\mathcal{T}_{\le k}$ be the set of Delaunay quadruples of P of weight at most k and similarly $\mathcal{T}_k$

denotes the set of Delaunay quadruples of weight exactly k. Similarly, a *Delaunay triple* is given by $(\{p, q\}, \{r\})$, where $p, q, r \in P$. Define its *weight*, denoted $W_{(\{p,q\},\{r\})}$, to be the number of points of P in $D_{pqr} \cup D_{pq\bar{r}}$. Let $\mathcal{S}_{\leq k}$ be the set of Delaunay triples of P of weight at most k, and $\mathcal{S}_k$ denotes the set of Delaunay triples of weight exactly k. One can upper bound the size of $\mathcal{T}_{\leq k}$, $\mathcal{S}_{\leq k}$ and using it, we derive an upper bound on the expected number of sub-problems with a certain number of points.

Lemma 2. *If $\epsilon n \geq 13$, $\mathbf{E}\left[|\{e \in \Xi \mid k_1 \epsilon n \leq |P_e| \leq k_2 \epsilon n\}|\right] \leq \frac{(3.1)c_0^3}{\epsilon e^{k_1 c_0}}(k_1^3 c_0 + 3.7 k_2^2)$.*

Lemma 3. *Algorithm* COMPUTE ϵ-NET *computes an ϵ-net of expected size* $13.4/\epsilon$.

Proof. First we show that the algorithm computes an ϵ-net. Take any disk D with center c containing ϵn points of P, and not hit by the initial random sample R. Increase its radius while keeping its center c fixed until it passes through a point, say p_1 of R. Now further expand the disk by moving c in the direction $\boldsymbol{p_1 c}$ until its boundary passes through a second point p_2 of R. The edge e defined by p_1 and p_2 belongs to Ξ, and the two extreme disks in the pencil of empty disks through p_1 and p_2 are the disks $D_{\Delta_e^1}$ and $D_{\Delta_e^2}$. Their union covers D, and so D contains ϵn points out of the set P_e. Then the net R_e computed for P_e must hit D, as $\epsilon n = (\epsilon n/|P_e|) \cdot |P_e|$.

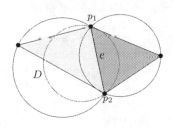

For the expected size, clearly, if $\epsilon n < 13$ then the returned set is an ϵ-net of size $\frac{13}{\epsilon}$. Otherwise we can calculate the expected number of points added to the ϵ-net during solving the sub-problems. We simply group them by the number of points in them. Set $E_i = \{e \mid 2^i \epsilon n \leq |P_e| < 2^{i+1} \epsilon n\}$, and let us denote the size of the ϵ-net returned by our algorithm with $f'(\epsilon)$. Then

$$\mathbf{E}\left[f'(\epsilon)\right] = \mathbf{E}[|R|] + \mathbf{E}\left[|\bigcup_{e \in \Xi} R_e|\right] = \frac{c_0}{\epsilon} + \mathbf{E}[|\{e \mid \epsilon n \leq |P_e| < 3\epsilon n/2\}|] \cdot f(2/3)$$

$$+ \mathbf{E}[|\{e \mid 3\epsilon n/2 \leq |P_e| < 2\epsilon n\}|] \cdot f\left(\frac{1}{2}\right) + \sum_{i=1} \mathbf{E}\left[\sum_{e \in E_i} f'\left(\frac{\epsilon n}{|P_e|}\right)\right]. \quad (1)$$

Noting that $\mathbf{E}[\sum_{e \in E_i} f'(\frac{\epsilon n}{|P_e|}) \mid |E_i| = t] \leq t\mathbf{E}[f'(1/2^{i+1})]$, we get

$$\mathbf{E}\left[\sum_{e \in E_i} f'\left(\frac{\epsilon n}{|P_e|}\right)\right] = \mathbf{E}\left[\mathbf{E}[\sum_{e \in E_i} f'\left(\frac{\epsilon n}{|P_e|}\right)|E_i]\right] \leq \mathbf{E}\left[|E_i| \cdot \mathbf{E}[f'(1/2^{i+1})]\right]$$

$$= \mathbf{E}[|E_i|] \cdot \mathbf{E}[f'(1/2^{i+1})] \quad (2)$$

as $|E_i|$ and $f'(\cdot)$ are independent. As $\epsilon' = \frac{\epsilon n}{|P_e|} > \epsilon$, by induction, assume $\mathbf{E}[f'(\epsilon')] \leq \frac{13.4}{\epsilon'}$. Then by using Lemma 1 and 2

$$\mathbf{E}\left[f'(\epsilon)\right] \leq \frac{c_0}{\epsilon} + \frac{(3.1) \cdot c_0^3(c_0 + 8.34)}{\epsilon e^{c_0}} \cdot 2 + \frac{(3.1) \cdot c_0^3((3/2)^3 c_0 + 14.8)}{\epsilon e^{3c_0/2}} \cdot 10$$

$$+ \sum_i \frac{(3.1) \cdot c_0^3(2^{3i} c_0 + 3.7 \cdot 2^{2i+2})}{\epsilon e^{c_0 2^i}} \cdot 13.4 \cdot 2^{i+1} \leq \frac{13.4}{\epsilon} \quad (3)$$

by setting $c_0 = 12$.   □

Lemma 4. *Algorithm* COMPUTE ϵ-NET *runs in expected time* $O(n \log n)$.

We have shown that the expected size of the returned ϵ-net is $13.4/\epsilon$. Furthermore, by Markov's inequality and repeatedly running the algorithm, an ϵ-net of size $(1 + \delta) \cdot 13.4/\epsilon$ is returned in exptected time $O(n/\delta \cdot \log n)$ for any constant $\delta > 0$. Setting δ small enough finishes the proof of Theorem 1.

3   Engineering the Agarwal-Pan Algorithm

The Agarwal-Pan (AP) algorithm (shown in Algorithm 2) uses an iterative reweighing strategy, where the idea is to assign a weight $w(\cdot)$ to each $p \in P$ such that the total weight of points contained in each $D \in \mathcal{D}$ is relatively high. It starts by setting $w(p) = 1$ for each $p \in P$. If there exists a disk D with small weight, it increases the weight of the points in D until their total weight exceeds a threshold of $cW/$OPT, where c is some constant and $W = \sum_{p \in P} w(p)$ is the current total weight. If after any iteration, all disks have weight above the threshold $\frac{cW}{2e\text{OPT}}$, return a $\frac{c}{2e\text{OPT}}$-net with respect to these weights, ensuring that every disk is hit.

For the purpose of analysis, Agarwal and Pan conceptually divide the reweighings into $O(\log n)$ phases, where each phase (except perhaps the last) performs $\Theta($OPT$)$ reweighings. The implementation of the AP algorithm requires two ingredients: **A)** a range reporting data structure and **B)** a dynamic approximate range counting data structure. The former is used to construct the set of points to be reweighed and the latter is required for figuring out whether a disk needs reweighing. As a pre-processing step, the AP algorithm first computes a $1/$OPT-net Q to be returned as part of the hitting set. This ensures that the remaining disks not hit by Q contain less than $n/$OPT points. Additionally they observe that in any iteration, if less than OPT disks are reweighed, then all disks have weight more than $\frac{cW}{2e\text{OPT}}$.

Algorithm 2. AP algorithm for computing hitting sets

Data: A point set P, a set of disks $\mathcal{D}$, a fixed constant c, and the value of OPT.

1 Compute a $(1/$OPT$)$-net, Q, of P and remove disks hit by Q
2 Set $w(p) = 1$ for all $p \in P$
3 **repeat**
4   **foreach** $D \in \mathcal{D}$ **do**
5   **if** $w(D) \leq cW/$OPT **then**
6   reweigh D repeatedly until the weight $w(D)$ exceeds $cW/$OPT

7   flag = false
8   **foreach** $D \in \mathcal{D}$ **do**
9   **if** $w(D) < (c/2e) \cdot W/$OPT **then** flag = true

10 **until** *flag = true*
11 **return** Q *along with a* $(c/2e$OPT$)$-*net of* P *with respect to* $w(\cdot)$

The AP algorithm is simple and has a clever theoretical analysis. Its main drawback is that the two data structures it uses are sophisticated with large constants in the running time. This unfortunately renders the AP algorithm impractical. Our goal is to find a method that avoids these sophisticated data structures and to develop additional heuristics which lead to not only a fast implementation but also one that generally gives an approximation ratio smaller than that guaranteed by the theoretical analysis of the AP algorithm. As part of the algorithm, we use the algorithm for constructing ϵ-nets described in the previous section, which already reduces the approximation factor significantly.

Removing A). Just as Agarwal and Pan do, we start by picking a c_1/OPT-net, for some constant c_1. The idea for getting rid of range-reporting data-structure is to observe that the very fact that a disk D is not hit by Q, *when Q is an ϵ-net*, makes it possible to use Q in a simple way to efficiently enumerate the points in D. We will show that D lies in the union of two Delaunay disks in the Delaunay triangulation of Q, which, as we show later, can be found by a simple binary search. The resulting algorithm still has worst-case near-linear running time.

Removing B). Our approach towards removing the dependence on dynamic approximate range counting data structure is the following: at the beginning of *each* phase we pick a c_2/OPT-net R, for some constant c_2. The set of disks that are not hit by R are then guaranteed to have weight at most $c_2 W/\text{OPT}$, which we can then reweigh during that phase. While this avoids having to use data-structure **B)**, there are two problems with this: $a)$ disks with small weight hit by R are not reweighed, and $b)$ a disk whose initial weight was less than $c_2 W/\text{OPT}$ could have its current weight more than $c_2 W/\text{OPT}$ in the middle of a phase, and so it is erroneously reweighed.

Towards solving these problems, the idea is to maintain an additional set S which is empty at the start of each phase. When a disk D is reweighed, we add a random point of D (sampled according to the probability distribution induced by $w(\cdot)$) to S. Additionally we maintain a nearest-neighbor structure for S, enabling us to only reweigh D if it is not hit by $R \cup S$. Now, if during a phase, there are $\Omega(\text{OPT})$ reweighings, then as in the Agarwal-Pan algorithm, we move on to the next phase, and $a)$ is not a problem. Otherwise, there have been less than OPT reweighings, which implies that less than OPT disks were not hit by R. Then we can return R together with the set S consisting of one point from each of these disks. This will still be a hitting set.

To remedy $b)$, before reweighing a disk, we compute the set of points inside D, and only reweigh if the total weight is at most $c_2 W/\text{OPT}$. Consequently we sometimes waste $O(n/\text{OPT})$ time to compute this list of points inside D without performing a reweighing. Due to this, the worst-case running time increases to $O(n^2/\text{OPT})$. In practice, this does not happen for the following reason: in contrast to the AP algorithm, our algorithm reweighs any disk *at most once* during a phase. Therefore if the weight of any disk D increases significantly, and yet D is not hit by S, the increase must have been due to the increase in weight of many disks intersected by D which were reweighed before D *and* for which the picked points (added to S) did not hit D. Reweighing in a random order makes these events very unlikely (in fact we suspect this gives an expected linear-time algorithm, though we have not been able to prove it).

Algorithm 3. Algorithm for computing small-sized hitting sets.

Data: A point set P, a set of disks $\mathcal{D}$, and the size of the optimal hitting set OPT.

1 Compute a (c_1/OPT)-net Q of P and the Delaunay triangulation $\Xi(Q)$ of Q.

2 **foreach** $q \in Q$ **do** construct $\Psi(Q)(q)$. **foreach** $D \in \mathcal{D}$ **do**

3 $\quad$ ⌊ **if** D *not hit by* Q **then** add D to $\mathcal{D}_1$. // using $\Xi(Q)$

4 $P_1 = P \setminus Q$.

5 **foreach** $p \in P_1$ **do** set $w(p) = 1$. **repeat**

6 $\quad$ Compute a (c_2/OPT)-net, R, of P_1 and the Delaunay triangulation $\Xi(R)$ of R.

7 $\quad$ Set $S = \emptyset$, $\Xi(S) = \emptyset$.

8 $\quad$ **foreach** $D \in \mathcal{D}_1$ *in a random order* **do**

9 $\quad\quad$ **if** D *not hit by* $R \cup S$ **then** // using $\Xi(R)$ and $\Xi(S)$

10 $\quad\quad\quad$ **foreach** $p \in D$ **do** set $w(p) = w(p) + c_3 w(p)$. // using $\Psi(Q)$ Add a
$\quad\quad\quad$ random point in D to S; update $\Xi(S)$.

11 **until** $|S| \le c_4 \text{OPT}$

12 **return** $\{Q \cup R \cup S\}$

See Algorithm 3 for the new algorithm (the data-structure $\Psi(Q)$ will be defined later).

Lemma 5. *The algorithm terminates, $Q \cup R \cup S$ is a hitting set, of size at most $(13.4 + \delta) \cdot$ OPT, for any $\delta > 0$.*

Proof. By construction, if the algorithm terminates, then $Q \cup R \cup S$ is a hitting-set. Set $c_1 = 13.4 \cdot 3/\delta$, $c_2 = 1/(1 + \delta/(13.4 \cdot 3))$, $c_3 = \delta/10000$ and $c_4 = \delta/3$. By the standard reweighing argument, we know that after t reweighings, we have:

$$\text{OPT}\,(1 + c_3)^{\frac{t}{\text{OPT}}} \le n \cdot (1 + \frac{c_2 c_3}{\text{OPT}})^t \tag{4}$$

which solves to $t = O(\frac{\text{OPT} \log n}{\delta})$. Each iteration of the repeat loop, except the last one, does at least $c_4 \text{OPT}$ reweighings. Then the repeat loop can run for at most $O(\frac{\text{OPT} \log n}{c_4 \text{OPT} \delta}) = O(\log n/\delta)$ times.

By Theorem 1, $|Q| \le (13.4/c_1)\text{OPT}$, $|R| \le (13.4/c_2)\text{OPT}$, and $|S| \le c_4 \text{OPT}$. Thus the overall size is $13.4 \text{OPT} \cdot \left(1/c_1 + 1/c_2 + c_4/13.4\right) \le (13.4 + \delta) \cdot \text{OPT}$.

Algorithmic details. Computing an ϵ-net takes $O(n \log n)$ time using Theorem 1. Checking if a disk D is hit by an ϵ-net (Q, R, or S) reduces to finding the closest point in the set to the center of D, again accomplished in $O(\log n)$ time using point-location in Delaunay/Voronoi diagrams $\Xi(\cdot)$. It remains to show how to compute, for a given disk $D \in \mathcal{D}_1$, the set of points of P contained in D:

Lemma 6. *Given a disk $D \in \mathcal{D}_1$, the set of points of P contained in D can be reported in time $O(n/\text{OPT} \log n)$.*

4   Implementation and Experimental Evaluation

In this section we present experimental results for our algorithms implemented in C++ and running on a machine equipped with an Intel Core i7 870 processor (2.93 GHz) and with 16 GB main memory. All our implementations are single-threaded, but we note that our hitting set algorithm can be easily multi-threaded. The source code can be obtained from the authors' website [1]. For nearest-neighbors and Delaunay triangulations, we use CGAL. It computes Delaunay triangulations in expected $O(n \log n)$ time. To calculate the optimal solution for the hitting set problem we use the IP solver SCIP (with the linear solver SoPlex). Creating the linear program is carried out efficiently by using the Delaunay triangulation of the points for efficient range searching.

Datasets. In order to empirically validate our algorithms we have utilized several real-world point sets. All our experiments' point sets are scaled to a unit square. The *World* dataset [3] contains locations of cities on Earth (except for the US) having around 10M records. For our experiments we use only the locations of cities in China having 1M records (the coordinates have been obtained from latitude and longitude data by applying the Miller cylindrical projection). The dataset *ForestFire* contains 700K locations of wildfire occurrences in the United States [2]. The *KDDCUP04Bio* dataset [1] (*KDDCU* for short) contains the first 2 dimensions of a protein dataset with 145K entries. We have also created a random data set *Gauss9* with 90K points sampled from 9 different Gaussian distributions with random mean and covariance matrices.

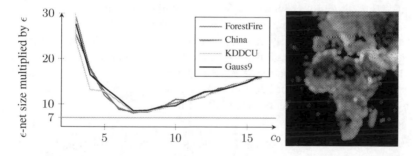

Fig. 1. ϵ-net size multiplied by ϵ for the datasets, $\epsilon = 0.01$ (left) and a subset of the ϵ-net for the *World* dataset (right).

Sizes of ϵ-nets. Setting the probability for random sampling to $\frac{12}{\epsilon \cdot n}$ results in approximately $\frac{12}{\epsilon}$ sized nets for nearly all datasets, as expected by our analysis. We note however, that in practice setting c_0 to 7 gives smaller size ϵ-nets, of size around $\frac{9}{\epsilon}$. See Figure 1 for the dependency of the net size on c_0 for $\epsilon = 0.01$. It also includes an ϵ-net calculated with our algorithm for a subset of the *World* data (red points denote the ϵ-net and each pixel's color is the logarithm of the number of disks it is contained in). See Table 1 for the ϵ-net sizes for different values of ϵ while c_0 is set to 7 and 12. This table also includes the size of the first random sample (R), which shows that the number of subproblems to solve increases as the random sample is more sparse.

Approximate Hitting Sets. For evaluating the practical usability of our approximate hitting set algorithm we compare it to the optimal solution. Our algorithm needs a guess

---

Table 1. The size of the ϵ-net multiplied by ϵ (left value in a column for a fixed ϵ) and the size of R, the first random sample multiplied by ϵ (right value in a column) for various point sets with $c_0 = 7$ or 12.

ϵ	$c_0 = 7$				$c_0 = 12$			
	0.2	0.1	0.01	0.001	0.2	0.1	0.01	0.001
China	7.8 6.6	8.3 6.1	8.28 6.80	8.426 7.090	14.2 14.2	10.6 10.6	12.33 12.33	12.152 12.138
ForestFire	7.4 7.4	8.3 7.3	8.46 7.46	8.522 6.892	13 13	11.6 11.6	12.01 12.01	12.103 12.077
KDDCU	7.4 7.4	8.4 7.4	8.31 7.29	8.343 6.989	10.2 10.2	9.8 9.8	11.65 11.57	12.006 11.978
Gauss9	7.4 5.8	7.8 7.6	8.00 7.18	8.100 6.882	9.8 9.8	12.0 12.0	11.61 11.43	11.969 11.965

for Opt, and so we run it with $O(\log n)$ guesses for the value of OPT. The parameters are set as follows: $c_0 = 10, c_1 = 30, c_2 = 12, c_3 = 2$ and $c_4 = 0.6$.

Our datasets only contain points and in order to create disks for the hitting set problem we have utilized two different strategies. In the first approach we create uniformly distributed disks in the unit square with uniformly distributed radius within the range $[0, r]$. Let us denote this test case as $RND(r)$. In the second approach we added disks centered at each point of the dataset with a fixed radius of 0.001. Let us denote this test case by $FIX(0.001)$. The results are shown in Table 2 for two values $r = 0.1$ and $r = 0.01$. Our algorithm provides a 1.3 approximation on average. With small radius the solver seems to outperform our algorithm but this is most likely due to the fact that the problems become relatively simpler and various branch-and-bound heuristics become efficient. With bigger radius and therefore more complex constraint matrix our algorithm clearly outperforms the IP solver. Our method obtains a hitting set for all point sets, while in some of the cases the IP solver was unable to compute a solution in reasonable time (we terminate the solver after 1 hour).

Table 2. Hitting sets. From top to bottom, $RND(0.1)$, $RND(0.01)$.

	# of points	# of disks	Q size	R size	S size	# of phases	IP solution	*dnet* solution	ap-prox.	IP time(s)	*dnet* time(s)
China	50K	50K	367	809	604	11	1185	1780	1.5	60	12
ForestFire	50K	16K	43	85	224	11	267	352	1.3	54.3	6.9
KDDCU	50K	22K	171	228	786	11	838	1185	1.4	40.9	9.8
Gauss9	50K	35K	322	724	1035	11	1493	2081	1.4	52.5	11.7
China	50K	49K	673	1145	4048	11	4732	5862	1.2	4.5	14.5
ForestFire	50K	25K	162	268	1021	11	1115	1451	1.3	6.2	9.5
KDDCU	50K	102K	1326	2492	6833	11	8604	10651	1.2	12.5	22.2
Gauss9	50K	185K	2737	6636	9867	11	15847	19239	1.2	22.4	36.0

In Table 3 we have included the memory consumption of both methods and statistics for range reporting. It is clear that the IP solver requires significantly more memory than our method. The statistics for range reporting includes the total number of range reportings (calculating the points inside a disk) and the number of range reportings when the algorithm doubles the weight of the points inside a disk (the doubling column in the table). It can be seen that only a fraction of the computations are wasted since the number of doublings is almost as high as the total number or range reportings. This in fact shows that the running time of our algorithm is near-linear in n.

Table 3. Memory usage in MB (left) and range reporting statistics (right).

	RND(0.01)		RND(0.1)		FIX(0.001)			RND(0.01)		RND(0.1)		FIX(0.001)	
	IP	dnet	IP	dnet	IP	dnet		total	doubling	total	doubling	total	doubling
China	243	21	4282	19	434	20	China	44014	43713	9406	9184	96335	95846
ForesFire	524	28	3059	18	5470	24	ForesFire	11167	11086	2767	2728	15648	15020
KDDCU	458	30	2999	23	175	22	KDDCU	75448	75016	8485	8364	173147	173044
Gauss9	569	33	3435	24	158	24	Gauss9	121168	120651	14133	13906	217048	217019

In order to test the scalability of our method compared to the IP solver we have used the *ForestFire* and *China* dataset with limiting the number of points to 10K, 20K, 30K... and repeating exactly the same experiments as above (while increasing the number of disks in a similar manner). In Figure 2 we plot the running time of the methods. The solid lines represent the case $RND(0.1)$ while the dashed ones denote $RND(0.01)$. One can see that as the number of points and disks increases our method becomes more efficient even though for small instances this might not hold. It can be seen that for the *China* dataset and $RND(0.01)$ the IP solver is faster than our method but after 500K points our method becomes faster. In Figure 2 the dotted line represents the running time of our algorithm for $FIX(0.001)$. In this case the IP running time is not shown because the solver was only able to solve the problem with 10K points within a reasonable time (for 20K and 30K points it took 15 and 21 hours respectively).

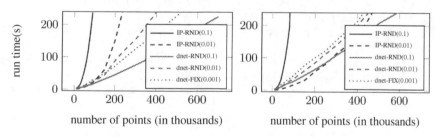

Fig. 2. Different point set sizes for the *ForestFire* (left) and *China* (right) datasets.

We have varied the radius of the disks for the fixed radius case to see how the algorithms behave. See Figure 3. With bigger radius the IP solver becomes very quickly unable to solve the problem (for radius 0.002 it was unable to finish within a day), showing that our method is more robust.

In order to test the extremes of our algorithm we have taken the *World* dataset containing 10M records. Our algorithm was able to calculate the solution of the $FIX(0.001)$ problem of size around 100K in 3.5 hours showing that the algorithm has the potential to calculate results even for extremely big datasets with a more optimized (e.g., multi-threaded) implementation.

914 N. Bus, N.H. Mustafa, and S. Ray

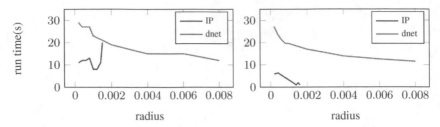

Fig. 3. Different radii settings for the *KDDCU* (left) and *China* (right) datasets.

References

1. Clustering datasets. http://cs.joensuu.fi/sipu/datasets/
2. Federal Wildland Fire Occurence Data, United States Geological Survey. http://wildfire.cr.usgs.gov/firehistory/data.html
3. Geographic Names Database, maintained by the National Geospatial-Intelligence Agency. http://geonames.nga.mil/gns/html/
4. Agarwal, P.K., Mustafa, N.H.: Independent set of intersection graphs of convex objects in 2d. Comput. Geom. 34(2), 83–95 (2006)
5. Agarwal, P.K., Pan, J.: Near-linear algorithms for geometric hitting sets and set covers. In: Symposium on Computational Geometry, p. 271 (2014)
6. Brönnimann, H., Goodrich, M.T.: Almost optimal set covers in finite vc-dimension. Discrete & Computational Geometry 14(4), 463–479 (1995)
7. Bus, N., Garg, S., Mustafa, N.H., Ray, S.: Improved local search for geometric hitting set. In: 32nd International Symposium on Theoretical Aspects of Computer Science (STACS 2015). LIPIcs, Schloss Dagstuhl - Leibniz-Zentrum fuer Informatik (2015)
8. Chazelle, B., Friedman, J.: A deterministic view of random sampling and its use in geometry. Combinatorica 10(3), 229–249 (1990)
9. Haussler, D., Welzl, E.: Epsilon-nets and simplex range queries. Discrete Comput. Geom. 2, 127–151 (1987)
10. Hochbaum, D.S., Maass, W.: Fast approximation algorithms for a nonconvex covering problem. J. Algorithms 8(3), 305–323 (1987)
11. Mustafa, N.H., Ray, S.: Improved results on geometric hitting set problems. Discrete & Computational Geometry 44(4), 883–895 (2010)
12. Pyrga, E., Ray, S.: New existence proofs for epsilon-nets. In: Proceedings of Symposium on Computational Geometry, pp. 199–207 (2008)

Efficient Computation
of Middle Levels Gray Codes

Torsten Mütze[1,*] and Jerri Nummenpalo[2]

[1] School of Mathematics, Georgia Institute of Technology, 30332 Atlanta GA, USA
muetze@math.gatech.edu
[2] Department of Computer Science, ETH Zürich, 8092 Zürich, Switzerland
njerri@inf.ethz.ch

Abstract. For any integer $n \geq 1$ a *middle levels Gray code* is a cyclic listing of all bitstrings of length $2n+1$ that have either n or $n+1$ entries equal to 1 such that any two consecutive bitstrings in the list differ in exactly one bit. The question whether such a Gray code exists for every $n \geq 1$ has been the subject of intensive research during the last 30 years, and has been answered affirmatively only recently [T. Mütze. Proof of the middle levels conjecture. *arXiv:1404.4442*, 2014]. In this work we provide the first efficient algorithm to compute a middle levels Gray code. For a given bitstring, our algorithm computes the next ℓ bitstrings in the Gray code in time $\mathcal{O}(n\ell(1 + \frac{n}{\ell}))$, which is $\mathcal{O}(n)$ on average per bitstring provided that $\ell = \Omega(n)$.

Keywords: Gray code, Middle levels conjecture.

1   Introduction

Efficiently generating all objects in a particular combinatorial class (e.g. permutations, combinations, partitions or trees) in such a way that each object is generated exactly once is one of the oldest and most fundamental problems in the area of combinatorial algorithms, and such generation algorithms are used as core building blocks in a wide range of practical applications (the survey [19] lists numerous references). A classical example is the so-called binary *Gray code*, which lists all 2^n bitstrings of length n such that any two consecutive bitstrings differ in exactly one bit. A straightforward implementation of this algorithm takes time $\mathcal{O}(n)$ to compute from a given bitstring the next one in the list (see Algorithm G in [13, Section 7.2.1.1]), which can be improved to $\mathcal{O}(1)$ [2] (see Algorithm L in [13, Section 7.2.1.1]). The space requirement of both algorithms is $\mathcal{O}(n)$. Similar minimum-change generation algorithms have been developed for various other combinatorial classes. We exemplarily cite four examples from the excellent survey [19] on this topic: (1) listing all permutations of $\{1, \ldots, n\}$

---

* The author was supported by a fellowship of the Swiss National Science Foundation. This work was initiated when the author was with the Department of Computer Science at ETH Zürich.

N. Bansal and I. Finocchi (Eds.): ESA 2015, LNCS 9294, pp. 915–927, 2015.
DOI: 10.1007/978-3-662-48350-3_76

so that consecutive permutations differ only by the swap of one pair of adjacent elements [11, 23], (2) listing all k-element subsets of an n-element set such that consecutive sets differ only by exchanging one element [2, 7, 18], (3) listing all binary trees with n vertices so that consecutive trees differ only by one rotation operation [14], (4) listing all spanning trees of a graph such that consecutive trees differ only by exchanging one edge [5].

Coming back to Gray codes, we say that a bitstring of length n has *weight* k, if it has exactly k entries equal to 1 (and $n - k$ entries equal to 0). Furthermore, for any integer $n \geq 1$ we define a *middle levels Gray code* as a cyclic listing of all bitstrings of length $2n + 1$ that have weight n or $n + 1$ such that any two consecutive bitstrings in the list differ in exactly one bit (the name 'middle levels' becomes clear when considering the relevant bitstrings as subsets in the Hasse diagram of the subset inclusion lattice). Clearly, a middle levels Gray code has to visit $N := \binom{2n+1}{n} + \binom{2n+1}{n+1} = 2\binom{2n+1}{n} = 2^{\Theta(n)}$ many bitstrings in total, and the weight of the bitstrings will alternate between n and $n + 1$ in every step. The existence of a middle levels Gray code for every value of n is asserted by the infamous *middle levels conjecture*, which originated probably with Havel [9] and Buck and Wiedemann [3], but has also been attributed to Dejter, Erdős, Trotter [12] and various others. It also appears as Exercise 56 in Knuth's book [13, Section 7.2.1.3]. It may come as a surprise that establishing the existence of a middle levels Gray code appears to be a difficult problem, given that by item (2) above one can easily come up with a listing of all bitstrings of length $2n + 1$ with weight exactly n (or exactly $n + 1$) such that any two consecutive bitstrings differ in *two* bits.

The middle levels conjecture has attracted considerable attention over the last 30 years (see e.g. [6, 8, 10, 12, 20]). Until recently, middle levels Gray codes had only been found with brute-force computer searches for $n \leq 19$ [21, 22] (for $n = 19$ this Gray code already consists of $N = 137.846.528.820$ bitstrings). A complete proof of the conjecture has only been announced very recently.

Theorem 1 ([15]). *A middle levels Gray code exists for every $n \geq 1$.*

Our Results. Even though the proof of Theorem 1 given in [15] is constructive, a straightforward implementation takes exponential (in n) time and space to compute for a given bitstring the next one in the middle levels Gray code (essentially, we need to compute and store the entire list of $N = 2^{\Theta(n)}$ bitstrings). The main contribution of this paper is a time- and space-efficient algorithm to compute a middle levels Gray code for every $n \geq 1$, which can be considered an algorithmic proof of Theorem 1. Specifically, given any bitstring of length $2n+1$ with weight n or $n+1$, our algorithm computes the next ℓ bitstrings in the Gray code in time $\mathcal{O}(n\ell(1 + \frac{n}{\ell}))$, which is $\mathcal{O}(n)$ on average per bitstring provided that $\ell = \Omega(n)$ (for most bitstrings the worst-case running time is $\mathcal{O}(n)$, for few it is $\mathcal{O}(n^2)$). Our algorithm requires $\mathcal{O}(n)$ space.

An implementation of this algorithm in C++ can be found on the authors' websites [1], and we invite the reader to experiment with this code. We used it to compute a middle levels Gray code for $n = 19$ in less than a day on an ordinary desktop computer. For comparison, the above-mentioned intelligent brute-force computation for $n = 19$ from [22] took 164 days (using comparable hardware).

Remark 1. Clearly, the ultimate goal would be a generation algorithm with a worst-case running time of $\mathcal{O}(1)$ per bitstring, but we believe this requires substantial new ideas that would in particular yield a much simpler proof of Theorem 1 than the one presented in [15].

Remark 2. It was shown in [15] that there are in fact double-exponentially (in n) many different middle levels Gray codes (which is easily seen to be best possible). This raises the question whether our algorithm can be parametrized to compute any of these Gray codes. While this is possible in principle, choosing between doubly-exponentially many different Gray codes would require a parameter of exponential size, spoiling the above-mentioned runtime and space bounds. Moreover, it would introduce a substantial amount of additional complexity in the description and correctness proof of the algorithm. To avoid all this, our algorithm computes only one particular 'canonical' middle levels Gray code.

Outline of This Paper. In Section 2 we present the pseudocode of our middle levels Gray code algorithm, and in Section 3 we discuss how to implement it to achieve the claimed runtime bound. The definition of one technical auxiliary function used in the algorithm and the correctness proof are omitted in this extended abstract due to the limited space. Complete proofs can be found in the preprint [16].

2   The Algorithm

It is convenient to reformulate our problem in graph-theoretic language: To this end we define the *middle levels graph*, denoted by $Q_{2n+1}(n, n+1)$, as the graph whose vertices are all bitstrings of length $2n+1$ that have weight n or $n+1$, with an edge between any two bitstrings that differ in exactly one bit (so bitstrings that may appear consecutively in the middle levels Gray code correspond to neighboring vertices in the middle levels graph). Clearly, computing a middle levels Gray code is equivalent to computing a Hamilton cycle in the middle levels graph (a Hamilton cycle is a cycle that visits every vertex exactly once). Throughout the rest of this paper we talk about middle levels Gray codes using this graph-theoretic terminology.

Our algorithm to compute a Hamilton cycle in the middle levels graph (i.e., to compute a middle levels Gray code) is inspired by the constructive proof of Theorem 1 in [15]. Efficiency is achieved by reformulating this inductive construction as a recursive algorithm. Even though the description of the algorithm is completely self-contained and illustrated with figures and examples that highlight the main ideas, the reader may find it useful to first read an informal overview of the proof of Theorem 1, which can be found in [15, Section 1.2].

Roughly speaking, our algorithm consists of a lower level function that computes sets of disjoint paths in the middle levels graph and several higher level functions that combine these paths to form a Hamilton cycle. In the following we explain these functions from bottom to top. Before doing so we introduce some notation that will be used throughout this paper.

2.1   Basic Definitions

Reversing/inverting and concatenating bitstrings. We define $\overline{0} := 1$ and $\overline{1} := 0$. For any bitstring x we let $\overline{\text{rev}}(x)$ denote the bitstring obtained from x by reversing the order of the bits and inverting every bit. Moreover, for any bitstring $x = (x_1, \ldots, x_{2n})$ we define $\pi(x) := (x_1, x_3, x_2, x_5, x_4, \ldots, x_{2n-1}, x_{2n-2}, x_{2n})$ (except the first and last bit, all adjacent pairs of bits are swapped). Note that the mappings $\overline{\text{rev}}$ and π are self-inverse. For two bitstrings x and y we denote by $x \circ y$ the concatenation of x and y. For any graph G whose vertices are bitstrings and any bitstring y we denote by $G \circ y$ the graph obtained from G by attaching y to every vertex of G.

Layers of the cube. For $n \geq 0$ and k with $0 \leq k \leq n$, we denote by $B_n(k)$ the set of all bitstrings of length n with weight k. For $n \geq 1$ and k with $0 \leq k \leq n-1$, we denote by $Q_n(k, k+1)$ the graph with vertex set $B_n(k) \cup B_n(k+1)$, with an edge between any two bitstrings that differ in exactly one bit.

Oriented paths, first/last vertices. An oriented path P in a graph is a path with a particular orientation, i.e., we distinguish its first and last vertex. For an oriented path $P = (v_1, v_2, \ldots, v_\ell)$ we define its first and last vertex as $F(P) := v_1$ and $L(P) := v_\ell$, respectively.

Bitstrings and Dyck paths. We often identify a bitstring x with a lattice path in the integer lattice $\mathbb{Z}^2$ as follows: Starting at the coordinate $(0,0)$, we read the bits of x from left to right and interpret every 1-bit as an upstep that changes the current coordinate by $(+1, +1)$ and every 0-bit as a downstep that changes the current coordinate by $(+1, -1)$. For any $n \geq 0$ and $k \geq 0$ we denote by $D_n(k)$ the set of lattice paths with k upsteps and $n - k$ downsteps (n steps in total) that never move below the line $y - 0$. For $n \geq 1$ we define $D_n^{>0}(k) \subseteq D_n(k)$ as the set of lattice paths that have no point of the form $(x, 0)$, $1 \leq x \leq n$, and $D_n^{=0}(k) \subseteq D_n(k)$ as the set of lattice paths that have at least one point of the form $(x, 0)$, $1 \leq x \leq n$. For $n = 0$ we define $D_0^{=0}(0) := \{()\}$, where $()$ denotes the empty lattice path, and $D_0^{>0}(0) := \emptyset$. We clearly have $D_n(k) = D_n^{=0}(k) \cup D_n^{>0}(k)$. Furthermore, for $n \geq 1$ we let $D_n^{-}(k)$ denote the set of lattice paths with k upsteps and $n - k$ downsteps (n steps in total) that have exactly one point of the form $(x, -1)$, $1 \leq x \leq n$. It is well known that in fact $|D_{2n}^{=0}(n)| = |D_{2n}^{-}(n)| = |D_{2n}^{>0}(n+1)|$ and that the size of these sets is given by the n-th Catalan number (see [4]). Observe that the mappings $\overline{\text{rev}}$ and π map each of the sets $D_{2n}^{=0}(n)$ and $D_{2n}^{-}(n)$ onto itself (see [15, Lemma 11]).

2.2   Computing Paths in $Q_{2n}(n, n+1)$

The algorithm PATHS() is at the core of our Hamilton cycle algorithm. For simplicity let us ignore for the moment the parameter flip $\in \{\text{true}, \text{false}\}$ and assume that it is set to false. Then for every n and k with $1 \leq n \leq k \leq 2n - 1$

the algorithm PATHS() defines a set of disjoint oriented paths $\mathcal{P}_{2n}(k, k+1)$ in the graph $Q_{2n}(k, k+1)$ in the following way: Given a vertex $x \in Q_{2n}(k, k+1)$ and the parameter dir $\in$ {prev, next}, the algorithm computes the neighbor of x on the path that contains the vertex x. The parameter dir controls the search direction, so for dir = prev we obtain the neighbor of x that is closer to the first vertex of the path, and for dir = next the neighbor that is closer to the last vertex of the path. If x is a first or last vertex on an oriented path from $\mathcal{P}_{2n}(k, k+1)$, then the result of a call to PATHS() with dir = prev or dir = next, respectively, is undefined (such calls will not be made).

Algorithm 1. PATHS(n, k, x, dir, flip)

Input: Integers n and k with $1 \leq n \leq k \leq 2n - 1$, a vertex $x \in Q_{2n}(k, k+1)$,
  parameters dir $\in$ {prev, next} and flip $\in$ {true, false}
Output: A neighbor of x in $Q_{2n}(k, k+1)$

P1 **if** $n = 1$ **then**   /* base cases */
P2 $\quad$ depending on x and dir, **return** previous/next neighbor of x on $\mathcal{P}_2(1, 2)$

P3 **else if** $n = k = 2$ and flip = true **then**
P4 $\quad$ depending on x and dir, **return** previous/next neighbor of x on $\widetilde{\mathcal{P}}_4(2, 3)$

P5 Split $x = (x_1, \ldots, x_{2n})$ into $x^- := (x_1, \ldots, x_{2n-2})$ and $x^+ := (x_{2n-1}, x_{2n})$
P6 **if** $k \geq n + 1$ **then**
P7 $\quad$ **return** PATHS($n - 1, k - x_{2n-1} - x_{2n}, x^-$, dir, flip) $\circ\, x^+$

P8 **else**   /* $k = n$ */
P9 $\quad$ **if** $x^+ = (1, 0)$ **then**
P10 $\quad\quad$ **return** PATHS($n - 1, n - 1, x^-$, dir, flip) $\circ\, x^+$

P11 $\quad$ **else if** $x^+ = (0, 0)$ **then**
P12 $\quad\quad$ **if** $x^- \in D_{2n-2}^{>0}(n)$ **then return** $x^- \circ (0, 1)$
P13 $\quad\quad$ **else return** PATHS($n - 1, n, x^-$, dir, flip) $\circ\, x^+$

P14 $\quad$ **else if** $x^+ = (0, 1)$ **then**
P15 $\quad\quad$ **if** $x^- \in D_{2n-2}^{=0}(n-1)$ **then return** $x^- \circ (1, 1)$
P16 $\quad\quad$ **else if** $x^- \in D_{2n-2}^-(n-1)$ and dir = next **then return** $x^- \circ (1, 1)$
P17 $\quad\quad$ **else if** $x^- \in D_{2n-2}^{>0}(n)$ and dir = prev **then return** $x^- \circ (0, 0)$
P18 $\quad\quad$ **else return** PATHS($n - 1, n - 1, x^-$, dir, false) $\circ\, x^+$

P19 $\quad$ **else if** $x^+ = (1, 1)$ **then**
P20 $\quad\quad$ **if** $x^- \in D_{2n-2}^{=0}(n-1)$ and dir = next **return** $x^- \circ (0, 1)$
P21 $\quad\quad$ **else if** $x^- \in D_{2n-2}^-(n-1)$ and dir = prev **then return** $x^- \circ (0, 1)$
P22 $\quad\quad$ **else return** $\overline{\mathrm{rev}}\left(\pi\left(\mathrm{PATHS}(n-1, n-1, \overline{\mathrm{rev}}^{-1}(\pi^{-1}(x^-)), \overline{\mathrm{dir}}, \mathrm{flip})\right)\right) \circ x^+$
    $\quad\quad$ where $\overline{\mathrm{dir}} :=$ prev if dir = next and $\overline{\mathrm{dir}} :=$ next otherwise

---

The algorithm PATHS() works recursively: For the base case of the recursion $n = 1$ (lines P1–P2) it computes neighbors for the set of paths $\mathcal{P}_2(1, 2) := \{((1, 0), (1, 1), (0, 1))\}$ (this set consists only of a single path on three vertices). E.g., the result of the call PATHS($1, 1, (1, 0)$, next, false) is $(1, 1)$ and the result of PATHS($1, 1, (1, 1)$, prev, false) is $(1, 0)$. For the recursion step the algorithm considers the last two bits of the current vertex x (see line P5) and, depending on their values (see lines P9, P11, P14 and P19), either flips one of these two bits (see lines P12, P15, P16, P17, P20, P21), or recurses to flip one of the first $2n-2$ bits instead, leaving the last two bits unchanged (see lines P7, P10, P13, P18, P22). As already mentioned, the algorithm PATHS() is a recursive formulation of the inductive construction of paths described in [15, Section 2.2], and the different cases in the algorithm reflect the different cases in this construction. The recursion step in line P22 is where the mappings $\overline{\text{rev}}$ and π introduced in Section 2.1 come into play (recall that $\overline{\text{rev}}^{-1} = \overline{\text{rev}}$ and $\pi^{-1} = \pi$).

It can be shown that the set of paths $\mathcal{P}_{2n}(n, n+1)$ computed by the algorithm PATHS() (with parameters $k = n$ and flip = false) has the following properties (we only state these properties here; a proof can be found in [16]):

(i) All paths in $\mathcal{P}_{2n}(n, n+1)$ are disjoint, and together they visit all vertices of the graph $Q_{2n}(n, n+1)$.

(ii) The sets of first and last vertices of the paths in $\mathcal{P}_{2n}(n, n+1)$ are $D_{2n}^{=0}(n)$ and $D_{2n}^{-}(n)$, respectively.

From (ii) we conclude that the number of paths in $\mathcal{P}_{2n}(n, n+1)$ equals the n-th Catalan number. E.g., the set of paths $\mathcal{P}_4(2,3)$ computed by PATHS() is $\mathcal{P}_4(2,3) = \{P, P'\}$ with $P := ((1,1,0,0), (1,1,0,1), (0,1,0,1), (0,1,1,1), (0,0,1,1), (1,0,1,1), (1,0,0,1))$ and $P' := ((1,0,1,0), (1,1,1,0), (0,1,1,0))$.

We will see in the next section how to combine the paths $\mathcal{P}_{2n}(n, n+1)$ in the graph $Q_{2n}(n, n+1)$ computed by our algorithm PATHS() (called with flip = false) to compute a Hamilton cycle in the middle levels graph $Q_{2n+1}(n, n+1)$. One crucial ingredient we need for this is another set of paths $\widetilde{\mathcal{P}}_{2n}(n, n+1)$, computed by calling the algorithm PATHS() with flip = true. We shall see that the core 'intelligence' of our Hamilton cycle algorithm consists of cleverly combining some paths from $\mathcal{P}_{2n}(n, n+1)$ and some paths from $\widetilde{\mathcal{P}}_{2n}(n, n+1)$ to a Hamilton cycle in the middle levels graph. We will see that from the point of view of all top-level routines that call the algorithm PATHS() only the paths $\mathcal{P}_{2n}(k, k+1)$ and $\widetilde{\mathcal{P}}_{2n}(k, k+1)$ for $k - n$ are used, but the recursion clearly needs to compute these sets also for all other values $k = n+1, n+2, \ldots, 2n-1$.

Specifically, calling the algorithm PATHS() with flip = true yields a set of paths $\widetilde{\mathcal{P}}_{2n}(k, k+1)$ that differs from $\mathcal{P}_{2n}(k, k+1)$ as follows: First note that for certain inputs of the algorithm the value of the parameter flip is irrelevant, and the computed paths are the same regardless of its value (this is true whenever the recursion ends in the base case in lines P1-P2). One such example is the path $((1,0,1,1,0,0), (1,1,1,1,0,0), (0,1,1,1,0,0))$ (computing previous or next neighbors for any vertex on this path yields the same result regardless of the value of flip). Such paths are not part of $\widetilde{\mathcal{P}}_{2n}(k, k+1)$, and we ignore them in the following. The remaining paths from $\mathcal{P}_{2n}(k, k+1)$ can be grouped into pairs and for every such pair (P, P') the set $\widetilde{\mathcal{P}}_{2n}(k, k+1)$ contains two paths R and R' that visit the same set of vertices as P and P', but that connect the end vertices of the paths the other way: Formally, denoting by $V(G)$ the vertex set of any graph G, we require that $V(P) \cup V(P') = V(R) \cup V(R')$ and $F(P) = F(R)$, $F(P') = F(R')$, $L(P) = L(R')$ and $L(P') = L(R)$. We refer to a pair (P, P') satisfying these conditions as a *flippable pair* of paths, and to the corresponding paths (R, R') as a *flipped pair*. This notion of flippable/flipped pairs is extremely valuable for designing our Hamilton cycle algorithm, as it allows the algorithm to decide *independently* for each flippable pair of paths, whether to follow one of the original paths or the flipped paths in the graph $Q_{2n}(n, n+1)$.

The base case for the recursive computation of $\widetilde{\mathcal{P}}_{2n}(k, k+1)$ is the set of paths $\widetilde{\mathcal{P}}_4(2,3) := \{R, R'\}$ (see lines P3–P4) that is defined by the two paths $R := ((1,1,0,0), (1,1,1,0), (0,1,1,0))$ and $R' := ((1,0,1,0), (1,0,1,1), (0,0,1,1), (0,1,1,1), (0,1,0,1), (1,1,0,1), (1,0,0,1))$ in $Q_4(2,3)$. Continuing the previous example, observe that the set of paths $\mathcal{P}_4(2,3) = \{P, P'\}$ computed by PATHS() mentioned before and this set of paths $\widetilde{\mathcal{P}}_4(2,3) = \{R, R'\}$ satisfy precisely the conditions for flippable/flipped pairs of paths, i.e., (P, P') is a flippable pair and (R, R') is a corresponding flipped pair of paths.

2.3 Computing a Hamilton Cycle in the Middle Levels Graph

The algorithm HAMCYCLEFLIP() uses the paths $\mathcal{P}_{2n}(n, n+1)$ and $\widetilde{\mathcal{P}}_{2n}(n, n+1)$ in the graph $Q_{2n}(n, n+1)$ computed by the algorithm PATHS() as described in the previous section to compute for a given vertex x the next vertex on a Hamilton cycle in the middle levels graph $Q_{2n+1}(n, n+1)$. To simplify the exposition of the algorithm, let us ignore for the moment the parameter $\texttt{flip} \in \{\texttt{true}, \texttt{false}\}$ and assume that it is set to $\texttt{false}$, and let us also ignore line F3 (so only paths from $\mathcal{P}_{2n}(n, n+1)$ are considered, and those from $\widetilde{\mathcal{P}}_{2n}(n, n+1)$ are ignored). With these simplifications the algorithm HAMCYCLEFLIP() will not compute a Hamilton cycle, but a set of several smaller cycles that together visit all vertices of the middle levels graph (this will be corrected later by setting $\texttt{flip}$ accordingly).

The algorithm HAMCYCLEFLIP() is based on the following decomposition of the middle levels graph $Q_{2n+1}(n, n+1)$ (see Figure 1): By partitioning the

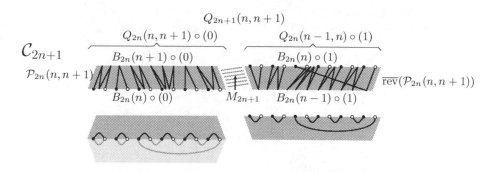

Fig. 1. The top part of the figure shows the decomposition of the middle levels graph and the definition (1). The 2-factor consists of three disjoint cycles that together visit all vertices of the graph. The first and last vertices of the paths are drawn in black and white, respectively. The bottom part of the figure shows a simplified drawing that helps analyzing the cycle structure of the 2-factor (it has two short cycles and one long cycle).

vertices (=bitstrings) of the graph $Q_{2n+1}(n, n+1)$ according to the value of the last bit, we observe that it consists of a copy of the graph $Q_{2n}(n, n+1) \circ (0)$ and a copy of the graph $Q_{2n}(n-1, n) \circ (1)$ plus the set of edges $M_{2n+1} = \{(x \circ (0), x \circ (1)) \mid x \in B_{2n}(n)\}$ along which the last bit is flipped. Observe furthermore that the graphs $Q_{2n}(n, n+1)$ and $Q_{2n}(n-1, n)$ are isomorphic, and the mapping $\overline{\mathrm{rev}}$ is an isomorphism between these graphs. It is easy to check that this isomorphism preserves the sets of end vertices of the paths $\mathcal{P}_{2n}(n, n+1)$: Using property (ii) mentioned in the previous section, we have $\overline{\mathrm{rev}}(D_{2n}^{=0}(n)) = D_{2n}^{=0}(n)$ and $\overline{\mathrm{rev}}(D_{2n}^-(n)) = D_{2n}^-(n)$ (in Figure 1, these sets are the black and white vertices). By property (i) of the paths $\mathcal{P}_{2n}(n, n+1)$, we conclude that

$$\mathcal{C}_{2n+1} := \mathcal{P}_{2n}(n, n+1) \circ (0) \ \cup \ \overline{\mathrm{rev}}(\mathcal{P}_{2n}(n, n+1)) \circ (1) \ \cup \ M'_{2n+1} \qquad (1)$$

with $M'_{2n+1} := \{(x \circ (0), x \circ (1)) \mid x \in D_{2n}^{=0}(n) \cup D_{2n}^-(n)\} \subseteq M_{2n+1}$ is a so-called 2-*factor* of the middle levels graph, i.e., a set of disjoint cycles that together visit all vertices of the graph. Note that along each of the cycles in the 2-factor, the paths from $\mathcal{P}_{2n}(n, n+1) \circ (0)$ are traversed in forward direction, and the paths from $\overline{\mathrm{rev}}(\mathcal{P}_{2n}(n, n+1)) \circ (1)$ in backward direction. The algorithm HAMCYCLEFLIP() (called with $\mathtt{flip} = \mathtt{false}$) computes exactly this 2-factor $\mathcal{C}_{2n+1}$: Given a vertex x of the middle levels graph, it computes the next vertex on one of the cycles from the 2-factor by checking the value of the last bit (line F2), and by returning either the next vertex on the corresponding path from $\mathcal{P}_{2n}(n, n+1) \circ (0)$ (line F5) or the previous vertex on the corresponding path from $\overline{\mathrm{rev}}(\mathcal{P}_{2n}(n, n+1)) \circ (1)$ (line F8; recall that $\overline{\mathrm{rev}}^{-1} = \overline{\mathrm{rev}}$). The cases that the next cycle edge is an edge from M'_{2n+1} receive special treatment: In these cases the last bit is flipped (lines F4 and F7).

Algorithm 2. HAMCYCLEFLIP$(n, x, \mathtt{flip})$

Input: An integer $n \geq 1$, a vertex $x \in Q_{2n+1}(n, n+1)$, state variable
$\qquad \mathtt{flip} \in \{\mathtt{true}, \mathtt{false}\}$
Output: Starting from x the next vertex on a Hamilton cycle in
$\qquad Q_{2n+1}(n, n+1)$, updated state variable $\mathtt{flip} \in \{\mathtt{true}, \mathtt{false}\}$

F1 Split $x = (x_1, \ldots, x_{2n+1})$ into $x^- := (x_1, \ldots, x_{2n})$ and the last bit x_{2n+1}
F2 **if** $x_{2n+1} = 0$ **then**
F3 $\quad$ **if** $x^- \in D_{2n}^{=0}(n)$ **then return** $\big(\mathrm{PATHS}(n, n, x^-, \mathtt{next}, a) \circ (x_{2n+1}), a\big)$ where
$\qquad a := \mathrm{ISFLIPVERTEX}(x^-)$
F4 $\quad$ **else if** $x^- \in D_{2n}^-(n)$ **then return** $(x^- \circ (1), \mathtt{false})$
F5 $\quad$ **else return** $\big(\mathrm{PATHS}(n, n, x^-, \mathtt{next}, \mathtt{flip}) \circ (x_{2n+1}), \mathtt{flip}\big)$
F6 **else** $\hfill$ /* $x_{2n+1} = 1$ */
F7 $\quad$ **if** $x^- \in D_{2n}^{=0}(n)$ **then return** $(x^- \circ (0), \mathtt{false})$
F8 $\quad$ **else return** $\big(\overline{\mathrm{rev}} \left(\mathrm{PATHS}(n, n, \overline{\mathrm{rev}}^{-1}(x^-), \mathtt{prev}, \mathtt{false}) \right) \circ (x_{2n+1}), \mathtt{false}\big)$

As mentioned before, calling the algorithm HAMCYCLEFLIP() with $\mathtt{flip} = \mathtt{false}$ yields the 2-factor in the middle levels graph defined in (1) that consists of more than one cycle. However, it can be shown that by replacing in (1) some of the flippable pairs of paths from $\mathcal{P}_{2n}(n, n+1) \circ (0)$ by the corresponding flipped paths from the set $\widetilde{\mathcal{P}}_{2n}(n, n+1) \circ (0)$ (which are computed by calling the algorithm PATHS() with $\mathtt{flip} = \mathtt{true}$) we obtain a 2-factor that has only one cycle, i.e., a Hamilton cycle. The key insight is that replacing a pair of flippable paths that are contained in two different cycles of the 2-factor C_{2n+1} by the corresponding flipped paths joins these two cycles to one cycle (this is an immediate consequence of the definition of flippable/flipped pairs of paths). This flipping is controlled by the parameter $\mathtt{flip}$ of the algorithm HAMCYCLEFLIP(): It decides whether to compute the next vertex on a path from $\mathcal{P}_{2n}(n, n+1) \circ (0)$ (if $\mathtt{flip} = \mathtt{false}$) or from $\widetilde{\mathcal{P}}_{2n}(n, n+1) \circ (0)$ (if $\mathtt{flip} = \mathtt{true}$). Note that these modifications do not affect the paths $\overline{\mathrm{rev}}(\mathcal{P}_{2n}(n, n+1)) \circ (1)$ in the union (1): For the corresponding instructions in lines F6–F8, the value of $\mathtt{flip}$ is irrelevant.

The decision whether to follow a path from $\mathcal{P}_{2n}(n, n+1) \circ (0)$ or from $\widetilde{\mathcal{P}}_{2n}(n, n+1) \circ (0)$ is computed at the vertices $x \circ (0)$, $x \in D_{2n}^{=0}(n)$, by calling the function ISFLIPVERTEX(x) (line F3). This decision is returned to the caller and maintained until the last vertex of the corresponding path in the graph $Q_{2n}(n, n+1) \circ (0)$ is reached. Recall that by the definition of flippable pairs of paths, this decision can be made *independently* for each flippable pair (of course it has to be consistent for both paths in a flippable pair: either both are flipped or none of them).

2.4   The Top-Level Algorithm

The algorithm HAMCYCLE(n, x, ℓ) takes as input a vertex x of the middle levels graph $Q_{2n+1}(n, n+1)$ and computes the next ℓ vertices that follow x on a Hamilton cycle in this graph (i.e., for $\ell \leq N = 2\binom{2n+1}{n}$, every vertex appears at most once in the output, and for $\ell = N$ every vertex appears exactly once and the vertex x comes last).

Algorithm 3. HAMCYCLE(n, x, ℓ)

Input: An integer $n \geq 1$, a vertex $x \in Q_{2n+1}(n, n+1)$, an integer $\ell \geq 1$
Output: Starting from x, the next ℓ vertices on a Hamilton cycle in
$\qquad Q_{2n+1}(n, n+1)$

H1 Split $x = (x_1, \ldots, x_{2n+1})$ into $x^- := (x_1, \ldots, x_{2n})$ and the last bit x_{2n+1}
H2 $\mathtt{flip} := \mathtt{false}$
H3 **if** $x_{2n+1} = 0$ **then**   /* initialize state variable flip */
H4 $\quad$ $y := x^-$
H5 $\quad$ **while** $y \notin D_{2n}^{=0}(n)$ **do**   /* move backwards to first path vertex */
H6 $\quad$ $\quad \lfloor \; y := \mathrm{PATHS}(n, n, y, \mathtt{prev}, \mathtt{false})$
H7 $\quad \lfloor$ $\mathtt{flip} := \mathrm{ISFLIPVERTEX}(y)$

H8 $y := x$
H9 **for** $i := 1$ **to** ℓ **do**   /* Hamilton cycle computation */
H10 $\quad$ $(y, \mathtt{flip}) := \mathrm{HAMCYCLEFLIP}(n, y, \mathtt{flip})$
H11 $\quad \lfloor$ **output** y

In terms of Gray codes, the algorithm takes a bitstring x of length $2n + 1$ that has weight n or $n + 1$ and outputs the ℓ subsequent bitstrings in a middle levels Gray code. A single call HAMCYCLE(n, x, ℓ) yields the same output as ℓ subsequent calls $x_{i+1} := $ HAMCYCLE$(n, x_i, 1)$, $i = 0, \ldots, \ell - 1$, with $x_0 := x$. However, with respect to running times, the former is faster than the latter.

The algorithm HAMCYCLE() consists of an initialization phase (lines H2–H7) in which the initial value of the state variable flip is computed. This is achieved by following the corresponding path from $\mathcal{P}_{2n}(n, n + 1) \circ (0)$ backwards to its first vertex (lines H5–H6) and by calling the function ISFLIPVERTEX() (line H7). The actual Hamilton cycle computation (lines H8–H11) repeatedly computes the subsequent cycle vertex and updates the state variable flip by calling the function HAMCYCLEFLIP() discussed in the previous section (line H10).

2.5 Flip Vertex Computation

To complete the description of our Hamilton cycle algorithm, it remains to specify the auxiliary function ISFLIPVERTEX(n, x). As mentioned before, this function decides for each vertex $x \in D_{2n}^{=0}(n)$ whether our Hamilton cycle algorithm should follow the path from $\mathcal{P}_{2n}(n, n + 1)$ that starts with this vertex (return value false) or the corresponding flipped path from $\widetilde{\mathcal{P}}_{2n}(n, n + 1)$ (return value true) in the graph $Q_{2n}(n, n + 1) \circ (0)$ (recall (1)). This function therefore nicely encapsulates the core 'intelligence' of our algorithm so that it produces a 2-factor consisting only of a single cycle and not of several smaller cycles in the middle levels graph. As the definition of this function is rather technical and unintuitive, it is omitted in this extended abstract. It can be found in [16, Section 2.5].

3 Runtime Analysis

A naive implementation of the function PATHS() takes time $\mathcal{O}(n^2)$: To see this observe that the membership tests whether x^- is contained in one of the sets $D_{2n-2}^{=0}(n-1)$, $D_{2n-2}^{-}(n-1)$ or $D_{2n-2}^{>0}(n)$ in lines P12, P15, P16, P17, P20 and P21 and the application of the mappings $\overline{\text{rev}}$ and π in line P22 take time $\mathcal{O}(n)$ (recall that $\overline{\text{rev}}^{-1} = \overline{\text{rev}}$ and $\pi^{-1} = \pi$), and that the value of n decreases by 1 with each recursive call. In the following we sketch how this can improved so that each call of PATHS() takes only time $\mathcal{O}(n)$ (more details can be found in the comments of our C++ implementation [1]): For this we maintain counters c_0 and c_1 for the number of zeros and ones of a given bitstring $x = (x_1, \ldots, x_{2n})$. Moreover, interpreting the bitstring x as a lattice path (as described in Section 2.1), we maintain vectors $c_{00}, c_{01}, c_{10}, c_{11}$ that count the number of occurences of pairs of consecutive bits (x_{2i}, x_{2i+1}), $i \in \{1, \ldots, n-1\}$, *per height level of the lattice path* for each of the four possible value combinations of x_{2i} and x_{2i+1}. E.g., for the bitstring $x = (1, 1, 0, 0, 0, 0, 1, 0, 1, 0)$ the vector c_{10} has a single nonzero entry 1 at height level (=index) 1 for the two bits $(x_2, x_3) = (1, 0)$, the vector c_{00} has a single nonzero entry 1 at height level (=index) 0 for the two bits $(x_4, x_5) = (0, 0)$, and the vector c_{01} has a single nonzero entry 2 at height level (=index) -1 for the pairs of bits $(x_6, x_7) = (x_8, x_9) = (0, 1)$. Using these counters, the three membership tests mentioned before can be performed in constant time. E.g., a bitstring $x = (x_1, \ldots, x_{2n})$ is contained in $D_{2n}^{=0}(n)$ if and only if $c_0 = c_1$ and $x_1 = 1$ and the entry of c_{00} at height level 0 equals 0 (i.e., the lattice path never moves below the line $y = 0$). Moreover, these counters can be updated in constant time when removing the last two bits of x and when applying the mappings $\overline{\text{rev}}$ and π: Note that $\overline{\text{rev}}$ simply swaps the roles of c_0 and c_1 and the roles of c_{00} and c_{11} (plus a possible index shift), and that π simply swaps the roles of c_{10} and c_{01}. To compute the applications of $\overline{\text{rev}}$ and π in line P22 in constant time, we do not modify x at all, but rather count the number of applications of $\overline{\text{rev}}$ and π and keep track of the middle range of bits of x that is still valid (when removing the last two bits of x, this range shrinks by 2 on one of the sides). Taking into account that multiple applications of $\overline{\text{rev}}$ and π cancel each other out, this allows us to compute the effect of applying those mappings *lazily* when certain bits are queried later on (when testing the values of the last two bits of some substring of x).

The auxiliary function IsFlipVertex() defined in [16, Section 2.5] can be implemented to run in time $\mathcal{O}(n^2)$ (see that paper for more details).

It was shown in [17, Eq. (25)] that the length of any path $P \in \mathcal{P}_{2n}(n, n+1)$ with a first vertex $F(P) =: x \in D_{2n}^{=0}(n)$ is given by the following simple formula: Considering the unique decomposition $x = (1) \circ x_\ell \circ (0) \circ x_r$ with $x_\ell \in D_{2k}^{=0}(k)$ for some $k \geq 0$, the length of P is given by $2|x_\ell| + 2 \leq 2(2n-2) + 2 = 4n - 2$. It follows that the while-loop in line H5 terminates after at most $\mathcal{O}(n)$ iterations, i.e., the initialization phase of HamCycle() (lines H2–H7) takes time $\mathcal{O}(n^2)$.

It was shown in [17, Theorem 10] that the distance between any two neighboring vertices of the form $x \circ (0)$, $x' \circ (0)$ with $x, x' \in D_{2n}^{=0}(n)$ (x and x' are first vertices of two paths $P, P' \in \mathcal{P}_{2n}(n, n+1)$) on a cycle in (1) is exactly $4n + 2$. Comparing the lengths of two paths from $\mathcal{P}_{2n}(n, n+1)$ that form a flippable pair with the lengths of the corresponding flipped paths from the set $\widetilde{\mathcal{P}}_{2n}(n, n+1)$, we observe that either the length of one the paths decreases by 4 and the length of the other increases by 4, or the lengths of the paths do not change (the paths P, P' and R, R' defined in Section 2.2 have exactly the length differences -4 and $+4$, and these differences only propagate through the first cases of the Paths() recursion, but not the last case in lines P19–P22). It follows that every call of HamCycleFlip() for which the condition in line F3 is satisfied and which therefore takes time $\mathcal{O}(n^2)$ due to the call of IsFlipVertex(), is followed by at least $4n - 3$ calls in which the condition is not satisfied, in which case HamCycleFlip() terminates in time $\mathcal{O}(n)$. Consequently, ℓ consecutive calls of HamCycleFlip() take time $\mathcal{O}(n^2 + n\ell)$.

Summing up the time $\mathcal{O}(n^2)$ spent for the initialization phase and $\mathcal{O}(n^2 + n\ell)$ for the actual Hamilton cycle computation, we conclude that the algorithm HamCycle(n, x, ℓ) runs in time $\mathcal{O}(n^2 + n\ell) = \mathcal{O}(n\ell(1 + \frac{n}{\ell}))$, as claimed.

Acknowledgements. We thank Günter Rote for persistently raising the question whether the proof of Theorem 1 could be turned into an efficient algorithm. We also thank the referees of this extended abstract for numerous valuable comments that helped improving the presentation of this work and the C++ code.

References

[1] Currently http://www.as.inf.ethz.ch/muetze and http://people.inf.ethz.ch/njerri

[2] Bitner, J., Ehrlich, G., Reingold, E.: Efficient generation of the binary reflected gray code and its applications. Commun. ACM 19(9), 517–521 (Sep 1976), http://doi.acm.org/10.1145/360336.360343

[3] Buck, M., Wiedemann, D.: Gray codes with restricted density. Discrete Math. 48(2-3), 163–171 (1984), http://dx.doi.org/10.1016/0012-365X(84)90179-1

[4] Chen, Y.: The Chung–Feller theorem revisited. Discrete Math. 308(7), 1328–1329 (2008), http://www.sciencedirect.com/science/article/pii/S0012365X07001811

[5] Cummins, R.: Hamilton circuits in tree graphs. IEEE Trans. Circuit Theory 13(1), 82–90 (Mar 1966)

[6] Duffus, D., Kierstead, H., Snevily, H.: An explicit 1-factorization in the middle of the Boolean lattice. J. Combin. Theory Ser. A 65(2), 334–342 (1994), http://dx.doi.org/10.1016/0097-3165(94)90030-2

[7] Eades, P., McKay, B.: An algorithm for generating subsets of fixed size with a strong minimal change property. Inf. Process. Lett. 19(3), 131–133 (1984), http://www.sciencedirect.com/science/article/pii/0020019084900917

[8] Felsner, S., Trotter, W.: Colorings of diagrams of interval orders and α-sequences of sets. Discrete Math. 144(1-3), 23–31 (1995), http://dx.doi.org/10.1016/0012-365X(94)00283-0, combinatorics of ordered sets (Oberwolfach, 1991)

[9] Havel, I.: Semipaths in directed cubes. In: Graphs and other combinatorial topics (Prague, 1982), Teubner-Texte Math., vol. 59, pp. 101–108. Teubner, Leipzig (1983)

[10] Johnson, R.: Long cycles in the middle two layers of the discrete cube. J. Combin. Theory Ser. A 105(2), 255–271 (2004), http://dx.doi.org/10.1016/j.jcta.2003.11.004

[11] Johnson, S.: Generation of permutations by adjacent transposition. Math. Comp. 17, 282–285 (1963)

[12] Kierstead, H., Trotter, W.: Explicit matchings in the middle levels of the Boolean lattice. Order 5(2), 163–171 (1988), http://dx.doi.org/10.1007/BF00337621

[13] Knuth, D.: The Art of Computer Programming, Volume 4A. Addison-Wesley (2011)

[14] Lucas, J., van Baronaigien, D.R., Ruskey, F.: On rotations and the generation of binary trees. J. Algorithms 15(3), 343–366 (1993), http://www.sciencedirect.com/science/article/pii/S019667748371045X

[15] Mütze, T.: Proof of the middle levels conjecture (Aug 2014), *arXiv:1404.4442*

[16] Mütze, T., Nummenpalo, J.: Efficient computation of middle levels Gray codes (Jun 2015), *arXiv:1506.07898*

[17] Mütze, T., Weber, F.: Construction of 2-factors in the middle layer of the discrete cube. J. Combin. Theory Ser. A 119(8), 1832–1855 (2012), http://www.sciencedirect.com/science/article/pii/S0097316512000969

[18] Ruskey, F.: Adjacent interchange generation of combinations. J. Algorithms 9(2), 162–180 (1988), http://www.sciencedirect.com/science/article/pii/0196677488900363

[19] Savage, C.: A survey of combinatorial Gray codes. SIAM Rev. 39(4), 605–629 (1997), http://dx.doi.org/10.1137/ S0036144595295272

[20] Savage, C., Winkler, P.: Monotone Gray codes and the middle levels problem. J. Combin. Theory Ser. A 70(2), 230–248 (1995), http://dx.doi.org/10.1016/0097-3165(95)90091-8

[21] Shields, I., Shields, B., Savage, C.: An update on the middle levels problem. Discrete Math. 309(17), 5271–5277 (2009), http://dx.doi.org/10.1016/j.disc.2007.11.010

[22] Shimada, M., Amano, K.: A note on the middle levels conjecture (Sep 2011), arXiv:0912.4564

[23] Trotter, H.: Algorithm 115: Perm. Commun. ACM 5(8), 434–435 (Aug 1962), http://doi.acm.org/10.1145/368637.368660

Computing the Similarity Between Moving Curves

Kevin Buchin, Tim Ophelders, and Bettina Speckmann

Eindhoven University of Technology
Eindhoven, The Netherlands
{k.a.buchin,t.a.e.ophelders,b.speckmann}@tue.nl

Abstract. In this paper we study similarity measures for moving curves which can, for example, model changing coastlines or glacier termini. Points on a moving curve have two parameters, namely the position along the curve as well as time. We therefore focus on similarity measures for surfaces, specifically the Fréchet distance between surfaces. While the Fréchet distance between surfaces is not even known to be computable, we show for variants arising in the context of moving curves that they are polynomial-time solvable or NP-complete depending on the restrictions imposed on how the moving curves are matched. We achieve the polynomial-time solutions by a novel approach for computing a surface in the so-called free-space diagram based on max-flow min-cut duality.

1 Introduction

Over the past years the availability of devices that can be used to track moving objects has increased dramatically, leading to an explosive growth in movement data. Naturally the goal is not only to track objects but also to extract information from the resulting data. Consequently recent years have seen a significant increase in the development of methods extracting knowledge from moving object data.

Tracking an object gives rise to data describing its movement. Often the scale at which the tracking takes place is such that the tracked objects can be viewed as point objects. Cars driving on a highway, birds foraging for food, or humans walking in a pedestrian zone: for many analysis tasks it is sufficient to consider them as moving points. Hence the most common data sets used in movement data processing are so-called trajectories: sequences of time-stamped points.

However, not all moving objects can be reasonably represented as points. A hurricane can be represented by the position of its eye, but a more accurate description is as a 2-dimensional region which represents the hurricanes extent. When studying shifting coastlines, reducing the coastline to a point is obviously unwanted: one is actually interested in how the whole coast line moves and changes shape over time. The same holds true when studying the terminus of a glacier. In such cases, the moving object is best represented as a polyline rather than by a single point. In this paper we hence go beyond the basic setting of

© Springer-Verlag Berlin Heidelberg 2015
N. Bansal and I. Finocchi (Eds.): ESA 2015, LNCS 9294, pp. 928–940, 2015.
DOI: 10.1007/978-3-662-48350-3_77

moving point objects and study moving complex, non-point objects. Specifically, we focus on similarity measures for moving curves, based on the Fréchet distance.

Definitions and Notation. The Fréchet distance is a well-studied distance measure for shapes, and is commonly used to determine the similarity between two curves A and $B : [0,1] \to \mathbb{R}^n$. A natural generalization to more complex shapes uses the definition of Equation 1 where the shapes A and B have type $X \to \mathbb{R}^n$.

$$D_{\mathrm{fd}}(A, B) = \inf_{\mu:X \to X} \sup_{x \in X} \|A(x) - B(\mu(x))\| \tag{1}$$

Here, $\| \cdot \| : \mathbb{R}^n \to \mathbb{R}$ is a norm such as the Euclidean norm (L^2) or the Manhattan norm (L^1). The *matching* μ ranges over orientation-preserving homeomorphisms (possibly with additional constraints) between the parameter spaces of the shapes compared; as such, it defines a correspondence between the points of the compared shapes. A matching between surfaces with parameters p and t is illustrated in Figure 1. Given one such matching we obtain a distance between A and B by taking the largest distance between any two

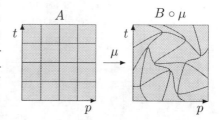

Fig. 1. A matching μ between surfaces A and B drawn as a homeomorphism between their parameter spaces.

corresponding points of A and B. The Fréchet distance is the infimum of these distances taken over all possible matchings. For moving points or static curves, we have as parameter space $X = [0, 1]$ and for moving curves or static surfaces, we have $X = [0, 1]^2$. We can define various similarity measures between shapes by imposing further restrictions on μ.

In practice a curve is generally represented by a sequence of $P + 1$ points. Assuming a linear interpolation between consecutive points, this results in a polyline with P segments. Analogously, a moving curve is a sequence of $T + 1$ polylines, each of P segments. We also interpolate the polylines linearly, yielding a bilinear interpolation, or a quadrilateral mesh of $P \times T$ quadrilaterals.

Related Work. The Fréchet distance or related measures are frequently used to evaluate the similarity between point trajectories [8,7,13]. The Fréchet distance is also used to match point trajectories to a street network [2,5]. The Fréchet distance between polygonal curves can be computed in near-quadratic time [3,6,9,17], and approximation algorithms [4,15] have been studied.

The natural generalization to moving (parameterized) curves is to interpret the curves as surfaces parameterized over time and over the curve parameter. The Fréchet distance between surfaces is NP-hard [16], even for terrains [10]. In terms of positive algorithmic results for general surfaces the Fréchet distance is only known to be semi-computable [1,12]. Polynomial-time algorithms have been given for the so called weak Fréchet distance [1] and for the Fréchet distance between simple polygons [11] and so called folded polygons [14].

When interpreting moving curves as surfaces it is important to take the different roles of the two parameters into account: the first is inherently linked to time and the other to space. This naturally leads to restricted versions of the Fréchet distance of surfaces. For curves, restricted versions of the Fréchet distance were considered [7,18]. For surfaces we are not aware of similar results.

1.1   Results

We refine the Fréchet distance between surfaces to meaningfully compare moving curves. To do so, we restrict matchings to be one of several suitable classes. Representative matchings for the considered classes together with the running times of our results are illustrated in Figure 2.

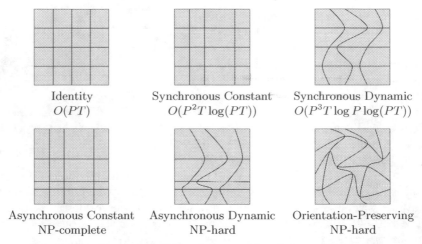

Identity
$O(PT)$

Synchronous Constant
$O(P^2T\log(PT))$

Synchronous Dynamic
$O(P^3T\log P\log(PT))$

Asynchronous Constant
NP-complete

Asynchronous Dynamic
NP-hard

Orientation-Preserving
NP-hard

Fig. 2. The time complexities of the considered classes of matchings.

The simplest class of matchings consists of a single predefined *identity* matching $\mu(p, t) = (p, t)$. Hence, to compute the *identity* Fréchet distance, we need only determine a pair of matched points that are furthest apart. It turns out that one of the points of a furthest pair is a vertex of a moving curve (i.e. quadrilateral mesh), allowing computation in $O(PT)$ time. See the full paper for more details.

We also discuss the *synchronous constant* Fréchet distance in the full paper. Here we assume that the matching of timestamps is known in advance, and the matching of positions is the same for each timestamp, so it remains constant. Our algorithm computes the positional matching that minimizes the Fréchet distance.

The *synchronous dynamic* Fréchet distance considered in Section 2 also assumes a predefined matching of timestamps, but does not have the constraint of the *synchronous constant* class that the matching of positions remains constant over time. Instead, the positional matching may change continuously over time.

Finally, in Section 3, we consider several cases where neither positional nor temporal matchings are predefined. The three considered cases are the *asynchronous constant*, *asynchronous dynamic*, and *orientation-preserving* Fréchet distance. The asynchronous constant class of matchings consists of a constant (but not predefined) matching of positions, as well as timestamps whereas in the asynchronous dynamic class of matchings, the positional matching may change continuously. In the orientation-preserving class (see the full paper), matchings range over orientation preserving homeomorphisms between the parameter spaces, given that the corners of the parameter space are aligned.

The last three classes are quite complex, and we give constructions proving that approximating the Fréchet distance within a factor 1.5 is NP-hard under these classes. For the asynchronous constant and asynchronous dynamic classes of matchings, this result holds even for moving curves embedded in $\mathbb{R}^1$ whereas the result for the orientation-preserving case holds for embeddings in $\mathbb{R}^2$.

Although we do not discuss classes where positional matchings are known in advance, these symmetric variants can be obtained by interchanging the time and position parameters for the discussed classes. Deciding which variant is appropriate for comparing two moving curves depends largely on how the data is obtained, as well as the use case for the comparison. For instance, the synchronous constant variant may be used on a sequence of satellite images which have associated timestamps. The synchronous dynamic Fréchet distance is better suited for sensors with different sampling frequencies, placed on curve-like moving objects.

2   Synchronous Dynamic Matchings

 Synchronous dynamic matchings align timestamps under the identity matching, but the matching of positions may change continuously over time. Specifically, the matching is defined as $\mu(p,t) = (\pi_t(p), t)$. Here, $\mu(p,t) : [0,P] \times [0,T] \to [0,P] \times [0,T]$ is continuous, and for any t the matching $\pi_t : [0,P] \to [0,P]$ between the two curves at that time is a nondecreasing surjection.

2.1   Freespace Partitions in 2D

The freespace diagram $\mathcal{F}_\varepsilon$ is the set pairs of points that are within distance ε of each other.

$$(x,y) \in \mathcal{F}_\varepsilon \Leftrightarrow \|A(x) - B(y)\| \leq \varepsilon$$

If A and B are curves with parameter space $[0,P]$, then their freespace diagram is two-dimensional, and the Fréchet distance is the minimum value of ε for which an xy-monotone path (representing μ) from $(0,0)$ to (P,P) through the freespace exists.

We use a variant of the max-flow min-cut duality to determine whether a matching through the freespace exists. Before we present the 3D variant for moving curves with synchronized timestamps, we illustrate the idea in the fictional 2D freespace of Figure 3. Here, any

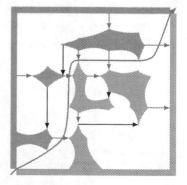

Fig. 3. A matching (green) in the 2D freespace (white).

matching—such as the green path—must be an x- and y-monotone path from
the bottom left to the top right corner and this matching must avoid all obstacles
(i.e. all points not in $\mathcal{F}_e$). Therefore each such matching divides the obstacles in
two sets: those above, and those below the matching.

Suppose we now draw a directed edge from an obstacle a to an obstacle b
if and only if any matching that goes over a must necessarily go over b. The
key observation is that a matching exists unless such edges can form a path
from the lower-right boundary to the upper-left boundary of the freespace. In
the example, a few trivial edges are drawn in black and gray. If all obstacles
were slightly larger, an edge could connect a blue obstacle with a red obstacle,
connecting the two boundaries by the edges drawn in black.

2.2 Freespace Partitions in 3D

In contrast to the 2D freespace where the matching is a path, matchings of the
form $\mu(p,t) = (\pi_t(p),t)$ form surfaces in the 3D freespace $\mathcal{F}_\varepsilon^{3D}$ (see Equation 2).
Such a surface again divides the obstacles in the freespace in two sets and can be
punctured by a path connecting two boundaries. We formalize this concept for
the 3D freespace and give an algorithm for deciding the existence of a matching.

$$(x,y,t) \in \mathcal{F}_\varepsilon^{3D} \text{ if and only if } \|A(x,t) - B(y,t)\| \le \varepsilon \qquad (2)$$

For $x,y,t \in \mathbb{N}$, the cell $C_{x,y,t}$ of the 3D freespace is the set $\mathcal{F}_\varepsilon^{3D} \cap ([x, x+1] \times [y, y+1] \times [t, t+1])$. The property of Lemma 1 holds for all such cells.

Lemma 1. *A cell $C_{x,y,t}$ of the freespace has a convex intersection with any line
parallel to the xy-plane or the t-axis.*

We divide the set of points not in $\mathcal{F}_\varepsilon^{3D}$ into a
set O of so-called obstacles, such that each indi-
vidual obstacle is a connected point set. Let u be
the open set of points representing the left and top
boundary of $\mathcal{F}_\varepsilon^{3D}$. Symmetrically, let d represent
the bottom and right boundary, see Figure 4. De-
note by $O' \subset O$ the obstacles between the boundaries.

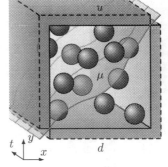

$O = \{u,d\} \cup O'$ with $\bigcup O' = ([0,P]^2 \times [0,T]) \setminus \mathcal{F}_\varepsilon^{3D}$;
$u = \{(x,y,t) \mid (x < 0 \wedge y > 0) \vee (x < P \wedge y > P)\}$;
$d = \{(x,y,t) \mid (x > 0 \wedge y < 0) \vee (x > P \wedge y < P)\}$.

Fig. 4. μ separates u and d.

Given a matching μ, let $D \subseteq O$ be the set of
obstacles below it, then $u \notin D$ and $d \in D$. Here, we use axes (x,y,t) and say
that a point is below some other point if it has a smaller y-coordinate. Because
each obstacle is a connected set and μ cannot intersect obstacles, a single obstacle
cannot lie on both sides of the same matching. Because all matchings have $u \notin D$
and $d \in D$, a matching exists if and only if $\neg(d \in D \Rightarrow u \in D)$.

We compute a relation $\triangleright$ of elementary dependencies between obstacles, such that its transitive closure $\oslash$ has $d \oslash u$ if and only if $d \in D \Rightarrow u \in D$. Let $a \triangleright b$ if and only if $a \cup b$ is connected (a touches b) or there exists some point $(x_a, y_a, t_a) \in a$ and $(x_b, y_b, t_b) \in b$ with $x_a \leq x_b$, $y_a \geq y_b$ and $t_a = t_b$. We prove in Lemmas 2 and 3 that this choice of $\triangleright$ satisfies the required properties and in Theorem 4 that we can use the transitive closure $\oslash$ of $\triangleright$ to solve the decision problem of the Fréchet distance.

Lemma 2. *If $a \oslash b$, then $a \in D \Rightarrow b \in D$.*

Proof. Assume that $a \triangleright b$, then either a touches b and no matching can separate them, or there exists some $(x_a, y_a, t) \in a$ and $(x_b, y_b, t) \in b$ with $x_a \leq x_b$, $y_a \geq y_b$. If there were some matching μ with $a \in D$, then $(x_a, y_\mu, t) \in \mu$ for some $y_\mu > y_a$. Similarly, if $b \notin D$, then $(x_b, y'_\mu, t) \in \mu$ for some $y'_\mu < y_b$. We can further deduce from $x_a \leq x_b$ and monotonicity of μ that we can pick y'_μ such that $y_a < y_\mu \leq y'_\mu < y_b$. However, this contradicts $y_a \geq y_b$, so such a matching does not exist. Hence, $a \in D \Rightarrow b \in D$ whenever $a \triangleright b$ and therefore whenever $a \oslash b$.   □

Lemma 3. *If $d \in D \Rightarrow u \in D$, then $d \oslash u$.*

Proof. Suppose $d \in D \Rightarrow u \in D$ but not $d \oslash u$. Then no matching exists, and no path from d to u exists in the directed graph $G = (O, \triangleright)$. Pick as D the set of obstacles reachable from d in G, then D does not contain u. Pick the *tightest* matching μ such that D lies below it, we define μ in terms of matchings $\pi_t \subseteq \mathbb{R}^2 \times \{t\}$ in the plane at each timestamp t.

$$(x, y, t) \in \pi_t \text{ if and only if } (x' > x \wedge y' < y) \Rightarrow \neg m(x', y', t) \wedge m(x, y, t) \text{ where}$$

$$m(x, y, t) \text{ if and only if } \{(x', y', t) \mid x' \leq x \wedge y' \geq y\} \cap \bigcup D = \emptyset$$

Because $u \notin D$, this defines a monotone path π_t from $(0,0)$ to (P, P) at each timestamp t. Suppose that π_t *properly* intersects some $o \in O$, such that some point of $(x_o, y_o, t) \in o$ lies below π_t. It follows from the definition of $\triangleright$ and $\neg m(x_o, y_o, t)$ that $d \triangleright o$ for some $d \in D$. However, such obstacle o cannot exist because D satisfies $\triangleright$. As a result, no path π_t intersects any obstacle and we can connect the paths π_t to obtain a continuous matching μ without intersecting any obstacles. So μ does not intersect obstacles in $O \setminus D$, contradicting $d \in D \Rightarrow u \in D$.   □

Theorem 4. *The Fréchet distance is greater than ε if and only if $d \oslash u$ for ε.*

Proof. We have for every matching that $u \notin D$ and $d \in D$. Therefore it follows from Lemma 2 that no matching exists if $d \oslash u$ for ε. In that case, the Fréchet distance is greater than ε. Conversely, if $\neg(d \oslash u)$ there is a set D satisfying $\oslash$ with $u \notin D$ and $d \in D$. In that case, a matching exists by Lemma 3, and the Fréchet distance is less than ε.   □

We choose the set of obstacles O' such that $\bigcup O' = ([0, P]^2 \times [0, T]) \setminus \mathcal{F}_\varepsilon$ and the relation $\triangleright$ is easily computable. Note that due to Lemma 1, each connected component contains a corner of a cell, so any cell in the freespace contains constantly many (up to eight) components of $\bigcup O'$. Moreover, we can index each obstacle in O' by a grid point $(x, y, t) \in \mathbb{N}^3$.

Let $o_{x,y,t} \subseteq ([0, P]^2 \times [0, T]) \cap ([x-1, x+1] \times [y-1, y+1] \times [t-1, t+1]) \setminus \mathcal{F}_\varepsilon$ be the maximal connected subset of the cells adjacent to (x, y, t), such that $o_{x,y,t}$ contains (x, y, t). Now, the obstacle $o_{x,y,t}$ is not well-defined if $(x, y, t) \in \mathcal{F}_\varepsilon$, in which case we define $o_{x,y,t}$ to be an empty (dummy) obstacle. We have $O' = \bigcup_{(x,y,t)} \{o_{x,y,t}\}$ and we remark that obstacles are not necessarily disjoint.

Each of the $O(P^2 T)$ obstacles is now defined by a constant number of vertices. We therefore assume that for each pair of obstacles $(a, b) \in O^2$, we can decide in constant time whether $a \triangleright b$; even though this decision procedure depends on the chosen distance metric. For each obstacle $a = o_{x,y,t}$, there are $O(P^2)$ obstacles $b = o_{x',y',t'}$ for which $a \triangleright b$, namely because $t - 2 \le t' \le t + 2$ if $a \triangleright b$. Furthermore, u and d contribute to $O(P^2 T)$ elements of the relation. Therefore we can compute the relation $\triangleright$ in $O(P^4 T)$ time.

Testing whether $d \otimes u$ is equivalent to testing whether a path from d to u exists in the directed graph $(O, \triangleright)$, which can be decided using a depth first search. So we can solve the decision problem for the Fréchet distance in $O(P^2 T + |\triangleright|) = O(P^4 T)$ time. However, the relation $\triangleright$ may yield many unnecessary edges. In Section 2.4 we show that a smaller set E of size $O(P^3 T)$ with the same transitive closure $\otimes$ is computable in $O(P^3 T \log P)$ time, so the decision algorithm takes only $O(P^3 T \log P)$ time.

2.3   Parametric Search

To give an idea of what the 3D freespace looks like, we have drawn the obstacles of the eight cells of the freespace between two quadrilateral meshes of size $P \times T = 2 \times 2$ in Figure 5. Cells of the 3D freespace lie within cubes, having six faces and twelve edges. We call such edges x-, y- or t-edges, depending on the axis to which they are parallel.

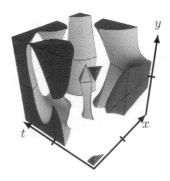

We are looking for the minimum value of ε for which a matching exists. When increasing the value of ε, the relation $\triangleright$ becomes sparser since obstacles shrink. Critical values of ε occur when $\triangleright$ changes. Due to Lemma 1, all critical values involve an edge or an xt-face or yt-face of a cell, but never the

Fig. 5. $[0, 2]^3 \setminus \mathcal{F}_\varepsilon^{3D}$

internal volume, so the following critical values cover all cases.

a) The minimal ε such that $(0, 0, t) \in \mathcal{F}_\varepsilon^{3D}$ and $(P, P, t) \in \mathcal{F}_\varepsilon^{3D}$ for all t.
b) An edge of $C_{x,y,t}$ becomes nonempty.
c) Endpoints of y-edges of $C_{x,y,t}$ and $C_{x+i,y,t}$ align in y-coordinate, or endpoints of x-edges of $C_{x,y,t}$ and $C_{x,y-j,t}$ align in x-coordinate.
d) Endpoints of a t-edge of $C_{x,y,t}$ and a t-edge of $C_{x+i,y-j,t}$ align in t-coordinate.

e) An obstacle in $C_{x,y,t}$ stops overlapping with an obstacle in $C_{x+i,y,t}$ or $C_{x,y-j,t}$ when projected along the x- or y-axis.

The endpoints involved in the critical values of type a), b), c) and d) can be captured in $O(P^2T)$ functions. We apply a parametric search for the minimum critical value ε_{abcd} of type a), b), c) or d) for which a matching exists. This takes $O((P^2T + time_{dec})\log(PT))$ time.

We illustrate the need for critical values of type e) in Figure 6, here obstacle a overlaps with both obstacles b and c while the overlap in edges does not contribute to $\triangleright$. It is unclear how critical values of type e) can be incorporated in the parametric search directly. Instead, we enumerate and sort the $O(P^3T)$ critical values of type e) in $O(P^3T\log(PT))$ time. Using $O(\log(PT))$ calls to the decision algorithm, we apply a binary search to find the minimum critical value ε_e of type e) for which a matching exists. Finding the critical value ε_e then takes $O((P^3T + time_{dec})\log(PT))$ time.

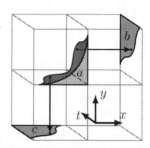

Fig. 6. $a \triangleright b$ and $a \triangleright c$

The synchronous dynamic Fréchet distance is then the minimum of ε_{abcd} and ε_e. This results in the following running time.

Theorem 5. *The synchronous dynamic Fréchet distance can be computed in $O((P^3T + time_{dec})\log(PT))$ time.*

Before stating the final running time, we present a faster algorithm for the decision algorithm.

2.4   A Faster Decision Algorithm

To speed up the decision procedure we distinguish the cases for which two obstacles may be related by $\triangleright$, these cases correspond to the five types of critical values of Section 2.3. Critical values of type a) and b) depend on obstacles in single cells, so there are at most $O(P^2T)$ elements of $\triangleright$ arising from type a) and b). Critical values of type c) and e) arise from pairs of obstacles in cells in the same row or column, so there are at most $O(P^3T)$ of them. In fact, we can enumerate the edges of type a), b), c), and e) of $\triangleright$ in $O(P^3T)$ time. On the other hand, edges of type d) arise between two cells with the same value of t, so there can be $O(P^4T)$ of them.

We compute a smaller directed graph (V, E) with $|E| = O(P^3T)$ that has a path from d to u if and only if $d \otimes u$. Let $V = O = \{u, d\} \cup O'$ be the vertices as before (we will include dummy obstacles for grid points in that lie in the freespace) and transfer the edges in $\triangleright$ except those of type d) to the smaller set of edges E. We must still induce edges of type d) in E, but instead of adding $O(P^4T)$ edges, we use only $O(P^3T)$ edges. The edges of type d) can actually be captured in the transitive closure of E using only $O(P)$ edges per obstacle in E.

Using an edge from $o_{x,y,t}$ to $o_{x+1,y,t}$ and to $o_{x,y-1,t}$, we construct a path from $o_{x,y,t}$ to any obstacle $o_{x+i,y-j,t}$. The sole purpose of the dummy obstacles is to construct these paths effectively. For obstacles whose gridpoints have the same t-coordinates, it then takes a total of $O(P^2T)$ edges to include the obstacles overlapping in t-coordinate related by type d), this is valid because $(x,y,t) \in o_{x,y,t}$ for non-dummy obstacles.

Denote by E_k^d the edges of type d) of the form $(a,b) = (o_{x,y,t_a}, o_{x+i,y-j,t_b})$ where $t_b = t_a + k$, then the set E_0^d of $O(P^2T)$ edges is the one we just constructed. Now it remains to induce paths with $t_a \neq t_b$, that still overlap in t-coordinates, i.e. the sets E_{-2}^d, E_{-1}^d, E_1^d and E_2^d. Denote by $t^-(a)$ and $t^+(a)$ the minimum and maximum t-coordinate over points in an obstacle a. For each obstacle, both the $t^-(a)$ and the $t^+(a)$ coordinates are an endpoint of a t-edge in a cell defining the obstacle due to Lemma 1, and therefore computable in constant time.

Our savings arise from the fact that if $o_{x,y,t} \triangleright o_{x+i,y-j,t+k}$ and $o_{x,y,t} \triangleright o_{x+i',y-j',t+k}$ with $i \leq i'$ and $j \leq j'$, then E_0^d induces a path from $o_{x+i,y-j,t+k}$ to $o_{x+i',y-j',t+k}$, so we do not need an additional edge to induce a path to the latter obstacle. To avoid degenerate cases, we start by exhaustively enumerating edges of E_k^d ($k \in \{-2,-1,1,2\}$) for which $i \leq 1$ or $j \leq 1$ in $O(P^3T)$ time so we need only consider edges with $i \geq 2 \wedge j \geq 2$.

For these remaining cases, we have $a \triangleright b$ if and only if $t^+(a) \geq t^-(b) \wedge t_b = t_a + k$, and $t^-(a) \leq t^+(b) \wedge t_b = t_a - k$ for positive k. From this we can derive the edges of E_k^d. Although for each a, there may be $O(P^2)$ obstacles b such that $a \triangleright b$ with $t_b = t_a + k$, the Pareto frontier of those obstacles b contains only $O(P)$ obstacles, see the grid of fictional values $t^-(b)$ in Figure 7. In the full paper, we show how to find these Pareto frontiers in $O(P \log P)$ time per obstacle a, using only $O(P^2T)$ preprocessing time for the complete freespace.

As a result, we can compute all $O(P^3T)$ edges of E_k^d in $O(P^3T \log P)$ time. By Theorem 6, the decision problem for the synchronous dynamic Fréchet distance is solvable in $O(P^3T \log P)$ time.

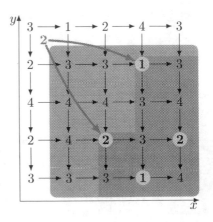

Fig. 7. Two edges (green) cover (red) all four obstacles b (green) within the query rectangle (blue) with values $t^-(b) \leq t^+(a) = 2$.

Theorem 6. *The decision problem for the synchronous dynamic Fréchet distance is solvable in $O(P^3T \log P)$ time.*

Proof. The edges E of types other than d) are enumerated in $O(P^3T)$ time, and using constantly many Pareto frontier queries for each obstacle, $O(P^3T)$ edges of type d) in E are computed in $O(P^3T \log P)$ time. Given the set E of edges, deciding whether a path between two vertices exists takes $O(|E|) = O(P^3T)$ time. The transitive closure of E equals $\oslash$, so a path from d to u exists in E

if and only if there was such a path in ▷. Since we compute E in $O(P^3 T \log P)$ time, the decision problem is solved in $O(P^3 T \log P)$ time.   □

The following immediately follows from Theorems 5 and 6.

Corollary 7. *The synchronous dynamic Fréchet distance can be computed in* $O(P^3 T \log P \log(PT))$ *time.*

3   Hardness

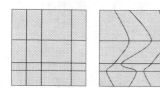

We extend the synchronous constant and synchronous dynamic classes of matchings to asynchronous ones. For this, we allow realignments of timestamps, giving rise to the asynchronous constant and asynchronous dynamic classes of matchings. The asynchronous constant class ranges over matchings of the form $\mu(p, t) = (\pi(p), \tau(t))$ where the π and τ are matchings of positions and timestamps. The asynchronous dynamic class of matchings has the form $\mu(p, t) = (\pi_t(p), \tau(t))$ for which the positional matching π_t changes over time. We first prove that the asynchronous constant Fréchet distance is in NP.

Theorem 8. *Computing the Fréchet distance is in NP for the asynchronous constant class of matchings.*

Proof. Given any matching $\mu(p, t) = (\pi(p), \tau(t))$ with a Fréchet distance of ε, we can derive—due to Lemma 1—a piecewise-linear matching τ^* in $O(T)$ time, such that a matching $\mu^*(p, t) = (\pi^*(p), \tau^*(t))$ with Fréchet distance at most ε exists. We can realign the quadrilateral meshes A and B under τ^* to obtain meshes A^* and B^* of polynomial size. Now the polynomial-time decision algorithm for synchronous constant matchings (see full paper) is applicable to A^* and B^*.   □

Due to critical values of type e), it is unclear whether each asynchronous dynamic matching admits a piecewise-linear matching τ^* of polynomial size, which would mean that the asynchronous dynamic Fréchet distance is also in NP.

We show that computing the Fréchet distance is NP-hard for both classes by a reduction from 3-SAT. The idea behind the construction is illustrated in the two height maps of Figure 8. These represent quadrilateral meshes embedded in $\mathbb{R}^1$ and correspond to a single clause of a 3-CNF formula of four variables.

We distinguish valleys (dark), peaks (white on A, yellow on B) and ridges (denoted X_i, F_i and T_i). An important observation is that in order to obtain a low Fréchet distance of $\varepsilon < 3$, the n-th valley of A must be matched with the n-th valley of B. Moreover, each ridge X_i must be matched with F_i or T_i and each peak of A must be matched to a peak of B. Note that even for asynchronous dynamic matchings, if X_i is matched to F_i, it cannot be matched to T_i and vice-versa because the (red) valley separating F_i and T_i has distance 3 from X_i.

The aforementioned properties are reflected more clearly in the 2D freespace between the curves at aligned timestamps t and $\tau(t)$. In Figure 8, we give a 2D slice (at $t_A = T/2$, $t_B = T/4$) of the 4D-freespace diagram with $\varepsilon = 2$ for the shown quadrilateral meshes. In this diagram with $\varepsilon = 2$, only 2^3 monotone paths exist (up to directed homotopy) whereas for $\varepsilon = 3$ there would be 2^4 monotone paths (one for each assignment of variables). For $\varepsilon = 2$, the peak of X_2 cannot be matched to F_2 at $t = T/4$ of B, corresponding to an assignment of $X_2 = true$.

Consider a 3-CNF formula with n variables and m clauses, then A and B consist of m clauses along the t-axis and n variables ($X_1 \ldots X_n$ and $F_1, T_1 \ldots F_n, T_n$) along the p-axis. The k-th clause of A is matched to the k-th clause of B due to the elevation pattern on the far left ($p = 0$). This means that the peaks of A are matched with peaks of the same clause on B and all these peaks have the same timestamp because $\tau(t)$ is constant (independent of p).

For each clause, there are three rows (timestamps) of B with peaks on the ridges. On each such timestamp, exactly one ridge (depending on the disjuncts of the clause) does not have a peak. Specifically, if a clause has X_i or $\neg X_i$ as its k-th disjunct, then the k-th row of that clause has no peak on ridge F_i or T_i, respectively. We use these properties in Theorem 11 where we prove that it is NP-hard to approximate the Fréchet distance within a factor 1.5.

Lemma 9. *The Fréchet distance between two such moving curves is at least* 3 *if the corresponding 3-CNF formula is unsatisfiable.*

Proof. Consider a matching yielding a Fréchet distance smaller than 3 given an unsatisfiable formula, then the peaks of A (of the k-th clause) are matched with peaks of B (of a single row of the k-th clause). Assign the value *true* to variable X_i if ridge X_i is matched with T_i and *false* if it is matched with F_i. Then for every clause ($V_i \vee V_j \vee V_k$) with $V_i \in \{X_i, \neg X_i\}$, there is a peak at $\pi(X_i)$, $\pi(X_j)$ or $\pi(X_k)$ for that clause. Such a matching cannot exist because then the 3-CNF formula would be satisfiable, so the Fréchet distance is at least 3. □

Lemma 10. *The Fréchet distance between two such moving curves is at most* 2 *if the corresponding 3-CNF formula is satisfiable.*

Proof. Consider a satisfying assignment to the 3-CNF formula. Match X_i with the center of F_i or T_i, if X_i is *false* or *true*, respectively. For every clause, the timestamp with peaks of A can be matched with a row of peaks on B. As was

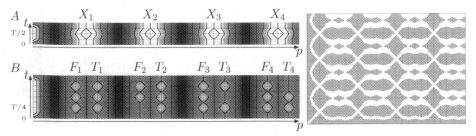

Fig. 8. Two quadrilateral meshes A and B embedded in $\mathbb{R}^1$ (indicated by color and isolines). Their Fréchet distance is 2 isolines if the clause ($X_2 \vee \neg X_3 \vee \neg X_4$) is satisfiable and 3 isolines otherwise. The freespace $\mathcal{F}_{\varepsilon=2}$ of (A, B) at times $(T/2, T/4)$ on the right.

already hinted at by Figure 8, the remaining parts of the curves can be matched with $\varepsilon = 2$. Therefore this yields a Fréchet distance of at most 2.   □

Theorem 11. *It is NP-hard to approximate the asynchronous constant or asynchronous dynamic Fréchet distance for moving curves in $\mathbb{R}^1$ within a factor 1.5.*

Proof. By Lemmas 9 and 10, the asynchronous constant or asynchronous dynamic Fréchet distance between two quadrilateral meshes embedded in $\mathbb{R}^1$ is at least 3 or at most 2, depending on whether a 3-CNF formula is satisfiable.   □

Acknowledgements. K. Buchin, T. Ophelders, and B. Speckmann are supported by the Netherlands Organisation for Scientific Research (NWO) under project no. 612.001.207 (K. Buchin) and no. 639.023.208 (T. Ophelders & B. Speckmann).

References

1. Alt, H., Buchin, M.: Can we compute the similarity between surfaces? Discrete Comput. Geom. 43(1), 78–99 (2010)
2. Alt, H., Efrat, A., Rote, G., Wenk, C.: Matching planar maps. In: Proc. 14th Sympos. Discrete Algorithms, pp. 589–598 (2003)
3. Alt, H., Godau, M.: Computing the Fréchet distance between two polygonal curves. Intern. J. Comput. Geom. Appl. 5(01n02), 75–91 (1995)
4. Aronov, B., Har-Peled, S., Knauer, C., Wang, Y., Wenk, C.: Fréchet Distances for Curves, Revisited. In: Azar, Y., Erlebach, T. (eds.) ESA 2006. LNCS, vol. 4168, pp. 52–63. Springer, Heidelberg (2006)
5. Brakatsoulas, S., Pfoser, D., Salas, R., Wenk, C.: On map-matching vehicle tracking data. In: Proc. 31st Intern. Conf. VLDB, pp. 853–864 (2005)
6. Bringmann, K.: Why walking the dog takes time: Fréchet distance has no strongly subquadratic algorithms unless SETH fails. In: Foundations of Computer Science, pp. 661–670 (2014)
7. Buchin, K., Buchin, M., Gudmundsson, J.: Constrained free space diagrams: a tool for trajectory analysis. Intern. J. GIS 24, 1101–1125 (2010)
8. Buchin, K., Buchin, M., Gudmundsson, J., Löffler, M., Luo, J.: Detecting commuting patterns by clustering subtrajectories. Intern. J. Comput. Geom. Appl. 21(3), 253–282 (2011)
9. Buchin, K., Buchin, M., Meulemans, W., Mulzer, W.: Four soviets walk the dog-with an application to Alt's conjecture. In: Proc. 25th Sympos. Discrete Algorithms, pp. 1399–1413 (2014)
10. Buchin, K., Buchin, M., Schulz, A.: Fréchet distance of surfaces: Some simple hard cases. In: de Berg, M., Meyer, U. (eds.) ESA 2010, Part II. LNCS, vol. 6347, pp. 63–74. Springer, Heidelberg (2010)
11. Buchin, K., Buchin, M., Wenk, C.: Computing the Fréchet distance between simple polygons. Comput. Geom. Theory Appl. 41(1–2), 2–20 (2008)
12. Buchin, M.: On the Computability of the Fréchet Distance Between Triangulated Surfaces. PhD thesis, Free University Berlin, Institute of Computer Science (2007)
13. Buchin, M., Dodge, S., Speckmann, B.: Context-aware similarity of trajectories. In: Proc. 6th Intern. Conf. GIS, pp. 43–56 (2012)

14. Cook IV, A.F., Driemel, A., Har-Peled, S., Sherette, J., Wenk, C.: Computing the Fréchet distance between folded polygons. In: Dehne, F., Iacono, J., Sack, J.-R. (eds.) WADS 2011. LNCS, vol. 6844, pp. 267–278. Springer, Heidelberg (2011)
15. Driemel, A., Har-Peled, S., Wenk, C.: Approximating the Fréchet distance for realistic curves in near linear time. Discrete Comput. Geom. 48(1), 94–127 (2012)
16. Godau, M.: On the complexity of measuring the similarity between geometric objects in higher dimensions. PhD thesis, Berlin, Freie Universität Berlin, Diss. (1998, 1999)
17. Har-Peled, S., Raichel, B.: The Fréchet distance revisited and extended. ACM Transactions on Algorithms 10(1), 3 (2014)
18. Maheshwari, A., Sack, J.-R., Shahbaz, K., Zarrabi-Zadeh, H.: Fréchet distance with speed limits. Comput. Geom. Theory Appl. 44(2), 110–120 (2011)

I/O-Efficient Similarity Join*

Rasmus Pagh[1], Ninh Pham[1], Francesco Silvestri[1]**, and Morten Stöckel[2],***

[1] IT University of Copenhagen, Denmark
{pagh,ndap,fras}@itu.dk
[2] University of Copenhagen, Denmark
most@di.ku.dk

Abstract. We present an I/O-efficient algorithm for computing similarity joins based on locality-sensitive hashing (LSH). In contrast to the filtering methods commonly suggested our method has provable subquadratic dependency on the data size. Further, in contrast to straightforward implementations of known LSH-based algorithms on external memory, our approach is able to take significant advantage of the available internal memory: Whereas the time complexity of classical algorithms includes a factor of N^ρ, where ρ is a parameter of the LSH used, the I/O complexity of our algorithm merely includes a factor $(N/M)^\rho$, where N is the data size and M is the size of internal memory. Our algorithm is randomized and outputs the correct result with high probability. It is a simple, recursive, cache-oblivious procedure, and we believe that it will be useful also in other computational settings such as parallel computation.

1 Introduction

The ability to handle noisy or imprecise data is becoming increasingly important in computing. In database settings this kind of capability is often achieved using similarity join primitives that replace equality predicates with a condition on similarity. To make this more precise consider a space $\mathbb{U}$ and a distance function $d : \mathbb{U} \times \mathbb{U} \to \mathbf{R}$. The *similarity join* of sets $R, S \subseteq \mathbb{U}$ is the following: Given a radius r, compute the set $R \bowtie_{\leq r} S = \{(x,y) \in R \times S \mid d(x,y) \leq r\}$. This problem occurs in numerous applications, such as web deduplication [3,15], document clustering [4], data cleaning [2,6]. As such applications arise in large-scale datasets, the problem of scaling up similarity join for different metric distances is getting more important and more challenging.

---

* The research leading to these results has received funding from the European Research Council under the EU 7th Framework Programme, ERC grant agreement no. 614331.
** In part supported by University of Padova project CPDA121378 and by MIUR of Italy project AMANDA while working at the University of Padova.
*** This work was done while at IT University of Copenhagen. Supported by the Danish National Research Foundation / Sapere Aude program and VILLUM FONDEN.

N. Bansal and I. Finocchi (Eds.): ESA 2015, LNCS 9294, pp. 941–952, 2015.
DOI: 10.1007/978-3-662-48350-3_78

Many known similarity join techniques (e.g., prefix filtering [2,6], positional filtering [15], inverted index-based filtering [3]) are based on *filtering* techniques that often, but not always, succeed in reducing computational costs. If we let $N = |R| + |S|$ these techniques generally require $\Omega(N^2)$ comparisons for worst-case data. Another approach is *locality-sensitive hashing* (LSH) where candidate output pairs are generated using collisions of carefully chosen hash functions. The LSH definition is as follows.

Definition 1. *Fix a distance function* $d : \mathbb{U} \times \mathbb{U} \rightarrow \mathbf{R}$. *For positive reals* r, c, p_1, p_2, *and* $p_1 > p_2, c > 1$, *a family of functions* $\mathcal{H}$ *is* (r, cr, p_1, p_2)-*sensitive if for uniformly chosen* $h \in \mathcal{H}$ *and all* $x, y \in \mathbb{U}$:

- *If* $d(x, y) \leq r$ *then* $Pr[h(x) = h(y)] \geq p_1$;
- *If* $d(x, y) > cr$ *then* $Pr[h(x) = h(y)] \leq p_2$.

We say that $\mathcal{H}$ *is monotone if* $Pr[h(x) = h(y)]$ *is a non-increasing function of the distance function* $d(x, y)$.

LSH is able to break the N^2 barrier in cases where for some constant $c > 1$ the number of pairs in $R \bowtie_{\leq cr} S$ is not too large. In other words, there should not be too many pairs that have distance within a factor c of the threshold, the reason being that such pairs are likely to become candidates, yet considering them does not contribute to the output. For notational simplicity, we will talk about *far* pairs at distance greater than cr (those that should not be reported), *near* pairs at distance at most r (those that should be reported), and *c-near* pairs at distance between r and cr (those that should not be reported but affect the I/O cost).

In this paper we study I/O-efficient similarity join methods based on LSH. That is, we are interested in minimizing the number of I/O operations where a block of B points from $\mathbb{U}$ is transferred between an external memory and an internal memory with capacity for M points from $\mathbb{U}$. Our main result is the first *cache-oblivious* algorithm for similarity join that has *provably* sub-quadratic dependency on the data size N and at the same time inverse polynomial dependency on M. In essence, where previous methods have an overhead factor of either N/M or $(N/B)^\rho$ we obtain an overhead of $(N/M)^\rho$, where $0 < \rho < 1$ is a parameter of the LSH employed, strictly improving both. We show:

Theorem 1. *Consider* $R, S \subseteq \mathbb{U}$, *let* $N = |R| + |S|$, *assume* $18 \log N + 3B \leq M < N$ *and that there exists a monotone* (r, cr, p_1, p_2)-*sensitive family of functions with respect to distance measure* d, *using space* B *and with* $p_2 < p_1 < 1/2$. *Let* $\rho = \log p_1 / \log p_2$. *Then there exists a cache-oblivious randomized algorithm computing* $R \bowtie_{\leq r} S$ *(wrt. d) with probability* $1 - \mathcal{O}(1/N)$ *using*

$$\tilde{\mathcal{O}} \left(\left(\frac{N}{M} \right)^\rho \left(\frac{N}{B} + \frac{|R \bowtie_{\leq r} S|}{MB} \right) + \frac{|R \bowtie_{\leq cr} S|}{MB} \right) \ I/Os.^{[1]}$$

---

[1] The $\tilde{\mathcal{O}}(\cdot)$-notation hides polylog($N$) factors.

At first, the given I/O complexity may look somewhat strange, but we argue that the bound is really quite natural. The first term matches the complexity of the standard LSH method if we set $M = 1$, and becomes essentially linear when $M = N$ (i.e., when we know that the I/O complexity is linear). Regarding the second term, we need to use $|R \bowtie_{\leq r} S|/(MB)$ I/Os to compute distances of $|R \bowtie_{\leq r} S|$ pairs since in a single I/O we can introduce at most MB pairs in memory. We get a factor $(N/M)^\rho$ because we may introduce some very near pairs this number of times in memory. We can make a similar argument for the last term. Because it is very hard to distinguish distance r from distance slightly above r we expect to have to get some, possibly all, pairs in $R \bowtie_{\leq cr} S$ into fast memory in order to conclude that they should not be output. A more detailed discussion and experimental evaluations can be seen in the full version [13].

It is worth noting that whereas most methods in the literature focus on a single (or a few) distance measure, our methods work for an arbitrary space and distance measure that allows LSH, e.g., Hamming, Manhattan (ℓ_1), Euclidean (ℓ_2), Jaccard, and angular metric distances. A primary technical hurdle in the paper is that we cannot use any kind of strong concentration bounds on the number of points having a particular value, since hash values of an LSH family may be correlated *by definition*. Another hurdle is *duplicate elimination* in the output stemming from pairs having multiple LSH collisions. However, in the context of I/O-efficient algorithms it is natural to not require the *listing* of all near pairs, but rather we simply require that the algorithm *enumerates* all such near pairs. More precisely, the algorithm calls for each near pair (x, y) a function $\mathtt{emit}(x, y)$. This is a natural assumption in external memory since it reduces the I/O complexity. In addition, it is desired in many applications where join results are intermediate results pipelined to a subsequent computation, and are not required to be stored on external memory. Our upper bound can be easily adapted to list all instances by increasing the I/O complexities of an *unavoidable* additive term of $\Theta(|R \bowtie_{\leq r} S|/B)$ I/Os.

The organization of the paper is as follows. In Section 2, we briefly review related work. Section 3 describes our algorithms including a warm-up cache-aware approach and the main results, a cache-oblivious solution, its analysis, and a randomized approach to remove duplicates. Section 4 concludes the paper.

2   Related Work

Because Locality-sensitive hashing (LSH) is a building block of our I/O-efficient similarity join, we briefly review LSH, the computational I/O model, and some state-of-the-art similarity join techniques.

Locality-Sensitive Hashing (LSH). LSH was originally introduced by Indyk and Motwani [12] for similarity search problem in high dimensional data. This technique obtains a sublinear (i.e., $\mathcal{O}(N^\rho)$) time complexity by increasing the gap of collision probability between near points and far points using the LSH family as defined in Definition 1. Such gap of collision probability is polynomial, with an exponent of $\rho = \log p_1/\log p_2$ dependent on c, and $0 < \rho < 1$.

In this work we will use LSH as a black box for the similarity join problem. It is worth noting that the standard LSHs for metric distances, including Hamming [12], ℓ_1 [7], ℓ_2 [1,7], Jaccard [4] and angular distances [5] are *monotone*. These common LSHs are space-efficient, and use space comparable to that required to store a point, except the LSH of [1] which requires space $N^{o(1)}$. We did not explicitly require the hash values themselves to be particularly small. However, using universal hashing we can always map to small bit strings while introducing no new collisions with high probability. Thus we assume that B hash values fit in one memory block.

Computational I/O Model. We study algorithms for similarity join in the *external memory model*, which has been widely adopted in the literature (see, e.g., the survey by Vitter [14]). The external memory model consists of an internal memory of M words and an external memory of unbounded size. The processor can only access data stored in the internal memory and move data between the two memories in blocks of size B. For simplicity we will here measure block and memory size in units of points from $\mathbb{U}$, such that a block can contain B points.

The *I/O complexity* of an algorithm is defined as the number of input/output blocks moved between the two memories by the algorithm. The *cache-aware* approach makes use of the parameter M explicitly to achieve its I/O complexity whereas the *cache-oblivious* one [8] does not explicitly use any model parameters. The latter is a desirable property as it implies optimality on all levels of the memory hierarchy and does not require parameter tuning when executed on different physical machines. The cache-oblivious model assumes that the internal memory is *ideal* in the sense that it has optimal cache-replacement policy that can evict the block that is used the farthest in the future, and also that a block can be placed anywhere in the cache (full associativity).

Similarity Join Techniques. We review some state-of-the-art of similarity join techniques most closely related to our work.

- **Index-based similarity join.** A popular approach is to make use of indexing techniques to build a data structure for one relation, and then perform queries using the points of the other relation. The indexes typically perform some kind of *filtering* to reduce the number of points that a given query point is compared to (see, e.g., [3,6]). Indexing can be space consuming, in particular for LSH, but in the context of similarity join this is not a big concern since we have many queries, and thus can afford to construct each hash table "on the fly". On the other hand, it is clear that index-based similarity join techniques will not be able to take significant advantage of internal memory when $N \gg M$. Indeed, the query complexity stated in [9] is $\mathcal{O}\left((N/B)^\rho\right)$ I/Os. Thus the I/O complexity of using indexing for similarity join will be high.

- **Sorting-based.** The indexing technique of [9] can be adapted to compute similarity joins more efficiently by using the fact that many points are being looked up in the hash tables. This means that all lookups can be done in a batched fashion using sorting. This results in a dependency on N that is $\tilde{\mathcal{O}}\left((N/B)^{1+\rho}\right)$ I/Os, where $\rho \in (0; 1)$ is a parameter of the LSH family.

- **Generic joins.** When N is close to M the I/O-complexity can be improved by using general join operators optimized for this case. It is easy to see that when N/M is an integer, a nested loop join requires $N^2/(MB)$ I/Os. Our cache-oblivious algorithm will make use of the following result on cache-oblivious nested loop joins:

Theorem 2. *(He and Luo [11]) For an arbitrary join condition, the join of relations R and S can be computed in $\mathcal{O}\left((|R| + |S|)/B + (|R||S|)/(MB)\right)$ I/Os by a cache-oblivious algorithm. This number of I/Os suffices to generate the result in memory, but may not suffice to write it to disk.*

3 Our Algorithms

In this section we describe our I/O efficient algorithms. We start in Section 3.1 with a warm-up cache-aware algorithm. It uses an LSH family where the value of the collision probability is set to be a function of the internal memory size. Section 3.2 presents our main result, a recursive and cache-oblivious algorithm which uses the LSH as a black-box approach and does not make any assumption on the value of collision probability. Section 3.3 describes the analysis and Section 3.4 shows how to reduce the expected number of times each near pair is emitted.

3.1 Cache-Aware Algorithm: ASimJoin

We will now describe a simple cache-aware algorithm called ASIMJOIN, which achieves the worst case I/O bounds as stated in Theorem 1. Due to the limit of space, we will sketch some intuitions of the algorithm and refer to [13] for the full discussion and omitted proof.

ASIMJOIN relies on an (r, cr, p_1', p_2')-sensitive family $\mathcal{H}'$ of hash functions with the following properties: $p_2' \leq M/N$ and $p_1' \geq (M/N)^\rho$, for a suitable value $0 < \rho < 1$. Given an arbitrary (r, cr, p_1, p_2)-sensitive family $\mathcal{H}$, the family $\mathcal{H}'$ can be built by concatenating $\lceil \log_{p_2}(M/N) \rceil$ hash functions from $\mathcal{H}$. For simplicity, we assume that $\log_{p_2}(M/N)$ is an integer and thus the probabilities p_1', p_2' can be exactly obtained. However, the algorithm and the analysis can be extended to the general case by increasing the I/O complexity by a factor at most p_1^{-1} in the worst case; in practical scenarios, this factor is a small constant [4,7,9].

Let R and S denote the input sets. The algorithm repeats $L = 1/p_1'$ times the following procedure. A hash function is randomly drawn from the (r, cr, p_1', p_2')-sensitive family, and it is used for partitioning the sets R and S into buckets of points with the same hash value. Then, the algorithm iterates through every hash value and, for each hash value v, it uses a double nested loop for generating all pairs of points in $R_v \times S_v$, where R_v and S_v denote the buckets respectively containing points of R and S with hash value v. A pair is emitted only if it is a near pair. For each input point $x \in R \cup S$, we maintain a counter that is increased every time a pair (x, y) is generated and y is far from x. The counter is maintained over all the L repetitions and keeps track of the number of collisions of x with its far points. As soon as the counter of a point exceeds $4LM$, the point is removed from the input set[2]. We refer to the pseudocode for more details.

By using the (r, cr, p_1', p_2')-sensitive family, ASIMJOIN guarantees that each point collides with at most M far points. Therefore, each point in $R_v \cup S_v$ is far from at most M points in $R_v \cup S_v$ (note that points in R and S are in the same universe and then collision probabilities apply independently of the belonging set). This implies that, if there are not too many near and c-near pairs, $R_v \bowtie_{\leq r} S_v$ can be efficiently computed with $\mathcal{O}(M/B)$ I/Os. On the other hand, if there are many near or c-near points, the I/O complexity can be upper bounded in an output sensitive way. In particular, we observe that a point cannot collide with too many far points since it is removed from the set after $4LM$ collisions with far points. Moreover, since each near pair has probability p_1' to be emitted by partitioning R and S with LSH, the process is repeated $L = 1/p_1'$ times.

Algorithm. ASimJoin(R, S): R, S are the input sets.

1   Associate to each point in S a counter initially set to 0;
2   **Repeat** $L = 1/p_1'$ times
3   Choose $h_i' \in \mathcal{H}'$ uniformly at random;
4   Use h_i' to partition (in-place) R and S in buckets R_v, S_v of points with hash value v;
5   **For** each hash value v generated in the previous step
6   /* For simplicity we assume that $|R_v| \leq |S_v|$ */
7   Split R_v and S_v into chunks $R_{i,v}$ and $S_{i,v}$ of size at most $M/2$;
8   **For** every chunk $R_{i,v}$ of R_v
9   Load in memory $R_{i,v}$;
10   **For** every chunk $S_{i,v}$ of S_v do
11   Load in memory $S_{i,v}$;
12   Compute $R_{i,v} \times S_{i,v}$ and emit all near pairs. For each far pair, increment the associated counters by 1;
13   Remove from $S_{i,v}$ and $R_{i,v}$ all points with the associated counter larger than $4LM$, and write $S_{i,v}$ back to external memory;
14   Write $R_{i,v}$ back to external memory;

3.2   Cache-Oblivious Algorithm: OSimJoin

The above cache-aware algorithm uses an (r, cr, p_1', p_2')-sensitive family of functions, with $p_1' \sim (M/N)^\rho$ and $p_2' \sim M/N$, for partitioning the initial sets into smaller buckets, which are then efficiently processed in the internal memory using the nested loop algorithm. As soon as the internal memory size M is known, this family of functions can be built by concatenating $\lceil \log_{p_2} p_2' \rceil$ hash functions from any given primitive (r, cr, p_1, p_2)-sensitive family. However, in the cache-oblivious settings the value of M is not known and such family cannot be built.

---

[2]   We observe that removing points that collide with at least $4LM$ points is only required for getting the claimed I/O complexity with high probability. The algorithm can be simplified by removing this operation, and yet obtaining the same I/O bound in expectation.

Therefore, we propose in this section an algorithm, named OSimJoin, that efficiently compute the similarity join even without knowing the values of the internal memory size M and the block length B, and uses as a black-box a given monotonic (r, cr, p_1, p_2)-sensitive family of functions[3]. The value of p_1 and p_2 can be considered constant in practical scenario.

As common in the cache-oblivious settings, we use a recursive approach for splitting the problem into smaller and smaller subproblems that at some point will fit the internal memory, although this point is not known in the algorithm. We first give a high level description of the cache-oblivious algorithm and an intuitive explanation. We then provide a more detailed description and analysis.

Algorithm. OSimJoin(R, S, ψ): R, S are the input sets, and ψ is the recursion depth.

1 **If** $|R| > |S|$, **then** swap (the references to) the sets such that $|R| \leq |S|$;
2 **If** $\psi = \Psi$ or $|R| \leq 1$, **then** compute $R \bowtie_{\leq r} S$ using the algorithm of Theorem 2 and return;
3 Pick a random sample S' of 18Δ points from S (or all points if $|S| < 18\Delta$);
4 Compute R' containing all points of R that have distance smaller than cr to at least half points in S'. Permute R such that points in R' are in the first positions;
5 Compute $R' \bowtie_{\leq r} S$ using the algorithm of Theorem 2;
6 **Repeat** $L = 1/p_1$ times
7   Choose $h \in \mathcal{H}$ uniformly at random;
8   Use h to partition (in-place) $R \backslash R'$ and S in buckets R_v, S_v of points with hash value v;
9   **For** each v where R_v and S_v are nonempty, recursively call OSimJoin $(R_v, S_v, \psi + 1)$;

OSimJoin receives in input the two sets R and S of similarity join, and a parameter ψ denoting the depth in the recursion tree (initially, $\psi = 0$) that is used for recognizing the base case. Let $|R| \leq |S|$, $N = |R| + |S|$, and denote with $\Delta = \log N$ and $\Psi = \lceil \log_{1/p_2} N \rceil$ two global values that are kept invariant in the recursive levels and computed using the initial input size N. For the sake of simplicity, we assume that $1/p_1$ and $1/p_2$ are integers, and further assume without loss of generality that the initial size N is a power of two. Note that, if $1/p_1$ is not integer, that the last iteration can be performed with probability $1/p_1 - \lfloor 1/p_1 \rfloor$, such that $L \in \{\lfloor 1/p_1 \rfloor, \lceil 1/p_1 \rceil\}$ and $\mathbb{E}[L] = 1/p_1$.

OSimJoin works as follows. If the problem is currently at recursive level $\Psi = \lceil \log_{1/p_2} N \rceil$ or $|R| \leq 1$, the recursion ends and the problem is solved using the cache-oblivious nested loop described in Theorem 2. Otherwise the following operations are executed. By exploiting sampling, the algorithm identifies a subset R' of R containing (almost) all points that are near or c-near to a constant fraction of points in S. More specifically, the set R' is computed by creating a random sample S' of S of size 18Δ and then adding to R' all points in R that have

---

[3] The monotonicity requirement can be relaxed to the following: $\Pr[h(x) = h(y)] \geq \Pr[h(x') = h(y')]$ for every two pairs (x, y) and (x', y') where $d(x, y) \leq r$ and $d(x', y') > r$. A monotone LSH family clearly satisfies this assumption.

distance at most cr to at least half points in S'. The join $R' \bowtie_{\leq r} S$ is computed by using the cache-oblivious nested-loop of Theorem 2 and then points in R' are removed from R. Subsequently, the algorithm repeats $L = 1/p_1$ times the following operations: a hash function is extracted from the (r, cr, p_1, p_2)-sensitive family and used for partitioning R and S into buckets, denoted with R_v and S_v with any hash value v; then, the join $R_v \bowtie_{\leq r} S_v$ is computed recursively.

The explanation of our approach is the following. By recursively partitioning input points with hash functions from an (r, cr, p_1, p_2)-sensitive family, the algorithm decreases the probability of collision between two far points. In particular, the collision probability of two far points is p_2^i at the i-th recursive level. On the other hand, by repeating the partitioning $1/p_1$ times in each level, the algorithm guarantees that a pair of near points is enumerated with constant probability since the probability that two near points collide is p_1^i at the i-th recursive level. It deserves to be noticed that the collision probability of far and near points at the recursive level $\log_{1/p_2}(N/M)$ is $\Theta(M/N)$ and $\Theta((M/N)^\rho)$, respectively, which are asymptotically equivalent to the values in the cache-aware algorithm. In other words, the partitioning of points at this level is equivalent to the one in the cache-aware algorithm, being the expected number of colliding far points is M. Finally, we observe that, when a point in R becomes close to many points in S, it is more efficient to detect and remove it, instead of propagating it down to the base cases. Indeed, it may happen that the collision probability of these points is large (close to 1) and the algorithm is not able to split them into subproblems that fit in memory.

3.3   I/O Complexity and Correctness of OSimJoin

Analysis of I/O Complexity. We will bound the *expected* number of I/Os of the algorithm rather than the worst case. This can be converted to a fixed time bound by a standard technique of restarting the computation when the expected number of I/Os is exceeded by a factor 2. To succeed with probability $1 - 1/N$ it suffices to do $\mathcal{O}(\log N)$ restarts to complete within twice the expected time bound, and the logarithmic factor is absorbed in the $\tilde{\mathcal{O}}$-notation. If the computation does not succeed within this bound we fail to produce an output, slightly increasing the error probability.

For notational simplicity, in this section we let R and S denote the initial input sets and let $\tilde{R}$ and $\tilde{S}$ denote the subsets given in input to a particular recursive subproblem (note that $\tilde{R}$ can be a subset of R but also of S; similarly for $\tilde{S}$). We also let $\tilde{S}'$ denote the sampling of $\tilde{S}$ in Step 3, and with $\tilde{R}'$ the subset of $\tilde{R}$ computed in Step 4. Our first lemma says that two properties of the choice of random sample in Step 3 are almost certain. The proof relies on Chernoff bounds on the choice of $\tilde{S}'$. See [13] for the omitted proof.

Lemma 1. *Consider a run of Steps 3 and 4 in a subproblem* $\text{OSimJoin}(\tilde{R}, \tilde{S}, \psi)$, *for any level* $0 \leq \psi \leq \Psi$. *Then with probability at least* $1 - \mathcal{O}(1/N)$ *over the choice of sample* $\tilde{S}'$ *we have:*

$$|\tilde{R}' \underset{\leq cr}{\bowtie} \tilde{S}| > \frac{|\tilde{R}'||\tilde{S}|}{6} \qquad and \qquad |(\tilde{R}\backslash\tilde{R}') \underset{>cr}{\bowtie} \tilde{S}| > \frac{5|\tilde{R}\backslash\tilde{R}'||\tilde{S}|}{6} .$$

In the remainder of the paper, we assume that Lemma 1 holds and refer to this event as $\mathcal{A}$. By the above, $\mathcal{A}$ holds with probability $1 - \mathcal{O}(1/N)$.

To analyze the number of I/Os for subproblems of size more than M we bound the cost in terms of different types of *collisions*, i.e., pairs in $R \times S$ that end up in the same subproblem of the recursion. We say that (x, y) *is in* a particular subproblem $\text{OSimJoin}(\tilde{R}, \tilde{S}, \psi)$ if $(x, y) \in (\tilde{R} \times \tilde{S}) \cup (\tilde{S} \times \tilde{R})$. Observe that a pair (x, y) is in a subproblem if and only if x and y have colliding hash values on every step of the call path from the initial invocation of OSimJoin.

Definition 2. *Given $Q \subseteq R \times S$ let $C_i(Q)$ be the number of times a pair in Q is in a call to OSimJoin at the ith level of recursion. We also let $C_{i,k}(Q)$, with $0 \leq k \leq \log M$ denote the number of times a pair in Q is in a call to OSimJoin at the ith level of recursion where the smallest input set has size in $[2^k, 2^{k+1})$ if $0 \leq k < \log M$, and in $[M, +\infty)$ if $k = \log M$. The count is over all pairs and with multiplicity, so if (x, y) is in several subproblems at the ith level, all these are counted.*

Next we bound the I/O complexity of OSimJoin in terms of $C_i(R \bowtie_{\leq cr} S)$ and $C_{i,k}(R \bowtie_{>cr} S)$, for any $0 \leq i < \Psi$. These quantities are then bounded in Lemma 3. Due to the space constraint, refer to [13] for the proof details.

Lemma 2. *Let $\ell = \lceil \log_{1/p_2}(N/M) \rceil$ and $M \geq 18 \log N + 3B$. Given that $\mathcal{A}$ holds, the I/O complexity of $\text{OSimJoin}(R, S, 0)$ is*

$$\tilde{\mathcal{O}}\left(\frac{NL^\ell}{B} + \sum_{i=0}^{\ell} \frac{C_i\left(R \underset{\leq cr}{\bowtie} S \right)}{MB} + \sum_{i=\ell}^{\Psi-1} \sum_{k=0}^{\log M} \frac{C_{i,k}\left(R \underset{>cr}{\bowtie} S \right) L}{B2^k} \right)$$

Proof. (Sketch) The proof of this lemma consists of bounding the I/O complexity of each step as a function of the number of c-near or far collisions. The first two terms give the cost of all subproblems at levels above ℓ: the first term is due to Step 5 and follows by expressing the cost in Theorem 2 in terms of c-near collision through Lemma 1; the second term follows from a simple analysis of Steps 7-8. The last term is the cost of levels below ℓ and follows by expressing the I/O complexity in terms of far collisions within subproblems of size in $[2^k, 2^{k+1})$ for any $k \geq 0$. The cost of level ℓ is asymptotically negligible compared to the other cases. □

We will now analyze the expected sizes of the terms in Lemma 2. Clearly each pair from $R \times S$ is in the top level call, so the number of collisions is $|R||S| < N^2$. But in lower levels we show that the expected number of times that a pair collides either decreases or increases geometrically, depending on whether the collision probability is smaller or larger than p_1 (or equivalently, depending on whether the distance is greater or smaller than the radius r). The lemma follows by expressing the number of collisions of the pairs at the ith recursive level as a *Galton-Watson branching process* [10]. See [13] for proof details.

Lemma 3. *Given that $\mathcal{A}$ holds, for each $0 \le i \le \Psi$ we have*

1. $\mathbb{E}\left[C_i\left(R \underset{>cr}{\bowtie} S\right)\right] \le |R \underset{>cr}{\bowtie} S| \, (p_2/p_1)^i$;

2. $\mathbb{E}\left[C_i\left(R \underset{>r,\le cr}{\bowtie} S\right)\right] \le |R \underset{>r,\le cr}{\bowtie} S|$;

3. $\mathbb{E}\left[C_i\left(R \underset{\le r}{\bowtie} S\right)\right] \le |R \underset{\le r}{\bowtie} S| \, L^i$;

4. $\mathbb{E}\left[C_{i,k}\left(R \underset{>cr}{\bowtie} S\right)\right] \le N 2^{k+1} (p_2/p_1)^i$, *for any* $0 \le k < \log M$.

We are now ready to prove the I/O complexity of OSIMJOIN as claimed in Theorem 1. By the linearity of expectation and Lemma 2, we get that the expected I/O complexity of OSIMJOIN is

$$
\tilde{O}\left(\frac{NL^\ell}{B} + \sum_{i=0}^{\ell} \frac{\mathbb{E}\left[C_i\left(R \underset{\le cr}{\bowtie} S\right)\right]}{MB} + \sum_{i=\ell}^{\Psi-1} \sum_{k=0}^{\log M} \frac{\mathbb{E}\left[C_{i,k}\left(R \underset{>cr}{\bowtie} S\right)\right] L}{B2^k} \right) ,
$$

where $\ell = \lceil \log_{1/p_2}(N/M) \rceil$. By noticing $C_{i,\log M}\left(R \bowtie_{>cr} S\right) \le C_i\left(R \bowtie_{>cr} S\right)$ we have $|R \bowtie_{>cr} S| \le N^2$ and $C_i\left(R \bowtie_{\le cr} S\right) = C_i\left(R \bowtie_{\le r} S\right) + C_i\left(R \bowtie_{>r,\le cr} S\right)$, and by plugging in the bounds on the expected number of collisions given in Lemma 3, we get the claimed result.

Analysis of Correctness. We now argue that a pair (x, y) with $d(x, y) \le r$ is output with good probability. Let $X_i = C_i((x, y))$ be the number of subproblems at level i containing (x, y). By applying Galton-Watson branching process, we get that $\mathbb{E}[X_i] = (\Pr[h(x) = h(y)]/p_1)^i$. If $\Pr[h(x) = h(y)]/p_1 > 1$ then in fact there is positive constant probability that (x, y) survives indefinitely, i.e., does not go extinct [10]. Since at every branch of the recursion we eventually compare points that collide under all hash functions on the path from the root call, this implies that (x, y) is reported with positive constant probability.

In the *critical case* where $\Pr[h(x) = h(y)]/p_1 = 1$ we need to consider the variance of X_i, which by [10, Theorem 5.1] is equal to $i\sigma^2$, where σ^2 is the variance of the number of children (hash collisions in recursive calls). If $1/p_1$ is integer the number of children in our branching process follows a binomial distribution with mean 1. This implies that $\sigma^2 < 1$. Also in the case where $1/p_1$ is not integer it is easy to see that the variance is bounded by 2. That is, we have $\text{Var}(X_i) \le 2i$, which by Chebychev's inequality means that for some integer $j^* = 2\sqrt{i} + \mathcal{O}(1)$:

$$
\sum_{j=j^*}^{\infty} \Pr[X_i \ge j] \le \sum_{j=j^*}^{\infty} \text{Var}(X_i)/j^2 \le 1/2 .
$$

Since we have $\mathbb{E}[X_i] = \sum_{j=1}^{\infty} \Pr[X_i \ge j] = 1$ then $\sum_{j=1}^{j^*-1} \Pr[X_i \ge j] > 1/2$, and since $\Pr[X_i \ge j]$ is non-increasing with j this implies that $\Pr[X_i \ge 1] \ge$

$1/(2j^*) = \Omega\left(1/\sqrt{i}\right)$. Since recursion depth is $\mathcal{O}\left(\log N\right)$ this implies the probability that a near pair is found is $\Omega\left(1/\sqrt{\log N}\right)$. Thus, by repeating $\mathcal{O}\left(\log^{3/2} N\right)$ times we can make the error probability $\mathcal{O}\left(1/N^3\right)$ for a particular pair and $\mathcal{O}\left(1/N\right)$ for the entire output by applying the union bound.

3.4 Removing Duplicates

The definition of LSH requires the probability $p(x,y) = \Pr\left[h(x) = h(y)\right]$ of two near points x and y of being hashed on the same value is at least p_1. If $p(x,y) \gg p_1$, our OSIMJOIN algorithm can emit (x,y) many times. As an example suppose that the algorithm ends in one recursive call: then, the pair (x,y) is expected to be in the same bucket for $p(x,y)L$ iterations of Step 6 and thus it is emitted $p(x,y)L \gg 1$ times in expectation. Moreover, if the pair is not emitted in the first recursive level, the expected number of emitted pairs increases as $(p(x,y)L)^i$ since the pair (x,y) is contained in $(p(x,y)L)^i$ subproblems at the ith recursive level. A simple solution requires to store all emitted near pairs on the external memory, and then using a cache-oblivious sorting algorithm [8] for removing repetitions. However, this approach requires $\tilde{\mathcal{O}}\left(\kappa|R \bowtie_{\leq_r} S|/B\right)$ I/Os, where κ is the expected average replication of each emitted pair, which can dominate the complexity of OSIMJOIN. A similar issue appears in the cache-aware algorithm ASIMJOIN as well: however, a near pair is emitted in this case at most $L' = (N/M)^\rho$ since there is no recursion and the partitioning of the two input sets is repeated only L' times.

If the collision probability $\Pr\left[h(x) = h(y)\right]$ can be explicitly computed in $\mathcal{O}\left(1\right)$ time and no I/Os for each pair (x,y), it is possible to emit each near pair once in expectation without storing near pairs on the external memory. We note that the collision probability can be computed for many metrics, including Hamming [12], ℓ_1 and ℓ_2 [7], Jaccard [4], and angular [5] distances. For the cache-oblivious algorithm, the approach is the following: for each near pair (x,y) that is found at the ith recursive level, with $i \geq 0$, the pair is emitted with probability $1/(p(x,y)L)^i$ and is ignored otherwise. For the cache-aware algorithm, the idea is the same but a near pair is emitted with probability $1/(p(x,y)L')$ with $L' = (N/M)^\rho$. The proof of the claim is provided in [13].

We observe that the proposed approach is equivalent to use an LSH where $p(x,y) = p_1$ for each near pair. Finally, we remark that this approach does not avoid replica of the same near pair when the algorithm is repeated for increasing the collision probability of near pairs.

4 Conclusion

In this paper we examine the problem of computing the similarity join of two relations in an external memory setting. Our new cache-aware algorithm of Section 3.1 and cache-oblivious algorithm of Section 3.2 improve upon current state of the art by around a factor of $(M/B)^\rho$ I/Os unless the number of c-near pairs is huge (more than NM). We believe this is the first cache-oblivious algorithm for similarity join, and more importantly the first subquadratic algorithm whose I/O performance improves significantly when the size of internal memory grows.

It would be interesting to investigate if our cache-oblivious approach is also practical — this might require adjusting parameters such as L. Our I/O bound is probably not easy to improve significantly, but interesting open problems are to remove the error probability of the algorithm and to improve the implicit dependence on dimension in B and M: In this paper we assume for simplicity that the unit of M and B is number of points, but in general we may get tighter bounds by taking into account the gap between the space required to store a point and the space for e.g., hash values. Also, the result in this paper is made with general spaces in mind and it is an interesting direction to examine if the dependence on dimension could be made explicit and improved in specific spaces.

References

1. Andoni, A., Indyk, P.: Near-optimal hashing algorithms for approximate nearest neighbor in high dimensions. In: Proceedings of FOCS 2006, pp. 459–468 (2006)
2. Arasu, A., Ganti, V., Kaushik, R.: Efficient exact set-similarity joins. In: Proceedings of VLDB 2006, pp. 918–929 (2006)
3. Bayardo, R.J., Ma, Y., Srikant, R.: Scaling up all pairs similarity search. In: Proceedings of WWW 2007, pp. 131–140 (2007)
4. Broder, A.Z., Glassman, S.C., Manasse, M.S., Zweig, G.: Syntactic clustering of the web. Computer Networks 29(8-13), 1157–1166 (1997)
5. Charikar, M.S.: Similarity estimation techniques from rounding algorithms. In: Proceedings of STOC 2002, pp. 380–388 (2002)
6. Chaudhuri, S., Ganti, V., Kaushik, R.: A primitive operator for similarity joins in data cleaning. In: Proceedings of ICDE 2006, p. 5 (2006)
7. Datar, M., Immorlica, N., Indyk, P., Mirrokni, V.S.: Locality-sensitive hashing scheme based on p-stable distributions. In: Proceedings of SOCG 2004, pp. 253–262 (2004)
8. Frigo, M., Leiserson, C.E., Prokop, H., Ramachandran, S.: Cache-oblivious algorithms. In: Proceedings of FOCS 1999, pp. 285–297 (1999)
9. Gionis, A., Indyk, P., Motwani, R.: Similarity search in high dimensions via hashing. In: Proceedings of VLDB 1999, pp. 518–529 (1999)
10. Harris, T.E.: The theory of branching processes. Courier Dover Publications (2002)
11. He, B., Luo, Q.: Cache-oblivious nested-loop joins. In: Proceedings of CIKM 2006, pp. 718–727 (2006)
12. Indyk, P., Motwani, R.: Approximate nearest neighbors: Towards removing the curse of dimensionality. In: Proceedings of STOC 1998, pp. 604–613 (1998)
13. Pagh, R., Pham, N., Silvestri, F., Stöckel, M.: I/O-efficient similarity join. Full version, arXiv:1507.00552 (2015)
14. Vitter, J.S.: Algorithms and Data Structures for External Memory. Now Publishers Inc. (2008)
15. Xiao, C., Wang, W., Lin, X., Yu, J.X.: Efficient similarity joins for near duplicate detection. In: Proceedings of WWW 2008, pp. 131–140 (2008)

Improved Approximation Algorithms
for Weighted 2-Path Partitions*

Amotz Bar-Noy[1], David Peleg[2], George Rabanca[1], and Ivo Vigan[1]

[1] Department of Computer Science, Graduate Center,
City University of New York, New York, United States
[2] Department of Computer Science and Applied Mathematics,
Weizmann Institute, Rehovot, Israel

Abstract. We investigate two NP-complete vertex partition problems on edge weighted complete graphs with $3k$ vertices. The first problem asks to partition the graph into k vertex disjoint paths of length 2 (referred to as 2-*paths*) such that the total weight of the paths is maximized. We present a cubic time approximation algorithm that computes a 2-path partition whose total weight is at least .5833 of the weight of an optimal partition; improving upon the $(.5265 - \epsilon)$-approximation algorithm of [26]. Restricting the input graph to have edge weights in $\{0, 1\}$, we present a .75 approximation algorithm improving upon the .55-approximation algorithm of [16].

Combining this algorithm with a previously known approximation algorithm for the 3-SET PACKING problem, we obtain a .6-approximation algorithm for the problem of partitioning a $\{0, 1\}$-edge-weighted graph into k vertex disjoint triangles of maximum total weight. The best known approximation algorithm for general weights achieves an approximation ratio of .5257 [4].

1 Introduction

Let $G = (V, E, \omega)$ be a complete graph on $3k$ vertices, having non-negative edge weights. The MAXIMUM WEIGHT 2-PATH PARTITION (M2PP) problem asks to compute a set of k vertex disjoint paths of length 2 (referred to as 2-paths) such that the sum of the weights of the paths is maximized. The MAXIMUM WEIGHT TRIANGLE PARTITION (MTP) problem asks to compute a set of k vertex disjoint cycles of length 3 (referred to as triangles) such that the sum of the edge weights of the k cycles is maximized.

---

* This work is supported by the Army Research Laboratory and was accomplished under Cooperative Agreement Number W911NF-09-2-0053, The views and conclusions contained in this document are those of the authors and should not be interpreted as representing the official policies, either expressed or implied, of the Army Research Laboratory or the U.S. Government. The U.S. Government is authorized to reproduce and distribute reprints for Government purposes notwithstanding any copyright notation here on.

© Springer-Verlag Berlin Heidelberg 2015
N. Bansal and I. Finocchi (Eds.): ESA 2015, LNCS 9294, pp. 953–964, 2015.
DOI: 10.1007/978-3-662-48350-3_79

Our main contribution is a 7/12-approximation algorithm for the M2PP problem. We also investigate $\{0, 1\}$-edge-weighted graphs – in which the weights of the edges of G are either 0 or 1 – and we present a .75-approximation algorithm for the M2PP problem and a .6-approximation algorithm for the MTP problem in this setting.

Both the M2PP and MTP problems have been studied before, mostly under the names MAXIMUM 2-PATH PACKING and MAXIMUM TRIANGLE PACKING respectively. Unfortunately, these names have also been used for the related but different problems defined below. Consequently, we use separate terminology to make a clear distinction between the packing and partitioning settings.

Given an unweighted graph H, a 2-PATH PACKING of H is a collection of vertex disjoint 2-paths and a TRIANGLE PACKING of H is a collection of vertex disjoint triangles. Such a collection is called *perfect* if it uses all the vertices of H. The MAXIMUM 2-PATH PACKING problem asks to find a 2-PATH PACKING of maximum cardinality. The MAXIMUM TRIANGLE PACKING problem is defined similarly.

The difference between packing and partitioning is that in the partitioning setting all edges exist (though some may have weight 0). Thus, any collection of disjoint sets of 3 vertices forms a valid partition of H into e.g. triangles, while such an arbitrary collection may not be a valid triangle packing.

Related Work. In their classic book [10], Garey and Johnson show that deciding whether a graph admits a perfect TRIANGLE PACKING or a perfect 2-PATH PACKING is NP-complete (p. 68 and 76 respectively). More general results on the NP-completeness of packing families of graphs into a given graph are provided in [17] and [22]. Both the MAXIMUM 2-PATH PACKING and MAXIMUM TRIANGLE PACKING problems are special cases of the unweighted 3-SET PACKING problem for which [18] (also see [12]) presents a local search algorithm that achieves a $\frac{2}{3} - \epsilon$ approximation (with $\epsilon > 0$).

In [19] it is shown that the MAXIMUM TRIANGLE PACKING is APX-hard even in graphs of maximum degree 4 and in [7] it is shown to be NP-hard to approximate within a factor of .9929. Moreover, [11] shows that the problem remains NP-complete even when restricted to the families of chordal, planar, line or total graphs. A .833-approximation algorithm for graphs with maximum degree 4 is presented in [23]. For the MAXIMUM 2-PATH PACKING problem, [24] presents a fixed parameter tractable algorithm and [2] presents an approximation algorithm for an edge weighted version of the problem.

The M2PP and MTP problems studied in this paper are special cases of the weighted 3-SET PACKING problem for which [1] presents a $\frac{1}{2} - \epsilon$ approximation algorithm. In [14], Hassin et al. observe that there exists a simple reduction from M2PP (respectively, MTP) to the problem of deciding whether a graph has a perfect 2-PATH PACKING (resp., TRIANGLE PACKING), implying that the two problems are NP-complete. Moreover, they present a randomized $\frac{35}{67} - \epsilon \approx .5222$ approximation algorithm for M2PP and in [26] this algorithm is refined and derandomized, leading to an improved approximation ratio of $.5265 - \epsilon$; a simpler analysis was presented in [27]. For the MTP problem, [14] (see also Erratum [15]) presents a $\frac{43}{83} - \epsilon \approx .518$ approximation algorithm and [5] (see also Erratum [6]) presents a randomized approximation algorithm which achieves a ratio of .5257. For $\{0, 1\}$-edge-weighted graphs, Hassin and Schneider [16] present a local search based .55-approximation algorithm for M2PP which runs in time $O(|V|^{10})$. In [13] the authors study the problem of partitioning a complete weighted graph into paths of length 3 and present a .75-approximation algorithm.

Contribution. In this paper we present a simple, matching based $7/12 \approx .583$ approximation algorithm for the M2PP problem on graphs with general non-negative weights, improving upon the $(.5265 - \epsilon)$-approximation algorithm of [26]. Besides improving the approximation ratio, our algorithm is significantly less computationally intensive: the algorithm of [26] runs in time exponential in $1/\epsilon$ while our algorithm runs in cubic time. Moreover, for $\{0, 1\}$-edge-weighted graphs we provide a .75-approximation algorithm for the M2PP problem improving upon the .55-approximation algorithm of [16]. The core idea of our algorithms is adapted from [13], where the authors show how to partition a complete weighted graph into paths of length 3. For a complete graph on $n = 3k$ vertices we prove the following two theorems in Section 2 and Section 3 respectively.

Theorem 1. *There exists a $7/12$-approximation algorithm for the M2PP problem running in time $O(k^3)$.*

Theorem 2. *For $\{0, 1\}$-edge-weighted graphs there exists a $3/4$-approximation algorithm for the M2PP problem running in time $O(k^3)$.*

In Section 4 we show how an approximation algorithm for M2PP can be combined with a 3-SET PACKING approximation algorithm to obtain an approximate solution for the MTP on $\{0, 1\}$-edge-weighted graphs. We are not aware of previous results for the MTP problem restricted to this case.

Theorem 3. *For $\{0, 1\}$-edge-weighted graphs there exists a $5/8$-approximation algorithm for the MTP problem running in time $O(k^3)$.*

Motivation. Besides being interesting variants of the MAXIMUM 2-PATH PACKING and MAXIMUM 2-PATH PACKING problems, M2PP and MTP are natural special cases of the TEAM FORMATION problem. Given a social network of experts, the TEAM FORMATION problem asks to find the most cohesive team. Some authors [9], [20], [21], [25] measure the cohesiveness of a team as the number of connections within a team, while others [20] consider only connections between a team leader and the other team members. Experimental evidence presented in [3] suggests that indeed, not all ties between team members are of equal importance, and maximizing the connection between the team leader and the rest of the group suffices. If, as in [20], we are interested in forming multiple teams, the TEAM FORMATION problem for teams of size 3 can be cast as either M2PP or MTP, depending on whether we are forming teams with or without a leader respectively .

2   M2PP in General Graphs

In this section we present and analyze the WEIGHTEDDOUBLEMATCHING approximation algorithm for the M2PP problem in graphs with general non-negative weights. For ease of presentation we restrict our attention to the case when k is even. The case when k is odd can be solved similarly.

Some of the graphs we argue about in this section may refer to non-simple graphs that have parallel edges and self-loops. To avoid confusion we begin by formally defining the notion of a graph as used below. A *graph* $G = (V, E, \gamma)$ is a set of vertices V together with a set of edges E and a function $\gamma : E \to \{\{u, v\} : u, v \in V\}$. For an edge $e \in E$ with $\gamma(e) = \{u, v\}$ we call the vertices u and v the *endpoints* of the edge e. Two edges $e_1, e_2 \in E$ are called *parallel* if $\gamma(e_1) = \gamma(e_2)$, and an edge $e \in E$ is called a *loop* if the two endpoints of e coincide. For clarity, we sometimes use $V(G)$ and $E(G)$ to denote the vertex and edge set of the graph G respectively. We say that the graph G is *complete* if for any vertices $u \neq v$ in $V(G)$ there exists an edge $e \in E(G)$ such that $\gamma(e) = \{u, v\}$, and we say that the graph G is *simple* if it contains no loops or parallel edges. In an edge weighted graph each edge $e \in E$ is associated with a weight $\omega(e)$. We slightly abuse notation by using $\omega(A)$ to denote the sum of the weights of the edges in an edge set A, and by using $V(A)$ to denote the set of endpoints of edges in A. Moreover, we use $\omega(G)$ to denote the sum of the weights of the edges in $E(G)$. For a set of 2-paths Π we use $E(\Pi)$ to denote the set of edges used by the 2-paths of Π and $\omega(\Pi)$ to denote the sum of the weights of the edges in $E(\Pi)$. For an edge $e \in E$, let V_e, E_e and γ_e be defined as follows:

- $V_e = (V \setminus \gamma(e)) \cup \{v_e\}$ for some new vertex $v_e \notin V$. We say that e is the edge of G *corresponding* to v_e, and vice versa.
- $E_e = E \setminus \{e\}$
- for any edge $f \in E_e$, $\gamma_e(f)$ is defined as:

$$\gamma_e(f) = \begin{cases} \gamma(f) & \text{if } \gamma(f) \cap \gamma(e) = \emptyset, \\ (\gamma(f) \setminus \gamma(e)) \cup \{v_e\} & \text{otherwise.} \end{cases}$$

We define $G/e = (V_e, E_e, \gamma_e)$, the graph resulting from *contracting* the edge e. For a set of edges A we denote by G/A the graph obtained from G by sequentially contracting all edges of A.

2.1   The WEIGHTEDDOUBLEMATCHING Algorithm

The WEIGHTEDDOUBLEMATCHING algorithm takes as input a weighted complete simple graph $G = (V, E, \omega_G, \gamma_G)$, with $|V| = 3k$, and in the first step it computes a perfect maximum weight matching M of G. If needed, M may use edges of weight 0 so that all vertices are matched. Therefore, the size of M is $3k/2$. In the second step, it contracts the edges of M to obtain a graph H and assigns to an edge e in H the weight

$$\omega_H(e) \equiv \omega_G(e) - \min\{\omega_G(a), \omega_G(b)\},$$

where a and b are the edges in M corresponding to the endpoints of e. In the next step, a maximum weight matching N of size exactly $k/2$ is computed in H based on the edge weight function ω_H. Observe that some of the edges in N may have negative weights in H. The matching N dictates how the edges of M are combined into the output 2-paths:

- for each edge $e \in N$, let $a, b \in M$ be the edges corresponding to the endpoints of e, with $\omega_G(a) \geq \omega_G(b)$; create a 2-path $\{a, e\}$, and call the vertex of b not incident to e a *residual vertex*.
- for each edge $a \in M$ corresponding to a vertex of H not matched by N, create a 2-path from a and an arbitrary residual vertex.

We denote by Π_1 the set of 2-paths created that contain an edge of N and the remaining 2-paths by Π_2. Notice that the size of both Π_1 and Π_2 is $k/2$. The algorithm outputs the set of 2-paths $\mathcal{A} = \Pi_1 \cup \Pi_2$. A more formal description of the algorithm is presented in the full paper.

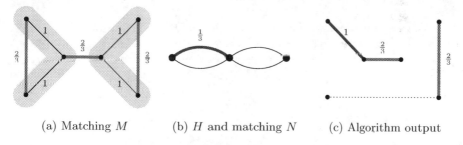

(a) Matching M (b) H and matching N (c) Algorithm output

Fig. 1. Illustration of a tight example.

Example. In Fig. 1 we illustrate the main steps of the **WEIGHTEDDOUBLEMA-TCHING** algorithm on a simple complete graph on six vertices (therefore $k = 2$). Fig. 1(a) shows the complete graph G with the zero weight edges omitted, and for each non-zero edge its corresponding weight. The optimal solution of weight 4 is drawn in light gray and a possible maximum weight matching M is displayed in bold (red). Fig. 1(b) shows the positive weight edges of the graph H. All the illustrated edges have weight 1 in the graph G and have weight $1 - 2/3$ in H. A maximum matching N of size $k/2 = 1$ is shown in bold (blue). Finally, Fig. 1(c) shows a possible output of the algorithm, with the dotted segment denoting an edge of zero weight. The weight of the output solution is $7/3$ which is a $7/12$ fraction of the weight of the optimal solution.

Algorithm Analysis. Since the most time consuming part of the **WEIGHTED-DOUBLEMATCHING** algorithm is computing the matchings M and N, the algorithm runs in time $O(k^3)$ [8].

We denote by OPT the set of 2-paths in a fixed optimal solution and in the remaining of this section show that the algorithm outputs a solution of weight at least $7/12 \cdot \omega_G(OPT)$. In Lemma 1 we show that the weight of the output solution is at least the weight of the matching M in G plus the weight of the matching N in H, i.e., $\omega_G(\mathcal{A}) \geq \omega_G(M) + \omega_H(N)$. In Lemma 2 we lower bound $\omega_H(E(OPT) \setminus M)$, the weight of the edges of the optimal solution in the graph H, by $\omega_G(E(OPT)) - 4/3 \cdot \omega_G(M)$; in Lemma 3 we show that the weight of the matching N in H is at least $\omega_H(E(OPT) \setminus M)/4$. Combining these two bounds it follows that $\omega_H(N) \geq \omega_G(OPT)/4 - \omega_G(M)/3$. Furthermore, it is easy to see that the weight of the matching M is at least half of the weight of the optimal solution since one can form a matching by selecting the heaviest edge of each 2-path in OPT and arbitrarily match the remaining vertices. Combining these results,

$$\omega_G(\mathcal{A}) \geq \omega_G(M) + \omega_H(N) \geq 2\omega_G(M)/3 + \omega_G(OPT)/4 \geq 7/12 \cdot \omega_G(OPT),$$

thus proving Theorem 1.

Lemma 1. $\omega_G(\mathcal{A}) \geq \omega_G(M) + \omega_H(N)$.

Proof sketch. First observe that all the edges of N are part of a 2-path in the output solution. Moreover, each edge $b \in M$ that does not appear in the output solution is adjacent to an edge $e \in N$ whose weight in H is $\omega_H(e) = \omega_G(e) - \omega_G(b)$. Therefore, the weight of e in H accounts for the absence of edge b in the output solution.

Lemma 2. $\omega_H(E(OPT) \setminus M) \geq \omega_G(E(OPT)) - 4/3 \cdot \omega_G(M)$.

In the full version of the paper we prove Lemma 2 by formalizing the intuition that, since the edges of OPT in H form a sparse graph, it is not possible that a small number of heavy edges of M correspond to the endpoints of many edges of OPT in H, which would make the total weight of OPT in H much smaller than the weight of OPT in G.

Lemma 3. *The graph H of the* **WEIGHTEDDOUBLEMATCHING** *algorithm has a matching of size $k/2$ of weight at least $\omega_H(E(OPT) \setminus M)/4$.*

Proof sketch. Let A be a set of k edges of $E(OPT)$ such that A is a matching of G and $A \cap M = \emptyset$. Then the edges of $A \cup M$ form a collection of disjoint alternating paths and cycles in G, which implies that the edges of A form a collection of disjoint paths and cycles in H. If all the cycles of A in H were of even length, then one could split A into two matchings of H, each of size $k/2$, and therefore show that there exists a matching of H with at least half the weight of A. However, although the cycles of $A \cup M$ in G have even length (since they are alternating), there is no guarantee that after contracting the edges of M the cycles formed by A are still of even length. In the full paper we show that there exists a matching A of G that "almost always" uses the heaviest edge from each optimal 2-path and moreover, the edges of $A \cup M$ do not contain any cycles and therefore form a collection of disjoint alternating paths.

3   M2PP on $\{0, 1\}$-Edge-Weighted Graphs

In this section we describe and analyze the DOUBLEMATCHING approximation algorithm for the M2PP problem in graphs in which the weights of the edges are either 0 or 1. The main difference to the algorithm described in section 2 is that the first matching M may not be perfect. The following steps are slightly more involved since after contracting the edges of M, one has to distinguish between those vertices that correspond to edges of M and those that don't.

The DOUBLEMATCHING algorithm operates in the following four phases. First, it finds a maximum weight matching M of G of minimum cardinality. In other words, M is a maximum weight matching of G that does not have any weight 0 edges. Next, it constructs a (complete) graph H by contracting the edges of M in G and colors the vertices of H white and black according to whether they are vertices from the original graph or they are obtained from contracted edges of M. Namely, $H = (V_W \cup V_B, E', \omega)$ is a $\{0,1\}$-edge weighted graph where V_W and V_B are the sets of white and black vertices respectively, and $|V_W| + 2|V_B| = 3k$.

We say that the vertex $\tilde{u} \in V(H)$ *corresponds* to the vertex $u \in V(G)$ if either u does not appear as a vertex in M and $\tilde{u} = u$ (in which case $\tilde{u}$ is white), or u is incident to some edge $e \in M$ and $\tilde{u}$ is the (black) vertex obtained by contracting e. We say that the edge $\{\tilde{u}, \tilde{v}\} \in E'$ *corresponds* to the edge $\{u, v\} \in G$ if $\tilde{u}$ corresponds to u and $\tilde{v}$ corresponds to v. (Note that an edge of H may correspond to up to 4 edges of G.) The edge $\{\tilde{u}, \tilde{v}\} \in E'$ has weight 1 iff at least one of the edges in G corresponding to it has weight 1. Hereafter, an edge of H between two black vertices is called a *black-black* edge; *white-black* and *white-white* edges are defined analogously.

In the third phase, the algorithm finds a maximum weight matching N of H among those having:

- $\min\{|V_B|, |V_W|\}$ white-black edges (the maximum possible).
- an equal number of black-black edges and unmatched black vertices.

In the fourth phase, the edges of the matching N dictate how the edges of M and the vertices not matched by M are combined into four collections $\Pi_1, \ldots, \Pi_4$ of 2-paths as follows:

- Π_1: 2-paths of maximum weight formed from the three vertices of G corresponding to each white-black edge in N.
- Π_2: 2-paths of maximum weight formed from each set of four vertices of G corresponding to a black-black edge in N. Each such set has an unused *residual* vertex. Denote by R be the set of all residual vertices.
- Π_3: 2-paths formed from the edge of G corresponding to a black vertex not matched by N, and an arbitrary vertex from the set R of residual vertices.
- Π_4: 2-paths of weight 0 formed by arbitrarily partitioning the remaining uncovered (white) vertices into triples.

The algorithm outputs the set of 2-paths $\mathcal{A} = \Pi_1 \cup \Pi_2 \cup \Pi_3 \cup \Pi_4$. Observe that when $|M| > k$ the set Π_4 is empty, and when $|M| \le k$ the sets Π_2 and Π_3 are empty.

Examples. We illustrate the main steps of the algorithm with two examples. Fig. 2(a) shows a complete $\{0, 1\}$-edge weighted graph on 12 vertices with its zero weight edges omitted. The matching $M = \{a, b, c\}$ is shown in bold (red). Fig. 2(b) shows the graph H obtained from contracting the edges of M, with the matching N shown in bold (red). Vertices v_a, v_b and v_c correspond to the edges a, b and c respectively. Fig. 2 (c) shows the resulting 2-paths with zero weight edges drawn as dotted segments. The algorithm outputs the sets $\Pi_1 = \{\pi_1, \pi_1', \pi_1''\}$ and $\Pi_4 = \{\pi_4\}$. Note that since G has a small matching number ($< k$), both Π_2 and Π_3 are empty.

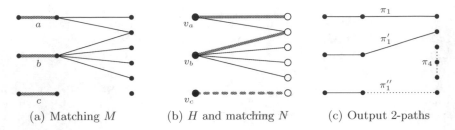

(a) Matching M (b) H and matching N (c) Output 2-paths

Fig. 2. DoubleMatching algorithm on a graph with small matching number.

Fig. 3(a) shows a complete $\{0, 1\}$-edge weighted graph G on 12 vertices with its zero weight edges omitted. A maximum matching $M = \{a, b, c, d, e\}$ is shown in bold (red). Fig. 3(b) shows the graph H obtained from G after the edges of M are contracted. The vertices v_a-v_e correspond to edges a-e respectively. The matching N is shown in bold (red). Fig. 3(c) shows the resulting 2-paths with zero weight edges drawn as dotted segments. The algorithm produces the sets $\Pi_1 = \{\pi_1, \pi_1'\}$, $\Pi_2 = \{\pi_2\}$ and $\Pi_3 = \{\pi_3\}$. Note that because G has a large matching number ($\geq k$), the set Π_4 is empty.

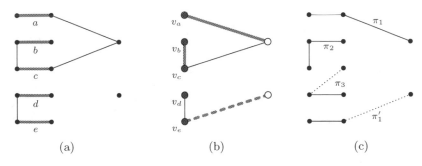

(a)   (b)   (c)

Fig. 3. DoubleMatching algorithm on a graph with large matching number.

Algorithm Analysis. In the full paper we show that the matching N can be found in $O(k^3)$ time by modifying the weights of the graph H and applying a standard maximum weight matching algorithm. Since contracting the edges of the matching M has quadratic time complexity and creating the 2-paths in the fourth phase takes linear time, the **DoubleMatching** algorithm has time complexity $O(k^3)$.

To prove the $3/4$-approximation guarantee, we split the analysis into two cases according to whether the size of M is large ($\geq k$) or small ($< k$). Let OPT be the set of 2-paths of a fixed optimal solution, and let Z_{opt} be the number of edges of weight 0 used by OPT. The weight of OPT is therefore $2k - Z_{opt}$. Moreover, we partition V_W, the set of white vertices of H, into two sets. Let $W^1 \subseteq V_W$ be the set of white vertices incident to at least one edge of OPT of weight 1 and let W^0 be the remaining white vertices. Let H_{OPT} denote the subgraph of H obtained by removing from H all edges not corresponding to edges of OPT. I.e., $H_{OPT} = (V_W \cup V_B, E'', \omega)$ where E'' consists of the H edges corresponding to the edges occurring in the 2-paths of OPT

The following lemma bounds the weight of the second matching N in the case when the size of the first matching is large.

Lemma 4. If $|M| \geq k$, then $\omega(N) \geq (2k - 3Z_{opt})/4$.

Proof sketch. By an argument similar to the one of Lemma 3, it can be shown that there exists a matching in H_{OPT} that has weight at least $(k - Z_{opt})/2$. However, this matching may not satisfy the two conditions imposed on the matching N, namely that it contains (a) the maximum number of white-black edges and (b) an equal number of black-black edges and unmatched black vertices. Let X be a matching of H_{OPT} of size and weight equal $(2k - 3Z_{opt})/4 \leq (k - Z_{opt})/2$. This matching can be modified (without changing its size or weight) to either contain $|W^1|/2$ or more white-black edges, or to not contain any black-black edges. Then, it is possible to add edges (possibly of weight 0) to this new matching, such that the conditions (a) and (b) imposed on the matching N are satisfied. Since N is the maximum weight matching satisfying (a) and (b), this proves the lemma.   □

The proof of the following lemma is based on the observation that every black vertex has at most two incident white-black edges of weight 1 in H_{OPT}.

Lemma 5. If $|M| < k$, then $\omega(N) \geq |W^1|/2$.

Proof. Let X be a maximum weight matching of H_{OPT} among those containing only white-black edges of weight 1. Let B_X be the set of black vertices that are incident to a white-black edge in X, and let W_X be the set of white vertices that are connected to a vertex in B_X by an edge of weight 1 in H_{OPT}. Since a black vertex cannot be incident to more than 2 white-black edges of weight 1, if $\omega(X) < |W^1|/2$ then $|B_X| < |W^1|/2$ and therefore $|W_X| < |W^1|$ implying that $W^1 \setminus W_X$ is not empty. Let u be a white vertex in $W^1 \setminus W_X$ and v a black vertex such that $\omega(\{u, v\}) = 1$. By adding $\{u, v\}$ to X we obtain a matching containing only white-black edges of weight 1 with weight larger than X, contradicting our assumption. Therefore, $\omega(N) \geq \omega(X) \geq |W^1|/2$, which proves the lemma.   □

Theorem 2. *For $\{0,1\}$-edge-weighted graphs there exists a $3/4$-approximation algorithm for the* M2PP *problem running in time $O(k^3)$.*

Proof sketch. It is easy to see that if $|M| \geq k$ the weight of the output solution is at least $\omega(M) + \omega(N) - (|M| - k)$. By using Lemma 4 to bound the weight of N we obtain:

$$\omega(\mathcal{A}) \geq |M| + \frac{2k - 3Z_{opt}}{4} - (|M| - k) = \frac{3}{4}(2k - Z_{opt}) = \frac{3}{4}\omega(OPT).$$

If $|M| < k$, the weight of the output solution is at least $\omega(M) + \omega(N) \geq \omega(M) + |W^1|/2$, where the inequality is due to Lemma 5. Since the vertices of G corresponding to vertices in $|W^0|$ are incident only to weight 0 edges of OPT, it is not difficult to see that $\omega(OPT) \leq 2 \cdot |V(M) \cup W^1|/3$ which equals $4/3(\omega(M) + |W^1|/2)$, and thus proves the claim. □

4   MTP on $\{0, 1\}$-Edge-Weighted Graphs

In this section we show how an approximation algorithm for the M2PP problem can be used in combination with a .5-approximation algorithm for the TRIANGLE PACKING problem to obtain an approximation algorithm for the MAXIMUM TRIANGLE PARTITIONING (MTP) problem on $\{0,1\}$-edge-weighted graphs. The most efficient .5-approximation algorithm for the TRIANGLE PACKING problem we are aware of is the depth 2 local search algorithm of [12, Theorem 3.4] (see also [18]) which works for general 3-SET PACKING. Our algorithm also relies on the existence of an α-approximate algorithm **2PP** for the M2PP problem on $\{0,1\}$-edge weighted graphs, with $\alpha \geq .5$ (e.g., our .75-approximation algorithm). The **TRIPART** algorithm is presented in Fig. 4.

We fix a $\{0,1\}$-edge-weighted graph $G = (V, E, \omega)$, and let OPT denote an optimal solution for MTP. Let A, B and C be the sets of triangles of OPT of weight 3, 2 and 1 respectively, and let $a = |A|$, $b = |B|$ and $c = |C|$. The details of the proofs in this section are omitted from this extended abstract.

By the definition of a, b and c, clearly $\omega(OPT) = 3a + 2b + c$. In Lemma 6 we show that the weight of the algorithm $\mathcal{A}_1$ is at least $2a + b + c$. It is interesting to note that this approximation bound cannot be improved by simply using an algorithm for the TRIANGLE PACKING problem with an approximation factor greater than .5.

Lemma 6. $\omega(\mathcal{A}_1) \geq 2a + b + c$.

It is easy to see that the weight of the algorithm $\mathcal{A}_2$ is at least $\alpha(2a + 2b + c)$ since a 2-path partition of weight $2a + 2b + c$ can be constructed from OPT. Proposition 1 below follows from the two bounds for $\mathcal{A}_1$ and $\mathcal{A}_2$ by applying standard techniques.

Proposition 1. $\max\{\omega(\mathcal{A}_1), \omega(\mathcal{A}_2)\} \geq \dfrac{2\alpha}{2\alpha + 1} \cdot \omega(OPT).$

Taking **DOUBLEMATCHING** as our algorithm **2PP**, with $\alpha = 3/4$, yields Theorem 3.

1. Let $G_1 = (V, E_1)$ be the unweighted graph formed by taking $E_1 = \{\{u, v\} \in E \mid \omega(\{u, v\}) = 1\}$.

2. Find a maximal triangle packing T in G_1 such that $|T|$ is at least half of the maximum triangle packing in G_1.

3. Let $G_{\overline{T}} = (V_{\overline{T}}, E_{\overline{T}}, \omega)$ be the subgraph of G, induced by the vertices of $V \setminus V(T)$.

4. Find a maximum weight matching M of size $|V_{\overline{T}}|/3$ in $G_{\overline{T}}$.

5. Complete M to a triangle partitioning T' in $G_{\overline{T}}$ by arbitrarily matching each edge $\{u, v\}$ in M to an unmatched vertex w in $V_{\overline{T}}$ and creating a triangle $\{u, v, w\}$.

6. Let $\mathcal{A}_1 = T \cup T'$.

7. Invoke algorithm **2PP** to get a 2-path partition Π of G.

8. Let $\mathcal{A}_2$ be the collection of triangles obtained by adding the missing edge to each 2-path of Π.

9. Return $\mathcal{A}$, the solution of larger weight among $\mathcal{A}_1$ and $\mathcal{A}_2$.

Fig. 4. The **TriPart** algorithm.

5 Conclusions

We presented two algorithms that significantly improve the approximation factor for the M2PP problem and for the MTP problem in $\{0, 1\}$-edge-weighted graphs. For graphs with arbitrary positive weights, the best known approximations for MTP only incrementally improve upon the approximation algorithms for the more general weighted 3-SET PACKING problem, while the only negative result is that the problem is NP-complete. Closing the approximation gap for any of these problems is an interesting open problem.

The triangle partition problem can be generalized to partitioning the graph into cliques of any fixed size, while the M2PP problem can be generalized to partitioning the graph into paths of fixed length or stars with a fixed number of leaves. Although some research has been done in this area (e.g. [13]), there are many open problems that deserve attention. Moreover, mirroring research in the graph packing literature, one might be interested in partitioning the graph into families of structures that maximize the weight of the chosen edges; for example, partitioning the graph into vertex disjoint edges and 2-Paths.

References

1. Arkin, E., Hassin, R.: On local search for weighted k-set packing. Math. Operations Research 23(3), 640–648 (1998)

2. Babenko, M., Gusakov, A.: New exact and approximation algorithms for the star packing problem in undirected graphs. In: STACS, pp. 519–530 (2011)

3. Bar-Noy, A., Basu, P., Baumer, B., Rabanca, G.: Star search: Effective subgroups in collaborative social networks (Unpublished)

4. Chen, Z.-Z., Tanahashi, R., Wang, L.: An improved randomized approximation algorithm for maximum triangle packing. In: Fleischer, R., Xu, J. (eds.) AAIM 2008. LNCS, vol. 5034, pp. 97–108. Springer, Heidelberg (2008)

5. Chen, Z.Z., Tanahashi, R., Wang, L.: An improved randomized approximation algorithm for maximum triangle packing. Discrete Applied Mathematics 157(7), 1640 (2009)

6. Chen, Z.Z., Tanahashi, R., Wang, L.: Erratum to, An improved randomized approximation algorithm for maximum triangle packing. Discrete Appl. Math. 157 (2009); Discrete Applied Mathematics 158(9), 1045–1047 (2010)
7. Chlebìk, M., Chlebìková, J.: Approximation hardness fo2653r small occurrence instances of NP-hard problems. In: Petreschi, R., Persiano, G., Silvestri, R. (eds.) ISAAC. LNCS, vol. 2653, pp. 152–164. Springer, Heidelberg (2003)
8. Gabow, H.: An efficient implementation of Edmonds' algorithm for maximum matching on graphs. J. ACM 23(2), 221–234 (1976)
9. Gajewar, A., Sarma, A.S.: Multi-skill collaborative teams based on densest subgraphs. In: SDM, pp. 165–176. SIAM/Omnipress (2012)
10. Garey, M.R., Johnson, D.S.: Computers and Intractability: a Guide to the Theory of NP-Completeness. Freeman, San Francisco (1979)
11. Guruswami, V., Pandu Rangan, C., Chang, M.S., Chang, G.J., Wong, C.K.: The vertex-disjoint triangles problem. In: Hromkovič, J., Sýkora, O. (eds.) WG 1998. LNCS, vol. 1517, pp. 26–37. Springer, Heidelberg (1998)
12. Halldórsson, M.: Approximating discrete collections via local improvements. In: SODA, pp. 160–169 (1995)
13. Hassin, R., Rubinstein, S.: An approximation algorithm for maximum packing of 3-edge paths. Information Processing Letters 63, 63–67 (1997)
14. Hassin, R., Rubinstein, S.: An approximation algorithm for maximum triangle packing. Discrete Applied Math. 154(6), 971–979 (2006)
15. Hassin, R., Rubinstein, S.: Erratum to "An approximation algorithm for maximum triangle packing". Discrete Applied Math. 154, 971–979 (2006); Discrete Applied Math. 154(18), 2620 (2006)
16. Hassin, R., Schneider, O.: A local search algorithm for binary maximum 2-path partitioning. Discrete Optimization 10(4), 333–360 (2013)
17. Hell, P., Kirkpatrick, D.C.: Packings by complete bipartite graphs. SIAM J. Algebraic Discrete Methods 7(2), 199–209 (1986)
18. Hurkens, C.A.J., Schrijver, A.: On the size of systems of sets every t of which have an sdr, with an application to the worst-case ratio of heuristics for packing problems. SIAM J. Discrete Math. 2(1), 68–72 (1989)
19. Kann, V.: Maximum bounded 3-dimensional matching is MAX SNP-complete. Information Processing Letters 37(1), 27–35 (1991)
20. Kargar, M., An, A.: Discovering top-k teams of experts with/without a leader in social networks. In: CIKM, pp. 985–994 (2011)
21. Li, C.T., Shan, M.K.: Team formation for generalized tasks in expertise social networks. In: SOCIALCOM, pp. 9–16 (2010)
22. Lonc, Z.: On the complexity of some edge-partition problems for graphs. Discrete Applied Mathematics 70(2), 177–183 (1996)
23. Manic, G., Wakabayashi, Y.: Packing triangles in low degree graphs and indifference graphs. Discrete Math. 308(8), 1455–1471 (2008)
24. Prieto, E., Sloper, C.: Looking at the stars. Theoretical Computer Science 351(3), 437–445 (2006)
25. Rangapuram, S., Bühler, T., Hein, M.: Towards realistic team formation in social networks based on densest subgraphs. In: WWW, pp. 1077–1088 (2013)
26. Tanahashi, R., Chen, Z.: A deterministic approximation algorithm for maximum 2-path packing. IEICE Tr. Inform. & Syst. E93-D(2), 241–249 (2010)
27. van Zuylen, A.: Multiplying pessimistic estimators: deterministic approximation of max TSP and maximum triangle packing. In: Thai, M.T., Sahni, S. (eds.) COCOON 2010. LNCS, vol. 6196, pp. 60–69. Springer, Heidelberg (2010)

A Multivariate Approach
for Weighted FPT Algorithms

Hadas Shachnai and Meirav Zehavi

Department of Computer Science, Technion, Haifa 32000, Israel
{hadas,meizeh}@cs.technion.ac.il

Abstract. We introduce a multivariate approach for solving weighted parameterized problems. Building on the *flexible* use of certain parameters, our approach defines a new general framework for applying the classic bounded search trees technique. In our model, given an instance of size n of a minimization/maximization problem, and a parameter $W \geq 1$, we seek a solution of weight at most/at least W. We demonstrate the wide applicability of our approach by solving the weighted variants of VERTEX COVER, 3-HITTING SET, EDGE DOMINATING SET and MAX INTERNAL OUT-BRANCHING. While the best known algorithms for these problems admit running times of the form $a^W n^{O(1)}$, for some constant $a > 1$, our approach yields running times of the form $b^s n^{O(1)}$, for some constant $b \leq a$, where $s \leq W$ is the minimum *size* of a solution of weight at most (at least) W. If no such solution exists, $s = \min\{W, m\}$, where m is the maximum size of a solution. Clearly, s can be substantially smaller than W. Moreover, we give an example for a problem whose polynomial-time solvability crucially relies on our flexible (in lieu of a strict) use of parameters.

We further show, among other results, that WEIGHTED VERTEX COVER and WEIGHTED EDGE DOMINATING SET are solvable in times $1.443^t n^{O(1)}$ and $3^t n^{O(1)}$, respectively, where $t \leq s$ is the minimum size of a solution.

1 Introduction

Many fundamental problems in graph theory are NP-hard already on *unweighted* graphs. This wide class includes, among others, VERTEX COVER, 3-HITTING SET, EDGE DOMINATING SET and MAX INTERNAL OUT-BRANCHING. Fast existing parameterized algorithms for these problems, which often exploit the structural properties of the underlying graph, cannot be naturally extended to handle weighted instances. Thus, solving efficiently weighted graph problems has remained among the outstanding open questions in parameterized complexity, as excellently phrased by Hajiaghayi [10]:

> "Most fixed-parameter algorithms for parameterized problems are inherently about *unweighted* graphs. Of course, we could add integer weights to the problem, but this can lead to a huge increase in the parameter. Can we devise fixed-parameter algorithms for weighted graphs that have less severe dependence on weights? Is there a nice framework for designing fixed-parameter algorithms on weighted graphs?"

© Springer-Verlag Berlin Heidelberg 2015
N. Bansal and I. Finocchi (Eds.): ESA 2015, LNCS 9294, pp. 965–976, 2015.
DOI: 10.1007/978-3-662-48350-3_80

We answer these questions affirmatively, by introducing a multivariate approach for solving weighted parameterized problems. We use this approach to obtain efficient algorithms for the following fundamental graph problems.

Weighted Vertex Cover (WVC): Given a graph $G = (V, E)$, a weight function $w : V \rightarrow \mathbb{R}^{\geq 1}$, and a parameter $W \in \mathbb{R}^{\geq 1}$, find a vertex cover $U \subseteq V$ (i.e., every edge in E has an endpoint in U) of weight at most W (if one exists).

Weighted 3-Hitting Set (W3HS): Given a 3-uniform hypergraph $G = (V, E)$, a weight function $w : V \rightarrow \mathbb{R}^{\geq 1}$, and a parameter $W \in \mathbb{R}^{\geq 1}$, find a hitting set $U \subseteq V$ (i.e., every hyperedge in E has an endpoint in U) of weight at most W (if one exists).

Weighted Edge Dominating Set (WEDS): Given a graph $G = (V, E)$, a weight function $w : E \rightarrow \mathbb{R}^{\geq 1}$, and a parameter $W \in \mathbb{R}^+$, find an edge dominating set $U \subseteq E$ (i.e., every edge in E touches an endpoint of an edge in U) of weight at most W (if one exists).

Weighted Max Internal Out-Branching (WIOB): Given a directed graph $G = (V, E)$, a weight function $w : V \rightarrow \mathbb{R}^{\geq 1}$, and a parameter $W \in \mathbb{R}^{\geq 1}$, find an out-branching of G (i.e., a spanning tree having exactly one vertex of in-degree 0) having internal vertices of total weight at least W (if one exists).

Parameterized algorithms solve NP-hard problems by confining the combinatorial explosion to a parameter k. More precisely, a problem is *fixed-parameter tractable (FPT)* with respect to a parameter k if it can be solved in time $O^*(f(k))$ for some function f, where O^* hides factors polynomial in the input size n. We note that it is *necessary* to assume that element weights are at least 1 in order to ensure fixed-parameter tractability with respect to W (see, e.g., [20]).

Existing FPT algorithms for the above problems have running times of the form $O^*(a^W)$, for some constant $a > 1$. Using our approach (described in Section 3), we obtain *faster* algorithms, whose running times are of the form $O^*(b^s)$, for some constant $b \leq a$, where $s \leq W$ is the minimum size of a solution of weight at most (at least) W. If no such solution exists, $s = \min\{W, m\}$, where m is the maximum size of a solution (for the unweighted version). Clearly, s can be significantly smaller than W. Moreover, in most of our results, $b < a$. We complement these results by developing algorithms for WEIGHTED VERTEX COVER and WEIGHTED EDGE DOMINATING SET parameterized by $t \leq s$, the minimum size of a solution (for the unweighted version). Any instance of the problems studied in this paper may satisfy $t < s$ (since if a solution A is smaller than a solution B, it is possible that the weight of A is larger than the weight of B).

1.1 Previous Work

Our problems are well known in graph theory and combinatorial optimization. They were also extensively studied in the area of parameterized complexity. We mention below known FPT results for their unweighted and weighted variants, parameterized by t and W, respectively.

Vertex Cover: VC is one of the first problems shown to be FPT. In the past two decades, it enjoyed a race towards obtaining the fastest FPT algorithm. The best FPT algorithm, due to Chen *et al.* [5], runs in time $O^*(1.274^t)$. In a similar race, focusing on graphs of bounded degree 3, the current winner is an algorithm of Issac *et al.* [17], whose running time is $O^*(1.153^t)$. For WVC, Niedermeier *et al.* [20] proposed an algorithm of $O^*(1.396^W)$ time and polynomial space, and an algorithm of $O^*(1.379^W)$ time and $O^*(1.363^W)$ space. Subsequently, Fomin *et al.* [13] presented an algorithm of $O^*(1.357^W)$ time and space. An alternative algorithm, using $O^*(1.381^W)$ time and $O^*(1.26^W)$ space, is given in [14].

3-Hitting Set: Several papers study FPT algorithms for 3HS. The best algorithm, by Wahlström [25], has running time $O^*(2.076^t)$. For W3HS, Fernau [12] gave an algorithm which runs in time $O^*(2.247^W)$ and uses polynomial space.

Edge Dominating Set: FPT algorithms for EDS, in general and bounded degree graphs, are given in several papers. The best known algorithm for general graphs, by Xiao *et al.* [26], runs in time $O^*(2.315^t)$, and for graphs of bounded degree 3, the current best algorithm, by Xiao *et al.* [27], runs in time $O^*(2.148^t)$. For WEDS in general graphs, the best algorithm, due to Binkele-Raible *et al.* [2], has running time $O^*(2.382^W)$ and uses polynomial space.

Max Internal Out-Branching: Although FPT algorithms for minimization problems are more common than those for maximization problems (see [11]), IOB was extensively studied in this area. The previous best algorithms run in time $O^*(6.855^t)$ [23], and in randomized time $O^*(4^t)$ [9,28] (faster algorithms are given in [29]). The weighted version, WIOB, was studied in the area of approximation algorithms (see [21,19]); however, to the best of our knowledge, its parameterized complexity is studied here for the first time.

We note that well-known tools, such as the color coding technique [1], can be used to obtain elegant FPT algorithms for some classic weighted graph problems (see, e.g., [15,16,24]). Recently, Cygan *et al.* [8] introduced a novel form of tree-decomposition to develop an FPT algorithm for minimum weighted graph bisection. Yet, for many other problems, these tools are not known to be useful. In fact, our flexible use of parameters may be essential for obtaining fast running times of the forms presented in this paper. We further elaborate in Section 3 on the limitations of known techniques in solving weighted graph problems.

1.2   Our Results

We introduce a novel multivariate approach for solving weighted parameterized problems. Our approach yields fast algorithms whose running times are of the form $O^*(c^s)$. We demonstrate its usefulness for the following problems.

- WVC: We give an algorithm that uses $O^*(1.381^s)$ time and polynomial space, or $O^*(1.363^s)$ time and space, complemented by an algorithm that uses $O^*(1.443^t)$ time and polynomial space. For graphs of bounded degree 3, this algorithm runs in time $O^*(1.415^t)$.[1]

---

[1] We also give an $O^*(1.347^W)$ time algorithm for WVC.

Table 1. Known results for WVC, W3HS, WEDS and WIOB, parameterized by t, W and s.

Problem	Unweighted	Parameter W	Parameter s	Parameter $(t+s)$	Comments
WVC	$O^*(1.274^t)$ [5]	$O^*(1.396^W)$ [20]	$\mathbf{O^*(1.381^s)}$	$\mathbf{O^*(1.443^t)}$	$O^*(1)$ space
	·	$O^*(1.357^W)$ [13]	$\mathbf{O^*(1.363^s)}$	·	
	$O^*(1.153^t)$ [17]	·	·	$\mathbf{O^*(1.415^t)}$	$\Delta = 3$
W3HS	$O^*(2.076^t)$ [25]	$O^*(2.247^W)$ [12]	$\mathbf{O^*(2.168^s)}$	$\mathbf{O^*(1.363^{s-t}2.363^t)}$	
WEDS	$O^*(2.315^t)$ [26]	$O^*(2.382^W)$ [2]	$\mathbf{O^*(2.315^s)}$	$\mathbf{O^*(3^t)}$	
	$O^*(2.148^t)$ [27]	·	·	·	$\Delta = 3$
WIOB	$O^*(6.855^t)$ [23]	—	$\mathbf{O^*(6.855^s)}$	—	

- W3HS: We develop an algorithm which uses $O^*(2.168^s)$ time and polynomial space, complemented by an algorithm which uses $O^*(1.381^{s-t}2.381^t)$ time and polynomial space, or $O^*(1.363^{s-t}2.363^t)$ time and $O^*(1.363^s)$ space.
- WEDS: We give an algorithm which uses $O^*(2.315^s)$ time and polynomial space, complemented by an $O^*(3^t)$ time and polynomial space algorithm.
- WIOB: We present an algorithm that has time and space complexities $O^*(6.855^s)$, or randomized time and space $O^*(4^sW)$.

Table 1 summarizes the known results for our problems. Results given in this paper are shown in boldface. Entries marked with · follow by inference from the first entry in the same cell. As shown in Table 1, our results imply that even if W is large, our problems can be solved efficiently, i.e., in times that are comparable to those required for solving their unweighted counterparts. Furthermore, most of the bases in our $O^*(c^s)$ running times are smaller than the bases in the corresponding known $O^*(c^W)$ running times. One may view such fast running times as somewhat surprising, since WVC, a key player in deriving our results, seems inherently more difficult than VC. Indeed, while VC admits a kernel of size $2t$, the smallest known kernel for WVC is of size $2W$ [4,6]. In fact, as shown in [18], WVC does not admit a polynomial kernel when parameterized by t.

Technical Contribution: A critical feature of our approach is that it allows an algorithm to "fail" in certain executions, e.g., to return NIL even if there exists a solution of weight at most (at least) W for the given input (see Section 3). We obtain improved running times for our algorithms by exploiting this feature, along with an array of sophisticated tools for tackling our problems. Specifically, in solving *minimization* problems, we show how the approach can be used to eliminate branching steps along the construction of bounded search trees, thus decreasing the overall running time. In solving WIOB, we reduce a given problem instance to an instance of an auxiliary problem, called WEIGHTED k-ITREE, for which we obtain an initial solution. This solution is then transformed into a solution for the original instance. Allowing "failures" for the algorithms simplifies the subroutine which solves WEIGHTED k-ITREE, since we do not need to ensure that the initial solution is not "too big". Again, this results in improved running times.

Furthermore, our approach makes non-standard use of the classic *bounded search trees* technique. Indeed, the analysis of an algorithm based on the technique relies on bounds attained by tracking the underlying input parameter, and the corresponding *branching vectors* of the algorithms (see Section 2). In deriving our results, we track the value of the *weight parameter* W, but analyze the branching vectors with respect to a *special size parameter* k. Our algorithms may base their output on the value of W only (i.e., ignore k; see, e.g., our algorithms for W3HS and WEDS in [22]), or may decrease k by *less* than its actual decrease in the instance (see, e.g., Rules 6 and 8 in Section 4, and Rule 5 of WVCnoW-Alg2 [22]; observe that when we decrease k by less than necessary, it is *not* guaranteed that we next decrease k by more than necessary).

We note that "discounting" analysis (namely, balancing bad cases against good cases that will inevitably follow) has some history in parameterized complexity. For example, an analysis of this type led to the results of Chen et al. [5]. Our approach (see Section 3) can both enhance and simplify "discounting" analysis. Indeed, in some of the algorithms we combine our approach with such analysis. Due to the general and intuitive nature of the approach, we believe it will find use in tackling the weighted variants of other classes of graph problems.

Organization: In Section 2, we give some definitions and notation, including an overview of the bounded search trees technique. Section 3 presents our general multivariate approach. In Section 4, we demonstrate the usefulness of our approach by developing an $O^*(1.381^s)$ time and polynomial space algorithm for WVC. Due to lack of space, all other applications are given in [22].

2 Preliminaries

Definitions and Notation: Given a (hyper)graph $G = (V, E)$ and a vertex $v \in V$, let $N(v)$ denote the set of neighbors of v; $E(v)$ denotes the set of edges adjacent to v. The *degree* of v is $|E(v)|$ (which, for hypergraphs, may not be equal to $|N(v)|$). Recall that a *leaf* is a degree-1 vertex. Given a subgraph H of G, let $V(H)$ and $E(H)$ denote its vertex set and edge set, respectively. For a subset $U \subseteq V$, let $N(U) = \bigcup_{v \in U} N(v)$, and $E(U) = \bigcup_{v \in U} E(v)$. Also, we denote by $G[U]$ the subgraph of G induced by U (if G is a hypergraph, $v, u \in U$ and $r \in V \setminus U$ such that $\{v, u, r\} \in E$, then $\{v, u\} \in E(G[U])$). Given a set S and a weight function $w : S \to \mathbb{R}$, the total weight of S is given by $w(S) = \sum_{s \in S} w(s)$. Finally, we say that a (hyper)edge $e \in E$ containing exactly d vertices is a *d-edge*.

Bounded Search Trees: The *bounded search trees* technique is fundamental in the design of recursive FPT algorithms (see, e.g., [11]). Informally, in applying this technique, one defines a list of rules. Each rule is of the form Rule X. [condition]action, where X is the number of the rule in the list. At each recursive call (i.e., a node in the search tree), the algorithm performs the action of the first rule whose condition is satisfied. If, by performing an action, the algorithm recursively calls itself at least twice, the rule is a *branching rule*; otherwise, it is a *reduction rule*. We only consider polynomial time actions that increase neither the parameter nor the size of the instance, and decrease at least one of them.

The running time of an algorithm that uses bounded search trees can be analyzed as follows. Suppose that the algorithm executes a branching rule which has ℓ branching options (each leading to a recursive call with the corresponding parameter value), such that in the ith branch option, the current value of the parameter decreases by b_i. Then, $(b_1, b_2, \ldots, b_\ell)$ is called the *branching vector* of this rule. We say that α is the *root* of $(b_1, b_2, \ldots, b_\ell)$ if it is the (unique) positive real root of $x^{b^*} = x^{b^* - b_1} + x^{b^* - b_2} + \ldots + x^{b^* - b_\ell}$, where $b^* = \max\{b_1, b_2, \ldots, b_\ell\}$. If $r > 0$ is the initial value of the parameter, and the algorithm (a) returns a result when (or before) the parameter is negative, and (b) only executes branching rules whose roots are bounded by a constant $c > 0$, then its running time is bounded by $O^*(c^r)$.

3   A General Multivariate Approach

In our approach, a problem parameterized by the solution weight is solved by adding a special size parameter. Formally, given a problem instance, and a weight parameter $W > 1$, we add an integer parameter $0 < k \leq W$. We then seek a solution of weight at most (at least) W. The crux of the approach is in allowing our algorithms to "fail" in certain cases. This enables to substantially improve running times, while maintaining the correctness of the returned solutions. Specifically, our algorithms satisfy the following properties. Given W and k,

(i) If there exists a solution of weight at most (at least) W, and size at most k, return a solution of weight at most (at least) W. The size of the returned solution may be larger than k.

(ii) Otherwise, return NIL, or a solution of weight at most (at least) W.

Clearly, the correctness of the solution can be maintained by iterating the above step, until we reach a value of k for which (i) is satisfied and the algorithm terminates with "success". Using our approach, we solve the following problems.

k-**WVC:** Given an instance of WVC, along with a parameter $k \in \mathbb{N}$, satisfy the following. If there is a vertex cover of weight at most W and size at most k, return a vertex cover of weight at most W; otherwise, return NIL, or a vertex cover of weight at most W.

k-**W3HS:** Given an instance of W3HS, along with a parameter $k \in \mathbb{N}$, satisfy the following. If there is a hitting set of weight at most W and size at most k, return a hitting set of weight at most W; otherwise, return NIL or a hitting set of weight at most W.

k-**WEDS:** Given an instance of WEDS, along with a parameter $k \in \mathbb{N}$, satisfy the following. If there is an edge dominating set of weight at most W and size at most k, return an edge dominating set of weight at most W; otherwise, return NIL or an edge dominating set of weight at most W.

k-**WIOB:** Given an instance of WIOB, along with a parameter $k < W$, satisfy the following. If there is an out-branching having a set of internal vertices of total weight at least W and cardinality at most k, return an out-branching with internal vertices of total weight at least W; otherwise, return NIL or an out-branching with internal vertices of total weight at least W.[2]

We develop FPT algorithms for the above variants, which are then used to solve the original problems. Initially, $k = 1$. We increase this value iteratively, until either $k = \min\{W, m\}$, or a solution of weight at most (at least) W is found, where m is the maximum size of a solution (for the unweighted version). Clearly, for WVC and W3HS, $m = |V|$; for WEDS, $m = |E|$; and for WIOB, m is the maximum number of internal vertices of a spanning tree of G. For WIOB, to ensure that $s \leq \min\{W, m\}$, we proceed as follows. Initially, we solve 1-WIOB. While the algorithm returns NIL, before incrementing the value of k, we solve IOB, in which we seek an out-branching having at least $(k+1)$ internal vertices (using [23,9]). If $k + 1 \geq W$, our algorithm returns the answer to IOB. Otherwise, our algorithm solves $(k + 1)$-WIOB (i.e., it increments k) only if the answer (to IOB) was not NIL.

We note that some weighted variants of parameterized problems were studied in the following *restricted* form. Given a problem instance, along with the parameters $W \geq 1$ and $k \in \mathbb{N}$, find a solution of weight at most (at least) W and size at most k; if such a solution does not exist, return NIL (see, e.g., [3,7]). Clearly, an algorithm for this variant can be used to obtain running time of the form $O^*(c^s)$, for some constant $c > 1$, for the original weighted instance. However, the efficiency of our algorithms crucially relies on the flexible use of the parameter k. In particular (as we show in the full version [22]), for some of the problems, the restricted form becomes NP-hard already on easy classes of graphs, as opposed to the above problems, which remain polynomial time solvable on such graphs.

4   An $O^*(1.381^s)$ Time Algorithm for WVC

In this section, we present our first algorithm, WVC-Alg. This algorithm employs the bounded search trees technique, described in Section 2. It builds upon rules used by the $O^*(1.396^W)$ time and polynomial space algorithm for WVC proposed in [20]. However, we also present new rules, including, among others, reduction rules that manipulate the weights of the vertices in the input graph. This allows us to easily and efficiently eliminate leaves and certain triangles (see, in particular, Rules 6 and 8). Furthermore, Rules 6 and 8 demonstrate the power of flexible use of parameters.[3] Thus, we obtain the following.

Theorem 1. WVC-Alg *solves* k-WVC *in* $O^*(1.381^k)$ *time and polynomial space.*

By the discussion in Section 3, this implies the desired result:

Corollary 2. WVC *can be solved in* $O^*(1.381^s)$ *time and polynomial space.*

---

[2] If $k \geq W$, assume that k-WIOB is simply WIOB.

[3] Excluding Rules 6 and 8, and the analysis of Rule 5, the rules in the extended abstract build upon the rules in [20].

Next, we present each rule within a call WVC-Alg($G = (V, E), w : V \to \mathbb{R}^{\geq 0}, W, k$). Recall that a rule can be performed only if its condition is true and the conditions of all of the rules preceding it are false. Initially, WVC-Alg is called with a weight function w, whose image lies in $\mathbb{R}^{\geq 1}$. After presenting a rule, we argue its correctness. For each branching rule, we also give the root of the corresponding branching vector (with respect to k). Since the largest root we shall get is bounded by 1.381, and the algorithm stops if $k < 0$, we have the desired running time.

Reduction Rule 1. [$\min\{W, k\} < 0$] Return NIL.

If $\min\{W, k\} < 0$, there is no vertex cover of weight at most W and size at most k.

Reduction Rule 2. [$E = \emptyset$] Return $\emptyset$.

Since $E = \emptyset$, an empty set is a vertex cover.

Reduction Rule 3. [There is a connected component H with at most one vertex of degree at least 3, where $|E(H)| \geq 1$] Use dynamic programming to compute a minimum-weight vertex cover U of H (see [20]). Return WVC-Alg($G[V \setminus V(H)], w, W - w(U), k - 1$) $\cup$ U.[4]

Since H is a connected component, any minimum-weight vertex cover of G consists of a vertex cover of H of weight $w(U)$, and a minimum-weight vertex cover of $G[V \setminus V(H)]$. Furthermore, any vertex cover of G contains a vertex cover of H of size at least 1. Therefore, we return a solution as required: if there is a solution of size at most k and weight at most W, we return a solution of weight at most W, and if there is no solution of weight at most W, we return NIL.

Reduction Rule 4. [There is a connected component H such that $|V(H)| \leq 100$ and $|E(H)| \geq 1$] Use brute-force to compute a minimum-weight vertex cover U of H. Return WVC-Alg($G[V \setminus V(H)], w, W - w(U), k - 1$) $\cup$ U.

The correctness of the rule follows from the same arguments as given for Rule 3. The next rule, among other rules, clarifies the necessity of Rule 4, and, in particular, the choice of the value 100.[5]

Branching Rule 5. [There is a vertex of degree at least 4, or all vertices have degree 3 or 0] Let v be a vertex of maximum degree.

1. If the result of WVC-Alg($G[V \setminus \{v\}], w, W - w(v), k - 1$) is not NIL: Return it along with v.
2. Else: Return WVC-Alg($G[V \setminus N(v)], w, W - w(N(v)), k - \max\{|N(v)|, 4\}$), along with $N(v)$.

---

[4] We assume that adding elements to NIL results in NIL.
[5] Choosing a smaller value is possible, but it is unnecessary and complicates the proof.

This branching is exhaustive. If the degree of v is at least 4, the rule is clearly correct; else, the degree of any vertex in G is 3 or 0. Then, we need to argue that decreasing k by 4 in the second branch, while $|N(v)| = 3$, leads to a correct solution.[6] Let C be the connected component that contains v. Since the previous rule did not apply, $|V(C)| > 100$. As we continue making recursive calls, as long as G contains edges from $E(C)$, it also contains at least one vertex of degree 1 or 2. For example, after removing v, it contains a neighbor of v whose degree is 1, and after removing $N(v)$, it contains a neighbor of a vertex in $N(v)$ whose degree is 1 or 2. Now, before we remove all the edges in $E(C)$, we encounter a recursive call where G contains a connected component of size at least 5, for which Rule 3 or 4 is applicable.[7] Therefore, if we apply Rule 5 again (and even if we do not, but there is a solution as required), we first apply Rule 3 or 4 which decrease k by 1, although the actual decrease is at least by 2. Indeed, 2 is the minimum size of any vertex cover of a connected component on at least 5 vertices and of maximum degree 3. Thus, it is possible to decrease k by 4 in Rule 5. By the definition of this rule, its branching vector is at least as good as $(1, 4)$, whose root is smaller than 1.381.

Reduction Rule 6. [There are $v, u \in V$ such that $N(v) = \{u\}$]

1. If $w(v) \geq w(u)$: Return WVC-Alg$(G[V \setminus \{v, u\}], w, W - w(u), k - 1) \cup \{u\}$.
2. Else if there is $r \in V$ such that $N(u) = \{v, r\}$:
   (a) Let w' be w, except for $w'(r) = w(r) - (w(u) - w(v))$.
   (b) If $w'(r) \leq 0$: Return WVC-Alg$(G[V \setminus \{v, u, r\}], w', W - w(v) - w(r), k - 1)$, along with $\{v, r\}$.
   (c) Else: Return WVC-Alg$(G[V \setminus \{v, u\}], w', W - w(u), k - 1)$, along with v if r is in the returned result, and else along with u.
3. Else: Let w' be w, except for $w'(u) = w(u) - w(v)$. Return WVC-Alg$(G[V \setminus \{v\}], w', W - w(v), k)$, along with v iff u is not in the returned result.

This rule, illustrated below, omits leaves (i.e., if there is a leaf, v, it is omitted from G in the recursive calls performed in this rule). Clearly, to obtain a solution, we should choose either u or $N(u)$. If $w(v) \geq w(u)$ (Case 1), we simply choose u (it is better to cover the only edge that touches v, $\{v, u\}$, by u).

Now, suppose that there is $r \in V$ such that $N(u) = \{v, r\}$. If $w'(r) \leq 0$ (Case 2b), it is better, in terms of weight, to choose $\{v, r\}$; yet, in terms of size, it might be better to choose u. In any case, k should be decreased by at least 1. Our flexible use of the parameter k allows us to decrease its value by 1, which is

---

[6] One cannot decrease k by 3, claiming that the case of a 3-regular connected component is negligible (this is done, e.g., in [20]), since during the execution of the algorithm, we can encounter many such components.

[7] The removal of $N(v) \cup \{v\}$ from C, which has maximum degree 3 and contains more than 100 vertices, generates at most 6 connected components; thus, it results in at least one component of at least $\lceil (101 - 4)/6 \rceil = 17$ vertices.

less than its actual decrease $(2 = |\{v, u\}|)$ in the instance.[8] Next, suppose that $w'(r) > 0$ (Case 2c). In $(G[V \setminus \{v, u\}], w', W - w(u), k - 1)$, choosing r reduces $W - w(u)$ to $W - w(u) - w'(r) = W - w(v) - w(r)$ and $k - 1$ to $k - 2$, which has the same effect as choosing $N(u)$ in the original instance. On the other hand, not choosing r has the same effect as choosing u in the original instance.

Finally, suppose that such r does not exist (Case 3). In $(G[V \setminus \{v\}], w', W - w(v), k)$, choosing u reduces $W - w(v)$ to $W - w(v) - w'(u) = W - w(u)$ and k to $k - 1$, which has the same effect as choosing u in the original instance. On the other hand, not choosing u has *almost* the same effect as choosing v in the original instance: the difference lies in the fact that we do not decrease k by 1. However, our flexible use of the parameter k allows us to decrease its value by less than necessary (as in Case 2b).[9]

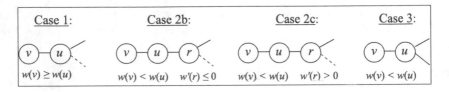

Fig. 1. Rule 6 of WVC-Alg.

Reduction Rule 7. [There are v, u such that $v \in N(u)$, $N(u) \setminus \{v\} \subseteq N(v) \setminus \{u\}$ and $w(v) \leq w(u)$] Return WVC-Alg$(G[V \setminus \{v\}], w, W - w(v), k - 1) \cup \{v\}$.

The vertices v and u are neighbors; thus, we should choose at least one of them. If we do not choose v, we need to choose $N(v)$, in which case we can replace u by v and obtain a vertex cover (since $N(u) \setminus \{v\} \subseteq N(v) \setminus \{u\}$) of the same or better weight (since $w(v) \leq w(u)$). Thus, in this rule, we choose v. Note that, if there is a triangle with two degree-2 vertices, or exactly one degree-2 vertex that is heavier than one of the other vertices in the triangle, this rule deletes a vertex of the triangle. Thus, in the following rules, such triangles do not exist.

Reduction Rule 8. [There are v, u, r such that $\{v, u\} = N(r)$, $\{v, r\} \subseteq N(u)$] Let w' be w, except $w'(v) = w(v) - w(r)$ and $w'(u) = w(u) - w(r)$. Return WVC-Alg$(G[V \setminus \{r\}], w', W - 2w(r), k)$, along with r iff not both v and u are in the returned result.

---

[8] In this manner, we may compute a vertex cover whose size is larger than k (since we decrease k only by 1), but we may not compute a vertex cover of weight larger than W. We note that if we decrease k by 2, we may overlook solutions: if there is a solution of size at most k and weight at most W that contains u, there is a solution of weight at most W that contains $\{v, r\}$, but there *might not* be a solution of size at most k and weight at most W that contains $\{v, r\}$.

[9] WVC-Alg is not called with $k-1$, as then choosing u overall decreases k by 2, which is more than required (thus we may overlook solutions, by reaching Rule 1 too soon).

The rule is illustrated below. First, note that $w'(v), w'(u) > 0$ (otherwise Rule 7 applies), and thus calling WVC-Alg with w' is possible.

We need to choose *exactly* two vertices from $\{v, u, r\}$: choosing less than two vertices does not cover all three edges of the triangle, and choosing r, if v and u are already chosen, is unnecessary. In WVC-Alg$(G[V \setminus \{r\}], w', W - 2w(r), k)$, choosing only v from $\{v, u\}$ reduces $W - 2w(r)$ to $W - 2w(r) - w'(v) = W - w(r) - w(v)$ and k to $k - 1$, which has *almost* the same effect as choosing r and v in the original instance, where the only difference lies in the fact that k is reduced by 1 (rather than 2). However, our flexible use of the parameter k allows us to decrease it by less than necessary. Symmetrically, choosing only u from $\{v, u\}$ has almost the same effect as choosing r and u in the original instance, and again, our flexible use of the parameter k allows us to decrease it by less than necessary. Finally, choosing both v and u reduces $W - 2w(r)$ to $W - 2w(r) - w'(v) - w'(u) = W - w(v) - w(u)$ and k to $k - 2$, which has the same effect as choosing v and u in the original instance. Thus, we have shown that each option of choosing exactly two vertices from $\{v, u, r\}$ is considered.

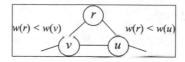

Fig. 2. Rule 8 of WVC-Alg.

From now on, since previous rules did not apply, there are no connected components on at most 100 vertices (by Rule 4), no leaves (by Rule 6), no vertices of degree at least 4 (by Rule 5), and no triangles that contain a degree-2 vertex (by Rules 7 and 8); also, there is a degree-2 vertex that is a neighbor of a degree-3 vertex (by Rules 3 and 5). We give the remaining rules in [22].

References

1. Alon, N., Yuster, R., Zwick, U.: Color coding. J. ACM 42(4), 844–856 (1995)
2. Binkele-Raible, D., Fernau, H.: Enumerate and measure: Improving parameter budget management. In: Raman, V., Saurabh, S. (eds.) IPEC 2010. LNCS, vol. 6478, pp. 38–49. Springer, Heidelberg (2010)
3. Chen, J., Fomin, F.V., Liu, Y., Lu, S., Villanger, Y.: Improved algorithms for feedback vertex set problems. J. Comput. Syst. Sci. 74(7), 1188–1198 (2008)
4. Chen, J., Kanj, I.A., Jia, W.: Vertex cover: further observations and further improvements. J. Algorithms 41(2), 280–301 (2001)
5. Chen, J., Kanj, I.A., Xia, G.: Improved upper bounds for vertex cover. Theor. Comput. Sci. 411(40-42), 3736–3756 (2010)
6. Chlebìk, M., Chlebìovà, J.: Crown reductions for the minimum weighted vertex cover problem. Discrete Appl. Math. 156(3), 292–312 (2008)

7. Cygan, M.: Deterministic parameterized connected vertex cover. In: Fomin, F.V., Kaski, P. (eds.) SWAT 2012. LNCS, vol. 7357, pp. 95–106. Springer, Heidelberg (2012)
8. Cygan, M., Lokshtanov, D., Pilipczuk, M., Pilipczuk, M., Saurabh, S.: Minimum bisection is fixed parameter tractable. In: STOC, pp. 323–332 (2014)
9. Daligault, J.: Combinatorial techniques for parameterized algorithms and kernels, with ppplications to multicut. PhD thesis, Universite Montpellier, France (2011)
10. Demaine, E.D., Hajiaghayi, M., Marx, D.: Open problems from Dagstuhl Seminar 09511 (2010). http://drops.dagstuhl.de/opus/volltexte/2010/2499/pdf/09511.SWM.Paper.2499.pdf
11. Downey, R.G., Fellows, M.R.: Fundamentals of parameterized complexity. Springer (2013)
12. Fernau, H.: Parameterized algorithms for d-hitting set: the weighted case. Theor. Comput. Sci. 411(16-18), 1698–1713 (2010)
13. Fomin, F.V., Gaspers, S., Saurabh, S.: Branching and treewidth based exact algorithms. In: Asano, T. (ed.) ISAAC 2006. LNCS, vol. 4288, pp. 16–25. Springer, Heidelberg (2006)
14. Fomin, F.V., Gaspers, S., Saurabh, S., Stepanov, A.A.: On two techniques of combining branching and treewidth. Algorithmica 54(2), 181–207 (2009)
15. Fomin, F.V., Lokshtanov, D., Saurabh, S.: Efficient computation of representative sets with applications in parameterized and exact agorithms. In: SODA, pp. 142–151 (2014)
16. Hüffner, F., Wernicke, S., Zichner, T.: Algorithm engineering for color-coding with applications to signaling pathway detection. Algorithmica 52(2), 114–132 (2008)
17. Issac, D., Jaiswal, R.: An $O^*(1.0821^n)$-time algorithm for computing maximum independent set in graphs with bounded degree 3. CoRR abs/1308.1351 (2013)
18. Jansen, B.M.P., Bodlaender, H.L.: Vertex cover kernelization revisited – upper and lower bounds for a refined parameter. Theory Comput. Syst. 53(2), 263–299 (2013)
19. Knauer, M., Spoerhase, J.: Better approximation algorithms for the maximum internal spanning tree problem. In: Dehne, F., Gavrilova, M., Sack, J.-R., Tóth, C.D. (eds.) WADS 2009. LNCS, vol. 5664, pp. 459–470. Springer, Heidelberg (2009)
20. Niedermeier, R., Rossmanith, P.: On efficient fixed-parameter algorithms for weighted vertex cover. J. Algorithms 47(2), 63–77 (2003)
21. Salamon, G.: Approximation algorithms for the maximum internal spanning tree problem. Theor. Comput. Sci. 410(50), 5273–5284 (2009)
22. Shachnai, H., Zehavi, M.: A multivariate framework for weighted FPT algorithms. CoRR abs/1407.2033 (2014)
23. Shachnai, H., Zehavi, M.: Representative families: a unified tradeoff-based approach. In: Schulz, A.S., Wagner, D. (eds.) ESA 2014. LNCS, vol. 8737, pp. 786–797. Springer, Heidelberg (2014)
24. Sharan, R., Dost, B., Shlomi, T., Gupta, N., Ruppin, E., Bafna, V.: QNet: a tool for querying protein interaction networks. J. Comput. Biol. 15(7), 913–925 (2008)
25. Wahlström, M.: Algorithms, measures and upper bounds for satisfiability and related problems. Ph.D. thesis Linköpings universitet, Sweden (2007)
26. Xiao, M., Kloks, T., Poon, S.H.: New parameterized algorithms for the edge dominating set problem. Theor. Comput. Sci. 511, 147–158 (2013)
27. Xiao, M., Nagamochi, H.: Parameterized edge dominating set in graphs with degree bounded by 3. Theor. Comput. Sci. 508, 2–15 (2013)
28. Zehavi, M.: Algorithms for k-internal out-branching. In: Gutin, G., Szeider, S. (eds.) IPEC 2013. LNCS, vol. 8246, pp. 361–373. Springer, Heidelberg (2013)
29. Zehavi, M.: Mixing color coding-related techniniques. In: ESA (2015, to appear)

Incidences with Curves in $\mathbb{R}^d$

Micha Sharir[1,*], Adam Sheffer[2], and Noam Solomon[1,**]

[1] Blavatnik School of Computer Science, Tel Aviv University, Tel Aviv, Israel
{michas,noamsolomon}@post.tau.ac.il
[2] Dept. of Mathematics, California Institute of Technology, Pasadena, CA, USA
adamsh@gmail.coml

Abstract. We prove that the number of incidences between m points and n bounded-degree curves with k degrees of freedom in $\mathbb{R}^d$ is $I(\mathcal{P}, \mathcal{C}) =$
$$O\left(m^{\frac{k}{dk-d+1}+\varepsilon} n^{\frac{dk-d}{dk-d+1}} + \sum_{j=2}^{d-1} m^{\frac{k}{jk-j+1}+\varepsilon} n^{\frac{d(j-1)(k-1)}{(d-1)(jk-j+1)}} q_j^{\frac{(d-j)(k-1)}{(d-1)(jk-j+1)}} \right.$$
$\left. +m + n\right),$ where the constant of proportionality depends on k, ε and d, for any $\varepsilon > 0$, provided that no j-dimensional surface of degree $c_j(k, d, j, \varepsilon)$, a constant parameter depending on k, d, j, and ε, contains more than q_j input curves, and that the q_j's satisfy certain mild conditions.

This bound generalizes a recent result of Sharir and Solomon [20] concerning point-line incidences in four dimensions (where $d = 4$ and $k = 2$), and partly generalizes a recent result of Guth [8] (as well as the earlier bound of Guth and Katz [10]) in three dimensions (Guth's three-dimensional bound has a better dependency on q). It also improves a recent d-dimensional general incidence bound by Fox, Pach, Sheffer, Suk, and Zahl [7], in the special case of incidences with algebraic curves. Our results are also related to recent works by Dvir and Gopi [4] and by Hablicsek and Scherr [11] concerning rich lines in high-dimensional spaces.

1 Introduction

Let $\mathcal{C}$ be a set of curves in $\mathbb{R}^d$. We say that $\mathcal{C}$ has k *degrees of freedom* with *multiplicity* s if (i) for every k points in $\mathbb{R}^d$ there are at most s curves of $\mathcal{C}$ that are incident to all k points, and (ii) every pair of curves of $\mathcal{C}$ intersect in at most s points. The bounds that we derive depend more significantly on k than on s—see below.

* Supported by Grant 2012/229 from the U.S.-Israel Binational Science Foundation, by Grant 892/13 from the Israel Science Foundation, by the Israeli Centers for Research Excellence (I-CORE) program (center no. 4/11), and by the Hermann Minkowski–MINERVA Center for Geometry at Tel Aviv University.
** Supported by Grant 892/13 from the Israel Science Foundation and the I-CORE program. Part of this research was performed while the authors were visiting the Institute for Pure and Applied Mathematics (IPAM), which is supported by the National Science Foundation.

© Springer-Verlag Berlin Heidelberg 2015
N. Bansal and I. Finocchi (Eds.): ESA 2015, LNCS 9294, pp. 977–988, 2015.
DOI: 10.1007/978-3-662-48350-3_81

In this paper we derive sharp upper bounds on the number of incidences between a set $\mathcal{P}$ of m points and a set $\mathcal{C}$ of n bounded-degree algebraic curves that have k degrees of freedom (with some multiplicity s). We denote the number of these incidences by $I(\mathcal{P}, \mathcal{C})$.

Before stating our results, let us put them in context. The basic and most studied case involves incidences between points and lines. In two dimensions, writing L for the given set of n lines, the classical Szemerédi–Trotter theorem [27] yields the worst-case tight bound

$$I(\mathcal{P}, L) = O\left(m^{2/3}n^{2/3} + m + n\right). \tag{1}$$

In three dimensions, in the 2010 groundbreaking paper of Guth and Katz [10], an improved bound has been derived for $I(\mathcal{P}, L)$, for a set $\mathcal{P}$ of m points and a set L of n lines in $\mathbb{R}^3$, provided that not too many lines of L lie in a common plane. Specifically, they showed:

Theorem 1 (Guth and Katz [10]). *Let $\mathcal{P}$ be a set of m distinct points and L a set of n distinct lines in $\mathbb{R}^3$, and let $q_2 \leq n$ be a parameter, such that no plane contains more than q_2 lines of L. Then*

$$I(P, L) = O\left(m^{1/2}n^{3/4} + m^{2/3}n^{1/3}q_2^{1/3} + m + n\right).$$

This bound was a major step in the derivation of the main result of [10], an almost-linear lower bound on the number of distinct distances determined by any set of n points in the plane, a classical problem posed by Erdős in 1946 [6]. Their proof uses several nontrivial tools from algebraic and differential geometry, most notably the Cayley–Salmon theorem on osculating lines to algebraic surfaces in $\mathbb{R}^3$, and additional properties of ruled surfaces. All this machinery comes on top of the main innovation of Guth and Katz, the introduction of the *polynomial partitioning technique*; see below.

In four dimensions, Sharir and Solomon [21] have obtained a still sharper bound:

Theorem 2 (Sharir and Solomon [21]). *Let $\mathcal{P}$ be a set of m distinct points and L a set of n distinct lines in $\mathbb{R}^4$, and let $q_2, q_3 \leq n$ be parameters, such that (i) each hyperplane or quadric contains at most q_3 lines of L, and (ii) each 2-flat contains at most q_2 lines of L. Then*

$$I(\mathcal{P}, L) \leq 2^{c\sqrt{\log m}}\left(m^{2/5}n^{4/5} + m\right) + A\left(m^{1/2}n^{1/2}q_3^{1/4} + m^{2/3}n^{1/3}q_2^{1/3} + n\right), \tag{2}$$

where A and c are suitable absolute constants. When $m \leq n^{6/7}$ or $m \geq n^{5/3}$, we get the sharper bound

$$I(\mathcal{P}, L) \leq A\left(m^{2/5}n^{4/5} + m + m^{1/2}n^{1/2}q_3^{1/4} + m^{2/3}n^{1/3}q_2^{1/3} + n\right). \tag{3}$$

In general, except for the factor $2^{c\sqrt{\log m}}$, the bound is tight in the worst case, for any values of m, n, with corresponding suitable ranges of q_2 and q_3.

This improves, in several aspects, an earlier treatment of this problem in Sharir and Solomon [20].

Another way to extend the Szemerédi–Trotter bound is for curves in the plane with k degrees of freedom (for lines, $k = 2$). This has been done by Pach and Sharir, who showed:[1]

Theorem 3 (Pach and Sharir [17]). *Let $\mathcal{P}$ be a set of m points in $\mathbb{R}^2$ and let $\mathcal{C}$ be a set of bounded-degree algebraic curves in $\mathbb{R}^2$ with k degrees of freedom and with multiplicity s. Then*

$$I(\mathcal{P},\mathcal{C}) = O\left(m^{\frac{k}{2k-1}} n^{\frac{2k-2}{2k-1}} + m + n\right),$$

where the constant of proportionality depends on k and s.

Several special cases of this result, such as the cases of unit circles and of arbitrary circles, have been considered separately [3,25].

Here too one can consider the extension of these bounds to higher dimensions. The literature here is rather scarce, and we only mention here the work of Sharir, Sheffer and Zahl [19] on incidences between points and circles in three dimensions; an earlier study of this problem by Aronov et al. [1] gives a different, dimension-independent bound.

The bounds given above include a "leading term" that depends only on m and n (like the term $m^{1/2}n^{3/4}$ in Theorem 1), and, except for the two-dimensional case, a series of "lower-dimensional" terms (like the term $m^{2/3}n^{1/3}q_2^{1/3}$ in Theorem 1 and the terms $m^{1/2}n^{1/2}q_3^{1/4}$ and $m^{2/3}n^{1/3}q_2^{1/3}$ in Theorem 2). The leading terms, in the case of lines, become smaller as d increases. Informally, by placing the lines in a higher-dimensional space, it should become harder to create many incidences on them.

Nevertheless, this is true only if the setup is "truly d-dimensional". This means that not too many lines or curves can lie in a common lower-dimensional space. The lower-dimensional terms handle incidences within such lower-dimensional spaces. There is such a term for every dimension $j = 2, \ldots, d - 1$, and the "j-dimensional" term handles incidences within j-dimensional subspaces (which, as the quadrics in the case of lines in four dimensions in Theorem 2, are not necessarily linear and might be algebraic of low constant degree). Comparing the bounds for lines in two, three, and four dimensions, we see that the j-dimensional term in d dimensions, for $j < d$, is a sharper variant of the leading term in j dimensions. More concretely, if that leading term is $m^a n^b$ then its counterpart in the d-dimensional bound is of the form $m^a n^t q_j^{b-t}$, where q_j is the maximum number of lines that can lie in a common j-dimensional flat or low-degree variety, and t depends on j and d.

Our Results. In this paper we consider a grand generalization of these results, to the case where $\mathcal{C}$ is a family of bounded-degree algebraic curves with k degrees of freedom (and some multiplicity s) in $\mathbb{R}^d$. This is a very ambitious and difficult

---

[1] Their result holds for more general families of curves, not necessarily algebraic, but, since algebraicity will be needed in higher dimensions, we assume it also in the plane.

project, and the challenges that it faces seem to be enormous. Here we make the first, and fairly significant, step in this direction, and obtain the following bounds. As the exponents in the bounds are rather cumbersome expressions in d, k, and j, we first state the special case of $d = 3$ (and prove it separately), and then give the general bound in d dimensions.

Theorem 4 (Curves in $\mathbb{R}^3$). *Let $k \geq 2$ be an integer, and let $\varepsilon > 0$. Then there exists a constant $c(k, \varepsilon)$ that depends on k and ε, such that the following holds. Let $\mathcal{P}$ be a set of m points and $\mathcal{C}$ a set of n irreducible algebraic curves of constant degree with k degrees of freedom (and some multiplicity s) in $\mathbb{R}^3$, such that every algebraic surface of degree at most $c(k, \varepsilon)$ contains at most q_2 curves of $\mathcal{C}$, for some given $q_2 \leq n$. Then*

$$I(\mathcal{P}, \mathcal{C}) = O\left(m^{\frac{k}{3k-2}+\varepsilon} n^{\frac{3k-3}{3k-2}} + m^{\frac{k}{2k-1}+\varepsilon} n^{\frac{3k-3}{4k-2}} q_2^{\frac{k-1}{4k-2}} + m + n \right),$$

where the constant of proportionality depends on k, s, and ε (and on the degree of the curves).

The corresponding result in d dimensions is as follows.

Theorem 5 (Curves in $\mathbb{R}^d$). *Let $d \geq 3$ and $k \geq 2$ be integers, and let $\varepsilon > 0$. Then there exist constants $c_j(k, d, \varepsilon)$, for $j = 2, \ldots, d-1$, that depend on k, d, j, and ε, such that the following holds. Let $\mathcal{P}$ be a set of m points and $\mathcal{C}$ a set of n irreducible algebraic curves of constant degree with k degrees of freedom (and some multiplicity s) in $\mathbb{R}^d$. Moreover, assume that, for $j = 2, \ldots, d-1$, every j-dimensional algebraic variety of degree at most $c_j(k, d, \varepsilon)$ contains at most q_j curves of $\mathcal{C}$, for given parameters $q_2 \leq \ldots \leq q_{d-1} \leq n$.*

$$I(\mathcal{P}, \mathcal{C}) = O\left(m^{\frac{k}{dk-d+1}+\varepsilon} n^{\frac{dk-d}{dk-d+1}} + \sum_{j=2}^{d-1} m^{\frac{k}{jk-j+1}+\varepsilon} n^{\frac{d(j-1)(k-1)}{(d-1)(jk-j+1)}} q_j^{\frac{(d-j)(k-1)}{(d-1)(jk-j+1)}} + m + n \right),$$

where the constant of proportionality depends on k, s, d, and ε (and on the degree of the curves), and provided that, for any $2 \leq j < l \leq d$, we have (with the convention that $q_d = n$)

$$q_j \geq \left(\frac{q_{l-1}}{q_l} \right)^{l(l-2)} q_{l-1}. \tag{4}$$

Discussion. The advantages of our results are obvious: They provide the first nontrivial bounds for the general case of curves with any number of degrees of freedom in any dimension (with the exception of one previous study of Fox et al. [7], in which weaker bounds are obtained, for arbitrary varieties instead of algebraic curves). Apart for the ε in the exponents, the leading term is "best possible," in the sense that (i) the polynomial partitioning technique [10] that our analysis employs (and that has been used in essentially all recent works on incidences in higher dimensions) yields a recurrence that solves to this bound, and, moreover, (ii) it is (nearly) worst-case tight for *lines* in two, three, and four dimensions (as shown in the respective works cited above), and in fact is likely to be tight for lines in higher dimensions too, using a suitable extension of a construction, due to Elekes and used in [10,21].

Nevertheless, our bounds are not perfect, and tightening them further is a major challenge for future research. Specifically:

(i) The bounds involve the factor m^ε. As the existing works indicate, getting rid of this factor is no small feat. Although the factor does not show up in the cases of lines in two and three dimensions, it already shows up (sort of) in four dimensions (Theorem 2), as well as in the case of circles in three dimensions [19]. (A recent study of Guth [8] also pays this factor for the case of lines in three dimensions, in order to simplify the analysis.) See the proofs and comments below for further elaboration of this issue.

(ii) The condition that no surface of degree $c_j(k,d,\varepsilon)$ contains too many curves of $\mathcal{C}$, for $j = 2,\ldots,d-1$, is very restrictive, especially since the actual values of these constants that arise in the proofs can be quite large. Again, earlier works also "suffer" from this handicap, such as Guth's work [8] mentioned above, as well as an earlier version of Sharir and Solomon's four-dimensional bound [20].

(iii) Finally, the lower-dimensional terms that we obtain are not best possible. For example, the bound that we get in Theorem 4 for the case of lines in $\mathbb{R}^3$ $(k-2)$ is $O(m^{1/2+\varepsilon}n^{3/4} + m^{2/3+\varepsilon}n^{1/2}q_2^{1/6} + m + n)$. When $q_2 \ll n$, the two-dimensional term $m^{2/3+\varepsilon}n^{1/2}q_2^{1/6}$ in that bound is worse than the corresponding term $m^{2/3}n^{1/3}q_2^{1/3}$ in Theorem 1 (even when ignoring the factor m^ε).

Our results are also related to recent works by Dvir and Gopi [4] and by Hablicsek and Scherr [11], that study rich lines in high dimensions. Specifically, let $\mathcal{P}$ be a set of n points in $\mathbb{R}^d$ and let L be a set of r-rich lines (each line of L contains at least r points of $\mathcal{P}$). If $|L| = \Omega(n^2/r^{d+1})$ then there exists a hyperplane containing $\Omega(n/r^{d-1})$ points of $\mathcal{P}$. Our bounds might be relevant for extending this result to rich curves. Concretely, for a set $\mathcal{P}$ of n points in $\mathbb{R}^d$ and a collection $\mathcal{C}$ of r-rich constant-degree algebraic curves, if $|\mathcal{C}|$ is too large then the incidence bound becomes larger than our "leading term", indicating that some lower-dimensional surface must contain many curves of $\mathcal{C}$, from which it might be possible to also deduce that such a surface has to contain many points of $\mathcal{P}$. While such an extension is not straightforward, we believe that it is doable, and plan to investigate it in our future work.

As in the classical work of Guth and Katz [10], and in the numerous follow-up studies of related problems, here too we use the polynomial partitioning method, as pioneered in [10]. The reason why our bounds suffer from the aforementioned handicaps is that we use a partitioning polynomial of (large but) constant degree. When using a polynomial of a larger, non-constant degree, we face the difficult task of bounding incidences between points and curves that are fully contained in the zero set of the polynomial, where the number of curves of this kind can be large, because the polynomial partitioning technique has no control over this value. We remark that for lines we have the classical Cayley–Salmon theorem (see, e.g., Guth and Katz [10]), which essentially bounds the number of lines that can be fully contained in an algebraic surface of a given degree, unless the surface is ruled by lines. However, such a property is not known for more general curves (see Nilov and Skopenkov [16] for an interesting exception involving circles in $\mathbb{R}^3$). Handling these incidences requires heavy-duty machinery from algebraic geometry, and leads to profound new problems in that domain that need to be tackled.

In contrast, using a polynomial of constant degree makes this part of the analysis much simpler, as can be seen below, but then handling incidences within the cells of the partition becomes non-trivial, and a naive approach yields a bound that is too large. To handle this part, one uses induction within each cell of the partitioning, and it is this induction process that is responsible for the weaker aspects of the resulting bound. Nevertheless, with these "sacrifices" we are able to obtain a "general purpose" bound that holds for a broad spectrum of instances. It is our hope that this study will motivate further research on this problem that would improve our results along the "handicaps" mentioned earlier. Recalling how inaccessible were these kinds of problems prior to Guth and Katz's breakthroughs seven and five years ago, it is quite gratifying that so much new ground can be gained in this area, including the progress made in this paper.

Background. Incidence problems have been a major topic in combinatorial and computational geometry for the past thirty years, starting with the aforementioned Szemerédi-Trotter bound [27] back in 1983. Several techniques, interesting in their own right, have been developed, or adapted, for the analysis of incidences, including the crossing-lemma technique of Székely [26], and the use of cuttings as a divide-and-conquer mechanism (e.g., see [3]). Connections with range searching and related algorithmic problems in computational geometry have also been noted and exploited, and studies of the Kakeya problem (see, e.g., [28]) indicate the connection between this problem and incidence problems. See Pach and Sharir [18] for a comprehensive (albeit a bit outdated) survey of the topic.

The landscape of incidence geometry has dramatically changed in the past seven years, due to the infusion, in two groundbreaking papers by Guth and Katz [9,10], of new tools and techniques drawn from algebraic geometry. Although their two direct goals have been to obtain a tight upper bound on the number of joints in a set of lines in three dimensions [9], and a near-linear lower bound for the classical distinct distances problem of Erdős [10], the new tools have quickly been recognized as useful for incidence bounds. See [5,13,14,19,24,29,30] for a sample of recent works on incidence problems that use the new algebraic machinery.

The present paper continues this line of research, and aims at extending the collection of instances where nontrivial incidence bounds in higher dimensions can be obtained.

2   The Three-Dimensional Case

Proof of Theorem 4. We prove by induction on $m + n$ that

$$I(\mathcal{P}, \mathcal{C}) \leq \alpha_1 \left(m^{\frac{k}{3k-2}+\varepsilon} n^{\frac{3k-3}{3k-2}} + m^{\frac{k}{2k-1}+\varepsilon} n^{\frac{3k-3}{4k-2}} q^{\frac{k-1}{4k-2}} \right) + \alpha_2(m+n), \qquad (5)$$

where α_1, α_2 are sufficiently large constants, α_1 depends on ε and k (and s), and α_2 depends on k (and s).

For the induction base, the case where m, n are sufficiently small can be handled by choosing sufficiently large values of α_1, α_2.

Since the incidence graph, as a subgraph of $\mathcal{P} \times \mathcal{C}$, does not contain $K_{k,s+1}$ as a subgraph, the Kővári-Sós-Turán theorem (e.g., see [15, Section 4.5]) implies that $I(\mathcal{P}, \mathcal{C}) = O(mn^{1-1/k} + n)$, where the constant of proportionality depends on k (and s). When $m = O(n^{1/k})$, this implies the bound $I(\mathcal{P}, \mathcal{C}) = O(n)$, which is subsumed in (5) if we choose α_2 sufficiently large. We may thus assume that $n \leq cm^k$, for some absolute constant c.

Applying the Polynomial Partitioning Technique. We construct an r-*partitioning polynomial* f for $\mathcal{P}$, for a sufficiently large constant r. That is, as established in Guth and Katz [10], f is of degree[2] $O(r^{1/3})$, and the complement of its zero set $Z(f)$ is partitioned into $u = O(r)$ open connected cells, each containing at most m/r points of $\mathcal{P}$. We impose the asymptotic relations $2^{1/\varepsilon} \ll r \ll \alpha_2 \ll \alpha_1$ between the various constants. Denote the (open) cells of the partition as $\tau_1, \ldots, \tau_u$. For each $i = 1, \ldots, u$, let $\mathcal{C}_i$ denote the set of curves of $\mathcal{C}$ that intersect τ_i and let $\mathcal{P}_i$ denote the set of points that are contained in τ_i. We set $m_i = |\mathcal{P}_i|$ and $n_i = |\mathcal{C}_i|$, for $i = 1, \ldots, u$, and $m' = \sum_i m_i$, and notice that $m_i \leq m/r$ for each i (and $m' \leq m$). An obvious property (which is a consequence of Bézout's theorem, see, e.g., [24, Theorem A.2] or [14]) is that every curve of $\mathcal{C}$ intersects $O(r^{1/3})$ cells of $\mathbb{R}^3 \setminus Z(f)$. Therefore, $\sum_i n_i \leq bnr^{1/3}$, for a suitable absolute constant $b > 1$ (that depends on the degree of the curves in $\mathcal{C}$). Using Hölder's inequality, we have

$$\sum_i n_i^{\frac{3k-3}{3k-2}} \leq \left(\sum_i n_i \right)^{\frac{3k-3}{3k-2}} \left(\sum_i 1 \right)^{\frac{1}{3k-2}} \leq b' \left(nr^{\frac{1}{3}} \right)^{\frac{3k-3}{3k-2}} r^{\frac{1}{3k-2}} = b' n^{\frac{3k-3}{3k-2}} r^{\frac{k}{3k-2}},$$

$$\sum_i n_i^{\frac{3k-3}{4k-2}} \leq \left(\sum_i n_i \right)^{\frac{3k-3}{4k-2}} \left(\sum_i 1 \right)^{\frac{k+1}{4k-2}} \leq b' \left(nr^{\frac{1}{3}} \right)^{\frac{3k-3}{4k-2}} r^{\frac{k+1}{4k-2}} = b' n^{\frac{3k-3}{4k-2}} r^{\frac{k}{2k-1}},$$

for another absolute constant b'. Combining the above with the induction hypothesis, applied within each cell of the partition, implies

$$\sum_i I(\mathcal{P}_i, \mathcal{C}_i) \leq \sum_i \left(\alpha_1 \left(m_i^{\frac{k}{3k-2}+\varepsilon} n_i^{\frac{3k-3}{3k-2}} + m_i^{\frac{k}{2k-1}+\varepsilon} n_i^{\frac{3k-3}{4k-2}} q_2^{\frac{k-1}{4k-2}} \right) + \alpha_2(m_i+n_i) \right)$$

$$\leq \alpha_1 \left(\frac{m^{\frac{k}{3k-2}+\varepsilon}}{r^{\frac{k}{3k-2}+\varepsilon}} \sum_i n_i^{\frac{3k-3}{3k-2}} + \frac{m^{\frac{k}{2k-1}+\varepsilon} q_2^{\frac{k-1}{4k-2}}}{r^{\frac{k}{2k-1}+\varepsilon}} \sum_i n_i^{\frac{3k-3}{4k-2}} \right) + \sum_i \alpha_2(m_i + n_i)$$

---

[2] The implied constants of proportionality in the $O(\cdot)$ notation are *absolute* constants. In contrast, r is a constant that depends on ε and on the other problem parameters.

$$\leq \alpha_1 b' \left(\frac{m^{\frac{k}{3k-2}+\varepsilon} n^{\frac{3k-3}{3k-2}}}{r^\varepsilon} + \frac{m^{\frac{k}{2k-1}+\varepsilon} n^{\frac{3k-3}{4k-2}} q_2^{\frac{k-1}{4k-2}}}{r^\varepsilon} \right) + \alpha_2 \left(m' + bnr^{1/3} \right).$$

Our assumption that $n = O(m^k)$ implies that $n = O\left(m^{\frac{k}{3k-2}} n^{\frac{3k-3}{3k-2}} \right)$ (with an absolute constant of proportionality). Thus, when α_1 is sufficiently large with respect to r, k, and α_2, we have

$$\sum_i I(\mathcal{P}_i, \mathcal{C}_i) \leq 2\alpha_1 b' \left(\frac{m^{\frac{k}{3k-2}+\varepsilon} n^{\frac{3k-3}{3k-2}}}{r^\varepsilon} + \frac{m^{\frac{k}{2k-1}+\varepsilon} n^{\frac{3k-3}{4k-2}} q_2^{\frac{k-1}{4k-2}}}{r^\varepsilon} \right) + \alpha_2 m'.$$

When r is sufficiently large, so that $r^\varepsilon \geq 6b'$, we have

$$\sum_i I(\mathcal{P}_i, \mathcal{C}_i) \leq \frac{\alpha_1}{3} \left(m^{\frac{k}{3k-2}+\varepsilon} n^{\frac{3k-3}{3k-2}} + m^{\frac{k}{2k-1}+\varepsilon} n^{\frac{3k-3}{4k-2}} q_2^{\frac{k-1}{4k-2}} \right) + \alpha_2 m'. \qquad (6)$$

Incidences on the Zero Set $Z(f)$. It remains to bound incidences with points that lie on $Z(f)$. Set $\mathcal{P}_0 := \mathcal{P} \cap Z(f)$ and $m_0 = |\mathcal{P}_0| = m - m'$. Let $\mathcal{C}_0$ denote the set of curves that are fully contained in $Z(f)$, and set $\mathcal{C}' := \mathcal{C} \setminus \mathcal{C}_0$, $n_0 := |\mathcal{C}_0|$, and $n' := |\mathcal{C}'| = n - n_0$. Since every curve of $\mathcal{C}'$ intersects $Z(f)$ in $O(r^{1/3})$ points, we have, taking α_1 to be sufficiently large, and arguing as above,

$$I(\mathcal{P}_0, \mathcal{C}') = O(nr^{1/3}) \leq \frac{\alpha_1}{3} m^{\frac{k}{3k-2}+\varepsilon} n^{\frac{3k-3}{3k-2}}. \qquad (7)$$

Finally, we consider the number of incidences between points of $\mathcal{P}_0$ and curves of $\mathcal{C}_0$. For this, we set $c(k, \varepsilon)$ to be the degree of f, which is $O(r^{1/3})$, and can be taken to be $O((6b')^{1/\varepsilon})$. Then, by the assumption of the theorem, we have $|\mathcal{C}_0| \leq q_2$. We consider a generic plane $\pi \subset \mathbb{R}^3$ and project $\mathcal{P}_0$ and $\mathcal{C}_0$ onto two respective sets $\mathcal{P}^*$ and $\mathcal{C}^*$ on π. Since π is chosen generically, we may assume that no two points of $\mathcal{P}_0$ project to the same point in π, and that no pair of distinct curves in $\mathcal{C}_0$ have overlapping projections in π. Moreover, the projected curves still have k degrees of freedom, in the sense that, given any k points on the projection γ^* of a curve $\gamma \in \mathcal{C}_0$, there are at most $s-1$ other projected curves that go through all these points. This is argued by lifting each point p back to the point $\bar{p}$ on γ in $\mathbb{R}^3$, and by exploiting the facts that the original curves have k degrees of freedom, and that, for a sufficiently generic projection, any curve that does not pass through $\bar{p}$ does not contain any point that projects to p. The number of intersection points between a pair of projected curves may increase but it must remain a constant since these are intersection points between constant-degree algebraic curves with no common components. By applying Theorem 3, we obtain

$$I(\mathcal{P}_0, \mathcal{C}_0) = I(\mathcal{P}^*, \mathcal{C}^*) = O(m_0^{\frac{k}{2k-1}} q_2^{\frac{2k-2}{2k-1}} + m_0 + q_2),$$

where the constant of proportionality depends on k (and s). Since $q_2 \leq n$ and $m_0 \leq m$, we have $m_0^{\frac{k}{2k-1}} q_2^{\frac{2k-2}{2k-1}} \leq m^{\frac{k}{2k-1}} n^{\frac{3k-3}{4k-2}} q_2^{\frac{k-1}{4k-2}}$. We thus get that $I(\mathcal{P}_0, \mathcal{C}_0)$ is at most

$$O\left(m^{\frac{k}{2k-1}}n^{\frac{3k-3}{4k-2}}q_2^{\frac{k-1}{4k-2}}+n+m_0\right) \leq \frac{\alpha_1}{3}m^{\frac{k}{2k-1}}n^{\frac{3k-3}{4k-2}}q_2^{\frac{k-1}{4k-2}}+b_2n+\alpha_2m_0, \quad (8)$$

for sufficiently large α_1 and α_2; the constant b_2 comes from Theorem 3, and is independent of ε and of the choices for α_1, α_2 made so far.

By combining (6), (7), and (8), including the case $m = O(n^{1/k})$, and choosing α_2 sufficiently large, we obtain

$$I(\mathcal{P},\mathcal{C}) \leq \alpha_1\left(m^{\frac{k}{3k-2}+\varepsilon}n^{\frac{3k-3}{3k-2}}+m^{\frac{k}{2k-1}+\varepsilon}n^{\frac{3k-3}{4k-2}}q_2^{\frac{k-1}{4k-2}}\right)+\alpha_2(m+n).$$

This completes the induction step and thus the proof of the theorem. $\square$

Example 1: The case of lines. Lines in $\mathbb{R}^3$ have $k = 2$ degrees of freedom, and we almost get the bound of Guth and Katz in Theorem 1. There are three differences that make this derivation somewhat inferior to that in Guth and Katz [10], as detailed in items (i)–(iii) in the discussion in the introduction. We also recall the two follow-up studies of point-line incidences in $\mathbb{R}^3$, of Guth [8] and of Sharir and Solomon [22]. Guth's bound suffers from weaknesses (i) and (ii), but avoids (iii), using a fairly sophisticated inductive argument. Sharir and Solomon's bound avoids (i) and (iii), and almost avoids (ii), in a sense that we do not make explicit here. In both cases, considerably more sophisticated machinery is needed to achieve these improvements.

Example 2: The case of circles. Circles in $\mathbb{R}^3$ have $k = 3$ degrees of freedom, and we get the bound

$$I(\mathcal{P},\mathcal{C}) = O\left(m^{3/7+\varepsilon}n^{6/7}+m^{3/5+\varepsilon}n^{3/5}q_2^{1/5}+m+n\right).$$

The leading term is the same as in Sharir et al. [19], but the second term is weaker, because it relies on the general bound of Pach and Sharir (Theorem 3), whereas the bound in [19] exploits an improved bound for point-circle incidences, due to Aronov et al. [1], which holds in any dimension. If we plug that bound into the above scheme, we obtain an exact reconstruction of the bound in [19]. In addition, considering the items (i)–(iii) discussed earlier, we note: (i) The requirements in [19] about the maximum number of circles on a surface are weaker, and are only for planes and spheres. (ii) The m^ε factors are present in both bounds. (iii) Even after the improvement noted above, the bounds still seem to be weak in terms of their dependence on q_2, and improving this aspect, both here and in [19], is a challenging open problem.

Theorem 4 can easily be restated as bounding the number of *rich points*.

Corollary 1. *For each $\varepsilon > 0$ there exists a parameter $c(k,\varepsilon)$ that depends on k and ε, such that the following holds. Let $\mathcal{C}$ be a set of n irreducible algebraic curves of constant degree and with k degrees of freedom (with some multiplicity s) in $\mathbb{R}^3$. Moreover, assume that every surface of degree at most $c(k,\varepsilon)$ contains at most q_2 curves of $\mathcal{C}$. Then, there exists some constant $r_0(k,\varepsilon)$ depending on ε, k (and s), such that for any $r \geq r_0(k,\varepsilon)$, the number of points that are incident to at least r curves of $\mathcal{C}$ (so-called r-rich points), is*

$$O\left(\frac{n^{3/2+\varepsilon}}{r^{\frac{3k-2}{2k-2}+\varepsilon}} + \frac{n^{3/2+\varepsilon}q_2^{1/2+\varepsilon}}{r^{\frac{2k-1}{k-1}+\varepsilon}} + \frac{n}{r}\right), \text{ where the constant of proportionality de-}$$

pends on k, s and ε.

Proof. Denoting by m_r the number of r-rich points, the corollary is obtained by combining the upper bound in Theorem 4 with the lower bound rm_r.   □

3   Incidences in Higher Dimensions

Due to lack of space, we omit the proof of Theorem 5, which is provided in the full version of the paper. It follows the proof of Theorem 4 rather closely, with appropriate modifications, and exploits the special assumption (4) on the values q_j to carry out the induction process successfully.

As a consequence of Theorem 5, we have:

Example: Incidences between Points and Lines in $\mathbb{R}^4$. In the earlier version [20] of our study of point-line incidences in four dimensions, we have obtained the following weaker version of Theorem 2.

Theorem 6. *For each $\varepsilon > 0$, there exists an integer c_ε, so that the following holds. Let P be a set of m distinct points and L a set of n distinct lines in $\mathbb{R}^4$, and let $q, s \leq n$ be parameters, such that (i) for any polynomial $f \in \mathbb{R}[x, y, z, w]$ of degree $\leq c_\varepsilon$, its zero set $Z(f)$ does not contain more than q lines of L, and (ii) no 2-plane contains more than s lines of L. Then,*

$$I(P, L) \leq A_\varepsilon\left(m^{2/5+\varepsilon}n^{4/5} + m^{1/2+\varepsilon}n^{2/3}q^{1/12} + m^{2/3+\varepsilon}n^{4/9}s^{2/9}\right) + A(m+n),$$

where A_ε depends on ε, and A is an absolute constant.

This result follows from our main Theorem 5, if we impose Equation (4) on $q_2 = s$, $q_3 = q$, and n, which in this case is equivalent to $s \leq q \leq n$ and $\frac{q^9}{n^8} < s$. This illustrates how the general theory developed in this paper extends similar results obtained earlier for "isolated" instances. Nevertheless, as already mentioned earlier, the bound for lines in $\mathbb{R}^4$ has been improved in Theorem 2 of [21], in its lower-dimensional terms.

Discussion. We first notice that similarly to the three-dimensional case, Theorem 5 implies an upper bound on the number of k-rich points in d dimensions (see Corollary 1 in three dimensions), and the proof thereof applies verbatim.

Second, we note that Theorems 4 and 5 have several weaknesses. The obvious ones are the items (i)–(iii) discussed in the introduction. Another, less obvious weakness, which has to do with the way in which the q_j-dependent terms in the bounds are derived. Specifically, these terms facilitate the induction step, when the constraining parameter q is passed *unchanged* to the inductive subproblems. Informally, since the overall number of lines in a subproblem goes down, one would expect q to decrease too, but so far we do not have a clean mechanism for doing so. This weakness is manifested, e.g., in Corollary 1, where one would like to replace the second term by one with a smaller exponent of n and a larger one of q. Specifically, for lines in $\mathbb{R}^3$, one would like to get a term close to $O(nq_2/k^3)$. This would yield $O(n^{3/2}/k^3)$ for the important special case $q_2 = O(n^{1/2})$ considered in [10]; the present bound is weaker.

A final remark concerns the relationships between the q_j, as set forth in Equation (4). These conditions are forced upon us by the induction process. As noted above, for incidences between points and lines in $\mathbb{R}^4$, the bound derived in our main theorem 5 is (asymptotically) the same as that of the main result of Sharir and Solomon in [20]. The difference is that there, no restrictions on the q_j are imposed. Their proof is facilitated by the so called "second partitioning polynomial" (see [13,20]). Recently, Basu and Sombra [2] proved the existence of a third partitioning polynomial (see [2, Theorem 3.1]), and conjectured the existence of a k-partitioning polynomial for general $k > 3$ (see [2, Conjecture 3.4] for an exact formulation); for completeness we refer also to [7, Theorem 4.1], where a weaker version of this conjecture is proved. Building upon the work of Basu and Sombra [2], the proof of Sharir and Solomon [21] is likely to extend and yield the same bound as in our main theorem 5, for the more general case of incidences between points and bounded degree algebraic curves in dimensions at most five, and, if [2, Conjecture 3.4] holds, in every dimension, without any conditions on the q_j.

References

1. Aronov, B., Koltun, V., Sharir, M.: Incidences between points and circles in three and higher dimensions. Discrete Comput. Geom. 33, 185–206 (2005)
2. Basu, S., Sombra, M.: Polynomial partitioning on varieties of codimension two and point-hypersurface incidences in four dimensions. arXiv:1406.2144
3. Clarkson, K., Edelsbrunner, H., Guibas, L., Sharir, M., Welzl, E.: Combinatorial complexity bounds for arrangements of curves and spheres. Discrete Comput. Geom. 5, 99–160 (1990)
4. Dvir, Z., Gopi, S.: On the number of rich lines in truly high dimensional sets. In: Proc. 31st Annu. Sympos. Comput. Geom. (2015, to appear)
5. Elekes, G., Kaplan, H., Sharir, M.: On lines, joints, and incidences in three dimensions. J. Combinat. Theory, Ser. A 118, 962–977 (2011); Also in arXiv:0905.1583
6. Erdős, P.: On sets of distances of n points. Amer. Math. Monthly 53, 248–250 (1946)
7. Fox, J., Pach, J., Sheffer, A., Suk, A., Zahl, J.: A semi-algebraic version of Zarankiewicz's problem. arXiv:1407.5705
8. Guth, L.: Distinct distance estimates and low-degree polynomial partitioning. Discrete Comput. Geom. 53, 428–444 (2015); Also in arXiv:1404.2321
9. Guth, L., Katz, N.H.: Algebraic methods in discrete analogs of the Kakeya problem. Advances Math. 225, 2828–2839 (2010); Also in arXiv:0812.1043
10. Guth, L., Katz, N.H.: On the Erdős distinct distances problem in the plane. Annals Math. 181, 155–190 (2015); Also in arXiv:1011.4105
11. Hablicsek, M., Scherr, Z.: On the number of rich lines in high dimensional real vector spaces. arXiv:1412.7025

12. Harris, J.: Algebraic Geometry: A First Course, vol. 133. Springer, New York (1992)
13. Kaplan, H., Matoušek, J., Safernová, Z., Sharir, M.: Unit distances in three dimensions. Combinat. Probab. Comput. 21, 597–610 (2012); Also in arXiv:1107.1077
14. Kaplan, H., Matoušek, J., Sharir, M.: Simple proofs of classical theorems in discrete geometry via the Guth–Katz polynomial partitioning technique. Discrete Comput. Geom. 48, 499–517 (2012); Also in arXiv:1102.5391
15. Matoušek, J.: Lectures on Discrete Geometry. Springer, Heidelberg (2002)
16. Nilov, F., Skopenkov, M.: A surface containing a line and a circle through each point is a quadric. arXiv:1110:2338
17. Pach, J., Sharir, M.: On the number of incidences between points and curves. Combinat. Probab. Comput. 7, 121–127 (1998)
18. Pach, J., Sharir, M.: Geometric incidences. In: Pach, J. (ed.) Towards a Theory of Geometric Graphs. Contemporary Mathematics, vol. 342, pp. 185–223. Amer. Math. Soc., Providence (2004)
19. Sharir, M., Sheffer, A., Zahl, J.: Improved bounds for incidences between points and circles. Combinat. Probab. Comput. 24, 490–520 (2015); Also in Proc. 29th ACM Symp. on Computational Geometry, 97–106, arXiv:1208.0053 (2013)
20. Sharir, M., Solomon, N.: Incidences between points and lines in four dimensions. In: Proc. 30th ACM Sympos. on Computational Geometry, pp. 189–197 (2014)
21. Sharir, M., Solomon, N.: Incidences between points and lines in four dimensions. J. AMS, arXiv:1411.0777 (submitted)
22. Sharir, M., Solomon, N.: Incidences between points and lines in three dimensions. In: Proc. 31st ACM Sympos. on Computational Geometry (2015, to appear); Also Discrete Comput. Geom. arXiv:1501.02544 (submitted)
23. Sharir, M., Solomon, N.: Incidences between points and lines on a two-dimensional variety in three dimensions. Combinatorica, arXiv:1501.01670 (submitted)
24. Solymosi, J., Tao, T.: An incidence theorem in higher dimensions. Discrete Comput. Geom. 48, 255–280 (2012)
25. Spencer, J., Szemerédi, E., Trotter, W.T.: Unit distances in the Euclidean plane. In: Bollobás, B. (ed.) Graph Theory and Combinatorics, pp. 293–303. Academic Press, New York (1984)
26. Székely, L.: Crossing numbers and hard Erdős problems in discrete geometry. Combinat. Probab. Comput. 6, 353–358 (1997)
27. Szemerédi, E., Trotter, W.T.: Extremal problems in discrete geometry. Combinatorica 3, 381–392 (1983)
28. Tao, T.: From rotating needles to stability of waves: Emerging connections between combinatorics, analysis, and PDE. Notices AMS 48(3), 294–303 (2001)
29. Zahl, J.: An improved bound on the number of point-surface incidences in three dimensions. Contrib. Discrete Math. 8(1) (2013); Also in arXiv:1104.4987
30. Zahl, J.: A Szemerédi-Trotter type theorem in $\mathbb{R}^4$, arXiv:1203.4600

D^3-Tree: A Dynamic Deterministic Decentralized Structure

Spyros Sioutas[3], Efrosini Sourla[1], Kostas Tsichlas[4], and Christos Zaroliagis[1,2]

[1] Dept of Computer Eng. & Informatics, Univ. of Patras, 26504 Patras, Greece
{sourla,zaro}@ceid.upatras.gr
[2] Computer Technology Institute & Press "Diophantus", N. Kazantzaki Str.,
Patras University Campus, 26504 Patras, Greece
[3] Dept of Informatics, Ionian University, 49100 Corfu, Greece
sioutas@ionio.gr
[4] Dept of Informatics, Aristotle Univ. of Thessaloniki, 54124 Thessaloniki, Greece
tsichlas@csd.auth.gr

Abstract. We present D^3-Tree, a dynamic deterministic structure for data management in decentralized networks, by engineering and further extending an existing decentralized structure. D^3-Tree achieves $O(\log N)$ worst-case search cost (N is the number of nodes in the network), $O(\log N)$ amortized load-balancing cost, and it is highly fault-tolerant. A particular strength of D^3-Tree is that it achieves $O(\log N)$ amortized search cost under massive node failures. We conduct an extensive experimental study verifying that D^3-Tree outperforms other well-known structures and that it achieves a significant success rate in element queries in case of massive node failures.

1 Introduction

Decentralized systems are ubiquitous and are encountered in various forms and structures. They are widely used for sharing resources and store very large data sets, using systems of small computers instead of large costly servers. Typical examples include cloud computing environments, peer-to-peer systems and the internet. In decentralized systems, data are stored at the network nodes and the most crucial operations are data search and data updates. A decentralized system is typically represented by a graph, a logical *overlay network*, where its N nodes correspond to the network nodes, while its edges may not correspond to existing communication links, but to communication paths. The complexity (cost) of an operation is measured in terms of the number of messages issued during its execution (internal computations at nodes are considered insignificant).

With respect to its *structure*, the overlay supports the operations *Join* (of a new node v; v communicates with an existing node u in order to be inserted into the overlay) and *Departure* (of an existing node u; u leaves the overlay announcing its intent to other nodes of the overlay). Moreover, the overlay implements an *indexing scheme* for the stored data, supporting the operations *Insert* (a new element), *Delete* (an existing element), *Search* (for an element), and *Range Query* (for elements in a specific range).

© Springer-Verlag Berlin Heidelberg 2015
N. Bansal and I. Finocchi (Eds.): ESA 2015, LNCS 9294, pp. 989–1000, 2015.
DOI: 10.1007/978-3-662-48350-3_82

Related Work. Considerable work has been done recently in order to build efficient decentralized systems with effective distributed search and update operations. In general, decentralized networks can be classified into two broad categories: distributed hash table (DHT)-based systems and tree-based systems. Examples of the former, which constitute the majority, include Chord, CAN, Pastry, Symphony, Tapestry (see [7] for an overview), Pagoda [1], SHELL [9] and P-Ring [3]. In general, DHT-based systems support exact match queries well and use (successfully) probabilistic methods to distribute the workload among nodes equally. Since hashing destroys the ordering on keys, DHT-based systems typically do not possess the functionality to support straightforwardly range queries, or more complex queries based on data ordering (e.g., nearest-neighbour and string prefix queries). Some efforts towards addressing range queries have been made in [4,8], getting however approximate answers and also making exact searching highly inefficient. Pagoda [1] achieves constant node degree but has polylogarithmic complexity for the majority of operations. SHELL [9] maintains large routing tables of $O(\log^2 N)$ space complexity, but achieves constant amortized cost for the majority of operations. Both are complicated hybrid structures and their practicality (especially concerning fault tolerant operations) is questionable. The most recent effort towards range queries is the P-Ring [3], a fully distributed and fault-tolerant system that supports both exact match and range queries, achieving $O(\log_d N + k)$ range search performance (d is the *order*[1] of the ring and k is the answer size). It also provides load-balancing by maintaining a load imbalance factor of at most $2 + \epsilon$ of a stable system, for any given constant $\epsilon > 0$, and has a stabilization process for fixing inconsistencies caused by node failures and updates, achieving an $O(d \cdot \log_d N)$ performance for load-balancing.

Tree-based systems are based on hierarchical structures. They support range queries more naturally and efficiently as well as a wider range of operations, since they maintain the ordering of data. On the other hand, they lack the simplicity of DHT-based systems, and they do not always guarantee data locality and load balancing in the whole system. Important examples of such systems include Family Trees [7], BATON [6], BATON* [5] and Skip List-based schemes like Skip Graphs (SG), NoN SG, SkipNet (SN), Bucket SG, Skip Webs, Rainbow Skip Graphs (RSG) and Strong RSG [7] that use randomized techniques to create and maintain the hierarchical structure. We should emphasize that w.r.t. load-balancing, the solutions provided in the literature are either heuristics, or provide expected bounds under certain assumptions, or amortized bounds but at the expense of increasing the memory size per node. In particular, in BATON [6], a decentralized overlay is provided with load-balancing based on data migration. However, their $O(\log N)$ amortized bound is valid only subject to a probabilistic assumption about the number of nodes taking part in the data migration process, and thus it is in fact an amortized expected bound. Moreover, its successor BATON*[5], exploits the advantages of higher *fanout* (number of children per node), to achieve reduced search cost of $O(\log_m N)$, where m is the fanout.

---

[1] Maximum fanout of the hierarchical structure on top of the ring. At the lowest level of the hierarchy, each node maintains a list of its first d successors in the ring.

However, the higher fanout leads to a higher update and load-balancing cost of $O(m \cdot \log_m N)$. Recently, a deterministic decentralized tree structure, called D^2-Tree [2], was presented that overcomes many of the aforementioned weaknesses of tree-based systems. In particular, D^2-Tree achieves $O(\log N)$ searching cost, amortized $O(\log N)$ update cost both for element updates and for node joins and departures, and deterministic amortized $O(\log N)$ bound for load-balancing. Its practicality, however, has not been tested so far.

Regarding fault tolerance, P-Ring [3] is considered highly fault-tolerant, using the Chord's Fault Tolerant Algorithms [11]. BATON [6] maintains vertical and horizontal routing information not only for efficient search, but to offer a large number of alternative paths between two nodes. In BATON* [5], fault tolerance is greatly improved due to higher fanout. D^2-Tree can tolerate the failure of a few nodes, but cannot afford a massive number of $O(N)$ node failures.

Our Contribution. In this work, we focus on hierarchical tree-based decentralized systems and introduce D^3-Tree (cf. Section 2), a dynamic deterministic decentralized structure. D^3-Tree is an extension of D^2-Tree [2] that adopts all of its strengths and extends it in two respects: it introduces an enhanced fault-tolerant mechanism and it is able to answer efficiently search queries when massive node failures occur. D^3-Tree achieves the same deterministic (worst-case or amortized) bounds as D^2-Tree for search, update and load-balancing operations, and answers search queries in $O(\log N)$ amortized cost under massive node failures. A comparison of D^3-Tree with state-of-the-art decentralized structures is given in Table 1. Note that all previous structures provided only empirical evidence of their capability to deal with massive node failures; no previous structure provided a theoretical guarantee for searching in such a case.

Our second contribution is an implementation of the D^3-Tree and a subsequent comparative experimental evaluation (cf. Section 3) with its main competitors BATON, BATON*, and P-Ring. Our experimental study verified the theoretical results (as well as those of the D^2-Tree) and showed that D^3-Tree outperforms other state-of-the-art hierarchical tree-based structures. Our experiments demonstrated

Table 1. Comparison of BATON, BATON*, P-Ring, D^2-Tree, and D^3-Tree.

Structures	Search	Search with massive failures		Node Updates (updating rout. tables)	Element Updates (load balancing)
		Theor.	Exp.		
BATON	$O(\log N)$	—	Yes	$\overline{O}(\log N)$	$\overline{O}(\log N)$
BATON*	$O(\log_m N)$	—	Yes	$\overline{O}(m \cdot \log_m N)$	$\overline{O}(m \cdot \log_m N)$
P-Ring	$O(\log_d N)$	—	Yes	$\widetilde{O}(d \cdot \log_d N)$	$\widetilde{O}(d \cdot \log_d N)$
D^2-Tree	$O(\log N)$	—	No	$\widetilde{O}(\log N)$	$\widetilde{O}(\log N)$
D^3-**Tree**	$O(\log N)$	$\widetilde{O}(\log N)$	Yes	$\widetilde{O}(\log N)$	$\widetilde{O}(\log N)$

N: number of nodes; m: fanout; d: order of the ring; $\widetilde{O}$: amortized bound; $\overline{O}$: expected amortized bound; Theor: theoretical bound; Exp: empirical evidence.

that D^3-Tree has a significantly small redistribution rate (structure redistributions after node joins or departures), while element load-balancing is rarely necessary. We also investigated the structure's fault tolerance in case of massive node failures and show that it achieves a significant success rate in element queries. Omitted details can be found in [10].

2   The D^3-Tree

In this section, we present D^3-Tree. A key feature is the weight-based mechanism (adopted from [2]), used for node redistribution after node updates and data load-balancing after element updates. The main idea is the almost equal distribution of elements among nodes, using *weights*, a metric showing how uneven is the load among nodes. The mechanism lazily updates the weight information on nodes, so load-balancing is performed only when it is absolutely necessary.

The new features of D^3-Tree are its enhanced fault-tolerant and search mechanisms, in case of node failures. The enhanced search operation is successful even when a considerable number of nodes fails. D^3-Tree is highly fault tolerant, since it supports a procedure of *node withdrawal* when a node is found unreachable, regardless of its position (internal node, leaf, bucket node). The success of these two operations is due to a small number of additional links a node maintains, through which it can reconstruct the routing table of a failed node.

2.1   The Structure

Let N be the number of nodes present in the network and let n denote the size of data elements residing in the nodes ($N \ll n$). The structure consists of two levels. The upper level is a Perfect Binary Tree (PBT) of height $O(\log N)$. The leaves of this tree are *representatives* of the *buckets* that constitute the lower level of the D^3-Tree. Each bucket is a set of $O(\log N)$ nodes which are structured as a doubly linked list.

Each node v of the D^3-Tree maintains an additional set of links (described below) to other nodes apart from the standard links which form the tree. The first four sets are inherited from the D^2-Tree, while the fifth set is a new one that contributes in establishing a better fault-tolerance mechanism.

1. Links to its father and its children.
2. Links to its adjacent nodes based on an in-order traversal of the tree.
3. Links to nodes at the same level as v. The links are distributed in exponential steps; the first link points to a node (if there is one) 2^0 positions to the left (right), the second 2^1 positions to the left (right), and the i-th link 2^{i-1} positions to the left (right). These links constitute the *routing table* of v and require $O(\log N)$ space per node.
4. Links to leftmost and rightmost leaf of its subtree. These links accelerate the search process and contribute to the structure's fault tolerance when a considerable number of nodes fail.

5. For leaf nodes only, links to the buckets of the nodes in their routing tables. The first link points to a bucket 2^0 positions left (right), the second 2^1 positions to the left (right) and the i-th link 2^{i-1} positions to the left (right). These links require $O(\log N)$ space per node and keep the structure fault tolerant, since each bucket has multiple links to the PBT.

The next lemma [2] captures some important properties of the routing tables.

Lemma 1. *(i) If a node v contains a link to node u in its routing table, then the parent of v also contains a link to the parent of u, unless u and v have the same father. (ii) If a node v contains a link to node u in its routing table, then the left (right) sibling of v also contains a link to the left (right) sibling of u, unless there are no such nodes. (iii) Every non-leaf node has two adjacent nodes in the in-order traversal, which are leaves.*

Regarding the index structure of the D^3-Tree, the range of all values stored in it is partitioned into sub-ranges each one of which is assigned to a node of the overlay. An internal node v with range $[x_v, x'_v]$ may have a left child u and a right child w with ranges $[x_u, x'_u]$ and $[x_w, x'_w]$ respectively such that $x_u < x'_u < x_v < x'_v < x_w < x'_w$. Ranges are dynamic in the sense that they depend on the values maintained by the node.

2.2   Node Joins and Departures

When a node z makes a join request to v, v forwards the request to an adjacent leaf u. If v is a PBT node, the request is forwarded to the left adjacent node, w.r.t. the in-order traversal, which is definitely a leaf (unless v is a leaf itself). In case v is a bucket node, the request is forwarded to the bucket representative, which is leaf. Then, node z is added to the doubly linked list of the bucket represented by u. In node joins, we make the simplification that the new node is clear of elements and we place it after the most loaded node of the bucket. Thus, the load is shared and the new node stores half of the elements of the most loaded node.

When a node v leaves the network, it is replaced by an existing node, so as to preserve the in-order adjacency. All navigation data are copied from the departing node v to the replacement node, along with the elements of v. If v is an internal PBT node, then it is replaced by its right adjacent node, which is a leaf and which in turn is replaced by the first node z in its bucket. If v is a leaf, then it is directly replaced by z. Then v is free to depart.

After a node join or departure, the modified weight-based mechanism [2] is activated and updates the sizes by ± 1 on the path from the leaf u to the root. Afterwards, the mechanism traverses the path from u to the root, in order to find the first unbalanced node (if such a node exists) and performs a *redistribution* in its subtree. The redistribution guarantees that if there are x nodes in total in the y buckets of the subtree of v, then after the redistribution each bucket maintains either $\lfloor x/y \rfloor$ or $\lfloor x/y \rfloor + 1$ nodes. The redistribution cost is $O(\log N)$ [2], which is indeed verified by our experiments.

The redistribution of nodes in the subtree of v starts from the rightmost bucket b and it is performed in an in-order fashion so that elements in the nodes are not affected. The transfer of nodes is accomplished by maintaining a link, called *dest*, to the bucket representative b' in which nodes should be put or taken from. In case b has q extra nodes, the nodes are removed from b and are added to b'. Finally, bucket b informs b' to take over and the same procedure applies again with b' as the source bucket. The case where q nodes must be transferred to bucket b from bucket b' is completely symmetric.

Throughout joins and departures of nodes, the size of buckets can increase undesirably or can decrease so much that some buckets may become empty. The structure guarantees that each bucket contains $O(\log N)$ nodes, throughout joins or departures of nodes, by employing two operations on the PBT, the *contraction* and the *extension*.

2.3   Single and Range Queries

The search for an element a may be initiated from any node v at level l. If v is a bucket node, then if its range contains a the search terminates, otherwise the search is forwarded to the bucket representative, which is a binary node. If v is a PBT node, then let z be the node with range of values containing a, $a \in [x_z, x'_z]$ and assume w.l.o.g. that $x'_v < a$. The case where $x_v > a$ is completely symmetric. First, we perform a horizontal binary search at the level l of v using the routing tables, searching for a node u with right sibling w (if there is such sibling) such that $x'_u < a$ and $x_w > a$.

Having located nodes u and w, the horizontal search is terminated and a vertical search is initiated. Node z will either be the common ancestor of u and w, or it will be in the right subtree rooted at u, or in the left subtree rooted at w. Node u contacts the rightmost leaf y of its subtree. If $x_y > a$ then an ordinary top down search from node u will suffice to find z. Otherwise, node z is in the bucket of y, or in its right in-order adjacent (this is also the common ancestor of u and w), or in the subtree of w.

When z is located, if a is found in z then the search was successful, otherwise a is not stored in the structure. The search for an element a is carried out in $O(\log N)$ steps [2], and it is indeed verified by our experiments.

A range query $[a, b]$ initiated at node v, invokes a search operation for element a. Node z that contains a returns to v all elements in its range. If all elements of u are reported then the range query is forwarded to the right adjacent node (in-order traversal) and continues until an element larger than b is reached for the first time.

2.4   Element Insertions and Deletions

Assume that an update operation (insertion/deletion) is initiated at node v involving element a. By invoking a search operation, node u with range containing element a is located and the update operation is performed on u.

In order to apply the weight-based mechanism for load balancing, the element should be inserted in a bucket node (similar to node joins) or in a leaf. If u is an internal node of the PBT, then element a is inserted in u and then the first element of u (note that elements into nodes are sorted) is removed from u and it is inserted into node q, which is the last node of the bucket of the left adjacent of u, in order to preserve the sequence of elements in the in-order traversal. This way, the insertion has been shifted to a bucket node. The case of element deletion is similar.

After an element update in leaf u or in its bucket, the weight-based mechanism is activated and updates the weights by ± 1 on the path from leaf u to the root. Afterwards, the mechanism traverses the path from leaf u to the root, in order to find the first node (if such a node exists) which is unbalanced and performs a load-balancing in its subtree.

The load-balancing mechanism guarantees that if there are $w(v)$ elements in total in the subtree of v of size $|v|$ (total number of nodes in the subtree of v including v), then after load-balancing each node stores either $\left\lfloor \frac{w(v)}{|v|} \right\rfloor$ or $\left\lfloor \frac{w(v)}{|v|} \right\rfloor + 1$ elements. The load-balancing cost is $O(\log N)$ [2], which is indeed verified by our experiments. The load-balancing mechanism is similar to the redistribution mechanism described above, so its description is omitted.

2.5   Fault Tolerance

Searches and updates in the D^3-Tree do not tend to favour any node, and in particular nodes near the root. However, a single node can be easily disconnected from the overlay, when all nodes with which it is connected fail. This means that 4 failures (two adjacent nodes and two children) are enough to disconnect the root. The most easily disconnected nodes are those which are near the root, since their routing tables are small in size.

When a node w discovers that v is unreachable, the network initiates a *node withdrawal* procedure by reconstructing the routing tables of v, in order for v to be removed smoothly, as if v was departing. If v belongs to a bucket, it is removed from the structure and the links of its adjacent nodes are updated. In case v is an internal binary node, its right adjacent node u is first located, making use of Lemma 1, in order to replace v.

If v is a leaf, then it should be replaced by the first node u in its bucket. In the D^2-Tree, if a leaf was found unreachable, contacting its bucket would be infeasible, since the only link between v and its bucket would have been lost. This weakness was eliminated in the D^3-Tree, by maintaining multiple links towards each bucket, distributed in exponential steps (in the same way as the horizontal adjacency links). This way, when w is unable to contact v, it contacts directly the first node of its bucket u and u replaces v. Regardless of node's v position in the structure, the elements stored in v are lost.

2.6   Single Queries with Node Failures

In a network with node failures, an unsuccessful search for element a refers to the cases where either z (the node with range of values containing a, i.e., $a \in [x_z, x'_z]$) is unreachable, or there is a path to z but the search algorithm can not follow it to locate z due to failures of intermediate nodes. The D^2-Tree provides a preliminary fault-tolerant mechanism that succeeds only in the case of a few node failures. That mechanism cannot deal with massive node failures (also known as churn), i.e., its search algorithm may fail to locate a. In the following, we present the key features of our D^3-Tree efficient search algorithm in case of massive node failures.

The search procedure is similar to the simple search described in Section 2.3. One difference in horizontal search lies in the fact that if the most distant right adjacent of v is unreachable, v keeps contacting its right adjacent nodes by decreasing the step by 1, until it finds node q which is reachable.

In case $x'_q < a$ the search continues to the right using the most distant right adjacent of q, otherwise the search continues to the left and q contacts its most distant left adjacent p which is in the right of v. If p is unreachable, q doesn't decrease the travelling step by 1, but contacts directly its nearest left adjacent (at step = 0) and asks it to search to the left. This improvement reduces the number of messages that are meant to fail, because of the exponential positions of nodes in routing tables and the nature of binary horizontal search. For example, in Fig. 1, the search starts from v_0 and v_8 contacts v_7, since v_4 has failed. No node contacts v_4 from then onwars and the number of messages is reduced by 2.

A vertical search to locate z is always initiated between two siblings u and w, which are either both active, or one of them is unreachable, as shown in Fig. 2 where the left sibling u is active and w, the right one, is unreachable. In both cases, first we search into the subtree of the active sibling, then we contact the common ancestor and then, if the other sibling is unreachable, the active sibling tries to contact its corresponding child (right child for left sibling and left child for right sibling). When the child is found the search is forwarded to its subtree.

In general, when node u wants to contact the left (right) child of unreachable node w, the contact is accomplished through the routing table of its own left

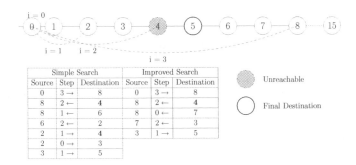

Simple Search			Improved Search		
Source	Step	Destination	Source	Step	Destination
0	3 →	8	0	3 →	8
8	2 ←	4	8	2 ←	4
8	1 ←	6	8	0 ←	7
6	2 ←	2	7	2 ←	3
2	1 →	4	3	1 →	5
2	0 →	3			
3	1 →	5			

Fig. 1. Example of binary horizontal search with node failures

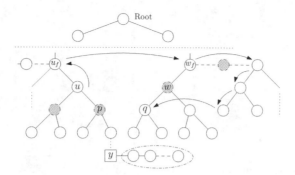

Fig. 2. Example of vertical search between u and unreachable w

(right) child. If its child is unreachable (Fig. 2), then u contacts its father u_f and u_f contacts the father of w, w_f, using Lemma 1(i). Then w_f, using Lemma 1(ii) twice in succession, contacts its grandchild through its left and right adjacents and their grandchildren.

In case initial node v is a bucket node, then if its range contains a the search terminates, otherwise the search is forwarded to the bucket representative. If the bucket representative has failed, the bucket contacts its other representatives right or left, until it finds a representative that is reachable. The procedure continues as described above for the case of a binary node.

The following lemma gives the amortized upper bound for the search cost in case of massive failures of $O(N)$ nodes.

Lemma 2. *The amortized search cost in case of massive node failures is $O(\log N)$.*

3 Experimental Study

We have built a simulator[2] with a user friendly interface and a graphical representation of the structure, to evaluate the performance of D^3-Tree. To evaluate the cost of operations, we ran experiments with different number of nodes N from 1,000 to 10,000, in order to be directly compared to BATON, BATON* and P-Ring. BATON* is a state-of-the-art decentralized architecture and P-Ring outperforms DHT-based structures in range queries and achieves a slightly better load-balancing performance compared to BATON*. For a structure of N nodes, 1000 x N elements where inserted. We used the number of passing messages to measure the performance of the system.

Cost of Node Joins/Departures: To measure the network performance for the operation of node updates, in a network of N initial nodes, we performed $2N$ node updates. In a preliminary set of experiments with mixed operations (joins/departures), we observed that redistributions rarely occurred, thus leading in negligible node update costs. Hence, we decided to perform only one type

---

[2] Our simulator is a standalone desktop application, developed in Visual Studio 2010, available in https://github.com/sourlaef/d3-tree-sim

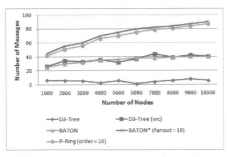
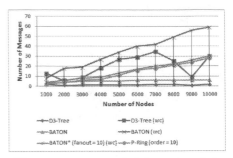

(a) average messages of node updates   (b) average messages of element updates

Fig. 3. Node and Element Update operations

of updates, $2N$ joins, that are expected to cause several redistributions. Fig. 3a
shows average case (nodes where joins occur are chosen randomly) and worst case
(joins occur only in the leftmost leaf), and in both cases the curves represent
the average amortized redistribution cost.

We observed that even for the worst case scenario, the D^3-Tree node update
and redistribution mechanism achieves a better amortized redistribution cost,
compared to that of BATON, BATON* and P-Ring. In the average case, during node joins, redistribution is rarely necessary (about 3% of join operations
lead to redistributions). However, in the worst case, during node joins, a great
number of nodes are accumulated into the bucket of the leftmost leaf, leaving
the other buckets unchanged. This naturally leads to more frequent and costly
redistributions (about 9% of join operations lead to redistributions).

Cost of Element Insertions/Deletions: To measure the network performance
for the operation of element updates, in a network of N nodes and n elements, we
performed n element updates. In a preliminary set of experiments with mixed
operations (insertions/deletions), we observed that load-balancing operations
rarely occurred, thus leading in negligible node update costs. Hence, we decided
to perform only one type of updates, n insertions. Fig. 3b shows average case
(element insertions occur at nodes chosen randomly) and worst case (element
insertions occur only in the leftmost leaf), and in both cases the curves represent
the average amortized load-balancing cost.

Conducting experiments, we observed that in the average case, the D^3-Tree
outperforms BATON, BATON* and P-Ring. However, in D^3-Tree's worst case,
the load-balancing performance is degraded compared to BATON* of *fanout* =
10 and P-Ring. In the average case, during element insertions, load-balancing
is rarely necessary (about 15% of insertions lead to load-balancing operations).
However, in worst case, a great number of element insertions take place into the
bucket of the leftmost leaf, leaving the other nodes unaffected, thus rendering the
subtree imbalanced very often. This leads to more frequent and costly operations
of load-balancing (about 50% of insertions evoke load-balancing).

Cost of Element Search with/without Node Failures. To measure the network
performance for the operation of single queries, we conducted experiments in

which for each N, we performed $2M$ (M is the number of binary nodes) searches. The search cost is depicted in Fig. 4a. An interesting observation here was that although the cost of search in D^3-Tree doesn't exceed $2 \cdot \log N$, it is higher that the cost of BATON, BATON* and P-Ring. This is due to the fact that when the target node is a *Bucket* node, the search algorithm, after locating the correct leaf, performs a serial search into its bucket to locate it.

To measure the network performance for the operation of element search with node failures, we conducted experiments for different percentages of node failures: 10%, 20%, 30%, 50% and 75%. For each N and node failure percentage, we performed $2M$ searches divided into 4 groups, each of $M/2$ searches. In order to get a better estimation of the search cost, we forced a different set of nodes to fail in each group. Fig. 4b depicts the increase in search cost when massive node failures take place in D^3-Tree, BATON, different fanouts of BATON* and P-Ring. We observe that D^3-Tree maintains low search cost, compared to the other structures, even for a failure percentage $\geq 30\%$.

Describing the effect of the enhanced search mechanism of D^3-Tree in case of massive failures in more detail, we must note that when the node failure percentage is small (10% to 15%), the majority of single queries that fail are the ones whose elements belong to failed nodes. When the number of failed nodes increases, single queries are not always successful, since the search mechanism fails to find a path to the target node although the node is reachable. However, even for the significant node failure percentage of 30%, our search algorithm is 85% successful, confirming thus our claim about the structure's fault tolerance.

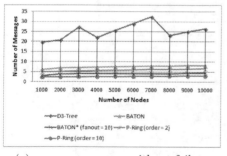
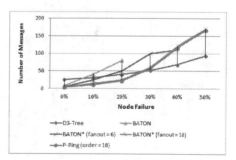

(a) average messages without failures   (b) effect of massive failures

Fig. 4. Single Queries without/with node failures

4 Conclusions

We presented D^3-Tree, a dynamic distributed deterministic structure, that turns out to be very efficient in practice and outperforms other state-of-the-art structures. Our experimental study showed (among others) that the $O(\log N)$ amortized bound for load balancing (the most costly operation) is achieved even for the worst case scenario. Moreover, investigating the structure's fault tolerance, we showed that D^3-Tree is highly fault tolerant, since even for a substantial amount of 30% node failures it achieves a significant success rate of 85% in element search, without increasing the search cost considerably.

Acknowledgments. This research has been co-financed by the European Union (European Social Fund - ESF) and Greek national funds through the Operational Program "Education and Lifelong Learning" of the National Strategic Reference Framework (NSRF) – Research Funding Programs Thales & Heracletus II, Investing in knowledge society through the European Social Fund.

References

1. Bhargava, A., Kothapalli, K., Riley, C., Scheideler, C., Thober, M.: Pagoda: A dynamic overlay network for routing, data management, and multicasting. In: ACM SPAA 2004, pp. 170–179 (2004)
2. Brodal, G., Sioutas, S., Tsichlas, K., Zaroliagis, C.: D^2-tree: A new overlay with deterministic bounds. Algorithmica 72(3), 860–883 (2015)
3. Crainiceanu, A., Linga, P., Machanavajjhala, A., Gehrke, J., Shanmugasundaram, J.: Load balancing and range queries in P2P systems using P-Ring. ACM Trans. Internet Technol. 10(4), Art.16, 1–16 (2011)
4. Gupta, A., Agrawal, D., Abbadi, A.E.: Approximate range selection queries in peer-to-peer systems. In: Proc. 1st Biennial Conference on Innovative Data Systems Research – CIDR (2003)
5. Jagadish, H.V., Ooi, B.C., Tan, K., Vu, Q.H., Zhang, R.: Speeding up search in P2P networks with a multi-way tree structure. ACM SIGMOD 2006, 1–12 (2006)
6. Jagadish, H.V., Ooi, B.C., Vu, Q.H.: Baton: a balanced tree structure for peer-to-peer networks. In: VLDB 2005, pp. 661–672 (2005)
7. Ozsu, M.T., Valduriez, P.: Principles of Distributed Database Systems. Springer (2011)
8. Sahin, O., Gupta, A., Agrawal, D., Abbadi, A.E.: A peer-to-peer framework for caching range queries. In: ICDE 2004, pp. 165–176 (2004)
9. Scheideler, C., Schmid, S.: A distributed and oblivious heap. In: Albers, S., Marchetti-Spaccamela, A., Matias, Y., Nikoletseas, S., Thomas, W. (eds.) ICALP 2009, Part II. LNCS, vol. 5556, pp. 571–582. Springer, Heidelberg (2009)
10. Sourla, E., Sioutas, S., Tsichlas, K., Zaroliagis, C.: D^3-tree: A dynamic distributed deterministic load–balancer for decentralized tree structures. Tech. Rep. ArXiv:1503.07905, ACM CoRR (March 2015)
11. Stoica, I., Morris, R., Karger, D., Kaashoek, M.F., Balakrishnan, H.: Chord: A scalable peer-to-peer lookup service for internet applications. SIGCOMM Comput. Commun. Rev. 31(4), 149–160 (2001)

Ignorant vs. Anonymous Recommendations

Jara Uitto and Roger Wattenhofer

ETH Zürich
{juitto,wattenhofer}@ethz.ch

Abstract. We start with an unknown binary $n \times m$ matrix, where the entries correspond to the preferences of n users on m items. The goal is to find at least one item per user that the user likes, with as few queries as possible. Since there are matrices where any algorithm performs badly without any preliminary knowledge of the input matrix, we reveal an anonymized version of the input matrix to the algorithm in the beginning of the execution. The input matrix is anonymized by shuffling the rows according to a randomly chosen hidden permutation. We observe that this anonymous recommendation problem can be seen as an adaptive variant of the Min Sum Set Cover problem and show that the greedy solution for the original version of the problem provides a constant approximation for the adaptive version.

1 Introduction

Algorithmic research studies a variety of models that are beyond the traditional input-output paradigm, where the whole input is given to the algorithm. Examples of such unconventional models are distributed, streaming, and multiparty algorithms (where the input is distributed in space), or regret, stopping, and online algorithms (where the input is distributed in time). Not having all the input initially is a drawback, and one will generally not be able to produce the optimal result. Instead, the algorithm designer often compares the result of the restricted online/distributed algorithm with the result of the best offline/centralized algorithm by means of competitive analysis.

However, it turns out, this is sometimes not possible. In particular, if the hidden input contains more information than we can learn within the execution of the algorithm, we might be in trouble. This is generally an issue in the domain of recommendation and active learning algorithms. In this paper, we study the following example. We are given an unknown binary matrix, where the entries correspond to preferences of n users on m items, i.e., entry (i, j) corresponds to whether user i likes item j. Thus, row i in the matrix can be seen as the *taste* vector of user i. In each round, the algorithm is allowed to reveal one entry in the matrix, i.e., query one user about one specific preference. The goal is to find at least one 1-entry in each row with a minimum number of queries. We call this problem the *ignorant* recommendation problem, since initially, the algorithm knows nothing about the taste matrix, and only over time (hopefully) learns about the taste of the users.

© Springer-Verlag Berlin Heidelberg 2015
N. Bansal and I. Finocchi (Eds.): ESA 2015, LNCS 9294, pp. 1001–1012, 2015.
DOI: 10.1007/978-3-662-48350-3_83

In the domain of recommendation and active learning, competitive analysis is still waiting to make its outburst. The approach is often to assume certain properties about the tastes of people, e.g., the users are partitioned into a small number of classes with very similar taste or, more abstractly, that the underlying taste matrix features certain algebraic properties such as low rank. Competitive analysis seems to be out of reach, exactly because a ignorant algorithm cannot compete against an algorithm that knows everything about the taste of the users.

Since we do not want to change the ignorant recommendation problem, our only hope is to make the competition weaker. What is the strongest model for the adversary that allows reasonable (or non-trivial) results? In this paper, we propose an *anonymous* version of the problem. In the anonymized problem, the adversarial algorithm knows the whole taste matrix, but the users are anonymous, i.e., the rows of the taste matrix have been permuted arbitrarily. We call this the anonymous recommendation problem.

We build on two previous results: First, a result that studies a *oblivious* version of the problem, called Min Sum Set Cover (**mssc**) [5]. The input for **mssc** is a collection of elements and a set of subsets of these elements, similarly to the classical Set Cover problem. The output is a linear order of these sets where the ordinal of the set that first covers an element e induces a cost $f(e)$ for e. The goal is to minimize the sum $\sum_e f(e)$ of the costs of the elements.

The **mssc** problem is oblivious in the sense that the strategy of an algorithm solving the **mssc** problem is independent of the recommendation history. In other words, the algorithm chooses an ordering of the items in the beginning of the execution and each user is recommended items according to this ordering. Our setting on the other hand allows the algorithm to be *adaptive* and change the strategy after each recommendation. In the oblivious setting, it is known that the greedy algorithm is a 4-approximation [5]. The second previous result compares the ignorant problem to the oblivious problem [19]. We strengthen this result by showing that the bounds hold (asymptotically) even when comparing the ignorant problem to the anonymous problem, instead of the oblivious problem. The relations between the aforementioned problems are illustrated in Figure 1.

The core of this paper hence deals with the power of anonymous input, showing that a good solution for the oblivious problem also yields a good result for the anonymous problem. In particular, we show that the greedy algorithm for

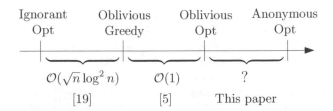

Fig. 1. We show that the greedy algorithm, and thus also the optimal algorithm, for the oblivious recommendation problem provide a constant approximation to the anonymous recommendation problem.

mssc yields a constant approximation to the anonymous problem. In this sense, our problem is an anonymous and adaptive variant of the **mssc** problem.

2 Related Work

The classic result related to the Min Sum Set Cover problem is that the greedy algorithm provides a constant approximation. It was shown by Bar-Noy, Bellare, Halldórsson, Shachnai, and Tamir [5] that the greedy solution is a 4-approximation. Feige, Lovász, and Tetali gave a simpler proof for this result and showed that getting an approximation ratio of $4 - \epsilon$ for any $\epsilon > 0$ is NP-hard [9]. Our work extends their results from the oblivious and offline setting into an adaptive and online setting, where the algorithm is allowed to change its strategy during the execution but is not given full information in the beginning. Online variants of the **mssc** problem have been studied before for example by Munagala et al. [17], who showed that even if the elements contained in the sets are hidden from the algorithm, one can achieve an $\mathcal{O}(\log n)$-approximation.

Also other variations of **mssc** have been considered. As an example, Azar and Gamzu studied ranking problems, where the goal is to maintain an adaptive ranking, i.e., an ordering of the sets that can change over time, while learning in an active manner [4]. They provided an $\mathcal{O}(\log(1/\epsilon))$-approximation algorithm for ranking problems, where the cost functions have submodular valuations. Golovin and Krause [12] studied problems with submodular cost functions further and in particular, they considered them in an adaptive environment. Furthermore, **mssc** is not the only classic optimization problem studied in an active or an adaptive environment. There exists work on adaptive and active versions of, for example, the well-known Set Cover [10,16], Knapsack [7], and Traveling Salesman [14] problems.

The input for the anonymous recommendation problem is a binary relation. Learning binary relations has been studied for example by Goldman et al [11]. They studied four different learning models: adversarial, random, a helpful teacher and similar to ours, a setting where the learner can choose which entry to look at. They studied upper and lower bounds on the number of mistakes the learner makes when predicting the entries in the input matrix. As a byproduct, they showed there are inputs where any algorithm can be forced to make $\Omega(k \cdot m)$ mistakes given a matrix with k different row types. In our setting, different row types correspond to users with different taste vectors and therefore, their bound immediately gives us a corresponding lower bound for any recommendation algorithm without any preliminary knowledge of the input, including **mssc** based solutions.

A common way to overcome such lower bounds is to perform a competitive analysis. However, an offline algorithm that sees the whole input for the ignorant recommendation problem can always solve the problem with 1 query per user. It was shown recently that if an online algorithm for the ignorant recommendation problem is compared to the optimal solution for **mssc**, the analysis becomes non-trivial [19]. Our results extend this work by showing that the *quasi*-competitive ratio stays asymptotically the same if we compare the online solution to the optimal anonymous algorithm. We also improve the results from previous work by allowing the recommendation algorithms to choose the sequence according to which the users are picked. In the previous work, the sequence was chosen uniformly at random. Another way to relax the competitive analysis is to give the online algorithm more power. For example the list update and bin packing problems have been studied under more powerful online algorithms [1,13].

The task we are considering can also be seen as a relaxed version of learning the identities of the users, that is, we wish to classify the unknown users into groups according to their preferences. Since the users are determined by their preferences, this can further be seen as finding a matching between the users and the preferences. The matching has to be perfect, i.e., in the end every user has to be matched to a unique preference. A similar setting was studied in economics, where the basic idea is that each buyer and seller have a hidden valuation on the goods that they are buying or selling and the valuations are learned during the execution. Then the goal is to find a perfect matching between a set of buyers and a set of sellers, where an edge in the matching indicates a purchase between the corresponding agents [6,15].

The main motivation for our work comes from the world of recommendations and the main interpretation of our variation of **mssc** is an online recommendation problem. Models of recommendation systems close to ours were studied by Drineas et al. [8], Awerbuch et al. [3] and Nisgav and Patt-Shamir [18], where the recommendation system wishes to find users that have interests in common or good items for users. One of their real life examples are social networks. In their works, the users are assumed to have preferences in common, whereas we study an arbitrary feasible input. With an arbitrary input, Alon et al. [2] showed that one can learn the whole preference matrix with minimal error in polylogarithmic time in a distributed setting.

3   Model

The input for the anonymous recommendation problem is a pair (U, V) consisting of a set of users $U = \{u_1, \ldots, u_n\}$ and a set of preference vectors $V = \{v_1, \ldots, v_n\}$ of length m, where preference vector $v \in V$ corresponds to the (binary) preferences of some user $u \in U$ on m items. Each user u_i is assigned exactly one preference vector v_j according to a hidden bijective mapping $\pi : U \to V$. By identifying the users with the preference vectors, π is a permutation of the users. The permutation π is chosen uniformly at random from the set of all possible permutations.

The execution of a recommendation algorithm works in rounds. In each round, a recommendation algorithm first picks a user $u \in U$ and then recommends some item b to this user. Recommending item b to user u is equivalent to checking whether user u likes b or not, i.e., the corresponding entry is revealed to the algorithm immediately after the recommendation. A recommendation algorithm is allowed to pick the user and the item at random.

We say that user u is *satisfied* after she has been recommended an item that she likes. The goal is to satisfy all users which corresponds to finding at least one 1-entry from each preference vector. The algorithm terminates when all users are satisfied. The runtime of a recommendation algorithm is measured as the expected number of queries. In other words, the runtime corresponds to the number of rounds until all users are satisfied. Therefore, the trivial upper and lower bound for the runtime are $n \cdot m$ and n, respectively, since it takes $n \cdot m$ queries to learn every element of every preference vector and n queries to learn one entry from each preference vector.

We assume that for any user u, there is always at least one item that she likes but we do not make any further assumptions on the input. Let OPT be the optimal recommendation algorithm for the anonymous recommendation problem. We measure the quality of a recommendation algorithm A by its approximation ratio, i.e., the maximum ratio between the expected number of queries by A and by OPT for any input I.

An important concept throughout the paper is the *popularity* of an item. The popularity of an item corresponds to the number of users that like it.

Definition 1. *Let b be an item. The popularity $|b|$ of item b is the number of users that like this item, i.e., $|b| = |\{v \in V \mid v(b) = 1\}|$.*

4   Anonymous Recommendations

The main goal of this paper is to show that from an asymptotic perspective, the anonymous and the oblivious recommendation problems are equally hard. Recall that in the oblivious setting, the algorithm sees a probability distribution D over the set of possible preference vectors for the users and must fix an ordering O of the items before the first query. Then, each user is recommended items according to O until she is satisfied. Otherwise, the oblivious model is similar to the anonymous model. To achieve our goal, we first observe that solving the oblivious recommendation problem takes at least as much time as solving the anonymous recommendation problem for any instance of preferences selected according to D. Clearly, an anonymous algorithm that chooses the best fixed ordering of books is at least as fast as any oblivious algorithm for this instance.

Then we show that the greedy algorithm for the oblivious recommendation problem is a 20-approximation to the anonymous recommendation problem. We follow the general ideas of the analysis of the greedy algorithm for **mssc** by Feige et al. [9], where they show that the greedy algorithm provides a 4-approximation. The fundamental difference between our analysis and theirs comes from bounding from below the number of recommendations needed to satisfy a given set $U' \subseteq U$ of users. In the oblivious setting, it is easy to get a lower bound on the number of recommendations needed per user. Given the most popular item b^* among users in U', $\Omega(|U'|^2/|b^*|)$ recommendations are needed, since one item can satisfy at most $|b^*|$ users, and each item is recommended to all unsatisfied users. In the adaptive setting, this is not necessarily the case.

We first give a lower bound on the amount of queries that are needed to satisfy any group of users as a function of the best item within this group. In essence, we show that from an asymptotic and amortized perspective, any adaptive algorithm also needs $\Omega(|U'|^2/|b^*|)$ rounds to satisfy all users in U'. Then, we utilize our lower bound and get that the greedy algorithm for **mssc** yields a constant approximation to the anonymous recommendation problem.

4.1   Consistency Graph

We identify the users with their preference vectors, which indicates that each user $u \in U$ corresponds to an (initially) unknown binary preference vector of length m. Therefore, each user can be seen as a preference vector that denotes the information we have gained about user u. We also identify the items with their corresponding indices, i.e., for an item b that has been recommended to u, $u(b)$ denotes the entry in the preference vector of user u that corresponds to whether u likes b or not. We call $u \in U$ and $v \in V$ *consistent*, if $u(i) = v(i)$ for every revealed entry $u(i)$.

Let OPT be the optimal anonymous recommendation algorithm. We model the state of an execution of OPT as a bipartite graph $G = (U \cup V, E)$, where $(u, v) \in E$ iff $u \in U$ and $v \in V$ are consistent. We refer to G as the *consistency graph*. The purpose of the consistency graph is to model the uncertainty that OPT has on the preferences of the users. To simplify our analysis, we provide OPT with the following advantage. Whenever OPT recommends user u an item b that u likes, we identify $u \in U$ with $v \in V$ in permutation π, i.e., OPT learns that $\pi(u) = v$. We note that this advantage can only improve the runtime of OPT, i.e., if we prove a lower bound for the performance of this "stronger" version of OPT, the same bound immediately holds for the optimal anonymous recommendation algorithm.

Now since the connection is revealed after finding a 1-entry and thus, the complete preference vector of u is learned, nothing further can be learned by recommending u more items. Therefore, we can simply ignore $u \in U$ and $\pi(u) \in V$ for the rest of the execution. Thus, upon recommending user u an item that she likes, we simply remove u from U and the corresponding preference vector $\pi(u)$ from V. The modification also implies that the termination condition, i.e., all users being satisfied, is equivalent to the sets U and V becoming empty.

The construction of G is illustrated in Figure 2. We emphasize that the graph G changes over time and we denote the state of G in round $r \geq 0$ by $G_r = (U_r \cup V_r, E_r)$, where U_r and V_r are the users and their preference vectors remaining in round r, respectively, and E_r is the set of edges between consistent nodes in round r. Notice that $G_0 = (U_0 \cup V_0, E_0)$ is a complete bipartite graph. We omit the index from the consistency graph whenever the actual round number is irrelevant. In addition, we note that even if the identity of a certain user u is clear (see user u_3 in Figure 2 for an example), the edges connected to u and $\pi(u)$ are only removed from G when the corresponding users and preference vectors become inconsistent.

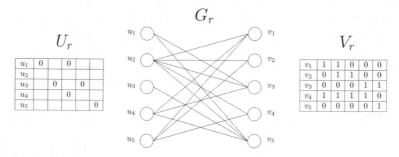

Fig. 2. A matrix representation of the unknown and the known entries of an input are given on the left and right, respectively. The consistency graph constructed based on V_r and the state of U_r is denoted by G_r. Nodes u_i and v_j are connected if and only if the corresponding rows are consistent.

5   Learning the Preferences

The goal of this section is to quantify the amount of knowledge OPT can gain per round. The intuition behind modeling the anonymous recommendation problem as a bipartite graph is that the number of remaining edges correlates with the amount of uncertainty OPT has. In other words, by querying the preferences of the users OPT can exclude inconsistent edges in G. When a 0-entry is discovered by recommending item b to user u, at most $|b|$ preference vectors can become inconsistent with u since there are $|b|$ vectors $v \in V$ such that $v(b) = 1$. On the other hand, when discovering a 1-entry, up to $2|U|$ edges might get removed due to removing u, $\pi(u)$ and all edges adjacent to them from G. Note that the consistency graph is simply a representation of the knowledge OPT has about the preference vectors, i.e., excluding the edges from the consistency graph only happens implicitly according to the revealed entries.

We employ an amortized scheme, where we pay in advance for edges that get removed by discovering 1-entries. Consider the case where a 0-entry is revealed from $u(b)$. Now all the edges $(u, v) \in E$, where $u(b) \neq v(b)$, are removed. For every edge (u, v) removed this way, we give both node u and node v two units of money that can be used later when their corresponding connections are revealed. Since at most $|b|$ edges are removed, we pay at most $2|b| + 2|b| = 4|b|$ units of money in total in a round where OPT discovers a 0-entry.

As an example, consider the graph illustrated in Figure 2 and assume that a 0 is revealed from $u_4(5)$. Now u_4 becomes inconsistent with v_3 and v_5, and the corresponding edges are removed. Upon removing these edges, we give two units of money to v_3, two units of money to v_5 and four units of money to u_4.

5.1   Finding a 1-Entry

Now we look at the case of discovering a 1-entry. In the following, we consider the consistency graph G_r for an arbitrary round r but omit the index when it is irrelevant for the proofs. Consider user $u \in U$ and let $\pi(u) = v$. Upon discovering the 1-entry from user u, we reveal that $\pi(u) = v$ and all the edges adjacent to u and v are removed. We divide the analysis into two cases. Consider first the case where $|\Gamma(u)| + |\Gamma(v)| \leq 4|U|/3$, where $\Gamma(u)$ denotes the exclusive neighborhood of u. The exclusive neighborhood of node $u \in U$ in graph $G = (U \cup V, E)$ contains all the nodes in the 1-hop neighborhood of u except node u itself, i.e., $\Gamma(u) = \{v \in V \mid (u, v) \in E\}$ and analogously for $v \in V$.

Note that since satisfied users are removed from U, $G = (U \cup V, E)$ is a complete bipartite graph if there are no revealed 0-entries. Therefore, any edge $(u, v) \notin E$, where $u \in U = U_r, v \in V = V_r$, was removed by revealing a 0-entry. Given that $|\Gamma(u)| + |\Gamma(v)| \leq 4|U|/3$, we know that at least $2|U| - 4|U|/3$ of the edges adjacent to u or v were removed by revealing 0-entries. We pay two units of money to either u or v, for every edge removed from the set of edges adjacent to nodes in $\Gamma(u) \cup \Gamma(v)$ and therefore, the combined money that the nodes have is at least $2(2|U| - 4|U|/3) = 4|U|/3$. Therefore, the money "pays" for all edges that are removed due to revealing the connection between u and v.

We use the rest of this section to study the second case, that considers the case where the sum of degrees of nodes u and v is high, i.e., larger than $4|U|/3$. The aim is to show that it is unlikely that v is the preference vector of u, since there are many consistent nodes with u and v and thus, there has to be considerably more valid permutations π', where $\pi'(u) \neq v$, than permutations, where $\pi'(u) = v$. This in turn implies that a randomly chosen permutation is likely not to have u connected to v.

We call a matching σ *compatible* with an edge $e = (u, v)$ if $(u, v) \in \sigma$ and *incompatible* with e otherwise. In the following lemma, we bound the number of perfect matchings that are compatible with a given edge e in G. Note that every perfect matching in G corresponds to some permutation of the users.

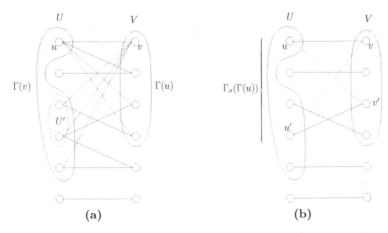

Fig. 3. The consistency graph is illustrated on the left. On the right, we show a perfect matching compatible with (u, v) with the solid lines. For every node $u' \in U'$, we have a valid perfect matching that is incompatible with (u, v) if we use edges (u, v') and (u', v) instead of (u, v) and (u', v').

Lemma 1. *Assume that $|\Gamma(u)| + |\Gamma(v)| > 4|U|/3$ for some nodes u and v in G. Let h be the total number of perfect matchings in G. Then there are at most $3h/|U|$ perfect matchings that are compatible with (u, v).*

Proof. Let σ be a perfect matching that is compatible with (u, v) in G. Let

$$U' = \{u' \in \Gamma(v) \mid (u', v') \in \sigma \text{ and } v' \in \Gamma(u)\} \setminus \{u\} \,,$$

$|\Gamma(u)| = k$ and let $\Gamma_\sigma(\Gamma(u)) = \{u' \in U \mid (u', v') \in \sigma \text{ and } v' \in \Gamma(u)\}$ be the set of nodes matched to $\Gamma(u)$ by σ. See Figure 3 for an illustration. Since σ is a matching, we get that $k = |\Gamma(u)| = |\Gamma_\sigma(\Gamma(u))|$. Also, we know that $|\Gamma(v)| + |\Gamma(u)| > 4|U|/3$ and therefore $|\Gamma(v)| \geq 4|U|/3 - k + 1$.

By taking a closer look at the definition of U', we see that $U' = \Gamma(v) \cap \Gamma_\sigma(\Gamma(u)) \setminus \{u\}$ and by re-writing, we get that $U' = \Gamma(v) \setminus (U \setminus \Gamma_\sigma(\Gamma(u))) \setminus \{u\}$. From the equations above, it follows that

$$|U'| \geq |\Gamma(v)| - (|U| - k) - 1 \geq \frac{4|U|}{3} - k + 1 - (|U| - k) - 1 = \frac{|U|}{3} \,.$$

For each node $u' \in U'$, we have a perfect matching $\sigma_{u'}$ that is incompatible with (u, v) in G, where (u, v) and $(u', v') \in \sigma$ are replaced by (u, v') and (u', v). In addition, the incompatible perfect matching $\sigma_{u'}$ is different for every $u' \in U'$, since $(u', v) \notin \sigma_z$ for any $u' \neq z \in U'$. Therefore, we have at least $|U|/3$ perfect matchings incompatible with (u, v) for every perfect matching that is compatible with (u, v). Note that no matchings are counted twice. The claim follows. $\square$

In the beginning of the execution, the probability of user $u \in U$ to be matched to vector $v \in V$ is simply $1/n$. When revealing the unknown entries, these probabilities change. The next step is to bound the probability of user $u \in U$ to be matched to vector $v \in V$ given the state of the consistency graph G. We identify the randomly chosen permutation π with a perfect matching σ_π where $(u, v) \in \sigma_\pi$ iff $\pi(u) = v$. Since the permutation π was chosen uniformly at random, any valid permutation, i.e., a permutation that does not contradict the revealed entries, is equally likely to be σ_π. Therefore, the probability that an edge (u, v) is in matching σ_π corresponds to the ratio of perfect matchings in G that are compatible with (u, v) and the number of all perfect matchings in G. We denote the event that edge (u, v) is in σ_π by $A(u, v)$ and the probability of $A(u, v)$ given G by $\mathbb{P}[A(u, v) \mid G]$.

Lemma 2. *(Proof deferred to the full version) Let $G = (U \cup V, E)$ be the consistency graph. For any nodes $u \in U$ and $v \in \Gamma(u)$, such that $|\Gamma(u)| + |\Gamma(v)| > 4|U|/3$, $\mathbb{P}[A(u, v) \mid G] \leq \frac{4}{|\Gamma(u)| + |\Gamma(v)|}$.*

5.2   Progress

Now, we define the *progress* $c(u, b, r)$ for recommending item b to user u in round $r \geq 0$. Informally, the idea of the progress value is to employ the money paid during the execution to bound the expected number of edges removed per round.

Consider any round r and let $w_r(z)$ denote the *wealth* of node $z \in U_r \cup V_r$, where wealth refers to the amount of money z has in round r. In the case of revealing a 0-entry, the progress indicates the number of removed edges and the money that is paid to the nodes adjacent to the removed edges. When finding a 1-entry and revealing the connection between u and $\pi(u)$, the progress indicates the number of removed edges minus the money already paid to u and $\pi(u)$. Let $\Gamma_r(u) = \{v \in V_r \mid (u,v) \in E_r\}$ and $\Gamma_r^b(u)$ denote the neighbors of u that like item b, i.e., $\Gamma_r^b(u) = \{v \in V_r \mid (u,v) \in E_r \wedge v(b) = 1\}$. Then, for entry $u(b)$ revealed in round r, the progress is given by

$$c(u,b,r) = \begin{cases} \sum_{v \in \Gamma_r^b(u)} 5 & \text{if } u(b) = 0 \\ |\Gamma_r(u)| + |\Gamma_r(\pi(u))| - (w_r(u) + w_r(\pi(u))) & \text{if } u(b) = 1 . \end{cases}$$

An illustration of the wealth and progress concepts is given in Figure 4. In the example given in Figure 4, revealing entry $u_2(1) = 1$, denoted by x, has a progress value of $|\Gamma(u_2)| + |\Gamma(v_4)| - (w(u_2) + w(v_4)) = 5 + 2 - 0 - 6 = 1$ and revealing entry $u_4(5) = 0$, denoted by y, yields a progress of $\sum_{v_3,v_5} 5 = 10$.

Next, we show that the total progress value counted from the first round up to any round r is never smaller than the number of edges removed from G within the first r rounds. We denote an execution of an algorithm until round r by $\mathcal{E}_r = (u^1, b^1), (u^2, b^2), \ldots, (u^{r-1}, b^{r-1})$, where u^i corresponds to the user selected in round $i < r$ and similarly for item b^i.

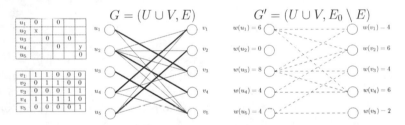

Fig. 4. In graph G' the dashed lines indicate the edges removed from the consistency graph during the execution of an algorithm. The wealth of each node is illustrated next to the corresponding node. The bold lines denote the underlying permutation π that connects users in U with their preference vectors in V.

Lemma 3. *(Proof deferred to the full version) For any round r and execution $\mathcal{E}_r$, it holds that $\sum_{i=0}^{r-1} c(u^i, b^i, i) \geq |E_0| - |E_r|$.*

The last thing we need to show is an upper bound on the expected progress per round for any round r. For ease of notation, we omit the round index for the rest of the section. In addition, we identify $c(u,b)$ with a random variable that equals to the progress gained by revealing entry (u,b).

Lemma 4. *Let b^* be the most popular item among the set U of unsatisfied users. Then for any user $u \in U$ and item b, $\mathbb{E}[c(u,b)] \leq 5|b^*|$.*

Proof. Consider any user $u \in U$ and any item b. We partition the event space into three disjoint parts according to the outcome of querying user u for item b and show that for each part, $\mathbb{E}[c(u,b)] \leq 5|b^*|$. First, consider the case where a 0-entry is revealed. By definition, we get $\mathbb{E}[c(u,b) \mid u(b) = 0] \leq 5|b^*|$. Furthermore,

$$\mathbb{E}\big[c(u,b) \mid (u(b) = 1) \wedge \big(|\Gamma(u)| + |\Gamma(\pi(u))| \leq 4|U|/3\big)\big] \leq 0 \ ,$$

since $w(u) + w(\pi(u)) \geq 4|U|/3$ given that $|\Gamma(u)| + |\Gamma(\pi(u))| \leq 4|U|/3$.

Let us then consider the third part of the event space, where $u(b) = 1$ and $|\Gamma(u)| + |\Gamma(\pi(u))| > 4|U|/3$ and let us denote this event by B. Let

$$E' = \{(u,v) \in E \mid |\Gamma(u)| + |\Gamma(\pi(u))| > 4|U|/3)]\}$$

and $\hat{\Gamma}(u) = \{v \in \Gamma(u) \mid (u,v) \in E' \wedge v(b) = u(b) = 1\}$. Then, by Lemma 2,

$$\mathbb{E}[c(u,b) \mid B] \leq \sum_{v \in \hat{\Gamma}(u)} (|\Gamma(u)| + |\Gamma(v)|) \cdot \mathbb{P}[A(u,v) \mid G]$$

$$\leq \sum_{v \in \hat{\Gamma}(u)} (|\Gamma(u)| + |\Gamma(v)|) \frac{4}{|\Gamma(u)| + |\Gamma(v)|} = \sum_{v \in \hat{\Gamma}(u)} 4 \leq 4|b^*| \ ,$$

where $|b^*| \geq |b| \geq |\hat{\Gamma}(u)|$, since b^* is the most popular item. Since the three aforementioned parts span the whole probability space, $\mathbb{E}[c(u,b)]$ is bounded by the maximum of the three cases and thus, the claim follows. $\square$

Theorem 1. *(Proof deferred to the full version) Let $R \subseteq U_0$ be a set of users and b^* the most popular item among these users. Any algorithm requires at least $|R|^2/(5|b^*|)$ queries to users in R in expectation to satisfy all users in R.*

As the last step, we establish that the greedy algorithm for the **mssc** problem provides an $\mathcal{O}(1)$-approximation for the anonymous recommendation problem. Due to the space constraints, we defer the proof of the theorem to the full version.

Theorem 2. *(Proof deferred to the full version) The greedy **mssc** algorithm provides a 20-approximation for the anonymous recommendation problem.*

References

1. Susanne Albers. A Competitive Analysis of the List Update Problem with Lookahead. *Theoretical Computer Science*, 197:95–109, 1998.
2. Noga Alon, Baruch Awerbuch, Yossi Azar, and Boaz Patt-Shamir. Tell Me Who I Am: an Interactive Recommendation System. In *Proceedings of the 18th Symposium on Parallelism in Algorithms and Architectures (SPAA)*, pages 261–279, 2006.
3. Baruch Awerbuch, Boaz Patt-Shamir, David Peleg, and Mark R. Tuttle. Improved Recommendation Systems. In *Proceedings of the 16th Symposium on Discrete Algorithms (SODA)*, pages 1174–1183, 2005.

4. Yossi Azar and Iftah Gamzu. Ranking with Submodular Valuations. In *Proceedings of the 22nd Symposium on Discrete Algorithms (SODA)*, pages 1070–1079, 2011.
5. Amotz Bar-Noy, Mihir Bellare, Magnús M. Halldórsson, Hadas Shachnai, and Tami Tamir. On Chromatic Sums and Distributed Resource Allocation. *Information and Computation*, 140(2):183–202, 1998.
6. Sushil Bikhchandani, Sven de Vries, James Schummer, and Rakesh Vohra. An Ascending Vickrey Auction for Selling Bases of a Matroid. *Operations Research*, 59(2):400–413, 2011.
7. Brian Dean, Michel Goemans, and Jan Vondrák. Approximating the Stochastic Knapsack Problem: The Benefit of Adaptivity. *Mathematics of Operations Research*, 33:945–964, 2008.
8. Petros Drineas, Iordanis Kerenidis, and Prabhakar Raghavan. Competitive Recommendation Systems. In *Proceedings of the 34th Symposium on Theory of Computing (STOC)*, pages 82–90, 2002.
9. Uriel Feige, László Lovász, and Prasad Tetali. Approximating Min Sum Set Cover. *Algorithmica*, 40:219–234, 2004.
10. Michel Goemans and Jan Vondrák. Stochastic Covering and Adaptivity. In *Proceedings of the 7th Latin American Conference on Theoretical Informatics (LATIN)*, pages 532–543, 2006.
11. Sally A. Goldman, Robert E. Schapire, and Ronald L. Rivest. Learning Binary Relations and Total Orders. *SIAM Journal of Computing*, 20(3):245 – 271, 1993.
12. Daniel Golovin and Andreas Krause. Adaptive Submodularity: Theory and Applications in Active Learning and Stochastic Optimization. *Journal of Artificial Intelligence Research (JAIR)*, 42:427–486, 2011.
13. Edward Grove. Online Bin Packing with Lookahead. In *Proceedings of the 6th Symposium on Discrete Algorithms (SODA)*, pages 430–436, 1995.
14. Anupam Gupta, Viswanath Nagarajan, and R. Ravi. Approximation Algorithms for Optimal Decision Trees and Adaptive TSP Problems. In *Proceedings of the 37th International Colloquium on Automata, Languages and Programming (ICALP)*, pages 690–701. Springer-Verlag, 2010.
15. Rachel Kranton and Deborah Minehart. A Theory of Buyer-Seller Networks. *American Economic Review*, 91:485–508, 2001.
16. Zhen Liu, Srinivasan Parthasarathy, Anand Ranganathan, and Hao Yang. Near-Optimal Algorithms for Shared Filter Evaluation in Data Stream Systems. In *Proceedings of the 2008 ACM SIGMOD International Conference on Management of Data*, pages 133–146, 2008.
17. Kamesh Munagala, Shivnath Babu, Rajeev Motwani, and Jennifer Widom. The Pipelined Set Cover Problem. In *Proceedings of the 10th International Conference on Database Theory (ICDT)*, pages 83–98, 2005.
18. Aviv Nisgav and Boaz Patt-Shamir. Finding Similar Users in Social Networks: Extended Abstract. In *Proceedings of the 21st Annual Symposium on Parallelism in Algorithms and Architectures (SPAA)*, 2009.
19. Jara Uitto and Roger Wattenhofer. On Competitive Recommendations. In *Proceedings of the 24th International Conference on Algorithmic Learning Theory (ALT)*, pages 83–97., 2013. Invited to a special issue of Theoretical Computer Science.

Lower Bounds in the Preprocessing and Query Phases of Routing Algorithms

Colin White

Carnegie Mellon University
crwhite@cs.cmu.edu
http://cs.cmu.edu/~crwhite

Abstract. In the last decade, there has been a substantial amount of research in finding routing algorithms designed specifically to run on real-world graphs. In 2010, Abraham et al. showed upper bounds on the query time in terms of a graph's *highway dimension* and diameter for the current fastest routing algorithms, including CONTRACTION HIERARCHIES, TRANSIT NODE ROUTING, and HUB LABELING. In this paper, we show corresponding lower bounds for the same three algorithms. We also show how to improve a result by Milosavljević which lower bounds the number of shortcuts added in the preprocessing stage for CONTRACTION HIERARCHIES. We relax the assumption of an optimal contraction order (which is NP-hard to compute), allowing the result to be applicable to real-world instances. Finally, we give a proof that optimal preprocessing for HUB LABELING is NP-hard. Hardness of optimal preprocessing is known for most routing algorithms, and was suspected to be true for HUB LABELING.

1 Introduction

The problem of finding shortest paths in road networks has been well-studied in the last decade, motivated by the application of computing driving directions. Although Dijkstra's algorithm runs in small polynomial time, for applications involving continental-sized road networks, Dijkstra's algorithm is simply not fast enough. There have been many different approaches to find algorithms that specifically run fast on real-world graphs.

Most recent innovations involve a two-stage algorithm: a preprocessing stage and a query stage. The preprocessing stage runs once and can spend hours calculating data. Then the query stage uses this data to find shortest paths very fast, often several orders of magnitude faster than Dijkstra's algorithm for a continental query. Once the preprocessing stage is completed, the users can run as many queries as they want. For a query between two nodes s and t (an s–t query), the algorithm returns $\text{dist}(s, t)$, the cost of the shortest path between s and t. Most algorithms can also return the vertices on the shortest path using an extra data structure.

The current fastest routing algorithm on real-world graphs is HUB LABELING [2], which achieves a speedup of six orders of magnitude over Dijkstra's algorithm. The TRANSIT NODE ROUTING algorithm is second-fastest, and requires

© Springer-Verlag Berlin Heidelberg 2015
N. Bansal and I. Finocchi (Eds.): ESA 2015, LNCS 9294, pp. 1013–1024, 2015.
DOI: 10.1007/978-3-662-48350-3_84

an order of magnitude less space than HUB LABELING. CONTRACTION HIERAR-
CHIES is also a fast routing algorithm, which was state of the art in 2008. For a
comprehensive overview of the best routing algorithms, see [6].

Until recently, it was known that these algorithms performed very well on
real-world maps, but there were no theoretical guarantees. In fact, it is not hard
to construct specific graphs for which these algorithms perform no faster than
Dijkstra's algorithm. So, an interesting theoretical question is to find properties
present in all real-life graphs that explain why these algorithms work so well.

With this motivation in mind, Abraham et al. defined the notion of *highway
dimension* [1], intuitively, the extent to which all shortest paths are hit by at
least one of a small set of *access nodes*. Although it is too computationally
intensive to calculate the exact highway dimension for a continental road map,
there is evidence that the highway dimension h is at most polylogarithmic in
the number of vertices. It is conjectured that real-world routing networks always
have low highway dimension, based on experimental evidence [3]. Abraham et
al. were able to prove strong upper bounds on the query times in terms of
highway dimension and diameter d for four of the fastest routing algorithms: HUB
LABELING, CONTRACTION HIERARCHIES, TRANSIT NODE ROUTING, and REACH.

1.1   Our Results

In this paper, we are interested in finding lower bounds for the current state-of-
the-art routing algorithms. We show tight or near-tight bounds on the runtime
for HUB LABELING, CONTRACTION HIERARCHIES, and TRANSIT NODE ROUTING.

Our lower bounds may facilitate proving better guarantees of these algorithms,
or provide intuition for new routing algorithms, if one can find differences between the graphs we use and real world instances. For example, the graphs we
use have low highway dimension, but they do not have small separators and are
nonplanar, so perhaps there is a way to modify HUB LABELING to take this into
account.

We show a tight lower bound for HUB LABELING, the fastest routing algorithm
to date [6]. For CONTRACTION HIERARCHIES and TRANSIT NODE ROUTING, the
definition of highway dimension in the lower bound versus upper bound is slightly
different (because of a recent redefinition by Abraham et al.), so we cannot quite
say the bounds are tight.

We can also use our analysis to generalize a known result by Milosavljević,
which lower bounds the number of shortcut edges in the preprocessing stage
of CONTRACTION HIERARCHIES [12]. This result assumes an optimal contraction
order which is NP-hard to compute [7]. So for real-world instances, we rely on using contraction orders based on heuristics. We show how to relax the assumption
about the contraction order, which means the result can be applied to real-world
instances.

We also contribute a hardness result for optimal preprocessing of HUB LA-
BELING. In 2010, Bauer et al. established hardness for optimal preprocessing for
a variety of the best routing algorithms, including CONTRACTION HIERARCHIES
and TRANSIT NODE ROUTING. In this paper, we show that in HUB LABELING

preprocessing, the problem of minimizing the maximum label size over all vertices is NP-hard.

This paper will proceed as follows. Section 2 will provide preliminary information, specifically about highway dimension, and also the graph construction used in our main theorems. In Section 3, we show a lower bound on the query time of the HUB LABELING algorithm, and prove that optimal preprocessing is NP-hard. In Section 4, we establish a lower bound on the query time for CONTRACTION HIERARCHIES, and generalize a lower bound on the number of shortcut edges added in the preprocessing phase. Section 5 establishes a lower bound on the query time of TRANSIT NODE ROUTING. We conclude and discuss future directions in Section 6.

Due to space constraints, we provide proof sketches. See the full version of this paper on arxiv for the complete proofs.

2 Preliminaries

In this paper, we assume nonnegative integral edge lengths and unique shortest paths. We will also assume graphs are undirected in all sections except for the hardness result. These are standard assumptions to make when proving bounds on routing algorithms, for example, [3] and [12].

$B_r(v)$ represents all nodes u such that $\text{dist}(u, v) < r$. We say a set of nodes *covers* a set of paths if each path has at least one of its vertices in the set of nodes.

2.1 Highway Dimension

Now we will formally define the notion of highway dimension.

The *highway dimension* of a graph $G = (V, E)$ is the smallest h such that for all $r > 0$ and for all $B_{4r}(v)$, there exists a set $H \subseteq V$, such that $|H| \leq h$ and H covers all shortest paths of length $\geq r$ in $B_{4r}(v)$.

Highway dimension was specifically designed to explain why the best routing algorithms perform well on real-world graphs but do not perform well on arbitrary graphs. Although it is too computationally intensive to calculate the exact highway dimension of a continental-sized road network, it is conjectured that the highway dimension of real-world graphs is at most polylogarithmic in the number of vertices [3].

Abraham et al. introduced a slightly refined version of the original highway dimension in 2013 [1].

The difference in the new definition versus the old one is that instead of having to hit all local shortest paths of length $\geq r$, we have to hit all paths P where there is a shortest path P' with endpoints s and t such that $l(P') > r$, $P \subseteq P'$, and $P' \setminus P \in \{\emptyset, \{s\}, \{t\}, \{s, t\}\}$. That is, we have to hit all paths that can be obtained by removing zero, one, or both endpoints of a shortest path with length $> r$. We will refer to a graph's highway dimension as h for the first definition, and $\hat{h}$ for the second definition.

The two definitions of highway dimension are very similar but have a few key differences. Most notably, the new definition bounds the degree of the graph, which was not true before [3]. The new definition of highway dimension allowed Abraham et al. to improve their results on the runtime of routing algorithms.

2.2   Definition of $G_{t,k,q}$

Now we will define the family of graphs $G_{t,k,q}$ that will be used in many of our proofs. $G_{t,k,q}$ was designed to by Milosavljević to show a lower bound on the number of shortcuts created during the preprocessing stage of CONTRACTION HIERARCHIES [12].

Consider a complete t-ary tree of height k for integers $t, k \geq 2$. Let $\lambda(v)$ denote the height of node v, and let $\lambda(u, v)$ denote the height of the lowest common ancestor between two nodes u and v.

Now define the edges as follows: for all nodes v and w such that w is a proper ancestor of v, there is an edge between v and w with length $16^{\lambda(w)-1}$. This means the edge length from a node w to one of its descendants v is independent of $\lambda(v)$. Furthermore, edge lengths increase for nodes higher up in the tree.

Denote this graph by $G_{t,k} = (V_{t,k}, E_{t,k})$. See Figure 1 for an example. For convenience, we will still refer to this graph as a tree, even though the additional edges create cycles.

Now we will define $G_{t,k,q} = (V_{t,k,q}, E_{t,k,q})$ by taking q copies of $G_{t,k}$, and naming them $G_{t,k}^{(a)} = (V_{t,k}^{(a)}, E_{t,k}^{(a)})$ for $a = 1, 2, ..., q$. The copy of a node $v \in G_{t,k}$ in $G_{t,k}^{(a)}$ is denoted $v^{(a)}$.

For all $v \in G_{t,k}$ and $a \neq b$, we add edge $v^{(a)} - v^{(b)}$ to $E_{t,k,q}$ with length $2^{\lambda(v)-k-1}$. This ensures that switching copies has a low penalty ($2^{\lambda(v)-k-1}$ is always less than 1), and it is always cheaper to switch among copies lower down in the tree. See Figure 1 for an example.

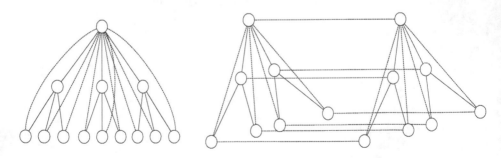

Fig. 1. *The left graph is $G_{3,3}$, and the right graph is $G_{2,3,2}$*

2.3   Properties of $G_{t,k,q}$

We will now discuss properties of $G_{t,k,q}$. The following three lemmas are proven in [12].

Lemma 1. *Given $s, t \in V_{t,k}$ with lowest common ancestor w, the unique shortest s–t path is s–w–t.*

Lemma 2. *Given $s^{(a)}$ and $t^{(b)}$ in $G_{t,k,q}$, let w be the lowest common ancestor between s and t. Then the shortest $s^{(a)}$–$t^{(b)}$ paths are:*
   $s^{(a)}$–$s^{(b)}$–$w^{(b)}$–$t^{(b)}$*, if $\lambda(s) \leq \lambda(t)$, and/or*
   $s^{(a)}$–$w^{(a)}$–$t^{(a)}$–$t^{(b)}$*, if $\lambda(t) \leq \lambda(s)$.*

Lemma 3. *The highway dimension h of $G_{t,k,q}$ is equal to q, the diameter D is $\Theta(16^k)$, and $|V_{t,k,q}| = \Theta(qt^k)$.*

It is worth noting that at the start we assumed graphs have unique shortest paths, but now many shortest paths in our main family of graphs are not unique. However, this is a common assumption in routing algorithm proofs because it is not hard to perturb the input to make all shortest paths unique while maintaining the validity of the proofs.

Additionally, integrality of edge lengths is violated. Since the smallest edge is 2^{-k} (and all edge lengths are multiples of this), all of the edge weights can be multiplied by 2^k to create integral lengths. This will increase D by a factor of k, doubling $\log D$, which will not affect our results.

3   Hub Labeling

The HUB LABELING algorithm was first devised in 2004 by Gavoille et al. [10], and further studied by Cohen et al. [8]. However, the algorithm was not practical for continental routing queries until 2011, when Abraham et al. came up with an efficient way to perform the preprocessing and query phases, which made it the fastest routing algorithm to date [2].

In this section, we will first give an introduction to the HUB LABELING algorithm. Then we will present a lower bound on the query time. Finally, we will show the preprocessing phase is NP-hard to optimize.

3.1   The Algorithm

HUB LABELING relies on the concept of labeling. Each node stores information about its shortest paths that allows us to reconstruct the shortest path during a query. This idea is used in a clever way to make queries run very fast.

In the HUB LABELING algorithm, we give each node $v \in V$ a *label* consisting of other nodes (the *hubs* of v), and we store the shortest distances to the hubs from v. We define a *labeling* L as the set of labels $L(v)$ for all $v \in V$.

We construct the labeling in such a way that for any pair of nodes s and t, $L(s) \cap L(t)$ contains at least one node on the shortest path from s to t. When satisfied, this is called the *cover property*. Then in order to perform an s–t query, we only need to find the $v \in L(s) \cap L(t)$ that minimizes $\text{dist}(s,v) + \text{dist}(v,t)$. This can be made to take $O(|L(s)| + |L(t)|)$ time if the labels are sorted with some arbitrary node order. This process returns $\text{dist}(s,t)$. To return the nodes on this shortest path, we need to add another data structure in the preprocessing stage, which does not increase the space complexity by more than a constant factor [2].

In Section 2, we will show that it is NP-hard to find the labeling that minimizes the maximum label size for all vertices. This was suspected to be true. Therefore, in practice we must rely on heuristics in the preprocessing stage.

Abraham et al. showed that the query time of HUB LABELING is $O(\hat{h} \log D)$, using a specific labeling [1]. The proof did not use any properties of $\hat{h}$ that are different from h, so we can also say that the query time is $O(h \log D)$.

It is not known how to construct the labeling used in their proof in polynomial time, so they showed a corollary that uses a polynomial preprocessing algorithm and permits queries to be handled in $O(h \log h \log D)$ time.

3.2   Lower Bounding the Query Time

We cannot prove a lower bound on the minimum query time, since labelings can be constructed to make any one query run in constant time. Instead, we will prove a bound on the average query time by bounding the sum of all label sizes.

Theorem 1. *For all h, D, n, there is a graph $G = (V, E)$ with highway dimension h, diameter $\Theta(D)$, and $|V| \geq n$, such that for any choice of labeling L, the average query requires $\Omega(h \log D)$ time.*

Proof (sketch). It suffices to show that for any h, D, and n, there exists a graph that fits the parameters of the theorem, and given any labeling L, the sum of the label size for all vertices is $\Omega(h|V| \log D)$.

We use $G_{t,k,q}$ for this task, where $k = O(\log D)$, $q = h$, and t is large enough such that $|V| \geq n$.

Leaf-leaf queries make up a constant fraction of all queries, so we limit our analysis to these queries.

The idea is to group leaf-leaf shortest paths by their lowest common ancestor: For $0 \leq i \leq k$, let $P_i = \{s\text{–}t \mid s \text{ and } t \text{ are leaves, and } \lambda(s,t) = i\}$.

Then we show that for $\frac{k}{2} \leq i \leq k$, the shortest paths in P_i contribute $\Omega(q^2 t^k)$ distinct nodes to the total sum of label sizes. We establish this by showing that for a leaf-leaf shortest path s–t in P_i for $i \geq \frac{k}{2}$, it is always best for s and t to choose their common ancestor as a hub, because this hub hits the greatest amount of shortest paths stemming from s and t. Despite this overlap, we are able to show $\Omega(q^2 t^k)$ distinct hubs are needed for each set P_i, which proves the theorem.   □

With this theorem, the upper bound presented in [1] becomes tight.

3.3   Hardness of Preprocessing

In 2010, Bauer et al. established hardness for optimal preprocessing for a variety of the best routing algorithms, including CONTRACTION HIERARCHIES and TRANSIT NODE ROUTING [7]. We provide hardness for optimal preprocessing in HUB LABELING which was suspected to be true [3]. By optimal preprocessing, we mean minimizing the maximum hub size over all vertices. Babenko et al. very recently established hardness for nearly the same problem, but they defined optimal preprocessing as minimizing over the *total* label size [5]. Our definition of optimal corresponds to minimizing the maximum query time, whereas the other definition corresponds to minimizing the average query time. We believe our definition is more motivated because minimizing the average query time does not take into account outlier queries that run for a long time.

We will switch to directed graphs, which was the original setting of HUB LABELING [2]. The main difference is that each node v has a forward label $L_f(v)$ and a reverse label $L_r(v)$, and the cover property states that for a directed s–t query, $L_f(s) \cap L_r(t)$ is not empty.

Now we formally define the problem MINIMUM HUB LABELING (MHL) as follows:

Problem (MHL). Given a directed graph $G = (V, A)$ and an integer k, find a labeling L satisfying the cover property such that $\max_{v \in V}(\max(|L_f(v)|, |L_r(v)|)) \leq k$.

Theorem 2. *Minimum hub-labeling is NP-hard.*

Here is a proof outline. We show a reduction from a classical NP-hard problem, *exact cover by 3-sets (X3C)*. In an X3C instance (U, C), U is a set of elements, 3 divides $|U|$, and C is a set of triples of U. The problem is whether there exists a set $C' \subseteq C$, $|C'| = \frac{|U|}{3}$ such that C' covers U (an exact 3-covering of U).

Given an X3C instance (U, C), we create an MHL instance (G, k) where $G = (V, E)$, $U \cup C \subseteq V$ and for $c \in C$, $u \in U$, c–$u \in E$ iff $u \in c$.

We also add a clique of vertices $\{b_1, \ldots, b_{2k-1}\} = B$ with arcs to nodes in U, whose sole purpose is to fill up the reverse labels of nodes in U. Finally, we add two vertices $\{a_1, a_2\} = A$ with arcs to every node in C.

By filling up the reverse labels of nodes $u \in U$, we force the nodes $a \in A$ to use nodes in C or U for the hubs of a–u shortest paths. And it is too inefficient to use nodes in U for the hubs, so nodes in C must act as the hubs. Then in order for A's label size to stay $\leq k$, there must be an exact cover for U.

4   Contraction Hierarchies

CONTRACTION HIERARCHIES [11] is a shortcut-based algorithm, making it fundamentally different from HUB LABELING. It works by running bidirectional Dijkstra search, pruning the searches based on a node's importance.

In this section, we explain how the CONTRACTION HIERARCHIES algorithm works, prove a lower bound on the query time, and then generalize a result about the number of shortcut edges added in the preprocessing phase.

4.1   The Algorithm

In the preprocessing stage for CONTRACTION HIERARCHIES, we iteratively *contract* nodes using a predefined ordering, called a *contraction ordering*. The contraction operation called on v first deletes v from the graph, and then may add edges between v's neighbors if they are needed to preserve the shortest path lengths. Any such edge is put into a set E^+. We contract every node in the graph based on the ordering, and we are left with the set E^+ of "shortcut edges".

To run an s–t query, run bidirectional Dijkstra search from s and t on the graph $G^+ = (V, E \cup E^+)$, except at node v, only consider edges v–w in which w was contracted after v. When there are no more nodes to consider in either direction, find the node v that minimizes the sum of its distances to s and to t.

In [11], it is proven that v is guaranteed to be on the shortest path between s and t, which means that $\text{dist}(s, t) = \text{dist}(s, v) + \text{dist}(v, t)$, so the query returns the shortest s–t path.

Note that any contraction ordering will give correct queries, but a better contraction ordering will make $|E^+|$ small, decreasing time and space requirements. Finding the optimal ordering is NP-hard [7], but there are fast heuristics that make $|E^+|$ within $\log h$ of optimal [1].

Abraham et al. showed an upper bound on the query time of CONTRACTION HIERARCHIES that depends on Δ: $O((\Delta + h \log D)(h \log D))$ [3]. Using the new definition of highway dimension, Abraham et al. achieved the better bound of $O((\hat{h} \log D)^2)$ time. Both of these assume optimal preprocessing. If a polynomial time preprocessing algorithm is required, the bounds are modified to $O((\hat{h} \log \hat{h} \log D)^2)$ and $O((\Delta + h \log h \log D)(h \log h \log D))$.

4.2   Lower Bounding the Query Time

We show a lower bound using the old definition of highway dimension.

Theorem 3. *For all h, D, n, there is a graph $G = (V, E)$ with highway dimension h, diameter $\Theta(D)$, and $|V| \geq n$ such that the average query time is $\Omega((h \log D)^2)$ for CONTRACTION HIERARCHIES.*

Our strategy will be to find a lower bound assuming Abraham et al.'s (optimal) ordering, and then show that modifying the ordering can only increase the runtime.

[12] provided a criterion for shortcut paths in the optimal ordering: the path $s^{(a)}$–$s^{(b)}$–$w^{(b)}$–$t^{(b)}$ is shortcut if and only if $a \neq b$, w is a proper ancestor of s, and $s^{(b)}$ is contracted before $s^{(a)}$.

Proof (sketch). Again we will use $G_{t,k,q}$, setting $k \in O(\log D)$, $q = h$, and t large enough such that $|V| \geq n$, and we limit our analysis to leaf-leaf queries, which make up the majority of all queries.

First we prove the theorem assuming Abraham et al.'s contraction order. For $G_{t,k,q}$, this means nodes are contracted based on their height in the tree.

Then we show that veering away from this ordering will only increase the number of shortcut edges produced, slowing the algorithm down.

The first claim can be established without too much work. We show that in the forward search of a leaf-leaf query $s^{(a)}$–$t^{(b)}$, the only nodes we may visit are ancestors $v^{(c)}$ of s such that $v^{(c)}$ is contracted after $v^{(a)}$. Then half of these nodes will have lower contraction order than the other half, and so it can be shown that the shortcut criterion guarantees $\Omega(q^2k^2)$ edges will be created along half of the forward searches.

The case of a general ordering is more technical. The main idea is to carefully examine the effects of contracting a node higher up in the tree, before all of its descendants were contracted. Although contracting a higher node v decreases some of the paths from any descendant u to v, it creates shortcuts between all pairs of descendants which have not yet been contracted, which could cause an exponential number of extra edges to be created. There is an overall net loss in time in the algorithm, which finishes the proof.   □

4.3   Lower Bounding the Size of E^+

Abraham et al.ś upper bound of $O((\hat{h} \log D)^2)$ on the query time involves proving that $|E^+| \in O(n\hat{h} \log D)$. The latter bound was proven tight in [1?] However, the proof assumes the contraction order from the algorithm in Abraham et al. which is thought to be NP-hard to compute. We show a new proof of this lower bound generalized to any contraction order.

Theorem 4. *For all h, D, n, there is a graph $G = (V, E)$ with highway dimension h, diameter $\Theta(D)$, and $|V| \geq n$ such that for any contraction ordering, $|E^+| \in \Omega(h|V| \log D)$.*

Due to space constraints, we refer the reader to the full version of this paper (on arxiv) for the proof.

5   Transit Node Routing

TRANSIT NODE ROUTING [4] was devised in 2007 by Bast et al., and it (and variants) remain the second-fastest family of routing algorithms, behind HUB LABELING [6]. However, TRANSIT NODE ROUTING requires about an order of magnitude less space than HUB LABELING. In this section, we first review the TRANSIT NODE ROUTING algorithm, and then we give a lower bound on the query time.

The algorithm works by picking a set $T \subset V$ of *transit nodes* that hits many long-distance shortest paths. $|T|$ is often chosen to be in $\Theta(\sqrt{|V|})$, which makes the algorithm run fastest while maintaining that additional memory requirements are bounded by the input graph size. Usually, the contraction order is used to pick T (since contraction order essentially seeks to measure a node's importance with respect to shortest paths), which works well in practice.

Next, given any node v, $A(v) \subset T$ is the set of that node's *access nodes*, which are chosen to hit the long-distance queries stemming from v. This usually means that we want to pick nodes in T that are close to v.

The distances between all pairs of transit nodes are computed and stored, as well as the distances between a node v and each of its access nodes. A query is called a *global query* if $\min(\text{dist}(s,u) + \text{dist}(u,v) + \text{dist}(v,t) \mid u \in A(s), \ v \in A(t)) = \text{dist}(s,t)$. Otherwise, it is a *local query*. To run an s–t query, first run a quick locality filter that determines whether the query is local. This filter is allowed to make one-sided errors; it can misclassify a global query as local, but not the other way around. Locality filters are historically calculated using the coordinates of the vertices. If it is a global query, calculate the minimum $\text{dist}(s,u) + \text{dist}(u,v) + \text{dist}(v,t)$ by trying all combinations of access nodes from $A(s)$ and $A(t)$. Local queries are handled by a fast local search such as CON-TRACTION HIERARCHIES.

Abraham et al. use a choice of T based on multiscale shortest-path covers to prove that access nodes are bounded in size by $O(\hat{h})$, from which it follows that global queries can be handled in $O(\hat{h}^2)$ time. Local queries done using CONTRACTION HIERARCHIES can be handled in $O((\hat{h} \log D)^2)$ time as we saw in the previous section (however, local queries tend to be small, making the queries run much faster than the average CONTRACTION HIERARCHIES query).

This bound is not possible without the new definition of highway dimension. Again, if we want polynomial time preprocessing, the query time bound for global queries increases to $O((\hat{h} \log \hat{h})^2)$.

5.1   Lower Bounding the Query Time

While the upper bound for TRANSIT NODE ROUTING was for global queries only, our lower bound will include both local and global searches. We will use CON-TRACTION HIERARCHIES for local queries.

Theorem 5. *For all h, D, n, there is a graph $G = (V, E)$ with highway dimension h, diameter $\Theta(D)$, and $|V| \geq n$ such that for any choice of transit nodes T and access nodes A, the average query time is $\Omega(h^2)$.*

Proof (sketch). First, we define a leaf-leaf shortest path as *regular* if the shortest path is global, and neither endpoint is a transit node. We are able to exclude irregular shortest paths from our analysis, for the following reasons. First, short-est paths with a transit node do not make up a constant fraction of all shortest paths. Second, if a constant fraction of all shortest paths were local, then we can use Theorem 4.2 and we are done.

So, we can assume a constant fraction of all leaf-leaf shortest paths are regular. Again we will use $G_{t,k,q}$ such that $k \in O(\log D)$, $q = h$, and t large enough so that $|V| \geq n$.

There is a simple intuition for the rest of the proof. Given a leaf-leaf shortest path $s^{(a)}$–$t^{(b)}$, either $s^{(a)}$ or $t^{(b)}$ must have an access node in the other's copy, since the non-endpoint vertices on the shortest path all come from one copy.

From here, the proof gets technical because we must show that a constant fraction of leaves have a large amount of access nodes in distinct copies and subtrees. But we are able to show that a constant fraction of the nodes need $\Omega(q^2)$ access nodes, and the proof follows.   □

6   Conclusions and Future Work

We proved lower bounds on the query time of HUB LABELING, CONTRACTION HIERARCHIES, and TRANSIT NODE ROUTING. The proofs are all quite different, despite using the same family of graphs for each proof. We also generalized a lower bound on the size of E^+ in CONTRACTION HIERARCHIES preprocessing, and established hardness for optimal preprocessing in HUB LABELING.

Although we have proven lower bounds for the query times of three state-of-the-art algorithms, the graphs used in the arguments are not representative of real-world graphs. For instance, the graphs do not have small separators and are not planar. This implies it may be possible to circumvent this lower bound using different properties that better capture the structure of real-world graphs.

Another way to work with more realistic road networks is to use the idea of multiscale dispersed graphs, defined in [9], as a new model for graphs that simulate real-world graphs. One may be able to obtain better bounds on the query time with this model.

Throughout this paper, we assumed undirected graphs, so future work could extend these results to the directed case. Furthermore, apart from HUB LABELING, the upper and lower bounds are not tight because of the different definitions of highway dimension. Ideally, we would find a way to prove the lower bounds using the more recent definition of highway dimension. However, we cannot use $G_{t,k,q}$ for this task. Under the new definition, $G_{t,k,q}$ has highway dimension at least $q + k$, since the new definition guarantees a graph's degree is bounded by its highway dimension.

Acknowledgments. The results in this paper are from the senior honors thesis of the author, written under the direction of Prof. Lyle McGeoch, at Amherst College. We would like to give a huge thanks to Lyle McGeoch for helpful discussions and suggestions throughout the writing process. We are grateful for the Post-Baccalaureate Summer Research Fellowship program at Amherst College, which supported the writing of this paper.

References

1. Abraham, I., Delling, D., Fiat, A., Goldberg, A.V., Werneck, R.F.: Highway dimension and provably efficient shortest path algorithms. Technical Report MSR-TR-2013-91, Microsoft Research (2013)
2. Abraham, I., Delling, D., Goldberg, A.V., Werneck, R.F.: A hub-based labeling algorithm for shortest paths on road networks. In: Pardalos, P.M., Rebennack, S. (eds.) SEA 2011. LNCS, vol. 6630, pp. 230–241. Springer, Heidelberg (2011)
3. Abraham, I., Fiat, A., Goldberg, A.V., Werneck, R.F.: Highway dimension, shortest paths, and provably efficient algorithms. In: Proceedings of the Twenty-First Annual ACM-SIAM Symposium on Discrete Algorithms (SODA 2010), pp. 782–793. SIAM (2010)
4. Arz, J., Luxen, D., Sanders, P.: Transit node routing reconsidered. In: Bonifaci, V., Demetrescu, C., Marchetti-Spaccamela, A. (eds.) SEA 2013. LNCS, vol. 7933, pp. 55–66. Springer, Heidelberg (2013)

5. Babenko, M.A., Goldberg, A.V., Kaplan, H., Savchenko, R., Weller, M.: On the complexity of hub labeling. CoRR, abs/1501.02492 (2015)

6. Bast, H., Delling, D., Goldberg, A., Müller-Hannemann, M., Pajor, T., Sanders, P., Wagner, D., Werneck, R.: Route planning in transportation networks. Technical report, Microsoft Research (2014)

7. Bauer, R., Columbus, T., Katz, B., Krug, M., Wagner, D.: Preprocessing speed-up techniques is hard. In: Calamoneri, T., Diaz, J. (eds.) CIAC 2010. LNCS, vol. 6078, pp. 359–370. Springer, Heidelberg (2010)

8. Cohen, E., Halperin, E., Kaplan, H., Zwick, U.: Reachability and distance queries via 2-hop labels. SIAM Journal on Computing 32(5), 1338–1355 (2003)

9. Eppstein, D., Goodrich, M.T.: Studying (non-planar) road networks through an algorithmic lens. In: Proceedings of the 16th ACM SIGSPATIAL International Conference on Advances in Geographic Information Systems (GIS 2008), pp. 1–10. ACM Press (2008)

10. Gavoille, C., Peleg, D., Pérennes, S., Raz, R.: Distance labeling in graphs. In: Proceedings of the Twelfth Annual ACM-SIAM Symposium on Discrete Algorithms, SODA 2001, pp. 210–219. Society for Industrial and Applied Mathematics, Philadelphia (2001)

11. Geisberger, R., Sanders, P., Schultes, D., Vetter, C.: Exact routing in large road networks using contraction hierarchies. Transportation Science 46(3), 388–404 (2012)

12. Milosavljević, N.: On optimal preprocessing for contraction hierarchies. In: Proceedings of the 5th ACM SIGSPATIAL International Workshop on Computational Transportation Science, pp. 33–38. ACM Press (2012)

Trip-Based Public Transit Routing

Sascha Witt

Karlsruhe Institute of Technology (KIT)
Karlsruhe, Germany
sascha.witt@kit.edu

Abstract. We study the problem of computing all Pareto-optimal journeys in a public transit network regarding the two criteria of arrival time
and number of transfers taken. We take a novel approach, focusing on
trips and transfers between them, allowing fine-grained modeling. Our
experiments on the metropolitan network of London show that the algorithm computes full 24-hour profiles in 70 ms after a preprocessing phase
of 30 s, allowing fast queries in dynamic scenarios.

1 Introduction

Recent years have seen great advances in route planning on continent-sized road
networks [2]. Unfortunately, adapting these algorithms to public transit networks is harder than expected [4]. On road networks, one is usually interested
in the shortest path between two points, according to some criterion. On public
transit networks, several variants of point-to-point queries exist. The simplest is
the *earliest arrival query*, which takes a departure time as an additional input
and returns a journey that arrives as early as possible. A natural extension is
the *multi-criteria* problem of minimizing both arrival time and the number of
transfers, resulting in a set of journeys. A *profile query* determines all optimal
journeys departing during a given period of time.

In the past, these problems have been solved by modeling the timetable information as a graph and running Dijkstra's algorithm or variants thereof on
that graph. Traditional graph models include the time-expanded and the time-
dependent model [14]. More recently, algorithms such as RAPTOR [10] and
Connection Scan [11] have eschewed the use of graphs (and priority queues) in
favor of working directly on the timetable.

In this work, we present a new algorithm that uses trips (vehicles) and the
transfers between them as its fundamental building blocks. Unlike existing algorithms, it does not assign labels to stops. Instead, trips are labeled with the
stops at which they are boarded. Then, a precomputed list of transfers to other
trips is scanned and newly reached trips are labeled. When a trip reaches the
destination, a journey is added to the result set. The algorithm terminates when
all optimal journeys have been found.

A motivating observation behind this is the fact that labeling stops with
arrival (or departure) times is not sufficient once minimum change times are
introduced. Some additional information is required to track which trips can

N. Bansal and I. Finocchi (Eds.): ESA 2015, LNCS 9294, pp. 1025–1036, 2015.
DOI: 10.1007/978-3-662-48350-3_85

be reached. For example, the realistic time-expanded model of Pyrga et al. [16] introduces additional nodes to deal with minimum change times, while Connection Scan [11] uses additional labels for trips. In contrast, once we know passengers boarded a trip at a certain stop, their further options are fully defined: Either they transfer to another trip using one of the precomputed transfers, or their current trip reaches the destination, in which case we can look up the arrival time in the timetable. In either case, there is no need to explicitly track arrival times at intermediary stops.

The core of the algorithm is similar to a breadth-first search, where levels correspond to the number of transfers taken so far. As a result, it is inherently multi-criterial, similar to RAPTOR [10]. Although a graph-like structure is used, there is no need for a priority queue. A preprocessing step is required to compute transfers, but can be parallelized trivially and only takes a few minutes, even on large networks (Section 4). By omitting unnecessary transfers, both space usage and query times can be improved at the cost of increased preprocessing time.

Section 2 introduces necessary notations and definitions, before Section 3 describes the algorithm and its variants. Section 4 presents the experimental evaluation. Finally, Section 5 concludes the paper.

2   Preliminaries

2.1   Notation

We consider public transit networks defined by an aperiodic *timetable*, consisting of a set of stops, a set of footpaths and a set of trips. A *stop* p represents a physical location where passengers can enter or exit a vehicle, such as a train station or a bus stop. Changing vehicles at a stop p may require a certain amount of time $\Delta\tau_{\mathrm{ch}}(p)$ (for example, in order to change platforms).[1] *Footpaths* allow travelers to walk between two stops. We denote the time required to walk from stop p_1 to p_2 by $\Delta\tau_{\mathrm{fp}}(p_1, p_2)$ and define $\Delta\tau_{\mathrm{fp}}(p, p) = \Delta\tau_{\mathrm{ch}}(p)$ to simplify some algorithms. A *trip* t corresponds to a vehicle traveling along a sequence of stops $\boldsymbol{p}(t) = \langle p_t^0, p_t^1, \ldots \rangle$. Note that stops may occur multiple times in a sequence. For each stop p_t^i, the timetable contains the arrival time $\tau_{\mathrm{arr}}(t, i)$ and the departure time $\tau_{\mathrm{dep}}(t, i)$ of the trip at this stop. Additionally, we group trips with identical stop sequences into *lines*[2] such that all trips t and u that share a line can be totally ordered by

$$t \preceq u \iff \forall i \in [0, |\boldsymbol{p}(t)|) : \tau_{\mathrm{arr}}(t, i) \le \tau_{\mathrm{arr}}(u, i) \tag{1}$$

and define

$$t \prec u \iff t \preceq u \wedge \exists i \in [0, |\boldsymbol{p}(t)|) : \tau_{\mathrm{arr}}(t, i) < \tau_{\mathrm{arr}}(u, i). \tag{2}$$

---

[1] More fine-grained models, such as different change times for specific platforms, can be used without affecting query times, since minimum change times are only relevant during preprocessing (Section 3.1).

[2] *Line* and *route* have both been previously used for this concept; we opted for *line* to avoid confusion with *routing* and the usage of *route* in the context of road networks.

If two trips have the same stop sequence, but cannot be ordered (because one overtakes the other), we assign them to different lines. We denote the line of a trip t by L_t and define $\boldsymbol{p}(L_t) = \boldsymbol{p}(t)$. We also define the set of lines at stop p as

$$\boldsymbol{L}(p) = \left\{(L, i) \mid p = p_L^i \text{ where } L \text{ is a line and } \boldsymbol{p}(L) = \langle p_L^0, p_L^1, \ldots \rangle \right\}. \quad (3)$$

A *trip segment* $p_t^b \to p_t^e$ represents a trip t traveling from stop p_t^b to stop p_t^e. A *transfer* between trips t and u $(t \neq u)$ is denoted by $p_t^e \to p_u^b$, where passengers exit t at the eth stop and board u at the bth. For all transfers,

$$p_t^e \to p_u^b \implies \tau_{\mathrm{arr}}(t, e) + \Delta\tau_{\mathrm{fp}}\left(p_t^e, p_u^b\right) \leq \tau_{\mathrm{dep}}(u, b) \quad (4)$$

must hold. Finally, a *journey* is a sequence of alternating trip segments and transfers, with optional footpaths at the beginning and end. Each leg of a journey must begin at the stop where the previous one ended.

We consider two well-known problems. Since both of them are multi-criteria problems, the results are *Pareto sets* representing non-dominated journeys. A journey dominates another if it is no worse in any criterion; if they are equal in every criterion, we break ties arbitrarily. Although multi-criteria Pareto optimization is NP-hard in general, it is efficiently tractable for natural criteria in public transit networks [15]. In the *earliest arrival problem*, we are given a source stop p_{src}, a target stop p_{tgt}, and a departure time τ. The result is a Pareto set of tuples $(\tau_{\mathrm{jarr}}, n)$ of arrival time and number of transfers taken during non-dominated journeys from p_{src} to p_{tgt} that leave no earlier than τ. For the *profile problem*, we are given source stop p_{src}, target stop p_{tgt}, an earliest departure time τ_{edt}, and a latest departure time τ_{ldt}. Here, we are asked to compute a Pareto set of tuples $(\tau_{\mathrm{jdep}}, \tau_{\mathrm{jarr}}, n)$ representing non-dominated journeys between p_{src} and p_{tgt} with $\tau_{\mathrm{edt}} \leq \tau_{\mathrm{jdep}} \leq \tau_{\mathrm{ldt}}$. Note that for Pareto-optimality, later departure times are considered to be better than earlier ones.

2.2   Related Work

Some existing approaches solve these problems by modeling timetable information as a graph, using either the *time-expanded* or the *time-dependent* model. In the (simple) time-expanded model, a node is introduced for each event, such as a train departing or arriving at a station. Edges are then added to connect nodes on the same trip, as well as between nodes belonging to the same stop (corresponding to a passenger waiting for the next train). To model minimum change times, additional nodes and edges are required [16]. One advantage of this model is that all edge weights are constant, which allows the use of speedup techniques developed for road networks, such as contraction. Unfortunately, it turns out that due to different network structures, these techniques do not perform as well for public transit networks [4]. Also, time-expanded graphs are rather large.

The time-dependent approach produces much smaller graphs in comparison. In the simple model, nodes correspond to stops. Edges no longer have constant weight, but are instead associated with (piecewise linear) travel time functions, which map departure times to travel times (or, equivalently, arrival times).

The weight then depends on the time at which this function is evaluated. This model can be extended to allow for minimum change times by adding a node for each line at each stop [16]. Some speedup techniques have been applied successfully to time-dependent graphs, such as ALT [6] and Contraction [12], although not for multi-criteria problems. For these, several extensions to Dijkstra's algorithm exist, among them the *Multicriteria Label-Setting* [13], the *Multi-Label Correcting* [7], the *Layered Dijkstra* [5], and the *Self-Pruning Connection Setting* [9] algorithms. However, as Dijkstra-variants, each of them has to perform rather costly priority queue operations.

Other approaches do not use graphs at all. *RAPTOR* (Round-bAsed Public Transit Optimized Router) [10] is a dynamic program. In each round, it computes earliest arrival times for journeys with n transfers, where n is the current round number. It does this by scanning along lines and, at each stop, checking for the earliest trip of that line that can be reached. It outperforms Dijkstra-based approaches in practice. The *Connection Scan Algorithm* [11] operates on *elementary connections* (trip segments of length 1). It orders them by departure time into a single array. During queries, this array is then scanned once, which is very fast in practice due to the linear memory access pattern.

A number of speedup techniques have been developed for public transit routing. *Transfer Patterns* [1,3] is based on the observation that for many optimal journeys, the sequence of stops where transfers occur is the same. By precomputing these transfer patterns, journeys can be computed very quickly at query time. *Public Transit Labeling* [8] applies recent advances in hub labeling to public transit networks, resulting in very fast query times. Another example is the *Accelerated Connection Scan Algorithm* [17], which combines CSA with multilevel overlay graphs to speed up queries on large networks. The algorithm presented in this work, however, is a new base algorithm; development of further speedup techniques is a subject for future research.

3   Algorithm

3.1   Preprocessing

We precompute transfers so they can be looked up quickly during queries. A key observation is that the majority of possible transfers is not needed in order to find Pareto-optimal journeys, and can be safely discarded. Preprocessing is divided into several steps: Initial computation and reduction. Initial computation of transfers is relatively straightforward. For each trip t and each stop p_t^i of that trip, we examine p_t^i and all stops reachable via (direct) footpaths from p_t^i. For each of these stops q, we iterate over $(L, j) \in \boldsymbol{L}(q)$ and find the first trip u of line L such that a valid transfer $p_t^i \to p_u^j$ satisfying (4) exists. Since, by definition, trips do not overtake other trips of the same line, we can discard any transfers to later trips of line L. Additionally, we do not add any transfers from the first stop $(i = 0)$ or to the last stop $(j = |\boldsymbol{p}(L)| - 1)$ of a trip. Furthermore, transfers to trips of the same line are only kept if either $u \prec t$ or $j < i$; otherwise, it is better to simply remain in the current trip.

After initial computation is complete, we perform a number of reduction steps, where we discard transfers that are not necessary to find Pareto-optimal journeys. First, we discard any transfers $p_t^i \to p_u^j$ where $p_u^{j+1} = p_t^{i-1}$ (we call these *U-turn transfers*) as long as

$$\tau_{\mathrm{arr}}(t, i-1) + \Delta\tau_{\mathrm{ch}}\big(p_t^{i-1}\big) \leq \tau_{\mathrm{dep}}(u, j+1) \tag{5}$$

holds. In this case, we can already reach u from t at the previous stop, and because

$$
\begin{aligned}
\tau_{\mathrm{arr}}(t, i-1) &\leq & \tau_{\mathrm{dep}}(t, i-1) &\leq & \tau_{\mathrm{arr}}(t, i) \\
\leq \tau_{\mathrm{dep}}(u, j) & & \leq \tau_{\mathrm{arr}}(u, j+1) &\leq & \tau_{\mathrm{dep}}(u, j+1),
\end{aligned}
\tag{6}
$$

all trips that can reach t at the previous stop can also reach u, and all trips reachable from u are also reachable from t. Equation (5) may not hold if the stops in question have different minimum change times.

Next, we further reduce the number of transfers by analyzing which transfers lead to improved arrival times. We do this by moving backwards along a trip, keeping track of where and when passengers in that trip can arrive, either by simply exiting the trip or by transferring to another trip reachable from their current position. Again, we iterate over all trips t. For each trip, we maintain two mappings τ_A and τ_C from stops to arrival time and earliest change time, respectively. Initially, they are set to ∞ for all stops. During execution of the algorithm, they are updated to reflect when passengers arrive (τ_A) or can board the next trip (τ_C) at each stop.[3] We then iterate over stops p_t^i of trip t in decreasing index order, meaning we examine later stops first. At each stop, we update the arrival time and change time for that stop if they are improved:

$$
\begin{aligned}
\tau_A\big(p_t^i\big) &\leftarrow \min\big(\tau_A\big(p_t^i\big), \tau_{\mathrm{arr}}(t, i)\big) \quad \text{and} \\
\tau_C\big(p_t^i\big) &\leftarrow \min\big(\tau_C\big(p_t^i\big), \tau_{\mathrm{arr}}(t, i) + \Delta\tau_{\mathrm{ch}}\big(p_t^i\big)\big).
\end{aligned}
$$

Similarly, we update τ_A and τ_C for all stops q reachable via footpaths from p_t^i:

$$
\begin{aligned}
\tau_A(q) &\leftarrow \min\big(\tau_A(q), \tau_{\mathrm{arr}}(t, i) + \Delta\tau_{\mathrm{fp}}\big(p_t^i, q\big)\big) \quad \text{and} \\
\tau_C(q) &\leftarrow \min\big(\tau_C(q), \tau_{\mathrm{arr}}(t, i) + \Delta\tau_{\mathrm{fp}}\big(p_t^i, q\big)\big).
\end{aligned}
$$

We then determine, for each transfer $p_t^i \to p_u^j$ from t at that stop, if u improves arrival and/or change times for any stop. To do this, we iterate over all stops p_u^k of u with $k > j$ and perform the same updates to τ_A and τ_C as we did above, this time for p_u^k and all stops reachable via footpaths from p_u^k. If this results in any improvements to either τ_A or τ_C, we keep the transfer, otherwise we discard it. Discarded transfers are not required for Pareto-optimal journeys, since we have shown that (a) taking later transfers (or simply remaining in the current trip) leads to equal or better arrival times (τ_A), and (b) all trips reachable via that transfer can also be reached via those later transfers (τ_C).

All these algorithms are trivially parallelized, since each trip is processed independently. Also, there is no need to perform them as separate steps; they

---

[3] If there are no minimum change times, then $\tau_A = \tau_C$ and we only maintain τ_A.

can easily be merged into one. We decided to keep them distinct to showcase the separation of concerns. Furthermore, more complex reduction steps are possible, where there are dependencies between trips. For example, to minimize the size of the transfer set, one could compute full profiles between all stops (all-to-all), then keep only those transfers required for optimal journeys. However, that would be computationally expensive. In contrast, the comparatively simple computations presented here can be performed within minutes, even for large networks, while still resulting in a greatly reduced transfer set (see Section 4 for details).

Note that this explicit representation of transfers allows fine-grained control over them. For instance, one can easily introduce transfers between specific trips that would otherwise violate the minimum change time or footpath restrictions, or remove transfers from certain trips. Transfer preferences are another example. If two trips travel in parallel (for part of their stop sequence), there may be multiple possible transfers between them. The algorithm described above discards all but the last of them; by modifying it, preference could be given to transfers that are more accessible, for instance. Since this only has to be considered during preprocessing, query times are unaffected.

3.2   Earliest Arrival Query

As a reminder, the input to an earliest arrival query consists of the source stop p_{src}, the target stop p_{tgt}, and the (earliest) departure time τ, and the objective is to calculate a Pareto set of (τ_{jarr}, n) tuples representing Pareto-optimal journeys arriving at time τ_{jarr} after n transfers. During the algorithm, we remember which parts of each trip t have already been processed by maintaining the index $R(t)$ of the first reached stop, initialized to $R(t) \leftarrow \infty$ for all trips. We also use a number of queues Q_n of trip segments reached after n transfers and a set $\mathcal{F}$ of tuples $(L, i, \Delta\tau)$. The latter indicates lines reaching the target stop p_{tgt}, and is computed by

$$\mathcal{F} = \{(L, i, 0) \mid (L, i) \in \boldsymbol{L}(p_{\text{tgt}})\}$$
$$\cup \{(L, i, \Delta\tau_{\text{fp}}(q, p_{\text{tgt}})) \mid (L, i) \in \boldsymbol{L}(q) \land \exists \text{ a footpath from } q \text{ to } p_{\text{tgt}}\}$$

We start by identifying the trips travelers can reach from p_{src} at time τ. For this, we examine p_{src} and all stops reachable via footpaths from p_{src}. For each of these stops q, we iterate over $(L, i) \in \boldsymbol{L}(q)$ and find the first trip t of line L such that

$$\tau_{\text{dep}}(t, i) \geq \begin{cases} \tau & \text{if } q = p_{\text{src}}, \\ \tau + \Delta\tau_{\text{fp}}(p_{\text{src}}, q) & \text{otherwise.} \end{cases}$$

For each of those trips, if $i < R(t)$, we add the trip segment $p_t^i \to p_t^{R(t)}$ to queue Q_0 and then update $R(u) \leftarrow \min(R(u), i)$ where $t \preceq u \land L_t = L_u$, meaning we update the first reached stop for t and all later trips of the same line. Due to the way $\preceq$ is defined in (1), none of these later trips u can improve upon t. By marking them as reached, we eliminate them from the search and avoid redundant work.

After the initial trips have been found, we operate on the trip segments in $Q_0, Q_1, \ldots$ until there are no more unprocessed elements. For each trip segment $p_t^b \to p_t^e \in Q_n$, we perform the following three steps. First, we check if this trip reaches the target stop. For each $(L_t, i, \Delta\tau) \in \mathcal{F}$ with $i > b$, we generate a tuple $(\tau_{\mathrm{arr}}(t, i) + \Delta\tau, n)$ and add it to the result set, maintaining the Pareto property. Second, we check if this trip should be pruned because it cannot lead to a non-dominated journey. This is the case if we already found a journey with $\tau_{\mathrm{jarr}} < \tau_{\mathrm{arr}}(t, b+1)$. Third, if the trip is not pruned, we examine its transfers. For each transfer $p_t^i \to p_u^j$ with $b < i \le e$, we check if $j < R(u)$. If so, we add $p_u^j \to p_u^{R(u)}$ to Q_{n+1} and update $R(v) \leftarrow \min(R(v), j)$ for all v with $u \preceq v \wedge L_u = L_v$. Otherwise, we already reached u or an earlier trip of the same line at j or an earlier stop, and we skip the transfer.

The main loop is similar to a breadth-first search: First, all trips reachable directly from the source stop are examined, then all trips reached after a transfer from those, etc. Therefore, we find journeys with the least number of transfers first. Any non-dominated journey discovered later cannot have a lower number of transfers and must therefore arrive earlier. This property enables the pruning in step two, which prevents us from having to examine all reachable trips regardless of the target. However, it also means that the journey with the earliest arrival time is the last one discovered, and all journeys with less transfers are found beforehand. This is why we only consider the multi-criteria problem variants.

3.3   Profile Query

We perform profile queries by running the main loop of an earliest arrival query for each distinct departure time in the given interval, preserving labels between runs to avoid redundant work. Later journeys dominate earlier journeys, provided arrival time and number of transfers are equal or better, while earlier journeys never dominate later ones. Therefore, we process later departures first. However, in order to reuse labels across multiple runs, we need to keep multiple labels for each trip, consisting of the index of the first reached stop and the number of transfers required to reach it. Since the number of transfers is limited in practice, we use $R_n(t)$ to denote the first stop reached on trip t after at most n transfers and update $R_{n+1}(t)$ (and following) whenever we update $R_n(t)$. To decide if a trip segment should be queued while processing Q_n, we compare against and update $R_{n+1}(t)$. We also change the pruning step so we compare against the minimum arrival time of journeys with no more than $n+1$ transfers.

To see why labels can be reused, consider two runs with departure times τ_1 and τ_2, where $\tau_1 < \tau_2$, which both reach trip t at stop i after n transfers. Continuing from this point, both will reach the destination at the same time and after the same number of transfers. However, since $\tau_1 < \tau_2$, the journeys departing at τ_2 dominate the journeys departing at τ_1. Knowing this, we can avoid computing them in the first place by computing τ_2 first and keeping the labels.

3.4   Implementation

We improve the performance of the algorithm by taking advantage of SIMD (single instruction, multiple data) instructions, avoiding dynamic memory allocations and increasing locality of reference (reducing cache misses). In our data instances, all lines have less than 200 stops. Also, none of our tests found Pareto-optimal journeys with 16 or more transfers. Thus, we set the maximum number of transfers to 15. During profile queries, we can then update $R_0(t)$ to $R_{15}(t)$ using a single 128-bit vector minimum operation.

To avoid memory allocations during query execution, we replace the n queues with a single, preallocated array. To see why this is possible, note that the maximum number of trip segments queued is bounded by the number of elementary connections. We use pointers to keep track of the current element, the end of the queue, and the level boundaries (where the number of transfers n is increased).

We improve locality of reference by splitting the steps of the inner loop into three separate loops. Thus, we iterate three times over each level, each time updating the elements in the "queue", before increasing n and moving on to the next level. In the first iteration, we look up $\tau_{arr}(t, b + 1)$ and store it next to the trip segment into the queue. Additionally, we check $\mathcal{F}$ to see if the trip reaches the destination, and update arrival times as necessary. In the second iteration, we perform the pruning step by comparing the time stored in the queue with the arrival time at the destination. If the element is not pruned, we replace it with two indices into the array of transfers, indicating the transfers corresponding to the trip segment. If the element is pruned, we set both indices to 0, resulting in an empty interval. Finally, in the third iteration, we examine this list of transfers and add new trip segments to the queue as necessary. Thus, arrival times $\tau_{arr}(\cdot, \cdot)$ are required only in the first loop, transfer indices only in the second loop, and transfers and reached stops $R_n(\cdot)$ only in the final loop. This leads to reduced cache pressure and therefore to less cache misses, which in turn results in improved performance (see Section 4).

3.5   Journey Descriptions

So far, we only described how to compute arrival time and number of transfers of journeys, which is enough for many applications. However, we can retrieve the full sequence of trip segments as follows. Whenever a trip segment is queued, we store with it a pointer to the currently processed trip segment. Since we replaced the queue with a preallocated array, all entries are preserved until the end of the query. Therefore, when we find a journey reaching the destination, we simply follow this chain of pointers to reconstruct the sequence of trip segments. If required, the appropriate transfers between the trips can be found by rescanning the list of transfers.

4   Experiments

We ran experiments on a dual 8-core Intel Xeon E5-2650 v2 processor clocked at 2.6 GHz, with 128 GB of DDR3-1600 RAM and 20 MB of L3 cache. Our code

Table 1. Instances used for experiments

	London	Germany
Stops	20 764	249 724
Trips	129 263	2 389 253
Connections	4 991 130	46 116 453
Footpaths	45 624	100 470
Lines (Routes)	2 161	232 644
Transfers (full)	121 339 213	1 826 424 894
Transfers (reduced)	19 502 791	186 296 771
Space consumption	115.5 MiB	1 140.9 MiB

Table 2. Preprocessing times for transfer computation and reduction

	London 1 thread	London 16 threads	Germany 1 thread	Germany 16 threads
Computation	18 s	3 s	177 s	37 s
Reduction	357 s	27 s	2 174 s	183 s
Total	375 s	30 s	2 351 s	220 s

was compiled using g++ 4.9.2 with optimizations enabled. We used two test instances, summarized in Table 1. The first, available at data.london.gov.uk, covers Greater London and includes data for underground, bus, and Docklands Light Railway services for one day. The second consists of data used by bahn.de during winter 2011/2012, containing European long distance trains, German local trains, and many buses over two days.

Table 1 also reports the number of transfers before and after reduction, as well as the total space consumption (for the reduced transfers and all timetable data). Reduction eliminates about 84% of transfers for London, and almost 90% for Germany. The times required for preprocessing can be found in Table 2.

Running times reported for queries are averages over 10 000 queries with source and target stops selected uniformly at random. For profile queries, the departure time range is the first day covered by the timetable; for earliest arrival queries, the departure time is selected uniformly at random from that range. We do not compute full journey descriptions (Section 3.5).

We evaluated the optimizations described in Section 3.4, as well as the effect of transfer reduction, on the London instance (Table 3). SIMD instructions are only used in profile queries and enabling them has no effect on earliest arrival queries. With all optimizations, running time for profile queries is improved by a factor of 2. Transfer reduction improves running times by a factor of 3.

We compare our new algorithm to the state of the art in Table 4. We distinguish between algorithms which optimize arrival time only (○) and those that compute Pareto sets optimizing arrival time and number of transfers (●), and between earliest arrival (○) and profile (●) queries. We report the average number

Table 3. Evaluation of optimizations in Section 3.4, using the London instance

variant	earliest arr. (ms)	profile (ms)
Basic, without SIMD	1.7	145.9
Basic, with SIMD	1.7	113.2
Optimized	1.2	70.0
Optimized, all transfers	3.5	226.0

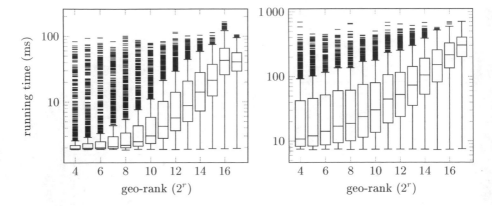

Fig. 1. Earliest arrival query times by geo-rank on Germany

Fig. 2. Profile query times by geo-rank on Germany

of label comparisons per stop[4], where available, and the average running time. Direct comparison with the Accelerated Connection Scan Algorithm (ACSA) [17] and Contraction Hierarchies (CH) [12] is difficult, since they do not support bicriteria queries.[5] We have faster query times than CSA [11] and RAPTOR [10], at the cost of a few minutes of preprocessing time. Transfer Patterns (TP) [1,3] and Public Transit Labeling (PTL) [8] have faster query times (especially on larger instances), however, their preprocessing times are several orders of magnitude above ours.

To examine query times further, we ran 1 000 geo-rank queries [17]. A geo-rank query picks a stop uniformly at random and orders all other stops by geographical distance. Queries are run from the source stop to the 2^r-th stop, where r is the geo-rank. Results for the Germany instance are reported in Figure 1 (earliest arrival queries) and Figure 2 (profile queries). Note the logarithmic scale on both axes. Query times for the maximum geo-rank are about the same as the average query time when selecting source and target uniformly at random, since randomly selected stops are unlikely to be near each other. Local queries, which

---

[4] Note that in our algorithm, labels are not associated with stops, but with trips instead. For better comparison with previously published work, we divided the total number of label comparisons by the number of stops.

[5] ACSA uses transfers to break ties between journeys with equal arrival times.

Table 4. Comparison with the state of the art. Results taken from [2,3,8,17]. Bicriteria algorithms computing a set of Pareto-optimal journeys regarding arrival time and number of transfers are marked in column "tr." (others only optimize arrival time). Profile queries are marked in column "pr.".

algorithm	instance	stops ($\cdot10^3$)	conn. ($\cdot10^6$)	tr.	pr.	prep. (h)	comp. /stop	query (ms)
TripBased	London	20.8	5.0	●	○	< 0.1	23.3	1.2
TP [2]	Madrid	4.6	4.8	●	○	185.0	n/a	3.1
PTL [8]	London	20.8	5.1	●	○	49.3	n/a	0.03
RAPTOR [10]	London	20.8	5.1	●	○	—	10.9	5.4
CSA [11]	London	20.8	4.9	○	○	—	26.6	1.8
CH [12]	Europe (LD)	30.5	1.7	○	○	< 0.1	n/a	0.3
TripBased	Germany	249.7	46.1	●	○	< 0.1	41.4	40.8
TP [3]	Germany	248.4	13.9	●	○	372.0	n/a	0.3
CSA [17]	Germany	252.4	46.2	○	○	—	n/a	298.6
ACSA [17]	Germany	252.4	46.2	○	○	0.2	n/a	8.7
TripBased	London	20.8	5.0	●	●	< 0.1	1 061.7	70.0
TP [2]	Madrid	4.6	4.8	●	●	185.0	n/a	3.1
rRAPTOR [10]	London	20.8	5.1	●	●	—	1 634.0	922.0
CSA [11]	London	20.8	4.9	●	●	—	3 824.9	466.0
TripBased	Germany	249.7	46.1	●	●	< 0.1	228.0	301.7
TP [3]	Germany	248.4	13.9	●	●	372.0	n/a	5.0
ACSA [17]	Germany	252.4	46.2	○	●	0.2	n/a	171.0

are often more relevant in practice, are generally much faster (by an order of magnitude), although there is a significant number of outliers, since physically close locations do not necessarily have direct or fast connections.

5 Conclusion

We presented a novel algorithm for route planning in public transit networks. By focusing on trips and transfers between them, we computed multi-criteria profiles optimizing arrival time and number of transfers on a metropolitan network in 70 ms with a preprocessing time of just 30 s, occupying a Pareto-optimal spot among current state of the art algorithms. The explicit representation of transfers allows fine-grained modeling, while the short preprocessing time allows the use in dynamic scenarios. In addition, localized changes (such as trip delays or cancellations) do not necessitate a full rerun of the preprocessing phase. Instead, only a subset of the data needs to be updated. Development of suitable algorithms is a subject of future studies. Future work also includes efficiently extending the covered period of time by exploiting periodicity in timetables, making the algorithm more scalable by using network decomposition, and extending it to support more generic criteria such as fare zones or walking distance.

References

1. Bast, H., Carlsson, E., Eigenwillig, A., Geisberger, R., Harrelson, C., Raychev, V., Viger, F.: Fast Routing in Very Large Public Transportation Networks Using Transfer Patterns. In: de Berg, M., Meyer, U. (eds.) ESA 2010, Part I. LNCS, vol. 6346, pp. 290–301. Springer, Heidelberg (2010)
2. Bast, H., Delling, D., Goldberg, A., Müller-Hannemann, M., Pajor, T., Sanders, P., Wagner, D., Werneck, R.F.: Route Planning in Transportation Networks. ArXiv e-prints arXiv:1504.05140 [cs.DS] (Apr 2015)
3. Bast, H., Storandt, S.: Frequency-based Search for Public Transit. In: SIGSPA-TIAL, pp. 13–22. ACM, New York (2014)
4. Berger, A., Delling, D., Gebhardt, A., Müller-Hannemann, M.: Accelerating Time-Dependent Multi-Criteria Timetable Information is Harder Than Expected. In: ATMOS 2009. OASIcs (2009)
5. Brodal, G.S., Jacob, R.: Time-dependent networks as models to achieve fast exact time-table queries. Electronic Notes in Theor. Computer Science 92, 3–15 (2004)
6. Cionini, A., D'Angelo, G., D'Emidio, M., Frigioni, D., Giannakopoulou, K., Paraskevopoulos, A., Zaroliagis, C.: Engineering Graph-Based Models for Dynamic Timetable Information Systems. In: ATMOS 2014. OASIcs (2014)
7. Dean, B.C.: Continuous-Time Dynamic Shortest Path Algorithms. Master's thesis, Massachusetts Institute of Technology (1999)
8. Delling, D., Dibbelt, J., Pajor, T., Werneck, R.F.: Public Transit Labeling. In: Bampis, E. (ed.) SEA 2015. LNCS, vol. 9125, pp. 273–285. Springer, Heidelberg (2015)
9. Delling, D., Katz, B., Pajor, T.: Parallel Computation of Best Connections in Public Transportation Networks. JEA 17, 4.4:4.1–4.4:4.26 (2012)
10. Delling, D., Pajor, T., Werneck, R.F.: Round-Based Public Transit Routing. Transportation Science, advance online publication (2012)
11. Dibbelt, J., Pajor, T., Strasser, B., Wagner, D.: Intriguingly Simple and Fast Transit Routing. In: Bonifaci, V., Demetrescu, C., Marchetti-Spaccamela, A. (eds.) SEA 2013. LNCS, vol. 7933, pp. 43–54. Springer, Heidelberg (2013)
12. Geisberger, R.: Contraction of Timetable Networks with Realistic Transfers. In: Festa, P. (ed.) SEA 2010. LNCS, vol. 6049, pp. 71–82. Springer, Heidelberg (2010)
13. Hansen, P.: Bicriterion Path Problems. In: Multiple Criteria Decision Making Theory and Application. LNEMS, vol. 177, pp. 109–127. Springer, Heidelberg (1980)
14. Müller-Hannemann, M., Schulz, F., Wagner, D., Zaroliagis, C.: Timetable Information: Models and Algorithms. In: Geraets, F., Kroon, L.G., Schoebel, A., Wagner, D., Zaroliagis, C.D. (eds.) Railway Optimization 2004. LNCS, vol. 4359, pp. 67–90. Springer, Heidelberg (2007)
15. Müller-Hannemann, M., Weihe, K.: On the cardinality of the Pareto set in bicriteria shortest path problems. Annals of Operations Research 147(1), 269–286 (2006)
16. Pyrga, E., Schulz, F., Wagner, D., Zaroliagis, C.: Efficient Models for Timetable Information in Public Transportation Systems. JEA 12, 2.4:1–2.4:39 (2008)
17. Strasser, B., Wagner, D.: Connection Scan Accelerated. In: ALENEX 2014, pp. 125–137 (2014)

Mixing Color Coding-Related Techniques

Meirav Zehavi

Department of Computer Science, Technion IIT, Haifa 32000, Israel
meizeh@cs.technion.ac.il

Abstract. Narrow sieves, representative sets and divide-and-color are three breakthrough techniques related to color coding, which led to the design of extremely fast parameterized algorithms. We present a novel family of strategies for applying mixtures of them. This includes: (a) a mix of representative sets and narrow sieves; (b) a faster computation of representative sets under certain separateness conditions, mixed with divide-and-color and a new technique, called "balanced cutting"; (c) two mixtures of representative sets and a new technique, called "unbalanced cutting". We demonstrate our strategies by obtaining, among other results, significantly faster algorithms for k-INTERNAL OUT-BRANCHING and WEIGHTED 3-SET k-PACKING, and a general framework for speeding-up the previous best deterministic algorithms for k-PATH, k-TREE, r-DIMENSIONAL k-MATCHING, GRAPH MOTIF and PARTIAL COVER.

1 Introduction

A problem is *fixed-parameter tractable (FPT)* with respect to a parameter k if it can be solved in time $O^*(f(k))$ for some function f, where O^* hides factors polynomial in the input size. The color coding technique, introduced by Alon et al. [1], led to the discovery of the first single exponential time FPT algorithms for many subcases of SUBGRAPH ISOMORPHISM. In the past decade, three breakthrough techniques improved upon it, and led to the development of extremely fast FPT algorithms for many fundamental problems. This includes the combinatorial divide-and-color technique [7], the algebraic multilinear detection technique [20,21,35] (which was later improved to the more powerful narrow sieves technique [2,3]), and the combinatorial representative sets technique [16].

Divide-and-color was the first technique that resulted in (both randomized and deterministic) FPT algorithms for weighted problems that are faster than those relying on color coding. Later, representative sets led to the design of deterministic FPT algorithms for weighted problems that are faster than the randomized ones based on divide-and-color. The fastest FPT algorithms, however, rely on narrow sieves. Unfortunately, narrow sieves is only known to be relevant to the design of *randomized* algorithms for *unweighted* problems.[1]

We present novel strategies for applying these techniques, combining the following elements (see Section 3).

---

[1] More precisely, when used to solve weighted problems, the running times of the resulting algorithms have *exponential* dependencies on the length of the input weights.

© Springer-Verlag Berlin Heidelberg 2015
N. Bansal and I. Finocchi (Eds.): ESA 2015, LNCS 9294, pp. 1037–1049, 2015.
DOI: 10.1007/978-3-662-48350-3_86

- Mixing narrow sieves and representative sets, previously considered to be two *independent* color coding-related techniques.
- Under certain "separateness conditions", speeding-up the best known computation of representative sets.
- Mixing divide-and-color-based preprocessing with the computation in the previous item, speeding-up standard representative sets-based algorithms.
- Cutting the universe into small pieces in two special manners, one used in the mix in the previous item, and the other mixed with a non-standard representative sets-based algorithm to improve its running time (by decreasing the size of the partial solutions it computes).

To demonstrate our strategies, we consider the following well-studied problems.

k-Internal Out-Branching (k-IOB): Given a directed graph $G = (V, E)$ and a parameter $k \in \mathbb{N}$, decide if G has an out-branching (i.e., a spanning tree with exactly one node of in-degree 0) with at least k internal nodes.

Weighted k-Path: Given a directed graph $G = (V, E)$, a weight function $w : E \to \mathbb{R}$, $W \in \mathbb{R}$ and a parameter $k \in \mathbb{N}$, decide if G has a simple directed path on exactly k nodes and of weight at most W.

Weighted 3-Set k-Packing $((3, k)$-WSP): Given a universe U, a family $\mathcal{S}$ of subsets of size 3 of U, a weight function $w : \mathcal{S} \to \mathbb{R}$, $W \in \mathbb{R}$ and a parameter $k \in \mathbb{N}$, decide if there is a family $\mathcal{S}' \subseteq \mathcal{S}$ of k disjoint sets and weight at least W.

The k-IOB problem is NP-hard since it generalizes HAMILTONIAN PATH. It is of interest, for example, in database systems [10], and for connecting cities with water pipes [31]. Many FPT algorithms were developed for k-IOB and related variants (see, e.g., [8,9,13,14,18,22,30,32,36]). We solve it in deterministic time $O^*(5.139^k)$ and randomized time $O^*(3.617^k)$, improving upon the previous best deterministic time $O^*(6.855^k)$ [32] and randomized time $O^*(4^k)$ [9,36]. To this end, we establish a relation between certain directed trees and paths on 2 nodes. This shows how certain partial solutions to k-IOB can be completed efficiently via a computation of a maximum matching in the underlying undirected graph.

We also present a unified approach for speeding-up standard representative sets-based algorithms. It can be used to modify the previous best deterministic algorithms (that already rely on the best known computation of representative sets) for k-PATH, k-TREE, r-DIMENSIONAL k-MATCHING $((r, k)$-DM), GRAPH MOTIF WITH DELETIONS (GM$_D$) and PARTIAL COVER (PC), including their weighted variants, which run in times $O^*(2.6181^k)$ [15,32], $O^*(2.6181^k)$ [15,32], $O^*(2.6181^{(r-1)k})$ [17], $O^*(2.6181^{2k})$ [29] and $O^*(2.6181^k)$ [32], to run in times $O^*(2.5961^k)$, $O^*(2.5961^k)$, $O^*(2.5961^{(r-1)k})$, $O^*(2.5961^{2k})$ and $O^*(2.5961^k)$, respectively. To demonstrate our approach, we use WEIGHTED k-PATH.

In the past decade, $(3, k)$-WSP and $(3, k)$-SP enjoyed a race towards obtaining the fastest FPT algorithms (see [3,5,6,7,11,12,19,20,23,24,33,34,37]). We solve $(3, k)$-WSP in deterministic time $O^*(8.097^k)$, improving upon $O^*(12.155^k)$, which is both the previous best running time of an algorithm for $(3, k)$-WSP and the previous best running time of a deterministic algorithm for $(3, k)$-SP [37]. The full version [38] also solves P_2-PACKING, a special case of $(3, k)$-WSP.

2   Color Coding-Related Techniques

In this paper, we use a known algorithm based on narrow sieves as a black box; thus, we avoid describing this technique. We proceed by giving a brief description of divide-and-color, followed by a more detailed one of representative sets.

Divide-and-Color: Divide-and-color is based on recursion; at each step, we color elements randomly or deterministically. In our strategies, we are interested in applying only one step, which can be viewed as using color coding with only two colors. In such a step, we have a set A of n elements, and we seek a certain subset A^* of k elements in A. We partition A into two (disjoint) sets, B and C, by coloring its elements. Thus, we get the problem of finding a subset $B^* \subseteq A^*$ in B, and another problem of finding the subset $C^* = A^* \setminus B^*$ in C. The partition should be done in a manner that is efficient and results in easier problems, which does not necessarily mean that we get two independent problems (of finding B^* in B and C^* in C). Deterministic applications of divide-and-color often use a tool called an (n, k)-universal set [28]. We need its following generalization:

Definition 1. *Let $\mathcal{F}$ be a set of functions $f : \{1, 2, \ldots, n\} \to \{0, 1\}$. We say that $\mathcal{F}$ is an (n, k, p)-universal set if for every subset $I \subseteq \{1, 2, \ldots, n\}$ of size k and a function $f' : I \to \{0, 1\}$ that assigns '1' to exactly p indices, there is a function $f \in \mathcal{F}$ such that for all $i \in I$, $f(i) = f'(i)$.*

The next result (of [16]) asserts that small universal sets can be computed fast.

Theorem 1. *There is an algorithm that, given integers n, k and p, computes an (n, k, p)-universal set $\mathcal{F}$ of size $O^*(\binom{k}{p} 2^{o(k)})$ in deterministic time $O^*(\binom{k}{p} 2^{o(k)})$.*

Representative Sets: We first give the definition of a representative family, and then discuss its relevance to the design of FPT algorithms. We note that a more general definition, not relevant to this paper, is given in [16,26].

Definition 2. *Given a universe E, a family $\mathcal{S}$ of subsets of size p of E, a function $w : \mathcal{S} \to \mathbb{R}$ and $k \in \mathbb{N}$, we say that a subfamily $\widehat{\mathcal{S}} \subseteq \mathcal{S}$ max (min) $(k-p)$-represents $\mathcal{S}$ if for any pair $X \in \mathcal{S}$ and $Y \subseteq E \setminus X$ such that $|Y| \leq k - p$, there is $\widehat{X} \in \widehat{\mathcal{S}}$ disjoint from Y such that $w(\widehat{X}) \geq w(X)$ $(w(\widehat{X}) \leq w(X))$.*

Informally, Definition 2 implies that if a set Y can be extended to a set of size at most k by adding a set $X \in \mathcal{S}$, then it can also be extended to a set of the same size by adding a set $\widehat{X} \in \widehat{\mathcal{S}}$ that is at least as good as X. The special case where the sets in $\mathcal{S}$ have the same weight is the unweighted version Definition 2.

Many FPT algorithms are based on dynamic programming, where after each stage, the algorithm computes a family $\mathcal{S}$ of sets that are partial solutions. At this point, we can compute a subfamily $\widehat{\mathcal{S}} \subseteq \mathcal{S}$ that represents $\mathcal{S}$. Then, each reference to $\mathcal{S}$ can be replaced by a reference to $\widehat{\mathcal{S}}$. The representative family $\widehat{\mathcal{S}}$ contains "enough" sets from $\mathcal{S}$; therefore, such replacement preserves the correctness of the algorithm. Thus, if we can efficiently compute representative families that are small enough, we can substantially improve the running time of the algorithm.

The *Two Families Theorem* of Bollobás [4] implies that for any universe E, a family $\mathcal{S}$ of subsets of size p of E and a parameter k ($\geq p$), there is a subfamily $\widehat{\mathcal{S}} \subseteq \mathcal{S}$ of size $\binom{k}{p}$ that $(k-p)$-represents $\mathcal{S}$. Monien [27] computed representative families of size $\sum_{i=0}^{k-p} p^i$ in time $O(|\mathcal{S}|p(k-p)\sum_{i=0}^{k-p} p^i)$, and Marx [25] computed such families of size $\binom{k}{p}$ in time $O(|\mathcal{S}|^2 p^{k-p})$. Recently, Fomin et al. [16] introduced a powerful technique that enables to compute representative families of size $\binom{k}{p}2^{o(k)}\log|E|$ in time $O(|\mathcal{S}|(k/(k-p))^{k-p}2^{o(k)}\log|E|)$. We need the following tradeoff-based generalization of their computation, given in [15,32]:

Theorem 2. *Given a fixed $c \geq 1$, a universe E, a family $\mathcal{S}$ of subsets of size p of E, a function $w : \mathcal{S} \to \mathbb{R}$ and a parameter $k \in \mathbb{N}$, a family $\widehat{\mathcal{S}} \subseteq \mathcal{S}$ of size $\frac{(ck)^k}{p^p(ck-p)^{k-p}}2^{o(k)}\log|E|$ that max (min) $(k-p)$-represents $\mathcal{S}$ can be found in time $O(|\mathcal{S}|(ck/(ck-p))^{k-p}2^{o(k)}\log|E| + |\mathcal{S}|\log|\mathcal{S}|\log W)$, where $W = \max_{S\in\mathcal{S}} |w(S)|$.*

3   Our Mixing Strategies

We give an overview of our mixing strategies in the context of the problems solved in this paper. The technical details, as well as another strategy (used to solve P_2-PACKING), are given in [38]. The first strategy builds upon the approach used in [16] to solve LONG DIRECTED CYCLE, and it does not involve a mixture of techniques, but it is relevant to this paper since the second strategy builds upon it; the other strategies are novel and involve mixtures of techniques.

3.1   Strategies I and II

Strategy I: Our deterministic k-IOB algorithm follows Strategy I in Fig. 1. The first reduction (of [8]) allows to focus on finding a small out-tree (i.e., a directed tree with exactly one node of in-degree 0) rather than an out-branching, while the second allows to focus on finding an even smaller out-tree, but we need to find it along with a set of paths on 2 nodes. The second reduction might be of independent interest. We then use a representative sets-based procedure in a manner that does not directly solve the problem, but returns a family of partial solutions that are trees. We try to extend the trees to solutions in polynomial time via computations of maximum matchings, which leads to the result below.

Theorem 3. *k-IOB is solvable in deterministic time $O^*(5.139^k)$.*

Generally, Strategy I is relevant to the following scenario. We are given a problem which can be reduced to two suproblems whose solutions should be disjoint. If we could disregard the disjointness condition, the first subproblem is solvable using representative sets, and the second subproblem is solvable in polynomial time. Then, we modify the representative sets-based procedure to compute a family of solutions (rather than one solution), such that for any set that might be a solution to the second subproblem, there is a solution in the family that is disjoint

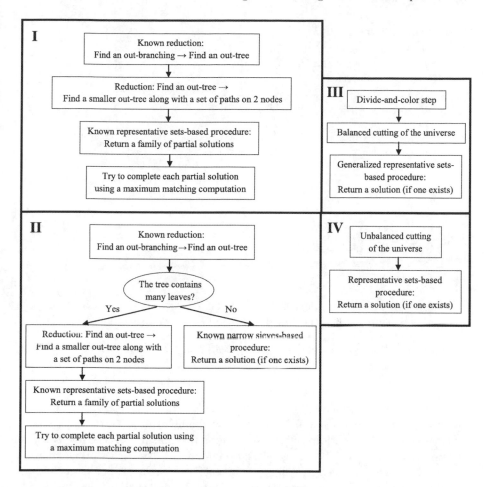

Fig. 1. Strategies for mixing color coding-related techniques, used to develop a deterministic algorithm for k-IOB (I), a randomized algorithm for k-IOB (II), deterministic algorithms for k-PATH, k-TREE, (r, k)-DM, GM_D and PC, including their weighted versions (III), and a deterministic algorithm for $(3, k)$-WSP (IV).

from this set. Thus, we can iterate over every solution A in the family, remove its elements from the input, and attempt to find a solution B to the second subproblem in the remaining part of the input—if we succeed, the combination of A and B should solve the original problem.

Strategy II: Our second result, a randomized FPT algorithm for k-IOB, builds upon our first algorithm and follows Strategy II in Fig. 1. This strategy shows the usefulness of mixing narrow sieves and representative sets, previously considered to be two independent tools for developing FPT algorithms. Strategy II indicates that the representative sets technique is relevant to the design of fast randomized

---

Algorithm 1. Strategy II ($\mathcal{I}$)

1: Let MIN and MAX be the minimum and maximum possible sizes of a solution.
2: **for** $size = MIN, MIN + 1, \ldots, MAX$ **do**
3:    **if** $size$ is "small" **then**
4:        Call a narrow sieves-based procedure, NarSie, with the input $(\mathcal{I}, size)$.
5:        **if** NarSie accepts the input **then** accept. **end if**
6:    **else**
7:        Call a representative sets-based procedure, RepSet, with the input $(\mathcal{I}, size)$, to compute a family $\mathcal{F}$ of partial solutions.
8:        **for all** $F \in \mathcal{F}$ **do**
9:            Call a polynomial-time procedure, PolTim, with the input $(\mathcal{I}, size, F)$.
10:           **if** PolTim accepts the input **then** accept. **end if**
11:       **end for**
12:   **end if**
13: **end for**
14: Reject.

FPT algorithms, even for unweighted problems. We thus obtain the following theorem, which breaks, for the first time, the $O^*(4^k)$-time barrier for k-IOB.[2]

Theorem 4. *There is a randomized algorithm for k-IOB that runs in time $O^*(3.617^k)$. The algorithm never decides that a no-instance is a yes-instance, but with probability $\leq \frac{1}{3}$, it decides that a yes-instance is a no-instance.*[3]

Generally, this strategy may be relevant to the scenario of Strategy I (in the context of *randomized* algorithms). Now, when we handle this scenario, we "guess" what fraction of the solution solves the harder subproblem (i.e., the one that we need to solve using a color coding-related technique). If this fraction is small, it is advised to follow Strategy I (since representative sets are efficient in finding a family of partial solutions that are significantly smaller than the size of the entire solution); otherwise, we need to rely on a condition stating that if this fraction is large, then the size of the entire solution is actually small—then, we can efficiently find it by using a narrow sieves-based procedure. Given a problem instance $\mathcal{I}$, Algorithm 1 summarizes the framework underlying Strategy II.

3.2   Strategy III

Our third result, a deterministic FPT algorithm for WEIGHTED k-PATH, follows Strategy III in Fig. 1. This strategy can be used to speed-up algorithms for other problems based on a standard application of representative sets, such as the k-TREE, (r, k)-DM, GM_D and PC algorithms of [15,17,29,32], including their

---

[2] Previous algorithms obtained solutions in time $O^*(2^k)$ to a problem generalizing (directed) k-PATH, and could thus solve k-IOB in time $O^*(2^{2k})$. Thus, it seemed that an $O^*((4 - \epsilon)^k)$ time algorithm for k-IOB would imply an $O^*((2 - \epsilon)^k)$ time algorithm for (directed) k-PATH, which is a problem that is open for several decades.

[3] Clearly, the success probability can be increased through multiple runs.

weighted variants, where: (1) elements are never deleted from partial solutions (this is not the case in our $(3, k)$-WSP algorithm); (2) the solution is one of the computed representative sets (this is not the case in our k-IOB and P_2-PACKING algorithms). We rely on a generalization of Definition 2 and Theorem 2:

Definition 3. *Let $E_1, E_2, \ldots, E_t$ be disjoint universes, $p_1, p_2, \ldots, p_t \in \mathbb{N}$, and $\mathcal{S}$ be a family of subsets of $(\bigcup_{i=1}^{t} E_i)$ such that $[\forall S \in \mathcal{S}, i \in \{1, 2, \ldots, t\} : |S \cap E_i| = p_i]$. Given a function $w : \mathcal{S} \to \mathbb{R}$ and parameters $k_1, k_2, \ldots, k_t \in \mathbb{N}$, we say that a subfamily $\widehat{\mathcal{S}} \subseteq \mathcal{S}$ max (min) $(k_1 - p_1, k_2 - p_2, \ldots, k_t - p_t)$-represents $\mathcal{S}$ if for any pair $X \in \mathcal{S}$ and $Y \subseteq (\bigcup_{i=1}^{t} E_i) \setminus X$ such that $[\forall i \in \{1, 2, \ldots, t\} : |Y \cap E_i| \leq k_i - p_i]$, there is $\widehat{X} \in \widehat{\mathcal{S}}$ disjoint from Y such that $w(\widehat{X}) \geq w(X)$ $(w(\widehat{X}) \leq w(X))$.*

Theorem 5. *Given fixed $c_1, c_2, \ldots, c_t \geq 1$, and $E_1, \ldots, E_t$, $p_1, \ldots, p_t$, $\mathcal{S}$, w and $k_1, \ldots, k_t$ as in Definition 3, a family $\widehat{\mathcal{S}} \subseteq \mathcal{S}$ of size $\prod_{i=1}^{t} (\frac{(c_i k_i)^{k_i}}{p_i{}^{p_i} (c_i k_i - p_i)^{k_i - p_i}}$ $2^{o(k_i)} \log |E_i|)$ that max (min) $(k_1 - p_1, k_2 - p_2, \ldots, k_t - p_t)$-represents $\mathcal{S}$ can be found in time $O(|\mathcal{S}| \prod_{i=1}^{t} ((c_i k_i / (c_i k_i - p_i))^{k_i - p_i} 2^{o(k_i)} \log |E_i|) + |\mathcal{S}| \log |\mathcal{S}| \log W)$, where $W = \max_{S \in \mathcal{S}} |w(S)|$.*

Relying on the proof of Theorem 2, we construct a separate data structure for each universe E_i; then, we combine information stored in these structures to compute a representative family. The details are given in [38]. We use Theorem 5 to efficiently solve a subcase of WEIGHTED k-PATH. We translate WEIGHTED k-PATH to this subcase via divide-and-color preprocessing, mixed with a technique that we call *balanced cutting*. The intuition underlying the translation process is explained below (the full version [38] also contains illustrations).

Consider some solution, which is a path $P = (V_P, E_P)$ (of minimal weight) on k nodes. Suppose that we have a set $E_1 \subseteq V$ such that when we look at P, starting from its first node (i.e., the node that does not have an ingoing neighbor), we count exactly $(i - 1)$ nodes that belong to $E_2 = V \setminus E_1$, then one node that belongs to E_1, then $(i - 1)$ nodes that belong to E_2, and so on. The computation of E_1 is discussed later. We call E_1 (E_2) the blue (red) part of the universe, and $E_1 \cap V_P$ ($E_2 \cap V_P$) the blue (red) part of P.

Suppose we try to find P by using a simple dynamic programming-based procedure, Simple, embedded with (standard) computations of representative sets. That is, at each stage, for every node $v \in V$, we have a family $\mathcal{S}$ of partial solutions such that each of them is a path that ends at v, and all of these paths have the same size (between 1 and k, depending on the stage). We decrease the size of $\mathcal{S}$ by computing a family that represents it. To advance to the next stage, we try to extend each of the remaining partial solutions, which is a path, by using every possible neighbor of its last node (that does not already belong to the path). Since i is large, to obtain an efficient algorithm, it is necessary to progress by adding nodes one-by-one. However, for the sake of clarity (of this explanation), suppose that we add i nodes at once (at each stage).

Now, consider a partition of the blue part of the universe into two sets, L and R, whose computation is discussed later. Visualize the nodes in L (R) as colored in dark (light) blue. We assume that if there is a solution, then there is also a solution, say P, such that L captures the first $\frac{k}{2i}$ elements of the blue part of P and R captures the remaining $\frac{k}{2i}$ elements of the blue part of P. Thus, when we look at a partial solution in the first half of the execution (of Simple), it should only contain *dark* blue and red elements. Moreover, we need to ensure that we have a partial solution that does not contain a certain set (using which it might be completed to a solution) of only *dark* blue and red elements—we do not need to ensure that it does not contain a certain set of light blue, dark blue and red elements, since we simply do not use any light blue element up to this point. When we reach the second half of the execution, we attempt to add to our partial solutions only *light* blue and red elements. Thus, we can ignore all of the dark blue elements in our partial solutions (we will not encounter them again), and we need to ensure that we have a partial solution that does not contain a certain set of only *light* blue and red elements.

We have shown that L (R) should be considered only in the first (second) half of the execution. Thus, in the first half of the execution, we compute representative families faster, since there are less elements in the sets to which we need to ensure separateness; in the second half of the execution, we also compute them faster, since we have smaller partial solutions. However, the time saved by faster computations of representative families does not justify the time necessary to obtain L and R. Fortunately, we save more time by using *generalized* representative families (Theorem 5). The slowest computation of a representative family for some family $\mathcal{S}$ (by [15,32]) occurs when each set in $\mathcal{S}$ contains slightly more than $k/2$ elements. We benefit from the usage of generalized representative sets, since the worst running times in the context of L and R actually occur at a point that is good with respect to the red part, and the worst running time in the context of the red part actually occurs at a point that is good with respect to L and R. For example, when we look at a partial solution that contains slightly more than 50%, say 55%, of the number of dark blue elements in a solution, it only contains $(55/2)\% = 27.5\%$ of the number of red elements in a solution; also, when we look at a partial solution that contains 55% of the number of light blue elements in a solution, it already contains $50 + (55/2)\% = 77.5\%$ of the number of red elements in a solution; finally, when we look at a partial solution that contains 55% of the number of red elements in a solution, it already contains 100% of the number of dark blue elements in a solution, and only $2 \cdot (55 - 50)\% = 10\%$ of the number of light blue elements in a solution. The computation in Theorem 5 is significantly faster under such distortion.

Still, letting L and R have the same size is not good enough (recall that the slowest computation of a representative family for some family $\mathcal{S}$ occurs when each set in $\mathcal{S}$ contains slightly *more* than $k/2$ elements). We will let the red part be significantly larger than the blue part (otherwise the computation of L and R is inefficient); thus, it seems reasonable that the separation between L and R should take place at the point corresponding to the slowest computation of a representative family with respect to the red part. In the algorithm, the choice is more complicated, since L, R and E_2 have different tradeoff parameters c (in fact, E_2 has two "main" tradeoff parameters, as well as a set of "transition" tradeoff parameters), and the choice effects the time required to compute L and R. Having E_1 (the blue part of the universe), it is easy to compute L and R by using a single divide-and-color step (see Section 2). That is, we compute many pairs (L, R), and ensure that at least one of them is good (i.e., captures the blue elements of a solution as described earlier). More precisely, assuming that we have some arbitrary association between indices and elements, we compute a $(|V|, |L \cup R|, |R|)$-universal set (see Theorem 1); for each function, we let L (R) contain the elements corresponding to indices to which the function assigns '0' ('1'). To ensure that the size of the universal set is not too large (which results in a slow running time), we let $|L \cup R|$ be significantly smaller than k.

Our computation of E_1 actually results in a set that may be very different from the nicely organized blue part described earlier (since obtaining such a part is inefficient). We only demand that E_1 contains exactly $|L \cup R|$ elements from some solution (if one exists). This can be accomplished in polynomial time: we define an arbitrary order on V, say $v_1, v_2, \ldots, v_{|V|}$, and for each node v_i, we let the (current) blue part contain all the elements that are greater or equal to v_i. We thus encounter at least one blue part that has the above mentioned desired property. Next, assume that we have such a blue part, E_1.

The blue part may be congested at some locations, and sparse in others, along the solution (since it contains the desired number of elements from a solution, regardless of their location in the path that is a solution). This is problematic with respect to *generalized* representative sets. For example, if the blue part is congested slightly after the middle of the solution, the point corresponding to the worst computation time for L (or R) is also the point corresponding to the worst computation time with respect to E_2. We do not need to ensure that E_1 is exactly as nicely ordered as described earlier, but we still need to ensure that congestions of the dark (light) blue part are as close as possible to the beginning (end) of the solution. Observe that the more the dark (light) blue part is congested at the beginning (end) of the solution, we gain more from our computation of generalized representative sets. We will only ensure that at worst, the blue part is *approximately* balanced along the solution. To this end, we cut the solution into a *fixed* number of small pieces of the same size (apart from one piece, which for the sake of clarity, this explanation ignores). We reorder the pieces, so that pieces that contain many blue elements are located closer to the beginning and the end of the solution. Now, we can benefit from the usage of generalized representative sets—of course, we no longer seek a directed path, but a collection of small directed paths (which is not a problem, since we can use their original order to obtain a directed path). Using the divide-and-color step described earlier, the dark blue part should appear only in "early" pieces, and the light blue part should appear only in the latter pieces.

Since we do not know the solution in advance, we cannot explicitly cut it into small pieces (and order them). Moreover, we cannot explicitly partition V into the set of elements that should be considered in the first piece, the set of elements that should be considered in the second piece, and so on, since this is inefficient. However, we can *implicitly* cut the universe as follows. We "guess" which are the

---

Algorithm 2. Strategy III ($\mathcal{I}$)

1: Let $e_1, e_2, \ldots, e_n$ define an order on the elements of the universe E.
2: Compute an $(n, |L \cup R|, |R|)$-universal set $\mathcal{F}$ by using Theorem 1.
3: **for** $i = 1, 2, \ldots, n$ **do**
4:     **for all** $f \in \mathcal{F}$ **do**
5:        Let $L = \{e_j \in E : j \geq i, f(e_j) = 0\}$ and $R = \{e_j \in E : j \geq i, f(e_j) = 1\}$.
6:        **for all** functions ℓ_1, ℓ_2, r_1, r_2 that implicitly cut the universe E **do**
7:           Associate a "distorted" problem instance, $\mathcal{I}'$, with $(\mathcal{I}, L, R, \ell_1, \ell_2, r_1, r_2)$.
8:           Call a *generalized* representative sets-based procedure, GenRepSet, with the input $\mathcal{I}'$.
9:           **if** GenRepSet accepts the input **then** accept. **end if**
10:        **end for**
11:     **end for**
12: **end for**
13: Reject.

---

nodes that are the first and last in every piece, according to the desired order of the pieces (more precisely, we iterate over every option of choosing these nodes—there is a polynomial number of options since there is a fixed number of pieces). The algorithm uses functions, ℓ_1, ℓ_2, r_1 and r_2, which indicate (via their images) which are the nodes that we guessed; the functions ℓ_1 and ℓ_2 belong to the first half of the execution, and for an index j, ℓ_1 (ℓ_2) assigns the first (last) node in the path that we are currently seeking; the functions r_1 and r_2 are similar, relating to the second half of the execution. Each node (besides two nodes) should appear exactly twice, both as a node that should start a piece and as a node that should end a piece (to ensure that the pieces form a path); observe that the pieces should be disjoint excluding the nodes connecting them, and that the node that ends piece j does not necessarily begins piece $j + 1$.

Algorithm 2 summarizes the framework underlying Strategy III. Overall, we obtain the following result.

Theorem 6. WEIGHTED k-PATH *is solvable in deterministic time* $O(2.59606^k \cdot poly(|V|) \cdot \log \widetilde{W}) = O^*(2.59606^k)$, *where* $\widetilde{W} = \max_{e \in E} |w(e)|$.

3.3   Strategy IV

Our fourth result, a deterministic FPT algorithm for $(3, k)$-WSP, follows Strategy IV in Fig. 1. Here we also cut the universe into small parts, though in a different manner, which allows us to delete more elements from partial solutions than [37]. We call this technique *unbalanced cutting*. Roughly speaking, unbalanced cutting *explicitly* partitions the universe into small pieces, using which it orders the entire universe (the order is only partially arbitrary), such that at certain points during the computation, we are "given" an element e that implies that from now on, we should not try to add (to partial solutions) elements that are "smaller" than e—thus, we can delete (from our partial solutions) all the elements that are smaller than e (since we will not encounter them again). Since we are handling smaller partial solutions, we get a better running time. The number of elements that we can delete at each point where we are "given" an element is computed by defining a recursive formula.

Due to lack of space, further explanations are omitted, and can be found in the full version [38]. Overall, we obtain the following result.

Theorem 7. $(3, k)$-WSP *is solvable in deterministic time* $O(8.097^k \cdot poly(|\mathcal{S}|) \cdot \log \widetilde{W}) = O^*(8.097^k)$, *where* $\widetilde{W} = \max_{S \in \mathcal{S}} |w(S)|$.

Generally, this strategy may be relevant to cases where we can isolate a layer of elements in a partial solution (in the case of k-WSP, this layer consists of the smallest element in each set of 3 elements in the partial solution) such that as the computation progresses, we can remove (from partial solutions) elements from this layer. Strategy IV attempts to allow us to delete elements not only from the isolated layer, but also from the other layers.

References

1. Alon, N., Yuster, R., Zwick, U.: Color coding. J. ACM 42(4), 844–856 (1995)
2. Björklund, A.: Determinant sums for undirected hamiltonicity. In: FOCS, pp. 173–182 (2010)
3. Björklund, A., Husfeldt, T., Kaski, P., Koivisto, M.: Narrow sieves for parameterized paths and packings. CoRR abs/1007.1161 (2010)
4. Bollobás, B.: On generalized graphs. Acta Math. Aca. Sci. Hun. 16, 447–452 (1965)
5. Chen, J., Feng, Q., Liu, Y., Lu, S., Wang, J.: Improved deterministic algorithms for weighted matching and packing problems. Theor. Comput. Sci. 412(23), 2503–2512 (2011)
6. Chen, J., Friesen, D., Jia, W., Kanj, I.: Using nondeterminism to design effcient deterministic algorithms. Algorithmica 40(2), 83–97 (2004)
7. Chen, J., Kneis, J., Lu, S., Molle, D., Richter, S., Rossmanith, P., Sze, S.H., Zhang, F.: Randomized divide-and-conquer: Improved path, matching, and packing algorithms. SIAM J. on Computing 38(6), 2526–2547 (2009)
8. Cohen, N., Fomin, F.V., Gutin, G., Kim, E.J., Saurabh, S., Yeo, A.: Algorithm for finding k-vertex out-trees and its application to k-internal out-branching problem. J. Comput. Syst. Sci. 76(7), 650–662 (2010)
9. Daligault, J.: Combinatorial techniques for parameterized algorithms and kernels, with applicationsto multicut. PhD thesis, Universite Montpellier, France (2011)
10. Demers, A., Downing, A.: Minimum leaf spanning tree. US Patent no. 6,105,018 (August 2013)
11. Downey, R., Fellows, M.: Parameterized Complexity. Springer, New York (1999)
12. Fellows, M., Knauer, C., Nishimura, N., Ragde, P., Rosamond, F., Stege, U., Thilikos, D., Whitesides, S.: Faster fixed-parameter tractable algorithms for matching and packing problems. Algorithmica 52(2), 167–176 (2008)
13. Fomin, F.V., Gaspers, S., Saurabh, S., Thomassé, S.: A linear vertex kernel for maximum internal spanning tree. J. Comput. Syst. Sci. 79(1), 1–6 (2013)
14. Fomin, F.V., Grandoni, F., Lokshtanov, D., Saurabh, S.: Sharp separation and applications to exact and parameterized algorithms. Algorithmica 63(3), 692–706 (2012)

15. Fomin, F.V., Lokshtanov, D., Panolan, F., Saurabh, S.: Representative sets of product families. In: Schulz, A.S., Wagner, D. (eds.) ESA 2014. LNCS, vol. 8737, pp. 443–454. Springer, Heidelberg (2014)

16. Fomin, F.V., Lokshtanov, D., Saurabh, S.: Efficient computation of representative sets with applications in parameterized and exact agorithms. In: SODA, pp. 142–151 (2014)

17. Goyal, P., Misra, N., Panolan, F.: Faster deterministic algorithms for r-dimensional matching using representative sets. In: FSTTCS, pp. 237–248 (2013)

18. Gutin, G., Razgon, I., Kim, E.J.: Minimum leaf out-branching and related problems. Theor. Comput. Sci. 410(45), 4571–4579 (2009)

19. Koutis, I.: A faster parameterized algorithm for set packing. Inf. Process. Lett. 94(1), 7–9 (2005)

20. Koutis, I.: Faster algebraic algorithms for path and packing problems. In: Aceto, L., Damgård, I., Goldberg, L.A., Halldórsson, M.M., Ingólfsdóttir, A., Walukiewicz, I. (eds.) ICALP 2008, Part I. LNCS, vol. 5125, pp. 575–586. Springer, Heidelberg (2008)

21. Koutis, I., Williams, R.: Limits and applications of group algebras for parameterized problems. In: Albers, S., Marchetti-Spaccamela, A., Matias, Y., Nikoletseas, S., Thomas, W. (eds.) ICALP 2009, Part I. LNCS, vol. 5555, pp. 653–664. Springer, Heidelberg (2009)

22. Li, W., Wang, J., Chen, J., Cao, Y.: A $2k$-vertex kernel for maximum internal spanning tree. CoRR abs/1412.8296 (2014)

23. Liu, Y., Chen, J., Wang, J.: On efficient FPT algorithms for weighted matching and packing problems. In: TAMC, pp. 575–586 (2007)

24. Liu, Y., Lu, S., Chen, J., Sze, S.H.: Greedy localization and color-coding: improved matching and packing algorithms. In: IWPEC, pp. 84–95 (2006)

25. Marx, D.: Parameterized coloring problems on chordal graphs. Theor. Comput. Sci. 351, 407–424 (2006)

26. Marx, D.: A parameterized view on matroid optimization problems. Theor. Comput. Sci. 410, 4471–4479 (2009)

27. Monien, B.: How to find long paths efficiently. Annals Disc. Math. 25, 239–254 (1985)

28. Naor, M., Schulman, J.L., Srinivasan, A.: Splitters and near-optimal derandomization. In: FOCS, pp. 182–191 (1995)

29. Pinter, R.Y., Shachnai, H., Zehavi, M.: Deterministic parameterized algorithms for the graph motif problem. In: Csuhaj-Varjú, E., Dietzfelbinger, M., Ésik, Z. (eds.) MFCS 2014, Part II. LNCS, vol. 8635, pp. 589–600. Springer, Heidelberg (2014)

30. Prieto, E., Sloper, C.: Reducing to independent set structure – the case of k-internal spanning tree. Nord. J. Comput. 12(3), 308–318 (2005)

31. Raible, D., Fernau, H., Gaspers, D., Liedloff, M.: Exact and parameterized algorithms for max internal spanning tree. Algorithmica 65(1), 95–128 (2013)

32. Shachnai, H., Zehavi, M.: Representative families: A unified tradeoff-based approach. In: Schulz, A.S., Wagner, D. (eds.) ESA 2014. LNCS, vol. 8737, pp. 786–797. Springer, Heidelberg (2014)

33. Wang, J., Feng, Q.: Improved parameterized algorithms for weighted 3-set packing. In: Hu, X., Wang, J. (eds.) COCOON 2008. LNCS, vol. 5092, pp. 130–139. Springer, Heidelberg (2008)
34. Wang, J., Feng, Q.: An $O^*(3.523^k)$ parameterized algorithm for 3-set packing. In: Agrawal, M., Du, D.-Z., Duan, Z., Li, A. (eds.) TAMC 2008. LNCS, vol. 4978, pp. 82–93. Springer, Heidelberg (2008)
35. Williams, R.: Finding paths of length k in $O^*(2^k)$ time. Inf. Process. Lett. 109(6), 315–318 (2009)
36. Zehavi, M.: Algorithms for k-internal out-branching. In: Gutin, G., Szeider, S. (eds.) IPEC 2013. LNCS, vol. 8246, pp. 361–373. Springer, Heidelberg (2013)
37. Zehavi, M.: Deterministic parameterized algorithms for matching and packing problems. CoRR abs/1311.0484 (2013)
38. Zehavi, M.: Mixing color coding-related techniques. CoRR abs/1410.5062 (2015)

Author Index

Printed in the United States
By Bookmasters